environment
THE SCIENCE BEHIND THE STORIES
CANADIAN EDITION

environment THE SCIENCE BEHIND THE STORIES

CANADIAN EDITION

JAY WITHGOTT

SCOTT BRENNAN

BARBARA MURCK
University of Toronto at Mississauga

Pearson Canada
Toronto

Library and Archives Canada Cataloguing in Publication

Withgott, Jay
    Environment: the science behind the stories / Jay Withgott, Scott Brennan, Barbara Murck. — Canadian ed.

Includes bibliographical references and index.
ISBN 978-0-321-31533-5

1. Environmental sciences. I. Murck, Barbara W. (Barbara Winifred), 1954– II. Brennan, Scott R. III. Title.

GE105.W58 2010   363.7   C2008-906038-5

Copyright © 2010 Pearson Canada Inc., Toronto, Ontario.

Pearson Benjamin Cummings. All rights reserved. This publication is protected by copyright and permission should be obtained from the publisher prior to any prohibited reproduction, storage in a retrieval system, or transmission in any form or by any means, electronic, mechanical, photocopying, recording, or likewise. For information regarding permission, write to the Permissions Department.

Original edition, entitled *Environment: The Science Behind the Stories*, Third Edition, published by Pearson Education, Inc., publishing as Pearson Benjamin Cummings. Copyright © 2008 Pearson Education, Inc. This edition is authorized for sale only in Canada.

ISBN-13: 978-0-321-31533-5
ISBN-10: 0-321-31533-2

Vice President, Editorial Director: Gary Bennett
Acquisitions Editor: Michelle Sartor
Marketing Manager: Kimberly Ukrainec
Senior Developmental Editor: Eleanor MacKay
Production Editor: Patricia Jones
Copy Editor: Laurel Sparrow
Proofreader: Dawn Hunter
Production Coordinator: Janis Raisen
Compositor: Macmillan Publishing Solutions
Photo Researcher: Lisa Brant
Permissions Researcher: Lisa Brant; The Editing Company
Art Director: Julia Hall
Cover Designer: Miriam Blier
Interior Designer: Jennifer Stimson
Cover Image: Image(s) copyright Edward Burtynsky, courtesy Nicholas Metivier Gallery, Toronto

For permission to reproduce copyrighted material, the publisher gratefully acknowledges the copyright holders listed on pages CR-1–CR-4, which are considered an extension of this copyright page.

Statistics Canada information is used with the permission of Statistics Canada. Users are forbidden to copy the data and redisseminate them, in an original or modified form, for commercial purposes, without permission from Statistics Canada. Information on the availability of the wide range of data from Statistics Canada can be obtained from Statistics Canada's Regional Offices, its World Wide Web site at http://www.statcan.ca, and its toll-free access number 1-800-263-1136.

3 4 5   13 12 11 10 09

Printed and bound in United States of America.

This book is printed on recycled paper

Brief Contents

Preface   xxvii
Acknowledgments   xxxi
Reviewers   xxxiii
About the Authors   xxxv

PART ONE
FOUNDATIONS OF ENVIRONMENTAL SCIENCE

1. An Introduction to Environmental Science   2
2. From Chemistry to Energy to Life   31
3. Evolution, Biodiversity, and Population Ecology   59
4. Species Interactions and Community Ecology   91
5. Earth Systems and Ecosystem Ecology   125

PART TWO
THE HUMAN ELEMENT IN THE NATURAL SYSTEM

6. Human Population   160
7. Soils and Agriculture   188
8. Biotechnology and Food Resources   219
9. Biodiversity and Conservation Biology   253
10. Forests and Land Management   287
11. Freshwater Systems and Water Resources   316
12. Marine and Coastal Systems and Fisheries   351
13. Atmospheric Science and Air Pollution   384
14. Global Climate Change   422
15. Fossil Fuels: Energy and Impacts   461

PART THREE
THE SEARCH FOR SOLUTIONS

16. Conventional Energy Alternatives   496
17. New Renewable Energy Alternatives   526
18. Managing Our Waste   556
19. Environmental Health and Risk Management   587
20. The Urban Environment: Creating Liveable Cities   624
21. Environmental Ethics and Economics: Values and Choices   653
22. Environmental Policy: Decision Making and Problem Solving   690
23. Sustainable Solutions   724

APPENDICES

Appendix A   Some Basics on Graphs   A-2
Appendix B   Units and Conversions   A-7
Appendix C   Periodic Table of the Elements   A-8

GLOSSARY   G-I
CREDITS   CR-I
SELECTED SOURCES AND REFERENCES FOR FURTHER READING   R-I
INDEX   I-I

Contents

Preface xxvii

Acknowledgments xxxi

Reviewers xxxiii

About the Authors xxxv

PART ONE
FOUNDATIONS OF ENVIRONMENTAL SCIENCE

1 AN INTRODUCTION TO ENVIRONMENTAL SCIENCE 2

Our Island, Earth 4

    Our environment is more than just our surroundings 4

    Environmental science explores interactions between humans and the physical and biological world 5

    Natural resources are vital to our survival 6

    Human population growth has shaped our resource use 7

    Resource consumption exerts social and environmental impacts 8

    Environmental science can help us avoid mistakes made in the past 10

The Nature of Environmental Science 10

    Environmental science is an interdisciplinary pursuit 10

    THE SCIENCE BEHIND THE STORY: The Lesson of Rapa Nui 11

    People differ in their perception of environmental problems 12

    Environmental science is not the same as environmentalism 13

The Nature of Science 14

    Scientists test ideas by critically examining evidence 14

    The scientific method is a key element of science 15

    There are different ways to test hypotheses 17

    The scientific process does not stop with the scientific method 18

    Science may go through "paradigm shifts" 19

Sustainability and the Future of Our World 20

    Population and consumption lie at the root of many environmental impacts 20

    We face many environmental challenges 20

    THE SCIENCE BEHIND THE STORY: Mission to Planet Earth: Monitoring Environmental Change 21

    Solutions to environmental problems must be global and sustainable 22

    Our energy choices will influence our future immensely 22

    Fortunately, potential solutions abound 23

    Are things getting better or worse? 23

    Sustainability is a goal for the future 24

    Sustainable development involves environmental protection, economic welfare, and social equity 24

    CANADIAN ENVIRONMENTAL PERSPECTIVES: David Suzuki 25

Conclusion 26

Reviewing Objectives 26

Testing Your Comprehension 27

Seeking Solutions 27

Interpreting Graphs and Data 28

Calculating Footprints 29

Take It Further 30

Chapter Endnotes 30

2 FROM CHEMISTRY TO ENERGY TO LIFE 31

Chemistry and the Environment 33

Chemistry can help solve environmental problems 33

Atoms and elements are chemical building blocks 34

THE SCIENCE BEHIND THE STORY: Letting Plants Do the Dirty Work 35

Atoms bond to form molecules and compounds 37

THE SCIENCE BEHIND THE STORY: How Isotopes Reveal Secrets of Earth and Life 38

The chemical structure of the water molecule facilitates life 38

Hydrogen ions determine acidity 40

Matter is composed of organic and inorganic compounds 41

Macromolecules are building blocks of life 42

We create synthetic polymers 44

Organisms use cells to compartmentalize macromolecules 44

Energy Fundamentals 45

Energy is always conserved . . . 46

. . . but energy can change inform and quality 46

Light energy from the Sun powers most living systems 47

Photosynthesis produces food for plants and animals 48

Cellular respiration releases chemical energy 49

Geothermal energy also powers Earth's systems 49

The Origin of Life 50

Early Earth was a very different place 50

Several hypotheses have been proposed to explain life's origin 51

The fossil record has taught us much about life's history 52

Present day organisms and their genes also help us decipher life's history 53

Conclusion 54

CANADIAN ENVIRONMENTAL PERSPECTIVES: Praveen Saxena 55

Reviewing Objectives 55

Testing Your Comprehension 56

Seeking Solutions 56

Interpreting Graphs and Data 57

Calculating Footprints 57

Take It Further 58

Chapter Endnotes 58

3 EVOLUTION, BIODIVERSITY, AND POPULATION ECOLOGY 59

Evolution as the Wellspring of Earth's Biodiversity 61

Natural selection shapes organisms and diversity 61

Natural selection acts on genetic variation 62

Evidence of natural selection is all around us 64

Evolution generates biological diversity 64

Speciation produces new types of organisms 66

Populations can be separated in many ways 67

Life's diversification results from numerous speciation events 67

Speciation and extinction together determine Earth's biodiversity 69

Some species are more vulnerable to extinction than others 69

Earth has seen several episodes of mass extinction 70

The sixth mass extinction has started 70

Levels of Ecological Organization   71

   Ecology is studied at several levels   71

      THE SCIENCE BEHIND THE STORY: The K-T Mass Extinction   72

   Habitat, niche, and degree of specialization are important in organismal ecology   73

Population Ecology   74

   Populations exhibit characteristics that help predict their dynamics   74

   Populations may grow, shrink, or remain stable   77

   Unregulated populations increase by exponential growth   78

   Limiting factors restrain population growth   79

   Carrying capacities can change   80

   The influence of some factors depends on population density   81

   Biotic potential and reproductive strategies vary from species to species   81

   Changes in populations influence the composition of communities   82

The Conservation of Biodiversity   82

   Social and economic factors affect species and communities   82

   Costa Rica took steps to protect its environment   83

      THE SCIENCE BEHIND THE STORY: Climate Change and Its Effects on Monteverde   84

      CANADIAN ENVIRONMENTAL PERSPECTIVES: Maydianne Andrade   86

Conclusion   86

Reviewing Objectives   86

Testing Your Comprehension   87

Seeking Solutions   87

Interpreting Graphs and Data   88

Calculating Footprints   89

Take It Further   89

Chapter Endnotes   90

4 SPECIES INTERACTIONS AND COMMUNITY ECOLOGY   91

Species Interactions   93

   Competition can occur when resources are limited   94

   Several types of interactions are exploitative   95

   Predators kill and consume prey   95

   Parasites exploit living hosts   96

   Herbivores exploit plants   97

   Mutualists help one another   98

   Some interactions have no effect on some participants   99

Ecological Communities   99

   Energy passes among trophic levels   99

   Energy, biomass, and numbers decrease at higher trophic levels   100

   Food webs show feeding relationships and energy flow   101

   Some organisms play bigger roles in communities than others   101

      THE SCIENCE BEHIND THE STORY: Assessing the Ecological Impacts of Zebra Mussels   103

   Succession follows severe disturbance   104

      THE SCIENCE BEHIND THE STORY: Otters, Urchins, Kelp, and a Whale of a Chain Reaction   106

   Invasive species pose new threats to community stability   107

   Some altered communities can be restored to their former condition   109

Earth's Biomes   110

   Biomes are groupings of communities that cover large geographic areas   111

   We can divide the world into roughly 10 terrestrial biomes   112

   Altitude creates patterns analogous to latitude   118

      CANADIAN ENVIRONMENTAL PERSPECTIVES: Zoe Lucas   120

   Aquatic and coastal systems also show biome-like patterns   120

CONTENTS

Conclusion 121
Reviewing Objectives 121
Testing Your Comprehension 122
Seeking Solutions 122
Interpreting Graphs and Data 123
Calculating Footprints 123
Take It Further 124
Chapter Endnotes 124

5 EARTH SYSTEMS AND ECOSYSTEM ECOLOGY 125

Earth's Environmental Systems 127
  Systems show several defining properties 128
  Understanding a complex system requires considering multiple subsystems 129
  Environmental systems may be perceived in various ways 130
Geologic Systems: How Earth Works 131
  The rock cycle is a fundamental environmental system 131
  Plate tectonics shapes Earth's geography 133
Ecosystems 135
  Ecosystems are systems of interacting living and nonliving entities 135
  Energy is converted to biomass 135
  Nutrients can limit productivity 136
  Ecosystems are integrated spatially 138
  Landscape ecologists study geographic areas with multiple ecosystems 139
Biogeochemical Cycles 139
  THE SCIENCE BEHIND THE STORY: Biodiversity Portrait of the St. Lawrence 140
  Remote sensing helps us apply landscape ecology 141
  Nutrients and other materials circulate via biogeochemical cycles 141
  The hydrologic cycle influences all other cycles 143

Our impacts on the hydrologic cycle are extensive 144
The carbon cycle circulates a vital organic nutrient 145
We are shifting carbon from the lithosphere to the atmosphere 146
The nitrogen cycle involves specialized bacteria 146
We have greatly influenced the nitrogen cycle 148
  THE SCIENCE BEHIND THE STORY: The Gulf of Mexico's "Dead Zone" 150
The phosphorus cycle involves mainly lithosphere and ocean 150
  CANADIAN ENVIRONMENTAL PERSPECTIVES: Robert Bateman 152
We affect the phosphorus cycle 152
Conclusion 153
Reviewing Objectives 154
Testing Your Comprehension 154
Seeking Solutions 155
Interpreting Graphs and Data 156
Calculating Footprints 156
Take It Further 157
Chapter Endnotes 157

PART TWO
THE HUMAN ELEMENT IN THE NATURAL SYSTEM

6 HUMAN POPULATION 160

Human Population: Approaching 7 Billion 162
  The human population is growing nearly as fast as ever 162
  Perspectives on human population have changed over time 163
  Is population growth really a "problem" today? 164
  Population is one of several factors that affect the environment 165

Demography   167

　The environment has a carrying capacity for humans   167

　Demography is the study of human population   168

　Population growth depends on rates of birth, death, immigration, and emigration   172

　Total fertility rate influences population growth   172

　Some nations have experienced the demographic transition   173

　Is the demographic transition a universal process?   175

Population and Society   175

　The status of women greatly affects population growth rates   175

　　THE SCIENCE BEHIND THE STORY: Fertility Decline in Bangladesh   176

　Population policies and family planning programs are working around the globe   176

　Poverty is strongly correlated with population growth   178

　Consumption from affluence creates environmental impacts   179

　The wealth gap and population growth contribute to conflict   180

　HIV/AIDS is exerting major impacts on African populations   180

　Severe demographic changes have social, political, economic, and environmental repercussions   181

　The U.N. has articulated sustainable development goals for humanity   182

Conclusion   182

　　CANADIAN ENVIRONMENTAL PERSPECTIVES: William Rees   183

Reviewing Objectives   184

Testing Your Comprehension   185

Seeking Solutions   185

Interpreting Graphs and Data   186

Calculating Footprints   186

Take It Further   187

Chapter Endnotes   187

7 SOILS AND AGRICULTURE   188

Soil as a System   190

　Soil formation is slow and complex   191

　A soil profile consists of layers known as horizons   192

　Soil can be characterized by colour, texture, structure, and pH   193

　Cation exchange is vital for plant growth   196

Soil: The Foundation for Feeding a Growing Population   196

　As population and consumption increase, soils are being degraded   196

　Agriculture began to appear around 10 000 years ago   196

　Industrialized agriculture is newer still   199

Soil Degradation: Problems and Solutions   199

　Regional differences in soil traits can affect agriculture   199

　Erosion can degrade ecosystems and agriculture   200

　Soil erodes by several mechanisms   201

　Soil erosion is a global problem   202

　Desertification reduces productivity of arid lands   203

　The Dust Bowl was a monumental event in North America   204

　The Soil Conservation Council emerged from the experience of drought   204

　Farmers can protect soil against degradation in various ways   205

　　THE SCIENCE BEHIND THE STORY: No-Till Agriculture in Southern Brazil   208

　Erosion control practices protect and restore plant cover   209

　Irrigation boosts productivity but can cause long-term soil problems   209

Salinization is easier to prevent than to correct 210

Agricultural fertilizers boost crop yields but can be overapplied 210

THE SCIENCE BEHIND THE STORY: Nitrate Contamination of the Abbotsford Aquifer 212

Grazing practices and policies can contribute to soil degradation 212

CANADIAN ENVIRONMENTAL PERSPECTIVES: Loretta Ford 214

Conclusion 214

Reviewing Objectives 215

Testing Your Comprehension 215

Seeking Solutions 216

Interpreting Graphs and Data 216

Calculating Footprints 217

Take It Further 218

Chapter Endnotes 218

8 BIOTECHNOLOGY AND FOOD RESOURCES 219

The Race to Feed the World 222

We are producing more food per person 222

We face undernourishment, overnutrition, and malnutrition 222

The "Green Revolution" led to dramatic increases in agricultural production 224

The Green Revolution has caused the environment both benefit and harm 224

Biofuels are having a significant impact on food availability 225

Pests and Pollinators 226

Many thousands of chemical pesticides have been developed 226

Pests evolve resistance to pesticides 227

Biological control pits one organism against another 228

Biological control agents themselves may become pests 229

Integrated pest management combines biocontrol and chemical methods 229

We depend on insects to pollinate crops 229

Conservation of pollinators is vital 230

THE SCIENCE BEHIND THE STORY: The Alfalfa and the Leafcutter 231

Genetically Modified Food 231

Genetic modification of organisms depends on recombinant DNA 232

Genetic engineering is like, and unlike, traditional agricultural breeding 233

Biotechnology is transforming the products around us 233

What are the impacts of GM crops? 234

Debate over GM foods involves more than science 235

Preserving Crop Diversity 237

Crop diversity provides insurance against failure 237

Seed banks are living museums for seeds 238

Raising Animals for Food 239

Consumption of animal products is growing 239

High consumption has led to feedlot agriculture 239

Our food choices are also energy choices 240

We also raise fish on "farms" 241

Aquaculture has some benefits 242

Aquaculture has some negative impacts, too 242

Sustainable Agriculture 243

Organic agriculture is on the increase 243

Locally supported agriculture is growing 244

THE SCIENCE BEHIND THE STORY: Organic Farming Put to the Test 245

Organic agriculture can even succeed in cities 245

CANADIAN ENVIRONMENTAL PERSPECTIVES: Alisa Smith and James MacKinnon 246

Conclusion 247

Reviewing Objectives 248

Testing Your Comprehension 249

Seeking Solutions 249

Interpreting Graphs and Data   250

Calculating Footprints   250

Take It Further   251

Chapter Endnotes   251

9 BIODIVERSITY AND CONSERVATION BIOLOGY   253

Our Planet of Life   255

Biodiversity encompasses several levels   255

Some groups hold more species than others   258

Measuring biodiversity is not easy   258

Biodiversity is unevenly distributed   259

Biodiversity Loss and Species Extinction   261

Extinction is a natural process   262

Earth has experienced five previous mass extinction episodes   262

Humans set the sixth mass extinction in motion years ago   262

Current extinction rates are much higher than normal   263

Biodiversity loss involves more than extinction   263

There are several major causes of biodiversity loss   264

THE SCIENCE BEHIND THE STORY: Amphibian Diversity and Decline   265

Benefits of Biodiversity   269

Biodiversity provides ecosystem services free of charge   270

Biodiversity helps maintain ecosystem function   270

Biodiversity enhances food security   271

Organisms provide drugs and medicines   271

Biodiversity provides economic benefits through tourism and recreation   272

People value and seek out connections with nature   272

Do we have ethical obligations toward other species?   273

Conservation Biology: The Search for Solutions   274

Conservation biology arose in response to biodiversity loss   274

Conservation biologists work at multiple levels   274

Island biogeography theory is a key component of conservation biology   275

Captive breeding, reintroduction, and cloning are single-species approaches   276

Some species act as "umbrellas" for protecting habitat and communities   278

Both national and international conservation efforts are widely supported   278

THE SCIENCE BEHIND THE STORY: The Swift Fox Returns   279

Other approaches highlight areas of high diversity   280

Community-based conservation is increasingly popular   280

Innovative economic strategies are being employed   281

CANADIAN ENVIRONMENTAL PERSPECTIVES: Biruté Mary Galdikas   282

Conclusion   282

Reviewing Objectives   283

Testing Your Comprehension   283

Seeking Solutions   284

Interpreting Graphs and Data   284

Calculating Footprints   285

Take It Further   286

Chapter Endnotes   286

10 FORESTS AND LAND MANAGEMENT   287

Forest Ecosystems and Their Status Worldwide   290

There are three major groups of forest biomes   290

Open-canopy wooded lands and grasslands are drier terrestrial ecosystems   291

Canada is a steward for much of the world's forest   292

Forests are ecologically valuable   293

Forests are a crucial link in nutrient and water cycles   293

Forest products are economically valued   293

Land Conversion   294

The growth of Canada and the United States was fuelled by land clearing and logging   295

Agriculture is the major cause of conversion of forests and grasslands   295

THE SCIENCE BEHIND THE STORY: Changing Climate and the Spruce Budworm on Vancouver Island   297

Livestock graze one-fourth of Earth's land surface   297

Demands for land and wood have led to some bad practices and deforestation   298

Deforestation is proceeding rapidly in many developing nations   298

Forest Management Principles   299

THE SCIENCE BEHIND THE STORY: Surveying Earth's Forests   300

Public forests in Canada are managed for many purposes   300

Today, many managers practise ecosystem-based management   302

Adaptive management evolves and improves   302

Plantation forestry has grown in North America   303

Timber is harvested by several methods   304

Fire policy has stirred controversy   305

Sustainable forestry is gaining ground   306

Parks and Reserves   307

Why have we created parks and reserves?   307

Not everyone supports land set-asides   308

Nonfederal entities also protect land   308

Parks and reserves are increasing internationally   309

The design of parks and reserves has consequences for biodiversity   310

CANADIAN ENVIRONMENTAL PERSPECTIVES: Tzeporah Berman   311

Conclusion   311

Reviewing Objectives   312

Testing Your Comprehension   312

Seeking Solutions   313

Interpreting Graphs and Data   313

Calculating Footprints   314

Take It Further   314

Chapter Endnotes   315

11 FRESHWATER SYSTEMS AND WATER RESOURCES   316

Freshwater Systems   318

Rivers and streams wind through landscapes   319

Wetlands include marshes, swamps, and bogs   320

Lakes and ponds are ecologically diverse systems   320

Groundwater plays key roles in the hydrologic cycle   321

Water is unequally distributed across Earth's surface   323

Climate change will cause water problems and shortages   323

How We Use Water   324

Water supplies our households, agriculture, and industry   325

We have erected thousands of dams   325

China's Three Gorges Dam is the world's largest   325

Some dams are now being removed   328

Dikes and levees are meant to control floods   328

We divert—and deplete—surface water to suit our needs   329

Inefficient irrigation wastes water   329

Wetlands have been drained for a variety of reasons   331

We are depleting groundwater   332

Our thirst for bottled water seems unquenchable   333

Will we see a future of water wars?   334

Solutions to Depletion of Freshwater   334

Solutions can address supply or demand   334

Desalination "makes" more water   334

Agricultural demand can be reduced   335

We can lessen residential and industrial water use in many ways   335

Economic approaches to water conservation are being debated   335

Freshwater Pollution and Its Control   336

Water pollution takes many forms   336

Water pollution comes from point and non-point sources   338

Scientists use several indicators of water quality   338

Groundwater pollution is a serious problem   339

There are many sources of groundwater pollution, including some natural sources   339

THE SCIENCE BEHIND THE STORY: Arsenic in the Waters of Bangladesh   340

Legislative and regulatory efforts have helped reduce pollution   341

We treat our drinking water   341

THE SCIENCE BEHIND THE STORY: When Water Turns Deadly: The Walkerton Tragedy   342

It is better to prevent pollution than to mitigate the impacts after it occurs   342

Wastewater and Its Treatment   343

Municipal wastewater treatment involves several steps   343

CANADIAN ENVIRONMENTAL PERSPECTIVES: David Schindler   345

Artificial wetlands can aid treatment   346

Conclusion   346

Reviewing Objectives   346

Testing Your Comprehension   347

Seeking Solutions   347

Interpreting Graphs and Data   348

Calculating Footprints   348

Take It Further   349

Chapter Endnotes   350

12 MARINE AND COASTAL SYSTEMS AND FISHERIES   351

The Oceans   354

Oceans cover most of Earth's surface   354

The oceans contain more than water   354

Ocean water is vertically structured   355

Ocean water flows vertically and horizontally, influencing climate   356

La Niña and El Niño demonstrate the atmosphere–ocean connection   357

Seafloor topography can be rugged and complex   358

Marine and Coastal Ecosystems   359

Open-ocean ecosystems vary in their biological diversity   359

Kelp forests harbour many organisms   360

Coral reefs are treasure troves of biodiversity   360

Intertidal zones undergo constant change   362

Salt marshes occur widely along temperate shorelines   363

Mangrove forests line coasts in the tropics and subtropics   363

Freshwater meets saltwater in estuaries   364

Human Use and Impact   364

Oceans provide transportation routes   364

We extract energy and minerals   364

Marine pollution threatens resources   365

Nets and plastic debris endanger marine life   366

Oil pollution comes from spills of all sizes   366

Pollutants can contaminate seafood 367

Excess nutrients cause algal blooms 367

Emptying the Oceans 368

We have long overfished 368

THE SCIENCE BEHIND THE STORY: China's Fisheries Data 369

Fishing has become industrialized 369

Fishing practices kill nontarget animals and damage ecosystems 370

Modern fishing fleets deplete marine life rapidly 371

Several factors mask declines 371

We are "fishing down the food chain" 372

Aquaculture can partly make up for the loss of capture fisheries 373

Aquaculture has both benefits and drawbacks 373

Consumer choice can influence fishing practices 374

Marine biodiversity loss erodes ecosystem services 374

Marine Conservation 375

Fisheries management has been based on maximum sustainable yield 375

We can protect areas in the ocean 375

THE SCIENCE BEHIND THE STORY: Do Marine Reserves Work? 376

Reserves can work for both fish and fishers 376

How should reserves be designed? 377

CANADIAN ENVIRONMENTAL PERSPECTIVES: Farley Mowat 378

Conclusion 379

Reviewing Objectives 379

Testing Your Comprehension 380

Seeking Solutions 381

Interpreting Graphs and Data 381

Calculating Footprints 382

Take It Further 382

Chapter Endnotes 383

13 ATMOSPHERIC SCIENCE AND AIR POLLUTION 384

The Atmosphere and Weather 386

The atmosphere is layered 387

Atmospheric properties include temperature, pressure, and humidity 388

Solar energy heats the atmosphere, helps create seasons, and causes air to circulate 388

The atmosphere drives weather and climate 390

Air masses interact to produce weather 390

Large-scale circulation systems produce global climate patterns 391

Outdoor Air Pollution 393

Natural sources can pollute 393

We create various types of outdoor air pollution 393

CEPA identifies harmful airborne substances 394

Government agencies share in dealing with air pollution 397

Monitoring shows that many forms of air pollution have decreased 398

Canada is attempting to "turn the corner" on air pollution 401

Smog is the most common, widespread air quality problem 401

Photochemical smog is produced by a complex series of reactions 403

THE SCIENCE BEHIND THE STORY: Identifying CFCs as the Main Cause of Ozone Depletion 404

Air quality is a rural issue, too 404

Industrializing nations are suffering increasing air pollution 405

Synthetic chemicals deplete stratospheric ozone 406

There are still many questions to be resolved about ozone depletion 406

The Montreal Protocol addressed ozone depletion 407

Acidic deposition is another transboundary pollution problem   408

Acid deposition has not been reduced as much as scientists had hoped   409

Indoor Air Pollution   410

Indoor air pollution in the developing world arises from fuelwood burning   411

THE SCIENCE BEHIND THE STORY: Acid Rain at Hubbard Brook Research Forest   412

Tobacco smoke and radon are the most dangerous indoor pollutants in the developed world   414

Many VOCs pollute indoor air   414

Living organisms can pollute indoor spaces   416

CANADIAN ENVIRONMENTAL PERSPECTIVES: David Phillips   417

We can reduce indoor air pollution   417

Conclusion   418
Reviewing Objectives   418
Testing Your Comprehension   419
Seeking Solutions   419
Interpreting Graphs and Data   420
Calculating Footprints   420
Take It Further   421
Chapter Endnotes   421

14 GLOBAL CLIMATE CHANGE   422

Our Dynamic Climate   424

What is climate change?   424

The Sun and atmosphere keep Earth warm   424

Greenhouse gases warm the lower atmosphere   425

Carbon dioxide is the anthropogenic greenhouse gas of primary concern   426

Human activity has released carbon from sequestration in long-term reservoirs   426

Other greenhouse gases contribute to warming   427

There are many feedback cycles in the climate system   427

Radiative forcing expresses change in energy input over time   428

The atmosphere is not the only factor that influences climate   428

The Science of Climate Change   430

Proxy indicators tell us about the past   430

Stable isotope geochemistry is a powerful tool for the study of paleoclimate   432

Direct atmospheric sampling tells us about the present   432

Models help us understand climate   432

THE SCIENCE BEHIND THE STORY: Reading History in the World's Longest Ice Core   434

Current and Future Trends and Impacts   435

The IPCC summarizes evidence of climate change and predicts future impacts   436

Temperature increases will continue   438

Changes in precipitation will vary by region   439

Melting ice and snow have far-reaching effects   440

The Arctic is changing dramatically   441

Rising sea levels will affect hundreds of millions of people and coastal zones   442

THE SCIENCE BEHIND THE STORY: Greenland's Glaciers Race to the Sea   444

Climate change affects organisms and ecosystems   445

Climate change exerts societal impacts—and vice versa   446

Are we responsible for climate change?   448

Responding to Climate Change   449

Shall we pursue mitigation, or adaptation?   449

We can look more closely at our lifestyle   451

Transportation is a significant source of greenhouse gases   452

We can reduce emissions in other ways as well   452

We will need to follow multiple strategies to reduce emissions   453

We began tackling climate change by international treaty   453

The Kyoto Protocol seeks to limit emissions   453

Market mechanisms are being used to address climate change   454

Carbon offsets are in vogue   455

You can reduce your own carbon footprint   455

CANADIAN ENVIRONMENTAL PERSPECTIVES: Sheila Watt-Cloutier   456

Conclusion   456

Reviewing Objectives   457

Testing Your Comprehension   457

Seeking Solutions   458

Interpreting Graphs and Data   458

Calculating Footprints   459

Take It Further   460

Chapter Endnotes   460

15 FOSSIL FUELS: ENERGY AND IMPACTS   461

Sources of Energy   464

We use a variety of energy sources   464

Fossil fuels are indeed fuels created from "fossils"   466

Fossil fuel reserves are unevenly distributed   466

Developed nations consume more energy than developing nations   466

It takes energy to make energy   467

Coal, Natural Gas, and Oil   468

Coal is the world's most abundant fossil fuel   468

Coal use has a long history   469

Coal is mined from the surface and from below ground   469

THE SCIENCE BEHIND THE STORY: Clean Coal for Electricity Generation   470

Coal varies in its qualities   470

Natural gas is the fastest-growing fossil fuel in use today   472

Natural gas is formed in two ways   472

Natural gas has only recently been widely used   472

Natural gas extraction becomes more challenging with time   473

Offshore drilling produces much of our gas and oil   473

Oil is the world's most-used fuel   474

Heat and pressure underground form petroleum   474

Petroleum geologists infer the location and size of deposits   474

We drill to extract oil   475

Petroleum products have many uses   475

THE SCIENCE BEHIND THE STORY: How Crude Oil Is Refined   476

We may have already depleted half our oil reserves   476

"Unconventional" Fossil Fuels   479

Canada owns massive deposits of oil sands   479

Oil shale is abundant in the American West   480

Methane hydrate shows potential   480

These alternative fossil fuels have downsides   480

Environmental Impacts of Fossil Fuel Use   481

Fossil fuel emissions cause pollution and drive climate change   481

Some emissions from fossil fuel burning can be "captured"   482

Coal mining affects the environment   483

Oil and gas extraction can alter the environment   484

Political, Social, and Economic Aspects   485

Nations can become dependent on foreign energy   485

Oil supply and prices affect the economies of nations   485

Residents may or may not benefit from their fossil fuel reserves   486

How will we convert to renewable energy?   487

Energy conservation has followed economic need   487

Personal choice and increased efficiency are two routes to conservation   488

> CANADIAN ENVIRONMENTAL PERSPECTIVES: Mary Griffiths   489

Both conservation and renewable energy are needed   489

Conclusion   490

Reviewing Objectives   490

Testing Your Comprehension   491

Seeking Solutions   491

Interpreting Graphs and Data   492

Calculating Footprints   492

Take It Further   493

Chapter Endnotes   493

PART THREE
THE SEARCH FOR SOLUTIONS

16 CONVENTIONAL ENERGY ALTERNATIVES   496

Alternatives to Fossil Fuels   498

Hydropower, nuclear power, and biomass energy are conventional alternatives   499

Conventional alternatives provide some of our energy and much of our electricity   499

Hydroelectric Power   499

Modern hydropower uses two approaches   499

Hydroelectric power is widely used   500

Hydropower is clean and renewable   501

Hydropower has negative impacts, too   502

Hydropower may not expand much more   503

Nuclear Power   503

Fission releases nuclear energy   504

Nuclear energy comes from processed and enriched uranium   504

Fission in reactors generates electricity in nuclear power plants   505

Breeder reactors make better use of fuel but have raised safety concerns   505

Fusion remains a dream   506

Nuclear power delivers energy more cleanly than fossil fuels   507

> THE SCIENCE BEHIND THE STORY: Assessing Emissions from Power Sources   508

Nuclear power poses small risks of large accidents   508

Chernobyl saw the worst nuclear accident yet   509

Radioactive waste disposal remains a problem   511

> THE SCIENCE BEHIND THE STORY: Health Impacts of Chernobyl   512

Multiple dilemmas have slowed nuclear power's growth   513

Biomass Energy   514

Traditional biomass sources are widely used in the developing world   515

New biomass strategies are being developed in industrialized countries   516

Biofuels can power automobiles   516

With biopower we generate electricity from biomass   518

Biomass energy brings environmental and economic benefits   520

Biomass energy also brings drawbacks   520

> CANADIAN ENVIRONMENTAL PERSPECTIVES: Gráinne Ryder   521

Conclusion   522

Reviewing Objectives   522

Testing Your Comprehension   523

Seeking Solutions   523

Interpreting Graphs and Data   524

Calculating Footprints   525

Take It Further   525

Chapter Endnotes   525

17 NEW RENEWABLE ENERGY ALTERNATIVES   526

"New" Renewable Energy Sources   529

The "new" renewables currently provide little of our power   529

The new renewables are growing quickly   530

THE SCIENCE BEHIND THE STORY: Energy from Landfill Gas at Beare Road   531

Our transition must begin soon   532

Solar Energy   533

Passive solar heating is simple and effective   533

Active solar energy collection can heat air and water in buildings   534

Concentrating solar rays magnifies energy   534

Photovoltaic cells generate electricity directly   534

Solar power is little used but fast growing   535

Solar power offers many benefits   536

Location and cost can be drawbacks   536

Wind Energy   538

Wind has long been used for energy   538

Modern wind turbines convert kinetic energy to electrical energy   538

Wind power is the fastest-growing energy sector   539

Offshore and high-elevation sites can be promising   539

Wind power has many benefits   541

Wind power has some downsides—but not many   542

Geothermal Energy   542

We can harness geothermal energy for heating and electricity   543

THE SCIENCE BEHIND THE STORY: Water and Earth Energy for Heating and Cooling in Toronto and Ottawa   544

Use of geothermal power is growing   545

Geothermal power has benefits and limitations   545

Ocean Energy   545

We can harness energy from tides, waves, and currents   545

The ocean stores thermal energy   547

Hydrogen   547

Hydrogen fuel may be produced from water or from other matter   548

CANADIAN ENVIRONMENTAL PERSPECTIVES: David Keith   550

Fuel cells produce electricity by joining hydrogen and oxygen   550

Hydrogen and fuel cells have many benefits   551

Conclusion   551

Reviewing Objectives   551

Testing Your Comprehension   552

Seeking Solutions   553

Interpreting Graphs and Data   553

Calculating Footprints   554

Take It Further   554

Chapter Endnotes   554

18 MANAGING OUR WASTE   556

Approaches to Waste Management   559

We have several aims in managing waste   559

Municipal Solid Waste   559

Patterns in the municipal solid waste stream vary from place to place   560

Waste generation is rising in all nations   561

Open dumping of the past has given way to improved disposal methods   562

Waste disposal is regulated by three levels of government   562

Sanitary landfills are engineered to minimize leakage of contaminants   563

Landfills can be transformed after closure   564

Landfills have drawbacks   565

Incinerating trash reduces pressure on landfills   565

> THE SCIENCE BEHIND THE STORY: Digging Garbage: The Archeology of Solid Waste   566

Many incinerators burn waste to create energy   567

Landfills can produce gas for energy   567

Reducing waste is a better option   568

Reuse is one main strategy for waste reduction   568

Composting recovers organic waste   569

Recycling consists of three steps   569

Recycling has grown rapidly and can expand further   570

Financial incentives can help address waste   571

Edmonton showcases reduction and recycling   571

Industrial Solid Waste   572

Regulation and economics each influence industrial waste generation   572

Industrial ecology seeks to make industry more sustainable   573

Businesses are adopting industrial ecology   573

Waste exchanges are an offshoot of industrial ecology   574

Hazardous Waste   574

Hazardous wastes have diverse sources   575

Organic compounds and heavy metals can be hazardous   575

> THE SCIENCE BEHIND THE STORY: Testing the Toxicity of "e-Waste"   576

"E-waste" is a new and growing problem   576

Several steps precede the disposal of hazardous waste   577

There are three disposal methods for hazardous waste   578

Radioactive waste is especially hazardous   579

Contaminated sites are being cleaned up, slowly   580

> CANADIAN ENVIRONMENTAL PERSPECTIVES: Brennain Lloyd   581

Conclusion   582

Reviewing Objectives   582

Testing Your Comprehension   583

Seeking Solutions   583

Interpreting Graphs and Data   584

Calculating Footprints   584

Take It Further   585

Chapter Endnotes   585

19 ENVIRONMENTAL HEALTH AND RISK MANAGEMENT   587

Environmental Health   590

Environmental hazards can be physical, chemical, biological, or cultural   590

Disease is a major focus of environmental health   593

Environmental health hazards exist indoors as well as outdoors   596

Toxicology is the study of poisonous substances   597

Toxic Agents in the Environment   598

Synthetic chemicals are ubiquitous in our environment   598

Silent Spring began the public debate over synthetic chemicals   598

Toxicants come in several different types   599

Endocrine disruption may be widespread   600

Endocrine disruption research has generated debate   602

> THE SCIENCE BEHIND THE STORY: Is Bisphenol-A Safe?   603

Toxicants may concentrate in water   604

Airborne toxicants can travel widely   605

Some toxicants persist for a long time 605

Toxicants may accumulate over time and up the food chain 605

Not all toxicants are synthetic 606

Studying the Effects of Hazards 607

Wildlife studies use careful observations in the field and lab 607

Human studies rely on case histories, epidemiology, and animal testing 607

THE SCIENCE BEHIND THE STORY: Pesticides and Child Development in Mexico's Yaqui Valley 608

Dose–response analysis is a mainstay of toxicology 608

Individuals vary in their responses to hazards 610

The type of exposure can affect the response 611

Mixes may be more than the sum of their parts 611

Risk Assessment and Risk Management 611

Risk is expressed in terms of probability 612

Our perception of risk may not match reality 612

Risk assessment analyzes risk quantitatively 613

Risk management combines science and other social factors 613

Philosophical and Policy Approaches 614

Two approaches exist for determining safety 614

Philosophical approaches are reflected in policy 614

Toxicants are also regulated internationally 616

CANADIAN ENVIRONMENTAL PERSPECTIVES: Wendy Mesley 617

Conclusion 618

Reviewing Objectives 618

Testing Your Comprehension 619

Seeking Solutions 619

Interpreting Graphs and Data 620

Calculating Footprints 621

Take It Further 621

Chapter Endnotes 622

20 THE URBAN ENVIRONMENT: CREATING LIVEABLE CITIES 624

Our Urbanizing World 627

Industrialization has driven the move to urban centres 627

Today's urban centres are unprecedented in scale and rate of growth 628

Various factors influence the geography of urban areas 630

People have moved to suburbs 630

Sprawl 631

Today's urban areas spread outward 631

Sprawl has several causes 633

THE SCIENCE BEHIND THE STORY: Measuring the Causes and Impacts of Sprawl 634

What is wrong with sprawl? 634

Creating Liveable Cities 636

City and regional planning are means for creating liveable urban areas 636

Zoning is a key tool for planning 637

Urban growth boundaries and greenbelts are now widely used 638

"Smart growth" aims to counter sprawl 639

The "new urbanism" and "liveable cities" are now in vogue 639

Transportation options are vital to liveable cities 640

Parks and open space are key elements of liveable cities 641

City parks were widely established at the turn of the last century 641

THE SCIENCE BEHIND THE STORY: Assessing the Benefits of Rail Transit 642

Smaller public spaces are also important 643

Urban Sustainability 644

Urban resource consumption brings a mix of environmental impacts 644

Urban intensification preserves land 645

Urban centres suffer and export pollution 646

Urban centres foster innovation and offer cultural resources 646

Some seek sustainability for cities 646

Conclusion 647

CANADIAN ENVIRONMENTAL PERSPECTIVES: Nola-Kate Seymoar 648

Reviewing Objectives 648

Testing Your Comprehension 649

Seeking Solutions 649

Interpreting Graphs and Data 650

Calculating Footprints 650

Take It Further 651

Chapter Endnotes 651

21 ENVIRONMENTAL ETHICS AND ECONOMICS: VALUES AND CHOICES 653

Culture, World View, and the Environment 656

Ethics and economics involve values 656

Culture and world view influence our perception of the environment 656

THE SCIENCE BEHIND THE STORY: The Mirrar Clan Confronts the Jabiluka Uranium Mine 657

Many factors shape our world views and perception of the environment 658

There are many ways to understand the environment 659

Environmental Ethics 660

Environmental ethics pertains to humans and the environment 661

We have extended ethical consideration to more entities through time 661

Environmental ethics has ancient roots 663

The Industrial Revolution inspired environmental philosophers 663

Conservation and preservation arose at the start of the twentieth century 664

The land ethic and deep ecology enlarged the boundaries of the ethical community 665

Ecofeminism recognizes connections between the oppression of nature and of women 666

Environmental justice seeks equitable access to resources and protection from environmental degradation 667

Economics: Approaches and Environmental Implications 669

Is there a trade-off between economics and the environment? 670

Economics studies the allocation of scarce resources 670

Several types of economies exist today 670

Environment and economy are intricately linked 670

Classical economics promoted the free market 672

Neoclassical economics incorporates human psychology 673

Cost–benefit analysis is a useful tool 673

Aspects of neoclassical economics have profound implications for the environment 674

THE SCIENCE BEHIND THE STORY: Ethics in Economics: Discounting and Global Climate Change 676

Is the growth paradigm good for us? 677

Economists disagree on whether economic growth is sustainable 678

A steady-state economy is a revolutionary alternative to growth 678

We can measure economic progress differently 679

We can give ecosystem goods and services monetary values 680

Markets can fail 683

Corporations are responding to sustainability concerns 683

CANADIAN ENVIRONMENTAL PERSPECTIVES: Matthew Coon Come 684

Conclusion   685

Reviewing Objectives   685

Testing Your Comprehension   686

Seeking Solutions   686

Interpreting Graphs and Data   687

Calculating Footprints   688

Take It Further   688

Chapter Endnotes   689

22 ENVIRONMENTAL POLICY: DECISION MAKING AND PROBLEM SOLVING   690

Environmental Policy   693

Environmental policy addresses issues of equity and resource use   694

Many factors hinder implementation of environmental policy   695

Canadian Environmental Law and Policy   696

Canada's environmental policies are influenced by our neighbour   696

Several legal instruments are used to ensure that environmental goals are achieved   696

Environmental goals and best practices can be promoted by voluntary initiatives   697

Canadian environmental policy arises from all three levels of government   697

Government and nongovernmental agencies work together to resolve environmental issues   701

Different environmental media require different regulatory approaches   701

Environmental policy has changed with the society and the economy   702

The social context for environmental policy changes over time   704

The concept of sustainable development now guides environmental policy   705

Scientific monitoring and reporting helps with environmental policy decisions   706

SOER presents organizational challenges   707

International Environmental Law and Policy   708

International law includes conventional and customary law   709

Several organizations shape international environmental policy   709

THE SCIENCE BEHIND THE STORY: The Environment in NAFTA and NAAEC   711

Exploring Approaches to Environmental Policy   714

Science plays a role in policy, but it can be politicized   714

Command-and-control policy has improved our lives, but it is not perfect   714

Economic tools also can be used to achieve environmental goals   714

Market incentives are being tried widely on the local level   717

Ecolabelling gives some choice back to the consumer   717

CANADIAN ENVIRONMENTAL PERSPECTIVES: Maude Barlow   718

Conclusion   719

Reviewing Objectives   719

Testing Your Comprehension   720

Seeking Solutions   720

Interpreting Graphs and Data   721

Calculating Footprints   721

Take It Further   722

Chapter Endnotes   722

23 SUSTAINABLE SOLUTIONS   724

Sustainability on Campus   727

Why strive for campus sustainability?   727

THE SCIENCE BEHIND THE STORY: A Campus Ecological Footprint Calculator   728

Campus efforts may begin with an audit   728

Recycling and waste reduction are common campus efforts   729

Green building design is a key to sustainable campuses   730

Efficient water use is important   732

Energy conservation is achievable   732

Students can promote renewable energy   732

Carbon neutrality is a new goal   733

Dining services and campus gardens let students eat sustainably   733

Institutional purchasing matters   734

Transportation alternatives are many   734

Campuses are restoring native plants, habitats, and landscapes   735

Sustainability efforts include curricular changes   736

Organizations are available to assist campus efforts   737

Sustainability and Sustainable Development   737

Sustainable development aims to achieve a triple bottom line   737

Environmental protection can enhance economic opportunity   738

What accounts for the perceived economy-versus-environment divide?   739

Humans are not separate from the environment   739

Strategies for Sustainability   740

We can refine our ideas about economic growth and quality of life   740

We can consume less   741

Population growth must eventually cease   742

Technology can help us toward sustainability   743

Industry can mimic natural systems   744

We can think in the long term   744

We can promote local self-sufficiency and embrace some aspects of globalization   744

Citizens exert political influence   745

Consumers vote with their wallets   745

THE SCIENCE BEHIND THE STORY: Rating the Environmental Performance of Nations   746

Promoting research and education is vital   746

Precious Time   746

We must pass through the environmental bottleneck   747

CANADIAN ENVIRONMENTAL PERSPECTIVES: Keleigh Annau   748

We must think of Earth as an island   748

Conclusion   749

Reviewing Objectives   749

Testing Your Comprehension   750

Seeking Solutions   750

Interpreting Graphs and Data   751

Calculating Footprints   752

Take It Further   752

Chapter Endnotes   753

APPENDICES

Appendix A   Some Basics on Graphs   A-2

Appendix B   Units and Conversions   A-7

Appendix C   Periodic Table of the Elements   A-8

GLOSSARY   G-1

CREDITS   CR-1

SELECTED SOURCES AND REFERENCES FOR FURTHER READING   R-1

INDEX   I-1

Preface

We live in extraordinary times. Human impact on our environment has never been so intensive or so far-reaching. The future of our society and the future of Earth's systems depend more critically than ever on the way we interact with the world around us. Fundamental aspects of climate, atmospheric composition, nutrient cycling, and biological diversity are being altered at dizzying speeds. Yet thanks to environmental science, we now understand better than ever how our planet's systems function and how we influence these systems. Environmental science helps us to characterize the problems we create, and it also illuminates the tremendous opportunities we have before us for effecting positive change.

The field of environmental science captures the very essence of this unique moment in history. This interdisciplinary pursuit stands at the vanguard of the current need to synthesize academic disciplines and to incorporate their contributions into a big-picture understanding of the world and our place within it.

The Canadian Edition

For the Canadian Edition of *Environment*, we wanted to do much, much more than just change the units, the spellings, and a few of the case studies and examples. Those of you who have used the wonderful U.S. 3rd edition of the book by Withgott and Brennan will know that, as strong as it is, it does have a fundamentally American focus that makes it challenging to use in a Canadian context. We wanted to produce a book that presents a truly Canadian perspective on environmental science. We wanted to tell students about the great, sometimes groundbreaking work being done by Canadian environmental scientists in many different disciplines. We wanted to celebrate our environmental heroes, achievements, and history, and to familiarize students with the people, locations, and events of that history, with examples from coast to coast to coast.

You will notice many changes between the U.S. 3rd edition and the Canadian edition, but the fundamental features that make *Environment: The Science Behind the Stories* such a powerful teaching and learning tool are still in place.

So, what's different? First, you will notice that the order of the chapters has been changed, and they are now organized into three parts:

The scientific fundamentals are now right up front in chapters 1 through 5, which form *Part I: Foundations of Environmental Science*. Here we introduce students to the chemistry, geology, ecology, and other scientific basics that they will need to move successfully through the rest of the book. (And don't forget to look at the extremely useful *Appendix A: Some Basics on Graphs*.)

In chapters 6 through 15 of *Part II: The Human Element in the Natural System*, we bring people into the mix, examining how human presence and activities affect each segment of the environment. This part of the book is not all doom-and-gloom, though; in every chapter there are stories and examples of approaches that have been taken and tools that are being developed to mitigate negative impacts. Of course the main focus is on the contribution of science and environmental scientists to the development of these tools.

In *Part III: The Search for Solutions*, chapters 16 through 23, we turn our attention to the future, examining how we can continue to develop new energy systems, better waste management processes, more liveable urban centres, and new economic and social tools, to create a healthier environment for us and for all who share this planetary system with us. In this part of the book you will find, in chapters 21 and 22, a comprehensive look at the social, political, and economic underpinnings of environmental management—now completely reworked to reflect a Canadian perspective and history. The focus on campus sustainability in Chapter 23—a unique feature of the U.S. 3rd edition—now provides inspiring examples of initiatives at Canadian colleges and universities from across the nation. In case you thought today's students were apathetic about environmental issues, this chapter will set you straight.

A completely new feature of the Canadian Edition is our *Canadian Environmental Perspectives*. Through these profiles we have tried to present the work of a wide variety of Canadians who contribute to our understanding of the environment. Some of those profiled are famous; others are regular people making their everyday contributions to environmental science. Some are controversial. Some work in the private sector, some in the public sector, some in non-profits, and some in academia. Our goal in this feature is to inspire young environmental scientists to realize that there are *many* different ways of approaching the study of the environment. We also wanted to explore more fully the differences between environmentalism and environmental science, to reveal to students that these *are* fundamentally different approaches, but they can and do often inform one another.

Meanwhile, the content and pedagogical features that have made *Environment: The Science Behind the Story* so successful in environmental science courses throughout North America are still in place, now enhanced with a Canadian focus.

Integrated Central Case Studies

Telling compelling stories about real people and real places is the best way to capture students' interest. Narratives with concrete detail also help teach abstract concepts, because they give students a tangible framework with which to incorporate new ideas. Many textbooks these days serve up case studies in isolated boxes, but we have chosen to integrate each chapter's central case study into the main text, weaving information and elaboration throughout the chapter. In this way, the concrete realities of the people and places of the central case study help to illustrate the topics we cover. Students and instructors using the book have consistently applauded this approach, and we hope it can continue to bring about a new level of effectiveness in environmental science education. In the Canadian Edition you will find that 17 of the 23 Central Cases now have a centrally Canadian focus. These cases present the stories that our students *need* to know if we are to move forward in our search for environmental solutions, rather than repeating the mistakes of the past.

The Science behind the Story

Our goal is not simply to present students with facts, but to engage them in the scientific process of testing and discovery. To do this, we feature in each chapter *The Science behind the Story* boxes, which elaborate on particular studies, guiding readers through details of the research. In this way we show not merely *what* scientists discovered, but *how* they discovered it. Instructors and students have confirmed that this feature enhances comprehension of chapter material and deepens understanding of the scientific process itself—a key component of effective citizenship in today's science-driven world. In this edition, the original *Science Behind the Story* features have been updated or replaced with examples of exciting and innovative work by Canadian scientists.

Weighing the Issues

 The multifaceted issues in environmental science often lack black-and-white answers, so students need critical-thinking skills to help navigate the grey areas at the juncture of science, policy, and ethics. We aim to help develop these skills with the *Weighing the Issues* questions dispersed through each chapter. These serve as stopping points for students to absorb and reflect upon what they have read, wrestle with some of the complex dilemmas in environmental science, and engage in spirited classroom discussion.

End-of-chapter features rich in options

The five features that conclude each chapter are targeted at particular student needs. *Reviewing Objectives* summarizes each chapter's main points and relates them to the learning objectives presented at the chapter's opening, enabling students to review concepts and to confirm that they have understood the most crucial ideas. *Testing Your Comprehension* questions provide concise study questions targeted to main topics in each chapter, while *Seeking Solutions* questions encourage broader creative thinking aimed at finding solutions. The *Think It Through* questions place students in a scenario and empower them to make decisions to resolve problems. *Interpreting Graphs and Data* uses figures from recent scientific studies to help students build quantitative and analytical skills in reading graphs and making sense of data. And *Calculating Footprints* enables students to calculate the environmental impacts of their own choices and then see how individual impacts scale up to impacts at the societal level.

An emphasis on solutions

The complaint we most frequently hear from students in environmental science courses is that the deluge of environmental problems can seem overwhelming. In the face of so many problems, students often come to feel that there is no hope or that there is little they can personally do to make a difference. We have aimed to counter this impression by drawing out innovative solutions that have worked, are being implemented, or can be tried in the future. While we do not paint an unrealistically rosy picture of the challenges that lie ahead, we portray dilemmas as opportunities and we try to instil hope and encourage action. Indeed, for every problem that human carelessness has managed to create, human ingenuity can devise one—and likely multiple—solutions.

To recognize the efforts of faculty and students toward encouraging sustainable practices on campus and in the community, Pearson Education Canada will be providing a grant of $1,000. Half of this will go to the student-staffed Campus Ecological Footprint research project at the University of Toronto Mississauga, profiled in *Science Behind the Story: A Campus Ecological Footprint Calculator*, Chapter 23. The other half will go to Lights Out Canada, the initiative of young environmental activist Keleigh Annau, who is profiled in *Canadian Environmental Perspectives*, Chapter 23.

Both the original *Environment: The Science Behind the Stories* and the Canadian edition of the book have grown directly from our professional experiences in teaching, research, and writing. Jay Withgott has synthesized and presented science to a wide readership. His experience in distilling and making accessible the fruits of scientific inquiry have shaped the book's content and the presentation of its material. Scott Brennan has taught environmental science to thousands of undergraduates and has developed an intimate feeling for what works in the classroom. His knowledge and experience have shaped the book's pedagogical approach. Barbara Murck brings to this strong team more than twenty years' experience teaching environmental science and Earth science to undergraduates at the University of Toronto, and 11 books published in related fields. She is an award-winning lecturer, who has worked on finding practical solutions to environmental problems from Toronto to Niger to Vietnam.

We have been guided in our efforts by extensive input from instructors across Canada who have served as reviewers and advisors for the Canadian edition. The participation of so many learned and thoughtful experts has improved this volume in countless ways.

We sincerely hope that our efforts will come close to being worthy of the immense importance of our subject matter. We invite you, students and instructors alike, to let us know how well we have achieved our goals and where you feel we have fallen short. We are committed to continual improvement, and value your feedback. Please feel free to write and send comments or suggestions to the Canadian Edition author, Barbara Murck, at barbara.murck@utoronto.ca.

At this most historic time to study environmental science, we are honoured to serve as your guides in the quest to better understand our world and ourselves.

Jay Withgott and Scott Brennan
Barbara Murck

The Teaching and Learning Package

We have prepared an excellent supplements package to accompany the text. This package includes the traditional supplements that students and professors have come to expect from authors and publishers, as well as some new kinds of supplements that involve electronic media.

For the Student

Companion Website
This site created specifically for the text, contains numerous review exercises (from which students get immediate feedback), exercises to expand one's understanding of environmental science, and resources for further exploration. It provides an excellent platform from which to start using the Internet for the study of environmental science. Please visit the site at www.myenvironmentplace.ca—a 24/7 personal study portal; it allows students to access the online materials they need to succeed in their courses and includes an **eBook** for this text. You will need the access code that has been packaged with your copy of the text to register and log on to the site.

For the Professor

Instructor's Resource CD ROM
This aid provides quick and easy access to a wealth of valuable teaching tools, including and Instructor's Manual, Computerized Test Item File in TestGen, PowerPoint Slides, and an Image Library in downloadable format.

Instructor's Manual
The Instructor's Manual includes lecture outlines, teaching notes that integrate material from the chapter, discussions of the Science Behind the Story features, suggestions for supplementary print and on-line resource material, and solutions to end-of-chapter questions and problems.

Computerized Test Item File in TestGen Test-Generating Software
Pearson TestGen is a special computerized test item file that enables instructors to view and edit existing questions, add questions, generate tests, and print tests in a variety of formats. This test bank contains approximately 1400 questions, and includes multiple choice, short-answer, graphing, and scenario-based items. We identify a suggested answer, an associated learning objective, and a difficulty level of easy, moderate, or difficult for all questions. The Pearson TestGen is compatible with IBM or MacIntosh systems.

PowerPoint Slides
Our colourful electronic slides are available in Microsoft PowerPoint®. The slides highlight, illuminate, and build on key concepts in the text.

Image Library
The Image Library is an impressive resource to help instructors create vibrant lecture presentations. Almost every figure and table from the text is provided in electronic format and is organized by chapter for convenience. These images can be imported easily into Microsoft PowerPoint to create new presentations or to add to existing ones.

Blackboard Premium for Environment: The Science Behind the Stories, Canadian Edition

Technology Specialists
Pearson's Technology Specialists work with faculty and campus course designers to ensure that Pearson technology products, assessment tools, and online course materials are tailored to meet your specific needs. This highly qualified team is dedicated to helping schools take full advantage of a wide range of educational resources, by assisting in the integration of a variety of instructional materials and media formats. Your local Pearson Education sales representative can provide you with more details on this service program.

CourseSmart
CourseSmart is a new way for instructors and students to access textbooks online anytime from anywhere. With thousands of titles across hundreds of courses, CourseSmart helps instructors choose the best textbook for their class and give their students a new option for buying the assigned textbook as a lower cost eTextbook. For more information, visit www.coursesmart.com.

Acknowledgments

A textbook is the product of *many* more minds and hearts than one might guess from the names on the cover. The three of us have been exceedingly fortunate to be supported and guided by a tremendous publishing team and by a small army of experts in environmental science who have generously shared their time and expertise. Although we alone, as authors, bear responsibility for any inaccuracies, the strengths of this book result from the collective labour and dedication of innumerable people.

The authors, particularly of the Canadian edition, are extremely grateful to the team at Pearson Education Canada for their support, advice, and professionalism throughout the development of this first Canadian edition. Special thanks go to Michelle Sartor, Executive Acquisitions Editor, for her enthusiasm about this project. I wish to acknowledge the contribution of Kim Ukrainec, Marketing Manager, whose boundless enthusiasm permeates this project as we move forward to get it into the hands of instructors and students across Canada. Many thanks go to Sherry Zweig, Senior Publishing and Editorial Representative, just for being herself, and for being the one who got me involved with this project and with Pearson Education Canada.

I want to thank Eleanor MacKay, Senior Developmental Editor, for her expert advice, feedback, and encouragement throughout the writing and development of this Canadian edition. And I wish to extend a special thanks to Patricia Jones, Production Editor; Janis Raisen, Production Coordinator; Laurel Sparrow, Copy Editor; and Lisa Brant, Beth McAuley, and Dayle Furlong, Researchers, for their dedication to quality and overseeing this Canadian edition's journey into print. Finally, Marnie Branfireun deserves many thanks for teaching my very large Environmental Science course so that I could devote myself to this project, and for her valuable contributions on the supplements.

In the lists that follow, we acknowledge the instructors and outside experts who have helped us maximize the quality and accuracy of our presentation through their chapter reviews, feature reviews, class tests, or other services. Their input made the Canadian edition a much stronger book. If the thoughtfulness and thoroughness of these reviewers are any indication, we feel confident that the teaching of environmental science is in excellent hands!

Lastly, for the Canadian Edition I would like to thank my ever-patient family, who lived with the innards of a book-in-progress strewn across the dining room table for more than a year, as well as my colleagues, friends, and students at the University of Toronto Mississauga for their faithful support. We dedicate this book to today's students, who will shape tomorrow's world.

Jay Withgott, Scott Brennan, and Barbara Murck

Reviewers

Don Alexander, Malaspina University College
Jason Bailey, Memorial University of Newfoundland
Sarah Boon, University of Lethbridge
Steven Cooke, Carleton University
David Kemp, Lakehead University
Keith P. Lewis, Memorial University of Newfoundland
Linda Lusby, Acadia University

Paul McMillan, Capilano College
Robert McLeman, University of Ottawa
Maren Oelbermann, University of Waterloo
Roxanne Razavi, Queen's University
Lawton Shaw, Athabasca University
Susan Vojoczki, McMaster University
Frank Williams, Langara College

About the Authors

Jay H. Withgott is a science and environmental writer with a background in scientific research and teaching. He holds degrees from Yale University, the University of Arkansas, and the University of Arizona. As a researcher, he has published scientific papers on topics in ecology, evolution, animal behaviour, and conservation biology in journals including *Proceedings of the National Academy of Sciences*, *Proceedings of the Royal Society of London B*, *Evolution*, and *Animal Behavior*. He has taught university-level laboratory courses in ecology, ornithology, vertebrate diversity, anatomy, and general biology.

As a science writer, Jay has authored articles for a variety of journals and magazines including *Science*, *New Scientist*, *BioScience*, *Smithsonian*, *Current Biology*, *Conservation in Practice*, and *Natural History*. He combines his scientific expertise with his past experience as a reporter and editor for daily newspapers to make science accessible and engaging for general audiences.

Jay lives with his wife, biologist Susan Masta, in Portland, Oregon.

Scott Brennan has taught environmental science, ecology, resource policy, and journalism at Western Washington University and at Walla Walla Community College. He has also worked as a journalist, photographer, and consultant.

Scott has cultivated his expertise in environmental science and public policy by serving as Executive Vice President and Chief Operating Officer for Alaskans for Responsible Mining, as Executive Conservation Fellow of the National Parks Conservation Association in Washington, D.C., and as a consultant to the U.S. Department of Defense Environmental Security Office at the Pentagon.

When not at work, Scott is likely to be found exploring the Chugach Mountains and the Bristol Bay drainages in southwest Alaska. He lives with his wife, Angela, and their dogs Raven and Hatcher, in south central Alaska's Chester Creek Watershed.

Barbara Murck has taught environmental and Earth science at the University of Toronto Mississauga for more than twenty years. Her academic background is in geology, with degrees from Princeton University and the University of Toronto. Barb has worked on a wide variety of environmental management projects in the developing world, from Africa to Asia, mainly as an expert on training and curriculum development. She has published numerous books on topics ranging from physical geology to environmental science to sustainability. She is an award-winning lecturer, who appreciates having had the opportunity to influence the lives and learning of thousands of students over the years.

Barb lives in a 100-year-old house in a heritage neighbourhood of Mississauga, Ontario with husband, Jack, children Eliza and Riley, and cat Zephyr.

PART ONE

This is Combers Beach in Pacific Rim National Park Reserve, British Columbia.

FOUNDATIONS OF ENVIRONMENTAL SCIENCE

An Introduction to Environmental Science

1

Earth is like an island.

Upon successfully completing this chapter, you will be able to

- Define the term *environment*
- Describe natural resources and explain their importance to human life
- Characterize the interdisciplinary nature of environmental science
- Understand the scientific method and how science operates
- Diagnose and illustrate some of the pressures on the global environment
- Articulate the concepts of sustainability and sustainable development

This photo of a crescent Earth is the first one taken of the planet as a whole, by automated camera on the unmanned *Apollo 4* spacecraft in November 1967.

CENTRAL CASE:
EARTH FROM SPACE: THE POWER OF AN IMAGE

"The two-word definition of sustainability is 'one planet.'"

—MATHIS WACKERNAGEL, ECOLOGICAL ECONOMIST AND CO-DEVELOPER OF THE ECOLOGICAL FOOTPRINT CONCEPT

"We're not the first to discover this, but we'd like to confirm, from the crew of *Apollo 17*, that the world is round."

—EUGENE CERNAN, *APOLLO 17* COMMANDER

Consider the following: Prior to November 9, 1967, no one had *ever seen* a photograph of the whole planet Earth, because no such thing existed.

Those of us who were alive back in 1967 were not completely clueless. We knew that Earth is a planet, surrounded by space. We knew that Earth is round (although visual confirmation of this fact still made a considerable impact on *Apollo 17* astronauts a few years later). Clearly we were familiar with the surface of this planet. Yet a simple photograph of Earth—floating in space, blue and shining and covered by clouds, vegetation, and a whole lot of water—managed to take everyone by surprise and changed both society and history in the process.

Actually, those very first photographs of the whole Earth taken in 1967 were not the ones that eventually caught the imagination of the general public. The 1967 photographs were taken by automated camera from the unmanned *Apollo 4* spacecraft, the first spacecraft to get far enough away from Earth to photograph the entire planet. Only part of the planet was in sunlight though, so the photographs show only a "crescent" Earth (see photo). Not long after, on December 24, 1968, *Apollo 8* astronauts took the first hand-held photographs showing Earth rising over the horizon of the Moon. The crew did a live radio broadcast that day, during which astronaut James Lovell commented,

"The Earth from here is a grand oasis in the big vastness of space."[1]

It was not until 1972 that the *Apollo 17* mission put astronauts in a position to photograph the entire *illuminated* planet Earth. The result was the famous Blue Marble[2] image, a version of which opens this chapter. The photograph was beautiful, its impact stunning, even unsettling. The original image was oriented with Antarctica at the top of the globe and an "upside-down" Africa in the middle. The unfamiliar perspective caused consternation among those who had never stopped to consider that the convention of orienting maps with north at the top is completely arbitrary.

The Blue Marble photograph is widely credited with kick-starting the modern environmental movement. Just five years elapsed between the first whole-Earth photographs in 1967 and the last ones to be recorded by human hands. (Since 1972, no manned space flight has been far enough away for the planet to be photographed in its entirety by astronauts.) In that five-year period was the summer of love, and war—the Vietnam War, the Six Days War, the Cold War. The Beatles sang on the first live international satellite television production. Canada celebrated the hundredth year of confederation. Neil Armstrong became the first person to walk on the Moon. Civil rights activist Martin Luther King, Jr., died; so did J. Robert Oppenheimer, the "father of the atomic bomb." The first hand-held calculator was sold (for almost $400).

Society changed dramatically during those five years, and it was a period of dawning awareness and public involvement in environmental issues. The first major oil spill happened in 1967, when the *Torrey Canyon* ran aground near England with 120 000 tonnes of crude oil on board. The first hints of trouble began to surface (literally) from hazardous chemicals stored underground at Love Canal, New York. Within a few years the site would be infamous, leading to the first declaration of an environmental state of emergency in the United States and making a grassroots hero of local activist Lois Gibbs. Books on environmental topics began to appear on bestseller lists, including *Limits to Growth*,[3] *The Population Bomb*,[4] *Small Is Beautiful*,[5] and their predecessor, *Silent Spring*.[6] The year 1970 opened with the signing of the first federal environmental legislation, the United States' *Environmental Protection Act*. The first Earth Day was held (1970). Greenpeace was founded (1971). The *United Nations Environment Programme* was established (1972).

British astronomer Sir Frederick Hoyle is reputed to have said, in 1948, "Once a photograph of the Earth, taken from outside, is available—once the sheer isolation of the Earth becomes known—a new idea as powerful as any in history will be let loose." To what extent were these and subsequent milestones in environmental history descended from the first glimpses of our planet from space, with all of its fragility and limitations? We will never know with certainty, but the Blue Marble is considered to be one of the most influential photographs in history—possibly the most widely distributed image of all time—and it remains an iconic symbol of the modern environmental movement.

Our Island, Earth

Viewed from space, our home planet resembles a blue marble suspended against a vast inky-black backdrop. Although few of us will ever witness that sight directly, photographs taken from space convey a sense that Earth is small, isolated, and fragile. From an astronaut's perspective it is apparent that Earth and its natural systems are not unlimited. As our population, our technological powers, and our consumption of resources increase, so does our ability to alter our planet and damage the very systems that keep us alive.

Our environment is more than just our surroundings

A photograph of Earth reveals a great deal, but it does not adequately convey the complexity of our environment. Our **environment** (compare the French word *environner*, "to surround") is more than water, land, and air; it is the sum total of our surroundings. It includes all of Earth's **biotic** components, or living things, as well as the **abiotic** components, or nonliving things, with which we interact. Our environment has abiotic physical constituents—the continents, oceans, clouds, rivers, and icecaps that you can see in the photo of Earth from space. It also has biotic constituents—the animals, plants, forests, soils, and people that occupy the landscape. In a more inclusive sense, it also encompasses the built environment—the structures, urban centres, and living spaces humans have created. In its *most* inclusive sense, our environment includes the complex webs of scientific, ethical, political, economic, and social relationships and institutions that shape our daily lives.

People commonly use the term *environment* in a narrower sense—of a nonhuman or "natural" world apart from human society. This connotation is unfortunate, because it masks the important fact that humans exist within the environment and are a part of the interactions that characterize it. As one of many species, we share with others a fundamental dependence on a healthy, functioning planet. The limitations of language make it all too easy to speak of "people and nature," or "society and the environment," as though they are separate and do not interact. However, the fundamental insight of environmental science is that we are part of the natural world, and our interactions with its other parts matter a great deal.

Why is it important that we give careful consideration to the meaning of the term *environment*? Back in 1970, when the federal government passed Canada's first environmental legislation, the environmental awareness of most North Americans was limited. If they thought about it at all, most people would have equated *environment* with *wilderness*, although this oversimplification was changing as public consciousness of environmental issues grew. Wilderness preservation is still an important concern, but our understanding of the environment, our impacts on it, and its role in our health and daily lives has broadened dramatically.

Today our definition of *environment* must be sufficiently comprehensive to include its legal, social, economic, and scientific aspects. Consequently, the mandate of **Environment Canada** is equally comprehensive:[7] to preserve and enhance the quality of Canada's natural environment, conserve our renewable resources, and protect our water resources. International relations, politics, ethics, business management, economics, social equity, engineering, law enforcement—all of these now play a role in managing and protecting the environment. In Chapter 22 we will look more closely at policies and decision making in relation to environmental problem solving, in the public, private, and nonprofit sectors.

To accomplish all this, our environmental leaders and policy makers need to know what they are talking about. As a community, we must constantly improve and refine our basic scientific understanding of water, air, land and soils, wildlife, weather and climate, and the dynamic interactions among all the components of which ecosystems are composed. That is where *environmental science*—the central focus of this book—comes in.

Environmental science explores interactions between humans and the physical and biological world

Appreciating how we interact with our environment is crucial for a well-informed view of our place in the world and for a mature awareness that we are one species among many on a planet full of life. Understanding our relationship with the environment is vital because we are altering the natural systems we need, in ways we do not yet fully comprehend.

We depend utterly on our environment for air, water, food, shelter, and everything else essential for living. However, our actions modify our environment, whether we intend them to or not. Many of these actions have enriched our lives, bringing us longer life spans, better health, and greater material wealth, mobility, and leisure time; however, many of them have damaged the natural systems that sustain us. Such impacts as air and water pollution, soil erosion, and species extinction compromise the well-being of all living organisms, pose risks to human life, and threaten our ability to build a society that will survive and thrive in the long term. The natural environment was functioning long before the human species appeared, and we would be wise to do our best to maintain its integrity and keep its key elements in place.

Environmental science is the study of how the natural world works, how our environment affects us, and how we affect our environment. We need to understand our interactions with—and our role in—the environment. Such knowledge is the essential first step toward devising solutions to our most pressing environmental problems. Many environmental scientists are taking this step, trying to apply their knowledge to develop solutions to the many environmental challenges we face. Chapters 2 through 5 in Part 1 of this book provide an introduction to the abiotic and biotic components of our environment, and to the basic concepts and principles of science as applied to the study of the environment.

It can be daunting to reflect on the magnitude of environmental dilemmas that confront us today. We will examine these challenges and issues in Part 2, Chapters 6 through 15, starting with a look at the human population itself and how it has grown and changed over time (Chapter 6).

Fortunately, with these problems also come countless opportunities for devising creative solutions. Right now, global conditions are changing more quickly than ever. Right now, through science, we as a civilization are gaining knowledge more rapidly than ever. And right now, the window of opportunity for acting to solve problems is still open. With such bountiful challenges and opportunities, this particular moment in history is an exciting time to be studying environmental science. Part 3 of the book, Chapters 16 through 23, will show you how to apply your knowledge of environmental science to begin an exploration of solutions for our current challenges.

Natural resources are vital to our survival

Islands are finite, and their inhabitants must cope with limitations in the material resources. On our island, Earth, human beings, like all living things, ultimately face environmental constraints. Specifically, there are limits to many of our **natural resources**, the various substances and energy sources we need to survive. We can view the renewability of natural resources as a continuum (**FIGURE 1.1**).

Natural resources that are replenishable over short periods are known as **renewable natural resources**. Some renewable resources, such as sunlight, wind, and wave energy, are perpetually renewed and essentially inexhaustible. Others renew themselves more slowly, and they may become nonrenewable if we use them at a rate that exceeds the rate at which they are renewed or replenished. Populations of animals and plants that we harvest from the wild may be renewable if we do not overharvest them but may vanish if we do.

Renewable resources, like groundwater and soil, can be harvested according to principles similar to those that govern living resources, like fish and trees. However, the rate of regeneration of such resources is limited by the rates of physical processes, such as the infiltration of groundwater to replenish an aquifer, or the physical and chemical weathering of rock to produce soil. Because these rates can be quite slow—it can take up to 10 000 years for soil formation to occur in cold climates like those of northern Canada, for example—it may take a very long time for these resources, once damaged or depleted, to be replenished.

Resource management is strategic decision making and planning aimed at balancing the use of a resource with its protection and preservation. The basic premise of renewable resource management—both living and non-living—is to balance the rate of *withdrawal* from the stock with the rate of *renewal* or *regeneration*. The **stock** is the harvestable portion of the resource. If the stock is being harvested or withdrawn at a faster rate than it can be replenished—faster than trees can be seeded and grow to maturity or faster than fish can be born and grow to a harvestable age or faster than precipitation can infiltrate to replenish the groundwater—then the stock will eventually be depleted. Renewable resources are sometimes called **stock-and-flow resources**, highlighting the importance of this balance in the their management.

In contrast, **nonrenewable natural resources,** like fossil fuels and mineral deposits, are in finite supply and are depletable, because they are formed *much* more slowly than we use them; it can take 100 million years for natural geological processes to form an ore deposit or a petroleum deposit. Once we use them up, they are no longer available because they will not be replenished on a humanly accessible time scale. Simply by withdrawing from the stock we are depleting the resource. These resources lie at the other end of the continuum in **FIGURE 1.1**.

Our civilization depends on numerous minerals: Iron is mined and processed to make steel. Copper is used in pipes, electrical wires, and a variety of other applications. Aluminum is extracted via bauxite ore and used in packaging and other end products. Lead is used in batteries, to shield medical patients from radiation, and in many other ways. Zinc, tungsten, phosphate, uranium,

FIGURE 1.1
Natural resources lie along a continuum from perpetually renewable to nonrenewable. Perpetually renewable or inexhaustible resources, such as sunlight, will always be there for us. Nonrenewable resources, such as oil and coal, exist in limited amounts that cannot be renewed on a humanly accessible time scale and could one day be gone. Other resources, such as timber, groundwater, soils, and food crops, can be renewed if we are careful not to deplete them or damage them.

Renewable natural resources ←——————→ **Nonrenewable natural resources**

- Sunlight
- Wind energy
- Wave energy
- Geothermal energy

- Agricultural crops
- Fresh water
- Forest products
- Soils

- Crude oil
- Natural gas
- Coal
- Copper, aluminum, and other metals

gold, silver—the list goes on and on. Although we rely on these resources, we do not manage their extraction in the way we manage renewable natural resources. Like fossil fuels, minerals are nonrenewable resources that are *mined* rather than *harvested*. Therefore, the mining industry benefits by extracting as much as it can as fast as it can and then, once extraction becomes too inefficient to be profitable, moving on to new sites. From a consumer's perspective, the management of nonrenewable mineral resources demands conservation, reuse, and recycling, all of which we will examine more closely in Chapter 18.

Still other resources are *truly* nonrenewable and nonreplenishable: once an atom has been split to release its nuclear energy, it will never return to its original state; once a species has become extinct, it will never return to life.

We need to manage the resources we take from the natural world carefully and effectively, because many of them are limited or may become so. Resource managers are guided in their decision making by available research in the natural sciences, but their decisions are also often influenced by political, economic, and social factors. A key question in managing resources is whether to focus narrowly on the resource of interest or to look more broadly at the environmental system of which the resource is a part. Taking a broader view can often help avoid damaging the system and can thereby help sustain the availability of the resource in the long term.

Preserving natural resources is an important consideration for the future, but it also speaks to the past and to our shared history as Canadians. Our economy, our identity, and even our national symbols have always been closely linked to the abundant physical resources of our environment. In recent years, however, the consumption of natural resources has increased greatly—in Canada and throughout the world—driven by rising affluence and the growth of the largest global human population in history (Chapter 6).

Human population growth has shaped our resource use

For nearly all of human history, only a few million people populated Earth at any one time. Although past populations cannot be calculated precisely, **FIGURE 1.2** gives some idea of how recently and suddenly our population has grown, surpassing 6 billion people just before the start of the twenty-first century.[8]

Four significant periods of societal change appear to have triggered remarkable increases in population size, concomitant with greatly increased environmental impacts (as discussed in greater detail in Chapter 6). The first happened as many as 2.5 million years ago during the *paleolithic* (or *Old Stone Age*) *period*, when early humans gained control of fire and began to shape and use stones as tools with which to modify their environment.

The second was the transition from a nomadic, hunter-gatherer lifestyle to a settled, agricultural way of life. This change began to occur around 10 000 to 12 000 years ago, and it is formally known as the *neolithic period* or *Agricultural Revolution*.

The third major societal change, known as the *Industrial Revolution*, began in the mid-1700s and entailed a shift from rural life, animal-powered agriculture, and manufacturing by craftspeople, to an urban society powered by fossil fuels (Chapter 15). Life improved in many ways as a result of the Industrial

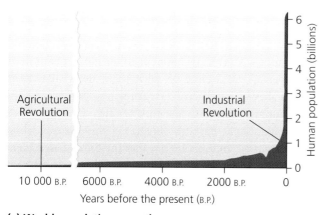

(a) **World population growth**

(b) **Urban society**

FIGURE 1.2
For almost all of human history, the world's population was low and relatively stable. It increased significantly as a result of the Agricultural Revolution and then as a result of the Industrial Revolution **(a)**. Our skyrocketing population has given rise to congested urban areas, such as this city in Java, Indonesia **(b)**.

Revolution, but it also marked the beginning of industrial-scale pollution and many other environmental and social problems that had not previously been experienced. Air quality declined dramatically as a result of the new reliance on coal. Water quality declined, and so did the urban landscape, as a result of the gathering of people into densely populated city centres (Chapter 6). Workplace health and safety, too, underwent a dramatic decline as factories were hastily erected and expanded. In many respects, the modern environmental movement had its roots in the efforts taken by concerned citizens during the Industrial Revolution to ensure a cleaner, safer environment for working and living.

Today we are in the midst of a fourth transition, which some have labelled the modern *Medical–Technological Revolution*. Advances in medicine and sanitation, the explosion of communication technologies, and the shift to modern agricultural practices collectively known as the Green Revolution have allowed more people to live longer, healthier lives. However, as in the Industrial Revolution, we are facing new environmental challenges as a result of the technological advancements. For example, in Chapter 8 we will look at the impacts of biotechnology on food production and the possibility that these new technologies could bring an end to hunger, as well as the potential for environmental and health impacts that are beyond our current understanding.

Each major societal transition introduced technological advancements that made life easier and resources more available, effectively increasing the carrying capacity of the environment for humans and allowing the human population to increase dramatically. The modern Medical–Technological Revolution is still ongoing, and the ultimate impacts on population and the environment are as yet unknown. We will explore the dynamics and implications of human population growth, especially the explosion of population in the past few decades, in Chapter 6. Then, in Chapter 20, we will examine the urban environment and the effort to create sustainable, liveable cities for growing human populations around the world.

Resource consumption exerts social and environmental impacts

Population growth affects resource use and availability, and it is unquestionably at the root of many environmental problems. However, patterns and habits of resource consumption are also to blame. The Industrial Revolution enhanced the material affluence of many of the world's people, raising standards of living by raising consumption. It led to an increase in population, but it also caused pressures on the environment to increase as a result of new technologies (e.g., coal-fired steam engines) and increased levels of consumption. We can expect that the same will be true of the Medical–Technological Revolution.

One approach to this relationship represents our total impact (I) on the environment as the product of population (P), affluence (A), and technology (T), as follows:

$$I = P \times A \times T$$

This "IPAT" model shows that impact is a function not only of population but also of affluence (which stands in for "level of consumption") and technology. An increase in the *number of people* (P) has impacts on the global environment, but we must also concern ourselves with the impacts of *increased consumption* of natural resources and manufactured goods by the world's people (A), and the impacts of *new* technologies (T) on the environment, sometimes in ways that we can just barely imagine. During the Industrial Revolution, the poor air quality caused by factories that belched dirty smoke was an entirely new phenomenon. Similarly, we are only just beginning to understand the impacts of genetic engineering on natural populations (Chapter 8). We will examine the IPAT model in greater detail in Chapter 6.

Carrying capacity and the "tragedy of the commons" When we think about Earth's limited resources and the capacity of the planet to support a growing human population, it is useful to consider the idea of carrying capacity. **Carrying capacity** refers to the biological productivity of a system; it is a measure of the ability of a system to support life. Environmental scientists quantify carrying capacity in terms of the number of individuals of a particular species that can be sustained by the biological productivity of a given area of land (in hectares or square kilometres). When the carrying capacity of the land (or water) system is exceeded—that is, when there are simply too many individuals for the system to support—one of two things will typically happen: either the population of that species will decline or collapse, or the system itself will be altered, damaged, or depleted.

Ecologist Garrett Hardin of the University of California, Santa Barbara, illustrated this process while disputing the economic theory that the unregulated exercise of individual self-interest serves the public good. According to Hardin's best-known essay, "The Tragedy of the Commons," published in the journal *Science* in 1968, resources that are open to unregulated exploitation inevitably become overused and, as a result, are damaged or depleted.

Hardin based his argument on the scenario of a public pasture, or "commons" that is open to unregulated grazing.

He argued that each person who puts animals to graze on the commons will be motivated by selfish interests to increase the number of his or her animals in the pasture. Because no single person owns the pasture, no one has the incentive to limit the number of grazing animals or to expend money or effort to care for the pasture. This is known as the **tragedy of the commons**: each individual withdraws whatever benefits are available from the common property as quickly as possible, until the resource becomes overused and depleted. Ultimately, the carrying capacity of the pasture will be exceeded, and its food production capacity will collapse.

In some situations, private ownership may address this problem. In China, for example, private land ownership—illegal for many decades, under Communism—has recently become possible in some rural areas. These limited experiments with private ownership have shown that landowners tend to be better environmental stewards than are short-term tenants, primarily because they are willing to make long-term investments in land management. In other cases, people who share a common resource may voluntarily organize and cooperate in enforcing its responsible use. In other cases the dilemma may require government regulation of the use of resources held in common by the public, from forests to air to freshwater.

---

weighing the issues

THE TRAGEDY OF THE COMMONS

Imagine you make your living by fishing. You are free to boat anywhere and set out as many traps as you like, and fish have been abundant. Limits and regulations are rarely enforced. However, the fishing grounds are getting crowded. Catches begin to decline, leaving you and all the others with catches too meagre to support your families. Some call for dividing the waters and selling access to individuals plot by plot. Others urge the fishers to team up, set quotas among themselves, and prevent newcomers from entering the market. Still others implore the government to get involved and pass laws regulating the size of the catch.

What do you think is the best way to combat this tragedy of the commons and save the fishery?

Calculating our ecological footprint As global affluence has increased, human society has consumed more and more of the planet's limited resources. We can quantify resource consumption by using the concept of the "ecological footprint," developed in the 1990s by environmental scientists Mathis Wackernagel and William Rees, working together at the University of British Columbia. The **ecological footprint** is a tool that can be used to express the environmental impact of an individual or a population. It is calculated in terms of the amount of biologically productive land and water required to provide the raw materials that person or population consumes and to absorb or recycle the waste the person or population produces. The footprint calculation gives the surface area "used" by a given person or population, after all the direct and indirect impacts are totalled. The ecological footprint is essentially the *inverse* of carrying capacity—it is a measure of the land (and water) required to sustain an individual, rather than the number of individuals that can be sustained by an area of land (or water).

Researchers calculate that our species is now using 39% more resources than are available on a sustainable basis from all the land on the planet. That is, we are not only exceeding the carrying capacity of the planet for the human species, but we are also depleting renewable resources 39% faster than they are being replenished. This is like drawing the principal out of a bank account, rather than living off the interest. Furthermore, people from wealthy nations have much larger ecological footprints than do people from poorer nations. The ecological footprint of an average Canadian is approximately 7.6 hectares—roughly two to four city blocks.[9] Yet, if we could divide up all the productive, habitable land of this planet among the 6.7 billion people who are now alive, each person would receive less than one city block. If all of the world's people consumed resources at the rate of North Americans, we would need the equivalent of more than two additional planet Earths to meet our resource needs.

Footprint calculations vary dramatically—you will probably even find some variations among the footprint calculation exercises in this book. This is because the calculation depends heavily on how certain components are defined. For example, different approaches to the ecological footprint calculation use different methodologies to account for the surface area of the oceans (which clearly does not have the same significance as land area does for humans, in terms of "living space" or even "biologically productive space"). Sometimes these differences can become political; for example, various energy sources—fossil fuels, nuclear energy, hydroelectric power—have very different environmental impacts, as you will learn in Chapters 15 and 16. How should the impacts of different energy sources be accounted for in footprint calculations? When is one impact "more negative" than another? The Global

Footprint Network[10] is an international nongovernmental organization that is working toward standardizing ecological footprint calculations worldwide. This should make the calculations more robust and their application to questions of environmental impact and sustainability more effective.

Environmental science can help us avoid mistakes made in the past

There is historical evidence that civilizations can crumble when pressures from population and consumption overwhelm resource availability. Perhaps the most intriguing of these is Easter Island (see "The Science Behind the Story: The Lesson of Rapa Nui," • p. 11), but many great civilizations have fallen after depleting resources or damaging their environment. The Greek and Roman empires show evidence of this, as do the civilizations of the Maya, the Anasazi, and other New World peoples. Plato wrote of the deforestation and environmental degradation accompanying ancient Greek cities, and today further evidence is accumulating from research by archeologists, historians, and paleoecologists, who study past societies and landscapes. The arid deserts of today's Near Eastern and Middle Eastern countries were far more lushly vegetated when the great ancient civilizations thrived there.

Researchers have now learned enough about ancient civilizations and their demise that scientist and author Jared Diamond—in his 2005 book, *Collapse*—could hypothesize why civilizations succeed and persist, or fail and collapse. Diamond identified five critical factors that determine the survival of civilizations: climate change, hostile neighbours, trade partners, environmental problems, and, finally, the society's response to environmental problems. It is interesting to note that only one of these factors—the response to environmental problems—is wholly controllable, and it is this factor that has been the crucial determinant of survival. Success and persistence, it turns out, depend largely on how societies interact with their environments.

Today we are confronted with news and predictions of environmental catastrophes on a regular basis, but it can be difficult to assess the reliability of such reports. It is even harder to evaluate the causes and effects of environmental change. Perhaps most difficult is to devise solutions to environmental problems. Studying environmental science will outfit you with the tools to evaluate information on environmental change and think critically and creatively about possible actions to take in response. Let us examine this broad field we call environmental science, and then explore the process and methods of science in general.

The Nature of Environmental Science

Environmental scientists aim to comprehend how Earth's natural systems function, how humans are influenced by those systems, and how we are influencing those systems. In addition, many environmental scientists are motivated by a desire to develop solutions to environmental quandaries. The solutions themselves (such as new technologies, policy decisions, or resource management strategies) are applications of environmental science. However, the study of such applications and their consequences is, in turn, also part of environmental science.

Environmental science is an interdisciplinary pursuit

Studying and addressing environmental problems is a complex endeavour that requires expertise from many disciplines, including ecology, Earth science, chemistry, biology, economics, political science, demography, ethics, and many others. Environmental science is thus an **interdisciplinary field**—one that employs concepts and techniques from numerous disciplines and brings research results from these disciplines together into a broad synthesis (**FIGURE 1.3**). Traditional disciplines are valuable because their scholars delve deeply into topics, uncovering new knowledge and developing expertise in particular areas. Interdisciplinary fields are valuable because their practitioners take specialized knowledge from different disciplines, consolidate it, synthesize it, and apply it in a broad context to serve the multifaceted interests of society.

Environmental science is especially broad because it encompasses not only the **natural sciences** (disciplines that study the natural world) but also the **social sciences** (disciplines that study human interactions and institutions). The natural sciences provide us with the means to gain accurate information about our environment and to interpret it reasonably. Addressing environmental problems, however, also involves weighing values and understanding human behaviour, and this requires the

THE SCIENCE BEHIND THE STORY

The Lesson of Rapa Nui

These immense moai (statues) are on Easter Island.

Rapa Nui (Easter Island) is one of the most remote islands on the globe. When European explorers reached the island in 1722, they found a barren landscape populated by fewer than 2000 people, living a marginal existence in caves. The desolate island featured gigantic statues of carved stone, evidence that a sophisticated civilization had once lived there. How could people without wheels or ropes, on an island without trees, have moved statues 10 m high, weighing 80 metric tons? The answer lies in the fact that the island did not always lack trees, and its people were not always without rope.

Scientists have determined that the island was once lushly forested, supporting a prosperous society of 6000 to 30 000 people. This once-flourishing civilization exceeded the carrying capacity of the island by overusing resources and cutting down trees, destroying itself in a downward spiral of starvation and conflict. Today, Rapa Nui stands as a demonstration of what can happen when a population consumes too much of the limited resources that support it.

To solve the mystery of the island's past, scientists have used various methods. British scientist John Flenley excavated sediments from the bottoms of the island's volcanic crater lakes, examining ancient grains of pollen to reconstruct changes in vegetation over time. Flenley and other researchers found that when Polynesian people first arrived (between 300 c.e. and 900 c.e.), the island was covered with a species of palm tree related to the tall, thick-trunked Chilean wine palm.

The palms would have provided fuelwood, building material for houses and canoes, fruit, and fibre—and, presumably, logs to move the stone statues. Scientists have tested hypotheses about how the islanders moved their monoliths, recreating the feat by using great quantities of rope, with tree trunks as rollers or sleds. The only likely source of rope on the island is the fibrous inner bark of the *hauhau* tree, a species that today is near extinction.

At least 21 other species of plants (including trees) that were once common on the island are now completely gone. Around 750 c.e., tree populations began to decline, and ferns and grasses became more common. By 950 c.e., the trees were largely gone. Around 1400 c.e., pollen levels plummeted, indicating a dearth of vegetation. The same sequence occurred two centuries later at two other sites, more remote from village areas. Evidence now supports the hypothesis that people gradually denuded the island.

With the trees gone, soil would have eroded away—confirmed by sediment that accumulated in Rapa Nui's lakes. Faster runoff of rainwater would have meant less freshwater available for drinking. Erosion would have degraded agricultural lands, lowering crop yields. Reduced agricultural production would have led to starvation and population decline.

Analyses by ornithologist David Steadman show that at least six species of land birds and 25 species of seabirds nested on Rapa Nui and were eaten by islanders. Today, no native land birds and only one type of seabird are left. Early islanders also feasted on porpoises, fish, sharks, turtles, octopi, and shellfish. Analyses of islanders' diets in the later years indicate that little seafood was consumed. With the trees gone, islanders could not build the great double-canoes their ancestors used for fishing. Europeans who visited in the eighteenth century observed only a few old small canoes and flimsy rafts made of reeds.

As resources declined, the islanders' main domesticated food animal, the chicken, became more valuable. Archeologists found that later islanders kept chickens in stone fortresses designed to prevent theft. The once prosperous and peaceful civilization fell into clan warfare, revealed by unearthed weapons made of hard volcanic rock and skeletons with head wounds.

The haunting statues of Rapa Nui (Easter Island) were erected by a sophisticated civilization that collapsed after depleting its resource base and devastating its island environment.

Canadian economists Scott Taylor and James Brander took a different approach to investigating what happened at Rapa Nui.[11] They developed a computer model of the interplay between renewable resources and population. The model is based on standard ecological *predator–prey models* (Chapter 4), with people in the role of predator, and resources as their prey. This scenario generates "feast-and-famine" cycles of rising and falling population and resource stocks. The researchers speculate that such cycles may account for the decline and eventual collapse of other civilizations as a result of rapid population growth and consequent resource degradation.

Is the story of Rapa Nui as unique and isolated as the island itself, or does it hold lessons for our world today? Earth may be vastly larger and richer in resources than was Rapa Nui, but the human population is also much larger. The islanders must have seen that they were depleting their resources, but they could not stop. Perhaps we can learn from them, and act wisely to conserve the resources on our island, Earth.

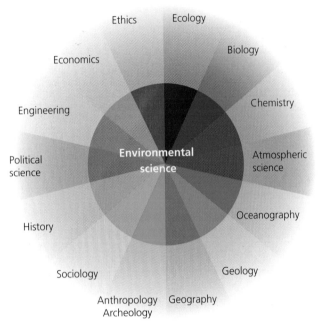

FIGURE 1.3
Environmental science is a highly interdisciplinary pursuit, involving input from many different established fields of study across the natural sciences and social sciences.

acidic solution and how it interacts with rocks and soils. Someone skilled at management would also be helpful, to act as a liaison between the scientists and the mine management team. Canadian mining companies routinely make use of teams like this in their efforts to control acid drainage.

IF YOU WERE ASKED... 1–2

If you were an environmental scientist, and you were asked to provide an assessment of a complex environmental situation, what kinds of experts would you call upon? Think about the various specialists and team members who might be needed to contribute their expertise to the following situations: the construction of a new hydroelectric dam; the proposed draining of a wetland to build a new subdivision; a proposal to permit bear hunting in a national park; or the management of a large oil spill just offshore from a pristine beach.

social sciences. Most environmental science programs focus predominantly on the natural sciences as they pertain to environmental issues. In contrast, programs heavily incorporating the social sciences often prefer the use of the term **environmental studies** to describe their academic umbrella. Whichever approach we take, these fields reflect many diverse perspectives and sources of knowledge.

Just as an interdisciplinary approach to studying issues can help us better understand them, an integrated approach to addressing problems can produce effective and lasting solutions. For example, consider how the Canadian mining industry is approaching the problem of *acid drainage*, which can occur wherever sulphur is present at a mine site. Sulphur is a very common constituent of coal and metal ores, both of which are important to the Canadian economy. If sulphur-bearing waste rock at a mine site interacts with rain or surface water, sulphuric acid is formed; if it is not contained, the acid can enter local streams, where it is devastating to affected ecosystems. To solve a problem involving acid drainage, a mining company would need to consult a biologist or an ecologist regarding the impacts of the acid on local plants and animals. A hydrologist would be helpful, to understand the flow of water at the site. A mining engineer could help decide how best to contain and isolate the waste rock piles. The company would want to consult with a chemist about the nature and behaviour of the

People differ in their perception of environmental problems

Environmental science arose in the latter half of the twentieth century, as people sought to better understand environmental problems and their origins. An *environmental problem*, stated simply, is any undesirable change in the environment. However, the perception of what constitutes an undesirable change may vary from one person or group of people to another, or from one context or situation to another. A person's age, gender, class, race, nationality, employment, and educational background can all affect whether he or she considers a given environmental change to be a "problem."

For instance, people today are more likely to view the spraying of the pesticide DDT as a problem than people did in the 1950s, because today more is known about the health risks from pesticides (**FIGURE 1.4**). However, a person living today in a malaria-infested village in Africa or India may still welcome the use of DDT if it kills mosquitoes that transmit malaria, because malaria is viewed as a more immediate health threat. Thus, an African and a North American who have each knowledgeably assessed the pros and cons may, because of differences in their circumstances, differ in their judgment of the severity of DDT as an environmental problem.

FIGURE 1.4
How a person or society defines an environmental problem can vary with time and circumstance. In 1945, health hazards from the pesticide DDT were not yet known, so children were doused with the chemical to treat head lice. Today, knowing of its toxicity to people and wildlife, developed nations have banned DDT. However, in some countries where malaria is a threat, DDT is still used as an effective means of eradicating mosquitoes, which transmit the disease.

People also vary in their awareness of problems. For example, in many cultures women are responsible for collecting water and fuelwood. As a result, they are often the first to perceive environmental degradation affecting these resources. In most societies, information about environmental health risks tends to reach wealthy people more readily than poor people. Thus, who you are, where you live, what you do, your income, your gender, and your socioeconomic status can have a huge effect on how you perceive your environment, how you perceive and react to change, and what impact those changes may have on how you live your life. In Chapter 21, we will examine the diversity of human values and philosophies and consider their effects on how we define environmental problems.

Environmental science is not the same as environmentalism

Although many environmental scientists are interested in solving problems, it is incorrect to confuse environmental science with environmentalism or environmental activism. They are *not* the same. Environmental science

FIGURE 1.5
Environmental scientists play roles very different from those of the environmental activists shown here. Some scientists do become activists to promote what they feel are workable solutions to environmental problems. However, those who do generally try hard to keep their advocacy separate from their pursuit of objective scientific work. This photograph shows Greenpeace activists protesting on Parliament Hill in Ottawa. Greenpeace is an international organization of environmental activists that was founded in Vancouver in 1971.

is the pursuit of knowledge about the workings of the environment and our interactions with it. **Environmentalism** is a social movement dedicated to protecting the natural world—and, by extension, humans—from undesirable changes brought about by human choices (**FIGURE 1.5**). Although environmental scientists study many of the same issues environmentalists care about, as scientists they attempt to maintain an objective approach in their work. Remaining as free as possible from personal or ideological bias—and open to whatever conclusions the data demand—is a hallmark of the effective scientist.

In each chapter of this book you will find *Canadian Environmental Perspectives*, offering brief profiles of Canadian environmental scientists as well as individuals in non-scientific professions who contribute to the understanding, protection, management, and sustainable use of the natural environment. These people play a wide range of roles, from policy maker to activist, artist, journalist, hunter, or animal rescuer. Many of them are scientists *and* writers, or scientists *and* filmmakers or gardeners or politicians or musicians—and, yes, many of them are also environmentalists. All these people have made a difference, one way or another. All of them rely on and contribute to our knowledge, as well as to our intuitive appreciation of the natural environment. Environmental science is *distinct* from philosophy, law, commerce, religion, politics, art, and activism, but is it necessarily *exclusive* of these human undertakings? You will have to judge this for yourself, but you can count on this book to help you make a more informed judgment of what you read, hear, and experience in your encounters with the natural environment.

The *Canadian Environmental Perspective* presented in this chapter highlights David Suzuki—scientist, activist, and environmentalist. Admired by many, but controversial for his activist stance on environmental issues, Suzuki exemplifies what a multifaceted undertaking it can be to work on behalf of the environment in today's world.

The Nature of Science

Science (from the Latin *scire*, "to know") is a systematic process for learning about the world and testing our understanding of it. The term *science* is also commonly used to refer to the accumulated body of knowledge that arises from this dynamic process of observation, testing, and discovery.

Knowledge gained from science can be applied to societal problems. Among the most important applications of science are its use in developing new technologies, and its use in informing policy and management decisions (**FIGURE 1.6**). These pragmatic applications in themselves are not science, but they must be informed by science in order to be effective. Many scientists are motivated simply by a desire to know how the world works, and others are motivated by the potential for developing useful applications and solutions to problems.

Environmental science is a dynamic yet systematic means of studying the world, and it is also the body of knowledge accumulated from this process. Like science in general, environmental science informs its practical applications and often is motivated by them.

Why does science matter? The late American astronomer Carl Sagan wrote the following in his 1995 treatise *The Demon-Haunted World: Science as a Candle in the Dark:*

> We've arranged a global civilization in which the most crucial elements—transportation, communications, and all other industries; agriculture, medicine, education, entertainment, protecting the environment; and even the key democratic institution of voting—profoundly depend on science and technology.

Sagan and many other thinkers before and since have argued that science is essential if we hope to develop solutions to the problems—environmental and otherwise—that we face today. We might go a step further and suggest that the *democratization* of science—making the science of our world accessible and understandable to as many people as possible—is also essential if we are to make informed decisions about the management of this planet.

Scientists test ideas by critically examining evidence

How can we tell whether warnings of impending environmental catastrophes—or any other claims, for that matter—are based on scientific thinking? Scientists examine ideas about how the world works by designing tests to determine whether these ideas are supported by evidence. If a particular statement or explanation is testable and resists repeated attempts to disprove it, scientists are likely to accept it as a useful and true explanation. Scientific inquiry thus consists of an incremental approach to the truth.

FIGURE 1.6
Scientific knowledge can be applied in policy and management decisions and in technology. Prescribed burning is a management practice to restore healthy forests and is informed by scientific research into forest ecology.

The scientific method is a key element of science

Scientists generally follow a process called the **scientific method**. A technique for testing ideas with observations, it involves several assumptions and a series of interrelated steps. There is nothing mysterious about the scientific method; it is merely a formalized version of the procedure any of us might naturally take, using common sense, to answer a question.

The scientific method is a theme with variations, however, and scientists pursue their work in many different ways. Because science is an active, creative, imaginative process, an innovative scientist may find good reason to stray from the traditional scientific method when a particular situation demands it. Moreover, scientists from different fields approach their work differently because they deal with dissimilar types of information. A natural scientist, such as a chemist, will conduct research quite differently from a social scientist, such as a sociologist. Because environmental science includes both natural and social sciences, in our discussion here we use the term *science* in its broad sense, to include both. Despite their many differences, scientists of all persuasions broadly agree on fundamental elements of the process of scientific inquiry.

The scientific method relies on the following assumptions:

- The universe functions in accordance with fixed natural laws that do not change from time to time or from place to place.
- All events arise from some cause and, in turn, lead to other events.
- We can use our senses and reasoning abilities to detect and describe natural laws that underlie the cause-and-effect relationships we observe in nature.

As practised by individual researchers or research teams, the scientific method (**FIGURE 1.7**) typically consists of the steps outlined below.

Make observations Advances in science typically begin with the observation of a phenomenon that the scientist wants to explain. Observations set the scientific method in motion and function throughout the process.

Ask questions Scientists are naturally curious about the world and love to ask questions. Why are certain plants or animals less common today than they once were? Why are storms becoming more severe, or flooding more frequent? What causes excessive growth of algae in local ponds? Do the impacts of pesticides on fish or frogs

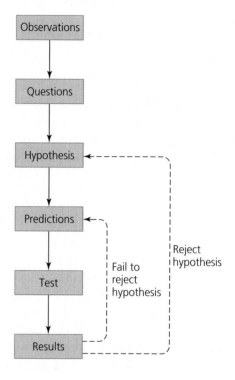

FIGURE 1.7
The scientific method is the observation-based hypothesis-testing approach that scientists use to learn how the world works. This diagram is a simplified generalization that, although useful for instructive purposes, cannot convey the true dynamic and creative nature of science. Moreover, researchers from different disciplines may pursue their work in ways that legitimately vary from this model.

mean that people could be affected in the same ways? All of these are questions environmental scientists have asked and attempted to answer.

Develop a hypothesis Scientists attempt to answer their questions by devising explanations that can be tested. A **hypothesis** is an educated guess that explains a phenomenon or answers a scientific question. For example, a scientist investigating the question of why algae are growing excessively in local ponds might observe chemical fertilizers being applied on farm fields nearby. The scientist might then state a hypothesis as follows: "Agricultural fertilizers running into ponds cause the algae in the ponds to increase." Sometimes this takes the form of a *null hypothesis*, a statement that the scientist is expecting no relationship between variables, such as between fertilizer and algal growth in a pond.

Make predictions The scientist next uses the hypothesis to generate **predictions**, which are specific statements that can be directly and unequivocally tested. In our algae example, a prediction might be: "If agricultural fertilizers are added to a pond, the quantity of algae in the pond will increase." A null hypothesis can also lead to predictions; for example, the scientist might predict that adding agricultural fertilizer to a pond will cause no change in the amount of algae growing in the pond.

Test the predictions Predictions are tested one at a time by gathering evidence that could potentially refute the prediction and thus disprove the hypothesis. The strongest form of evidence comes from experimentation. An **experiment** is an activity designed to test the validity of a hypothesis. It involves manipulating **variables**, conditions that can change. For example, a scientist could test the hypothesis linking algal growth to fertilizer by selecting two identical ponds and adding fertilizer to one while leaving the other in its natural state. In this example, fertilizer input is an **independent variable**, a variable the scientist manipulates, whereas the quantity of algae that results is the **dependent variable**, one that depends on the fertilizer input. If the two ponds are identical except for a single independent variable (fertilizer input), then any differences that arise between the ponds can be attributed to that variable. Such an experiment is known as a **controlled experiment** because the scientist controls for the effects of all variables except the one being tested—the dependent variable. In our example, the pond left unfertilized serves as a **control**, an unmanipulated point of comparison for the manipulated or treated pond. Whenever possible, it is best to replicate an experiment, that is, to stage multiple tests of the same comparison of control and treatment. Our scientist could perform a replicated experiment on, say, 10 pairs of ponds, adding fertilizer to one of the ponds in each pair.

Experiments can establish causal relationships, showing that changes in an independent variable cause changes in a dependent variable. However, experiments are not the only means of testing a hypothesis. Sometimes a hypothesis can be convincingly addressed through **correlation**, that is, searching for relationships and patterns among variables.

Suppose our scientist surveys 50 ponds, 20 of which are fed by fertilizer runoff from nearby farm fields and 30 of which are not. Let us also say he or she finds seven times as much algal growth in the fertilized ponds as in the unfertilized ponds. The scientist would conclude that algal growth is correlated with fertilizer input; that is, that one tends to increase along with the other.

Although this type of evidence is weaker than the causal demonstration that controlled experiments can provide, sometimes it is the best approach, or the only feasible one. For example, in studying the effects of global climate change (Chapter 14), we could hardly run an experiment that involved adding carbon dioxide to 10 treatment planets and comparing the result to 10 control planets.

Analyze and interpret results Scientists record **data**, or information, from their studies. They particularly value *quantitative* data, information expressed by using numbers, because numbers provide precision and are easy to compare. The scientist running the fertilization experiment, for instance, might quantify the area of water surface covered by algae in each pond, or measure the dry weight of algae in a certain volume of water taken from each pond. Even with the precision that numbers provide, however, a scientist's results may not be clear-cut. Experimental data may differ from control data only slightly, or different replicates may yield different results. The scientist must therefore analyze the data by using statistical tests. With these mathematical methods, scientists can determine objectively and precisely the strength and reliability of patterns they find. If the results are unreliable or cannot be replicated, it may be necessary to attempt a different kind of test.

Some research, especially in the social sciences, involves information that is *qualitative*, or not expressible in terms of numbers. Research involving historical texts, personal interviews, surveys, detailed examination of case studies, or descriptive observations of behaviour can include qualitative data on which statistical analyses may not be possible. Such studies are still scientific in the broad sense, because their data can be interpreted systematically by using other accepted methods of analysis.

If experiments disprove a hypothesis, the scientist will reject the hypothesis and may develop a new one to replace it. If experiments fail to disprove the hypothesis, this outcome lends support to the hypothesis but does not *prove* it is correct. The scientist may choose to generate new predictions to test the hypothesis in a different way and further assess its likelihood of being true. Thus, the scientific method loops back on itself, often giving rise to repeated rounds of hypothesis revision, prediction, and testing (see **FIGURE 1.7**).

If repeated tests fail to disprove a particular hypothesis, and evidence in its favour is accumulating, the researcher may eventually conclude that the idea is well supported. Ideally, a scientist would want to test all possible explanations for the question of interest. For instance, our scientist might propose an additional hypothesis that algae increase in fertilized ponds because numbers of fish or invertebrate animals that eat algae decrease. It is possible, of course, that both hypotheses could be correct and that each may explain some portion of the initial observation that local ponds were experiencing algal blooms.

There are different ways to test hypotheses

An experiment in which the researcher actively chooses and manipulates the independent variable is known as a **manipulative experiment** (FIGURE 1.8A). A manipulative experiment provides the strongest type of evidence a scientist can obtain. In practice, however, some modes of scientific inquiry are more amenable to manipulative experimentation than others. Physics and chemistry tend to involve manipulative experiments, but many other fields deal with entities less easily manipulated than are physical forces and chemical reagents. This is true of *historical sciences,* such as cosmology, which deals with the history of the universe, and paleontology, which explores the history of past life. It is difficult to experimentally manipulate a star thousands of light years away, or the tooth from a mastodon that lived 15 000 years ago. Moreover, many of the most interesting questions in these fields centre on the causes and consequences of particular historical events, rather than the behaviour of general constants.

Disciplines that do not quite fit the so-called physics model of science sometimes rely on **natural experiments** rather than manipulative ones (FIGURE 1.8B). For instance, an evolutionary biologist might want to test whether animal species isolated on oceanic islands tend to evolve large body sizes over time. The biologist cannot run a manipulative experiment by placing animals on islands and continents and waiting long enough for evolution to do its work. However, this is exactly what nature has already done. The biologist might test the idea by comparing pairs of closely related species, in which one of each pair lives on an island and the other on a continental mainland, or one is a modern species and the other an ancient, fossilized relative. The experiment has in essence been conducted naturally, and it is up to the scientist to interpret the results.

In ecology, both manipulative and natural experimentation are used. The science of **ecology** deals with the distribution and abundance of organisms (living things), the interactions among them, and the interactions between organisms and their abiotic environments. When possible, ecologists try to run manipulative experiments. An ecologist wanting to measure the importance of a certain insect in pollinating the flowers of a given crop plant might, for example, fit some flowers with a device to keep the insects out while leaving other flowers accessible, and later measure the fruit output of each group. Other questions that involve large spatial scales or long time scales may instead require natural experiments.

The social sciences generally involve less experimentation than the natural sciences, depending more on careful observation and statistical interpretation of patterns in data. For example, a sociologist studying how people from different cultures conceive of the notion of wilderness might conduct a survey and analyze responses to the questions, looking for similarities and differences among respondents. Such analyses may be either quantitative or qualitative, depending on the nature of the data and the researchers' particular questions and approaches.

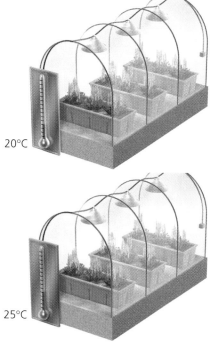

(a) Manipulative experiment

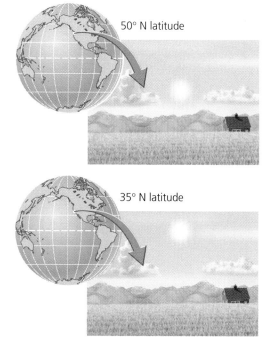

(b) Natural experiment, or correlational study

FIGURE 1.8
A researcher wanting to test how temperature affects the growth of a crop might run a manipulative experiment in which the crop is grown in two identical greenhouses, one kept at 20°C and the other kept at 25°C **(a)**. Alternatively, the researcher might run a natural experiment in which he or she compares the growth of the crop in two fields at different latitudes: a cool northerly location and a warm southerly one **(b)**. Because it would be difficult to hold all variables besides temperature constant, the researcher might want to collect data on a number of northern and southern fields and correlate temperature and crop growth using statistical methods.

Descriptive observational studies and natural experiments can show correlation between variables, but they cannot demonstrate that one variable *causes* change in another, as manipulative experiments can. Not all variables are controlled for in a natural experiment, so a single result could give rise to several interpretations. However, correlative studies, when done well, can make for very convincing science, and they preserve the real-world complexity that manipulative experiments often sacrifice. Moreover, sometimes correlation is all we have. Because large-scale manipulations are difficult, some of the most important questions in environmental science tend to be addressed with correlative data.

The large scale and complexity of many questions in environmental science also mean that few studies, manipulative or correlative, come up with neat and clean results. As such, scientists are not always able to give policy makers and society definitive answers to questions. Even when science is able to provide answers, deciding upon the optimal social response to a problem can still be very difficult.

The scientific process does not stop with the scientific method

Individual researchers or teams of researchers follow the scientific method as they investigate questions that interest them. However, scientific work takes place within the context of a community of peers, and to have any impact, a researcher's work must be published and made accessible to this community. Thus, the scientific method is embedded within a larger process that takes place at the level of the scientific community as a whole (**FIGURE 1.9**).

Peer review When a researcher's work is done and the results have been analyzed, he or she writes up the findings and submits them to a journal for publication. Several other scientists specializing in the topic of the paper examine the manuscript, provide comments and criticism (generally anonymously), and judge whether the work merits publication. This procedure, known as **peer review**, is an essential part of the scientific process. Peer review is a valuable guard against faulty science contaminating the literature on which all scientists rely. However, because scientists are human and may have their own personal biases and agendas, politics can sometimes creep into the review process. Fortunately, just as the vast majority of individual scientists strive to remain accurate and objective in conducting their research, the scientific community does its best to ensure fair review of all work.

Conference presentations Scientists frequently present their work at professional conferences, where they

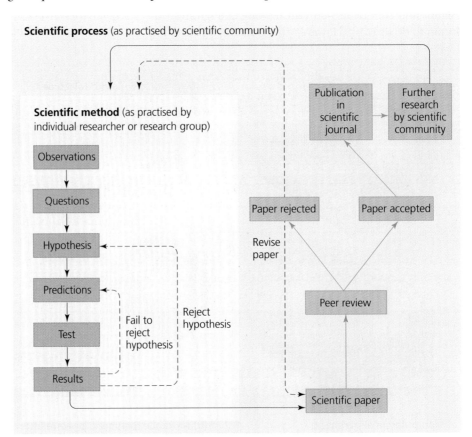

FIGURE 1.9
The scientific method (inner box) followed by individual researchers or research teams exists within the context of the overall process of science at the level of the scientific community (outer box). This process includes peer review and publication of research, acquisition of funding, and the development of theory through the cumulative work of many researchers.

interact with colleagues and receive comments informally on their research. Feedback from colleagues can help improve the quality of a scientist's work before it is submitted for publication.

Grants and funding Research scientists spend large portions of their time writing grant applications requesting money to fund their research from private foundations or government agencies, such as the Natural Sciences and Engineering Research Council or the Social Sciences and Humanities Research Council. Grant applications undergo peer review just as scientific papers do, and competition for funding is often intense. Scientists' reliance on funding sources can also lead to potential conflicts of interest. A scientist who obtains data showing his or her funding source in an unfavourable light may be reluctant to publish the results for fear of losing funding—or, worse yet, may be tempted to doctor the results. This situation can arise, for instance, when an industry funds research to test its products for safety or environmental impact. Most scientists do not succumb to these temptations, but some funding sources have been known to pressure their scientists for certain results. This is why as a student or an informed citizen, when critically assessing a scientific study, you should always try to find out where the researchers obtained the funding.

Repeatability Sound science is based on doubt rather than certainty and on repeatability rather than one-time occurrences. Even when a hypothesis appears to explain observed phenomena, scientists are inherently wary of accepting it. The careful scientist will test a hypothesis repeatedly in various ways before submitting the findings for publication. Following publication, other scientists usually will attempt to reproduce the results in their own experiments and analyses.

weighing the issues: FOLLOW THE MONEY 1–3

Let us say that you are a research scientist, and you are interested in studying the impacts of chemicals released by pulp-and-paper mills on nearby freshwater lakes. Obtaining research funding has been difficult; then a representative from a large pulp-and-paper company contacts you. The company also is interested in the impacts of its chemical effluents on nearby water bodies, and it would like to fund your research project. What are the pros and cons of this offer?

Theories If a hypothesis survives repeated testing by numerous research teams and continues to predict experimental outcomes and observations accurately, it may potentially be incorporated into a theory. A **theory** is a widely accepted, well-tested explanation of one or more cause-and-effect relationships, which has been extensively validated by extensive research. Whereas a hypothesis is a simple explanatory statement that may be refuted by a single experiment, a theory consolidates many related hypotheses that have been tested and supported by a large body of experimental and observational data.

Note that scientific use of the word *theory* differs from popular usage of the word. In everyday language when we say something is "just a theory," we are suggesting it is a speculative idea without much substance. Scientists, however, mean just the opposite when they use the term; to them, a theory is a conceptual framework that effectively explains a phenomenon and has undergone extensive and rigorous testing, such that confidence in it is extremely strong. For example, Darwin's theory of evolution by natural selection has been supported and elaborated upon by many thousands of studies over 150 years of intensive research. Such research has shown repeatedly and in great detail how plants and animals change over generations, or evolve, to express characteristics that best promote survival and reproduction. Because of its strong support and explanatory power, evolutionary theory is the central unifying principle of modern biology.

Science may go through "paradigm shifts"

Results obtained by the scientific method may sometimes later be reinterpreted to show that earlier interpretations were incorrect. Thomas Kuhn's 1962 book *The Structure of Scientific Revolutions* argued that science goes through periodic revolutions, or dramatic upheavals in thought, in which one scientific **paradigm**, or dominant view, is abandoned for another. For example, before the sixteenth century, scientists believed that Earth was at the centre of the universe. They made accurate measurements of the movement of the planets, then applied elaborate corrections that seemed to be needed in order to explain their measurements from a geocentric (Earth-centred) viewpoint. Nicolaus Copernicus eventually disproved the geocentric model of the solar system by demonstrating that placing the Sun at the centre of the solar system explained the planetary data and observations much better. A similar paradigm shift occurred in the 1960s, when geologists accepted the theory of plate tectonics because evidence for the movement of continents and the action of tectonic plates had accumulated and had become overwhelmingly convincing.

Understanding how science works is vital to assessing how scientific ideas and interpretations change through time, with new information. This process is especially relevant in environmental science, a young field that is changing rapidly as we gather vast amounts of new information, as human impacts on the planet multiply, and as lessons from the consequences of our actions become apparent. Because so much remains unstudied and undone, and because so many issues we cannot foresee are likely to arise in the future, environmental science will remain an exciting frontier for you to explore as a student and as an informed citizen throughout your life.

Sustainability and the Future of Our World

Throughout this book you will see examples of environmental scientists asking questions, developing hypotheses, conducting experiments, gathering and analyzing data, and drawing conclusions about environmental processes and the causes and consequences of environmental change. Environmental scientists who aim to understand the condition of our environment and the consequences of our impacts are studying the most centrally important issues of our time. New technologies and new scientific approaches have made it easier than ever before to monitor these changes (see "The Science Behind the Story: Mission to Planet Earth," • p. 21). You will also see, throughout this book, profiles of environmental leaders who are attempting to translate the scientific understanding of environmental change into political, social, and economic actions to protect our environment.

Population and consumption lie at the root of many environmental impacts

We modify our environment in diverse ways, but the steep and sudden rise in human population has amplified nearly all our impacts (Chapter 6). Our numbers have nearly quadrupled in the past 100 years, passing 6 billion in 1999 and 6.7 billion in 2007. We add about 78 million people to the planet each year—more than 200 000 per day. Today, the rate of population growth is slowing, but our absolute numbers continue to increase and shape our interactions with one another and with our environment.

Our consumption of resources has risen even faster than our population growth. The rise in affluence has been a positive development for humanity, and our conversion of the planet's natural capital has made life more pleasant for us so far. However, like rising population, rising per capita consumption amplifies the demands we make on our environment.

Moreover, affluence and consumption have not grown equally for all the world's citizens. Today, the 20 wealthiest nations boast 40 times the income of the 20 poorest nations—twice the gap that existed four decades earlier. The ecological footprint of the average citizen of a developed nation, such as Canada or the United States, is considerably larger than that of the average resident of a developing country (**FIGURE 1.10**).

We face many environmental challenges

The dramatic growth in human population and consumption is due in part to our successful efforts to expand and intensify the production of food (Chapters 7 and 8). Since

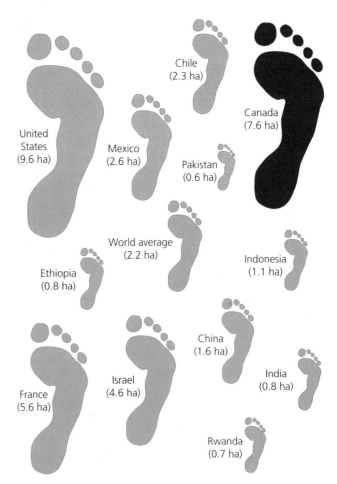

FIGURE 1.10
The citizens of some nations have larger ecological footprints than the citizens of others. United States residents consume more resources—and thus use more land—than residents of any other nation. Shown here are ecological footprints for average citizens of several developed and developing nations, as of 2003. Data from Global Footprint Network, 2006.

THE SCIENCE BEHIND THE STORY

Mission to Planet Earth: Monitoring Environmental Change

Ellesmere Island National Park Reserve is the most northerly park on Earth.

Science begins with observation. One of the most significant legacies of the space program is the ability to observe Earth from afar. The development of this ability changed our perspective on this planet permanently, leading to a new sense of respect, care, and concern for the environment that continues to grow as a social priority.

Since the last hand-held photograph of the whole planet was taken in 1972, scientists and technologists have dramatically improved their ability to capture and interpret a wide variety of images of this planet. Hand-held photography continues to provide an important part of the data returned by manned space missions (you can see many of these photographs archived at NASA's website, www.nasa.gov). Today, satellites and the sophisticated instrumentation they carry provide us with the opportunity to observe, study, monitor change, and gather an unprecedented amount of information about the planet, through technologies and processes that are collectively referred to as *remote sensing*.

By the early 1980s, NASA scientists and administrators had faced the reality that they were unlikely to obtain the necessary funding to return to the Moon any time in the near future. Instead, they began to turn their attention to another nearby planetary object: Earth. They called the new scientific approach *Earth system science* and defined as its goal "to obtain a scientific understanding of the entire Earth system on a global scale by describing how its component parts and their interactions have evolved, how they function, and how they may be expected to continue to evolve on all timescales."[12] The observation and interpretation of environmental change was fundamental to this new scientific approach, right from the beginning.

To achieve the goals of Earth system science, scientists began to use technologies that had been developed for other purposes—space exploration, communications, even warfare—to observe Earth and its component parts over time, and they called the new endeavour *Mission to Planet Earth*. The scientific results of this mission have been and continue to be spectacular. Satellites and remote detection and measuring technologies have provided a massive amount of information about the environment, how it changes over time, and how human activity affects it—information with a depth and breadth that would have been unimaginable back when those first whole-planet photographs were taken. The photo of Ellesmere Island National Park Reserve (top left), acquired in August 2003 by the Advanced Spaceborne Thermal Emission and Reflection Radiometer (ASTER) aboard NASA's *Terra* satellite, shows a tidewater glacier in Greely Fjord, situated in the southwestern corner of the Reserve. Icebergs floating in the fjord are chunks that have broken off the glacier. The corrugated surface on the glacier near its terminus (or end) is a network of crevasses. The strong linear features running through the glacier are flow lines. The dark blue features are melt ponds; they are darker than the ice and absorb more sunlight, thus melting more ice. A great deal of what environmental scientists now know about this planet as a coherent system, how its various parts interact, and how it is changing and evolving has been based on information derived from the Mission to Planet Earth.

This satellite image shows Lake Athabasca, the dark, irregular patch straddling the border between Alberta (west) and Saskatchewan (east), and numerous active summer forest fires (indicated by red dots). A large smoke plume stretches across Saskatchewan and into Manitoba, to the east. This image was acquired by the Moderate Resolution Imaging Spectroradiometer aboard the *Terra* satellite in July 2002.

the origins of agriculture and the Industrial Revolution, new technologies have enabled us to grow increasingly more food per unit of land. These advances in agriculture must be counted as one of humanity's great achievements, but they have come at some cost. We have converted nearly half the planet's land surface for agriculture; our extensive use of chemical fertilizers and pesticides poisons organisms and alters natural systems; and erosion, climate change, and poorly managed irrigation are destroying 5 million to 7 million hectares of productive cropland each year.

Meanwhile, pollution from our farms, industries, households, and individual actions dirties our land, water, and air (**FIGURE 1.11**). Outdoor air pollution, indoor air pollution, and water pollution contribute to the deaths of millions of people each year (Chapters 11 and 13). Environmental toxicologists are chronicling the impacts on people and wildlife of the many synthetic chemicals and other pollutants we emit into the environment (Chapter 19).

Perhaps our most pressing pollution challenge may be to address the looming spectre of global climate change (Chapter 14). Scientists have firmly concluded that human activity is altering the composition of the atmosphere and that these changes are affecting Earth's climate. Since the start of the Industrial Revolution, atmospheric carbon dioxide concentrations have risen by 37%, to a level not present in at least 650 000 years. This increase results from our reliance on burning fossil fuels to power our civilization. Carbon dioxide and several other gases absorb heat and warm Earth's surface, which is likely responsible for glacial melting, sea-level rise, impacts on wildlife and crops, and increased episodes of destructive weather.

The combined impact of human actions, such as climate change, overharvesting, pollution, the introduction of non-native species, and particularly habitat alteration, has driven many aquatic and terrestrial species out of large parts of their ranges and toward the brink of extinction (Chapter 9). Today Earth's biological diversity, or **biodiversity**, the cumulative number and diversity of living things (Chapter 3), is declining dramatically. Many biologists say we are already in a mass extinction event comparable to only five others documented in all of Earth's history. Biologist Edward O. Wilson has warned that the loss of biodiversity is our most serious and threatening environmental dilemma, because it is not the kind of problem that responsible human action can remedy. The extinction of species is irreversible; once a species has become extinct, it is lost forever.

Solutions to environmental problems must be global and sustainable

The nature of virtually all these environmental issues is being changed by the set of ongoing phenomena commonly dubbed *globalization*. Our increased global interconnectedness in trade, politics, and the movement of people and of other species poses many challenging problems, but it also sets the stage for novel and effective solutions.

The most comprehensive scientific assessment of the present condition of the world's ecological systems and their ability to continue supporting our civilization was completed in 2005, when more than 2000 of the world's leading environmental scientists from nearly 100 nations completed the **Millennium Ecosystem Assessment**. The four main findings of this exhaustive project are summarized in Table 1.1. The Assessment makes clear that our degradation of the world's environmental systems is having negative impacts on all of us but that with care and diligence we can still turn many of these trends around.

Our energy choices will influence our future immensely

Our reliance on fossil fuels to power our civilization has intensified virtually every negative impact we have on the environment, from habitat alteration to air pollution to climate change. Fossil fuels have also brought us the material affluence we enjoy. By taking advantage of the richly concentrated energy in coal, oil, and natural gas, we have been able to power the machinery of the Industrial Revolution, produce the chemicals that boosted agricultural yields, run the vehicles and transportation networks of our mobile society, and manufacture and distribute our countless consumer products. It is little exaggeration

FIGURE 1.11
Indoor and outdoor air pollution contribute to millions of premature deaths each year, and environmental scientists and policy makers are working to reduce these problems in a variety of ways.

Table 1.1 Main Findings of the Millennium Ecosystem Assessment

- Over the past 50 years, humans have changed ecosystems more rapidly and extensively than in any comparable period of time in human history, largely to meet rapidly growing demands for food, freshwater, timber, fibre, and fuel. This has resulted in a substantial and largely irreversible loss in the diversity of life on Earth.
- The changes made to ecosystems have contributed to substantial net gains in human well-being and economic development, but these gains have been achieved at growing costs. These costs include the degradation of ecosystems and the services they provide for us, and the exacerbation of poverty for some groups of people.
- This degradation could grow significantly worse during the first half of this century.
- The challenge of reversing the degradation of ecosystems while meeting increasing demands for their services can be partially overcome, but doing so will involve significantly changing many policies, institutions, and practices.

Adapted from *Millennium Ecosystem Assessment*, Synthesis Report, 2005.

to say that the lives we live today are a result of the availability of fossil fuels (Chapter 15).

However, in extracting fossil fuels, we are splurging on a one-time bonanza. Scientists calculate that we have depleted half the world's oil supplies and that we are in for a rude awakening very soon, once the supply begins to decline while the demand continues to rise. We are also approaching the peak production of natural gas, and coal is also nonrenewable and in limited supply. The search is now on for alternative sources of energy that will allow us to maintain an acceptable standard of living while minimizing the environmental impacts of energy use (Chapters 16 and 17). How we handle the imminent crisis of fossil fuel depletion and the search for replacements will largely determine the nature of our lives in the twenty-first century.

Fortunately, potential solutions abound

We cannot, of course, live without exerting *any* impact on Earth's systems. We face trade-offs with many environmental issues, and the challenge is to develop solutions that increase our quality of life while minimizing harm to the environment that supports us. Fortunately, many workable solutions are at hand, and we can achieve many more potential solutions with further effort.

In response to agricultural problems, scientists and others have developed and promoted soil conservation, high-efficiency irrigation, and organic agriculture. In addition, technological advances and new laws have greatly reduced the pollution emitted by industry and automobiles in wealthier countries. Canadian scientists have been at the forefront of many of these technological advances and have made fundamental contributions to global environmental management theories and to our current understanding of the human–environment relationship. Amid ample reasons for concern about the state of global biodiversity, advances in conservation biology are enabling scientists and policy makers in many cases to work together to protect habitat, slow extinction, and safeguard endangered species (**FIGURE 1.12**).

Recycling (Chapter 18) is helping to relieve our waste disposal problems, and alternative renewable energy sources (Chapters 16 and 17) are being developed to take the place of fossil fuels (**FIGURE 1.13**). These are but a few of the many solutions we will explore in the course of this book. In Chapter 23, we will examine some of the new structures, programs, processes, and technologies that are emerging in support of these solutions.

Are things getting better or worse?

Despite the myriad challenges we face, some people maintain that the general conditions of human life and the environment are, in fact, getting better, not worse. A recent

FIGURE 1.12
Human activities are pushing many organisms toward extinction. Efforts to save endangered species and reduce biodiversity loss include many approaches, but all require that adequate areas of appropriate habitat be preserved in the wild. The habitat of polar bears is increasingly threatened by global climate change.

FIGURE 1.13
Our dependence on fossil fuels has caused a wide array of environmental impacts. Although fossil fuels have powered our civilization since the Industrial Revolution, many renewable energy sources exist, such as solar energy, which can be collected with panels like these. Such alternative energy sources could be further developed for sustainable use now and in the future.

proponent of this view, Danish statistician Bjorn Lomborg, wrote in his book *The Skeptical Environmentalist*:

> We are not running out of energy or natural resources. There will be more and more food per head of the world's population. Fewer and fewer people are starving. In 1900 we lived for an average of 30 years; today we live for 67.... The air and water around us are becoming less and less polluted. Mankind's lot has actually improved in terms of practically every measurable indicator.

Furthermore, some people maintain that we will find ways to make Earth's natural resources meet all our needs indefinitely and that human ingenuity will see us through any difficulty. Such views are sometimes characterized as *cornucopian*. In Greek mythology, *cornucopia*—literally "horn of plenty"—is the name for a magical goat's horn that overflowed with grain, fruit, and flowers. In contrast, people who predict doom and disaster for the world because of our impact upon it have been called *Cassandras*, after the mythical princess of Troy with the gift of prophecy, whose dire predictions were not believed.

At least three questions are worth asking each time you are confronted with seemingly conflicting statements from Cassandras and cornucopians:

1. Do the impacts being debated pertain only to humans or also to other organisms and natural systems?
2. Are the debaters thinking in the short term or the long term?
3. Are they considering all costs and benefits relevant for the question at hand, or only some?

As you proceed through this book and encounter countless contentious issues, consider how a person's perception of them may be influenced by these three factors.

Sustainability is a goal for the future

The primary challenge in our increasingly populated world is how to live within our planet's means, such that Earth and its resources can sustain us, and the rest of Earth's biota, for the foreseeable future. This is the challenge of **sustainability**, a guiding principle of modern environmental science. Sustainability means leaving our children and grandchildren a world as rich and full as the world we live in now. It means not depleting Earth's natural capital, so that after we are gone our descendants will enjoy the use of resources, as we have. It means developing solutions that are able to work in the long term. Sustainability requires maintaining fully functioning ecological systems, because we cannot sustain human civilization without sustaining the natural systems that nourish it. Our final chapter (Chapter 23) takes a wide-ranging look at emerging sustainable solutions—on college and university campuses, and in the world at large.

Sustainable development involves environmental protection, economic welfare, and social equity

Environmental protection is often cast as being in opposition to the economic and social needs of human society, but environmental scientists have long recognized that our civilization cannot exist without a functional natural environment. In recent years, people of all persuasions have increasingly realized the connection between environmental quality and human quality of life. Moreover, we now recognize that often it is society's poorer people who suffer the most from environmental degradation. This realization has led advocates of environmental protection, economic development, and social justice to begin working together toward common goals. This cooperative approach has given rise to the modern drive for sustainable development.

Economists employ the term **development** to describe the use of natural resources for economic advancement (as opposed to simple subsistence, or survival). Construction of homes, schools, hospitals, power plants, factories, and transportation networks are all examples of activities in support of development. **Sustainable**

development is the use of renewable and nonrenewable resources in a manner that satisfies our current needs without compromising future availability of resources. The United Nations has defined sustainable development as development that " ... meets the needs of the present without sacrificing the ability of future generations to meet their needs." This definition is taken from the United Nations–sponsored *Brundtland Commission* (named after its chair, Norwegian prime minister Gro Harlem Brundtland), which published an influential 1987 report entitled *Our Common Future*.

Prior to the Brundtland Report, most people aware of human impact on the environment might have thought *sustainable development* to be an oxymoron—a phrase that contradicts itself. Although development involves making purposeful changes intended to improve the quality of human life, environmental advocates have long pointed out that development often so degrades the

---

CANADIAN ENVIRONMENTAL PERSPECTIVES

David Suzuki is an environmentalist, but he was trained as a scientist.

- **Zoologist, geneticist, and professor** University of British Columbia, Sustainable Development Research Institute
- **Environmentalist** and **activist**
- **Writer** *The Sacred Balance: Rediscovering Our Place in Nature*
- **Radio and TV broadcast journalist** *Quirks & Quarks* and *The Nature of Things*

What remains to be said about a man who has been called "Canada's environmental conscience"? He has a long list of honorary degrees. He is a Companion of the Order of Canada, placed fifth in the Canadian Broadcasting Corporation's "Greatest Canadian" contest, and has received scores of major environmental and journalism awards. We see him on TV every day, exhorting us not to air condition our homes to the point at which penguins would be comfortable.

What we sometimes forget is that before he was a journalist, writer, activist, or

David Suzuki

TV broadcaster, David Suzuki was a scientist.[13] He completed a degree in biology in 1958, spent that summer working as a fish biologist for the Department of Lands and Forests in Ontario, and then went on to graduate school in zoology at the University of Chicago. By 1962 Suzuki had accepted an appointment in the Department of Genetics at the University of Alberta, moving a year later to a faculty position at the University of British Columbia.

When Suzuki started out as a young professor teaching genetics—a science he adored for its precision and the promise it held for society—he encountered questions at the intersection between science and ethics. Specifically, he learned how genetics had been used by the Nazis and others with an interest in institutionalizing racism.[14] Suzuki's own family had had property seized and been placed in internment camps by the Government of Canada during the Second World War. "Once a Jap, always a Jap," a Member of Parliament had said at the time,[15] echoing the belief that treachery and untrustworthiness were genetically encoded into anyone of Japanese descent.

This history makes Suzuki a perfect starting point for a discussion about the difference between environmentalism and environmental science. They are different—but not entirely separate. Ideally, science informs and responds to political and social influences, without being overly influenced by them. David Suzuki has consciously given up "doing" science on an everyday basis, freeing himself to focus on the more political side of science. Today he is much better known for his activism and journalism than for his science. On the rapidly developing field of biotechnology, so close to his own scientific background, he comments:

There is absolutely no reason to suppose that biologists know enough to anticipate the ecological and health ramifications of a revolutionary technology such as genetic engineering. Governments must resist the economic pressures and show leadership and concern for the long-term health of people and nature. And scientists involved in this exciting area should learn from history and welcome free and open discussion about ecological, health, and social implications of their work.[16]

Science is a human endeavour; it can never be entirely free of political or social influence. We want our leaders to incorporate scientific understanding into their social decisions, but there is no foolproof way to ensure that science is not misused to serve political ends. By becoming aware of the complex relationships among science, society, and politics, we can work to ensure that an appropriate balance is maintained.

"We must reinvent a future free of blinders so that we can choose from real options." —David Suzuki

Thinking About Environmental Perspectives

David Suzuki is no stranger to controversy. Throughout much of his career he has faced criticism for speaking out on environmental issues. Is it acceptable, in your view, for a scientist also to be an activist, or a spokesperson, for a cause? To what extent, if at all, does environmental activism compromise a person's ability to function as a scientist? If a scientist uncovers something in the course of doing research that may be of importance to the general public, is it appropriate to speak out about this discovery? Or does the act of speaking out compromise objectivity as a scientist?

natural environment that it threatens the very improvements for human life that were intended. Conversely, many people remain under the impression that protecting the environment is incompatible with serving people's economic needs.

Fortunately, sustainable development efforts by governments, businesses, industries, organizations, and individuals everywhere—from students on campus to international representatives at the United Nations—are beginning to alter these perceptions. These efforts are generating sustainable solutions that meet environmental, economic, and social goals simultaneously, satisfying the so-called "triple bottom line."

Sustainability and the triple bottom line demand that our current human population limit its environmental impact while also promoting economic well-being and social equity. These aims require us to make an ethical commitment to our fellow citizens and to future generations. They also require that we apply knowledge from the sciences to help us devise ways to limit our impact and maintain the functioning environmental systems on which all life depends.

"Will we develop in a sustainable way?" may well be the single most important question in the world today. Environmental science holds one crucial key to addressing it: Because so much remains unstudied and undone, and because it is so central to our modern world, environmental science will remain an exciting frontier for you to explore as a student and as an informed citizen throughout your life.

Conclusion

Finding effective ways of living peacefully, healthfully, and sustainably on our diverse and complex planet will require a thorough scientific understanding of both natural and social systems. Environmental science helps us understand our intricate relationship with the environment and informs our attempts to solve and prevent environmental problems.

Identifying a problem is the first step in devising a solution to it. Many of the trends detailed in this book may cause us worry, but others give us reason to hope. One often-heard criticism of environmental science courses and books is that they emphasize the negative. Recognizing the validity of this criticism, in this book we attempt to balance the discussion of environmental problems with a corresponding focus on potential solutions. Solving environmental problems can move us toward health, longevity, peace, and prosperity. Science in general, and environmental science in particular, can aid us in our efforts to develop balanced and workable solutions to the many environmental dilemmas we face today and to create a better world for ourselves and our children.

REVIEWING OBJECTIVES

You should now be able to:

Define the term *environment*

- Our environment consists of everything around us, including living and nonliving things.
- Humans are a part of the environment and are not separate from it.

Describe natural resources and explain their importance to human life

- Resources from nature are essential to human life and civilization.
- Some resources are inexhaustible or perpetually renewable, others are nonrenewable, and still others are renewable if we are careful not to exploit them at too fast a rate.
- Hardin articulated the concept of carrying capacity, the number of individuals that can be sustained by a given area of productive land. Wackernagel and Rees pioneered the idea of the ecological footprint, a measure of the amount of productive land it would take to support an individual at a certain level of consumption.

Characterize the interdisciplinary nature of environmental science

- Environmental science uses the approaches and insights of numerous disciplines from the natural sciences and the social sciences.

Understand the scientific method and how science operates

- Science is a process of using observations to test ideas.
- The scientific method consists of a series of steps, including making observations, formulating questions, stating a hypothesis, generating predictions, testing predictions, and analyzing the results obtained from the tests.

- The scientific method has many variations, and there are many different ways to test questions scientifically.
- Scientific research occurs within a larger process that includes peer review of work, journal publication, and interaction with colleagues.

Diagnose and illustrate some of the pressures on the global environment

- The increasing human population and increasing per capita consumption exacerbate human impacts on the environment.
- Human activities, such as industrial agriculture and the use of fossil fuels for energy, are having diverse environmental impacts, including resource depletion, air and water pollution, habitat destruction, and the diminishment of biodiversity.

Articulate the concepts of sustainability and sustainable development

- Sustainability means living within the planet's means, such that Earth's resources can sustain us—and other species—for the foreseeable future.
- Sustainable development means pursuing environmental, economic, and social goals in a coordinated way, and it is the most important pursuit in our society today.

TESTING YOUR COMPREHENSION

1. What do renewable resources and nonrenewable resources have in common? How are they different? Identify two renewable and two nonrenewable resources.
2. How did the Agricultural Revolution affect human population size and the environment? How did the Industrial Revolution affect human population size and the environment? Explain your answers.
3. What is *the tragedy of the commons*? Explain how the concept might apply to an unregulated industry that is a source of water pollution.
4. What is *environmental science*? Name several disciplines involved in environmental science.
5. What are the two meanings of *science*? Name three applications of science.
6. Describe the scientific method. What is the typical sequence of steps?
7. Explain the difference between a manipulative experiment and a natural experiment.
8. What needs to occur before a researcher's results are published? Why is this important?
9. Give examples of three major environmental problems in the world today, along with their causes.
10. What is sustainable development?

SEEKING SOLUTIONS

1. Many resources are renewable if we use them in moderation but can become nonrenewable if we overexploit them. Order the following resources on a continuum of renewability (see Figure 1.1), from most renewable to least renewable: soils, timber, freshwater, food crops, and biodiversity. What factors influenced your decision? For each of these resources, what might constitute overexploitation, and what might constitute sustainable use?
2. Why do you think the inhabitants of Rapa Nui did not or could not stop themselves from stripping their island of all its trees? Do you see similarities between the history of Rapa Nui and the modern history of our society? Why, or why not?
3. What environmental problem do *you* feel most acutely yourself? Do you think there are people in the world who do not view your issue as an environmental problem? Who might they be, and why might they take a different view?
4. If the human population were to stabilize tomorrow and never surpass 7 billion people, would that solve our environmental problems? Which types of problems might be alleviated, and which might continue to become worse?
5. Consider the historic expansion of agriculture and our ability to feed increasing numbers of people, as described in this chapter. Now ask yourself, "Are things getting better or worse?" Ask this question from four points of view: (1) the human perspective, (2) the perspective of other organisms, (3) a short-term perspective, and (4) a long-term perspective. Do your answers to this question change? If so, how?

6. **THINK IT THROUGH** You have become the head of a major funding agency that disburses funding to researchers pursuing work in environmental science. You must give your staff several priorities to determine what types of scientific research to fund. What environmental problems would you most like to see addressed with research? Describe the research you think would need to be completed so that workable solutions to these problems could be developed. Would more than science be needed to develop sustainable solutions?

INTERPRETING GRAPHS AND DATA

Environmental scientists study phenomena that range in size from individual molecules (Chapter 2) to the entire Earth (Chapter 5) and that occur over time periods lasting from fractions of a second to billions of years. To simultaneously and meaningfully represent data covering so many orders of magnitude, scientists have devised a variety of mathematical and graphical techniques, such as exponential notation and logarithmic scales. At the right are two graphical representations *of the same data,* representing the growth of a hypothetical population from an initial size of 10 individuals at a rate of increase of approximately 2.3% per generation. The graph in part (a) uses a conventional linear scale for the population size; the graph in part (b) uses a logarithmic scale.

1. Using the graph in part (a), what would you say was the population size after 200 generations? After 400? After 600? After 800? How would you answer the same questions by using the graph in part (b)? What impression does the graph in part (a) give about population change for the first 600 generations? What impression does the graph in part (b) give?

2. Compare these graphs to Figure 1.2a. What does the human population appear to be doing between 10 000 b.p. and 2000 b.p.?

3. The size of a population that is growing by a constant rate of increase will plot as a straight line on a logarithmically scaled graph like the one in part (b), but if the annual rate of increase changes, the line will curve. Do you think the data for the human population over the past 12 000 years would plot as a straight line on a logarithmically scaled graph? If not, when and why do you think the line would bend?

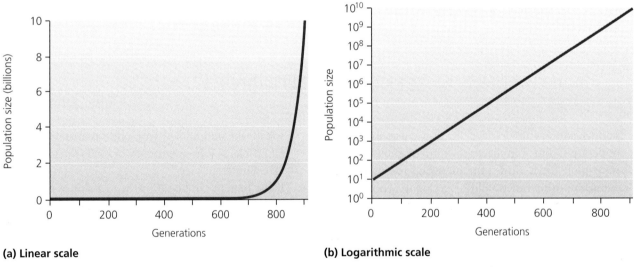

(a) Linear scale

(b) Logarithmic scale

Hypothetical population growth curves **(a)** and **(b)**, assuming an initial size of 10 and a constant rate of increase of approximately 2.3% per generation.

CALCULATING FOOTPRINTS

Mathis Wackernagel and his many colleagues at the Global Footprint Network have continued to refine the method of calculating ecological footprints—the amount of biologically productive land and water required to produce the energy and natural resources we consume and to absorb the wastes we generate. According to their 2006 report, there are less than 1.8 hectares available for every person in the world, yet we use on average more than 2.2 hectares per person, creating a global ecological deficit or overshoot of about 25%.

Compare the ecological footprints of each of the countries listed in the table. Calculate their proportional relationships to the world population's average ecological footprint and to the land available globally to meet our ecological demands.

1. Why is the ecological footprint for people in Bangladesh so low?
2. Why is it so high in the United States?
3. The population of the United States is expected to grow to 349 million (from the current 302 million) by 2025. What impact, if any, do you think this growth will have on the average global ecological footprint? In comparison, what would be the impact of adding an equivalent number of Bangladeshis? What about Canadians?
4. Based on the data in the table, what impacts do you think average family income has on ecological footprints?
5. Go to an online footprint calculator, such as the one at www.myfootprint.org or http://ecofoot.org, and take the test to determine your own personal ecological footprint. Enter the value you obtain in the table, and calculate the other values as you did for each nation. How does your footprint compare with those of people in Canada and in other nations? Name three actions you could take to reduce your footprint. (*Note:* Save this number—you will calculate your footprint again in Chapter 23 at the end of your course.)

Country	Ecological Footprint (Hectares per Person)	Proportion Relative to World Average Footprint	Proportion Relative to World Area Available
Bangladesh	0.5		7.6 0.73 (1.3 ÷ 1.78)
Canada			
Colombia	1.3		
Mexico	2.6		
Sweden	6.1		
Thailand	1.4		
United States	9.6		
World average	2.23	1.0 (2.23 ÷ 2.23)	1.25 (2.23 ÷ 1.78)
Your personal footprint (see Question 5)			

Data from Living Planet Report 2006, WWF International, Zoological Society of London, and Global Footprint Network.

TAKE IT FURTHER

 Go to www.myenvironmentplace.ca where you will find

- Suggested answers to end-of-chapter questions
- Quizzes, animations, and flashcards to help you study
- *Research Navigator*™ database of credible and reliable sources to assist you with your research projects
- Tutorials to help you master how to interpret graphs
- Current news articles that link the topics that you study to case studies from your region and around the world

- **ECO Occupational Profiles:** If you found this chapter especially interesting, you might want to learn more about the following jobs by visiting the Occupational Profiles website of the Environmental Careers Organization. Go to www.eco.ca and check out the following careers:
  - Environmental educator
  - Environmental manager
  - Environmental reporter
  - Environmental communications officer

CHAPTER ENDNOTES

1. NASA, The *Apollo 8* Christmas Eve Broadcast, http://nssdc.gsfc.nasa.gov/planetary/lunar/apollo8_xmas.html.
2. Information about the Blue Marble and other photographs of Earth from space can be obtained from NASA's Earth Observatory website, http://earthobservatory.nasa.gov/Newsroom/BlueMarble/BlueMarble_history.html.
3. Meadows, Donella, Dennis L. Meadows, Jørgen Randers, and William W. Behrens III (1972) *Limits to Growth*, New York, Universe Books.
4. Ehrlich, Paul R. (1968) *The Population Bomb*, Sierra Club-Ballantine Books.
5. Schumacher, E. F. (1973) *Small Is Beautiful: Economics as if People Mattered*, New York, Harper & Row.
6. Carson, Rachel (1962) *Silent Spring*, Houghton Mifflin.
7. Environment Canada, www.ec.gc.ca
8. The United Nations officially marked the "Day of Six Billion" on October 12, 1999.
9. Based on the assumption that an average city block is roughly 2–4 ha.
10. Global Footprint Network, www.footprintstandards.org.
11. Brander, James A., M. Scott Taylor (1998) The Simple Economics of Easter Island: A Ricardo-Malthus Model of Renewable Resource Use, *The American Economic Review*, Vol. 88, No. 1, pp. 119–138.
12. Earth System Science Committee (1986) *Earth System Science: A Program for Global Change*, NASA Advisory Council.
13. Based partly on information from the website of the David Suzuki Foundation, www.davidsuzuki.org/About_us/Dr_David_Suzuki/.
14. Suzuki, David, Biotechnology: A Geneticist's Personal Perspective, www.davidsuzukifoundation.org/files/General/DTSbiotech.pdf.
15. Ibid, p. 5.
16. Ibid, p. 23.

2 From Chemistry to Energy to Life

Beautiful Pavilion Lake, British Columbia, hides scientific secrets beneath its surface.

Upon successfully completing this chapter, you will be able to

- Explain the fundamentals of chemistry and apply them to real-world environmental situations
- Describe the molecular building blocks of living organisms
- Differentiate among the types of energy and recite the basics of energy flow
- Distinguish photosynthesis, respiration, and chemosynthesis, and summarize their importance to living things
- Itemize and evaluate the major hypotheses for the origin of life on Earth
- Outline our knowledge regarding early life and give supporting evidence for each major concept

This is a sample of a microbialite from Pavilion Lake.

CENTRAL CASE:
THE UNUSUAL MICROBIALITES OF PAVILION LAKE

"Life exists in the universe only because the carbon atom possesses certain exceptional properties."
—SIR JAMES JEANS, ASTRONOMER, PHYSICIST, AND MATHEMATICIAN

"Science never gives up searching for truth, since it never claims to have achieved it."
—JOHN CHARLES POLANYI, CHEMISTRY NOBEL PRIZE WINNER

Pavilion Lake (see photo) is a beautiful, clear blue-green lake that is protected as part of the Marble Canyon Provincial Park system in the interior of British Columbia. The lake holds traditional spiritual significance for the Ts'kw'aylaxw First Nation. In recent years, a group of scientists from NASA's Ames Research Center also has taken a particular interest in some of the unique characteristics of the lake. The feature of greatest scientific interest is the presence in the lake of a variety of *microbialites*—reeflike sedimentary structures composed of calcium carbonate (a combination of calcium, carbon, and oxygen) of organic derivation. The light-blue patches visible within the lake in the photo are some of these structures. Microbialites that occur in freshwater environments are less well known scientifically than the much more common oceanic reefs, which are also composed of calcium carbonate.

Darlene Lim and her colleagues from the University of British Columbia and NASA's Ames Research Center are interested in learning about the biological origins of the Pavilion Lake freshwater microbialites and the geobiological conditions that control their *morphology* (that is, their physical form). The widely varying morphologies of the Pavilion Lake microbialites, which range from chimney-shaped to cone-, leaf-, and dome-shaped, vary with depth; the structures occur in water depths ranging from 5 cm to 60 m, and the pressure and light levels at these depths differ substantially. These differences in the physical environment must

Geobiologist Darlene Lim collects a sediment sample from a core.

have influenced the biological processes through which the structures are formed.

Understanding how life-supporting environments originated on Earth depends to a great extent on understanding the formation of carbonate rocks. Carbonate rocks contain carbon dioxide and are extremely important long-term storage reservoirs for carbon that would otherwise remain in the atmosphere. The formation of carbonates early in Earth's history was a crucial step in the chemical evolution of the terrestrial atmosphere and an integral part of the history of chemical interaction among the atmosphere, hydrosphere, biosphere, and lithosphere.

Scientific models predict that if Mars once had a thick CO_2-rich atmosphere, like that of early Earth, there should be massive carbonates present at or near the surface of the planet, as there are on Earth. However, scientists have detected only trace levels of carbonate in Martian soil, and no massive carbonate outcrops have been identified on the surface. Where, then, are the missing carbonates?

The study of terrestrial carbonates, like the microbialites of Pavilion Lake, may help scientists answer questions like these. In the process, they hope to develop a deeper understanding of early environments and life on Earth, as well.[1]

Chemistry and the Environment

Examine any environmental issue—whether it is a question of basic science or an application to human–environment interactions—and you will likely discover chemistry playing a central role. Chemistry is crucial to understanding how gases, such as carbon dioxide and methane, contribute to global climate change, and to understanding how the climate, atmosphere, and life on Earth came to be as they now are. Scientists are using chemistry in new ways to help us understand life and biological processes on this planet. Chemistry is central to understanding how pollutants, such as sulphur dioxide and nitric oxide, cause acid rain, and how pesticides and other artificial compounds we release into the environment affect the health of wildlife and people. Chemistry is also essential in understanding water pollution and sewage treatment, atmospheric ozone depletion, hazardous waste and its disposal, and just about any energy issue.

Chemistry can help solve environmental problems

Scientists use chemistry to develop new solutions to environmental problems. **Bioremediation**—the use of naturally occurring microbial organisms to accelerate the cleanup of chemicals at polluted sites—is one illustration of this. Hydrocarbon-consuming bacteria and fungi are used to clean up the soil beneath leaky gasoline tanks that threaten drinking water supplies. Other kinds of microbes are used to degrade chemical pesticide residues in soil.

When the *Exxon Valdez* oil tanker ran aground in Alaska's Prince William Sound in 1989, it spilled 42 million litres of crude oil, coating 2100 km of Alaskan coastline. The largest oil spill in North American history, it killed thousands of seabirds, sea otters, and harbour seals and countless fish, smothered intertidal plants and animals, and defiled the area's relatively pristine environment.

Thousands of workers and volunteers launched a cleanup effort of unprecedented scope. The cleanup crews corralled the oil with booms, skimmed it from the water, soaked it up with absorbent materials, and dispersed it with chemicals. They pressure-washed the beaches (**FIGURE 2.1**), removed contaminated sand with backhoes, and even tried burning the oil. Scientists used the opportunity to test a new bioremediation strategy in which they stimulated naturally occurring bacteria to *biodegrade*, or break down, the oil. About 5% of the single-celled microbes naturally present on Alaskan beaches feed on hydrocarbons produced by the region's conifer trees. Hydrocarbons from conifers are chemically similar to those that make up crude oil, so scientists predicted that the microbes might also be able to degrade oil.

Today many wildlife populations at the site of the *Exxon Valdez* spill have recovered, but some have not, and pockets of oil remain. The results of the bioremediation experiments were interpreted differently by different researchers. Some felt that the new approach had increased the rate of remediation fivefold; others felt that the benefits had been minimal. Part of the reason for the unpredictable behaviour of spilled oil and its response to remediation efforts is the widely variable chemistry of crude oil. It is the chemistry of oil that causes it to gum up birds' feathers and mammals' fur, impairing their insulating abilities and causing hypothermia. It is the chemistry of oil that causes it to float on water, or clump together and sink to the bottom, accumulate on beaches, or be washed away by the rain. It is the chemistry of oil—so challenging to deal with when released into a natural setting—that provides the energy to power our remarkable civilization and modern way of life (Chapter 15).

Plants also have been pressed into service in environmental cleanups. Plants, such as wheat, tobacco, water hyacinths, chrysanthemums, and cattails, have been used to clean up toxic materials from soils, in an approach called **phytoremediation**. Some types of plants draw up heavy metals, such as lead and cadmium, through their roots and stomata, thus removing the toxins from the soil and concentrating them in plant tissue, which can later be harvested and disposed of properly (see "The Science Behind the Story: Letting Plants Do the Dirty Work," • p. 35).

Environmental chemists are excited about the countless future applications of chemistry that may help us address environmental problems. To appreciate the complex chemistry involved in environmental science, and to understand how Earth's system supports life, we must begin with a grasp of the fundamentals.

Atoms and elements are chemical building blocks

All material in the universe that has mass and occupies space is called **matter**. Matter can be transformed from one form into another, but it cannot be created or destroyed. This principle is referred to as the **law of conservation of matter**. In environmental science, this principle helps us understand that the amount of matter stays constant as it is recycled in nutrient cycles and ecosystems. It also makes clear that we cannot simply wish away the matter (such as waste and pollution) that we want to get rid of. Every drop of oil spilled in a pristine bay will end up somewhere, whether it sinks into the sediment, coats a bird's feathers, or is consumed by bacteria

FIGURE 2.1
Workers spray fertilizer on an oil-coated Alaskan beach, in an effort to balance the chemistry of the site and stimulate naturally occurring bacteria to consume and biodegrade the spilled oil.

CALCULATING FOOTPRINTS

Mathis Wackernagel and his many colleagues at the Global Footprint Network have continued to refine the method of calculating ecological footprints—the amount of biologically productive land and water required to produce the energy and natural resources we consume and to absorb the wastes we generate. According to their 2006 report, there are less than 1.8 hectares available for every person in the world, yet we use on average more than 2.2 hectares per person, creating a global ecological deficit or overshoot of about 25%.

Compare the ecological footprints of each of the countries listed in the table. Calculate their proportional relationships to the world population's average ecological footprint and to the land available globally to meet our ecological demands.

1. Why is the ecological footprint for people in Bangladesh so low?
2. Why is it so high in the United States?
3. The population of the United States is expected to grow to 349 million (from the current 302 million) by 2025. What impact, if any, do you think this growth will have on the average global ecological footprint? In comparison, what would be the impact of adding an equivalent number of Bangladeshis? What about Canadians?
4. Based on the data in the table, what impacts do you think average family income has on ecological footprints?
5. Go to an online footprint calculator, such as the one at www.myfootprint.org or http://ecofoot.org, and take the test to determine your own personal ecological footprint. Enter the value you obtain in the table, and calculate the other values as you did for each nation. How does your footprint compare with those of people in Canada and in other nations? Name three actions you could take to reduce your footprint. (*Note:* Save this number—you will calculate your footprint again in Chapter 23 at the end of your course.)

Country	Ecological Footprint (Hectares per Person)	Proportion Relative to World Average Footprint	Proportion Relative to World Area Available
Bangladesh	0.5		7.6 0.73 (1.3 ÷ 1.78)
Canada			
Colombia	1.3		
Mexico	2.6		
Sweden	6.1		
Thailand	1.4		
United States	9.6		
World average	2.23	1.0 (2.23 ÷ 2.23)	1.25 (2.23 ÷ 1.78)
Your personal footprint (see Question 5)			

Data from Living Planet Report 2006, WWF International, Zoological Society of London, and Global Footprint Network.

TAKE IT FURTHER

 Go to www.myenvironmentplace.ca where you will find

- Suggested answers to end-of-chapter questions
- Quizzes, animations, and flashcards to help you study
- *Research Navigator*™ database of credible and reliable sources to assist you with your research projects
- Tutorials to help you master how to interpret graphs
- Current news articles that link the topics that you study to case studies from your region and around the world

- **ECO Occupational Profiles:** If you found this chapter especially interesting, you might want to learn more about the following jobs by visiting the Occupational Profiles website of the Environmental Careers Organization. Go to www.eco.ca and check out the following careers:
  - Environmental educator
  - Environmental manager
  - Environmental reporter
  - Environmental communications officer

CHAPTER ENDNOTES

1. NASA, The *Apollo 8* Christmas Eve Broadcast, http://nssdc.gsfc.nasa.gov/planetary/lunar/apollo8_xmas.html.
2. Information about the Blue Marble and other photographs of Earth from space can be obtained from NASA's Earth Observatory website, http://earthobservatory.nasa.gov/Newsroom/BlueMarble/BlueMarble_history.html.
3. Meadows, Donella, Dennis L. Meadows, Jørgen Randers, and William W. Behrens III (1972) *Limits to Growth*, New York, Universe Books.
4. Ehrlich, Paul R. (1968) *The Population Bomb*, Sierra Club-Ballantine Books.
5. Schumacher, E. F. (1973) *Small Is Beautiful: Economics as if People Mattered*, New York, Harper & Row.
6. Carson, Rachel (1962) *Silent Spring*, Houghton Mifflin.
7. Environment Canada, www.ec.gc.ca
8. The United Nations officially marked the "Day of Six Billion" on October 12, 1999.
9. Based on the assumption that an average city block is roughly 2–4 ha.
10. Global Footprint Network, www.footprintstandards.org.
11. Brander, James A., M. Scott Taylor (1998) The Simple Economics of Easter Island: A Ricardo-Malthus Model of Renewable Resource Use, *The American Economic Review*, Vol. 88, No. 1, pp. 119–138.
12. Earth System Science Committee (1986) *Earth System Science: A Program for Global Change*, NASA Advisory Council.
13. Based partly on information from the website of the David Suzuki Foundation, www.davidsuzuki.org/About_us/Dr_David_Suzuki/.
14. Suzuki, David, Biotechnology: A Geneticist's Personal Perspective, www.davidsuzukifoundation.org/files/General/DTSbiotech.pdf.
15. Ibid, p. 5.
16. Ibid, p. 23.

THE SCIENCE BEHIND THE STORY

Letting Plants Do the Dirty Work

The lemon-scented geranium has the potential for use in phytoremediation.

When soil is contaminated with heavy metals from mining, manufacturing, oil extraction, or military facilities, the standard solution is to dig up tons of soil and pile it into a hazardous waste dump. Bulldozing so much dirt can release toxic chemicals into the air and costs up to $7.5 million per hectare. As an alternative, scientists are developing methods of *phytoremediation*, using plants (*phyto* means "plant") to remediate, or detoxify, contaminated soils.

For example, researchers at the University of Guelph in Ontario have recently shown that the lemon-scented geranium (*Pelargonium*, see photo) has a natural ability to absorb heavy metals, such as copper, mercury, lead, and nickel, from contaminated soil. The plants absorb metal ions through their roots and store them in their leaves and shoots. Normally, heavy metals are not readily accessible to plants, because they are tied up in the soil in the form of compounds that do not dissolve easily in water. But chemicals called chelating agents can bind to the metals, making them more water-soluble, and thus more accessible to plant roots.

Dr. Praveen Saxena, a professor of horticulture at the University of Guelph, says his research team has found that the geranium is a more powerful accumulator of toxic metals than any plant that has yet been tested. The geraniums could be available for commercial use in phytoremediation within three to five years.

The use of plants to remove hazardous materials from contaminated soil or groundwater is a relatively new technology, which has been developed commercially only in the past 10 years or so. Before then, plants that were known to absorb metals from the environment had been used in prospecting for valuable ores. Samples of the plants were collected and analyzed in the lab; if higher-than-normal concentrations of the metal were found, a more detailed investigation would be undertaken to determine the ore-bearing potential of the area. Yet despite this natural ability of plants to absorb contaminants, research into their use for the remediation of contaminated soil is still relatively new, and very few plants have yet demonstrated commercial success.

However, the lemon-scented geranium has many characteristics of a good phytoremediation agent. It grows quickly, and its large volume—it is dense and bushy, and grows to a height of one metre—allows it to accumulate lots of metal ions. Unlike many plants, the geranium can also take up a wide variety of metals and can grow in a variety of soils with relatively low requirements for water and nutrients. Dr. Saxena says that lemon-scented geraniums have survived as long as four months in lab conditions that would kill a less hardy plant.

By using a process called embryogenesis, Saxena's research team produces hundreds of geraniums from a small piece of plant tissue. The researchers are also testing different genes that may be inserted into the embryos to enhance the plants' ability to absorb metals. Such results are beginning to be applied at contaminated sites. Once the plants have accumulated metals from contaminated soil, they can be harvested and burned in a closed chamber, then put through a smelting procedure to recover the metals. Alternatively, the plants can be dried and disposed of at a hazardous waste site.

The lemon-scented geranium has a strong natural fragrance, but this quality does not contribute to the plant's ability to absorb toxins and is not affected by the toxins taken up. After accumulating heavy metals from the soil, the lemon-scented geranium still smells nice. Because metals and oils do not mix, the aromatic oil of the lemon-scented geranium, called citronella, can still be extracted and sold.[2]

Phytoremediation is a new pursuit, and it faces some hurdles. One is time; individual plants can take up only so much of a substance, and 5 to 20 years of repeated plantings may be needed to reduce a soil's metal content to an acceptable level. Metals also need to be in a water-soluble form. In addition, cleanup is limited to the depth of soil that plants' roots reach. Finally, plants that accumulate toxins can potentially harm insects that eat the plants and, in turn, animals that eat the insects.

Despite such obstacles, phytoremediation is catching on. Many market analysts predict success for these new technologies. The petroleum, mining, and smelting industries have many heavy metal-contaminated sites, and there should be no shortage of potential customers for the lemon-scented geranium and other hard-working plants.

(**FIGURE 2.2**). Every piece of garbage or billow of smokestack pollution or canister of nuclear waste we dispose of will not simply disappear; instead, we will need to take responsible initiatives to mitigate its impacts.

The carbon, calcium, and oxygen of which the Pavilion Lake microbialites are composed are elements. An **element** is a fundamental type of matter, a chemical substance with a given set of properties, which cannot be broken down into substances with other properties. Chemists currently recognize 92 elements occurring in nature, as well as more than 20 others that have been artificially created. Elements that are especially abundant in living organisms include **carbon**, **nitrogen**, **hydrogen**, and **oxygen** (**Table 2.1**). Each element has its own abbreviation, or chemical symbol. The *periodic table of the elements* summarizes information on the elements in a comprehensive and elegant way. (Please see Appendix C on the Companion Website at www.myenvironmentplace.ca.)

Elements are composed of **atoms**, the smallest components that maintain the chemical properties of the

FIGURE 2.2
Matter can never just disappear or go away, as much as we might sometimes wish it. In this case, oil that was spilled in a pristine bay when the *Exxon Valdez* ran aground in Alaska moved around and changed its form, but all of it ended up somewhere—in the rocks and sediments at the bottom of the bay, dispersed in the water, or clotting the feathers of thousands of seabirds, as shown here.

element (see **FIGURE 2.3**). Every atom has a nucleus consisting of **protons** (positively charged particles) and **neutrons** (particles lacking any electric charge). The atoms of each element have a defined number of protons, referred to as the element's **atomic number**. (Elemental carbon, for instance, has 6 protons in its nucleus; thus, its atomic number is 6.) An atom's nucleus is surrounded by negatively charged particles known as **electrons**, which balance the positive charge of the protons.

Isotopes Although all atoms of a given element contain the same number of protons, they do not necessarily contain the same number of neutrons. Atoms of the same element with differing numbers of neutrons are referred to as **isotopes** (**FIGURE 2.3A**). Isotopes are denoted by their elemental symbol, preceded by the **mass number**, or combined number of protons and neutrons in the atom. For example, ^{14}C (carbon-14) is an isotope of carbon with 8 neutrons (and 6 protons) in the nucleus rather than the normal 6 neutrons and 6 protons of ^{12}C (carbon-12). The atomic number of carbon is 6, therefore all the isotopes of carbon have 6 protons (if not, they would not be carbon atoms), but different numbers of neutrons are possible, leading to different mass numbers for the various isotopes of carbon.

Because they differ slightly in mass, isotopes of an element differ slightly in their behaviour. This fact has turned out to be very useful for researchers. Scientists have been able to use isotopes to study a number of phenomena that help illuminate the history of Earth's physical environment. Researchers also have used them to study the flow of nutrients within and among organisms, and the movement of organisms from one geographic location to another (see "The Science Behind the Story: How Isotopes Reveal Secrets of Earth and Life," • pp. 38–39).

Some isotopes are **radioactive** and therefore "decay" spontaneously, changing their chemical identity as they shed subatomic particles and emit high-energy radiation. **Radioisotopes** decay into lighter radioisotopes, until they become *stable isotopes,* which are not radioactive. Each radioisotope decays at a rate determined by that isotope's **half-life**, the amount of time it takes for one-half the atoms to give off radiation and decay. Each radioisotope has its own characteristic half-life, and the half-lives of various radioisotopes range from fractions of a second to billions of years. For example, the naturally occurring radioisotope uranium-235 (^{235}U) is the principal source of energy for commercial nuclear power. It decays into a series of daughter isotopes, eventually forming lead-207 (^{207}Pb), and has a half-life of about 700 million years.

Ions Atoms can also gain or lose electrons, thereby becoming **ions**, electrically charged atoms or combinations of atoms (**FIGURE 2.3B**). Ions are denoted by their elemental symbol followed by their ionic charge. For

Table 2.1 Earth's Most Abundant Chemical Elements, by Mass			
Earth's crust	**Oceans**	**Air**	**Organisms**
Oxygen (O), 49.5%	Oxygen (O), 85.8%	Nitrogen (N), 78.1%	Oxygen (O), 65.0%
Silicon (Si), 25.7%	Hydrogen (H), 10.8%	Oxygen (O), 21.0%	Carbon (C), 18.5%
Aluminum (Al), 7.4%	Chlorine (Cl), 1.9%	Argon (Ar), 0.9%	Hydrogen (H), 9.5%
Iron (Fe), 4.7%	Sodium (Na), 1.1%	Other, < 0.1%	Nitrogen (N), 3.3%
Calcium (Ca), 3.6%		Other, 0.4%	Calcium (Ca), 1.5%
Sodium (Na), 2.8%			Phosphorus (P), 1.0%
Potassium (K), 2.6%			Potassium (K), 0.4%
Magnesium (Mg), 2.1%			Sulphur (S), 0.3%
Other, 1.6%			Other, 0.5%

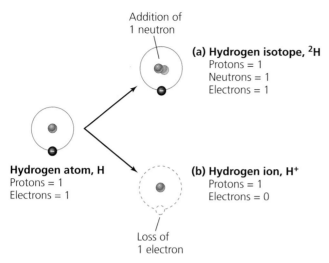

FIGURE 2.3
The mass number of hydrogen is 1 (1 proton + 0 neutrons). Deuterium, an isotope of hydrogen **(a)**, contains a neutron as well as a proton and thus it has greater mass than a typical hydrogen atom. Another isotope of hydrogen, called tritium (not shown here), has a mass number of 3 (1 proton + 2 neutrons). Note that all these variants have 1 proton—hence, they are all still "hydrogen," chemically, but they have different mass numbers because of the different numbers of neutrons in their nuclei. Shown in **(b)** is the hydrogen ion, H^+. By losing its electron, it gains a positive charge.

instance, a common ion used by mussels and clams to form shells is Ca^{2+}, a calcium atom that has lost two electrons and so has a positive charge of 2. Ions that form as a result of the loss of electrons, and which therefore carry a positive charge, are called *cations*. Ions that form as a result of gaining electrons, and which therefore carry a negative charge, are called *anions*.

Atoms bond to form molecules and compounds

Atoms can bond together to form **molecules**, combinations of two or more atoms. Molecules can contain one element or several. Common molecules containing only a single element include those of oxygen gas (O_2) and nitrogen gas (N_2), both of which are abundant in air. A molecule composed of atoms of two or more different elements is called a **compound**. Water is a compound; it is composed of two hydrogen atoms bonded to one oxygen atom, and denoted by the chemical formula H_2O. Another compound is carbon dioxide, consisting of one carbon atom bonded to two oxygen atoms; its chemical formula is CO_2.

Bonding Atoms **bond** or combine chemically because of an attraction for one another's electrons. Because the strength of this attraction varies among elements, atoms may be held together in different ways, according to whether and how they share or transfer electrons. When atoms in a molecule share electrons, they generate a *covalent bond*. For instance, two atoms of hydrogen bond to form hydrogen gas, H_2, by sharing electrons equally. Atoms in a covalent bond can also share electrons unequally, with one atom exerting a greater pull. Such is the case with water, in which oxygen attracts electrons more strongly than hydrogen does, forming what are termed *polar* covalent bonds. If the strength of attraction is sufficiently unequal, an electron may be transferred from one atom to another. Such a transfer creates oppositely charged ions that are said to form *ionic bonds*. These associations are called *ionic compounds*, or *salts*. Table salt (NaCl) contains ionic bonds between positively charged sodium cations (Na^+), each of which donated an electron, and negatively charged chloride anions (Cl^-), each of which received an electron.

Redox reactions The loss of an electron by a molecule, an atom, or an ion is an example of a process called *oxidation*, whereas the addition of an electron is an example of *reduction*. The oxidation state or oxidation number of an element is a measure of its charge in standard chemical conditions. The oxidation state of a free, uncombined element is zero. In reactions that involve oxidation and reduction (abbreviated **redox**), a change in oxidation state occurs, typically (but not always) accompanied by a transfer of electrons. Substances that induce oxidation are called *oxidizing agents*; they are electron acceptors. Substances that induce reduction are called *reducing agents*; they are electron donors. Redox reactions are very common in the environment. Photosynthesis and respiration, discussed below, are both examples of redox reactions; so is the combustion of wood, gasoline, or oil. Another common example is the formation of rust, in which iron oxidizes to Fe^{3+} and oxygen in the surrounding air receives electrons.

Mixtures and solutions Elements, molecules, and compounds can also come together in mixtures without chemically bonding or reacting. A physical mixture of two or more substances is called a **solution**, a term most often applied to liquids but also applicable to some gases and solids. Air in the atmosphere is a solution formed of constituents, such as nitrogen, oxygen, water, carbon dioxide, methane (CH_4), and ozone (O_3). Human blood, ocean water, mud, plant sap, and metal alloys, such as brass, are all solutions. Crude oil at high pressure may carry natural gas in solution and often contains other substances distributed unevenly. It is a heavy liquid mixture of many kinds of molecules consisting primarily of carbon and hydrogen atoms. Its physical properties vary with temperature, pressure, and composition.

THE SCIENCE BEHIND THE STORY

How Isotopes Reveal Secrets of Earth and Life

Dr. Keith Hobson works with Environment Canada and the University of Saskatchewan.

Isotopes have become one of the most powerful instruments in the environmental scientist's toolkit. These alternative versions of chemical elements enable scientists to date ancient materials, reconstruct past climates, and study the lifestyles of prehistoric humans. They also allow researchers to work out photosynthetic pathways, measure animals' diets and health, and trace nutrient flows through organisms and ecosystems.

Researchers studying the past often use **radiocarbon dating**. Carbon's most abundant isotope is ^{12}C, but ^{13}C and ^{14}C also occur in nature. Carbon-14 is radioactive and occurs in organisms at the same low concentration that it occurs in the atmosphere. Once an organism dies, no new ^{14}C is incorporated into its structure, and the radioactive decay process gradually reduces its store of ^{14}C, converting these atoms to ^{14}N (nitrogen-14).

The decay is slow and steady, acting like a clock. Scientists can date ancient organic materials by measuring the percentage of carbon that is ^{14}C and matching this value against the clock-like progression of decay. In this way, archeologists and paleontologists have dated prehistoric human remains; charcoal, grain, and shells found at ancient campfires; and bones and frozen tissues of recently extinct animals, such as mammoths. The most recent ice age has been dated from ^{14}C analysis of trees overrun by glacial ice sheets.

Because the half-life of ^{14}C is 5730 years, radiocarbon dating is not useful for items more than 50 000 years old; too little ^{14}C would remain to permit accurate analysis. For older items, scientists use other isotopes. Uranium-238 (with a half-life of 4.5 billion years) has been used to date very early fossils. For dating geological formations, potassium–argon dating is useful (potassium-40 decays to argon-40). Oxygen-18 has been widely used to measure changes in climate and sea level.

Researchers interested in present-day processes can use **stable isotope** analysis. Unlike radioactive isotopes, stable isotopes occur in nature in constant ratios. For instance, nitrogen occurs as 99.63% nitrogen-14 and 0.37% nitrogen-15. Ratios of isotopes are called *isotopic signatures*. The isotopic signatures of various environmental materials and processes are diagnostic—almost like fingerprints. For example, organisms tend to retain ^{15}N in their tissues but readily excrete ^{14}N. As a result, animals higher in the food chain show isotopic signatures biased toward ^{15}N, as do animals that are starving. Keith Hobson—an ecologist with the Prairie and Northern Wildlife Research Centre, Canadian Wildlife Service of Environment Canada, and the University of Saskatchewan—used nitrogen signatures to analyze the diets of seabirds and marine mammals, to show that geese fast while nesting and to trace artificial contaminants in food chains.

Hobson and other scientists have also used stable carbon isotopes for ecological studies. Plants produce food through one of three photosynthetic pathways, and the isotopic signature of carbon in plants varies among these pathways. Grasses have higher ratios of ^{13}C to ^{12}C than oak trees do, for instance, whereas cacti have intermediate ratios. When animals eat plants,

The chemical structure of the water molecule facilitates life

Water dominates Earth's surface, covering more than 70% of the globe, and its abundance is a primary reason Earth is hospitable to life. Scientists think life originated in water and stayed there for 3 billion years before moving onto land. Today every land-dwelling creature remains critically tied to water for its existence.

The water molecule's amazing capacity to support life results from its unique chemical properties. As just mentioned, the oxygen atom attracts electrons more strongly than do the two hydrogen atoms in a water molecule, resulting in a polar molecule in which the oxygen end has a partial negative charge and the hydrogen end has a partial positive charge. Because of this configuration, water molecules can adhere to one another in a special type of interaction called a *hydrogen bond,* in which the oxygen atom of one water molecule is weakly attracted to one or two hydrogen atoms of another (**FIGURE 2.4**). The weak electrical attraction of hydrogen bonding can also occur between hydrogen and certain other atoms, such as nitrogen. In water, hydrogen bonds are most stable in ice, somewhat stable in liquid water, and broken in water vapour.

These loose connections among molecules give water several properties important in supporting life and stabilizing Earth's climate:

- Water remains liquid over a wide range of temperatures. At Earth's surface, water exists in liquid form from 0°C all the way to 100°C. This means that water-based biological processes can occur in a very wide range of environmental conditions.
- Water exhibits strong cohesion. (Think of how water holds together in drops, and how drops on a surface join together when you touch them to one another.) This cohesion facilitates the transport of chemicals,

they incorporate the plants' isotopic signatures into their own tissues, and this signature passes up the food chain. As a result, carbon isotope studies can tell ecologists what an animal has been eating. Similarly, archeologists have used isotopic signatures in human bone to determine when ancient people switched from a hunter–gatherer diet to an agricultural one.

Isotopic data can even tell a scientist where an animal has been. For example, nectar-feeding bats have been shown to move seasonally between communities dominated by cacti and communities dominated by trees. Such movements have been inferred for migrating warblers, for elephants hunted for ivory, and for the oceanic movements of seals and salmon.

Recently, researchers used isotopes to track the movements of birds and other animals that migrate thousands of kilometres. This is possible because the isotopic signature of hydrogen in rainfall varies systematically across large geographic regions.

This signature gets passed from rainwater to plants, and from plants to animals, leaving a fingerprint of geographic origin in an animal's tissues. Hobson and colleagues used a combination of isotopic data from hydrogen and carbon to pinpoint the geographic origins of monarch butterflies that had migrated to communal roosts in Mexico, providing important information for their conservation (see map).

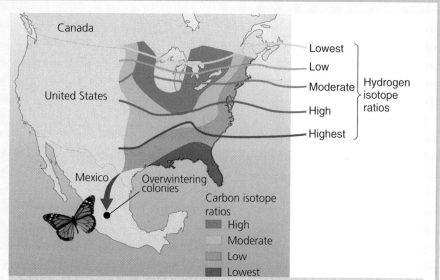

Plants in different geographic areas show different isotopic ratios for elements, such as carbon and hydrogen. Caterpillars of monarch butterflies incorporate into their tissues carbon and hydrogen in the isotopic ratios present in the plants they eat. When these caterpillars metamorphose into butterflies and migrate, they carry these isotopic signatures with them, providing scientists with clues to their origin. Shown is a map of isotopic ratios across eastern North America produced from measurements of monarchs in the summer. The four coloured bands show decreasing ratios of ^{13}C to ^{12}C from north to south. The five grey lines show increasing ratios of ^2H (heavy hydrogen or deuterium) to ^1H from north to south. By measuring carbon and hydrogen isotope ratios in monarchs wintering in Mexico, and matching the combination of these numbers against this map, researchers were able to pinpoint the geographic origin of many of the butterflies.

Source: Wassenaar, L. I., and K. A. Hobson. 1998. Proceedings of the National Academy of Sciences of the USA 95:15436–15439.

Stable isotope analysis is used to study many different environmental materials and processes today. Other elements show patterns of natural variation that have not yet been used or even discovered, researchers say, so there remains much more we can learn from the use of these subtle chemical clues.

such as nutrients and waste, in plants and animals and in the physical environment.
- Water has a high heat capacity. Initial heating weakens hydrogen bonds between molecules but does not speed molecular motion. As a result, water can absorb a large amount of heat with only small changes in its temperature. This quality helps stabilize systems against change, whether those systems are organisms, ponds, lakes, or climate systems.
- Water molecules in ice are farther apart than in liquid form (**FIGURE 2.5A**), so ice is less dense than liquid water—the reverse pattern of most other compounds, which become denser as they freeze. This is why ice floats on liquid water. Floating ice has an insulating effect that can prevent water bodies from freezing solid in winter.
- Water molecules bond well with other polar molecules, because the positive end of one molecule bonds readily to the negative end of another. As a result,

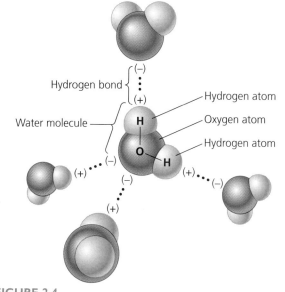

FIGURE 2.4
Water is a unique compound that has several properties crucial for life. Hydrogen bonds give water cohesion by enabling water molecules to adhere loosely to one another.

(a) Why ice floats on water

FIGURE 2.5
(a) Ice floats on water because solid ice is less dense than liquid water. This is an unusual property of H_2O—it is far more common for the solid form of a material to be denser than the liquid form. In ice, each molecule is connected to neighbouring molecules by stable hydrogen bonds, forming a spacious crystal lattice. In liquid water, hydrogen bonds frequently break and reform, and the molecules are closer together and less well organized. **(b)** Water is often called the "universal solvent" because it can dissolve so many chemicals, especially polar and ionic compounds. Seawater holds sodium and chloride ions, among others, in solution.

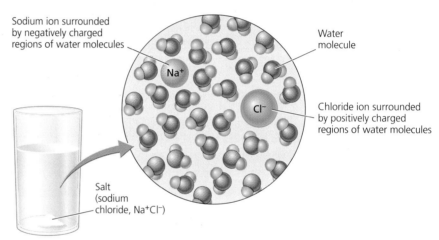

(b) Water as a solvent; how water dissolves salt

water can hold in solution, or dissolve, many other molecules, including chemicals necessary for life (**FIGURE 2.5B**). It follows that most biologically important solutions involve water.

weighing the issues — WATER'S PROPERTIES FOR LIFE — 2–1

Water has several special properties that make it accommodating to life. Think about the process of a plant taking up water through its root system; how might the chemical properties of water influence this process, or cause other materials to be taken up by the plant along with the water?

Hydrogen ions determine acidity

In any *aqueous solution* (a solution in which water is the solvent), a small number of water molecules *dissociate* or split apart, each forming a hydrogen ion (H^+) and a hydroxide ion (OH^-). The product of hydrogen and hydroxide ion concentrations is always the same; as the concentration of one increases, the concentration of the other decreases, and the product of their concentrations remains constant. Pure water contains equal numbers of these ions, and we call this water *neutral*. Most aqueous solutions, however, contain different concentrations of these two ions. Solutions in which the H^+ concentration is greater than the OH^- concentration are **acidic**; the stronger the acid, the more readily dissociation occurs and H^+ ions are released. Solutions in which the OH^- concentration is greater than the H^+ concentration are **basic**.

The **pH** scale (**FIGURE 2.6**) quantifies the acidity or basicity of solutions. The scale runs from 0 to 14; pure water has a hydrogen ion concentration of 10^{-7} and has a pH of 7. Solutions with pH less than 7 are acidic, those with a pH greater than 7 are basic, and those with a pH of 7 are neutral. Because the pH scale is logarithmic, each step on the scale represents a tenfold difference in hydrogen ion concentration. Thus, a substance with a pH of 6 contains 10 times as many hydrogen ions as a substance with a pH of 7, and a substance with a pH of 5 contains 100 times as many hydrogen ions as one with a pH of 7. **FIGURE 2.6** shows pH for a number of common substances. Industrial air pollution has intensified the acidity of precipitation, which is naturally slightly acidic, such that the pH of rain in parts of the northeastern and midwestern United States and south–central Canada now frequently dips to 4 or lower.

Matter is composed of organic and inorganic compounds

Beyond their need for water, living things also depend on organic compounds, which they create and of which they are created. **Organic compounds** consist of carbon atoms (and generally hydrogen atoms) joined by covalent bonds, often with other elements, such as nitrogen, oxygen, sulphur, and phosphorus. Carbon's unusual ability to build elaborate molecules has resulted in millions of different organic compounds, many of which are highly complex.

Chemists differentiate organic compounds from **inorganic compounds**, which also are important in the support of life. Some inorganic compounds contain carbon as a constituent, but they lack the carbon–carbon bonds that are characteristic of organic compounds. It is important to remember that in scientific terminology *organic* does not mean "natural" or

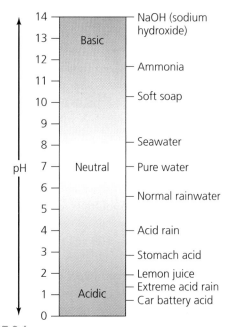

FIGURE 2.6
The pH scale measures how acidic or basic a solution is. The pH of pure water is 7, the midpoint of the scale. Acidic solutions have higher hydrogen ion concentrations and lower pH, whereas basic solutions have lower hydrogen ion concentrations and higher pH.

"environmentally friendly" or "pesticide free," as we have come to use the word in everyday language. The term *organic* does not even imply that a compound is or was once alive—it simply refers to the presence in the chemical compound of carbon-based molecules with carbon–carbon bonding.

Most biological materials, including crude oil and petroleum products, are made up of organic compounds called hydrocarbons. **Hydrocarbons** consist primarily of atoms of carbon and hydrogen (although other elements may enter the compounds, typically as impurities). The simplest hydrocarbon is **methane** (CH_4), the key component of natural gas; it has one carbon atom bonded to four hydrogen atoms (**FIGURE 2.7A**). Adding another carbon atom and two more hydrogen atoms gives us

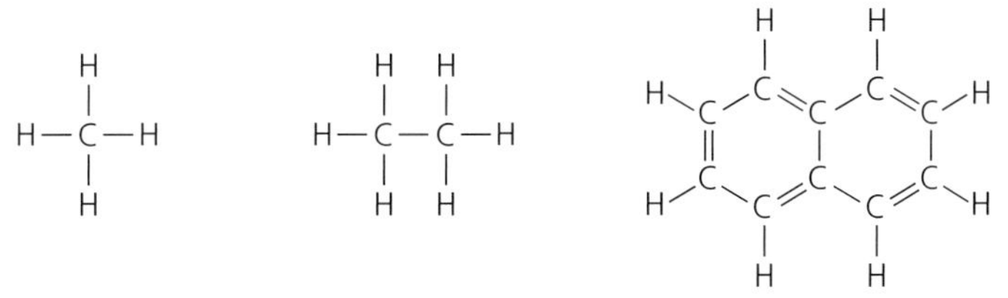

(a) Methane, CH_4 (b) Ethane, C_2H_6 (c) Naphthalene, $C_{10}H_8$ (a polycyclic aromatic hydrocarbon)

FIGURE 2.7
Hydrocarbons are a major class of organic compound, and mixtures of them make up fossil fuels, such as crude oil. The simplest hydrocarbon is methane **(a)**. Many hydrocarbons consist of linear chains of carbon atoms with hydrogen atoms attached; the shortest of these is ethane **(b)**. Volatile hydrocarbons with multiple rings, such as naphthalene **(c)**, are called polycyclic aromatic hydrocarbons (PAHs).

ethane (C_2H_6), the next-simplest hydrocarbon (FIGURE 2.7B). The smallest (and therefore lightest-weight) hydrocarbons (those consisting of four or fewer carbon atoms) exist in a gaseous state at normal temperatures and pressures. Larger (and therefore heavier) hydrocarbons are liquids, and those consisting of more than 20 carbon atoms are normally solids.

Some hydrocarbons from petroleum pose health hazards to wildlife and people, as you will see in Chapters 11, 13, 15, and 19. For example, *polycyclic aromatic hydrocarbons*, or *PAHs* (FIGURE 2.7C), which are volatile molecules with a structure of multiple carbon rings, can evaporate from spilled oil and gasoline and can mix with water. The eggs and young of fish and other aquatic creatures are often most at risk. PAHs also occur in particulate form in various combustion products, including cigarette smoke, wood smoke, and charred meat.

Macromolecules are building blocks of life

Just as the carbon atoms in hydrocarbons may be strung together in chains, other organic compounds sometimes combine to form long chains of repeated molecules. Some of these chains, called **polymers**, play key roles as the building blocks of life. Three types of polymers are essential to life: proteins, nucleic acids, and carbohydrates. Lipids are not considered polymers but are also fundamental to life. These four types of molecules are referred to as *macromolecules* because of their large size.

Proteins **Proteins** consist of long chains of organic molecules called amino acids. **Amino acids** are organic molecules in which a central carbon atom is linked to a hydrogen atom, an acidic carboxyl group (—COOH), a basic amine group (—NH_2), and an organic side chain unique to each type of amino acid (FIGURE 2.8A). Organisms combine up to 20 different types of amino acids into long chains to build proteins (FIGURE 2.8B). A protein's identity is determined by its particular sequence of amino acids and by the shape the protein molecule assumes as it folds. Protein molecules typically have highly convoluted shapes, with certain parts of the chain exposed and others hidden inside the folds (FIGURE 2.8C). A protein's folding pattern affects its function, because the position of each chemical group helps determine how it interacts with cell surfaces and with other molecules.

Proteins serve many functions. Some help produce tissues and provide structural support for the organism. For example, animals use proteins to generate skin, hair, muscles, and tendons. Some proteins help store energy, and others transport substances. Some function as components of the immune system, defending the organism against foreign attackers. Still others act as hormones, molecules that serve as chemical messengers within an organism. Finally, proteins can serve as **enzymes**, molecules that catalyze, or promote, certain chemical reactions. For example, bacteria used for bioremediation use specialized enzymes to break down hydrocarbons, just as we use enzymes to digest our food.

Nucleic acids Protein production is directed by **nucleic acids**. The two nucleic acids—**deoxyribonucleic acid (DNA)** and **ribonucleic acid (RNA)**—carry the hereditary information for organisms and are responsible for passing traits from parents to offspring. Nucleic acids are composed of series of nucleotides, each of which contains a sugar molecule, a phosphate group, and a nitrogenous base (FIGURE 2.9A). The double strands of

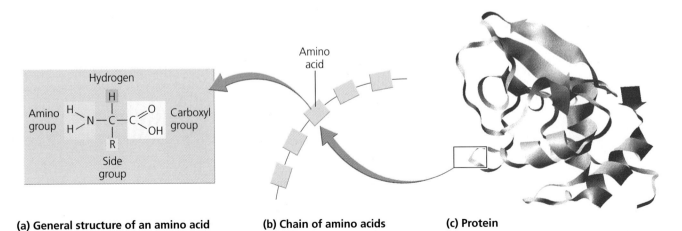

(a) General structure of an amino acid   **(b)** Chain of amino acids   **(c)** Protein

FIGURE 2.8
Proteins are polymers that are vital for life. They are made up of long chains of amino acids **(a, b)** and fold up into complex convoluted shapes **(c)** that help determine their functions.

CHAPTER TWO   FROM CHEMISTRY TO ENERGY TO LIFE   43

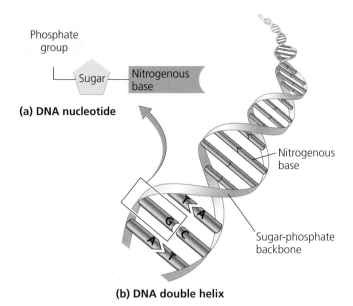

FIGURE 2.9
Nucleic acids encode genetic information in the sequence of nucleotides **(a)**, small molecules that pair together like rungs of a ladder. DNA includes four types of nucleotides, each with a different nitrogenous base: adenine (A), guanine (G), cytosine (C), and thymine (T). Adenine pairs with thymine, and cytosine pairs with guanine. In RNA, thymine is replaced by uracil (U). DNA twists in the shape of a double helix **(b)**.

DNA can be pictured like a ladder twisted into a spiral, giving the entire molecule a shape called a double helix (**FIGURE 2.9B**). RNA is similar, but it is generally single-stranded and uses ribose (instead of deoxyribose) as its sugar group.

Hereditary information encoded in the nucleotide sequence of DNA is rewritten to a molecule of RNA. RNA then directs the order in which amino acids assemble to build proteins, which go on to influence the structure, growth, and maintenance of the organism. Genetic information from DNA is passed from one generation to another as the strands replicate during cell division and egg or sperm formation. Regions of DNA coding for particular proteins that perform particular functions are called **genes**. In most organisms, the *genome*—the set of all an organism's genes—is divided among specialized areas called *chromosomes*, which act as the carriers of genetic information. Different types of organisms have different numbers of genes and chromosomes. Most bacteria have a single circular chromosome, for instance, whereas humans have 46 linear ones.

Carbohydrates   A third type of biologically vital polymer, **carbohydrates** consist of atoms of carbon, hydrogen, and oxygen. Simple carbohydrates, called *sugars* or *monosaccharides*, have structures that are three to seven carbon atoms long, and formulas that are some multiple of CH_2O. Glucose ($C_6H_{12}O_6$) is one of the most common and important sugars, providing energy that fuels plant and animal cells (**FIGURE 2.10A**). Glucose also serves as a building block for complex carbohydrates, or *polysaccharides*. Plants use *starch*, a glucose-based polysaccharide, to store energy, and animals eat plants to acquire starch (**FIGURE 2.10B**). In addition, both plants and animals use complex carbohydrates to

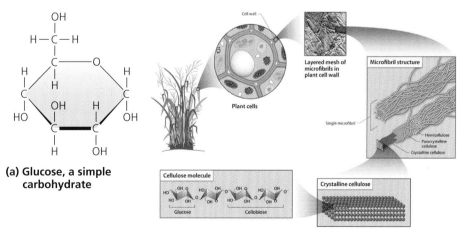

FIGURE 2.10
The monosaccharide glucose **(a)** is the simplest and most abundant carbohydrate and is a vital energy source for organisms. Linked glucose molecules form starch **(b)**, an important polysaccharide. Cellulose is an insoluble, fibrous polysaccharide that gives strength to the stems, trunks, and cell walls of plants.

build structure. Insects and crustaceans form hard shells from the carbohydrate *chitin*. *Cellulose*, the most abundant organic compound on Earth, is a complex carbohydrate found in the cell walls of leaves, bark, stems, and roots. Like starch, it is composed of glucose molecules, but bound together in a different way (**FIGURE 2.10B**). Cellulose is an insoluble, indigestible fibrous material that gives strength to plant structures and is used in the manufacture of many products that incorporate fibres, such as papers and textiles.

Lipids A fourth type of macromolecule includes a chemically diverse group of compounds called **lipids**, which are classified together because they do not dissolve in water. These include the following:

- *Fats* and *oils*, which are convenient forms of energy storage, especially for mobile animals. Their hydrocarbon structures somewhat resemble gasoline, a similarity echoed in their function: to effectively store energy and release it when burned.
- *Phospholipids*, which are similar to fats but consist of one water-repellent (or *hydrophobic*) side and one water-attracting (or *hydrophilic*) side. This characteristic allows them, when arranged in a double layer, to make up the primary component of animal cell membranes.
- *Waxes*, which are lipids that are digestible by some but not all organisms. They can play structural roles (for instance, beeswax in beehives).
- *Steroids*, which are used in animal cell membranes and in the production of hormones, including the sex hormones estrogen and androgen, which are vital to sexual maturation.

We create synthetic polymers

The polymers in nature that are so vital to our survival have inspired chemists. These scientists have taken the polymer concept and run with it, creating innumerable types of **synthetic** (human-made) polymers, which we call **plastics**. Polyethylene, polypropylene, polyurethane, and polystyrene are just a few of the many synthetic polymers in our manufactured products today (we often know them by their brand names, such as Nylon, Teflon, and Kevlar). Plastics, many of them derived from hydrocarbons in petroleum, are all around us in our everyday lives, from furniture to food containers to fibre optics to fleece jackets.

We value synthetic polymers because they resist chemical breakdown. Although plastics make our lives easier, the waste and pollution they create when we discard them is long-lasting as well. In later chapters we will see how pollutants that resist breakdown can cause problems for wildlife and human health, for water quality, for marine animals, and for waste management. Fortunately, chemists, policy makers, and citizens are finding more ways to design and use less-polluting substances and to recycle materials effectively.

Organisms use cells to compartmentalize macromolecules

Natural polymers and macromolecules help to build **cells**, the most basic unit of life's organization. All living things are composed of cells, and organisms range in their complexity from single-celled bacteria to plants and animals that contain millions of cells. Cells vary greatly in size, shape, and function.

Biologists classify organisms into two groups based on the structure of their cells. **Eukaryotes** include plants, animals, fungi, and protists. The cells of eukaryotes (**FIGURE 2.11A**) consist of an outer membrane of lipids and an inner fluid-filled chamber containing **organelles**, internal structures that perform specific functions. These internal structures include (among others) ribosomes, which are organelles that synthesize proteins, and mitochondria, where the last step in the extraction of energy from sugars and fats occurs. Eukaryotes also have within each of their cells a membrane-enclosed nucleus that houses DNA. Eukaryotic organisms generally have many cells.

Prokaryotic organisms are much simpler and probably existed on Earth for many millions of years before the emergence of the more complicated eukaryotic cells. **Prokaryotes** are generally single-celled, and their cells lack membrane-bound organelles and a nucleus (**FIGURE 2.11B**). All bacteria are prokaryotes, as are the lesser-known microorganisms called *archaea*. Bacteria are diverse and are ubiquitous in the environment, and, of course, they do far more than attack oil spills. Many types of bacteria perform functions vital to human life—for instance, aiding in digestion and preventing the buildup of harmful wastes.

In eukaryotes, cells specialize in different roles and are organized into collections of cells performing the same function, called *tissues*. Tissues make up *organs,* and organisms are composed of *organ systems*. We have now completed a (very quick!) review of the hierarchy in which matter is organized in living things on Earth (**FIGURE 2.12**). Over the next three chapters, we will explore the levels of this hierarchy above the organismal level, as we study the science of ecology. But first we will examine energy, something that underlies every process in environmental science.

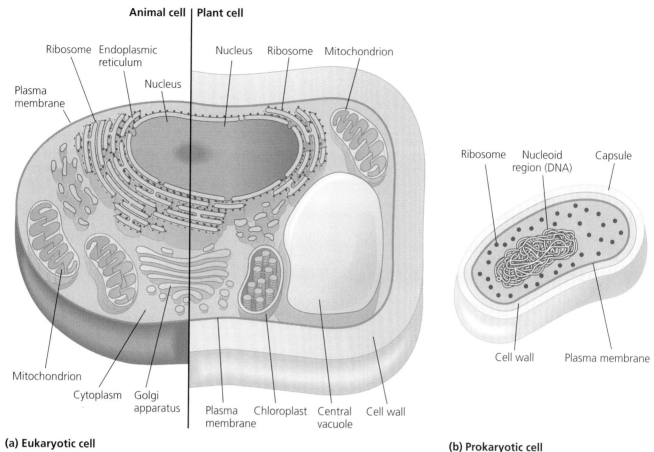

(a) Eukaryotic cell

(b) Prokaryotic cell

FIGURE 2.11
Cells are the smallest unit of life that can function independently. Eukaryotic cells **(a)** contain organelles, such as mitochondria and chloroplasts, as well as a membrane-enclosed nucleus that contains DNA. Plant cells (right) have rigid cell walls of cellulose, whereas animal cells (left) have more flexible cell membranes. Prokaryotic cells **(b)** are simpler, lacking membrane-bound organelles and an enclosed nucleus.

Energy Fundamentals

Creating and maintaining organized complexity, whether of a cell or an organism or an ecological system, requires energy. Energy is needed to power the geological forces that shape our planet; to organize matter into complex forms, such as biological polymers; to build and maintain cellular structure; and to power the interactions that take place among species. Indeed, energy is somehow involved in nearly every biological, chemical, and physical event. A sparrow in flight expends energy to propel its body through the air. When the sparrow lays an egg, its body uses energy to create the calcium-based eggshell and colour it with pigment. The sparrow sitting on its nest transfers energy from its body in heating the developing chicks inside its eggs. Some of the most dramatic releases of energy in nature do not involve living things; think of volcanoes erupting or tornadoes sweeping across the plains.

But what, exactly, is energy? Although intangible, **energy** can change the position, physical composition, or temperature of matter. Scientists differentiate between two types of energy: **potential energy**, energy of position; and **kinetic energy**, energy of motion. Consider river water held behind a dam. By preventing water from moving downstream, the dam causes the water to accumulate potential energy. When the dam gates are opened, the potential energy is converted to kinetic energy, in the form of water's motion as it rushes downstream.

Such energy transfers take place at the atomic level every time a chemical bond is broken or formed. **Chemical energy** is potential energy held in the bonds between atoms. Bonds differ in their amounts of chemical energy, depending on the atoms they hold together. Converting a molecule with high-energy bonds (such as the carbon–carbon bonds of petroleum products) into molecules with lower-energy bonds (such as the bonds in water or carbon dioxide) releases energy by changing potential energy into kinetic energy and produces motion, action, or heat. Just as our automobile engines split the hydrocarbons of gasoline to release chemical energy and generate movement, our bodies split glucose molecules in our food for the same purpose (**FIGURE 2.13**).

Hierarchy of Matter Within Organisms		
	Organism	An individual living thing
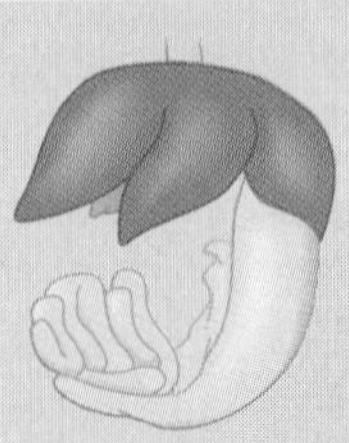	Organ system	An integrated system of organs whose action is coordinated for a particular function
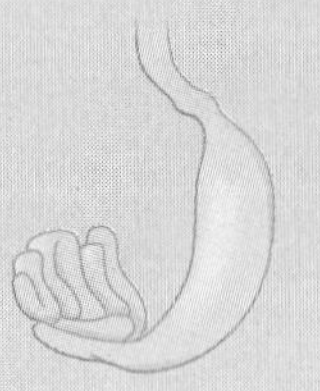	Organ	A structure in an organism composed of several types of tissues and specialized for some particular function
	Tissue	A group of cells with common structure and function
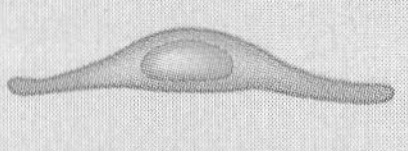	Cell	The smallest unit of living matter able to function independently, enclosed in a semi-permeable membrane
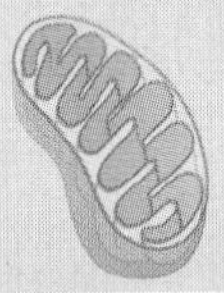	Organelle	A structure inside a eukaryotic cell that performs a particular function
	Macro-molecule	A large organic molecule (includes proteins, nucleic acids, carbohydrates, and lipids)
	Molecule	A combination of two or more atoms chemically bonded together
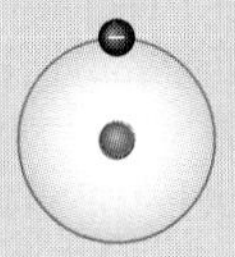	Atom	The smallest component of an element that maintains the element's chemical properties

FIGURE 2.12
Within an organism, matter is organized in a hierarchy of levels, from atoms through cells through organ systems.

In addition to occurring in the form of chemical energy, potential energy can occur as nuclear energy, the energy that holds atomic nuclei together and is released when an atom is split. It can also occur as stored mechanical energy, such as the energy in a compressed spring or a tree that bends in the wind. Kinetic energy also can take a variety of forms, though all are typically expressed through movement of electrons, atoms, molecules, or objects. In addition to movement, such as wind or running water, these include radiant or **electromagnetic energy**, which travels in photons of light; electrical energy, the movement of electrons, of which lightning is a natural example; thermal energy, or heat, expressed in the vibrational movement of atoms and molecules; and sound, which results when something causes an object to vibrate.

Energy is always conserved ...

Although energy can change from one form to another, it cannot be created or destroyed. The total energy in the universe remains constant and thus is said to be *conserved*. Scientists have dubbed this principle the **first law of thermodynamics**. The potential energy of the water behind a dam will equal the kinetic energy of its eventual movement down the riverbed. Similarly, burning converts the potential energy in a log of firewood to an equal amount of energy produced as heat and light. We obtain energy from the food we eat, which we expend in exercise, put toward the body's maintenance, or store as fat. We do not somehow create additional energy or end up with less than the food gives us. Any individual system can temporarily increase or decrease in energy, but the total amount in the universe remains constant.

... but energy can change in form and quality

Although the first law of thermodynamics requires that the overall amount of energy be conserved in any process of energy transfer, the **second law of thermodynamics** states that the nature of energy will change from a more-ordered state to a less-ordered state, if no force counteracts this tendency. That is, systems tend to move toward increasing disorder, or *entropy*. For instance, after death every organism undergoes decomposition and loses its structure. A log of firewood—the highly organized and structurally complex product of many years of slow tree growth—transforms in the campfire to a residue of carbon ash, smoke, and gases, such as carbon dioxide and

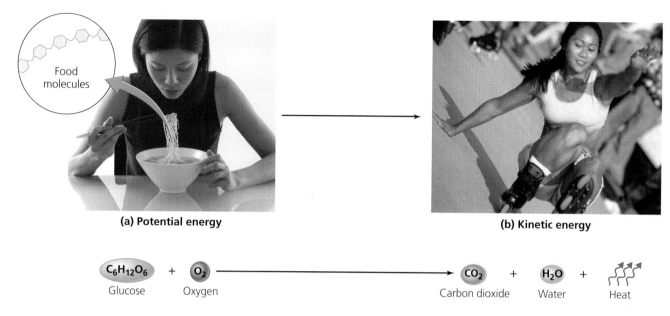

FIGURE 2.13
Energy is released when potential energy is converted to kinetic energy. Potential energy stored in sugars, such as glucose, in the food we eat **(a)**, combined with oxygen, becomes kinetic energy when we exercise **(b)**, releasing carbon dioxide and water as by-products.

water vapour, as well as the light and the heat of the flame. With the help of oxygen, the complex biological polymers that make up the wood are converted into a disorganized assortment of rudimentary molecules and heat and light energy.

The nature of any given energy source helps determine how easily humans can harness it. Such sources as petroleum products and high-voltage electricity contain concentrated energy that is easily released. It is relatively easy for us to gain large amounts of energy efficiently from these high-quality sources. In contrast, sunlight and the heat stored in ocean water are considered low-quality energy sources. Each and every day the world's oceans absorb heat energy from the Sun equivalent to that of 250 billion barrels (roughly 40 trillion litres) of oil—more than 3000 times as much as our global society uses in a year. But because this energy is spread out across such vast spaces, it is diffuse and difficult to harness.

In every transfer of energy, some portion usable to us is lost. The inefficiency of some of the most common energy conversions that power our society can be surprising. When we burn gasoline in an automobile engine, only about 16% of the energy released is used to power the automobile's movement. The rest of the energy is converted to heat. Incandescent light bulbs are worse; only 5% of their energy is converted to the light that we use them for, while the rest escapes as heat. Viewed in this context, the 15% efficiency of much current solar technology looks pretty good.

 ENERGY AVAILABILITY AND ENERGY POLICY 2–2

Contrast the ease of harnessing concentrated energy, such as that of petroleum, with the ease of harnessing highly diffuse energy, such as that of heat from the oceans. How do you think these differences have affected our society's energy policy and energy sources through the years?

Although the second law of thermodynamics specifies that systems tend to move toward disorder, the order of an object or a system can be increased through the input of additional energy from outside the system. This is precisely what living organisms do. Organisms maintain their structure and function by consuming energy. They represent a constant struggle to maintain order and combat the natural tendency toward disorder.

Light energy from the Sun powers most living systems

The energy that powers Earth's ecological systems comes primarily from the Sun. The Sun releases radiation from large portions of the electromagnetic spectrum, although our atmosphere filters much of this out, and we can see only some of this radiation as visible light (**FIGURE 2.14**).

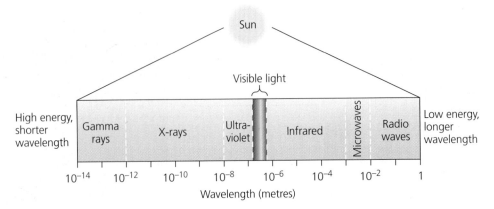

FIGURE 2.14
The Sun emits radiation from many portions of the electromagnetic spectrum. Visible light makes up only a small portion of this energy. Some radiation that reaches our planet is reflected back; some is absorbed by air, land, and water; and a small amount powers photosynthesis.

Most of the Sun's energy is reflected, or else absorbed and reemitted, by the atmosphere, land, or water (see Chapter 14). **Solar energy** drives our weather and climate patterns, including winds and ocean currents. A small amount (less than 1% of the total) powers plant growth, and a still smaller amount flows from plants into the organisms that eat them and the organisms that decompose dead organic matter. A minuscule amount of energy, relatively speaking, is eventually deposited below ground in the form of the chemical bonds in fossil fuels.

The Sun's light energy is used directly by some organisms to produce their own food. Such organisms, called **autotrophs** or **primary producers**, include green plants, algae, and cyanobacteria (a type of bacteria named for their characteristic blue-green, or cyan, colour.) Autotrophs turn light energy from the Sun into chemical energy via a process called photosynthesis (**FIGURE 2.15**). In **photosynthesis**, sunlight powers a series of chemical reactions that convert carbon dioxide and water into sugars, transforming low-quality energy from the Sun into high-quality energy the organism can use. This is called **primary production**.

Photosynthesis produces food for plants and animals

Photosynthesis occurs within cell organelles called **chloroplasts**, where the light-absorbing pigment **chlorophyll** (which is what makes plants green) uses solar energy to initiate a series of light-dependent chemical reactions. During these reactions, water molecules are split, and they react to form hydrogen ions (H^+) and molecular oxygen (O_2), thus creating the oxygen that we breathe. The light-dependent reactions also produce small, high-energy molecules that are used to fuel reactions in the **Calvin cycle**. In these reactions, carbon atoms from carbon dioxide are linked together to manufacture sugars.

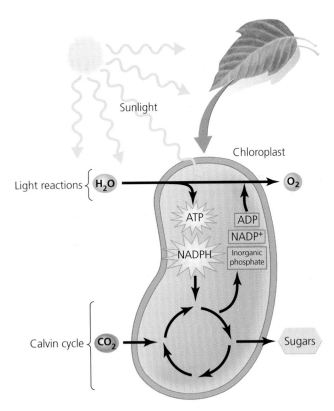

FIGURE 2.15
In photosynthesis, autotrophs, such as plants, algae, and cyanobacteria, use sunlight to convert carbon dioxide and water into sugars and oxygen. This schematic diagram summarizes the complex sets of chemical reactions that take place within chloroplasts. In the light reactions, water is converted to oxygen in the presence of sunlight, creating high-energy molecules (ATP and NADPH) that help drive reactions in the Calvin cycle, in which carbon dioxide is used to produce sugars. Molecules of ADP, $NADP^+$, and inorganic phosphate created in the Calvin cycle in turn help power the light reactions, creating an endless loop.

Photosynthesis is a complex process, but the overall reaction can be summarized with the following equation:

$$6CO_2 + 12H_2O + \text{ENERGY FROM THE SUN} \rightarrow C_6H_{12}O_6 \text{ (SUGAR)} + 6O_2 + 6H_2O$$

The numbers preceding each molecular formula indicate how many molecules of each type are involved in the reaction. Note that the sums of the numbers on each side of the equation for each element are equal; that is, there are 6 C, 24 H, and 24 O on each side. This illustrates how chemical equations are balanced, with each atom recycled and matter conserved. No atoms are lost; they are simply rearranged among molecules. Note also that water appears on both sides of the equation. The reason is that for every 12 water molecules that are input and dissociated in the process, 6 water molecules are newly created. We can streamline the photosynthesis equation by showing only the net loss of 6 water molecules:

$$6CO_2 + 6H_2O + \text{ENERGY FROM THE SUN} \rightarrow C_6H_{12}O_6 \text{ (SUGAR)} + 6O_2$$

Thus, in photosynthesis, water, carbon dioxide, and light energy from the Sun are transformed to produce sugar (glucose) and oxygen. Thus, photosynthesis is a redox reaction: carbon dioxide is reduced in the formation of sugar, and water is oxidized in the formation of molecular oxygen. To accomplish this, green plants draw up water from the ground through their roots, absorb carbon dioxide from the air through their leaves, and harness sunlight. With these ingredients, they create sugars for their growth and maintenance, and they release oxygen as a by-product. Animals, in turn, depend on the sugars and oxygen from photosynthesis. Animals survive by being **consumers** or **heterotrophs**, organisms that gain their energy by feeding on other organisms. The energy generated by consumers is called **secondary production**. They eat plants (thus becoming **primary consumers**), or animals that have eaten plants (thus becoming **secondary consumers**), and they take in oxygen. In fact, it is thought that animals appeared on Earth's surface only after the planet's atmosphere had been supplied with oxygen by cyanobacteria, the earliest autotrophs.

Cellular respiration releases chemical energy

Organisms make use of the chemical energy created by photosynthesis in a process called **cellular respiration**. To release the chemical energy of glucose, cells use the reactivity of oxygen to convert glucose back into its original starting materials; in other words, respiration oxidizes glucose to produce carbon dioxide and water. The energy released during this process is used to form chemical bonds or to perform other tasks within cells. The net equation for cellular respiration is thus the opposite of that for photosynthesis:

$$C_6H_{12}O_6 \text{ (SUGAR)} + 6O_2 \rightarrow 6CO_2 + 6H_2O + \text{ENERGY}$$

However, the energy gained per glucose molecule in respiration is only two-thirds of the energy input per glucose molecule in photosynthesis—a prime example of the second law of thermodynamics in action. Respiration occurs in autotrophs that create glucose and also in heterotrophs. Most animals are heterotrophs, including the fungi and microbes that decompose organic matter. In most ecological systems, plants, algae, or cyanobacteria form the base of a food chain through which energy passes to heterotrophs.

Geothermal energy also powers Earth's systems

Although the Sun is Earth's primary power source, it is not the only one. A minor additional source is the gravitational pull of the Moon, which causes ocean tides. A more significant additional energy source is heat emanating from inside Earth, powered primarily by radioactivity. When we think of radioactivity, nuclear power plants and atomic weapons may come to mind, but radioactivity is a natural phenomenon.

As discussed earlier, radioactivity is the release of high-energy rays or particles by radioisotopes as their nuclei spontaneously decay. Radiation from naturally occurring radioisotopes deep inside Earth heats the inside of the planet, and this heat gradually makes its way to the surface. This internal heat energy drives plate tectonics, heats magma that erupts from volcanoes, and warms groundwater (FIGURE 2.16). Called **geothermal energy**, this heat from deep within the planet is now being harnessed for commercial power in some locations where it is particularly concentrated at the surface.

Long before humans came along, geothermal energy was powering other biological communities. On the floor of the ocean, jets of geothermally heated water—essentially underwater geysers—gush into the icy-cold depths. One of the amazing scientific discoveries of recent decades was the realization that these *hydrothermal vents* can host entire communities of organisms that thrive in the extreme high-temperature, high-pressure conditions. Gigantic clams, immense tubeworms, and odd mussels, shrimps, crabs, and fish all flourish in the seemingly hostile environment near scalding water that

FIGURE 2.16
These thermal pools at Rabbitkettle Hot Springs in Nahanni National Park Reserve, Northwest Territories, are heated year-round by geothermal energy from deep below ground. The bright colours of the rocks are from colonies of bacteria that thrive in the hot mineral-laden water. The calcium carbonate deposits that form the edges of the pools were made primarily by inorganic precipitation processes; contrast this with the calcium carbonate microbialite structures from Pavilion Lake, B.C., discussed in the Central Case.

(a) Hydrothermal vent

(b) Giant tubeworms

FIGURE 2.17
Hydrothermal vents on the ocean floor **(a)** send spouts of hot mineral-rich water into the cold blackness of the deep sea. Amazingly, specialized biological communities thrive in these unusual conditions. Odd creatures, such as these giant tubeworms **(b)**, survive thanks to bacteria that produce food from hydrogen sulphide by the process of chemosynthesis.

shoots out of tall chimneys of encrusted minerals (**FIGURE 2.17**).

These locations are so deep underwater that they completely lack sunlight, so their communities cannot fuel themselves through photosynthesis. Instead, bacteria in deep-sea vents use the chemical-bond energy of hydrogen sulphide (H_2S) to transform inorganic carbon into organic carbon compounds in a process called **chemosynthesis**:

$$6CO_2 + 6H_2O + 3H_2S \rightarrow C_6H_{12}O_6 \text{ (SUGAR)} + 3 H_2SO_4$$

There are many different types of chemosynthesis, but note how this particular reaction for chemosynthesis closely resembles the photosynthesis reaction. These two processes use different energy sources, but each uses water and carbon dioxide to produce sugar and a by-product, and each produces potential energy that is later released during respiration. Energy from chemosynthesis passes through the deep-sea-vent animal community, as heterotrophs, such as clams, mussels, and shrimp, gain nutrition from chemoautotrophic bacteria. Hydrothermal vent communities excited scientists not only because they were novel and unexpected, but also because some researchers believe they may help us understand how life itself originated.

The Origin of Life

How and where life originated is one of the most centrally important—and intensely debated—questions in modern science. In searching for the answer, scientists have learned a great deal about the history of life on Earth and about what early Earth was like. That scientific interest has extended to other planets, such as Mars, which may prove to have once harboured life. We study the geological and chemical environments of both planets to learn more about what this part of the solar system was like billions of years ago, when life first took hold on this planet.

Early Earth was a very different place

Earth formed about 4.5 billion years ago in the same way as the other planets of our solar system: dispersed bits of material whirling through space around our Sun were drawn by gravity into one another, coalescing into a series of spheres. For several hundred million years after the planets formed, there remained enough stray material in the solar system that Earth and the other young planets were regularly bombarded by large chunks of debris in the form of asteroids, meteorites, and comets. The largest impacts were probably so explosive that they vaporized the newly formed oceans. Add to this the severe volcanic and tectonic activity and the intense ultraviolet radiation, and it is clear that early Earth was a pretty hostile place (**FIGURE 2.18**). Any life that emerged during this "bombardment stage" might easily have been killed off. Only after most debris was cleared from the solar system was life able to gain a foothold.

Earth's early atmosphere was very different from our atmosphere today. It was chemically reducing, and free (uncombined) oxygen was largely lacking until

FIGURE 2.18
The young Earth on which life originated was a very different place from our planet today. Microbial life first evolved amid sulphur-spewing volcanoes, intense ultraviolet radiation, frequent extraterrestrial impacts, and an atmosphere containing ammonia.

photosynthesizing microbes started producing it. Whereas today's atmosphere is dominated by nitrogen and oxygen (see Table 2.1), Earth's early atmosphere is thought to have contained large amounts of hydrogen, ammonia (NH_3), methane, carbon dioxide, carbon monoxide (CO), and water vapour. Figuring out how Earth's atmosphere evolved into its current state is an interesting and challenging area of research. We know where some of the constituents that were so abundant early in Earth's history have gone; for example, much of the carbon dioxide from the early atmosphere is now bound up in thick sequences of carbonate rocks—limestones. This is a good thing; if the carbon dioxide were released from carbonate rocks, we would have an atmosphere of 95% carbon dioxide, similar to that of Venus, and that would definitely not be conducive to life as we know it.

Several hypotheses have been proposed to explain life's origin

Most scientists interested in life's origin think that life must have begun when inorganic chemicals linked themselves into small molecules and formed organic compounds. Some of these compounds gained the ability to replicate, or reproduce themselves, whereas others found ways to group together into proto-cells. There is much debate and ongoing research, however, on the details of this process, especially concerning the location of the first chemical reactions and the energy source(s) that powered them.

Primordial soup: The heterotrophic hypothesis
The hypothesis traditionally favoured is that life evolved from a "primordial soup" of simple inorganic chemicals—carbon dioxide, oxygen, and nitrogen—dissolved in the ocean's surface waters or tidal shallows. Scientists since the 1930s have suggested how simple amino acids might have formed under these conditions and how more complex organic compounds could have followed, including simple ribonucleic acids that could replicate themselves. This hypothesis is termed *heterotrophic* because it proposes that the first life forms used organic compounds from their environment as an energy source.

Lab experiments have provided evidence that such a process can work. In 1953, biochemists Stanley Miller and Harold Urey passed electricity through a mixture of water vapour, hydrogen, ammonia, and methane, which was believed at that time to represent the early atmosphere. They were able to produce many organic compounds, including amino acids. Subsequent experiments confirmed these findings, but scientists since then have modified their ideas about early atmospheric conditions, so these experiments seem less likely to represent what actually happened.

"Seeds" from space: The panspermia hypothesis
Another hypothesis proposes that microbes from elsewhere in the solar system travelled on meteorites that crashed to Earth, seeding our planet with life. Scientists had long rejected this idea, believing that even if amino acids or bacteria were to exist in space, the searing temperatures that comets and meteors attain as they enter our atmosphere should destroy them before they reach the surface.

However, the Murchison meteorite, which fell in Australia in 1969, was found to contain many amino acids, suggesting that amino acids within rock can survive impact. Since then, experiments simulating impact conditions have shown that organic compounds and some bacteria can withstand a surprising amount of abuse. Furthermore, planetary scientists have shown that large asteroid impacts on one planet, such as Mars, can throw up so much material that some eventually may make its way to other planets, such as Earth. And recent astrobiology research suggests that comets have brought large amounts of water, and possibly organic compounds, to Earth throughout its history. As a result of such findings, long-distance travel of microbes through space and into our atmosphere now seems more plausible than previously thought.

Life from the depths: The chemoautotrophic hypothesis
In the 1970s and 1980s, several scientists proposed that life originated at deep-sea hydrothermal vents, like those in **FIGURE 2.17**, where sulphur was abundant. In this scenario, the first organisms were chemoautotrophs, creating their own food from hydrogen

sulphide. A related hypothesis suggests that life originated in the hot, moist environment of thermal pools and hot springs, like those in **FIGURE 2.16**—an environment that is presently favoured by specifically adapted types of bacteria. Current research on *extremophiles*—organisms that are adjusted to conditions of extreme heat, cold, pressure, acidity, or salinity—by scientists like Darlene Lim and her colleagues is helping to further our understanding of the earliest life forms on Earth and the environmental conditions in which they survived.

Genetic analysis of the relationships of present-day organisms suggests that some of the most ancient ancestors of today's life forms lived in extremely hot, wet environments. The extreme heat of hydrothermal vents could act to speed up chemical reactions that link atoms together into long molecules, a necessary early step in life's formation. Scientists have shown experimentally that it is possible to form amino acids and begin a chain of steps that might potentially lead to the formation of life under high-temperature, high-pressure conditions similar to those of hydrothermal vents.

weighing the issues 2–3
HYPOTHESES ON LIFE'S ORIGIN

Which lines of evidence in the debate over the origin of life strike you as the most convincing, and why? Which strike you as the least convincing, and why? Can you think of any further scientific research that could be done to address the question of how life originated?

The fossil record has taught us much about life's history

Whether the first life arose in deep-sea vents, tidal pools, or comet craters, we know that life diversified into countless forms over Earth's long history. The earliest evidence of life on Earth comes from rocks about 3.5 billion years old. Although these earliest traces are controversial, there is ample evidence that simple forms of life, such as single-celled bacteria, were present on Earth well over 3 billion years ago. Remains of these microscopic life forms (and their chemical by-products, in the form of the isotopic signatures of biological processes) have been preserved in the rock record, just as have later, much larger creatures,

such as dinosaurs. Scientists also learn about the biological processes that formed these ancient structures by studying similar processes in operation today, in places like Pavilion Lake. **FIGURE 2.19** compares modern-day structures built by algae with the ancient remains of similar structures built by algae hundreds of millions of years ago.

Ancient life forms are preserved in the rock record by the process of fossilization. As organisms die, some are buried by sediment. Under the right conditions, the hard parts of their bodies—such as bones, shells, and teeth—may be preserved as the sediments are compressed into rock. Minerals replace the organic material, leaving behind a **fossil**, a remnant or an imprint, preserved in

(a) Modern-day stromatolites, Australia

(b) Remains of ancient stromatolites, Ottawa River

FIGURE 2.19
These modern-day algal structures **(a)**, called stromatolites, shown here in a protected salt-water environment in Australia, are probably very similar to the structures built by the first bacterial life forms on Earth. Compare the modern stromatolites to the ancient remnants of algal structures from 400 million to 500 million years ago **(b)** preserved near the Champlain Bridge, Ottawa River, and to the freshwater microbialite structures currently forming at Pavilion Lake in British Columbia.

FIGURE 2.20
The fossil record has helped reveal the history of life on Earth. The numerous fossils of trilobites suggest that these animals, now extinct, were abundant in the oceans from roughly 540 million to 250 million years ago.

stone, of the dead organism (**FIGURE 2.20**). Geological processes over millions of years have buried sedimentary rock layers and later brought them to the surface, revealing assemblages of fossils representing plants and animals from different time periods. The cumulative body of fossils worldwide is known as the **fossil record**. Paleontologists study the fossil record to infer the history of past life on Earth. The rocks in which these fossils were preserved are also scientifically significant; they give us abundant information about the environments in which the organisms lived, including the chemistry, topography, other organisms, and even climate.

The fossil record clearly shows that:

- The species living today are but a tiny fraction of all the species that ever lived; the vast majority of Earth's species are long extinct.
- Earlier types of organisms changed, or evolved, into later ones.
- The number of species existing at any one time has increased through history.
- There have been several episodes of *mass extinction*, or simultaneous loss of great numbers of species in Earth's history.

The fossil record also tells us that for most of life's history, microbes, like the bacteria that consume hydrocarbons or the cyanobacteria that produce oxygen, were the only life on Earth. It was not until about 600 million years ago that large and complex organisms, such as animals, land plants, and fungi, appeared.

The crude oil with which we power modern society is itself a kind of fossil. Plant and animal matter can be preserved as it sinks to the seafloor and is buried in the absence of oxygen; eventually it becomes compressed and turns into the amorphous mixes of hydrocarbons we call fossil fuels. Coal, oil, and natural gas are the *fossil fuels* we use to power our civilization. When we drive a car, ignite a stove, or flick on a light switch, we are using energy from life that died and was buried millions of years ago and preserved in the rock record.

Present day organisms and their genes also help us decipher life's history

Besides fossils, biologists also use present-day organisms to infer how evolution proceeded in the past. By comparing the genes, the external characteristics of organisms, or both, scientists can create branching trees, similar to family genealogies, which show the relationships among organisms and thus their history of divergence through time. As you follow such a tree from its trunk to the tips of its branches, you proceed forward through time, tracing the history of life.

A major advance was made in recent years as scientists discovered an entire new domain of life, the archaea, single-celled prokaryotes that are genetically very different from bacteria. Today most biologists view the tree of life as a three-pronged edifice consisting of the bacteria, the archaea, and the eukaryotes (**FIGURE 2.21**).

We will examine the flowering of the diversity of life on our planet in Chapter 3 and further in Chapter 9. The relationships among organisms, and those between organisms and their environments, form the basis for ecology, a discipline of primary importance to environmental science.

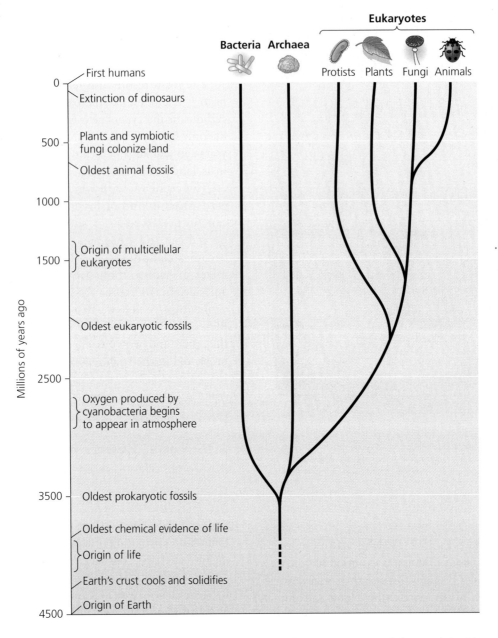

FIGURE 2.21
The fossil record and the analysis of present-day organisms and their genes allow scientists to reconstruct evolutionary relationships among organisms and to build a "tree of life." As you progress upward from the trunk of the tree to the tips of its branches, you are moving forward in time. Each fork denotes the divergence of major groups of organisms, each group of which in this greatly simplified diagram includes many thousands of species. The tree of life as understood by scientists today consists of three main groups: the bacteria, the recently discovered archaea, and the diverse eukaryotes. Protists and ancestral eukaryotes are poorly known, and their future study may well produce discoveries that further revise our understanding of life's history.

Conclusion

Life has flourished on Earth for more than 3 billion years, stemming from an origin that scientists are eagerly attempting to understand. Deciphering how life originated depends on understanding energy, energy flow, and chemistry. Knowledge in these areas also enhances our understanding of how present-day organisms interact with one another, how they relate to their nonliving environment, and how environmental systems function. Energy and chemistry are in some way tied to nearly every significant process involved in environmental science.

Chemistry can also be a tool for finding solutions to environmental problems. Cleaning up chemical pollution through bioremediation with microbes or plants is just one example. Knowledge of chemistry can be a powerful ally, whether you are interested in analyzing agricultural practices, managing water resources, reforming energy policy, conducting toxicological studies, or finding ways to mitigate global climate change.

CANADIAN ENVIRONMENTAL PERSPECTIVES

Praveen Saxena

Praveen Saxena studies traditional plants, including breadfruit, in his lab at the University of Guelph.

- **Professor** of physiology and developmental biology
- **Horticulturalist** and herbal medicine expert
- **Researcher**, Plant Cell Technology Laboratory, University of Guelph
- **Conservationist** of rare and endangered plants

Praveen Saxena is interested in plants—mainly, he is interested in plants that have been used as herbal remedies for many generations. As he points out, plants naturally produce thousands of chemicals. Some of these are potentially useful in the treatment of human ailments, but others may be seriously harmful. In addition to their natural chemical constituents, many plants bioconcentrate chemicals—some of them potentially harmful, such as heavy metals—from the surrounding environment. (This is the basis for their potential usefulness for phytoremediation; see "The Science Behind the Story: Letting Plants Do the Dirty Work.").

When we use plants as herbal remedies, we risk ingesting these harmful chemicals. Yet herbal remedies are poorly controlled in comparison with synthetic pharmaceuticals. There are wide variations in the chemical compositions of the plants used in various herbal mixtures, and gaps in the scientific understanding of how plants concentrate certain elements. One of the projects in Dr. Saxena's lab involves technologies for growing plants *in vitro* (that is, in test tubes), so their chemical characteristics can be more closely monitored and controlled, and their use as medicines rendered safer.

Dr. Saxena is also a plant conservationist. He is particularly interested in the conservation of unique, rare, and endangered medicinal, ornamental, and food crop plants. One of his main interests is in breadfruit, a nutritious food crop that is important in traditional agro-forestry systems in many parts of the world. A number of breadfruit *cultivars* are becoming rare and endangered as a result of cultural and environmental changes. Dr. Saxena and his colleagues have developed techniques for the *in vitro* conservation and multiplication of breadfruit cultivars. This technology has the potential to promote sustainable agriculture and food security in the tropics, where breadfruit is a multipurpose life-supporting crop. The scientists hope to extend their work on breadfruit to other important food and medicinal crops.[3]

"Plants have been used since ancient times to heal and cure disease, yet only now are they being grown in controlled conditions and measured for how much medicine they actually contain."
—*Praveen Saxena*

Thinking About Environmental Perspectives

We tend to think of herbal medicines as being "natural," or inherently safe. However, as Dr. Saxena points out, this is not always the case. Herbal remedies are *much* less actively studied, screened, regulated, and monitored than conventional pharmaceuticals. Do you think herbal remedies should be more tightly regulated? How might scientific research such as that of Dr. Saxena and his colleagues contribute to the safety of herbal medicines?

REVIEWING OBJECTIVES

You should now be able to:

Explain the fundamentals of chemistry and apply them to real-world environmental situations

- Understanding chemistry provides a powerful tool for developing solutions to many environmental problems.
- Atoms form molecules, and changes at the atomic level can result in alternative forms of elements, such as ions and isotopes.
- Characteristics of the water molecule help facilitate life.
- Living things depend on organic compounds, which are carbon based.

Describe the molecular building blocks of living organisms

- Proteins, nucleic acids, carbohydrates, and lipids are key building blocks of life.
- Organisms use cells to compartmentalize their component parts.

Differentiate among the types of energy and recite the basics of energy flow

- Energy can be either potential (stored energy, or energy of position) or kinetic (energy of motion). Chemical energy is an example of potential energy stored in the bonds between atoms.

- The total amount of energy in the universe is conserved; energy cannot be created or destroyed.
- Systems tend to increase in entropy, or disorder, unless energy is added to build or maintain order and complexity.
- Earth's systems are powered mainly by radiation from the Sun, as well as by geothermal heating from the planet's core, and by tidal interactions among Earth, the Sun, and the Moon.

Distinguish photosynthesis, respiration, and chemosynthesis, and summarize their importance to living things

- In photosynthesis, autotrophs use carbon dioxide, water, and solar energy to produce the sugars they need, as well as oxygen.
- In respiration, organisms extract energy from sugars by converting them in the presence of oxygen into carbon dioxide and water.
- In chemosynthesis, specialized autotrophs use carbon dioxide, water, and chemical energy from minerals to produce the sugars they need.

Itemize and evaluate the major hypotheses for the origin of life on Earth

- The heterotrophic hypothesis proposes that life arose from chemical reactions in surface or shallow waters of the ocean.
- The panspermia hypothesis proposes that substances needed for life's origin on Earth arrived from space.
- The chemoautotrophic hypothesis proposes that life arose from chemical reactions near deep-sea hydrothermal vents.

Outline our knowledge regarding early life and give supporting evidence for each major concept

- The fossil record has revealed many patterns in the history of life, including that species evolve, most species are extinct, and species numbers on Earth have increased.
- By comparing modern-day organisms scientists can infer genetic relationships among them and understand their evolutionary history.

TESTING YOUR COMPREHENSION

1. What are the basic building blocks of matter? Provide examples by using chemicals common in living organisms.
2. Name four ways in which the chemical nature of the water molecule facilitates life.
3. What is a redox reaction? Give an example of a redox reaction that occurs in nature.
4. What are the three classes of biological polymers, and what are their functions?
5. Describe the two major forms of energy, and give examples of each.
6. State the first law of thermodynamics, and describe some of its implications.
7. What are the three major sources of energy that power Earth's environmental systems?
8. What substances are produced by photosynthesis? By cellular respiration? By chemosynthesis?
9. Compare and contrast three competing hypotheses for the origin of life.
10. Name three things scientists have learned from the fossil record.

SEEKING SOLUTIONS

1. Under what types of conditions might bioremediation be a successful strategy, and when might it not be?
2. Can you think of an example of an environmental problem not mentioned in this chapter that a good knowledge of chemistry could help us solve?
3. Describe an example of energy transformation from one form to another that is not mentioned in this chapter.
4. Give three examples of ways in which the input of energy can impede the tendency toward disorder that the second law of thermodynamics describes.
5. Referring to the chemical reactions for photosynthesis and respiration, provide an argument for why increasing amounts of carbon dioxide in the atmosphere because of global climate change might potentially increase amounts of oxygen in the atmosphere. Give an argument for why it might potentially decrease amounts of atmospheric oxygen. What would you need to know to determine which of these two outcomes might occur?
6. **THINK IT THROUGH** The ministry of the environment has put you in charge of cleaning up an old

industrial site so that it meets safety standards and does not contaminate drinking water supplies. Your staff's initial inspection of surface soil at the site shows that it is contaminated with oil and with lead, a toxic heavy metal. Your job allows you to engage experts in bioremediation and phytoremediation at local universities, as well as environmental engineers in the private sector. You have a budget of several million dollars and five years to get the job done. What steps will you take to get the site cleaned up? Describe scientific research you would commission, economic questions you would ask, and engineering options you might consider.

INTERPRETING GRAPHS AND DATA

In phytoremediation, plants are used to clean up soil or water contaminated by heavy metals, such as lead (Pb), arsenic (As), zinc (Zn), and cadmium (Cd). For plants to absorb these metals from soil, the metals must be dissolved in soil water. For any given instance, all metal can be accounted for as either remaining bound to soil particles, being dissolved in soil water, or being stored in the plant.

In a study on the effectiveness of alpine penny-cress (*Thlaspi caerulescens*) for phytoremediation, Enzo Lombi and his colleagues grew crops of this small perennial plant for approximately one year in pots of soil from contaminated sites. They then measured the amount of zinc and cadmium in the soil and in the plants when they were harvested.

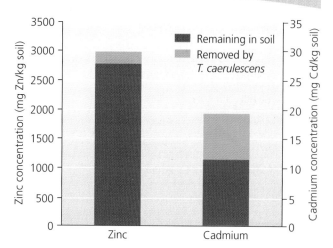

Removal of zinc and cadmium from contaminated soil by alpine penny-cress, *Thlaspi caerulescens*. Data from Lombi, E., et al. 2001. Phytoremediation of heavy metal-contaminated soils: natural hyperaccumulation versus chemically enhanced phytoextraction. *Journal of Environmental Quality* 30:1919–1926.

1. What were the zinc and cadmium concentrations in the soil prior to phytoremediation? What were the zinc and cadmium concentrations in the soil after one year of phytoremediation?
2. How much zinc and cadmium were removed from the soil? If the plants continued to remove zinc and cadmium from the soil at the rates shown above, approximately how long would it take to remove all the zinc and cadmium?
3. Alpine penny-cress is one of many plants that produce natural chelating agents (see "The Science Behind the Story: Letting Plants Do the Dirty Work," that increase the solubility of metals in soil water. If these dissolved metals were *not* subsequently taken up by the plants, what might be an unintended consequence of having increased their solubility?

CALCULATING FOOTPRINTS

In ecological systems, a rough rule of thumb is that when energy is transferred from plants to plant-eaters or from prey to predator, the efficiency is only about 10%. Much of this inefficiency is a consequence of the second law of thermodynamics. Another way to think of this is that eating 10 calories of plant material is the ecological equivalent of eating one calorie of material from an animal.

Humans are considered omnivores because we can eat both plants and animals. The choices we make about what to eat have significant ecological impacts. With this in mind, calculate the ecological energy requirements for four different diets, each of which provides a total of 2000 dietary calories per day.

1. How many ecologically equivalent calories would it take to support you for a year on each of the four diets listed?
2. What is the relative ecological impact of including as little as 10% of your calories from animal sources (e.g., milk, dairy products, eggs, and meat)? What is

the ecological impact of a strictly carnivorous diet compared with a strict vegetarian diet?

3. What percentages of the calories in your own diet do you think come from plant versus animal sources? Estimate the ecological impact of your diet, relative to a strictly vegetarian one.

4. Describe some challenges of providing food for the growing human population, especially as people in many poorer nations develop a taste for an American-style diet rich in animal protein and fat.

Diet	Source of Calories	Number of Calories Consumed By Source	Ecologically Equivalent Calories By Source	Total Ecologically Equivalent Calories Per Day
100% plant	Plant			0.00
0% animal	Animal			
90% plant	Plant	1800	1800	3800
10% animal	Animal	200	2000	
50% plant	Plant			
50% animal	Animal			
0% plant	Plant			
100% animal	Animal			

TAKE IT FURTHER

Go to www.myenvironmentplace.ca where you will find

- Suggested answers to end-of-chapter questions
- Quizzes, animations, and flashcards to help you study
- *Research Navigator*™ database of credible and reliable sources to assist you with your research projects
- Tutorials to help you master how to interpret graphs
- Current news articles that link the topics that you study to case studies from your region and around the world

ECO Occupational Profiles: If you found this chapter especially interesting, you might want to learn more about the following jobs by visiting the Occupational Profiles website of the Environmental Careers Organization. Go to www.eco.ca and check out the following careers:
- Analytical chemist
- Biochemist
- Chemical technician
- Environmental chemist
- Microbiologist

CHAPTER ENDNOTES

1. Pavilion Lake Research Project: Relevance to Astrobiology and Space Exploration, www.pavilionlake.com.
2. Based on Fragrant Geraniums Clean Pollutants from Soil, www.carleton.ca/jmc/cnews/30031998/story1.html
3. University of Guelph, Department of Plant Agriculture, www.plant.uoguelph.ca/faculty/psaxena/, and Plant Cell Technology Laboratory, www.plant.uoguelph.ca/research/cellculture/index.html, among other sources.

3 Evolution, Biodiversity, and Population Ecology

The Monteverde cloud forest in Costa Rica is a hotspot of biodiversity.

Upon successfully completing this chapter, you will be able to

- Explain the process of natural selection, and cite evidence for this process
- Describe the ways in which evolution results in biodiversity
- Discuss reasons for species extinction and mass extinction events
- List the levels of ecological organization
- Outline the characteristics of populations that help predict population growth
- Assess logistic growth, carrying capacity, limiting factors, and other fundamental concepts of population ecology
- Identify efforts and challenges involved in the conservation of biodiversity

The golden toads of Monteverde have not been observed since 1989 and are believed to be extinct.

CENTRAL CASE:
STRIKING GOLD IN A COSTA RICAN CLOUD FOREST

"What a terrible feeling to realize that within my own lifetime, a species of such unusual beauty, one that I had discovered, should disappear from our planet."
—DR. JAY M. SAVAGE, DESCRIBING THE GOLDEN TOAD IN 1998

"To keep every cog and wheel is the first precaution of intelligent tinkering."
—ALDO LEOPOLD

During a 1963 visit to Central America, biologist Jay Savage heard rumours of a previously undocumented toad living in Costa Rica's mountainous Monteverde region. The elusive amphibian, according to local residents, was best known for its colour: a brilliant golden yellow-orange. Savage was told that the toad was hard to find because it appeared only during the early part of the region's rainy season.

Monteverde means "green mountain" in Spanish, and the name could not be more appropriate. The settlement of Monteverde sits beneath the verdant slopes of the Cordillera de Tilarán, mountains that receive more than 400 cm of annual rainfall. Some of the lush forests above Monteverde, which begin at an altitude of around 1600 m, are known as *lower montane rainforests*. They are also known as *cloud forests* because much of the moisture they receive arrives in the form of low-moving clouds that blow inland from the Caribbean Sea. Monteverde's cloud forest was not fully explored at the time of Savage's first visit, and researchers who had been there described the area as pristine, with a rich bounty of ferns, liverworts, mosses, clinging vines, orchids, and other organisms that thrive in cool, misty environments. Savage knew that such conditions create ideal habitat for many toads and other amphibians.

In May of 1964, Savage organized an expedition into the muddy mountains above Monteverde to try to document the existence of the previously unknown toad species in its natural habitat. Late in the afternoon of May 14, he and his colleagues found what they were looking for. Approaching the mountain's crest, they spotted bright orange patches on the forest's black floor. In one area that was only 5 m in diameter, they counted 200 golden toads.

The discovery received international attention, making a celebrity of the tiny toad—which Savage named *Bufo periglenes* (literally, "brilliant toad")—and making a travel destination of its mountain home. At the time, no one knew that the Monteverde ecosystem was about to be transformed. No one foresaw that the oceans and atmosphere would begin warming because of global climate change and cause Monteverde's moisture-bearing clouds to rise, drying the forest. No one could guess that this newly discovered species of toad would become extinct in less than 25 years. By 1988 the annual count of golden toads had dropped to one individual; the count was the same in 1989, and not a single golden toad has been observed anywhere since then, in spite of extensive searches. Listed previously as "endangered," the toad's extinction was declared final in 2004 on the IUCN (International Union of Conservation Network) *Red List of Threatened Species* by Dr. Jay Savage, who first identified the new species in 1966.[1]

Evolution as the Wellspring of Earth's Biodiversity

The golden toad was new to science, and countless species still await discovery, but scientists understand quite well how the world became populated with the remarkable diversity of organisms we see today. We know that the process of biological evolution has brought us from a stark planet inhabited solely by microbes to a lush world of 1.5 million (and likely millions more) species (**FIGURE 3.1**).

Perceiving how organisms adapt to their environments and change over time is crucial for understanding the history of life. Understanding evolution is also vital for appreciating ecology, a central component of environmental science. Evolutionary processes are relevant to many aspects of environmental science, including pesticide resistance, agriculture, medicine, and environmental health.

The term *evolution* in the broad sense means change over time, but scientists most often use the term to refer specifically to biological evolution. Biological **evolution** consists of genetic change in organisms across generations. This genetic change often leads to modifications in the appearance, functioning, or behaviour of organisms through time. Biological evolution results from random genetic changes and may proceed randomly or may be directed by natural selection.

Natural selection is the process by which traits that enhance survival and reproduction are passed on more frequently to future generations than those traits that do not, altering the genetic makeup of populations through time. The theory of evolution by natural selection is one of the best supported and most illuminating concepts in all of science. Recall from Chapter 1 that to be designated as a theory, a scientific hypothesis must be widely accepted and extensively validated, surviving repeated testing by numerous research teams and successfully and accurately predicting experimental outcomes and observations. From a scientific standpoint, evolutionary theory is indispensable, because it is the foundation of modern biology.

Natural selection shapes organisms and diversity

In 1858, Charles Darwin and Alfred Russell Wallace each independently proposed the concept of natural selection as a mechanism for evolution and as a way to explain the great variety of living things. Darwin and Wallace were English naturalists who had studied plants and animals in such exotic locales as the Galápagos Islands and the Malay Archipelago. Both men recognized that organisms face a constant struggle to gain sufficient resources to survive and reproduce. Both were influenced by the writings of Thomas Malthus, who hypothesized that human population growth would outstrip resource availability and lead to widespread death and social upheaval. Darwin and Wallace observed that organisms produce more offspring than can possibly survive, and they realized that some offspring may be more likely than others to survive and reproduce. Furthermore, they recognized that whichever characteristics give certain individuals an advantage in surviving and reproducing might be inherited by their offspring. These characteristics, Darwin and Wallace realized, would tend to become more prevalent in the population in future generations.

Natural selection is a simple concept that offers an astonishingly powerful explanation for patterns apparent in nature. The idea of natural selection follows logically from a few straightforward premises (Table 3.1). One is that

FIGURE 3.1
Much of our planet's biological diversity resides in tropical rainforests. Monteverde's cloud-forest community includes organisms such as these: **(a)** resplendent quetzal (*Pharomachrus mocinno*), **(b)** puffball mushroom (*Calostoma cinnabarina*), **(c)** harlequin frog (*Atelopus varius*), and **(d)** scutellerid bug (*Pachycoris torridus*).

individuals of the same species vary in their characteristics. Although not known in Darwin and Wallace's time, we now know that variation is due to differences in genes, the environments within which genes are expressed, and the interactions between genes and environment. As a result of this variation, some individuals within a species will happen to be better suited to their particular environment than others and thus will be able to survive longer and/or reproduce more. (Note, however, that the same genes that make an individual particularly well suited to one environment might actually be disadvantageous in a different environment.)

Table 3.1   The Logic of Natural Selection

- Organisms produce more offspring than can survive.
- Individuals vary in their characteristics.
- Many characteristics are inherited by offspring from parents.

Therefore,

- Some individuals will be better suited to their environment than other individuals and thus will be able to survive longer and/or reproduce more
- By producing more offspring and/or offspring of higher quality, better suited individuals transmit more genes to future generations than poorly suited individuals
- Future generations will contain more genes, and thus more characteristics, of the better-suited individuals; as a result, characteristics evolve across generations through time

Many, many scientists since the time of Darwin and Wallace have contributed to our understanding of evolution and have helped to refine it and move it from the status of a hypothesis to an accepted theory. As with most scientific theories, however, our understanding of evolution grows and evolves with the continued contributions of modern scientists.

Perhaps no modern scientist has made more important contributions to the understanding of evolution than Stephen Jay Gould, a Harvard University professor of paleontology. Gould (with Niles Eldridge) developed a refinement of the theory of evolution, which stated that evolutionary change tends to occur relatively quickly, in short bursts separated by long periods of stability. This is referred to as *punctuated equilibrium*, in contrast to the idea that evolution is an extremely slow, gradual, and continuous process. This idea continues to stir discussion among evolutionary biologists, although Gould himself did not see it as being in any way inconsistent with prior understandings of the evolutionary process.

Natural selection acts on genetic variation

Many characteristics are passed from parent to offspring through the genes, and a parent that is long-lived, robust, and produces many offspring will pass on genes to more offspring than a weaker, shorter-lived individual that produces only a few offspring. In the next generation, therefore, the genes of better adapted individuals will be more prevalent than those of less well adapted individuals. From one generation to another through time, species will evolve to possess characteristics that lead to better and better success in a given environment. A trait that promotes success is called an **adaptive trait**, or an *adaptation*. A trait that reduces success is said to be *maladaptive*.

For an organism to pass a trait along to future generations—that is, for the trait to be *heritable*—genes in the organism's DNA must code for the trait. Accidental alterations that arise during DNA replication give rise to genetic variation among individuals. In an organism's lifetime, its DNA will be copied millions of times by millions of cells. In all this copying and recopying, sometimes a mistake is made. Accidental changes in DNA, called **mutations**, can range in magnitude from the addition, deletion, or substitution of single nucleotides to the insertion or deletion of large sections of DNA. If a mutation occurs in a sperm or egg cell, it may be passed on to the next generation. Although most mutations have little effect, some can be deadly, whereas others can be beneficial. Those that are not lethal provide the genetic variation on which natural selection acts.

Sexual reproduction also generates variation. In sexual organisms, genetic material is mixed, or recombined, so that a portion of each parent's genome is included in the genome of the offspring. This process of *recombination* produces novel combinations of genes, generating variation among individuals.

Genetic variation can lead to variation in organismal-level traits. We can visualize how traits vary by using distribution graphs, which help us see that selection can alter the characteristics of organisms through time in three main ways (**FIGURE 3.2**). Selection that drives a feature in one direction rather than another—for example, toward larger or smaller, faster or slower—is called **directional selection**. In contrast, **stabilizing selection** produces intermediate traits, in essence preserving the status quo. Under **disruptive selection**, traits diverge from their starting condition in two or more directions.

An organism's environment determines what pressures natural selection will exert on the organism. Sometimes two completely separate and distinct species evolve similar

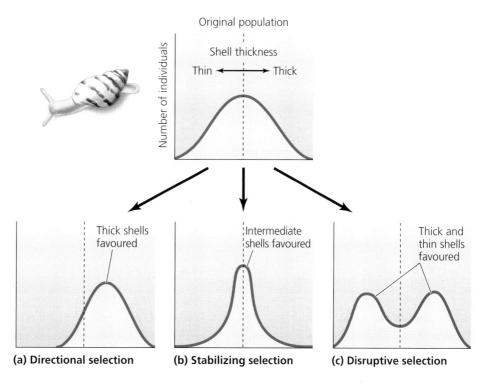

FIGURE 3.2
Selection can act in three ways. Assume that there is a population of snails with shells of different thicknesses (top graph). Because shells protect snails against predators, snails with thick shells may be favoured over those with thin shells, through *directional selection* **(a)**. Alternatively, suppose that a shell that is too thin breaks easily, whereas a shell that is too thick wastes resources that are better used for feeding or reproduction. In such as case, *stabilizing selection* **(b)** could act to favour snails with shells that are neither too thick nor too thin. Under *disruptive selection* **(c)**, extreme traits are favoured. For example, perhaps thin-shelled snails are so resource-efficient that they can outreproduce intermediate-shelled snails, whereas thick-shelled snails are so well protected from predators that they also outreproduce intermediate-shelled snails. In such a case, each of the "extreme" strategies works more effectively than a compromise between the two, and natural selection increases the relative numbers of thin- *and* thick-shelled snails, while reducing the number of intermediate-shelled ones.

traits, as a result of adapting to selective pressures from similar environments; this is called **convergent evolution**. For example, many desert plants have independently evolved similar characteristics in response to the rigours of life in an arid environment; these adaptations include the absence of large leaf surfaces, to minimize evaporation, and thickened stems to enhance water storage.

Organisms respond, over time, to the selective pressures exerted by the physical characteristics of the environment in which they live by adapting and evolving the specific traits that allow them to be successful in that environment. In turn, once an organism has become specifically adapted to a particular habitat, it may become highly dependent on the characteristics and resources offered by that environment. If this environment is very limited in scope—like the cloud forest pools required by the golden toads—this may threaten the species' chances for long-term survival. Golden toads that were well adapted to the moist conditions of Monteverde's cloud forest did not persist after Monteverde's climate became drier starting 25 years ago.

In all these ways, variable genes and variable environments interact as organisms engage in a perpetual process of adapting to the changing conditions around them. Natural selection does not simply weed out individuals that are ill-suited to the environment; it also allows for the development and diversification of traits that may lead to the emergence of new species and completely new types of organisms. A burst of species diversification that occurs in response to environmental change is called an **adaptive radiation**.

---

weighing the issues — ARTIFICIAL SELECTION 3-1

Consider the pets and farm animals that humans have domesticated through artificial selection, such as horses, dogs, and cats. In what ways do they differ from their wild relatives? For thousands of years artificial selection has allowed plant and animal breeders to create new breeds by the careful selection of valued traits, expressed and reinforced through numerous generations of reproduction. If breeders could generate a new, prize-winning or useful breed of horse or tomato or chicken or corn or rose *without* having to wait several generations for the desired traits to appear, do you think they should make use of it? (This question will become more than just academic in Chapter 8, where we will look at new technologies for the genetic modification of crops.)

Evidence of natural selection is all around us

The results of natural selection are all around us, visible in every adaptation of every organism (**FIGURE 3.3**). In addition, countless lab experiments (mostly with fast-reproducing organisms, such as bacteria and fruit flies) have demonstrated rapid evolution of traits. The evidence for selection that may be most familiar to us is that which Darwin himself cited prominently in his work 150 years ago: our breeding of domestic animals. In our dogs, our cats, and our livestock, we have conducted our own version of selection. We have chosen animals with traits we like and bred them together, while not breeding those with variants we do not like. Through such **selective breeding**, we have been able to exaggerate particular traits we prefer.

Consider the great diversity of dog breeds, all of which are variations on a single subspecies, *Canis lupus familiaris*. From Great Dane to Chihuahua, they can interbreed freely and produce viable offspring, yet breeders maintain the striking differences between them by allowing only like individuals to breed with like. This process of selection conducted under human direction is termed **artificial selection**.

Artificial selection has also given us the many crop plants we depend on for food, all of which were domesticated from wild ancestors and carefully bred over years, centuries, or millennia. Through selective breeding, we have created corn with larger, sweeter kernels; wheat and rice with larger and more numerous grains; and apples, pears, and oranges with better taste. We have diversified single types into many, for instance, breeding variants of the plant *Brassica oleracea* to create broccoli, cauliflower, cabbage, and Brussels sprouts. Our entire agricultural system is based on artificial selection.

Evolution generates biological diversity

When Charles Darwin wrote about the wonders of a world full of diverse animals and plants, he conjured up the vision of a "tangled bank" of vegetation harbouring all kinds of creatures. Such a vision fits well with the arching vines, dripping leaves, and mossy slopes of the tropical cloud forest of Monteverde. Indeed, tropical forests worldwide teem with life and harbour immense biological diversity (see **FIGURE 3.1**).

Biological diversity, or **biodiversity**, refers to the sum total of all organisms in an area, taking into account the diversity of species, their genes, their populations, and their communities. A **species** is a particular type of

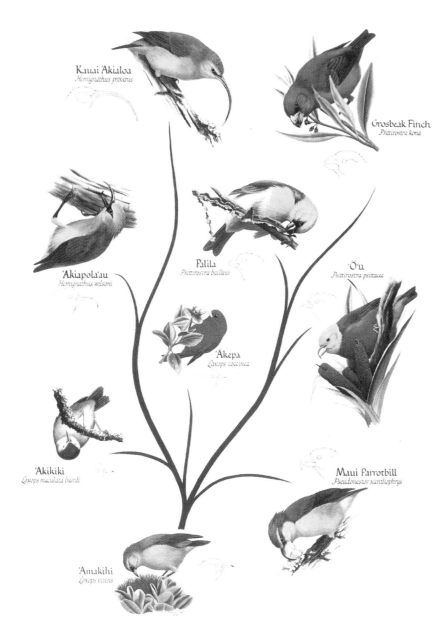

FIGURE 3.3
Natural selection has produced tremendous diversity among organisms in the wild. In the group of birds known as Hawaiian honeycreepers, closely related species have adapted to different food resources, physical environments, or ways of life, as indicated by the diversity in their plumage colours and the shapes of their bills. Such a burst of species formation caused by natural selection is known as an *adaptive radiation*.

organism or, more precisely, a population or group of populations whose members share certain characteristics and can freely breed with one another and produce fertile offspring. A **population** is a group of individuals of a particular species that live in the same area. We have already discussed genes, and we will introduce communities shortly (Chapter 4).

Scientists have described between 1.5 million and 1.8 million species, but many more remain undiscovered or unnamed. Estimates for the total number of species in the world range up to 100 million, with many of them thought to occur in tropical forests. In this light, the discovery of a new toad species in Costa Rica in 1964 seems far less surprising. Although Costa Rica covers a tiny fraction (0.01%) of Earth's surface area, it is home to 5%–6% of all species known to scientists. And of the 500 000 species that scientists estimate exist in the country, only 87 000 (17.4%) have been inventoried and described.

Tropical rainforests, such as Costa Rica's, are by no means the only places rich in biodiversity. Step outside anywhere on Earth, even in a major city, and you will find numerous species within easy reach. They may not always be large and conspicuous, like polar bears or blue whales or elephants, but they will be there. Plants poke up from cracks in asphalt in every city in the world, and even Antarctic ice harbours microbes. In a handful of backyard soil there may exist an entire miniature world of life, including several insect species, several types of mites, a millipede or two, many nematode worms, a few plant seeds, countless fungi, and millions upon millions of bacteria. We will examine Earth's biodiversity in detail in Chapter 9.

Speciation produces new types of organisms

How did Earth come to have so many species? Whether there are 1.5 million or 100 million, such large numbers require scientific explanation. The process by which new species are generated is termed **speciation**. Speciation can occur in a number of ways, but most biologists consider the main mode of species formation to be **allopatric speciation**, the emergence of a new species as a result of the physical separation of populations over some geographic distance. To understand allopatric speciation, begin by picturing a population of organisms. Individuals within the population possess many similarities that unify them as a species, because they are able to reproduce with one another and share genetic information. However, if the population is broken up into two or more populations that become isolated from one another, individuals from one population cannot reproduce with individuals from the others.

When mutations or natural variations are present in the DNA of an organism in one of these isolated populations, it cannot spread to the other populations. Over time, each population will independently accumulate its own set of variations. Eventually, the populations may diverge, or grow different enough, that their members can no longer mate with one another; this process is referred to as **divergent evolution**. Individuals from the two differing populations may no longer recognize one another as being the same species because they have diverged so much in appearance or behaviour, or they may simply become genetically incapable of producing viable offspring.

Once this has happened, there is no going back; the two populations cannot interbreed, and they have embarked on their own independent evolutionary trajectories as separate species (**FIGURE 3.4**). The populations will continue diverging in their characteristics as chance variations accumulate that confer traits, causing the populations to become different in random ways. If environmental conditions happen to be different for the two

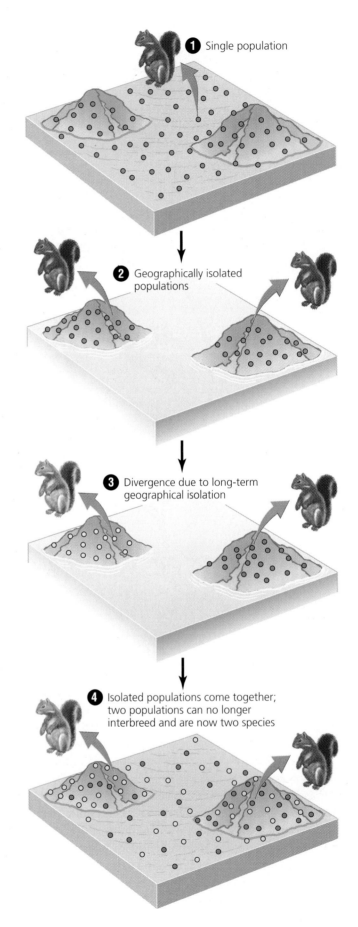

FIGURE 3.4
Allopatric speciation has generated much of Earth's diversity. In this process, some geographic barrier splits a population. In this diagram, two mountaintops (**1**) are turned into islands by the rising sea level (**2**), isolating populations of squirrels. Each isolated population accumulates its own independent set of genetic changes over time, until individuals become genetically distinct and unable to breed with individuals from the other population (**3**). The two populations now represent separate species and will remain so even if the geographic barrier is removed and the new species intermix (**4**).

populations, then natural selection may accelerate the divergence. Through the speciation process, single species can generate multiple species, each of which can in turn generate more.

Populations can be separated in many ways

The long-term geographic isolation of populations can occur in various ways (Table 3.2). Glacial ice sheets may move across continents during ice ages and split populations in two. Major rivers may change course and do the same. Mountain ranges may rise and divide regions and their organisms. Drying climate may partially evaporate lakes, subdividing them into multiple smaller bodies of water. Warming or cooling temperatures may cause whole plant communities to move northward or southward, or upslope or downslope, creating new patterns of plant and animal distribution. Regardless of the mechanism of separation, in order for speciation to occur, populations must remain isolated for a long time, generally thousands of generations.

If the geologic or climatic process that has isolated populations reverses itself—the glacier recedes or the river returns to its old course or warm temperatures turn cool again—then the populations can come back together. This is the moment of truth for speciation. If the populations have not diverged enough, their members will begin interbreeding and reestablish gene flow, mixing the variations that each population accrued while isolated. However, if the populations have diverged sufficiently, they will not interbreed; two species will have been formed, each fated to continue on its own evolutionary path.

Although allopatric speciation has long been considered the main mode of species formation, speciation appears to occur in other ways as well. **Sympatric speciation** occurs when species form from populations that become reproductively isolated, occupying a new ecological niche within the same geographic area. For example, populations of some insects may become isolated by feeding and mating exclusively on different types of plants. Or they may mate during different seasons, isolating themselves in time rather than space. In some plants, speciation apparently has occurred as a result of hybridization between species. In others, it seems to have resulted from natural variations, or from mutations that changed the numbers of chromosomes, creating plants that could not mate with plants with the original number of chromosomes. Garnering solid evidence for speciation mechanisms is difficult, so biologists still actively debate the relative importance of each of these modes of speciation.

Life's diversification results from numerous speciation events

Repeated speciation events have generated complex patterns of diversity at levels above the species level. Such patterns are studied by evolutionary biologists, who examine how groups of organisms arose and evolved the characteristics they show. For instance, how did we end up with plants as different as mosses, palm trees, daisies, and redwoods? Why do fish swim, snakes slither, and sparrows sing? How and why did the ability to fly evolve independently in birds, bats, and insects? To address such questions, scientists need to know how the major groups diverged from one another, and this pattern ultimately results from the history of individual speciation events.

We saw in Chapter 2 how the history of divergence can be represented in a treelike diagram (**FIGURE 2.21**). Such branching diagrams, called *cladograms*, or **phylogenetic trees**, illustrate scientists' hypotheses as to how divergence took place (**FIGURE 3.5**). Phylogenetic trees can show relationships among species, among major groups of species, among populations within a species, or even among individuals. In addition, by mapping traits onto a tree according to which organisms possess them, we can trace how the traits themselves may have evolved. For instance, the tree of life shows that birds, bats, and insects are distantly related, with many other flightless groups between them. So, it is far simpler to conclude that the three groups evolved flight independently than to conclude that the many flightless groups all lost an ancestral ability to fly. Because phylogenetic trees help biologists make such inferences about so many traits, they have become one of the modern biologist's most powerful tools.

Table 3.2 Natural Mechanisms of Population Isolation That Can Give Rise to Allopatric Speciation

- Glacial ice sheets advance.
- Mountain chains are uplifted.
- Major rivers change course.
- Sea level rises, creating islands (see **FIGURE 3.4**).
- Climate warms, pushing vegetation up mountain slopes and fragmenting it.
- Climate dries, dividing large single lakes into multiple smaller lakes.
- Ocean current patterns shift.
- Islands are formed in the sea by volcanism.

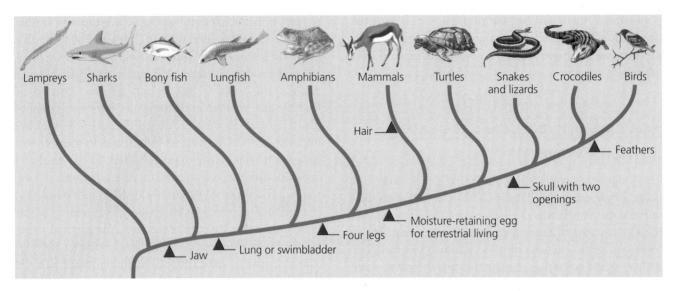

FIGURE 3.5
Phylogenetic trees show the history of life's divergence. Similar to family genealogies, these trees illustrate relationships among groups of organisms, as inferred from the study of similarities and differences among present-day creatures. The diagram here is a greatly simplified representation of relationships among a few major groups—one small portion of the huge and complex "tree of life." Each branch results from a speciation event, and time proceeds upward from bottom to top. By mapping traits onto phylogenetic trees, biologists can study how traits have evolved over time. In this diagram, several major traits are mapped, using triangular arrows indicating the point at which they originated. For instance, all vertebrates "above" the point at which jaws are indicated have jaws, whereas lampreys diverged before jaws originated and thus lack them.

Life's history, as revealed by phylogenetic trees and by the fossil record is complex indeed, but a few big-picture trends are apparent. Life, in its 3.5 billion years, has evolved many complex structures from simple ones, and large sizes from small ones. However, these are only generalizations. Many organisms have evolved to become simpler or smaller when natural selection favoured it. Many very complex life forms have disappeared altogether (**FIGURE 3.6**), and it is easy to argue that Earth still belongs to the bacteria and other microbes, some of them little changed over eons.

Even fans of microbes, however, must marvel at some of the exquisite adaptations that animals, plants, and fungi have evolved:

- The heart that beats so reliably for an animal's entire lifetime that we take it for granted
- The complex organ system of which the heart is a part
- The stunning plumage of a peacock in full display
- The ability of each and every plant on the planet to lift water and nutrients from the soil, gather light from the sun, and turn it into food
- The staggering diversity of beetles and other insects
- The human brain and its ability to reason

FIGURE 3.6
Life has not always progressed from simple to complex during evolution. Many complex organisms have gone extinct, taking their designs and innovations with them. For example, the strange creatures portrayed in this painting were found fossilized in the Burgess Shale of the Canadian Rockies in British Columbia. They lived in marine environments 530 million years ago and vanished without leaving descendants. Painting by Marella J. Sibbick.
Source: The National History Museum, London.

All these and more have resulted from the process of evolution as it has generated new species and completely new branches on the tree of life.

Speciation and extinction together determine Earth's biodiversity

Although speciation generates Earth's biodiversity, it is only one part of the equation—for, as you will recall from Chapter 2, the vast majority of species that once lived are now gone. The disappearance of a species from Earth is called **extinction**. From studying the fossil record, paleontologists calculate that the average time a species spends on Earth is 1 million to 10 million years. The number of species in existence at any one time is equal to the number added through speciation minus the number removed by extinction.

It is crucial to understand that extinction is a natural process, but human impact can profoundly affect the rate at which it occurs (**FIGURE 3.7**). The apparent extinction of the golden toad made headlines worldwide, but unfortunately it was not such an unusual occurrence. As we will see in Chapter 9, the biological diversity that makes Earth such a unique planet is being lost at an astounding rate. This loss affects humans directly, because other organisms provide us with life's necessities—food, fibre, medicine, and ecosystem services. Species extinction brought about by human impact may well be the single biggest environmental problem we face, because the loss of a species is irreversible.

Some species are more vulnerable to extinction than others

In general, extinction occurs when environmental conditions change rapidly or severely enough that a species cannot adapt genetically to the change; natural selection simply does not have enough time to work. All manner of environmental events can cause extinction, from climate change to the rise and fall of sea level, to the arrival of new harmful species, to severe weather events, such as extended droughts. In general, small populations and species narrowly specialized on some particular resource or way of life are most vulnerable to extinction from environmental change.

The golden toad was a prime example of a vulnerable species. It was **endemic** to the Monteverde cloud forest, meaning that it occurred nowhere else on the planet. Endemic species face relatively high risks of extinction because all their members belong to a single, sometimes small, population. At the time of its discovery, the golden toad was found in only a 4 km² area of Monteverde. It also required very specific conditions to breed successfully. During the spring at Monteverde, water collects in shallow pools within the network of roots that spans the cloud forest's floor. The golden toad gathered to breed in these rootbound reservoirs, and it was here that Jay Savage and his companions collected their specimens in 1964. Monteverde provided an ideal living environment for the golden toad, but the minuscule size of that environment meant that any stresses that deprived the toad of

FIGURE 3.7
Until 10 000 years ago, the North American continent teemed with a variety of large mammals, including mammoths, camels, giant ground sloths, lions, sabre-toothed cats, and various types of horses, antelope, bears, and others. Nearly all of this megafauna went extinct suddenly about the time that humans first arrived on the continent. Similar extinctions occurred in other areas simultaneously with human arrival, suggesting to many scientists that overhunting or other human impacts were responsible.
Source: National Museum of Natural History, Smithsonian Institution.

the resources it needed to survive might doom the entire world population of the species.

Partly because of their very specialized requirements, many amphibian species around the globe are at a high level of risk of extinction. The recent Global Amphibian Assessment ranked nearly one-third (32%) of the world's amphibians as "threatened," compared with 23% of mammal species and 12% of bird species. Of Canada's 45 amphibian species, 16 (35%) are currently designated as "at risk" or "sensitive."[2] (We will discuss the specific meanings of terms like *at risk*, *endangered*, and *threatened* in Chapter 9.)

However, vulnerability resulting from a restricted or specialized habitat, lifestyle, or resource requirements is certainly not limited to amphibians. Consider the Vancouver Island marmot (*Marmota vancouverensis*), one of the rarest mammals in North America and one of only five species of mammals whose natural range is entirely within Canada. Marmots prefer small patches of south- to southwest-facing, steeply sloping, boulder-filled meadows, between 800 m and 1500 m in altitude. The marmot's Vancouver Island range has been drastically reduced over the past few decades, primarily by habitat alteration resulting from logging. There may now be fewer than 100 individuals of this species alive in the wild,[3] with only a handful of colonies in a few locations in southern Vancouver Island (**FIGURE 3.8**).

Earth has seen several episodes of mass extinction

Most extinction occurs gradually, one species at a time. The rate at which this type of extinction occurs is referred to as the *background extinction rate*. However, Earth has seen five events of staggering proportions that killed off massive numbers of species at once. These episodes, called **mass extinction** events, have occurred at widely spaced intervals in Earth's history and have wiped out 50%–95% of our planet's species each time. The best known mass extinction occurred 65 million years ago and brought an end to the age of dinosaurs (although birds are modern descendants of dinosaurs). Evidence suggests that the impact of a gigantic asteroid caused this event, called the Cretaceous–Tertiary, or K-T, event (see "The Science Behind the Story: The K-T Mass Extinction)".

The sixth mass extinction has started

Many biologists have concluded that Earth is currently entering its sixth mass extinction event—and that we are the cause. Changes to Earth's natural systems set in motion by human population growth, development, and resource depletion have made many species extinct and are threatening countless more. The alteration and outright destruction of natural habitats, the hunting and harvesting of species, and the introduction of invasive species from one place to another where they can harm native species—these processes and many more have combined to threaten Earth's biodiversity.

When we look around us, it may not appear as though a human version of an asteroid impact is taking place, but we cannot judge such things on a human timescale. On the geologic timescale, extinction over 100 years or over 10 000 years appears every bit as instantaneous as extinction over a few days.

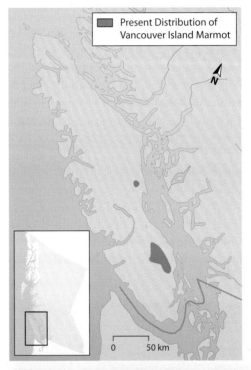

FIGURE 3.8
An extremely restricted natural range with very specific habitat requirements can leave some species highly vulnerable, especially if local environmental changes occur. The Vancouver Island marmot, seen here (bottom), is one of the rarest mammals in North America and one of only five species of mammals that are found only in Canada. This marmot's range (top, shown in purple) has been drastically reduced to just a few locations with a handful of colonies, primarily as a result of logging.

weighing the issues

WHY SHOULD WE CARE ABOUT EXTINCTION? 3–2

Many scientists say biodiversity loss is our biggest environmental problem today. Can you elaborate on some of the specific reasons why we should be concerned about loss of biodiversity and extinction of species? We discussed the Vancouver Island marmot, on the brink of extinction as a result of habitat alteration; what would be lost if this animal were to become extinct? Other endangered species in Canada include the northern abalone, the whooping crane, and the Atlantic halibut—each with its own appeal, certainly, but marmots are cute and furry; does that make them more compelling as examples of the value or importance of biodiversity?

Another of Canada's endangered species is the blue whale (*Balaenoptera musculus*), the largest animal that has *ever lived*. Is the blue whale more worthy of preservation than the Vancouver Island marmot or the whooping crane or the northern abalone? How should we define the criteria for what is "worthy" of preservation? Should we be concerned about the value of one species, or is the true importance that these species are part of a larger whole? (We will consider such questions in greater detail in Chapter 21.)

Levels of Ecological Organization

The extinction of species, their generation through speciation, and other evolutionary mechanisms and patterns have substantial influence on ecology. Moreover, it is often said that ecology provides the stage on which the play of evolution unfolds. The two, it is clear, are tightly intertwined in many ways. As we discussed in Chapter 1, ecology is the study of interactions among organisms and between organisms and their environments.

Ecology is studied at several levels

Life occurs in a hierarchy of levels. The atoms, molecules, and cells we reviewed in Chapter 2 represent the lowest levels in this hierarchy (see **FIGURE 2.12**). Aggregations of cells of particular types form tissues, and tissues form organs, all housed within an individual living organism. Ecologists study relationships on the higher levels of this hierarchy (**FIGURE 3.9**), namely on the organismal, population, community, and ecosystem levels. **Communities** are made up of multiple interacting species that live in the same area. A population of golden toads, a population of resplendent quetzals, and populations of ferns and mosses, together with all the other interacting plant, animal, fungal, and microbial populations in the Monteverde cloud forest, would be considered a community. **Ecosystems** encompass communities and the abiotic (nonliving) material and forces with which their members interact. Monteverde's cloud forest ecosystem consists of the community plus the air, water, soil, nutrients, and energy the community's organisms use.

At the organismal level, the science of ecology describes relationships between organisms and their physical environments. It helps us understand, for example, what aspects of the golden toad's environment were important to it and why.

Population ecology investigates the quantitative dynamics of how individuals within a species interact with one another. It helps us understand why populations of some species (such as the golden toad or the Vancouver Island marmot) decline, while populations of others (such as humans) increase.

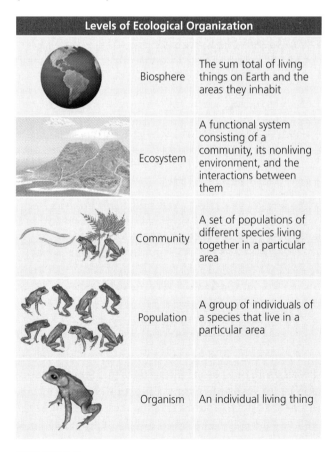

FIGURE 3.9
Life exists in a hierarchy of levels. Ecology includes the study of the organismal, population, community, and ecosystem levels, and, increasingly, the level of the biosphere. Levels below the organismal level are illustrated in **FIGURE 2.12**.

THE SCIENCE BEHIND THE STORY

The K-T Mass Extinction

Dr. Luis Alvarez (left) and Dr. Walter Alvarez (right) examine samples of the iridium layer.

On five occasions in the history of life, huge numbers of species went extinct in a geologic instant. The last mass extinction occurred 65 million years ago, at the dividing line between the Cretaceous and Tertiary periods, called the *K-T boundary*. About 70% of the species then living, including the dinosaurs, disappeared.

When he first started working at Bottaccione Gorge in Italy, geologist Walter Alvarez had no idea he might soon help discover what killed off the dinosaurs. Alvarez was developing a new method to determine the age of sedimentary rocks, and he chose Bottaccione Gorge because it presented an ideal geologic archive. Its 400 m walls are stacked like layer cake with beds of rose-coloured limestone that formed between 100 million and 50 million years ago from dust that had settled to the bottom of an ancient sea.

While analyzing these layers, Alvarez noticed a band of reddish clay 1 cm thick sandwiched between two layers of limestone. The older layer just below it was packed with fossils of *Globotruncana*, a sand grain-sized animal that lived in the late Cretaceous period. The newer layer just above it contained just a few scattered fossils of a cousin of *Globotruncana*, typical of sedimentary rock formed in the early Tertiary period. What Alvarez found interesting was that the K-T boundary clay layer, which had formed just as dinosaurs were going extinct, had no fossils at all.

To get an idea of how quickly the K-T mass extinction event occurred, Walter and his father, physicist Luis Alvarez, analyzed the boundary clay layer, by using the rare metal iridium as a kind of clock, to see how long the layer had taken to form. Almost all the iridium on Earth's surface comes from the dust of tiny meteorites (fragments of asteroids, or shooting stars) that enter and then burn up in the atmosphere. A smaller amount comes from internal terrestrial sources, reaching the atmosphere via volcanism and then on the surface being deposited along with the iridium of extraterrestrial origin. The rate of delivery of iridium to the surface by meteorite dust and by volcanism is approximately constant over time. Therefore, an analysis of the iridium content of the clay layer should have indicated how many have indicated years' worth of iridium had accumulated.

Iridium levels in the underlying and overlying limestone units were typical for sedimentary rocks, about 0.3 parts per billion. In the clay layer, however, the Alvarezes were surprised to find levels 30 times as high. To make sure the finding was not unique to Bottaccione Gorge, they checked the K-T clay layer at a Danish sea cliff in which the same sequence of rock layers was exposed. The clay layer there had 160 times as much iridium as the surrounding rock. The Alvarezes hypothesized that the excess iridium had come from a massive asteroid that had smashed into Earth, causing a global environmental catastrophe that wiped out the dinosaurs.

To convince themselves and the science community that an asteroid impact did cause the K-T event, the Alvarezes had to rule out other possible explanations. For example, the extra iridium could have come from seawater of an unusual composition. Calculations proved, however, that seawater could not contain enough iridium to account for the high levels in the clay layers. An alternative mechanism would be to bring the excess iridium to the surface through an extended period of unusually active and violent volcanic activity; many scientists who work on the K-T extinction event still prefer this hypothesis to explain the event.

How reasonable is it to postulate that an asteroid impact could have caused the K-T extinction? An asteroid 10 km wide strikes Earth, on average, every 100 million years. Such an impact would unleash an explosion 1000 times as forceful as the 1883 eruption of the Indonesian volcano Krakatau, which scattered so much dust around the world that sunsets were intense and summers were unusually cool for two years after. An asteroid impact 65 million years ago, scientists suggested, could have kicked up enough dust to blot out the sun for several years (see picture). This would have inhibited photosynthesis, causing plants to die off, food webs to collapse, and most animals, including dinosaurs, to die of starvation. Only a few smaller animals survived, feeding on rotting vegetation. When sunlight returned, plants sprouted from dormant seeds, and a long recovery began.

Published in the journal *Science* in 1980, the Alvarezes' explanation was immediately attacked by other geologists, who claimed that there were other, more

---

Community ecology focuses on interactions among species, from one-to-one interactions to complex interrelationships involving entire communities of organisms. In the case of Monteverde, it allows us to study how the golden toad and many other species of its cloud forest community interact.

Finally, **ecosystem ecology** reveals patterns, such as energy and nutrient flow, by studying living and nonliving components of systems in conjunction. As you will see, changing climate has had a strong influence on the organisms of Monteverde's cloud forest ecosystem.

As improved technologies allow scientists to learn more about the complex operations of natural systems on a global scale, ecologists are increasingly expanding their horizons beyond ecosystems to the biosphere as a whole. In this chapter we explore ecology up through the population level. In Chapter 4 we examine the community level, and in Chapter 5 we explore the ecosystem and biosphere levels.

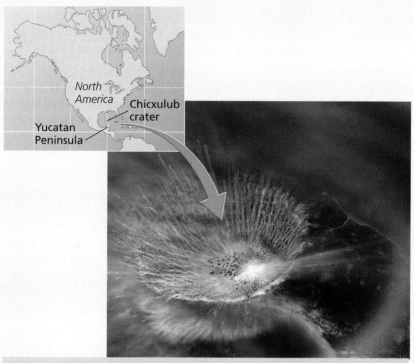

A colossal asteroid impact 65 million years ago is thought to have caused the Cretaceous–Tertiary mass extinction.

great extinctions in geologic history. The greatest mass extinction ever recorded in the rock record ended the Permian period 251 million years ago (called the Permian–Triassic or P-Tr extinction). This mass extinction involved the demise of 96% of all marine species and 70% of terrestrial species, but its cause remains controversial. Several large impact craters have been investigated as possible evidence of a meteorite impact cause, including the Manicouagan Crater in Quebec (see photo), but none of them is of appropriate age to account for the P-Tr extinction. Other hypotheses include massive volcanism, shifts in the chemistry and/or depth of ocean waters, sudden atmospheric changes, or a combination of many factors. So far, none of the hypotheses has accumulated conclusive evidentiary support.

Manicouagan Crater in Quebec is the remnant of a meteorite impact that may have been large enough to cause a major extinction (the crater, now somewhat eroded and marked by a ring lake, was originally about 100 km in diameter). However, the date of the crater's formation, between 214 million and 206 million years ago, is inconsistent with the timing of the P-Tr extinction, the cause of which remains controversial.

likely explanations for the high iridium levels in Bottaccione Gorge. Some scientists also disputed the timescale (and some still do), suggesting that the K-T extinction likely took place over thousands of years—a vanishingly short period in geologic terms, but much, much longer than the almost instantaneous events associated with a meteorite impact.

Throughout the 1980s, scientists kept finding evidence supporting the asteroid-impact hypothesis. Iridium-enriched clay turned up in K-T layers around the world, as did minerals called shocked quartz and stishovite, which form only under the extreme pressure of thermonuclear explosions and asteroid impacts. Models of the impact event predicted that global wildfires would have ensued; the clay layers were found to contain an abundance of tiny soot particles, perhaps from the smoke of such fires.

The Alvarezes' hypothesis became widely accepted after 1991, once scientists pinpointed a 65 million-year-old crater of the expected size in the ocean off the coast of Mexico (see map). Today there is still much scientific discussion and controversy about the details of the K-T extinction events, but scientists broadly agree that an asteroid impact 65 million years ago either caused or contributed to our planet's most recent mass extinction.

There is much weaker consensus about the causes of some of the other

Habitat, niche, and degree of specialization are important in organismal ecology

On the organismal level, each organism relates to its environment in ways that tend to maximize its survival and reproduction. One key relationship involves the specific environment in which an organism lives—its **habitat**. A species' habitat consists of both living and nonliving elements—of rock, soil, leaf litter, and humidity, as well as the other organisms around it. The golden toad lived in a habitat of cloud forest—more specifically, on the moist forest floor, using seasonal pools for breeding and burrows for shelter. Plants known as *epiphytes* use other plants as habitat; they grow on trees for physical support, obtaining water from the air and nutrients from organic debris that collects among their leaves. Epiphytes thrive in cloud forests because they require a habitat with high humidity. Monteverde hosts more than 330 species of epiphytes, mostly ferns, orchids, and bromeliads (pineapple relatives). By collecting pools of rainwater and pockets of leaf litter, epiphytes create habitat for many other organisms, including many invertebrates and even frogs that lay their eggs in the rainwater pools.

Habitats are scale dependent. A tiny soil mite may require only a few square centimetres of soil as its habitat. A vulture, an elephant, or a whale, in contrast, may require many square kilometres of habitat to traverse by air, land, or water.

Each organism thrives in certain habitats and not in others, leading to nonrandom patterns of **habitat use**. Mobile or *motile* organisms (those that are able to move about freely) actively select habitats from within the range of options they encounter, a process called **habitat selection**. In the case of plants and *sessile* animals (those that are not freely mobile and whose progeny disperse passively), patterns of habitat use result from success in some habitats and failure in others.

The criteria by which organisms favour some habitats over others vary greatly. To a soil mite, the important characteristics of a habitat might be the chemistry, moisture, and compactness of the soil, and the percentage and type of organic matter. A vulture may ignore not only soil but also topography and vegetation, focusing solely on the abundance of dead animals in the area that it scavenges for food. To a whale, water temperature and salinity, light level, and abundance of marine microorganisms might be the critical characteristics. Every species assesses habitats differently because every species has different needs.

Habitat selection is important because the availability and quality of habitat are crucial to an organism's well-being. Indeed, because habitats provide everything an organism needs, including nutrition, shelter, breeding sites, and mates, the organism's very survival depends on the availability of suitable habitats. Often this engenders conflict with people who want to alter or develop a habitat for their own purposes.

Another way in which an organism relates to its environment is through its niche. A species' **niche** reflects its use of resources and its functional role in a community. This includes its habitat use, its consumption of certain foods, its role in the flow of energy and matter, and its interactions with other organisms. The niche is a multidimensional concept, a kind of summary of everything an organism does. Eugene Odum, who pioneered the science of ecology and the concept of the ecosystem, once wrote that "habitat is the organism's address, and the niche is its profession." We will examine the niche concept further in Chapter 4.

Organisms vary in the breadth of their niche. Species with narrow breadth, and thus very specific requirements, are said to be **specialists**. Those with broad tolerances, able to use a wide array of habitats or resources, are **generalists**. For example, in a study of eight Costa Rican bird species that feed from epiphytes, ornithologist T. Scott Sillett found that four were generalists. The other four were specialists on the insect resources the epiphytes provided and spent more than 75% of their foraging efforts feeding only from epiphytes.

Specialist and generalist strategies both have advantages and disadvantages. Specialists can be successful over evolutionary time by being extremely good at the things they do, but they are vulnerable when conditions change and threaten the habitat or resource on which they have specialized. Generalists meet with success by being able to live in many different places and weather variable conditions, but they may not thrive in any one situation to the degree that a specialist does. An organism's habitat, niche, and degree of specialization each reflect the adaptations of the species and are products of natural selection.

Population Ecology

Individuals of the same species inhabiting a particular area make up a population. Species may consist of multiple populations that are geographically isolated from one another. This is the case with a species characteristic of Monteverde—the resplendent quetzal (*Pharomachrus mocinno*), considered one of the world's most spectacular birds (see **FIGURE 3.1A**). Although it ranges from southernmost Mexico to Panama, the resplendent quetzal lives only in high-elevation tropical forests and is absent from low-elevation areas. Human development has destroyed much of its forest habitat; thus, the species today exists in many separate populations scattered across Central America.

Humans have become more mobile than any other species and have spread nearly everywhere on the planet. As a result, it is difficult to define a distinct human population on anything less than the global scale. Some would maintain that in the ecological sense of the word, all 6.7 billion of us compose one population.

Populations exhibit characteristics that help predict their dynamics

Whether one is considering humans or whales or marmots or golden toads, all populations have characteristics that help population ecologists predict the future dynamics of that population. Such attributes as density, distribution, sex ratio, age structure, and birth and death rates all

(a) Passenger pigeon

(b) 19th-century lithograph of pigeon hunting in Iowa

FIGURE 3.10
The passenger pigeon **(a)** was once North America's most numerous bird, and its flocks literally darkened the skies when millions of birds passed overhead **(b)**. However, human cutting of forests and hunting drove the species to extinction within a few decades.

help the ecologist understand how a population may grow or decline. The ability to predict growth or decline is especially useful in monitoring and managing threatened and endangered species. It is also vital in its application to human populations (Chapter 6). Understanding human population dynamics, their causes, and their consequences is one of the central elements of environmental science and one of the prime challenges for our society today.

Population size Expressed as the number of individual organisms present at a given time, **population size** may increase, decrease, undergo cyclical change, or remain the same over time. Extinctions are generally preceded by dramatic population declines. As late as 1987, scientists documented a golden toad population at Monteverde in excess of 1500 individuals, but in 1988 and 1989 scientists sighted only a single toad. By 1990, the species had disappeared.

The passenger pigeon (*Ectopistes migratorius*), also now extinct, illustrates the extremes of population size (**FIGURE 3.10**). This was once the most abundant bird in North America; flocks of passenger pigeons literally darkened the skies, nesting in gigantic colonies in the forests of the American Midwest and southern Canada. Once people began cutting the forests, the birds' great concentration made them easy targets for hunters, who gunned down thousands and shipped them to market by the wagonload. By the end of the nineteenth century, the passenger pigeon population had declined to such a low number that they could not form the large colonies they needed to breed effectively. In 1914, the last passenger pigeon on Earth died in the Cincinnati Zoo, bringing the continent's most numerous bird species to extinction within just a few decades.

Population density The flocks and breeding colonies of passenger pigeons showed high population density, another attribute that ecologists assess to better understand populations. **Population density** describes the number of individuals within a population, per unit area. For instance, the 1500 golden toads counted in 1987 within 4 km^2 indicated a density of 375 toads/km^2. In general, larger organisms tend to have lower population densities because they require more resources to survive.

High population density can make it easier for organisms to group together and find mates, but it can also lead to conflict in the form of competition if space, food, or mates are in limited supply. Overcrowded organisms may also become more vulnerable to the predators that feed on them, and close contact among individuals can increase the transmission of infectious disease. For these reasons, organisms sometimes leave an area when densities become too high. In contrast, at low population densities, organisms benefit from more space and resources but may find it harder to locate mates and companions.

High population densities in small remnants of habitat may have doomed Monteverde's harlequin frog (*Atelopus varius*; see **FIGURE 3.11C**), an amphibian that disappeared from the cloud forest around the same time as the golden toad. The harlequin frog is a specialist, favouring "splash zones," areas alongside rivers and streams that receive spray from waterfalls and rapids. As Monteverde's climate grew warmer and drier in the 1980s and 1990s, water flow decreased, and many streams dried up. Splash zones grew smaller and fewer, and harlequin frogs were forced to cluster together in what remained of the habitat. Researchers J. Alan Pounds and Martha Crump recorded frog population densities up to 4.4 times as high as normal, with more than two frogs per metre of stream. Such overcrowding likely made the frogs more vulnerable to disease transmission, predator attack, and assault from parasitic flies. The researchers concluded that these factors led to the harlequin frog's disappearance from Monteverde.

Thankfully, a new population of harlequin frogs was found in 2003 on a private reserve elsewhere in Costa Rica, so there is still hope that the species may survive. The frog was rediscovered by University of Delaware student Justin Yeager, who was doing field research during his study abroad trip in Costa Rica that summer.

Population distribution It was not simply the harlequin frog's density but also its distribution in space that led to its demise at Monteverde. **Population distribution**, or *population dispersion*, describes the spatial arrangement of organisms within an area. Ecologists define three distribution types: random, uniform, and clumped (**FIGURE 3.11**). In a *random distribution*, individuals are located haphazardly in space in no particular pattern. This type of distribution can occur when the resources an organism needs are found throughout an area, and other organisms do not strongly influence where members of a population settle.

A *uniform distribution* is one in which individuals are evenly spaced. This can occur when individuals hold territories or otherwise compete for space. For instance, in a desert where there is little water, each plant may need a certain amount of space for its roots to gather adequate moisture. As a result, each individual plant may be equidistant from others.

In a *clumped distribution*, the pattern most common in nature, organisms arrange themselves according to the availability of the resources they need to survive. Many desert plants grow in patches around isolated springs or along arroyos that flow with water after rainstorms. During their mating season, golden toads were found clumped at seasonal breeding pools. Humans, too, exhibit clumped distribution; people frequently aggregate together in urban centres. Clumped distributions often arise from habitat selection. Distributions can depend on the scale at which one measures them. At very large scales, all organisms show clumped or patchy distributions, because some parts of the total area they inhabit are bound to be more hospitable than others.

Sex ratios For organisms that reproduce sexually and have distinct male and female individuals, the sex ratio

(a) Random

(b) Uniform

(c) Clumped

FIGURE 3.11
Individuals in a population can be spatially distributed over a landscape in three fundamental ways. In a random distribution **(a)**, organisms are dispersed at random through the environment. In a uniform distribution **(b)**, individuals are spaced evenly, at equal distances from one another. Territoriality can result in such a pattern. In a clumped distribution **(c)**, individuals occur in patches, concentrated more heavily in some areas than in others. Habitat selection or flocking to avoid predators can result in such a pattern.

of a population can help determine whether it will increase or decrease in size over time. A population's **sex ratio** is its proportion of males to females. In monogamous species (in which each sex takes a single mate), a 50/50 sex ratio maximizes population growth, whereas an unbalanced ratio leaves many individuals of one sex without mates.

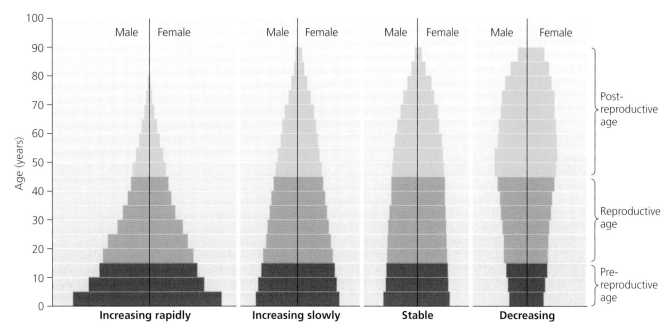

FIGURE 3.12
Age structure diagrams show the relative frequencies of individuals of different age classes, by gender, in a population. In this example for humans, populations heavily weighted toward young age classes (at left) grow most quickly, whereas those weighted heavily toward old age classes (at right) decline.

Age structure Populations most often consist of individuals of different ages. **Age distribution**, or **age structure**, describes the relative numbers of organisms of each age within a population. Like sex ratio, age distribution can have strong effects on rates of population growth or decline. A population made up mostly of individuals past reproductive age will tend to decline over time. In contrast, a population with many individuals of reproductive age or soon to be of reproductive age is likely to increase. A population with an even age distribution will likely remain stable as births keep pace with deaths.

Age structure diagrams, often called **age pyramids**, are visual tools scientists use to show the age structure of populations (**FIGURE 3.12**). The width of each horizontal bar represents the relative size of each age class. A pyramid with a wide base has a relatively large age class that has not yet reached its reproductive stage, indicating a population much more capable of rapid growth. In this respect, a wide base of an age pyramid is like an oversized engine on a rocket—the bigger the booster, the faster the increase. We will examine age pyramids further in Chapter 6 in reference to human populations.

Birth and death rates All the preceding factors can influence the rates at which individuals within a population are born and die. A convenient way to express birth and death rates is to measure the number of births and deaths per 1000 individuals for a given time period. Such a rate is termed a *crude birth rate* or *crude death rate*.

Just as individuals of different ages have different abilities to reproduce, individuals of different ages show different probabilities of dying. For instance, people are more likely to die at old ages than young ages; if you were to follow 1000 10-year-olds and 1000 80-year-olds for a year, you would find that at year's end more 80-year-olds had died than 10-year-olds. However, this pattern does not hold for all organisms. Amphibians, such as the golden toad, produce large numbers of young, which suffer high death rates. For a toad, death is less likely (and survival more likely) at an older age than at a very young age.

To show how the likelihood of death can vary with age, ecologists use graphs called **survivorship curves** (**FIGURE 3.13**). There are three fundamental types of survivorship curves. Humans, with higher death rates at older ages, show a *type I* survivorship curve. Toads, with higher death rates at younger ages, show a *type III* survivorship curve. A *type II* survivorship curve is intermediate and indicates equal rates of death at all ages. Many birds can be characterized with type II curves.

Populations may grow, shrink, or remain stable

Now that we have outlined some key attributes of populations, we are ready to take a quantitative view of population change by examining some simple mathematical concepts used by population ecologists and *demographers* (those who study human populations, Chapter 6). Population growth, or decline, is determined by four factors:

1. Births within the population, or **natality**
2. Deaths within the population, or **mortality**
3. **Immigration**, the arrival of individuals from outside the population

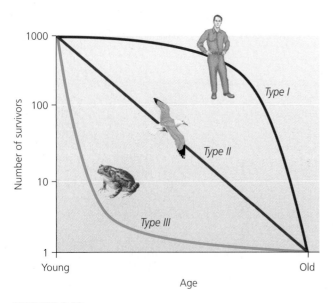

FIGURE 3.13
In a type I survivorship curve, survival rates are high when individuals are young and decrease sharply when individuals are old. In a type II survivorship curve, survival rates are equivalent regardless of an individual's age. In a type III survivorship curve, most mortality takes place at young ages, and survival rates are greater at older ages. Some examples include humans (type I), birds (type II), and amphibians (type III).

4. **Emigration**, the departure of individuals from the population

To understand how a population changes, we measure its **growth rate**, which can be calculated as the crude birth rate plus the immigration rate, minus the crude death rate plus the emigration rate, each expressed as the number per 1000 individuals per year:

$$(\text{Crude birth rate} + \text{Immigration rate})$$
$$- (\text{Crude death rate} + \text{Emigration rate})$$
$$= \text{Growth rate}$$

The resulting number tells us the net change in a population's size per 1000 individuals per year. For example, a population with a crude birth rate of 18 per 1000/yr, a crude death rate of 10 per 1000/yr, an immigration rate of 5 per 1000/yr, and an emigration rate of 7 per 1000/yr would have a growth rate of 6 per 1000/yr:

$$(18/1000/\text{yr} + 5/1000/\text{yr}) - (10/1000/\text{yr} + 7/1000/\text{yr})$$
$$= 6/1000/\text{yr}$$

Thus, a population of 1000 with these birth, death, immigration, and emigration rates will reach 1006 after one year. If the population is 1 000 000, it will reach 1 006 000 after one year. These population increases are often expressed as percentages, which we can calculate using the following formula:

$$\text{Growth rate} \times 100\%$$

Thus, a growth rate of 6/1000/yr would be expressed as

$$6/1000/\text{yr} \times 100\% = 0.6\%/\text{yr}$$

By measuring population growth in terms of percentages, scientists can compare increases and decreases in species that have far different population sizes. They can also project changes that will occur in the population over longer periods, much like you might calculate the amount of interest your savings account will earn over time.

Unregulated populations increase by exponential growth

When a population, or anything else, increases by a fixed percentage each year, it is said to undergo geometric growth or **exponential growth**. A savings account is a familiar frame of reference for describing exponential growth. If at the time of your birth your parents had invested $1000 in a savings account earning 5% interest compounded each year, you would have only $1629 by age 10, and $2653 by age 20, and you would have more than $30 000 when you turn 70. If you could wait just 10 years more, that figure would rise to nearly $50 000. Only $629 was added during your first decade, but approximately $19 000 was added during the decade between ages 70 and 80. The reason is that a fixed percentage of a small number makes for a small increase, but that same percentage of a large number produces a large increase. Thus, as savings accounts (or populations) become larger, each incremental increase likewise gets larger. Such acceleration is a fundamental characteristic of exponential growth.

In contrast, if your parents only added a fixed amount to your savings account each year—say, $1000 per year—your savings would still grow, but it would be arithmetic or **linear growth**. If both accounts (linear growth, with interest at a fixed *amount* per year, and exponential growth, with interest at a fixed *percentage* per year) were allowed to proceed unchecked, the account with the exponential growth would necessarily outstrip the linear growth account eventually. This will be the case, even if the balance in the linear growth account is higher for the first few years.

This fundamental difference between linear and exponential growth was the basis for the work of Thomas Malthus, who suggested that the human population (with exponential growth) would eventually exceed available food resources (with linear growth, or so he thought at the time), to the likely distress of all humanity. The J-shaped curve in **FIGURE 3.14** shows an example of an exponential population increase. As Malthus realized, populations of organisms increase exponentially *unless they meet constraints*. Each organism reproduces by a certain amount, and as populations get larger, more individuals reproduce by that amount. If there were no external limits on growth, ecologists theoretically would expect exponential growth to occur as a

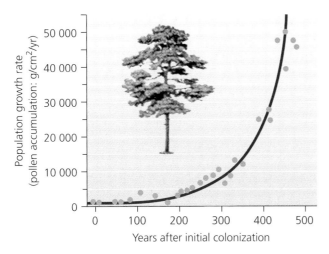

FIGURE 3.14
Although no species can maintain exponential growth indefinitely, some may grow exponentially for a time when colonizing an unoccupied environment or exploiting an unused resource. Scientists have used pollen records to determine that the Scots pine (*Pinus sylvestris*) increased exponentially after the retreat of glaciers following the last ice age around 9500 years ago. Data from Bennett, K. D. 1983. Postglacial population expansion of forest trees in Norfolk, U.K. *Nature* 303:164–167.

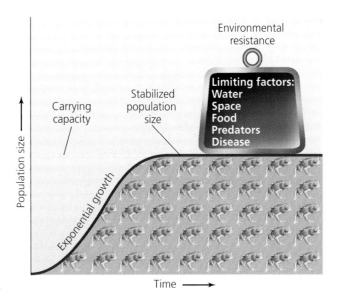

FIGURE 3.15
The logistic growth curve shows how population size may increase rapidly at first, then grow more slowly, and finally stabilize at a carrying capacity. Carrying capacity is determined both by the biotic potential of the organism and by various external limiting factors, collectively termed *environmental resistance*.

result of a population fulfilling its *biotic potential*, which we will discuss in further detail below.

Exponential growth usually occurs in nature when a population is small, competition is minimal, and environmental conditions are ideal for the organism in question. Most often, these conditions occur when organisms are introduced to a new environment. Mould growing on a piece of bread or fruit, and bacteria colonizing a recently dead animal, are cases in point. But species of any size may show exponential growth under the right conditions. A population of the Scots pine, *Pinus sylvestris*, grew exponentially when it began colonizing the British Isles after the end of the last ice age (see **FIGURE 3.14**). Receding glaciers had left conditions ideal for its exponential expansion.

Limiting factors restrain population growth

Exponential growth, however, rarely lasts long. If even a single species in Earth's history had increased exponentially for very many generations, it would have blanketed the planet's surface, and nothing else could have survived. Instead, every population eventually is constrained by **limiting factors**—physical, chemical, and biological characteristics of the environment that restrain population growth. The interaction of these factors is what ultimately determines the carrying capacity, the maximum population size of a species that a given environment can sustain.

Ecologists use the curve in **FIGURE 3.15** to show how an initial exponential increase is slowed and finally brought to a standstill by limiting factors. Called a sigmoidal ("S-shaped") or **logistic growth curve**, it rises sharply at first but then begins to level off as the effects of limiting factors become stronger. Eventually the force of these factors—which taken together are termed **environmental resistance**—stabilizes the population size at its carrying capacity.

The logistic curve is a simplified model; real populations can behave quite differently. Some may cycle indefinitely above and below the carrying capacity. Some may show cycles that become less extreme and approach the carrying capacity. Others may overshoot the carrying capacity and then crash, fated either for extinction or recovery (**FIGURE 3.16**).

Many factors contribute to environmental resistance and influence a population's growth rate and carrying capacity. Space is one factor that limits the number of individuals a given environment can support; if there is no physical room for additional individuals, they are unlikely to survive. Other limiting factors for animals in a terrestrial environment can include the availability of food, water, mates, shelter, and suitable breeding sites; temperature extremes; prevalence of disease; and abundance of predators. Plants are often limited by amounts of sunlight and moisture and the type of soil chemistry, in addition to disease and attack from plant-eating animals. In aquatic systems, limiting factors include salinity, sunlight, temperature, dissolved oxygen, fertilizers, and pollutants.

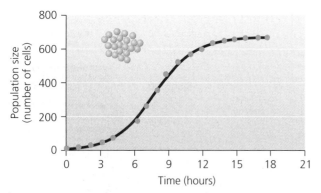

(a) Yeast cells, *Saccharomyces cerevisiae*

(c) Stored-product beetle, *Callosobruchus maculatus*

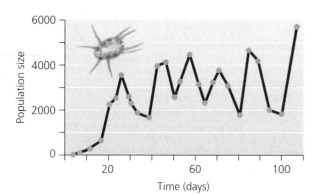

(b) Mite, *Eotetranychus sexmaculatus*

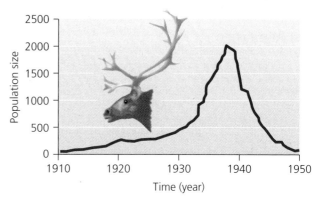

(d) St. Paul reindeer, *Rangifer tarandus*

FIGURE 3.16
Population growth in nature often departs from the stereotypical logistic growth curve, and it can do so in several fundamental ways. Yeast cells from an early lab experiment show logistic growth **(a)** which, like that of the Scots pine shown in Figure 3.14, closely matches the theoretical model. Some organisms, like the mite **(b)**, show cycles in which population fluctuates indefinitely above and below the carrying capacity. Population oscillations can also dampen, lessening in intensity and eventually stabilizing at carrying capacity **(c)**, as in a lab experiment with the stored-product beetle. Populations that rise too fast and deplete resources may crash just as suddenly **(d)**, like the population of reindeer introduced to the Bering Sea island of St. Paul. Data for (a) from Pearl, R. 1927. The growth of populations. *Quarterly Review of Biology* 2:532–548; for (b) from Huffaker, C. B. 1958. Experimental studies on predation: Dispersion factors and predator–prey oscillations, *Hilgardia* 27:343–383; for (c) from Utida, S. 1967. Damped oscillation of population density at equilibrium, *Researches on Population Ecology* 9:1–9; for (d) from Scheffer, V. C. 1951. Rise and fall of a reindeer herd, *Scientific Monthly* 73:356–362.

Sometimes one limiting factor may outweigh all others and restrict population growth. For example, scientists hypothesize that Monteverde's population of golden toads had plenty of space, food, and shelter but lacked adequate moisture. If moisture were the primary limiting factor, then increasing moisture would have increased the carrying capacity of the habitat for the toads. Indeed, to determine limiting factors, ecologists often conduct experiments in which they increase or decrease a hypothesized limiting factor to observe its effects on population size. Unfortunately in the case of the golden toad, such experiments were not conducted before its disappearance.

Carrying capacities can change

Because limiting factors can be numerous, and because environments are complex and changeable, carrying capacity can vary constantly. The human species illustrates another reason that carrying capacity is not necessarily a fixed entity. Although all organisms are subject to environmental resistance, some are capable of altering their environment to reduce this resistance. Our own species has proven to be particularly effective at this. When our ancestors began to build shelters and use fire for heating and cooking, they reduced the environmental resistance of areas with cold climates and were able to expand into new territory. As limiting factors are overcome (either through the development of new technologies or through natural environmental change), the carrying capacity for a species increases. We humans have managed so far to increase the planet's carrying capacity for ourselves, through agricultural, medical, and other technologies. Unfortunately for the golden toad, which lacked the capacity to alter its own environment, the limiting factors on its population growth exerted ever-increasing pressure during the late 1980s.

weighing the issues

3-3 CARRYING CAPACITY AND HUMAN POPULATION GROWTH

As we saw in Chapter 1, the global human population has risen from fewer than 1 billion people 200 years ago to 6.7 billion today, and we have far exceeded our historic carrying capacity. In fact, some demographers argue that Earth's true carrying capacity for the human species is only about 10 million (the estimated human population at the time when the human species had expanded to cover most of the globe). What factors increased Earth's carrying capacity for people? Are there limiting factors for the human population? What might they be? Do you think we can keep raising Earth's carrying capacity in the future? Are there any factors that might cause Earth's carrying capacity for the human species to decrease?

The influence of some factors depends on population density

Just as carrying capacity is not a fixed entity, the influence of limiting factors can vary with changing conditions. In particular, the density of a population can increase or decrease the impact of certain factors on that population. Recall that high population density can help organisms find mates but can also increase competition and the risk of predation and disease. Such factors are said to be *density-dependent* factors, because their influence waxes and wanes according to population density. The logistic growth curve in **FIGURE 3.15** represents the effects of density dependence. The more population size increases, the more environmental resistance kicks in.

Density-independent factors are limiting factors whose influence is not affected by population density. Temperature extremes and catastrophic events, such as floods, fires, and landslides, are examples of density-independent factors, because they can eliminate large numbers of individuals without regard to population density.

Biotic potential and reproductive strategies vary from species to species

Limiting factors from an organism's environment provide only half the story of population regulation. The other half comes from the attributes of the organism itself. For example, organisms differ in their **biotic potential**, or maximum capacity to produce offspring under ideal environmental conditions. A fish with a short gestation period that lays thousands of eggs at a time has high biotic potential, whereas a whale with a long gestation period that gives birth to a single calf at a time has low biotic potential. The interaction between an organism's biotic potential and the environmental resistance to its population growth helps determine the fate of its population.

Giraffes, elephants, humans, and other large animals with low biotic potential produce a relatively small number of offspring and take a long time to gestate and raise each of their young. Species that take this approach to reproduction compensate by devoting large amounts of energy and resources to caring for and protecting the relatively few offspring they produce during their lifetimes. Such species are said to be **K-selected** (or they are called *K-strategists*). K-selected species are so named because their populations tend to stabilize over time at or near their carrying capacity, and *K* is a commonly used abbreviation for carrying capacity. Because their populations stay close to carrying capacity, these organisms must be good competitors, able to hold their own in a crowded world. Thus in these species, natural selection favours individuals that invest in producing offspring of high quality that can be good competitors.

In contrast, species that are **r-selected** focus on quantity, not quality. Species considered to be r-selected (or called *r-strategists*) have high biotic potential and devote their energy and resources to producing as many offspring as possible in a relatively short time. Their offspring do not require parental care after birth, so r-strategists simply leave their survival to chance. The abbreviation *r* denotes the rate at which a population increases in the absence of limiting factors. Populations of r-selected species fluctuate greatly, such that they are often well below carrying capacity. This is why natural selection in these species favours traits that lead to rapid population growth. Many fish, plants, frogs, insects, and others are r-selected. The golden toad was one example. Each adult female laid 200 to 400 eggs, and her tadpoles spent five weeks unsupervised in the breeding pools metamorphosing into adults.

Table 3.3 summarizes stereotypical traits of r-selected and K-selected species. However, it is important to note that these are two extremes on a continuum and that most species fall somewhere between these endpoints. Moreover, some organisms show combinations of traits that do not clearly correspond to a place on the continuum. A redwood tree (*Sequoia sempervirens*), for instance, is large and long-lived, yet it produces many small seeds and offers no parental care.

Table 3.3 Typical Characteristics of r-Selected and K-Selected Species

r-selected species	K-selected species
Small size	Large size
Fast development	Slow development
Short-lived	Long-lived
Reproduction early in life	Reproduction later in life
Many small offspring	Few large offspring
Fast population growth rate	Slow population growth rate
No parental care	Parental care
Weak competitive ability	Strong competitive ability
Variable population size, often well below carrying capacity	Constant population size, close to carrying capacity
Variable and unpredictable mortality	More constant and predictable mortality

Populations of K-selected species are generally regulated by density-dependent factors, such as disease, predation, and food limitation. In contrast, density-independent factors tend to regulate populations of r-selected species, whose success or failure is often determined by large-scale environmental change. Many r-selected species frequently experience large swings in population size, such as rapid increases during the breeding season and rapid declines soon after, when unfit and unlucky young are removed from the population. For this reason, scientists often have difficulty determining whether steep population declines are a part of natural cycles or a sign of serious trouble. For years, scientists debated the golden toad's apparent extinction. Now that it has failed to reappear for more than 15 years, most agree that the toad's population crash was not part of a normal, repeating cycle.

Changes in populations influence the composition of communities

In the late 1980s, the golden toad and the harlequin frog were the most diligently studied species that had been affected by changing environmental conditions in the Costa Rican cloud forest. However, once scientists began looking at populations of other species at Monteverde, they began to notice more troubling changes. By the early 1990s, not only had golden toads, harlequin frogs, and other organisms been pushed from their cloud forest habitat into apparent extinction, but many species from lower, drier habitats also had begun to appear at Monteverde. These immigrants included species tolerant of drier conditions, such as blue-crowned motmots (*Momotus momota*) and brown jays (*Cyanocorax morio*).

By the year 2000, 15 dry forest species had moved into the cloud forest and begun to breed. Meanwhile, population sizes of several cloud forest bird species had declined. After 1987, 20 of 50 frog species vanished from one part of Monteverde, and ecologists later reported more disappearances, including those of two lizards native to the cloud forest. Scientists hypothesized that the warming, drying trends that researchers were documenting (see "The Science Behind the Story: Climate Change and Its Effects on Monteverde") were causing population fluctuations and unleashing changes in the composition of the community.

The Conservation of Biodiversity

Changes in populations and communities have been taking place naturally as long as life has existed, but today human development, resource extraction, and population pressure are speeding the rate of change and altering the types of change. The ways we modify our environment cannot be fully understood in a scientific vacuum, however. The actions that threaten biodiversity have complex social, economic, and political roots, and environmental scientists appreciate that we must understand these aspects if we are to develop solutions.

Fortunately, people can do things to forestall population declines of species threatened with extinction. Millions of people around the world are already taking action to safeguard the biodiversity and ecological and evolutionary processes that make Earth a unique place (Chapter 9). Costa Ricans have been confronting the challenges to their nation's biodiversity, and their actions so far show what even a small country of modest means can do.

Social and economic factors affect species and communities

Many of the threats to Costa Rica's species and ecological communities result from past economic and social forces whose influences are still evident. European immigrants and their descendants viewed Costa Rica's lush forests as an obstacle to agricultural development, and timber companies saw them simply as a source of wood products. Costa Rica's leading agricultural products have long included beef and bananas, whose production and cultivation require extensive environmental modification.

Between 1945 and 1995, the country's population grew from 860 000 to 3.34 million, and the percentage of land devoted to pasture increased from 12% to 33%. With much of the formerly forested land converted to agriculture, the proportion of the country covered by forest decreased from 80% to 25%. In 1991, Costa Rica was losing its forests faster than any other country in the world—nearly 140 ha per day. As a result, populations of innumerable species were declining, and some were becoming endangered (**FIGURE 3.17**). Few people foresaw the need to conserve biological resources until it became clear that they were rapidly being lost.

Costa Rica took steps to protect its environment

In 1970, the Costa Rican government and international representatives came together to create the country's first national parks and protected areas. This was an enormously important step for this developing country; until the early 1980s, the rate of deforestation in Costa Rica was one of the highest in the world, and as much as 60% of the country was cleared, mainly to accommodate ranching, in just a few decades.

The first parks in Costa Rica centred on areas of spectacular scenery, such as the Poás Volcano National Park. Santa Rosa National Park encompassed valuable tropical dry forest; Tortuguero National Park contained essential nesting beaches for the green turtle (*Chelonia mydas;* see **FIGURE 3.17B**); and Cahuita National Park was established to protect a prominent coral reef system. In 1972 the efforts of local residents, along with contributions from international conservation organizations, provided the beginnings of what is today the Monteverde Cloud Forest Biological Reserve. This privately managed 10 500 ha reserve was established to protect the forest and its populations of 2500 plant species, 400 bird species, 500 butterfly species, 100 mammal species, and 120 reptile and amphibian species, including the golden toad.

Initially the government gave the parks little real support. According to Costa Rican conservationist Mario Boza, in their early years the parks were granted only five guards, one vehicle, and no funding. Today government support for protected areas in Costa Rica is much stronger. Fully 12% of the nation's area is contained in national parks, and a further 16% is devoted to other types of wildlife and conservation reserves. (In comparison, 6.3% of Canada's land area is protected as a nature reserve or wilderness area, with a total of only 10.4% for all protected areas of any type.[4]) Costa Ricans, along with international biologists, are working to protect endangered species and recover their populations.

(a) **Golden-cheeked warbler**

(b) **Green sea turtle**

(c) **Red-backed squirrel monkey**

FIGURE 3.17
Costa Rica is home to a number of species classified as globally threatened or endangered. The golden-cheeked warbler, *Dendroica chrysoparia* **(a)**, winters in Central America but breeds in Texas, where its habitat is being rapidly lost to housing development. The green sea turtle, *Chelonia mydas* **(b)**, is widely distributed throughout the world's oceans and lays eggs on beaches in Costa Rica and elsewhere, but it has undergone steep population declines. The red-backed squirrel monkey, *Saimiri oerstedii* **(c)**, is endemic to a tiny area in Costa Rica and is vulnerable to forest loss because of its small geographic range. These vertebrate species receive attention from scientists and the media and are the focus of recovery efforts, but many more plants, insects, and other less-celebrated species also are declining in number.

THE SCIENCE BEHIND THE STORY

Climate Change and Its Effects on Monteverde

Dr. J. Alan Pounds (left) and Dr. Luis Coloma look for harlequin frogs.

Soon after the golden toad's disappearance, scientists began to investigate the role of climate change in driving cloud forest species toward extinction. They noted that the period from July 1986 to June 1987 was the driest on record in Monteverde, with unusually high temperatures and record-low stream flows. These conditions caused the golden toad's breeding pools to dry up shortly after they filled in the spring of 1987, killing nearly all of the eggs and tadpoles present in the pools.

Scientists began reviewing weather data and found that the number of dry days and dry periods each winter in the Monteverde region had increased. Biologists knew that such local climate trends were bad news for amphibians like the golden toad and harlequin frog. Because amphibians breathe and absorb moisture through their skin, they are susceptible to dry conditions, high temperatures, acid rain, and pollutants concentrated by reduced water levels. Based on these facts, herpetologists J. Alan Pounds and Martha Crump hypothesized that hot, dry conditions were to blame for increased adult mortality and breeding problems among golden toads and other amphibians.

Throughout this period, scientists worldwide were realizing that the oceans and atmosphere were warming because of human release of carbon dioxide and other gases into the atmosphere. Global climate change (Chapter 14), experts were learning, could produce varying effects on climate at regional and local levels.

With this in mind, Pounds and others concerned about Monteverde's changing conditions reviewed the scientific literature on ocean and atmospheric science to analyze the effects on Monteverde's local climate of warming patterns in the ocean around Costa Rica.

By 1997 these researchers had determined that Monteverde's cloud forest was becoming drier. The clouds that had given the forest its name and much of its moisture now passed by at higher elevations, where they were no longer in contact with the trees. The primary factor determining the clouds' altitude, the researchers found, is nearby ocean temperatures; as ocean temperatures increase, clouds pass over Monteverde at higher elevations. Once the cloud forest's water supply was pushed upward, out of reach of the mountaintops, the cloud forest began to dry out.

In a paper in the journal *Nature*, Pounds and two colleagues reported these findings. Their conclusion—that broad-scale climate modification was causing local changes at the species, population, and community levels—explained a great number of events occurring at Monteverde. Rising cloud levels and decreasing moisture could explain not only the disappearance of the golden toad and harlequin frog but also the concurrent population crashes in 1987 and subsequent disappearance of 20 species of frogs and toads from the Monteverde region. Amphibians that survived underwent population crashes in each of the region's three driest years.

Pounds and his co-workers further described "a constellation of demographic changes that have altered communities of birds, reptiles and amphibians" in the area as likely additional consequences of this shift in moisture availability. As these mountaintop forests dried out, dry-tolerant species crept in, and moisture-dependent species were stranded at the mountaintops by a rising tide of dryness.

Although organisms may in general be driven from one area to another by changing environmental conditions, if a species has nowhere to go, then extinction may result.

Costa Rica and its citizens are now reaping the benefits of their conservation efforts—not only ecological benefits but also economic ones. Because of its parks and its reputation for conservation, tourists from around the world now visit Costa Rica, a phenomenon called **ecotourism** (**FIGURE 3.18**). The ecotourism industry draws more than 1 million visitors to Costa Rica each year, provides thousands of jobs to Costa Ricans, and is a major contributor to the country's economy. Today's Costa Rican economy is fuelled in large part by commerce and tourism, whose contributions (40%) outweigh those of industry (22%) and agriculture (13%) combined.

It remains to be seen how effectively ecotourism can help preserve natural systems in Costa Rica in the long

FIGURE 3.18
Costa Rica has protected a wide array of its diverse natural areas. This protection has stimulated the nation's economy through ecotourism. Here, visitors experience a walkway through the forest canopy in one of the nation's parks.

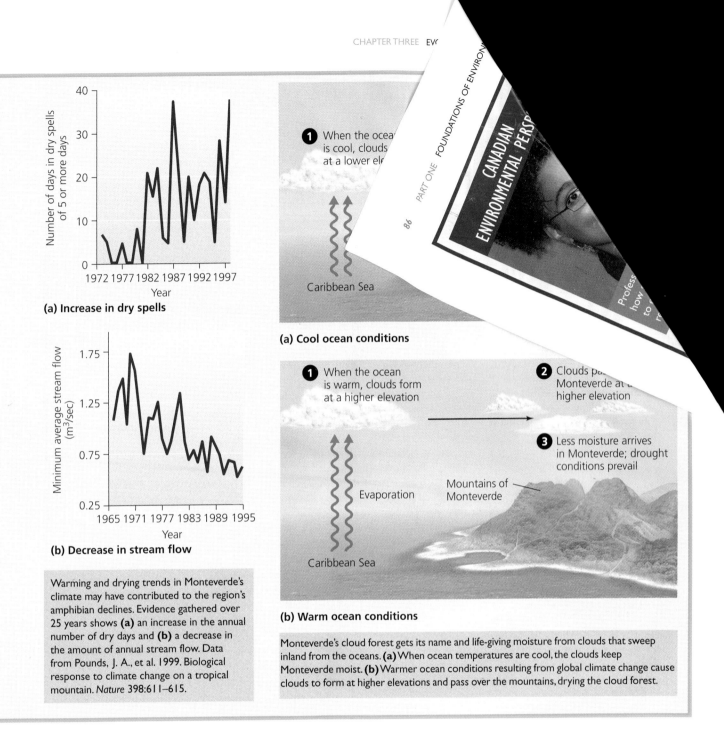

(a) Increase in dry spells

(b) Decrease in stream flow

Warming and drying trends in Monteverde's climate may have contributed to the region's amphibian declines. Evidence gathered over 25 years shows (a) an increase in the annual number of dry days and (b) a decrease in the amount of annual stream flow. Data from Pounds, J. A., et al. 1999. Biological response to climate change on a tropical mountain. *Nature* 398:611–615.

(a) Cool ocean conditions

(b) Warm ocean conditions

Monteverde's cloud forest gets its name and life-giving moisture from clouds that sweep inland from the oceans. (a) When ocean temperatures are cool, the clouds keep Monteverde moist. (b) Warmer ocean conditions resulting from global climate change cause clouds to form at higher elevations and pass over the mountains, drying the cloud forest.

term. As forests outside the parks disappear, the parks are beginning to suffer from illegal hunting and timber extraction. Conservationists say the parks are still under-protected and underfunded. Ecotourism will likely need to generate still more money to preserve habitat, protect endangered species, and restore altered communities to their former condition. Restoration is being carried out in Costa Rica's Guanacaste Province, for instance, where scientists are restoring dry tropical forest from grazed pasture. Restoration of ecological communities is one phenomenon we will examine in our next chapter, as we move from populations to communities.

weighing the issues

HOW BEST TO CONSERVE BIODIVERSITY?

3–4

Most people view national parks and ecotourism as excellent means of keeping ecological systems intact. Yet the golden toad went extinct despite living within a reserve established to protect it, and climate change does not pay attention to park boundaries. What lessons can we learn from this, about the conservation of biodiversity?

Maydianne Andrade

Professor Maydianne Andrade studies how sex and cannibalism come together to ensure reproductive success for the redback spider.

- **Professor** of ecology and evolutionary biology at the University of Toronto
- **Entomologist, arachnologist**, and "Spider Woman"
- One of *Popular Science*'s **"Brilliant 10"** Scientists for 2005

Maydianne Andrade is interested in sex but not just any kind of sex. She studies how sexual selection, social behaviour, and ecological conditions interact to affect the evolution of mating systems in the redback spider, *Latrodectus hasselti*.[5]

Andrade is an associate professor in the Department of Ecology and Evolutionary Biology at the University of Toronto's Scarborough campus. She completed her B.Sc. at Simon Fraser University, her M.Sc. at the University of Toronto, and her Ph.D. at Cornell University. The animals she studies—the redback spider and its close relative, the black widow spider—are unusual because the males of the species make an extreme investment—actually, a "terminal investment"—in seeking success in the mating process.

This means pretty much what it sounds like: the male sacrifices himself to ensure his success in mating. He actually does this by offering himself to the female as "dinner." Yes, female redback spiders are cannibals. This may seem counterintuitive; how can it be a beneficial adaptation for a male spider to sacrifice himself to achieve success in mating? Does this not mean that he gives up all possibility of any future mating opportunities?

Andrade has discovered some interesting features of the species. For one thing, the physiology of the male redback spider is such that he can transfer sperm to the female even as he is being consumed. Furthermore, the self-sacrifice gives such an enormous advantage to the male redback spider—which would likely only mate once anyway, even in the absence of cannibalism—that it is worth his while to give up future mating opportunities for success just this one time.

As Andrade points out,

This extreme form of male mating investment provides a unique opportunity to test sexual selection and life history theory because males are under strong selection to succeed in their single mating opportunity, but are constrained by ecology and physically dominant females.

We have a good understanding of factors affecting the strength and natural selection on redback males, and can manipulate cues indicating the strength of selection in the field and laboratory.[6]

Andrade's current work—which has ramifications not only for the study of spiders but for the study of natural selection and evolutionary theory, too—is focused on furthering our understanding of the effects of factors such as diet on the ratio of males to females, and how male competition and female choice affect the mating process.

"When we talk about 'thinking outside the box,' that's what thinking like a scientist is all about." —**Maydianne Andrade**

Thinking About Environmental Perspectives

How do researchers like Maydianne Andrade advance our *general* understanding of evolutionary theory through the study of *specific*, unusual, or extreme mating habits like those of the redback spider? Visit Andrade's website and read some of her research papers to find out more about this (www.utsc.utoronto.ca/mandrade).

Conclusion

The golden toad and other organisms of the Monteverde cloud forest have helped illuminate the fundamentals of evolution and population ecology that are integral to environmental science. The evolutionary processes of natural selection, speciation, and extinction help determine Earth's biodiversity. Understanding how ecological processes work at the population level is crucial to protecting biodiversity threatened by the mass extinction event that many biologists maintain is already underway.

REVIEWING OBJECTIVES

You should now be able to:

Explain the process of natural selection, and cite evidence for this process

- Because organisms produce excess young, individuals vary in their traits, and many traits are inherited, some individuals will prove better at surviving and reproducing. Their genes will be passed on and become more prominent in future generations.
- Mutations and recombination provide the genetic variation for natural selection.
- We have produced our pets, farm animals, and crop plants through artificial selection.

Describe the ways in which evolution results in biodiversity

- Natural selection can act as a diversifying force as organisms adapt to their environments in myriad ways.
- Speciation (by geographic isolation and other means) produces new species.
- Once they have diverged, lineages continue diverging, a process represented in a phylogenetic tree.

Discuss reasons for species extinction and mass extinction events

- Extinction often occurs when species that are highly specialized or that have small populations encounter rapid environmental change.
- Earth's life has experienced five known episodes of mass extinction, because of an asteroid impact and possibly volcanism and other factors.

List the levels of ecological organization

- Ecologists study phenomena on the organismal, population, community, and ecosystem levels—and, increasingly, on the biosphere level.

Outline the characteristics of populations that help predict population growth

- Populations are characterized by population size, population density, population distribution, sex ratio, age structure, and birth and death rates.
- Immigration and emigration, as well as birth and death rates, determine how a population will grow or decline.

Assess logistic growth, carrying capacity, limiting factors, and other fundamental concepts of population ecology

- Populations unrestrained by limiting factors will undergo exponential growth until they meet environmental resistance.
- Logistic growth describes the effects of density dependence; exponential growth slows as population size increases, and population size levels off at a carrying capacity.
- K-selection and r-selection describe theoretical extremes in how organisms can allocate growth and reproduction.

Identify efforts and challenges involved in the conservation of biodiversity

- Social and economic factors influence our impacts on natural systems.
- Extensive efforts to protect and restore species and habitats will be needed to prevent further erosion of biodiversity.

TESTING YOUR COMPREHENSION

1. Explain the premises and logic that support the concept of natural selection.
2. How does allopatric speciation occur?
3. Name two examples of evidence for natural selection.
4. Name three organisms that have become extinct, and give a probable reason for each extinction.
5. What is the difference between a species and a population? Between a population and a community?
6. Contrast the concepts of habitat and niche.
7. List and describe each of the five major population characteristics discussed in this chapter. Explain how each shapes population dynamics.
8. Could any species undergo exponential growth forever? Explain your answer.
9. Describe how limiting factors relate to carrying capacity.
10. Explain the difference between K-selected species and r-selected species. Can you think of examples of each that were not mentioned in the chapter?

SEEKING SOLUTIONS

1. In what ways has artificial selection changed people's quality of life? Give examples. Can you imagine a way in which artificial selection could be used to improve our quality of life further? Can you imagine a way it could be used to lessen our environmental impact?

2. What types of species are most vulnerable to extinction, and what kinds of factors threaten them? Can you think of any species in your region that are threatened with extinction today? What reasons lie behind their endangerment?
3. Do you think the human species can continue raising its global carrying capacity? How so, or why not? Do you think we *should* try to keep raising our carrying capacity? Why or why not?
4. Describe the evidence suggesting that changes in temperature and precipitation led to the extinction of the golden toad and to population crashes for other amphibians at Monteverde. What do you think could be done to help make similar future declines less likely?
5. What are the advantages of ecotourism for a country like Costa Rica? Can you think of any disadvantages? What would you recommend that Costa Rica do to prevent the loss of its biodiversity?
6. **THINK IT THROUGH** You are a population ecologist studying animals in a national park, and the government is asking for advice on how to focus its limited conservation funds. How would you rate the following three species, from most vulnerable (and thus most in need of attention) to least vulnerable? Give reasons for your choices.

- A bird with an even sex ratio that is a habitat generalist
- A salamander endemic to the park that lives in high-elevation forest
- A fish that specializes on a few types of invertebrate prey and has a high population size

INTERPRETING GRAPHS AND DATA

Amphibians are sensitive biological indicators of climate change because their reproduction and survival are so closely tied to water. One way in which drier conditions may affect amphibians is by reducing the depth of the pools of water in which their eggs develop. Shallower pools offer less protection from UV-B (ultraviolet) radiation, which some scientists maintain may kill embryos directly or make them more susceptible to disease.

Herpetologist Joseph Kiesecker and colleagues conducted a field study of the relationships among water depth, UV-B radiation, and survivorship of western toad (*Bufo boreas*) embryos in the Pacific Northwest. In manipulative experiments, the researchers placed toad embryos in mesh enclosures at three different depths of water. The researchers placed protective filters that blocked all UV-B radiation over some of these embryos, while leaving other embryos unprotected without the filters. Some of the study's results are presented in the accompanying graph.

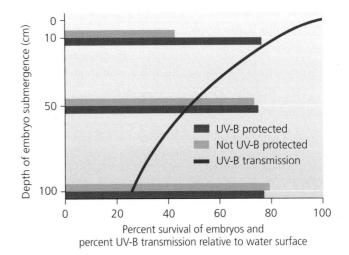

Embryo survivorship in western toads (*Bufo boreas*) at different water depths and UV-B light intensities. Red bars indicate embryos protected under a filter that blocked UV-B light; orange bars indicate unprotected embryos. The blue line indicates the amount of UV-B light reaching different depths in the water column, expressed as a percentage of the UV-B radiation at the water surface. Data from Kiesecker, J. M., et al. 2001. Complex causes of amphibian population declines. *Nature* 410:681–684.

1. If the UV-B radiation at the surface has an intensity of 0.27 watts/m², approximately what is its intensity at depths of 10 cm, 50 cm, and 100 cm?
2. Approximately how much did survival rates at the 10 cm depth differ between the protected and unprotected treatments? Why do you think survival rates differed significantly at the 10 cm depth but not at the other depths?
3. What do you think would be the effect of drier-than-average years on the western toad population, if the average depth of pools available for toad spawning dropped? How do the data above address your hypothesis? Do they support cause-and-effect relationships among water depth, UV-B exposure, disease, and toad mortality?

CALCULATING FOOTPRINTS

Canadians love their coffee. In 2004, coffee imports to Canada topped 223 million kilograms (out of about 7360 million kilograms produced globally per year, as of 2006, according to the International Coffee Organization).[7] Next to petroleum, coffee is the most valuable (legal) commodity on the world market, and Canadians are among the most enthusiastic consumers on a per capita basis. Most coffee is produced in large tropical plantations, where coffee is the only tree species and is grown in full sun. However, approximately 2% of coffee is produced in small groves where coffee trees and other species are intermingled. These *shade-grown* coffee forests maintain greater habitat diversity for tropical rainforest wildlife. Given the information above, estimate the coffee consumption rates in the table at right. Then repeat the calculations based on American coffee consumption, given that the population of the United States is approximately 301 million, and coffee imports to the United States are approximately 1391 million kilograms per year, as of 2005.

1. What percentage of global coffee production is imported to Canada? If only shade-grown coffee were imported to Canada, how much would shade-grown production need to increase to meet that demand?

2. How much extra would you be willing to pay for a kilogram of shade-grown coffee, if you knew that your money would help to prevent habitat loss or extinction for animals, such as Sumatran tigers, rhinoceroses, and the many songbirds that migrate between Latin America and North America each year?

3. If everyone in Canada were willing to pay as much extra per kilogram for shade-grown coffee as you are, how much additional money would that provide for the conservation of biodiversity in the tropics each year?

	Population	Kilograms of Coffee Per Day	Kilograms of Coffee Per Year
You	1	0.0185	6.75
Your hometown			
Your province			
Canada			

Data from International Coffee Organization, www.ico.org/index.asp.

TAKE IT FURTHER

Go to www.myenvironmentplace.ca where you will find

- Suggested answers to end-of-chapter questions
- Quizzes, animations, and flashcards to help you study
- *Research Navigator*tm database of credible and reliable sources to assist you with your research projects
- Tutorials to help you master how to interpret graphs
- Current news articles that link the topics that you study to case studies from your region and around the world

- **ECO Occupational Profiles:** If you found this chapter especially interesting, you might want to learn more about the following jobs by visiting the Occupational Profiles website of the Environmental Careers Organization. Go to www.eco.ca and check out the following careers:
  - Ecologist
  - Entomologist
  - Ornithologist
  - Wildlife biologist

CHAPTER ENDNOTES

1. Pounds, A., and J. Savage, 2004. *Bufo periglenes*. In: IUCN 2007. *2007 IUCN Red List of Threatened Species*. www.iucnredlist.org, downloaded September 24, 2007.
2. Environment Canada Environmental Monitoring and Assessment Network (EMAN), *Status of Amphibian and Reptile Species in Canada*, www.eman-rese.ca/eman/reports/publications/2004/amph_rept_status/.
3. *Wildlife in British Columbia at Risk: Vancouver Island Marmot*, BC Ministry of Environment, Lands, and Parks, www.env.gov.bc.ca/wld/documents/marmot.pdf.
4. Environment Canada, *State of the Environment Infobase*, www.ec.gc.ca/soer-ree/English/Indicator_series/techs.cfm?tech_id=1&issue_id=2.
5. Andrade Lab Research, Department of Ecology and Evolutionary Biology, University of Toronto Scarborough, www.scar.utoronto.ca/~mandrade.
6. Andrade Lab Research, Department of Ecology and Evolutionary Biology, University of Toronto Scarborough, www.scar.utoronto.ca/~mandrade.
7. International Coffee Organization, www.ico.org/index.asp, accessed September 2007.

4 Species Interactions and Community Ecology

This marsh–wetland community is one of the few remaining along the shore of Lake Ontario.

Upon successfully completing this chapter, you will be able to

- Compare and contrast the major types of species interactions
- Characterize feeding relationships and energy flow, using them to construct trophic pyramids and food webs
- Distinguish characteristics of a keystone species
- Characterize the process of succession and the debate over the nature of communities
- Perceive and predict the potential impacts of invasive species on communities
- Explain the goals and methods of ecological restoration
- Describe and illustrate the terrestrial biomes of the world

This is a typical aggregation of zebra mussels, like those that have invaded the Great Lakes.

CENTRAL CASE:
BLACK AND WHITE AND SPREAD ALL OVER: ZEBRA MUSSELS INVADE THE GREAT LAKES

"The zebra mussel is helping us understand what makes a good invader."
—ANTHONY RICCIARDI, MCGILL UNIVERSITY

"The zebra mussel has altered aquatic ecosystems beyond recognition."
—MICHAEL BARDWAJ, *CANADIAN GEOGRAPHIC*[1]

As if the Great Lakes had not been through enough, the last thing they needed was the zebra mussel. The pollution-fouled waters of Lake Erie and the other Great Lakes had become gradually cleaner in the years following the establishment of the International Joint Commission and the signing of the Great Lakes Water Quality Agreement between Canada and the United States in 1972. As these international efforts brought industrial discharges under control, people once again began to use the lakes for recreation, and populations of fish rebounded.

Then the zebra mussel arrived. Black-and-white-striped shellfish the size of a dime (see photo), zebra mussels (*Dreissena polymorpha*) attach to hard surfaces and feed on algae by filtering water through their gills. This mollusc is native to the Caspian Sea, Black Sea, and Azov Sea in western Asia and eastern Europe. It made its North American debut in 1988 when it was discovered in Canadian waters at Lake St. Clair, which connects Lake Erie with Lake Huron. Evidently ships arriving from Europe had discharged ballast water containing the mussels or their larvae into the Great Lakes.

Within two years of their discovery in Lake St. Clair, zebra mussels had reached all five of the Great Lakes. The next year, they entered New York's Hudson River to the east and the Illinois River at Chicago to the west. From the Illinois River and its canals, they soon reached the Mississippi River, giving them access to a vast watershed covering 40% of the United States. By 2007, zebra mussel colonies and sightings had been reported in Ontario, Manitoba, Quebec, and 24 U.S. states.

How could a mussel spread so quickly? The zebra mussel's larval stage is well adapted for long-distance dispersal. Its tiny larvae drift freely for several weeks, travelling as far as the currents take them. Adults that attach themselves to boats and ships may be transported from one place to another, even to isolated lakes and ponds well away from major rivers. They can survive out of the water for several days and are known to have been transported overland to many locations. In North America the mussels encountered none of the particular species of predators, competitors, and parasites that had evolved to limit their population growth in the Old World.

Why the fuss? Zebra mussels are best known for clogging up water intake pipes at factories, power plants, municipal water supplies, and wastewater treatment facilities. At one power plant, workers counted 700 000 mussels per square metre of pipe surface. Great densities of these organisms can damage boat engines, degrade docks, foul fishing gear, and sink buoys that ships use for navigation. Through such impacts, it is estimated that zebra mussels cost hundreds of millions of dollars each year. Over the first 10 years of the zebra mussel invasion, the total cost to Great Lakes economies is estimated to have reached $5 billion, with ongoing annual costs of $20 000 to $350 000 per industrial facility. These figures include only costs to industry, not to individuals or cottagers, who also suffer costs such as clogged water pipes, ruined motorboats, and fouled beaches.[2]

Zebra mussels also have severe impacts on the ecological systems they invade. They eat **phytoplankton**, microscopic algae that drift in open water. Because each mussel filters a litre or more of water every day, they consume so much phytoplankton that they can deplete populations. Phytoplankton is the foundation of the Great Lakes food web, so its depletion is bad news for **zooplankton**, the tiny aquatic animals that eat phytoplankton—and for the fish that eat both. Water bodies with zebra mussels have fewer zooplankton and open-water fish than water bodies without them, researchers are finding.

However, zebra mussels also provide benefits to some bottom-feeding invertebrates and fish. By filtering algae and organic matter from open water and depositing nutrients in their feces, they shift the community's nutrient balance to the bottom and benefit the species that feed there. Once they have cleared the water, sunlight penetrates deeper, spurring the growth of large-leafed underwater plants and algae. Such changes have ripple effects throughout the community that scientists are only beginning to understand.

In the past several years, scientists have noticed a surprising twist: One invader is being displaced by another. The quagga mussel (*Dreissena buensis*), a close relative of the zebra mussel, is spreading through the Great Lakes, replacing the zebra mussel in many locations. What consequences this may have for ecological communities, scientists are only beginning to understand.

Species Interactions

By interacting with many species in a variety of ways, zebra mussels have set in motion an array of changes in the ecological communities they have invaded. Interactions among species are the threads in the fabric of communities, holding them together and determining their nature. Ecologists have organized species interactions into several fundamental categories. Most prominent are competition, predation, parasitism, herbivory, and mutualism. Table 4.1 summarizes the positive (+) and negative (−) impacts of each type of interaction for

Table 4.1 Effects of Species Interactions on Their Participants		
Type of interaction	Effect on Species 1	Effect on Species 2
Mutualism	+	+
Commensalism	+	0
Predation, parasitism, herbivory	+	−
Neutralism	0	0
Amensalism	−	0
Competition	−	−

"+" denotes a positive effect; "−" denotes a negative effect; "0" denotes no effect.

each participant. An interaction with no impact is shown by a "0" in the table.

Competition can occur when resources are limited

When multiple organisms seek the same limited resource, their relationship is said to be one of **competition**. Competing organisms do not usually fight with one another directly and physically. Competition is commonly more subtle and indirect, involving the consequences of one organism's ability to match or outdo others in procuring resources. The resources for which organisms compete can include just about anything an organism might need to survive, including food, water, space, shelter, mates, sunlight, and more. Competitive interactions can take place among members of the same species (**intraspecific competition**) or among members of two or more different species (**interspecific competition**).

We have already discussed intraspecific competition in Chapter 3, without naming it as such. Recall that density dependence limits the growth of a population; individuals of the same species compete with one another for limited resources, such that competition is more acute when there are more individuals per unit area (denser populations). Thus, intraspecific competition is really a population-level phenomenon.

In contrast, interspecific competition can have substantial effects on the composition of communities. If one species is a very effective competitor, it may exclude another species from resource use entirely. This outcome, called **competitive exclusion**, occurred in Lake St. Clair and western Lake Erie as the zebra mussel outcompeted a native mussel species.

Alternatively, if neither competing species fully excludes the other, the species may live side by side at a certain ratio of population sizes. This result, called **species coexistence**, may produce a stable point of equilibrium, in which the population size of each remains fairly constant through time.

Coexisting species that use the same resources tend to adjust to their competitors to minimize competition with them. Individuals can do this by changing their behaviour so as to use only a portion of the total array of resources they are capable of using. In such cases, individuals are not fulfilling their entire *niche,* or ecological role. The full niche of a species is called its **fundamental niche** (**FIGURE 4.1A**). An individual that plays only part of its role because of competition or other species interactions is said to be displaying a **realized niche** (**FIGURE 4.1B**), the portion of its fundamental niche that is actually filled, or realized.

Species make similar adjustments over evolutionary time. They adapt to competition by evolving to use slightly different resources or to use their shared resources in different ways. If two bird species eat the same type of seeds, one might come to specialize on larger seeds and the other to specialize on smaller seeds. Or one bird might become more active in the morning and the other more active in the evening, thus avoiding direct interference. This process is called **resource partitioning**, because the species divide, or partition, the resource they use in common by specializing in different ways (**FIGURE 4.2**).

Resource partitioning can lead to *character displacement,* which occurs when competing species evolve

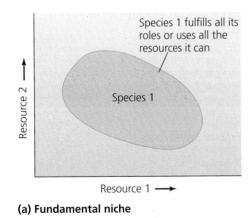

(a) Fundamental niche

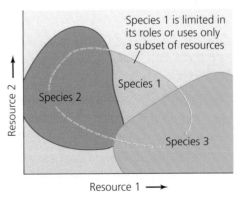

(b) Realized niche

FIGURE 4.1
An organism facing competition may be forced to play a lesser ecological role or use fewer resources than it would in the absence of its competitor. With no competitors, an organism can exploit its full fundamental niche **(a)**. But when competitors restrict what an organism can do or what resources it can use, the organism is limited to a realized niche **(b)**, which covers only a subset of its fundamental niche. In considering niches, ecologists have traditionally focused on competition, but they now recognize that other species interactions are also influential.

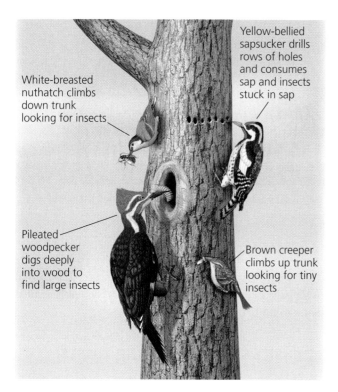

FIGURE 4.2
When species compete, they tend to partition resources, each specializing on a slightly different resource or way of attaining a shared resource. A number of types of birds—including the woodpeckers, creeper, and nuthatch shown here—feed on insects from tree trunks, but they use different portions of the trunk, seeking different foods in different ways.

physical characteristics that reflect their reliance on the portion of the resource they use. By becoming more different from one another, two species reduce their competition. Through natural selection, birds that specialize on larger seeds may evolve larger bills that enable them to make best use of the resource, whereas birds specializing on smaller seeds may evolve smaller bills. This is precisely what extensive research has revealed about the finches from the Galápagos Islands that were first described by Charles Darwin.

Several types of interactions are exploitative

In competitive interactions, each participant has a negative effect on other participants, because each takes resources the others could have used. This is reflected in the two minus signs shown for competition in Table 4.1. In other types of interactions, some participants benefit while others are harmed (note the +/− interactions in Table 4.1). We can think of interactions in which one member exploits another for its own gain as exploitative

interactions. Such interactions include predation, parasitism, herbivory, and related concepts, outlined below.

Predators kill and consume prey

Every living thing needs to procure food and, for most animals, that means eating other living organisms. **Predation** is the process by which individuals of one species—the **predator**—hunt, capture, kill, and consume individuals of another species, the **prey** (**FIGURE 4.3**). Along with competition, predation has traditionally been viewed as one of the primary organizing forces in community ecology. Interactions between predators and prey structure the food webs that we will examine shortly, and they influence community composition by helping determine the relative abundance of predators and prey.

Zebra mussel predation on phytoplankton has reduced phytoplankton populations by up to 90%, according to many studies in the Great Lakes and Hudson River. Zebra mussels also consume the smaller types of zooplankton. This predation, combined with the competition mentioned above, has caused zooplankton population sizes and biomass to decline by up to 70% in Lake Erie and the Hudson River since zebra mussels arrived. Meanwhile, the mussels do not eat some cyanobacteria, so concentrations of these cyanobacteria rise in lakes with zebra mussels. Most predators are also prey, however, and zebra mussels have become a food source for a number of North American species since their introduction. These include diving ducks, muskrats, crayfish, flounder, sturgeon, eels, and several types of fish with grinding teeth, such as carp and freshwater drum.

FIGURE 4.3
Predator–prey interactions have ecological and evolutionary consequences for both prey and predator. Here, a fire-bellied snake (*Liophis epinephalus*) devours a frog in the Monteverde cloud forest we studied in Chapter 3.

FIGURE 4.4
Predator–prey systems sometimes show paired cycles, in which increases and decreases in one organism apparently drive increases and decreases in the other. Although such cycles are predicted by theory and are seen in lab experiments, they are very difficult to document conclusively in natural systems. Data from Maclulich, D. A. 1937. *Fluctuation in the numbers of varying hare (Lepus americanus)*. Univ. Toronto Stud. Biol. Ser. No. 43, Toronto: University of Toronto Press.

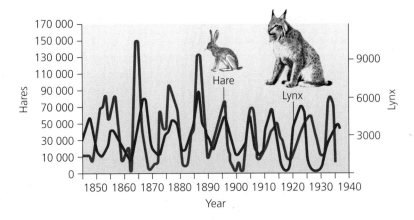

Predation can sometimes drive population dynamics by causing cycles in population sizes. An increase in the population size of prey creates more food for predators, which may survive and reproduce more effectively as a result. As the predator population rises, additional predation drives down the population of prey. Fewer prey in turn causes some predators to starve, so that the predator population declines. This allows the prey population to begin rising again, starting the cycle anew. Most natural systems involve so many factors that such cycles do not last long, but in some cases we see extended cycles (**FIGURE 4.4**).

Predation also has evolutionary ramifications. Individual predators that are more adept at capturing prey will likely live longer, healthier lives and be better able to provide for their offspring than will less adept individuals. Thus, natural selection on individuals within a predator species leads to the evolution of adaptations that make them better hunters. Prey are faced with an even stronger selective pressure—the risk of immediate death. For this reason, predation pressure has caused organisms to evolve an elaborate array of defences against being eaten (**FIGURE 4.5**).

Parasites exploit living hosts

Organisms can exploit other organisms without killing them. **Parasitism** is a relationship in which one organism, the **parasite**, depends on another, the **host**, for nourishment or some other benefit while simultaneously doing the host harm. Unlike predation, parasitism usually does not result in an organism's immediate death, although it sometimes contributes to the host's eventual death.

Many parasites live in close contact with their hosts. These parasites include disease pathogens, such as the protists that cause malaria and dysentery, as well as animals, such as tapeworms, that live in the digestive tracts of their hosts. Other parasites live on the exterior of their

(a) Cryptic coloration

(b) Warning coloration

(c) Mimicry

FIGURE 4.5
Natural selection to avoid predation has resulted in many fabulous adaptations. Some prey hide from predators by *crypsis*, or camouflage, such as this gecko on tree bark **(a)**. Other prey are brightly coloured to warn predators that they are toxic or distasteful, such as this monarch butterfly **(b)**. Still others fool predators with mimicry. Some, like walking sticks imitating twigs, mimic for crypsis. Others mimic toxic, distasteful, or dangerous organisms, like this caterpillar **(c)**; when it is disturbed, the caterpillar swells and curves its tail end and shows eyespots, to look like a snake's head.

hosts, such as the ticks that attach themselves to their hosts' skin, and the sea lamprey (*Petromyzon marinus*), another invader of the Great Lakes (**FIGURE 4.6A**). Sea lampreys are tube-shaped vertebrates that grasp the bodies of fish by using a suction-cup mouth and a rasping tongue, sucking blood from the fish for days or weeks. Sea lampreys invaded the Great Lakes from the Atlantic Ocean after people dug canals to connect the lakes for shipping, and the lampreys soon devastated economically important fisheries of chub, lake herring, whitefish, and lake trout. Since the 1950s, Great Lakes fisheries managers have reduced lamprey populations by applying chemicals that selectively kill lamprey larvae.

Other types of parasites are free-living and come into contact with their hosts only infrequently. For example, the cuckoos of Eurasia and the cowbirds of the Americas parasitize other birds by laying eggs in their nests and letting the host bird raise the parasite's young.

Some parasites cause little harm, but others may kill their hosts. Many insects parasitize other insects, often killing them in the process, and are called *parasitoids*. Various species of parasitoid wasps lay eggs on caterpillars. When the eggs hatch, the wasp larvae burrow into the caterpillar's tissues and slowly consume them. The wasp larvae metamorphose into adults and fly from the body of the dying caterpillar.

Just as predators and prey evolve in response to one another, so do parasites and hosts, in a process termed *coevolution*. Hosts and parasites can become locked in a duel of escalating adaptations, a situation sometimes referred to as an "evolutionary arms race." Like rival nations racing to stay ahead of one another in military technology, host and parasite may repeatedly evolve new responses to the other's latest advance. In the long run, though, it may not be in a parasite's best interests to become too harmful to its host. Instead, a parasite might leave more offspring in the next generation—and thus be favoured by natural selection—if it allows its host to live a longer time, or even to thrive.

(a) **Sea lamprey**

(b) **Fungus *Cordyceps***

FIGURE 4.6
Parasites harm their host organism in some way. With its suction-like mouth and rasping tongue, the sea lamprey (*Petromyzon marinus*) attaches itself to fish and sucks the fish's blood for days or weeks, sometimes killing the fish **(a)**. Sea lampreys wreaked havoc on Great Lakes fisheries after entering the lakes through human-built canals. In **(b)**, an ant (*Pachycondyla*) is infected by a fungus (*Cordyceps*) that will eventually kill it. In the meantime, fruiting bodies of the fungus are sprouting from the ant's head. The fungus will soon alter the ant's behaviour, causing it to climb to the highest branches of a nearby plant, so its spores will attain the broadest possible distribution.

Herbivores exploit plants

One of the most common types of exploitation is **herbivory**, which occurs when animals feed on the tissues of plants. Insects that feed on plants are the most widespread type of herbivore; just about every plant in the world is attacked by some type of insect (**FIGURE 4.7**). In most cases, herbivory does not kill a plant outright, but may affect its growth and reproduction.

Like animal prey, plants have evolved a wide array of defences against the animals that feed on them. Many plants produce chemicals that are toxic or distasteful to herbivores. Others arm themselves with thorns, spines, or irritating hairs. In response, herbivores may evolve ways to overcome these defences, and the plant and the animal may embark on an evolutionary arms race.

Some plants go a step further and recruit certain animals as allies to assist in their defence. Many such plants encourage ants to take up residence by providing thorns or swelled stems for the ants to nest in or nectar-bearing structures for the ants to feed from. These ants protect the plant in return by attacking other insects that land or crawl on it. Other plants respond to herbivory by releasing volatile chemicals when they are bitten or pierced. The airborne chemicals attract predatory insects that may attack the herbivore. Such cooperative strategies as trading defence for food are examples of our next type of species interaction, mutualism.

FIGURE 4.7
Herbivory is a common way to make a living. The world holds many thousands, and perhaps millions, of species of plant-eating insects, such as this larva (caterpillar) of the death's head hawk moth (*Acherontia atropos*) from western Europe.

Mutualists help one another

Mutualism is a relationship in which two or more species benefit from interaction with one another. Generally each partner provides some resource or service that the other needs.

Many mutualistic relationships—like many parasitic relationships—occur between organisms that live in close physical contact. (Indeed, biologists hypothesize that many mutualistic associations evolved from parasitic ones.) Such physically close association is called **symbiosis**. Thousands of terrestrial plant species depend on mutualisms with fungi; plant roots and some fungi together form symbiotic associations called mycorrhizae. In these symbioses, the plant provides energy and protection to the fungus, while the fungus assists the plant in absorbing nutrients from the soil. In the ocean, coral polyps, the tiny animals that build coral reefs, share beneficial arrangements with algae known as zooxanthellae. The coral provide housing and nutrients for the algae in exchange for a steady supply of food—90% of their nutritional requirements.

You, too, are part of a symbiotic association. Your digestive tract is filled with microbes that help you digest food—microbes for which you are providing a place to live. Indeed, we may owe our very existence to symbiotic mutualisms. It is now widely accepted that the eukaryotic cell originated after certain prokaryotic cells engulfed other prokaryotic cells and established mutualistic symbioses. Scientists have inferred that some of the engulfed cells eventually evolved into cell organelles.

Not all mutualists live in close proximity. One of the most important mutualisms in environmental science involves free-living organisms that may encounter each other only once in their lifetimes. This is *pollination* (**FIGURE 4.8**), an interaction of key significance to agriculture and our food supply. Bees, birds, bats, and other creatures transfer pollen (male sex cells) from one flower to the ova (female cells) of another, fertilizing the female egg, which subsequently grows into a fruit. The pollinating animals visit flowers for their nectar, a reward the plant uses to entice them. The pollinators receive food, and the plants are pollinated and reproduce. Various types of bees alone pollinate 73% of our crops, one expert has estimated—from soybeans to potatoes to tomatoes to beans to cabbage to oranges.

FIGURE 4.8
In mutualism, organisms of different species benefit one another. An important mutualistic interaction for environmental science is pollination. This hummingbird visits flowers to gather nectar and in the process transfers pollen between flowers, helping the plant reproduce. Pollination is of key importance to agriculture, ensuring the reproduction of many crop plants.

Some interactions have no effect on some participants

Two other types of species interaction get far less attention. **Amensalism** is a relationship in which one organism is harmed and the other is unaffected. In **commensalism**, one species benefits and the other is unaffected. Amensalism has been difficult to pin down, because it is hard to prove that the organism doing the harm is not in fact besting a competitor for a resource. For instance, some plants release poisonous chemicals that harm nearby plants (a phenomenon called *allelopathy*), and some experts have suggested that this is an example of amensalism. However, allelopathy can also be viewed as one plant investing in chemicals to outcompete others for space.

One association commonly cited as an example of commensalism occurs when the conditions created by one plant happen to make it easier for another plant to establish and grow. For instance, palo verde trees in the Sonoran Desert create shade and leaf litter that allow the soil beneath them to hold moisture longer, creating an area that is cooler and moister than the surrounding sun-baked ground. Young plants find it easier to germinate and grow in these conditions, so seedling cacti and other desert plants generally grow up directly beneath "nurse" trees, such as palo verde. This phenomenon, called *facilitation*, influences the structure and composition of communities and how they change through time.

Ecological Communities

In Chapter 3 we defined a *community* as a group of populations of organisms that live in the same place at the same time. The members of a community interact with one another in the ways described above, and the direct interactions among species often have indirect effects that ripple outward to affect other community members. The strength of interactions also varies, and together species' interactions determine the species composition, structure, and function of communities. *Community ecologists* are interested in which species coexist, how they relate to one another, how communities change through time, and why these patterns exist.

Energy passes among trophic levels

The interactions among members of a community are many and varied, but some of the most important involve who eats whom. As we saw in Chapter 2, the energy that drives such interactions in most systems comes ultimately from the sun via photosynthesis. As organisms feed on one another, this energy moves through the community, from one rank in the feeding hierarchy, or **trophic level**, to another (**FIGURE 4.9**).

Producers *Producers*, or *autotrophs* ("self-feeders," as defined in Chapter 2), compose the first trophic level, as we saw in Chapter 2. Terrestrial green plants, cyanobacteria, and algae capture solar energy and use photosynthesis to produce sugars. The chemosynthetic bacteria of hot springs and deep-sea hydrothermal vents use geothermal energy in a similar way to produce food.

Consumers Organisms that consume producers are known as *primary consumers* and compose the second trophic level. Grazing animals, such as deer and grasshoppers, are primary consumers. The third trophic level consists of *secondary consumers,* which prey on primary consumers. Wolves that prey on deer are considered secondary consumers, as are rodents and birds that prey on grasshoppers. Predators that feed at even higher trophic levels are known as *tertiary consumers.* Examples of tertiary consumers include hawks and owls that eat rodents that have eaten grasshoppers. Note that most primary consumers are **herbivores** because they consume plants, whereas secondary and tertiary consumers are **carnivores** because they eat animals. Animals that eat both plant and animal food are referred to as **omnivores**.

Detritivores and decomposers Detritivores and decomposers consume nonliving organic matter. **Detritivores**, such as millipedes and soil insects, scavenge the waste products or the dead bodies of other community members. **Decomposers**, such as fungi and bacteria, break down leaf litter and other nonliving matter further into simpler constituents that can then be taken up and used by plants. These organisms play an essential role as the community's recyclers, making nutrients from organic matter available for reuse by living members of the community.

In Great Lakes communities, phytoplankton are the main producers, floating freely and photosynthesizing with sunlight that penetrates the upper layer of the water. Zooplankton are primary consumers, feeding on the phytoplankton. Phytoplankton-eating fish are primary consumers, and zooplankton-eating fish are secondary consumers. At higher trophic levels are tertiary consumers, such as larger fish and birds that feed on plankton-eating fish. Zebra mussels, by eating both phytoplankton and zooplankton, function on multiple trophic levels. When any of these organisms dies and sinks to the bottom, detritivores scavenge its tissues and microbial decomposers recycle its nutrients.

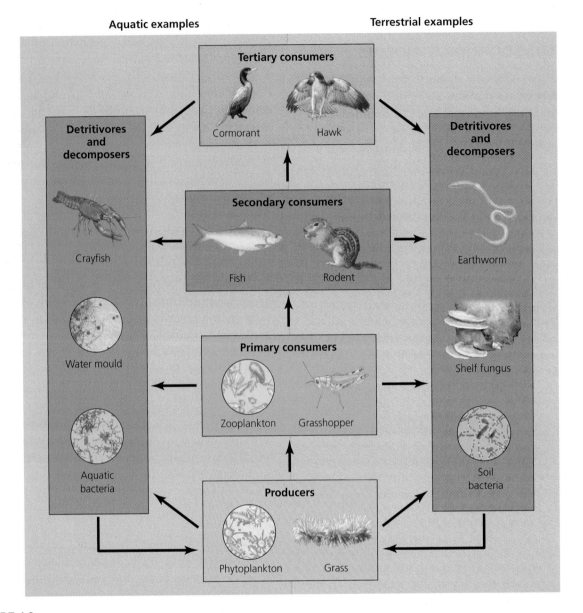

FIGURE 4.9
Ecologists organize species hierarchically by their feeding rank, or trophic level. The diagram shows aquatic (left) and terrestrial (right) examples at each level. Arrows indicate the direction of energy flow. Producers produce food by photosynthesis, primary consumers (herbivores) feed on producers, secondary consumers eat primary consumers, and tertiary consumers eat secondary consumers. Communities can have more or fewer trophic levels than in this example. Detritivores and decomposers feed on nonliving organic matter and the remains of dead organisms from all trophic levels, and they "close the loop" by returning nutrients to the soil or the water column for use by producers.

Energy, biomass, and numbers decrease at higher trophic levels

At each trophic level, most of the energy that organisms use is lost through respiration. Only a small amount of the energy is transferred to the next trophic level through predation, herbivory, or parasitism. The first trophic level (producers) contains a large amount of energy, but the second (primary consumers) contains less energy—only that amount gained from consuming producers. The third trophic level (secondary consumers) contains still less energy, and higher trophic levels (tertiary consumers) contain the least. A general rule of thumb is that each trophic level contains just 10% of the energy of the trophic level below it, although the actual proportion can vary greatly.

This pattern, which can be visualized as a **trophic pyramid**, generally also holds for the numbers of organisms at each trophic level (**FIGURE 4.10**). Generally, fewer organisms exist at higher trophic levels than at lower trophic levels. A grasshopper eats many plants in its lifetime, a rodent eats many grasshoppers, and a hawk eats

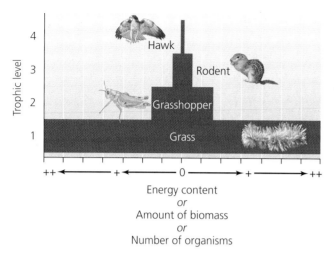

FIGURE 4.10
An atrophic pyramid illustrates a rough rule of thumb for the way ecological communities are structured. Organisms at lower tropic levels generally exist in far greater numbers, with greater energy content and greater biomass, than organisms at higher tropic levels. When one organism consumes another, most energy gets used up in respiration rather than in building new tissue. The example shown here is generalized; the actual shape of any given pyramid may be different.

many rodents. Thus, for every hawk in a community there must be many rodents, still more grasshoppers, and an immense number of plants. Because the difference in numbers of organisms among trophic levels tends to be large, the same pyramid-like relationship also often holds true for biomass. Even though rodents are larger than grasshoppers, and hawks larger than rodents, the sheer number of prey relative to the predators means that prey biomass will likely be greater overall.

Food webs show feeding relationships and energy flow

As energy is transferred from species on one trophic level to species on other trophic levels, it is said to pass up a **food chain**. Plant, grasshopper, rodent, and hawk make up a food chain (as in **FIGURE 4.10**), a linear series of feeding relationships. Thinking in terms of food chains is conceptually useful, but in reality ecological systems are far more complex than simple linear chains. A more accurate representation of the feeding relationships in a community is a **food web**, a visual map of feeding relationships and energy flow, showing the many paths by which energy passes among organisms as they consume one another.

FIGURE 4.11 shows a food web from a temperate deciduous forest of eastern North America. Like virtually all diagrams of ecological systems, it is greatly simplified, leaving out the vast majority of species and interactions that occur. Note, however, that even within this simplified diagram, we can pick out a number of different food chains involving different sets of species.

A Great Lakes food web would involve the phytoplankton and cyanobacteria that photosynthesize near the water's surface, the zooplankton that eat them, the fish that eat all these, the larger fish that eat the smaller fish, and the lampreys that parasitize the fish. It would include a number of native mussels and clams and, since 1988, the zebra and quagga mussels that are crowding them out. It would include diving ducks that used to feed on native bivalves and now are preying on mussels.

This food web would also show that an array of bottom-dwelling invertebrates feed from the refuse of the exotic mussels. These waste products promote bacterial growth and disease pathogens that harm native bivalves, but they also provide nutrients that nourish crayfish and many smaller *benthic* (bottom-dwelling) invertebrate animals. Finally, the food web would include underwater plants and macroscopic algae, whose growth is promoted by the non-native mussels. The mussels clarify the water by filtering out phytoplankton, and sunlight penetrates deeper into the water column, spurring photosynthesis and plant growth. Thus, zebra and quagga mussels alter this food web essentially by shifting productivity from the open-water regions to the benthic and *littoral* (nearshore) regions. In so doing, the mussels affect fish indirectly, helping benthic and littoral fishes and making life harder for open-water fishes (see "The Science Behind the Story: Assessing the Ecological Impacts of Zebra Mussels").

Some organisms play bigger roles in communities than others

"Some animals are more equal than others," George Orwell wrote in his 1945 book *Animal Farm*. Although Orwell was making wry sociopolitical commentary, his remark hints at a truth in ecology. In communities, ecologists have found, some species exert greater influence than do others. A species that has a particularly strong or far-reaching impact is often called a **keystone species**. A keystone is the wedge-shaped stone at the top of an arch that is vital for holding the structure together; remove the keystone, and the arch will collapse. In an ecological community, removal of a keystone species will have substantial ripple effects and will alter a large portion of the food web.

Some keystone species have been removed from their natural communities with unintended consequences, in what are essentially uncontrolled large-scale experiments. A well-known example is the elimination of wolves (intentionally, for the most part) from many parts of North America. Wolves are voracious predators; when

FIGURE 4.11
Food webs are conceptual representations of feeding relationships in a community. This food web pertains to eastern North America's temperate deciduous forest and includes organisms on several trophic levels. In a food web diagram, arrows are drawn from one organism to another to indicate the direction of energy flow as a result of predation, parasitism, or herbivory. For example, an arrow leads from the grass to the cottontail rabbit to indicate that cottontails consume grasses. The arrow from the cottontail to the tick indicates that parasitic ticks derive nourishment from cottontails. Communities include so many species and are complex enough, however, that most food web diagrams are bound to be gross simplifications.

they are eliminated from an area, the populations of large herbivores, such as elk and moose, can grow out of control. This can have far-reaching impacts on vegetation and, consequently, on all other animals in the area.

Ecologists also have verified the keystone species concept by careful observation and controlled experiments. For example, classic work by marine biologist Robert Paine established that the predatory starfish *Pisaster ochraceus* has great influence on the community composition of intertidal organisms on the Pacific coast of North America. When *Pisaster* is present in this community, species diversity is high, with several types of barnacles, mussels, and algae. When *Pisaster* is removed, the mussels it preys on become numerous and displace other species, suppressing species diversity. More recent work off the Atlantic coast published in 2007 suggests that the reduction of shark populations by commercial fishing has allowed populations of certain skates and rays to increase, which has depressed numbers of bay scallops and other bivalves they eat.

THE SCIENCE BEHIND THE STORY

Assessing the Ecological Impacts of Zebra Mussels

Echinogammarus ischnus is a small (8–10 mm) amphipod, a freshwater shrimp that has invaded the Great Lakes and is replacing the local amphipod species, *Gammarus fasciatus*.

When zebra mussels appeared in the Great Lakes, people feared for sport fisheries and estimated that fish population declines could cost billions of dollars. The mussels would deplete the phytoplankton and zooplankton that fish depended on, people reasoned, and many fewer fish would survive. Now, 20 years after the first appearance of *Dreissena polymorpha* in the Great Lakes, scientists are using a variety of approaches to assess the true long-term impacts of these invaders on local ecosystems.

Dr. Anthony Ricciardi is an aquatic ecologist at McGill University School of Environment. He and his graduate students are working to understand *Dreissena polymorpha* and other aquatic invasive species, through a combination of field studies, replicated experiments, modelling, and analysis of published data.

Graduate student Jessica Ward has taken the approach of *meta-analysis*—analyzing and drawing new conclusions from data that have already been published. One interesting and challenging aspect of zebra mussel infestations is that the impacts differ from one part of an ecosystem to another. Zebra mussels may crowd out some native species and decrease the availability of phytoplankton as food for other species. However, the resulting increase in waste products may enhance the availability of nutrients for some *benthic* (bottom-dwelling) organisms. In other words, zebra mussels have some negative and some positive impacts. Understanding and quantifying these impacts is crucial for predicting how infestations will affect ecosystems.

Ward examined published data from 47 sites and developed statistical models of the impacts of zebra mussels on benthic invertebrates. She confirmed that the introduction of zebra mussels is generally associated with an increase in benthic invertebrate diversity across a broad range of habitats and environmental conditions.[3]

Student Michelle Palmer took a different approach. Palmer wanted to know why an invasion can lead to the local extinction of a native species in one part of an ecosystem, while the two species (invader and native) may coexist in another part of the ecosystem. Her hypothesis was that the physical–chemical characteristics and existing biota of the local community can help or hinder the invading species, thus either weakening or exacerbating its impacts. Palmer specifically investigated *Echingammarus ischnus* (an *amphipod*, or small freshwater shrimp, see photo), an invasive species that is replacing the common native amphipod *Gammarus fasciatus* in some parts of the Great Lakes–St. Lawrence system but not in others.

It had been suggested by other researchers that the replacement of the native amphipod by the invader might be more successful in environments that had previously been invaded by *Dreissena polymorpha*. Zebra mussels alter the physical environment by changing the water quality and by attaching to and dominating all hard or rocky surfaces. The presence of zebra mussel colonies alters the complexity of the habitat, greatly increasing the number of places where smaller organisms like amphipods could hide from predators; such hiding places are called *refugia*. Perhaps *E. ischnus* was more skillful than *G. fasciatus* at taking advantage of the refugia offered by *D. polymorpha* colonies. If so, then the presence of zebra mussels could enhance the invasive impacts of *E. ischnus*, allowing it to be more successful at replacing native shrimps.

To investigate, Palmer carried out field surveys, documenting the physical and chemical characteristics of the environments in which the species were coexisting. She designed *in situ* (on-site) experiments to test the susceptibility of the amphipods to predators by using hard surfaces (bricks) that were bare, half-covered, or fully covered with zebra mussel colonies. Some of the experimental set-ups were designed to allow access by the amphipods' main predators, and some were designed to exclude predators. Ten replicates of each set-up were investigated.

Palmer found that the presence of zebra mussels correlated with a higher population density of amphipods of both types, confirming that the refugia offered by the zebra mussel colonies were beneficial to the amphipods. She also confirmed that overall numbers of amphipods were lower in the experimental set-ups that allowed predation, compared with those in which predators were excluded. However, Palmer found that the native *G. fasciatus* was actually *more* successful than the invader *E. ischnus* in the zebra mussel-infested locations that were available to predation.

One reason for this, she suggested, is that *G. fasciatus* is less active, burrowing into the refugia among the zebra mussel shells, compared with *E. ischnus*, which exposes itself to predation by moving about more actively. In other words, Palmer's results suggested that the presence of zebra mussels levelled the playing field, rendering the native species less susceptible to predation, and contributing to the coexistence of the two species.[4]

To understand the significance of one or two large species in an ecosystem, and the delicacy of the balance, we can consider the case of moose and wolves in the boreal forest of Cape Breton Highlands National Park. Moose were rare by 1900 and had completely disappeared from the area by 1924, because of excessive hunting and habitat destruction. Parks Canada reintroduced moose to the park during 1947 and 1948, by importing and releasing 18 animals from Elk Island National Park. The reintroduction was highly successful, and moose are currently plentiful in the park, possibly too plentiful.[5] In the absence of the moose's natural predator, the wolf, which disappeared from the area as early as the mid-1800s, there are few natural controls on the moose population within the park. Moose are selective eaters; their preferred winter food is the balsam fir. By browsing heavily on certain types of food but not others, they can alter the forest landscape, leading to overall changes in the ecosystem. The success of the reintroduced moose population in Cape Breton Highlands National Park may be leading to changes in the composition of the boreal forest there.

Animals at high trophic levels, such as wolves, starfish, and sea otters, are most often seen as keystone species. Other species attain keystone status as "ecosystem engineers" by physically modifying the environment shared by community members. Beavers build dams and turn streams into ponds, flooding vast expanses of dry land and turning it to swamp. Prairie dogs dig burrows that aerate the soil and serve as homes for other animals. Bees are absolutely crucial, for example, even for human food security, because they are *pollinators*, moving pollen from male to female plants to facilitate the plants' sexual reproduction, as well as ensuring plant genetic diversity.

Less conspicuous organisms and those toward the bottoms of food chains can potentially be viewed as keystone species, too. Remove the fungi that decompose dead matter or the insects that control plant growth or the phytoplankton that are the base of the marine food chain and a community may change very rapidly indeed. Because there are usually more species at lower trophic levels, however, it is less likely that any one of them alone might have wide influence; if one species is removed, other species that remain may be able to perform many of its functions.

Identifying keystone species is no simple task, and there is no cut-and-dried definition of the term to help us. Community dynamics are complex, species interactions differ in their strength, and the strength of species interactions can vary through time and space. "The Science Behind the Story: Otters, Urchins, Kelp, and a Whale of a Chain Reaction" gives an idea of the surprises that are sometimes in store for ecologists studying these interactions.

Communities respond to disturbance in different ways. The removal of a keystone species is just one of many types of disturbance that can modify the composition, structure, or function of an ecological community. Over time, any given community may experience natural disturbances ranging from gradual phenomena, such as climate change, to sudden events, such as hurricanes, floods, or avalanches.

weighing the issues 4–1
KEYSTONE SPECIES AND CONSERVATION

Imagine the government is gathering citizen input on three development options. Detailed studies assess the potential effects on the environment that would likely result from each of the development projects. A report of the results of these studies, an **environmental impact statement (EIS)**, says that option 1 would likely result in the extermination of bobcats. Option 2, the EIS predicts, would probably kill off a species of pocket mouse, a primary consumer that is common in the community. Option 3 would likely eliminate a species of lupine, a plant that covers a large percentage of the ground in the present community.

You are a citizen desiring minimal change in the natural community so that your children grow up in an area like the one in which you grew up. What questions would you ask of an ecologist about the bobcat, pocket mouse, and lupine so that you could decide which might most likely be a keystone species? If you had to provide input without further information, what would you advise the government?

Communities are dynamic systems and may respond to disturbance in several ways. A community that resists change and remains stable despite disturbance is said to show **resistance** to the disturbance. Alternatively, a community may show **resilience**, meaning that it changes in response to disturbance but later returns to its original state. Or a community may be modified by disturbance permanently and may never return to its original state.

Succession follows severe disturbance

If a disturbance is severe enough to eliminate all or most of the species in a community, the affected site will undergo a somewhat predictable series of changes that ecologists call **succession**. In the traditional view of this process, ecologists described two types of succession.

Primary succession follows a disturbance so severe that no vegetation or soil life remains from the community that occupied the site. In primary succession, a biotic community is built essentially from scratch. In contrast, **secondary succession** begins when a disturbance dramatically alters an existing community but does not destroy all living things or all organic matter in the

FIGURE 4.12
As this glacier retreats, small pioneer plants (foreground) begin the process of primary succession.

The pioneers best suited to colonizing bare rock are the mutualistic aggregates of fungi and algae known as **lichens**. Lichens succeed because their algal component provides food and energy via photosynthesis while the fungal component takes a firm hold on rock and captures the moisture that both organisms need to survive. As lichens grow, they secrete acids that break down the rock surface. The resulting waste material forms the beginnings of soil, and once soil begins to form, small plants, insects, and worms find the rocky outcrops more hospitable. As new organisms arrive, they provide more nutrients and habitat for future arrivals. As time passes, larger plants establish themselves, the amount of vegetation increases, and species diversity rises.

Secondary succession on land begins when a fire, a hurricane, logging, or farming removes much of the biotic community. Consider a farmed field in eastern North America that has been abandoned (**FIGURE 4.13**). In the first few years after farming ends, the site will be colonized by pioneer species of grasses, herbs, and forbs that were already in the vicinity and that disperse effectively. As time passes, shrubs and fast-growing trees, such as aspens, rise from the field. Pine trees subsequently rise above the aspens and shrubs, forming a pine-dominated forest. This pine forest develops an understory of hardwood trees, because pine seedlings do not grow well under mature pines, whereas some hardwood seedlings do. Eventually the hardwoods outgrow the pines, creating a hardwood forest (see **FIGURE 4.13**).

soil. In secondary succession, vestiges of the previous community remain, and these building blocks help shape the process.

At terrestrial sites, primary succession takes place after a bare expanse of rock, sand, or sediment becomes newly exposed to the atmosphere. This can occur when glaciers retreat, lakes dry up, or volcanic lava flows spread across the landscape (**FIGURE 4.12**). Species that arrive first and colonize the new substrate are referred to as **pioneer species**. Pioneer species are well adapted for colonization, having such traits as spores or seeds that can travel long distances.

Processes of succession occur in many diverse ecological systems, from ponds to rocky intertidal areas to the carcasses of animals. A lake or pond that originates as nothing but water on a lifeless substrate begins to undergo succession as it is colonized by algae, microbes, plants, and zooplankton. As these organisms grow, reproduce, and die, the water body slowly fills with organic matter. The lake or pond acquires further organic matter and sediments from the water it receives from rivers, streams,

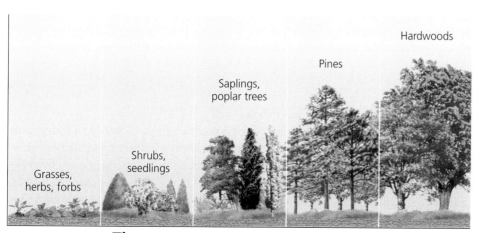

Time ⟶

FIGURE 4.13
Secondary succession occurs after a disturbance, such as fire, landslides, or farming, removes most of the vegetation from an area. Shown here is a typical series of changes in a plant community of eastern North America following the abandonment of a farmed field.

THE SCIENCE BEHIND THE STORY

Otters, Urchins, Kelp, and a Whale of a Chain Reaction

Dr. James Estes, University of California at Santa Cruz, works at sea in Alaska.

Even after ecologists thought they understood the relationship among sea otters, sea urchins, and kelp, some surprises were in store.

Sea otters (*Enhydra lutris*) live in coastal waters of the Pacific Ocean. These mammals float on their backs amid the waves, feasting on sea urchins (*Strongylocentrotus* spp.) that they pry from the ocean bottom. Once abundant, sea otters were hunted nearly to extinction for their fur. Protection by international treaty in 1911 allowed their numbers to grow. Otters returned to high densities in some regions but failed to return in others.

Biologists noted that regions with abundant otters hosted dense "forests" of kelp, a brown alga (seaweed) that anchors to the seafloor, growing up to 60 m high toward the sunlit surface. Kelp forests provide complex physical structures in which diverse communities of fish and invertebrates find shelter and food. In regions without sea otters, scientists found kelp forests absent. In the absence of otters, urchins become so numerous that they eat every last bit of kelp, creating empty seafloors called "urchin barrens" that are relatively devoid of life.

Ecologists determined that otters were largely responsible for the presence of the kelp forests, simply by keeping urchin numbers in check through predation. This research—mostly by James Estes of the University of California at Santa Cruz and his colleagues—established sea otters as a prime example of a keystone species. Jane Watson investigated 60 randomly selected sites in the shallow rocky communities off the northwest coast of Vancouver Island. Sea otters were present in 40 sites and absent from 20. In the locations where sea otters were absent, urchins occupied 84% of the sea bottom. In contrast, areas with sea otters hosted large populations of kelp, and urchins occupied only 1.8% of the bottom.

But the story did not end there. In the 1990s, otter populations dropped precipitously near Alaska and the Aleutians. No one knew why. Estes and his co-workers placed radio tags on Aleutian otters and studied them at sea. Their first hypothesis was that fertility rates had dropped, but radio-tracking observations showed that females were raising pups without problem. Their second hypothesis was that the otters were simply moving to other locations. But the radio tracking showed no unusual dispersals.

They were left with only one viable hypothesis: increased mortality. Then one day in 1991, Estes's team witnessed something never seen before. They watched as a sea otter was killed and eaten by an orca, or killer whale (*Orcinus orca*). These striking black-and-white predators grow up to 10 m long, hunt in groups, and usually attack larger prey. A sea otter to them is a mere snack. Yet over the following years, Estes's team saw more cases of orca predation on otters. Could killer whales be killing off the otters?

The researchers compared a bay where otters were vulnerable to orcas with a lagoon where they were protected. Otter numbers in the lagoon remained stable over four years, whereas those in the bay dropped by 76%. Radio tracking showed no movement between these locations. Using data on otter birth rates, death rates, and population age structure, they estimated that to account for the otter decline, 6788 orca attacks per year would have had to occur in their study area. This *expected* rate of observed attacks matched their *actual* number of observed attacks. These lines of evidence led the researchers to propose that predation by orcas was eliminating otters.

As otters declined in the Aleutians, urchins increased, and kelp density fell dramatically (see figure). These changes supported the idea that otters were a keystone species, but now it seemed that one keystone species was being controlled by another.

---

and surface runoff. Eventually, the water body fills in, becoming a bog or even a terrestrial ecosystem.

In this traditional view of succession that we have described, the transitions between stages of succession eventually lead to a **climax community**, which remains in place, with little modification, until some disturbance restarts succession. Early ecologists felt that each region had its own characteristic climax community, determined by the region's climate.

Today, ecologists recognize that succession is far more variable and less predictable than originally thought. The trajectory of succession can vary greatly according to chance factors, such as which particular species happen to gain an early foothold. The stages of succession blur into one another and vary from place to place, and some stages may sometimes be skipped completely. In addition, climax communities are not predetermined solely by climate but may vary with other conditions from one time or place to another.

Once a climax community is disturbed and succession is set in motion, there is no guarantee that the community will ever return to that climax state. Many communities disturbed by human impact have not returned to their former conditions. This is the case with vast areas of the Middle East that once were fertile enough to support productive farming but now are deserts.

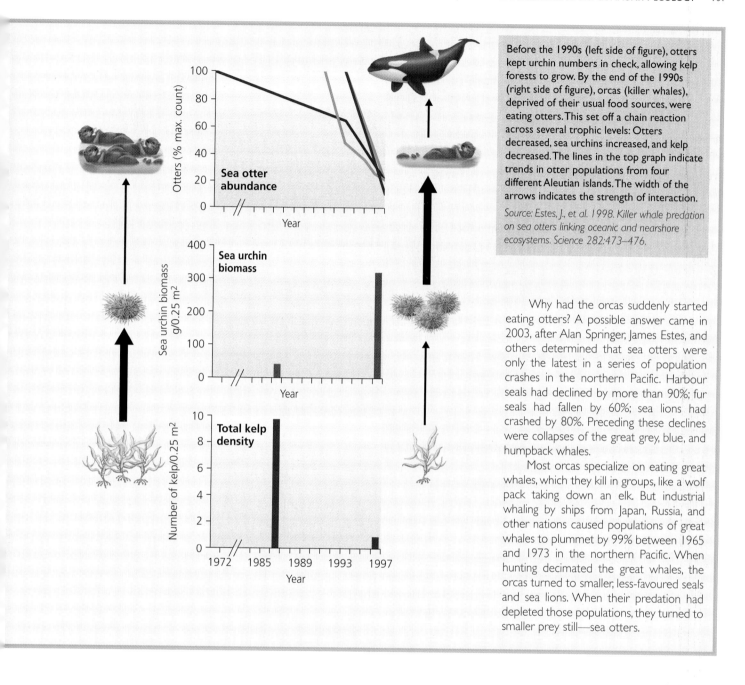

Before the 1990s (left side of figure), otters kept urchin numbers in check, allowing kelp forests to grow. By the end of the 1990s (right side of figure), orcas (killer whales), deprived of their usual food sources, were eating otters. This set off a chain reaction across several trophic levels: Otters decreased, sea urchins increased, and kelp decreased. The lines in the top graph indicate trends in otter populations from four different Aleutian islands. The width of the arrows indicates the strength of interaction.

Source: Estes, J., et al. 1998. Killer whale predation on sea otters linking oceanic and nearshore ecosystems. Science 282:473–476.

Why had the orcas suddenly started eating otters? A possible answer came in 2003, after Alan Springer, James Estes, and others determined that sea otters were only the latest in a series of population crashes in the northern Pacific. Harbour seals had declined by more than 90%; fur seals had fallen by 60%; sea lions had crashed by 80%. Preceding these declines were collapses of the great grey, blue, and humpback whales.

Most orcas specialize on eating great whales, which they kill in groups, like a wolf pack taking down an elk. But industrial whaling by ships from Japan, Russia, and other nations caused populations of great whales to plummet by 99% between 1965 and 1973 in the northern Pacific. When hunting decimated the great whales, the orcas turned to smaller, less-favoured seals and sea lions. When their predation had depleted those populations, they turned to smaller prey still—sea otters.

Invasive species pose new threats to community stability

Traditional concepts of communities and successions involve sets of organisms understood to be native to an area. But what if a new organism arrives from elsewhere? And what if this non-native (also called *alien* or **exotic**) organism turns *invasive*, spreads widely, and becomes dominant? Such **invasive species** can alter a community substantially and are one of the central ecological forces in today's world.

Most often, invasive species are non-native species that people have introduced, intentionally or by accident, from elsewhere in the world. Any non-native organism introduced into an ecosystem will require adjustments, but species become invasive **pests** when the negative impacts outweigh the benefits, especially when the limiting factors that might regulate their population growth are absent. Thus, the main characteristics of *problematic* invasive species include the ability to spread rapidly, and unimpeded, in the new environment, and the ability to have a negative impact on native species and ecosystems into which it has been introduced.

Plants and animals brought to one area from another may leave their predators, parasites, and competitors behind, freeing them from natural constraints on their

FIGURE 4.14
The zebra mussel is a prime example of a biological invader that has modified an ecological community. By filtering phytoplankton and small zooplankton from open water, it generates a number of impacts on other species, both negative (red downward arrows) and positive (green upward arrows) **(a)**. This map **(b)** shows the range of zebra and quagga mussels in North America as of 2007. In less than three decades the zebra mussel has spread to Ontario, Manitoba, Quebec, and 24 U.S. states, mainly by boats but in some cases by overland transport. The quagga mussel is rapidly following the zebra mussel's spread.
Source for (b): U.S. Geological Survey.[6]

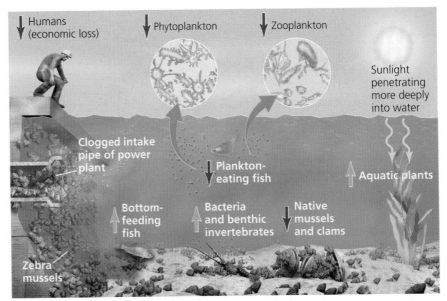

(a) Impacts of zebra mussels on members of a Great Lakes nearshore community

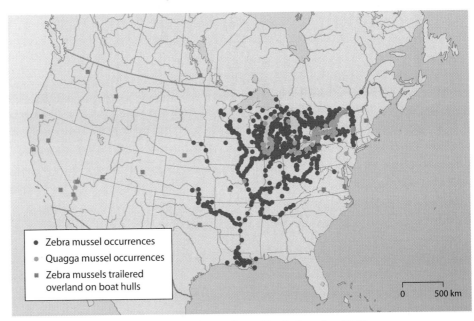

(b) Distribution of zebra and quagga mussels on North America, 2007

population growth (**FIGURE 4.14**). If there happen to be few organisms in the new environment that can act as predators, parasites, or competitors, the introduced species may do very well. As it proliferates, it may exert diverse influences on its fellow community members. An example is the chestnut blight, an Asian fungus that killed nearly every mature American chestnut (*Castanea dentata*), the dominant tree species of many forests of eastern North America, in the quarter-century preceding 1930. Asian trees had evolved defences against the fungus over long millennia of coevolution, but the American chestnut had not.

In other cases, a species may be considered a pest even in regions where it is native. An example is the Asian long-horned beetle (*Anoplophora glabripennis*), a voracious wood-eating insect that is native to China but is also a pest, with few natural predators. The Asian long-horned beetle feeds on many different species of temperate hardwood trees, including maple, birch, horse chestnut, poplar, willow, elm, ash, and black locust. The beetle, introduced to North America from China via packaging materials, has the potential to cause widespread destruction. To date, the only approach to controlling the spread of the beetle is to cut down and burn any infested trees.[7]

So far we have considered examples of the impacts of a non-native insect, a fungus, and a mollusc, but virtually any type of organism can become an invasive pest, given circumstances that facilitate its spread. Introduced grasses and shrubs, such as the Scotch broom (*Cytisus scoparius*), have had dramatic impacts on the Garry oak (*Quercus garryana*)-based ecosystem of southeastern Vancouver Island and the Gulf Islands, a disappearing ecosystem that supports many rare plant species.[8] Fish introduced into streams purposely for sport or accidentally via shipping compete with and exclude native fish.

A particularly troublesome example of a native invasive species is the sea lamprey (*Petromyzon marinus*), an eel-like fish that is native to the Atlantic Ocean and was probably introduced to the Great Lakes as early as the 1830s by oceangoing vessels. The sea lamprey is a pretty unpleasant creature to begin with, as it lives by attaching itself to the flanks of other fish and feeding parasitically on their blood. But lampreys have had tremendous success as they have spread—mainly via shipping channels and canals—into the Great Lakes, devastating some local fish populations.

Hundreds of island-dwelling animals and plants worldwide have been driven extinct by the goats, pigs, and rats intentionally introduced by human colonists. The cane toad (*Bufo marinus*), introduced to Australia to control insects in sugar cane fields (which it never did very successfully), is poisonous to just about anything that tries to eat it and has been extremely damaging to a wide variety of indigenous animal populations.

We will examine more examples of introduced species in our discussion of biodiversity in Chapter 9. The impact of invasive species on native species and ecological communities is severe already, and it is growing year by year with the increasing mobility of humans and the globalization of our society. Global trade helped spread zebra mussels, which were unintentionally transported in the ballast water of cargo ships. To maintain stability at sea, ships take water into their hulls as they begin their voyage and discharge that water at their destination. Decades of unregulated exchange of ballast water have ferried hundreds of species across the oceans.

In North America, zebra mussels—and the media attention they generated—helped put invasive species on the map as a major environmental and economic problem. Scientific research into introduced species has proliferated, and many ecologists view invasive species as the second-greatest threat to species and natural systems, behind only habitat destruction.

Funding has now become more widely available for the control and eradication of invasive species, and control mechanisms are widely researched and shared across jurisdictional boundaries. Managers at all levels of government have been trying a variety of techniques to control the spread of zebra and quagga mussels—removing them manually, applying toxic chemicals, drying them out, depriving them of oxygen, introducing predators and diseases, and stressing them with heat, sound, electricity, carbon dioxide, and ultraviolet light. However, most of these are localized and short-term fixes that are not capable of making a dent in the huge populations at large in the environment. In case after case, managers are finding that controlling and eradicating invasive species are so difficult and expensive that preventive measures (such as ballast water regulations) represent a much better investment.

weighing the issues 4–2
ARE INVASIVE SPECIES ALL BAD?

Some ethicists have questioned the notion that all invasive species should automatically be considered bad. If we introduce a non-native species to a community and it greatly modifies the community, do you think that is a bad thing? What if it drives another species extinct? What if the invasive species arrived on its own, rather than through human intervention? What ethical standard(s) would you apply to determine whether an invasive species should be battled or accepted?

Some altered communities can be restored to their former condition

Invasive species are adding to the tremendous transformations that humans have already forced on natural landscapes through habitat alteration, deforestation, hunting of keystone species, pollution, and other impacts. With so much of Earth's landscape altered by human impact, it is impossible to find areas that are truly pristine. This realization has given rise to the conservation effort known as **ecological restoration**. The practice of ecological restoration is informed by the science of **restoration ecology**. Restoration ecologists research the historical conditions of ecological communities as they existed before our industrialized civilization altered them. They then try to devise ways to restore some of these areas to an earlier condition, often to a natural "presettlement" condition.

For example, activities underway at Chatterton Hill Park in Victoria, British Columbia, are aimed at the ecological restoration of Garry oak–associated ecosystems (**FIGURE 4.15**). As mentioned above, these delicate, complex ecosystems, which occur almost exclusively in south Vancouver Island and the Gulf Islands, are being invaded

FIGURE 4.15
Garry oak ecosystems in south Vancouver Island and the Gulf Islands of British Columbia, like the one shown here, are being invaded by aggressive exotic species. Some have been targeted for possible ecological restoration.

by aggressive exotic species that suppress the growth of native plants. Ecological restoration activities at Chatterton Hill, which began in 2002, include active restoration, removal of invasive plants, natural feature inventories (soil, vegetation, and animals), monitoring, and educational activities. In 2006, local restoration ecologists reevaluated the site, comparing it with a similar nearby site where no restoration activities had taken place; they concluded that biodiversity at the restoration site was increasing. At the comparison site, the tree layer was similar but the understory vegetation was still dominated by grasses and the invasive exotic, Scotch broom.[9]

Perhaps the world's largest restoration project is the ongoing effort to restore parts of the Florida Everglades. The Everglades, a unique 7500 km^2 ecosystem of interconnected marshes and seasonally flooded grasslands, has been drying out for decades because the water that feeds it has been heavily managed for flood control and overdrawn for irrigation and development. The water management system inadvertently caused extensive degradation of the environment, resulting in the loss of more than half of the Everglades and the elimination of whole classes of ecosystems. Populations of wading birds have dropped 90%–95%, and economically important fisheries have suffered greatly. Extensive engineering of river channels also led to a loss of aesthetic appeal, problems with stagnancy and pollution of water in the channels, and a host of other problems. The 30-year, $7.8 billion restoration project intends to restore water by undoing damming and diversions of 1600 km of canals, 1150 km of levees, and 200 water control structures.

Ecosystem restoration is almost always expensive and only sometimes, or partially, successful. Regardless, the more our population grows and development spreads, the more ecological restoration will become a vital conservation strategy for the future.

weighing the issues — RESTORING "NATURAL" COMMUNITIES 4–3

Practitioners of ecological restoration in North America aim to restore communities to their natural state. But what is meant by "natural"? Does it mean the state of the community before industrialization? Before Europeans came to the New World? Before any people laid eyes on the community? Let us say Aborigines altered a forest community 8000 years ago by burning the underbrush regularly to improve hunting, and continued doing so until Europeans arrived 400 years ago and cut down the forest for farming. Today, the area's inhabitants want to restore the land to its "natural" forested state. Should restorationists try to recreate the forest of Aboriginal time, or the forest that existed even before the Aboriginal peoples arrived? What are some advantages and disadvantages of each approach?

Earth's Biomes

Across the world, each portion of each continent has different sets of species, leading to endless variety in community composition. However, communities in far-flung places often share strong similarities in their structure and function. This allows us to classify communities into broad types. A **biome** is a major regional complex of similar communities—a large ecological unit recognized primarily by its dominant plant type and vegetation structure. The world contains a number of biomes, each covering large contiguous geographic areas (**FIGURE 4.16**).

A term that is often used interchangeably with *biome*, but probably should not be, is *ecoregion*. An **ecoregion** is defined by the World Wildlife Fund as a large area of land or water that contains a geographically distinct assemblage of natural communities that share a large majority of their species and ecological dynamics, share similar environmental conditions, and interact ecologically in ways that are critical for their long-term persistence.[10] A particular ecoregion—such as the short grasslands of the Canadian Prairies of southern Alberta and Saskatchewan, for example—is thus a representative of a biome (such as the "temperate grasslands" biome) that is broader in scope and occurs in numerous localities around the world.

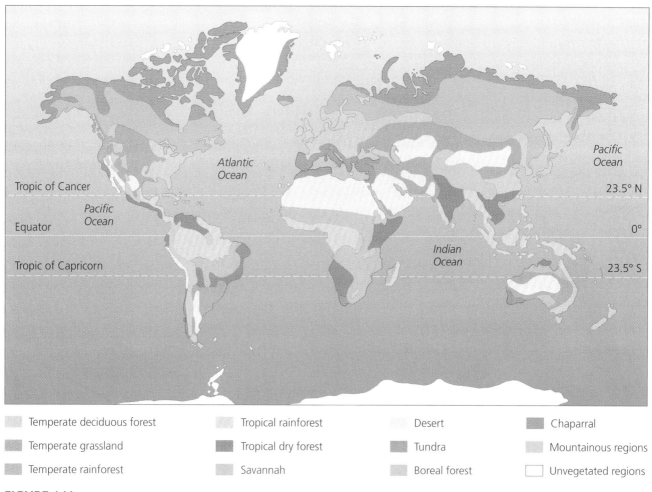

FIGURE 4.16
Biomes are distributed around the world according to temperature, precipitation, atmospheric and oceanic circulation patterns, and other factors.

The difference between an ecoregion and a biome might be easiest to grasp for a highly distinctive environment, such as a desert. The desert biome has certain climatic and ecological characteristics in common (see below), notably the lack of precipitation. There are many representatives of the desert biome around the world, including the Mojave Desert, the Gobi Desert, and the Sahara Desert, among many others. Each of these individual desert environments constitutes or is part of an ecoregion, with its own particular characteristics and flora and fauna, but it is still consistent with the broad characteristics that define the desert biome.

Biomes are groupings of communities that cover large geographic areas

Which biome covers any particular portion of the planet depends on a variety of abiotic factors, including temperature, precipitation, atmospheric circulation, and soil characteristics. Among these factors, temperature and precipitation exert the greatest influence (**FIGURE 4.17**). Because biome type is largely a function of climate, and because average monthly temperature and precipitation are among the best indicators of an area's climate, scientists often use climate diagrams, or **climatographs**, to depict such information. Global climate patterns cause biomes to occur in large patches in different parts of the world. For instance, temperate deciduous forest occurs in eastern North America, north–central Europe, and eastern China. Note how patches representing the same biome tend to occur at similar latitudes. This is due to the north–south gradient in temperature and to atmospheric circulation patterns.

Each biome encompasses a variety of communities that share similarities. For example, the eastern United States and the southernmost part of eastern Canada support part of the temperate deciduous forest biome. From New Hampshire to the Great Lakes to eastern Texas, precipitation and temperature are similar enough that most of the region's natural plant cover consists of broad-leafed

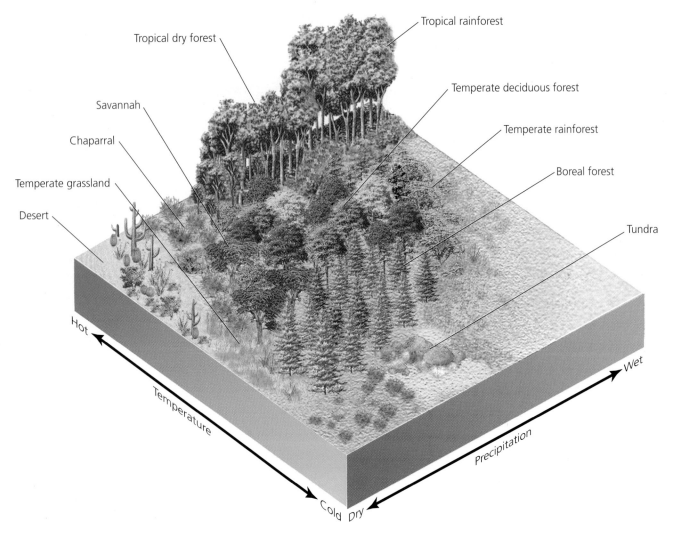

FIGURE 4.17
As precipitation increases, vegetation generally becomes taller and more luxuriant. As temperature increases, types of plant communities change. Together, temperature and precipitation are the main factors determining which biome occurs in a given area. For instance, deserts occur in dry regions, tropical rainforests occur in warm, wet regions, and tundra occurs in the coldest regions.

trees that lose their leaves in winter. Within this region, however, exist many different types of temperate deciduous forest, such as oak–hickory, beech–maple, and pine–oak forests, each sufficiently different to be designated a separate community.

We can divide the world into roughly 10 terrestrial biomes

Let us look briefly at the characteristics that define the world's 10 major terrestrial biomes.

Tundra The **tundra** (**FIGURE 4.18**) is a dry biome—nearly as dry as a desert—but located at very high latitudes along the northern edges of Russia, Canada, and Scandinavia. Extremely cold winters with little daylight and moderately cool summers with lengthy days characterize this landscape of lichens and low, scrubby vegetation without trees. The great seasonal variation in temperature and day length results from this biome's high-latitude location, angled toward the sun in the summer and away from the sun in the winter.

Because of the cold climate, underground soil remains more or less permanently frozen and is called *permafrost*. During the long, cold winters, the surface soils freeze as well; then, when the weather warms, they melt and produce seasonal accumulations of surface water that make ideal habitat for mosquitoes and other biting insects. The swarms of insects benefit bird species that migrate long distances to breed during the brief but productive summer. Caribou also migrate to the tundra to breed, and then leave for the winter. Only a few large animals, such as polar bears (*Ursus maritimus*) and musk

(a) Tundra

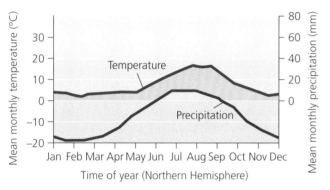

(b) Typical location: Cumberland Sound, Baffin Island, Nunavut

FIGURE 4.18
Tundra is a cold, dry biome found near the poles and atop high mountains at lower latitudes. Scientists use climate diagrams to illustrate an area's average monthly precipitation and temperature.[11] Typically in these diagrams, the x-axis marks months of the year (beginning in January for regions in the Northern Hemisphere and in July for regions in the Southern Hemisphere). Paired y-axes denote average monthly temperature and average monthly precipitation. The twin curves plotted on a climate diagram indicate trends in precipitation (blue) and in temperature (red) from month to month.

oxen (*Ovibos moschatus*), can survive year-round in this extreme climate. Tundra also occurs as **alpine tundra** at the tops of high mountains in temperate and tropical regions.

Because of the extreme climate in Canada's North, Alaska, Russia, and Scandinavia, much of the tundra biome remains intact and relatively unaltered by direct human occupation and interference. For example, the World Wildlife Fund reports the tundra ecoregions of North America to be 95%–98% intact. However, much of the tundra is unprotected by national or provincial or territorial legislation, rendering it potentially susceptible to alteration through human activities. Furthermore, the indirect effects of human modification of the global environment, especially the climate, are increasingly evident in the tundra. One problem that is of external origin is atmospheric fallout, which results in the deposition of heavy metals, and pesticide pollution.[12] In many areas of the tundra, seasonal ice connects the many islands with the mainland; with climatic warming and the associated melting of sea ice, these habitats and the animals that depend on them will be in increasing peril.

Boreal forest The northern coniferous forest, or **boreal forest**, also called **taiga** (**FIGURE 4.19**), stretches in a broad band across much of Canada, Alaska, Russia, and Scandinavia. In Canada the boreal forest encompasses nearly 6 million km². A few species of **coniferous** or evergreen trees, which create seed cones, have needle-like leaves, and remain green year-round, dominate large stretches of the boreal forest, interspersed with bogs and lakes. The black spruce (*Picea mariana*) is a common evergreen species.

(a) Boreal forest

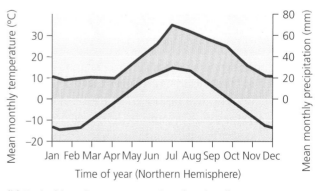

(b) Typical location: Jasper National Park, Alberta

FIGURE 4.19
Boreal forest is defined by long, cold winters, relatively cool summers, and moderate precipitation.

The boreal forest's uniformity over huge areas reflects the climate common to this latitudinal band of the globe: These forests develop in cooler, drier regions than do temperate forests, and they experience long, cold winters and short, cool summers. Soils are typically nutrient poor and somewhat acidic. As a result of the strong seasonal variation in day length, temperature, and precipitation, many organisms compress a year's worth of feeding, breeding, and rearing of young into a few warm, wet months. Year-round residents of boreal forest include mammals, such as moose (*Alces alces*), wolves (*Canis lupus*), bears, lynx (*Felis lynx*), and many burrowing rodents. This biome also hosts many insect-eating birds that migrate from the tropics to breed during the brief, intensely productive summer season.

The Boreal Forest Conservation Framework puts Canada's stewardship responsibilities with regard to the boreal forest in perspective, stating in part that

> *Canada's boreal region contains one-quarter of the world's remaining original forests. The largest intact forest ecosystem left on Earth, Canada's boreal is home to a rich array of wildlife including migratory songbirds, waterfowl, bears, wolves and some of the world's largest woodland caribou herds. The boreal region's natural wealth sustains many of Canada's Aboriginal communities, who have lived in harmony with the boreal for thousands of years. It also supports thousands of jobs and contributes billions to the Canadian economy.*
>
> *The Boreal Forest Conservation Framework is based on a shared vision to sustain the ecological and cultural integrity of the Canadian boreal forest region, in perpetuity. The Framework's goal is to conserve the cultural, sustainable economic and natural values of the entire Canadian boreal region by employing the principles of conservation biology to protect at least 50% of the region in a network of large interconnected protected areas, and to support sustainable communities, world-leading ecosystem-based resource management practices, and state-of-the-art stewardship practices in the remaining landscape.*[13]

The members of the Boreal Leadership Council are described as "historically unlikely partners" and include representatives from industry and finance; nongovernmental organizations, nonprofits, and environmental groups; and Aboriginal organizations and governing bodies. This type of collaborative, cross-disciplinary, cross-sectoral management process is typical of Canada's historical approach to the management of complex and sometimes thorny environmental issues, and it is one of the reasons that Canada is considered to be a world leader in environmental management.

Temperate deciduous forest The **temperate deciduous forest** (FIGURE 4.20) that dominates the landscape around the central and southern Great Lakes is characterized by broad-leafed trees that are **deciduous**, meaning that they lose their leaves each fall and remain dormant during winter, when hard freezes would endanger leaves. These mid-latitude forests occur in much of Europe and eastern China as well as in eastern North America—all areas in which precipitation is spread relatively evenly throughout the year. Although soils of the temperate deciduous forest are relatively fertile, the biome generally consists of far fewer tree species than are found in tropical rainforests. Oaks, beeches, and maples are a few of the most abundant types of trees in these forests. A sampling of typical animals of the temperate deciduous forest of eastern North America is shown in FIGURE 4.11.

(a) Temperate deciduous forest

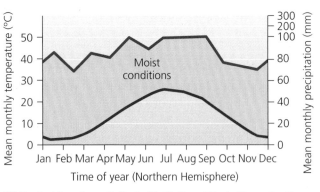

(b) Typical location: Kejimkujik National Park, Nova Scotia

FIGURE 4.20
Temperate deciduous forests experience relatively stable seasonal precipitation but more variation in seasonal temperatures. When the precipitation curve falls well above the temperature curve, as shown here, the region experiences relatively moist conditions, indicated here by the green shading.

Much of the temperate deciduous or broad-leafed forest in North America has been greatly altered since European settlement; for example, it has been estimated that only about 5% of New England–Acadian mixed broad-leaf forest in Canada remains intact, with about 50% of the habitat in this region described as "heavily altered" by human activity.[14] However, forest cover has been making a comeback in the New England–Acadia region in recent decades, thanks to changing land use and conservation efforts.

Temperate grassland Moving westward from the Great Lakes, we find **temperate grasslands** (FIGURE 4.21). This is because temperature differences between winter and summer become more extreme, and rainfall diminishes. The limited amount of precipitation in the Prairies and the Great Plains region can support grasses more easily than trees. Also known as *steppes* or *prairies*, temperate grasslands were once widespread throughout parts of North and South America and much of central Asia.

Today people have converted most of the world's grasslands for agriculture, greatly reducing the abundance of native plants and animals. Both the tallgrass prairies that characterize the midwestern United States and the shortgrass prairies of southern Alberta and Saskatchewan are described by the World Wildlife Fund as having been "virtually converted," mainly for wheat production and grazing, with only small undisturbed patches remaining (less than 2% remaining "intact" in Canada). However, restoration ecology efforts in this biome have seen some success, with efforts to reestablish populations of native species, such as the black-footed ferret (*Mustela nigripes*) and bison (*Bison bison*), well underway.[15] Other characteristic vertebrate animals of the North American grasslands include prairie dogs, pronghorn antelope (*Antilocapra americana*), and ground-nesting birds, such as meadowlarks.

Temperate rainforest Moving still further west in North America, the topography becomes more varied, and biome types are intermixed. The coastal Pacific region, with its heavy rainfall, features **temperate rainforest** (FIGURE 4.22), a forest type known for its potential to produce large volumes of commercially important forest products, such as lumber and paper. Coniferous trees such as cedars, spruces, hemlocks, and Douglas fir (*Pseudotsuga menziesii*) grow very tall in the temperate rainforest, so the forest interior is shaded and damp. In the Queen Charlotte Islands, for example, moisture-loving animals, such as the bright yellow banana slug (*Ariolimax columbianus*) are common, and old-growth conifer stands host the endangered spotted owl (*Strix occidentalis*). The soils of temperate rainforests are usually quite fertile but are susceptible to landslides and erosion if forests are cleared.

Temperate rainforests have been the focus of controversy in Pacific coastal regions, where overharvesting has driven some species toward extinction. Clear-cut logging and road building into forested areas remain the greatest threats to temperate rainforest habitat in these areas (see Chapter 10, "Central Case: Battling Over the Last Big Trees at Clayoquot Sound".

Tropical rainforest In tropical regions we see the same pattern found in temperate regions: Areas of high rainfall grow rainforests, areas of intermediate rainfall host dry or deciduous forests, and areas of lower rainfall become dominated by grasses. However, tropical biomes differ from their temperate counterparts in other ways because they are closer to the equator and therefore

(a) Temperate grassland

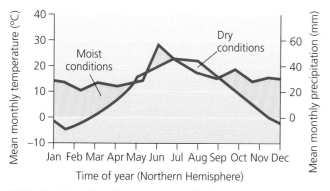

(b) Typical location: Moose Jaw, Saskatchewan

FIGURE 4.21
Temperate grasslands experience temperature variations throughout the year and too little precipitation for many trees to grow. This climatograph indicates both "moist" (green) and "dry" (yellow) climate conditions. When the temperature curve is above the precipitation curve, as is the case in May and mid-June through September, the climate conditions are "dry."

(a) Temperate rainforest

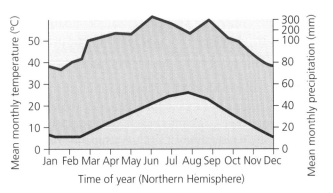

(b) Typical location: Queen Charlotte Islands and Vancouver, BC.

FIGURE 4.22
Temperate rainforests receive a great deal of precipitation and feature moist, mossy interiors.

(a) Tropical rainforest

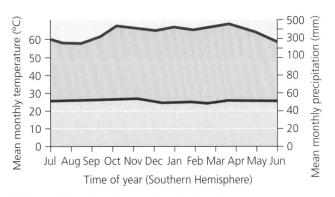

(b) Typical location: Bogor, Java, Indonesia

FIGURE 4.23
Tropical rainforests, famed for their biodiversity, grow under constant, warm temperatures and a great deal of rain.

warmer on average year-round. For one thing, they hold far greater biodiversity.

The **tropical rainforest** biome (**FIGURE 4.23**) is found in Central America, South America, Southeast Asia, West Africa, and other tropical regions, and is characterized by year-round rain and uniformly warm temperatures. Tropical rainforests have dark, damp interiors, lush vegetation, and highly diverse biotic communities, with greater numbers of species of insects, birds, amphibians, and various other animals than any other biome. These forests are not dominated by single species of trees, as are forests closer to the poles, but instead consist of very high numbers of tree species intermixed, each at a low density. Any given tree may be draped with vines, enveloped by strangler figs, and loaded with *epiphytes* (orchids and other plants that grow in trees), such that trees occasionally collapse under the weight of all the life they support. Despite this profusion of life, tropical rainforests have very poor, acidic soils that are low in organic matter. Nearly all nutrients present in this biome are contained in the trees, vines, and other plants—not in the soil. An unfortunate consequence is that once tropical rainforests are cleared, the nutrient-poor soil can support agriculture for only a short time. As a result, farmed areas are abandoned quickly, and the soil and forest vegetation recover very slowly.

Tropical dry forest Tropical areas that are warm year-round but where rainfall is lower overall and highly seasonal give rise to **tropical dry forest**, or tropical deciduous forest (**FIGURE 4.24**), a biome widespread in India, Africa, South America, and northern Australia. Wet and dry seasons each span about half a year in tropical dry forest. Rains during the wet season can be extremely heavy and, coupled with erosion-prone soils, can lead to severe soil loss when forest clearing occurs over large areas. Across the globe, much tropical dry forest has been converted to agriculture. Clearing for farming or ranching is made easier by the fact that vegetation heights are much lower and canopies less dense than in tropical rainforest. Organisms that inhabit tropical dry forest have

(a) Tropical dry forest

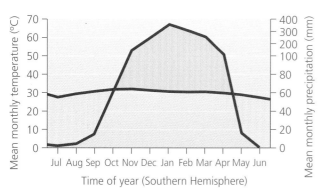

(b) Typical location: Darwin, Australia

FIGURE 4.24
Tropical dry forests experience significant seasonal variations in precipitation and relatively stable warm temperatures.

(a) Savannah

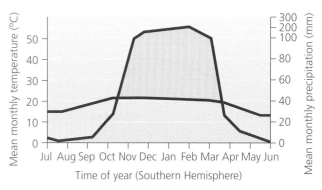

(b) Typical location: Harare, Zimbabwe

FIGURE 4.25
Savannahs are grasslands with clusters of trees. They experience slight seasonal variation in temperature but significant variation in rainfall.

adapted to seasonal fluctuations in precipitation and temperature. For instance, plants are deciduous and often leaf out and grow profusely with the rains, then drop their leaves during the driest times of year.

Savannah Drier tropical regions give rise to **savannah** (**FIGURE 4.25**), tropical grassland interspersed with clusters of acacias or other trees. The savannah biome is found today across stretches of Africa (the ancestral home of our species), South America, Australia, India, and other dry tropical regions. Precipitation in savannahs usually arrives during distinct rainy seasons and concentrates grazing animals near widely spaced water holes. Common herbivores on the African savannah include zebras, gazelles, and giraffes, and the predators of these grazers include lions, hyenas, and other highly mobile carnivores.

Desert Where rainfall is very sparse, **desert** (**FIGURE 4.26**) forms. This is the driest biome on Earth; most deserts receive well under 25 cm of precipitation per year,

much of it during isolated storms months or years apart. Depending on rainfall, deserts vary greatly in the amount of vegetation they support. Some, like the Sahara and Namib deserts of Africa, are mostly bare sand dunes; others, like the Sonoran Desert of Arizona and northwest Mexico, are quite heavily vegetated. Deserts are not always hot; the high desert of the western United States is one example. Because deserts have low humidity and relatively little vegetation to insulate them from temperature extremes, sunlight readily heats them in the daytime, but daytime heat is quickly lost at night. As a result, temperatures vary widely from day to night and across seasons of the year. Desert soils can often be quite saline and are sometimes known as lithosols, or stone soils, for their high mineral and low organic-matter content.

Desert animals and plants have evolved many adaptations to deal with the harsh climatic conditions. Most reptiles and mammals, such as rattlesnakes and kangaroo mice, are active in the cool of night, and many Australian desert birds are nomadic, wandering long distances

(a) Desert

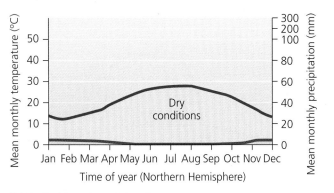

(b) Typical location: Cairo, Egypt

FIGURE 4.26
Deserts are dry year-round, but they are not always hot. Precipitation can arrive in intense, widely spaced storm events. The photograph, from the Sonoran Desert in Arizona, shows the maximum amount of vegetation a desert can support.

(a) Chaparral

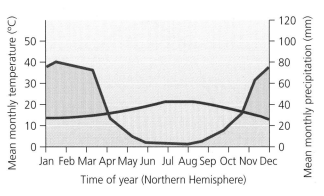

(b) Typical location: Baja Peninsula, California, USA

FIGURE 4.27
Chaparral is a highly seasonal biome dominated by shrubs, influenced by marine weather, and dependent on fire.

to find areas of recent rainfall and plant growth. Many desert plants have thick leathery leaves to reduce water loss, or green trunks so that the plant can photosynthesize without leaves, which would lose water. The spines of cacti and many other desert plants guard those plants from being eaten by herbivores desperate for the precious water they hold. These are examples of convergent evolution of plants and animals, adapting separately to the dry conditions that are characteristic of the desert biome.

Chaparral In contrast to the boreal forest's broad, continuous distribution, **chaparral** or *Mediterranean "scrub" woodland* (**FIGURE 4.27**) is limited to fairly small patches widely flung around the globe. Scrub woodland consists mostly of evergreen shrubs and is densely thicketed. This biome is also highly seasonal, with mild, wet winters and warm, dry summers. This type of climate is induced by oceanic influences; in addition to ringing the Mediterranean Sea, chaparral occurs along the coasts of California, Chile, and southern Australia. In Europe it is called *maquis*; in Chile, *matorral*; and in Australia, *mallee*. Chaparral communities experience frequent fire, and their plant species are adapted to resist fire or even to depend on it for germination of their seeds.

Altitude creates patterns analogous to latitude

As any hiker or skier knows, climbing in elevation causes a much more rapid change in climate than moving the same distance toward the poles. Vegetative communities change along mountain slopes in correspondence with this small-scale climate variation (**FIGURE 4.28**). These changes with altitude define the *alpine* biome. *Altitudinal zonation*, as it is referred to, is independent of latitude—the variation of temperature, precipitation, and ecological communities with altitude can occur anywhere from the equator to the Rockies. A hiker ascending one of southern Arizona's higher mountains, for example, would begin in

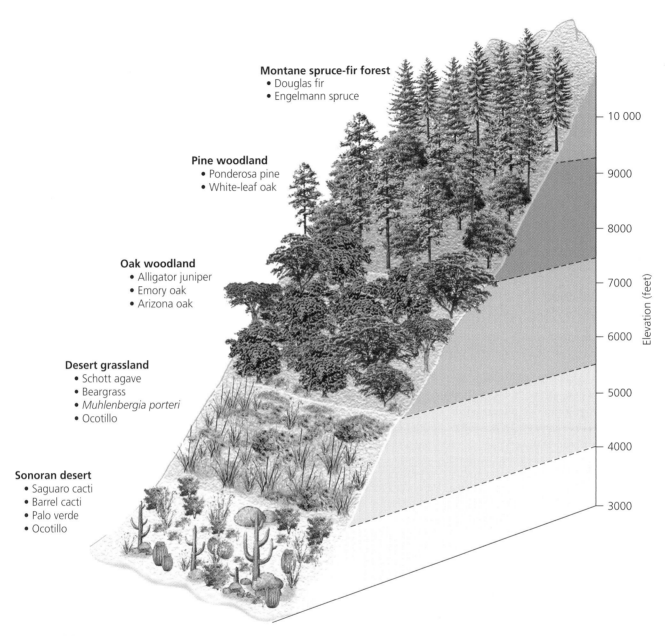

FIGURE 4.28
As altitude increases, vegetation changes in ways similar to how it changes as one moves toward the poles, taking a hiker through the local equivalent of several biomes.

Sonoran Desert or desert grassland and proceed through oak woodland, pine forest, and finally spruce–fir forest—the equivalent of passing through several biomes, without any change in latitude. A hiker scaling one of the great peaks of the Andes in Ecuador, near the equator, could begin in tropical rainforest and end amid glaciers in alpine tundra.

Characteristics that are typical of the alpine biome on a high-mountain peak, compared with lowlands or foothills that surround it, include lower temperatures, lower atmospheric pressures (and less oxygen), higher exposure to ultraviolet radiation, and higher precipitation. The foothills of the Canadian Rockies, for example, are characterized by meadows, grasslands, and riparian (riverside) woodlands, which grade upward through boreal forests and into alpine tundra and glaciers on the mountain peaks; precipitation on the peaks can be as much as twice that of the foothill areas. The zonation of ecological communities on mountain slopes is also influenced by secondary climatic effects related to topographic relief, such as rainshadow effects and exposure to or shelter from the prevailing winds and sunlight.[16] For example, when moisture-laden air ascends a steep mountain, it releases precipitation as it cools; by the time it flows over the top of the mountain and down the other side, it can be very dry, creating a *rainshadow desert*.

CANADIAN ENVIRONMENTAL PERSPECTIVES

Zoe Lucas

Zoe Lucas works with the wild horses of Sable Island.

- **Independent researcher** on the biology and ecology of wild horses
- **Advocate** for Sable Island and its unique ecology
- **Photographer** and **documenter** of life (and death) on Sable Island

As a student at the Nova Scotia School of Art and Design in Halifax, Zoe Lucas had not even considered environmental science as an option. Then Lucas had the opportunity to visit Sable Island, a 41 km patch of sand located about 300 km southeast of Halifax. Sable is known for its remote, windswept landscapes and bands of wild horses, as well as grey seals, harbour seals, and several types of rare and endangered migratory birds. It also boasts a long marine history involving frequent shipwrecks. After a couple of days on Sable Island, Lucas was captivated and became driven to find her way back there. Completing her master of fine arts degree, Lucas eventually started teaching and opened her own studio, but her desire to return to Sable remained strong. She describes it as a combination of esthetics ("Sable is an extremely beautiful place; it's like living in a watercolour painting") and being highly activated by being outdoors.

Lucas eventually made it back to Sable Island as a volunteer cook with a university research group. Still driven largely by the esthetics of the place, she found that the longer she spent on the island, the more curious she became about its nature (in both senses of the word). Eventually she began to participate in scientific research and environmental monitoring projects. She has now authored or contributed to numerous reports and scientific studies on the island's horses, vegetation, and seals, and collaborates with researchers at various universities.

Lucas carries out a complete search of the island's shoreline once every four to five weeks, looking for oiled birds, stranded cetaceans, and marine litter. Between beach surveys she records field data on the feral horses, including their range, reproductive activities, and band structure. She also works on short-term projects, such as collecting invertebrate or plant specimens requested by researchers.

Lucas says that, in addition to the benefits of working in such a unique and beautiful place, the work of environmental monitoring appeals to her because there are endless questions. "Hardly a day passes without learning something new—sometimes just a snippet of a detail, sometimes a big aha! Being able to do this work in such a wonderful and occasionally challenging environment is very compelling."

The island is administered by the Canadian Coast Guard, and landing there without permission (except in the case of an emergency) is against the law. Scientific research on the island is carried out with logistical support provided by the Sable Island Station, originally set up by the Meteorological Service of Canada (a branch of Environment Canada) as a meteorological station but now also used as a base for year-round environmental stewardship. There has been some discussion about initiating ecotourism trips to the island, but this would require significant care and planning to ensure preservation of the island's fragile ecosystems.

"I can't imagine anyone not wanting to live and work here." —Zoe Lucas, about Sable Island

Thinking About Environmental Perspectives

Sable Island is a remote wilderness. Great care has been taken to preserve the fragile ecosystems and unique wild populations that live there. However, few people will ever visit Sable Island. What do you think about this? What is the value of wilderness? Do you think more people should be able to visit, even if their presence would be disruptive and potentially damaging to the natural environment?

Aquatic and coastal systems also show biome-like patterns

In our discussion of biomes, we have focused exclusively on terrestrial systems, because the biome concept, as traditionally developed and applied, has been limited to terrestrial systems. Areas equivalent to biomes also exist in the oceans, but their geographic shapes would look very different from those of terrestrial biomes if plotted on a world map. We can consider the thin strips along the world's coastlines to represent one aquatic system, the continental shelves another, and the open ocean, deep sea, coral reefs, and kelp forests as still other distinct sets of communities. There are also many coastal systems that straddle the line between terrestrial and aquatic, such as salt marshes, rocky intertidal communities, mangrove forests, and estuaries. And, of course, there are freshwater systems, such as those of the Great Lakes.

Unlike terrestrial biomes, aquatic systems are shaped not by air temperature and precipitation but by such factors as water temperature, salinity, dissolved nutrients, wave action, currents, depth, and type of substrate (e.g., sandy, muddy, or rocky bottom). Light levels can play a role, too. Shallow-water environments tend to be saltier (because of the evaporation of fresh water from the surface), with higher light levels and warmer temperatures, in comparison with deep-water environments. Marine communities are also more clearly delineated by their animal life than by their plant life. We will examine freshwater, marine, and coastal systems in the greater detail they deserve in Chapters 11 and 12.

Conclusion

The natural world is so complex that we can visualize it in many ways and at various scales. Dividing the world's communities into major types, or biomes, is informative at the broadest geographic scales. Understanding how communities function at more local scales requires understanding how species interact with one another. Species interactions, such as predation, parasitism, competition, and mutualism, give rise to effects that are both weak and strong, direct and indirect. Feeding relationships can be represented by the concepts of trophic levels and food webs, and particularly influential species are sometimes called keystone species. Increasingly humans are altering communities, in part by introducing non-native species that may turn invasive. But increasingly, through ecological restoration, we are also attempting to undo the changes we have caused.

REVIEWING OBJECTIVES

You should now be able to:

Compare and contrast the major types of species interactions

- Competition results when individuals or species vie for limited resources. It can occur within or among species and can result in coexistence or exclusion. It also can lead to realized niches, resource partitioning, and character displacement.
- In predation, one species kills and consumes another. It is the basis of food webs and can influence population dynamics and community composition.
- In parasitism, one species derives benefit by harming (but usually not killing) another.
- Herbivory is an exploitative interaction whereby an animal feeds on a plant.
- In mutualism, species benefit from one another. Some mutualists are symbiotic, whereas other mutualists are free-living.

Characterize feeding relationships and energy flow, using them to construct trophic pyramids and food webs

- Energy is transferred in food chains among trophic levels.
- Lower trophic levels generally contain more energy, biomass, and numbers of individuals than higher trophic levels.
- Food webs illustrate feeding relationships and energy flow among species in a community.

Distinguish characteristics of a keystone species

- Keystone species have impacts on communities that are far out of proportion to their abundance.
- Top predators are frequently considered keystone species, but other organisms may be thought of as keystones for other reasons.

Characterize the process of succession and the debate over the nature of communities

- Succession is a stereotypical pattern of change within a community through time.
- Primary succession begins with an area devoid of life. Secondary succession begins with an area that has been severely disturbed.

Perceive and predict the potential impacts of invasive species on communities

- Invasive species, such as the zebra mussel, have altered the composition, structure, and function of communities.
- Humans are the cause of most modern species invasions, but we can also respond to invasions with prevention and control measures.

Explain the goals and methods of ecological restoration

- Ecological restoration aims to restore communities to a more "natural" state, variously defined as before human or industrial interference.
- Restoration efforts in the field are informed by the growing science of restoration ecology.

Describe and illustrate the terrestrial biomes of the world

- Biomes represent major classes of communities spanning large geographic areas.
- The distribution of biomes is determined by temperature, precipitation, and other factors.
- The biome concept by tradition refers to terrestrial systems. Aquatic systems can be classified in similar ways, determined by different factors.

TESTING YOUR COMPREHENSION

1. How does competition lead to a realized niche? How does it promote resource partitioning?
2. Contrast the several types of exploitation. How do predation, parasitism, and herbivory differ?
3. Give examples of symbiotic and nonsymbiotic mutualisms. Describe at least one way in which mutualisms affect your daily life.
4. Explain how trophic levels, food chains, and food webs are related.
5. Name several ways in which a species could be considered a keystone species.
6. Explain and contrast primary and secondary terrestrial succession.
7. Name five changes to Great Lakes communities that have occurred since the invasion of the zebra mussel.
8. What is restoration ecology?
9. What factors most strongly influence the type of biome that forms in a particular place on land? What factors determine the type of aquatic system that may form in a given location?
10. Draw climate diagrams for a boreal forest and for a desert. Label all parts of the diagram, and describe all the types of information an ecologist could glean from such a diagram.

SEEKING SOLUTIONS

1. Imagine that you spot two species of birds feeding side by side, eating seeds from the same plant, and that you begin to wonder whether competition is at work. Describe how you might design scientific research to address this question. What observations would you try to make at the outset? Would you try to manipulate the system to test your hypothesis that the two birds are competing? If so, how?
2. Spend some time outside on your campus or in your yard or in the nearest park or natural area. Find at least 10 species of organisms, and observe them long enough to watch them feed or to make an educated guess about what they feed on. Now, using Figure 4.11 as a model, draw a simple food web involving all the organisms you observed.
3. Can you think of one organism not mentioned in this chapter as a keystone species that you believe may be a keystone species? For what reasons do you suspect this? How could you experimentally test whether an organism is a keystone species?
4. Why do scientists consider invasive species to be a problem? What makes a species "invasive," and what ecological effects can invasive species have?
5. From year to year, biomes are stable entities, and our map of world biomes appears to be a permanent record of patterns across the planet. But are the locations and identities of biomes permanent, or could they change over time? Provide reasons for your answers.
6. Can you devise possible responses to the zebra mussel invasion? What strategies would you consider if you were put in charge of the effort to control this species' spread and reduce its impacts? Name some advantages of each of your ideas, and identify some obstacles it might face in being implemented.
7. **THINK IT THROUGH** Consider this real-life example of invasive-species management and restoration ecology, and then answer the questions that follow, in your new role as an environmental land manager: The Garry oak meadows on Trinity Western University's Crow's Nest Ecological Research Area are experiencing encroachment by the Douglas fir (*Pseudotsuga menziesii*), which threatens the health and survival of the meadow. Traditionally, Aborigines maintained the Garry oak meadows with frequent burnings. However, with European settlement and the resulting control and suppression of fires, Douglas firs have been increasingly successful at invading the meadows. Some Garry oak restoration projects have undertaken to remove Douglas firs physically and/or using controlled fires. However, when the Douglas firs are removed, the resulting disturbance of the soil apparently facilitates the establishment of other invasive species.[17]
   (a) How would you design a comprehensive scientific study to determine the best ways of removing Douglas firs without enhancing conditions for encroachment by other invasive species?
   (b) Removing Douglas firs from the Garry oak meadows would not return the meadows to their "natural," pristine, or pre-human state; instead, it would

represent a return to an earlier phase of human (Aboriginal) ecosystem management. Is this the best approach? Do you think the Douglas firs should be allowed to advance naturally, without interference by fire or other removal techniques? Or is it worthwhile to preserve the meadows—which are disappearing rapidly, and host a variety of unusual and threatened species—by returning to an earlier land management approach?

(c) Can you think of some general guidelines that might be used to make ecosystem management decisions in other cases of this type?

INTERPRETING GRAPHS AND DATA

The grey wolf (*Canis lupus*) is a keystone species in Yellowstone National Park's ecosystem. Wolf packs hunt elk, gorge themselves on the kill, and leave the carcass as carrion for scavenger species, such as ravens, magpies, eagles, coyotes, and bears. As the global climate has warmed, winters in Yellowstone have become shorter over the past 55 years. Fewer elk weaken and die in milder weather, and so less carrion is available to scavengers during warmer, shorter winters. Biologists Christopher Wilmers and Wayne Getz studied the links among climate change, wolves, elk, and scavenger populations in Yellowstone. They used empirical field data on wolf predation rates and elk carrion availability recorded over 55 years to develop a model that estimated carrion availability with and without wolves for each winter month. Some of their findings are presented in the graph.

1. How much less carrion is available in April than in November when wolves are present? When wolves are not present?
2. Wolves were hunted nearly to extinction in the 1930s and were reintroduced to Yellowstone only in 1995. How, would you suspect, has their reintroduction affected scavenger populations since then? Why?

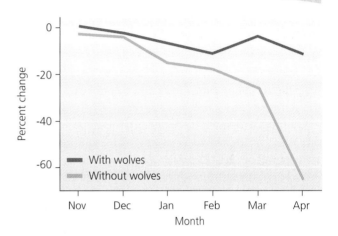

Changes in amount of winter carrion available to scavengers in Yellowstone National Park, with and without wolves, according to the model. Differences in March and April are statistically significant.
Data from Wilmers, C. C., and W. M. Getz. 2005. Gray wolves as climate change buffers in Yellowstone. PLoS Biology 3(4):e92.

3. Predict what effect continued shorter, warmer winters would have on scavenger populations. Why? Are the predicted effects of wolf reintroduction and of climate change compounded, or do they tend to cancel one another out?

CALCULATING FOOTPRINTS

Species appearing in a new area are generally called "invasive" if they increase markedly in population and also increase in their impacts on the biotic communities and landscapes around them. By these measures, are human beings an invasive species? The table below shows human population and per capita energy consumption for Canada, the United States, and the world for 1990 and 2003. Total energy consumption gives us a very rough

	Annual per capita energy consumption, kilogram of oil equivalent (kgoe)[18]		Population (000s)		Total energy consumption (calculated)	
	1990	2003	1990	2003	1990	2003
Canada	7558.4	8300.7	27 701	31 676		
United States	7543.4	7794.8	248 710	294 043		
World	1633.3	1674.4	5 278 640	6 302 309		

measure of total environmental impact. Calculate total energy consumption in the table by multiplying the population and per capita consumption values.

1. In percentage terms, how much did total energy consumption, as calculated here, increase between 1990 and 2003 for Canada, the United States, and the world?

2. What effects on biotic communities do you think this change in total energy consumption has had? Speculate on overall effects, and give several specific known or likely examples.

3. After considering these data, do you consider yourself to be a member of an invasive species? Why or why not?

TAKE IT FURTHER

Go to www.myenvironmentplace.ca where you will find
- Suggested answers to end-of-chapter questions
- Quizzes, animations, and flashcards to help you study
- *Research Navigator*™ database of credible and reliable sources to assist you with your research projects
- Tutorials to help you master how to interpret graphs
- Current news articles that link the topics that you study to case studies from your region and around the world

- **ECO Occupational Profiles:** If you found this chapter especially interesting, you might want to learn more about the following jobs by visiting the Occupational Profiles website of the Environmental Careers Organization. Go to www.eco.ca and check out the following careers:
  - Ecologist
  - Entomologist
  - Ornithologist
  - Wildlife technician
  - Zoologist

CHAPTER ENDNOTES

1. "'Musseling' in on an Ecosystem: Tracing the Natural History of the Zebra Mussel in Lake Erie," *Canadian Geographic*, September 2003.
2. Environment Canada Backgrounder: Lake Ontario Invasion Begins, 1999, www.on.ec.gc.ca/press/goby-invasion.html.
3. Ward, Jessica M., and Anthony Ricciardi (2007) Impacts of *Dreissena* invasions on benthic macroinvertebrate communities: A Metaanalysis, *Diversity and Distributions*, Vol. 13, pp. 155–165.
4. Palmer, M. E., and Anthony Ricciardi (2005) Community interactions affecting the relative abundances of native and invasive amphipods in the St. Lawrence River, *Canadian Journal of Fisheries and Aquatic Science*, Vol. 62, pp. 1111–1118.
5. Parks Canada Fact Sheet, 2007, "Moose a-plenty," Cape Breton Highlands National Park of Canada, www.pc.gc.ca/canada/pn-tfn/itm2-/2007/2007-04-30_e.asp.
6. Map showing 2007 distribution of zebra mussel sightings, "Zebra Mussels Cause Economic and Ecological Problems in the Great Lakes," U.S. Geological Survey.
7. Danoff-Burg, James, ed., *Introduced Species Summary Project*, Columbia University, Centre for Environmental Research and Conservation, and other sources.
8. Haber, Erich, *Guide to Monitoring Exotic and Invasive Plants*, Environment Canada Ecological Monitoring and Assessment Network (EMAN).
9. Carder, Judith E. W. *Restoration of Natural Systems Program*, Garry Oak Ecosystems Recovery Team, Research Colloquium 2006, University of Victoria, BC; and Garry Oak Restoration Project (GORP), www.gorpsaanich.com
10. World Wildlife Fund, "Ecoregions," www.worldwildlife.org/science/ecoregions
11. Climatographs adapted from Breckle, S. W. (1999) *Walter's vegetation of the Earth: The ecological systems of the geo-biosphere*, 4th ed., Berlin: Springer-Verlag.
12. World Wildlife Fund, "Ecoregions," www.worldwildlife.org/science/ecoregions/nearctic.cfm.
13. Canadian Boreal Institute, Boreal Leadership Council, www.borealcanada.ca
14. World Wildlife Fund, "Ecoregions," www.worldwildlife.org/science/ecoregions/.
15. World Wildlife Fund, "Ecoregions," www.worldwildlife.org/science/ecoregions/nearctic.cfm.
16. World Wildlife Fund, "Ecoregions," www.worldwildlife.org/science/ecoregions/nearctic.cfm,
17. Roberts, Kimberly, Stephanie Koole, and David Clements, Garry Oak Ecosystems Recovery Team—Research Colloquium 2006, Trinity Western University, Langley, B.C.
18. World Resources Institute, www.earthtrends.org.

5 Earth Systems and Ecosystem Ecology

This is the estuary of the St. Lawrence River.

Upon completing this chapter, you will be able to

- Describe the nature of environmental systems
- Explain how plate tectonics and the rock cycle shape the landscape around us and Earth beneath our feet
- Define ecosystems and evaluate how living and nonliving entities interact in ecosystem-level ecology
- Outline the fundamentals of landscape ecology
- Compare and contrast how water, carbon, nitrogen, and phosphorus cycle through the environment

Beluga whales in the St. Lawrence estuary are suffering from pollution-related health problems.

CENTRAL CASE:
THE PLIGHT OF THE ST. LAWRENCE BELUGAS

"In nature there is no 'above' or 'below,' and there are no hierarchies. There are only networks nesting within other networks."
—FRITJOF CAPRA, THEORETICAL PHYSICIST

"The concept of a subtly interconnected world, of a whispering pond in and through which we are intimately linked to each other and to the universe ... is part of humanity's response to the challenges that we now face in common."
—ERVIN LASZLO, SYSTEMS THEORIST

The St. Lawrence is one of the great river systems of Canada. From its origin in Lake Ontario (see map) it flows approximately 1200 km to the Gulf of St. Lawrence, the world's largest estuary, where fresh water meets salt water. The cold Labrador current brings an abundant food source of plankton and fish from the Atlantic Ocean into the estuary, supporting a small population of beluga whales. The playful, sociable beluga or white whale, *Delphinapterus leucas*, prefers cold saltwater estuaries as its habitat.

Health problems have plagued the St. Lawrence belugas for several decades, causing their population to decrease to fewer than 700. This is much smaller than the population of many thousands that occupied the estuary at the beginning of the twentieth century. The belugas appear to be dying of cancer, at the rate of about 14 to 15 per year.

Daniel Martineau and a team of veterinarians from the University of Montreal carried out autopsies on more than 100 dead whales from the St. Lawrence. They found that 27% of the adults and 17% of the young belugas had died of cancer, mainly gastrointestinal. "In dolphins and terrestrial animals, the figure is closer to 2%," says Dr. Martineau. Toxicological studies showed that the whales had been exposed to organochloride pollutants, notably polycyclic aromatic hydrocarbons or PAHs.[1]

PAHs come from the burning of fossil fuels and other combustion sources. They do not break down easily and have become one of the most widespread contaminants in aquatic and marine environments. Once deposited, they are carried to waterways by runoff and accumulate in the bottom sediments of rivers and near shorelines. Belugas feed on organisms that live in these sediments. The concentration of contaminants increases in animals that eat higher up the food chain, through *biomagnification*. PAHs are also *lipophilic*, or "fat-loving," compounds, so they combine easily with fats and accumulate over time in the blubber of the belugas, through *bioaccumulation*.

In addition to the cancers, the St. Lawrence belugas have very low reproductive rates and other health issues, including cysts and bacterial infections. It is not clear to what extent these health problems may have been caused by exposure to pollutants in the estuary.

Where do the PAHs come from, and what does this have to do with the belugas' estuarine habitat? The answer can be found by looking at the Great Lakes and St. Lawrence River as a single great connected system. Any changes upstream in the system will be felt downstream. This includes the deposition of pollutants—some generated as far away as the Golden Horseshoe industrial zone in Ontario—that eventually make their way to the estuary.

PAHs are not the only problem. Other organochloride compounds and heavy metals from airborne sources are concentrated in the waters of the estuary. Agricultural development along the St. Lawrence River has contributed pesticides to the system. Some of these contaminants find their way into the thousands of tons of fish harvested from St. Lawrence commercial fisheries each year. Excess organic matter from fertilizer runoff and animal waste also has contributed to a sharp drop in oxygen concentration in the deepest waters of the estuary, where dissolved oxygen levels have declined by half since the 1930s.

The lack of dissolved oxygen reflects a condition known as **hypoxia**, which is not uncommon in estuarine and coastal waters around the world (see "The Science Behind the Story: The Gulf of Mexico's 'Dead Zone'"). It is most likely caused by the increased nutrient availability from fertilizer runoff into the deep water of the estuary, where decomposition of the nutrients consumes the available oxygen. Nutrient overenrichment can lead to algal overgrowths, called *blooms*, and subsequent ecosystem degradation. This process, called **eutrophication**, can happen in both freshwater and saltwater systems. Another possible contributor to hypoxia and eutrophication in the Gulf of St. Lawrence is an influx of warm, oxygen-depleted water from the Gulf Stream, which displaces the cold, oxygen-rich water of the Labrador Current.

The St. Lawrence beluga population is classified as "vulnerable" on the IUCN Red List of Endangered Species. The health of the whales is an indicator of the overall health of the estuary, and a reminder of the interconnectedness of the Great Lakes–St. Lawrence ecosystem. Recognizing this interconnectedness, scientists are striving to understand the ecosystem as a whole and develop strategies for protecting the estuary and the organisms that inhabit it.[2]

Earth's Environmental Systems

Our planet's environment consists of complex networks of interlinked systems. In the realm of community ecology (Chapter 4), these systems include the ecological webs of relationships among species. At the ecosystem level, they include the interaction of living species with the nonliving entities around them. Earth's systems also include **cycles** that shape the landscapes around us and guide the flow of key chemical elements and compounds that support life and regulate climate. We depend on these systems for our very survival.

Assessing questions holistically by taking a "systems approach" is helpful in environmental science, because so many issues are multifaceted and complex. However, systems often show behaviour that is difficult to understand and predict. Environmental scientists are rising to the challenge of studying systems broadly and are beginning to find comprehensive solutions to problems, such as pollution and hypoxia in the estuary of the St. Lawrence.

Systems show several defining properties

A **system** is a network of relationships among parts, elements, or components that interact with and influence one another through the exchange of energy, matter, or information. Systems receive inputs of energy, matter, or information, process these inputs, and produce outputs. Systems that receive inputs of both energy and matter and produce outputs of both are called **open systems**. Systems that receive inputs and produce outputs of energy, but not matter, are called **closed systems**. In a closed system, matter cycles among the various parts of the system but does not leave or enter the system. Although it is scientifically more straightforward to deal with closed systems, in nature no system is truly, perfectly closed.

Energy inputs to Earth's environmental systems include solar radiation as well as heat released by geothermal activity, organismal metabolism, and human activities, such as fossil fuel combustion. Information inputs can come in the form of sensory cues from visual, olfactory (chemical), magnetic, or thermal signals. Inputs of matter occur when chemicals or physical material moves among systems, such as when seeds are dispersed long distances, migratory animals deposit waste far from where they consumed food, or plants convert carbon in the air to living tissue by photosynthesis.

As a system, the Gulf of St. Lawrence receives inputs of freshwater, sediments, nutrients, and pollutants from the St. Lawrence and other rivers. Fishers and large animals, like the belugas, harvest some of the system's output: matter and energy in the form of fish and plankton. This output subsequently becomes input to the human economic system and to the digestive systems of the people and whales who consume seafood from the St. Lawrence.

Sometimes a system's output can serve as input to that same system, a circular process described as a **feedback loop**. Feedback loops are of two types, negative and positive.

In a **negative feedback loop** (**FIGURE 5.1A**), output that results from a system moving in one direction acts as

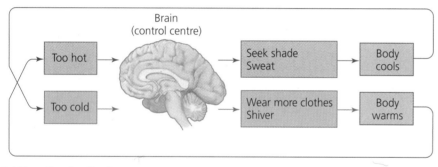

(a) Negative feedback

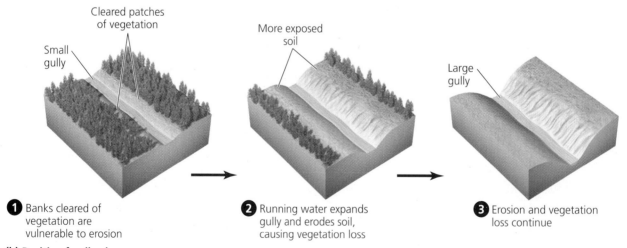

(b) Positive feedback

FIGURE 5.1

(a) exert a stabilizing influence on systems and are common in nature. The human body's response to heat and cold involves a negative feedback loop. Positive feedback loops **(b)** have a destabilizing effect on systems and push them toward extremes. For example, the clearing of vegetated land can lead to a runaway process of soil erosion: water flowing through an eroded gully may expand the gully and lead to further erosion. Rare in nature, positive feedback loops are common in natural systems altered by human impact.

input that moves the system in the other direction. Input and output essentially neutralize one another's effects, stabilizing the system. A thermostat, for instance, stabilizes a room's temperature by turning on the furnace when the room gets cold and shutting it off when the room gets warm. Similarly, negative feedback regulates our body temperature. If we get too hot, our sweat glands pump out moisture that evaporates to cool us down, or we may move from sun to shade. If we get too cold, we shiver, creating heat, or we move into the sun or put on more clothing. Another example of negative feedback is a predator–prey system in which predator and prey populations rise and fall in response to each other (see **FIGURE 4.4**). Most systems in nature involve negative feedback loops. Negative feedback loops enhance stability and, in the long run, only those systems that are stable will persist.

Positive feedback loops have the opposite effect. Rather than stabilizing a system, they drive it further toward one extreme or another. Exponential growth in human population provides an example. The more people who are born, the more there are to give birth to further people; increased output leads to increased input, leading to further increased output. Another example is the spread of cancer; as cells multiply out of control, the process is self-accelerating. Positive feedback can also occur with the process of erosion, the removal of soil by water or wind. Once vegetation has been cleared to expose soil, erosion may become progressively more severe if the forces of water or wind surpass the rate of vegetative regrowth (**FIGURE 5.1B**). Positive feedback can alter a system substantially. Positive feedback loops are rare in nature, but they are common in natural systems altered by human impact.

The inputs and outputs of complex natural systems usually occur simultaneously, keeping the system constantly active. Earth's climate system, for instance, does not ever stop. When processes within a system move in opposing directions at equivalent rates so that their effects balance out, the process is said to be in **dynamic equilibrium**. The term *dynamic* is used to indicate that even though the system is in a state of balance or equilibrium, it is an ever-changing, ever-adjusting balance, not static or unchanging.

Processes in dynamic equilibrium can contribute to **homeostasis**, the tendency of a system to maintain constant or stable internal conditions. The properties of resilience and resistance, discussed in Chapter 4, are related to homeostasis. *Resistance* refers to the strength of the system's tendency to remain constant; *resilience* is a measure of how readily the system will return to its original state, once it has been disturbed. Homeostatic systems are often said to be in a stable or **steady state**; however, the state itself may change over time, even while the system maintains its ability to stabilize conditions internally. For instance, organisms grow and mature. Similarly, Earth has experienced a gradual increase in atmospheric oxygen over its history, yet life has adapted, and Earth remains, by most definitions, a homeostatic system.

It is difficult to understand systems fully by focusing on their individual components because systems can show **emergent properties**, characteristics not evident in the components alone. Stating that systems possess emergent properties is a lot like saying "the whole is more than the sum of its parts." For example, if you were to reduce a tree to its component parts (leaves, branches, trunk, bark, roots, fruit, and so on), you would not be able to predict the whole tree's emergent properties, which include the role the tree plays as habitat for birds, insects, parasitic vines, and other organisms (**FIGURE 5.2**). You could analyze the tree's chloroplasts (photosynthetic cell organelles), diagram its branch structure, and evaluate its fruit's nutritional content, but you would still be unable to understand the tree as habitat, as part of a forest landscape, or as a reservoir for carbon storage.

Systems seldom have well-defined boundaries, so deciding where one system ends and another begins can be difficult. Consider a desktop computer system. It is certainly a network of parts that interact and exchange energy and information, but what are its boundaries? Is the system what arrives in a packing crate and sits on top of your desk? Or does it include the network you connect it to at school, home, or work? What about the energy grid you plug it into, with its distant power plants and transmission lines? And what of the internet? Browsing the web, you are drawing in digitized text, light, and sound from around the world.

No matter how we attempt to isolate or define a system, we soon see that it has many connections to systems larger and smaller than itself. Systems may exchange energy, matter, and information with other systems, and they may contain or be contained within other systems—so where we draw boundaries may depend on the spatial or temporal scale at which we choose to focus.

Understanding a complex system requires considering multiple subsystems

The Great Lakes, St. Lawrence River, and Atlantic Ocean are systems that interact with one another. On a map, the river appears as a branched network of water channels.

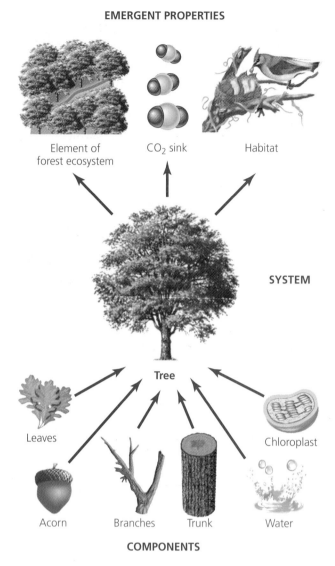

FIGURE 5.2
A system's emergent properties are not evident when we break the system down into its component parts. For example, a tree serves as wildlife habitat and plays roles in forest ecology and global climate regulation, but you would not know that from considering the tree only as a collection of leaves, branches, and chloroplasts. If we try to understand systems solely by breaking them into component parts, we will miss much of what makes them important.

But where are this system's boundaries? You might argue that the river consists primarily of water, originates in Ontario, and ends in the Atlantic Ocean.

But what about the rivers that feed it and the farms, cities, and forests that line its banks (**FIGURE 5.3**)? Major rivers like the Ottawa and Saguenay Rivers flow into the St. Lawrence. Hundreds of smaller tributaries drain vast expanses of farmland, woodland, fields, cities, towns, and industrial areas before their water joins the St. Lawrence. These waterways carry with them millions of tons of sediment, hundreds of species of plants and animals, and numerous pollutants. The St. Lawrence River system is also intimately interconnected with the entire Great Lakes system—together they constitute an integrated hydrological system.

Therefore, for an environmental scientist interested in runoff and the flow of water, sediment, or pollutants, it may make the best sense to view the Great Lakes–St. Lawrence River *watershed* as a system. One must consider the entire area of land a river drains to comprehend and solve problems of river pollution. However, for a scientist interested in the estuary's hypoxic zone, it may make the best sense to view the river together with the Gulf and coastal waters of the Atlantic as the system of interest, because their interaction is central to the problem. In environmental science, one's delineation of a system can and should depend on the questions one is addressing.

Environmental systems may be perceived in various ways

There are many ways to delineate natural systems, and your choice will depend on the particular issues in which you are interested. Categorizing environmental systems can help make Earth's dazzling complexity comprehensible to the human brain and accessible to problem solving.

Scientists divide Earth's components into broad structural systems. The **lithosphere** is the rock and sediment beneath our feet, in the planet's uppermost layers. The **atmosphere** is composed of the air surrounding our planet. The **hydrosphere** encompasses all water—salt or fresh, liquid, ice, or vapour—in surface bodies, underground, and in the atmosphere. The **biosphere** consists of the total of all the planet's living organisms and the abiotic (nonliving) portions of the environment with which they interact.

Although these categories can be useful, their boundaries overlap, so the systems interact. Picture a robin plucking an earthworm from the ground after a rain. You are witnessing an organism (the robin) consuming another organism (the earthworm) by removing it from part of the lithosphere (the soil) that the earthworm had been modifying—all this made possible because rain (from the hydrosphere) recently wet the ground. The robin might then fly through the air (the atmosphere) to a tree (an organism), in the process respiring (combining oxygen from the atmosphere with glucose from the organism, and adding water to the hydrosphere and carbon dioxide and heat to the atmosphere). Finally, the bird might defecate, adding nutrients to the lithosphere below. The study of such interactions among living and nonliving things is a key part of ecology at the ecosystem level, as scientists become more inclined to approach Earth's systems holistically.

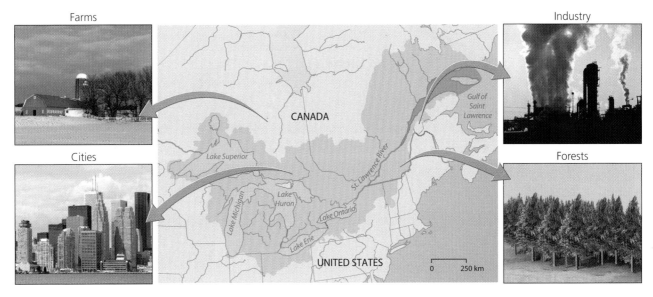

FIGURE 5.3
The watershed of the Great Lakes and St. Lawrence covers much of Ontario and Quebec, and extends into the northern United States. Runoff from the land into the river carries water, sediment, and pollutants from a variety of sources downstream to the Gulf of St. Lawrence, where pollution has given rise to a hypoxic zone and other environmental problems. Farms, cities, and industry are all contributors; so are natural sources, such as forests and soils.

Geologic Systems: How Earth Works

The physical processes of geology determine Earth's landscape and form the foundation for the biotic patterns that overlay the landscape. Let us start with these physical systems, which occur primarily in the lithosphere.

The rock cycle is a fundamental environmental system

We tend to think of rock as pretty solid stuff. Yet in the long run, over geologic time, rocks do change. **Rocks** and the **minerals** (naturally occurring, inorganic crystalline solids) that compose them are heated, melted, cooled, broken down, and reassembled in a very slow process called the **rock cycle** (FIGURE 5.4). The type of rock in a given region helps determine soil chemistry and thereby influences the biotic components of the region's ecosystems. Understanding the rock cycle enables us to appreciate more clearly the formation and conservation of soils, mineral resources, fossil fuels, and other natural resources, as well as the movement and storage of groundwater in the subsurface.

Igneous rock All rocks can melt. At high enough temperatures, rock will enter a molten, liquid state called **magma**. If magma is released (as in a volcanic eruption), it may flow or spatter across Earth's surface as **lava**. Rock that forms when magma cools is called **igneous rock** (from the Latin *ignis*, meaning "fire").

Igneous rock comes in several different types, because magma can solidify in different ways. Magma that cools slowly and solidifies while it is still well below Earth's surface is known as **intrusive** or *plutonic* rock (FIGURE 5.5A). The Coast Mountains in the Pacific Cordillera, British Columbia, were formed in this way and later exposed at the surface (FIGURE 5.5A). Granite is the best-known type of intrusive rock. A slow cooling process allows minerals of different types to grow into the larger crystals that give granite its multicoloured, coarse-grained appearance. In contrast, when magma is ejected from a volcano, it cools very quickly, so minerals have little time to grow into coarser crystals. This kind of igneous rock is called **extrusive** or *volcanic* rock, and its most common representative is basalt, the principal rock type of the Hawaiian Islands (FIGURE 5.5B).

Sedimentary rock All rock weathers away with time. The relentless forces of wind, water, freezing, and thawing eat away at rocks, stripping off one tiny grain (or large chunk) after another. Particles of rock blown by wind or washed away by water finally come to rest downhill, downstream, or downwind from their sources, forming **sediments**. These eroded remains of rocks usually are deposited very slowly, but floods can accelerate the process. Floods that sweep down the St. Lawrence River and its tributaries deposit sediments and nutrients across the floodplain, where they enrich soils, and at the river's mouth. Sediment layers accumulate over time, causing the weight and pressure of overlying layers to increase.

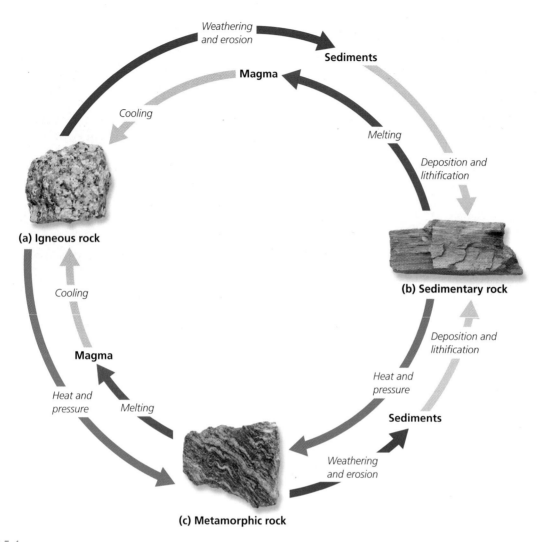

FIGURE 5.4
In the rock cycle, igneous rock **(a)** is formed when rock melts to form magma and the magma then cools. Sedimentary rock **(b)** is formed when rock is weathered and eroded and the resulting sediments are compressed to form new rock. Metamorphic rock **(c)** is formed when rock is subjected to intense heat and pressure underground. Through these processes (shown by different-coloured arrows), each type of rock can be converted into either of the other two types.

Sedimentary rock (**FIGURE 5.4B**) is formed when dissolved minerals seep through sediment layers and act as a kind of glue, called *cement*, which crystallizes and binds the sediment particles together. The formation of rock through these processes of compaction, cementation, and crystallization is termed **lithification**. Similar processes of physical compaction and chemical transformation preserve the fossil remnants of organisms, and create the fossil fuels we use for energy.

Like igneous rock, the several types of sedimentary rock are classified by the way they form and the size and composition of the particles they contain. Such rocks as limestone (**FIGURE 5.5C**) and rock salt form by chemical means when rocks dissolve and their components recrystallize to form new rocks. A second type of sedimentary rock forms when layers of sediment compress and become physically cemented to one another. Examples include conglomerate, made up of large pebbles and cobbles; sandstone, made of cemented sand particles; and shale, composed of still smaller mud particles. The cliffs of the Alberta Badlands and Dinosaur Provincial Park are made of layer upon layer of sedimentary rock (**FIGURE 5.5D**).

Metamorphic rock Geologic forces may bend, uplift, compress, or stretch rock. When great heat or pressure is exerted on rock, the rock may change its form, becoming **metamorphic rock** (**FIGURE 5.4C**). The forces that metamorphose rock occur at temperatures lower than the rock's melting point but high enough to reshape the crystals within the rock and change its appearance and physical properties. Common types of metamorphic rock include marble, formed when limestone is heated and pressurized, strengthening its structure; and slate, formed when shale is heated and metamorphosed. A more coarsely layered metamorphic

(a) Intrusive igneous rock—granite

(b) Extrusive igneous rock—basalt

(c) Sedimentary rock—limestone

(d) Sedimentary rock—sandstone

(d) Metamorphic rock—gneiss

FIGURE 5.5
The Coast Range Mountains of British Columbia **(a)** are made of granite, an intrusive igneous rock. In contrast, the volcanic islands of Hawaii **(b)** are built of basalt, an extrusive igneous rock. These classic cliffs in Ha Long Bay, Vietnam, are made of the sedimentary rock limestone **(c)**. These cliffs in Dinosaur Provincial Park, Alberta, are beautiful examples of sedimentary layers **(d)**. The Canadian Shield, which makes up the core of the North American continent, is built of very ancient metamorphic rocks, mainly gneiss **(e)**.

rock, called gneiss, makes up the Canadian Shield, which forms the core of the North American continent (**FIGURE 5.5E**).

Plate tectonics shapes Earth's geography

The rock cycle takes place within the broader context of **plate tectonics**, a process that underlies earthquakes and volcanoes and that determines the geography of Earth's surface. Earth's surface consists of a lightweight thin **crust** of rock floating atop a malleable **mantle**, which in turn surrounds a molten heavy **core** made mostly of iron. Earth's internal heat drives convection currents that flow in loops in the mantle, pushing the mantle's soft rock cyclically upward (as it warms) and downward (as it cools), like a gigantic conveyor belt. As the mantle material moves, it drags large plates of crust along its surface edge.

Earth's surface consists of about 15 major tectonic *plates*, most including some combination of ocean and continent (**FIGURE 5.6**). Imagine peeling an orange and putting the pieces of peel back onto the fruit; the ragged pieces of peel are like the plates of crust riding atop

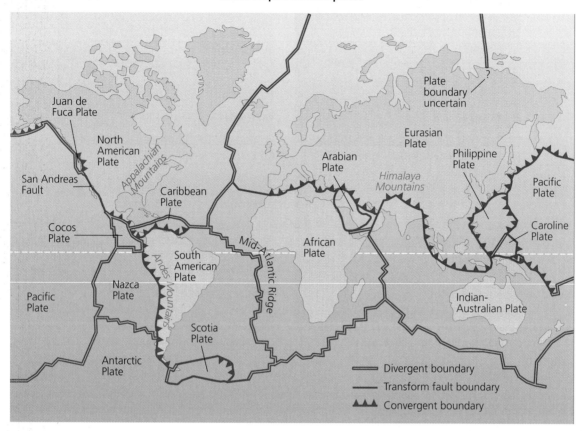

FIGURE 5.6
Earth's crust consists of roughly 15 major plates that move very slowly by the process of plate tectonics.

Earth's surface. These plates move at rates of roughly 2–15 cm per year. This movement has influenced Earth's climate and life's evolution throughout our planet's history as the continents combined, separated, and recombined in various configurations. By studying ancient rock formations throughout the world, geologists have determined that land masses that were joined together have split and moved apart numerous times.

At **divergent plate boundaries**, magma surging upward to the surface divides plates and pushes them apart, creating new crust as it cools and spreads (**FIGURE 5.7A**). A prime example is the Mid-Atlantic Ridge,

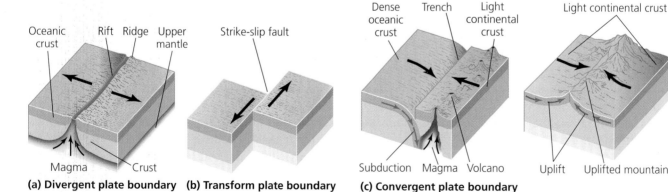

FIGURE 5.7
Different types of boundaries between tectonic plates result in different geologic processes. At a divergent plate boundary, such as a midocean ridge on the seafloor **(a)**, magma extrudes from beneath the crust, and the two plates move gradually away from the boundary in the manner of conveyor belts. At a transform plate boundary **(b)**, two plates slide alongside each other, creating friction that leads to earthquakes. Where plates collide at a convergent plate boundary **(c)**, one plate may be subducted beneath another, leading to volcanism, or both plates may be uplifted, causing mountain ranges to form.

part of a 74 000 km system of magmatic extrusion cutting across the seafloor. Plates expanding outward from divergent plate boundaries at midocean ridges bump against other plates, creating different types of plate boundaries.

When two plates meet, they may slip and grind alongside one another, forming a **transform plate boundary** (**FIGURE 5.7B**) and creating friction that spawns earthquakes along slip–strike faults. The Pacific Plate and the North American Plate rub against each other along California's San Andreas Fault. Southern California is slowly creeping toward northern California along this fault.

When plates collide at **convergent plate boundaries**, either of two consequences may result (**FIGURE 5.7C**). First, one plate of crust may slide beneath another in a process called **subduction**. The subducted crust is heated as it dives into the mantle, and it may send up magma that erupts through the surface in volcanoes. Mount St. Helens in Washington, which erupted violently in 1980 and renewed its activity in 2004, is fuelled by magma from subduction. When denser ocean crust slides beneath lighter continental crust, volcanic mountain ranges are formed that parallel coastlines. Examples include the Cascades (which include Mount St. Helens) and South America's Andes Mountains (where the Nazca Plate slides below the South American Plate).

When one plate of oceanic crust is subducted beneath another, the resulting volcanism may form arcs of islands, such as Japan and the Aleutians. In addition, deep trenches may be created, such as the Mariana Trench, the planet's deepest abyss. The Juan da Fuca hydrothermal vents, off the west coast of Vancouver Island, are in a marine protected area that lies atop an active subduction zone.

Alternatively, two colliding plates of continental crust may slowly lift material from both plates. The Himalayas, the world's highest mountains, are the result of the Indian–Australian Plate's collision with the Eurasian Plate 40 million to 50 million years ago; these mountains are still being uplifted today. The Appalachian Mountains, once the world's highest mountains themselves, resulted from a much earlier collision with the edge of what is today Africa.

The lithosphere and the tectonic processes that operate within it form the foundation for all of the biological and ecological processes on Earth. They are responsible for shaping the land surface and determining the positions of the continents and ocean basins, which, in turn, influence climate, weather, and the resulting distribution of biomes. Let us go on now to look more closely at the biological and ecological processes that are superimposed on this physical backdrop.

Ecosystems

An **ecosystem** consists of all organisms and nonliving entities that occur and interact in a particular area at the same time. The ecosystem concept builds on the idea of the biological community (Chapter 4), but ecosystems include abiotic components as well as biotic ones. In ecosystems, energy flows and matter cycles among these components.

Ecosystems are systems of interacting living and nonliving entities

The idea of ecosystems originated early in the twentieth century with such scientists as British ecologist Arthur Tansley, who saw that biological entities are tightly intertwined with chemical and physical entities. Tansley and others felt that there was so much interaction and feedback between organisms and their abiotic environments that it made the most sense to view living and nonliving elements together. For instance, the flow of water, sediment, and nutrients from the St. Lawrence River plays a key role in the nearshore ecosystem of the Gulf of St. Lawrence. The input of moisture from clouds plays a key role in the Monteverde cloud forest ecosystem we visited in Chapter 3. Abiotic factors, such as temperature, precipitation, latitude, and elevation, have substantial influence over which biomes exist in a given locality.

Ecologists began analyzing ecosystems as an engineer might analyze the operation of a machine. In this view, ecosystems are systems that receive inputs of energy, process and transform that energy while cycling matter internally, and produce a variety of outputs (such as heat, water flow, and animal waste products) that can move into other ecosystems. Energy flows in one direction through ecosystems; most arrives as radiation from the Sun, powers the system, and exits in the form of heat (**FIGURE 5.8A**). Matter, in contrast, is generally recycled within ecosystems (**FIGURE 5.8B**). We saw in Chapter 4 how energy and matter are passed among organisms (producers, consumers, and decomposers) through food web relationships. Matter is recycled because when organisms die and decay their nutrients remain in the system. In contrast, most energy that organisms take in is later lost through respiration.

Energy is converted to biomass

Energy flow in most ecosystems begins with radiation from the Sun. In Chapter 2 we explored how autotrophs, such as green plants and phytoplankton, use photosynthesis to

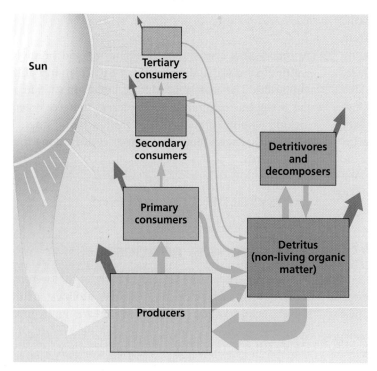
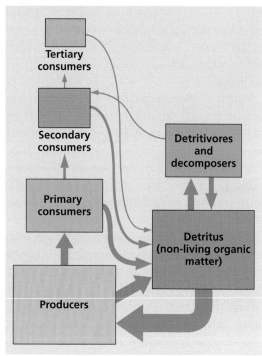

(a) Energy flowing through an ecosystem

(b) Matter cycling within an ecosystem

FIGURE 5.8
Energy enters, flows through, and exits an ecosystem. In **(a)**, light energy from the Sun (yellow arrow) drives photosynthesis in producers, which begins the transfer of chemical energy (green arrows) among trophic levels and detritus. Energy exits the system through respiration in the form of heat (red arrows). In contrast, matter cycles within an ecosystem. In **(b)**, blue arrows show the movement of nutrients among trophic levels and detritus. In both diagrams, box sizes represent relative magnitudes of energy or matter content, and arrow widths represent relative magnitudes of energy or matter transfer. Such magnitudes may vary tremendously from one ecosystem to another. Some abiotic components (such as water, air, and inorganic soil content) of ecosystems are omitted from these schematic diagrams.

capture the Sun's energy and produce food. We then saw in Chapter 4 how organisms at higher trophic levels consume organisms at lower trophic levels, transferring energy. The result of this process is the production of **biomass**, organic material of which living organisms are formed.

As autotrophs convert solar energy to the energy of chemical bonds in sugars, they perform *primary production*. Specifically, the assimilation of energy by autotrophs is termed **gross primary production**. Autotrophs use a portion of this production to power their own metabolism by respiration. The energy that remains after respiration, which is used to generate biomass, ecologists call **net primary production**. Thus, net primary production equals gross primary production minus respiration. Net primary production can be measured by the energy or the organic matter stored by plants after they have metabolized enough for their own maintenance.

Another way to think of net primary production is that it represents the energy or biomass available for consumption by heterotrophs. Plant matter not eaten by herbivores becomes fodder for detritivores and decomposers once the plant dies or drops its leaves. Heterotrophs use the energy they gain from plants for their own metabolism, growth, and reproduction. The total biomass that heterotrophs generate by consuming autotrophs is termed *secondary production*.

Ecosystems vary in the rate at which plants convert energy to biomass. The rate at which production occurs is termed **productivity**, and ecosystems in which plants convert solar energy to biomass rapidly are said to have high **net primary productivity**.

Freshwater wetlands, tropical forests, coral reefs, and algal beds tend to have the highest net primary productivities, whereas deserts, tundra, and open ocean tend to have the lowest (**FIGURE 5.9A**). Variation in net primary productivity among ecosystems and biomes results in geographic patterns of variation across the globe (**FIGURE 5.9B**). In terrestrial ecosystems, net primary productivity tends to increase with temperature and precipitation. In aquatic ecosystems, net primary productivity tends to rise with light and the availability of nutrients.

Nutrients can limit productivity

Nutrients are elements and compounds that organisms consume and require for survival. Organisms need several dozen naturally occurring chemical elements to survive.

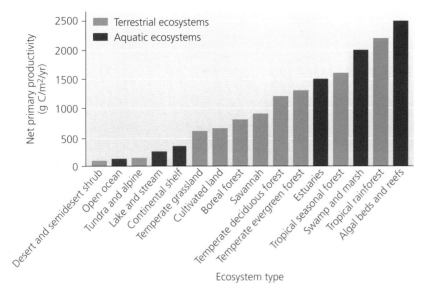

(a) Net primary productivity for major ecosystem types

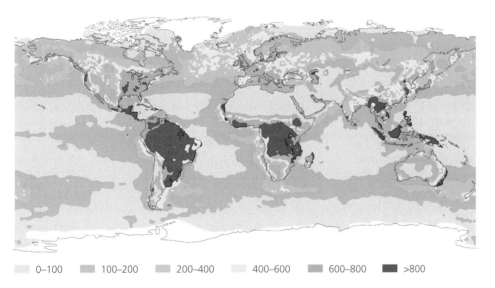

0–100   100–200   200–400   400–600   600–800   >800

(b) Global map of net primary productivity

FIGURE 5.9
(a) Freshwater wetlands, tropical forests, coral reefs, and algal beds show high net primary productivities on average, whereas deserts, tundra, and the open ocean show low values. (b) On land, net primary production varies geographically with temperature and precipitation, which also influence the locations of biomes. In the world's oceans, net primary production is highest around the margins of continents, where nutrients (of both natural and human origin) run off from land.
Data in (a) from Whittaker, R. H. 1975. Communities and ecosystems, 2nd ed. New York: MacMillan. Map in (b) from satellite data presented by Field, C. B., et al. 1998. Primary production of the biosphere: Integrating terrestrial and oceanic components. Science 281:237–240.

Elements and compounds required in relatively large amounts are called **macronutrients** and include nitrogen, carbon, and phosphorus. Nutrients needed in small amounts are called **micronutrients**.

Nutrients stimulate production by plants, and lack of nutrients can limit production. The availability of nitrogen or phosphorus frequently is a *limiting factor* for plant or algal growth. When these nutrients are added to a system, producers show the greatest response to whichever nutrient has been in shortest supply. Phosphorus tends to be limiting in freshwater systems, and nitrogen in marine systems. Thus, marine hypoxic zones result primarily from excess nitrogen, whereas freshwater ponds and lakes tend to suffer eutrophication when they contain too much phosphorus.

Canadian ecologist David Schindler and others demonstrated the effects of phosphorus on freshwater systems in the 1970s by experimentally manipulating entire lakes. In one experiment, the researchers divided a 16 ha lake in Ontario in half with a plastic barrier. To one half they added carbon, nitrate, and phosphate; to the other they added only carbon and nitrate. Soon after the experiment began, they saw a dramatic increase in algae in the half of the lake that received phosphate, whereas the other half hosted algal levels typical for lakes in the region (**FIGURE 5.10**). This difference held until shortly after they stopped fertilizing seven years later, when algae decreased to normal levels in the half that had previously received phosphate. Such experiments showed clearly that phosphorus addition can markedly increase primary productivity in lakes.

Similar experiments in coastal ocean waters show nitrogen to be the more important limiting factor for primary productivity. In experiments in the 1980s and

FIGURE 5.10
A portion of this lake in Ontario was experimentally treated with the addition of phosphate. This treated portion experienced an immediate, dramatic, and prolonged algal bloom, visible in the opaque water in the topmost part of this photo.

FIGURE 5.11
Satellite images, like this one, have helped scientists track runoff and phytoplankton blooms. Shown is the Northumberland Strait in the southern part of the Gulf of St. Lawrence. Greenish brown at the bottom left is land; the large island is Prince Edward Island. The deep blue to the top and right is the water of the Gulf and Atlantic Ocean. Along the coast, the light brown plumes are sediment washing into the Gulf's waters via rivers and surface runoff. The shades of turquoise represent phytoplankton blooms that result from nutrient inputs. Image from the SeaWiFS Project, NASA/Goddard Space Flight Center.

1990s, Swedish ecologist Edna Granéli took samples of ocean water from the Baltic Sea and added phosphate, nitrate, or nothing. Chlorophyll and phytoplankton increased greatly in the flasks with nitrate, whereas those with phosphate did not differ from the controls. Experiments in Long Island Sound by other researchers show similar results. For open ocean waters far from shore, research indicates that iron is a highly effective nutrient.

Because nutrients run off from land into the Baltic Sea, Long Island Sound, the Gulf of St. Lawrence, the Gulf of Mexico, and worldwide, primary productivity in the oceans tends to be greatest in nearshore waters, which receive nutrient runoff from land, and lowest in open ocean areas far from land (see **FIGURE 5.9**). Satellite imaging technology that reveals phytoplankton densities has given scientists an improved view of productivity at regional and global scales, which has helped them track blooms of algae that contribute to coastal hypoxic zones (**FIGURE 5.11**).

The number of known hypoxic zones is increasing globally, with about 200 documented so far. Most are located off the coasts of Europe and eastern North America. Specific causes vary from place to place, but most result from rising nutrient pollution from farms, cities, and industry. In North America, the Gulf of Mexico and Chesapeake Bay may be the most severely affected. Decades of pollution and human impact have devastated fisheries and greatly altered the ecology of these water bodies.

Good news of a sort comes from the Black Sea, which borders Ukraine, Russia, Turkey, and Eastern Europe. This immense inland sea had long suffered one of the world's worst hypoxic zones, which destroyed many of its fisheries and seagrass beds. Then in the 1990s, after the Soviet Union collapsed, industrial agriculture in the region declined drastically. With fewer fertilizers running off into it, the Black Sea began to recover, and today fish and mussel beds are reviving. However, agricultural collapse is not a strategy anyone would choose to alleviate hypoxia. Rather, scientists are proposing a variety of innovative and economically acceptable ways to reduce nutrient runoff.

Ecosystems are integrated spatially

We can conceptualize ecosystems at different scales. An ecosystem can be as small as an ephemeral puddle of water where brine shrimp and tadpoles feed on algae and detritus with mad abandon as the pool dries up. Or an ecosystem might be as large as a bay, lake, or forest. For some purposes, scientists even view the entire biosphere as a single all-encompassing ecosystem. The term is most often used, however, to refer to systems of moderate geographic extent that are somewhat self-contained. For example, the salt marshes that line the outer part of the St. Lawrence estuary, where its waters mix with those of the Atlantic Ocean, may be classified as ecosystems.

Adjacent ecosystems may share components and interact extensively. For instance, a pond ecosystem is very different from a forest ecosystem that surrounds it, but salamanders that develop in the pond live their adult lives under logs on the forest floor until returning to the pond to breed. Rainwater that nourishes forest plants may eventually make its way to the pond, carrying with it nutrients from the forest's leaf litter. Likewise, coastal dunes, the ocean, and a lagoon or salt marsh all may interact, as do forests and prairie where they converge. Areas in which ecosystems meet may consist of transitional zones called **ecotones**, in which elements of both ecosystems mix.

> **weighing the issues**   **ECOSYSTEMS WHERE YOU LIVE**   5-1
>
> Think about the area where you live. How would you describe that area's ecosystems? How do these systems interact with one another? If one ecosystem were greatly disturbed (say, if a wetland or forest were replaced by a shopping mall), what impacts might that have on nearby natural systems?

Landscape ecologists study geographic areas with multiple ecosystems

Because components of different ecosystems may intermix, ecologists often find it useful to view these systems on a larger geographic scale that encompasses multiple ecosystems. For instance, if you are studying large mammals, such as black bears, which move seasonally from mountains to valleys or between mountain ranges, you had better consider the overall landscape that includes all these areas. If you study fish, such as salmon, which move between marine and freshwater ecosystems, you need to know how these systems interact.

In such a broad-scale approach, called **landscape ecology**, scientists study how landscape structure affects the abundance, distribution, and interaction of organisms. A landscape-level approach is also proving useful for scientists, citizens, planners, and policy makers to plan for sustainable regional development (see "The Science Behind the Story: Biodiversity Portrait of the St. Lawrence").

For a landscape ecologist, a landscape is made up of a spatial array of *patches*. Depending on the researcher's perspective, patches may be ecosystems, or communities, or areas of habitat for a particular organism. Patches are spread spatially over a landscape in a *mosaic*. This metaphor reflects how natural systems often are arrayed across landscapes in complex patterns, like an intricate work of art. Thus, a forest ecologist may refer to a mosaic of forested patches remaining in an agricultural landscape, whereas a butterfly biologist might speak of a mosaic of patches of grassland habitat for a particular species of butterfly.

In referring to a *landscape*, ecologists generally imply a spatial scale larger than an ecosystem but smaller than a biome. However, one can view a landscape at different scales. **FIGURE 5.12** illustrates a landscape consisting of four ecosystem types, with ecotones along their borders indicated by thick red lines. At this scale, we perceive a mosaic consisting of four patches plus a river. However, the inset shows a magnified view of an ecotone. At this finer resolution, we see that the ecotone consists of patches of forest and grassland in a complex arrangement. The scale at which an ecologist focuses will depend on the questions being explored, or on the organisms being studied.

FIGURE 5.13 illustrates in a simplified way how different data sets of a GIS are combined, layer upon layer, to form a composite map. GIS has become a valuable tool used by geographers, landscape ecologists, resource managers, and conservation biologists. GIS technology also brings insights that affect planning and land-use decisions.

Some conservation groups, such as the Nature Conservancy land trust, now apply a landscape ecology approach widely in their land acquisition and management strategies. Principles of landscape ecology, and tools, such as GIS, are increasingly used in local and regional planning processes.

Biogeochemical Cycles

Materials move through the environment in complex and fascinating ways. Energy enters an ecosystem from the Sun, flows from one organism to another, and is dissipated to the atmosphere as heat; however, the physical matter of an ecosystem is circulated through natural systems over and over again.

Every organism has specific habitat needs, so when its habitat is distributed in patches across a landscape, individuals may need to expend energy and risk predation travelling from one patch to another. If the patches are distant enough, the organism's population may become divided into subpopulations, each occupying a different patch in the mosaic. Such a network of subpopulations, most of whose members stay within their respective

THE SCIENCE BEHIND THE STORY

Biodiversity Portrait of the St. Lawrence

The expanse of the St. Lawrence River at Quebec City.

A major interdisciplinary team of scientists, led by Jean-Luc DesGranges of the Canadian Wildlife Service and Jean-Pierre Ducruc of the Ministère de l'environnement du Québec, has undertaken to produce a comprehensive conservation plan to document and ultimately protect the biodiversity of the St. Lawrence system. The project is called, simply enough, The Biodiversity Portrait of the St. Lawrence.

But the St. Lawrence ecosystem is far from simple—it is an enormous and highly complex system. The team's approach aims to address and integrate as much of that complexity as possible (see figure). They are taking an integrated, holistic approach to the task, using the principles of landscape ecology. The vast territory of the ecosystem has been subdivided into smaller tracts, representing each major landscape type. Scientific studies and species inventories have been undertaken, along with extensive compendiums and compilations of existing information.

The team also is making full use of satellite data. In particular, satellite images at a variety of resolutions will assist in the assessment of anthropogenic impacts on the system, in eight categories: agriculture, urbanization, transportation, shoreline modifications, loss of wetlands, acidification, toxic substances, and commercial marine fisheries.

One of the most daunting tasks has been to reconcile the enormous variety of data, derived from many different sources and in widely varying formats, into a meaningful and accessible catalogue that can provide the basis for an integrated management system. This has been accomplished through the use of a comprehensive georeferenced database, also called a geographic information system (GIS).

Until the Biodiversity Portrait was undertaken, information sources on the physical and biotic characteristics of the St. Lawrence were scattered far and wide throughout the scientific literature and in noncentralized databases. This prevented scientists and managers from improving their understanding by developing a comprehensive picture of the richness, structure, and interrelationships

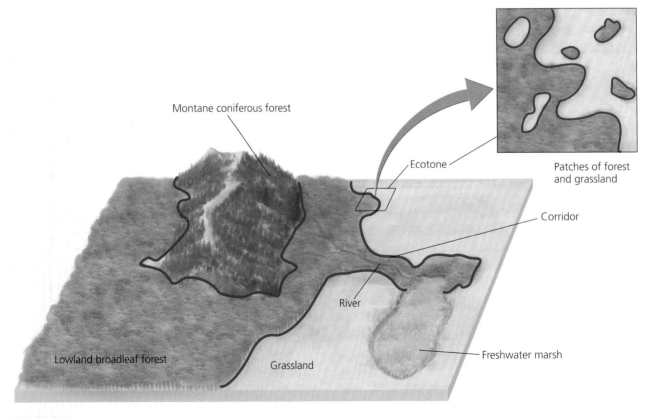

FIGURE 5.12
Landscape ecology deals with spatial patterns above the ecosystem level. This generalized diagram of a landscape shows a mosaic of patches of five ecosystem types (three terrestrial types, a marsh, and a river). Thick red lines indicate ecotones. A stretch of lowland broad-leaf forest running along the river serves as a corridor connecting the large region of forest on the left to the smaller patch of forest alongside the marsh. The inset shows a magnified view of the forest–grassland ecotone and how it consists of patches on a smaller scale.

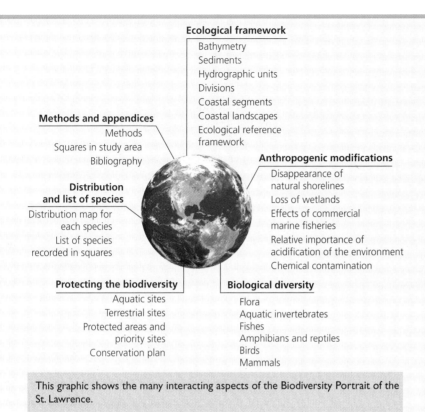

This graphic shows the many interacting aspects of the Biodiversity Portrait of the St. Lawrence.

of the components of this vast system. By collecting and consolidating much of the data collected along the St. Lawrence over the past 30 or so years into a structured and standardized georeferenced database, the Biodiversity Portrait has gone a long way toward overcoming this obstacle.

The georeferenced database now serves as a tool for describing the terrestrial, coastal, and marine environments of the St. Lawrence, and for studying their ecological and biological diversity and the anthropogenic pressures acting on them. Synthesizing all this information gave rise to offshoots, such as small-scale ecological maps of the St. Lawrence, thematic maps of the natural environment, and a cartographic atlas showing the key sites for the biodiversity of the St. Lawrence, where urgent conservation action is required.

For the first time in the long history of human interaction with the St. Lawrence, scientists and managers have a powerful, modern tool with which to move toward more integrated, effective, science-based management of this important ecosystem.[3]

patches but some of whom move among patches or mate with members of other patches, is called a *metapopulation*. When patches are still more isolated from one another, individuals may not be able to travel between them at all. In such a case, smaller subpopulations may be at risk of extinction.

Because of this extinction risk, metapopulations and landscape ecology are of great interest to scientists specializing in **conservation biology**, who study the loss, protection, and restoration of biodiversity. Of particular concern is the fragmentation of habitat into small and isolated patches—something that often results from human impact.

Establishing corridors of habitat (see **FIGURE 5.12**) to link patches is one approach that conservation biologists pursue as they attempt to maintain biodiversity in the face of human impact. We will return to these issues when we study conservation biology and habitat fragmentation in more detail in Chapters 9 and 10.

Remote sensing helps us apply landscape ecology

More and more scientists are taking a landscape perspective these days, and remote sensing technologies are improving our ability to do so. Satellites orbiting Earth are sending us more and better data than ever before on what the surface of our planet looks like (see **FIGURE 5.11**, for example). By helping us monitor our planet from above, satellite imagery has become a vital tool in modern environmental science.

A common tool for research in landscape ecology is the **geographic information system (GIS)**. A GIS consists of computer software that takes multiple types of data (for instance, on geology, hydrology, vegetation, animal species, and human development) and combines them on a common set of geographic coordinates. The idea is to create a complete picture of a landscape and to analyze how elements of the different data sets are arrayed spatially and how they may be correlated. See **FIGURE 5.13**.

Nutrients and other materials circulate via biogeochemical cycles

Nutrients move through ecosystems in **nutrient cycles** or **biogeochemical cycles**. In these cycles, materials travel through the atmosphere, hydrosphere, and lithosphere, and from one organism to another, in dynamic equilibrium. A carbon atom in your fingernail today might have

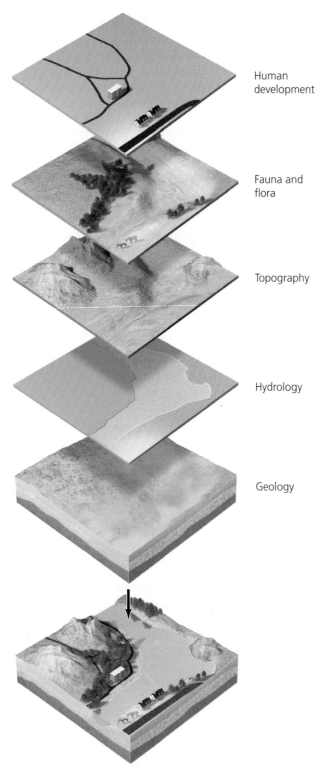

FIGURE 5.13
Geographic information systems (GIS) allow us to layer different types of data on natural landscape features and human land uses so as to produce maps integrating this information. GIS can be used to explore correlations among these data sets and to help in regional planning.

helped compose the muscle of a cow a year ago, may have resided in a blade of grass a month before that, and may have been part of a dinosaur's tooth 100 million years ago. After we die, the nutrients in our bodies will spread widely through the environment, eventually being incorporated by an untold number of organisms far into the future.

Nutrients and other materials (including toxins, as you will see in subsequent chapters) move from one **pool**, or **reservoir**, to another, remaining for varying amounts of time—the **residence time**—in each reservoir. The dinosaur, the grass, the cow, and you are each reservoirs for carbon atoms. The average residence time for an atom of carbon in your body will be much longer than the average residence time for an atom of carbon in a blade of grass, which has a short life span and will soon die, releasing its carbon back to the surrounding environment.

The movement of materials among reservoirs is termed a **flux** (**FIGURE 5.14**). Fluxes are rates, so they are stated in terms of mass or volume of material moving among reservoirs *per unit of time*. The flux of a material between reservoirs can change over time. Human activity has influenced the fluxes of certain materials. For example, we have increased the flux of nitrogen from the atmosphere to terrestrial reservoirs and have shifted the flux of carbon in the opposite direction.

Reservoirs that release more nutrients (or any other material of interest) than they accept are called **sources**; reservoirs that accept more nutrients than they release are called **sinks**. Carbon sinks are of particular importance today, as we struggle to lower the rate at which carbon is released into the atmosphere, potentially affecting our global climate system (Chapter 14).

The time it would take for all the atoms (or particles) of a particular material to be flushed through a particular reservoir is called the **turnover time**. Turnover time represents a balance between fluxes *into* the reservoir (from a source) and fluxes *out of* the reservoir (to a sink). If we stop all new sources of material coming into the reservoir, then the turnover time is the amount of time it would take for the material to be completely flushed through and out of the system.

Turnover time also depends on processes that influence the residence time of the material, including any processes that might hold or bind the material within the reservoir, or (alternatively) cause it to be flushed through more quickly. For example, let us say that we are interested in mercury in the water of a particular lake. The turnover time for mercury in the lake will depend on how quickly fresh water runs into the lake, and how quickly the mercury-laden water leaves the lake. It will also depend on other sources of mercury—for example, mercury that is deposited onto the lake's surface from the air or taken up from mercury-bearing rocks in the lakebed are two other potential sources. Now, what if some of the mercury in the water sinks to the bottom and binds to bottom sediments? This would greatly increase both the residence time (how long the mercury stays in the lake) and the

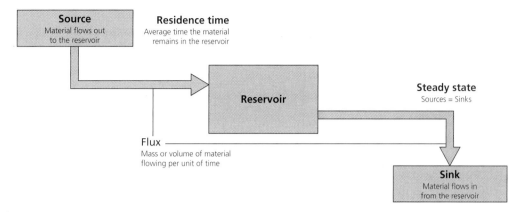

FIGURE 5.14
The properties of reservoirs and how cyclical fluxes move materials into and out of reservoirs are fundamentally important concepts in environmental science today.

turnover time (how quickly the mercury would be flushed through, if all the sources of new mercury into the lake were stopped).

These are extremely important concepts in environmental science today. For example, we are concerned about the presence in the atmosphere of substances that damage the stratospheric ozone layer (Chapter 13). Many of these substances have very long residence times in the atmosphere. Even if we stop producing them altogether—which we have almost accomplished as a result of the Montreal Protocol—the turnover time for these materials to be cleared out of the atmosphere by natural processes will be measured in decades. As we discuss biogeochemical cycles, think about how they involve negative feedback loops that promote dynamic equilibrium, and consider how human actions can influence fluxes and generate positive feedback loops. Let us start with a look at the most fundamental of the biogeochemical cycles: the water cycle.

The hydrologic cycle influences all other cycles

Water is so integral to life that we frequently take it for granted. The essential medium for all manner of biochemical reactions, water plays key roles in nearly every environmental system, including each of the nutrient cycles we are about to discuss. Water carries nutrients and sediments from the continents to the oceans via rivers, streams, and surface runoff, and it distributes sediments onward in ocean currents. Increasingly, water also distributes artificial pollutants.

The *water cycle*, or **hydrologic cycle** (**FIGURE 5.15**), summarizes how water—in liquid, gaseous, and solid forms—flows through our environment. A brief introduction to the hydrologic cycle here sets the stage for our more in-depth discussion of freshwater and marine systems in Chapters 11 and 12.

The oceans (Chapter 12) are the main reservoir in the hydrologic cycle, holding 97% of all water on Earth. The freshwater we depend on for our survival (Chapter 11) accounts for less than 3%, and two-thirds of this small amount is tied up in glaciers, snowfields, and icecaps. Thus, considerably less than 1% of the planet's water is in a form that we can readily use—groundwater, surface freshwater, and rain from atmospheric water vapour.

Evaporation and transpiration Water moves from oceans, lakes, ponds, rivers, and moist soil into the atmosphere by **evaporation**, the conversion of a liquid to gaseous form. Warm temperatures and strong winds speed rates of evaporation. A greater degree of exposure has the same effect; an area logged of its forest or converted to agriculture or residential use will lose water more readily than a comparable area that remains vegetated. Water also enters the atmosphere by **transpiration**, the release of water vapour by plants through their leaves. Transpiration and evaporation act as natural processes of distillation, effectively creating pure water by filtering out minerals carried in solution.

Precipitation, runoff, and surface water Water returns from the atmosphere to Earth's surface as **precipitation** when water vapour condenses and falls as rain or snow. Precipitation may be taken up by plants and used by animals, but much of it flows as **runoff** into streams, rivers, lakes, ponds, and oceans. Amounts of precipitation vary greatly from region to region globally, helping give rise to the variety of biomes.

Groundwater Some precipitation and surface water soaks down through soil and rock to recharge underground reservoirs known as **aquifers**. Aquifers are sponge-like regions of rock and soil that hold **groundwater**, water found underground beneath layers of soil. The upper limit

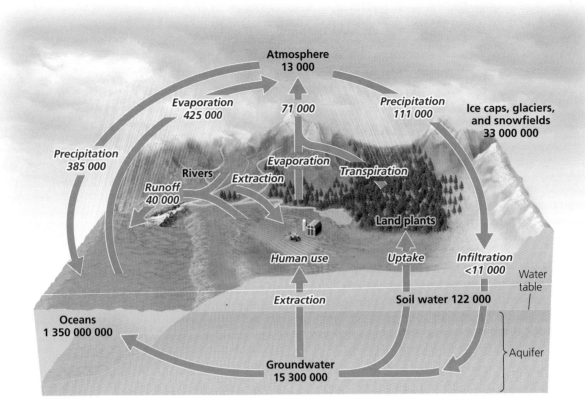

FIGURE 5.15
The hydrologic cycle summarizes the many routes that water molecules take as they move through the environment. Grey arrows represent fluxes among reservoirs for water. The hydrologic cycle is a system unto itself but also plays key roles in other biogeochemical cycles. Oceans hold 97% of our planet's water, whereas most freshwater resides in groundwater and icecaps. Water vapour in the atmosphere condenses and falls to the surface as precipitation, and then evaporates from land and transpires from plants to return to the atmosphere. Water flows downhill into rivers, eventually reaching the oceans. In the figure, reservoir names are printed in black type, and numbers in black type represent reservoir sizes expressed in units of cubic kilometres (km^3). Processes, printed in italic red type, give rise to fluxes, printed in italic red type and expressed in cubic kilometres per year. Data from Schlesinger, W. H. 1997. *Biogeochemistry: An analysis of global change*, 2nd ed. London: Academic Press.

of groundwater held in an aquifer is referred to as the **water table**. Aquifers may hold groundwater for long periods, so the water may be quite ancient. In some cases groundwater can take hundreds or even thousands of years to recharge fully after being depleted. Groundwater becomes exposed to the air where the water table reaches the surface, and the exposed water can run off toward the ocean or evaporate into the atmosphere.

weighing the issues **YOUR WATER** 5–2

Are you aware of any water shortages or conflicts over water use in your region? What is the quality of your water, and what pollution threats does it face? Given your knowledge of the hydrologic cycle, what solutions would you propose for water problems in your region?

Our impacts on the hydrologic cycle are extensive

Human activity affects every aspect of the water cycle. By damming rivers to create reservoirs, we increase evaporation and, in some cases, infiltration of surface water into aquifers. By altering Earth's surface and its vegetation, we increase surface runoff and erosion. By spreading water on agricultural fields, we can deplete rivers, lakes, and streams and can increase evaporation. By removing forests and other vegetation, we reduce transpiration and may lower water tables. By emitting into the atmosphere pollutants that dissolve in water droplets, we change the chemical nature of precipitation, in effect sabotaging the natural distillation process that evaporation and transpiration provide. Perhaps most threatening to our future, we are overdrawing groundwater to the surface for drinking, irrigation, and industrial use and have thereby begun to deplete groundwater resources. Water shortages have already given rise to numerous conflicts worldwide, and many people think this situation will worsen (see Chapter 11 "Central Case: Turning the Tap: The Prospect of Canadian Bulk Water Exports".

The carbon cycle circulates a vital organic nutrient

As the definitive component of organic molecules, **carbon** (C) is an ingredient in carbohydrates, fats, and proteins and in the bones, cartilage, and shells of all living things. From fossil fuels to DNA, from plastics to pharmaceuticals, carbon atoms are everywhere. The **carbon cycle** describes the routes that carbon atoms take through the environment (**FIGURE 5.16**).

Photosynthesis, respiration, and food webs
Producers, including terrestrial and aquatic plants, algae, and cyanobacteria, pull carbon dioxide out of the atmosphere and out of surface water to use in photosynthesis. Photosynthesis breaks the bonds in carbon dioxide (CO_2) and water (H_2O) to produce oxygen (O_2) and carbohydrates (e.g., glucose, $C_6H_{12}O_6$). Autotrophs use some of the carbohydrates to fuel their own respiration, thereby

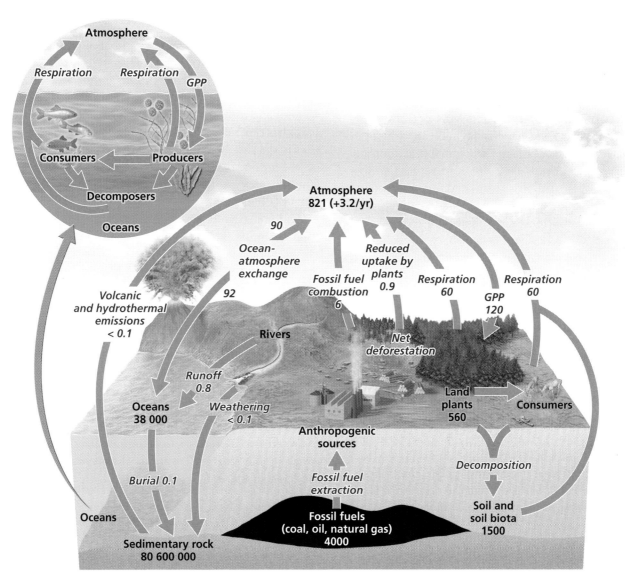

FIGURE 5.16
The carbon cycle summarizes the many routes that carbon atoms take as they move through the environment. Grey arrows represent fluxes among carbon reservoirs. Plants use carbon dioxide from the atmosphere for photosynthesis (gross primary production, or "GPP" in the figure). Carbon dioxide is returned to the atmosphere through respiration by plants, their consumers, and decomposers. The oceans sequester carbon in their water and in deep sediments. The vast majority of the planet's carbon is stored in sedimentary rock. In the figure, reservoir names are printed in black type, and numbers in black type represent reservoir sizes expressed in petagrams (units of 10^{15} g) of C. Processes, printed in italic red type, give rise to fluxes, printed in italic red type and expressed in petagrams of C per year. Data from Schlesinger, W. H. 1997. *Biogeochemistry: An analysis of global change*, 2nd ed. London: Academic Press.

releasing some of the carbon back into the atmosphere and oceans as CO_2. When producers are eaten by primary consumers, which in turn are eaten by secondary and tertiary consumers, more carbohydrates are broken down in respiration, producing carbon dioxide and water. The same process occurs when decomposers consume waste and dead organic matter. Respiration from all these organisms releases carbon back into the atmosphere and oceans.

All organisms use carbon for structural growth, so a portion of the carbon an organism takes in becomes incorporated into its tissues. The abundance of plants and the fact that they take in so much carbon dioxide for photosynthesis makes plants a major reservoir for carbon. Because CO_2 is a greenhouse gas of primary concern, much research on global climate change is directed toward measuring the amount of CO_2 that plants tie up. Scientists are working toward understanding exactly how much this portion of the carbon cycle influences Earth's climate.

Sediment storage of carbon As organisms die, their remains may settle in sediments in ocean basins or in freshwater wetlands. As layers of sediment accumulate, older layers are buried more deeply, experiencing high pressure over long periods. These conditions can convert soft tissues into fossil fuels—coal, oil, and natural gas—and shells and skeletons into sedimentary rock, such as limestone, as discussed above. Sedimentary rock composes the largest single reservoir in the carbon cycle. Although any given carbon atom spends a relatively short time in the atmosphere, carbon trapped in sedimentary rock may reside there for hundreds of millions of years.

Carbon trapped in sediments and fossil fuel deposits may eventually be released into the oceans or atmosphere by geologic processes, such as uplift, erosion, and volcanic eruptions. It also reenters the atmosphere when we extract and burn fossil fuels (Chapter 15).

The oceans The world's oceans are the second-largest reservoir in the carbon cycle. They absorb carbon-containing compounds from the atmosphere, from terrestrial runoff, from undersea volcanoes, and from the waste products and detritus of marine organisms. Some carbon atoms absorbed by the oceans—in the form of carbon dioxide, carbonate ions (CO_3^{2-}), and bicarbonate ions (HCO_3^-)—combine with calcium ions (Ca^{2+}) to form calcium carbonate ($CaCO_3$), an essential ingredient in the skeletons and shells of microscopic marine organisms. As these organisms die, their calcium carbonate shells sink to the ocean floor and begin to form sedimentary rock. The rates at which the oceans absorb and release carbon depend on many factors, including temperature and the numbers of marine organisms converting CO_2 into carbohydrates and carbonates.

We are shifting carbon from the lithosphere to the atmosphere

By mining fossil fuel deposits, we are essentially removing carbon from an underground reservoir with a residence time of millions of years. By combusting fossil fuels in our automobiles, homes, and industries, we release carbon dioxide and greatly increase the flux of carbon from the ground to the air. Since the mid-eighteenth century, our fossil fuel combustion has added about 250 billion metric tons of carbon to the atmosphere. Meanwhile, the movements of CO_2 *from* the atmosphere *back* to the hydrosphere, lithosphere, and biosphere (that is, the sinks) have not kept pace.

In addition, cutting down forests and burning fields removes carbon from the reservoir of vegetation and releases it to the air. And if less vegetation is left on the surface, there are fewer plants to draw CO_2 back out of the atmosphere. As a result, scientists estimate that today's atmospheric carbon dioxide reservoir is the largest that Earth has experienced in the past 650 000 years and perhaps in the past 20 million years. The anthropogenic flux of carbon out of the fossil fuel reservoir and into the atmosphere is a driving force behind global climate change (Chapter 14).

Our understanding of the carbon cycle is not yet complete. Scientists have long been baffled by the so-called *missing carbon sink*. Scientists have measured how much of the carbon dioxide we emit by fossil fuel combustion and deforestation goes into the atmosphere and oceans, but there remain roughly 1 billion to 2 billion metric tons unaccounted for. Many researchers think this is taken up by plants or soils of the northern temperate and boreal forests. But they would like to know for sure—because if certain forests are acting as a major sink for carbon (and thus restraining global climate change), we would like to keep it that way (see Chapter 7 "Central Case: Mer Bleue: A Bog of International Significance"). If forests that today are sinks were to turn into sources and begin releasing the "missing" carbon, climate change could accelerate drastically.

The nitrogen cycle involves specialized bacteria

Nitrogen (N) makes up 78% of our atmosphere by mass and is the sixth most abundant element on Earth. It is an essential ingredient in the proteins, DNA, and RNA that build our bodies. Nitrogen is an essential nutrient for plant growth. Thus the **nitrogen cycle** (FIGURE 5.17) is of vital importance to us and to all other organisms. Despite its abundance in the air, nitrogen gas (N_2) is chemically inert

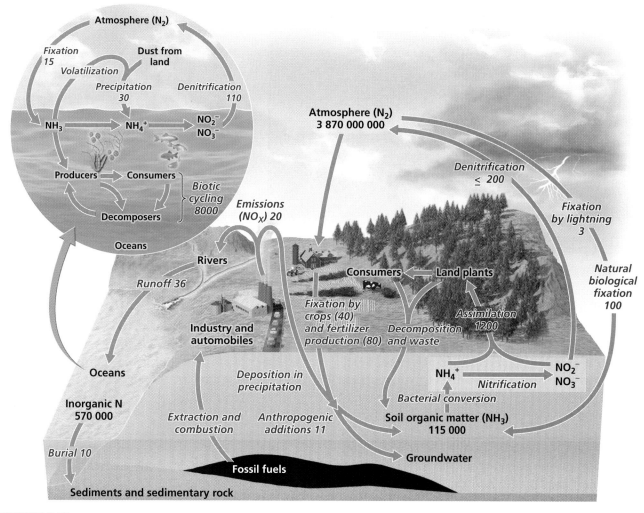

FIGURE 5.17
The nitrogen cycle summarizes the many routes that nitrogen atoms take as they move through the environment. Grey arrows represent fluxes among reservoirs for nitrogen. In the nitrogen cycle, specialized bacteria play key roles in "fixing" atmospheric nitrogen and converting it to chemical forms that plants can use. Other types of bacteria convert nitrogen compounds back to the atmospheric gas N_2. In the oceans, inorganic nitrogen is buried in sediments, whereas nitrogen compounds are cycled through food webs as they are on land. In the figure, reservoir names are printed in black type, and numbers in black type represent reservoir sizes expressed in teragrams (units of 10^{12} g) of N. Processes, printed in italic red type, give rise to fluxes, printed in italic red type and expressed in teragrams of N per year. Data from Schlesinger, W. H. 1997. *Biogeochemistry: An analysis of global change*, 2nd ed. London: Academic Press.

and cannot cycle out of the atmosphere and into living organisms without assistance from lightning, highly specialized bacteria, or human intervention. For this reason, the element is relatively scarce in the lithosphere and hydrosphere and in organisms. However, once nitrogen undergoes the right kind of chemical change, it becomes biologically active and available to the organisms that need it, and it can act as a potent fertilizer. Its scarcity makes biologically active nitrogen a limiting factor for plant growth.

Nitrogen fixation To become biologically available, inert nitrogen gas (N_2) must be "fixed," or combined with hydrogen in nature to form ammonia (NH_3), whose water-soluble ions of ammonium (NH_4^+) can be taken up by plants. **Nitrogen fixation** can be accomplished in two ways: by the intense energy of lightning strikes, or when air in the top layer of soil comes in contact with particular types of **nitrogen-fixing** bacteria. These bacteria live in a mutualistic relationship with many types of plants, including soybeans and other *legumes*, providing them with nutrients by converting nitrogen to a usable form. As we will see in Chapter 7, farmers have long nourished their soils by planting crops that host nitrogen-fixing bacteria among their roots (**FIGURE 5.18**).

Nitrification and denitrification Other types of specialized bacteria then perform a process known as **nitrification**. In this process, ammonium ions are first converted into nitrite ions (NO_2^-), and then into nitrate ions (NO_3^-). Plants can take up these ions, which also become available after atmospheric deposition on soils or in water or after application of nitrate-based fertilizer.

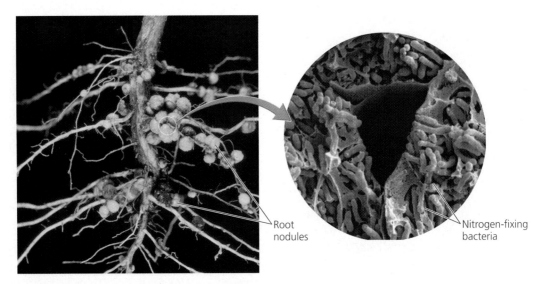

FIGURE 5.18
Specialized bacteria live in nodules on the roots of leguminous plants. In the process of nitrogen fixation, the bacteria convert nitrogen to a form that the plant can take up into its roots.

Animals obtain the nitrogen they need by consuming plants or other animals. Recall from Chapter 2 ("The Science Behind the Story: How Isotopes Reveal Secrets of Earth and Life") how scientists use stable isotopes of nitrogen to study the trophic level and nutritional condition of animals. Decomposers obtain nitrogen from dead and decaying plant and animal matter and from animal urine and feces. Once decomposers process the nitrogen-rich compounds they take in, they release ammonium ions, making these available to nitrifying bacteria to convert again to nitrites and nitrates.

The next step in the nitrogen cycle occurs when **denitrifying bacteria** convert nitrates in soil or water to gaseous nitrogen via a multistep process. Denitrification thereby completes the cycle by releasing nitrogen back into the atmosphere as a gas.

We have greatly influenced the nitrogen cycle

The impacts of excess nitrogen from agriculture and other human activities in the Great Lakes and St. Lawrence system have had a negative impact on both water quality and the health of marine organisms in downstream areas, as discussed in "Central Case: The Plight of the St. Lawrence Belugas." Similar impacts in the Mississippi River watershed have become painfully evident to shrimpers and scientists with an interest in the Gulf of Mexico (see "The Science Behind the Story: The Gulf of Mexico's 'Dead Zone'").

But hypoxia in the Gulf of Mexico, the estuary of the St. Lawrence, and other coastal locations around the world is hardly the only problem resulting from human manipulation of the nitrogen cycle.

Historically, nitrogen fixation was a *bottleneck*, a step that limited the flux of nitrogen out of the atmosphere. This changed when the research of two German chemists enabled us to fix nitrogen on an industrial scale. Fritz Haber worked in the German army's chemical weapons program during the First World War. Shortly before the war, Haber found a way to combine nitrogen and hydrogen gases to synthesize ammonia, a key ingredient in modern explosives and agricultural fertilizers. Several years later, Carl Bosch built on Haber's work and devised methods to produce ammonia on an industrial scale.

The work of these two scientists enabled people to overcome the limits on productivity long imposed by nitrogen scarcity in nature. The widespread application of their findings has enhanced agriculture and thereby contributed to the enormous increase in human population over the past 90 years. Farmers, golf course managers, and home owners have all taken advantage of the fertilizers made possible by the **Haber–Bosch process**.

These developments have led to a dramatic alteration of the nitrogen cycle. Today, by using the Haber–Bosch process, our species is fixing at least as much nitrogen artificially as is being fixed naturally. We have effectively doubled the natural rate of nitrogen fixation on Earth.

By fixing atmospheric nitrogen, we increase its flux out of the atmosphere and into other reservoirs. In addition, we have affected fluxes in other parts of the cycle. When we burn forests and fields, we force nitrogen out of soils and vegetation and into the atmosphere. When we burn fossil fuels, we increase the rate at which nitric oxide (NO) enters the atmosphere and reacts to form nitrogen dioxide (NO_2). This compound is a precursor to nitric acid (HNO_3), a key component of acid precipitation. We introduce another nitrogen-containing gas, nitrous oxide (N_2O), by allowing anaerobic bacteria to break down the tremendous volume of animal waste produced in agricultural feedlots. We have also accelerated the introduction of nitrogen-rich compounds into terrestrial and aquatic systems by destroying wetlands and cultivating more legume crops that host nitrogen-fixing bacteria in their roots.

These activities increase amounts of nitrogen available to aquatic plants and algae, boosting their growth (**FIGURE 5.19**). Algal populations soon outstrip the availability of other required nutrients and begin to die and decompose. As in the St. Lawrence estuary and Gulf of Mexico hypoxic zones, this large-scale decomposition can rob other aquatic organisms of oxygen, leading to shellfish die-offs and other significant impacts on ecosystems. Nitrate pollution, common in the groundwater of agricultural areas, can also lead to human health effects.

Researchers report that human activities have

- Doubled the rate at which fixed nitrogen enters terrestrial ecosystems (and the rate is still increasing)
- Increased atmospheric concentrations of the greenhouse gas N_2O and of other oxides of nitrogen that produce smog
- Depleted essential nutrients (such as calcium and potassium) from soils, because fertilizer helps flush them out
- Acidified surface water and soils
- Greatly increased transfer of nitrogen from rivers to oceans
- Encouraged plant growth, causing more carbon to be stored within terrestrial ecosystems
- Reduced biological diversity, especially plants adapted to low nitrogen concentrations
- Changed the composition and function of estuaries and coastal ecosystems
- Harmed many coastal marine fisheries

weighing the issues
NITROGEN POLLUTION AND ITS FINANCIAL IMPACTS 5–3

Most nitrate that enters the Great Lakes and the St. Lawrence River originates from farms and other sources in Ontario and Quebec, yet many of its negative impacts are borne by downstream users, such as fishers and organisms in the estuary and the Gulf of St. Lawrence. Who should be responsible for addressing this problem? How would you deal with the fact that the watershed straddles the international border between Canada and the United States?

Many of the impacts associated with excess nitrogen, including eutrophication, also occur when phosphorus—another plant nutrient that boosts the growth of phytoplankton—is present in excess in aquatic systems. Let us look briefly at the phosphorus cycle, the fourth major biogeochemical cycle.

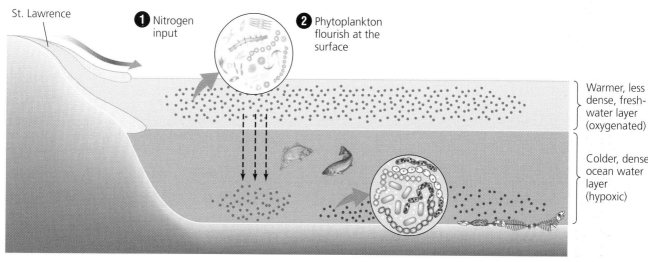

FIGURE 5.19
Excess nitrogen causes eutrophication in marine and freshwater systems, such as the St. Lawrence estuary. Coupled with stratification (layering) of water, eutrophication can severely deplete dissolved oxygen. Nitrogen from river water (1) boosts growth of phytoplankton (2), which die and are decomposed at the bottom by bacteria (3). Stability of the surface layer prevents deeper water from absorbing oxygen to replace oxygen consumed by decomposers (4), and the oxygen depletion suffocates or drives away bottom-dwelling marine life (5), giving rise to hypoxic zones.

THE SCIENCE BEHIND THE STORY

The Gulf of Mexico's "Dead Zone"

This is the Mississippi River as it enters the Gulf of Mexico.

Louisiana fishers have long hauled in more seafood than those of any other U.S. state except Alaska. Each year they have harvested nearly 600 million kilograms of shrimp, fish, and shellfish from the rich waters of the northern Gulf of Mexico. Then in 2005, Hurricane Katrina and Hurricane Rita pummelled the Gulf Coast and left Louisiana's fisheries in ruin.

But for years before the hurricanes hit, fishing had become increasingly difficult. The reason? Billions of organisms were suffocating in the Gulf's "dead zone," water so depleted of oxygen that marine organisms are killed or driven away. The dead zone appears each spring and grows through the summer and fall, starting near the mouths of the Mississippi and Atchafalaya Rivers off the Louisiana coast. At the time of writing the size of the dead zone in 2008 was on track to surpass the previous record of 22 000 km^2, set in 2002.

Hypoxia has a negative impact on aquatic animals, which asphyxiate if deprived of oxygen. Fully oxygenated water contains up to 10 parts per million (ppm) of oxygen; when concentrations drop below 2 ppm, creatures that can leave an affected area will do so. Below 1.5 ppm, most marine organisms die. In the Gulf of Mexico's hypoxic zone, oxygen concentrations frequently drop below these levels.

Scientists have identified human impacts hundreds of kilometres away as the cause of the oxygen starvation. The Gulf is being excessively enriched by nitrogen and phosphorus flushed down the Mississippi River. This nutrient pollution comes from fertilizers used on farms far upriver, and from urban runoff, industrial discharges, fossil fuel emissions, and municipal sewage outflow. The rivers draining into the Gulf carry excess nutrients that feed algal blooms, whose decomposition then depletes the oxygen in wide stretches of ocean water.

In 1985, Dr. Nancy Rabalais and researchers from the Louisiana Universities Marine Consortium started tracking oxygen levels at nine sites in the Gulf every month and continued those measurements for five years. At dozens of other spots near the shore and in deep water, they took less frequent oxygen readings. The team also collected and tested hundreds of coastal and Gulf water samples, and donned scuba gear to view the condition of shrimp, fish, and other sea life. Such a range of long-term data allowed the scientists to build a detailed "map" of the dead zone.

In 1991, Dr. Rabalais made that map public, earning immediate headlines. That year, her group mapped the size of the zone at more than 10 000 km^2. Bottom-dwelling shrimp were stretching out of their burrows, straining for oxygen. Many fish had fled. The bottom waters, infused with sulphur from bacterial decomposition, smelled of rotten eggs.

Years of continuous tracking allowed scientists to explain the dead zone's predictable emergence. As rivers rose each spring (and fertilizers were applied in Midwestern farm states), oxygen would start to disappear in the northern Gulf. The hypoxia would last through the summer or fall, until seasonal storms mixed oxygen into hypoxic areas. Over time, monitoring linked the dead zone's size to the volume of river flow and its nutrient load; the 1993 flooding of the Mississippi created a zone much larger than the year before. Conversely, a drought in 2000 brought lower river flows, lower nutrient loads, and a smaller dead zone (see figure).

In 2000, an assessment involving dozens of scientists blamed the dead zone on nutrients from fertilizers and other sources. The government acted on the findings, proposing that farmers in Ohio, Iowa, and Illinois cut down on fertilizer use. Farmers' advocates protest that farmers are being singled out while urban pollution sources are ignored. Meanwhile, scientists have documented coastal dead zones in 200 other areas throughout the world.

In 2004, Environmental Protection Agency water quality scientist Howard Marshall suggested that to alleviate the dead zone, we would be best off reducing phosphorus pollution from industry and sewage treatment. His reasoning: Phytoplankton need both nitrogen and phosphorus, but the Gulf now has so much nitrogen that phosphorus has become the limiting factor on phytoplankton growth.

Many scientists now propose that nitrogen and phosphorus should be managed jointly. Further research indicates that a federally mandated 30% reduction in nitrogen in the river will not be enough. Some scientists suggest that restoration of wetlands along the river and at the delta would filter pollutants before they reach the Gulf. All this research is guiding a federal plan to reduce runoff, clean up the Mississippi, restore wetlands, and shrink the dead zone—and it has led to a better understanding of hypoxic zones around the world.

The phosphorus cycle involves mainly lithosphere and ocean

The element **phosphorus** (P) is a key component of cell membranes and of several molecules vital for life, including DNA, RNA, ATP, and ADP. Although phosphorus is indispensable for life, the amount of phosphorus in organisms is dwarfed by the vast amounts in rocks, soil, sediments, and the oceans. Unlike the carbon and nitrogen cycles, the **phosphorus cycle** (FIGURE 5.20) has no appreciable atmospheric component besides the transport of tiny amounts of windblown dust and seaspray.

Geology and phosphorus availability The vast majority of Earth's phosphorus is contained within rocks and is released only by weathering, which releases

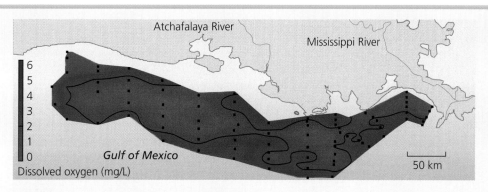

(a) Dissolved oxygen at bottom, July 2006

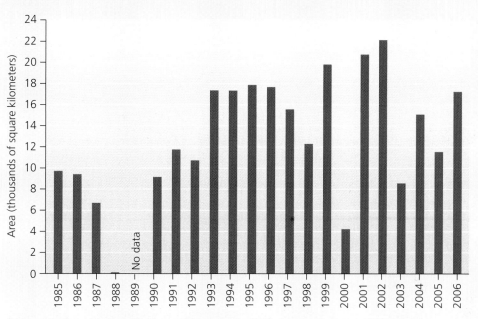

(b) Area of hypoxic zone in the northern Gulf of Mexico

The map in (a) shows dissolved oxygen concentrations in bottom waters of the Gulf of Mexico off the Louisiana coast from the July 2006 survey. Areas in red indicate the lowest oxygen levels, and areas in blue indicate the highest levels. Regions considered hypoxic (<2 mg/L) are encircled with a black line. Black dots indicate sampling points. The size of the Gulf's hypoxic zone varies (b) as a result of several factors. Floods increase its size by bringing additional runoff (as with the Mississippi River floods of 1993), whereas tropical storms decrease its size by mixing oxygen-rich water into the dead zone (as in 2003). Between 1985 and 2006, the hypoxic zone averaged 13 000 km² in size. Data from Nancy Rabalais, LUMCON.

phosphate ions (PO_4^{3-}) into water. Phosphates dissolved in lakes or in the oceans precipitate into solid form, settle to the bottom, and reenter the lithosphere's phosphorus reservoir in sediments. Because most phosphorus is bound up in rock and only slowly released, environmental concentrations of phosphorus available to organisms tend to be very low. This relative rarity explains why phosphorus is frequently a limiting factor for plant growth, and why an artificial influx of phosphorus can produce immediate and dramatic effects.

Food webs   Plants can take up phosphorus through their roots only when phosphate is dissolved in water. Primary consumers acquire phosphorus from water and plants and pass it on to secondary and tertiary consumers. Consumers also pass phosphorus to the soil through the

CANADIAN ENVIRONMENTAL PERSPECTIVES

Robert Bateman

Robert Bateman is one of Canada's leading naturalist artists.

- Award-winning and internationally recognized **artist**
- **Naturalist** and **environmentalist**
- **Teacher**

It is hard to think of a painter with broader popular appeal than Robert Bateman. His works are so realistic, so close to the natural subjects he portrays, that we can almost smell the hot breath of the bear or feel the dry, choking dust stirred by the buffalo's hooves.

Bateman started life as both an artist and a naturalist when, as a young boy in Toronto, he undertook to draw all of the birds in the neighbouring ravine. He describes his life since then as having been "immersed in nature."[4] Although he claims to have always been an artist at heart, there was a time when Bateman thought he could not make a living as an artist, so he chose to become a teacher. But he kept painting, travelling, and observing the natural world, and eventually his work began to receive wider notice.

Although he experimented with other painting styles, such as impressionism and even cubism, Bateman has settled into a meticulously realistic style that pays homage to each tiny detail of the ecological communities he portrays. He was inspired in this by American painter Andrew Wyeth, another great realist interpreter of the natural world.

Having spent his life devoted to the detailed interpretation and representation of the natural world, Bateman is a committed environmentalist. He was recently filmed painting a black "oil slick" over one of his orca images, to dramatize the possibility of tanker spills in the pristine Douglas Channel of British Columbia. He wishes that students could spend at least half of their time in the wilderness, stating, "In outdoor education you not only learn about nature, you learn about yourself, your limits and your relationship with others."[5]

In 2005 Bateman volunteered to be a test subject for Toxic Nation, an initiative of Environmental Defence, a Canadian environmental organization. For this study, children and adults—celebrities like Bateman, as well as ordinary citizens—were tested for 88 toxic chemicals. The tests on Bateman detected 32 carcinogens, 19 hormone disruptors, 16 respiratory toxicants, and 42 reproductive/developmental toxicants.[6] (You will learn more about these various categories of toxins in subsequent chapters, especially Chapter 19.)

Bateman viewed his participation in this study as a way to convince decision makers to take action on behalf of the natural environment and the health of humans.

"I can't conceive of anything being more varied and rich and handsome than the planet Earth. And its crowning beauty is the natural world. I want to soak it up, to understand it as well as I can, and to absorb it." —Robert Bateman

Thinking About Environmental Perspectives

1. Robert Bateman spends a lot of time observing and studying the natural world. It is extremely important to him to portray the plant and animal communities in his paintings with as much scientific accuracy as possible. However, he has been known to say that "art wins" if he ever comes up against a conflict between science and art. Do you think this compromises the value of his works from a naturalist's perspective? Or does it add power to his environmental message?

---

excretion of waste. Decomposers break down phosphorus-rich organisms and their wastes and, in so doing, return phosphorus to the soil.

We affect the phosphorus cycle

Humans influence the phosphorus cycle in several ways. We mine rocks containing phosphorus to extract this nutrient for the inorganic fertilizers we use on crops and lawns. Our wastewater discharge also tends to be rich in phosphates. Phosphates that run off into waterways can boost algal growth and cause eutrophication, leading to murkier waters and altering the structure and function of aquatic ecosystems. Phosphates are also present in detergents, so one way each of us can reduce phosphorus input into the environment is to purchase phosphate-free detergents.

In the 1970s Lake Erie began to exhibit signs of hypoxia (see Chapter 11 "Central Case: The Death and Rebirth of Lake Erie"). As in the Gulf of Mexico and St. Lawrence examples that we have touched on, the problem was traced to inputs of phosphate from fertilizers, detergents, and municipal sewage. The sources were land based, but runoff carried the phosphate-bearing materials into the lake from surrounding farms and towns in the United States and Canada.

The International Joint Commission (IJC), set up under the International Boundary Waters Treaty of 1909, adopted the Great Lakes Water Quality Agreement (GLWQA) in 1972 to begin to address the problem of hypoxia, eutrophication, and phosphate runoff into Lake Erie. The broader goal of the agreement was to promote an integrated, cooperative, scientific, and ecosystem-based approach to the management of the international waters of the Great Lakes and adjacent transboundary waters. Initially, IJC activities under the GLWQA focused on reducing phosphate from detergents and municipal

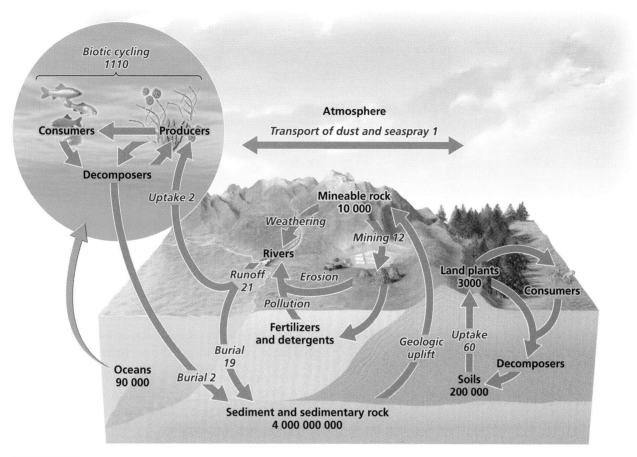

FIGURE 5.20
The phosphorus cycle summarizes the many routes that phosphorus atoms take as they move through the environment. Grey arrows represent fluxes among reservoirs for phosphorus. Most phosphorus resides underground in rock and sediment, but the phosphorus cycle moves this element through the soil, the oceans, and freshwater and terrestrial ecosystems. Rocks containing phosphorus are uplifted geologically and weathered away in this slow process, and small amounts of phosphorus cycle through food webs, where this nutrient is often a limiting factor for plant growth. In the figure, reservoir names are printed in black type, and numbers in black type represent reservoir sizes expressed in teragrams (units of 10^{12} g) of P. Processes, printed in italic red type, give rise to fluxes, printed in italic red type and expressed in teragrams of P per year. Data from Schlesinger, W. H. 1997. *Biogeochemistry: An analysis of global change*, 2nd ed. London: Academic Press.

sewage, as these were the best understood pollutants from a scientific perspective.

Since then, the agreement has grown in both strength and scope, and now deals with all threats to water quality in the Great Lakes system from chemical, physical, and biological factors, as well as from other causes, such as habitat destruction. Phosphate runoff into the lakes has been controlled to a degree, and Lake Erie has recovered from its most severely oxygen-depleted state. However, runoff of phosphates and many other pollutants into the Great Lakes continues to be problematic, and all the lakes still face periodic episodes of oxygen depletion.

Conclusion

Earth hosts many interacting systems, and the way one perceives them depends on the questions in which one is interested. Physical systems and processes, such as the hydrologic cycle, the rock cycle, and plate tectonics, lay the groundwork for the ways in which life spreads itself across the planet. Life interacts with its abiotic environment in ecosystems, systems through which energy flows and materials are recycled. Understanding the biogeochemical cycles that describe the movement of nutrients within and among ecosystems is crucial, because human activities are causing significant changes in the ways those cycles function.

Thinking in terms of systems is crucial to understanding how Earth works, so that we may learn how to avoid disrupting its processes and how to mitigate any disruptions we cause. By studying the environment from a systems perspective and by integrating scientific findings with the policy process, people who care about the St. Lawrence River and its estuarine ecosystem are working today to address the environmental issues there from an integrated, holistic perspective.

Unperturbed ecosystems use renewable solar energy, recycle nutrients, exhibit dynamic equilibrium, and involve negative feedback loops. The environmental systems we see on Earth today are those that have survived the test of time. Our industrialized civilization is young in comparison. Might we not be able to take a few lessons about sustainability from a careful look at the natural systems of our planet?

REVIEWING OBJECTIVES

You should now be able to:

Describe the nature of environmental systems

- Systems are networks of interacting components that generally involve feedback loops, show dynamic equilibrium, and result in emergent properties.
- Earth's natural systems are complex, so environmental scientists often take a holistic approach to studying environmental systems.
- Because environmental systems interact and overlap, one's delineation of systems depends on the questions in which one is interested.

Explain how plate tectonics and the rock cycle shape the landscape around us and Earth beneath our feet

- Matter is cycled within the lithosphere, and rocks transform from one type to another.
- Plate tectonics is a fundamental system that produces earthquakes and volcanoes and guides Earth's physical geography.

Define ecosystems and evaluate how living and nonliving entities interact in ecosystem-level ecology

- Ecosystems consist of all organisms and nonliving entities that occur and interact in a particular area at the same time.
- Energy flows in one direction through ecosystems, whereas matter is recycled.
- Energy is converted to biomass, and ecosystems vary in their productivity.
- Input of nutrients can boost productivity, but an excess of nutrients can alter ecosystems in ways that cause severe ecological and economic consequences.

Outline the fundamentals of landscape ecology

- Landscape ecology studies how landscape structure influences organisms.
- Landscapes consist of patches spatially arrayed in a mosaic. Organisms dependent on certain types of patches may occur in metapopulations.
- Remote sensing technology and GIS are assisting the use of landscape ecology in conservation and regional planning.

Compare and contrast how water, carbon, nitrogen, and phosphorus cycle through the environment

- Water moves widely through the environment in the hydrologic cycle.
- Most carbon is contained in sedimentary rock.
- Substantial amounts of carbon also occur in the oceans and in soil.
- Carbon flux between organisms and the atmosphere occurs via photosynthesis and respiration.
- Nitrogen is a vital nutrient for plant growth. Most nitrogen is in the atmosphere, so it must be "fixed" by specialized bacteria or lightning before plants can use it.
- Phosphorus is most abundant in sedimentary rock, with substantial amounts in soil and the oceans. Phosphorus has no appreciable atmospheric pool. It is a key nutrient for plant growth.
- Humans are causing substantial impacts to Earth's biogeochemical cycles. These impacts include shifting carbon from fossil fuel reservoirs into the atmosphere, shifting nitrogen from the atmosphere to the planet's surface, and depleting groundwater supplies, among many others.

TESTING YOUR COMPREHENSION

1. Which type of feedback loop is most common in nature, and which more commonly results from human action? How might the emergence of a positive feedback loop affect a system in homeostasis?

2. Describe how hypoxic conditions can develop in coastal marine ecosystems, such as the Gulf of St. Lawrence.

3. What is the difference between an ecosystem and a community?
4. Describe the typical movement of energy through an ecosystem. Describe the typical movement of matter through an ecosystem.
5. What role does each of the following play in the carbon cycle?
   - Cars
   - Photosynthesis
   - Oceans
   - Earth's crust
6. Contrast the function performed by nitrogen-fixing bacteria with that performed by denitrifying bacteria.
7. How has human activity altered the carbon cycle? The phosphorus cycle? The nitrogen cycle? To what environmental problems have these changes given rise?
8. What is the difference between evaporation and transpiration? Give examples of how the hydrologic cycle interacts with the carbon, phosphorus, and nitrogen cycles.
9. Name the three main types of rocks, and describe how each type may be converted to the others via the rock cycle.
10. How does plate tectonics account for mountains? For volcanoes? For earthquakes? Why do you think it took so long for scientists to discover such a fundamental environmental system as plate tectonics?

SEEKING SOLUTIONS

1. As global warming (Chapter 14) melts icecaps in the Arctic, it exposes darker-coloured surfaces—and dark surfaces absorb more sunlight and heat than light surfaces. Would you expect this to result in a feedback process? If so, which type—negative or positive? Explain your answer.
2. Consider the ecosystem(s) that surround(s) your campus. How do some of the principles from our discussion on ecosystems apply to the ecosystem(s) around your campus?
3. For a conservation biologist interested in sustaining populations of each organism listed below, why would it be helpful to take a landscape ecology perspective? Explain your answer in each case.
   - A forest-breeding warbler that suffers poor nesting success in small fragmented forest patches
   - A bighorn sheep that must move seasonally between mountains and lowlands
   - A toad that lives in upland areas but travels cross-country to breed in localized pools each spring
4. A simple change in the flux rate between just two reservoirs in a single nutrient cycle can potentially have major consequences for ecosystems and, indeed, for the globe. Explain how this can be, using one example from the carbon cycle and one example from the nitrogen cycle.
5. How do you think we might solve the problems of PAH contamination and hypoxia in the Gulf of St. Lawrence? Assess several possible solutions for each problem, your reasons for believing they might work, and the likely hurdles we might face. Explain who should be responsible for implementing solutions, and why.
6. **THINK IT THROUGH** Imagine that you are a fisher in the St. Lawrence estuary and that your income is decreasing because nutrient pollution is causing algal blooms that affect the quality of your catch. One day, your Member of Parliament comes to town, and you have a one-minute audience with her. What steps would you urge her to take in Ottawa, to try to help alleviate the nutrient pollution in the St. Lawrence and help maintain the quality of the fishery?

   Now imagine that you are a farmer in rural Quebec who has learned that the government is insisting that you use 30% less fertilizer on your crops each year. You know that in good growing years you could do without that fertilizer, and you would be glad not to have to pay for it. But in bad growing years, you need the fertilizer to ensure a harvest so that you can continue making a living. And you must apply the fertilizer each spring before you know whether it will be a good or bad year. What would you tell your Member of Parliament when she comes to town?

INTERPRETING GRAPHS AND DATA

Scientists are debating what effects global climate change (Chapter 14) may have on nutrient cycles. As soil becomes warmer, especially at far northern latitudes, nutrients in the soil should become more available to plants, stimulating plant growth. One hypothesis is that more carbon will end up stored in the soil as a result, because plants will pull carbon from the atmosphere and transfer it to the soil reservoir as they shed leaves or die.[1] Under this hypothesis, increased flux of carbon from the atmosphere to the soil would act as negative feedback counteracting climate warming, because less carbon in the atmosphere would lead to less warming.

To test whether the carbon flux actually changes in this way when nutrients are made more available in a tundra ecosystem, researchers are conducting a long-term study in Alaska.[2] For 20 years, they have added fertilizer to treatment plots while leaving control plots unfertilized. Recently, they estimated amounts of carbon by measuring biomass aboveground and belowground in both sets of plots. Aboveground biomass consists of living plant material, whereas belowground biomass consists mostly of nonliving organic material stored in the soil and not yet decomposed. Some of the research team's results are presented in the graph at right.

1. Calculate the sizes of the aboveground, belowground, and total carbon pools (in g C/m^2) for the Control and Fertilized treatment groups.
2. What do the aboveground data indicate about the effect of fertilizer on plant growth? What do the belowground data indicate about the effect of fertilizer on organic material stored in the soil?
3. Do the data support the hypothesis that the net effect of increased nutrient availability will be to remove carbon from the atmosphere and store it in the soil?

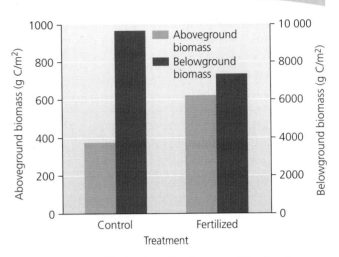

Effects of 20 years of fertilization (10 g nitrogen and 5 g phosphorus per square metre per year) on carbon pools in tundra near Toolik Lake, Alaska. Differences are statistically significant for the aboveground, belowground, and total carbon pools.

Using FIGURE 5.16 as a reference, can you suggest a different hypothesis that might explain the data better? Based on this data, would you predict that the warming of tundra soil will decrease the atmospheric concentration of CO_2 and act as negative feedback to climate change, or increase it and act as positive feedback?

Endnotes

1. Hobbie, S. E., et al. (2002) A synthesis: The role of nutrients as constraints on carbon balances in boreal and arctic regions. *Plant and Soil*, Vol. 242, pp. 163–170.
2. Mack, M. C., et al. (2004) Ecosystem carbon storage in arctic tundra reduced by long-term nitrogen fertilization. *Nature*, Vol. 431, pp. 440–443.

CALCULATING FOOTPRINTS

A common dream of young people is to own a home in the suburbs, surrounded by a weed-free green lawn. But lawns must be managed, and traditional lawn care requires the application of fertilizers and pesticides, both of which raise environmental concerns. More than 140 municipalities in Canada, including the entire province of Quebec, have now placed restrictions on the cosmetic use of synthetic lawn pesticides as a result of health and environmental concerns. In the United States, where lawns are particularly popular, it has recently been estimated on the basis of satellite data that there are more than 10 million hectares of lawn grass, making it the nation's largest single crop!

Assuming that all the populations indicated in the following table have lawns and will fertilize them at a typical fertilizer application rate of 60 kg of nitrogen per hectare, calculate the total amount of nitrogen that will be applied to their lawns. When estimating the number of lawns, assume that the typical household includes three people.

1. Where does all of this nitrogen come from? Where does it go?
2. What other environmental impacts are caused by fertilizer production, transport, and application?
3. What steps could you take to reduce nitrogen pollution?

Fertilizer Application	Number of Lawns	Amount of Nitrogen Applied (kg)
To your 0.25 ha lawn	1	15
To the lawns of your classmates		
To all the lawns in your home town		
To all the lawns in your province		
To all the lawns in Canada	Approx. 5 000 000	

TAKE IT FURTHER

Go to www.myenvironmentplace.ca where you will find

- Suggested answers to end-of-chapter questions
- Quizzes, animations, and flashcards to help you study
- *Research Navigator*™ database of credible and reliable sources to assist you with your research projects
- Tutorials to help you interpret graphs
- Current news articles that link the topics that you study to case studies from your region and around the world

- **ECO Occupational Profiles**: If you found this chapter especially interesting, you might want to learn more about the following jobs by visiting the Occupational Profiles website of the Environmental Careers Organization. Go to www.eco.ca and check out the following careers:
  - Botanist
  - Environmental geologist
  - Environmental geophysicist
  - Geographer
  - Zoologist

CHAPTER ENDNOTES

1. Environment Canada *EnviroZine*, Great Lakes–St. Lawrence Basin: A Freshwater Giant, 2006, www.ec.gc.ca/EnviroZine/english/issues/61/feature2_e.cfm.
2. This piece is based on information from IUCN Red List of Endangered Species, 2007, www.iucnredlist.org/search/details.php/6335/summ; Université de Montréal Faculté de Médecine Vétérinaire, www.medvet.umontreal.ca/pathologie_microbiologie/beluga/anglais/default_ang.asp; Minister of the Environment, Canada, *The Contribution of Agriculture to the Deterioration of the St. Lawrence River*, 1999, www.slv2000.qc.ec.gc.ca/communiques/phase3/enjeu_agricoles_a.pdf; Environment Canada EnviroZine, Great Lakes–St. Lawrence Basin: A Freshwater Giant, 2006, www.ec.gc.ca/EnviroZine/english/issues/61/feature2_e.cfm; and Environment Canada St. Lawrence Centre, www.qc.ec.gc.ca/csl/inf/inf069_e.html.

3. Based on information from DesGranges, J.-L. 2000. Protecting the biodiversity of the St. Lawrence: Conservation plan. In DesGranges, J.-L. and J.-P. Ducruc (eds.). *Biodiversity Portrait of the St. Lawrence*. Canadian Wildlife Service, Environment Canada, Quebec Region and the Direction du patrimoine écologique, Ministère de l'Environnement du Québec. Internet version, www.qc.ec.gc.ca/faune/biodiv.

4. Robert Bateman on Outdoor Education, www.batemanideas.com/outdooreducation.html.

5. Robert Bateman on Education, www.batemanideas.com/education.html.

6. Environmental Defence, Toxic Nation Reports, *Robert Bateman*, www.toxicnation.ca/toxicnation-studies/pollution-in-adults/Robert.

Even in Nunavut, Canada's far north, the formerly pristine landscape shows the impact of human activity.

PART TWO

THE HUMAN ELEMENT IN THE NATURAL SYSTEM

Human Population 6

This crowded street is in Guangzhou, one of China's largest cities.

Upon successfully completing this chapter, you will be able to

- Assess the scope and historical patterns of human population growth
- Evaluate how human population, affluence, and technology affect the environment
- Explain and apply the fundamental concepts of demography
- Outline the concept of demographic transition
- Describe how wealth and poverty, the status of women, and family planning programs and policies affect population growth
- Characterize the dimensions of the HIV/AIDS pandemic and its impacts on population, the environment, and sustainability

This billboard in Chengdu promotes China's one-child policy.

CENTRAL CASE:
CHINA'S ONE-CHILD POLICY

"As you improve health in a society, population growth goes down. You know, I thought it was ... before I learned about it, I thought it was paradoxical."
—BILL GATES, CHAIR, MICROSOFT CORP.

"Population growth is analogous to a plague of locusts. What we have on this earth today is a plague of people."
—TED TURNER, MEDIA MAGNATE AND SUPPORTER OF THE UNITED NATIONS POPULATION FUND

"There is no population problem."
—SHELDON RICHMAN, SENIOR EDITOR, CATO INSTITUTE

The People's Republic of China is the world's most populous nation, home to one-fifth of the 6.7[1] billion people living on Earth as of 2008.

The first significant increases in China's population in the past 2000 years of the nation's history resulted from enhanced agricultural production and a powerful government during the Qing (Manchu) Dynasty in the 1800s. Population growth began to outstrip food supplies by the mid-1850s, and quality of life for the average Chinese peasant began to decline. From the mid-1800s (an era of increased European intervention in China) until 1949, China's population grew very slowly, at about 0.3% per year. This slow population growth was due, in part, to food shortages and political instability, which caused a decline in birth rates. Population growth rates rose again following the establishment of the People's Republic and have declined once again since the establishment of the one-child policy.

When Mao Zedong founded the country's current regime in 1949, roughly 540 million people lived in a mostly rural, war-torn, impoverished nation. Mao believed population growth was desirable, and under his leadership China grew and changed. By 1970, China's population had grown to approximately 790 million people. At that time, the average Chinese woman gave birth to 5.8 children in her lifetime.

Unfortunately, the country's burgeoning population and its industrial and agricultural development were eroding the nation's soils, depleting its water, levelling its forests, and polluting its air. Chinese leaders realized that the nation might not be able to feed its people if their numbers grew much larger. They saw that continued population growth could exhaust resources and threaten the stability and economic progress of Chinese society. The government decided to institute a population control program that prohibited most Chinese couples from having more than one child.

The program began with education and outreach efforts encouraging people to marry later and have fewer children. Along with these efforts, the Chinese government increased the accessibility of contraceptives and abortion. By 1975, China's annual population growth rate had dropped from 2.8% to 1.8%. To further decrease the birth rate, in 1979 the government took the more drastic step of instituting a system of rewards and punishments to enforce a one-child limit. One-child families received better access to schools, medical care, housing, and government jobs, and mothers with only one child were given longer maternity leaves. Families with more than one child, meanwhile, were subjected to social scorn and ridicule, employment discrimination, and monetary fines. In some cases, the fines exceeded half the offending couple's annual income.

Beginning in 1984, the one-child policy was loosened, strengthened, and then loosened again as government leaders sought to maximize population control while minimizing public opposition. Today the one-child program applies mostly to urban couples, whereas many rural farmers and ethnic minorities are exempt.

In enforcing these policies, China has, in effect, been conducting one of the largest and most controversial social experiments in history. In purely quantitative terms, the experiment has been a major success; the nation's growth rate is now down to 0.6%, making it easier for the country to deal with its many social, economic, and environmental challenges.

However, China's population control policies have also produced unintended consequences, such as widespread killing of female infants, an unbalanced sex ratio, and a black market trade in teenaged girls. It is expected to lead to further problems in the future, including an ageing population and shrinking workforce. Moreover, the policies have elicited intense criticism from those who oppose government intrusion into personal reproductive choices.

As other nations become more and more crowded, might their governments also feel forced to turn to drastic policies that restrict individual freedoms? In this chapter, we examine human population dynamics worldwide, consider their causes, and assess their consequences for the environment and our society.

Human Population: Approaching 7 Billion

While China works to slow its population growth and speed its economic growth, populations continue to rise in most nations of the world. Most of this growth is occurring in poverty-stricken nations that are ill equipped to handle it. India (**FIGURE 6.1**) is currently on course to surpass China as the world's most populous nation. Although the *rate* of global growth is slowing, we are still increasing in absolute numbers, and today more than 6.7 billion of us inhabit the planet.

Just how much is 6.7 billion? We often have trouble conceptualizing huge numbers like a billion; 1 billion is 1000 times greater than 1 million. If you were to count once each second without ever sleeping, it would take more than 30 years to reach 1 billion. To travel 1 billion kilometres, you would have to drive the entire length of the Trans-Canada Highway 128 000 times, or fly to the Moon 2500 times.

The human population is growing nearly as fast as ever

As you learned in Chapter 1, the human population has been growing at a tremendous rate. The population has doubled since 1966 and is growing by roughly 80 million people annually (nearly 2.6 people are added—that is,

FIGURE 6.1
Population growth in developing countries is leading our global population toward 7 billion and possibly beyond. India is on course to surpass China soon as the world's most populous nation. In this photo, Indian women wait in line for immunizations for their babies.

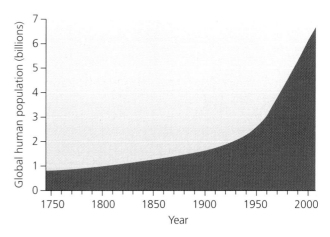

FIGURE 6.2
The global human population has grown exponentially, rising from less than 1 billion in 1800 to approximately 6.7 billion today. Data from U.S Bureau of the Census.

births minus deaths—every *second*). This is the equivalent of adding two and a half times the population of Canada to the world each year. It took until after 1800—most of human history—for our population to reach 1 billion. Yet we reached 2 billion by 1927, and 3 billion in less than 30 more years, in 1960. Our population added its next billion in just 14 years (1974), its next billion in a mere 13 years (1987), and the most recent billion in another 12 years (**FIGURE 6.2**). Think about when you were born and how many people have been added to the planet since that time. No previous generations have ever lived amid so many other people.

What accounts for such unprecedented growth? We saw in Chapter 3 how exponential growth—the increase in a quantity by a fixed percentage per unit time—accelerates the absolute increase of population size over time, just as compound interest accrues in a savings account. The reason, you will recall, is that a given percentage of a large number is a greater quantity than the same percentage of a small number. Thus, even if the growth *rate* remains steady, population *size* will increase by greater increments with each successive generation.

In fact, the world's population growth rate has not remained steady. During much of the twentieth century the growth rate actually rose from year to year. It peaked at 2.1% per year during the 1960s and has declined to 1.2% per year since then. Although 1.2% may sound small, exponential growth endows small numbers with large consequences. For instance, a hypothetical population starting with one man and one woman, growing at 1.2% per year, gives rise to a population of 2939 after 40 generations and 112 695 after 60 generations. In today's world, rates of annual growth vary greatly from region to region. **FIGURE 6.3** maps this variation.

At a 2.1% annual growth rate, a population doubles in only 33 years. For low rates of increase, we can estimate doubling times with a handy rule of thumb. Just take the number 70, and divide it by the annual percentage growth rate: 70 ÷ 2.1 = 33.3. Had China not instituted its one-child policy—that is, had its growth rate remained unchecked at 2.8%—it would have taken only 25 years to double in size (70 ÷ 2.8 = 25). Had population growth continued at this rate, China's population would have exceeded 2 billion people by 2004; based on current projections, that milestone will not be reached, as China's population is expected to stabilize at around 1.4 billion.

Perspectives on human population have changed over time

At the outset of the Industrial Revolution in England of the 1700s, population growth was regarded as a good thing. For parents, a high birth rate meant more children to support them in old age. For society, it meant a greater pool of labour for factory work.

Population growth and environmental scarcity

British economist Thomas Malthus (1766–1834) had a different opinion. Malthus claimed that unless population growth was limited by laws or other social controls, the number of people would outgrow the available food supply until starvation, war, or disease arose and reduced the population (**FIGURE 6.4**). Malthus's most influential work, *An Essay on the Principle of Population*, published in 1798, argued that a growing population would eventually be checked either by limits on births or increases in

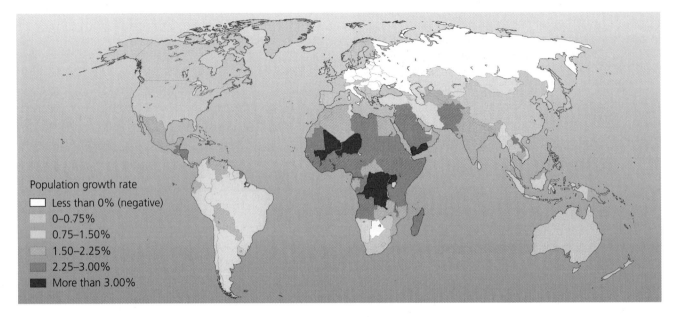

FIGURE 6.3
Population growth rates vary greatly from place to place. Population is growing fastest in poorer nations of the tropics and subtropics but is now beginning to decrease in some northern industrialized nations. Shown are natural rates of population change as of 2006. Data from the Population Reference Bureau, *2006 World Population Data Sheet*.

(a) 18th-century London, England

(b) Thomas Malthus

FIGURE 6.4
The England of Thomas Malthus's era (1766–1834), shown in this engraving **(a)**, favoured population growth as society industrialized. Malthus **(b)** argued that the pressure of population growth on the availability of resources could lead to disaster.

deaths. If limits on births (such as abstinence and contraception) were not implemented soon enough, Malthus wrote, deaths would increase.

Malthus's thinking was specifically shaped by the rapid urbanization and industrialization he witnessed during the early years of the Industrial Revolution. More recently, biologist Paul Ehrlich of Stanford University has been called a "neo-Malthusian" because—like Malthus— he warned that population growth would have disastrous effects on the environment and human welfare.

In his 1968 book, *The Population Bomb*, Ehrlich predicted that the rapidly increasing human population would unleash widespread famine and conflict that would consume civilization by the end of the twentieth century. Ehrlich and other neo-Malthusians have argued that population is growing much faster than our ability to produce and distribute food, and that population control is the only way to prevent massive starvation, environmental degradation, and civil strife.

Is population growth really a "problem" today?

The world's ongoing population growth has been made possible by technological innovations, improved sanitation, better medical care, increased agricultural output, and other factors that have led to a decline in death rates, particularly a drop in rates of infant mortality. Birth rates have not declined as much, so births have outpaced deaths for many years now. Thus, the population "problem" actually arises from a very good thing—our ability to keep more people alive longer.

The mainstream view in Malthus's day held that population increase was a good thing, and today there are still many people who argue that population growth poses no problems. Even though human population has nearly quadrupled in the past 100 years—the fastest it has ever grown—Malthusian and neo-Malthusian predictions have not materialized on a devastating scale. This is due, in part, to enormous increases in crop yields and advances in agricultural technology associated with the Green Revolution. Increasing material prosperity has also helped bring down birth rates—something Malthus and Ehrlich did not foresee.

Some cornucopian thinkers believe that resource depletion caused by population increase is not a problem if new resources can be found to replace depleted resources. Libertarian writer Sheldon Richman expressed this view as follows:

> *The idea of carrying capacity doesn't apply to the human world because humans aren't passive with respect to their environment. Human beings create resources. We find potential stuff and human intelligence turns it into resources. The computer revolution is based on sand; human intelligence turned that common stuff into the main component [silicon] of an amazing technology.*

In contrast to the idea that humankind will always be able to save itself with a "technological fix," environmental scientists recognize that few resources are actually "created" by humans, and that not all resources can be replaced or reinvented once they have been depleted. For example, once species have gone extinct, we cannot replicate their exact function in ecosystems, or know what medicines or other practical applications we might have obtained from them, or regain the educational and esthetic value of observing them. Another irreplaceable resource is land; we cannot expand Earth like a balloon to increase its surface area.

Even if resource substitution could enable population growth to continue indefinitely, could we maintain the *quality* of life that we desire for our descendants and ourselves? Surely some of today's resources are easier or cheaper to use, and less environmentally destructive to harvest or mine, than the resources that might replace them. Unless resource availability keeps pace with population growth, the average person in the future will have less space in which to live, less food to eat, and less material wealth than the average person does today. Thus population increases are indeed a problem if they create stress on resources, social systems, or the natural environment, such that our quality of life declines.

In today's world, population growth is much more strongly correlated with poverty than with wealth. In spite of this, many governments have found it difficult to let go of the notion that population growth increases a nation's economic, political, or military strength. Many national governments still offer financial and social incentives to encourage their citizens to produce more children. Governments of countries currently experiencing population declines (such as many in Europe) feel especially uneasy. According to the Population Reference Bureau, more than three of every five European national governments now take the view that their birth rates are too low, and none states that its rate is too high. However, outside Europe, 56% of national governments feel their birth rates are too high, and only 8% feel they are too low.

Researchers in the Malthusian tradition are still making important contributions to our understanding of population and the environment today. Research has shown that environmental degradation and scarcity, particularly in situations of overcrowding in sensitive environments, can lead to migrantism, refugeeism, and even armed conflict. For example, political scientist Thomas Homer-Dixon of the University of Toronto investigated the linkages between agricultural land scarcity and ethnic tensions in initiating the violent armed conflicts that occurred in Rwanda in the mid-1990s. In other active conflicts we can see that global population growth and resulting environmental scarcity have played a central role in causing famine, disease, and social and political conflict around the world.

The dire predictions of Malthus, Ehrlich, and other neo-Malthusians were not accurate; however, the idea that a "technological fix" will always emerge in time to rescue humans from problems brought on by population growth is equally unsatisfactory. Our current understanding of the limitations of the resources of this planet suggests that the real answer lies somewhere between these extremes.

Population is one of several factors that affect the environment

The extent to which population increase can be considered a problem involves more than just numbers of people. One widely used formula gives us a handy way to think about factors that affect the environment. Nicknamed the **IPAT model**, it is a variation of a formula proposed in 1974 by Paul Ehrlich and John Holdren, a professor of environmental policy at Harvard University. The IPAT model represents how our total impact (I) on the environment results from the interaction among population (P), affluence (A), and technology (T):

$$I = P \times A \times T$$

Increased population intensifies impact on the environment as more individuals take up space, use natural resources, and generate waste. Increased affluence magnifies environmental impact through the greater per capita resource consumption that generally has accompanied enhanced wealth. Changes in technology may either decrease or increase human impact on the environment. Technology that enhances our abilities to exploit minerals, fossil fuels, old-growth forests, or ocean fisheries generally increases impact, but technology to reduce smokestack emissions, harness renewable energy, or improve manufacturing efficiency can decrease impact.

We might also add a sensitivity factor (S) to the equation to denote how sensitive a given environment is to human pressures:

$$I = P \times A \times T \times S$$

For instance, the arid lands of western China are more sensitive to human disturbance than the moist regions of southeastern China. Plants grow more slowly in the arid west, making deforestation and soil degradation more likely. Thus, adding an additional person to western China should have more environmental impact than adding one to southeastern China (all other conditions being equal).

Various population researchers have refined the IPAT equation by adding terms for the effects of social institutions, such as education, laws, and their enforcement; stable and cohesive societies; and ethical standards that promote environmental well-being. Factors like these affect how population, affluence, and technology translate into environmental impact.

Impact can be thought of in various ways, but it generally boils down to either pollution or resource consumption. Pollution became a problem in the modern world once our population grew large enough that we produced great quantities of waste. The depletion of resources by larger and hungrier populations has been a focus of scientists and philosophers since before Malthus's time. Recall how the people of Rapa Nui brought down their own civilization by depleting their most important limited resource, trees. History offers other cases in which resource depletion helped end civilizations, from the Mayans to the Mesopotamians. Some environmental scientists have predicted similar problems for our global society in the near future if we do not manage to embark on a path toward sustainability (**FIGURE 6.5**).

As discussed above, Malthus and the "neo-Malthusians" have not yet seen their direst predictions come true. The reason is that we have developed technology—the T in the IPAT equation—time and again to alleviate our strain on resources and allow us to further

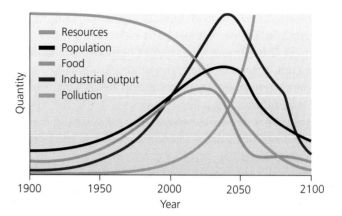

(a) Projection based on status quo policies

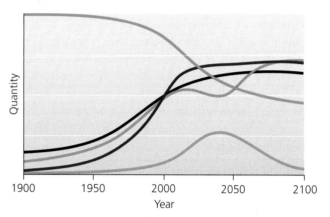

(b) Projection based on policies for sustainability

FIGURE 6.5
Environmental scientist Donella Meadows and her colleagues used computer simulations to generate a series of projections of trends in human population, resource availability, food production, industrial output, and pollution. The projections are based on data from the past century and current scientific understanding of the environment's biophysical limits. Shown in **(a)** is their projection for a world in which "society proceeds in a traditional manner without any major deviation from the policies pursued during most of the twentieth century." In this projection, population and production increase until declining nonrenewable resources make further growth impossible, causing population and production to decline rather suddenly. The researchers also ran their simulations with different parameters to examine possible alternative futures. Under a scenario with policies aimed at sustainability **(b)**, population levelled off at 8 billion, production and resource availability levelled off at medium-high levels, and pollution declined to low levels. Data from Meadows, D., et al. 2004. *Limits to Growth: The 30-Year Update*. White River Junction, VT: Chelsea Green Publishing.

expand our population. For instance, we have employed technological advances to increase global agricultural production faster than our population has risen.

Modern-day China shows how all elements of the IPAT formula can combine to cause tremendous environmental impact in very little time. While millions of Chinese are increasing their material wealth and consumption of resources, the country is battling unprecedented environmental challenges brought about by its extremely rapid economic development. Intensive agriculture has expanded westward out of the country's historic moist rice-growing areas, causing farmland to erode and literally blow away, much like the Dust Bowl tragedy that befell the agricultural heartland of North America in the 1930s. China has overpumped many of its aquifers and has drawn so much water for irrigation from the Huang He (Yellow River) that the once-mighty waterway now dries up in many stretches. Although China has been reducing its air pollution from industry and charcoal-burning homes, the country faces new urban pollution and congestion threats from rapidly increasing numbers of automobiles.

As the world's developing countries try to attain the level of material prosperity that industrialized nations enjoy, China is a window on what much of the rest of the world could soon become.

Demography

It is a fallacy to think of people as being somehow outside nature. Humans exist within their environment as one species out of many. As such, the principles of population ecology we outlined in Chapter 3 that apply to toads, frogs, and passenger pigeons apply to humans as well. The application of population ecology principles to the study of statistical change in human populations is the focus of the social science of **demography**.

The environment has a carrying capacity for humans

Environmental factors set limits on our population growth, and the environment has a carrying capacity for our species, as it does for every other. We happen to be a particularly successful organism, however—one that has repeatedly increased the carrying capacity of the environment by developing technology to overcome natural limits on population growth. As mentioned in Chapter 1, four significant periods of societal change appear to have fundamentally altered the human relationship with the environment and increased the carrying capacity, triggering remarkable increases in population size (**FIGURE 6.6**).

The first transition happened in the *paleolithic period* (or *Old Stone Age*), when early humans gained control of fire (as much as 1.5 million years ago) and began to shape and use stones (as much as 2.5 million years ago) as tools with which to modify their environment. We can speculate that this transition made life so much easier and the environment so much more manageable for our ancestors that their population grew substantially, although we have little direct evidence about world population dating from that period.

The second major change was the transition from a nomadic hunter–gatherer lifestyle to a settled agricultural way of life. This change began to occur around 10 000 to 12 000 years ago and is known as the **Agricultural Revolution**, in what is known as the *neolithic* (or *New Stone Age*) *period*. This agriculture-based lifestyle was much more intensive and manipulative in the production of resources from the land. As people began to grow their own crops, raise domestic animals, and live settled, sedentary lives in villages, they found it easier to meet their nutritional needs. As a result, they began to live longer and to produce more children who survived to adulthood. The Agricultural Revolution initiated a permanent change in the way humans relate to the natural environment.

The third major societal change, known as the **Industrial Revolution**, began in the mid-1700s. It

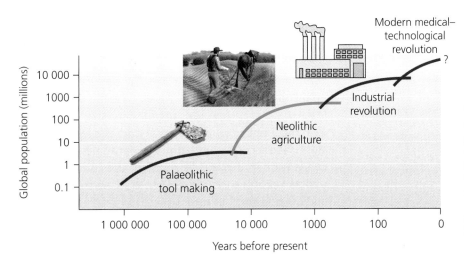

FIGURE 6.6
Tool making, agriculture, and industrialization each allowed our species to increase the global carrying capacity. The logarithmic scale of the axes makes it easier to visualize this pattern. We are currently in the midst of a fourth transition, involving the globalization of modern medical–technological advances; its impacts on population and the environment are as yet unknown. Data from Goudie, A. 2000. *The Human Impact*. Cambridge, MA: MIT Press.

entailed a shift from rural life, animal-powered agriculture, and manufacturing by craftsmen, to an urban society powered by fossil fuels (Chapter 15). The Industrial Revolution introduced improvements in sanitation and medical technology. Another very important aspect was the impact on agriculture and animal husbandry. Agricultural production was greatly enhanced during the Industrial Revolution by the introduction of fossil fuel powered equipment, steam engines, and synthetic fertilizers, along with advances in plant and animal breeding.

We are currently in the midst of a fourth major transition, involving the globalization of modern medical and technological advancements. The **Medical–Technological Revolution** is marked by developments in medicine, sanitation, and pharmaceuticals; the explosion of communication technologies; and the shift to modern agricultural practices known as the *Green Revolution* that have collectively allowed more people to live longer, healthier lives. This transition is still in progress, and the long-term implications for the human population, individual health, and the environment are unknown. Perhaps this will also be a period during which human society makes the transition to more sustainable, renewable energy sources and away from dependence on fossil fuels.

Environmental scientists who have tried to quantify the human carrying capacity of this planet have come up with wildly differing estimates. Estimates range from 1 billion to 2 billion people living prosperously in a healthy environment to 33 billion living in extreme poverty in a degraded world of intensive cultivation without natural areas. As our population climbs toward 7 billion and beyond, we may yet continue to find ways to increase carrying capacity. Given our knowledge of population ecology, however, we have no reason to presume that human numbers can go on growing indefinitely. Indeed, as we have seen (see **FIGURE 3.16D**), populations that exceed their carrying capacity can crash.

Demography is the study of human population

The field of demography developed along with and partly preceded population ecology, and the disciplines have influenced and borrowed from one another. Demographic data help us understand how differences in population characteristics and related phenomena (for instance, decisions about reproduction) affect human communities and their environments. **Demographers** study population size, density, distribution, age structure, sex ratio,

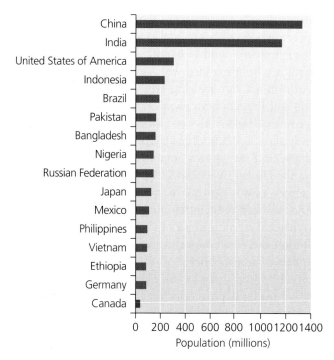

FIGURE 6.7
The world's nations range in population from several thousand (on some South Pacific islands) to China's 1.3 billion. Shown here are the 2006 populations for the world's most populous 15 countries, and Canada. Land area can also play an important role in determining the density and environmental impacts of population; for example, the current population of almost 33 million people in Canada contrasts with a population of almost 85 million in Vietnam, but Canada's land mass is more than 30 times as large. Data from the United Nations Population Division, 2007, *World Population Prospects: The 2006 Revision*.

and rates of birth, death, immigration, and emigration of humans, just as population ecologists study these characteristics in other organisms. Each of these characteristics is useful for predicting population dynamics and potential environmental impacts.

Population size The global human population of 6.7 billion comprises more than 200 nations, with populations ranging from China's 1.3 billion and India's 1.2 billion (**FIGURE 6.7**) to a number of island nations with populations below 100 000. The size that our global population will eventually reach remains to be seen (**FIGURE 6.8**). However, **population size** alone—the absolute number of individuals—does not tell the whole story. Rather, a population's environmental impact depends on its density, distribution, and composition, as well as on affluence, technology, level of consumption, and other factors outlined earlier.

Population density and distribution People are distributed very unevenly over the globe. In ecological

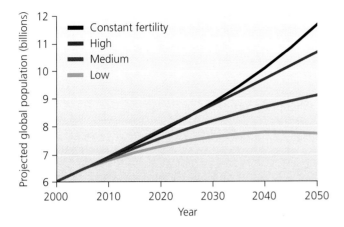

FIGURE 6.8
The United Nations predicts trajectories of world population growth, presenting its estimates in several scenarios based on different assumptions of fertility rates. In this 2006 projection, population was estimated to reach almost 12 billion in the year 2050 if fertility rates remained constant at 2006 levels (top line on graph). However, U.N. demographers expect fertility rates to continue falling, so they arrived at a best guess (*medium* scenario) of 9.2 billion for the human population in 2050. In the *high* scenario, if women on average have 0.5 children more than in the medium scenario, population will reach 10.8 billion in 2050. In the *low* scenario, if women have 0.5 children less than in the medium scenario, the world will contain 7.8 billion people in 2050. Data from United Nations Population Division, 2007, *World Population Prospects: The 2006 Revision*.

terms, our distribution is clumped at all spatial scales. At the global scale (**FIGURE 6.9**), **population density**—the number of people per unit of land area—is particularly high in regions with temperate, subtropical, and tropical climates, such as China, Europe, Mexico, southern Africa, and India. Population density is low in regions with extreme-climate biomes, such as desert, deep rainforest, and tundra. Dense along seacoasts and rivers, human population is less dense at locations far from water. At intermediate scales, we cluster together in cities and suburbs and are spread more sparsely across rural areas. At small scales, we cluster in certain neighbourhoods and in individual households.

This uneven distribution means that certain areas bear far more environmental impact than others. Just as the Huang He has experienced intense pressure from millions of Chinese farmers, the world's other major rivers—from the Nile to the Danube, the Ganges, and the Mississippi—have all received more than their share of human impact. Urbanization (Chapter 20) entails the packaging and transport of goods, intensive fossil fuel consumption, and hotspots of pollution. However, the concentration of people in cities increases efficiency and economies of scale, and relieves pressure on ecosystems in less-populated areas by releasing some of them from some human development.

At the same time, areas with low population density are often vulnerable to environmental impacts, because the reason they have low populations in the first place is that they are sensitive and cannot support many people (a high S value in our revised IPAT model). Deserts, for instance, are easily affected by development that commandeers a substantial share of available water. Grasslands can be turned to deserts if they are farmed too intensively, as has happened across vast stretches of

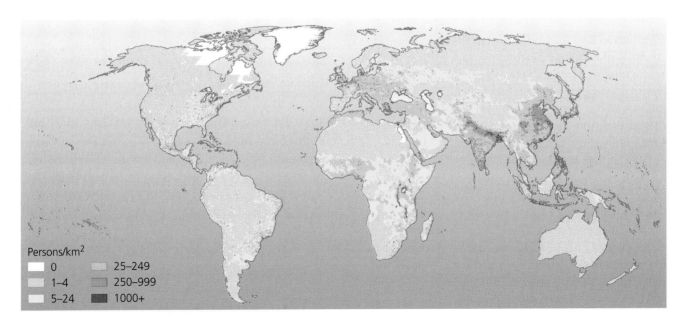

FIGURE 6.9
Human population density varies tremendously from one region to another. Arctic and desert regions have the lowest population densities, whereas areas of India, Bangladesh, and eastern China have the densest populations. Data are for 2000, from the Center for International Earth Science Information Network (CIESIN), Columbia University; and Centro Internacional de Agricultura Tropical (CIAT), 2004.

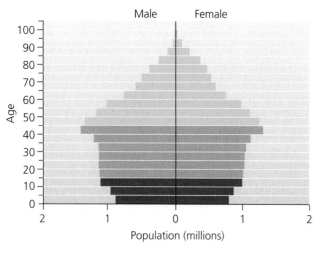
(a) Age pyramid of Canada in 2005

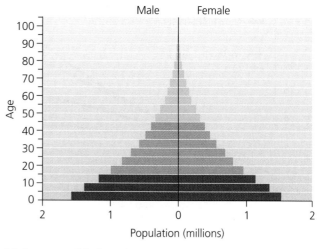
(b) Age pyramid of Madagascar in 2005

FIGURE 6.10
Canada **(a)** shows a balanced age structure, with relatively even numbers of individuals in various age classes. Madagascar **(b)** shows an age distribution heavily weighted toward young people. Madagascar's population growth rate is nine times that of Canada. The post-Second World War "baby boom" is visible as a "bump" in the age pyramid for Canada, between the ages of 40 and 50. In future years the nation will experience an ageing population, as baby boomers grow older. Go to **GRAPHit!** at www.myenvironmentplace.ca. The Statistics Canada website features animated population pyramids for Canada from 1900 through 2006. With these animations you can track the baby boom "bump" from its beginning after the Second World War to the present day, as baby boomers age.

the Sahel region bordering Africa's Sahara Desert, in the Middle East, and in parts of China. The Arctic tundra, in the northern circumpolar region, is another environment that is highly sensitive to environmental change and human impacts. For example, a disturbance of vegetation—something as simple as a set of car tracks—can cause deep melting of permafrost and collapse of soil, the scars of which may last for years or even decades.

Age structure Data on the age structure or age distribution of human populations are especially valuable to demographers trying to predict future dynamics of populations. As we saw in Chapter 3, large proportions of individuals in young age groups portend a great deal of reproduction and, thus, rapid population growth. Examine the age structure diagrams, or age pyramids, for Canada and Madagascar (**FIGURE 6.10**). Not surprisingly, Madagascar has the greater population growth rate. In fact, its annual growth rate, 2.7%, is nine times that of Canada, 0.3%.

By causing dramatic reductions in the number of children born since 1970, China virtually guaranteed that its population age structure would change (**FIGURE 6.11**). In 1995 the median age in China was 27; by 2030 it will be 39. In 1997 there were 125 children under age five for every 100 people aged 65 or older in China, but by 2030 there will be only 32. The number of people older than 65 will rise from 100 million in 2005 to 236 million in 2030.

This dramatic shift in age structure will challenge China's economy, health care systems, families, and military forces because fewer working-age people will be available to support social programs that assist the increasing number of older people. However, the shift in age structure also reduces the proportion of dependent children. The reduced number of young adults may mean

weighing the issues 6-1
CHINA'S REPRODUCTIVE POLICY

Consider the benefits as well as the problems associated with a reproductive policy, such as China's. Do you think a government should be able to enforce strict penalties for citizens who fail to abide by such a policy? If you disagree with China's policy, what alternatives can you suggest for dealing with the resource demands of a quickly growing population?

This pattern of ageing in the population is occurring in many countries, including Canada (see **FIGURE 6.10A**). Older populations will present new challenges for many nations, as increasing numbers of older people require the care and financial assistance of relatively fewer working-age citizens.

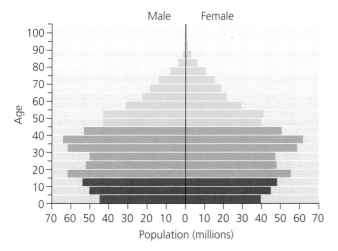

(a) Age pyramid of China in 2005

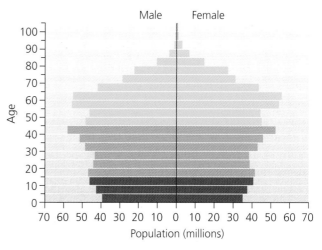

(b) Projected age pyramid of China in 2030

FIGURE 6.11
As China's population ages, older people will outnumber the young. Age pyramids show the predicted greying of the Chinese population between 2005 **(a)** and 2030 **(b)**. Today's children may, as working-age adults **(c)**, face pressures to support greater numbers of older citizens than has any previous generation. Data from U.N. Population Division.

(c) Young female factory workers in Hong Kong

a decrease in the crime rate. Moreover, older people are often productive members of society, contributing volunteer activities and services to their children and grandchildren. Clearly, in terms of both benefits and drawbacks, life in China will continue to be profoundly affected by the particular approach its government has taken to population control.

Sex ratios The ratio of males to females also can affect population dynamics. Imagine two islands, one populated by 99 men and 1 woman and the other by 50 men and 50 women. Where would we be likely to see the greatest population increase over time? Of course, the island with an equal number of men and women would have a greater number of potential mothers and thus a greater potential for population growth.

The naturally occurring sex ratio in human populations at birth features a slight preponderance of males; for every 100 female infants born, 105 to 106 male infants are born. This phenomenon may be an evolutionary adaptation to the fact that males are slightly more prone to death during any given year of life. It usually ensures that the ratio of men to women is approximately equal at the time people reach reproductive age; women then begin to predominate as the population ages, generally leading to a ratio of males to females that is slightly less than one-to-one in the population as a whole. Thus, a slightly uneven sex ratio at birth may be beneficial. However, a greatly distorted ratio can lead to problems.

In recent years, demographers have noted an unsettling trend in China, also observed in other nations where there is a strong traditional preference for boys: The ratio of newborn boys to girls has become skewed. In the 2000 census, 120 boys were reported born for every 100 girls. Some provinces reported sex ratios as high as 138 boys for every 100 girls. The overall ratio of males per 100 females for the Chinese population as a whole is currently around 107, but that will change as the younger, male-dominated portion of the population ages, possibly ballooning to 135. The leading hypothesis for these unusual sex ratios is that many parents, having learned the sex of their fetuses by ultrasound, are selectively aborting female fetuses.

Traditionally, Chinese culture has valued sons because they can carry on the family name, assist with farm labour in rural areas, and care for ageing parents. Daughters, in contrast, will most likely marry and leave their parents, as the culture dictates. As a result, they will not provide the same benefits to their parents as will sons. Sociologists hold that this cultural gender preference, combined with the government's one-child policy, has led some couples to abort female fetuses or to abandon or kill female infants. The Chinese government reinforced this gender discrimination when in 1984 it exempted rural peasants from the one-child policy if their first child was a girl, but not if the first child was a boy.

China is, of course, not the only nation in the world to experience the phenomenon of a skewed sex ratio. According to the U.N. Population Division, the United Arab Emirates and Qatar both have population sex ratios of more than 200, meaning that two boys survive to adulthood for every one girl, and Oman, Bahrain, and Kuwait are all more than 125.[2]

The unbalanced sex ratio in China and elsewhere may have the effect of further lowering population growth rates, with the ageing of today's children under 15. However, it has already proven tragic for some of the "missing girls." It is also beginning to have the undesirable social consequence of leaving many Chinese men single. This, in turn, has resulted in a grim new phenomenon: In parts of rural China, teenaged girls are being kidnapped and sold to families in other parts of the country as brides for single men.

Population growth depends on rates of birth, death, immigration, and emigration

Rates of birth, death, immigration, and emigration help determine whether a human population grows, shrinks, or remains stable. The formula for measuring population growth that we used in Chapter 3 also pertains to humans: Birth and immigration add individuals to a population, whereas death and emigration remove individuals. As discussed in Chapter 3, it is convenient to express birth and death rates as the number of births and deaths per 1000 individuals for a given period—the **crude birth rate** and **crude death rate**.

Technological advances have led to a dramatic decline in human death rates, widening the gap between crude birth rates and crude death rates and resulting in the global human population expansion. Just as individuals of different ages have different abilities to reproduce, individuals of different ages show different probabilities of dying. For instance, people are more likely to die at old ages than young ages; if you were to follow 1000 10-year-olds and 1000 80-year-olds for a year, you would find that at year's end more 80-year-olds had died than 10-year-olds.

In today's ever-more-crowded world, immigration and emigration are playing increasingly large roles. Refugees, people forced to flee their home country or region, have become more numerous in recent decades as a result of war, civil strife, and environmental degradation. The United Nations puts the number of refugees who flee to escape poor environmental conditions in the millions per year. It is also widely acknowledged that environmental degradation and resource shortages often contribute to other causes of refugeeism, including internal and international conflicts.

FIGURE 6.12
The flight of refugees from Rwanda into the Democratic Republic of Congo in 1994 following the Rwandan genocide caused tremendous hardship for the refugees and tremendous stress on the environment into which they moved.

The movement of refugees also causes significant environmental problems in receiving regions, as desperate victims try to eke out an existence with no livelihood and no cultural or economic attachment to the land or incentive to conserve its resources. The millions who fled Rwanda following the genocide there in the mid-1990s, for example, inadvertently destroyed large areas of forest while trying to obtain fuelwood, food, and shelter to stay alive once they reached the Democratic Republic of Congo (**FIGURE 6.12**).

Since 1970, growth rates in many countries have been declining, even without population control policies, and the global growth rate has declined (**FIGURE 6.13**). This decline has come about, in part, from a steep drop in birth rates. Note, however, that this is the *rate of growth* that is slowing, while the *absolute size* of the population continues to increase.

Total fertility rate influences population growth

One key statistic demographers calculate to examine a population's potential for growth is the **total fertility rate (TFR),** or the average number of children born per female member of a population during her lifetime. **Replacement fertility** is the TFR that keeps the size of a population stable. For humans, replacement fertility is equal to a TFR of 2.1. When the TFR drops below 2.1,

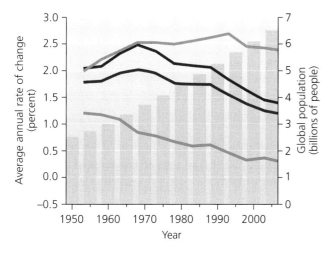

FIGURE 6.13
The annual growth rate of the global human population (dark blue line) peaked in the 1960s and has declined since then. Growth rates of industrialized nations (green line) have fallen since 1950, while those of developing nations (red line) have fallen since the global peak in the late 1960s. For the world's least-developed nations (orange line), growth rates began to fall in the 1990s. Although growth *rates* are declining, global population size (grey bars) is still growing by about the same amount each year, because smaller percentage increases of ever-larger numbers produce roughly equivalent additional amounts. Data from United Nations Population Division, 2007, *World Population Prospects: The 2006 Revision.*

Table 6.1	Total Fertility Rates (TFR) for Major Continental Regions	
Region	TFR 1970–1975	TFR 2005–2010
Africa	6.7	4.7
Latin America and Caribbean	5.0	2.4
Asia	5.0	2.3
Oceania	3.2	2.3
North America	2.0	2.0
Europe	2.2	1.4

Data from United Nations Population Division, 2007, *World Population Prospects: The 2006 Revision.*

population size, in the absence of immigration, will shrink.

Various factors influence TFR and have acted to drive it downward in many countries in recent years. Historically, people tended to conceive many children, which helped ensure that at least some would survive. Lower infant mortality rates have made this less necessary. Increasing urbanization has also driven TFR down; whereas rural families need children to contribute to farm labour, in urban areas children are usually excluded from the labour market, are required to go to school, and impose economic costs on their families. If a government provides some form of social security, as most do these days, parents need fewer children to support them in their old age when they can no longer work. Finally, with greater education and changing roles in society, women tend to shift into the labour force, putting less emphasis on child rearing.

All these factors have come together in Europe, where TFR has dropped from 2.6 to 1.4 in the past half-century. Every European nation now has a fertility rate below the replacement level, and populations are declining in 28 of 44 European nations. In 2006, Europe's overall annual **natural rate of population change** (change caused by birth and death rates alone, excluding migration) was 0.1%. Worldwide by 2006, a total of 74 countries had fallen below the replacement fertility rate of 2.1. These countries made up roughly 45% of the world's population and included China (with a TFR of 1.6). Table 6.1 shows the TFRs of major continental regions.

weighing the issues: CONSEQUENCES OF LOW FERTILITY?   6-2

In Canada, the United States, and every European nation, the total fertility rate has now dipped below the replacement fertility rate. What economic, social, or environmental consequences—positive or negative—do you think might result from below-replacement fertility rates?

Some nations have experienced the demographic transition

Many nations that have lowered their birth rates and TFRs have been going through a similar set of interrelated changes. In countries with good sanitation, good health care, and reliable food supplies, more people than ever before are living long lives. As a result, over the past 50 years the life expectancy for the average person has increased from 46 to 67 years as the global crude death rate has dropped from 20 deaths per 1000 people to 9 deaths per 1000 people. Strictly speaking, **life expectancy** is the average number of years that an individual in a particular age group is likely to continue to live, but often people use this term to refer to the average number of years a person can expect to live from birth. Much of the increase in life expectancy is due to reduced rates of infant mortality. Societies going through these changes are mostly the ones that have undergone urbanization and industrialization and have been able to generate personal wealth for their citizens.

To make sense of these trends, demographers developed a concept called the **demographic transition**. This is a model of economic and cultural change proposed in the 1940s and 1950s by demographer Frank Notestein and elaborated on by others to explain the declining death rates and birth rates that have occurred in Western nations as they became industrialized. Notestein observed that nations tend to move from a stable pre-industrial state of high birth and death rates to a stable post-industrial state of low birth and death rates. Industrialization, he proposed, caused these rates to fall naturally by first decreasing mortality and then lessening the need for large families. Parents would thereafter choose to invest in quality of life rather than quantity of children. Because death rates fall before birth rates fall, a period of net population growth results. Thus, under the demographic transition model, population growth is seen as a temporary phenomenon that occurs as societies move from one stage of development to another.

The pre-industrial stage The first stage of the demographic transition model (**FIGURE 6.14**) is the **pre-industrial stage**, characterized by conditions that have defined most of human history. In pre-industrial societies, both death rates and birth rates are high. Death rates are high because disease is widespread, medical care rudimentary, and food supplies unreliable and difficult to obtain. Birth rates are high because people must compensate for high mortality rates in infants and young children by having several children. In this stage, children are valuable as additional workers who can help meet a family's basic needs. Populations within the pre-industrial stage are not likely to experience much growth, which is why the human population was relatively stable from neolithic times until the industrial revolution.

Industrialization and falling death rates Industrialization initiates the second stage of the demographic transition, known as the **transitional stage**. This transition from the pre-industrial stage to the industrial stage is generally characterized by declining death rates because of increased food production and improved medical care. Birth rates in the transitional stage remain high, however, because people have not yet grown used to the new economic and social conditions. As a result, population growth surges.

The industrial stage and falling birth rates The third stage in the demographic transition is the **industrial stage**. Industrialization increases opportunities for employment outside the home, particularly for women. Children become less valuable, in economic terms, because they do not help meet family food needs as they did in the pre-industrial stage. If couples are aware of this, and if they have access to birth control, they may choose to have fewer children. Birth rates fall, closing the gap with death rates and reducing the rate of population growth.

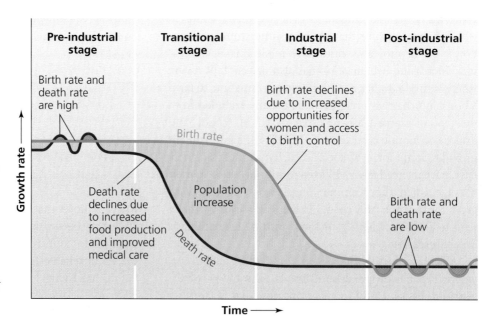

FIGURE 6.14
The demographic transition is an idealized process that has taken some populations from a pre-industrial state of high birth rates and high death rates to a post-industrial state of low birth rates and low death rates. In this diagram, the wide green area between the two curves illustrates the gap between birth and death rates that causes rapid population growth during the middle portion of this process. Data from Kent, M. M., and K. A. Crews. 1990. *World Population: Fundamentals of Growth*. Washington, DC: Population Reference Bureau.

The post-industrial stage   In the final stage, the **post-industrial stage**, both birth and death rates have fallen to low levels. Population sizes stabilize or decline slightly. The society enjoys the fruits of industrialization without the threat of runaway population growth.

Is the demographic transition a universal process?

The demographic transition has occurred in many European countries, the United States, Canada, Japan, and several other developed nations over the past 200 to 300 years. Nonetheless, it is a model that may not apply to all developing nations as they industrialize now and in the future. Some social scientists doubt that it will apply; they point out that population dynamics may be different for developing nations that adopt the Western world's industrial model rather than devising their own. Some demographers assert that the transition will fail in cultures that place greater value on childbirth or grant women fewer freedoms.

Moreover, natural scientists warn that there are not enough resources in the world to enable all countries to attain the standard of living that developed countries now enjoy. It has been estimated that for people of all nations to have the quality of life that Canadians do, we would need the natural resources of two to three more Earths. Whether developing nations, which include the vast majority of the planet's people, will pass through the demographic transition as developed nations have is one of the most important and far-reaching questions for the future of our civilization and Earth's environment.

Population and Society

Demographic transition theory links the statistical study of human populations with various societal factors that influence, and are influenced by, population dynamics. Let us now examine a few of these major societal factors more closely.

The status of women greatly affects population growth rates

Many demographers had long believed that fertility rates were influenced largely by degrees of wealth or poverty. However, affluence alone cannot determine the total fertility rate, because a number of developing countries now have fertility rates lower than that of Canada. Instead, recent research is highlighting factors pertaining to the social empowerment of women. Drops in TFR have been

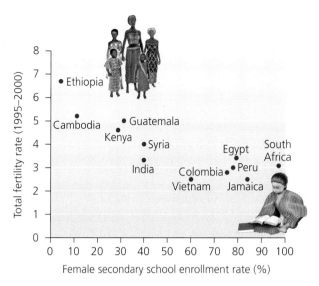

FIGURE 6.15
Increasing female literacy is strongly associated with reduced birth rates in many nations. Data from McDonald, M., and D. Nierenberg. 2003. Linking population, women, and biodiversity. *State of the World 2003*. Washington, DC: Worldwatch Institute.

most noticeable in countries where women have gained access to contraceptives and education, particularly family planning education (see "The Science Behind the Story: Fertility Decline in Bangladesh," and **FIGURE 6.15**).

In 2007, 54% of married women worldwide (aged 15–49) reported using some modern method of contraception to plan or prevent pregnancy. China, at 86%, had the highest rate of contraceptive use of any nation. Five European nations showed rates of contraceptive use higher than 70%, as did Canada, Costa Rica, Cuba, New Zealand, Australia, Brazil, and Thailand. At the other end of the spectrum, 26 African nations had rates of 10% or lower. These low rates of contraceptive use contribute to high fertility rates in sub-Saharan Africa, where the region's TFR is 5.5 children per woman. By comparison, in Asia, where the TFR in 1950 was 5.9, it is 2.4 today—in part a legacy of the population control policies of China and some other Asian countries.

These data clearly demonstrate that in societies where women have little power, substantial numbers of pregnancies are unintended. Studies show that when women are free to decide whether and when to have children, fertility rates have fallen and the resulting children are better cared for, healthier, and better educated. Unfortunately, in many societies, by tradition men restrict women's decision-making abilities, including decisions about how many children they will bear.

The gap between the power held by men and by women is just as obvious at the highest levels of government. Worldwide, only 17% of elected government officials in national legislatures are women. Canada (at about 21%)

THE SCIENCE BEHIND THE STORY

Fertility Decline in Bangladesh

This woman is in Matlab, Bangladesh.

Research in developing countries indicates that poverty and overpopulation can create a vicious cycle, in which poverty encourages high fertility and high fertility obstructs economic development. Are there policy steps that such countries can take to bring down fertility rates? Scientific analysis of family planning programs in the South Asian nation of Bangladesh suggests that there are.

Bangladesh is one of the poorest, most densely populated countries on the planet. Its 145 million people live in an area about the size of Newfoundland and Labrador, and 45% of them live below the poverty line. With few natural resources and 1000 people per square kilometre (Newfoundland and Labrador, for comparison, has about 4.5 people per square kilometre), limiting population growth is critically important. As Bangladeshi president Ziaur Rahman declared in 1976, "If we cannot do something about population, nothing else that we accomplish will matter much."

Since then, Bangladesh has made striking progress in controlling population growth. Despite stagnant economic development, low literacy rates, poor health care, and limited rights for women, the nation's total fertility rate (TFR) has dropped markedly. In the 1970s, the average woman in Bangladesh gave birth to more than six children over the course of her life. Today, the TFR is 3.2.

Researchers hypothesized that family planning programs were responsible for Bangladesh's rapid reduction in TFR. Because conducting an experiment to test such a hypothesis is difficult, some researchers took advantage of a natural experiment. By comparing Bangladesh to countries that are socioeconomically similar but have had less success in lowering TFR, like Pakistan, researchers concluded that Bangladesh succeeded because of aggressive, well-funded outreach efforts that were sensitive to the values of its traditional society.

However, because no two countries are identical, it is difficult to draw firm conclusions from such broad-scale studies. This is why the Matlab Family Planning and Health Services Project, in the isolated rural area of Matlab, Bangladesh, has become one of the best-known experiments in family planning in developing countries.

The Matlab Project was an intensive outreach program run collaboratively by the Bangladeshi government and international aid organizations. Each household in the project area received biweekly visits from local women offering counselling, education, and free contraceptives. Compared with a similar government-run program in a nearby area, the Matlab Project featured more training, more services, and more frequent visits. In both areas, a highly organized health surveillance system gave researchers detailed information about births, deaths, and health-related behaviours, such as contraceptive use. The result was an experiment comparing the Matlab Project with the government-run area.

When Matlab Project director James Phillips and colleagues reviewed a decade's worth of data in 1988, they found that fertility rates had declined in both areas. The decline appeared to be due almost entirely to a rise in contraceptive use, because other factors—such as the average age of marriage—remained the same. Phillips and his colleagues found that the declines had been significantly greater in the Matlab area than in the government-run area. These findings suggested that high-intensity outreach efforts could affect fertility rates even in the absence of significant improvements in women's status, education, or economic development.

Why was the outreach program successful? One hypothesis was that visits

---

lags behind not only Europe (e.g., Sweden at 48%) but also many developing nations (e.g., Rwanda at 49% and Argentina at 36%) in the proportion of women in positions of power in government.[3] As more women win positions of power, it will have environmental consequences, for when women have economic and political power and access to education, they gain the option, and often the motivation, to limit the number of children they bear.

Population policies and family planning programs are working around the globe

Data show that funding and policies that encourage family planning have been effective in lowering population growth rates in all types of nations, even those that are least industrialized. No nation has pursued a population control program as extreme as China's, but other rapidly growing nations have implemented less-restrictive programs.

When policy makers in India introduced the idea of forced sterilization as a means of population control in the 1970s, the resulting outcry brought down the government. Since then, India's efforts have been more modest and far less coercive, focusing on family planning and reproductive health care. A number of Indian states also run programs of incentives and disincentives promoting a "two-child norm," and current debate centres on whether this is a just and effective approach. Regardless, unless India strengthens its efforts to slow population growth, it seems set to overtake China and become the world's most populous nation by about the year 2030.

In the Matlab Project, Bangladeshi households received visits from local women offering counselling, education, and free contraceptives.

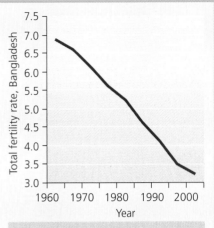

Total fertility rate has declined markedly in Bangladesh in the past 40 years, in part because of family planning programs. Data from United Nations Population Division.

from health care workers had helped convince local women that small families are desirable. However, in 1999, Mary Arends-Kuenning, a graduate student in economics at the University of Michigan, and her colleagues reported that there was no relationship between women's perception of the ideal family size and the number of visits made by outreach workers, either in Matlab or nearby comparison areas. Ideal family size declined equally in all areas. Instead of creating new demand for birth control, the Matlab Project appears to have helped women convert an already existing desire for fewer children into behaviours, such as contraceptive use, that reduce fertility.

Bangladesh's ability to rein in fertility rates despite unfavourable social and economic conditions bodes well for impoverished nations facing explosive population growth. However, significant challenges remain. If rates fail to decline further, the country's population could double to 290 million—nine times the current population of Canada—within 40 years. Scientific research has helped illuminate the impact of family planning programs on fertility, but further reductions may require fundamental social, political, and economic changes that are difficult to implement in resource-strapped countries like Bangladesh. Nonetheless, scientific evidence collected at Matlab has played an important role in informing population control efforts in Bangladesh and elsewhere.

The government of Thailand relies on an education-based approach to family planning that has reduced birth rates and slowed population growth. In the 1960s, Thailand's growth rate was 2.3%, but today it stands at 0.7%. This decline was achieved without a one-child policy. It has resulted, in large part, from government-sponsored programs devoted to family planning education and increased availability of contraceptives. Brazil, Mexico, Iran, Cuba, and many other developing countries have instituted active programs to reduce their population growth. These programs entail setting targets and providing incentives, education, contraception, and reproductive health care.

Many of these programs are working. The data shown in **FIGURE 6.15** are not the only cases in which family planning programs have helped lower fertility rates. One study in 2000 examined four different pairs of nations located in the same parts of the world, with one country in each pair having a stronger program: Thailand and the Philippines, Pakistan and Bangladesh, Tunisia and Algeria, and Zimbabwe and Zambia (**FIGURE 6.16**). The demographers concluded that in all four cases, the country with the stronger program (Thailand, Bangladesh, Tunisia, and Zimbabwe) initiated or accelerated a decline in fertility with its policies. In the case of Thailand and the Philippines, the researchers also concluded that the Catholic Church's strong presence in the Philippines held back the success of family planning there.

In 1994, the United Nations hosted the milestone *International Conference on Population and Development* in Cairo, Egypt, at which 179 nations endorsed a platform calling on all governments to offer universal access to reproductive health care within 20 years. The conference

FIGURE 6.16
Data from four pairs of neighbouring countries demonstrate the effectiveness of family planning in reducing fertility rates. In each case, the nation that invested in family planning and (in some cases) made other reproductive rights, education, and health care more available to women (blue lines) reduced its total fertility rate (TFR) far more dramatically than its neighbour (red lines). Data from United Nation Population Division; and Harrison, P., and F. Pearce. 2000. *AAAS Atlas of Population and Environment*. Berkeley, CA: University of California Press.

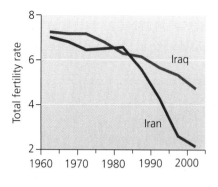
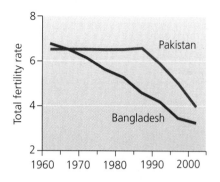
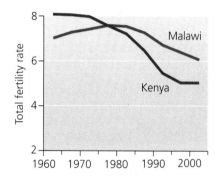
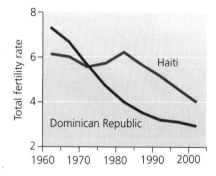

marked a turning away from older notions of command-and-control population policy geared toward pushing contraception and lowering population to preset targets. Instead, it urged governments to offer better education and health care and to address social needs that bear indirectly on population (such as alleviating poverty, disease, and sexism). Unfortunately, worldwide funding for family planning programs fell by at least a third in the decade following the Cairo conference.

weighing the issues 6–3
ABSTAINING FROM THE REAL WORLD? FUNDS FOR INTERNATIONAL FAMILY PLANNING

In the past, the U.S. government has withheld funds from the United Nations Population Fund (UNFPA) because its programs provide education in family planning, HIV/AIDS prevention, and teen pregnancy prevention in many nations. Canada and the European Union offered additional funding to UNFPA to offset the loss of U.S. contributions. What do you think of the U.S. decision? What conditions, if any, should Canada place on the use of funds for international aid in the area of family planning?

Poverty is strongly correlated with population growth

The alleviation of poverty was a prime target of the 1994 Cairo conference, because poorer societies tend to show higher population growth rates than do wealthier societies. This pattern is consistent with demographic transition theory. Table 6.2 shows that poorer nations tend to have higher fertility and growth rates, along with higher birth and infant mortality rates and lower rates of contraceptive use.

Such trends as these have affected the distribution of people on the planet. In 1960, 70% of all people lived in developing nations. By 2007, 82% of the world's population was living in these countries. Moreover, fully 99% of the next billion people to be added to the global population will be born in these poor, less developed regions.

This is unfortunate from a social standpoint, because these people will be added to the countries that are least able to provide for them. It is also unfortunate from an environmental standpoint, because poverty often results in environmental degradation. People dependent on agriculture in an area of poor farmland, for instance, may need to try to farm even if doing so degrades the soil and is not sustainable. This is largely why Africa's once-productive Sahel region, like many regions of western China, is turning to desert (**FIGURE 6.17**). Poverty also drives the hunting of many large mammals in Africa's forests, including the great apes that are now disappearing as local settlers and miners kill them for their "bush meat."

Table 6.2 Per Capita Wealth, with Rates of Fertility, Population Growth, and Contraceptive Use, for Selected Nations

Nation	Per capita GNI PPP (U.S. $)*	Rate of natural population increase (% per year)	Children born per woman (TFR)	Population density (per km^2)	Infant mortality (per 1000)	Couples using birth control (%)
Tanzania	740	2.6	5.4	41	78	20
Niger	830	3.4	7.1	11	126	5
Ethiopia	1 190	2.5	5.4	70	77	14
Haiti	1 490	1.8	4.0	323	57	25
Pakistan	2 500	2.3	4.1	213	78	22
India	3 800	1.6	2.9	344	58	49
Syria	3 920	2.5	3.5	108	19	35
China	7 730	0.5	1.6	138	27	86
Brazil	8 800	1.4	2.3	22	27	70
Romania	9 820	−0.2	1.3	90	14	34
Mexico	11 330	1.7	2.4	54	21	59
Spain	28 420	0.3	1.4	90	4	53
Japan	33 730	0.0	1.3	338	3	48
Canada	34 610	0.3	1.5	3	5	73
United Kingdom	35 690	0.3	1.8	251	5	79
United States	44 260	0.6	2.1	31	7	68

*GNI PPP is "gross national income in purchasing power parity," a measure that standardizes income and makes it comparable among nations, by converting income to "international" dollars by using a conversion factor. International dollars indicate the amount of goods and services one could buy with a given amount of money. Data from Population Reference Bureau. 2007. World Population Data Sheet 2007.

FIGURE 6.17
In the semi-arid Sahel region of Africa, where population is increasing beyond the land's ability to handle it, dependence on grazing agriculture has led to environmental degradation.

Consumption from affluence creates environmental impacts

Poverty can lead people into environmentally destructive behaviour, but wealth can produce even more severe and far-reaching environmental impacts. The affluence that characterizes such societies as Canada, the United States, Japan, or the Netherlands is built on massive and unprecedented levels of resource consumption. Much of this chapter has dealt with numbers of people rather than on the amount of resources each member of the population consumes or the amount of waste each member produces. The environmental impact of human activities, however, depends not only on the number of people involved but also on the way those people live. Recall the A for affluence in the IPAT equation. Affluence and consumption are spread unevenly across the world, and affluent societies generally consume resources from other localities as well as from their own.

In Chapter 1, we introduced the concept of the *ecological footprint*, the cumulative amount of Earth's surface area required to provide the raw materials a person or population consumes and to dispose of or recycle the waste that they produce. Individuals from affluent societies leave a considerably larger per capita ecological footprint (see **FIGURE 1.10**). In this sense, the addition of 1 Canadian to the world has as much environmental impact as the addition of 6 Chinese, or 12 Indians or Ethiopians, or 40 Somalians. This fact should remind us that the "population problem" does not lie entirely with the developing world.

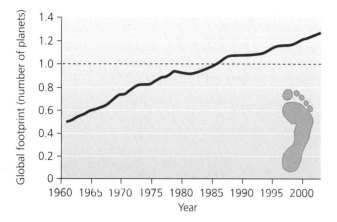

FIGURE 6.18
The global ecological footprint of the human population is more than 2.5 times what it was in 1961 and now exceeds what Earth can bear in the long run, scientists have calculated. The estimate shown here indicates that we have already overshot our carrying capacity by 25%; that is, we are using renewable resources 25% faster than they are being replenished. Another way to think of this is that we would require 1.25 Earths to support humanity in the long term, at current levels of population and consumption. Data from WWF Worldwide Fund for Nature. 2006. *Living Planet Report*. Gland, Switzerland: WWF.

Indeed, just as population is rising, so is consumption, and some environmental scientists have calculated that we are already living beyond the planet's means to support us sustainably. One recent analysis concluded that humanity's global ecological footprint surpassed Earth's capacity to support us in 1987 and that our species is now living as much as 39% beyond its means (**FIGURE 6.18**). The rising consumption that is accompanying the rapid industrialization of China, India, and other populous nations makes it all the more urgent for us to find a path to global sustainability.

The wealth gap and population growth contribute to conflict

The stark contrast between affluent and poor societies in today's world is, of course, the cause of social as well as environmental stress. More than half the world's people live below the internationally defined poverty line of $2 per day. The richest one-fifth of the world's people possesses more than 80 times the income of the poorest one-fifth (**FIGURE 6.19**). The richest one-fifth also uses 86% of the world's resources. That leaves only 14% of global resources—energy, food, water, and other essentials—for the remaining four-fifths of the world's population to share. As the gap between rich and poor grows wider and as the sheer numbers of those living in poverty continue to increase, it seems reasonable to predict increasing tensions between the "haves" and the "have-nots."

(a) A family living in North America

(b) A family living in Egypt

FIGURE 6.19
A typical North American family **(a)** may own a large house, keep numerous material possessions, and have enough money to afford luxuries, such as vacation travel. A typical family in a developing nation, such as Egypt **(b)**, may live in a small, sparsely furnished dwelling with few material possessions and little money or time for luxuries.

HIV/AIDS is exerting major impacts on African populations

The rising material wealth and falling fertility rates of many industrialized nations today are slowing population growth in accordance with the demographic transition model. Some other nations, however, are not following Notestein's script. Instead, in these countries mortality is beginning to rise, presenting a scenario more akin to Malthus's fears. This is especially the case in countries where the HIV/AIDS epidemic has taken hold (**FIGURE 6.20**). African nations are being hit hardest. Of the 40 million people in the world infected with HIV/AIDS as of 2006, 27 million live in the nations of sub-Saharan Africa. The low rate of use of contraceptives, which

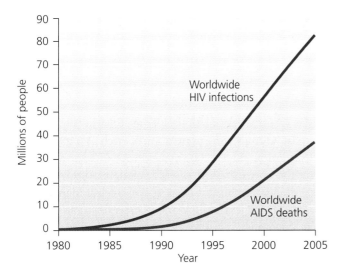

FIGURE 6.20
AIDS cases are increasing rapidly in much of the world. As of 2006, total cumulative HIV infections since 1980 were estimated at nearly 83 million, and 37 million people are estimated to have died from the disease so far. Data from UNAIDS; and *Vital Signs 2006–2007*. Washington, DC: Worldwatch Institute.

contributes to this region's high fertility rate, also fuels the expansion of AIDS. One in every 16 people aged 15–49 in sub-Saharan Africa is infected with HIV, and for southern African nations, the figure is more than one in five.

The AIDS pandemic is having the greatest impact on human populations of any disease since the Black Death killed roughly one of three people in fourteenth-century Europe, and since smallpox brought by Europeans to the New World wiped out perhaps millions of native people. As AIDS takes roughly 6000 lives in Africa every day, the pandemic is unleashing a variety of demographic changes. Infant mortality in sub-Saharan Africa has risen to 9 deaths out of 100 live births—15 times the rate in the developed world. The high numbers of infant deaths and premature deaths of young adults has caused life expectancy in parts of southern Africa to fall from a high of close to 59 years in the early 1990s back down to less than 40 years, where it stood in the early 1950s. AIDS is also leaving behind millions of orphans.

weighing the issues
HIV/AIDS AND POPULATION 6–4

What sorts of problems would you predict might occur in the surviving population after a major disease, such as AIDS, kills a high percentage of the population?

Africa is not the only region with reason to worry. HIV is well established in the Caribbean and in Southeast Asia, and it is spreading quickly in eastern Europe and central Asia. In China, where injected drug use is fuelling its spread, the World Health Organization estimated that 10 million people could become infected between 2005 and 2010. As of 2006, an estimated 14 million children under the age of 15 worldwide had lost one or both parents to the disease.

Severe demographic changes have social, political, economic, and environmental repercussions

Everywhere in sub-Saharan Africa, AIDS is undermining the ability of developing countries to make the transition to modern technologies because it is removing many of the youngest and most productive members of society. In 1999 Zambia lost 1300 teachers to AIDS, and only 300 new teachers graduated to replace them. South Africa loses an estimated $7 billion per year to declines in its labour force as AIDS patients fill the nation's hospitals. The loss of productive household members to AIDS causes families and communities to break down as income and food production decline. Valuable environmental and farming knowledge is being lost as an entire generation of Africans is decimated. (See "The Science Behind the Story: Is AIDS an Environmental Threat?" available online at www.myenvironmentplace.ca.)

These problems are hitting many countries at a time when their governments are already experiencing what has been called *demographic fatigue*. Demographically fatigued governments face overwhelming challenges related to population growth, including educating and finding jobs for their swelling ranks of young people. With the added stress of HIV/AIDS, these governments face so many demands that they are stretched beyond their capabilities to address problems. As a result, the problems grow worse, and citizens lose faith in their governments' abilities to help them.

If nations in sub-Saharan Africa—and other regions where the disease is spreading fast, such as India and Southeast Asia—do not take aggressive steps soon, and if the rest of the world does not try to help, these countries could fail to advance through the demographic transition. Instead, their rising death rates could push birth rates back up, potentially causing these countries to fall back to the pre-industrial stage of the demographic transition model. Such an outcome would lead to greater population growth while economic and social conditions worsen. It would be a profoundly negative outcome, both for human welfare and for the well-being of the environment.

The U.N. has articulated sustainable development goals for humanity

In 2000, world leaders came together to adopt the *Millennium Declaration*, which set out a framework of basic goals for humanity over the next decade and a half. The **Millennium Development Goals** set an aggressive target date of 2015 to achieve many of the fundamental goals for sustainable development that aid organizations have worked so hard to achieve over the past few decades (Table 6.3). Each of the broad goals has several specific underlying targets. You can find out more about these targets, and about our global progress toward them, at the United Nations Millennium Development Goals website, www.un.org/millenniumgoals/global.shtml.

Interestingly, population control is *not* one of the Millennium Development Goals. However, the interconnections we have discussed in this chapter should make it clear that in order to achieve the *other* goals, both population growth and resource consumption levels will need to be addressed.

If humanity's overarching goal is to generate a high standard of living and quality of life for all the world's people, then developing nations must find ways to reduce their population growth. However, those of us living in the industrialized world must also be willing to reduce our consumption, otherwise the goal of achieving environmental sustainability will elude us. Earth does not hold enough resources to sustain all 6.7 billion of us at the current North American standard of living, nor can we go out and find extra planets. We must make the best of the one place that supports us all.

Conclusion

Today's human population is larger than at any time in the past. Our growing population, as well as our growing consumption, affects the environment and our ability to meet the needs of all the world's people. Approximately 90% of children born today are likely to live their lives in conditions far less healthy and prosperous than most of us in the industrialized world are accustomed to.

However, there are at least two major reasons to be encouraged. First, although global population is still rising, the *rate* of growth has decreased nearly everywhere, and some countries are even seeing population declines. Most developed nations have passed through the demographic transition, showing that it is possible to lower death rates while stabilizing population and creating more prosperous societies. Second, progress has been made in expanding rights for women worldwide. Although there is still a long way to go, women are slowly being treated more fairly, receiving better education, obtaining more economic independence, and gaining more ability to control their reproductive decisions. Aside from the clear ethical progress these developments entail, they are helping to slow population growth; where aggressive programs to control population growth have failed in many countries, educating girls and giving reproductive rights to women are succeeding.

Human population cannot continue to rise forever. The question, however, is how will it stop rising: through the gentle and benign process of the demographic transition, through restrictive governmental intervention, such as China's one-child policy, or through the miserable Malthusian checks of disease and social conflict caused by overcrowding and competition for scarce resources? Moreover, sustainability demands a further challenge—that we stabilize our population size in time to avoid destroying the natural systems that support our economies and societies. We are indeed a special species. We are the only one to come to such dominance as to fundamentally change so much of Earth's landscape and even its climate system. We are also the only species with the intelligence needed to turn around an increase in our own numbers before we destroy the very systems on which we depend.

Table 6.3 United Nations Millennium Development Goals for 2015
■ Eradicate extreme poverty and hunger
■ Achieve universal primary education
■ Promote gender equality and empower women
■ Reduce child mortality
■ Improve maternal health
■ Combat HIV/AIDS, malaria and other diseases
■ Ensure environmental sustainability
■ Develop a global partnership for development

Source: United Nations, End Poverty 2015, Millennium Development Goals, www.un.org/millenniumgoals/global.shtml.

CANADIAN ENVIRONMENTAL PERSPECTIVES

William Rees

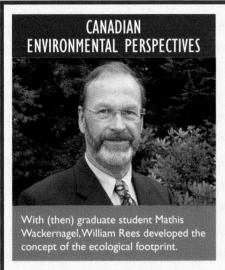

With (then) graduate student Mathis Wackernagel, William Rees developed the concept of the ecological footprint.

- **Professor** and **researcher**
- **Bioecologist**
- **Ecological economist**

As a young professor, William Rees spent his time calculating the carrying capacity of the land on which the University of British Columbia was located. One day, he turned the calculation upside down—literally—and ended up with one of the most useful tools we now have for understanding human impacts on the environment. Rees is a professor in the School of Community and Regional Planning at UBC, where he has taught since 1969. Rees teaches and carries out research on a variety of topics related to sustainability, but he is best known as the co-inventor (with Mathis Wackernagel, now Executive Director of the Global Footprint Network, his graduate student at the time) of the ecological footprint concept.

Rees grew up on a farm in southern Ontario, and his connection to the land through farming led to his interest in sustainability. Rees describes a warm summer day at the farm when he was about 9 or 10: "We were in my grandmother's country porch, 13 of us or so, having lunch after a hard morning's work in the field. I happened to glance down at my plate full of young new carrots, little potatoes, fresh lettuce, and so on, and to make a long story short, I realized that there wasn't a single thing on the plate that I hadn't had a hand in growing . . . I was so excited by that notion, I don't think I was able to eat my lunch."[4] He credits this early emotional connection to the land with his later decisions to study zoology and ecology.

The idea that humans—like all other species—rely on nature for sustenance and resources followed Rees throughout university and a Ph.D. in population ecology at the University of Toronto. As a young professor, Rees investigated the relationship between ecological economics and human ecology. Economic thought in the 1970s still relied on the idea that resource shortages could be resolved through technology, or by extensification or intensification of resource extraction. This did not sit well with Rees; surely this could not work indefinitely. While attempting to compute the carrying capacity of the Lower Fraser Valley (Lower Mainland) in BC, he had the idea of turning the calculation upside down—instead of computing the carrying capacity of the land, why not figure out how much land or ecosystem capacity would be needed to support a human population of a given size and consumption level? The ecological footprint concept was born.

Since the 1996 publication of *Our Ecological Footprint: Reducing Human Impact on the Earth*, both Rees and Wackernagel have continued their work on the ecological footprint as an indicator of human impact. Rees's current research is focused on environmental change and determining the necessary ecological conditions for biodiversity preservation and sustainable socioeconomic development. His current book asks the question, "Is humanity inherently unsustainable?"

"The whole thrust toward sustainability has been to treat it as an economic problem or a technological problem . . . What we really ought to be thinking about is what makes for people's welfare.[5] *—William Rees*

Thinking About Environmental Perspectives

Bill Rees points to his childhood on a farm in southern Ontario as the origin of his interest in sustainability and human impacts on the environment. Did you have any childhood experiences that created in you a deep connection with the land or the natural environment? If so, did these influence your later decisions about what to study?

REVIEWING OBJECTIVES

You should now be able to:

Assess the scope and historical patterns of human population growth

- The current global population of 6.7 billion people adds about 80 million people per year (2.6 people every second).
- The world's population growth rate peaked at 2.1% in the 1960s and now stands at 1.2%. Growth rates vary among regions of the world.
- Attitudes toward the "population problem" have changed over time. The Malthusian perspective holds that population is a problem to the extent that it depletes resources, intensifies pollution, stresses social systems, or degrades ecosystems, such that the natural environment or our quality of life declines.

Evaluate how human population, affluence, and technology affect the environment

- The IPAT model summarizes how environmental impact (I) results from interactions among population size (P), affluence (A), and technology (T).
- Rising population and rising affluence (leading to greater consumption) each increase environmental impact. Technological advances have frequently exacerbated environmental degradation, but they can also help mitigate our impact.
- Four major societal transitions (paleolithic tool-use; neolithic development of agriculture; Industrial Revolution switch to fossil fuels and mechanization; and the modern Medical–Technological and Green Revolutions) have fundamentally altered the way the human population interacts with the environment.

Explain and apply the fundamental concepts of demography

- Demography applies principles of population ecology to the statistical study of human populations.
- Demographers study size, density, distribution, age structure, and sex ratios of populations, as well as rates of birth, death, immigration, and emigration.
- Total fertility rate (TFR) contributes greatly to change in a population's size.

Outline the concept of demographic transition

- The demographic transition model explains why population growth has slowed in industrialized nations. Industrialization and urbanization have reduced the economic need for children, while education and the empowerment of women have decreased unwanted pregnancies. Parents in developed nations choose to invest in quality of life rather than quantity of children.
- The demographic transition may or may not proceed to completion in all of today's developing nations. Whether it does is of immense importance in the quest for population stabilization and sustainability.

Describe how wealth and poverty, the status of women, and family planning programs and policies affect population growth

- When women are empowered and achieve equality with men, fertility rates fall, and children tend to be better cared for, healthier, and better educated.
- Family planning programs and reproductive education have successfully reduced population growth in many nations.
- Poorer societies tend to have higher population growth rates than do wealthier societies.
- The high consumption rates of affluent societies may make their ecological impact greater than that of poorer nations with larger populations.

Characterize the dimensions of the HIV/AIDS pandemic and its impacts on population, the environment, and sustainability

- About 38 million people worldwide are infected with HIV/AIDS, of which 25 million live in sub-Saharan Africa.
- Epidemics that claim large numbers of young and productive members of society influence population dynamics, and can have severe social, political, and environmental ramifications, particularly for traditional food production.

TESTING YOUR COMPREHENSION

1. What is the approximate current human global population? How many people are being added to the population each day? How many have been added since you were born?
2. Why has the human population continued to grow in spite of environmental limitations?
3. Contrast the views of environmental scientists with those of libertarian writer Sheldon Richman and similar-thinking economists over whether population growth is a problem. Why does Richman think the concept of carrying capacity does not apply to human populations?
4. Explain the IPAT model. How can technology either increase or decrease environmental impact? Provide at least two examples.
5. What characteristics and measures do demographers use to study human populations? Which of these help determine the impact of human population on the environment?
6. What is the total fertility rate (TFR)? Can you explain why the replacement fertility for humans is approximately 2.1? Why would it not be exactly 2.0? How is Europe's TFR affecting its natural rate of population change?
7. Why have fertility rates fallen in many countries?
8. In the demographic transition model, why is the pre-industrial stage characterized by high birth and death rates, and the industrial stage by falling birth and death rates?
9. How does the demographic transition model explain the increase in population growth rates in recent centuries? How does it explain the decrease in population growth rates in recent decades?
10. Why do poorer societies have higher population growth rates than wealthier societies? How does poverty affect the environment? How does affluence affect the environment?

SEEKING SOLUTIONS

1. China's reduction in birth rates is leading to significant change in the nation's age structure. Review Figure 6.11, which portrays the projected change. You can see that the population is growing older, based on the top-heavy age pyramid for the year 2030. What sorts of effects might this ultimately have on Chinese society? Explain your answer.
2. The World Bank estimates that more than half the world's people survive on less than the equivalent of $2 per day. What effect would you expect this situation to have on the political stability of the world? Explain your answer.
3. Apply the IPAT model to the example of China provided in the chapter. How do population, affluence, technology, and ecological sensitivity affect China's environment? Now consider your own country, region, province, or territory. How do population, affluence, technology, and ecological sensitivity affect your environment? How can we regulate the relationship between population and its effects on the environment?
4. Do you think that all of today's developing nations will complete the demographic transition and come to enjoy a permanent state of low birth and death rates? Why or why not? What steps might we as a global society take to help ensure that they do? Now think about developed nations, including Canada. Do you think these nations will continue to lower their birth and death rates in a state of prosperity? What factors might affect whether they do so?
5. **THINK IT THROUGH** India's prime minister has put you in charge of that nation's population policy. India has a population growth rate of 1.7% per year, a TFR of 2.9, a 46% rate of contraceptive use, and a population that is 72% rural. What policy steps would you recommend, and why?
6. **THINK IT THROUGH** Now imagine that you have been tapped to design population policy for Germany. Germany is losing population at an annual rate of 0.2%, has a TFR of 1.3, a 72% rate of contraceptive use, and a population that is 88% urban. What policy steps would you recommend, and why?

INTERPRETING GRAPHS AND DATA

At right are graphed data representing the economic condition of the world's population. The *y*-axis indicates the per capita income for each country or region expressed as purchasing power (termed *gross national income in purchasing power parity*, or *GNI PPP*; see Table 6.2 for a more detailed explanation). The *x*-axis indicates the cumulative percentage of the world population whose per capita GNI PPP is equal to or greater than that country's or region's per capita GNI PPP. The horizontal dotted line indicates the global average per capita GNI PPP.

1. What percentage of the world population lives at or below the global average per capita GNI PPP? What percentage lives at or below one-half of the global average per capita GNI PPP? What percentage lives at or above twice the global average per capita GNI PPP?
2. Given a global average per capita GNI PPP of $9190 and a world population of 6 700 000 000 people, what is the total global GNI PPP? What would the global GNI PPP be if everyone lived at the level of affluence of Canada?
3. How do you personally resolve the ethical conflict between the desirable goal of raising the standard of living of the billions of desperately poor people in the world and the likelihood that increasing their affluence (A in the equation I = PAT) will have a negative impact on the environment?

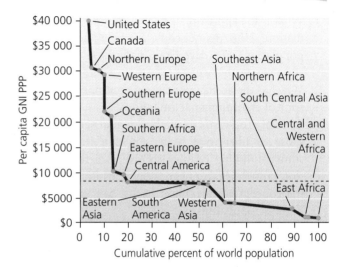

Percentage of world population at various income levels. *Data source: Population Reference Bureau. 2006. World Population Data Sheet 2006.*

CALCULATING FOOTPRINTS

The equation I = PAT (Impact = Population × Affluence × Technology) suggests that a population's size and affluence are not the only determinants of its ecological impact; its technological choices also have an effect. Technologies can be either efficient or wasteful. One way of gauging the relative value of T is to calculate a per capita value of I/A (equivalent to I divided by A divided by P). The table presents per capita values of I (estimated ecological footprints, based on "Footprint 2.0" from Redefining Progress) and A (income, expressed as GNI PPP). Calculate the relative values of T by completing the blank column.

1. If the world average value of T were decreased (improved) to that of Canada, what per capita GNI PPP could be supported at the current average per capita ecological footprint of 21.9 hectares?
2. What value of T would enable the world's population to live at its current affluence within the 15.7 hectares per capita of biological capacity that is estimated to be available to humans? Do you think this is achievable? (Data from Redefining Progress, *Ecological Footprints of Nations: 2005 Update.*)
3. Which country's technological choices would you choose to study if you were interested in learning how to maximize your standard of living while minimizing your ecological impact? Using the value of T for this country and the mid-2007 world population of 6 700 000 000, calculate the following:

    (a) The number of people the world could support at the current per capita impact of 21.9 hectares and affluence level of $9190
    (b) The number of people the world could support sustainably on the available 15.7 global hectares per capita and at an affluence level of $9190
    (c) The per capita GNI PPP that the world's current population could achieve with a footprint of 21.9 hectares per capita
    (d) The per capita GNI PPP that the world's current population could achieve with a footprint of 15.7 hectares per capita

Nation	Impact (ecological footprint, in hectares per capita)	Affluence (per capita income, GNI PPP)	Technology (I/A) (footprint per $1000 income)
Bangladesh	2.3	$2 090	1.1
Colombia	11.2	$7 420	
Mexico	23.1	$10 030	
Sweden	66.8	$31 420	
Thailand	15.9	$8 440	
United States	108.9	$41 950	
Canada	83.0	$32 220	
World	21.9	$9 190	

Data sources: *Population Reference Bureau. 2006.* World Population Data Sheet 2006; *and Redefining Progress. February 2005.* Ecological Footprints of Nations—2005 Update. *The organization Redefining Progress provides up-to-date calculations of the ecological footprints of nations. Its methodology has changed over the past few years, from "Footprint 1.0," which only counted one-third of Earth's surface area as being "biologically productive," to "Footprint 2.0," which accounts for the biological productivity of the entire surface of the planet. Other changes include reserving a small portion of Earth's biological productivity in support of other species, accounting for the biological productivity of Earth's oceans, and allocating a greater footprint to energy-related impacts. Where Footprint 1.0 suggested that humanity's overall ecological footprint is 1.18 (necessitating an Earth 18% larger than this planet), Footprint 2.0 suggests that we may have overshot Earth's productive capacity even more and would require 1.39 Earths, or an Earth 39% larger than this planet, to sustain humanity at present levels of consumption and population. (Humanity's overall per capita ecological footprint 21.9 ÷ the available biological capacity of Earth 15.7 hectares per capita = 1.39.)*

TAKE IT FURTHER

Go to www.myenvironmentplace.ca where you will find

- Suggested answers to end-of-chapter questions
- Quizzes, animations, and flashcards to help you study
- *Research Navigator*™ database of credible and reliable sources to assist you with your research projects
- Tutorials to help you master how to interpret graphs
- Current news articles that link the topics that you study to case studies from your region and around the world

- **ECO Occupational Profiles:** If you found this chapter especially interesting, you might want to learn more about the following jobs by visiting the Occupational Profiles website of the Environmental Careers Organization. Go to www.eco.ca and check out the following careers:
  - Cartographer
  - Geographer
  - Biometrician

CHAPTER ENDNOTES

1. Unless otherwise noted, all population data in this chapter are from (or calculated from) the United Nations Department of Economic and Social Affairs, Population Division (2007) *World Population Prospects: The 2006 Revision.*
2. Data from United Nations Population Division (2007) *World Population Prospects: The 2006 Revision,* Table A.1.
3. Data from Inter-Parliamentary Union (April 2007) *Women in National Parliaments,* www.ipu.org/wmne/classif.htm.
4. Quoted from Gismondi, M. (2000) Dr. William Rees Interviewed by Dr. Michael Gismondi. *Aurora Online,* http://aurora.icaap.org/index.php/aurora/article/view/18/29.
5. Quoted from Gismondi, M. (2000) Dr. William Rees Interviewed by Dr. Michael Gismondi. *Aurora Online,* http://aurora.icaap.org/index.php/aurora/article/view/18/29.

Soils and Agriculture

7

This is the Mer Bleue provincial wetland, near Ottawa, Ontario.

Upon successfully completing this chapter, you will be able to

- Delineate the fundamentals of soil science, including soil formation and the properties of soil
- Explain the importance of soils to agriculture, and describe the impacts of agriculture on soils
- Outline major historical developments in agriculture
- State the causes and predict the consequences of soil erosion and soil degradation
- Describe the history and explain the principles of soil conservation

This satellite image shows the ancient channel of the Ottawa River that is now occupied by the Mer Bleue wetland.

CENTRAL CASE:
MER BLEUE: A BOG OF INTERNATIONAL SIGNIFICANCE

"This soil of ours, this precious heritage, what an unobtrusive existence it leads! To the rich soil let us give the credit due. The soil is the reservoir of life."
—J.A. TOOGOOD, CANADIAN SOIL SCIENTIST

"The nation that destroys its soil destroys itself."
—FRANKLIN D. ROOSEVELT, FORMER U.S. PRESIDENT

The Mer Bleue Conservation Area is a 35 km² provincially protected wetland situated just east of Ottawa, Ontario. It is located in an ancient, now-abandoned channel of the Ottawa River (see satellite photo), and hosts a number of unusual plant species that are specially adapted to moist, boggy, acidic conditions, including *Sphagnum* moss, bog rosemary, blueberry, cottongrass, cattails, and tamarack (see photo). The area, classified as an open bog, has been recognized under the Ramsar Convention as a wetland site of international importance.

The Mer Bleue wetland provides an example of a specific type of soil—peat—that has been accumulating in many northern areas since the end of the last ice age. Canada has some of the most extensive peatlands in the world, covering 14% of the land area, and the peat deposits in some parts of the Mer Bleue bog, which formed over the past 8000 years, are up to 6 m thick!

Northern peatlands are extremely important storage reservoirs for carbon, and they are thought to hold about one-third of all the carbon stored in soils. Through decomposition, peat produces soil gases, such as CO_2 and CH_4, which function as greenhouse gases in the atmosphere. Thus, understanding the potential reaction of these very sensitive soils to climate change, particularly changes in water content and temperature, is of great interest to scientists. For example, if warming leads northern peat soils to have faster decomposition rates and thus to release more soil gases, there could be a major increase in the concentration of these carbon-based gases in the atmosphere. This could set up positive feedback and have a reinforcing influence on global warming.

The storage of carbon in peat depends on the balance between net primary production and decomposition, with plants storing or sequestering carbon as a result of photosynthesis, then contributing the stored carbon to the peat soil where it accumulates in the form of plant litter. Temperature and light levels are of obvious importance to this balance, since the process of photosynthesis is involved. Moisture is another very important factor. When water levels are high, CH_4, the by-product of anaerobic (reduced) decomposition, is produced in large quantities; when conditions are drier, on the other hand, respiration tends to be aerobic (oxidized).

The Peatland Carbon Study (PCARS) was initiated by a group of Canadian scientists in 1997. Researchers involved in the project, who are linked by an interest in ecosystem structure and function, include a soil scientist, two microclimatologists, a hydrologist, a palynologist (who studies ancient pollen), a plant ecologist, and graduate students from fields as varied as geochemistry, botany, and microbial ecology. The work continues today as part of the Fluxnet Canada–Canadian Carbon Project research networks, to measure and model the influence of climatic and seasonal changes on the carbon balance of a peatland. Scientific activities at the Mer Bleue site use an instrument tower equipped for meteorological measurements, including energy balance, water vapour, and carbon dioxide and methane fluxes, combined with field investigations on plant growth and decomposition, hydrology, and experimental manipulations, such as drainage and the addition of nutrients (see photo).

On the basis of these ongoing measurements, now one of the longest-standing continuous sets of measurements of a northern peatland, scientists are developing a series of comprehensive ecosystem models for peatlands.[1] By studying this most typical of Canadian ecosystems, scientists are contributing to our understanding of how soils may behave in a global context, in response to the phenomenon of climate change.

These structures are part of the scientific instrumentation at Mer Bleue, measuring fluxes of soil gases.

Soil as a System

We generally overlook the startling complexity of soils. In everyday language we tend to equate the word *soil* with the word *dirt*. **Soil**, however, is not merely loose material derived from rock; it is a complex plant-supporting system consisting of disintegrated rock, organic matter, water, gases, nutrients, and microorganisms (**FIGURE 7.1**). Soil is also fundamental to the support of life on this planet and the provision of food for the growing human population. As a resource it is renewable if managed carefully, but it is currently at risk in many locations around the world.

Soil consists very roughly of half mineral matter with varying proportions of organic matter, with the rest of the pore space being taken up by air, water, and other soil gases. The organic matter in soil includes living and dead microorganisms as well as decaying material derived from plants and animals. A single teaspoonful of soil can contain 100 million bacteria, 500 000 fungi, 100 000 algae, and 50 000 protists. Soil also provides habitat for earthworms, insects, mites, millipedes, centipedes, nematodes, sow bugs, and other invertebrates, as well as burrowing mammals, amphibians, and reptiles. The composition of a region's soil can have as much influence on the region's ecosystems as do the climate, latitude, and elevation. In fact, because soil is composed of living and nonliving

CHAPTER SEVEN SOILS AND AGRICULTURE   191

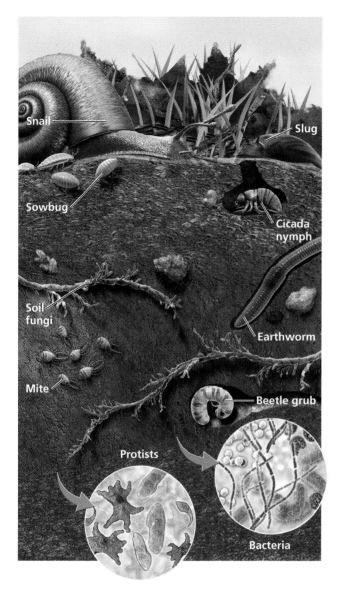

FIGURE 7.1
Soil is a complex mixture of organic and inorganic components and is full of living organisms whose actions help keep it fertile. In fact, entire ecosystems exist in soil. Most soil organisms, from bacteria to fungi to insects to earthworms, decompose organic matter. Many, such as earthworms, also help to aerate the soil.

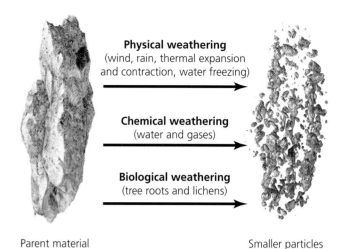

FIGURE 7.2
The weathering of parent material is the first step in soil formation. Rock is broken down into finer particles by physical, chemical, or biological means.

components that interact in complex ways, soil itself meets the definition of an ecosystem.

Soil formation is slow and complex

The formation of soil plays a key role in terrestrial primary succession, which begins when the lithosphere's parent material is exposed to the effects of the atmosphere, hydrosphere, and biosphere. **Parent material** is the base geological material in a particular location. It can include lava or volcanic ash; rock or sediment deposited by glaciers; wind-blown dunes; sediments deposited by rivers, in lakes, or in the ocean; or **bedrock**, the continuous mass of solid rock that makes up Earth's crust.

The processes most responsible for soil formation are weathering, erosion, and the deposition and decomposition of organic matter. **Weathering** describes the physical, chemical, and biological processes that break down rocks and minerals, turning large particles into smaller particles (**FIGURE 7.2**). These small particles of mineral matter—called *regolith*—are the precursors of soils.

Physical weathering (or **mechanical weathering**) breaks rocks down without triggering a chemical change in the parent material. Wind and rain are two main forces of physical weathering. Daily and seasonal temperature variation aids their action by causing the thermal expansion and contraction of parent material. Areas with extreme temperature fluctuations experience rapid rates of physical weathering. Water freezing and expanding in cracks in rock also causes physical weathering.

Chemical weathering results when water or other substances chemically interact with parent material. Warm, wet conditions usually accelerate chemical weathering. Circumstances where precipitation or groundwater is unusually acidic, such as in bogs, also promote chemical weathering.

Biological weathering occurs when living things break down parent material by physical or chemical means. For instance, lichens initiate primary terrestrial succession by producing acid, which chemically weathers rock. A tree may accelerate weathering through the physical action of its roots as they grow and rub against rock. It may also accelerate weathering chemically through the decomposition of its leaves and branches or with chemicals it releases from its roots.

weighing the issues

EARTH'S SOIL RESOURCES

7-1

It can take anywhere from 500 to 10 000 years to produce 1 cm of natural topsoil, depending on local conditions of temperature, moisture, and the other factors that influence soil formation. Consider that much of Canada's land area was scraped free of soil by the passage of huge ice masses during the last glaciation. The glaciers retreated about 10 000 years ago, but even today much of interior and northern Canada is not covered by a great thickness of soil. Given this very long renewal time, is soil a renewable resource? How should the long renewal time influence soil management?

Weathering produces fine particles and is the first step in soil formation. Another process often involved is **erosion**, the movement of soil from one area to another. Erosion may sometimes help form soil in one locality by depositing material it has depleted from another. The transport process itself also can promote physical weathering, as the transported particles collide and scrape against one another. Erosion is particularly prevalent when soil is denuded of vegetation, leaving the surface exposed to water and wind that may wash or blow it away. Although erosion can help build new soil in the long term, on the timescale of human lifetimes and for the natural systems on which we depend, erosion is generally perceived as a destructive process that reduces the amount of life that a given area of land can support.

Biological activity contributes to soil formation through the deposition, decomposition, and accumulation of organic matter. As plants, animals, and microbes die or deposit waste, this material is incorporated into the substrate, mixing with minerals. The deciduous trees of temperate forests, for example, drop their leaves each fall, making leaf litter available to the detritivores and decomposers that break it down and incorporate its nutrients into the soil. In decomposition, complex organic molecules are broken down into simpler ones, including those that plants can take up through their roots.

Partial decomposition of organic matter creates **humus**, a dark, spongy, crumbly mass of material made up of complex organic compounds. Soils with high humus content hold moisture well and are productive for plant life. Soils that are dominated by partially decayed, compressed organic material—like the soil at Mer Bleue—are called **peat**. Peat is characteristic of northern climates

Table 7.1 The Five Factors That Influence Soil Formation

Factor	Effects
Climate	Soil forms faster in warm, wet climates. Heat speeds chemical reactions and accelerates weathering, decomposition, and biological growth. Moisture is required for many biological processes and can speed weathering.
Organisms	Earthworms and other burrowing animals mix and aerate soil, add organic matter, and facilitate microbial decomposition. Plants add organic matter and affect a soil's composition and structure.
Topographical relief	Hills and valleys affect exposure to sun, wind, and water, and they influence where and how soil moves. Steeper slopes result in more runoff and erosion and in less leaching, accumulation of organic matter, and differentiation of soil layers.
Parent material	Chemical and physical attributes of the parent material influence properties of the resulting soil.
Time	Soil formation takes decades, centuries, or millennia. The four factors above change over time, so the soil we see today may be the result of multiple sets of factors.

because cool temperatures slow the decay process, allowing great thicknesses of organic material to accumulate.

Weathering, erosion, the accumulation and transformation of organic matter, and other processes that contribute to soil formation are all influenced by outside factors. Soil scientists cite five primary factors that influence the formation of soil (Table 7.1).

A soil profile consists of layers known as horizons

Once weathering has produced an abundance of small mineral particles, wind, water, and organisms begin to move and sort them. Eventually, distinct layers develop. Each layer of soil is known as a **horizon**, and the cross-section as a whole, from surface to bedrock, is known as a **soil profile**. Soil scientists subdivide the layers according to their characteristics and the processes that take place within them. For our purposes we will discuss five major horizons, known as the O, A, B, C, and R horizons (**FIGURE 7.3**). Soils from different locations vary, and few soil profiles contain all these horizons, but any given soil contains at least some of them. Generally, the degree of

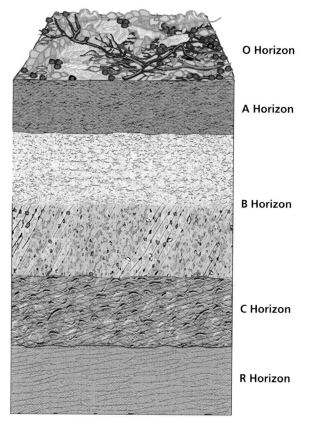

FIGURE 7.3
Mature soil consists of layers, or horizons, that have different compositions and characteristics. The number and depth of horizons vary from place to place and from soil type to soil type, producing different soil profiles. In general, organic matter and the degree of weathering decrease as one moves downward in a soil profile. The O horizon consists mostly of organic matter deposited by organisms. The A horizon, or topsoil, consists of some organic material mixed with mineral components. Minerals tend to be leached from the A horizon down into the B horizon. The C horizon consists largely of weathered parent material, which may overlie an R horizon of pure parent material.

weathering and the concentration of organic matter decrease as one moves downward in the soil profile.

Many soil profiles include an uppermost layer consisting mostly of organic matter, such as decomposing branches, leaves, and animal waste. This thin layer is designated the **O horizon** (O for *organic*) or litter layer. Distinctions also are made among *L, F,* and *H horizons,* which are organic horizons derived from the accumulation of forest litter in various stages of decomposition.[2]

Just below the organic horizon lies the **A horizon**, consisting of inorganic mineral components, with organic matter and humus from above mixed in. The A horizon is often referred to as **topsoil**, that portion of the soil that is most nutritive for plants and therefore most vital to ecosystems and agriculture. Topsoil takes its loose texture and dark colour from its humus content. The O and A horizons are home to most of the countless organisms that give life to soil.

Generally, the degree of weathering and the concentration of organic matter decrease as one moves downward in a soil profile from the surface. Minerals are carried downward as a result of **leaching**, the process whereby solid particles suspended or dissolved in liquid are transported to another location. Soil that undergoes leaching is like coffee grounds in a drip filter. When it rains, water infiltrates the soil (just as it infiltrates coffee grounds), dissolves some of its components, and carries them downward into the deeper horizons. Minerals that are commonly leached include iron, aluminum, and silicate clay. In some soils, minerals may be leached so rapidly that plants are deprived of nutrients. Minerals that leach rapidly from soils may be carried into groundwater and can pose human health threats when the water is extracted.

Minerals and organic matter that are leached from the topsoil move down into the **B horizon**, or **subsoil**, where they accumulate. The **C horizon**, if present, is located below the B horizon and consists of parent material unaltered or only slightly altered by the processes of soil formation. It therefore contains rock particles that are larger and less weathered than the layers above. The C horizon sits directly above the **R horizon**, or parent material (*R* stands for *rock*). Finally, certain soils are characterized by the presence of a distinct layer of water, called a *W horizon*. For example, some arctic soils contain a segregated layer of perennially frozen ice, or **permafrost**.

Soil can be characterized by colour, texture, structure, and pH

The horizons presented above depict an idealized, "typical" soil, but soils display great variety. Canadian soil scientists classify soils into 10 major groups, based largely on the processes thought to form them (Table 7.2). Within these 10 *orders*, there are dozens of *great groups*, hundreds of *subgroups*, and thousands of soils belonging to lower categories, all arranged in a hierarchical system. Scientists classify soils into these various categories by using properties such as colour, texture, structure, and pH.

Soil colour The colour of soil (**FIGURE 7.4**) can indicate its composition and sometimes its fertility. For example, the famously red colour of soils on Prince Edward Island is a result of the high iron content of the soil. Black or dark brown soils are usually rich in organic matter, whereas a pale grey to white colour often indicates leaching or low organic content. This colour variation

Table 7.2 The Canadian System of Soil Classification

a. Categories or "Taxa" in the Canadian System of Soil Classification

Taxa	Principles used	Number of classes
Order	Dominant soil-forming process	10
Great group	Strength of soil-forming process	31
Subgroup	Kind and arrangement of horizons	231
Family	Parent material characteristics	About 10 000
Series	Detailed features of the soil	About 100 000

b. Orders in the Canadian System of Soil Classification

Brunisolic	Poorly developed soils (i.e., lacking in horizon development, sometimes only lightly weathered, but slightly more developed than regosols, see below) that typically form under boreal forests
Chernozemic	Well-drained to imperfectly drained soils with surface horizons darkened by the accumulation of organic matter from the decomposition of grasses; typical of the Interior Plains of Western Canada
Gleysolic	Soils that are mottled (i.e., patchy) in colour, as a result of intermittent or continuous saturation with water and reducing (i.e., non-oxygenated) conditions; saturation may result from either a high groundwater table or temporary accumulation of water, or both
Cryosolic	Soils that form in either mineral or organic materials that have permafrost within 1 m of the surface; occupying much of the northern third of Canada
Luvisolic	Soils with light-coloured eluvial (E, or leached) horizons, and B horizons in which clay has accumulated; characteristic of well-drained to imperfectly drained sites, in sandy loam to clay, base-saturated parent materials under forest vegetation in subhumid to humid, mild to very cold climates, from the southern extremity of Ontario to the zone of permafrost and from Newfoundland to British Columbia
Organic	Soils that are composed largely of organic materials, including soils commonly known as peat, muck, or bog and fen soils; commonly saturated with water for prolonged periods
Regosolic	Weakly developed soils, lacking a B horizon
Podzolic	Soils with a B horizon in which the dominant accumulation product is amorphous material composed mainly of humified organic matter combined in varying degrees with aluminum and iron; typically form under forest or heath vegetation in cool to very cold climates
Solonetzic	Soils with a B horizon that is very hard when dry, and swells to a sticky mass of very low permeability when wet; occur on saline parent materials in some areas of the Interior Plains in association with chernozemic soils, mostly associated with a vegetative cover of grasses
Vertisolic	Disturbed soils with high clay contents, characterized by shrinking—swelling or wetting—drying cycles that either disrupt or inhibit the formation of soil horizons

Source: Table based on Soil Classification Working Group. 1998. The Canadian System of Soil Classification, 3rd ed. Ottawa: Agriculture and Agri-Food Canada. Publication 1646, 187 pp. (available online through the National Land and Water Information Service, http://sis2.agr.gc.ca/cansis/taxa/cssc3/index.html).

---

occurs among soil horizons in any given location and also among soils from different geographic locations. Long before modern analytical tests of soil content were developed, the colour of topsoil provided farmers and ranchers with information about a region's potential to support crops and provide forage for livestock.

Soil texture **Soil texture** is determined by the size of particles and is the basis on which soils are assigned to one of three general categories (**FIGURE 7.5**). **Clay** consists of particles less than 0.002 mm in diameter, **silt** of particles 0.002–0.05 mm, and **sand** of particles 0.05–2 mm. Sand grains, as any beachgoer knows, are large enough to see individually and do not adhere to one another. Clay particles, in contrast, readily adhere to one another and give clay a sticky feeling when moist. Soil with a relatively even mixture of the three particle sizes is known as **loam**.

For a farmer, soil texture influences a soil's *workability*, its relative ease or difficulty of cultivation. This is because texture influences the soil's **porosity**, a measure of the relative volume of spaces within the material, as well as its **permeability**, a measure of the interconnectedness of the spaces and the ease with which fluids can move around in the material. In general, the finer the particles in a sediment or soil, the smaller the spaces between them. The smaller the spaces, the harder it is for water and air to travel through the soil, slowing infiltration and reducing the amount of oxygen available to soil biota.

FIGURE 7.4
The colour of soil may vary drastically from one location to another. A soil's composition affects its colour. For instance, soils high in organic matter tend to be dark brown or black.

It is possible for a material to have a fairly high porosity but low permeability. This is typical of soils with high clay content. Conversely, soils with large particles tend to have larger spaces that are highly interconnected, allowing water to pass through (and beyond the reach of plant roots) too quickly. Thus, crops planted in sandy soils require frequent irrigation. For this reason, silty soils with medium-sized pores, or loamy soils with mixtures of pore sizes, are generally best for plant growth and crop agriculture.

Soil structure **Soil structure** is a measure of the organization or "clumpiness" of soil. Some degree of structure encourages soil productivity, and biological activity helps promote this structure. However, soil clumps that are too large can discourage plant roots from establishing if soil particles are compacted too tightly together. Repeated tilling can compact soil and make it less able to absorb water. When farmers repeatedly till the same field at the same depth, they may end up forming *ploughpan* or *hardpan*, a hard layer that resists the infiltration of water and the penetration of roots.

Soil pH The degree of acidity or alkalinity influences a soil's ability to support plant growth. Plants can die in soils that are too acidic (low pH) or alkaline (high pH), but even a moderate variation can influence the

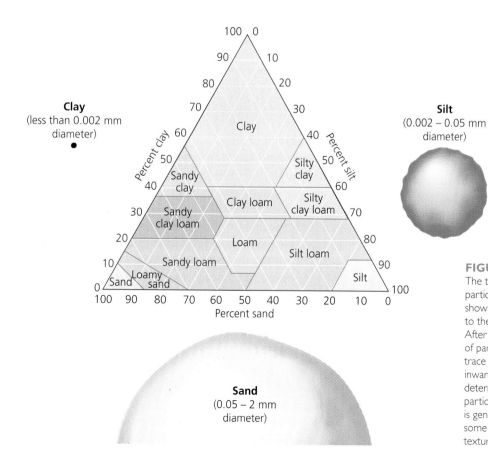

FIGURE 7.5
The texture of soil depends on its mix of particle sizes. Using the triangular diagram shown, scientists classify soil texture according to the relative proportions of sand, silt, and clay. After measuring the percentage of each type of particle size in a soil sample, a scientist can trace the appropriate white lines extending inward from each side of the triangle to determine what type of soil texture that particular combination of values creates. Loam is generally the best for plant growth, although some types of plants grow better in other textures of soil.

availability of nutrients for plants' roots. During leaching, for instance, acids from organic matter may remove some nutrients from the sites of exchange between plant roots and soil particles, and water carries these nutrients deeper.

Cation exchange is vital for plant growth

The characteristics of soil affect its ability to provide plants with nutrients. Plants gain many nutrients through a process called **cation exchange**. Soil particle surfaces that are negatively charged hold cations, or positively charged ions, such as those of calcium, magnesium, and potassium. In cation exchange, plant roots donate hydrogen ions to the soil in exchange for these nutrient ions, which the soil particles then replenish by exchange with soil water. *Cation exchange capacity* expresses a soil's ability to hold cations (preventing them from leaching and thus making them available to plants) and is a useful measure of soil fertility. Soils with fine texture (e.g., clay) and soils rich in organic matter have the greatest cation exchange capacity. As soil pH becomes lower (more acidic), cation exchange capacity diminishes, nutrients leach away, and soil instead may supply plants with harmful aluminum ions. This is one way in which acidic precipitation can harm soils and plant communities.

Soil: The Foundation for Feeding a Growing Population

Healthy soil is vital for agriculture, for forests (Chapter 10), and for the functioning of Earth's natural systems. Productive soil is a renewable resource, but if we abuse it through careless or uninformed practices, we can greatly reduce its productivity. Like other renewable resources, if soil is degraded or washed away at a rate that is faster than the rate at which it can be renewed, it effectively becomes nonrenewable because the supply, or stock, of the resource is being depleted.

As the human population has increased, so have the amounts of land and resources we devote to agriculture, which currently covers 38% of Earth's land surface. We can define **agriculture** as the practice of raising crops and livestock for human use and consumption. We obtain most of our food and fibre from **cropland**, land used to raise plants for human use, and **rangeland**, land used for grazing livestock.

As population and consumption increase, soils are being degraded

If we are to feed the world's rising human population, we will need to change our diet patterns or increase agricultural production—and do so sustainably, without degrading the environment and reducing its ability to support agriculture. We cannot simply keep expanding agriculture into new areas—the spreading or **extensification** of resource extraction—because land suitable and available for farming is running out. Instead, we must find ways to improve the efficiency of food production in areas that are already in agricultural use.

Today many lands unsuitable for farming are being farmed, causing considerable environmental damage. Mismanaged agriculture has turned grasslands into deserts and has removed ecologically precious forests. It has extracted nutrients from soils and added them to water bodies, harming both systems. It has diminished biodiversity, encouraged invasive species, and polluted soil, air, and water with toxic chemicals. Poor agricultural practices have allowed countless tons of fertile soil to be blown and washed away.

As our planet gains more than 80 million people each year, we lose 5 million to 7 million hectares of productive cropland annually. Throughout the world, especially in drier regions, it has gotten more difficult to raise crops and graze livestock as soils have become eroded and degraded (**FIGURE 7.6**). **Soil degradation**, damage to or loss of soil, around the globe has resulted from roughly equal parts of forest removal, cropland agriculture, and overgrazing of livestock, with a much smaller (though still significant) contribution from industrial contamination (**FIGURE 7.7**).

Soil degradation has direct impacts on agricultural production. Scientists estimate that over the past 50 years soil degradation has reduced potential rates of global grain production by 13% on cropland and 4% on rangeland. By the middle of the twenty-first century, there will likely be 3 billion more mouths to feed. For these reasons, it is imperative that we learn to farm in sustainable ways that are gentler on the land and that maintain the integrity of soil.

Agriculture began to appear around 10 000 years ago

During most of our species' 200 000-year existence,[3] we were hunter–gatherers, depending on wild plants and animals. Then about 10 000 years ago, as the climate

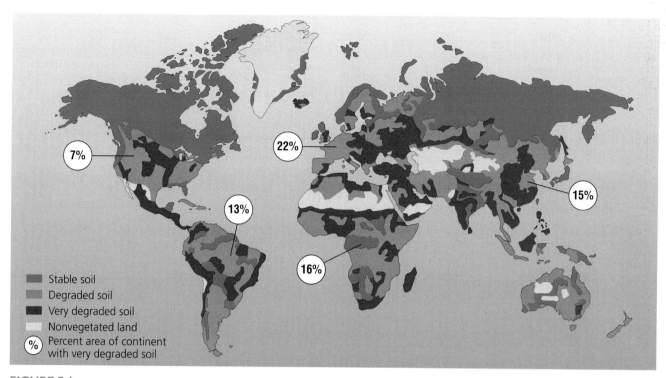

FIGURE 7.6
Soils are becoming degraded in many areas worldwide. Europe currently has a higher proportion of degraded land than other continents because of its long history of intensive agriculture, but degradation is rising quickly in developing countries in Africa and Asia. Data from United Nations Environment Programme (UNEP). 2002. *Global environmental outlook 3*. London: UNEP and Earthscan Publications Limited.

warmed following a period of glaciation, people in some cultures began to raise plants from seed and to domesticate animals.

Agriculture most likely began as hunter–gatherers brought back to their encampments wild fruits, grains, and nuts. Some of these foods fell to the ground, were thrown away, or were eaten and survived passage through the digestive system. The plants that grew from these seeds near human encampments likely produced fruits that were on average larger and tastier than those in the wild—they sprang from seeds of fruits selected by people because they were especially large and delicious. As these plants bred with others nearby that shared their characteristics, they gave rise to subsequent generations of plants with large and flavourful fruits.

Eventually, people realized that they could guide this selective process, and they began intentionally planting seeds from the plants whose produce was most desirable. This is, of course, artificial selection at work. This practice of selective breeding continues to the present day and has produced the many hundreds of crops we enjoy, all of which are artificially selected versions of wild plants. People followed the same process of selective breeding with animals, creating livestock from wild species—by accident at first, then by intention.

Once our ancestors learned to cultivate crops and raise animals, they began to settle in more permanent camps and villages near water sources. Agriculture and a sedentary lifestyle likely reinforced each other. The need to harvest crops kept people sedentary, and once they were sedentary, it made sense to plant more crops; this is a positive feedback cycle. Population increase resulted from these developments and further promoted them. This is a simple consequence of the biological

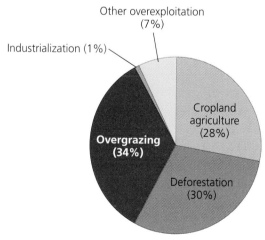

FIGURE 7.7
Most of the world's soil degradation results from cropland agriculture, overgrazing by livestock, and deforestation. Data from Wali, M. K., et al., 1999. Assessing terrestrial ecosystem sustainability: Usefulness of regional carbon and nitrogen models. *Nature and Resources* 35:21–33.

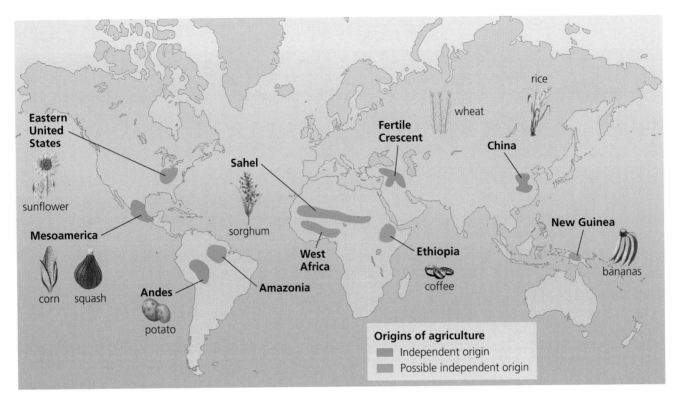

FIGURE 7.8
Agriculture appears to have originated independently in multiple locations throughout the world, as different cultures domesticated certain plants and animals from wild species living in their environments. This depiction summarizes conclusions from diverse sources of research on evidence for early agriculture. Areas where people are thought to have independently invented agriculture are coloured green. (China may represent two independent origins.) Areas coloured blue represent regions where people either invented agriculture independently or obtained the idea from cultures of other regions. A few of the many crop plants domesticated in each region are shown. Data from syntheses in Diamond, J., 1997. *Guns, Germs, and Steel*. New York: W.W. Norton; and Goudie, A. 2000. *The Human Impact*, 5th ed. Cambridge, MA: MIT Press.

productivity of land, and its relationship to carrying capacity.

Agriculture is thus a form of **intensification**—a way to increase the productivity of a given unit of land. Intensification can increase the carrying capacity of a land area (up to a point). A hunter–gatherer lifestyle requires a very large land area to support a given population; switching to a sedentary lifestyle based on agriculture allowed for larger groups to be supported on much smaller areas of land. In human history, the development of agriculture represents a huge technological advancement that permitted—or possibly even caused—a sudden dramatic increase in population. The ability to grow excess farm produce enabled some people to leave farming and live off the food that others produced. This led to the development of professional specialties, commerce, technology, densely populated urban centres, social stratification, and politically powerful elites. For better or worse, the advent of agriculture eventually brought us the civilization we have today.

Evidence from archeology and paleoecology suggests that agriculture was invented independently by different cultures in at least five areas of the world and possibly 10 or more (**FIGURE 7.8**). The earliest widely accepted archeological evidence for plant domestication is from the "Fertile Crescent" region of the Middle East about 10 500 years ago, and the earliest evidence for animal domestication also is from that region, just 500 years later. Crop remains have been dated by using radiocarbon dating and similar methods. Wheat and barley originated in the Fertile Crescent, as did rye, peas, lentils, onions, garlic, carrots, grapes, and other food plants familiar to us today. The people of this region also domesticated goats and sheep. Meanwhile, in China, domestication began as early as 9500 years ago, leading eventually to the rice, millet, and pigs we know today. Agriculture in Africa (coffee, yams, sorghum, and more) and the Americas (corn, beans, squash, potatoes, llamas, and more) developed later in several areas, 4500–7000 years ago.

For thousands of years, the work of cultivating, harvesting, storing, and distributing crops was performed by human and animal muscle power, along with hand tools and simple machines (**FIGURE 7.9**). This biologically powered agriculture is known as **traditional agriculture**. In the oldest form of traditional agriculture, known as **subsistence agriculture**, farming families produce only

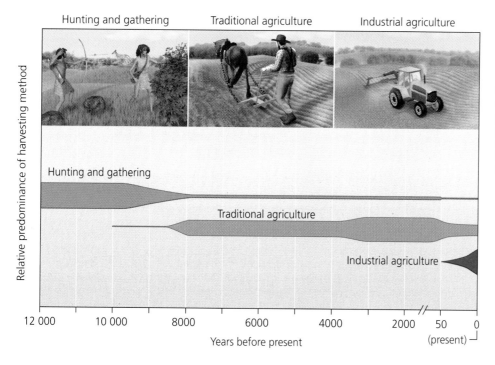

FIGURE 7.9
Hunting and gathering composed the predominant human lifestyle until the onset of agriculture and sedentary living, which centred on farms, villages, and cities, beginning nearly 10 000 years ago. Over the millennia, societies practising traditional agriculture gradually replaced hunter–gatherer cultures. Only within the past century has industrialized agriculture spread, replacing much traditional agriculture.

enough food for themselves and do not make use of large-scale irrigation, fertilizer, or teams of labouring animals. Intensive traditional agriculture sometimes uses draft animals and employs significant quantities of irrigation water and fertilizer, but stops short of using fossil fuels. This type of agriculture aims to produce food for the farming family, as well as excess food to sell in the market.

Industrialized agriculture is newer still

The Industrial Revolution introduced large-scale mechanization and fossil fuel combustion to agriculture just as it did to industry, enabling farmers to replace horses and oxen with faster and more powerful means of cultivating, harvesting, transporting, and processing crops. Other advances facilitated irrigation and fertilizing, while the invention of chemical pesticides reduced competition from weeds and herbivory by insects and other crop pests.

To be efficient, **industrialized agriculture** demands that vast fields be planted with single types of crops. The uniform planting of a single crop, termed **monoculture**, is distinct from the *polyculture* approach of much traditional agriculture, such as Native American farming systems that mixed maize, beans, squash, and peppers in the same fields. Today, industrialized agriculture occupies about 25% of the world's cropland.

Industrialized agriculture spread from developed nations to developing nations with the advent of the **Green Revolution**, a phenomenon we will explore in Chapter 8. Beginning around 1950, the Green Revolution introduced new technology, crop varieties, and farming practices to the developing world. These advances dramatically increased yields per hectare of cropland and helped millions avoid starvation. But despite its successes, the Green Revolution is exacting a high price. The intensive cultivation of farmland is creating new problems and exacerbating old ones. Many of these problems pertain to the integrity of soil, which is the very foundation of our terrestrial food supply.

Soil Degradation: Problems and Solutions

Scientists' studies of soil and the practical experience of farmers have shown that the most desirable soil for agriculture is a loamy mixture with a pH close to neutral that is workable and capable of holding nutrients. Many soils deviate from this ideal and prevent land from being arable or limit the productivity of arable land. Increasingly, limits to productivity are being set by human impact that has degraded many once-excellent soils. Common problems affecting soil productivity include erosion, desertification, salinization, waterlogging, nutrient depletion, structural breakdown, and pollution.

Regional differences in soil traits can affect agriculture

The characteristics of soil and soil profiles can vary from place to place. One example that bears on agriculture is

FIGURE 7.10
In tropical forested areas, the traditional form of farming is swidden agriculture, as seen here in Surinam. In this practice, forest is cut, the plot is farmed for one to a few years, and the farmer then moves on to clear another plot, leaving the first to regrow into forest. This frequent movement is necessary because tropical soils are nutrient-poor, with nearly all nutrients held in the vegetation. Burning the cut vegetation adds nutrients to the soil, which is why this practice is often called "slash-and-burn" agriculture. At low population densities, this form of farming had little large-scale impact on forests, but at today's high population densities, it is a leading cause of deforestation.

the difference between soils of tropical rainforests and those of temperate grasslands. Although rainforest ecosystems have high primary productivity, most of their nutrients are tied up in plant tissues and not in the soil. For example, the soil of the Amazonian rainforest is much less productive than the soil of the grasslands in Saskatchewan.

To understand how this can be, consider the main differences between the two regions: temperature and rainfall. The enormous amount of rain that falls in the Amazon readily leaches minerals and nutrients out of the topsoil. Those not captured by plants are taken quickly down to the water table, out of reach of most plants' roots. High temperatures speed the decomposition of leaf litter and the uptake of nutrients by plants, so amounts of humus remain small, and the topsoil layer remains thin.

Thus when forest is cleared for farming, cultivation quickly depletes the soil's fertility. This is why the traditional form of agriculture in tropical forested areas is **swidden** agriculture, in which the farmer cultivates a plot for one to a few years and then moves on to clear another plot, leaving the first to grow back to forest (**FIGURE 7.10**). This method may work well at low population densities, but with today's high human populations, soils may not be allowed enough time to regenerate. As a result, intensive agriculture has ruined the soils and forests of many tropical areas.

In temperate grassland areas, such as the Saskatchewan prairies, in contrast, rainfall is low enough that leaching is reduced and nutrients remain high in the soil profile, within reach of plants' roots. Plants take up nutrients and then return them to the topsoil when they die; this cycle maintains the soil's fertility. The thick, rich topsoil of temperate grasslands can be farmed repeatedly with minimal loss of fertility if proper farming techniques are used. However, growing and harvesting crops without returning adequate organic matter to the soil gradually depletes organic material, and leaving soil exposed to the elements increases erosion of topsoil. It is such consequences that farmers in many locations around the world have sought to forestall through the use of reduced tillage and other location-appropriate agricultural approaches.

Erosion can degrade ecosystems and agriculture

Erosion, as we have noted, is the removal of material from one place and its transport toward another by the action of wind or water. **Deposition** is the arrival of eroded material at its new location. Erosion and deposition are natural processes that in the long run can help create soil. Flowing water can deposit eroded sediment in river valleys and deltas, producing rich and productive soils. This is why floodplains are excellent for farming and why flood-control measures can decrease the productivity of agriculture in the long run.

However, erosion often becomes a problem locally for ecosystems and agriculture because it nearly always takes place much more quickly than soil is formed. Furthermore, erosion tends to remove topsoil, the most valuable soil layer for living things. People have increased the vulnerability of fertile lands to erosion through three widespread practices:

1. Overcultivating fields through poor planning or excessive ploughing, disking, or harrowing
2. Overgrazing rangelands with more livestock than the land can support
3. Clearing forested areas on steep slopes or with large clear-cuts

Erosion can be gradual and hard to detect. For example, an erosion rate of 12 tonnes per hectare per year removes only a penny's thickness of soil per year. In many parts of the world, scientists, farmers, and extension agents are measuring erosion rates in hopes of identifying areas in danger of serious degradation before they become too badly damaged.

Soil erodes by several mechanisms

Several types of erosion can occur, including wind erosion and four principal kinds of water erosion (**FIGURE 7.11**).

Research indicates that rill erosion has the greatest potential to move topsoil, followed by sheet erosion and splash erosion, respectively. All types of water erosion—particularly gully erosion—are more likely to occur where

(a) Splash erosion

(b) Sheet erosion

(c) Rill erosion

(d) Gully erosion

FIGURE 7.11
Splash erosion **(a)** occurs as raindrops strike the ground with enough force to dislodge small amounts of soil. *Sheet erosion* **(b)** results when thin layers of water traverse broad expanses of sloping land. *Rill erosion* **(c)** leaves small pathways along the surface where water has carried topsoil away. *Gully erosion* **(d)** cuts deep into soil, leaving large gullies that can expand as erosion proceeds.

Table 7.3 Universal Soil Loss Equation

$A = R \times K \times LS \times C \times P$

A	Predicted soil loss caused by water erosion	Results in tonnes per hectare per year
R	Erosivity factor	Quantifies the erosive force of rainfall and runoff; takes into account both total amount of rainfall and its intensity
K	Soil erodibility factor	Represents the ease with which a soil is eroded; based on the cohesiveness of a soil and its resistance to detachment and transport
LS	Slope length and steepness factor	Steeper slopes lead to higher flow velocities; longer plots accumulate runoff from larger areas and result in higher flow velocities
C	Vegetative cover and management factor	Considers the type and density of vegetative cover as well as all related management practices, such as tillage, fertilization, and irrigation
P	Erosion control practices factor	Influence of conservation practices, such as contour planting, strip cropping, grassed waterways, and terracing relative to the erosion potential of simple up–down slope cultivation

Source: Table based on material from University of BC, SoilWeb, www.landfood.ubc.ca/soil200/soil_mgmt/soil_erosion.htm.

Table 7.4 Wind Erosion Prediction Equation

$E = f(I\ C\ K\ L\ V)$

E	Predicted soil loss caused by wind erosion	Results in tonnes per acre per year, or tonnes per hectare per year
I	Soil erodibility factor	Represents the resistance of the soil to the abrasive action of wind-carried particles; dependent upon a number of factors, including soil texture and aggregation
C	Local wind erosion climate factor	Takes into account moisture, wind speed, and wind direction; wind erosion is most common in arid and semi-arid regions
K	Roughness factor	Describes the surface roughness of the soil; greater roughness indicates greater resistance to erosion
L	Length of field factor	A measure of the unsheltered length of the field; a longer field will have higher wind velocities and thus greater erosion potential
V	Vegetative cover factor	Accounts for cover type and density, including cover from crop residues

slopes are steeper. In general, steeper slopes, greater precipitation intensities, and sparser vegetative cover all lead to greater water erosion.

The *Universal Soil Loss Equation (USLE)* was developed as a tool for estimating erosion losses by water from cultivated fields, and to show how different soil and management factors influence soil erosion (Table 7.3).

Wind, like water, is a moving fluid. Wind often flows very quickly over the surface, but it typically does not have the same ability to pick up and transport large particles that water has. Nevertheless, wind can be a highly effective agent of erosion. Wind erosion, also called *aeolian erosion*, operates mainly by *deflation*, whereby all loose, fine-grained material is picked up from the surface. Another important mechanism of wind erosion is *abrasion*, whereby wind-transported particles become "projectiles," striking other rocks at the surface and causing them to break up. The *Wind Erosion Prediction Equation* shows how wind erosion is a function of five factors and their interactions (Table 7.4).

Grasslands, forests, and other plant communities protect soil from both wind and water erosion. Vegetation breaks the wind and slows water flow, while plant roots hold soil in place and take up water. Removing plant cover will nearly always accelerate erosion.

Soil erosion is a global problem

In today's world, humans are the primary cause of erosion, and we have accelerated it to unnaturally high rates. In a 2004 study, geologist Bruce Wilkinson analyzed prehistoric erosion rates from the geologic record and compared these with modern rates. He concluded that humans are more than 10 times as influential at moving soil than are all other natural processes on the surface of the planet combined.

More than 1.9 billion hectares of the world's croplands suffer from erosion and other forms of soil degradation resulting from human activities. Between 1957 and 1990, China lost as much arable farmland as exists in Denmark, France, Germany, and the Netherlands combined. In Kazakhstan, central Asia's largest nation, industrial cropland agriculture imposed on land better suited for grazing caused tens of millions of hectares to be degraded by wind erosion. For Africa, soil degradation over the next 40 years could reduce crop yields by half. Couple these declines in soil quality and crop yields with the rapid population growth occurring in many of these areas, and we begin to see why some observers describe the future of agriculture as a crisis situation.

Erosion of agricultural soil has been a significant concern in Canada for the past 25 years or more, but improvements are occurring. Soil researchers from Agriculture and Agri-Food Canada have determined the on-farm cost of agricultural land degradation in Canada to be almost $670 million per year.[4] However, the same report showed significant reductions in the area of cropland at risk of erosion. As of 1996, approximately 85% of cropland area in Canada was deemed to be in the "tolerable" range of risk for water-related erosion, a 22% improvement over the situation 15 years earlier. The situation for wind erosion risk improved by 59% over the same period. A combination of reduced tillage, less intensive crop production, and removal of *marginal land* (that is, land ill-suited to agriculture) from production were cited as contributing to lower erosion rates.[5] Other sources[6] also point to changes in farming techniques since the recognition of a soil erosion crisis in the early 1980s, with the result of overall improvement in the health and stability of agricultural soils in Canada.

Desertification reduces productivity of arid lands

Much of the world's population lives and farms in arid environments, where **desertification** is a concern. This term describes a loss of more than 10% of productivity because of erosion, soil compaction, forest removal, overgrazing, drought, salinization, climate change, depletion of water sources, and other factors. The terms *desertification* and *degradation* are often confused, and the United Nations is careful to distinguish between them.[7] Desertification is a type of land degradation that occurs in arid and semi-arid areas and can result from various factors, including climatic variations and human activities. Land degradation is, in turn, defined as the reduction or loss of the biological or economic productivity of land.

Severe desertification can result in the expansion of desert areas or creation of new deserts in areas that once supported fertile land. This process has occurred in many areas of the Middle East that have been inhabited, farmed, and grazed for long periods. To appreciate the cumulative impact of centuries of traditional agriculture, we need only look at the present desertified state of that portion of the Middle East where agriculture originated, nicknamed the "Fertile Crescent." These arid lands—in present-day Iraq, Syria, Turkey, Lebanon, and Israel—are not so fertile anymore.

Arid and semi-arid lands are prone to desertification because their precipitation is too meagre to meet the demand for water from growing human populations. According to the United Nations Environment Programme (UNEP), 40% of Earth's land surface can be classified as drylands, arid areas that are particularly subject to degradation. Declines of soil quality in these areas have endangered the food supply or well-being of more than 1 billion people around the world. Of the affected lands, most degradation results from wind and water erosion. In recent years gigantic dust storms from denuded land in China have blown across the Pacific Ocean to North America, and dust storms from Africa's Sahara Desert have blown across the Atlantic Ocean to the Caribbean Sea. Such massive dust storms occurred in the Canadian Prairies and the Great Plains of the United States during the Dust Bowl days of the early twentieth century, when desertification shook North American agriculture and society to their very roots (**FIGURE 7.12**).

It has been estimated that desertification affects fully one-third of the planet's land area, impinging on the lives of 250 million people[8] and costing tens of billions of dollars in lost income per year. China alone loses $6.5 billion annually from desertification. In its western reaches, desert areas are expanding and joining one another because of overgrazing from more than 400 million goats, sheep, and cattle. In the Sistan Basin along the border of Iran and Afghanistan, an oasis that supported a million livestock recently turned barren in just five years, and windblown sand buried more than 100 villages. In Kenya, overgrazing and deforestation fuelled by rapid population growth has left 80% of its land vulnerable to desertification. In a positive feedback cycle, the soil degradation forces ranchers to crowd onto more marginal land and farmers to reduce fallow periods, both of which further exacerbate soil degradation.

FIGURE 7.12
Canada is not exempt from the impacts of wind erosion, as shown in this photo from the Dust Bowl of the 1930s.

A 2007 United Nations report estimated that desertification, worsened by climate change, could displace 50 million people in 10 years. The report suggested that industrialized nations fund reforestation projects in dryland areas of the developing world to slow desertification while gaining carbon credits in emissions trading programs.

The Dust Bowl was a monumental event in North America

Prior to large-scale cultivation of the Prairies and the Great Plains, native prairie grasses of this temperate grassland region held erosion-prone soils in place. In the late nineteenth and early twentieth centuries, however, many homesteading settlers arrived with hopes of making a living there as farmers. Between 1879 and 1929, cultivated area in the region soared, driven primarily by rapid increases in the price of wheat. Farmers in the region grew abundant wheat, and ranchers grazed many thousands of cattle, sometimes expanding onto unsuitable land. Both types of agriculture contributed to erosion by removing native grasses and breaking down soil structure.

At the end of 1929 the stock market crashed, sending the price of wheat lower than the price of seed; the Great Depression began and with it an inexorable cycle of poverty and land degradation that would last most of the decade. Starting in the early 1930s, a prolonged period of drought in the region exacerbated the ongoing human impacts on the soil from overly intensive agricultural practices. Strong winds began to carry away millions of tonnes of topsoil, and often newly planted seed, as well. Dust storms travelled up to 2000 km, blackening skies and coating the skins of farm workers. Some areas lost as much as 10 cm of topsoil in a few short years (**FIGURE 7.12**).

The affected region in the Prairies and Great Plains became known as the *Dust Bowl*, a term now also used for the historical event itself. The "black blizzards" of the Dust Bowl (see **FIGURE 7.12**) destroyed livelihoods and caused many people to suffer a type of chronic lung irritation and degradation known as dust pneumonia, similar to the silicosis that afflicts coal miners exposed to high concentrations of coal dust. Large numbers of farmers were forced off their land; those who stayed faced infestations of grasshoppers so thick that they clogged car radiators and made the roads slippery. Chickens and turkeys ate the grasshoppers, which caused their meat and eggs to develop a bad taste. There were no pesticides (chemical pesticides had not yet been invented) and no way to control the grasshoppers.

By 1937, Dust Bowl conditions had peaked. Since the price of wheat was so low, farmers began to plant alternative crops, such as oats, rye, flax, peas, and alfalfa. They adapted to the dry weather with reduced tilling, crop rotations (to allow the soil a chance to replenish itself), and fertilizer applications. By the end of the decade, the slow recovery from the Dust Bowl had begun.[9]

The Soil Conservation Council emerged from the experience of drought

In 1935 the Prairie Farm Rehabilitation Administration (PFRA) was set up; interestingly, however, it was not until the early 1980s that the issue of soil erosion and degradation of agricultural soils really took centre stage in Canada on a nationwide basis. This partly occurred as a result of another serious drought in the late 1970s, followed by the publication in 1984 of a book by the federal government entitled *Soil at Risk: Canada's Eroding Future*.[10] The following year saw the establishment of the first National Soil Conservation Week, and the establishment of the National Soil Conservation Program.

The Soil Conservation Council of Canada was established in 1987, with the following goals:

- *To develop a national spirit to foster a feeling of unity among those who are concerned for soil conservation;*
- *To improve the level of understanding and awareness about the causes of soil degradation among all Canadians and to increase their support of soil conservation goals;*
- *To facilitate communication among soil conservation groups, governments and industry relating to soil conservation needs, programs and policies;*
- *To communicate to the general public policies, programs or activities that affect the sustainable use of this country's soil resources;*
- *To encourage the development of policies, production methods and management systems for agriculture, forestry and land use which enables sustainable use of our soil and related resources.*[11]

Today a number of government agencies provide services to assist farmers with all aspects of soil management and conservation in Canada. Internationally, the United Nations promotes soil conservation and sustainable agriculture through a variety of programs of its Food and Agriculture Organization (FAO). The FAO's Farmer-Centred Agricultural Resource Management (FARM) Programme is a project that supports innovative approaches to resource management and sustainable agriculture in China, Thailand, Vietnam, Indonesia, Sri Lanka, Nepal, the Philippines, and India. The program studies agricultural success stories and tries to help other farmers duplicate the successful efforts. Rather than following a top-down, government-controlled approach, the FARM program relies on the creativity of local communities to educate and encourage farmers throughout Asia to conserve soils and secure their food supply.

Farmers can protect soil against degradation in various ways

Several farming techniques can reduce the impacts of conventional cultivation on soils. Such measures have been widely shared and applied in many places around the world (**FIGURE 7.13**), and some have been practised by traditional farmers for centuries.

Crop rotation The practice of alternating the kinds of crops grown in a particular field from one season or year to the next is called **crop rotation** (**FIGURE 7.14A**). Rotating crops can return nutrients to the soil, break cycles of disease associated with continuous cropping, and minimize the erosion that can come from letting fields lie fallow. Farmers in Alberta, Prince Edward Island, and elsewhere have returned to an earlier approach, in which they plant alternating swaths of land each year but leave the field stubble from the previous year to protect newly tilled fields from exposure to wind and water erosion. Many farmers rotate between wheat or corn and soybeans from one year to the next. Soybeans are *legumes*, which have specialized bacteria on their roots that can fix nitrogen. Soybeans revitalize soil that a previous crop has partially depleted of nutrients. Crop rotation also reduces insect pests; if an insect is adapted to feed and lay eggs on one particular crop, planting a different crop will leave its offspring with nothing to eat.

Contour farming Water running down a hillside can easily carry soil away, particularly if there is too little vegetative cover to hold the soil in place. Thus, sloped agricultural land is especially vulnerable to erosion. Several methods have been developed for farming on slopes. **Contour farming** (**FIGURE 7.14B**) consists of ploughing furrows sideways across a hillside, perpendicular to its slope, to help prevent formation of rills and gullies. The technique is so named because the furrows follow the natural contours of the land. In contour farming, the downhill side of each furrow acts as a small dam that slows

FIGURE 7.13
Government agricultural extension agents assist farmers worldwide, by providing information on the newest research and techniques that can help them farm productively while minimizing damage to the land. Here, an extension agent from Colombia's Instituto Colombiano Agropecuario inspects yucca plants grown by farmer Pedro Gomez on a farm in Valle del Cauca.

runoff and catches soil before it is carried away. Contour farming is most effective on gradually sloping land with crops that grow well in rows.

Intercropping and agroforestry Farmers may also gain protection against erosion by **intercropping**, planting different types of crops in alternating bands or other spatially mixed arrangements (**FIGURE 7.14C**). Intercropping helps slow erosion by providing more complete ground cover than does a single crop. Like crop rotation, intercropping offers the additional benefits of reducing vulnerability to insect and disease incidence, and, when a nitrogen-fixing legume is one of the crops, of replenishing the soil. Cover crops can be physically mixed with primary food crops, which include maize, soybeans, wheat, onions, cassava, grapes, tomatoes, tobacco, and orchard fruit.

FIGURE 7.14
The world's farmers have adopted various strategies to conserve soil. **(a)** Rotating crops, such as soybeans and corn, helps restore soil nutrients and reduce impacts of crop pests. **(b)** Contour farming reduces erosion on hillsides. **(c)** Terracing minimizes erosion in steep mountainous areas. **(d)** Intercropping can reduce soil loss while maintaining soil fertility. **(e)** Shelterbelts protect against wind erosion. **(f)** In no-till farming, corn grows up from amid the remnants of a "cover crop."

When crops are interplanted with trees, in a practice called **agroforestry**, even more benefits can be realized (see also Chapter 10). Trees draw nutrients and water from deep in the soil through their root systems, cycling them into the shallower layers of soil. They also contribute organic material to the topsoil, in the form of tree *litter*, or fallen branches and leaves. Trees can also provide partial shade for crops, although light levels must be managed by appropriate pruning and spacing of the trees. In many agricultural regions, especially in the developing world where small-scale agriculture is still practised, agroforestry provides a low-cost way to close the nutrient cycle and rehabilitate soils, while providing other sustainably harvested forest products, such as fruits, nuts, and timber.

Terracing   On extremely steep terrain, terracing (FIGURE 7.14D) is the most effective method for preventing erosion. Terraces are level platforms, sometimes with raised edges, that are cut into steep hillsides to contain water from irrigation and precipitation. **Terracing** transforms slopes into series of steps, like a staircase, enabling farmers to cultivate hilly land without losing huge amounts of soil to water erosion. Terracing is common in ruggedly mountainous regions, such as the foothills of the Himalayas and the Andes, and has been used for centuries by farming communities in such areas. Terracing is labour-intensive to establish but in the long term is likely the only sustainable way to farm in mountainous terrain.

Shelterbelts   A widespread technique to reduce erosion from wind is to establish **shelterbelts** or **windbreaks** (FIGURE 7.14E). These are rows of trees or other tall, perennial plants that are planted along the edges of fields to slow the wind. Shelterbelts have been widely planted across the Prairies and the Great Plains, where fast-growing species, such as poplars, are often used. Shelterbelts have also been combined with intercropping in a practice known as **alley cropping**. In this approach, fields planted in rows of mixed crops are surrounded by or interspersed with rows of trees that provide fruit, wood, or protection from wind. Agroforestry and alley cropping methods have been widely used in India, Africa, China, and Brazil, where coffee growers near a national conservation area have established farming systems combining farming and forestry.

Reduced tillage   To plant by using the reduced-tillage or **no-till** method (FIGURE 7.14F), a tractor pulls a drill that cuts long, shallow furrows through the O horizon of dead weeds and crop residue and the upper levels of the A horizon. The device drops seeds into the furrow and closes the furrow over the seeds. Often a localized dose of fertilizer is added to the soil along with the seeds. By increasing organic matter and soil biota while reducing erosion, no-till and reduced tillage farming can build soil up, restore it, and improve it. Proponents of no-till farming claim that the practice offers a number of benefits (Table 7.5). See "The Science Behind the Story: No-Till Agriculture in Southern Brazil".

Table 7.5   No-Till Farming

Direct benefits of no-till farming
- Conserves biodiversity in soil and in terrestrial and aquatic ecosystems
- Produces sustainable, high crop yields
- Heightens environmental awareness among farmers
- Provides shelter and winter food for animals
- Reduces irrigation demands by 10%–20%
- Crop residues act as a sink for carbon (one metric ton per hectare)
- Reduces fossil fuel use by 40%–70%
- Enhances food security by increasing drought resistance
- Reduces erosion by 90%

Indirect benefits arising from the reduction in erosion
- Reduces silt deposition in reservoirs
- Reduces water pollution from chemicals
- Increases groundwater recharge and lessens flooding
- Increases sustained crop yields and lowers food prices
- Lowers costs of treating drinking water
- Reduces costs of maintaining dirt roads
- Eliminates dust storms in towns and cities
- Increases efficiency in use of fertilizer and machinery

Source: Modified from Shaxson, T. F. 1999. The roots of sustainability: Concepts and practice: Zero tillage in Brazil, ABLH Newsletter ENABLE; World Association for Soil and Water Conservation (WASWC) Newsletter.

In Argentina, the area under no-till farming exploded from 100 000 ha in 1990 to 7.3 million ha in 1999, covering 30% of all arable land in the country. The results included increased crop yields, reduced erosion, enhanced soils, and a healthier environment. Maize yields grew by 37% and soybean yields by 11%, while costs to farmers fell by 40%–57%. Erosion, pesticide use, and water pollution declined. The techniques spread largely because of the actions of farmers themselves and their national no-till farmers' organization.

The no-till or reduced-tillage practice has been widely adopted in Canada but not with unmitigated success. The Soil Conservation Council of Canada estimates that as much as 50% of some crops in Ontario are now grown by using a no-till approach, including wheat and soybeans. However, some crops—notably corn—are not amenable to reduced tillage, which tends to keep the soil colder and moister for longer than conventional tillage methods.[12] Critics of no-till and reduced-tillage farming also note that these techniques often require substantial use of chemical herbicides (because weeds are not physically removed from fields) and synthetic fertilizer (because other plants take up a significant portion of the soil's nutrients).

THE SCIENCE BEHIND THE STORY

No-Till Agriculture in Southern Brazil

Wheat is grown extensively in Paraná, Brazil.

In southernmost Brazil, hundreds of thousands of people make their living by farming. The warm climate and rich soils of this region's rolling highlands and coastal plains have historically made for bountiful harvests (see photo). However, repeated cycles of ploughing and planting over many decades diminished the productivity of the soil. More and more topsoil—the valuable surface layer of soil richest in organic matter and nutrients—was being eroded away by water and wind. Meanwhile, the synthetic fertilizers used to restore nutrients were polluting area waterways. Yields were falling, and by 1990 farmers were looking for help.

As a result, many of southern Brazil's farmers abandoned the conventional practice of tilling the soil after harvests. In its place, they turned to *no-tillage* farming, otherwise known as *zero-tillage*, *no-till*, or in Brazil, *plantio direto*. Turning the earth by tilling (ploughing, disking, harrowing, or chiselling) aerates the soil and works weeds and old crop residue into the soil to nourish it. Tilling, however, also leaves the surface bare of vegetation for a period of time, during which erosion by wind and water can remove precious topsoil. Although tilling historically boosted the productivity of agriculture in Europe, many experts now think it is less appropriate for soils in subtropical regions, such as southern Brazil. The heavy rainfall of tropical and subtropical regions results in greater rates of erosion, causing tilled soils to lose organic matter and nutrients.

Working with agricultural scientists and government extension agents, southern Brazil's farmers began leaving crop residues on their fields after harvesting and planting "cover crops" to keep soil protected during periods when they were not raising a commercial crop. When they went to plant the next crop, they merely cut a thin, shallow groove into the soil surface, dropped in seeds, and covered them. They did not invert the soil as they had when tilling, and the soil stayed covered with plants or their residues at all times, reducing erosion by 90%.

With less soil eroding away, and more organic material being added to it, the soil held more water and was better able to support crops. The improved soil quality meant better plant growth and greater crop production. In the state of Santa Catarina, maize yields per hectare increased by 47% between 1991 and 1999, wheat yields rose by 82%, and soybean yields grew by 83%, according to local farmers, extension agents, and international scientists. In the states of Paraná and Rio Grande do Sul, maize yields were up 67% over 10 years, and soybean yields rose by 68%.

Besides boosting yields, no-till farming methods reduced farmers' costs, because farmers now used less labour and less fuel. No-till agriculture spread quickly in the region as farmers saw their neighbours' successes and traded information through "Friends of the Land" clubs organized on local, municipal, regional, and statewide levels. In Paraná and Rio Grande do Sul, the area being farmed with no-till methods shot up from 700 000 ha in 1990 to 10.5 million ha in 1999, when it involved 200 000 farmers. In Santa Catarina, where farms are generally smaller, more than 100 000 farmers now apply no-till methods to 880 000 ha of farmland. No-till farming is now spreading northward into Brazil's tropical regions and to other parts of Latin America.

By enhancing soil conditions and reducing erosion, no-till techniques have benefited southern Brazil's society and environment as well; its air, waterways, and ecosystems are less polluted. Similar effects are being felt elsewhere in the world where no-till and reduced-tillage methods are being applied.

Reduced tillage is certainly not a panacea for all areas of the world. In general, tropical areas benefit more than temperate regions, because erosion is greater in the tropics and hot weather can overheat tilled soil. The benefits and drawbacks of different tillage approaches vary with location, soil characteristics, and type of crop. In regions suitable for reduced tillage, proponents say these approaches can help make agriculture sustainable. We will need sustainable agriculture if we are to feed the world's human population while protecting the natural environment, including the soils that vitally support our production of food.

Southern Brazil's no-till farmers have departed somewhat from the industrialized model by relying more heavily on *green manures* (dead plants as fertilizer) and by rotating fields with cover crops, including nitrogen-fixing legumes. The manures and legumes nourish the soil, and cover crops also reduce weeds by taking up space the weeds might occupy. Critics maintain, however, that green manures are generally not practical for large-scale intensive agriculture. Certainly, reduced tillage methods work well in some areas but not in others, and they work better with some crops than with others. Farmers will do best by educating themselves on the options and doing what is best for their particular crops on their own land.

weighing the issues

HOW WOULD YOU FARM?

7–2

Let us say that you are a farmer who owns land on both sides of a steep ridge. You want to plant a sun-loving crop on the sunny, but very windy, south slope of the ridge and a crop that needs a great deal of irrigation on the north slope. What types of farming techniques might maximize conservation of your soil? What other factors might you want to know about before you decide to commit to one or more methods?

FIGURE 7.15
Vast swaths of countryside in western China have been planted with fast-growing poplar trees. These reforestation efforts do not create ecologically functional forests—the plantations are too biologically simple—but they greatly slow soil erosion.

Erosion control practices protect and restore plant cover

Farming methods to control erosion make use of the general principle that maximizing vegetative cover will protect soils, and this principle has been applied widely beyond farming. It is common throughout the developed world to stabilize eroding banks along creeks and roadsides by planting vegetation to anchor the soil. In areas with severe and widespread erosion, some nations have introduced vast plantations of fast-growing trees. China has embarked on the world's largest tree-planting program to slow its soil loss (**FIGURE 7.15**). Although such reforestation efforts do help to slow erosion, they do not at the same time produce ecologically functioning forests, because tree species are selected only for their fast growth and are planted in monocultures.

Irrigation boosts productivity but can cause long-term soil problems

Erosion is not the only threat to the health and integrity of soils. Soil degradation can result from other factors as well, such as impacts caused by the application of water to crops. The artificial provision of water to support agriculture is known as **irrigation**. Some crops, such as rice and cotton, require large amounts of water, whereas others, such as beans and wheat, require relatively little. Other factors influencing the amount of water required for growth include the rate of evaporation, as determined by climate, and the soil's ability to hold water and make it available to plants' roots.

If the climate is too dry or too much water evaporates or runs off before it can be absorbed into the soil, crops may require irrigation. By irrigating crops, people have managed to turn previously dry and unproductive regions into fertile farmland. Irrigation accounts for 70% of all freshwater withdrawn by people. Irrigated land area has increased dramatically around the world, reaching almost 400 million hectares in 2007,[13] greater than the entire area of Mexico and Central America.

If some water is good for plants and soil, it might seem that more must be better. But this is not necessarily the case; there is indeed such a thing as too much water. Overirrigation in poorly drained areas can cause or exacerbate certain soil problems. Soils too saturated with water may become waterlogged. When **waterlogging** occurs, the water table is raised to the point that water bathes plant roots, depriving them of access to gases and essentially suffocating them. If it lasts long enough, waterlogging can damage or kill plants.

An even more common problem is **salinization** (or **salination**), the buildup of salts in surface soil layers. In dryland areas where precipitation is minimal and evaporation rates are high, water evaporating from the A horizon may pull water from lower horizons upward by capillary action. As this water rises through the soil, it carries dissolved salts, and when it evaporates at the surface, those salts precipitate and are left at the surface. Irrigation in arid areas generally hastens salinization, because it provides repeated doses of moderate amounts of water, which dissolve salts in the soil and gradually raise them to the surface. Moreover, because irrigation water often contains some dissolved salt in the first place, irrigation introduces new sources of salt to the soil. Overirrigation and waterlogging can worsen salinization problems, and in many areas of farmland, soil is turning whitish with encrusted salt.

Salinization now inhibits agricultural production on one-fifth of all irrigated cropland globally, costing more than $11 billion annually. As of 2000, the Food and Agriculture Organization of the United Nations estimated that the total area of salinized soil is 397 million hectares. Of 230 million hectares of irrigated land, 45 million hectares (or 19.5%) of soil was salt-affected, and of 1500 million hectares of dryland agriculture, 32 million hectares (or 2.1%) of soil was affected by salization caused by human activities.[14]

Salinization is easier to prevent than to correct

The remedies for mitigating salinization once it has occurred are more expensive and difficult to implement than the techniques for preventing it in the first place. The best way to prevent salinization is to avoid planting crops that require a great deal of water in areas that are prone to the problem. A second way is to irrigate with water that is as low as possible in salt content. A third way is to irrigate efficiently, supplying no more water than the crop requires, thus minimizing the amount of water that evaporates and hence the amount of salt that accumulates in the topsoil.

Currently, irrigation efficiency worldwide is low; only 43% of the water applied actually gets used by plants. *Drip irrigation* systems (**FIGURE 7.16**) that target water directly to plants are one solution to the problem. These systems allow more control over where water is aimed and waste far less water. Once considered expensive to install, they are becoming less costly, such that more farmers in developing countries will be able to afford them.

If salinization has occurred, one potential way to mitigate it would be to stop irrigating and wait for rain to flush salts from the soil. However, this solution is unrealistic because salinization generally becomes a problem in dryland areas where precipitation is never adequate to flush soils. A better option may be to plant salt-tolerant plants, such as barley, that can be used as food or pasture. A third option is to bring in large quantities of less-saline water with which to flush the soil. However, using too much water may cause waterlogging. As is the case with many environmental problems, preventing salinization is easier than correcting it after the fact.

(a) Conventional irrigation

(b) Drip irrigation

FIGURE 7.16
Currently, plants take up less than half the water we apply in irrigation. Conventional methods lose a great deal of water to evaporation **(a)**. In more efficient drip irrigation approaches, such as this one watering grape vines **(b)**, hoses are arranged so that water drips from holes in the hoses directly onto the plants that need the water.

weighing the issues: MEASURING AND REGULATING SOIL QUALITY 7–3

The government of Canada has adopted comprehensive measures to control air and water quality, and has set legal standards for allowable levels of various pollutants in air and water. Could such standards be developed for soil quality? If so, what properties should be measured to inform the standards?

Agricultural fertilizers boost crop yields but can be overapplied

Salinization is not the only source of chemical damage to soil. Overapplying fertilizers can also chemically damage soils. Plants grow through photosynthesis, requiring sunlight, water, and carbon dioxide, but they also require nitrogen, phosphorus, and potassium, as well as smaller amounts of more than a dozen other nutrients. Plants remove these nutrients from soil as they grow, and leaching likewise removes nutrients. If agricultural soils come to contain too few nutrients, crop yields decline. Therefore, a great deal of effort is aimed at enhancing nutrient-limited soils by adding **fertilizer**, any of various substances that contain essential nutrients.

There are two main types of fertilizers. **Inorganic fertilizers** are mined or synthetically manufactured mineral

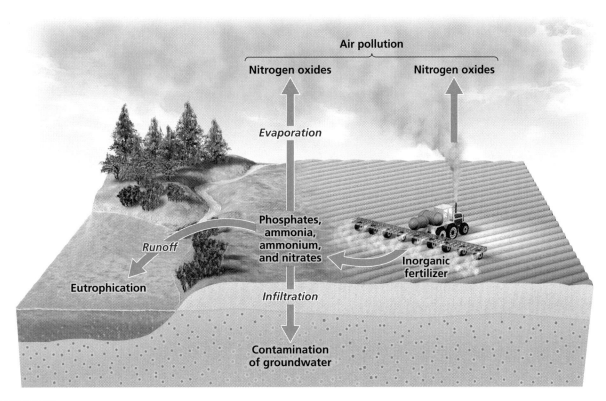

FIGURE 7.17
The overapplication of inorganic (or organic) fertilizers can have effects beyond the farm field, because nutrients that are not taken up by plants may end up in other places. Nitrates can leach into groundwater, where they can pose a threat to human health in drinking water. Phosphates and some nitrogen compounds can run off into surface waterways and alter the ecology of streams, rivers, ponds, and lakes through eutrophication. Some compounds, such as nitrogen oxides, can even enter and pollute the air. Anthropogenic inputs of nitrogen have greatly modified the nitrogen cycle, and now account for one-half the total nitrogen flux on Earth.

supplements. **Organic fertilizers** consist of natural materials (largely the remains or wastes of organisms) and include animal manure, crop residues, fresh vegetation (or "green manure"), and **compost**, a mixture produced when decomposers break down organic matter, including food and crop waste, in a controlled environment. Organic fertilizers can provide some benefits that inorganic fertilizers cannot. The proper use of compost improves soil structure, nutrient retention, and water-retaining capacity, helping to prevent erosion. As a form of recycling, composting reduces the amount of waste consigned to landfills and incinerators.

However, organic fertilizers are no panacea. For instance, manure, when applied in amounts needed to supply sufficient nitrogen for a crop, may introduce excess phosphorus that can run off into waterways. Inorganic fertilizers are generally more susceptible than are organic fertilizers to leaching and runoff, and they are somewhat more likely to cause unintended off-site impacts. Inorganic fertilizer use is growing globally; unfortunately, its mismanagement is causing increasingly severe pollution problems.

Applying substantial amounts of fertilizer to croplands has impacts far beyond the boundaries of the fields (**FIGURE 7.17**). Nitrogen and phosphorus runoff from farms and other sources can lead to phytoplankton blooms, creating an oxygen-depleted "dead zone." Such eutrophication occurs at many river mouths, lakes, and ponds throughout the world. Moreover, nitrates readily leach through soil and contaminate groundwater, and components of some nitrogen fertilizers can even volatilize (evaporate) into the air.

Through these processes, unnatural amounts of nitrates and phosphates spread through ecosystems and pose human health risks, including cancer and methemoglobinemia, or blue baby syndrome (see "The Science Behind the Story: Nitrate Contamination of the Abbotsford Aquifer"). Health Canada and the U.S. Environmental Protection Agency have both determined that nitrate concentrations in excess of 10 mg/L for adults and 5 mg/L for infants in drinking water are unsafe, yet many sources around the world exceed even the looser standard of 50 mg/L set by the World Health Organization.

THE SCIENCE BEHIND THE STORY

Nitrate Contamination of the Abbotsford Aquifer

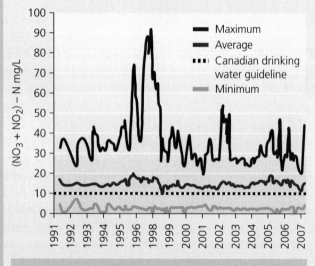

This map shows the location of the Abbotsford Aquifer in British Columbia.

The Abbotsford Aquifer is a sand and gravel deposit that covers approximately 100 km² in the Lower Fraser River Valley, British Columbia, and a similar area in adjacent Washington State, U.S.A. The *aquifer*, or water-bearing unit, is largely unprotected from surface runoff and infiltration, rendering it vulnerable to contamination.

This graph shows average, minimum, and maximum nitrate concentrations in samples of Abbotsford Aquifer groundwater over time. Most sites were intentionally chosen in areas near intense agricultural activities.

Source: Environment Canada, Pacific and Yukon Region, Vancouver, B.C., 2004.

Nitrate contamination of groundwater in the aquifer has been observed since the early 1950s. From an assessment of 2757 groundwater samples collected from monitoring wells in the region since 1992, 71% have exceeded the 10 mg/L Guideline for Canadian Drinking Water Quality set by Health Canada, with individual values up to 91.9 mg/L (see graph).

The land that overlies the aquifer is experiencing population growth and industrial, commercial, and agricultural development. This land use change brings many potential sources of nitrate contamination, which can originate from manure and chemical fertilizer use, sewage and septic tank discharges, and airport de-icing chemicals, among other sources. However, recent studies suggest that nitrate concentrations in the aquifer are primarily from agricultural sources.

Agriculture and Agri-Foods Canada estimates that the application of nitrogen fertilizer in the area of the Abbotsford Aquifer exceeds the actual uptake of nitrogen by crops, by as much as 50%. The main crop grown in the area has changed from grass hay to raspberries, which require much less nitrogen. There has also been a significant increase in poultry production (and thus excess manure) over the same period. The increased poultry production and shift to crops that require less nitrogen have resulted in excess nitrogen concentrated within a relatively small geographical area.

The Abbotsford Aquifer provides water to more than 100 000 people in the City of Abbotsford and northern Washington State. The direction of groundwater flow is generally southward, and some Washington counties whose drinking water wells draw from areas with high nitrate levels have expressed concerns about the contamination on the Canadian side.

Nitrates can affect human health by reducing the ability of blood to carry oxygen. Infants are particularly at risk from drinking water with high nitrate concentrations, and a potentially fatal condition called methemoglobinemia or "blue baby syndrome" can result. Nitrate is also a major nutrient for aquatic vegetation; excessive amounts can lead to uncontrolled growth of algae and other aquatic plants, a process known as eutrophication.

Environment Canada is conducting research on the aquifer's water quality and quantity. With other federal and provincial agencies, the Fraser River Action Program has undertaken several studies on groundwater in the Lower Fraser Valley, including studies to identify the sources of aquifer contamination, and agricultural nutrient management and manure management studies. Environment Canada has also initiated and implemented projects to educate the public on groundwater stewardship. In 1992 the Abbotsford–Sumas Aquifer International Task Force was created, in an attempt to coordinate cross-border efforts to manage this important aquifer.[15]

Grazing practices and policies can contribute to soil degradation

We have focused in this chapter largely on the cultivation of crops as a source of impacts on soils and ecosystems, but raising livestock also has such impacts. When sheep, goats, cattle, or other livestock graze on the open range, they feed primarily on grasses. As long as livestock populations do not exceed a range's carrying capacity and do not consume grasses faster than grasses can be replaced, grazing may be sustainable. However, when too many animals eat too much of the plant cover, impeding plant regrowth and preventing the replacement of biomass, the result is **overgrazing**.

Rangeland scientists have shown that overgrazing causes a number of impacts, some of which give rise to

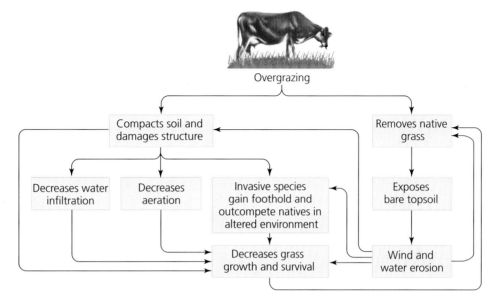

FIGURE 7.18
When grazing by livestock exceeds the carrying capacity of rangelands and their soil, overgrazing can set in motion a series of consequences and positive feedback loops that degrade soils and grassland ecosystems.

positive feedback cycles that exacerbate damage to soils, natural communities, and the land's productivity for grazing (**FIGURE 7.18**). When livestock removes too much of an area's plant cover, more soil surface is exposed and made vulnerable to erosion. Soil erosion makes it difficult for vegetation to regrow, perpetuating the lack of cover and giving rise to more erosion. Moreover, non-native weedy plants may invade denuded soils (**FIGURE 7.19**). These invasive plants are usually less palatable to livestock and can outcompete native vegetation in the new, modified environment, further decreasing native plant cover.

FIGURE 7.19
The effects of overgrazing can be striking, as shown in this photo along a fence line separating a grazed plot (right) from an ungrazed plot (left) in the Konza Prairie Reserve in Kansas. The overgrazed plot has lost much of its native grass and has been invaded by weedy plants that compete more effectively in conditions of degraded soil.

Overgrazing can also compact soils and alter their structure. Soil **compaction** makes it harder for water to infiltrate, harder for soils to be aerated, harder for plants' roots to expand, and harder for roots to conduct cellular respiration. All these effects further decrease the growth and survival of native plants. Soil compaction also can be caused by overtilling and by clear-cut logging (which we will look at in Chapter 10) and the rapid withdrawal of groundwater (which we will look at in Chapter 11).

As a cause of soil degradation, overgrazing is equal to cropland agriculture, and it is a greater cause of desertification. Humans keep a total of 3.4 billion cattle, sheep, and goats. Rangeland classified as degraded now adds up to 680 million hectares, although some estimates put the number as high as 2.4 billion hectares, fully 70% of the world's rangeland area. Rangeland degradation is estimated to cost $23.3 billion per year. Grazing exceeds the sustainable supply of grass in India by 30% and in parts of China by up to 50%. To relieve pressure on rangelands, both nations are now beginning to feed crop residues to livestock.

Range managers do their best to assess the carrying capacity of rangelands and inform livestock owners of these limits, so that herds are rotated from site to site as needed to conserve grass cover and soil integrity. Managers can also establish and enforce limits on grazing on publicly owned land when necessary, circumventing the tragedy of the commons, which we saw in Chapter 1. Today, increasing numbers of ranchers are working cooperatively with government agencies, environmental scientists, and even environmental advocates to find ways to ranch more sustainably and safeguard the health of the land.

CANADIAN ENVIRONMENTAL PERSPECTIVES

Loretta Ford

Loretta Ford is a soil scientist who specializes in remediation at mine sites.

- **Soil scientist** and **agrologist**
- **Soil remediation specialist** and **reclamation planner**, and assistant director of environmental affairs for Northern Dynasty Minerals
- **Kayaking and camping enthusiast**

Loretta Ford knew back in high school that she wanted to be an environmental scientist. She loved the outdoors, biking, kayaking, hiking, and camping. When she saw a promotional poster for a program in conservation and reclamation technology, she knew it was meant for her. Her two years at Lakeland College taught her, among other things, that she most enjoyed working with soils and plants. "I was fascinated with how soil developed under different climatic and vegetative conditions. The whole nutrient cycling process was interesting." Ford decided to specialize and went on to study soil science at the University of Alberta.

While working toward her B.Sc. degree, Ford focused her studies on practical applications where she could use her knowledge of soils. This led her eventually to an M.Sc. degree in soil reclamation. Her research project involved growing plants on reclaimed soil from a mine site. Today her main role in her work is to provide environmental expertise to mining companies, working through Hunter Dickinson Services, which manages nine mineral exploration and mining companies around the world. Her main focus is on mining in the North, and she is currently working on the Pebble Project, a proposed copper–gold mine in Alaska.

Ford says that what she really enjoys about her work is being able to apply her knowledge of soil science to find practical solutions for problems in the everyday world. Her advice is used in the engineering design of mines before they are even built, all the way to the development of plans for closure and reclamation of the site at the end of the mine's productive life span. If there are negative impacts that cannot be avoided, Ford helps mining companies come up with innovative ways to repair or compensate for the damage. As she says, "Just imagine being able to take a natural soil, dig it up, churn it about, store it for years, then place it back on top of disturbed land and have it grow a plant community again. The work I do is about taking something that could be detrimental and with a few careful steps, having results that make the land beautiful and productive again. Mining really can be a temporary use of the land and I work toward achieving that goal." The idea of working from *within* industry to protect the environment is at the core of Loretta Ford's working philosophy. She received an award from the Soil and Water Conservation Society for her work.

"In my role, I can help mining companies do the right thing by minimizing the impacts to the environment." —Loretta Ford

Thinking About Environmental Perspectives

Do you think it would be difficult to be an environmental scientist working in a sector that, like mining, necessarily involves significant disturbances on the environment? How do people like Loretta Ford balance the desire of mining companies to manage costs against the need to take steps to protect the environment from permanent damage?

Conclusion

Many of the policies enacted and the practices developed to combat soil degradation in Canada and worldwide have been quite successful, particularly in reducing the erosion of topsoil. Despite these successes, however, soil is still being degraded at a rate that calls into question the sustainability of industrial agriculture.

Our species has enjoyed a 10 000-year history with agriculture, yet despite all we have learned about soil degradation and conservation, many challenges remain. It is clear that even the best-conceived soil conservation programs require research, education, funding, and commitment from both farmers and governments if they are to fulfill their potential. In light of continued population growth, we will likely need better technology and wider adoption of soil conservation techniques if we hope to be able to feed the 9 billion people expected to crowd our planet in mid-century.

REVIEWING OBJECTIVES

You should now be able to:

Delineate the fundamentals of soil science, including soil formation and the properties of soil

- Soil includes diverse biotic communities that decompose organic matter.
- Climate, organisms, relief, parent material, and time are factors influencing soil formation.
- Soil profiles consist of distinct horizons with characteristic properties.
- Soil can be characterized by colour, texture, structure, and pH.
- Soil properties affect the potential for plant growth and agriculture in any given location.

Explain the importance of soils to agriculture, and describe the impacts of agriculture on soils

- Successful agriculture requires healthy soil.
- As the human population grows and consumption increases, pressures from agriculture are degrading Earth's soil, and we are losing 5 million to 7 million hectares of productive cropland annually.

Outline major historical developments in agriculture

- About 10 000 years ago, people began breeding crop plants and domesticating animals.
- Domestication took place through the process of selective breeding, or artificial selection.
- Agriculture is thought to have originated independently in different cultures around the world.
- Industrial agriculture is gradually replacing traditional agriculture, which, in turn, largely replaced hunting and gathering.

State the causes and predict the consequences of soil erosion and soil degradation

- Some agricultural practices have resulted in high rates of erosion across the world, lowering crop yields.
- Desertification affects a large portion of the world's soils, especially those in arid regions.
- Overirrigation can cause salinization and waterlogging, which lower crop yields and are difficult to mitigate.
- Overapplying fertilizers can cause pollution problems that affect ecosystems and human health.
- Overgrazing, overtilling, and careless forestry practices can cause soil degradation and have diverse impacts on native ecosystems.

Describe the history and explain the principles of soil conservation

- The Dust Bowl in North America and similar events elsewhere have encouraged scientists and farmers to develop ways of better protecting and conserving topsoil.
- Farming techniques such as crop rotation, contour farming, intercropping, terracing, shelterbelts, and reduced tillage, enable farmers to reduce soil erosion and boost crop yields.
- In Canada and across the world, governments are devising innovative policies and programs to deal with the problems of soil degradation.

TESTING YOUR COMPREHENSION

1. How did the practices of selective breeding and human agriculture begin roughly 10 000 years ago? Summarize the influence of agriculture on the development and organization of human communities.
2. Describe the methods used in traditional agriculture, and contrast subsistence agriculture with intensive traditional agriculture. What makes industrialized agriculture different from traditional agriculture?
3. What processes are most responsible for the formation of soil? Describe the three types of weathering that may contribute to the process of soil formation.
4. Name the five primary factors thought to influence soil formation, and describe one effect of each.
5. How are soil horizons created? What is the general pattern of distribution of organic matter in a typical soil profile?

6. Why is erosion generally considered a destructive process? Name three human activities that can promote soil erosion. Describe four kinds of soil erosion by water. What factors affect the intensity of water erosion?
7. List the farming techniques that can help reduce the risk of erosion caused by conventional cultivation methods.
8. How does terracing effectively turn very steep and mountainous areas into arable land? Explain the method of no-till farming. Why does this method reduce soil erosion?
9. How do fertilizers boost crop growth? How can large amounts of fertilizer added to soil also end up in water supplies and the atmosphere?
10. What policies can be linked to the practice of overgrazing? Describe the effects of overgrazing on soil. What conditions characterize sustainable grazing practices?

SEEKING SOLUTIONS

1. How do you think a farmer can best help to conserve soil? How do you think a scientist can best help to conserve soil? How do you think a national government can best help to conserve soil?
2. How and why might actual soils differ from the idealized five-horizon soil profile presented in the chapter? How might departures from the idealized profile indicate the impact of human activities? Provide at least three examples.
3. What method of farming would you choose to employ on a gradual slope with the threat of natural erosion? What kinds of plants might you use to prevent erosion, and why?
4. Discuss how the methods of no-till or reduced tillage farming can enhance soil quality. What drawbacks or negative effects might no-till or reduced-tillage practices have on crops and soil quality, and how might these be prevented?
5. **THINK IT THROUGH** You are a land manager with your provincial government and you have just been put in charge of 200 000 ha of public lands that have been degraded by decades of overgrazing. Soil is eroding, creating large gullies. Shrubs are encroaching on grassland areas because fire was suppressed. Environmentalists want an end to ranching on the land. Ranchers want continued grazing, but they are concerned about the land's condition and are willing to entertain new ideas. What steps would you take to assess the land's condition and begin restoring its soil and vegetation? Would you allow grazing, and if so, would you set limits on it?
6. **THINK IT THROUGH** You are the head of an international granting agency that helps farmers with soil conservation and sustainable agriculture. You have $10 million to disburse. Your agency's staff has decided that the funding should go to (1) farmers in an arid area of Africa prone to salinization, (2) farmers in a fast-growing area of Indonesia where swidden agriculture is practised, (3) farmers in southern Brazil practising no-till agriculture, and (4) farmers in a dryland area of Mongolia undergoing desertification. What types of projects would you recommend funding in each of these areas, how would you apportion your funding among them, and why?

INTERPRETING GRAPHS AND DATA

Kishor Atreya and his colleagues at Kathmandu University in Nepal conducted a field experiment to test the effects of reduced tillage versus conventional tillage on erosion and nutrient loss in the Himalayan Mountains in central Nepal. The region in which they worked has extremely steep terrain (with an average slope of 18%) and receives more than 138 cm of rain per year, with 90% of it coming during the monsoon season from May to September. Atreya's team measured the amounts of soil, organic carbon, and nitrogen lost from the research plots (which were unterraced) during one year. Some of their results are presented in the graph.

1. Under the conditions of the study reported above, how much soil, organic carbon, and nitrogen would be saved annually in fields with reduced tillage relative to fields with conventional tillage? Express your answers both in absolute units and as percentages.
2. Given that annual crop yields in the study plots were approximately four metric tons per hectare, what is the ratio of soil lost to crop yield under conventional tillage? Under reduced tillage?
3. Is reduced tillage a sustainable management practice for Nepalese farmers? If so, what data from the study above would you cite in support of your answer? If not, or if you cannot say, then what concerns raised by the data above would still need to be addressed, or what additional data would be needed, to answer the question?

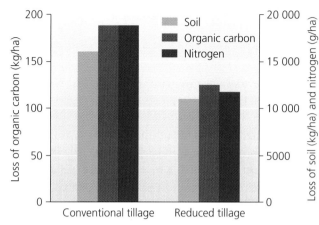

Annual soil and nutrient losses in plots under conventional and reduced tillage systems. All reduced tillage values are significantly different from their conventional tillage counterparts.
Data from Atreya, K., et al. 2005. Applications of reduced tillage in hills of central Nepal. Soil & Tillage Research 88(1–2):16–29.

CALCULATING FOOTPRINTS

UNEP's Global Assessment of Human-Induced Soil Degradation (GLASOD) was the most comprehensive assessment of the state of the world's soils ever undertaken (in 1991, updated in 2001, but only for Asia).[16] The study had flaws and limitations, but even today it represents one of the most comprehensive data sets available on global soil resources. GLASOD estimated that about 2 billion hectares of land (562 million hectares of the world's cropland) are susceptible to soil erosion and degradation. The study further estimated that about 6 million to 7 million hectares of agricultural land is lost globally through erosion each year (some estimates are higher), and another 1.5 million hectares as a result of salinization and alkalinization. (These estimates vary widely; try researching online to find out the methodologies used by various researchers to estimate soil degradation rates.) The Food and Agriculture Organization of the United Nations (FAO) estimated in 2006 that there are 1402 million hectares of arable land in the world.[17]

Given these figures, calculate the following:

1. The FAO has estimated that soil erosion (7 million hectares per year) accounts for about 40% of soil loss. How much would the total be, if this is true? How much would it be per capita?
2. What percentage of total arable land does this represent each year?
3. If 7 million hectares of agricultural land are lost per year to erosion, how much does this amount to, per person?
4. If 1.5 million hectares of agricultural land are lost per year to salinization and alkalinization, how much does this amount to, per person?

Now fill in the missing information for the following table. (Note that the *actual* rates of soil loss vary widely from one region to another, depending upon climate and other natural factors, as well as human activities, notably agricultural practices. Other factors, such as population density, also play a role.)

	Total world soil loss per year per person	Total world soil lost to erosion per year per person	Total world soil lost to salinization per year per person
World			
Canada			
Your province			
You			

TAKE IT FURTHER

 Go to www.myenvironmentplace.ca where you will find

- Suggested answers to end-of-chapter questions
- Quizzes, animations, and flashcards to help you study
- *Research Navigator*™ database of credible and reliable sources to assist you with your research projects
- Tutorials to help you master how to interpret graphs
- Current news articles that link the topics that you study to case studies from your region and around the world
- **ECO Occupational Profiles:** If you found this chapter especially interesting, you might want to learn more about the following jobs by visiting the Occupational Profiles website of the Environmental Careers Organization. Go to www.eco.ca and check out the following careers:
  - Agriculture specialist
  - Agrologist
  - Environmental geologist
  - Remediation specialist
  - Soil conservationist
  - Soil scientist

CHAPTER ENDNOTES

1. Based on information from Trent University, The Blue Lab: Mer Bleue Research project, www.trentu.ca/academic/bluelab/research_merbleue.html and personal communications with Nigel Roulet, Tim Moore, and Nathan Basiliko, August, 2008.
2. Soil Classification Working Group (1998) *The Canadian System of Soil Classification*, 3rd ed. Ottawa: Agriculture and Agri-Food Canada, Publication 1646, 187 pp. (available online through the National Land and Water Information Service, http://sis.agr.gc.ca/cansis/references/1998sc_a.html).
3. The species *Homo sapiens* is thought to have originated in Africa approximately 200 000 years B.P. (before present). The modern human subspecies, *H. sapiens sapiens*, which is the only subspecies of *H. sapiens* that survives today, probably originated about 130 000 B.P. Smithsonian Institute Human Origins Project, www.mnh.si.edu/anthro/humanorigins/.
4. Agriculture and Agri-Food Canada (1997) *Profile of Production Trends and Environmental Issues in Canada's Agriculture and Agri-food Sector*, www4.agr.gc.ca/resources/prod/doc/policy/environment/pdfs/sds/profil_e.pdf.
5. van Vlietl, L. J. P., G. A. Padbury, and D. A. Lobb (2003) Soil Erosion Risk Indicators Used in Canada. Paper presented at the OECD Expert Meeting on Soil Erosion and Biodiversity, Rome, Italy, March 2003.
6. Agriculture and Agri-Food Canada (1997) *Profile of Production Trends and Environmental Issues in Canada's Agriculture and Agri-food Sector*. Ottawa: Author.
7. From the U.N. Convention to Combat Desertification, as cited in the Millennium Ecosystem Assessment, *Ecosystems and Human Well-Being: Desertification Synthesis*, 2005, available online at www.millenniumassessment.org/documents/document.355.aspx.pdf.
8. United Nations Convention to Combat Desertification, www.unccd.int.
9. CBC News Online (2004) In-depth: Agriculture, 1930s Drought, www.cbc.ca/news/background/agriculture/drought1930s.html.
10. Based on Government of Canada (1984). *Soil at Risk: Canada's Eroding Future*. Ottawa: Author.
11. Soil Conservation Council of Canada, About Us: Overview, www.soilcc.ca/about-us.htm.
12. Soil Conservation Council of Canada, www.soilcc.ca.
13. International Water Management Institute, Global Irrigated Area Mapping Project, www.iwmigiam.org/stats/.
14. FAO Land and Plant Nutrition Management Service, Global Network on Integrated Soil Management for Sustainable Use of Salt-Affected Soils, http://193.43.36.103/ag/AGL/agll/spush/intro.htm.
15. Environment Canada, Nitrate Levels in the Abbotsford Aquifer: An Indicator of Groundwater Contamination in the Lower Fraser Valley, www.ecoinfo.org/env_ind/region/nitrate/nitrate_e.cfm.
16. Oldeman, L. R., R. T. A. Hakkeling, and W. G. Sombroek (1991) *GLASOD, World Map of the Status of Human-Induced Soil Degradation: A Brief Explanatory Note*. The Netherlands: International Soil Reference and Information Centre and UNEP.
17. UNFAO, 2006, Food and Agriculture Statistics Global Outlook, http://faostat.fao.org/Portals/_Faostat/documents/pdf/world.pdf.

8 Biotechnology and Food Resources

In order for his farm to be considered organic, this farmer must adhere to strict regulations.

Upon successfully completing this chapter, you will be able to

- Explain the challenge of feeding a growing human population
- Identify the goals, methods, and environmental impacts of the "Green Revolution"
- Categorize the strategies of pest management
- Discuss the importance of pollination
- Describe the science behind genetically modified food
- Evaluate controversies and the debate over genetically modified food
- Ascertain approaches for preserving crop diversity
- Assess feedlot agriculture for livestock and poultry
- Weigh approaches in aquaculture
- Evaluate sustainable agriculture

This farmer grows maize in Oaxaca, Mexico

CENTRAL CASE:
GM MAIZE AND ROUNDUP READY CANOLA

"Worrying about starving future generations won't feed them. Food biotechnology will ... At Monsanto, we now believe food biotechnology is a better way forward."
— MONSANTO COMPANY ADVERTISEMENT

"I never put those plants on my land. The question is, where do Monsanto's rights end and mine begin?"
— PERCY SCHMEISER

Corn is a staple grain of the world's food supply. We can trace its ancestry back roughly 5500 years, when people in the highland valleys of what is now the state of Oaxaca in southern Mexico first domesticated that region's wild maize plants. The corn we eat today arose from some of the many varieties that evolved from the early selective crop breeding conducted by the people of this region.

Today Oaxaca remains a world centre of biodiversity for maize, with many native varieties growing in the rich, well-watered soil (see photo). Preserving such varieties of crops in their ancestral homelands is important for securing the future of our food supply, scientists maintain, because these varieties serve as reservoirs of genetic diversity—reservoirs we may need to draw on to sustain or advance our agriculture.

In 2001, Mexican government scientists conducting routine genetic tests of Oaxacan farmers' maize announced that they had turned up DNA that matched *transgenes*, that is, genes transferred from genetically modified (GM) corn, even though Mexico had banned its cultivation since 1998. Corn is one of many crops that scientists have genetically engineered to express desirable traits, such as large size, fast growth, and resistance to insect pests.

Activists opposed to GM food trumpeted the disturbing news and urged a ban on imports of transgenic crops from producer countries, such as the United States, into developing nations. The agrobiotech industry defended the safety of its crops and questioned the validity of the research. Further research by Mexican government scientists confirmed the presence of transgenes in Mexican maize. Those studies were controversial and still have not been definitively verified, although the findings were accepted by a special commission of experts convened under the North American Free Trade Agreement (NAFTA). The commission concluded that corn imported from the United States was the source for the transgenes, which spread by wind pollination and interbreeding with native cultivars once in Mexico.

Meanwhile, back in Canada, GM producer Monsanto was engaged in a highly public battle with 74-year-old Saskatchewan farmer Percy Schmeiser (see photo) over Schmeiser's canola crop. Canola is an edible oil derived from rapeseed. It is widely grown throughout the Canadian prairies, and 80% of the rapeseed grown in Canada is genetically modified.

Schmeiser maintained that pollen from Monsanto's Roundup Ready canola had blown from a neighbouring farm onto his land and pollinated his non-GM canola. Schmeiser had never purchased the patented seed and did not want the crossbreeding. Monsanto investigators took seed samples from his plants and charged him with violating a Canadian law that makes it illegal for farmers to reuse or grow patented seed without a contract. The courts sided with Monsanto, ordering the farmer to pay the corporation $238 000. Schmeiser appealed to the Supreme Court of Canada, which ruled that Monsanto's patent had indeed been violated but acknowledged that the farmer had not benefited from the GM seeds and had not intended their use, and exempted him from paying any fines or fees to the company.

In spite of the Supreme Court loss, Schmeiser received wide public support. He and his wife, Louise, were given the 2007 Right Livelihood Award in recognition of his struggle. A government committee sought a revision in the patent law, and the National Farmers Union of Canada called for a moratorium on GM food.

Saskatchewan farmer Percy Schmeiser was accused by the Monsanto Company of planting its patented canola without a contract with the company. Schmeiser said his non-GM plants were contaminated with Monsanto's transgenes from neighbouring farms. Monsanto won the David-and-Goliath case in Canada's Supreme Court, but Schmeiser became a hero to small farmers and anti-GM food activists worldwide.

Schmeiser says that the most difficult part of the entire saga has been the loss, through contamination by transgenes, of the local variety of rapeseed that he had planted throughout his 60-year farming career.[1]

The larger question of the legality of holding patents on living organisms remains unresolved. In 2002 Canada became the first nation in the industrialized world to prohibit the holding of patents on higher organisms (a genetically modified mouse was the organism in question). Monsanto continues to demand that farmers heed patent laws and is continuing with several similar lawsuits. In a fascinating turn of events, Percy Schmeiser reached an out-of-court settlement with Monsanto in March 2008, in which the company not only admitted that contamination had occurred but also agreed to pay all costs of remediating it.

The discussion about genetically modified organisms, their patenting and use, and their potential benefits and impacts, is far from over. In this chapter, we take a widely ranging view of the ways people have devised to increase agricultural output, the environmental effects of these efforts, and the implications for food resources of the future.

The Race to Feed the World

Although human population growth has slowed, we can still expect our numbers to swell to 9 billion by the middle of this century. For every two people living today, there will be three in 2050. Feeding 50% more mouths half a century from now while protecting the integrity of soil, water, and ecosystems will require sustainable agriculture. This could involve approaches as diverse as organic farming and the genetically modified crops that are eliciting so much controversy in Oaxaca, Saskatchewan, and elsewhere.

We are producing more food per person

Over the past half-century, our ability to produce food has grown even more quickly than global population (FIGURE 8.1). However, largely because of political obstacles and inefficiencies in distribution, today 850 million people in developing countries still do not have enough to eat. Every five seconds, somewhere in the world, a child starves to death. Agricultural scientists and policy makers pursue a goal of **food security**, the guarantee of an adequate, reliable, and available food supply to all people at all times. Making a food supply sustainable depends on maintaining healthy soil, water, and biodiversity. As we saw in Chapter 7, careless expansion of agriculture can have devastating effects on the environment and the long-term ability of the world's soils to continue supporting crops and livestock.

Starting in the 1960s, a number of scientists predicted widespread starvation and a catastrophic failure of agricultural systems, arguing that the human population could not continue to grow without outstripping its food supply. However, the human population has continued to increase well past the scientists' predictions. Although it is tragic that 850 million people are hungry today, this number is smaller than the 960 million who lacked reliable and sufficient food in 1970. In percentage terms, we have reduced hunger by half, from 26% of the population in 1970 to 13% today.

We have achieved these advances in part by increasing our ability to produce food. We have increased food production by devoting more energy (especially fossil fuel energy) to agriculture; by planting and harvesting more frequently; by greatly increasing the use of irrigation, fertilizers, and pesticides; by increasing the amount of cultivated land; and by developing (through crossbreeding and genetic engineering) more productive crop and livestock varieties.

Although agricultural production has so far outpaced population growth, there is no guarantee that it will continue to do so. Already with grain crops, the world's staple foods, we are producing less food per person each year. Since 1985, world grain production per person has fallen by 9%. Moreover, the world's soils are in decline, and nearly all the planet's arable land has already been claimed.

We face undernourishment, overnutrition, and malnutrition

Although many people lack access to adequate food, others are affluent enough to consume more than is healthy. People suffering from **undernourishment**, who receive less than 90% of their daily caloric needs, mostly live in the developing world. Meanwhile, in the developed world, many people suffer from **overnutrition**, receiving

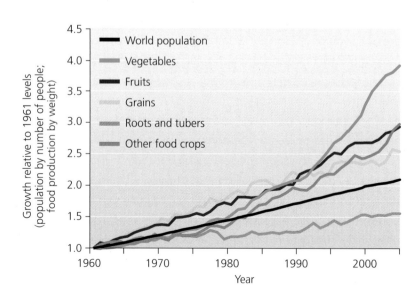

FIGURE 8.1
Global agricultural food production rose more than two-and-a-half times in the past four decades, growing at a faster rate than world population. Production of all types of foods, particularly vegetables, increased from 1961 to 2005. Trend lines show cumulative increases relative to 1961 levels. Data from the Food and Agriculture Organization of the United Nations.

too many calories each day. In Canada, where food is available in abundance and people tend to lead sedentary lives with little exercise, 48% of adults exceed healthy weight standards, and 14% are obese (according to World Health Organization standards).[2]

For most people who are undernourished, the reasons are economic. One-fifth of the world's people live on less than $1 per day, and more than half live on less than $2 per day, the World Bank estimates. Hunger is a problem even in Canada, where more than 750 000 people used the services of a food bank during a typical month in 2006, according to the Canadian Association of Food Banks; more than 41% of them were children.[3] This measure underrepresents the true situation, particularly for rural areas that lack access to food banks. The National Population Health Survey has estimated that *food insecurity*—the inability to procure sufficient food when needed—is a factor in 10.2% of Canadian households, affecting well over 3 million people.[4]

In a wealthy food-producing nation like Canada, if you thought that hunger must be caused by more than just a lack of available food, you would certainly be right. Many factors coalesce to cause food insecurity around the world, and these factors have as much to do with poverty and the weaknesses of our food delivery systems as with the abundance and availability of food. Some people say that food security is a function of five "A's":

- *Availability.* Food must be produced in sufficient quantity and not diverted for other uses, such as the production of biofuels.
- *Affordability.* Poverty is the reason for most food bank visits in Canada.
- *Accessibility.* It doesn't help if the food is available but you can't get to it, perhaps because you lack the means of transportation, or because there is unrest in your region.
- *Acceptability.* Do you think you would eat *anything* if you were really hungry? What if you were Muslim and the only thing available was pork? What if insect grubs were the only available food?
- *Adequacy.* The nutritional quality of food, as well as its abundance, must be sufficient to maintain a person's health, or they will slowly waste away.

Just as the *quantity* of food a person eats is important for health, so is the *quality* of food. **Malnutrition**, a shortage of nutrients the body needs, including a complete complement of vitamins and minerals, can occur in both undernourished and overnourished individuals. Malnutrition can lead to disease (**FIGURE 8.2**). When people eat a high-starch diet but not enough protein or

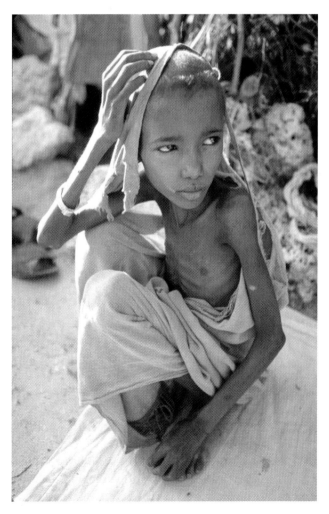

FIGURE 8.2
Millions of children, including this child in Somalia, suffer from forms of malnutrition, such as kwashiorkor and marasmus.

weighing the issues

POVERTY AND FOOD INSECURITY   8-1

Think about what it would truly mean to live on $1 per day, like one-fifth of the world's people do. This doesn't just mean leaving your house with a loonie in your pocket for the day. It means that *everything* in your life must be paid for with just $1: housing, furniture, food, energy, clothing, water (if it is not freely available where you live), toys (if you are a child), medicine (if you are ill), books (if you are a student), transportation (if you need to get somewhere, such as to a job site). And this is the case day after day. In this list of daily priorities, where would nutritious food be for you?

essential amino acids, *kwashiorkor* results. Children who have recently stopped breastfeeding are most at risk for developing kwashiorkor, which causes bloating of the abdomen, deterioration and discoloration of hair, mental disability, immune suppression, developmental delays, anemia, and reduced growth. Protein deficiency together with a lack of calories can lead to *marasmus*, which causes wasting or shrivelling among millions of children in the developing world.

The "Green Revolution" led to dramatic increases in agricultural production

The desire for greater quantity and quality of food for the growing human population led in the mid- and late twentieth century to the **Green Revolution**. Realizing that farmers could not go on indefinitely cultivating more and more land to increase crop output, agricultural scientists created methods and technologies to increase crop output per unit area of existing cultivated land. Industrialized nations had been dramatically increasing their per-area yields during the first part of the twentieth century. Many people saw growth in production and efficiency as key to ending starvation in developing nations. Note that this switches the focus of agricultural production from *extensification*—increasing resource productivity by simply bringing more land into production—to *intensification*, in which new technologies permit greater resource productivity from each unit of land. In the end, the Green Revolution was characterized by *both* extensification and intensification of agricultural production.

The transfer of technology to the developing world that marked the Green Revolution began in the 1940s, when U.S. agricultural scientist Norman Borlaug introduced Mexico's farmers to a specially bred type of wheat (**FIGURE 8.3**). This strain of wheat produced large seed heads, was short in stature to resist wind, was resistant to diseases, and produced high yields. Within two decades of planting and harvesting this specially bred crop, Mexico tripled its wheat production and began exporting wheat.

The stunning success of this program inspired others. Borlaug—who won the Nobel Peace Prize for his work—took his wheat to India and Pakistan and helped transform agriculture there. Soon many developing countries were increasing their crop yields by using selectively bred strains of wheat, rice, corn, and other crops from developed nations. Some varieties yielded three or four times as much per hectare as did their predecessors.

FIGURE 8.3
Norman Borlaug holds examples of the wheat variety he bred that helped launch the Green Revolution. The high-yielding disease-resistant wheat helped increase agricultural productivity in many developing countries. Borlaug won the Nobel Peace Prize for his work on food production.

The Green Revolution has caused the environment both benefit and harm

Along with the new grains, developing nations imported the methods of modern **industrialized agriculture**. They began applying large amounts of synthetic fertilizers and chemical pesticides to their fields, irrigating crops with generous amounts of water, and using heavy equipment powered by fossil fuels. This high-input agriculture succeeded dramatically in allowing farmers to harvest more corn, wheat, rice, and soybeans from each hectare of land. From 1900 to 2000, humans expanded the world's total cultivated area by 33%, yet increased energy inputs into agriculture by 800%. Intensive agriculture saved millions in India from starvation in the 1970s and eventually turned that nation into a net exporter of grain.

These developments had mixed effects on the environment. On the positive side, the intensified use of already cultivated land reduced pressures to convert additional natural lands for new cultivation. Between 1961 and 2002, food production rose 150% and population rose 100%, while area converted for agriculture increased only 10% (see "Interpreting Graphs and Data"). For this reason, the Green Revolution prevented some degree of deforestation and habitat conversion in many countries at the very time those countries were experiencing their fastest population growth rates. In this sense, the Green Revolution was beneficial for biodiversity and natural ecosystems.

FIGURE 8.4
Most agricultural production in industrialized countries comes from monocultures—large stands of single types of crop plant, such as this wheat field. Clustering crop types in uniform fields on large scales greatly improves the efficiency of planting and harvesting, but it also decreases biodiversity and makes crops susceptible to outbreaks of pests that specialize on particular crops. Armyworms (inset) are major agricultural pests whose outbreaks can substantially reduce crop yields. The caterpillars of these moths defoliate a wide variety of crops, including wheat, corn, cotton, alfalfa, canola, and beets.

---

weighing the issues

THE GREEN REVOLUTION AND POPULATION 8–2

In the 1960s, India's population was growing at an unprecedented rate, and traditional agriculture was not producing enough food to support the growth. By adopting Green Revolution agriculture, India sidestepped mass starvation. In the years since intensifying its agriculture, India has added several hundred million more people and continues to suffer widespread poverty and hunger.

Can we call the Green Revolution a success? Has it solved problems or delayed our resolution of problems or just created new ones? How sustainable are Green Revolution approaches? Norman Borlaug hoped that the Green Revolution would give us "breathing room" in which to deal with what he called "the Population Monster." Have we been dealing effectively with the "population monster" during the breathing room that the Green Revolution has bought for us?

However, the intensive use of water, fossil fuels, and chemical fertilizers and pesticides had extensive negative impacts on the environment in terms of pollution, salinization, and desertification. The impacts of the Green Revolution on small-scale farmers in the developing world—many of whom are disadvantaged by lack of income, lack of education, or both, in terms of accessing or benefiting from these technologies—are particularly controversial.

One key aspect of Green Revolution techniques has had particularly negative consequences for biodiversity and mixed consequences for crop yields. The planting of crops in **monocultures**, large expanses of single crop types (**FIGURE 8.4**), has made planting and harvesting more efficient and has thereby increased output. However, monocultural planting has reduced biodiversity over huge areas, because many fewer wild organisms are able to live in monocultures than in native habitats or in traditional small-scale polycultures. Moreover, when all plants in a field are genetically similar, as in monocultures, all will be equally susceptible to viral diseases, fungal pathogens, or insect pests that can spread quickly from plant to plant (**FIGURE 8.4**). For this reason, monocultures bring significant risks of catastrophic failure.

Monocultures have also contributed to a narrowing of the human diet. Globally, 90% of the food we consume now comes from only 15 crop species and eight livestock species—a drastic reduction in diversity from earlier times. The nutritional dangers of such dietary restriction have been alleviated by the fact that expanded global trade has provided many people access to a wider diversity of foods from different locations around the world. However, this effect has benefited wealthy people far more than poor people. One reason that farmers and scientists were so concerned about transgenic contamination of Oaxaca's native maize is that Oaxacan maize varieties serve as a valuable source of genetic variation in a world where so much variation is being lost to monocultural practices.

Biofuels are having a significant impact on food availability

The Green Revolution was all about inputs. Enormous energy inputs (both chemical and mechanical) allowed for the huge increases in production that characterized the transition to modern industrialized agriculture. Nowhere is the relationship between energy and food more starkly apparent than in the current controversy over *biofuels,* organic materials that are converted into liquid or gaseous fuels for use in internal combustion engines as replacements for oil and natural gas. Notable for its recent enormous surge is corn-derived ethanol, an efficient alcohol fuel that can be mixed in various proportions with normal gasoline.

The principle of using renewable energy sources to replace limited, nonrenewable fuels, like petroleum, is founded in the principles of sustainability. The production of corn ethanol is now soaring in many countries, but most notably in the United States—where production was 38% higher in 2007 than in the previous year—in response to federal government goals and incentives for renewable energy development. Some states have even begun to require a certain proportion of alcohol content in all gasoline.

However, it has rapidly become apparent that there is something fundamentally wrong in taking a crop that provides food for both people and livestock, and burning it in cars, trucks, and machines. The U.N. Food and Agriculture Organization has called it a "conflict of interests." The greatly increased demand for corn has led to much higher prices and scarcities worldwide among all of the basic grains, not just corn. Land that was once devoted to growing food crops now produces fuel. Ultimately this has contributed to accelerated deforestation in agricultural nations, such as Indonesia, where more land has had to be pressed into service for the production of food grains. We will look more closely at the pros and cons of biofuels in Chapter 17.

Pests and Pollinators

Throughout the history of agriculture, the insects, fungi, viruses, rodents, and weeds that eat or compete with our crops have taken advantage of the ways we cluster food plants into agricultural fields. These organisms, in making a living for themselves, cut crop yields and make it harder for farmers to make a living. As just one example of thousands, various species of moth caterpillars known as armyworms (see **FIGURE 8.4**) decrease yields of everything from beets to sorghum to millet to canola to pasture grasses. Pests and weeds have always posed problems for traditional agriculture, and they pose an even greater threat in a monoculture situation, where a pest adapted to specialize on that particular crop can easily move from one individual plant to many others of the same type.

What humans term a **pest** is any organism that damages crops that are valuable to us. What we term a **weed** is any plant that competes with our crops. These are subjective categories that we define entirely by our own economic interests. (American writer and philosopher Ralph Waldo Emerson wrote, "What is a weed? A plant whose virtues have not yet been discovered.") There is nothing inherently malevolent in the behaviour of a pest or a weed. These organisms are simply trying to survive and reproduce. From the viewpoint of an insect that happens to be adapted to feed on corn, grapes, or apples, a grain field, vineyard, or orchard represents an endless buffet.

---

weighing the issues   **WHAT A PEST!**   8–3

Compare the concept of a pest or weed species with that of alien and invasive species, discussed in Chapter 4. At what point should a species be considered a pest? Does it have to cause damage to human interests? What if it causes harm only to natural ecosystems? How should pest species be managed?

Many thousands of chemical pesticides have been developed

To prevent pest outbreaks and to limit competition with weeds, people have developed thousands of artificial chemicals to kill insects (**insecticides**), plants (**herbicides**), and fungi (**fungicides**). Poisons that target pest organisms are collectively termed **pesticides**. In Canada today, more than 7000 pesticides are registered for use. Many of the more than 500 active ingredients in these pesticides have not been evaluated for health or environmental impacts for many years—more than 150 were approved for use in Canada prior to 1960. The new *Pest Control Products Act*, which came into effect as federal legislation in Canada in 2006, will require products to be reevaluated 15 years after they are initially approved for use, among other provisions designed to improve the safety and minimize the environmental impacts of pesticide use.[5] Table 8.1 shows the main categories of chemical pesticides used in Canada.

Today more than $32 billion is expended annually on pesticides, as much as $1.5 billion of it in Canada. Most pesticides used in Canada are for agricultural purposes (91% of sales), while the remaining nonagricultural pest management products—primarily for domestic use but also used in the forestry and industrial sectors and for the management of golf courses and other landscapes—represent 9% of total sales. Interestingly, 85% of the total pesticides sold in Canada are herbicides, followed by fungicides (7%), insecticides (4%), and other specialty pest management chemicals, such as rodenticides (4%).[6]

Table 8.1	Categories of Pesticides Used in Canada				
Class of chemical pesticide	**First used**	**Examples**	**Types**	**Current status**	**Effects**
Organochlorines	1942	aldrin; chlordane; dieldrin; endrin; heptachlor; lindane; methoxychlor; toxaphene; HCB; PCP; DDT	Mostly insecticides	Some registered in Canada; others (such as DDT) discontinued in Canada but still used in developing nations	Persistent; bioaccumulative; affect ability to reproduce, develop, and withstand environmental stress
Organophosphates	Very early 1940s	schradan; parathion; malathion	Insecticides	Schradan discontinued in 1964, resulting in a move toward less toxic groups (malathion, parathion)	Nonpersistent; systemic; not very selective; toxic to humans
Carbamates	First appeared in 1930; large-scale use 1950s	carbaryl; methomyl; propoxur; aldicarb	Fungicides, insecticides	Aldicarb discontinued in 1964; the others are registered in Canada	Nonpersistent; not very selective; toxic to birds and fish
Phenoxy	1946	2,4-D; 2,4,5-T	Herbicide	2,4-D is widely used; 2,4,5-T banned in Canada	Selective effects on humans and mammals are not well known; some potential to cause cancer in laboratory animals
Pyrethroids	1980	fenpropanthrin; deltamethrin; cypermethrin	Insecticide	Fenpropanthrin is not registered in Canada, unlike the two other pesticides	Target-specific: more selective than organophosphates or carbamates; not acutely toxic to birds or mammals, but particularly toxic to aquatic species

Source: Based on Government of Canada. 2000. Pesticides: Making the Right Choice for Health and the Environment. Report of the Standing Committee on Environment and Sustainable Development, Appendix 3.2, http://cmte.parl.gc.ca/cmte/CommitteePublication.aspx?COM=173&Lang=1&SourceId=36396

Noting that pesticides are, by definition, designed to be toxic to organisms and that the toxic effects may not be limited to the target organisms, we will address the health consequences of synthetic pesticides for humans and other organisms (the "Effects" column of Table 8.1) in greater detail in Chapter 19.

Pests evolve resistance to pesticides

Despite the toxicity of these chemicals, their usefulness tends to decline with time as pests evolve resistance to them. Recall from our discussion of natural selection that organisms within populations vary in their traits. Because most insects and microbes occur in huge numbers, it is likely that a small fraction of individuals may by chance have genes that confer some degree of immunity to a given pesticide. Even if a pesticide application kills 99.99% of the insects in a field, 1 in 10 000 survives. If an insect survives by being genetically resistant to a pesticide, and if it mates with other resistant individuals of the same species, the insect population may grow. This new population will consist of individuals that are genetically resistant to the pesticide. As a result, pesticide applications will cease to be effective (**FIGURE 8.5**).

In many cases, industrial chemists are caught up in an "evolutionary arms race" with the pests they battle, racing to increase or retarget the toxicity of their chemicals while the armies of pests evolve ever-stronger resistance to their efforts. The number of species known to have evolved resistance to pesticides has grown over the decades. As of 2007, there were more than 2700 known cases of resistance by 550 species to more than 300 pesticides, and some species have evolved resistance to multiple pesticides. Resistant pests can take a significant economic toll on crops. In Canada, resistance to herbicides is of great concern (recall that 85% of pesticide sales in Canada are for herbicides); both the number of herbicide-resistant weeds and the area of land covered by such weeds are increasing.[7]

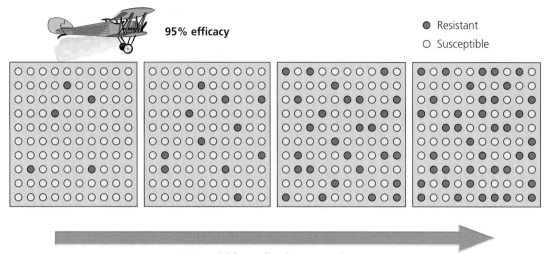

FIGURE 8.5
Through the process of natural selection, crop pests frequently evolve resistance to the poisons we apply to kill them. This simplified diagram shows that when a pesticide is applied to an outbreak of insect pests, it may kill virtually all individuals except those few with an innate immunity to the poison. Those surviving individuals may found a population with genes for resistance to the poison. Future applications of the pesticide may then be ineffective, forcing us to develop a more potent poison or an alternative means of pest control.

Biological control pits one organism against another

Because of pesticide resistance and the health risks of some synthetic chemicals, agricultural scientists increasingly battle pests and weeds with organisms that eat or infect them. This strategy, called **biological control**, or **biocontrol** for short, operates on the principle that "the enemy of one's enemy is one's friend." For example, parasitoid wasps are natural enemies of many caterpillars. These wasps lay eggs on a caterpillar, and the larvae that hatch from the eggs feed on the caterpillar, eventually killing it. Parasitoid wasps have been used as biocontrol agents in many situations. Some such efforts have succeeded at pest control and have led to steep reductions in chemical pesticide use.

One classic case of successful biological control is the introduction of the cactus moth, *Cactoblastis cactorum*, from Argentina to Australia in the 1920s to control invasive prickly pear cactus that was overrunning rangeland (**FIGURE 8.6**). Within just a few years, the moth managed to free millions of hectares of rangeland from the cactus.

(a) Before cactus moth introduction

(b) After cactus moth introduction

FIGURE 8.6
In one of the classic cases of biocontrol, larvae of the cactus moth (*Cactoblastis cactorum*) were used to clear non-native prickly pear cactus from millions of hectares of rangeland in Queensland, Australia. These photos from the 1920s show an Australian ranch before **(a)** and after **(b)** introduction of the moth.

A widespread modern biocontrol effort has been the use of **Bacillus thuringiensis** (**Bt**), a naturally occurring soil bacterium that produces a protein that kills many caterpillars and the larvae of some flies and beetles. Farmers have used the natural pesticidal activity of this bacterium to their advantage by spraying spores of this bacterium on their crops. If used correctly, Bt can protect crops from pest-related losses.

Biological control agents themselves may become pests

In most cases, biological control involves introducing an animal or a microbe into a foreign ecosystem, often on another continent. Such relocation helps ensure that the target pest has not already evolved ways to deal with the biocontrol agent, but it also introduces risks. In some cases, biocontrol has produced unintended consequences once the biocontrol agent became invasive and began affecting nontarget organisms. Following the cactus moth's success in Australia, for example, it was introduced in other countries to control prickly pear; however, it is now feared that the moth larvae could decimate native and economically important species of prickly pear and other cacti.

Scientists debate the relative benefits and risks of biocontrol measures. If biocontrol works as planned, it can be a permanent solution that requires no further maintenance and is environmentally benign. However, if the agent has nontarget effects, the harm done may become permanent, because removing the agent from the system once it is established is far more difficult than simply stopping a chemical pesticide application. The potential impacts of releasing a biocontrol agent into the natural environment are basically the same as for any alien or non-native species.

Because of concerns about unintended impacts, researchers now study biocontrol proposals carefully before putting them into action, and government regulators must approve these efforts. Canada has been a world leader in this regard. However, there will never be a foolproof way of knowing in advance whether a given biocontrol program will work as planned.

Integrated pest management combines biocontrol and chemical methods

As it became clear that both chemical and biocontrol approaches have their drawbacks, many agricultural scientists and farmers developed a more sophisticated strategy, trying to combine the best attributes of both approaches.

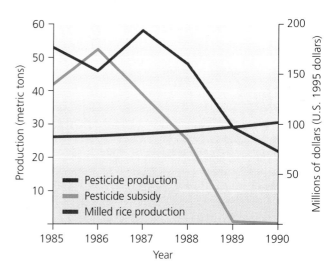

FIGURE 8.7
The Indonesian government threw its weight behind integrated pest management starting in 1986. Within just a few years, pesticide production and pesticide imports were down drastically, pesticide subsidies were phased out, and yields of rice increased.

In **integrated pest management (IPM)**, numerous techniques are integrated to achieve long-term suppression of pests, including biocontrol, use of chemicals, close monitoring of populations, habitat alteration, crop rotation, transgenic crops, alternative tillage methods, and mechanical pest removal. IPM is defined broadly enough that it encompasses a wide variety of strategies.

In recent decades, IPM has become popular in many parts of the world. Indonesia stands as an exemplary case (**FIGURE 8.7**). The nation had subsidized pesticide use heavily for years, but its scientists came to understand that pesticides were actually making pest problems worse. They were killing the natural enemies of the brown planthopper, which began to devastate rice fields as its populations exploded. Concluding that pesticide subsidies were costing money, causing pollution, and apparently decreasing yields, the Indonesian government in 1986 banned the importation of 57 pesticides, slashed pesticide subsidies, and encouraged IPM. Within four years, pesticide production fell to below half its 1986 level, imports fell to one-third, and subsidies were phased out (saving $179 million annually). Rice yields rose 13%.

We depend on insects to pollinate crops

Managing insect pests is such a major issue in agriculture that many people fall into a habit of thinking of all insects as somehow bad or threatening. But in fact, most insects are harmless to agriculture, and some are absolutely essential. The insects that pollinate agricultural crops are one of the most vital, yet least understood and least

FIGURE 8.8
Many agricultural crops depend on insects or other animals to pollinate them. Our food supply, therefore, depends partly on conservation of these vital organisms. These apple blossoms are being visited by a European bee. Flowers use colours and sweet smells to advertise nectar and pollen, enticements that attract pollinators.

appreciated, factors in cropland agriculture. **Pollination** is the process by which male sex cells of a plant (pollen) fertilize female sex cells of another plant; it is the botanical version of sexual intercourse. Without pollination, no plants could reproduce sexually, and no plant species would persist for long. Many plants achieve pollination by wind distribution. Millions of minuscule pollen grains are blown long distances, and by chance a small number land on the female parts of other plants of their species. The many kinds of plants that sport showy flowers, however, typically are pollinated by animals, such as hummingbirds, bats, and insects (**FIGURE 8.8**; see also **FIGURE 4.8**). Flowers are, in fact, evolutionary adaptations that function to attract pollinators. The sugary nectar and protein-rich pollen in flowers serve as rewards to lure these sexual intermediaries, and the sweet smells and bright colours of flowers are signals that advertise these rewards.

Although our staple grain crops are derived from grasses and are wind-pollinated, many other crops depend on insects for pollination. This environmental service has economic—not just ecological—value; the estimated value of pollination services rendered by insects in Canada each year is $1.2 billion.[8] A comprehensive survey by tropical bee biologist Dave Roubik documented 800 species of cultivated plants that rely on bees and other insects for pollination. An estimated 73% of cultivars are pollinated, at least in part, by bees; 19% by flies; 5% by wasps; 5% by beetles; and 4% by moths and butterflies. In addition, bats pollinate 6.5% and birds 4%. According to Seeds of Diversity and Environment Canada:

Animals pollinate three-quarters of the world's staple crops, 80% of all flowering plants in temperate climates, and 90% globally. One mouthful in three requires insect pollination. Since 25% of all birds eat seeds or fruit, they are also dependent on pollinators. Quite simply, without pollinators, the entire terrestrial globe would look entirely different and would not be able to support the number of people that it currently does.[9]

Many pollinating insects are at risk from the same pressures that are threatening other species, including habitat loss, land degradation, habitat fragmentation, pesticide use, invasive species, and climate change. These are not strictly new phenomena. For example, habitat destruction in the 1930s caused the failure of native leafcutter bee populations in Manitoba; this, in turn, led to the collapse of alfalfa seed production in the Canadian Prairies (see "The Science Behind the Story: The Alfalfa and the Leafcutter"). However, increases in the rate and extent of environmental change now put these ecological relationships at greater risk. The flowering plant–insect pollinator relationship is so specific and fragile (the precise shape of an insect's appendages can determine its suitability for pollinating a particular plant species) that even minor changes can have devastating impacts.

Conservation of pollinators is vital

Preserving the biodiversity of native pollinators is especially important today because the domesticated workhorse of pollination, the European honeybee (*Apis apis*), is being devastated by parasites. North American farmers regularly hire beekeepers to bring colonies of this introduced honeybee to their fields when it is time to pollinate crops (**FIGURE 8.9**). In recent years, certain parasitic mites have swept through honeybee populations, decimating hives and pushing many beekeepers toward financial ruin. Moreover, research indicates that honeybees are sometimes less effective pollinators than many native species, and often outcompete them, keeping the native species away from the plants.

Farmers and homeowners alike can help maintain populations of pollinating insects by reducing or eliminating pesticide use. All insect pollinators, including honeybees, are vulnerable to the vast arsenal of insecticides that modern industrial agriculture applies to crops and that many homeowners apply to lawns and gardens. Some insecticides are designed to specifically target certain types of insects, but many are not. Without full and detailed information on the effects of pesticides, farmers and homeowners trying to control the "bad" bugs

THE SCIENCE BEHIND THE STORY

The Alfalfa and the Leafcutter

This leafcutter bee is pollinating an alfalfa flower.

In the first half of the twentieth century, land clearing for agriculture destroyed many nesting sites of native leafcutter bees in the Canadian Prairies. As a result, Canadian alfalfa seed production decreased dramatically, virtually collapsing by mid-century. Alfalfa is an important forage crop for livestock, and it is used to control moisture and nutrient levels in agricultural fields. Honeybees are ineffective alfalfa pollinators because they can steal nectar without "tripping" the alfalfa flower, a process that uncovers the plant's stigma and is required for the pollination to be successful. With the loss of the most effective native pollinators, by 1950 Canada was importing alfalfa seed to meet 95% of its domestic needs.

In response to this crisis, the European alfalfa leafcutter bee (Megachile rotundata) was introduced to Canada in 1961 (see photo). Scientists, beekeepers, and seed growers worked together to develop a management system for the new bees, which eventually resulted in a sixfold increase in alfalfa yields. By the end of the twentieth century, Canada was back to meeting or exceeding its demand for alfalfa seed, thanks to the alfalfa leafcutter bee. The bees are now being used by blueberry producers in Eastern Canada and to pollinate buckwheat and hybrid canola in the Prairies.

The importation of the leafcutter bees also led to a new kind of beekeeper, one who sells bee larvae to other growers for pollination. The leafcutter bee is gentler than the honeybee, and typically will sting only if it is squeezed. Leafcutters are solitary (rather than gregarious, like other kinds of bees), and less likely to wander than honeybees. These characteristics make leafcutters easier to manage and handle than other bees. Management of leafcutters, which do not build colonies or store honey, involves building large nesting arrays of layered, grooved materials. These arrays mimic the sites in which the bees would naturally build their individual nests, in cracks and grooves in soft, decaying wood.

By controlling temperature, the emergence of the adult bees can be synchronized with the alfalfa bloom. The bees are kept dormant at 4°C until about three weeks before the expected crop bloom, when the temperature is turned up to 29°C to trigger the development of adult leafcutter bees. This management system has made Canada the leading producer of alfalfa leafcutters, currently producing 4 billion bees for pollination of domestic and international crops each year.

Interestingly (but not surprisingly, given their name) leafcutters can also become a pest species. To line their nests, the bees carefully incise small, round pieces of leaves and carry them back to the nesting site, which can potentially damage the leaf. The damage is usually minor, unless there is an unusually large population of feral (escaped domesticated) or native leafcutter bees in a small area.[10]

FIGURE 8.9
European honeybees are widely used to pollinate crop plants, and beekeepers transport hives of bees to crops when it is time for flowers to be pollinated. However, honeybees have recently suffered devastating epidemics of parasitism, making it increasingly important for us to conserve native species of pollinators.

that threaten the plants they value all too often kill the "good" insects as well.

Homeowners, even in the middle of a city, can encourage populations of pollinating insects by planting gardens of flowering plants that nourish pollinating insects and by providing nesting sites for bees. By allowing noncrop flowering plants (such as clover) to grow around the edges of their fields, farmers can maintain a diverse community of insects—some of which will pollinate their crops.

Genetically Modified Food

The Green Revolution enabled us to feed a greater number and proportion of the world's people, but relentless population growth demands still more. A new set of potential solutions began to arise in the 1980s and 1990s as advances in genetics enabled scientists to directly alter the genes of organisms, including crop plants and livestock. The genetic modification of organisms that provide us with food holds promise for increasing nutrition and the efficiency of agriculture, while lessening the impacts of agriculture on the planet's environmental systems.

However, genetic modification may pose risks that are not yet well understood, which has given rise to protest around the globe from consumer advocates, small farmers, opponents of big business, and environmental activists. Because genetically modified foods have generated so much emotion and controversy, it is vital at the outset to clear up the terminology and clarify exactly what the techniques involve.

Genetic modification of organisms depends on recombinant DNA

The genetic modification of crops and livestock is one type of **genetic engineering**, any process whereby scientists directly manipulate an organism's genetic material in the lab, by adding, deleting, or changing segments of its DNA. To genetically engineer organisms, scientists extract genes from the DNA of one organism and transfer them into the DNA of another to create a **genetically modified (GM) organism**. The technique is called **recombinant DNA** technology, referring to DNA that has been patched together from the DNA of multiple organisms. In this process, scientists break up DNA from multiple organisms and then splice segments together, trying to place genes that produce certain proteins and code for certain desirable traits (such as rapid growth, disease and pest resistance, or higher nutritional content) into the genetic makeup, or **genomes**, of organisms lacking those traits.

Recombinant DNA technology was developed in the 1970s by scientists studying the bacterium *Escherichia coli*. As shown in **FIGURE 8.10**, scientists first isolate *plasmids*, small, circular DNA molecules, from a bacterial culture. At the same time, DNA containing a gene of interest is removed from the cells of another organism. Scientists insert the gene of interest into the plasmid to form recombinant DNA. This recombinant DNA enters new bacteria, which then reproduce, generating many copies of the desired gene.

When scientists use recombinant DNA technology to develop new varieties of crops, they can often introduce

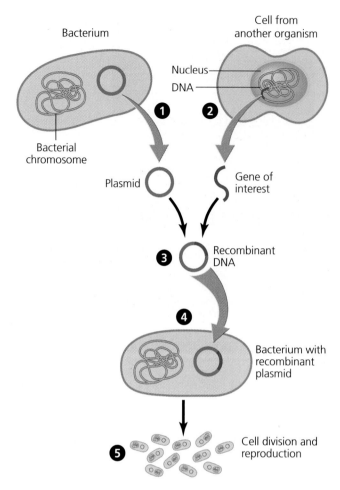

FIGURE 8.10
To create recombinant DNA, a gene of interest is excised from the DNA of one type of organism and is inserted into a stretch of bacterial DNA called a plasmid. The plasmid is then introduced into cells of the organism to be modified. If all goes as planned, the new gene will be expressed in the GM organism as a desirable trait, such as rapid growth or high nutritional content in a food crop.

the recombinant DNA directly into a plant cell and regenerate an entire plant from that single cell. Some plants, including many grains, are not receptive to plasmids, in which case scientists may use a "gene gun" to shoot DNA directly into plant cells. An organism that contains DNA from another species is called a **transgenic** organism, and the genes that have moved between them are called **transgenes**.

The creation of transgenic organisms is one type of **biotechnology**, the material application of biological science to create products derived from organisms. Recombinant DNA and other types of biotechnology have helped us develop medicines, clean up pollution, understand the causes of cancer and other diseases, dissolve blood clots after heart attacks, and make better beer and cheese. FIGURE 8.11 details several of the most notable developments in GM foods. These examples and the stories behind them illustrate both the promises and the pitfalls of food biotechnology.

Genetic engineering is like, and unlike, traditional agricultural breeding

The genetic alteration of plants and animals by humans is nothing new; we have been influencing the genetic make-up of livestock and crop plants for thousands of years. As we saw in Chapter 7, our ancestors altered the gene pools of domesticated plants and animals through selective breeding by preferentially mating individuals with favoured traits so that offspring would inherit the traits. Early farmers selected plants and animals that grew faster, were more resistant to disease and drought, and produced large amounts of fruit, grain, or meat.

Proponents of GM crops often stress this continuity with our past and say there is little reason to expect that today's GM food will be less safe than selectively bred food. However, as biotech's critics are quick to point out, the techniques that geneticists use to create GM organisms differ from traditional selective breeding in important ways. For one, selective breeding generally mixes genes of individuals of the same species, whereas with recombinant DNA technology, scientists mix genes of different species, even those as different as viruses and crops, or spiders and goats. For another, selective breeding deals with whole organisms living in the field, whereas genetic engineering involves lab experiments dealing with genetic material apart from the organism. And whereas traditional breeding selects from among combinations of genes that come together on their own, genetic engineering creates novel combinations directly. Thus, traditional breeding changes organisms through the process of selection, whereas genetic engineering is more akin to the process of mutation.

Biotechnology is transforming the products around us

In just three decades, GM foods have gone from science fiction to big business. As recombinant DNA technology first developed in the 1970s, scientists debated among themselves whether the new methods were safe. They collectively regulated and monitored their own research until most scientists were satisfied that reassembling genes in bacteria did not create dangerous superbacteria. Once the scientific community declared itself confident that the technique was safe in the 1980s, industry leaped at the chance to develop hundreds of applications, from improved medicines (such as a vaccine for hepatitis B and insulin for diabetes) to designer plants and animals.

Most GM crops today are engineered to resist herbicides, so that farmers can apply herbicides to kill weeds without having to worry about killing their crops. Other crops are engineered to resist insect attacks. Some are modified for both types of resistance. Resistance to herbicides and pests makes it efficient, and in some cases more economical, for large-scale commercial farmers to do their jobs. As a result, sales of GM seeds to these farmers have risen quickly.

Today more than half of the world's soybean harvest is transgenic, as is one of every four cotton plants, one of every five canola plants, and one of every six corn plants. Globally in 2006, it was estimated that GM crops grew on 106 million hectares of farmland,[11] but the number of producing nations and crop types, as well as the variety of genetically controlled traits, remain low. Just six countries (Canada, the United States, Argentina, Brazil, China, and South Africa), four crops (soybean, cotton, maize, and canola), and two controlled traits (herbicide tolerance and insect resistance) account for more than 99% of the area devoted to the production of transgenic crops worldwide (FIGURE 8.12).[12]

Because these nations are major food exporters, much of the produce on the world market is transgenic, from such crops as soybeans, corn, cotton, and canola. The global area planted in GM crops has jumped by more than 10% annually every year since 1996, and more than half the world's people now live in nations in which GM crops are grown. The market value of GM crops in 2006 was estimated at $6.15 billion.

weighing the issues — GM FOODS AND YOU — 8–4

Do you think you have ever eaten a food product that contained genetically modified organisms? Hint: If you live in North America, your answer should almost certainly be "yes." Perhaps as much as 70% of the food products on shelves in North American grocery stores contain at least some GM ingredients. Check your kitchen cupboards for any foods that contain products or ingredients made from corn, soy, or canola. The probability that some of those ingredients came from genetically modified plants is very high.

Several Notable Examples of Genetically Modified Food Technology	
Food	Development
Golden rice 	Millions of people in the developing world get too little vitamin A in their diets, causing diarrhea, blindness, immune suppression, and even death. The problem is worst with children in east Asia, where the staple grain, white rice, contains no vitamin A. Researchers took genes from plants that produce vitamin A and spliced the genes into rice DNA to create more-nutritious "golden rice" (the vitamin precursor gives it a golden colour). Critics charged that biotech companies hyped their product, which contains only small amounts of the nutrient and may not be the best way to combat vitamin A deficiency. India's foremost critic of GM food, Vandana Shiva, charged that "vitamin A rice is a hoax . . . a very effective strategy for corporate takeover of rice production, using the public sector as a Trojan horse." Backers of the technology counter that the nutritive value can be further improved and could enhance the health of millions of people.
Flavr Savr tomato 	By reversing the function of a normal tomato gene, the Calgene Corporation created the Flavr Savr tomato, which Calgene maintained would ripen longer on the vine, taste better, stay firm during shipping, and last longer in the produce department. The U.S. Food and Drug Administration approved the Flavr Savr tomato for sale in the United States in 1994. Calgene stopped selling the Flavr Savr in 1996, however, for several reasons, including problems with the technique and public safety concerns.
Ice-minus strawberries 	University of California–Berkeley researcher Steven Lindow removed a gene that facilitated the formation of ice crystals from the DNA of a particular bacterium, *Pseudomonas syringae*. The modified, frost-resistant bacteria could then serve as a kind of antifreeze when sprayed on the surface of frost-sensitive crops such as strawberries. The multiplying bacteria would coat the berries, protecting them from frost damage. However, early news coverage of this technique showed scientists spraying plants while wearing face masks and protective clothing, an image that caused public alarm.
Bt crops 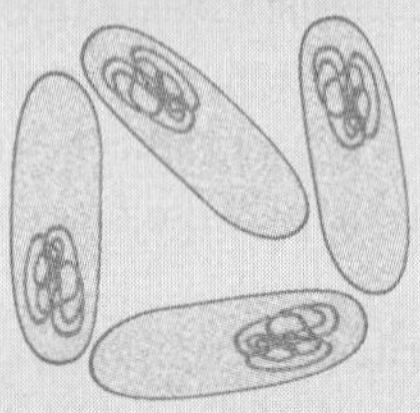	By equipping plants with the ability to produce their own pesticides, scientists hoped to boost crop yields by reducing losses to insects. By the late 1980s, scientists working with *Bacillus thuringiensis* (Bt) had pinpointed the genes responsible for producing that bacterium's toxic effects on insects, and had managed to insert the genes into the DNA of crops. The USDA and EPA approved Bt versions of 18 crops for field testing, from apples to broccoli to cranberries. Corn and cotton are the most widely planted Bt crops today. Proponents say Bt crops reduce the need for chemical pesticides. However, critics worry that the continuous presence of Bt in the environment will induce insects to evolve resistance to the toxins and that Bt crops might cause allergic reactions in humans. Another concern is that the crops may harm nontarget species. A 1999 study reported that pollen from Bt corn can kill the larvae of monarch butterflies, a nontarget species, when corn pollen drifts onto milkweed plants that monarchs eat. Another study that year showed that the Bt toxin could leach from corn roots and poison the soil.

FIGURE 8.11
The early development of genetically modified foods has been marked by a number of cases in which these products ran into trouble in the marketplace or were opposed by activists. A selection of these cases serves to illustrate some of the issues that proponents and opponents of GM foods have being debating.

What are the impacts of GM crops?

As GM crops were adopted, as research proceeded, and as biotech business expanded, many citizens, scientists, and policy makers became concerned. Some feared the new foods might be dangerous for people to eat—what if there were unexpected health consequences, such as unanticipated allergic reactions to transgenes in GM foods? Others were concerned that transgenes might escape, pollute ecosystems, and damage nontarget organisms. Still others worried that pests would evolve resistance to the supercrops and become "superpests" or that transgenes would be transferred from crops to other plants and turn them into "superweeds." Some worried that transgenes might ruin the integrity of native ancestral races of crops.

Because the technology is new and its large-scale introduction into the environment is newer still, there remains a lot that scientists do not know about how transgenic crops behave in the field. Certainly, millions of North Americans eat GM foods every day without any obvious signs of harm, and evidence for negative ecological effects is limited so far. However, it is still too early to dismiss all concerns without further scientific research. There are numerous mechanisms whereby transgenes can "escape" from the confines of the organism into which

Several Notable Examples of Genetically Modified Food Technology	
Food	Development
StarLink corn	StarLink corn, a variety of Bt corn, had been approved and used in the United States for animal feed but not for human consumption. In 2000, StarLink corn DNA was discovered in taco shells and other corn products, causing fears that the corn might cause allergic reactions. No such health effects have been confirmed, but the corn's French manufacturer, Aventis CropScience, chose to withdraw the product from the market. Although StarLink corn was grown on only a tiny portion of U.S. farmland, its transgene apparently spread widely to other corn through cross-pollination. This episode cost U.S. taxpayers, because the U.S. government spent $20 million to purchase contaminated corn and remove it from the food supply.
Sunflowers and superweeds	Sunflowers have also been engineered to express the Bt toxin. Research on Bt sunflowers suggests that their transgenes might spread to other plants and turn them into vigorous weeds that compete with the crop. This is most likely to happen with crops like squash, canola, and sunflowers that can breed with their wild relatives. In 2002, Ohio State University researcher Allison Snow and colleagues bred wild sunflowers with Bt sunflowers and found that hybrids with the Bt gene produced more seeds and suffered less herbivory than hybrids without it. They concluded that if Bt sunflowers were planted commercially, the Bt gene would spread into wild sunflowers, potentially turning them into superweeds. Researcher Norman Ellstrand of the University of California–Riverside, meanwhile, had found that transgenes from radishes were transferred to wild relatives 1 km away and that hybrids produced more seeds, so the gene could be expected to spread in wild populations. He found the same results with sorghum and its weedy relative, johnsongrass. Such results suggest that transgenic crops can potentially create superweeds that can compete with crops and harm nontarget organisms.
Roundup Ready crops 	The Monsanto Company manufactures a widely used herbicide called Roundup. Roundup kills weeds, but kills crops too, so farmers must apply it carefully. Thus, Monsanto engineered Roundup Ready crops—including soybeans, corn, cotton, and canola—that are immune to the effects of its herbicide. With these variants, farmers can spray Roundup on their fields without killing their crops, in theory making the farmer's life easier. Of course, this also creates an incentive for farmers to use Monsanto's Roundup herbicide rather than a competing brand. Unfortunately, Roundup is not completely benign; its active ingredient, glyphosate, is the third-leading cause of illness for California farm workers. It also harms nitrogen-fixing bacteria and desirable fungi in soils that are essential for crop production. Biotech proponents have argued that GM crops are good for the environment because they reduce pesticide use. This may often be the case, but some studies have shown that farmers apply more herbicide when they use Roundup Ready crops.
Terminator seeds	In the late 1990s the USDA worked with Delta and Pine Land Company to engineer a line of crop plants that can kill their own seeds. This so-called "terminator" technology would ensure that farmers buy seeds from seed companies every year rather than planting seeds saved from the previous year's harvest. Because GM crops require a great deal of research and development, seed companies reason that they need to charge farmers annually for seeds in order to recoup their investment. Critics worried that pollen from terminator plants might fertilize normal plants, damaging the crops of farmers who save seeds from year to year. Some nations, like India and Zimbabwe, banned terminator seeds. These countries saw the efforts of biotech seed companies to sell them terminator seeds as a ploy to make poor farmers dependent on multinational corporations for seeds. In the face of this opposition, in 1999 agrobiotech companies Monsanto and AstraZeneca announced that they would not bring their terminator technologies to market.

FIGURE 8.11
(Continued)

they have been implanted and move out into native populations, as well as locations where it has happened. Therefore, critics argue that we should adopt the **precautionary principle**, the idea that one should not proceed until the ramifications of an action are well understood.

The British government, in considering whether to allow the planting of GM crops, commissioned three large-scale studies between 2003 and 2005. The first study, on economics, found that GM crops could produce long-term financial benefits for Britain, although short-term benefits would be minor. The second study addressed health risks and found little to no evidence of harm to human health, but it noted that effects on wildlife and ecosystems should be tested before crops are approved. The third study looked at effects on bird and invertebrate populations from four GM crops modified for herbicide resistance. Results showed that fields of GM beets and GM spring oilseed rape supported less biodiversity than fields of their non-GM counterparts. Fields of GM maize supported more, however, and fields of winter oilseed rape showed mixed results. Policy makers had hoped that the biodiversity study would end the debate, but the science showed that the impacts of GM crops are complex.

Debate over GM foods involves more than science

Much more than science has been involved in the debate over GM foods. Ethical issues have played a large role. For

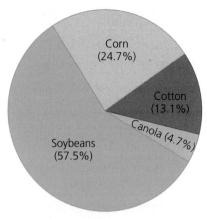

(a) GM crops by type

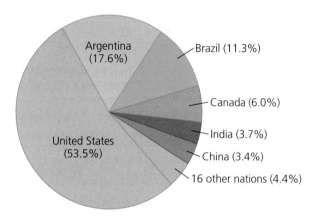

(b) GM crops by nation

FIGURE 8.12
At 58.8% of the global total, the United States leads the world in land area dedicated to genetically modified crops. Following far behind the United States is Argentina at 20.0%, and Canada is third at 6.7%. Data from International Service for the Acquisition of Agri-Biotech Application (ISAAA), 2004.

many people, the idea of "tinkering" with the food supply seems dangerous or morally wrong. Even though our agricultural produce is the highly artificial product of thousands of years of selective breeding, people tend to think of food as natural. Furthermore, because every person relies on food for survival and cannot choose *not* to eat, the genetic modification of dietary staples, such as corn, wheat, and rice, essentially forces people to consume GM products or to go to special effort to avoid them.

The perceived lack of control over one's own food has driven widespread concern about domination of the global food supply by a few large businesses. Gigantic agrobiotech companies—among them Monsanto, Syngenta, Bayer CropScience, Dow, DuPont, and BASF—create GM technologies. Many activists say these multi-national corporations threaten the independence and well-being of the small farmer. This perceived loss of democratic local control is a driving force in the opposition to GM foods, especially in Europe and the developing world. Critics of biotechnology also voice concern that much of the research into the safety of GM organisms is funded, overseen, or conducted by the corporations that stand to profit if their transgenic crops are approved for human consumption, animal feed, or ingredients in other products.

So far, GM crops have not lived up to their promise of feeding the world's hungry—perhaps because they haven't been allowed to. Nearly all commercially available GM crops have been engineered to express either pesticidal properties (e.g., Bt crops) or herbicide tolerance. Often, these GM crops are tolerant to herbicides that the same company manufactures and profits from (e.g., Monsanto's Roundup Ready crops). Crops with traits that might benefit poor small-scale farmers of developing countries (such as increased nutrition, drought tolerance, and salinity tolerance) have not been widely commercialized, perhaps because corporations have less economic incentive to do so. Similarly, such crops as the infamous "golden rice"—engineered with a higher than usual content of vitamin A, and proposed as the solution to vitamin A deficiencies throughout the developing world—have met with limited success. Whereas the Green Revolution was a largely public venture, the "gene revolution" promised by GM crops has mainly been driven by market considerations of companies selling proprietary products.

When the U.S.-based Monsanto Company began developing GM products in the mid-1980s, it foresaw public anxiety and worked hard to inform, reassure, and work with environmental and consumer advocates, who the company feared would otherwise oppose the technology. Monsanto even lobbied the U.S. government to regulate the industry so the public would feel safer about it. These efforts were undermined, however, when the

---

weighing the issues   **EARLY HURDLES FOR GM FOODS**   8–5

As the vignettes in **FIGURE 8.11** illustrate, a number of GM foods have run into difficulties. Do you think this reveals an underlying problem with the approach, or are these simply examples of unavoidable glitches that are bound to occur during the early development of any new technology? Or are politics and hysteria simply getting in the way of the successful implementation of these new technologies? Do you expect that debate between proponents and opponents of GM organisms will subside in the future, or not?

company's first GM product to market, a growth hormone to spur milk production in cows, alarmed consumers concerned about children's health. Then, when the company went through a leadership change, its new head changed tactics and pushed new products aggressively without first reaching out to opponents. Opposition built, and the company lost the public's trust, especially in Europe and in the developing world. David-and-Goliath battles that pitted giant Monsanto against lone farmers, like Percy Schmeiser, have not helped to repair the company's tarnished public image.

Given such developments, the future of GM foods seems likely to hinge on social, economic, legal, and political factors, as well as scientific ones. European consumers have been particularly vocal in expressing their unease about the possible risks of GM technologies. Opposition in nations of the European Union resulted in a *de facto* moratorium on GM foods from 1998 to 2003, blocking the importation of hundreds of millions of dollars in agricultural products. This prompted the United States to bring a case before the World Trade Organization in 2003, complaining that Europe's resistance was hindering free trade.

Europeans now widely demand that GM foods be labelled as such and criticize the United States for not joining 100 other nations in signing the Cartagena Protocol on Biosafety (part of the United Nations Convention on Biodiversity), a treaty that lays out guidelines for open information about exported crops. Canada has been a party to the Convention on Biodiversity since 1992, but has never ratified the Cartagena Protocol.[13]

Transnational spats will surely affect the future direction of agriculture, but consumers and the governments of the world's developing nations could exert the most influence in the end. Recent decisions by the governments of India and Brazil to approve GM crops (following long and divisive debates) are already adding greatly to the world's transgenic agriculture. Moreover, China is aggressively expanding its use of transgenic crops.

A counterexample is Zambia, one of several African nations that refused U.S. food aid meant to relieve starvation during a drought in late 2002. The governments of these nations worried that their farmers would plant some of the GM corn seed that was meant to be eaten, and that GM corn would thereby establish itself in their countries. They viewed this as undesirable because African economies depend on exporting food to Europe, which has put severe restrictions on GM food. In the end, Zambia's neighbours accepted the grain after it had been milled (so none could be planted), but Zambia held out. Citing health and environmental risks, uncertain science, and the precautionary principle, the Zambian government declined the aid, despite the fact that 2 million to

FIGURE 8.13
Debate over GM foods reached a dramatic climax in Zambia in 2002, when the government refused U.S. shipments of GM corn that were intended to relieve starvation caused by drought. Here a Zambian mother with her child waits in a line for food assistance.

3 million of its people were at risk of starvation (**FIGURE 8.13**). Intense debate followed within the country and around the world. Eventually the United Nations delivered non-GM grain, and in April 2003 the Zambian government announced a plan to coordinate a comprehensive long-term policy on GM foods.

The Zambian experience demonstrates some of the ethical, economic, and political dilemmas modern nations face. The corporate manufacturers of GM crops naturally aim to maximize their profits, but they also aim to develop products that can boost yields, increase food security, and reduce hunger. Although industry, activists, policy makers, and scientists all agree that hunger and malnutrition are problems and that agriculture should be made environmentally safer, they often disagree about the solutions to these dilemmas and the risks that each proposed solution presents.

Preserving Crop Diversity

As the controversy over transgenic crops in Oaxaca and Saskatchewan demonstrates, one concern many people harbour is that transgenes might move, by pollination, into local native races of crop plants. There is certainly now abundant evidence that this is possible or, indeed, likely.

Crop diversity provides insurance against failure

Preserving the integrity of native variants gives us a bulwark against commercial crop failure. The regions where crops first were domesticated generally remain important repositories of crop biodiversity. Although modern industrial agriculture relies on a small number of plant types, its foundation lies in the diverse varieties that still exist in places like Oaxaca. These varieties contain genes that, through conventional crossbreeding or genetic engineering, might confer resistance to disease, pests, inbreeding, and other pressures that challenge modern agriculture. Monocultures essentially place all our eggs in one basket, such that any single catastrophic cause could potentially wipe out entire crops. Having available the domesticated varieties, or **cultivars**, and the wild relatives of crop plants gives us the genetic diversity that may include ready-made solutions to unforeseen problems.

Because accidental interbreeding can decrease the diversity of local variants, many scientists argue that we need to protect areas like Oaxaca. For this reason, the Mexican government helped create the Sierra de Manantlán Biosphere Reserve around an area harbouring the localized plant thought to be the direct ancestor of maize. For this reason, too, it imposed a national moratorium in 1998 on the planting of transgenic corn (although that ban was lifted in 2005).

We have lost a great deal of genetic diversity in our crop plants already. The number of wheat varieties in China is estimated to have dropped from 10 000 in 1949 to 1000 by the 1970s, and Mexico's famed maize varieties now number only 30% of what was extant in the 1930s. In the United States, many fruits and vegetables have decreased in diversity by 90% in less than a century. Note, however, that the number of varieties that exist is not, on its own, indicative of the robustness of biodiversity. For example, in recent years the number of wheat varieties in China has actually increased, but the genetic diversity among those varieties has narrowed.[14]

A primary cause of the loss of crop diversity is that market forces have discouraged diversity in the appearance of fruits and vegetables. Commercial food transporters and processors prefer items to be similar in size and shape, for convenience. Consumers, for their part, have shown preferences for uniform, standardized food products over the years. Now that local organic agriculture is growing in affluent societies, however, consumer preferences for diversity are increasing.

Seed banks are living museums for seeds

Protecting areas with high crop diversity is one way to preserve genetic assets for our agricultural systems. Another is to collect and store seeds from crop varieties and periodically plant and harvest them to maintain a diversity of cultivars. This is the work of **seed banks** or

(a) Traditional food plants of the Desert Southwest

(b) Pollination by hand

FIGURE 8.14
Seed banks preserve genetic diversity of traditional crop plants. Native Seeds/SEARCH of Tucson, Arizona, preserves seeds of food plants important to traditional diets of Native Americans of Arizona, New Mexico, and northwestern Mexico. Beans, chilies, squashes, gourds, maize, cotton, and lentils are all in its collections, as well as less-known plants, such as amaranth, lemon basil, and devil's claw **(a)**. Traditional foods, like mesquite flour, prickly pear pads, chia seeds, tepary beans, and cholla cactus buds, help fight the diabetes that Native Americans frequently suffer from having adopted a Western diet. At the farm where seeds are grown, care is taken to pollinate varieties by hand **(b)** to protect their genetic distinctiveness.

gene banks, institutions that preserve seed types as a kind of living museum of genetic diversity (**FIGURE 8.14**). In total, these facilities hold roughly 6 million seed samples, keeping them in cold, dry conditions to encourage long-term viability. The $300 million in global funding for these facilities is not adequate for proper storage and for the labour of growing out the seed periodically to renew the stocks. Therefore, it is questionable how many of these 6 million seeds are actually preserved.

The Royal Botanic Gardens' Millennium Seed Bank in Britain holds more than 1 billion seeds and aims to bank seed from 10% of the world's plants by 2010. In Arctic

Norway, construction has begun on a "doomsday vault" seed bank, intended to hold seeds from around the world as a safeguard against global agricultural calamity. Other major efforts include large seed banks, such as the U.S. National Seed Storage Laboratory at Colorado State University, Seed Savers Exchange in Iowa, Plant Gene Resources of Canada, and the Wheat and Maize Improvement Centre (CIMMYT) in Mexico.

Raising Animals for Food

Food from cropland agriculture makes up a large portion of the human diet, but most people also eat animal products. People don't *need* to eat meat or other animal products to live full, active, healthy lives, but for most people it is difficult to obtain a balanced diet without incorporating animal products. Many of us do eat animal products, and this choice has environmental, social, agricultural, and economic impacts.

Consumption of animal products is growing

As wealth and global commerce have increased, so has our consumption of meat, milk, eggs, fish, and other animal products (FIGURE 8.15). The world population of **domesticated animals** and animals raised in captivity for food rose from 7.3 billion animals to 20.6 billion animals between 1961 and 2000. Most of these animals are chickens, although the most-eaten meat per unit of weight is pork. Global meat production has increased fivefold since 1950, and per capita meat consumption has nearly doubled.

Like other domesticated species, livestock and other farm animals can be at risk of biodiversity loss, and even extinction. For example, the FAO's Global Databank for Animal Genetic Resources for Food and Agriculture contains information on a total of 7616 livestock breeds, of which 20% are classified as "at risk." During the first six years of the twenty-first century, 62 livestock breeds became extinct, which amounts to a loss of almost one breed per month.[15]

High consumption has led to feedlot agriculture

In traditional agriculture, livestock were kept by farming families near their homes or were grazed on open grasslands by nomadic herders or sedentary ranchers. These traditions have survived, but the advent of industrial agriculture, responding to the pressure of global population growth, has added a new method. **Feedlots**, also known as *factory farms* or **concentrated animal feeding operations** (**CAFOs**), are essentially huge warehouses or pens designed to deliver energy-rich food to animals living at extremely high densities (FIGURE 8.16). Today, more than half of the world's pork and poultry comes from feedlots, as does much of its beef.

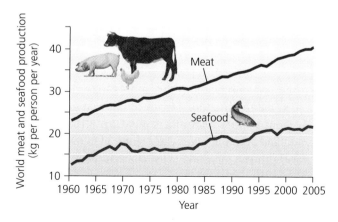

FIGURE 8.15
Per capita consumption of meat from farm animals has increased steadily worldwide over the past few decades, as has per capita consumption of seafood (marine and freshwater, harvested and farmed). Data from Food and Agriculture Organization of the United Nations.

FIGURE 8.16
These chickens at a factory farm are housed several to a cage and have been "debeaked," the tips of their beaks cut off, to prevent them from pecking one another. The hens cannot leave the cages and essentially spend their lives eating, defecating, and laying eggs, which roll down slanted floors to collection trays. The largest chicken farms house hundreds of thousands of individuals.

Feedlot operations allow for greater production of food and are probably necessary to keep up with the levels of meat consumption in Canada and the United States. Feedlots have one overarching benefit for environmental quality: Taking cattle, sheep, goats, and other livestock off the land and concentrating them in feedlots reduces the impact they would otherwise exert on large portions of the landscape. In Chapter 7, you learned that overgrazing can degrade soils and vegetation and that hundreds of millions of hectares of land are considered overgrazed. Animals that are densely concentrated in feedlots will not contribute to overgrazing and soil degradation.

However, feedlots are not without impact, and many environmental advocates have attacked them for their contributions to water and air pollution. Waste from feedlots can emit strong odours and can pollute surface water and groundwater, because livestock produce prodigious amounts of feces and urine. One dairy cow can produce about 20 400 kg of waste in a single year. Greeley, Colorado, is home to North America's largest meatpacking plant and two adjacent feedlots, all of which are owned by the agribusiness firm ConAgra. Each feedlot has room for 100 000 cattle that are fed surplus grain and injected with anabolic steroids to stimulate growth. During its stay at the feedlot, a typical steer will eat 1360 kg of grain, gain 180 kg in body weight, and generate 23 kg of manure each day. Poor waste containment practices at feedlots have been linked to outbreaks of disease, including virulent strains of *Pfiesteria*, a microbe that poisons fish. In 2000 in Walkerton, Ontario, a deadly strain of *E. coli* bacteria, thought to have originated from the contamination of municipal water wells by runoff from factory farms, caused the deaths of seven people and serious illness in hundreds of others (see Chapter 11).

The crowded and dirty conditions under which animals are often kept at factory farms necessitate the use of antibiotics to control disease. These chemicals can be transferred up the food chain, and their overuse can cause microbes to evolve resistance to them. Crowded conditions also can exacerbate outbreaks of diseases, such as avian influenza ("bird flu") and bovine spongiform encephalitis (BSE, or "mad cow" disease), which are now known to be transferable to humans in serious and even deadly forms.

Feedlot impacts can be minimized when properly managed, and both the federal and provincial or territorial governments regulate feedlots in Canada. Most feedlot manure is applied to farm fields as fertilizer, reducing the need for chemical fertilizers. Manure in liquid form can be injected into the ground where plants need it, and farmers can conduct tests to determine amounts that are appropriate to apply.

weighing the issues 8–6

FEEDLOTS AND ANIMAL RIGHTS

Animal rights activists decry factory farming because they say it mistreats animals. Chickens, pigs, and cattle are kept crowded together in small pens, fattened up, and slaughtered. Do you think animal rights concerns should be given weight as we determine how best to raise our food? Should we concern ourselves with the quality of life—and death—of the animals that constitute part of our diet?

Our food choices are also energy choices

What we choose to eat has significant ramifications for how we use energy and the land that supports agriculture. Recall our discussions of thermodynamics and trophic levels. Whenever energy moves from one trophic level to the next, as much as 90% is lost. For example, if we feed grain to a cow and then eat beef from the cow, we lose a great deal of the grain's energy to the cow's digestion and metabolism. Energy is used up when the cow converts the grain to tissue as it grows and as the cow uses its muscle mass on a daily basis to maintain itself. For this reason, eating meat is far less energy-efficient than relying on a vegetarian diet. The lower in the food chain from which we take our food sources, the greater the proportion of the Sun's energy we put to use as food and the more people Earth can support.

Some animals convert grain feed into milk, eggs, or meat more efficiently than others (**FIGURE 8.17**). Scientists have calculated relative energy conversion efficiencies for different types of animals. Such energy efficiencies have ramifications for land use—land and water are required to raise food for the animals, and some animals require more than others. **FIGURE 8.18** shows the area of land and weight of water required to produce 1 kg of food protein for milk, eggs, chicken, pork, and beef. Producing eggs and chicken meat requires the least space and water, whereas producing beef requires the most. Such differences make clear that when we choose what to eat, we are also indirectly choosing how to make use of resources such as land and water.

In 1900, we fed about 10% of global grain production to animals. In 1950, this number had reached 20%, and

CHAPTER EIGHT BIOTECHNOLOGY AND FOOD RESOURCES 241

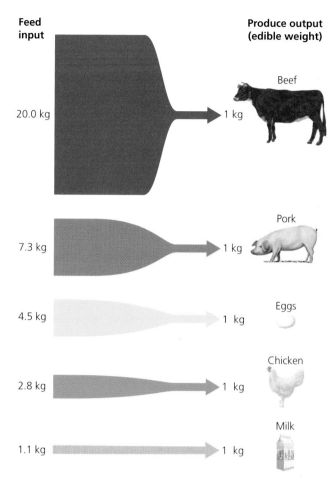

FIGURE 8.17
Different animal food products require different amounts of input of animal feed. Chickens must be fed 2.8 kg of feed for each 1 kg of resulting chicken meat, for instance, whereas 20 kg of feed must be provided to cattle to produce 1 kg of beef. Data from Smil, V. 2001. *Feeding the World: A Challenge for the Twenty-First Century.* Cambridge, MA: MIT Press.

to the deforestation of tropical rainforests and their conversion into rangelands. We will discuss these connections in greater detail in Chapter 11.

We also raise fish on "farms"

In addition to plants grown in croplands and animals raised on rangelands and in feedlots, we rely on aquatic organisms for food. Wild fish populations are plummeting throughout the world's oceans as increased demand and new technologies have led us to overharvest most marine fisheries. This means that raising fish and shellfish on "fish farms" may be the only way to meet the growing demand for these foods.

We call the raising of aquatic organisms for food in controlled environments **aquaculture**. Many aquatic

by the beginning of the twenty-first century, we were feeding 45% of global grain production to animals. Although much of the grain fed to animals is not of a quality suitable for human consumption, the resources required to grow it could have instead been applied toward growing food for people. One partial solution is to feed livestock crop residues—plant matter, such as stems and stalks, that we would not consume anyway—and this is increasingly being done.

An additional environmental problem associated with meat production is that the plants required to feed the livestock must be grown on large ranches, which are basically domesticated grasslands. The growth in meat consumption thus requires that forested land worldwide be converted to rangelands in support of livestock production. In the 1970s, this was first dubbed "the Hamburger Connection"—by purchasing a hamburger made from South American beef, one was unwittingly contributing

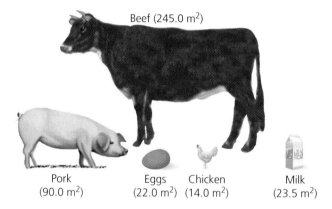

(a) Land required to produce 1 kg of protein

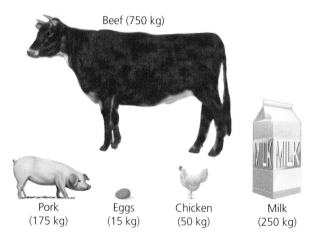

(b) Water required to produce 1 kg of protein

FIGURE 8.18
Producing different types of animal products requires different amounts of land and water. Raising cattle for beef requires by far the most land (a) and water (b) of all animal products. Data from Smil, V. 2001. *Feeding the World: A Challenge for the Twenty-First Century.* Cambridge, MA: MIT Press.

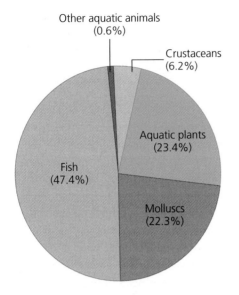

(a) World aquaculture production by groups

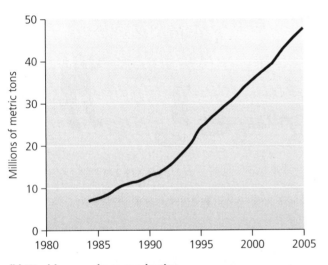

(b) World aquaculture production

FIGURE 8.19
Aquaculture involves a wide diversity of marine and freshwater organisms **(a)**. Global production of meat (all species other than plants) from aquaculture has risen steeply in the past two decades **(b)**. Data in (a) from FAO. 2004. *The State of World Fisheries and Aquaculture.* Rome, Italy: Author; and (b) FAO. 1995. *Aquaculture Production Statistics, 1984–1993, and Fishery Statistics: Aquaculture Production.* Rome, Italy: Author.

species are grown in open water in large, floating netpens. Others are raised in land-based ponds or holding tanks. People pursue both freshwater and marine aquaculture. Aquaculture is the fastest-growing type of food production; in the past 20 years, global output has increased sevenfold (**FIGURE 8.19**). Aquaculture today provides a third of the world's fish for human consumption, is most common in Asia, and involves more than 220 species. Some, such as carp, are grown for local consumption, whereas others, such as salmon and shrimp, are exported to affluent countries.

Aquaculture has some benefits

When conducted on a small scale by families or villages, as in China and much of the developing world, aquaculture helps ensure people a reliable protein source. This type of small-scale aquaculture can be sustainable, and it is compatible with other activities. For instance, uneaten fish scraps make excellent fertilizers for crops. Aquaculture on larger scales can help improve a region's or nation's food security by increasing overall amounts of fish available.

Aquaculture on any scale has the benefit of reducing fishing pressure on overharvested and declining wild stocks, as well as providing employment for fishers who can no longer fish from depleted natural stocks. Reducing fishing pressure also reduces *by-catch*, the unintended catch of nontarget organisms that results from commercial fishing. Furthermore, aquaculture relies far less on fossil fuels than do fishing vessels and provides a safer work environment. Fish farming can also be remarkably energy-efficient, producing as much as 10 times as much fish per unit area as is harvested from oceanic waters on the continental shelf and up to 1000 times as much as is harvested from the open ocean.

Aquaculture has some negative impacts, too

Along with its benefits, aquaculture has disadvantages. Dense concentrations of farmed animals can increase the incidence of disease, which reduces food security, necessitates antibiotic treatment, and results in additional expense. A virus outbreak wiped out half a billion dollars in shrimp in Ecuador in 1999, for instance. If farmed aquatic organisms escape into ecosystems where they are not native, they may spread disease to native stocks or may outcompete native organisms for food or habitat. The opposite has also occurred—recent research suggests that wild Pacific salmon swimming near aquaculture pens have passed on parasites, which then are able to spread rapidly as a result of the high densities of organisms in the pens.

The possibility of competition also arises when the farmed animals have been genetically modified. Like the transgenic corn that has influenced Mexican maize, transgenic fish have become a part of the food production system in recent years. Genetic engineering of Pacific salmon has produced transgenic fish that weigh up to 11 times as much as nontransgenic ones. Transgenic Atlantic salmon

FIGURE 8.20
Efforts to genetically modify important food fish have resulted in the creation of transgenic salmon (top), which can be considerably larger than wild salmon of the same species.

raised in Scotland have been engineered to grow to 5–50 times the normal size for their species (**FIGURE 8.20**). Such GM fish as these could outcompete their non-GM wild cousins. They may also interbreed with native and hatchery-raised fish and weaken already troubled stocks. Researchers have concluded that under certain circumstances, escaped transgenic salmon may increase the extinction risk that native populations of their species face, in part because the larger male fish have better odds of mating successfully.

We will revisit aquaculture in Chapter 12, in the context of our discussion of marine and coastal systems and fisheries.

Sustainable Agriculture

Post–Green Revolution industrialized agriculture has allowed food production to keep pace with a growing population, but it involves many adverse environmental impacts. These range from the degradation of soils (Chapter 7) to reliance on fossil fuels (Chapter 15) to problems arising from pesticide use, genetic modification, and intensive feedlot and aquaculture operations. Although developments in intensive commercial agriculture have alleviated some environmental pressures, they have exacerbated others. Industrial agriculture in some form seems necessary to feed our planet's 6.7 billion people, but many feel we will be better off in the long run by practising less-intensive methods of raising animals and crops.

Farmers and researchers have made great advances toward sustainable agriculture in recent years. **Sustainable agriculture** is agriculture that does not deplete soils faster than they form. It is farming and ranching that does not reduce the amount of healthy soil, clean water, and genetic diversity essential to long-term crop and livestock production. It is, simply, agriculture that can be practised in the same way far into the future. For example, the no-till agriculture practised in southern Brazil that we examined in Chapter 7 appears to fit the notion of sustainable agriculture, as does the traditional Chinese practice of carp aquaculture in small ponds. Sustainable agriculture is closely related to *low-input agriculture*, agriculture that uses smaller amounts of pesticides, fertilizers, growth hormones, water, and fossil fuel energy than are currently used in industrial agriculture. Food growing practices that use no synthetic fertilizers, insecticides, fungicides, or herbicides—but instead rely on biological approaches, such as composting and biocontrol—are termed **organic agriculture**.

Organic agriculture is on the increase

Citizens, government officials, farmers, and agricultural industry representatives have debated the meaning of the word *organic* for many years. Experimental organic gardens began to appear in North America in the 1940s, but in Canada the federal *Organic Products Regulations* came into effect only in December 2006, as a new part of the *Canadian Agricultural Products Act*. This law established national standards for organic products and facilitates the labelling, quality, and sale of organic food. The organic certification logo is permitted only on food products that meet specific Canadian standards for organic production, such as using natural fertilizers and raising animals in conditions that mimic nature as much as possible.

On June 30, 2009, the Organic Products Regulations will come into force and will form the basis by which the Canadian Food Inspection Agency (CFIA), as competent authority, will regulate the use of the organic agricultural product legend (legend). This regulatory regime will facilitate international market access, providing protection to consumers against deceptive and misleading labelling practices.

Products represented as organic in interprovincial and international trade will require certification to the Canadian Organic Standards by a CFIA accredited Certification Body. To bear the legend, products must also have an organic content greater than 95%. The use of the legend is voluntary. The legend may be either printed in colour or in black and white. Imported products bearing the legend must also bear the statement "Product of" immediately preceding the name of the country of origin or the statement "Imported", in close proximity to the legend. Also, products bearing the legend must also bear the name of the organization that has certified the product.

Long viewed as a small niche market, the market for organic foods is on the increase. Although they still account for only a small percentage of food expenditures in Canada, sales of organic products are increasing by about 20% annually. In 2001, 3%–5% (close to $10 billion) of Europe's food market was organic. Although 3%–5% of the market may not seem like much, organic agriculture in Europe grew by about 30% per year between 1985 and 2001, effectively doubling every two to three years.

Production is increasing along with demand. Although organic agriculture takes up less than 1% of cultivated land worldwide (24 million hectares in 2004), this area is rapidly expanding. In North America, the amount of land used in organic agriculture has recently increased 15%–20% each year. Today more than 500 000 ha are used to grow organic products in Canada, by almost 4000 producers,[16] and farmers in more than 130 nations practise organic farming commercially to some extent.

Two motivating forces have fuelled these trends. Many consumers favour organic products because of concern that consuming produce grown with the use of pesticides may pose risks to their health. Consumers also buy organic produce out of a desire to improve environmental quality by reducing chemical pollution and soil degradation (see "The Science Behind the Story: Organic Farming Put to the Test"). Many other consumers, however, will not buy organic produce because it usually is more expensive and often looks less uniform and esthetically appealing in the supermarket aisle compared with the standard produce of high-input agriculture.

weighing the issues 8-7
HOW MUCH INFORMATION DO YOU WANT ON FOOD LABELS?

The Canadian Food Inspection Agency issues labels to certify that produce claiming to be organic has met the government's organic standards. Increasingly, critics of GM products want them to be labelled as well. Given that 70% of processed food currently contains GM ingredients, labelling would cause added—and, many people think, unnecessary—costs. The European Union currently labels such foods. Do you want your food to be labelled? Would you choose among foods based on whether they were organic or genetically modified? Do you feel your food choices have environmental impacts, good or bad?

Overall, though, enough consumers are willing to pay more for organic meat, fruit, and vegetables that businesses are making such foods more widely available. In early 2000, one of Britain's largest supermarket brands announced that it would sell only organic food—and that the new organic products would cost customers no more than had nonorganic products. In addition to food products, many textile makers (among them The Gap, Levi's, and Patagonia) are increasing their use of organic cotton. (We tend to think of cotton as a "natural" fibre, but it is actually a highly *erosive* crop, meaning that it has an intensive impact on agricultural land. It takes about 0.5 kg of pesticides and fertilizers to grow, by conventional agricultural means, the cotton required to manufacture one T-shirt.[17]) Roots, a company founded in Canada, was one of the first major clothing manufacturers to begin experimenting with large-scale use of organic cotton, in 1989.[18]

Organic agriculture succeeds in part because it alleviates many problems introduced by high-input agriculture, even while passing up many of the benefits. For instance, although in many cases more insect pests attack organic crops because of the lack of chemical pesticides, biocontrol methods can often keep these pests in check. Moreover, the lack of synthetic chemicals maintains soil quality and encourages helpful pollinating insects. In the end, consumer choice will determine the future of organic agriculture. Falling prices and wider availability suggest that organic agriculture will continue to increase. In addition, sustainable agriculture, whether organic or not, will sooner or later need to become the rule rather than the exception.

Government initiatives have also spurred the growth of organic farming. For example, several million hectares of land have undergone conversion from conventional to organic farming in Europe since the European Union adopted a policy in 1993 to support farmers financially during the first years of conversion. Such support is important, because conversion often means a temporary loss in income for farmers. More and more studies, however, suggest that reduced inputs and higher market prices can, in the long run, make organic farming more profitable for the farmer than conventional methods.

Locally supported agriculture is growing

Increasing numbers of farmers and consumers are also supporting local small-scale agriculture. Farmers' markets (**FIGURE 8.22**) are becoming more numerous as consumers rediscover the joys of fresh, locally grown produce. The average food product sold in North American supermarkets travels at least 2300 km between the farm and the shelf, and supermarket produce is often chemically treated to preserve freshness and colour. At farmers' markets, consumers can buy fresh produce in

THE SCIENCE BEHIND THE STORY

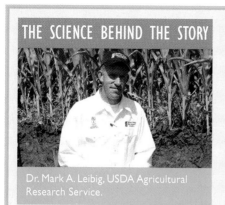

Dr. Mark A. Leibig, USDA Agricultural Research Service.

Organic Farming Put to the Test

Fields of wheat and potatoes, some grown organically and some cultivated with the synthetic chemicals favoured by industrialized agriculture, stand side by side on an experimental farm in Switzerland. Although conventionally farmed fields receive up to 50% more fertilizer, they produce only 20% more food than organically farmed fields. How are organic fields able to produce decent yields without synthetic agricultural chemicals? The answer, scientists have found, lies in the soil.

Swiss researchers at the Research Institute of Organic Agriculture have been comparing organic and conventional fields since 1978 by using a series of growing areas that feature four different farming systems. The first group of plots mirrors conventional farms, in which large amounts of chemical pesticides, herbicides, and fertilizers are applied to soil and plants. The second set is treated with a mixed approach of conventional and organic practices, including chemical additives, synthetic sprays, and livestock manure as fertilizer. The third set, the organic plots, use only manure, mechanical weeding machines, and plant extracts to control pests. A fourth group of plots follows organic practices but also uses extra natural boosts, such as adding herbal extracts to compost. The two organic plots receive about 35%–50% less fertilizer than the conventional fields and 97% fewer pesticides.

Over more than 20 years of monitoring, the organic fields yielded 80% of what the conventional fields produced, researchers reported in the journal *Science* in 2002. Organic crops of winter wheat yielded about 90% of the conventional wheat crop yield. Organic potato crops averaged about 68% of the conventional potato yields. The comparatively low potato yield was due to nutrient deficiency and a fungus-caused potato blight.

Scientists have hypothesized that organic farms keep their yields high because organic agricultural practices better conserve soil quality, keeping soil fertile over the long term. Soil scientist Paul Voroney of the University of Guelph in Ontario has been looking at the impact of organic methods. By using adjacent fields in nine different locations, he compared the impact of conventional practices with organic practices. On average, organic matter levels were 15% higher in the organically managed fields. They also had better soil structure and a 20% increase in the number of living soil microbes in the top 10 cm of the soil profile.[19]

Other researchers have shown that organic farming produced soils that contained more naturally occurring nutrients, held greater quantities of water, and had higher concentrations of microbial life than conventionally farmed soil. Organic farms also have deeper nutrient-rich topsoil and greater earthworm activity—all signs of soils healthy enough to produce crops without help from synthetic chemicals. Increasingly, researchers are concluding that organically managed soil

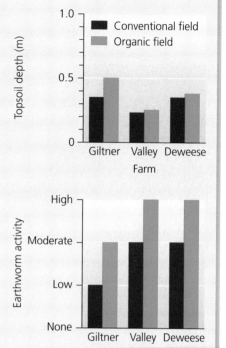

Researchers demonstrated that organic farming at three sites increased topsoil depth moderately and activity of earthworms dramatically. Data from Liebig, M.A., and J.W. Doran, 1999. Impact of organic production practices on soil quality indicators, *Journal of Environmental Quality* 28:1601–1609.

supports a more diverse range of microbial and plant life, which translates into increased biodiversity, self-sustaining fields, and strong crop yields. Such findings may be pivotal as large growers increasingly debate whether to turn to organic farming.

---

season from local farmers and often have a wide choice of organic items and unique local varieties.

Some consumers are partnering with local farmers in a phenomenon called **community-supported agriculture (CSA)**. In this practice, consumers pay farmers in advance for a share of their yield, usually in the form of weekly deliveries of produce. Consumers get fresh seasonal produce, while farmers get a guaranteed income stream up front to invest in their crops—an alternative to taking out loans and being at the mercy of the weather. As of 2005, the CSA network in Quebec alone counted almost 100 farm members, supplying 20 000 people with local organic meats and produce.[20]

Organic agriculture can even succeed in cities

One surprising place that organic agriculture is making inroads is within cities. Many urban areas now offer

CANADIAN ENVIRONMENTAL PERSPECTIVES

Alisa Smith and James MacKinnon

Alisa Smith and J. B. MacKinnon invented the "100-mile diet" concept.

- **Environmentalists**
- **Writers** and **bloggers**
- **Local food enthusiasts** and **inventors** of the 100-Mile Diet

One night Alisa Smith and James MacKinnon were expecting company, but found that they were out of food. They scrounged around their cabin in northern B.C., and came up with a trout, wild mushrooms, potatoes, garlic, dandelion leaves, apples, sour cherries, and rose hips. Everything tasted so good that the two embarked upon a venture to learn more about the food they ate—specifically, where it comes from and how far it has to travel to get to the dinner table. After learning that food travels almost 2000 km to reach the average North American home, Smith and MacKinnon pledged to spend one year eating only foods that were grown and produced within 100 miles (160 km) of their Vancouver apartment, and they blogged their way through the entire experience. The result was a book (*The 100-Mile Diet: A Year of Local Eating*, by Alisa Smith and J. B. MacKinnon) and a new concept in eating.

They were surprised both by the benefits of the 100-Mile Diet ("it rolls off the tongue easier than '160 km Diet'"), and by the enthusiastic response they received from people around the world. Some of the benefits—on top of the environmental benefits of cutting back on the long-distance transportation of food products—included fresher taste, more fruits and vegetables in a generally healthier diet, seasonal awareness of food, and support of the local economy.[21]

However, there were many challenges. It took a while to figure out how to find some local products, and how to tell the provenance of ingredients (which usually don't specify their origin, even if they are listed on food labels). Among the items they could not find in local production were sugar, rice, lemons, ketchup, olive oil, peanut butter, and orange juice.[22] One of the only exceptions or "cheats" they allowed themselves during that first year was the occasional beer (it may be brewed locally, but the ingredients come from elsewhere).

The 100-Mile Diet is just one aspect—though perhaps the best known—of the growing movement of local and "slow" food enthusiasts; some are now calling themselves "locavores." MacKinnon and Smith, for their part, are still eating about 85% locally, by their reckoning, and don't believe they will ever abandon this way of life.

They firmly believe that local eating will continue to grow in popularity, especially with rising prices for grains and other basic food products, and the cost of transporting them. In fact, they believe that local eating will fundamentally transform our approach to food in the coming decades.

"Alisa and I are increasingly convinced that much larger numbers of people could be fed from smaller landscapes than we think; we'd just have to put food closer to the centre of daily life (where it belongs!)."
—James MacKinnon[23]

Thinking About Environmental Perspectives

What foods would you have to do without, if you decided to try to eat locally? Remember to think about where the individual ingredients come from, as well as where the item is produced. Next time you are in the supermarket, read the labels on items that you normally buy, and try to determine their place of origin. (You might have to do some additional research; does this suggest anything to you about how food is labelled?) What are some local substitutions you could make for the "missing" items on your dinner table? What are some of the local specialties that you would be able to enjoy, and how would they vary from season to season?

FIGURE 8.22
Farmers' markets, like this one in Toronto, have become more widespread as consumers have rediscovered the benefits of buying fresh, locally grown produce. There has been farmers' market activity at this particular site, the St. Lawrence Market, since 1803.

community gardens in which residents can grow small plots of fruits and vegetables (Chapter 20).

For example, organic agriculture is deeply entrenched in both the cities and the rural areas of Cuba. Long a close ally of the former Soviet Union, Cuba suffered economic and agricultural upheaval following the Soviet Union's dissolution. In 1989, as the USSR was breaking up, Cuba lost 75% of its total imports, 53% of its oil imports, and 80% of its fertilizer and pesticide imports. Faced with such losses, Cuba's farmers had little choice but to "go organic."

Because far less oil was available to fuel Cuba's transportation system, farmers began growing food closer to cities and even within them. By 1998 the Cuban government's Urban Agriculture Department had encouraged the development of more than 8000 gardens in the capital city of Havana (**FIGURE 8.23**). More than 30 000 people, including farmers, government workers, and private citizens, worked in these gardens, which covered 30% of the city's available land. Cuba has also taken steps to compensate for the loss of fossil fuels, fertilizers, and pesticides by, for example, using oxen instead of tractors, using integrated pest management, encouraging people to live outside urban areas and to remain involved in agriculture, and establishing centres to breed organisms for biological pest control.

Cuba's agriculture likely requires more human labour per unit output than do intensive commercial farms of developed nations, and Cuba's economic and agricultural policies are guided by tight top-down control in a rigid state socialist system. Nevertheless, Cuba's low input farming has produced some positive achievements. The practices have led to the complete control of the sweet-potato borer, a significant pest insect, and in the 1996–1997 growing season the Cuban people produced record yields for 10 crops. Although Cuba's move toward organic agriculture was involuntary, its response to its economic and agricultural crisis illustrates how other nations might, by choice, begin to farm in ways that rely less on enormous inputs of fossil fuels and synthetic chemicals.

Conclusion

Many of the intensive commercial agricultural practices we have discussed have substantial negative environmental impacts. At the same time, it is important to realize that many aspects of industrialized agriculture have had positive environmental effects by relieving certain pressures on land or resources. Whether Earth's natural systems would be under more pressure from 6.7 billion people practising traditional agriculture or from 6.7 billion people living under the industrialized agriculture model is a very complicated question.

What is certain is that if our planet is to support 9 billion people by mid-century without further degradation of the soil, water, pollinators, and other ecosystem services that support our food production, we must find ways to shift to sustainable agriculture. Such approaches as biological pest control, organic agriculture, pollinator conservation, preservation of native crop diversity, sustainable aquaculture, and likely some degree of careful and responsible genetic modification of food may all be parts of the game plan we will need to set in motion. What remains to be seen is the extent to which individuals, governments, and corporations will be able to put their own interests and agendas in perspective to work together toward a sustainable future.

FIGURE 8.23
Organic gardening takes place within the city limits of Havana, Cuba, out of necessity. With little money to pay for the large amounts of fertilizers and pesticides required for industrialized agriculture, Cubans get much of their food from local agriculture without these inputs.

REVIEWING OBJECTIVES

You should now be able to:

Explain the challenge of feeding a growing human population

- Our food production has outpaced the growth of our population, yet there are still 850 million hungry people in the world.

Identify the goals, methods, and environmental impacts of the "Green Revolution"

- The goal of the Green Revolution was to increase agricultural productivity per unit area of land to feed the world's hungry without further degrading natural lands.
- Agricultural scientists used selective breeding to develop strains of crops that grew quickly, were more nutritious, or were resistant to disease or drought.
- The expanded use of fossil fuels and chemical fertilizers and pesticides has increased pollution. However, the increased efficiency of production has reduced the amount of natural land converted for farming.

Categorize the strategies of pest management

- Most "pests" and "weeds" are killed with synthetic chemicals that also can pollute the environment and pose health hazards.
- Pests tend to evolve resistance to chemical pesticides, forcing chemists to design ever more toxic poisons.
- Natural enemies of pests can be employed against them in the practice of biological control.
- Integrated pest management includes a combination of techniques, and attempts to minimize use of synthetic chemicals.

Discuss the importance of pollination

- Insects and other organisms are essential for ensuring the reproduction of many of our crop plants.
- Conservation of native pollinating insects is vitally important to our food supply.

Describe the science behind genetically modified food

- Genetic modification depends on the technology of recombinant DNA. Genes containing desirable traits are moved from one type of organism into another.
- Modification through genetic engineering is both like and unlike traditional selective breeding.
- GM crops may have ecological impacts, including the spread of transgenes, the creation of "superweeds," and indirect impacts on biodiversity. More research is needed to determine how widespread or severe these impacts may be.

Evaluate controversies and the debate over genetically modified food

- Little evidence exists so far for human health impacts from GM foods, but anxiety over health impacts inspires wide opposition to GM foods.
- Many people have ethical qualms about altering the food we eat through genetic engineering.
- Opponents of GM foods view multinational biotechnology corporations as a threat to the independence of small farmers.

Ascertain approaches for preserving crop diversity

- Protecting regions of diversity of native crop varieties, such as Oaxaca, can provide insurance against failure of major commercial crops.
- Seed banks preserve rare and local varieties of seed, acting as storehouses for genetic diversity.

Assess feedlot agriculture for livestock and poultry

- Increased consumption of animal products has driven the development of high-density feedlots.
- Feedlots create tremendous amounts of waste and other environmental impacts, but they also relieve pressure on lands that could otherwise be overgrazed.

Weigh approaches in aquaculture

- Aquaculture provides economic benefits and food security, can relieve pressures on wild fish stocks, and can be sustainable.
- Aquaculture also creates pollution, causes habitat loss, and has other environmental impacts.

Evaluate sustainable agriculture

- Organic agriculture has fewer environmental impacts than industrial agriculture. It is a small part of the market but is growing rapidly.
- Locally supported agriculture, as shown by farmers' markets and community-supported agriculture, is also growing.

TESTING YOUR COMPREHENSION

1. What kinds of techniques have people employed to increase agricultural food production? How did agricultural scientist Norman Borlaug help inaugurate the Green Revolution?
2. Explain how pesticide resistance occurs.
3. Explain the concept of biocontrol. List several components of a system of integrated pest management (IPM).
4. About how many and what types of cultivated plants are known to rely on insects for pollination? Why is it important to preserve the biodiversity of native pollinators?
5. What is recombinant DNA? How is a transgenic organism created? How is genetic engineering different from traditional agricultural breeding? How is it similar?
6. Describe several reasons why many people support the development of genetically modified organisms, and name several uses of such organisms that have been developed so far.
7. Describe the scientific concerns of those opposed to genetically modified crops. Describe some of the other concerns.
8. Name several positive and negative environmental effects of feedlot operations. Why is beef an inefficient food from the perspective of energy consumption?
9. What are some economic benefits of aquaculture? What are some negative environmental impacts?
10. What are the objectives of sustainable agriculture? What factors are causing organic agriculture to expand?

SEEKING SOLUTIONS

1. Assess several ways in which high-input agriculture can be beneficial for the environment and several ways in which it can be detrimental to the environment. Now suggest several ways in which we might modify industrial agriculture to lessen its environmental impact.
2. What factors make for an effective biological control strategy of pest management? What risks are involved in biocontrol? If you had to decide whether to use biocontrol against a particular pest, what questions would you want to have answered before you decide?
3. From what you have learned in this chapter about the staple crop corn, how would you choose to farm corn, if you had to do so for a living? Would you choose to grow genetically modified corn? How would you manage pests and weeds? Would you grow corn for people, livestock, or both? Think of the various ways corn is grown, purchased, and valued in different places—such as Canada, the United States, Europe, Oaxaca, and Zambia—as you formulate your answer.
4. Those who view GM foods as solutions to world hunger and pesticide overuse often want to speed their development and approval. Others adhere to the precautionary principle and want extensive testing for health and environmental safety. How much caution do you think is warranted before a new GM crop is introduced?
5. **THINK IT THROUGH** Imagine it is your job to make the regulatory decision as to whether to allow the planting of a new genetically modified strain of cabbage that produces its own pesticide and has twice the vitamin content of regular cabbage. What questions would you ask of scientists before deciding whether to approve the new crop? What scientific data would you want to see, and how much would be enough? Would you also consult nonscientists or take ethical, economic, and social factors into consideration?
6. **THINK IT THROUGH** Cuba adopted low-input organic agriculture out of necessity. If the country were to become economically prosperous once more, do you think Cubans would maintain this form of agriculture, or do you think they would turn to intensive, high-input farming instead? What path do you think they should pursue, and why?

INTERPRETING GRAPHS AND DATA

In the year 2000, more than 80 million metric tons of nitrogen fertilizer was used in producing food for the world's 6 billion people. Food production, the use of nitrogen fertilizers, and world population all had grown over the preceding 40 years, but at somewhat different rates. Food production grew slightly faster than population, while relatively little additional land was converted to agricultural use during this time. Fertilizer use grew most rapidly.

1. Express the 2002 values of the four graphed indices as percentages of the value of each index in 1961.
2. Calculate the ratio of the food production index to the nitrogen fertilizer use index in 1961 and in 2002. What does comparing these two ratios tell you about how the efficiency of nitrogen use in agriculture has changed? Is this an example of the law of diminishing returns?
3. As world population has grown, so has the demand for food, yet little additional land has been devoted to food production. Calculate the ratio of the agricultural land index to the population index for 1961 and 2002. What does comparing these two ratios tell you about how the per capita demand on agricultural land has changed over the years? To what factors can you attribute this change?

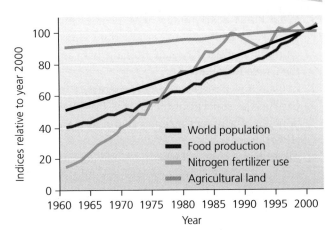

Global food production, nitrogen fertilizer use, human population, and land converted to agriculture, 1961–2002, relative to 2000 levels (2000 = 100). Data from Food and Agriculture Organization of the United Nations.

CALCULATING FOOTPRINTS

As food production became more industrialized during the twentieth century, several trends emerged. One trend, documented in this chapter, was a loss in the number of varieties of crops grown. A second trend was the increasing amount of energy expended to store food and ship it to market. In North America today, food travels an average of more than 2000 km from the field to your table. The price you pay for the food covers the cost of this long-distance transportation, which is approximately $0.70 per tonne per kilometre, although this obviously varies widely.[24] Assuming that the average person eats around 4 kg of food per day, calculate the food transportation costs for each category in the table.

1. What specific challenges to environmental sustainability are imposed by a food production and distribution system that relies on long-range transportation to bring food to market?
2. A study by Pirog[25] and Benjamin noted that locally produced food travelled only 75 km or so to market, thus saving 96% of the economic and environmental costs of transportation, as well as supporting local farmers. Locally grown foods may be fresher and cause less environmental impact as they are brought to market, but what are the disadvantages to you as a consumer in relying on local food production? Do you think the advantages of a "100-Mile Diet" outweigh those disadvantages?
3. What has happened to gasoline prices recently? Would future increases in the price of gas affect your answers to the preceding questions?

Consumer	Daily Cost	Annual Cost
You		
Your class		
Your town		
Your province		
Canada		

TAKE IT FURTHER

 Go to www.myenvironmentplace.ca where you will find

- Suggested answers to end-of-chapter questions
- Quizzes, animations, and flashcards to help you study
- *Research Navigator*™ database of credible and reliable sources to assist you with your research projects
- Tutorials to help you master how to interpret graphs
- Current news articles that link the topics that you study to case studies from your region and around the world

- **ECO Occupational Profiles:** If you found this chapter especially interesting, you might want to learn more about the following jobs by visiting the Occupational Profiles website of the Environmental Careers Organization. Go to www.eco.ca and check out the following careers:
  - Agricultural engineer
  - Agriculture technician
  - Agronomist
  - Aquaculturist
  - Biotechnologist
  - Crop and livestock producer
  - Microbiologist

CHAPTER ENDNOTES

1. Monsanto vs Schmeiser, www.percyschmeiser.com.
2. Canadian Institute for Health Information, 2004, *Overweight and Obesity in Canada: A Population Health Perspective,* http://secure.cihi.ca/cihiweb/dispPage.jsp?cw_page=GR_1130_E.
3. Canadian Association of Food Banks, *Hunger Count 2006,* www.cafb-acba.ca/documents/HungerCount_20061.pdf.
4. Rainville, B., and S. Brink (2001) Food Insecurity in Canada, 1998–1999. Research paper R-01-2E. Ottawa: Applied Research Branch, Human Resources Development Canada.
5. Health Canada, Pest Management Regulatory Agency, www.pmra-arla.gc.ca/english/legis/pcpa-e.html.
6. Government of Canada, 2000, *Pesticides: Making the Right Choice for Health and the Environment. Report of the Standing Committee on Environment and Sustainable Development,* http://cmte.parl.gc.ca/cmte/CommitteePublication.aspx?COM=173&Lang=1&SourceId=36396.
7. McEwan, K., and W. A. Deen (1997) *Review of Agricultural Pesticide Pricing and Availability in Canada.* Prepared for Saskatchewan Agriculture and Food; Ontario Ministry of Agriculture, Food and Rural Affairs; and Agriculture and Agri-Food Canada. Guelph, ON: Ridgetown College Research Report, University of Guelph.
8. Based on information from Kevan, P. G., E. A. Clark, and V. G. Thomas (1990) Pollinators and sustainable agriculture. *American Journal of Alternative Agriculture,* Vol. 5, No. 1, pp. 13–22.
9. Dyer, J. S. (2006) *Raising Awareness Among Canadians About Plant Pollinators and the Importance of Monitoring and Conserving Them.* Published electronically by Seeds of Diversity Canada (SoDC) for the Ecological Monitoring and Assessment Network Coordinating Office (EMAN CO) of Environment Canada, www.pollinationcanada.ca/lit/Pollinator%20Awareness%20Paper.pdf.
10. Kevan, P. G., E. A. Clark, and V. G. Thomas (1990) Pollinators and sustainable agriculture. *American Journal of Alternative Agriculture,* Vol. 5, No. 1, pp. 13–22; and other sources.
11. International Service for the Acquisition of Agri-Biotech Applications, ISAAA Brief 35-2006 Executive Summary Global Status of Commercialized Biotech/GM Crops: 2006, www.isaaa.org/resources/publications/briefs/35/executivesummary/default.html.
12. Traxler, G. (2004) The economic impacts of biotechnology-based technological innovations, ESA Working Paper No. 04-08, May, Agricultural and Development Economics Division, The Food and Agriculture Organization of the United Nations, www.fao.org/es/esa.
13. Convention on Biological Diversity, www.cbd.int/default.shtml.

14. Eaton, D., J. Windig, S. J. Hiemstra, and M. van Veller (2006) Indicators for livestock and crop diversity, *North-South Policy Brief,* Vol. 2006-1, pp. 1–4. Programme International Cooperation, Wageningen International, The Netherlands, www.wi.wur.nl/NR/rdonlyres/FADBC382-F7C9-4D2E-B39D-E209AF3C32D2/42280/Policybrief20061.pdf.
15. *The State of the World's Animal Genetic Resources for Food and Agriculture* (2007). Commission on Genetic Resources for Food and Agriculture, Food and Agriculture Organization of the United Nations.
16. Canadian Organic Growers, *Quick Facts About Canada's Organic Sector,* www.cog.ca/orgquickfacts.htm.
17. Sustainable Cotton Project, www.sustainablecotton.org.
18. Roots, Organic Cotton, www.roots.com.
19. Carter, Jeffrey (2006) Researchers show soil microbes increase in organic fields—*Timing of cover crop planting is important to good performance, depending on variety, Ontario Farmer,* February 7, as reported by Ontario Agriculture Centre of Canada, www.organicagcentre.ca/ResearchDatabase/res_microbes_covercrop_of.asp.
20. Equiterre, www.equiterre.org/en/agriculture/paniersBios/index.php.
21. Smith, Alisa, and James MacKinnon, The 100-Mile Diet: Local Eating for Global Change, http://100milediet.org.
22. Smith, Alisa, and J. B. MacKinnon (2007) *The 100-Mile Diet: A Year of Local Eating.* Toronto: Random House Canada.
23. MacKinnon, James. The 100-Mile Diet: Blog, http://100milediet.org/category/the-latest.
24. Calculating the cost of a "food mile" is far from straightforward. For an overview and discussion of the caveats involved in these calculations, check out the study by Pirog and Benjamin cited in question #2, or read "How the myth of food miles hurts the planet," by Robin McKie, *The Observer,* Sunday, March 23, 2008, www.guardian.co.uk/environment/2008/mar/23/food.ethicalliving.
25. Data from Pirog, R., 2005. *Energy Efficiency as an Integral Part of Sustainable Agriculture: Food Miles and Fuel Usage in Food Transport, ACEEE Forum on Energy Efficiency in Agriculture.* Ames, IA: Leopold Center for Sustainable Agriculture, Iowa State University.

9 Biodiversity and Conservation Biology

This is where the Sikhote-Alin Mountains meet the Pacific Ocean.

Upon successfully completing this chapter, you will be able to

- Characterize the scope of biodiversity on Earth
- Describe ways to measure biodiversity
- Contrast background extinction rates and periods of mass extinction
- Evaluate the primary causes of biodiversity loss
- Specify the benefits of biodiversity
- Assess conservation biology and its practice
- Explain island biogeography theory and its application to conservation biology
- Compare and contrast traditional and more innovative biodiversity conservation efforts

Siberian tigers are rare, even in their native habitat in the Sikhote-Alin Mountains.

CENTRAL CASE:
SAVING THE SIBERIAN TIGER

"Future generations would be truly saddened that this century had so little foresight, so little compassion, such lack of generosity of spirit for the future that it would eliminate one of the most dramatic and beautiful animals this world has ever seen."
—GEORGE SCHALLER, WILDLIFE BIOLOGIST, ON THE TIGER

"Except in pockets of ignorance and malice, there is no longer an ideological war between conservationists and developers. Both share the perception that health and prosperity decline in a deteriorating environment. They also understand that useful products cannot be harvested from extinct species."
—EDWARD O. WILSON, HARVARD UNIVERSITY BIODIVERSITY EXPERT

Historically, tigers roamed widely across Asia from Turkey to northeast Russia to Indonesia. Within the past 200 years, however, people have driven the majestic striped cats from most of their historic range. Today, tigers are exceedingly rare and are creeping toward extinction.

Of the tigers that still survive, those of the subspecies known as the Siberian tiger are the largest cats in the world. Males reach more than 3.5 m in length and weigh up to 360 kg. Also named Amur tigers for the watershed they occupied along the Amur River, which divides Siberian Russia from Manchurian China, these cats now find their last refuge in the forests of the remote Sikhote-Alin Mountains of the Russian Far East.

For thousands of years the Siberian tiger coexisted with the region's native people and held a prominent place in native language and lore. These people referred to the tiger as "Old Man" or "Grandfather" and equated it with royalty or viewed it as a guardian of the mountains and forests. Indigenous people of the region rarely killed a tiger unless it had preyed on a person.

The Russians who moved into the region and exerted control in the early twentieth century had no such cultural traditions. They hunted tigers for sport and hides, and some Russians reported killing as many as 10 tigers in a single hunt. In addition, poachers began killing tigers to sell their body parts to China and other Asian countries, where they are used in traditional medicine and as aphrodisiacs. Meanwhile, road building, logging, and agriculture began to fragment tiger habitat and provide easy access for well-armed hunters.

The tiger population dipped to perhaps 20–30 animals. With a population this small, the concern is not only that the last few individuals might die out, but also that there might be insufficient genetic variation within the remaining tigers for the population to continue to survive and breed successfully.

International conservation groups began to get involved, working with Russian biologists to try to save the dwindling tiger population. One such group was the Hornocker Wildlife Institute, now part of the Wildlife Conservation Society. In 1991 the group helped launch the Siberian Tiger Project, devoted to studying the tiger and its habitat. The team put together a plan to protect the tiger, began educating people regarding the tiger's importance and value, and worked closely with those who live in proximity to the big cats.

Thanks to such efforts by conservation biologists, today Siberian tigers in the wild number roughly 330–370, and about 600 more survive in zoos and captive breeding programs around the world. The outlook for the species' survival is still challenging, but many people are trying to save these endangered animals. It is one of many efforts around the world today to stem the loss of our planet's priceless biological diversity.

Our Planet of Life

Growing human population and resource consumption are putting ever greater pressure on the flora and fauna of the planet, from tigers to tiger beetles. We are diminishing Earth's diversity of life, the very quality that makes our planet so special. In Chapter 3 we introduced the concept of **biological diversity**, or **biodiversity**, as the sum total of all organisms in an area, taking into account the diversity of species, their genes, their populations, and their communities. In this chapter we will refine this definition and examine current biodiversity trends and their relevance to our lives. We will then explore science-based solutions to biodiversity loss.

Biodiversity encompasses several levels

Biodiversity is a concept as multifaceted as life itself, and definitions of the term are plentiful. Different biologists use different working definitions according to their own aims, interests, and values. Nonetheless, there is broad agreement that the concept applies across several major levels in the organization of life (**FIGURE 9.1**). The level that is easiest to visualize and most commonly used is species diversity.

Species diversity As you recall from Chapter 3, a species is a distinct type of organism, a set of individuals that uniquely share certain characteristics and can breed with one another and produce fertile offspring. Biologists may use differing criteria to delineate species boundaries; some emphasize characteristics shared because of common ancestry, whereas others emphasize ability to interbreed. In practice, however, scientists broadly agree on species identities. We can express **species diversity** in terms of both the number and the variety of species in the world or in a particular region. One component of species diversity is **species richness**, the number of species. Another is **evenness** or **relative abundance**, the extent to which numbers of individuals of different species are equal or skewed.

As we have seen, speciation generates new species, adding to species richness, whereas extinction decreases species richness. Although immigration, emigration, and local extinction may increase or decrease species richness locally, only speciation and extinction change it globally.

Taxonomists, scientists who classify species, use an organism's physical appearance and genetic makeup to determine its species. Taxonomists also group species by their similarity into a hierarchy of categories meant to reflect evolutionary relationships. Related species are grouped together into genera (singular, genus), related genera are grouped into families, and so on (**FIGURE 9.2**). Every species is given a two-part Latin or Latinized scientific name denoting its genus and species. The tiger, *Panthera tigris*, differs from the world's other species of large cats, such as the jaguar (*Panthera onca*), the leopard (*Panthera pardus*), and the African lion (*Panthera leo*). These four species are closely related in evolutionary terms, as indicated by the genus name they share, *Panthera*. They

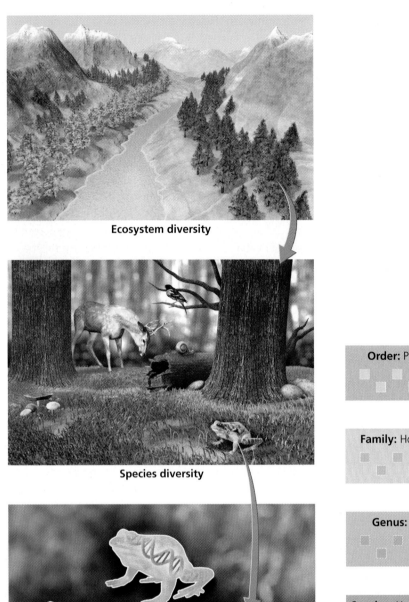

FIGURE 9.1
The concept of biodiversity encompasses several levels in the hierarchy of life. Species diversity (middle frame of figure) refers to the number or variety of species. Genetic diversity (bottom frame) refers to variety of genes among individuals within a given population or species. Ecosystem diversity (top frame) and related concepts refer to variety at levels above the species level, such as ecosystems, communities, habitats, or landscapes.

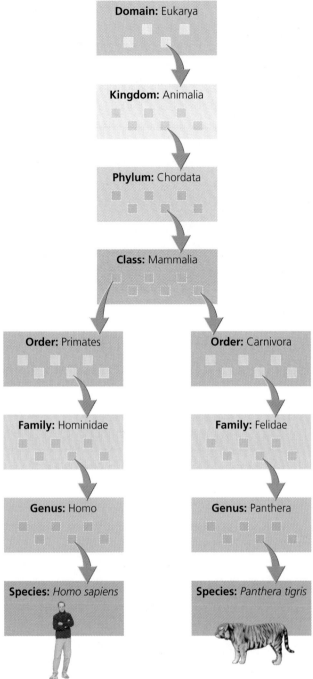

FIGURE 9.2
Taxonomists classify organisms by using a hierarchical system meant to reflect evolutionary relationships. Species that are similar in their appearance, behaviour, and genetics (because of recent common ancestry) are placed in the same genus. Organisms of similar genera are placed within the same family. Families are placed within orders, orders within classes, classes within phyla, phyla within kingdoms, and kingdoms within domains. For instance, humans (*Homo sapiens,* a species in the genus *Homo*) and tigers (*Panthera tigris,* a species in the genus *Panthera*) are both within the class *Mammalia*. However, the differences between our two species, which have evolved over millions of years, are great enough that we are placed in different orders and families.

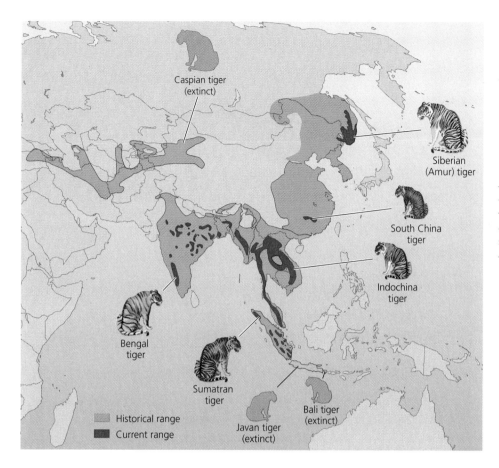

FIGURE 9.3
Three of the eight subspecies of tiger became extinct during the twentieth century. The Bali, Javan, and Caspian tigers are extinct. Today only the Siberian (Amur), Bengal, Indochina, Sumatran, and South China tigers persist, and the Chinese government estimates that fewer than 30 individuals of the South China tiger remain. Deforestation, hunting, and other pressures from people have caused tigers of all subspecies to disappear from most of the geographic range they historically occupied. This map contrasts the ranges of the eight subspecies in the years 1800 (orange) and 2000 (red). Data from the Tiger Information Center.

are more distantly related to cats in other genera, such as the cheetah (*Acinonyx jubatus*) and the bobcat (*Felis rufus*), although all cats are classified together in the family *Felidae*.

Biodiversity exists below the species level in the form of subspecies, populations of a species that occur in different geographic areas and differ from one another in some characteristics. Subspecies are formed by the same processes that drive speciation but result when divergence does not proceed far enough to create separate species. Scientists denote subspecies with a third part of the scientific name. The Siberian tiger, *Panthera tigris altaica*, is one of five subspecies of tiger still surviving (**FIGURE 9.3**). Tiger subspecies differ in colour, coat thickness, stripe patterns, and size. For example, *Panthera tigris altaica* is 5–10 cm taller at the shoulder than the Bengal tiger (*Panthera tigris tigris*) of India and Nepal, and it has a thicker coat and larger paws.

Genetic diversity Scientists designate subspecies when they recognize substantial genetically based differences among individuals from different populations of a species. However, all species consist of individuals that vary genetically from one another to some degree, and this genetic diversity is an important component of biodiversity. **Genetic diversity** encompasses the varieties in DNA present among individuals within species and populations.

Genetic diversity provides the raw material for adaptation to local conditions. According to Environment Canada, "genetic diversity is what enables a species to adapt to ecological change."[1] A diversity of genes for coat thickness in tigers allowed natural selection to favour genes for thin coats of fur in Bengal tigers living in warm regions, and genes for thick coats of fur for Siberian tigers living in cold regions. In the long term, populations with more genetic diversity may stand better chances of persisting, because their variation better enables them to cope with environmental change.

Populations with little genetic diversity are vulnerable to environmental change for which they are not genetically prepared. Populations with depressed genetic diversity may also be more vulnerable to disease and may suffer *inbreeding depression*, which occurs when genetically similar parents mate and produce weak or defective offspring. Scientists have sounded warnings over low genetic diversity in species that have dropped to low population sizes in the past, including cheetahs, bison, and elephant seals, but the full consequences of reduced diversity in these species remain to be seen. Diminishing genetic diversity in our crop plants also is a prime concern to humanity.

FIGURE 9.4
This illustration shows organisms scaled in size to the number of species known from each major taxonomic group. This gives a visual sense of the disparity in species richness among groups. However, because most species are not yet discovered or described, some groups (such as bacteria, archaea, insects, nematodes, protists, fungi, and others) may contain far more species than we now know of. Data from Groombridge, B., and M. D. Jenkins. 2002. *Global Biodiversity: Earth's Living Resources in the 21st Century.* UNEP-World Conservation Monitoring Centre. Cambridge, U.K.: Hoechst Foundation.

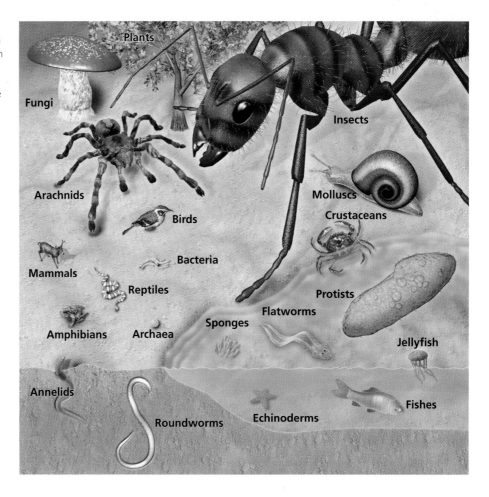

Ecosystem diversity Biodiversity also encompasses levels above the species level. **Ecosystem diversity** refers to the number and variety of ecosystems, but biologists may also refer to the diversity of biotic community types or habitats within some specified area. If the area is large, scientists may also consider the geographic arrangement of habitats, communities, or ecosystems at the landscape level, including the sizes, shapes, and interconnectedness of patches of these entities. Under any of these concepts, a coastal zone of rocky and sandy beaches, forested cliffs, offshore coral reefs, and ocean waters would hold far more biodiversity than the same acreage of a monocultural cornfield. A mountain slope whose vegetation changes from desert to hardwood forest to coniferous forest to alpine meadow would hold more biodiversity than an area the same size consisting of only desert, forest, or meadow.

Some groups hold more species than others

Species are not evenly distributed among taxonomic groups. In terms of number of species, insects show a staggering predominance over all other forms of life (**FIGURE 9.4** and **FIGURE 9.5**). Within insects, about 40% are beetles. Beetles outnumber all non-insect animals and all plants. No wonder the twentieth-century British biologist J. B. S. Haldane famously quipped that God must have had "an inordinate fondness for beetles."

Some groups have given rise to many species in a relatively short period of time through the process of adaptive radiation (see **FIGURE 3.3**). Others have diversified because of a tendency to become separated by barriers that promote allopatric speciation. Still other groups have accumulated species through time because of low rates of extinction.

Measuring biodiversity is not easy

Coming up with precise quantitative measurements to express a region's biodiversity is difficult. This is partly why scientists often express biodiversity in terms of its most easily measured component, species diversity, and in particular, species richness. Species richness is a good gauge for overall biodiversity, but we still are profoundly ignorant of the number of species that exist worldwide. So far, scientists have identified and described 1.7 million to

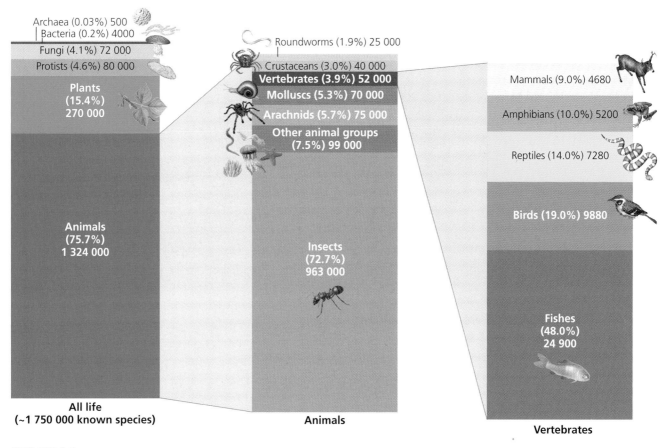

FIGURE 9.5
In the left portion of the figure, we see that three-quarters of known species are animals. The central portion subdivides animals, revealing that nearly three-quarters of animals are insects and that vertebrates compose only 3.9% of animals. Among vertebrates (right portion of figure), nearly half are fishes, and mammals compose only 9%. As noted, most species are not yet discovered or described, so some groups may contain far more species than we now know of. Data from Groombridge, B., and M. D. Jenkins. 2002. *Global Biodiversity: Earth's Living Resources in the 21st Century.* UNEP-World Conservation Monitoring Centre. Cambridge, U.K.: Hoechst Foundation.

2.0 million species of plants, animals, and microorganisms. However, estimates for the total number that actually exist range from 3 million to 100 million, with our best educated guesses ranging from 5 million to 30 million.

Our knowledge of species numbers is incomplete for several reasons. First, some areas of Earth remain little explored. We have barely sampled the ocean depths, hydrothermal vents, or the tree canopies and soils of tropical forests. Second, many species are tiny and easily overlooked. These inconspicuous organisms include bacteria, nematodes (roundworms), fungi, protists, and soil-dwelling arthropods. Third, many organisms are so difficult to identify that some thought to be identical sometimes turn out, once biologists look more closely, to be multiple species. This is frequently the case with microbes, fungi, and small insects, but also sometimes with organisms as large as birds, trees, and whales.

Smithsonian Institution entomologist Terry Erwin pioneered one method of estimating species numbers. In 1982, Erwin's crews fogged rainforest trees in Central America with clouds of insecticide and then collected insects, spiders, and other arthropods as they died and fell from the treetops. Erwin concluded that 163 beetle species specialized on the tree species *Luehea seemannii*.

If this were typical, he figured, then the world's 50 000 tropical tree species would hold 8 150 000 beetle species and—since beetles represent 40% of all arthropods—20 million arthropod species. If canopies hold two-thirds of all arthropods, then arthropod species in tropical forests alone would number 30 million. Many assumptions were involved in this calculation, and several follow-up studies have revised Erwin's estimate downward.

Biodiversity is unevenly distributed

Numbers of species tell only part of the story of Earth's biodiversity. Living things are distributed across our planet unevenly, and scientists have long sought to explain the distributional patterns they see.

FIGURE 9.6
For many types of organisms, number of species per unit area tends to increase as one moves toward the equator. This trend, the latitudinal gradient in species richness, is one of the most readily apparent—yet least understood—patterns in ecology. One example is bird species in North and Central America: In any one spot in arctic Canada and Alaska, 30 to 100 species can be counted; in areas of Costa Rica and Panama, the number rises to more than 600. Adapted from Cook, R. E. 1969. Variation in species density in North American birds. *Systematic Zoology* 18:63–84.

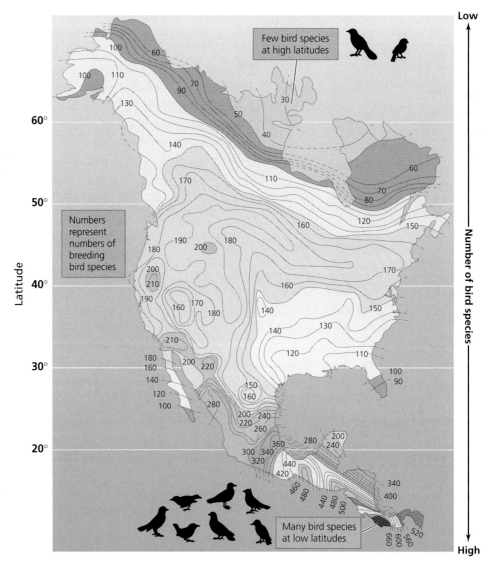

One of the most striking patterns of diversity is seen in the fact that species richness generally increases as one approaches the equator (**FIGURE 9.6**). This pattern of variation with latitude, called the **latitudinal gradient**, has been one of the most obvious patterns in ecology, but it also has been one of the most difficult for scientists to explain.

Hypotheses abound for the cause of the latitudinal gradient in species richness, but it seems likely that plant productivity and climate stability play key roles in the phenomenon (**FIGURE 9.7**). Greater amounts of solar energy, heat, and humidity at tropical latitudes lead to more plant growth, making areas nearer the equator more productive and able to support larger numbers of animals. In addition, the relatively stable climates of equatorial regions—their similar temperatures and rainfall from day to day and season to season—help ensure that single species won't dominate ecosystems and instead that numerous species can coexist. Whereas varying environmental conditions favour generalists—species that can deal with a wide range of circumstances but that do no single thing very well—stable conditions favour organisms with specialized niches that do particular things very well. In addition, polar and temperate regions may be relatively lacking in species because glaciation events repeatedly forced organisms out of these regions and toward more tropical latitudes.

The latitudinal gradient influences the species diversity of Earth's biomes. Tropical dry forests and rainforests tend to support far more species than tundra and boreal forests, for instance. Tropical biomes typically show more evenness as well, whereas in high-latitude biomes with low species richness, particular species may greatly outnumber others. For example, Canadian boreal forest is dominated by immense expanses of black spruce, whereas Panamanian tropical forest contains hundreds of tree species, no one of which greatly outnumbers the others.

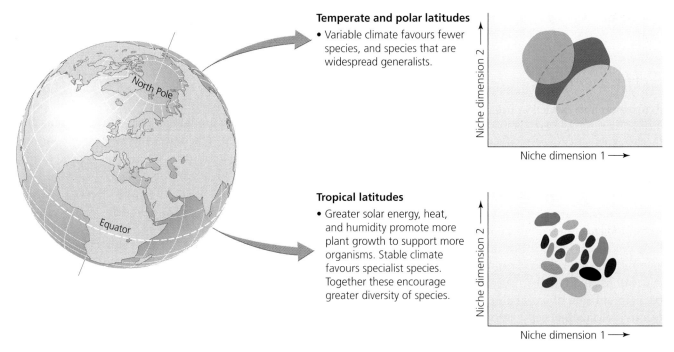

FIGURE 9.7
Ecologists have offered many hypotheses for the latitudinal gradient in species richness, and one set of ideas is summarized here. The variable climates (across days, seasons, and years) of polar and temperate latitudes favour organisms that can survive a wide range of conditions. Such generalist species have expansive niches; they can do many things well enough to survive, and they spread over large areas. In tropical latitudes, the abundant solar energy, heat, and humidity induce greater plant growth, which supports more organisms. The stable climates of equatorial regions favour specialist species, which have restricted niches but do certain things very well. Together these factors are thought to promote greater species richness in the tropics.

At smaller scales, diversity patterns vary with habitat type. Generally, habitats that are structurally diverse allow for more ecological niches and support greater species richness and evenness. For instance, forests generally support greater diversity than grasslands.

For any given geographic area, species diversity tends to increase with diversity of habitats, because each habitat supports a somewhat different community of organisms. Thus, ecotones, where habitats intermix, often have high biodiversity. Because human disturbance (such as clearing plots of forest) can sometimes increase habitat diversity (which ecologists call *habitat heterogeneity*), species diversity may be higher in disturbed areas. However, this is true only at local scales. At larger scales, human disturbance decreases diversity because species that rely on large unbroken expanses of single habitat will disappear.

Understanding such patterns of biodiversity is vital for landscape ecology, regional planning, and forest management. We will discuss further geographic patterns later in this chapter, when we explore solutions to the ongoing loss of biodiversity that our planet is currently experiencing.

Biodiversity Loss and Species Extinction

Biodiversity at all levels is being lost to human impact, most irretrievably in the extinction of species. Once vanished, a species can never return. **Extinction** occurs when the last member of a species dies and the species ceases to exist, as apparently was the case with Monteverde's golden toad. The disappearance of a particular population from a given area, but not the entire species globally, is referred to as **extirpation**. The tiger has been extirpated from most of its historic range, but it is not yet extinct. Although a species that is extirpated from one place may still exist in others, extirpation is an erosive process that can, over time, lead to extinction. A species that is in imminent danger of becoming extirpated or extinct is referred to as **endangered**, and one that is likely to become endangered in the near future is called **threatened**. These categories—threatened, endangered, and extirpated or extinct—are the main classifications used by the Canadian ***Species-At-Risk Act (SARA)***. Any species that is agreed to have fallen within one of these categories is considered to be *at risk*, and is listed on the *SARA* Public Registry, through a process that is described in greater detail later in this chapter.

Extinction is a natural process

Extirpation and extinction occur naturally. If organisms did not naturally go extinct, we would be up to our ears in dinosaurs, trilobites, ammonites, and the millions of other types of creatures that vanished from Earth long before humans appeared. Paleontologists estimate that roughly 99% of all species that have ever lived are now extinct. This means that the wealth of species on our planet today compose only about 1% of all species that have ever lived. Most extinctions preceding the appearance of humans have occurred one by one for independent reasons, at a rate that paleontologists refer to as the **background rate of extinction**. For example, the fossil record indicates that for mammals and marine animals, 1 species out of 1000 would typically become extinct every 1000 to 10 000 years. This translates to an annual rate of 1 extinction per 1 million to 10 million species.

Earth has experienced five previous mass extinction episodes

Extinction rates have risen far above this background rate during several mass extinction events in Earth's history. In the past 440 million years, our planet has experienced five major episodes of **mass extinction** (FIGURE 9.8). Each of these events has eliminated more than one-fifth of life's families and at least half its species (Table 9.1).

The most severe episode occurred at the end of the Permian period, 248 million years ago, when close to 54% of all families, 90% of all species, and 95% of marine species went extinct.

The best-known episode occurred at the end of the Cretaceous period, 65 million years ago, when an apparent asteroid impact brought an end to the dinosaurs and many other groups. In addition, there is evidence for further mass extinctions in the Cambrian period and earlier, more than half a billion years ago.

If the current trend continues, the modern era may see the extinction of more than half of all species. Although similar in scale to previous mass extinctions, today's ongoing mass extinction is different in two primary respects. First, humans are causing it. Second, humans will suffer as a result of it.

Humans set the sixth mass extinction in motion years ago

We have recorded many instances of human-induced species extinction over the past few hundred years. Sailors documented the extinction of the dodo on the Indian Ocean island of Mauritius in the seventeenth century, and we still have a few of the dodo's body parts in museums. Among North American birds in the past two centuries, we have driven into extinction the Carolina parakeet, great auk, Labrador duck, and passenger pigeon, and probably the Bachman's warbler and Eskimo curlew. Several more species—including the whooping crane, California condor, Kirtland's warbler, and ivory-billed woodpecker, recently rediscovered in the wooded swamps of Arkansas—teeter on the brink of extinction.

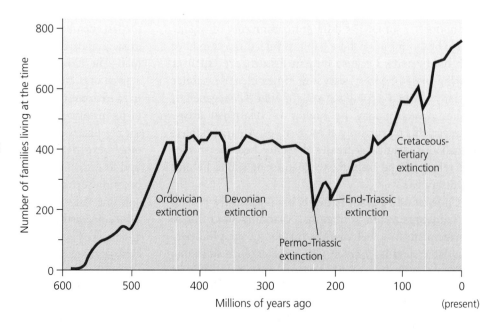

FIGURE 9.8
The fossil record shows evidence of five episodes of mass extinction during the past half-billion years of Earth's history. At the end of the Ordovician, Devonian, Permian, Triassic, and Cretaceous periods, 50%–95% of the world's species appear to have gone extinct. Each time, biodiversity later rebounded to equal or higher levels, but the rebound required millions of years in each case. Data from Raup, D. M., and J. J. Sepkoski. 1982. Mass extinctions in the marine fossil record. *Science* 215:1501–1503.

TABLE 9.1 Mass Extinctions

Event	Date (millions of years ago)	Cause	Types of life most affected	Percentage of life deepleted
Ordovician	440 mya	Unknown	Marine organisms; terrestrial record is unknown	>20% of families
Devonian	370 mya	Unknown	Marine organisms; terrestrial record is unknown	>20% of families
Permo-Triassic	250 mya	Possibly volcanism	Marine organisms; terrestrial record is unknown	>50% of families; 80%–95% of species
End-Triassic	202 mya	Unknown	Marine organisms; terrestrial record is unknown	20% of families; 50% of genera
Cretaceous-Tertiary*	665 mya	Asteroid impact	Marine organisms; terrestrial record is unknown	15% of families; >50% of species
Current	Beginning 0.01 mya	Human impact, through habitat destruction and other means	Large animals, specialized organisms, island organisms, and organisms hunted or harvested by humans	Ongoing

*Note that the term "Tertiary" is no longer in common use among geologists.

However, species extinctions caused by humans precede written history. Indeed, people may have been hunting species to extinction for thousands of years. Archeological evidence shows that in case after case, a wave of extinctions followed close on the heels of human arrival on islands and continents (FIGURE 9.9). After Polynesians reached Hawaii, half its birds went extinct. Birds, mammals, and reptiles vanished following human arrival on many other oceanic islands, including large island masses, such as New Zealand and Madagascar. The pattern appears to hold for at least two continents, as well. Dozens of species of large vertebrates died off in Australia after Aborigines arrived roughly 50 000 years ago, and North America lost 33 genera of large mammals after people arrived on the continent at least 10 000 years ago.

Current extinction rates are much higher than normal

Today, species loss is accelerating as our population growth and resource consumption put increasing strain on habitats and wildlife. In 2005, scientists with the Millennium Ecosystem Assessment calculated that the current global extinction rate is 100 to 1000 times as great as the background rate. They noted a decrease in genetic diversity as well as declining population sizes and numbers of species, accompanied by greater demands on ecosystem services in the past few decades. Moreover, they projected that the rate of species extinctions would increase tenfold or more in future decades.[2]

To keep track of the current status of endangered species, World Conservation Union, a nongovernmental organization, maintains the **Red List**, a regularly updated list of species facing high risks of extinction. The 2006 Red List reported that 23% (1093) of mammal species and 12% (1206) of bird species are threatened with extinction. Among other major groups (for which assessments are not fully complete), estimates of the percentage of species threatened ranged from 31% to 86%. Since 1970, less than one generation ago, at least 58 fish species, 9 bird species, and 1 mammal species have become extinct. For all these figures, the *actual* numbers of species extinct and threatened, like the actual number of total species in the world, are doubtless greater than the *known* numbers.

Among the 1093 mammals facing possible extinction on the Red List is the tiger, which despite—or perhaps because of—its tremendous size and reputation as a fierce predator, is one of the most endangered large animals on the planet. In 1950, eight tiger subspecies existed (see FIGURE 9.3). Today, three are extinct. The Bali tiger, *Panthera tigris balica*, went extinct in the 1940s; the Caspian tiger, *Panthera tigris virgata*, during the 1970s; and the Javan tiger, *Panthera tigris sondaica*, during the 1980s.

Biodiversity loss involves more than extinction

Extinction is only part of the story of biodiversity loss. The larger part of the story is the decline in population sizes of many organisms. Declines in numbers are accompanied by shrinkage of species' geographic ranges. Thus,

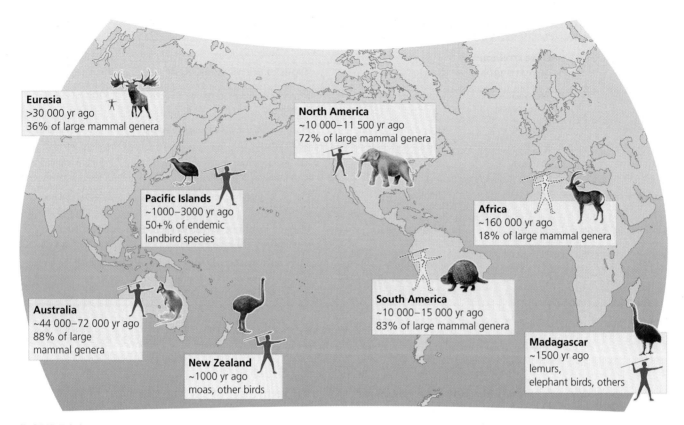

FIGURE 9.9
This map shows for each region the time of human arrival and the extent of the recent extinction wave. Illustrated are representative extinct megafauna from each region. The human hunter icons are sized according to the degree of evidence that human hunting was a cause of extinctions; larger icons indicate more certainty that humans (as opposed to climate change or other forces) were the cause. Data for South America and Africa are so far too sparse to be conclusive, and future archeological and paleontological research could well alter these interpretations. Adapted from Barnosky, A. D., et al. 2004. Assessing the causes of late Pleistocene extinctions on the continents. *Science* 306:70–75; and Wilson, E. O. 1992. *The Diversity of Life*. Cambridge, MA: Belknap Press.

many species today are less numerous and occupy less area than they once did. Tigers numbered well over 100 000 worldwide in the nineteenth century but number only about 5000 today.

To measure and quantify this degradation, scientists at the World Wildlife Fund and the United Nations Environment Programme (UNEP) developed a metric called the *Living Planet Index*. This index summarizes trends in the populations of 555 terrestrial species, 323 freshwater species, and 267 marine species that are well enough monitored to provide reliable data. Between 1970 and 2003, the Living Planet Index fell by roughly 40% (**FIGURE 9.10**).

There are several major causes of biodiversity loss

Reasons for the decline of any given species are often multifaceted and complex, so they can be difficult to determine. The current precipitous decline in populations of amphibians throughout the world provides an example. Frogs, toads, and salamanders worldwide are decreasing drastically in abundance. As we saw in Chapter 3 with the

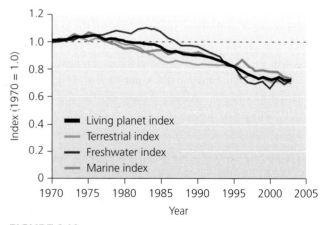

FIGURE 9.10
The Living Planet Index serves as an indicator of the state of global biodiversity. Index values summarize population trends for 1145 species. Between 1970 and 2000, the Living Planet Index fell by roughly 40%. The indices for terrestrial and marine species fell 30%, and the index for freshwater species fell 50%. Data from World Wide Fund for Nature and U.N. Environment Programme. 2006. *The Living Planet Report, 2006*. Gland, Switzerland: WWF.

golden toad, several have already gone extinct, and scientists are struggling to explain why. Recent studies have implicated a wide array of factors, and most scientists

THE SCIENCE BEHIND THE STORY

Amphibian Diversity and Decline

Dr. Madhava Meegaskumbura searches for frogs at night in Sri Lanka.

New species of most vertebrate classes are discovered at a rate of only a few per year, but the number of known amphibian species (frogs, salamanders, etc.) has jumped nearly 42% since 1985. At the same time, more than 200 amphibian species worldwide are in steep decline. Researchers feel they may be naming some species just before they go extinct, and losing others before they are even discovered. The recent Global Amphibian Assessment ranked nearly one-third (32%) of the world's amphibians as threatened, compared with 23% of mammals and 12% of bird species. At least 32 species of frogs, toads, and salamanders studied just years or decades ago, including the golden toad (Chapter 3), are now gone.

These losses are especially worrying because amphibians are regarded as "biological indicators" of the condition of an ecosystem. Amphibians rely on both aquatic and terrestrial environments, and may breathe and absorb water through their skin, so they are sensitive to environmental stresses. Studying the reasons for their decline can tell us much about the state of our environment. The major threats to amphibians worldwide include habitat loss and degradation, exotic species, overharvesting, increases in UV radiation, pollution, disease, and road mortality. Fungal and viral diseases have been implicated in some amphibian declines. Some of these causes of mortality are exacerbated by climate change.[3]

The status of amphibians in Canada is documented in Wild Species 2005, the most recent report from the National General Status Working Group (NGSWG).[4] NGSWG comprises representatives from the provinces and territories, and the three federal agencies whose mandate includes wildlife: Canadian Wildlife Service, Department of Fisheries and Oceans, and Parks Canada. The report concluded that habitat loss, especially draining of wetlands, is a leading threat to amphibians in Canada. Wetlands that remain within agricultural or urban landscapes may be polluted. Habitat fragmentation also reduces or prevents the movement of individuals among populations, leading to reduced population stability and reduced exchange of genes.[5]

In some parts of the world, scientific scrutiny has revealed amphibian "hot spots." In the 1990s, an international team of scientists set out to determine whether Sri Lanka, a large tropical island off the coast of India, held more than the 40 frog species that were already known. Researcher Madhava Meegaskumbura and his team combed through trees, rivers, ponds, and leaf litter for eight years, collecting more than 1400 frogs at 300 study sites. They analyzed the frogs' physical appearance, habitat use, and vocalizations, as well as their genes.

The studies led to the discovery of more than 100 previously unknown species of frogs, on this one island. Some of these unusual animals live on rocks and have leg fringes and markings that help disguise them as clumps of moss. Others are tree frogs that lay their eggs in baskets they construct. These discoveries, however, come against the backdrop of distressing amphibian declines worldwide. Most worrisome is that populations are vanishing even when no direct damage to habitat is apparent. Researchers surmise that a combination of factors might be at work.

In one study, researchers presented young frogs with two common dangers—pesticides and predators—to see how the mix affected their survival. Groups of tadpoles were put in different tubs of water.

The odd-looking purplish frog *Nasikabatrachus sahyadrensis*, of India, is one of many new amphibian species recently discovered.

Some tubs contained pure water, others contained varying levels of the pesticide carbaryl, and others contained a harmless solvent as a control. To some of the tubs, the researchers added a hungry predator—a young salamander. The salamander was caged and could not reach the tadpoles, but they were aware of its presence. Researchers watched to see how many tadpoles survived the different combinations of stress factors.

Their results revealed that tadpoles that withstood one type of stress might not survive two. All tadpoles in clean water with no predators survived, and all tadpoles exposed to high concentrations of carbaryl died within several days, regardless of predator presence. But when carbaryl levels were lower, the presence of the salamander made a noticeable difference. In one trial, 75% of tadpoles survived the pesticide if no predator was present, but in the presence of the salamander, survival rates dropped to 25%. Thus when both stresses were present, death rates increased by two to four times.

As scientists learn more about how factors combine to threaten amphibians, they are gaining a clearer picture of how the fate of these creatures may foreshadow the future for other organisms.

---

now suspect that such factors may be interacting synergistically (see "The Science Behind the Story: Amphibian Diversity and Decline").

Overall, scientists have identified four primary causes of population decline and species extinction: habitat alteration, invasive species, pollution, and overharvesting. Global climate change (Chapter 14) is becoming the fifth. Each of these factors is intensified by human population growth and by our increase in per capita consumption of resources.

Habitat alteration Nearly every human activity can alter the habitat of the organisms around us. Farming replaces diverse natural communities with simplified ones of only one or a few plant species. Grazing modifies the structure and species composition of grasslands. Either type of agriculture can lead to desertification. Clearing forests removes the food, shelter, and other resources that forest-dwelling organisms need to survive. Hydroelectric dams turn rivers into reservoirs upstream and thereby affect water conditions and floodplain communities downstream. Urbanization and suburban sprawl supplant diverse natural communities with simplified human-made ones, driving many species from their homes.

Because organisms are adapted to the habitats in which they live, any major change is likely to render their habitat less suitable for them. Of course, human-induced habitat change may benefit some species. Such animals as starlings, house sparrows, pigeons, and grey squirrels do very well in urban and suburban environments and benefit from our modification of natural habitats. However, the species that benefit are relatively few; for every species that gains, more lose. Furthermore, the species that do well in our midst tend to be weedy, cosmopolitan generalists that are in little danger of disappearing any time soon.

Habitat alteration is by far the greatest cause of biodiversity loss today. It is the primary source of population declines for 83% of threatened mammals and 85% of threatened birds, according to UNEP data. As just one example of thousands, the prairies native to central North America have been almost entirely converted to agriculture, especially south of the border in the United States, where the area of prairie habitat has been reduced by more than 99%. As a result, grassland bird populations have declined by an estimated 82%–99%. Many grassland species have been extirpated from large areas, and the two species of prairie chickens still persisting in pockets of the Great Plains could soon go extinct.

Habitat destruction has occurred widely in nearly every biome. More than half of temperate forests, grasslands, and shrublands had been converted by the year 1950 (mostly for agriculture). Across Asia, scientists estimate that 40% of the tiger's remaining habitat has disappeared just in the past decade. Today habitat is being lost most rapidly in tropical rainforests, tropical dry forests, and savannahs.

Invasive species Our introduction of non-native species to new environments, where some may become invasive (**FIGURE 9.11**), has also pushed native species toward extinction. Some introductions have been accidental. Examples include aquatic organisms (such as zebra mussels, Chapter 4) transported among continents in the ballast water of ships, animals that have escaped from the pet trade, and the weed seeds that cling to our socks as we travel from place to place. Other introductions have been intentional. People have brought with them food crops, domesticated animals, and other organisms as they colonized new places, generally unaware of the ecological consequences that could result.

Species native to islands are especially vulnerable to disruption from introduced species because the native species have been in isolation for so long with relatively few parasites, predators, and competitors. As a result, they have not evolved the defences necessary to resist invaders that are better adapted to these pressures.

Most organisms introduced to new areas perish, but the few types that survive may do very well, especially if they find themselves without the predators and parasites that attacked them back home or without the competitors that had limited their access to resources. Once released from the limiting factors of predation, parasitism, and competition, an introduced species may increase rapidly, spread, and displace native species. Moreover, invasive species cause billions of dollars in economic damage each year.

Pollution Pollution can harm organisms in many ways. Air pollution (Chapter 13) can degrade forest ecosystems; water pollution (Chapter 11) can adversely affect fish and amphibians; agricultural runoff (including fertilizers, pesticides, and sediments; Chapters 7 and 8) can harm many terrestrial and aquatic species. Heavy metals, PCBs, endocrine-disrupting compounds, and various other toxic chemicals can poison both people and wildlife (Chapter 19), and the effects of oil and chemical spills on wildlife are dramatic and well known.

Although pollution is a substantial threat, it tends to be less significant than public perception holds it to be. The damage to wildlife and ecosystems caused by pollution can be severe, but it tends to be less so than the damage caused by habitat alteration or invasive species.

Overharvesting For most species, a high intensity of hunting or harvesting by humans will not *in itself* pose a threat of extinction, but for some species it can. The Siberian tiger is one such species. Large in size, few in number, long-lived, and raising few young in its lifetime—a classic K-selected species—the Siberian tiger is just the type of animal to be vulnerable to population reduction by hunting. The advent of Russian hunting nearly drove the animal extinct, whereas decreased hunting during and after the Second World War contributed to a population increase. By the mid-1980s, the Siberian tiger population was likely up to 250 individuals.

The political freedom that came with the Soviet Union's breakup in 1989, however, brought with it a freedom to

Invasive Species			
Species	Native to...	Invasive in...	Effects
Mosquito fish (*Gambusia affinis*) 	North America	Africa, Asia, Europe, and Australia	Introduced to control mosquito populations, the mosquito fish outcompetes native fish, eats their eggs, and does no better than native species in controlling mosquitoes.
Zebra mussels (*Dreissenna polymorpha*) 	Caspian Sea	Freshwater ecosystems including the Great Lakes of Canada and the United States	Zebra mussels (Chapter 4) most likely made their way from their home by travelling in ballast water taken on by cargo ships. They compete with native species and clog water treatment facilities and power plant cooling systems.
Kudzu (*Pueraria montana*) 	Japan	Southeastern United States	Kudzu is a vine that can grow 30 m (100 ft) in a single season. The U.S. Soil Conservation Service introduced kudzu in the 1930s to help control erosion. Adaptable and extraordinarily fast-growing, kudzu has taken over thousands of hectares of forests, fields, and roadsides in the southeastern United States.
Asian long-horned beetles (*Anoplophora glabripennis*) 	Asia	United States	Having first arrived in the United States in imported lumber in the 1990s, these beetles burrow into hardwood trees and interfere with the trees' ability to absorb and process water and nutrients. They may wipe out the majority of hardwood trees in an area. Several U.S. cities, including Chicago in 1999 and Seattle in 2002, have cleared thousands of trees after detecting these invaders.
Rosy wolfsnail (*Euglandina rosea*) 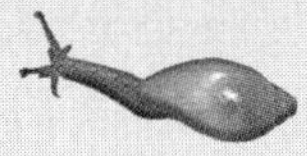	Southeastern United States and Latin America	Hawaii	In the 1950s, well-meaning scientists introduced the rosy wolfsnail to Hawaii to prey upon and reduce the population of another invasive species, the giant African land snail (*Achatina fulica*), which had been introduced early in the 20th century as an ornamental garden animal. Within a few decades, however, the carnivorous rosy wolfsnail had instead driven more than half of Hawaii's native species of banded tree snails to extinction.
Cane toad (*Bufo marinus*) 	Southern United States to tropical South America	Northern Australia and other locations	Since being introduced 70 years ago to control insects in sugarcane fields, the cane toad has wreaked havoc across northern Australia (and other locations). The skin of this tropical American toad can kill its predators, and the cane toad outcompetes native amphibians.
Bullfrog (*Rana catesbiana*) 	Eastern North America	Western North America	The bullfrog is contributing to amphibian and reptile declines in western North America. Bullfrog tadpoles grow large and can outcompete and prey on other tadpoles, but need to grow a long time in permanent water to do so. Historically most water bodies in the arid West dried up part of the year, making it impossible for bullfrogs to live there, but artificial impoundments—dams, farm ponds, canals—gave the bullfrogs bases from which they could spread.

FIGURE 9.11
Invasive species are species that thrive in areas where they are introduced, outcompeting, preying on, or otherwise harming native species. Of the many thousands of invasive species, this chart shows a few of the best known.

harvest Siberia's natural resources, the tiger included, without regulations or rules. This coincided with an economic expansion in many Asian countries, where tiger penises are traditionally used to try to boost human sexual performance and where tiger bones, claws, whiskers, and other body parts are used to treat a wide variety of maladies (**FIGURE 9.12**). Thus, the early 1990s brought a boom in poaching (poachers killed at least 180 Siberian tigers between 1991 and 1996), as well as a dramatic increase in logging of the Korean pine forests on which the tigers and their prey depend.

Over the past century, hunting has led to steep declines in the populations of many other K-selected animals. The Atlantic grey whale has gone extinct, and several other

AN ELEMENT IN THE NATURAL SYSTEM

	Invasive Species		
	Native to...	Invasive in...	Effects
 (Gypsy moth)	Eurasia	Northeastern United States	In the 1860s, a scientist introduced the gypsy moth to Massachusetts in the mistaken belief that it might be bred with others to produce a commercial-quality silk. The gypsy moth failed to start a silk industry, and instead spread through the northeastern United States and beyond, where its outbreaks defoliate trees over large regions every few years.
European starling (*Sturnus vulgaris*) 	Europe	North America	The bird was first introduced to New York City in the late 19th century by Shakespeare devotees intent on bringing every bird mentioned in Shakespeare's plays to the new continent. It only took 75 years for the birds to spread to the Pacific coast, Alaska, and Mexico, becoming one of the most abundant birds on the continent. Starlings are thought to outcompete native birds for nest sites.
Indian mongoose (*Herpestes auropunctatus*) 	Southeast Asia	Hawaii	Rats that had invaded the Hawaiian islands from ships in the 17th century were damaging sugarcane fields, so in 1883 the Indian mongoose was introduced to control rat populations. Unfortunately, the rats were active at night and the mongooses fed during the day, so the plan didn't work. Instead mongooses began preying on native species like ground-nesting seabirds and the now-endangered Nene or Hawaiian goose (*Branta sandvicensis*).
A green alga (*Caulerpa taxifolia*) 	Tropical oceans and seas	Mediterranean Sea	Dubbed the "killer algae," *Caulerpa taxifolia* has spread along the coasts of several Mediterranean countries since it apparently escaped from Monaco's aquarium in 1984. Creeping underwater over the sand and mud like a green shag carpet, it crowds out other plants, is inedible to most animals, and tangles boat propellers. It has been the focus of intense eradication efforts since arriving recently in Australia and California.
Cheatgrass (*Bromus tectorum*) 	Eurasia	Western United States	In just 30 years after its introduction to Washington state in the 1890s, cheatgrass has spread across much of the western United States. Its secret: fire. Its thick patches, which choke out other plants and use up the soil's nitrogen, burn readily. Fire kills many of the native plants, but not cheatgrass, which grows back even stronger amid the lack of competition.
Brown tree snake (*Boiga irregularis*) 	Southeast Asia	Guam	Nearly all native forest bird species on the South Pacific island of Guam have disappeared. The culprit is the brown tree snake. The snakes were likely brought to the island inadvertently as stowaways in cargo bays of military planes in the Second World War. Guam's birds had not evolved with tree snakes, and so had no defences against the snake's nighttime predation. The snakes also cause numerous power outages each year on Guam and have spread to other islands where they are repeating their ecological devastation. The arrival of this snake is the greatest fear of conservation biologists in Hawaii.

FIGURE 9.11
(Continued)

whales remain threatened or endangered. Gorillas and other primates that are killed for their meat may be facing extinction soon. Thousands of sharks are killed each year simply for their fins, which are used in soup. Today the oceans contain only 10% of the large animals they once did.

Climate change The preceding four types of human impacts affect biodiversity in discrete places and times. In contrast, our manipulation of Earth's climate system (Chapter 14) is beginning to have global impacts on both habitat and biodiversity. As we will explore in Chapter 14, our emissions of carbon dioxide and other "greenhouse

FIGURE 9.12
Body parts from tigers have long been used as medicines or aphrodisiacs in some traditional Asian cultures. Hunters and poachers have illegally killed countless tigers through the years to satisfy market demand for these items. Here a street vendor in northern China displays tiger penises and other body parts for sale.

gases" that trap heat in the atmosphere are causing average temperatures to warm worldwide, modifying global weather patterns and increasing the frequency of extreme weather events. Scientists foresee that these effects, together termed *global climate change*, will accelerate and become more severe in the years ahead until we find ways to reduce our emissions from fossil fuels.

Climate change is beginning to exert effects on plants and animals. Extreme weather events, such as droughts, put increased stress on populations, and warming temperatures are forcing species to move toward the poles and higher in altitude. Some species will be able to adapt, but others will not. Consider the cloud forest fauna from Monteverde that we examined in Chapter 3. Mountaintop organisms cannot move further upslope to escape warming temperatures, so they will likely perish. Trees may not be able to move poleward fast enough. Animals and plants may find themselves among different communities of prey, predators, and parasites to which they are not adapted.

In the Arctic, where warming has been greatest, the polar bear (**FIGURE 9.13**) has now been added to the U.S. endangered species list. Melting ice hinders the bear's ability to hunt seals. Overall, scientists now predict that a 1.5°C–2.5°C global temperature increase could put 20%–30% of the world's plants and animals at increased risk of extinction.

All five of these primary causes of population decline are intensified by human population growth and rising per capita consumption. More people and more consumption mean more habitat alteration, more invasive species, more pollution, more overharvesting, and more climate change. Growth in population and growth in consumption are the ultimate reasons behind the proximate threats to biodiversity. Just as we now have a solid scientific understanding of the causes of biodiversity loss, we are also coming to

FIGURE 9.13
The polar bear (*Ursus maritimus*) is thought to be threatened by climate change as Arctic warming melts the sea ice from which it hunts seals, forcing the bears to swim farther for food. Partly in response to a lawsuit from environmental groups, the U.S. Fish and Wildlife Service added the polar bear to the U.S. endangered species list in 2008. This will likely have an impact on how polar bear hunts are regulated and carried out by Aboriginal people in Canada's north.

appreciate its consequences as we begin to erode the many benefits that biodiversity brings us.

Benefits of Biodiversity

Scientists worldwide are presenting us with data that confirm what any naturalist who has watched the habitat change in his or her hometown already knows: From amphibians to tigers, biodiversity is being lost rapidly and visibly within our lifetimes. The loss of one species may or may not affect us as individuals in any discernible way, but it is important to consider biodiversity from a holistic perspective. A comparison has been made (perhaps first by Paul Ehrlich) to the rivets in an airplane wing. The loss of one rivet, or two, or three, will not cause the plane to crash. But at some point the structure will be compromised, and eventually the loss of just one more rivet will cause it to fail. If individual species are like the rivets in the airplane wing, then we might well ask how many more we can afford to lose before the structure is compromised.

This raises the question "Why does biodiversity matter?" There are many ways to answer this question, but we can begin by considering the ways that biodiversity benefits people. Scientists have offered a number of tangible, pragmatic reasons for preserving biodiversity, showing how biodiversity directly or indirectly supports human society. In addition, many people feel that organisms have an intrinsic right to exist and that ethical and esthetic dimensions to biodiversity preservation cannot be ignored. (We will consider these questions in greater detail in Chapter 21.)

Biodiversity provides ecosystem services free of charge

Contrary to popular opinion, some things in life can indeed be free, as long as we choose to protect the living systems that provide them. Intact forests provide clean air and buffer hydrologic systems against flooding and drought. Native crop varieties provide insurance against disease and drought. Abundant wildlife can attract tourists and boost the economies of developing nations. Intact ecosystems provide these and other valuable processes, known as *ecosystem services*, for all of us, free of charge.

Maintaining these ecosystem services is one clear benefit of protecting biodiversity. According to UNEP, biodiversity

- Provides food, fuel, and fibre
- Provides shelter and building materials
- Purifies air and water
- Detoxifies and decomposes wastes
- Stabilizes and moderates Earth's climate
- Moderates floods, droughts, wind, and temperature extremes
- Generates and renews soil fertility and cycles nutrients
- Pollinates plants, including many crops
- Controls pests and diseases
- Maintains genetic resources as inputs to crop varieties, livestock breeds, and medicines
- Provides cultural and esthetic benefits
- Gives us the means to adapt to change

Organisms and ecosystems support a vast number of vital processes that humans could not replicate or would need to pay for if nature did not provide them. As we will see in Chapter 21, the annual value of just 17 of these ecosystem services may be in the neighbourhood of $16 trillion to $54 trillion per year. The Millennium Ecosystem Assessment estimated that 60% of ecosystem services are being degraded or used unsustainably.[6]

Biodiversity helps maintain ecosystem function

Functioning ecosystems are vital, but does biodiversity really help them maintain their function? Ecologists have found that the answer appears to be yes. Research has demonstrated that high levels of biodiversity tend to increase the *stability* of communities and ecosystems. Research has also found that high biodiversity tends to increase the *resilience* of ecological systems—their ability to weather disturbance, bounce back from stresses, or adapt to change. Most of this research has dealt with species diversity, but new work is finding similar effects for genetic diversity. Thus, a decrease in biodiversity could diminish a natural system's ability to function and to provide services to our society.

What about the extinction of selected species, however? Skeptics have asked whether the loss of a few endangered species will really make much difference in an ecosystem's ability to function. Ecological research suggests that the answer to this question depends on which species are removed. Removing a species that can be functionally replaced by others may make little difference. Recall, however, our discussion of keystone species. Like the keystone that holds an arch together, a keystone species is one whose removal results in significant changes in an ecological system. If a keystone species is extirpated or driven extinct, other species may disappear or experience significant population changes as a result.

Top predators, such as tigers, are often considered keystone species. A single top predator may prey on many other carnivores, each of which may prey on many herbivores, each of which may consume many plants. Thus the removal of a single individual at the top of a food chain can have impacts that multiply as they cascade down the food chain. Moreover, top predators, such as tigers, wolves, and grizzly bears, are among the species most vulnerable to human impact. Large animals are frequently hunted and also need large areas of habitat, making them susceptible to habitat loss and fragmentation. Top predators are also vulnerable to the buildup of toxic pollutants in their tissues through the process of biomagnification, as seen in the example of the beluga whales in the St. Lawrence estuary.

The influence of "ecosystem engineers," such as ants and earthworms, can be every bit as far-reaching as those of keystone species. Ecosystems are complex, and it is difficult to predict which particular species may be important. Thus, many people prefer to apply the precautionary principle in the spirit of Aldo Leopold, who advised, "To keep every cog and wheel is the first precaution of intelligent tinkering."

Biodiversity enhances food security

Biodiversity benefits agriculture as well. As our discussion of native landraces of corn in Oaxaca, Mexico, in Chapter 8 showed, genetic diversity within crop species and their ancestors is enormously valuable. In 1995, Turkey's wheat crops received at least $50 billion worth of disease resistance from wild wheat strains. California's barley crops annually receive $160 million in disease resistance benefits from Ethiopian strains of barley. During the 1970s a researcher discovered a maize species in Mexico known as *Zea diploperennis*. This maize is highly resistant to disease, and it is a perennial, meaning it will grow back year after year without being replanted. At the time of its discovery, its entire range was limited to a 10 ha plot of land in the mountains of the Mexican state of Jalisco.

Other potentially important food crops await utilization (**FIGURE 9.14**). The babassu palm (*Orbignya phalerata*) of the Amazon produces more vegetable oil than any other plant. The serendipity berry (*Dioscoreophyllum cumminsii*) produces a sweetener that is 3000 times sweeter than table sugar. Several species of salt-tolerant grasses and trees are so hardy that farmers can irrigate them with saltwater. These same plants also produce animal feed, a substitute for conventional vegetable oil, and other economically important products. Such species could be immeasurably beneficial to areas undergoing soil salinization caused by poorly managed irrigation.

Organisms provide drugs and medicines

People have made medicines from plants for centuries, and many of today's widely used drugs were discovered by studying chemical compounds present in wild plants, animals, and microbes (**FIGURE 9.15**). Each year, pharmaceutical products owing their origin to wild species generate up to $150 billion in sales.

Food Security and Biodiversity: Potential new food sources		
Species	Native to...	Potential uses and benefits
Amaranths (three species of *Amaranthus*) 	Tropical and Andean America	Grain and leafy vegetable; livestock feed; rapid growth, drought resistant
Buriti palm (*Mauritia flexuosa*) 	Amazon lowlands	"Tree of life" to Amerindians; vitamin-rich fruit; pith as source for bread; palm heart from shoots
Maca (*Lepidium meyenii*) 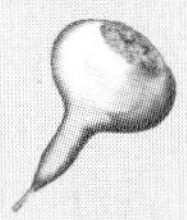	Andes Mountains	Cold-resistant root vegetable resembling radish, with distinctive flavour; near extinction
Tree tomato (*Cyphomandra betacea*) 	South America	Elongated fruit with sweet taste
Babirusa (*Babyrousa babyrussa*) 	Indonesia: Moluccas and Sulawesi	A deep-forest pig; thrives on vegetation high in cellulose and hence less dependent on grain
Capybara (*Hydrochoeris hydrochoeris*) 	South America	World's largest rodent; meat esteemed; easily ranched in open habitats near water
Vicuna (*Lama vicugna*) 	Central Andes	Threatened species related to llama; valuable source of meat, fur, and hides; can be profitably ranched
Chachalacas (*Ortalis*, many species) 	South and Central America	Birds, potentially tropical chickens; thrive in dense populations; adaptable to human habitations; fast-growing
Sand grouse (*Pterocles*, many species) 	Deserts of Africa and Asia	Pigeon-like birds adapted to harshest deserts; domestication a possibility

FIGURE 9.14
By protecting biodiversity, we can enhance food security. The wild species shown here are a tiny fraction of the many plants and animals that could someday supplement our food supply. Adapted from Wilson, E. O. 1992. *The Diversity of Life*. Cambridge, MA: Belknap Press.

It can truly be argued that every species that goes extinct represents one lost opportunity to find a cure for cancer or AIDS. The rosy periwinkle (*Catharanthus roseus*) produces compounds that treat Hodgkin's disease

Medicines and Biodiversity: Natural sources of pharmaceuticals		
Plant	Drug	Medical application
Pineapple (*Ananas comosus*) 	Bromelain	Controls tissue inflammation
Autumn crocus (*Colchicum autumnale*) 	Colchicine	Anticancer agent
Yellow cinchona (*Cinchona ledgeriana*) 	Quinine	Antimalarial
Common thyme (*Thymus vulgaris*) 	Thymol	Cures fungal infection
Pacific yew (*Taxus brevifolia*) 	Taxol	Anticancer (especially ovarian cancer)
Velvet bean (*Mucuna deeringiana*) 	L-Dopa	Parkinson's disease suppressant
Common foxglove (*Digitalis purpurea*) 	Digitoxin	Cardiac stimulant

FIGURE 9.15
By protecting biodiversity, we can enhance our ability to treat illness. Shown here are just a few of the plants that have so far been found to provide chemical compounds of medical benefit. Adapted from Wilson, E. O. 1992. *The Diversity of Life*. Cambridge, MA: Belknap Press.

and a particularly deadly form of leukemia. Had this native plant of Madagascar become extinct prior to its discovery by medical researchers, two deadly diseases would have claimed far more victims than they have to date. In Australia, where the government has placed high priority on research into products from rare and endangered species, a rare species of cork, *Duboisia leichhardtii*, now provides hyoscine, a compound that physicians use to treat cancer, stomach disorders, and motion sickness. Another Australian plant, *Tylophora*, provides a drug that treats lymphoid leukemia. Researchers are now exploring the potential of the compound prostaglandin E2 in treating gastric ulcers. This compound was first discovered in two frog species unique to the rainforest of Queensland, Australia. Scientists believe that both species are now extinct.

A compound that forms the basis for the anti-cancer drug Taxol is derived from the bark of the Pacific yew (genus *taxus*), native to British Columbia. At first, overharvesting threatened not only the very slow-growing yew but another endangered species, the spotted owl, which relies on the yew as part of its natural habitat. Today the basic ingredient for Taxol is still extracted from the bark of the Pacific yew, but the tree is cultivated specifically for this purpose.

Biodiversity provides economic benefits through tourism and recreation

Besides providing for our food and health, biodiversity can represent a direct source of income through tourism, particularly for developing countries in the tropics that have impressive species diversity. As we saw in Chapter 3 with Costa Rica, many people like to travel to experience protected natural areas, and in so doing they create economic opportunity for residents living near those natural areas. Visitors spend money at local businesses, hire local people as guides, and support the parks that employ local residents. Ecotourism thus can bring jobs and income to areas that otherwise might be poverty-stricken.

Ecotourism has become a vital source of income for such nations as Costa Rica, with its rainforests; Australia, with its Great Barrier Reef; Belize, with its reefs, caves, and rainforests; and Kenya and Tanzania, with their savannah wildlife. Canada, too, benefits from ecotourism; its national, provincial, and territorial parks draw millions of visitors domestically and from around the world. In Chapter 10 we will consider the case of logging versus ecotourism in Clayoquot Sound, British Columbia, the scene of the largest episode of civil disobedience in Canadian history. Ecotourism serves as a powerful financial incentive for nations, states, and local communities to preserve natural areas and reduce impacts on the landscape and on native species.

As ecotourism increases in popularity, however, critics have warned that too many visitors to natural areas can degrade the outdoor experience and disturb wildlife. Anyone who has been to Yellowstone Park on a crowded summer weekend can attest to this. Ecotourism's effects on species living in parks and reserves are much debated, and likely they vary enormously from one case to the next. As ecotourism continues to increase, so will debate over its costs and benefits for local communities and for biodiversity.

People value and seek out connections with nature

Not all the benefits of biodiversity to humans can be expressed in the hard numbers of economics or the day-to-day practicalities of food and medicine. Some scientists and philosophers argue that there is a deeper importance to biodiversity. E. O. Wilson (**FIGURE 9.16**) has described

FIGURE 9.16
Edward O. Wilson is the world's most recognized authority on biodiversity and its conservation, and he has inspired many people who study our planet's life. A Harvard professor and world-renowned expert on ants, Wilson has written more than 20 books and has won two Pulitzer prizes. His books *The Diversity of Life* and *The Future of Life* address the value of biodiversity and its outlook for the future.

a phenomenon he calls **biophilia**, "the connections that human beings subconsciously seek with the rest of life." Wilson and others have cited as evidence of biophilia our affinity for parks and wildlife, our keeping of pets, the high value of real estate with a view of natural landscapes, and our interest—despite being far removed from a hunter–gatherer lifestyle—in hiking, bird watching, fishing, hunting, backpacking, and similar outdoor pursuits.

weighing the issues 9–1
BIOPROSPECTING IN COSTA RICA

Bioprospectors search for organisms that can provide new drugs, foods, or other valuable products. Scientists working for pharmaceutical companies, for instance, scour biodiversity-rich countries for potential drugs and medicines. Many have been criticized for harvesting indigenous species to create commercial products that do not benefit the country of origin. To make sure it would not lose the benefits of its own biodiversity, Costa Rica reached an agreement with the Merck pharmaceutical company in 1991. The nonprofit National Biodiversity Institute of Costa Rica (INBio) allowed Merck to evaluate a limited number of Costa Rica's species for their commercial potential in return for $1.1 million, a small royalty rate on any products developed, and training for Costa Rican scientists.

Do you think both sides win in this agreement? What if Merck discovers a compound that could be turned into a billion-dollar drug? Does this provide a good model for other countries, and for other companies?

In a 2005 book, writer Richard Louv adds that as today's children are increasingly deprived of outdoor experiences and direct contact with wild organisms, they suffer what he calls "nature-deficit disorder." Although it is not a medical condition, this alienation from biodiversity and the natural environment, Louv argues, may damage childhood development and lie behind many of the emotional and physical problems young people in developed nations face today.

Do we have ethical obligations toward other species?

If Wilson, Louv, and others are right, then biophilia not only may affect ecotourism and real estate prices, but also may influence our ethics. When Maurice Hornocker and his associates first established the Siberian Tiger Project, he wrote: "Saving the most magnificent of all the cat

weighing the issues 9–2
BIOPHILIA AND NATURE-DEFICIT DISORDER

What do you think of the concepts of biophilia and "nature-deficit disorder"? Have you ever felt a connection to other living things that you couldn't explain in scientific or economic terms? Do you think that an affinity for other living things is innately human? How could you determine whether or not most people in your community feel this way?

species and one of the most endangered should be a global responsibility . . . If they aren't worthy of saving, then what are we all about? What is worth saving?"

We humans are part of nature, and like any other animal we need to use resources and consume other organisms to survive. In that sense, there is nothing immoral about our doing so. However, we have conscious reasoning ability and are able to control our actions and make conscious decisions. Our ethical sense has developed from this intelligence and ability to choose. As our society's sphere of ethical consideration has widened over time, and as more of us take up biocentric or ecocentric world views, more people have come to believe that other organisms have intrinsic value and an inherent right to exist.

Despite our ethical convictions, however, and despite biodiversity's many benefits—from the pragmatic and economic to the philosophical and spiritual—the future of biodiversity is far from secure. Even our protected areas and national parks are not big enough or well protected enough to ensure that biodiversity is fully safeguarded within their borders. The search for solutions to today's biodiversity crisis is an exciting and active one, and scientists are playing a leading role in developing innovative approaches to maintaining the diversity of life on Earth.

Conservation Biology: The Search for Solutions

Today, more and more scientists and citizens perceive a need to do something to stem the loss of biodiversity. In his 1994 autobiography, *Naturalist*, E. O. Wilson wrote:

> When the [twentieth] century began, people still thought of the planet as infinite in its bounty. The highest mountains were still unclimbed, the ocean depths never visited, and vast wildernesses stretched across the equatorial continents . . . In one lifetime exploding human populations have reduced wildernesses to threatened nature reserves. Ecosystems and species are vanishing at the fastest rate in 65 million years. Troubled by what we have wrought, we have begun to turn in our role from local conqueror to global steward.

Conservation biology arose in response to biodiversity loss

The urge to act as responsible stewards of natural systems, and to use science as a tool in that endeavour, helped spark the rise of conservation biology. **Conservation biology** is a scientific discipline devoted to understanding

FIGURE 9.17
Conservation biologists integrate lab and field research to develop solutions to biodiversity loss. Here, a conservation biologist checks on a jabiru stork nest in the Pantanal region of Brazil.

the factors, forces, and processes that influence the loss, protection, and restoration of biological diversity. It arose as biologists became increasingly alarmed at the degradation of the natural systems they had spent their lives studying.

Conservation biologists choose questions and pursue research with the aim of developing solutions to such problems as habitat degradation and species loss (**FIGURE 9.17**). Conservation biology is thus an applied and goal-oriented science, with implicit values and ethical standards. This perceived element of advocacy sparked some criticism of conservation biology in its early years. However, as scientists have come to recognize the scope of human impact on the planet, more of them have directed their work to addressing environmental problems. Today conservation biology is a thriving pursuit that is central to environmental science and to achieving a sustainable society.

Conservation biologists work at multiple levels

Conservation biologists integrate an understanding of evolution and extinction with ecology and the dynamic nature of environmental systems. They use field data, lab data, theory, and experiments to study the impacts of humans on other organisms. They also attempt to design, test, and implement ways to mitigate human impact.

These researchers address the challenges facing biological diversity at all levels, from genetic diversity to species diversity to ecosystem diversity. At the genetic level, *conservation geneticists* study genetic attributes of organisms, generally to infer the status of their populations. If two populations of a species are found to be genetically distinct enough to be considered subspecies, they may have different ecological needs and may require different types of management.

In addition, as a population dwindles, genetic variation is lost from the gene pool. Conservation geneticists ask how small a population can become and how much genetic variation it can lose before running into problems, such as inbreeding depression. By determining a minimum viable population size for a given population, conservation geneticists and population biologists provide wildlife managers with an indication of how vital it may be to increase the population.

Problems for populations and subspecies spell problems for species, because declines and local extirpation generally precede rangewide endangerment and extinction. As we will see, it is at the species level that much of the funding and resources for conservation biology exist.

Many efforts also revolve around habitats, communities, ecosystems, and landscapes—and these efforts are often informed by studies of genes, populations, and species. As we saw in our discussion of landscape ecology, organisms are sometimes distributed across a landscape as a *metapopulation*, or a network of subpopulations. Because small and isolated subpopulations are most vulnerable to extirpation, conservation biologists pay special attention to them. By examining how organisms disperse from one habitat patch to another, and how their genes flow among subpopulations, conservation biologists try to learn how likely a population is to persist or succumb in the face of habitat change or other threats.

Island biogeography theory is a key component of conservation biology

Safeguarding habitat for species and conserving communities and ecosystems requires thinking and working at the landscape level. One key conceptual tool for doing so is the **equilibrium theory of island biogeography**. This theory, introduced by E. O. Wilson and ecologist Robert MacArthur in 1963, explains how species come to be distributed among oceanic islands. Since then, researchers have also applied it to "habitat islands"—patches of one habitat type isolated within "seas" of others. The Sikhote-Alin Mountains, last refuge of the Siberian tiger, are a habitat island, isolated from other mountains by deforested regions, a seacoast, and populated lowlands.

Island biogeography theory explains how the number of species on an island results from an equilibrium balance between the number added by immigration and the number lost through extirpation. It predicts an island's species richness based on the island's size and its distance from the mainland:

- The farther an island is located from a continent, the fewer species tend to find and colonize it; this is called the **distance effect**. Thus, remote islands host fewer species because of lower immigration rates (**FIGURE 9.18A**).
- Large islands have higher immigration rates because they present fatter targets for wandering or dispersing organisms to encounter (**FIGURE 9.18B**).

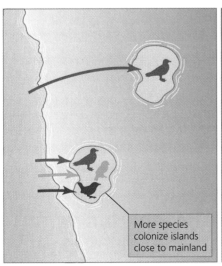
(a) Distance effect

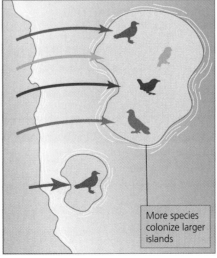
(b) Target size

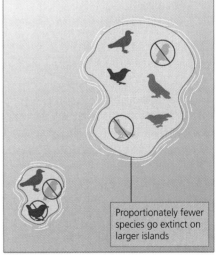
(c) Differential extinction

FIGURE 9.18
Islands located close to a continent receive more immigrants than islands that are distant **(a)**, so that near islands end up with more species.
Large islands present fatter targets for dispersing organisms to encounter **(b)**, so that more species immigrate to large islands than to small islands.
Large islands also experience lower extinction rates **(c)**, because their larger area allows for larger populations.

- Large islands have lower extinction rates because more space allows for larger populations, which are less vulnerable to dropping to zero by chance (**FIGURE 9.18C**).

Together, these last two trends give large islands more species at equilibrium than small islands—a phenomenon called the **area effect**. Large islands also tend to contain more species because they generally possess more habitats than smaller islands, providing suitable environments for a wider variety of arriving species. Very roughly, the number of species on an island is expected to double as island size increases tenfold. This effect can be illustrated with **species-area curves** (**FIGURE 9.19**).

These theoretical patterns have been widely supported by empirical data from the study of species on islands. The patterns hold up for terrestrial habitat islands, such as forests fragmented by logging and road building (**FIGURE 9.20**). Small islands of forest lose their diversity fastest, starting with large species that were few in number to begin with. In a landscape of fragmented habitat, species requiring the habitat will gradually disappear, winking out from one fragment after another over time. Fragmentation of forests and other habitats constitutes one of the prime threats to biodiversity. In response to this process of **habitat fragmentation**, conservation biologists have designed landscape-level strategies to try to optimize the arrangement of areas to be preserved. We will examine a few of these strategies in our discussion of parks and preserves in Chapter 10.

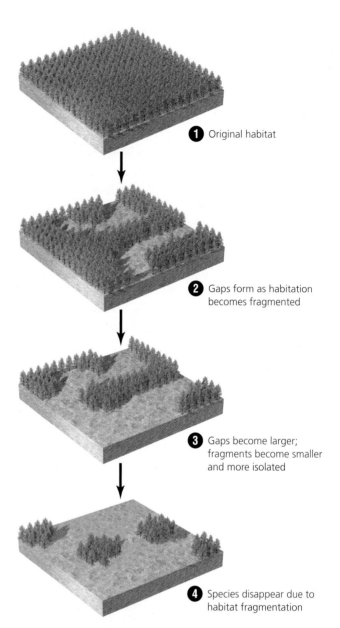

FIGURE 9.20
Forest clearing, farming, road building, and other types of human land use and development can fragment natural habitats. Habitat fragmentation usually begins when gaps are created within a natural habitat. As development proceeds, these gaps expand, join together, and eventually dominate the landscape, stranding islands of habitat in their midst. As habitat becomes fragmented, fewer populations can persist, and numbers of species in the fragments decrease with time.

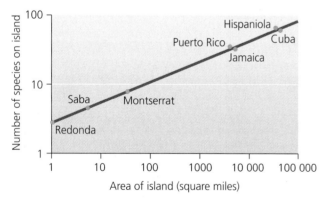

FIGURE 9.19
The larger the island, the greater the number of species—a prediction borne out by data from around the world. By plotting the number of amphibians and reptile species on Caribbean islands as a function of the areas of these islands, the species-area curve shows that species richness increases with area. The increase is not linear, but logarithmic; note the scales of the axes. Data from MacArthur, R. H., and E. O. Wilson. 1967. *The Theory of Island Biogeography*. Princeton University Press.

Captive breeding, reintroduction, and cloning are single-species approaches

In the effort to save threatened and endangered species, biologists are going to impressive lengths. Zoos and botanical gardens have become centres for the **captive breeding** of these species, so that individuals can be raised and

FIGURE 9.21
In efforts to save the California condor (*Gymnogyps californianus*) from extinction, biologists have raised hundreds of chicks in captivity with the help of hand puppets designed to look and feel like the heads of adult condors. Using these puppets, biologists feed the growing chicks in an enclosure and shield them from all contact with humans, so that when the chick is grown it does not feel an attachment to people.

reintroduced into the wild. One example is the program to save the California condor, North America's largest bird (**FIGURE 9.21**). Condors were persecuted in the early twentieth century, collided with electrical wires, and succumbed to lead poisoning from scavenging carcasses of animals killed with lead shot. By 1982, only 22 condors remained, and biologists decided to take all the birds into captivity, in hopes of boosting their numbers and then releasing them. The ongoing program is succeeding. So far, more than 100 of the 250 birds raised in captivity have been released into the wild at sites in California and Arizona, where a few pairs have begun nesting.

Other reintroduction programs have been more controversial. Reintroducing wolves to Yellowstone National Park—an effort that involved imported wolves from across the border in Canada—has proven popular with the public. However, reintroducing wolves to sites in Arizona and New Mexico met stiff resistance from ranchers who fear the wolves will attack their livestock. The program is making slow headway, and several of the wolves have been shot.

Some reintroduction programs require international cooperation (see "The Science Behind the Story: The Swift Fox Returns"). For example, the only naturally occurring wild population of whooping cranes nests and breeds each spring in Wood Buffalo Park, straddling the border between Alberta and the Northwest Territories. The flock then migrates and winters over in the Aransas National Wildlife Refuge on the gulf coast of Texas. The population of almost 2000 cranes in the late 1800s dipped to an all-time low of just 15 in 1941, after being decimated by hunting. In the past few years the population has slowly increased to about 180 birds as a result of careful reintroductions, led by the International Crane Recovery Team consisting of scientists from both Canada and the United States.[7]

9–3 FRAGMENTATION AND BIODIVERSITY

Suppose a critic of conservation tells you that human development increases biodiversity, pointing out that when a forest is fragmented, new habitats, such as grassy lots and gardens, may be introduced to an area and allow additional species to live there. How would you respond?

For the Siberian tiger, China is considering a similar reintroduction program. The Chinese government says it is preparing 600 captive Siberian tigers for release into the forests in the far northeastern portion of the country. However, critics note that the forests are so fragmented that efforts would be better focused on improving habitat first.

The newest idea for saving species from extinction is to create more individuals by cloning them. In this technique, DNA from an endangered species is inserted into a cultured egg without a nucleus, and the egg is implanted into a closely related species that can act as a surrogate mother. So far two Eurasian mammals have been cloned in this way. With future genetic technology, some scientists even talk of recreating extinct species from DNA recovered from preserved body parts. However, even if cloning can succeed from a technical standpoint, most biologists agree that such efforts are not an adequate response to biodiversity loss. Without ample habitat and protection in the wild, having cloned animals in a zoo does little good.

Some species act as "umbrellas" for protecting habitat and communities

Protecting habitat and conserving communities, ecosystems, and landscapes are the goals of many conservation biologists. Often, particular species are essentially used as tools to conserve communities and ecosystems. Species-specific legislation can provide legal justification and resources for species conservation, but no such laws exist for communities or ecosystems. Large species that roam great distances, such as the Siberian tiger and Asian and African elephants, require large areas of habitat. Meeting the habitat needs of these so-called *umbrella species* automatically helps meet those of thousands of less charismatic animals, plants, and fungi that would never elicit as much public interest.

Environmental advocacy organizations have found that using large and charismatic vertebrates as spearheads for biodiversity conservation has been an effective strategy. This approach of promoting particular "flagship species" is evident in the long-time symbol of the World Wide Fund for Nature (World Wildlife Fund in North America), the panda. The panda is a large endangered animal requiring sizeable stands of undisturbed bamboo forest. Its lovable appearance has made it a favourite with the public—and an effective tool for soliciting funding for conservation efforts that protect far more than just the panda.

---

weighing the issues

SINGLE-SPECIES CONSERVATION?   9–4

What would you say are some advantages of focusing on conserving single species, versus trying to conserve broader communities, ecosystems, or landscapes? What might be some of the disadvantages? Which do you think is the better approach, or should we use both?

At the same time, many conservation organizations today are moving beyond the single-species approach. The Nature Conservancy, for instance, has in recent years focused more on whole communities and landscapes. The most ambitious effort may be the Wildlands Project, a group proposing to restore huge amounts of North America's land to its presettlement state.

Both national and international conservation efforts are widely supported

Canada enacted its long-awaited endangered species law, the *Species at Risk Act* (*SARA*), in 2002. The federal government was careful to stress cooperation with landowners and provincial and territorial governments, rather than presenting the law as a decree from the national government. Canada's environment minister at the time, David Anderson, wanted to avoid the hostility unleashed by the "command-and-control" approach of the U.S. *Endangered Species Act* (*ESA*) some 30 years before. That hostility against *ESA* continues to this day in the United States, although the legislation can point to some significant successes in protecting endangered species. In 2007 the Centre for Biological Diversity published "100 Success Stories for Endangered Species Day," among which were the Virginia big-eared bat, California's southern sea otter, Florida's red wolf, and the Hawaiian goose, the state bird of Hawaii, which were brought back from the brink of extinction through *ESA* interventions.[8]

Environmentalists and many scientists in Canada have protested that *SARA* is too weak and fails to protect species and habitat adequately. One of the main objections it that the process of listing a species is not based principally on scientific information but is heavily influenced by both politics and economics. The process for listing a species through the *SARA* Registry, which qualifies it for legal protection under the Act, begins with monitoring and an assessment of the status of the species. This is a scientific process, carried out by the Committee on the Status of Endangered Wildlife in Canada (COSEWIC). A committee of experts produces a scientific report assigning a status to species that are thought to be at risk. The minister of the environment then responds to the assessment and designation. Between 2003 and 2006, COSEWIC recommended 186 plants and animals for *SARA* listing, of which 30 were turned down.

Arne Mooers and colleagues from Simon Fraser University have studied the final decisions made by the federal government on *SARA* listings recommended by COSEWIC and have found that science often takes a back seat to economics and politics. A 2006 study by the

THE SCIENCE BEHIND THE STORY

The Swift Fox Returns

The swift fox has been reintroduced to Canada.

The swift fox (*Vulpes velox*) is a small, slender tan-coloured fox whose natural habitat is the prairies of Manitoba, Saskatchewan, and Alberta, as well as the western grasslands of the United States. It is mainly nocturnal and is named for its speed (*velox*), as it races through the prairie grasslands at up to 60 km/h.

In the first part of the twentieth century, the swift fox disappeared entirely from the wild in Canada, mostly as a result of the loss of habitat, both in quality and in quantity. The last swift fox was captured in Canada in 1928; it was considered extirpated by 1938, and officially declared as such by COSEWIC in 1978.[9]

In 1971, Miles and Beryl Smeeton of the Wildlife Reserve of Western Canada imported two pairs of swift foxes from the United States, with the intention of starting a privately funded captive breeding and release program. Over the next decade or so the organization (which was later renamed the Cochrane Ecological Institute) developed partnerships, first with the University of Calgary, then with the Canadian Wildlife Service, and finally with the government of Saskatchewan and Fish and Wildlife Service of Alberta. Several masters' theses at the University of Calgary were undertaken to examine the logistics and plan the best approach for the reintroductions.

In 1983 the reintroduction program started to place swift foxes into parts of their native territory in Canada.[10] Between 1983 and 1997, 814 swift foxes were released into the wild. Many of the foxes that were reintroduced through the program came from wild populations in the United States; some were bred in captivity in Canada. The program was carried on with funding from a variety of sources over the years, including government agencies, nongovernmental environmental organizations, and private corporations, but in many years the funding for the program was precarious.

In the mid- to late 1980s, 155 reintroduced foxes were radio-collared and tracked by program-related researchers: 33 of the foxes had been relocated from the United States; 41 had been raised in captivity in Canada and released in the spring; and 81 had been raised in captivity and released in the fall. The researchers determined that the relocation strategy was more successful than captive breeding. The survival rate after the first year for the relocated foxes was 85%, compared with 25% for the captive-bred foxes.[11]

However, continuing to relocate wild foxes from the United States is not a sustainable option. Wild swift fox populations there are already small and facing similar challenges to those faced by the Canadian populations. Program scientists began to experiment with different release approaches for the captive-bred foxes. In a "soft" release, the young foxes are kept in an outdoor pen over the winter to acclimatize, and then released in the spring or winter. In a "hard" release, the young foxes are released straight into the wild.

The swift fox has been successfully reintroduced into parts of its natural range in Canada, where it had been declared extirpated.

Researchers determined that the hard release is as successful as the soft release, in terms of the survival rates for the animals, and much more cost-effective.[12]

According to Joel Nicholson, a biologist with Alberta Fish and Wildlife, "the swift fox reintroduction program has been one of the most successful canid reintroductions in the world."[13] A 2005–2006 survey of the swift fox population in Alberta, Saskatchewan, and Montana counted 1162 foxes, predominantly born in the wild. This is still a very low number, though. The swift fox is no longer considered extirpated, but it is still designated as an endangered species under *SARA* in Canada. Populations elsewhere in the world, however, are healthy and thriving.

---

researchers showed that animals that are of economic value (including fish) and animals from Canada's North have the least chance of being approved for legal protection under *SARA*. For example, none of the 10 species from Nunavut that were recommended for listing by COSEWIC was approved. The only marine fish that was approved for listing (of the 11 recommended) was the green sturgeon, which is considered inedible and is thus avoided by the commercial fishing industry.[14]

Today, a number of nations have laws protecting species, although they are not always well enforced. In Russia, the government issued *Decree 795* in 1995, creating a Siberian tiger conservation program and declaring the tiger one of the nation's most important natural and national treasures. However, funding from the state for tiger conservation is so meagre that the Wildlife Conservation Society feels it necessary to help pay for Russians to enforce their own anti-poaching laws.

At the international level, the United Nations has facilitated several treaties to protect biodiversity. The 1973 **Convention on International Trade in Endangered Species of Wild Fauna and Flora (CITES)** protects endangered species by banning the international transport of their body parts. When nations enforce it, CITES can protect the tiger and other rare species whose body parts are traded internationally.

In 1992, leaders of many nations agreed to the **Convention on Biological Diversity**. This treaty embodies three goals: to conserve biodiversity, to use biodiversity in a sustainable manner, and to ensure the fair distribution of biodiversity's benefits. The convention aims to help

- Provide incentives for biodiversity conservation
- Manage access to and use of genetic resources
- Transfer technology, including biotechnology
- Promote scientific cooperation
- Assess the effects of human actions on biodiversity
- Promote biodiversity education and awareness
- Provide funding for critical activities
- Encourage every nation to report regularly on its biodiversity conservation efforts

The treaty's many accomplishments so far include ensuring that Ugandan people share in the economic benefits of wildlife preserves, increasing global markets for "shade-grown" coffee and other crops grown without removing forests, and replacing pesticide-intensive farming practices with sustainable ones in some rice-producing Asian nations. As of 2007, 188 states had become parties to the Convention on Biological Diversity. Those choosing *not* to do so include Iraq, Somalia, the Vatican, and the United States. This decision is just one example of why the U.S. government is no longer widely regarded as a leader in biodiversity conservation efforts.

Other approaches highlight areas of high diversity

One international approach oriented around geographic regions, rather than single species, has been the effort to map **biodiversity hotspots**. The concept of biodiversity hotspots was introduced in 1988 by British ecologist Norman Myers as a way to prioritize regions that are most important globally for biodiversity conservation. A hotspot is an area that supports an especially great number of species that are **endemic** to the area, that is, found nowhere else in the world (**FIGURE 9.22**). To qualify as a hotspot, a location must harbour at least 1500 endemic plant species, or 0.5% of the world total. In addition, a hotspot must have already lost 70% of its habitat as a result of human impact and be in danger of losing more.

The nonprofit group Conservation International maintains a list of 34 biodiversity hotspots (**FIGURE 9.23**). The ecosystems of these areas together once covered 15.7% of the planet's land surface, but today, because of

FIGURE 9.22
The golden lion tamarin (*Leontopithecus rosalia*), a species endemic to Brazil's Atlantic rainforest, is one of the world's most endangered primates. Captive breeding programs have produced roughly 500 individuals in zoos, but the tamarin's habitat is quickly disappearing.

habitat loss, cover only 2.3%. This small amount of land is the exclusive home for 50% of the world's plant species and 42% of all terrestrial vertebrate species. The hotspot concept gives incentive to focus on these areas of endemism, where the greatest number of unique species can be protected with the least amount of effort.

The World Wide Fund for Nature (WWF) has organized its conservation efforts around the concept of the **ecoregion**, a large area of land or water with a geographically distinct assemblage of natural communities that share similar environmental conditions and ecological dynamics, and interact ecologically in ways that are critical for their long-term persistence.[15] The organization has identified a "Global 200" list of ecoregions that are priorities for conservation, including ecoregions (there are actually 238 of them on the list) from both terrestrial and marine settings.

Community-based conservation is increasingly popular

Taking a global perspective and prioritizing optimal locations to set aside as parks and reserves make good sense. However, setting aside land for preservation affects the people that live in and near these areas. In past decades,

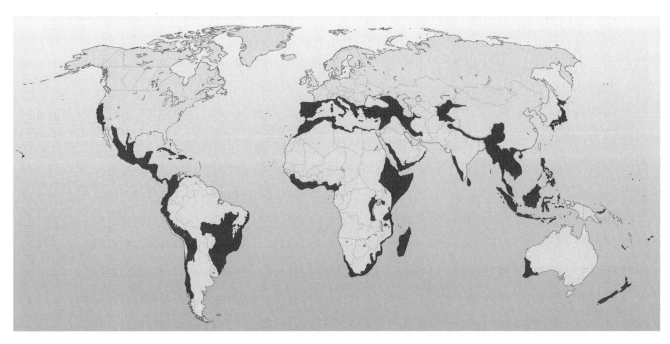

FIGURE 9.23
Some areas of the world possess exceptionally high numbers of species found nowhere else. Some conservation biologists have suggested prioritizing habitat preservation in these areas, dubbed *biodiversity hotspots*. Shown in red are the 34 biodiversity hotspots mapped by Conservation International. Data from Conservation International, 2005.

many conservationists from developed nations, in their zeal to preserve ecosystems in other nations, too often neglected the needs of people in the areas they wanted to protect. Many developing nations came to view this international environmentalism as a kind of neocolonialism.

Today this has largely changed, and many conservation biologists actively engage local people in efforts to protect land and wildlife in their own backyards, in an approach sometimes called *community-based conservation*. Setting aside land for preservation deprives local people of access to natural resources, but it can also guarantee that these resources will not be used up or sold to foreign corporations and can instead be sustainably managed. Moreover, parks and reserves draw ecotourism, which can support local economies.

In the small Central American country of Belize, conservation biologist Robert Horwich and his group, Community Conservation, Inc., have helped start a number of community-based conservation projects. The Community Baboon Sanctuary consists of tracts of riparian forest that farmers have agreed to leave intact, to serve as homes and travelling corridors for the black howler monkey, a centrepiece of ecotourism. The fact that the reserve uses the local nickname for the monkey signals respect for residents, and today a local women's cooperative is running the project. A museum was built, and residents receive income for guiding and housing visiting researchers and tourists. Community-based conservation has not always been so successful, but in a world of increasing human population, locally based management that meets people's needs sustainably will likely be essential.

Innovative economic strategies are being employed

As conservation moves from single-species approaches to the hotspot approach to community-based conservation, innovative economic strategies are also being attempted. One strategy is the *debt-for-nature swap*. In such a swap, a conservation organization raises money and offers to pay off a portion of a developing country's international debt in exchange for a promise by the country to set aside reserves, fund environmental education, and better manage protected areas.

A newer strategy that Conservation International has pioneered is the *conservation concession*. Nations often sell concessions to foreign multinational corporations, allowing them to extract resources from the nation's land. A nation can, for instance, earn money by selling to an international logging company the right to log its forests. Conservation International has stepped in and paid nations for concessions for conservation rather than resource extraction. The nation gets the money *and* keeps its natural resources intact. The South American country of Surinam, which still has extensive areas of pristine rainforest, entered into such an agreement and has virtually

CANADIAN ENVIRONMENTAL PERSPECTIVES

Biruté Mary Galdikas

Canadian Biruté Mary Galdikas is an advocate for endangered orangutans in the rainforests of Indonesia.

- **Biological anthropologist**
- **Field primatologist**
- **Conservationist** and **orangutan advocate**

They are sometimes called the "trimates"—three women who made such groundbreaking expansions upon our understanding of primate behaviour that they practically carved out their own private niche in the world of science. All three—Jane Goodall, who studies chimpanzees and works to conserve their habitat; Dian Fossey, who worked with mountain gorillas; and Biruté Mary Galdikas, whose research provided the baseline for what we know today about orangutans—were hand-picked and funded originally by Louis Leakey, the famous paleontologist whose work helped uncover the prehistory of our own species. Leakey was convinced that the women would make great strides in "field primatology," the observation of our closest genetic relatives in their own natural habitat, because of their comparative lack of scientific training (an "uncluttered mind") and what he felt was the greater patience and empathy of female researchers. For whatever reason, Leakey was right.

Galdikas was born in Germany but grew up as a naturalized Canadian in Toronto. She met Leakey in California in 1969, several years after Goodall and Fossey had begun to establish their names and unconventional research techniques. She convinced him upon their first meeting that she deserved a place among them, and he set about gathering the funds needed to sponsor her first research outing to the rainforests of Indonesia.[16] Once there, Galdikas had an even tougher time than her predecessors. The approach of all three was to allow the primates to gradually become accustomed to their presence, so their natural behaviour could be observed. It was a tall order for Goodall and Fossey, but even more so for Galdikas. Orangutans are more solitary than chimps or gorillas; not only did they not welcome Galdikas, but they actively tried to dissuade her from approaching by throwing things at her and even defecating on her. It took 12 long years for Galdikas to habituate one of the orangutans to her presence.

Her patience paid off in scientific results. In her Ph.D. thesis for UCLA in 1978, Galdikas documented a number of behaviours that had never before been witnessed in wild orangutans, including the observation that male and female orangutans form relationships that last for extended periods. She later made observations on tool use among orangutans that would come to be considered as classic research. For example, in the journal *Science* she documented her observation that wild orangutans spontaneously use tools for a number of different purposes. It had previously been thought that only captive orangutans—influenced by constant exposure to humans—were tool users.[17] This provided a basis for later research suggesting that some orangutan groups in the wild have developed basic "cultural" differences, such as having specific sounds for communication with members of the group.

Today Galdikas has part-time affiliations with Simon Fraser University in British Columbia and Universitas Nasional in Jakarta, Indonesia. Like Goodall and Fossey (until the latter's death in 1985), she devotes most of her time to advocacy for the critically endangered orangutans, and conservation of their rapidly disappearing habitat—the rainforests of Borneo and Sumatra. Her group, Orangutan Conservation International, runs a rescue and rehabilitation centre that saves young orangutans orphaned by fire, deforestation, or hunting, nurses them back to health, and returns them to the wild.[18]

"To follow them, you would have had to just jump in the swamp, which was neck-deep there, and that's when I thought, 'Gee, this is going to be really hard.'" —Biruté Mary Galdikas, on her first encounter with orangutans in the rainforest of Indonesia.

Thinking About Environmental Perspectives

In June of 2008, Spain's Parliament voted in favour of new legislation that would extend, for the first time ever, certain limited rights to great apes. The proposed laws make it illegal to kill, torture, or arbitrarily imprison apes, including their use in medical experimentation, circuses, and films. What do you think of this? Is it "about time" for a law of this type, to provide protection to our closest genetic relatives? Or is it an ill-conceived attempt to extend "human" rights to nonhumans?

---

halted logging while pulling in $15 million. It remains to be seen how large a role such strategies will play in the future protection of biodiversity.

Conclusion

The erosion of biological diversity on our planet threatens to result in a mass extinction event equivalent to the mass extinctions of the geological past. Human-induced habitat alteration, invasive species, pollution, and overharvesting of biotic resources are the primary causes of biodiversity loss. This loss matters, because human society could not function without biodiversity's pragmatic benefits. As a result, conservation biologists are rising to the challenge of conducting science aimed at saving endangered species, preserving their habitats, restoring populations, and keeping natural ecosystems intact. The innovative strategies of these scientists hold promise to slow the erosion of biodiversity that threatens life on Earth.

REVIEWING OBJECTIVES

You should now be able to:

Characterize the scope of biodiversity on Earth

- Biodiversity can be thought of at three levels, commonly called species diversity, genetic diversity, and ecosystem diversity.
- Roughly 1.7 million to 2.0 million species have been described so far, but scientists agree that the world holds millions more.
- Some taxonomic groups (such as insects) hold far more diversity than others.
- Diversity is unevenly spread across different habitats and areas of the world.

Describe ways to measure biodiversity

- Global estimates of biodiversity are based on extrapolations from scientific assessments in local areas and certain taxonomic groups.

Contrast background extinction rates and periods of mass extinction

- Species have gone extinct at a background rate of roughly 1 species per 1 million to 10 million species each year. Most species that have ever lived are now extinct.
- Earth's life has experienced five mass extinction events in the past 440 million years.
- Human impact is presently causing the beginnings of a sixth mass extinction.

Evaluate the primary causes of biodiversity loss

- Habitat alteration is the main cause of current biodiversity loss. Invasive species, pollution, and overharvesting are also important causes. Climate change threatens to become a major cause very soon.

Specify the benefits of biodiversity

- Biodiversity is vital for functioning ecosystems and the services they provide us.
- Wild species are sources of food, medicine, and economic development.
- Many people feel humans have a psychological need to connect with the natural world.

Assess conservation biology and its practice

- Conservation biology is an applied science that studies biodiversity loss and seeks ways to protect and restore biodiversity at all its levels.

Explain island biogeography theory and its application to conservation biology

- Island biogeography theory explains how size and distance influence the number of species occurring on islands.
- The theory applies to terrestrial islands of habitat in fragmented landscapes.

Compare and contrast traditional and more innovative biodiversity conservation efforts

- Most conservation efforts and laws so far have focused on threatened and endangered species. Efforts include captive breeding and reintroduction programs.
- Species that are charismatic and well known are often used as tools to conserve habitats and ecosystems. Increasingly, landscape-level conservation is being pursued in its own right.
- International conservation approaches include treaties, biodiversity hotspots, community-based conservation, debt-for-nature swaps, and conservation concessions.

TESTING YOUR COMPREHENSION

1. What is biodiversity? List and describe three levels of biodiversity.
2. What are the five primary causes of biodiversity loss? Can you give a specific example of each?
3. List and describe five invasive species and the adverse effects they have had.
4. Define the term *ecosystem services*. Give five examples of ecosystem services that humans would have a hard time replacing if their natural sources were eliminated.
5. What is the relationship between biodiversity and food security? Between biodiversity and pharmaceuticals? Give three examples of potential benefits of biodiversity conservation for food security and medicine.
6. Describe four reasons why people suggest biodiversity conservation is important.

7. What is the difference between an umbrella species and a keystone species? Could one species be both an umbrella species and a keystone species?
8. Explain the theory of island biogeography. Use the example of the Siberian tiger to describe how this theory can be applied to fragmented terrestrial landscapes.
9. What is a biodiversity hotspot?
10. Describe community-based conservation.

SEEKING SOLUTIONS

1. In one of the quotations that open this chapter, biologist E. O. Wilson argues that "Except in pockets of ignorance and malice, there is no longer an ideological war between conservationists and developers. Both share the perception that health and prosperity decline in a deteriorating environment." Do you agree or disagree? How do people in your community view biodiversity?
2. Many arguments have been advanced for the importance of preserving biodiversity. Which argument do you think is most compelling, and why? Which argument do you think is least compelling, and why?
3. Some people argue that we shouldn't worry about endangered species because extinction has always occurred. How would you respond to this view?
4. Compare the biodiversity hotspot approach with the approach of community-based conservation. What are the advantages and disadvantages of each? Can we—and should we—follow both approaches?
5. **THINK IT THROUGH** You are an influential legislator in a country that has no endangered species law, and you want to introduce legislation to protect your country's vanishing biodiversity. Consider the Canadian *Species at Risk Act*, as well as international efforts, such as CITES and the Convention on Biological Diversity. What strategies would you write into your legislation? How would your law be similar to and different from the existing Canadian and international efforts?
6. **THINK IT THROUGH** As a citizen and resident of your community, and a parent of two young children, you attend a town meeting called to discuss the proposed development of a shopping mall and condominium complex. The development would eliminate a 40 ha stand of forest, the last sizeable forest stand in the town. The developers say the forest's loss will not matter because plenty of smaller stands still exist scattered throughout town. One of the town's decision makers recognizes you as someone who hikes, bird watches, and fishes with your children in the forest, and she asks you to comment about the development's possible impacts on the community's biodiversity. What will you choose to tell your fellow citizens and the town's decision makers at this meeting?

INTERPRETING GRAPHS AND DATA

Habitat alteration is the primary cause of present-day biodiversity loss. Of all human activities, the one that has resulted in the most habitat alteration is agriculture. Between 1850 and 2000, 95% of the native grasslands of the Midwestern United States were converted to agricultural use. As a result, conventional farming practices replaced diverse natural communities with greatly simplified ones. The vast monocultures of industrialized agriculture produce bountiful harvests, but at substantial costs in lost ecosystem services.

Data from a recent study reviewing the scientific literature on the effects of organic farming practices on biodiversity are shown in the graph.

1. Overall, how many studies showed a positive effect of organic farming on biodiversity, relative to conventional farming? How many studies reported a negative effect? How many studies reported no effect?

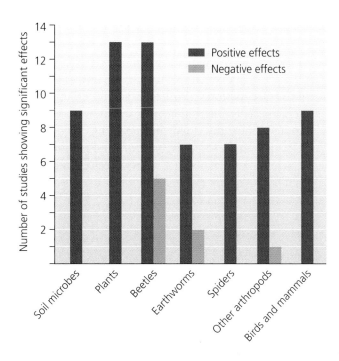

Numbers of scientific studies reporting negative or positive effects on biodiversity of organic agriculture versus conventional farming practices. In parentheses are numbers of studies reporting no effect. Data from Hole, D., et al. 2005. Does organic farming benefit biodiversity? *Biological Conservation* 122:113–130.

2. For which group or groups of organisms is evidence of positive effects the strongest? Reference the numbers to support your choice(s).
3. Recall the ecosystem services provided by biodiversity. What services do the groups you chose in Question 2 provide?

CALCULATING FOOTPRINTS

Of the five major causes of biodiversity loss discussed in this chapter, habitat alteration arguably has the greatest impact. In their 1996 book introducing the ecological footprint concept, authors Mathis Wackernagel and William Rees present a consumption/land-use matrix for an average North American. Each cell in the matrix lists the number of hectares of land of that type required to provide for the different categories of a person's consumption (food, housing, transportation, consumer goods, and services). Of the 4.27 ha required to support this average person, 0.59 ha are forest, with most (0.40 ha) being used to meet the housing demand. Using this information, calculate the missing values in the table.

1. Approximately two-thirds of the forests' productivity is consumed for housing. To what use(s) would you speculate that most of the other third is put?
2. If the harvesting of forest products exceeds the sustainable harvest rate, what will be the likely consequence for the forest?
3. What will be the impact of deforestation, or of the loss of old-growth forests and their replacement with plantations of young trees, on the species diversity of the forest community? In your answer, discuss the possibilities of both extirpation and extinction.

	Hectares of forest used for housing	Total forest hectares used
You	0.40	0.59
Your class		
Your province		
Canada		

Data from Wackernagel, M., and W. Rees. 1996. Our Ecological Footprint: Reducing Human Impact On The Earth. *British Columbia, Canada: New Society Publishers.*

TAKE IT FURTHER

 Go to www.myenvironmentplace.ca where you will find

- Suggested answers to end-of-chapter questions
- Quizzes, animations, and flashcards to help you study
- *Research Navigator*™ database of credible and reliable sources to assist you with your research projects
- Tutorials to help you master how to interpret graphs
- Current news articles that link the topics that you study to case studies from your region and around the world

- **ECO Occupational Profiles:** If you found this chapter especially interesting, you might want to learn more about the following jobs by visiting the Occupational Profiles website of the Environmental Careers Organization. Go to www.eco.ca and check out the following careers:
  - Conservation biologist
  - Conservation officer
  - Naturalist
  - Park interpreter

CHAPTER ENDNOTES

1. Environment Canada, 1996, Glossary of Selected Terms: genetic diversity, www.environment-canada.ca/soer-ree/English/SOER/1996report/Doc/1-10-1.cfm?StrLetter=G.
2. Millennium Ecosystem Assessment (2005) *Ecosystems and Human Well-Being: Synthesis.* Washington, DC: Island Press.
3. National General Status Working Group (2006) *Wild Species 2005: The General Status of Species in Canada,* www.wildspecies.ca/wildspecies2005/index.cfm?lang=e.
4. National General Status Working Group (2006) *Wild Species 2005: The General Status of Species in Canada,* www.wildspecies.ca/wildspecies2005/index.cfm?lang=e.
5. National General Status Working Group (2006) *Wild Species 2005: The General Status of Species in Canada,* www.wildspecies.ca/wildspecies2005/index.cfm?lang=e.
6. Millennium Ecosystem Assessment (2005) *Ecosystems and Human Well-Being: Synthesis.* Washington, DC: Island Press.
7. Whooping Crane Eastern Partnership, www.bringbackthecranes.org/back/proj-facts.htm.
8. Center for Biological Diversity, 2007, *The Road to Recovery: 100 Success Stories for Endangered Species Day,* www.esasuccess.org/reports/.
9. *Swift Fox,* Species at Risk Public Registry, Government of Canada, www.sararegistry.gc.ca/species/speciesDetails_e.cfm?sid=140.
10. Weagle, Ken, and Clio Smeeton (1997) *Captive breeding of Swift Fox for reintroduction: Final report to funding bodies 1994 to 1997.* Cochrane, AB: Cochrane Ecological Institute, www.ceinst.org/Final%20Report%20to%20Funders%2097.pdf.
11. Cotterill, S. E. (1997) Status of the Swift Fox (*Vulpes velox*) in Alberta. Alberta Environmental Protection, Wildlife Management Division, *Wildlife Status Report No. 7,* Edmonton, AB. 17 pp.
12. Cotterill, S. E. (1997) Status of the Swift Fox (*Vulpes velox*) in Alberta. Alberta Environmental Protection, Wildlife Management Division, *Wildlife Status Report No. 7,* Edmonton, AB. www.srd.gov.ab.ca/fishwildlife/status/swfox/cons.html.
13. Nature Conservancy of Canada, *The Endangered Swift Fox,* www.natureconservancy.ca/site/News2?abbr=ncc_work_&page=NewsArticle&id=5047.
14. Mooers, A. Ø., L. R. Prugh, M. Festa-Bianchet, and J. A. Hutchings (2006) Biases in legal listing under Canadian endangered species legislation, *Conservation Biology,* Vol. 21, No. 3, pp. 572–575.
15. Worldwide Fund for Nature: Ecoregions, www.worldwildlife.org/science/ecoregions/item1847.html.
16. Morell, Virginia (1993) Called 'trimates,' three bold women shaped their field, *Science,* Vol. 260, No. 5106 (April 16), pp. 420–425.
17. Galdikas, Biruté M. F. (1989) Orangutan tool use, *Science,* Vol. 243, No. 4888 (January 13), pp. 152–152.
18. Dan Ferber (2000) Orangutans face extinction in the wild, *Science,* Vol. 288, No. 5469 (May 19), pp. 1147, 1149–1150.

10 Forests and Land Management

Clear cutting is changing the visual landscape of much of British Columbia.

Upon successfully completing this chapter, you will be able to

- Identify the principles, goals, and approaches of forest resource management
- Summarize the ecological roles and economic contributions of forests, and outline the history and scale of forest loss
- Explain the fundamentals of forest management, and describe the major methods of harvesting timber
- Identify major federal land management agencies and the lands they manage
- Recognize types of parks and reserves, and evaluate issues involved in their design

Anti-logging protestors rallied at Clayoquot Sound in 1993.

CENTRAL CASE:
BATTLING OVER THE LAST BIG TREES AT CLAYOQUOT SOUND

"Clear-cutting... may be either desirable or undesirable, acceptable or unacceptable, according to the type of forest and the management objectives."
—DR. HAMISH KIMMINS, UNIVERSITY OF BRITISH COLUMBIA, 1992

"What we have is nothing less than an ecological Holocaust occurring right now in British Columbia."
—MARK WAREING, WESTERN CANADA WILDERNESS COMMITTEE, 1990

It was the largest act of civil disobedience in Canadian history, and it played out along a seacoast of majestic beauty, at the foot of some of the world's biggest trees. At Clayoquot Sound on the western coast of Vancouver Island, British Columbia, protestors blocked logging trucks, preventing them from entering stands of ancient temperate rainforest. The activists chanted slogans, sang songs, and chained themselves to trees.

Loggers complained that the protestors were keeping them from doing their jobs and making a living. In the end, 850 of the 12 000 protestors were arrested, and this remote, mist-enshrouded land of cedars and hemlocks became ground zero in the debate over how we manage forests.

That was in 1993, and the activists were opposing *clear-cutting*, the logging practice that removes all trees from an area. Most of Canada's old-growth temperate rainforest had already been cut, and the forests of Clayoquot Sound were among the largest undisturbed stands of temperate rainforest left on the planet.

Timber from **old-growth forests**—complex primary forests in which the trees are at least 150 years old—had long powered British Columbia's economy.

Historically, one in five jobs in B.C. depended on its $13 billion timber industry, and many small towns would have gone under without it. By 1993, however, the timber industry was cutting thousands of jobs a year because of mechanization, and the looming depletion of old growth threatened to slow the industry. Meanwhile, Greenpeace was convincing overseas customers to boycott products made from trees clear-cut by multinational timber company MacMillan Bloedel. Soon British Columbia's premier found himself trying to persuade European nations not to boycott his province's main export.

In 1995, the provincial government called for an end to clear-cutting at Clayoquot Sound, after its appointed scientific panel of experts submitted a new forestry plan for the region. The plan recommended reducing harvests, retaining 15%–70% of old-growth trees in each stand, decreasing the logging road network, designating forest reserve areas, and managing *riparian* (water's-edge) zones. Two years later, the provincial government reversed many of these regulations, and a new premier pronounced forest activists "enemies of British Columbia."

The antagonists struck a deal; wilderness advocates and MacMillan Bloedel agreed to log old growth in limited areas, using environmentally friendly practices. In 1998, First Nations people of the region formed a timber company, Iisaak Forest Resources, in agreement with MacMillan Bloedel's successor, Weyerhaeuser, and began logging in Clayoquot Sound in a more environmentally sensitive manner (see photo).

In the Nuu-chah-nulth language of First Nations people from the Clayoquot Sound area, *iisaak* (pronounced E-sock) means "respect," which became a guiding principle for forestry in Clayoquot Sound. The **variable retention harvesting** they applied—logging selectively with the goal of retaining a certain percentage and particular characteristics of the forest ecosystem—is more expensive than normal clear-cutting. Iisaak Forest Resources hoped to recoup some of the extra cost by achieving a premium price for the cut timber and through ecotourism and the sustainable exploitation of other forest resources.

Leaving most of the trees standing accomplished what forest advocates had predicted: People from all

Logging continues in some parts of the Clayoquot Sound region.

over the world—1 million each year—are now visiting Clayoquot Sound for its natural beauty and kayaking and whale watching in its waters. Ecotourism (along with fishing and aquaculture) has surpassed logging as a driver of local economies. The United Nations designated the site as a biosphere reserve in 2000, encouraging land protection and sustainable development. From the perspective of ecotourism, the trees appear to be worth more standing than cut down.

Tensions continue today, however. Logging has never completely stopped in Clayoquot Sound—even in areas near park and biosphere reserve boundaries—to the dismay of environmental activists.[1] Local forest advocates worry that the provincial government's new Working Forest Policy will increase logging, and the town of Tofino petitioned the province to exempt Clayoquot Sound's forests from the policy.

Ultimately, Iisaak Forest Resources found it difficult to make money doing sustainable forestry and entered into an agreement with the environmental organization Ecotrust Canada in 2006. Today logging is being done more sustainably, and at a profit, under this arrangement. The provincial government is considering new forestry plans that would shift logging out of old-growth forests and into younger forests that were already logged in the past. As long as our demand for lumber, paper, and forest products keeps increasing, pressures will keep building on the remaining forests on Vancouver Island and around the world.

Forest Ecosystems and Their Status Worldwide

Forests cover roughly 30% of Earth's land surface (**FIGURE 10.1**). They provide habitat for countless organisms and help maintain soil, air, and water quality. They play key roles in our planet's biogeochemical cycles, serving as one of the most important reservoirs in the carbon cycle. Forests have also long provided humanity with wood for fuel, construction, paper production, and more.

There are three major groups of forest biomes

There are three major types of forest biomes, corresponding roughly to the high, middle, and equatorial latitudes. However, there are *many* local variations, as well as altitudinal variations; some of these were discussed in Chapter 4, and we will look more closely at the forests of Canada specifically later in this chapter.

1. The *boreal forest* is a high-latitude forest type that is characterized by cold, relatively dry climates with short growing seasons. The boreal forest biome stretches across much of Canada, Russia, and Scandinavia. Farther to the north, the boreal forest grades into the more open *tundra* biome.

2. The *temperate forest* occurs in mid-latitude areas of seasonal climate, which typically experience a distinct winter season and summer growing season. Temperate forests occur throughout eastern North America, northeastern Asia, and western and central Europe. Temperate forests cover much less area globally than boreal forests, in part because people have already cleared so many of them.

3. *Tropical forests*, which host extremely diverse flora and fauna, occur in the wet, tropical climates of equatorial South and Central America, equatorial Africa, and Indonesia and Southeast Asia. For rainforests we tend to think first of the tropical environment, but in fact their central characteristic is not temperature but rainfall. Thus, variations include the evergreen rainforests of the cool, wet Pacific northwest. Similarly, there are tropical forests in which the climate is warm but not wet year-round, alternating instead between a rainy season and a dry season. As the dry season gets longer, the character of the forest changes and other wooded biome types emerge, such as the more open tropical dry forest and savannah biomes.

These broad descriptions of the major types of forest biomes should make it obvious that not all forests are completely dominated by trees. A **forest**, strictly speaking, is a land area with significant tree cover, in which the **canopy** (the upper level of leaves and branches defined by the treetops) is largely **closed**. A **woodland** is a wooded

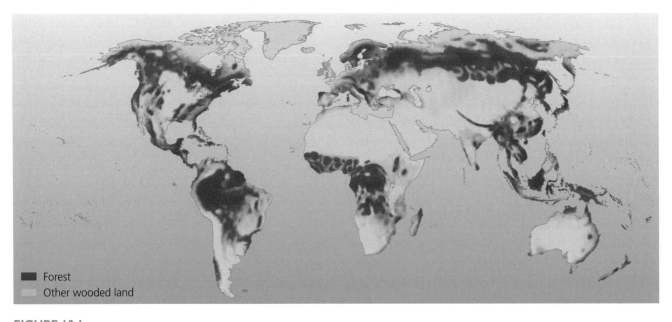

FIGURE 10.1
About 30% of Earth's land surface is covered by forest. Most of this consists of the boreal forests of the north and the tropical forests of South America and Africa. Other lands (including tundra, shrubland, and savannah) can be classified as "wooded land," implying a more open forest type that supports trees, but at sparser densities. Data from U.N. Food and Agriculture Organization (FAO). 2005. *Global forest resources assessment*.

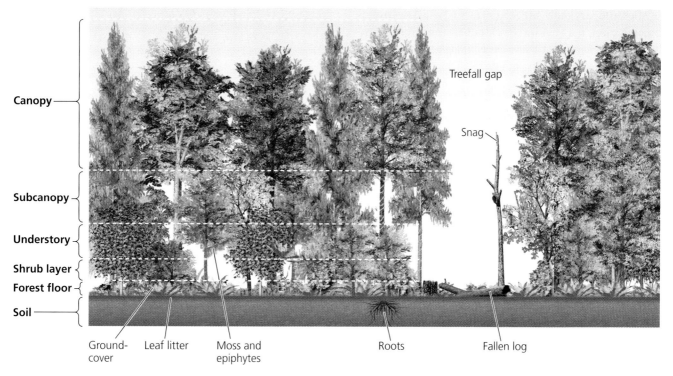

FIGURE 10.2
Mature forests are complex ecosystems. In this cross-section of a generalized mature forest, the crowns of the largest trees form the canopy, and smaller trees beneath them form the shaded subcanopy and understory. Shrubs and groundcover grow just above the forest floor, which may be covered in leaf litter rich with invertebrate animals. Vines, mosses, lichens, and epiphytes cover portions of trees and the forest floor. Snags (standing dead trees), whose wood can easily be hollowed out, provide food and nesting and roosting sites for woodpeckers and other animals. Fallen logs nourish the soil and young plants and provide habitat for countless invertebrates as the logs decompose. Treefall gaps caused by fallen trees let light through the canopy and create small openings in the forest, allowing early successional plants to grow in patches within the mature forest.

(treed) area in which the canopy is more **open**, that is, there are some openings between the trees that allow light to penetrate to the ground, or **floor**, of the forest (**FIGURE 10.2**).

Open-canopy wooded lands and grasslands are drier terrestrial ecosystems

At the drier end of the climatic spectrum (in both cold high-latitude regions and warm low-latitude regions), the canopies of wooded areas tend to be even more open. **Shrublands** are wooded areas that are covered by smaller, bushier trees, or *shrubs*, often interspersed with occasional taller trees. *Tundra* is a high-latitude (and high-altitude) version of shrubland. **Savannah** is an open area dominated by grasses, with widely scattered trees. Finally, **grasslands** are lands that are dominated by grasses and other non-woody vegetation. Again, all these basic biome types have many local variations (**FIGURE 10.3**).

Grasslands, savannahs, and even shrublands are not forests, strictly speaking, although any of them can be partially "wooded." It is common to group these biome types together under the category of **drylands**, emphasizing their central defining characteristic of low precipitation. This is a broad category that includes some areas with a relatively long dry season alternating with a rainy season; some semi-arid regions that experience low precipitation year-round; and—at the extreme end of the dryland spectrum—the arid *desert* biome.

FIGURE 10.3
Savannahs and grasslands are characterized by a relatively dry climate and dispersed trees.

Because drylands are characterized by low overall precipitation, they tend to be extremely sensitive to environmental change and are easily damaged if land use practices become overly intensive. Therefore, *desertification* and *land degradation* are major environmental issues in dryland management. Much of the world's grassland and dry woodland has been converted for the purpose of agriculture or rangeland. We will discuss this in greater detail below.

Canada is a steward for much of the world's forest

Canada's current 402 million hectares of forested and other wooded land (310 million hectares of which is "true" forest)[2] represents more than 10% of the world's forest cover, 40% of Canada's total land area, 25% of the world's natural (rather than planted) forest, 30% of the world's boreal forest, and 20% of the world's temperate rainforest, and includes some of the world's largest intact forest ecosystems.[3] About one-third of the nation's forests are in British Columbia, and 38% are in Quebec and Ontario. According to the National Forest Inventory, about 1.5 million hectares of the wooded land in Canada is grassland.[4]

According to the Food and Agricultural Organization of the United Nations, about 53.3% of Canada's primary forest remains more or less intact.[5] In comparison, of the original 4 million square kilometres (400 million hectares) of forested land in the United States, the vast majority was deforested by the late nineteenth century. (In the early 1900s, forest cover in the United States began to stabilize, however, and in the past few decades has actually increased.)

Canada clearly has an obligation to the rest of the world to manage its forests as effectively and sustainably as possible. Canada's forest biomes (also discussed in Chapter 4) include many regional variations, some of which are described next.

Forests of the north The boreal forest, or *taiga*, the largest forested region of Canada, stretches through all of the provinces and territories except Nova Scotia and Prince Edward Island (**FIGURE 10.4A**). White spruce, tamarack, jack pine, and lodgepole pine are the main coniferous species in the boreal forest, and white birch, aspen, and balsam poplar are the main deciduous trees. In the north the boreal forest merges with the tundra, an open woodland biome.

Forests of the west In the west is the *subalpine forest region* of the mountains of British Columbia and western Alberta, with characteristic Engelmann spruce, alpine fir, and lodgepole pine, and the *montane forest region* in

(a) Northern boreal forest

(b) Western ponderosa pine forest

(c) Eastern deciduous forest

FIGURE 10.4
Canada's forest biomes include many regional variations.

British Columbia's central plateau, with Rocky Mountain Douglas fir, lodgepole pine, trembling aspen, and ponderosa pine (**FIGURE 10.4B**). The *coast forest region*, found in Clayoquot Sound and elsewhere on Vancouver Island, is the temperate rainforest, characterized by western red cedar, western hemlock, Sitka spruce, yellow cypress, and deciduous big-leaf maple, red alder, cottonwood, Garry oak, and arbutus. The *Columbia forest region* of the Kootenay, Thompson, and Fraser river valleys includes species like western white pine, Engelmann spruce, western larch, and grand fir.

Forests of the east In the east, the *deciduous forest region* north of Lake Erie and Lake Ontario is the smallest forest region in Canada, characterized by deciduous species, such as sugar maple, beech, elm, and oak, and conifers like the eastern white pine and eastern hemlock (**FIGURE 10.4C**). In this region there are also pockets of "Carolinian" species, like the tulip tree and black gum, which are more common farther to the south in the eastern United States. The *Great Lakes–St. Lawrence forest region* extends from northwestern New Brunswick, through the St. Lawrence, Lac St. Jean, and Saguenay river valleys, over southern and central Ontario, and into Manitoba. Typical conifers include the eastern white and red pine and eastern hemlock. The characteristic deciduous species is yellow birch. Finally, the *Acadian forest region* of Nova Scotia, New Brunswick, and Prince Edward Island is typified by spruce and balsam fir, with common deciduous sugar maple, yellow birch, and beech.[6]

Forests are ecologically valuable

Because of their structural complexity and their ability to provide many niches for organisms, forests are some of the richest ecosystems for biodiversity. Trees furnish food and shelter for an immense diversity of vertebrate and invertebrate animals. Countless insects, birds, mammals, and other organisms subsist on the leaves, fruits, and seeds that trees produce.

Some animals are adapted for living in the dense treetop canopy, where beetles, caterpillars, and other leaf-eating insects abound, providing food for birds, such as tanagers and warblers, while arboreal mammals from squirrels to sloths to monkeys consume fruit and leaves. Other animals specialize on the *subcanopies* of trees, and still others utilize the bark, branches, and trunks. Cavities in trunks provide nest and shelter sites for a wide variety of vertebrates. Dead and dying trees are valuable for many species; these *snags* are decayed by insects that, in turn, are eaten by woodpeckers and other animals.

Meanwhile, the shrubs and groundcover plants of the **understory**, the forest floor and the lowest levels of growth, give a forest structural complexity and provide habitat for still more organisms. Moreover, the leaves, stems, and roots of forest plants are colonized by an extensive array of fungi and microbes, in both parasitic and mutualistic relationships. And much of a forest's diversity resides in the forest floor, where the soil is generally nourished by fallen leaves and branches, called *litter*. As we saw in Chapter 7, myriad soil organisms help decompose plant material and cycle nutrients.

In general, forests with a greater diversity of plants, such as tropical rainforests, host a greater diversity of organisms overall. And in general, fully mature forests, such as the undisturbed old-growth forests remaining at Clayoquot Sound, contain more biodiversity than younger forests. Older forests contain more structural diversity and thus more microhabitats and resources to support more species.

Forests are a crucial link in nutrient and water cycles

In addition to hosting a significant proportion of the world's biodiversity, forests provide all manner of vital ecosystem services. Forest vegetation stabilizes soil and prevents erosion. The principal direct cause of soil erosion and degradation is the removal of vegetation. This is especially true in tropical rainforests where, counterintuitively, soils are not particularly fertile because most of the biomass of the system resides in the trees and other forest plants. Once the trees have been removed, the thin soil is exposed to wind and water and can quickly erode.

Trees and other forest plants help regulate the hydrologic cycle, slowing runoff, lessening flooding, and purifying water as they take it in from the soil and release it to the atmosphere. Forests thus function as a link among the biogeochemical cycles of the atmosphere, hydrosphere, biosphere, and geosphere. They draw mineral nutrients and water from depth through their root systems, delivering them to near-surface soil layers, where they become available for other plants (**FIGURE 10.5**). They also deliver organic material to the topsoil, in the form of **litter**, the fallen branches and leaves of the trees. Forests also store carbon, release oxygen, and act as moderating influence on climate. By performing such ecological functions, forests are indispensable for our survival.

Forest products are economically valued

In addition to the immense value of their ecological services, forests also provide people with economically

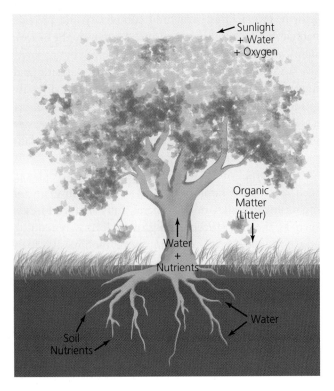

FIGURE 10.5
Trees perform many ecological services. One of the most important is drawing nutrients and water from depth through their root systems. In this way, they contribute to biogeochemical cycling.

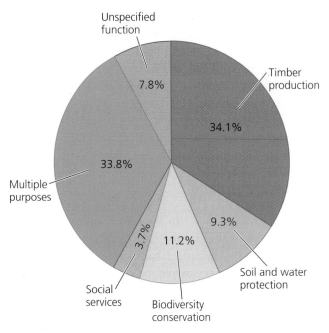

FIGURE 10.6
Worldwide, nations designate more than one-third of forests primarily for production of timber and other forest products. Smaller areas are designated for conservation of biodiversity, protection of soil and water quality, and "social services," such as recreation, tourism, education, and conservation of culturally important sites. About one-third of forests are designated for combinations of these functions. Data from U.N. Food and Agriculture Organization (FAO). 2005. *Global forest resources assessment.*

valuable wood products. For millennia, wood from forests has fuelled our fires, keeping people warm and well fed. It has housed people, keeping us sheltered. It built the ships that carried people and cultures from one continent to another. It allowed us to produce paper, the medium of the first information revolution.

In recent decades, industrial harvesting has allowed the extraction of more timber than ever before, supplying all these needs of a rapidly growing human population and its expanding economy. The exploitation of forest resources has been instrumental in helping our society achieve the standard of living we enjoy today. Indeed, without industrial timber harvesting, you would not be reading this book. Most commercial logging today takes place in Canada, Russia, and other nations that hold large expanses of boreal forest, and in tropical countries with large amounts of rainforest, such as Brazil and Indonesia. Timber harvested from coniferous trees (such as those that dominate the boreal forest) is called **softwood**, whereas timber that comes from deciduous trees is called **hardwood**. (The terms are not related to the actual hardness of the wood.) The softwood lumber industry is extremely important to Canada's economy.

Forests also supply non-wood products in abundance. Some of these *non-timber forest products* (*NTFPs*) include medicinal and herbal products, such as ginseng, echinacea, and St. John's wort; decorative products, such as wreaths and other greenery; and many edible products, including fruits, honey, truffles, and nuts. Many Aboriginal and indigenous people make their livelihoods by harvesting non-timber forest products. The *seringeiro* rubbertappers of the Brazilian Amazon are one example of a group of people whose lifestyle is adapted to the sustainable extraction of forest resources.

Nations maintain and use forests for all these economic and ecological reasons. An international survey in 2005 found that globally, more than one-third of forests were designated primarily for timber production. Others were designated for a variety of functions, including conservation of biodiversity, protection of soil and water quality, and "social services," such as recreation, tourism, education, and conservation of culturally important sites (**FIGURE 10.6**).

Land Conversion

The conversion of forested lands and grasslands for other purposes is nothing new. We all depend in some way on wood, and people have cleared forests for millennia to exploit timber resources. Historically, as agriculture emerged and some cultures began to adopt a *sedentary* or settled lifestyle, the clearing of forested land for farming

FIGURE 10.7
Huge trees were harvested in many locations in the United States and Canada in the first part of the nineteenth century, including the pines shown here in the Madawaska River Valley near Ottawa. Early timber harvesting practices in North America caused significant environmental impacts and removed virtually all the virgin timber from one region after another.

would have been one of the very first significant human-generated environmental impacts. Throughout human history, forests have been cleared to build settlements, harvest forest products, and establish farms. **Deforestation**, the loss of forested land, has by now altered the landscapes and ecosystems of much of our planet.

The growth of Canada and the United States was fuelled by land clearing and logging

When we think of deforestation today, we tend to think of other parts of the world. Historically, however, logging for timber and the clearing of land for farming propelled the growth of both Canada and the United States throughout the phenomenal expansion westward across the North American continent over the past 400 years. The vast deciduous forests of the eastern United States were virtually stripped of their trees by the mid-nineteenth century, making way for countless small farms. Timber from these forests built the cities of eastern North America.

As the farming economy shifted to an industrial one, wood was used to stoke the furnaces of industry. Once most of the mature trees were removed from the eastern hardwood forests, timber companies moved to the south and west, eventually harvesting some of the continent's biggest trees in the Rocky Mountains, the Sierra Nevada, and the Pacific Coast ranges (**FIGURE 10.7**).

By the early twentieth century, very little **primary forest**—natural forest uncut by people—was left in the United States. Today, the largest oaks and maples found in eastern North America, and even most redwoods of the California coast, are **second-growth** trees: trees that have sprouted and grown to partial maturity after old-growth timber has been cut. The size of the gargantuan trees they replaced can be seen in the enormous stumps that remain in the more recently logged areas of the Pacific coast. The scarcity of old-growth trees on the North American continent today explains the concern that scientists have for old-growth ecosystems and the passion with which environmental advocates have fought to preserve ancient forests in such areas as Clayoquot Sound.

In spite of vigorous historical logging, much of Canada remains forested (**FIGURE 10.8A**). Deforestation continues in Canada, but in 2005 it affected less than 0.02% (approximately 56 000 ha) of Canada's forests (**FIGURE 10.8B, 10.8C**). The principal cause of deforestation in Canada today is not logging but land clearing for agriculture. Urban development and, increasingly, outbreaks of parasites and other invasive pest species, such as the Asian longhorned beetle, mountain pine beetle, and spruce budworm, are also significant (see "The Science Behind the Story: Changing Climate and the Spruce Budworm on Vancouver Island").

Agriculture is the major cause of conversion of forests and grasslands

Agriculture now covers more of the planet's surface than does forest; 38% of Earth's terrestrial surface is devoted to agriculture—more than the area of North America and Africa combined. Of this land, 26% supports pasture, and 12% consists of crops and arable land. Agriculture is the most widespread type of human land use and the principal driver of land conversion today, causing tremendous impacts on land and ecosystems. Although agricultural methods, such as organic farming and no-till farming, can be sustainable, the majority of the world's cropland hosts either intensive traditional agriculture or monocultural industrial agriculture, involving heavy use of fertilizers, pesticides, and irrigation (Chapters 7 and 8).

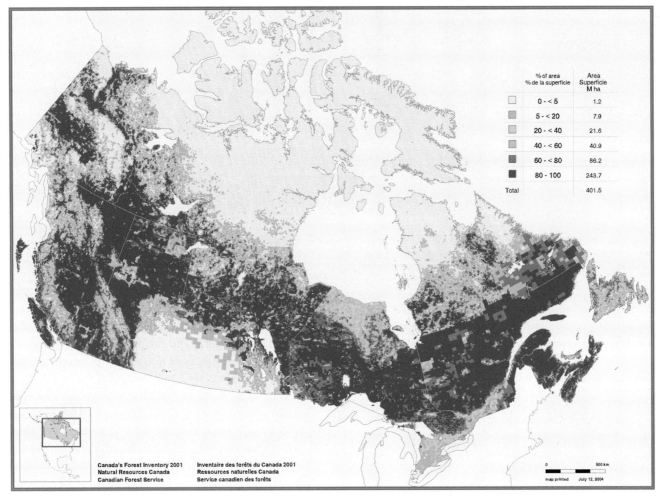

(a) Forested land in Canada

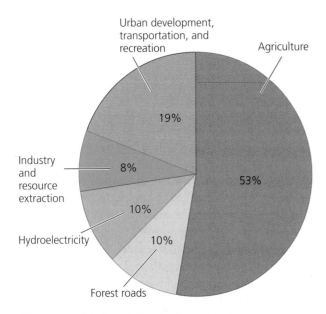

(b) Causes of deforestation in Canada today

(c) Land-clearing for agriculture

FIGURE 10.8
In spite of vigorous logging and deforestation during the westward expansion from the 1600s to the early 1900s, much of Canada remains forested today **(a)**. This is in contrast to the United States, where the majority of primary forest had been removed by the late 1800s. Today in Canada the principal causes of deforestation **(b)** are land clearing for agriculture **(c)**, as seen here in a photo of Quyon, Quebec, and urban development.

THE SCIENCE BEHIND THE STORY

Changing Climate and the Spruce Budworm on Vancouver Island

Spruce budworm larvae explore a Douglas fir.

As climate changes, ecosystems change too.[7] On southern Vancouver Island, the invasive parasitic western spruce budworm (see photo) is no longer the problem that it once was, as a result of a local increase in sea temperature over the past century that has limited the availability of the pest's food.

Ross Benton and Dr. Alan Thomson, research scientists at Natural Resources Canada's Pacific Forestry Centre in Victoria, have investigated the effects of changing climate at a local level. Their research shows that a 90-year increase in overall winter temperature on southern Vancouver Island has nearly eliminated western spruce budworm outbreaks in the area. "This pest will always be there at a very low level, but its population can't expand," says Benton. "The opportunity for the bug to infest its host, the Douglas fir, is less and less."

Although budworm outbreaks have been known in this area since 1909, it is unlikely to happen again, says Benton. During outbreaks in the first part of the twentieth century, the budworm defoliated a total area of 35 732 ha on southern Vancouver Island. However, the changing climate now means that food isn't available when the budworm larvae hatch. Budworm development is triggered by temperature, while Douglas fir development is triggered more by light and day length. As a result, the budworm larvae are emerging from their shelters earlier than a century ago, while Douglas fir bud development has remained constant. "The pest is starting to develop earlier and can't bore into the outer sheathing that protects new buds," says Benton. "The food is there, but not accessible," adds Thomson.

Although this may sound like good news, it may not be, according to Dr. Thomson. "Previous problems with budworm outbreaks are gone, but new species that weren't a problem before may become one." He points out that although spruce budworm outbreaks are

This B.C. forest has suffered an extensive attack by the mountain pine beetle.

no longer a problem in this part of the island, British Columbia's interior forests are experiencing large-scale outbreaks.

Meanwhile, mountain pine beetles have been one of the highest-profile forestry problems in British Columbia in the past decade. "Without the mountain pine beetle, the budworm would be big news," says Dr. Thomson. Since 1997, mountain pine beetles have infested more than 300 000 ha of lodgepole pine forests in the central interior of the province (see photo). In previous outbreaks, pine beetles have killed as many as 80 million trees over 450 000 ha, making them the second most important natural disturbance agent after fire in these forests.

In many parts of the developing world, forests are cleared for a traditional form of farming called **swidden** agriculture, in which a small area of forest is cleared (often by a *slash-and-burn* method) and crops are planted. After one or two seasons of planting, when the soil has been depleted of nutrients, the farmer moves on to clear another patch of forest, leaving the first clearing in a *fallow* or resting state, giving it time to replenish itself. This can actually be a sustainable practice, if the initial clearings are given sufficient time—often as much as seven years—with which to replenish the nutrient content of the soil. However, social and economic pressures in the developing world, including population pressure, have led to shorter and shorter fallow times, with the result that the cleared forest soils erode away, rather than regenerating. After that, the soil will no longer support either crops or forests.

In theory, the marketplace should discourage people from farming with intensive methods that degrade land they own if such practices are not profitable. But agriculture in many countries (including Canada) is supported by government subsidies, which amount to billions of dollars in some cases. For example, the Brazilian government provides financial incentives to farmers to clear areas of the Amazon rainforest for agriculture. Proponents of agricultural subsidies stress that the vagaries of weather make profits and losses from farming unpredictable from year to year. To persist, these proponents say, an agricultural system needs some way to compensate farmers for bad years. Opponents of subsidies argue that subsidization of environmentally destructive agricultural practices is unsustainable.

Livestock graze one-fourth of Earth's land surface

Most cattle in North America today are raised in feedlots, but they have traditionally been raised by grazing on open **rangelands**, grasslands or wooded areas converted for the

FIGURE 10.9
Livestock grazing covers a quarter of the earth's land surface.

purpose of supporting livestock (**FIGURE 10.9**). As we saw in Chapter 7, grazing can be sustainable, but overgrazing damages soils, waterways, and vegetative communities. Range managers are responsible for regulating ranching on public lands, and they advise ranchers on sustainable grazing practices.

Cropland agriculture uses less than half the land taken up by livestock grazing, which covers a quarter of the world's land surface. Human use of rangeland, however, does not necessarily exclude its use by wildlife or its continued functioning as a grassland ecosystem. Grazing can be sustainable if done carefully and at low intensity. In the West, some ranching proponents claim that cattle are merely taking the place of the vast herds of bison that once roamed the plains. Indeed, most of the world's grasslands have historically been home to large herds of grass-eating mammals, and grasses have adapted to herbivory.

Poorly managed grazing can have adverse impacts on soil and grassland ecosystems. In Central and South America, the conversion of forested land and grasslands to rangelands occurred with phenomenal rapidity during the decades from the 1940s to the 1970s. The dramatic loss of forested land has led a few countries, notably Costa Rica, to institute some tough new environmental restrictions.

Ranchers and environmentalists have traditionally been at loggerheads. In the past several years, however, they have been finding some common ground, teaming up to preserve ranchland against what both of them view as a threat—the encroaching housing developments of suburban sprawl. Although developers often pay high prices for ranchland, many ranchers do not want to see the loss of the wide-open spaces and the ranching lifestyle that they cherish.

Demands for land and wood have led to some bad practices and deforestation

Deforestation has caused soil degradation, population declines, and species extinctions, and, as we saw in the case of Easter Island (see "The Science Behind the Story: The Lesson of Rapa Nui"), it has helped bring whole civilizations to ruin. Timber resources can, in principle, be harvested sustainably, but unfortunately this hasn't always happened. Impacts are greatest in tropical areas because of the potentially massive loss of biodiversity, and in dryland regions because of the vulnerability of these lands to desertification. In addition, deforestation adds carbon dioxide (CO_2) to the atmosphere: CO_2 is released when plant matter is burned or decomposed, and thereafter less vegetation remains to soak up CO_2. Deforestation is thereby one contributor to global climate change (Chapter 14).

Globally, about 13 million hectares of forest are deforested each year—the area of Nova Scotia and New Brunswick combined. Today forests are being felled at the fastest rates in the tropical rainforests of Latin America and Africa (**FIGURE 10.10**). Developing countries in these regions are striving to expand areas of settlement for their burgeoning populations and to boost their economies by extracting natural resources and selling them abroad. Moreover, many people in these societies cut trees for fuelwood for their daily cooking and heating needs. In contrast, areas of Europe and eastern North America are slowly gaining forest cover as they recover from the severe deforestation of past decades and centuries. Overall, the world is losing its forests (see "The Science Behind the Story: Surveying Earth's Forests").

Land uses, such as grazing, farming, and timber harvesting, need not have strongly adverse impacts. It is not these activities *per se* that cause environmental problems, but rather the overexploitation of resources beyond what ecosystems can handle. Unfortunately, economic and social pressures, particularly in the developing world, often drive overexploitation and unsustainable practices.

Deforestation is proceeding rapidly in many developing nations

Uncut primary tropical forests still remain in many developing countries. These nations are in the position Canada faced a century or two ago: having a vast frontier that they

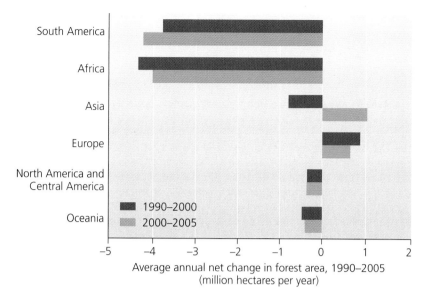

FIGURE 10.10
Nations in South America and Africa are experiencing rapid deforestation as they attempt to develop, extract resources, and provide new agricultural land for their growing populations. In Europe, meanwhile, forested area is slowly increasing as some formerly farmed areas are abandoned and allowed to grow back into forest. The data for North and Central America reflect a balance of forest regrowth in North America and forest loss in Central America. In Asia, natural forests are being lost, but the extensive planting of tree plantations (here counted as forests) in China has increased forest cover for Asia since 2000. Data from U.N. Food and Agriculture Organization (FAO). 2005. *Global forest resources assessment.*

can develop for human use. Today's advanced technology, however, has allowed these countries to exploit their resources and push back their frontiers even more quickly than occurred in North America. As a result, deforestation is extremely rapid in such places as Brazil, Indonesia, and parts of West Africa.

Developing nations are often desperate enough for economic development, and for foreign capital with which to maintain the interest payments on enormous national debt loads, that they impose few or no restrictions on logging. Often their timber is extracted by foreign multinational corporations, which have paid fees to the developing nation's government for a *concession*, or right to extract the resource. In such cases, the foreign corporation has little or no incentive to manage forest resources sustainably. Many of the short-term economic benefits are reaped not by local residents but by the corporations that log the timber and export it elsewhere. Local people may or may not receive temporary employment from the corporation, but once the timber is harvested they no longer have the forest and the ecosystem services it once provided.

In Sarawak, the Malaysian portion of the island of Borneo, foreign corporations that were granted logging concessions have deforested several million hectares of tropical rainforest since 1963 (**FIGURE 10.11A**). The clearing of this forest—one of the world's richest, hosting such organisms as orangutans and the world's largest flower, *Rafflesia arnoldii* (**FIGURE 10.11B**)—has had direct impacts on the 22 tribes of people who live as hunter–gatherers in Sarawak's rainforest. The Malaysian government did not consult the tribes about the logging, which decreased the wild game on which these people depended. Oil palm agriculture was established afterward, leading to pesticide and fertilizer runoff that killed fish in local streams. The tribes protested peacefully and finally began blockading logging roads. The government, which at first jailed them, now is negotiating, but it insists on converting the tribes to a farming way of life.

LOGGING HERE OR THERE 10-1

Imagine you are an environmental activist protesting a logging operation that is cutting old-growth trees near your hometown. If the protest is successful, the company will move to a developing country and cut its primary forest instead. Would you still protest the logging in your hometown? Would you pursue any other approaches?

Forest Management Principles

Professionals who manage forests through the practice of **forestry** (or **silviculture**) must balance the central importance of forests as ecosystems with civilization's demand for wood products. Sustainable forest management, like the management of other renewable natural resources, is based on maintaining equilibrium between *stocks* and

THE SCIENCE BEHIND THE STORY

Surveying Earth's Forests

This old-growth forest is in British Columbia.

In the time it takes you to read this sentence, two hectares of tropical forest will have been cleared.

Where do such numbers come from? How do we know how much forest our planet is losing—or how much forest there was to begin with?

Every five to 10 years since 1948, the United Nations Food and Agriculture Organization (FAO) has conducted a global inventory of forest resources. FAO researchers ask the world's national governments to provide data on forest area, types of forest, deforestation, regrowth, and other parameters. However, nations may not keep accurate data or may not care to share it.

In recent years, satellite technology has provided a way to confirm information from national surveys. With remote sensing data from satellites, we can measure and map forest cover from space.

Satellites that observe Earth's surface measure wavelengths of energy being emitted. This information is processed, and used to infer what materials are on the surface. Plants absorb most wavelengths of light but reflect infrared radiation and green light (which is why plants appear green). When satellite data shows green and infrared wavelengths reflected from the surface but an absence of other wavelengths, such as red and blue, the presence of vegetation can be inferred.

Researchers use data from a number of satellites (for instance, the Land Remote Sensing Satellite [Landsat], the Advanced Very High Resolution Radiometer [AVHRR], and the Moderate Resolution Imaging Spectroradiometer [MODIS]) to infer vegetative cover. By quantifying plant cover, researchers can compare one site to another and can measure how sites change across seasons or years.

A 2001 study of global forest cover by the United Nations Environment Programme (UNEP) used AVHRR data and loaded this into a geographic information system, or GIS. Researchers then added data on human population distribution, political boundaries, and land protected against development. The study found that forests in densely populated nations, such as India and Indonesia, are under pressure from expanding human settlement and require urgent conservation efforts.

In 2005, the FAO released its latest global accounting of forests, which superseded all previous efforts. In its *Global Forest Resources Assessment*, researchers combined remote sensing data, questionnaire responses, analysis from forestry experts, and statistical modelling, all to form a comprehensive picture of the state of the world's forests.

Major findings of the FAO assessment included the following:

- Forests cover 30% of the world's land area.
- Just 10 nations account for two-thirds of all forests.
- About 34% of forests are designated primarily for wood production.
- Forests store 283 billion metric tons of carbon in living tissue and more overall than the atmosphere.

---

flows. In principle, the removal or **harvesting** of material from the resource by logging should not occur at a rate that exceeds the capability of the resource to replenish or regenerate itself.

Public forests in Canada are managed for many purposes

Nearly 94% of Canada's forest is publicly owned; the majority of this is under provincial jurisdiction. Only about 6%–8% of forested land in Canada is privately owned, about 1.5% by logging companies, with the remainder under federal or territorial control.[8] An increasing amount of land in Canada, including forested land, is under Aboriginal jurisdiction as land claims are settled.[9]

In Canada, timber is extracted from both privately owned and publicly held forests by private timber companies. In fact, much of the resource extraction industry in Canada—both logging and mining—is carried out on Crown lands (mainly provincial). The provinces and territories and the federal government have different requirements governing the acquisition of rights to carry out resource extraction on Crown lands. Under the *Constitution*, provinces and territories own and regulate the natural resources within their boundaries.

The federal role in forestry is based on its responsibility for the national economy, trade, science and technology, the environment, and federal Crown lands and parks, as well as Aboriginal and treaty rights, which are constitutionally protected by the federal *Constitution Act of 1982*.

The Canadian Forest Service, part of Natural Resources Canada, was established in 1899 and given the responsibility to preserve timber on Dominion lands and

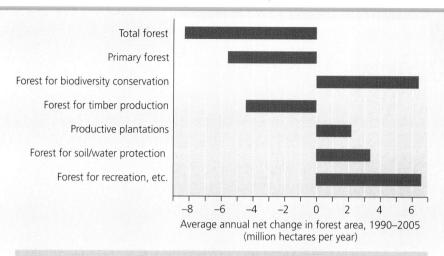

Progress toward sustainable forestry is mixed, as shown by 7 of 21 indicators assessed by the Food and Agriculture Organization (FAO). Forest uses are shifting toward conservation and away from production, but forest area is still shrinking. Data from U.N. Food and Agriculture Organization (FAO), 2005. *Global forest resources assessment.*

- Rare tree species valued for wood (such as teak and mahogany) are often in danger of vanishing.
- Some 84% of forests are publicly owned, but private ownership is increasing.
- Forestry employs 10 million people.

By comparing the results from 2005 with those of previous assessments, the FAO could also reveal trends through time. Major findings included the following:

- Globally, we deforest 13 million hectares per year.
- Today, 5.7 million hectares per year support regrowing forests.
- Net annual forest loss is 7.3 million hectares—about the area of New Brunswick.
- Regions vary in rates of forest loss or gain (see **FIGURE 10.6**).
- Primary forests are being lost more quickly than second-growth forests.
- Forest plantations make up only 3.8% of forests but are increasing.
- The world's forests hold 1.1 billion fewer tons of carbon each year.

The study examined how our society was progressing toward sustainable management of forests. Researchers examined a number of indicators (see the figure). Among other trends, they found the following:

- More and more forests are being managed for multiple uses.
- About 11% of forests are designated for biodiversity conservation.
- Some 9% of forests are designated for soil and water conservation.
- Use of forests for recreation and education is increasing rapidly.

These trends, however, are occurring against a backdrop of continued net loss of forests, especially primary forests. Overall, the report concludes that the outlook is mixed: "There are many good signs and positive trends, but many negative trends remain. While intensive forest plantation and conservation efforts are on the rise, primary forests continue to become degraded or converted to agriculture at alarming rates in some regions."

---

to develop policies to encourage tree culture.[10] Since then, the Canadian Forest Service has been involved in the scientific study and monitoring of Canada's forests and in managing the extraction of timber and non-timber forest products from national forests. The National Forest Strategies began to be officially developed in 1980. The current National Forest Strategy makes reference to Canada's obligation to the world to maintain and preserve the vast forests of which we are the stewards. The vision statement of this document reads, in part,

> *The health of the forest is directly linked to environmental processes on local, regional and international scales as well as to the social, cultural, spiritual and economic well-being of us all. As a result, we want to improve our understanding of how we are a part of and how we affect the forest—in short, to act on our increasingly informed understanding to become even better stewards of our forest resource.*[11]

For the past half-century, forest management throughout North America has nominally been guided by the policy of **multiple use**, meaning that forests were to be managed for recreation, wildlife habitat, mineral extraction, and various other uses. In reality, however, timber production was most often the primary use. In recent decades, increased awareness of the problems associated with logging has prompted many citizens to protest the way public forests are managed in Canada. These citizens have urged that provincial forests be managed for recreation, wildlife, and ecosystem integrity, rather than for timber. They want forests managed as ecologically functional entities, not as cropland for trees.

The National Forest Strategy for 2003–2008 includes the following objectives:

- Implementing ecosystem-based management (as described above)

(a) Logging in Borneo

(b) *Rafflesia arnoldii* **in bloom**

FIGURE 10.11
Logging, both illegal and legal **(a)** and associated deforestation are rampant in Borneo, Malaysia. The central habitat of *Rafflesia arnoldii*, the world's largest flower **(b)**, and that of many other important and threatened species, is in Borneo.

- Improving the environmental, social, and economic sustainability of forest communities through legislation and policies
- Recognizing the historical and legal rights of Aboriginal peoples and their fundamental connection to the forest ecosystem
- Stimulating the diversification of markets for forest products, benefits, and services
- Enhancing the skills and knowledge of forest practitioners, to promote innovation for competitiveness and sustainability
- Actively engaging Canadians in sustainability through the planning, maintenance, and management of urban forests
- Strengthening policies and services that support the contribution of private woodlots to forest sustainability
- Creating a comprehensive national forest reporting system that consolidates data, information, and knowledge for all valued features of the forest, both urban and rural[12]

Today, many managers practise ecosystem-based management

Increasing numbers of renewable resource managers today espouse ecosystem-based management. **Ecosystem-based management** attempts to manage the harvesting of resources in ways that minimize impact on the ecosystems and ecological processes of the forest. In Canada, ecosystem-based management aims to preserve forest health, structure, functions, composition, and biodiversity. This has been partly achieved by the establishment, over time, of a system of provincially, federally, and internationally protected areas (discussed below). An additional goal of ecosystem-based forest management is to maintain Canada's forests as viable reservoirs, or *sinks*, for atmospheric carbon.[13] This is crucial for the management of carbon emissions in the global effort to control climate change, which we will consider in Chapter 14.

As an example, the plan proposed in 1995 by the Scientific Panel for Sustainable Forest Practices on Clayoquot Sound and approved by British Columbia's government was essentially a plan for ecosystem-based management. By carefully managing ecologically important areas, such as riparian corridors, by considering patterns at the landscape level, and by affording protection to some forested areas, the plan aimed to allow continued timber harvesting at reduced levels while preserving the functional integrity of the ecosystem.

Although ecosystem-based management has gained a great deal of support in recent years, it is challenging for managers to determine how best to implement this type of management. Ecosystems are complex, and our understanding of how they operate is limited. Thus, ecosystem-based management has often come to mean different things to different people.

Adaptive management evolves and improves

Some management actions will succeed, and some will fail. A wise manager will try new approaches if old ones are not effective. **Adaptive management** involves systematically testing different management approaches and aiming to improve methods as time goes on. It entails monitoring the results of one's practices and continually adjusting them as needed, based on what is learned. This approach is intended as a true fusion and partnership of science and management, because hypotheses about how best to manage resources are explicitly tested. Adaptive management can be time-consuming and complicated, however. It has posed a challenge for many managers, because those who adopt new approaches must often overcome inertia and resistance to change from proponents of established practices.

Adaptive management has become a guiding principle for forest management in Canada. In British Columbia, the Ministry of Forests and Range has embraced adaptive management, promoting the approach through collaborative projects that range from testing alternative forestry practices, to monitoring whole watersheds, to evaluating the effectiveness of various land and resource management strategies.[14] Some examples of adaptive management projects that have been undertaken in B.C. forests include the following:

- The West Arm Demonstration Forest Experiments, which were designed to study a broad range of forest values and to evaluate the effectiveness and impacts of several new forest management approaches
- The Donna Creek Biodiversity Project, which tested whether maintenance of tall stumps and small residual tree islands benefit cavity-using animals within clear-cuts
- The Grizzly Bear Habitat Project, which was designed to assist in the development of forestry management systems to maintain grizzly habitat, while at the same time producing timber[15]

Plantation forestry has grown in North America

Logging today (in North America, at least) is largely offset by **reforestation**, the planting of trees after logging, and **afforestation**, the planting of trees where forested cover has not existed for some time. In Canada, as in some other economically developed nations, reforestation and afforestation have more than offset losses to deforestation in the past decade or so, leading to a (small) net increase in forested land. The North American timber industry is largely centred on production from plantations of fast-growing tree species that are single-species monocultures. Because all trees in a given stand are planted at the same time, the stands are **even-aged**, with all trees the same age (**FIGURE 10.12**). Stands are cut after a certain number of years (called the *rotation time*), and the land is replanted with seedlings.

It is important to acknowledge that planting new trees will not replace complex old-growth forests that may have taken hundreds of years to develop. Even when regrowth outpaces removal, the character of forests may still change. In North America and worldwide, primary forest continues to be lost and to be replaced by younger second-growth forest. Most ecologists and foresters view these plantations more as crop agriculture than as ecologically functional forests. Because there are few tree species and little variation in tree age, plantations do not offer many forest organisms the habitat they need.

The principle of **maximum sustainable yield**, a basic principle of renewable resource management, argues for cutting trees shortly after they have gone through their

FIGURE 10.12
Even-aged tree stand management is practised on tree plantations where all trees are of equal age, as seen in the stand in the foreground that is regrowing after clear-cutting. In uneven-aged tree stand management, harvests are designed to maintain a mix of tree ages, as seen in the more mature forest in the background. The increased structural diversity of uneven-aged stands provides superior habitat for most wild species and makes these stands more akin to ecologically functional forests.

fastest stage of growth, and trees often grow most quickly at intermediate ages. Thus, trees may be cut long before they have grown as large as they would in the absence of harvesting. Although this practice may maximize timber production over time, it can cause drastic changes in the ecology of a forest by eliminating habitat for species that depend on mature trees. However, some harvesting methods aim to maintain **uneven-aged** stands, where a mix of ages (and often a mix of tree species) makes the stand more similar to a natural forest.

Timber is harvested by several methods

When they harvest trees, timber companies use any of several methods. From the 1950s through the 1970s, many timber harvests were conducted by using the **clear-cutting** method, in which all trees in an area are cut, leaving only stumps. Clear-cutting is generally the most cost-efficient method in the short term, but it has the greatest impacts on forest ecosystems (**FIGURE 10.13**). In the best-case scenario, clear-cutting may mimic natural disturbance events, such as fires, tornadoes, or windstorms, that knock down trees across large areas. In the worst-case scenario, entire communities of organisms are destroyed or displaced, soil erodes, and the penetration of sunlight to ground level changes microclimatic conditions such that new types of plants replace those that had composed the native forest. Essentially, clear-cutting sets in motion an artificially driven process of succession in which the resulting climax community may turn out to be quite different from the original climax community.

FIGURE 10.13
Clear-cutting is the most cost-efficient method for timber companies, but it can have severe ecological consequences, including soil erosion and species turnover. Although certain species do use clear-cuts as they regrow, most people find these areas esthetically unappealing, and public reaction to clear-cutting has driven changes in forestry methods.

Clear-cutting occurred widely across North America at a time when public awareness of environmental problems was blossoming. The combination produced public outrage toward the timber industry and public forest managers. Eventually the industry integrated other harvesting methods (**FIGURE 10.14**). A set of approaches dubbed **new forestry** called for timber cuts that came closer to mimicking natural disturbances. For instance, "sloppy clear-cuts" that leave a variety of trees standing were intended to mimic the changes a forest might experience if hit by a severe windstorm.

Clear-cutting (**FIGURE 10.14A**) is still widely practised, but other methods involve cutting some trees and leaving some standing. In the *seed-tree* approach (**FIGURE 10.14B**), small numbers of mature and vigorous seed-producing trees are left standing so that they can reseed the logged area. In the *shelterwood* approach (also **FIGURE 10.14B**), small numbers of mature trees are left in place to provide shelter for seedlings as they grow. These three methods all lead to even-aged stands of trees.

Selection systems, in contrast, allow uneven-aged stand management. In selection systems (**FIGURE 10.14C**), like the variable retention harvest system practised on Clayoquot Sound, only some trees in a forest are cut at any one time. The stand's overall rotation time may be the same as in an even-aged approach, because multiple harvests are made, but the stand remains mostly intact between harvests. Selection systems include single-tree selection, in which widely spaced trees are cut one at a time, and group selection, in which small patches of trees are cut.

It was a form of selection harvesting that Iisaak Forest Services and other logging organizations pursued at Clayoquot Sound after old-growth advocates applied pressure and the scientific panel published its guidelines. Not wanting to bring a complete end to logging when so many local people depended on the industry for work, these activists and scientists instead promoted what they considered a more environmentally friendly method of timber removal.

However, selection systems are by no means ecologically harmless. Moving trucks and machinery over an extensive network of roads and trails to access individual trees compacts the soil and disturbs the forest floor. Selection methods are also unpopular with timber companies because they are expensive, and loggers dislike them because they are more dangerous than clear-cutting.

The bottom line, from an ecological perspective, is that all methods of logging result in habitat disturbance, which invariably affects the plants and animals inhabiting an area. All methods change forest structure and composition. Most methods increase soil erosion, leading to siltation of waterways, which can degrade habitat and affect drinking water quality. Most methods also speed runoff,

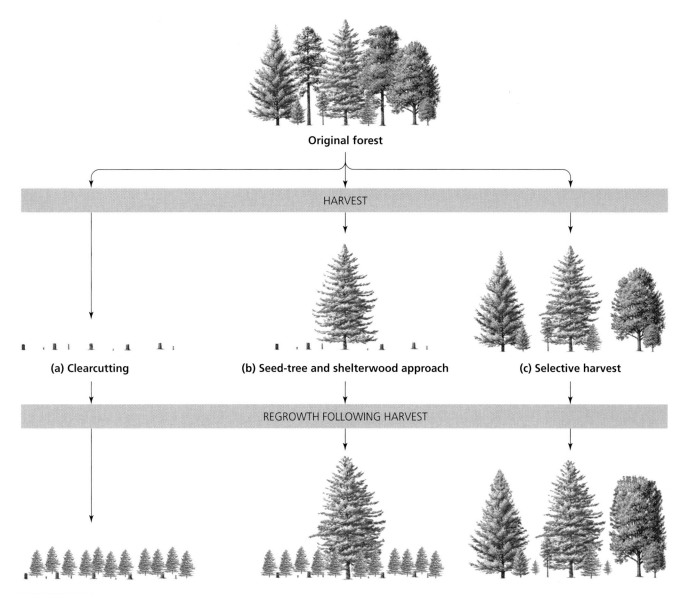

FIGURE 10.14
Foresters and timber companies have devised various methods to harvest timber from forests. In clear-cutting **(a)**, all trees in an area are cut, extracting a great deal of timber inexpensively but leaving a vastly altered landscape. In seed-tree systems and shelterwood systems **(b)**, small numbers of large trees are left in clear-cuts to help reseed the area or provide shelter for growing seedlings. In selection systems **(c)**, a minority of trees is removed at any one time, while most are left standing. These latter methods involve less environmental impact than clear-cutting, but all methods can cause significant changes to the structure and function of natural forest communities.

sometimes causing flooding. In extreme cases, as when steep hillsides are clear-cut, landslides can result.

Fire policy has stirred controversy

The management of fires is one of the most controversial aspects of forest management today. For more than a century, land management agencies throughout North America suppressed fire whenever and wherever it broke out (as per the warnings of "Smokey the Bear"). Yet ecological research now clearly shows that many ecosystems depend on fire. Certain plants have seeds that germinate only in response to fire, and researchers studying tree rings have documented that many ecosystems historically experienced frequent fire. Burn marks in a tree's rings reveal past fires, giving scientists an accurate history of fire events extending back hundreds or even thousands of years.

Many wooded dryland ecosystems also depend on fire. For example, researchers have found that North America's grasslands and open pine woodlands burn regularly in the natural system. Ecosystems dependent on fire are adversely affected by its suppression; pine woodlands become cluttered with hardwood understory that ordinarily would be cleared away by fire, for instance, and animal diversity and abundance decline.

FIGURE 10.15
Forest fires are natural phenomena to which many plants are adapted and which maintain many ecosystems. The suppression of fire by people over the past century has led to a buildup of leaf litter, woody debris, and young trees, which serve as fuel to increase the severity of fires when they do occur. As a result, catastrophic wildfires have become more common in recent years. To avoid these unnaturally severe fires, most fire ecologists suggest allowing natural fires to burn when possible and instituting controlled burns to reduce fuel loads and restore forest ecosystems.

In the long term, fire suppression can lead to catastrophic fires that truly damage forests, destroy human property, and threaten human lives. Fire suppression allows limbs, logs, sticks, and leaf litter to accumulate on the forest floor, effectively producing kindling for a catastrophic fire. Such fuel buildup helped cause the 1988 fires in Yellowstone National Park, the 2003 fires in southern California, the 2003 fires in British Columbia, and thousands of other wildfires across the continent each year (**FIGURE 10.15**). Fire suppression and fuel buildup have made catastrophic fires significantly greater problems than they were in the past. Now, global climate change is bringing drier weather to much of the Canadian Prairies, further worsening the wildfire risk. At the same time, increasing residential development on the edges of forested land is placing more homes in fire-prone situations.

To reduce fuel load and improve the health and safety of forests, forest management agencies have in recent years been burning areas of forest under carefully controlled conditions. These **prescribed burns**, or **controlled burns**, have worked effectively, but they have been implemented on only a relatively small amount of land.

Another significant and controversial aspect of fire management concerns what happens after a fire, which may include the physical removal of small trees, underbrush, and dead trees by timber companies. The removal of dead trees, or snags, following a natural disturbance is called **salvage logging**. From an economic standpoint, salvage logging may seem to make good sense. Proponents of salvage logging argue that forests regenerate best after a fire if they are logged and replanted with seedlings. Moreover, they maintain, salvage logging reduces future fire risk by removing woody debris that could serve as fuel for the next fire. However, ecologically, snags have immense value; the insects that decay them provide food for wildlife, and many birds, mammals, and reptiles depend on holes in snags for nesting and roosting sites. Removing timber from recently burned land can also cause severe erosion, collapse of streambanks, and soil damage.

weighing the issues 10–2
HOW TO HANDLE FIRE?

A century of fire suppression has left vast swaths of forested lands in North America in danger of catastrophic wildfires. Yet we will probably never have adequate resources to conduct careful prescribed burning over all these lands. Can you suggest any solutions to help protect people's homes near forests while improving the ecological condition of some forested lands? Do you think people should be allowed to build homes in fire-prone areas?

Sustainable forestry is gaining ground

Any company can claim that its timber-harvesting practices are sustainable, but how is the purchaser of wood products to know whether they really are? In the last several years, a consumer movement has grown that is making

weighing the issues 10–3
POST-FIRE MANAGEMENT

When a fire burns a forest, should the killed trees be cut and sold for timber? Proponents of salvage logging say yes: we should not let economically valuable wood go to waste. Opponents of post-fire logging say the burned wood is more valuable left in place—for erosion control, wildlife habitat (snags provide holes for cavity-dwelling animals, and food for insects and birds), and organic material to enhance the soil and nurse the growth of future trees. If you were in charge of a provincial forest and had to make a policy for post-fire management, what would you do?

FIGURE 10.16
A Brazilian woodcutter taking inventory marks timber harvested from a forest certified for sustainable management in Amazonian Brazil. A consumer movement centred on independent certification of sustainable wood products is allowing consumer choice to promote sustainable forestry practices.

informed consumer choice possible. Several organizations now examine the practices of timber companies and offer **sustainable forestry certification** to products produced by using methods they consider sustainable (**FIGURE 10.16**).

Such organizations as the International Organization for Standardization (ISO), the Sustainable Forestry Initiative (SFI) program, and the Forest Stewardship Council (FSC) have varying standards for certification. Consumers can look for the logos of these organizations on forest products they purchase. The FSC is widely perceived to have the strictest certification standards. In 2001, Iisaak, the Native-run timber company at Clayoquot Sound, became the first tree farm licence holder in British Columbia to receive FSC certification.

Consumer demand for sustainable wood has been great enough that Home Depot and other major retail businesses have begun selling sustainable wood. The decisions of such retailers are influencing the logging practices of many timber companies. In British Columbia, 70% of the province's annual harvest now is certified or meets ISO requirements. Sustainable forestry is more costly for the timber industry, but if certification standards can be kept adequately strong, then consumer choice in the marketplace can be a powerful driver for good forestry practices for the future.

Parks and Reserves

For forests, drylands, and other terrestrial ecosystems, debates continue over how best to use land and manage resources. Resource extraction from Crown lands in Canada has helped propel our economy. But as resources dwindle, as forests, grasslands, and soils are degraded, and as the landscape fills with more people, the arguments for conservation of resources—for their sustainable use—have grown stronger. Also growing stronger is the argument for preservation of land—setting aside tracts of relatively undisturbed land intended to remain forever undeveloped.

For ethical reasons (see Chapter 21) as well as pragmatic, ecological and economic ones, Canadian citizens and many other people worldwide have chosen to set aside tracts of land in perpetuity to be preserved and protected from development.

Why have we created parks and reserves?

Historian Alfred Runte has cited four traditional reasons that parks and protected areas have been established:

1. Enormous, beautiful, or unusual features, such as the Rocky Mountains and Clayoquot Sound, inspire people to protect them—an impulse termed *monumentalism* (**FIGURE 10.17**).
2. Protected areas offer recreational value to tourists, hikers, fishers, hunters, and others.
3. Protected areas offer utilitarian benefits. For example, undeveloped watersheds provide cities with clean drinking water and a buffer against floods.
4. Parks make use of sites lacking economically valuable material resources or that are hard to develop; land that holds little monetary value is easy to set aside.

FIGURE 10.17
The awe-inspiring beauty of some regions of Canada was one reason for the establishment of national parks. Scenic vistas such as this one near Banff, Alberta, have inspired millions of people to visit them.

To these four traditional reasons, a fifth has been added in recent years: parks can help preserve biodiversity. As we saw in Chapter 9, human impact alters habitats and has led to countless population declines and species extinctions. A park or reserve is widely viewed as an island of habitat that can, scientists hope, maintain species that might otherwise disappear.

Today there are 43 **national parks** in the Canadian parks system, covering a total of 27 million hectares, or 2.7% of the total land area of Canada.[16] Approximately 16 million people visit Canada's national parks each year. Yellowstone National Park in the United States was the very first national park, established in 1872 and followed soon after by Yosemite National Park. Canada's first national park was established in Banff, Alberta, in 1885.

At Clayoquot Sound, Pacific Rim National Park Reserve is a protected area designated for future national park status. The Clayoquot Sound region also encompasses several provincial parks. Provincial parks in Canada number in the hundreds and cover more area than national parks. Canada's parks system includes other types of protected areas as well, such as marine conservation areas and cultural, historic, and natural heritage sites.

The Canadian Wildlife Service, part of Environment Canada, contributes to the management and scientific understanding of wildlife and habitat management in Canada. Many sites in the parks system also serve as **wildlife refuges**, which are havens for the conservation of wildlife and habitat, as well as, in some cases, being available for hunting, fishing, wildlife observation, photography, environmental education, and other public uses.

Some wildlife advocates find it ironic that hunting is allowed at many parks and refuges, but hunters have long been in the forefront of the conservation movement and have traditionally supplied the bulk of funding for land acquisition and habitat management for the refuges. Ducks Unlimited Canada is an example of a nonprofit, nongovernmental organization founded by hunters but with the specific goal of conserving wetlands.

Not everyone supports land set-asides

The restriction of activities in some wilderness areas has helped generate opposition among those who seek to encourage resource extraction and development, as well as hunting and increased motor vehicle access, on protected Crown lands. The drive to extract more resources, secure local control of lands, and expand recreational access to public lands is epitomized by the *wise-use movement*, a loose confederation of individuals and groups that coalesced in the 1980s and 1990s in response to the increasing success of environmental advocacy. Wise-use advocates are dedicated to protecting private property rights; opposing government regulation; transferring federal lands to provincial, municipal, or private hands; and promoting motorized recreation on public lands. The wise-use movement, which has been described as "anti-conservation," includes many farmers, ranchers, trappers, and mineral prospectors, as well as groups representing the industries that extract timber, mineral, and fossil fuel resources.

Debate between mainstream environmental groups and wise-use spokespeople has been vitriolic. Each side claims to represent the will of the people and paints the other as the oppressive establishment. Wise-use advocates have played key roles in ongoing debates over policy issues, such as whether recreational activities that disturb wildlife should be allowed.

Nonfederal entities also protect land

Efforts to set aside land—and the debates over such efforts—at the federal level are paralleled at regional and local levels. Each Canadian province has agencies that manage resources on provincial Crown lands, as do many municipalities. Private nonprofit groups also preserve land. **Land trusts** are local or regional organizations that purchase land with the aim of preserving it in its natural condition. The Nature Conservancy can be considered the world's largest land trust, but smaller ones are springing up throughout North America. Probably the earliest private land trust in Canada was the Hamilton Naturalists Club in Ontario, which began to acquire land for conservation purposes in 1919.[17] **FIGURE 10.18** shows the dramatic increase in land trusts operating in Canada, particularly since the 1970s.

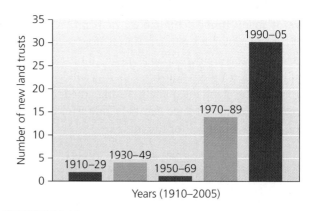

FIGURE 10.18
Land trusts that promote private land stewardship have increased dramatically in Canada since the first, the Hamilton Naturalists Club, in 1919. This figure is based on a sample of 49 land trust organizations.

Parks and reserves are increasing internationally

Many nations have established national park systems and are benefiting from ecotourism as a result—from Costa Rica (Chapter 3) to Ecuador to Thailand to Tanzania. The total worldwide area in protected parks and reserves increased more than fourfold from 1970 to 2000, and in 2003 the world's 38 536 protected areas covered 1.3 billion hectares, or 9.6% of the planet's land area. However, parks in developing countries do not always receive the funding, legal support, or enforcement support they need to manage resources, provide for recreation, and protect wildlife from poaching and timber from logging. Thus many of the world's protected areas are merely *paper parks*—protected on paper but not in reality.

Some types of protected areas fall under national sovereignty but are designated or partly managed internationally by the United Nations. **World heritage sites** are an example; currently more than 830 sites across 184 countries are listed for their natural or cultural value. One such site is Australia's Kakadu National Park (see Chapter 21). Another is the mountain gorilla reserve shared by three African countries. The gorilla reserve, which integrates national parklands of Rwanda, Uganda, and the Democratic Republic of Congo, is also an example of a *transboundary park*, an area of protected land overlapping national borders. Transboundary parks can be quite large, and they account for 10% of protected areas worldwide, involving more than 100 countries. A North American example is Waterton-Glacier National Parks on the Canadian–American border. Some transboundary reserves function as *peace parks*, helping ease tensions by acting as buffers between nations that have quarrelled over boundary disputes. This is the case with Peru and Ecuador, as well as Costa Rica and Panama, and many people hope that peace parks can also help resolve conflicts between Israel and its neighbours.

Another land protection vehicle that has increased worldwide over the past couple of decades is the **debt-for-nature swap**, conceived in the early 1980s by Thomas Lovejoy and pioneered by Conservation International more than 20 years ago with partners in Costa Rica. In debt-for-nature swaps, a nongovernmental environmental group or even a corporation takes on a portion of the debt of a developing country, in exchange for some form of environmental protection or conservation. Since national debt is one of the principal driving forces of habitat destruction in developing nations (because of the need to extract natural resources and generate foreign capital), debt-for-nature swaps hold the promise of protecting the environment while addressing one of its fundamental threats. So far, in spite of millions of dollars devoted to debt-for-nature swaps worldwide, results have been mixed.

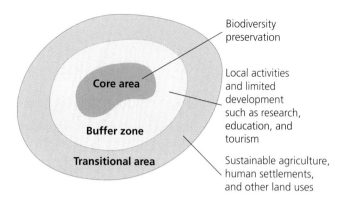

FIGURE 10.19
Biosphere reserves are international efforts that couple preservation with sustainable development to benefit local residents. Each reserve includes a core area that preserves biodiversity, a buffer zone that allows limited development, and a transition zone that permits various uses.

Biosphere reserves are tracts of land with exceptional biodiversity that couple preservation with sustainable development to benefit local people. They are designated by UNESCO (the United Nations Educational, Scientific, and Cultural Organization), following application by local stakeholders. Each biosphere reserve consists of (1) a **core** area that is isolated from the surroundings, and serves to preserve habitat and biodiversity, (2) a *buffer zone* that allows local activities and limited development that do not hinder the core area's function, and (3) an outer *transitional zone* in which agriculture, human settlement, and other land uses can be pursued in a sustainable way (**FIGURE 10.19**).

Research has demonstrated that wildlife adapted to forest core habitats suffer declines when they are forced to occupy **edge** habitats, in parts of the forest that are immediately adjacent to surrounding areas. Edge habitat—even if it is still forested—can be quite different in character from habitat in the forest core, particularly in terms of properties, such as light levels, density of vegetation, and moisture. The design of biosphere reserves seeks to address this concern and to reconcile human needs with the interests of wildlife and the goal of habitat preservation.

Clayoquot Sound was designated as Canada's 12th biosphere reserve in 2000 (**FIGURE 10.20**), in an attempt to help build cooperation among environmentalists, timber companies, Native people, and local residents and businesses. The core area consists of provincial parks and Pacific Rim National Park Reserve. Environmentalists hoped the designation would help promote stronger land preservation efforts. Local residents supported it because outside money was being offered for local development efforts. The timber industry did not stand in the way once it was clear that harvesting operations would not be affected. The designation has brought Clayoquot Sound

FIGURE 10.20
Clayoquot Sound was declared a UNESCO Biosphere Reserve in 2000.

more international attention, but it has not created new protected areas and has not altered land use policies.

The design of parks and reserves has consequences for biodiversity

Often it is not outright destruction of habitat that threatens species, but rather the *fragmentation* of habitat. Expanding agriculture, spreading cities, highways, logging, and many other impacts have chopped up large contiguous expanses of habitat into small, disconnected ones (**FIGURE 10.21**). When this happens, many species suffer. Bears, mountain lions, and other animals that need large ranges in which to roam may disappear. Bird species that thrive in the interior of forests may fail to reproduce when forced near the edge of a fragment. Their nests often are attacked by predators and parasites that favour open habitats surrounding the fragment or that travel along habitat edges. Avian ecologists judge forest fragmentation to be a main reason why populations of many songbirds of eastern North America are declining.

Because habitat fragmentation is such a central issue in biodiversity conservation, and because there are limits on how much land can be set aside, conservation biologists have argued heatedly about whether it is better to make reserves large in size and few in number, or many in number but small in size. Nicknamed the **SLOSS dilemma**, for "**S**ingle **L**arge **O**r **S**everal **S**mall," this debate is ongoing and complex, but it seems clear that large species that roam great distances, such as the elephant or the Siberian tiger (Chapter 9), benefit more from the "single large" approach to reserve design. In contrast, creatures such as insects, that live as larvae in small areas may do just fine in a number of small isolated reserves, if they can disperse as adults by flying from one reserve to another.

A related issue is whether **corridors** of protected land are important for allowing animals to travel between islands of protected habitat. In theory, connections between fragments provide animals with access to more habitat and help enable gene flow to maintain populations in the long term. Many land management agencies and environmental groups try, when possible, to join new reserves to existing reserves for these reasons. It is clear that we will need to think on the landscape level if we are to preserve a great deal of our natural heritage.

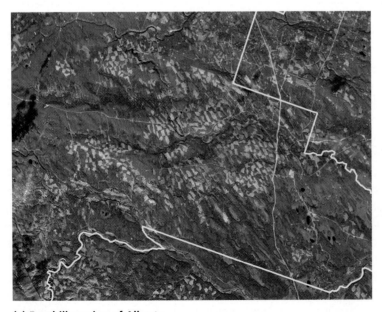

(a) Foothills region of Alberta

(b) Wood thrush

FIGURE 10.21
As human populations have grown and human impacts have increased, most large expanses of natural habitat have become fragmented into smaller disconnected areas. Forest fragmentation, for example, is evident in this satellite image of the foothills region of Alberta.

CANADIAN ENVIRONMENTAL PERSPECTIVES

Tzeporah Berman

Tzeporah Berman, environmental activist

- **Co-founder** of environmental non-profit organization **ForestEthics**
- **Anti-logging activist**
- **Market campaigner** and "**corporate re-educator**"

In the spring of 1992, 22-year-old Tzeporah Berman returned to Vancouver Island to continue her fieldwork as a volunteer for Western Canada Wilderness Committee and researchers at the University of Victoria, studying marbled murrelet nests. But she could not find the nesting area. The approach to the site had been logged, and with the landmarks obliterated, she could not get her bearings. Gradually, the reality dawned: this was the nesting site. She found a ring of stumps that had been the 70-metre-high Sitka spruce trees under which she had camped. She found a trickle of water that had been a waterfall and pool where she had swum. Eagles wheeled overhead, surveying their fallen nests.

Sitting on a stump, in tears, Berman reconsidered her summer and her future. She had planned to finish environmental studies and then go into law. But by the time she did that, she decided, there would be no marbled murrelets left. The next day, a van stopped on its way to a blockade in Clayoquot Sound. Berman climbed in.[18]

The rest, as they say, is history. Berman was right in the midst of the historic protests at Clayoquot Sound, described in the Central Case. Eventually she was arrested and dragged into court for "aiding and abetting" the commission of hundreds of criminal acts by the thousands of other Clayoquot protestors (the charges were dropped). She became a figurehead for the cause—sometimes hugged and sometimes attacked in public, she experienced death threats and even had her apartment burned by an unknown arsonist. The B.C. premier's reference to "enemies of the state" was aimed at Berman and her associates.

Working for Greenpeace, Berman began her transformation into "corporate re-educator," while helping to organize a huge (and very successful) public boycott against logging company MacMillan Bloedel. The boycott, or "market campaign," focused specifically on MacMillan Bloedel's customers. "Did Pacific Bell know, they would ask, that their phone books were made from the ancient rainforest of Clayoquot Sound? Surely the company would be alarmed at this, and if they weren't, certainly Pacific Bell's customers would be. Alarmed, Pacific Bell called MacMillan Bloedel; so did Scott Paper and the *New York Times*."[19] As the tide began to turn, Berman realized that money and the wishes of customers could be leveraged to convince large corporations that having environmentally responsible policies is not only the *right* thing to do, it is also the most *financially sound* thing to do.

Today ForestEthics, the nonprofit organization that Berman and others founded in 1994, collaborates with some of the same corporations it previously boycotted. Through its Corporate Action Program, ForestEthics has turned "corporate adversaries into allies."[20] ForestEthics works with companies, such as The Home Depot, Dell, Victoria's Secret, Estée Lauder, and Staples (the largest paper retailer in the world), to develop strong, financially sound, cutting-edge corporate environmental policies. By some accounts, Berman's actions and her work through ForestEthics have saved as many as 60 million hectares of forest from destruction in Canada and elsewhere in the world.

"The influence of the marketplace is integral to protecting the key areas of the natural forest that are left. Economic prosperity in the long term is tied to ecological prosperity now." —Tzeporah Berman[21]

Thinking About Environmental Perspectives

Tzeporah Berman and her colleagues used the threat of public dissatisfaction and boycotts to convince large corporations to adopt responsible policies regarding logging and product sources. Can you think of other cases where large, multinational corporations have been convinced to change their policies because of environmental or social justice concerns and campaigns on the part of their customers?

Conclusion

Forests and other terrestrial biomes provide crucial ecosystem services, supporting a vast diversity of species and providing goods that are of economic value to humans as well. Managing natural resources sustainably is particularly important for such resources as timber and soil, which otherwise can be carelessly exploited, degraded, or overharvested. Canada and many other nations have established various federal and regional agencies to oversee and manage publicly held land and the natural resources that are extracted from public land. Historically, forest management in North America has reflected general trends in land and resource management. Early emphasis on resource extraction evolved into policies on sustained yield and multiple use, a shift that occurred as land and resource availability declined and as the public became more aware of environmental degradation.

Public forests today are managed not only for timber production but also for recreation, wildlife habitat, and ecosystem integrity. Meanwhile, support for the preservation of natural lands has resulted in parks, wilderness areas, and other reserves, both in North America and abroad. These trends are positive ones, because the preservation and conservation of land and resources is essential if we want our society to be sustainable and to thrive in the future.

REVIEWING OBJECTIVES

You should now be able to:

Identify the principles, goals, and approaches of resource management

- Resource management enables us to sustain natural resources that are renewable if we are careful not to deplete them.
- Resource managers have increasingly focused not only on extraction but also on sustaining the ecological systems that make resources available.
- Resource managers have long managed for maximum sustainable yield and are beginning to implement ecosystem-based management and adaptive management.

Summarize the ecological roles and economic contributions of forests, and outline the history and scale of forest loss

- Forests not only provide us with economically important timber but also support biodiversity and contribute ecosystem services.
- Developed nations deforested much of their land as settlement, farming, and industrialization proceeded. Today deforestation is taking place most rapidly in developing nations.

Explain the fundamentals of forest management, and describe the major methods of harvesting timber

- Harvesting methods for timber include clear-cutting and other even-aged techniques, as well as selection strategies that maintain uneven-aged stands that more closely resemble natural forest.
- Foresters are beginning to manage for recreation, wildlife habitat, and ecosystem integrity, as well as timber production.
- Fire policy has been politically controversial, but scientists agree that we need to address the impacts of a century of fire suppression.
- Certification of sustainable forest products allows consumer choice in the marketplace to influence forestry techniques.

Analyze the scale and impacts of agricultural land use

- Agriculture has contributed greatly to deforestation and has had enormous impacts on landscapes and ecosystems worldwide.

Identify major federal land management agencies and the lands they manage

- The Canadian Forest Service, Parks Canada, Canadian Wildlife Service, and Environment Canada each play a role in managing Canada's forests and national parks at the federal level. In addition, Agriculture and Agri-Foods Canada oversees the management of rangelands.

Recognize types of parks and reserves, and evaluate issues involved in their design

- Public demand for preservation and recreation has led to the creation of parks, reserves, and wilderness areas in North America and across the world.
- Biosphere reserves are one of several types of internationally managed protected lands.
- Because habitat fragmentation affects wildlife, conservation biologists are working on how best to design parks and reserves.

TESTING YOUR COMPREHENSION

1. How do minerals differ from timber when it comes to resource management?
2. Compare and contrast maximum sustainable yield, adaptive management, and ecosystem-based management. Why may pursuing maximum sustainable yield sometimes conflict with what is ecologically desirable?
3. Name several major causes of deforestation. Where is deforestation most severe today?
4. Compare and contrast the major methods of timber harvesting.
5. Describe several ecological effects of logging. How has the Canadian Forest Service responded to public concern over the ecological effects of logging?

6. Are forest fires a bad thing? Explain your answer.
7. Approximately what percentage of Earth's land is used for agriculture? How has agriculture contributed to the loss of wetlands in Canada?
8. Name five reasons that people have created parks and reserves. What types of protected areas are there in Canada, aside from national parks?
9. Why do some people oppose federal land protection?
10. Roughly what percentage of Earth's land is protected? What types of protected areas have been established in countries outside North America?

SEEKING SOLUTIONS

1. Do you think maximum sustainable yield represents an appropriate policy for resource managers to follow? Why or why not?
2. People in developed countries are fond of warning people in developing countries to stop destroying rainforest. People of developing countries often respond that this is hypocritical, because the developed nations became wealthy by deforesting their land and exploiting its resources in the past. What would you say to the president of a developing nation, such as Brazil, that is seeking to clear much of its forest?
3. Can you think of a land use conflict that has occurred in your region? How was it resolved? If it is unresolved, then how could it be resolved?
4. What are some ecological effects of agricultural subsidies? Propose arguments for and against subsidies from an ecological point of view.
5. **THINK IT THROUGH** You have just become the supervisor of a national forest. Timber companies are requesting to cut as many trees as you will let them, and environmentalists want no logging at all. Your forest consists of 10% old-growth primary forest, and 90% secondary forest. Your forest managers are split among preferring maximum sustainable yield, ecosystem-based management, and adaptive management. What management approach(es) will you take? Will you allow logging of all, none, or some old-growth trees? Will you allow logging of secondary forest? If so, what harvesting strategies will you encourage? What would you ask your scientists before deciding on policies on fire management and salvage logging?
6. **THINK IT THROUGH** You have just been elected mayor of a town on Clayoquot Sound. A timber company that employs 20% of your town's residents wants to log a hillside above the town, and the provincial government is supportive of the harvest. But owners of ecotourism businesses that run whale watching excursions and rent kayaks to out-of-town visitors are complaining that the logging would destroy the area's esthetic appeal and devastate their businesses—and these businesses provide 40% of the tax base for your town. Greenpeace is organizing a demonstration in your town soon, and news reporters are beginning to call your office, asking what you will do. How will you proceed?

INTERPRETING GRAPHS AND DATA

The invention of the movable-type printing press by Johannes Gutenberg in 1450 stimulated a demand for paper that has only increased as the world population has grown.

The twentieth-century invention of the xerographic printing process used in photocopiers and laser printers has accelerated our demand for paper, with most raw fibre for paper production coming from wood pulp from forest trees.

1. How many millions of tons of paper and paperboard were consumed worldwide in 1970? 1980? 1990? 2000?

2. By what percentage did worldwide consumption of paper and paperboard increase from 1970 to 1980? From 1980 to 1990? From 1990 to 2000?
3. Explain three steps that your school could take to reduce its paper consumption.

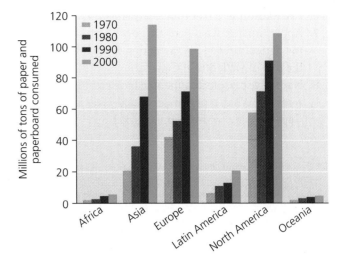

Global consumption of paper and paperboard, 1970–2000. Data from the Food and Agriculture Organization of the United Nations.

CALCULATING FOOTPRINTS

The average North American uses more than 300 kg of paper and paperboard per year. By using the estimates of paper and paperboard consumption for each region of the world (based on the year 2000)—as shown in the figure in the "Interpreting Graphs and Data" section—calculate the per capita consumption of paper and paperboard for each region of the world by using the population data in the table.

1. How much paper would North Americans save each year if we consumed paper at the rate of Europeans?
2. How much paper would be consumed if everyone in the world used as much paper as the average European? As the average North American?
3. Why do you think people in other regions consume less paper, per capita, than North Americans? Name three things you could do to reduce your paper consumption.

	Population* (millions)	Total paper consumed (millions of tonnes)	Per capita paper consumed (kilograms)
Africa	840	5.5	6.3
Asia	3 766		
Europe	728		
Latin America	531		
North America	319		
Oceania	32		
World	6 216		~50

*Data source: Population Reference Bureau.

TAKE IT FURTHER

 Go to www.myenvironmentplace.ca where you will find

- Suggested answers to end-of-chapter questions
- Quizzes, animations, and flashcards to help you study
- *Research Navigator*™ database of credible and reliable sources to assist you with your research projects
- Tutorials to help you master how to interpret graphs
- Current news articles that link the topics that you study to case studies from your region and around the world
- **ECO Occupational Profiles:** If you found this chapter especially interesting, you might want to learn more about the following jobs by visiting the Occupational Profiles website of the Environmental

Careers Organization. Go to www.eco.ca and check out the following careers:

- Arborist
- Cartographer
- Environmental manager
- Forester
- Forestry technician
- Geographer
- Geomatics technician
- GIS analyst
- Horticulturalist
- Park warden

CHAPTER ENDNOTES

1. *Friends of Clayoquot Sound Newsletter*, Fall 2007–Winter 2008.
2. Canada's National Forest Inventory, 2001, Canadian Forest Service, http://nfi.cfs.nrcan.gc.ca/canfi/facts_e.html.
3. National Forest Strategy Steering Committee, *National Forest Strategy 2003–2008: A Sustainable Forest: The Canadian Commitment*, http://nfsc.forest.ca/strategies/strategy5.html.
4. Canada's National Forest Inventory, *Area classification by forest region, 2001*, www.pfc.cfs.nrcan.gc.ca/monitoring/inventory/canfi/data/area-class-small_e.html.
5. Food and Agriculture Organization (of the United Nations) (2005) *Global Forest Resource Assessment 2005—Progress toward sustainable forest management*. FAO Forestry Paper. Rome, Italy, United Nations, Food and Agriculture Organization: p. 235, annex 3, table 9, ftp://ftp.fao.org/docrep/fao/008/A0400E/A0400E14.pdf. Primary forests are defined as "forests of native species, in which there are no clearly visible indications of human activity and ecological processes are not significantly disturbed" (p. 40).
6. Armson, Ken (1999) *Canadian Forests: A Primer*. Knowledge of the Environment for Youth (KEY) Environmental Literacy Series, The KEY Foundation.
7. Turner, Jennifer (2008) *Changing Climate Stops Pest Outbreaks on Vancouver Island*, Forest NewsTips, April, http://cfs.nrcan.gc.ca/news/585.
8. Canada's National Forest Inventory, 2001, Canadian Forest Service, http://nfi.cfs.nrcan.gc.ca/canfi/facts_e.html.
9. National Forest Strategy Steering Committee, *National Forest Strategy 2003–2008: A Sustainable Forest: The Canadian Commitment*, http://nfsc.forest.ca/strategies/strategy5.html.
10. NRCAN, Canadian Forest Service, *About the Canadian Forest Service: Our History*, http://cfs.nrcan.gc.ca/aboutus/organization/3.
11. National Forest Strategy Coalition, *National Forest Strategy 2003–2008: A Sustainable Forest: The Canadian Commitment*, http://nfsc.forest.ca/strategies/strategy5.html.
12. National Forest Strategy Coalition, *National Forest Strategy 2003–2008: A Sustainable Forest: The Canadian Commitment*, http://nfsc.forest.ca/strategies/strategy5.html.
13. National Forest Strategy Steering Committee, *National Forest Strategy 2003–2008: A Sustainable Forest: The Canadian Commitment*, http://nfsc.forest.ca/strategies/strategy5.html
14. B.C. Ministry of Forests and Range, Forest Practices Branch, *Adaptive Management Initiatives in the BC Forest Service*, www.for.gov.bc.ca/hfp/amhome/amhome.htm.
15. B.C. Ministry of Forests and Range, Forest Services Branch, www.for.gov.bc.ca/hfp/amhome/canadaprojects.htm.
16. Parks Canada, Canada's National Parks and National Reserves, www.pc.gc.ca/docs/v-g/nation/nation103_e.asp.
17. Campbell, L., and C. D. A. Rubec (2006) *Land Trusts in Canada: Building Momentum for the Future*. Ottawa: Wildlife Habitat Canada and the Stewardship Section, Canadian Wildlife Service, Environment Canada.
18. These two paragraphs are directly quoted from *The Clayoquot Women* by Bob Bossin, www3.telus.net/oldfolk/women.htm.
19. The preceding three sentences are quoted directly from *The Clayoquot Women* by Bob Bossin, www3.telus.net/oldfolk/women.htm.
20. ForestEthics, *About ForestEthics*, http://forestethics.org/article.php?list=type&type=9.
21. As quoted by Julia Dault, *Enviro Heroes: Tzeporah Berman*, Green Living Online, www.greenlivingonline.com/EnviroHeroes/enviro-heroes-tzeporah-berman.

Freshwater Systems and Water Resources

Canada has an abundance of fresh water, but much of it is at risk.

Upon successfully completing this chapter, you will be able to

- Explain the importance of water and the hydrologic cycle to ecosystems, human health, and economic pursuits
- Delineate the distribution of freshwater on Earth
- Describe major types of freshwater ecosystems
- Discuss how we use water and alter freshwater systems
- Assess problems of water supply and propose solutions to address depletion of freshwater
- Assess problems of water quality and propose solutions to address water pollution
- Explain how wastewater is treated

About 8% of Canada's surface is covered by water, more than any other nation, and 18% of the world's total available freshwater resides in the Great Lakes, seen here in a satellite image.

CENTRAL CASE:
TURNING THE TAP: THE PROSPECT OF CANADIAN BULK WATER EXPORTS

"Water promises to be to the twenty-first century what oil was to the twentieth century: the precious commodity that determines the wealth of nations."
—FORTUNE MAGAZINE, MAY 2000

"The wars of the twenty-first century will be fought over water."
—WORLD WATER COMMISSION CHAIRMAN ISMAIL SERAGELDIN

"I predict that the United States will be coming after our fresh water aggressively within three to five years. I hope that when the day comes, Canada will be ready."
—PETER LOUGHEED, FORMER PREMIER OF ALBERTA

There are few topics more emotional or more controversial for Canadians than the subject of freshwater.

Some argue that water is our legacy, our natural capital, and that its abundance defines us as Canadians. They fear that we will place our sovereignty at risk if we allow large-scale diversions of freshwater, or **bulk water exports**, from Canadian water bodies. Once bulk water exports are allowed to begin flowing to the thirsty southwestern United States, they maintain, they will be impossible to stop.

For others, access to freshwater is a fundamental human right. They argue that those who possess it in abundance have the moral duty to provide water to those who lack it. As the stewards of 25% of the world's wetlands, 7% of the world's renewable flowing water, and 18% of the world's surface freshwater (mainly in the Great Lakes),[1] Canada truly has abundant resources to manage.

For some, water is a valuable marketable commodity, which Canadians possess in surplus, and for which prices will continue to increase in the coming decades. Canada could be in an enviable position of economic and strategic strength in a world water market.

Others maintain that we should not even consider exporting our water to serve those who have mistreated, mismanaged, and depleted their own water supplies. Transporting large quantities of water from Canadian water bodies would cause massive changes, perhaps even permanent damage, to our natural ecosystems. And to what end, if those who would import our water have such a poor track record in appropriately managing this precious resource?

For example, desert areas of the southwestern United States are home to the enormous and rapidly increasing populations of cities like Los Angeles, Phoenix, and Las Vegas. The Imperial Valley of California—a natural desert (the sand dune scenes from *Star Wars: Return of the Jedi* were filmed there)—is one of the most fertile agricultural areas in the world, turning out water-intensive fruits like strawberries all winter. This is possible only because vast quantities of water are transported to the area from the Colorado River, via the 132 km All-American Canal. These deserts could never, under natural circumstances, sustainably support such large populations and water-dependent human activities.

Finally, there are those who reason that the very thirsty states of the American Southwest will be coming for our water before long anyway, and that perhaps we would be wise to sell it to them before they contemplate taking it by force.

As an environmental issue and a political issue it is confusing, complicated, and controversial, but it is crucially important for us as humans, as Canadians, and as citizens of North America and the world. Canada possesses some of the most enviable water resources in the world, but we also are some of the most wasteful users of water; per capita daily use of water in Canada (about 343 L/day) is surpassed only by the United States.

Water is already exported to the United States from Canada. The 66 bottlers of water in Canada produced more than 2.3 billion litres of bottled water in 2006; more than a third of it was exported, mostly to the United States.[2] Significant **interbasin transfers**—the transportation of water from one drainage basin to another—already occur between Canada and the United States. Most of this—about 97% of the volume—is for electrical power production.[3]

However, Canada has not yet approved the wholesale bulk export or massive diversion of water to the United States. The North American Free Trade Agreement (NAFTA) identifies water as a marketable and tradable commodity, which effectively means that Canada is prohibited from restricting water for use exclusively within its national boundaries. Interestingly, Simon Reisman, Canada's chief trade negotiator for NAFTA, was a director of the GRAND Canal Company, a private-sector proponent of bulk water diversion.[4] The "GRAND" (Great Recycling and Northern Development) Canal scheme would involve damming James Bay and diverting the 20 rivers that flow into it toward the south. Another large-scale bulk water export plan, NAWAPA (North American Water and Power Alliance), would divert the Yukon, Peace, and Liard rivers through an 800 km-long canal running along the Rocky Mountain Trench and into the United States.

For the moment, Canada's freshwater is protected against bulk exports by a watershed-based approach in which each province and territory individually prohibits bulk water exports under the *International Boundary Waters Treaty Act*. Because Canada's position on bulk water exports has been relatively firm, some American companies are eyeing the significant water resources in Alaska as an alternative source.

This discussion is ongoing, and it will continue in the coming decades. What is *your* position on bulk water exports? You may be called upon, before too long, to decide.

Freshwater Systems

"Water, water, everywhere, nor any drop to drink." The well known line from Coleridge's poem *The Rime of the Ancient Mariner* describes the situation on our planet quite well. Water may seem abundant to us, but water that we can drink is actually quite rare and limited (**FIGURE 11.1**). Roughly 97.5% of Earth's water resides in the oceans and is too salty to drink or use to water crops. Only 2.5% is considered **freshwater**, water that is relatively pure, with few dissolved salts. Because most freshwater is tied up in glaciers, icecaps, and underground aquifers, just over 1 part in 10 000 of Earth's water is easily accessible for human use.

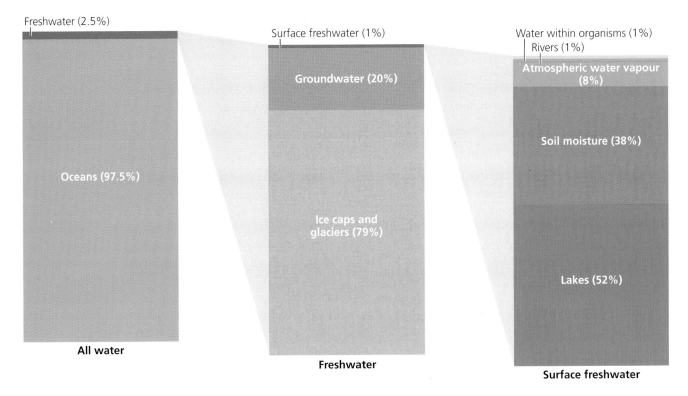

FIGURE 11.1
Only 2.5% of Earth's water is freshwater. Of that 2.5%, most is tied up in glaciers and icecaps. Of the 1% that is surface water, most is in lakes and soil moisture. Data from United Nations Environment Programme (UNEP) and World Resources Institute.

Water is constantly moving among the reservoirs specified in **FIGURE 11.1** via the *hydrologic cycle* (**FIGURE 5.15**). As water moves, it redistributes heat, erodes mountain ranges, builds river deltas, maintains organisms and ecosystems, shapes civilizations, and gives rise to political conflicts. Let us first examine the portions of the hydrologic cycle that are most conspicuous to us—surface water bodies—and take stock of the ecological systems they support.

Rivers and streams wind through landscapes

Water from rain, snowmelt, or springs runs downhill and converges where the land dips lowest, forming streams, creeks, or brooks. These watercourses merge into rivers, whose water eventually reaches the ocean (or sometimes ends in a landlocked water body). A smaller river flowing into a larger one is a **tributary**, and the area of land drained by a river and all its tributaries is that river's **drainage basin** or **watershed**.

Rivers shape the landscape through which they run. The force of water rounding a river's bend gradually eats away at the outer shore, eroding soil from the bank. Meanwhile, sediment is deposited along the inside of the bend, where water currents are weaker. In this way, over time, river bends become exaggerated in shape (**FIGURE 11.2**). Eventually, a bend may become such an extreme loop (called an *oxbow*) that water erodes a short-cut from one end of the loop to the other, pursuing a direct course. The bend is cut off and remains as an isolated U-shaped water body called an *oxbow lake*.

Over thousands or millions of years, a river may shift from one course to another, back and forth over a large area, carving out a flat valley. Areas nearest a river's course

FIGURE 11.2
Rivers and streams flow downhill, shaping landscapes, as shown by an oxbow of this meandering river in Alberta.

that are flooded periodically are said to be within the river's **floodplain**. Frequent deposition of silt from flooding makes floodplain soils especially fertile. As a result, agriculture thrives in floodplains, and **riparian** (riverside) forests are productive and species-rich.

The water of rivers and streams hosts diverse ecological communities. Algae and detritus support many types of invertebrates, from water beetles to crayfish. Insects as diverse as dragonflies, mayflies, and mosquitoes develop as larvae in streams and rivers before maturing into adults that take to the air. Fish consume aquatic insects, and birds, such as kingfishers, herons, and ospreys, dine on fish. Many amphibians spend their larval stages in streams, and some live their entire lives in streams. Salmon migrate from oceans up rivers and streams to spawn.

Wetlands include marshes, swamps, and bogs

Systems that combine elements of freshwater and dry land are enormously rich and productive. Often lumped under the term **wetlands**, such areas include different types of systems. In **freshwater marshes** (FIGURE 11.3), shallow water allows plants to grow above the water's surface. Cattails and bulrushes are plants typical of North American marshes. **Swamps** also consist of shallow water rich in vegetation, but they occur in forested areas. The cypress swamps of the southeastern United States, where cypress trees grow in standing water, are an example. Swamps are also created when beavers build dams across streams with limbs from trees they have cut, flooding wooded areas upstream. **Bogs** are ponds thoroughly covered with thick floating mats of vegetation and can represent a stage in aquatic succession; an example is the Mer Bleue Bog in Ontario, featured in the Central Case of Chapter 7.

Wetlands are extremely valuable as habitat for wildlife. They also provide important ecosystem services by slowing runoff, reducing flooding, recharging aquifers, and filtering pollutants. Despite these vital roles, people have drained and filled wetlands extensively, largely for agriculture. It is estimated that southeastern Canada has lost well over half of all wetlands since European colonization, with up to 90% loss in some areas. The Potholes region of the Canadian Prairies (and extending into the U.S. Midwest) is the most highly productive agricultural region of the country, as well as the host of about 4.5 million hectares of wetlands. The vast grasslands and small wetlands of the Potholes region have been widely converted to agricultural production, such that only about half of the wetlands that were present in the late 1700s (prior to European settlement) remain today.[5]

Lakes and ponds are ecologically diverse systems

Lakes and ponds are bodies of open standing water. Their physical conditions and the types of life within them vary with depth and the distance from shore. As a result, scientists have described several zones typical of lakes and ponds (FIGURE 11.4).

The region ringing the edge of a water body is named the **littoral zone**. Here the water is shallow enough that aquatic plants grow from the mud and reach above the water's surface. The nutrients and productive plant growth of the littoral zone make it rich in invertebrates—such as insect larvae, snails, and crayfish—on which fish, birds, turtles, and amphibians feed. The **benthic zone** extends along the bottom of the entire water body, from shore to the deepest point. Many invertebrates live in the mud on the bottom, feeding on detritus or preying on one another.

In the open portion of a lake or pond, away from shore, sunlight penetrates shallow waters of the **limnetic zone**. Because light enables photosynthesis and plant growth, the limnetic zone supports phytoplankton, which in turn support zooplankton, both of which are eaten by fish. Within the limnetic zone, sunlight intensity (and therefore water temperature) decreases with depth. The water's turbidity affects the depth of this zone; water that is clear allows sunlight to penetrate deeply, whereas turbid water does not. Below the limnetic zone is the **profundal zone**, the volume of open water that sunlight does not reach. This zone lacks plant life and thus is lower in dissolved oxygen and supports fewer animals. Aquatic

FIGURE 11.3
Shallow water bodies with ample vegetation are called wetlands and include swamps, bogs, and marshes, such as this one in Botswana, Africa.

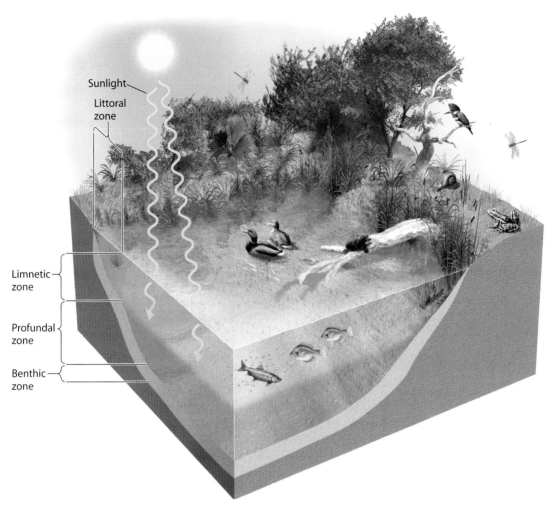

FIGURE 11.4
In lakes and ponds, emergent plants grow around the shoreline in the littoral zone. The limnetic zone is the layer of open, sunlit water, where photosynthesis takes place. Sunlight does not reach the deeper profundal zone. The benthic zone, which is the bottom of the water body, often is muddy, rich in detritus and nutrients, and low in oxygen.

animals rely on dissolved oxygen, and its concentration depends on the amount released by photosynthesis and the amount removed by animal and microbial respiration, among other factors.

Ponds and lakes change over time naturally as streams and runoff bring them sediment and nutrients. **Oligotrophic** lakes and ponds, which have low-nutrient and high-oxygen conditions, may slowly give way to the high-nutrient, low-oxygen conditions of **eutrophic** water bodies (jump ahead to see **FIGURE 11.20**). Eventually, water bodies may fill in completely by the process of aquatic succession. As lakes or ponds change over time, species of fish, plants, and invertebrates adapted to oligotrophic conditions may give way to those that thrive under eutrophic conditions.

Some lakes are so large that they differ substantially in their characteristics from small lakes. These large lakes are sometimes known as inland seas; the Great Lakes are prime examples. Because they hold so much water, most of their biota is adapted to open water. Major fish species of the Great Lakes include lake sturgeon, lake whitefish, northern pike, alewife, bass, walleye, and perch. Lake Baikal in Asia is the world's deepest lake, at 1637 m deep, and the Caspian Sea is the world's largest freshwater body, at 371 000 km^2.

Groundwater plays key roles in the hydrologic cycle

Any precipitation reaching Earth's land surface that does not evaporate, flow into waterways, or get taken up by organisms infiltrates the surface. Most percolates downward through the soil to become **groundwater** (**FIGURE 11.5**). Groundwater makes up one-fifth of Earth's freshwater supply and plays a key role in meeting human water needs.

Groundwater is contained within **aquifers**: porous formations of rock, sand, or gravel that hold water. An

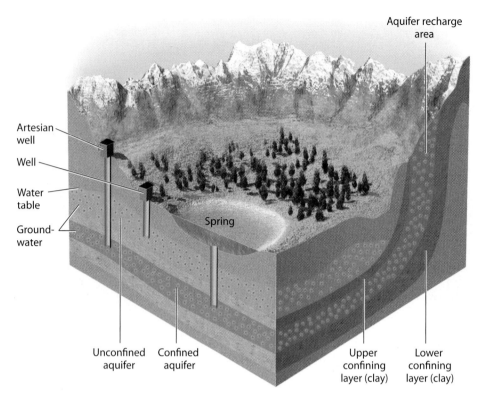

FIGURE 11.5
Groundwater may occur in unconfined aquifers above impermeable layers or in confined aquifers under pressure between impermeable layers. Water may rise naturally to the surface at springs and through the wells we dig. Artesian wells tap into confined aquifers to mine water under pressure.

aquifer's upper layer, or *zone of aeration*, contains pore spaces partly filled with water. In the lower layer, or *zone of saturation*, the spaces are completely filled with water. The boundary between these two zones is the **water table**. Picture a sponge resting partly submerged in a tray of water; the lower part of the sponge is completely saturated, whereas the upper portion may be moist but contains plenty of air in its pores. Any area where water infiltrates Earth's surface and reaches an aquifer below is known as an aquifer **recharge zone**.

There are two broad categories of aquifers. A **confined aquifer**, or **artesian aquifer**, exists when a water-bearing porous layer of rock, sand, or gravel is trapped between upper and lower layers of less permeable substrate (often clay). In such a situation, the water is under great pressure. In contrast, an **unconfined aquifer** has no such impermeable layer to confine it, so its water is under less pressure and can be readily recharged by surface water.

Just as surface water becomes groundwater by infiltration and percolation, groundwater becomes surface water through springs (and human-drilled wells), sometimes keeping streams flowing when surface conditions are otherwise dry. Groundwater flows downhill and from areas of high pressure to areas of low pressure, emerging to join surface water bodies at **discharge zones**. A typical rate of groundwater flow might be only about 1 m per day, so groundwater may remain in an aquifer for a long time. In fact, groundwater can be ancient. The average age of groundwater has been estimated at 1400 years, and some is tens of thousands of years old.

The world's largest known aquifer is the Ogallala Aquifer, which underlies the Great Plains of the United States. Water from this massive aquifer has enabled American farmers to create a bountiful grain-producing region. The volumes, the specific characteristics, and even the exact areal extents of the major aquifers in Canada are not yet thoroughly known; they are being studied under the Natural Resources Canada (NRCAN) Groundwater Mapping Program (**FIGURE 11.6**).[6]

Among the most significant aquifers in Canada—both in terms of extent or volume of water and in terms of the areas and populations they service—are the Paskapoo Formation, which covers more than 10 000 km^2 of southwestern Alberta; the Oak Ridges Moraine, a series of glacial deposits that cover 1900 km^2 and provide much of the Greater Toronto area with water; and the Annapolis-Cornwallis Valley Aquifers in Nova Scotia, covering a surface area of 2400 km^2 in a valley running parallel to the Bay of Fundy.[7] Groundwater is of particular importance in Prince Edward Island, which draws nearly all of its water from aquifers that underlie most of the island.

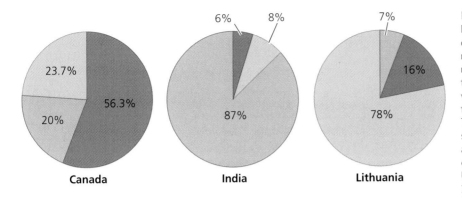

FIGURE 11.8
Nations apportion their freshwater consumption differently. Industry consumes most water used in Canada, agriculture uses the most in India, and most water in Lithuania goes toward domestic use. The largest category of water withdrawal in Canada (more than 60% of the total) is used in the production of electricity. This category includes only withdrawals, not in-stream use. Withdrawals for power production are not included in the data shown in this diagram. Data are based on information from U.N. Food and Agriculture Organization (FAO, 2000) and from Environment Canada, 2005, *The Management of Water: Water Use*.

the world's people are already affected by water scarcity (with less than 1000 m³ of water per person per year), according to a comprehensive global assessment presented in 2006.

Water supplies our households, agriculture, and industry

We all use water at home for drinking, cooking, and cleaning. Farmers and ranchers use water to irrigate crops and water livestock. Most manufacturing and industrial processes require water. The proportions of each of these three types of use—residential/municipal, agricultural, and industrial—vary dramatically among nations (**FIGURE 11.8**). Nations with arid climates tend to use more freshwater for agriculture, and heavily industrialized nations use a great deal for industry. Globally, we spend about 70% of our annual freshwater allotment on agriculture. Industry accounts for roughly 20%, and residential and municipal uses for only 10%.

When we remove water from an aquifer or surface water body and do not return it, this is called **consumptive use**. A large portion of agricultural irrigation and of many industrial and residential uses is consumptive. **Nonconsumptive use** of water does not remove, or only temporarily removes, water from an aquifer or surface water body. Using water to generate electricity at hydroelectric dams is an example of nonconsumptive use; water is taken in, passed through dam machinery to turn turbines, and released downstream.

We have erected thousands of dams

A **dam** is any obstruction placed in a river or stream to block the flow of water so that water can be stored in a reservoir. We build dams to prevent floods, provide drinking water, facilitate irrigation, and generate electricity (**FIGURE 11.9**; Table 11.1). Power generation with hydroelectric dams is discussed in Chapter 16.

Worldwide, we have erected more than 45 000 large dams (greater than 15 m high) across rivers in more than 140 nations, and tens of thousands of smaller dams. In the Prairie provinces alone there are almost 800 dams.[10] Only a few major rivers in the world remain undammed and free-flowing; these run through the tundra and taiga of Canada, Alaska, and Russia and in remote regions of Latin America and Africa.

Our largest dams are some of humanity's greatest engineering feats. The Gardiner Dam in Saskatchewan is the largest in Canada in terms of water-holding capacity; the Mica Dam in British Columbia is the tallest. The two behemoths in the United States are the Hoover and Glen Canyon dams. The Hoover Dam, probably the most recognizable icon of the great dam-building period in the United States, holds 35.2 km³ of water in a reservoir that is 177 km long and 152 m deep. The Glen Canyon's reservoir is almost as large; together they store four times as much water as flows in the river in an entire year.

China's Three Gorges Dam is the world's largest

The complex mix of benefits and costs that dams produce is exemplified by the world's largest dam project, the Three Gorges Dam on China's Yangtze River. (It is worth noting that the second-largest dam in the world is not a water dam but the Syncrude tar sand tailings dam in Alberta; see Chapter 15.) The Three Gorges Dam, 186 m high and 2 km wide, was completed in 2003 (**FIGURE 11.10A**). When completely filled in 2009–2010, its reservoir should

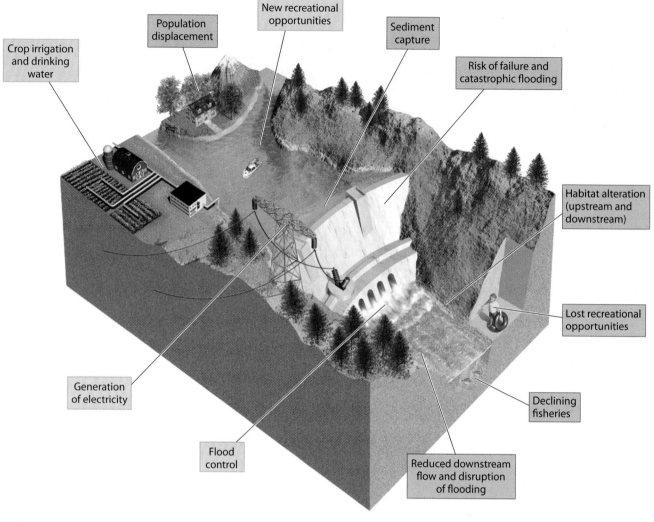

FIGURE 11.9
Damming rivers has diverse consequences for people and the environment. The generation of clean and renewable electricity is one of several major benefits (green boxes) of hydroelectric dams. Habitat alteration is one of several negative impacts (red boxes).

stretch for 616 km (as long as Lake Superior) and hold more than 38 trillion litres of water. This project will enable boats and barges to travel farther upstream, provide flood control, and generate enough hydroelectric power to replace dozens of large coal or nuclear plants.

However, the Three Gorges Dam has cost $25 billion to build, and its reservoir is flooding 22 cities and the homes of 1.13 million people, requiring the largest resettlement project in China's history (**FIGURE 11.10B**). The major earthquake in southern China in 2008 raised fears of potential damage to the great structure; the consequences of a collapse of a dam this large would be devastating.

The filling of the reservoir behind the dam is submerging 10 000-year-old archeological sites, productive farmlands, and wildlife habitat. Moreover, the reservoir slows the river's flow so that suspended sediment settles behind the dam. Indeed, the reservoir began accumulating

weighing the issues

IS THE THREE GORGES DAM WORTH IT? 11–1

Do you think the projected benefits from China's great Three Gorges Dam (energy and water for industry and municipalities) will be worth the risk to the environment, and the social disruption of displacing more than a million people?

---

sediment as soon as the dam was completed, and because the river downstream is deprived of sediment, the tidal marshes at the Yangtze's mouth are eroding away, leaving the city of Shanghai with a degraded coastal environment

Table 11.1 Major Benefits and Costs of Dams

Benefits

- **Power generation.** Hydroelectric dams (• pp. 595–598) provide inexpensive electricity.
- **Emissions reduction.** Hydroelectric power produces no greenhouse gases in its operation (although some are produced during construction and maintenance of infrastructure). By replacing fossil fuel combustion as an electricity source, hydropower reduces air pollution and climate change, and their health and environmental consequences.
- **Crop irrigation.** Reservoirs can release irrigation water when farmers most need it and can buffer regions against drought.
- **Drinking water.** Many reservoirs store plentiful, reliable, and clean water for municipal drinking water supplies, provided that watershed lands draining into the reservoir are not developed or polluted.
- **Flood control.** Dams can prevent floods by storing seasonal surges, such as those following snowmelt or heavy rain.
- **Shipping.** By replacing rocky river beds with deep placid pools, dams enable ships to transport goods over longer distances.
- **New recreational opportunities.** People can fish from boats and use personal watercraft on reservoirs in regions where such recreation was not possible before.

Costs

- **Habitat alteration.** Reservoirs flood riparian habitats and displace or kill riparian species. Dams modify rivers downstream. Shallow warm water downstream from a dam is periodically flushed with cold reservoir water, stressing or killing many fish.
- **Fisheries declines.** Salmon and other fish that migrate up rivers to spawn encounter dams as a barrier. Although "fish ladders" at many dams allow passage, most fish do not make it.
- **Population displacement.** Reservoirs generally flood fertile farmland and have flooded many human settlements. An estimated 40–80 million people globally have been displaced by dam projects over the past half century.
- **Sediment capture.** Sediment settles behind dams. Downstream floodplains and estuaries are no longer nourished, and reservoirs fill with silt.
- **Disruption of flooding.** Floods create productive farmland by depositing rich sediment. Without flooding, topsoil is lost, and farmland deteriorates.
- **Risk of failure.** There is always risk that a dam could fail, causing massive property damage, ecological damage, and loss of life.
- **Lost recreational opportunities.** Tubing, whitewater rafting, fly-fishing and kayaking opportunities are lost.

(a) The Three Gorges Dam in Yichang, China

(b) Displaced people in Sichuan Province, China

(c) Archival photo of the construction of the St. Lawrence Seaway

FIGURE 11.10
China's Three Gorges Dam, completed in 2003, is the world's largest dam **(a)**. Well over a million people were displaced, archeological treasures were lost forever, and whole cities were levelled for its construction, as shown here in Sichuan Province **(b)**. The reservoir began filling in 2003 and will continue filling for several years. The construction of the St. Lawrence Seaway **(c)** from 1954 to 1959 required the displacement of 6500 people, and caused the flooding of 14 000 hectares of land, seven villages, and 225 farms.

and less coastal land to develop. Many scientists worry that the Yangtze's many pollutants will also be trapped in the reservoir, making the water undrinkable. In fact, high levels of bacteria were found in the water as it began building up behind the dam. The Chinese government plans to sink $5 billion into building hundreds of sewage treatment and waste disposal facilities.

Lest you have the impression that large structures with extensive social and environmental impacts are only constructed in developing countries, consider the case of the St. Lawrence Seaway (**FIGURE 11.10C**). Although it took more than 50 years of discussion, once underway the St. Lawrence Seaway took only five years to complete; it opened in 1959. The Seaway is a series of canals that connects the Great Lakes to the Atlantic Ocean, following the route of the St. Lawrence River. Its construction required the displacement of 6500 people from farms and homes along the canal route; whole villages were flooded. The Lost Villages Historical Society website (http://lostvillages.ca/) documents the construction of the St. Lawrence Seaway and the farms and villages that were affected.

Some dams are now being removed

People who feel that the costs of some dams have outweighed their benefits are pushing for such dams to be dismantled. By removing dams and letting rivers flow freely, these people say, we can restore riparian ecosystems, reestablish economically valuable fisheries, and revive river recreation, such as fly-fishing and rafting. Another common reason for the decommissioning of dams is that many aging dams are in need of costly repairs or have outlived their economic usefulness.

Roughly 500 dams have been removed in the United States in recent years, and more will soon follow. In Canada only a handful of dams have been decommissioned so far, but the concept of dam removal for river restoration and ecological recovery is beginning to take hold. Dam removals also provide opportunities for scientific study of the response of the river and aquatic communities to changes in flow rate, water temperature, sedimentation, and other factors that accompany decommissioning.

Dikes and levees are meant to control floods

Flood prevention ranks high among reasons we control the movement of freshwater. People have always been attracted to riverbanks for their water supply and for the flat topography and fertile soil of floodplains. Flooding is a normal, natural process caused by snowmelt or heavy rain, and floodwaters spread nutrient-rich sediments over large areas, benefiting both natural systems and human agriculture.

In the short term, however, floods can do tremendous damage to the farms, homes, and property of people who choose to live in floodplains. To protect against floods, individuals and governments have built **dikes** and levees (long raised mounds of earth) along the banks of rivers to hold rising water in main channels. Many dikes are small and locally built, but some are massive. In Canada, the flood diversion wall that protects the City of Winnipeg from the Red River is a major example of a flood protection structure (**FIGURE 11.11**). In the United States the Army Corps of Engineers has constructed thousands of kilometres of massive levees along the banks of major waterways (those that failed in New Orleans after Hurricane Katrina are examples). Although these structures prevent flooding at most times and places, they can sometimes worsen flooding because they force water to stay in channels and accumulate, building up enormous energy and leading to occasional catastrophic overflow events.

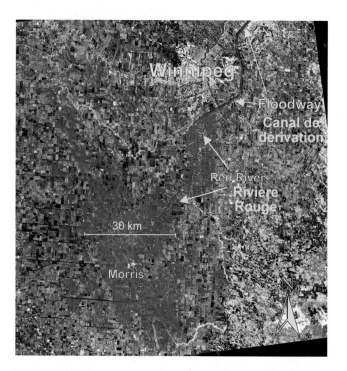

FIGURE 11.11
A flood diversion wall protected the City of Winnipeg during the 1997 Red River flood but may have exposed other villages to flooding. In this satellite image, the water appears blue, and the diversion wall can be seen in the upper righthand corner of the photograph, keeping the water from entering the core of the City of Winnipeg.

We divert—and deplete—surface water to suit our needs

People have long diverted water from rivers, streams, lakes, and ponds to farm fields, homes, and cities. *Diversion* refers to the process of removing water from its channel, or modifying its flow for the purpose of using it elsewhere.

The Colorado River in the southwestern United States is one of the world's classic examples of diversion and over-allocation of water from a major river (FIGURE 11.12). Early in its course, some Colorado River water is piped through a mountain tunnel and down the Rockies' eastern slope to supply the city of Denver. More is removed for Las Vegas and other cities and for farmland as the water proceeds downriver. When it reaches Parker Dam on the California–Arizona state line, large amounts are diverted into the Colorado River Aqueduct, which brings water to millions of people in the Los Angeles and San Diego areas via a long open-air canal. From Parker Dam, Arizona also draws water, transporting it in the large canals of the Central Arizona Project. Farther south at Imperial Dam, water is diverted into the Coachella and All-American Canals, destined for agriculture, mostly in California's Imperial Valley. To make this desert bloom, Imperial Valley farmers soak the soil with subsidized water for which they pay one penny per 795 L.

What water is left in the Colorado River after all the diversions is just a trickle making its way to the Gulf of California and Mexico. On some days, water does not reach the Gulf at all. This reduction in flow (FIGURE 11.12) has drastically altered the ecology of the lower river and the once-rich delta, changing plant communities, wiping out populations of fish and invertebrates, and devastating fisheries. It also led to a tense international incident between the United States and Mexico in the 1970s, necessitating high-level international talks and agreements concerning water reallocation, which improved the situation somewhat.

Nowhere are the effects of surface water depletion so evident as at the Aral Sea. Once the fourth-largest lake on Earth, just larger than Lake Huron, it has lost more than four-fifths of its volume in just 45 years (FIGURE 11.13). This dying inland sea, on the border of present-day Uzbekistan and Kazakhstan, is the victim of irrigation practices. The former Soviet Union instituted large-scale cotton farming in this region by flooding the dry land with water from the two rivers leading into the Aral Sea. For a few decades this action boosted Soviet cotton production, but it caused the Aral Sea to shrink, and the irrigated soil became waterlogged and salinized.

Today 60 000 fishing jobs are gone, winds blow pesticide-laden dust up from the dry lake bed, and what cotton grows on the blighted soil cannot bring the regional economy back. However, all may not be lost: Scientists, engineers, and local people struggling to save the northern portion of the Aral Sea and its damaged ecosystems may now have finally begun reversing its decline.

weighing the issues

FLOOD PROTECTION AT WHAT COST? 11–2

The Red River Floodway protected the City of Winnipeg during the massive flood of 1997. However, some have argued that this diversion of flood waters from the city's core put other farms and villages at risk of flooding that would not normally have been in the path of the flood, most notably the town of St. Agathe. Do you believe it is worthwhile, or even crucial, to protect the homes, businesses, and other costly resources, as well as the people of the City of Winnipeg, even if it means sacrificing some other areas to the flood? What would you do to protect or compensate people in those other areas?

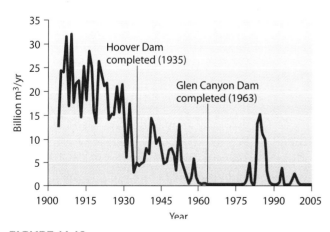

FIGURE 11.12
Flow at the mouth of the Colorado River has greatly decreased over the past century as a result of withdrawals, mostly for agriculture. The river now often runs dry at its mouth. Data from Postel, S. 2005. *Liquid assets: The critical need to safeguard freshwater ecosystems.* Worldwatch Paper 170. Washington, DC: Worldwatch Institute.

Inefficient irrigation wastes water

The Green Revolution required significant increases in irrigation, and 70% more water is withdrawn for irrigation today than in 1960. During this period, the amount of land under irrigation has doubled (FIGURE 11.14). Expansion of irrigated agriculture has kept pace with

(a) Ships stranded by the Aral Sea's fast-receding waters

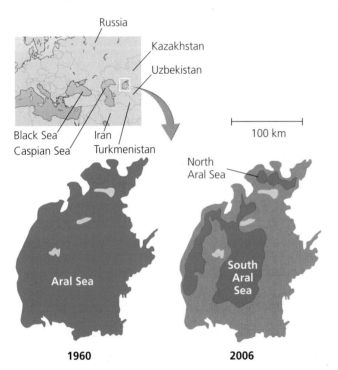

(c) The shrinking Aral Sea, then and now

(b) Satellite view of Aral Sea, 2002

FIGURE 11.13
Ships lie stranded in the sand **(a)** because the waters of Central Asia's Aral Sea have receded so far and so quickly **(b)**. The Aral Sea was once the world's fourth-largest lake. However, it has been shrinking for the past four decades **(c)** because of overwithdrawal of water to irrigate cotton crops. Today restoration efforts are beginning to reverse the decline in the northern portion of the sea, and waters there are slowly rising.

population growth; irrigated area per capita has remained stable for at least four decades at around 460 m².

Irrigation can more than double crop yields by allowing farmers to apply water when and where it is needed. The world's 274 million hectares of irrigated cropland make up only 18% of world farmland but yield fully 40% of world agricultural produce, including 60% of the global grain crop. Still, most irrigation remains highly inefficient. Only about 45% of the freshwater we use for irrigation actually is taken up by crops. Inefficient "flood and furrow" irrigation, in which fields are liberally flooded with water that may evaporate from standing pools, accounts for 90% of irrigation worldwide. Overirrigation leads to waterlogging and salinization, which affect one-fifth of farmland today and reduce world farming income by $11 billion.

Many national governments have subsidized irrigation to promote agricultural self-sufficiency. Unfortunately, inefficient irrigation methods in arid regions, such as the Middle East, are using up huge amounts of groundwater for little gain. Worldwide, roughly 15%–35% of water withdrawals for irrigation are thought to be

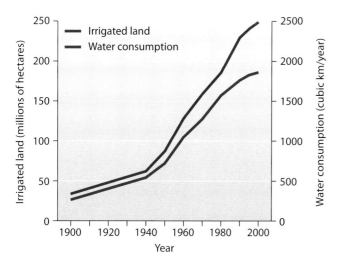

FIGURE 11.14
Throughout the twentieth century, overall global water consumption rose in tandem with the area of land irrigated for agriculture. Data from United Nations Sustainable Development Commission on Freshwater, 2002.

unsustainable. In areas where agriculture is demanding more freshwater than can be sustainably supplied, **water mining**—withdrawing water faster than it can be replenished—is taking place (**FIGURE 11.15**). In these areas, aquifers are being depleted or surface water is being piped in from other regions.

Wetlands have been drained for a variety of reasons

Throughout recent history, governments have encouraged laborious efforts to drain wetlands in order to promote settlement and farming. Many of today's crops grow on the sites of former wetlands—swamps, wooded marshes, bogs, and river floodplains—that people have drained and filled in (**FIGURE 11.16**). Wetlands also have historically been seen as "swamps"—insect-infested, smelly, and useless for any kind of industrial or agricultural development. In the 1930s the governor of Florida announced his intention to drain the entire Everglades, a unique cypress, mangrove, and tropical hardwood wetland that occupies much of the southern portion of the state. This attitude began to change dramatically in North America with the rise of the modern environmental movement in the 1970s and has continued to change as a result of research efforts to determine the economic value of services provided by wetlands (see Chapter 21).

In 1971, an international agreement was reached in Ramsar, Iran, concerning the documentation and protection of wetlands around the world. It is known as the *Ramsar Convention*, or more accurately, the *Convention on Wetlands of International Importance, Especially as*

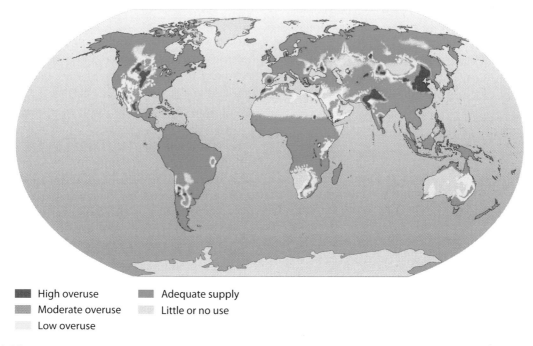

FIGURE 11.15
Irrigation for agriculture is the main contributor to unsustainable water use. Mapped are regions where overall use of freshwater (for agriculture, industry, and domestic use) exceeds the available supply, requiring groundwater depletion or diversion of water from other regions. The map understates the problem, because it does not reflect seasonal shortages. Data from UNESCO. 2006. *Water: A Shared Responsibility.* World Water Development Report 2. Paris and New York: UNESCO and Berghahn Books.

FIGURE 11.16
Most of North America's wetlands have been drained and filled, and the land converted to agricultural use. The northern Great Plains region of Canada was pockmarked with thousands of "prairie potholes," water-filled depressions that served as nesting sites for most of the continent's waterfowl. Today many of these wetlands have been lost; shown are farmlands encroaching on prairie potholes.

Waterfowl Habitat. The Ramsar Convention, to which Canada is a signatory, demonstrates the global concern regarding wetland loss and degradation. The mission of the treaty is the "conservation and wise use of all wetlands through local, regional, and national actions and international cooperation."[11] Today a large portion—perhaps 90%—of the original wetlands in southern Canada have been lost, but Canada as a whole still has the largest area of wetlands of any country in the world, according to Ramsar Convention data (**FIGURE 11.17**).[12]

Many people now have a different view of wetlands. Rather than seeing them as worthless swamps, science has made it clear that wetlands are valuable ecosystems. This scientific knowledge, along with a preservation ethic, has induced policy makers to develop regulations to safeguard remaining wetlands. Yet, because of loopholes, differing laws, development pressures—and even debate over the legal definition of wetlands—many of these vital ecosystems are still being lost.

We are depleting groundwater

Groundwater is more easily depleted than surface water because most aquifers recharge very slowly. If we compare an aquifer to a bank account, we are making more withdrawals than deposits, and the balance is shrinking. Today we are extracting 160 km^3 more water each year than is finding its way back into the ground. This is a major problem because one-third of Earth's human population—including 26% of the population of Canada—relies on groundwater for its needs. As aquifers are depleted, water tables drop. Groundwater becomes more difficult and expensive to extract, and eventually it may run out. In parts of Mexico, India, China, and other Asian and Middle Eastern nations, water tables are falling 1–3 m per year.

When groundwater is overpumped in coastal areas, saltwater can intrude into aquifers, making water undrinkable. This has occurred widely in the Middle East and in localities as varied as Florida, Turkey, and Bangkok. Moreover, as aquifers lose water, their substrate can become weaker and less capable of supporting overlying strata, and the land surface above may subside. For this reason, cities from Venice to Bangkok to Shanghai are slowly sinking. Mexico City's downtown has sunk over 10 m since the time of Spanish arrival; streets are buckled, old

FIGURE 11.17
The Columbia Wetlands, seen here, is Ramsar site number 1463—a wetland of international significance and the largest of its kind in British Columbia.

FIGURE 11.18
When too much groundwater is withdrawn too quickly, the land above it may collapse in sinkholes, sometimes bringing buildings down with it.

buildings lean at angles, and underground pipes break so often that 30% of the system's water is lost to leaks.

Sometimes land subsides suddenly in the form of **sinkholes**, areas where the ground gives way with little warning, occasionally swallowing people's homes (**FIGURE 11.18**). Once the ground subsides, soil can undergo *compaction*, becoming compressed and losing the porosity that enabled it to hold water. Recharging a depleted aquifer may thereafter become much more difficult.

Falling water tables also do vast ecological harm. Permanent wetlands exist where water tables are high enough to reach the surface, so when water tables drop, wetland ecosystems dry up. In Jordan, the Azraq Oasis covered 7500 ha and enabled migratory birds and other animals to find water in the desert. The water table beneath this oasis dropped 2.5–7 m during the 1980s because of increased well use by the city of Amman. As a result, the oasis dried up altogether during the 1990s. Today international donors are collaborating with the Jordanian government to try to find alternative sources of water and restore this oasis.

Our thirst for bottled water seems unquenchable

It seems to fit our busy lifestyle. Canadians' use of bottled water is surpassed only by that of Americans. Statistics Canada reports that almost 3 in 10 households in Canada today utilize bottled water as their main source for domestic drinking water. The proportion of households dependent on bottled water increases with household income, although the relationship between income and bottled water consumption is complex. Interestingly, university-educated households in the Statistics Canada 2008 study[13] were shown to be *less* likely to consume bottled water.

For the nation as a whole, some 820 million litres of water were bottled for Canadian consumption in 2000; by 2003 the amount had increased to almost 1.5 billion litres. This means that the average per capita consumption of 17.9 L of bottled water per year in 1995 had risen to 27.6 L by 2000, and then jumped to almost 50 L by 2003![14]

The interesting thing about bottled water is that much of it is just plain, ordinary tap water—in fact, it has been estimated that as much as a quarter of all bottled water comes straight from a municipal water tap, sometimes with additional filtering or other treatment. Advertising and labelling legislation prohibits bottling companies from misrepresenting what is in the bottle, of course, so if you read the label carefully you should be able to determine whether the water has come from a "natural" source, such as groundwater or a spring, or from a municipal supply.

But is natural source bottled water necessarily more healthful or safer for you? Although most people think so, in fact there are far fewer checks on bottled water and the bottling process than on municipal water supplies, which are rigorously monitored and regulated. Studies also have now shown that the small amount of bacteria present in water—even clean bottled water—rapidly grow when exposed to sunlight, especially if the bottle has been opened. Once someone drinks from the bottle, bacteria from the mouth enter the water and can multiply.

So the next time you reach for a bottle of water, consider what's in the bottle, where it came from, how it was extracted and monitored, and exactly what you are paying for. *Then* decide whether you think it is worth generating an empty plastic bottle.

weighing the issues: THE PRICE OF A LITRE 11-3

Do you drink bottled water? Why? (Do you think it is safer than municipal water? Do you prefer the taste? Is it more convenient?)

What do you pay for a litre of bottled water? What do you pay for a litre of gas at the pump? What do you think *should* be reflected in these prices? Are all of these things adequately reflected in the current prices for these two commodities? What price do you think was paid for the water by the company that bottled it?

What about the source of the water you consume—is it groundwater and, if so, is its source adequately protected? And what about the plastic waste that is generated?

Will we see a future of water wars?

Depletion of freshwater leads to shortages, and resource scarcity can lead to conflict, as shown by the research of Thomas Homer-Dixon and others. Many predict that water's role in regional conflicts will increase as human population continues to grow in water-poor areas and as climate change alters regional patterns of precipitation. A total of 261 major rivers, whose watersheds cover 45% of the world's land area, are **transboundary** waterways, that is, they cross or flow along national borders, and disagreements are common. Water is already a key element in the hostilities among Israel, the Palestinian people, and neighbouring nations.

On the positive side, many nations have cooperated with neighbours to resolve water disputes. India has struck cooperative agreements over management of transboundary rivers with Pakistan, Bangladesh, Bhutan, and Nepal. In Europe, international conventions have been signed by multiple nations along the Rhine and the Danube rivers. The international agreements between Canada and the United States, governing the Great Lakes and other water bodies that straddle the boundary between our two nations, have been examples of largely successful water management agreements. Such progress gives reason to hope that future water wars will be few and far between.

Solutions to Depletion of Freshwater

Human population growth, expansion of irrigated agriculture, and industrial development doubled our annual freshwater use between 1960 and 2000. We now use an amount equal to 10% of total global runoff. The hydrologic cycle makes freshwater a renewable resource, but if our usage exceeds what a lake, a river, or an aquifer can provide, we must either reduce our use, find another water source, or be prepared to run out of water.

Solutions can address supply or demand

To address depletion of freshwater, we can aim either to increase supply or to reduce demand. Strategies for reducing demand include conservation and efficiency measures. Lowering demand is more difficult politically in the short term but may be necessary in the long term. In the developing world, international aid agencies are increasingly funding demand-based solutions over supply-based solutions, because demand-based solutions offer better economic returns and cause less ecological and social damage.

To increase supply in a given area, people have transported water through pipes and aqueducts from areas where it is more plentiful or accessible. In many instances, water-poor regions have forcibly appropriated water from communities too weak to keep it for themselves. For instance, Los Angeles grew by using water it appropriated from the Owens Valley, Mono Lake, and other less-inhabited regions of California. In so doing, it desertified the environments of those areas, creating dust-bowls and destroying rural economies. Today Las Vegas is trying to win approval for a 450 km pipeline to import groundwater from sparsely populated eastern Nevada, where local residents and wildlife advocates oppose the diversion plan.

> ### weighing the issues
> **WEIGHING THE ISSUES:**
> **REACHING FOR WATER**
>
> In 1941, Los Angeles needed water and decided to divert streams feeding into Mono Lake, more than 565 km away in northern California. As the lake level fell 14 m over 40 years, salt concentrations doubled and aquatic communities suffered. Other desert cities in the American Southwest—such as Las Vegas, Phoenix, and Denver—are expected to double in population in the coming decades.
>
> What challenges might these cities face in trying to pipe in water from distant sources, including Canada? How will people living in the source areas be affected? Do you think such diversions are justified? How else could these cities meet their future water needs? Or should water shortage force an end to development in these areas, and oblige residents to adopt a more sustainable lifestyle with regard to water consumption? As a Canadian voter, you may one day need to decide whether Canada should sell bulk water to California, Nevada, and Arizona. What is your position?

Desalination "makes" more water

Another supply-side strategy is to develop technologies to find or "make" more water. The best known technological approach to generate freshwater is **desalination**, or *desalinization*, the removal of salt from seawater or other water of marginal quality. One method of desalination mimics the hydrologic cycle by hastening evaporation from allotments of ocean water with heat and then

FIGURE 11.19
The Jubail Desalinization Plant in Saudi Arabia is the largest facility in the world that turns saltwater into freshwater.

condensing the vapour—essentially *distilling* freshwater. Another method involves forcing water through membranes to filter out salts; the most common process of this type is called *reverse osmosis*.

More than 7500 desalination facilities are operating worldwide, most in the arid Middle East and some in small island nations that lack groundwater. The largest plant, in Saudi Arabia, produces 485 million litres of freshwater every day (**FIGURE 11.19**). However, desalination is expensive, requires large inputs of fossil fuel energy, and generates concentrated salty waste.

Agricultural demand can be reduced

Because most water is used for agriculture, it makes sense to look first to agriculture for ways to decrease demand. Farmers can improve efficiency by lining irrigation canals to prevent leaks, levelling fields to minimize runoff, and adopting efficient irrigation methods. Low-pressure spray irrigation sprays water downward toward plants, and drip irrigation systems target individual plants and introduce water directly onto the soil (see **FIGURE 7.16**). Both methods reduce water lost to evaporation and surface runoff. Low-pressure precision sprinklers in use in a number of arid localities have efficiencies of 80%–95% and have resulted in water savings of 25%–37%. Experts have estimated that drip irrigation, which has efficiencies as high as 90%, could cut water use in half while raising yields by 20%–90% and giving developing-world farmers $3 billion in extra annual income.

Choosing crops to match the land and climate in which they are being farmed can save huge amounts of water. Currently, crops that require a great deal of water, such as cotton, rice, and alfalfa, are often planted in arid areas with government-subsidized irrigation. As a result of the subsidies, the true cost of water is not part of the costs of growing the crop. Eliminating subsidies and growing crops in climates with adequate rainfall could greatly reduce water use. Finally, selective breeding and genetic modification can result in crop varieties that require less water.

We can lessen residential and industrial water use in many ways

We can each help reduce agricultural water use by decreasing the amount of meat we eat, because producing meat requires far greater water inputs than producing grain or vegetables. In households, we can reduce water use by installing low-flow faucets, showerheads, washing machines, and toilets. Automatic dishwashers, studies show, use less water than does washing dishes by hand. If your home has a lawn, it is best to water it at night, when water loss from evaporation is minimal. Better yet, you can replace a water-intensive lawn with native plants adapted to the region's natural precipitation patterns. An example is **xeriscaping**—landscaping with plants that are well adapted to a dry environment.

Industry and municipalities can take water-saving steps as well. Manufacturers are shifting to processes that use less water and in doing so are reducing their costs. Some cities are recycling municipal wastewater for irrigation and industrial uses. Governments in England and in Arizona are capturing excess surface runoff during their rainy seasons and pumping it into aquifers. Finding and patching leaks in pipes has saved some cities and companies large amounts of water—and money. A program of retrofitting enabled Massachusetts to reduce water demand by 31% and avoid an unpopular $500 million river diversion scheme.

Economic approaches to water conservation are being debated

Economists who want to use market-based strategies to achieve sustainable water use have suggested ending government subsidies of inefficient practices and letting water become a commodity whose price reflects the true costs of its extraction. Others worry that making water a fully priced commodity would make it less available to the world's poor and increase the gap between rich and poor. Because industrial use of water can be 70 times as profitable as agricultural use, market forces alone might favour uses that would benefit wealthy and industrialized people, companies, and nations at the expense of the poor and less industrialized.

Similar concerns surround another potential solution, the privatization of water supplies. During the 1990s, many public water systems were partially or wholly privatized, with their construction, maintenance, management, or ownership being transferred to private companies. This was done in the hope of increasing the systems' efficiency, but many firms have little incentive to allow equitable access to water for rich and poor alike. Already in some developing countries, rural residents without access to public water supplies, who are forced to buy water from private vendors, end up paying on average 12 times more than those connected to public supplies.

Other experiences indicate that decentralization of control over water, from the national level to the local level, may help conserve water. In Mexico, the effectiveness of irrigation systems improved dramatically once they were transferred from public ownership to the control of 386 local water user associations.

Regardless of how demand is addressed, the ongoing shift from supply-side to demand-side solutions is beginning to pay dividends. A new focus on demand (through government mandates and public education) has decreased public water consumption, and industries are becoming more water-efficient.

Freshwater Pollution and Its Control

The quantity and distribution of freshwater poses one set of environmental and social challenges. Safeguarding the *quality* of water involves another collection of environmental and human health dilemmas. To be safe for consumption by human beings and other organisms, water must be relatively free of disease-causing organisms and toxic substances.

Although developed nations have made admirable advances in cleaning up water pollution over the past few decades, the World Commission on Water recently concluded that more than half the world's major rivers are "seriously depleted and polluted, degrading and poisoning the surrounding ecosystems, threatening the health and livelihood of people who depend on them." The largely invisible pollution of groundwater, meanwhile, has been termed a "covert crisis."

Water pollution takes many forms

The term **pollution** describes the release into the environment of matter or energy that causes undesirable impacts on the health and well-being of humans or other organisms. Pollution can be physical, chemical, or biological and can affect water, air, or soil.

Water pollution comes in many forms and can cause diverse impacts on aquatic ecosystems and human health. We can categorize pollution into several types, including nutrient pollution, biological pollution by disease-causing organisms, toxic chemical pollution, physical pollution by sediment, and thermal pollution.

Nutrient pollution   We saw in Chapter 5 how nutrient pollution from fertilizers and other sources can lead to eutrophication and hypoxia in coastal marine areas (see **FIGURE 5.19**). Eutrophication proceeds in a similar fashion in freshwater systems, where phosphorus is usually the nutrient that spurs growth. When excess phosphorus enters surface waters, it fertilizes algae and aquatic plants, boosting their growth rates and populations. Although such growth provides oxygen and food for other organisms, algae can cover the water's surface, depriving deeper-water plants of sunlight. As algae die off, they provide food for decomposing bacteria. Decomposition requires oxygen, so the increased bacterial activity drives down levels of dissolved oxygen. These levels can drop too low to support fish and shellfish, leading to dramatic changes in aquatic ecosystems.

Eutrophication (**FIGURE 11.20**) is a natural process, but excess nutrient input from runoff from farms, golf courses, lawns, and sewage can dramatically increase the rate at which it occurs. We can reduce nutrient pollution by treating wastewater, reducing fertilizer application, planting vegetation to increase nutrient uptake, and purchasing phosphate-free detergents.

Pathogens and waterborne diseases   Disease-causing organisms (pathogenic viruses, protists, and bacteria) can enter drinking water supplies when these are contaminated with human waste from inadequately treated sewage or with animal waste from feedlots. Specialists monitoring water quality can tell when water has been contaminated by waste when they detect fecal coliform bacteria, which live in the intestinal tracts of people and other vertebrates. These bacteria are usually not pathogenic themselves, but they serve as indicators of fecal contamination, which may mean that the water holds other pathogens that can cause ailments, such as giardiasis, typhoid, or hepatitis A.

(a) Oligotrophic water body

(b) Eutrophic water body

FIGURE 11.20
An oligotrophic water body **(a)** with clear water and low nutrient content may eventually become a eutrophic water body **(b)** with abundant algae and high nutrient content. Pollution of freshwater bodies by excess nutrients accelerates the process of eutrophication.

Biological pollution by pathogens causes more human health problems than any other type of water pollution. A study of global water supply and sanitation issues by the World Health Organization (WHO) and the United Nations Children's Fund (UNICEF) in 2000 showed that despite advances in many parts of the world, major problems still existed. On the positive side, 4.9 billion people (82% of the population) had access to safe water as a result of some form of improvement in their water supply—an increase from 4.1 billion (79% of the population) in 1990. However, more than 1.1 billion people were still without safe water supplies. In addition, 2.4 billion people had no sewer or sanitation facilities. Most of these people were Asians and Africans, and four-fifths of the people without sanitation lived in rural areas. These conditions contribute to widespread health impacts and 5 million deaths per year.

Treating sewage constitutes one approach to reducing the risks that waterborne pathogens pose. Another is using chemical or other means to disinfect drinking water. Others include hygienic measures, such as public education to encourage personal hygiene and government enforcement of regulations to ensure the cleanliness of food production, processing, and distribution.

Toxic chemicals Our waterways have become polluted with toxic organic substances of our own making, including pesticides, petroleum products, and other synthetic chemicals. Many of these can poison animals and plants, alter aquatic ecosystems, and cause a wide array of human health problems, including cancer. In addition, toxic metals (such as arsenic, lead, and mercury) and acids (from acid precipitation and acid drainage from mining sites also cause negative impacts on human health and the environment. Health impacts of toxic chemicals are discussed in Chapter 19.

Legislating and enforcing more stringent regulations of industry can help reduce releases of these toxic inorganic chemicals. Better yet, we can modify our industrial processes and our purchasing decisions to rely less on these substances.

Sediment Although floods build fertile farmland, sediment that rivers transport can also impair aquatic ecosystems. Mining, clear-cutting, land clearing for housing development, and careless cultivation of farm fields all expose soil to wind and water erosion. Some water bodies, such as China's Yellow River, are naturally sediment-rich, but many others are not. When a clear-water river receives a heavy influx of eroded sediment, aquatic habitat can change dramatically, and fish adapted to clear-water environments may not be able to adjust. We can reduce sediment pollution by better managing farms and forests and avoiding large-scale disturbance of vegetation.

Thermal pollution Water's ability to hold dissolved oxygen decreases as temperature rises, so some aquatic organisms may not survive when human activities raise water temperatures. When we withdraw water from a river and use it to cool an industrial facility, we transfer heat energy from the facility back into the river where the water is returned. People also raise surface water temperatures by removing streamside vegetation that shades water.

Too little heat can also cause problems. On many dammed rivers, water at the bottoms of reservoirs is colder than water at the surface. When dam operators release water from the depths of a reservoir, downstream water temperatures drop suddenly. These low water temperatures may favour cold-loving invasive species over endangered native species.

Water pollution comes from point and non-point sources

Some water pollution is emitted from **point sources**—discrete locations, such as a factory or sewer pipe. In contrast, **non-point-source** pollution arises from multiple cumulative inputs over larger areas, such as farms, city streets, and residential neighbourhoods (**FIGURE 11.21**).

Many common activities give rise to non-point-source water pollution, such as applying fertilizers and pesticides to lawns, applying salt to roads in winter, and changing automobile oil. To minimize non-point-source pollution of drinking water, governments can limit development on watershed land surrounding reservoirs.

Scientists use several indicators of water quality

Most forms of water pollution are not very visible to the human eye, so scientists and technicians measure certain physical, chemical, and biological properties of water to characterize **water quality**. Biological properties include the presence of fecal coliform bacteria and other disease-causing organisms, as discussed above. Algae and aquatic

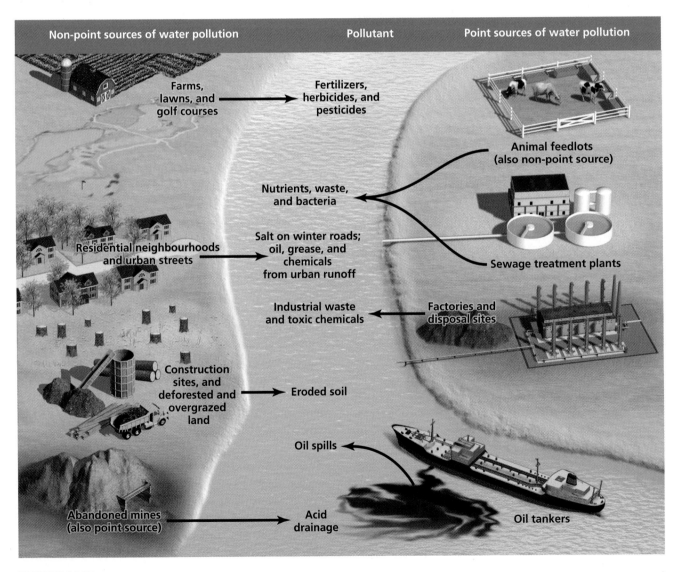

FIGURE 11.21
Point-source pollution comes from discrete facilities or locations, usually from single outflow pipes. Non-point-source pollution (such as runoff from streets, residential neighbourhoods, lawns, and farms) originates from numerous sources spread over large areas.

invertebrates are also commonly used as biological indicators of water quality.

Chemical properties include nutrient concentrations, pH, taste and odour, and hardness. "Hard" water contains naturally high concentrations of calcium and magnesium ions, which prevents soap from lathering and leaves chalky deposits behind when heated or boiled. An important chemical characteristic is dissolved oxygen content. **Dissolved oxygen** is an indicator of aquatic ecosystem health because surface waters low in dissolved oxygen are less capable of supporting aquatic life.

Among physical characteristics, **turbidity** measures the density of suspended particles in a water sample. If scientists can measure only one parameter, they will often choose turbidity, because it tends to correlate with many others and is thereby a good indicator of overall water quality. Fast-moving rivers that cut through arid or eroded landscapes, like the Yellow River, carry a great deal of sediment and are turbid and muddy looking as a result. The colour of the water can reveal particular substances present in a sample. Some forest streams run the colour of iced tea because of chemicals called tannins that occur naturally in decomposing leaf litter. Finally, temperature can be used to assess water quality. High temperatures can interfere with some biological processes, and warmer water holds less dissolved oxygen.

Groundwater pollution is a serious problem

Most efforts at pollution control have focused on surface water. Yet increasingly, groundwater sources once assumed to be pristine have been contaminated by pollution from industrial and agricultural practices. Groundwater pollution is largely hidden from view and is extremely difficult to monitor (**FIGURE 11.22**); it can be out-of-sight, out-of-mind for decades until widespread contamination of drinking supplies is discovered.

Groundwater pollution is also more difficult to manage than surface water pollution. Rivers flush their pollutants fairly quickly, but groundwater retains its contaminants until they decompose, which in the case of persistent pollutants can be many years or decades. The long-lived pesticide DDT, for instance, is still found widely in aquifers in North America, even though it was banned 40 years ago. Moreover, chemicals break down much more slowly in aquifers than in surface water or soils. Groundwater generally contains less dissolved oxygen, microbes, minerals, and organic matter, so decomposition is slower. For instance, concentrations of the herbicide alachlor decline by half after 20 days in soil, but in groundwater this takes almost four years.

FIGURE 11.22
Groundwater supplies must be closely monitored, especially in areas where the potential for contamination is high. It is much easier to prevent contamination of groundwater than to remediate after contamination has occurred.

There are many sources of groundwater pollution, including some natural sources

Some chemicals that are toxic at high concentrations, including aluminum, fluoride, nitrates, and sulphates, occur naturally in groundwater. After all, water that resides in aquifers is surrounded by rocks—the natural sources for many chemical compounds, both benign and toxic—for hundreds to tens of thousands of years. During that time of close contact many compounds are leached from the rocks into the groundwater (including the calcium and magnesium that lead to water "hardness," mentioned above). The poisoning of Bangladesh's wells by arsenic is one case of natural contamination (see "The Science Behind the Story: Arsenic in the Waters of Bangladesh").

However, there is no escaping the fact that groundwater pollution from human activity is widespread. Industrial, agricultural, and urban wastes—from heavy metals to petroleum products to industrial solvents to pesticides—can leach through soil and seep into aquifers. Pathogens and other pollutants can enter groundwater through improperly designed wells and from the pumping of liquid hazardous waste below ground.

Leakage from underground septic tanks, tanks of industrial chemicals, and tanks of oil and gas also pollutes groundwater (**FIGURE 11.23**). According to Environment Canada, without adequate corrosion protection, more than half of underground gasoline storage tanks can be expected to begin leaking by the time they are 15 years old, and just 1 L of gasoline can contaminate up to 1 000 000 L of groundwater.[15] Intercepting carcinogenic or otherwise toxic pollutants, such as chlorinated solvents

THE SCIENCE BEHIND THE STORY

Arsenic in the Waters of Bangladesh

These skin lesions were caused by arsenic poisoning in Bangladesh.

In the 1970s, UNICEF, with the help of environmental scientists at the British Geological Survey, launched a campaign to improve access to freshwater in Bangladesh. By digging thousands of small artesian wells, the designers of the program hoped to reduce Bangladeshis' dependence on disease-ridden surface waters. In the mid-1990s, however, scientists began to suspect that the wells dug to improve Bangladeshis' health were contaminated with arsenic, a poison that, if ingested frequently, can cause serious skin disorders and other illnesses, including cancer.

A medical doctor sounded the first alarm. In 1983, dermatologist K. C. Saha of the School of Tropical Medicine in Calcutta, India, saw the first of many patients from West Bengal, an area of India just west of Bangladesh, who showed signs of arsenic poisoning. Through a process of elimination, contaminated well water was identified as the likely cause of the poisoning. The hypothesis was confirmed by groundwater testing and by the work of epidemiologists, among them Dipankar Chakraborti of Calcutta's Jadavpur University.

However, it was not until the late 1990s that large-scale testing of Bangladesh's wells began. By 2001, when the British Geological Survey and the government of Bangladesh published their final report, 3524 wells had been tested. Of the shallow wells—those less than 150 m deep—46% exceeded the World Health Organization's maximum recommended level of 10 μg/L of arsenic. Extrapolating across all of Bangladesh, the scientists estimated that as many as 2.5 million wells serving 57 million people were contaminated. As the figure shows, arsenic contamination is most prevalent in southern Bangladesh, but localized hot spots are found in northern regions of the country.

Scientists have not yet reached consensus on the chemical processes by which Bangladesh's shallow aquifers became contaminated. All agree that the arsenic is of natural origin; what remains unclear is how the low levels of arsenic naturally present in soils were dissolved in the aquifers in elemental and highly toxic form. One initial explanation, suggested by Chakraborti and his colleagues, placed most of the blame on agricultural irrigation. By drawing large amounts of water out of aquifers during Bangladesh's dry season, they argued, irrigation had permitted oxygen to enter the aquifers and prompted the release of arsenic from pyrite, a common mineral.

Other scientists contend that pyrite oxidation cannot explain most cases of arsenic contamination. In a 1998 paper in the journal *Nature*, Ross Nickson of the British Geological Survey and his colleagues suggested that arsenic was being released from iron oxides carried into Bangladesh by the Ganges River. They pointed to results of a hydrochemical survey that measured the chemical composition of aquifers throughout Bangladesh. Contrary to the predictions of the pyrite oxidation hypothesis, the survey found that arsenic concentrations tended to increase with aquifer depth and to be inversely correlated with concentrations of sulphur, a component of pyrite. Nickson and his colleagues concluded that highly reducing chemical conditions created by buried organic matter, such as peat, had probably leached arsenic from iron oxides over thousands of years.

Recently, Massachusetts Institute of Technology hydrologist Charles Harvey and colleagues suggested that irrigation may contribute to the arsenic problem after all but not because of pyrite oxidation. In a 2002 paper in the journal *Science*, they described an experiment in which more than a dozen wells were dug near the capital city of Dhaka. Contrary to the pyrite oxidation hypothesis, and in agreement with Nickson and his colleagues, they found little evidence of a connection between sulphur or oxygen and arsenic.

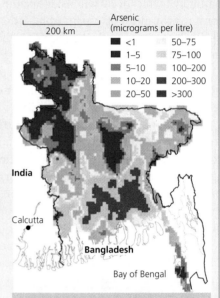

In a modern tragedy, thousands of wells dug for drinking water in Bangladesh at the urging of international aid workers turned out to be laced with arsenic. This map shows that arsenic concentrations are highest in the southern portion of the country.

Source: Kinniburgh, D. G., and P. L. Smedley, eds. 2001. Arsenic Contamination of Groundwater in Bangladesh. Department of Public Health Engineering Bangladesh, British Geological Survey Report.

However, they also found that they could increase arsenic concentrations by injecting organic matter, such as molasses, into their experimental wells. In the process of being metabolized by microbes, the molasses appeared to be freeing arsenic from iron oxides. A similar process might take place naturally, Harvey's team argued, when runoff from rice paddies, ponds, and rivers recharges aquifers that have been depleted by heavy pumping for irrigation. In support of this hypothesis, they found that much of the carbon in the shallow wells was of recent origin. Other scientists, however, have found arsenic in much older waters. This finding suggests that Bangladesh's arsenic problem may be caused by multiple hydrological and geologic factors.

FIGURE 11.23
Leaky underground storage tanks have been a major source of groundwater pollution. They are particularly problematic in the Atlantic provinces, where reliance on groundwater is very high.

and gasoline, before they reach aquifers is vital because once an aquifer is contaminated, it is extremely difficult to remediate.

Agriculture also contributes to groundwater pollution. Nitrate from fertilizers has leached into groundwater in agricultural areas throughout Canada and in 49 of the 50 U.S. states (see "The Science Behind the Story: Nitrate Contamination of the Abbotsford Aquifer" in Chapter 7). Nitrate in drinking water has been linked to cancers, miscarriages, and "blue baby syndrome," which reduces the oxygen carrying capacity of infants' blood. Agriculture can also contribute pathogens, primarily originating from animal wastes. In 2000, the groundwater supply of Walkerton, Ontario, became contaminated with the bacterium *Escherichia coli*, or *E. coli*. Two thousand people became ill, and seven died (see "The Science Behind the Story: When Water Turns Deadly: The Walkerton Tragedy").

Legislative and regulatory efforts have helped reduce pollution

As numerous as our freshwater pollution problems may seem, it is important to remember that many of them were worse a few decades ago, when Lake Erie was declared officially "dead" (see Chapter 22, "Central Case: The Death and Rebirth of Lake Erie").

Citizen activism and government response during the 1960s and 1970s resulted in fundamental changes in environmental practices and legislation that made it illegal to discharge pollution from a point source without a permit, set standards for industrial wastewater and for contaminant levels in surface waters, and funded construction of sewage treatment plants. In Canada most such legislation is enacted and enforced at the provincial level, although the federal government sets environmental guidelines through the *Canadian Environmental Protection Act* and other federal legislation, and regulates any interprovincial transfers of hazardous materials. However, probably the single most powerful act that serves to protect water quality in Canada is a federal law, the *Fisheries Act*, which makes it illegal to damage any water body that serves as a habitat for fish. Thanks to such legislation, point-source pollution has been reduced, and rivers and lakes in most parts of North America are cleaner than they have been in decades.

The Great Lakes represent a success story in fighting water pollution. Much of this work has been carried out through the International Joint Commission (IJC), the *Great Lakes Water Quality Agreement*, and the *International Boundary Waters Treaty Act*. In the 1970s the Great Lakes, which hold 18% of the world's surface freshwater, were badly polluted with wastewater, fertilizers, and toxic chemicals. Today, coordinated efforts of the Canadian and U.S. governments have paid off. Releases of toxic chemicals are down, and phosphorus runoff has decreased. Bird populations are rebounding, and Lake Erie is now home to the world's largest walleye fishery.

The Great Lakes' troubles are by no means over—sediment pollution is still heavy, PCBs and mercury still settle on the lakes from the air, and fish are not always safe to eat. However, the progress so far shows how conditions can improve when citizens push their governments to take action.

Other developed nations have also reduced water pollution. In Japan, Singapore, China, and South Korea, legislation, regulation, enforcement, and investment in wastewater treatment have brought striking water quality improvements. However, non-point-source pollution, eutrophication, and acid precipitation remain major challenges.

We treat our drinking water

Technological advances have also improved our ability to control pollution. The treatment of drinking water and the treatment of wastewater are mainstream practices in developed nations today. Health Canada publishes standards for drinking water contaminants, which local governments and water suppliers are obligated to meet. Categories for standards include microbiological parameters (viruses, bacteria, protozoa, turbidity); chemical and physical parameters (including both health and esthetic guidelines); and radiological parameters. More

THE SCIENCE BEHIND THE STORY

When Water Turns Deadly: The Walkerton Tragedy[16]

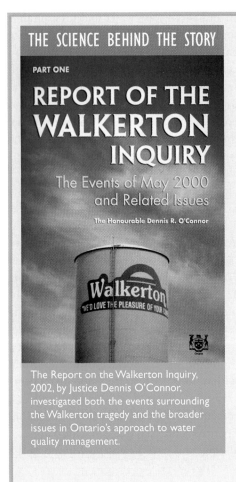

The Report on the Walkerton Inquiry, 2002, by Justice Dennis O'Connor, investigated both the events surrounding the Walkerton tragedy and the broader issues in Ontario's approach to water quality management.

Until May 2000, there was little to distinguish Walkerton from dozens of small towns in southern Ontario.[17]

That's how Justice Dennis O'Connor started his 2002 report entitled *Report of the Walkerton Inquiry: The Events of May 2000 and Related Incidents*. He continued with a matter-of-fact summary of events:

In May 2000, Walkerton's drinking water system became contaminated with deadly bacteria, primarily Escherichia coli O157:H7. Seven people died, and more than 2,300 became ill. The community was devastated. The losses were enormous. There were widespread feelings of frustration, anger, and insecurity.[18]

In response to these events the Ontario government ordered the public inquiry, which found that contaminants entered Walkerton's municipal water wells around May 12, via any of a number of pathways. The transport of contaminants and their eventual entry into three wells was probably facilitated by heavy rains. The main source for the contaminants was animal wastes from livestock on a nearby farm. The farmer had followed appropriate procedures, the inquiry concluded, and was not at fault.

The contaminants were *Escherichia coli* and, less importantly, *Campylobacter jejuni*. Both are short, curved, rod-shaped bacteria (see photo) that occur commonly in the feces of birds, bats, and other warm-blooded animals. Symptoms of infection in humans include bloody diarrhea and stomach pain.

Some strains of *E. coli*, including the O157:H7 strain found at Walkerton, can be deadly. *C. jejuni* also has been linked to the development of a debilitating and sometimes fatal condition, Guillain-Barré syndrome. These are seriously pathogenic organisms.

Events began to unfold in Walkerton about six days after the initial contamination. Twenty children missed school on May 18; two were admitted to hospital. Multiple cases of intestinal illness prompted the medical officer of health, Dr. Murray McQuigge, to inquire about the safety of the water supply. Three times he was assured of its safety by Stan Koebel, manager of the Walkerton Public

---

than 80 characteristics are considered in these guidelines; some have numerical standards associated with them, others do not. The guidelines are set by the Federal–Provincial–Territorial Committee on Drinking Water, which includes members from Health Canada, Environment Canada, and the Council of Environment Ministers, as well as the Canadian Advisory Council on Plumbing.

Before being sent to your tap, water from a reservoir or aquifer is treated with chemicals to remove particulate matter; passed through filters of sand, gravel, and charcoal; and/or disinfected with small amounts of an agent, such as chlorine.

It is better to prevent pollution than to mitigate the impacts after it occurs

In many cases, solutions to pollution will need to involve prevention, not simply **"end-of-pipe"** treatment and cleanup. With groundwater contamination, preventing pollution in the first place is by far the best strategy when one considers the other options for dealing with the problem: filtering groundwater before distributing it can be extremely expensive; pumping water out of an aquifer, treating it, then injecting it back in, repeatedly, takes an impracticably long time; and restricting pollutants on lands above selected aquifers would simply shift pollution elsewhere.

There are many things ordinary people can do to help minimize freshwater pollution. One is to exercise the power of consumer choice in the marketplace by purchasing phosphorus-free detergents and other "environmentally friendly" products. Another is to become involved in protecting local waterways. Locally based "riverwatch" groups or watershed associations enlist volunteers to collect data and help provincial and federal agencies safeguard the health of rivers and other water bodies. Such programs are proliferating as citizens and policy makers increasingly demand clean water.

Utilities Commission (PUC). On May 21, Dr. McQuigge issued a boil-water advisory, in spite of the PUC's assurances.

The inquiry later concluded that Ontario's procedures for water safety testing were faulty and inadequate, and that PUC operators were insufficiently trained, had not done the daily testing they were supposed to do, and falsified data entries to give the impression that testing had taken place.

The inquiry concluded that chlorine residual testing and turbidity monitoring would have allowed for the detection and resolution of the problem. Chlorine residual testing is a simple approach that indicates whether sufficient chlorine has been added to the water to inactivate all the bacteria. If no "leftover" or residual chlorine remains in the water, then all of the added chlorine was used up and it is possible that some bacteria remain active. *Turbidity* refers to lack of clarity in the water; sometimes this is caused by suspended sediment, but a concentration of microorganisms can also cause turbidity. If chlorine residual testing and turbidity monitoring had been done on a daily basis, the problems would have been detected earlier and the outbreak could have been contained.

If a problem had been detected, further testing would have been triggered to identify the cause. This would have included technologies that detect an enzyme called beta-glucuronidase, which is produced by almost all forms of *E. coli* and very few other bacteria.

The sample is placed in a medium that changes colour or fluoresces if it is exposed to the enzyme. Sometimes it is necessary to culture the bacteria for a few days to have a sufficient concentration in the sample.

Most tests reveal the presence of *E. coli* but neither the exact amount nor the strain. Professor Ulrich Krull and colleagues at the University of Toronto Mississauga have developed a detection technology based on DNA. These *biosensors* contain short sequences of single-stranded DNA or *ssDNA*, which link or *hybridize* with complementary sequences of ssDNA in the sample, if they are present. An indicator, such as a fluorescing gel, shows that hybridization has occurred. Biosensors are useful because they instantly reveal not only the presence of the bacteria, but the specific strain.

In his report, Justice O'Connor brought forward some pressing concerns

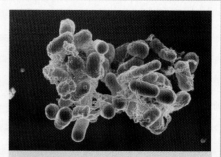

The contaminants found in the municipal water supply in Walkerton in 2000 included the bacteria *C. jejuni* and the deadly O157:H7 strain of *E. coli*, shown here.

about Ontario's approach to drinking water. Health Canada sets guidelines for drinking water quality; the provinces and territories have the responsibility to meet these guidelines and ensure drinking water safety. In some provinces this responsibility falls to the health ministry, in others the environment ministry. In 2002 Ontario passed the *Safe Drinking Water Act* to address the concerns raised by Justice O'Connor through the Walkerton Inquiry.

Wastewater and Its Treatment

Wastewater refers to water that has been used by people in some way. It includes water carrying sewage; water from showers, sinks, washing machines, and dishwashers; water used in manufacturing or cleaning processes by businesses and industries; and stormwater runoff.

Although natural systems can process moderate amounts of wastewater, the large and concentrated amounts generated by our densely populated areas can harm ecosystems and pose threats to human health. Thus, attempts are now widely made to treat wastewater before releasing it into the environment.

Municipal wastewater treatment involves several steps

In rural areas, **septic systems** are the most popular method of wastewater disposal. In a septic system, wastewater runs from the house to an underground septic tank, inside which solids and oils separate from water. The clarified water proceeds downhill to a drain field of perforated pipes laid horizontally in gravel-filled trenches underground. Microbes decompose the wastewater these pipes emit. Periodically, solid waste needs to be pumped from the septic tank and taken to a landfill.

In more densely populated areas, municipal sewer systems carry wastewater from homes and businesses to centralized treatment locations. There, pollutants in wastewater are removed by physical, chemical, and biological means (**FIGURE 11.24**).

At a treatment facility, **primary treatment**, the physical removal of contaminants in settling tanks or clarifiers, generally removes about 60% of suspended solids from wastewater. Wastewater then proceeds to **secondary treatment**, in which water is stirred and aerated so that aerobic bacteria degrade organic pollutants. Roughly 90% of suspended solids may be removed after secondary treatment. Finally, the clarified water is treated with chlorine, and sometimes ultraviolet light, to kill bacteria. Most often, the treated water, called **effluent**,

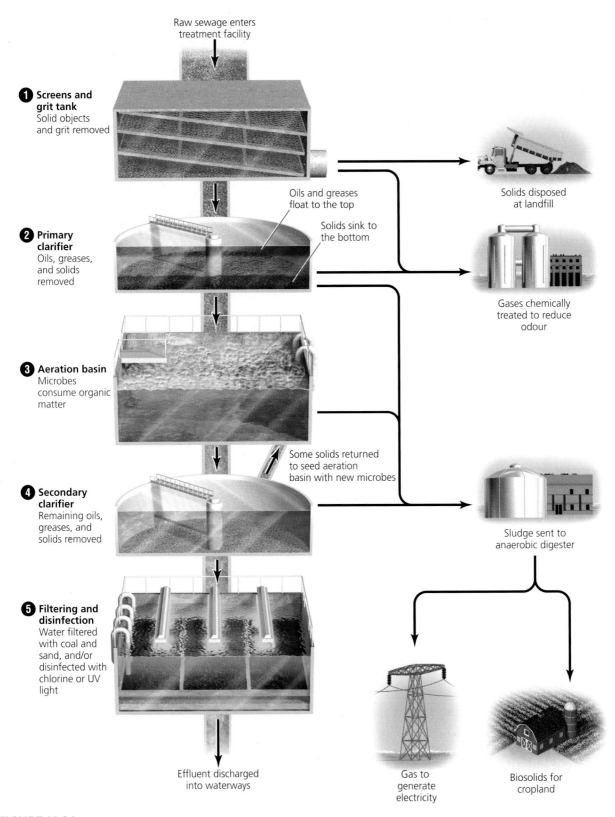

FIGURE 11.24

Shown here is a generalized process from a modern, environmentally sensitive wastewater treatment facility. Wastewater initially passes through screens to remove large debris and into grit tanks to let grit settle **(1)**. It then enters tanks called primary clarifiers **(2)**, in which solids settle to the bottom and oils and greases float to the top for removal. Clarified water then proceeds to aeration basins **(3)** that oxygenate the water to encourage decomposition by aerobic bacteria. Water then passes into secondary clarifier tanks **(4)** for removal of further solids and oils. Next, the water may be purified by chemical treatment with chlorine, passage through carbon filters, and/or exposure to ultraviolet light **(5)**. The treated water (called *effluent*) may then be piped into natural water bodies, used for urban irrigation, flowed through an artificial wetland, or used to recharge groundwater. In addition, most treatment facilities control odour in the early steps and use anaerobic bacteria to digest sludge removed from the wastewater. Sludge from digesters may be sent to farm fields as fertilizer, and gas from digestion may be used to generate electric power.

CANADIAN ENVIRONMENTAL PERSPECTIVES

David Schindler

University of Alberta ecologist David Schindler warns that Canadian freshwater resources are under stress.

- **Ecosystem ecologist** and Killam Memorial **Professor** of Biological Sciences at the University of Alberta
- **Limnologist** and **biogeochemist**
- **Freshwater advocate**

Although his is not a "household name," it is probable that no other environmental scientist in Canada today is better known or more highly regarded for his commitment to freshwater resources and ecosystems than the University of Alberta's David Schindler.

Schindler is an ecologist and a *limnologist*—a scientist who studies ponds, lakes, wetlands, and freshwater ecology. He received a D. Phil. in ecology from Oxford University, where he attended as a Rhodes Scholar in the late 1960s. After serving two years as an Assistant Professor at Trent University, Schindler joined the Fisheries Research Board of Canada, which later became the Department of Fisheries and Oceans, as the founding director of the Experimental Lakes Area (ELA), as one of the most successful and productive long-term research programs on freshwater systems. (You will learn more about the ELA in Chapter 13's "Central Case: The Rain and The Big Nickel," which looks at the impacts of acid rain on northern lakes.)

Through his work at ELA, Schindler contributed significantly to the scientific understanding of the effects of phosphorus and acid rain on lake ecosystems. This understanding helped lead to restrictions on the phosphorus content of detergents and sewage, as well as changes in air quality legislation in Canada, the United States, and the European Union. In recognition of his contributions to the scientific understanding of freshwater ecosystems, Schindler has received many prizes, awards, medals, and honorary degrees, including the first Stockholm Water Prize, the Volvo International Environment Prize (he is the only Canadian to win either prize), the Tyler Environmental Prize, and the Gerhard Herzberg Canada Gold Medal for Science and Engineering, which came with a research grant of $1 million.

Schindler has most recently turned his attention to the Alberta tar sand developments and their intensive use of water. In an article entitled *The Myth of Abundant Canadian Water* he wrote, "It is perhaps ironic that Alberta, the province most vociferously opposed to controlling greenhouse gases in order to protect its pampered petrochemical industries, will almost certainly be the first to suffer from freshwater shortages."[19] In the same article he continued, "To a water expert, looking ahead is like the view from a locomotive, 10 seconds before the train wreck. Sometime in the coming century, the increasing human demand for water, the increasing scarcity of water due to climate warming, and one of the long droughts of past centuries will collide, and Albertans will learn first-hand what water scarcity is all about."[20] Schindler is also now making important contributions to the scientific understanding of the effects of global warming and of stratospheric ozone depletion on northern lakes.

Dr. Schindler's instruction at the University of Alberta is somewhat out of the ordinary for a typical science professor. He co-teaches a graduate course entitled Limnology: The Philosophy, Sociology, and Politics of Science and Public Policy in Canada; Environmental Decision Making.[21] The course is consistent with what is perhaps his most important achievement, which goes beyond the science. It lies instead in his success at communicating the scientific message about freshwater vulnerability to those who most need to hear it. In a profile of Schindler's achievements, the Natural Sciences and Engineering Research Council of Canada noted that he ha "succeeded in conveying his knowledge and its importance to legislators in Canada and around the world and to the general public,"[22] and that through his work and teaching he has inspired many students to become involved with research and careers in environmental science. This contribution will have a lasting impact on Canada and Canadians.

"The time to make these decisions is now, not after our water is gone." —David Schindler

Thinking About Environmental Perspectives

David Schindler is widely recognized as a model for active, engaged science, and it is often noted that he has inspired many students to enter the field of environmental science. Is there a professor in your college or university who has inspired you to continue in your studies? How do you think Professor Schindler manages to balance his scientific research with his political activism on behalf of freshwater resources? Compare Schindler's story to the discussion of David Suzuki in Chapter 1, recalling that Suzuki chose to give up doing science on a daily basis in order to focus more exclusively on environmental activism.

---

is piped into rivers or the ocean following primary and secondary treatment. Sometimes, however, "reclaimed" water is used for lawns and golf courses, for irrigation, or for industrial purposes, such as cooling water in power plants.

As water is purified throughout the treatment process, solid material called **sludge** is removed. Sludge is sent to digesting vats, where microorganisms decompose much of the matter. The result, a wet solution of "biosolids," is then dried and either disposed of in a landfill, incinerated, or used as fertilizer on cropland. Methane-rich gas created by the decomposition process is sometimes burned to generate electricity, helping to offset the cost of the treatment facility.

Artificial wetlands can aid treatment

Natural wetlands already perform the ecosystem service of water purification, and wastewater treatment engineers are now manipulating wetlands and even constructing wetlands *de novo* to employ them as tools to cleanse wastewater. The practice of treating wastewater with so-called **constructed** (or **artificial**) **wetlands** is growing quickly. For example, the government of Nova Scotia and the Nova Scotia Agricultural College are constructing a test series of three artificial wetlands that will be used to filter agricultural wastewaters from livestock operations. Generally in this approach, wastewater that has gone through primary treatment at a conventional facility is pumped into the wetland, where microbes living amid the algae and aquatic plants decompose the remaining pollutants. Water cleansed in the wetland can then be released into waterways or allowed to percolate underground.

Constructed wetlands also serve as havens for wildlife and areas for human recreation. Restored and artificial wetlands in Ontario, Alberta, Nova Scotia, and many other locations in Canada are serving as wetland habitats for birds and wildlife, while helping to recharge depleted aquifers.

Conclusion

Citizen action, government legislation and regulation, new technologies, economic incentives, and public education are all enabling us to confront what will surely be one of the great environmental challenges of the new century: ensuring adequate quantity and quality of freshwater for ourselves and for the planet's ecosystems.

Accessible freshwater is only a minuscule percentage of the hydrosphere, but we generally take it for granted. With our expanding population and increasing water usage, we are approaching conditions of widespread scarcity. Water depletion and water pollution are already taking a toll on the health, economies, and societies of the developing world, and they are beginning to do so in arid areas of the developed world. There is reason to hope that we may yet attain sustainability in our water usage, however. Potential solutions are numerous, and the issue is too important to ignore.

REVIEWING OBJECTIVES

You should now be able to:

Explain the importance of water and the hydrologic cycle to ecosystems, human health, and economic pursuits

- We depend utterly on drinkable water, and a functioning hydrologic cycle is vital to maintaining ecosystems and our civilization.

Delineate the distribution of freshwater on Earth

- Of all the water on Earth, only about 1% is readily available for our use.
- Water availability varies in space and time, and regions vary greatly in the amounts they possess.

Describe major types of freshwater ecosystems

- The main types of freshwater ecosystems include rivers and streams, wetlands, and lakes and ponds.

Discuss how we use water and alter freshwater systems

- We use water for agriculture, industry, and residential use. The ratio of these uses varies among societies, but globally 70% is used for agriculture.
- Most of the world's rivers are dammed. Dams bring a diverse set of benefits and costs. Increasingly, people are proposing dam removal.
- We divert water with canals and irrigation ditches and attempt to control floods with dikes and levees.
- We pump water from aquifers and surface water bodies, sometimes at unsustainable rates.

Assess problems of water supply and propose solutions to address depletion of freshwater

- Water tables are dropping worldwide from unsustainable groundwater extraction. Surface water extraction has caused rivers to run dry and water bodies to shrink.
- Unequal water distribution amid shrinking supplies may heighten political tensions over water in the future.
- Solutions to expand supply, such as desalination, are worth pursuing, but not to the exclusion of finding ways to decrease demand.
- Solutions to reduce demand include technology, approaches, and consumer products that increase efficiency in agriculture, industry, and the home.

Assess problems of water quality and propose solutions to address water pollution

- Water pollutants include excessive nutrients, microbial pathogens, toxic chemicals, sediment, and thermal pollution.
- Water pollution stems from point sources and non-point sources.
- Scientists who monitor water quality use biological, chemical, and physical indicators.
- Groundwater pollution can be more persistent than surface water pollution.
- Legislation and regulation have improved water quality in developed nations in recent decades.
- Preventing water pollution is better than mitigation.

Explain how wastewater is treated

- Septic systems are used to treat wastewater in rural areas.
- Wastewater is treated physically, biologically, and chemically in a series of steps at municipal wastewater treatment facilities.
- Artificial wetlands enhance wastewater treatment while restoring habitat for wildlife.

TESTING YOUR COMPREHENSION

1. Define *groundwater*. What role does groundwater play in the hydrologic cycle?
2. Why are sources of freshwater unreliable for some people and plentiful for others?
3. Describe three benefits of damming rivers, and three costs. What particular environmental, health, and social concerns has China's Three Gorges Dam and its reservoir raised?
4. Why do the Colorado, Rio Grande, Nile, and Yellow rivers now slow to a trickle or run dry before reaching their deltas?
5. Why are water tables dropping around the world? What are some environmental costs of falling water tables?
6. Name three major types of water pollutants, and provide an example of each. List three properties of water that scientists use to determine water quality.
7. Why do many scientists consider groundwater pollution a greater problem than surface water pollution?
8. What are some anthropogenic (human) sources of groundwater pollution?
9. Describe how drinking water is treated. How does a septic system work?
10. Describe and explain the major steps in the process of wastewater treatment. How can artificial wetlands aid such treatment?

SEEKING SOLUTIONS

1. Discuss possible strategies for equalizing distribution of water throughout the world. Consider supply and transport issues. Have our methods of drawing, distributing, and storing water changed very much throughout history? How is the scale of our efforts affecting the availability of water supplies?
2. How can we lessen agricultural demand for water? Describe some ways in which we can reduce household water use. How can industrial uses of water be reduced?
3. Discuss some of the methods we can adopt, in addition to "end-of-pipe" solutions, to prevent contamination and ensure "water security."
4. How might desalination technology help "make" more water? Describe two methods of desalination. Where is this technology being used?
5. **THINK IT THROUGH** You have been put in charge of water policy for your region. The aquifer beneath your region has been overpumped, and many wells have already run dry. Agricultural production last year decreased for the first time in a generation, and farmers are clamouring for you to do something. Meanwhile, the region's largest city is growing so fast that more water is needed for its burgeoning urban population. What policies would you consider to restore your region's water supply? Would you try to take steps to increase supply, to decrease demand, or both? Explain why you would choose such policies.
6. **THINK IT THROUGH** Having solved the water depletion problem in your region, your next task is to deal with pollution of the groundwater that provides

the region's drinking water supply. Recent studies have shown that one-third of the province's groundwater has levels of pollutants that violate international standards for human health. Citizens are fearful for their safety, and the provincial government is threatening enforcement. What steps would you consider taking to safeguard the quality of your region's groundwater supply, and why?

INTERPRETING GRAPHS AND DATA

Close to 75% of the freshwater used by people is used in agriculture, and about 1 of every 14 people lives where water is scarce. By the year 2050, scientists project that two-thirds of the world's population will live in water-scarce areas, including most of Africa, the Middle East, India, and China. How much water is required to feed 7 billion people a basic dietary requirement of 2700 calories (11.3 kJ) per day? The answer depends on the efficiency with which we use water in agricultural production and on the type of diet we consume.

1. How many litres of water are needed to produce 2300 calories (9.6 kJ) of vegetable food? How many litres of water are needed to produce 400 calories (1.7 kJ) of animal food? How many litres of water are needed daily to provide this diet? Annually?
2. How many litres of water would be saved daily, compared to the diet in the graph, if the 2700 calories were provided entirely by vegetables? Annually?
3. Reflect on one of the quotes at the beginning of this chapter: "Water promises to be to the twenty-first century what oil was to the twentieth century: the precious commodity that determines the wealth of nations." How do you think the demographic pressure

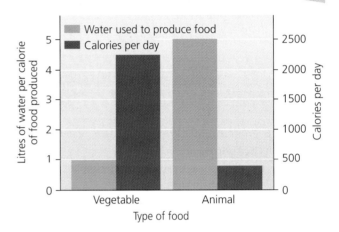

Amount of water needed to produce vegetable and animal food (orange), and global average calories per day consumed of vegetable and animal food (red) (1000 calories = 0.004184 kJ). Data from Wallace, J. S. 2000. Increasing agricultural water use efficiency to meet future food production. Agriculture, Ecosystems and Environment 82:105–119.

on the water supply could affect world trade, particularly trade of agricultural products? Do you think it could affect prospects for peace and stability in and among nations? How so?

CALCULATING FOOTPRINTS

In Canada, household water use averages almost 350 L per person per day; only the United States has higher per capita water use.[23] One of the single greatest personal uses of water is for showering (see figure).

Standard showerheads dispense 15 L of water per minute, but so-called low-flow showerheads dispense only 9 L per minute. Given an average daily shower time of 10 minutes, calculate the amounts of water used and saved over the course of a year with standard versus low-flow showerheads, and record your results in the table below.

1. What percentage of personal water consumption would you calculate is used for showering?
2. How much additional water would you be able to save by shortening your average shower time from 10 minutes to 8 minutes? To 5 minutes?
3. Can you think of any factors that are not being considered in this scenario of water savings? Explain.

Canada's watery lifestyle
In 2001, the average daily freshwater domestic use per capita was 335 litres, equal to more than 55 cases of standard-size bottled water.* Here's how the average Canadian used that much water.

Water goes down the drain faster than most of us realize. Of the 343 L of water that each Canadian now uses per day, on average, a total of 35% of it goes to bathing in one form or another.

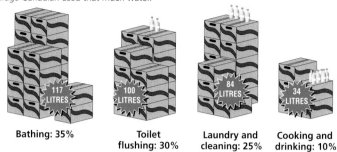

Bathing: 35% (117 LITRES)
Toilet flushing: 30% (100 LITRES)
Laundry and cleaning: 25% (84 LITRES)
Cooking and drinking: 10% (34 LITRES)

How can we be using *that* much water?
Water goes down the drain faster than most of us realize. Here's how some of our daily activities contribute to our total water usage.

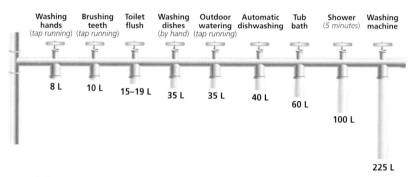

- Washing hands (tap running): 8 L
- Brushing teeth (tap running): 10 L
- Toilet flush: 15–19 L
- Washing dishes (by hand): 35 L
- Outdoor watering (tap running): 35 L
- Automatic dishwashing: 40 L
- Tub bath: 60 L
- Shower (5 minutes): 100 L
- Washing machine: 225 L

*A standard size container of bottled water is 500 mL

© Environment Canada, 2005

	Annual water use with standard showerheads (litres)	Annual water use with low-flow showerheads (litres)	Annual water savings with low-flow showerheads (litres)
You	54 750	32 850	21 900
Your class			
Your province			

Based on data from U.S. EPA. 1995. *Cleaner water through conservation: Chapter 1—How we use water in the United States*, EPA 841-B-95-002; and Environment Canada, Canada's Watery Lifestyle, EnviroZine, www.ec.gc.ca/EnviroZine/english/issues/52/any_questions_e.cfm.

TAKE IT FURTHER

 Go to www.myenvironmentplace.ca where you will find

- Suggested answers to end-of-chapter questions
- Quizzes, animations, and flashcards to help you study
- *Research Navigator*™ database of credible and reliable sources to assist you with your research projects
- Tutorials to help you master how to interpret graphs
- Current news articles that link the topics that you study to case studies from your region and around the world
- **ECO Occupational Profiles:** If you found this chapter especially interesting, you might want to learn more about the following jobs by visiting the Occupational Profiles website of the Environmental Careers Organization. Go to www.eco.ca and check out the following careers:

- Biological technician
- Environmental monitoring technician
- Fisheries technician
- Hydrologist
- Limnologist
- Microbiologist
- Wastewater collection and treatment operator
- Water and wastewater plant engineer
- Water quality technician
- Water treatment and distribution operator

CHAPTER ENDNOTES

1. Environment Canada, *Fresh Water: Quick Facts*, www.ec.gc.ca/water/en/e_quickfacts.htm.
2. Agriculture and Agri-Foods Canada, *The Canadian Bottled Water Industry*, www4.agr.gc.ca/AAFC-AAC/display-afficher.do?id=1171644581795&lang=e#sig.
3. Quinn, Frank (2007) *Water Diversion, Export, and Canada–U.S. Relations: A Brief History* (August). Program on Water Issues (POWI), Munk Centre for International Studies at the University of Toronto, www.powi.ca.
4. Quinn, Frank (2007) *Water Diversion, Export, and Canada–U.S. Relations: A Brief History* (August). Program on Water Issues (POWI), Munk Centre for International Studies at the University of Toronto, www.powi.ca.
5. Prairie Pothole Joint Venture, *Introduction to the Prairie Pothole Region, 2005 Implementation Plan Section I*, www.ppjv.org
6. Natural Resources Canada, *Groundwater Mapping Program: Overview*, http://ess.nrcan.gc.ca/gm-ces/overview_e.php.
7. Coté, François (2006) *Freshwater Management in Canada IV: Groundwater*. Library of Parliament, Parliamentary Information and Research Service, Science and Technology Division, February 2006, www.parl.gc.ca/information/library/PRBpubs/prb0554-e.html#aquifiers.
8. Environment Canada, *Water—Vulnerable to Climate Change*, www.ec.gc.ca/water/en/info/pubs/FS/e_FSA9.htm.
9. Environment Canada, Informational and Resources Services, *Water—Vulnerable to Climate Change*, www.ec.gc.ca/water/en/info/pubs/FS/e_FSA9.htm#supply.
10. Environment Canada, *Fresh Water, Bulk Removal of Water*, www.ec.gc.ca/water/en/manage/removal/e_remove.htm.
11. Ramsar Convention, www.ramsar.org.
12. Ramsar Convention, www.ramsar.org.
13. Statistics Canada (2008) Against the flow: Which households drink bottled water? *EnviroStats*, Summer, Vol. 2, No. 2, www.statcan.ca/Daily/English/080625/d080625c.htm.
14. Statistics Canada (2008) Against the flow: Which households drink bottled water? *EnviroStats*, Summer, Vol. 2, No. 2, www.statcan.ca/Daily/English/080625/d080625c.htm.
15. Environment Canada, *The Management of Water: Leaking Underground Storage Tanks and Pipelines*, www.ec.gc.ca/water/en/manage/poll/e_tanks.htm.
16. Based partly on information from O'Connor, Dennis R. (2002) *Report of the Walkerton Inquiry: The Events of May 2000 and Related Issues*. Toronto: Government of Ontario.
17. O'Connor, Dennis R. (2002) *Report of the Walkerton Inquiry: The Events of May 2000 and Related Issues*. Toronto: Government of Ontario.
18. O'Connor, Dennis R. (2002) *Report of the Walkerton Inquiry: The Events of May 2000 and Related Issues*. Toronto: Government of Ontario.
19. Schindler, David (2006) *The Myth of Abundant Canadian Water*, March, www.innovationcanada.ca/en/articles/the-myth-of-abundant-canadian-water.
20. Schindler, David (2006) *The Myth of Abundant Canadian Water*, March, www.innovationcanada.ca/en/articles/the-myth-of-abundant-canadian-water.
21. University of Alberta, Faculty of Science, Department of Biological Sciences, www.biology.ualberta.ca/faculty/david_schindler/?Page=1023.
22. NSERC News Releases, *David Schindler*, www.nserc.gc.ca/news/2000/aoe_schindler_e.htm.
23. NRCAN, *The Atlas of Canada: Domestic Water Consumption*, http://atlas.nrcan.gc.ca/site/english/maps/ freshwater/consumption/domestic/1.

12 Marine and Coastal Systems and Fisheries

Cod fishers haul in a dwindling catch

Upon successfully completing this chapter, you will be able to

- Identify physical, geographic, chemical, and biological aspects of the marine environment
- Describe major types of marine ecosystems
- Outline historic and current human uses of marine resources
- Assess human impacts on marine environments
- Review the current state of ocean fisheries and reasons for their decline
- Evaluate marine protected areas and reserves as innovative solutions

Cod fishing has been a way of life for generations in Newfoundland and Labrador. Monster cod like these are no longer there to be harvested.

CENTRAL CASE:
LESSONS LEARNED: THE COLLAPSE OF THE COD FISHERIES

"All of a sudden they just crashed."
—DONALD PAUL, NEWFOUNDLAND FISHER, IN 1997

"Either we have sustainable fisheries, or we have no fishery."
—CANADIAN FISHERIES MINISTER DAVID ANDERSON, IN 1998

No fish has had more impact on human civilization than the Atlantic cod. Europeans exploring the coasts of North America 500 years ago discovered that they could catch these abundant fish by dipping baskets over the railings of their ships, and the race to harvest this resource helped lead to the colonization of the New World. Starting in the early 1500s, schooners captured countless millions of cod, and the fish became a dietary staple on both sides of the Atlantic.

Since then, cod fishing has been the economic engine for hundreds of coastal communities in eastern Canada and New England. In many Canadian coastal villages, cod fishing has been a way of life for generations (see photo). So it came as a shock when the cod all but disappeared, and governments had to step in and close the fisheries.

The Atlantic cod (*Gadus morhua*) is a type of **groundfish**, a fish that lives or feeds along the bottom. People have long coveted groundfish, such as halibut, pollock, haddock, and flounder. Adult cod eat smaller fish and invertebrates, commonly grow to 60–70 cm long, and can live 20 years. A mature female cod can produce several million eggs. Atlantic cod inhabit cool ocean waters on both sides of the North Atlantic and occur in 24 discrete populations, called stocks. One stock inhabits the Grand Banks off the Newfoundland coast,

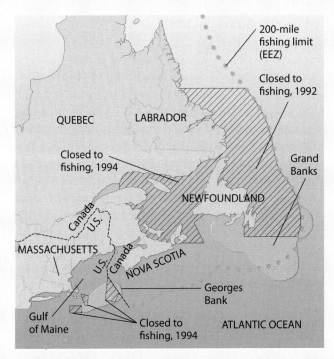

FIGURE 12.1
Ten stocks of Atlantic cod inhabit areas of the northwestern Atlantic Ocean, including the Grand Banks and Georges Bank, regions of shallow water that are especially productive for groundfish. Portions of these and other areas have been closed to fishing in recent years because cod populations have collapsed after being overfished.

and another lives on Georges Bank off Massachusetts (FIGURE 12.1).

The Grand Banks provided ample fish for centuries. With advancing technology, however, ships became larger and more effective at finding fish. By the 1960s, massive industrial trawlers from Europe were vacuuming up unprecedented numbers of groundfish. In 1977, Canada exercised its legal right to the waters 200 nautical miles from shore, the **Exclusive Economic Zone** established by the U.N. Convention on the Law of the Sea. Foreign fleets were expelled from most of the Grand Banks, and Canada's fishing industry was revved up like never before.

Catches began to dwindle in the 1980s. Too many fish had been taken, and trawling had destroyed much underwater habitat. Environmental factors also played a role. By 1992 the situation was dire: scientists reported that mature cod were at just 10% of their long-term abundance. On July 2, fisheries minister John Crosbie announced a two-year ban on commercial cod fishing off of Newfoundland and Labrador, where the $700 million fishery supplied income to 16% of the province's workforce. To compensate fishers, the government offered 10 weekly payments of $225, along with training for new job skills and incentives for early retirement. Over the next two years, 40 000 fishers and processing plant workers lost their jobs. Some coastal communities faced economic ruin; for generations, fishing had been their reason for being.

Cod stocks did not rebound by 1994, so the government extended the moratorium, enacted bans on all other major cod fisheries, and scrambled to offer more compensation, eventually spending more than $4 billion. In 1997–1998, Canada partially reopened some fisheries, but data soon confirmed that the stocks were not recovering. In April 2003, the cod fisheries were closed indefinitely, to recreational as well as commercial fishing. Today it is illegal for Canadians in these areas to catch even one cod. Fishers challenging the ban have been arrested, fined, and jailed.

Across the border in U.S. waters, cod stocks were collapsing in the Gulf of Maine and on Georges Bank. In 1994, the National Marine Fisheries Service (NMFS) closed three prime fishing areas. Over the next several years, NMFS designed a number of regulations, but these steps were too little, too late. A 2005 report revealed that the cod were not recovering as hoped, and scientists are struggling to explain why. Research suggests that once mature cod were eliminated, the species they preyed upon proliferated. Now those species compete with and prey on young cod, preventing the population from rebuilding.

There is some good news, however: the closures have allowed some other species to rebound. Seafloor invertebrates have begun to recover in the absence of trawling; spawning stocks of haddock and yellowtail flounder have risen; and sea scallops have increased in biomass fourteenfold in the Georges Bank fishery. Such recoveries in no-fishing areas show scientists, fishers, and policy makers that protecting areas of ocean can help save dwindling marine populations. In the Grand Banks, however, research has shown that the fundamental nature of the ecosystem has shifted from a bottom-oriented fauna to a shallow-water fauna. There are no signs that a complete recovery will ever be achievable.

The Oceans

It has been said that our planet Earth should more properly be named "Ocean." After all, ocean water covers the vast majority of our planet's surface. Moreover, the oceans strongly influence how our planet's systems work. They influence global climate, teem with biodiversity, facilitate transportation and commerce, and provide us with resources. Even landlocked areas far from the coasts are affected. The oceans provide fish for people to eat in Saskatchewan, supply oil to power cars in Ontario, and influence the weather in Manitoba. In this chapter you will learn something about **oceanography**, the scientific study of the physics, chemistry, biology, and geology of the oceans.

Oceans cover most of Earth's surface

Although we generally speak of the world's oceans (**FIGURE 12.2**) in the plural, giving each major basin a name—Pacific, Atlantic, Indian, Arctic, and Southern—all these oceans are connected, composing a single vast body of water. This one "world ocean" covers 71% of Earth's surface and contains 97.2% of its surface water. The oceans take up most of the hydrosphere, influence the atmosphere and lithosphere, and encompass much of the biosphere, including at least 250 000 species. The world's oceans touch and are touched by virtually every environmental system and every human endeavour.

The oceans contain more than water

Ocean water contains approximately 96.5% H_2O by mass; most of the remainder consists of ions from dissolved salts (**FIGURE 12.3**). Ocean water is salty primarily because ocean basins are the final repositories for water that runs off the land. Rivers carry sediment and dissolved salts from the continents into the ocean, as do winds. Evaporation from the ocean surface then removes pure water, leaving a higher concentration of salts. If we were able to evaporate all the water from the oceans, the empty basins would be covered with a layer of dried salt 63 m thick.

The **salinity**—basically, the saltiness—of ocean water generally ranges from 33 to 37 parts per thousand (ppt), varying from place to place because of differences in evaporation, precipitation, and freshwater runoff from land and glaciers. Salinity near the equator is low because this region has a great deal of precipitation, which is relatively salt free. In contrast, surface salinity is high at latitudes roughly 30–35 degrees north and south, where evaporation exceeds precipitation.

Besides the dissolved salts shown in **FIGURE 12.3**, nutrients, such as nitrogen and phosphorus, occur in seawater in trace amounts (well under one part per million) and play essential roles in nutrient cycling in marine ecosystems. Another aspect of ocean chemistry is dissolved gas content. Roughly 36% of the gas dissolved in seawater is oxygen, which is produced by photosynthetic

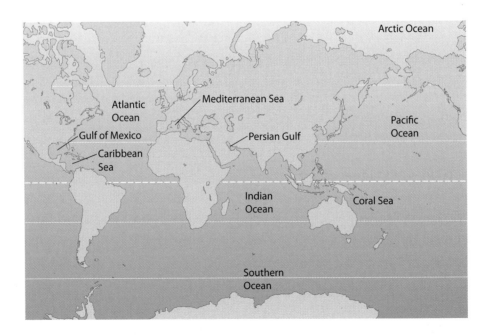

FIGURE 12.2
The world's oceans are connected in a single vast body of water but are given different names. The Pacific Ocean is the largest and, like the Atlantic and Indian Oceans, includes both tropical and temperate waters. The smaller Arctic and Southern Oceans include the waters in the north and south polar regions, respectively. Many smaller bodies of water are named as seas or gulfs; a selected few are shown here.

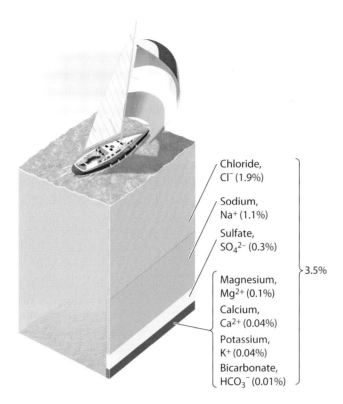

FIGURE 12.3
Ocean water consists of 3.5% salt, by mass, as shown by the proportionally thin coloured slices of the cube in this diagram. Most of this salt is NaCl in solution, so sodium and chloride ions are abundant. A number of other ions and trace elements are also present.

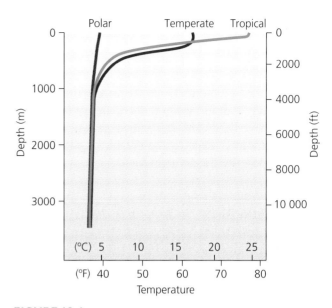

FIGURE 12.4
Ocean water varies in temperature with depth. Water temperatures near the surface are warmer because of daily heating by the sun, and within the top 1000 m they become rapidly colder with depth. This temperature differential is greatest in the tropics because of intense solar heating and is least in the polar regions. Deep water at all latitudes is equivalent in temperature.
Source: Garrison, T. 2005. Oceanography, 5th ed. Belmont, CA: Brooks/Cole.

plants, bacteria, and phytoplankton, and by diffusion from the atmosphere. Marine animals depend on this oxygen, and oxygen concentrations are highest in the upper layer of the ocean, reaching 13 mL/L of water.

Ocean water is vertically structured

Surface waters in tropical regions receive more solar radiation and therefore are warmer than surface waters in temperate or polar regions. In all regions, however, temperature declines with depth (**FIGURE 12.4**). Water density increases as salinity rises and as temperature falls. These relationships give rise to different layers of water; heavier (colder and saltier) water sinks, and lighter (warmer and less salty) water remains nearer the surface. Waters of the surface zone are heated by sunlight each day and are stirred by wind such that they are of similar density throughout, down to a depth of approximately 150 m. The salinity of surface water is influenced by a variety of interacting factors, some of which contribute to greater salinity (e.g., evaporation and sea ice formation, both of which remove freshwater from the surface layer) and some of which contribute to reduced salinity (e.g., precipitation and the influx of river water, both of which bring freshwater into the surface layer).

Below the zone of warm, salty surface water is the **thermocline**, a zone in which temperature decreases rapidly with depth, toward the much colder deep layer. The salinity of the water also changes, increasing with depth along the **halocline**. In response to the changes in temperature and salinity, the density of the water also changes rapidly with depth, increasing along the **pycnocline**. The transitional zone marked by the thermocline, halocline, and pycnocline contains about 18% of ocean water by volume, compared with the surface zone's 2%. The remaining 80% resides in the ocean's vast deep layer. The dense, cold water in this zone is sluggish and unaffected by winds, storms, sunlight, or daily temperature fluctuations.

Despite the daily heating and cooling of surface waters, ocean temperatures are much more stable than temperatures on land. Midlatitude oceans experience yearly temperature variation of only around 10°C, and tropical and polar oceans are still more stable. The reason for this stability is that water has a very high *heat capacity*, a measure of the heat required to increase temperature by a given amount. It takes much more heat energy to increase the temperature of water than it does to increase the temperature of air by the same amount. High heat capacity enables the oceans to absorb a tremendous amount of heat from the atmosphere. In fact, the heat content of the entire atmosphere is equal to that of just the top 2.6 m of the oceans. By absorbing heat and releasing it to the atmosphere, the oceans help regulate Earth's climate (Chapter 14).

Also influencing climate are the ocean's circulation systems, both deep and shallow. These circulation systems move both vertically and horizontally, affecting marine ecosystems and connecting the ocean's transitional and surface zones with the deeper waters.

Ocean water flows vertically and horizontally, influencing climate

Surface winds and heating in seawater create huge vertical flows of water, or **currents**. **Upwelling**, the vertical flow of cold, deep water toward the surface, occurs where horizontal currents diverge, or flow away from one another. Because upwelled water is rich in nutrients from the bottom, upwellings are often sites of high primary productivity and lucrative fisheries. Upwellings also occur where strong winds blow away from or parallel to coastlines (**FIGURE 12.5**). An example is the western coast of Vancouver Island, where north winds and the Coriolis effect move surface waters away from the shore, raising nutrient-rich water from below and creating a biologically rich region. The cold water also chills the air along the coast, giving Vancouver Island its famous cool, rainy summers.

In areas where surface currents converge, or come together, surface water sinks, a process called **downwelling**. Downwelling transports warm water rich in dissolved gases, providing an influx of oxygen for deep-water life. Vertical currents also occur in the deep zone, where differences in water density can lead to rising and falling convection currents, such as those seen in molten rock and in air.

The **thermohaline circulation** is the global oceanic circulation system of upwelling and downwelling currents. It connects surface water flows to deeper water flows, with far-reaching effects on global climate. The term *thermohaline* comes from root words that mean "heat" and "salinity." In this worldwide circulatory system, warmer, fresher water moves along the surface and water in the deep zones, which is colder, saltier, and denser, circulates far beneath the surface. In the Atlantic Ocean, warm surface water flows northward from the equator in the Gulf Stream, carrying heat to high latitudes and keeping Europe warmer than it would otherwise be. As the surface water of this conveyor belt system releases heat energy and cools, it becomes denser and sinks, creating **North Atlantic Deep Water** (**NADW**). The sinking of this cold, dense water keeps the northern part of the Atlantic basin connected to the global thermohaline circulation system; without it, the climate in areas bordering the North Atlantic would be very different.

Ocean water also flows horizontally, in vast riverlike flows (**FIGURE 12.6**), driven mainly by wind systems and differences in air pressure. These surface currents flow within the upper 400 m of water, horizontally and for great distances. These long-lasting patterns influence global climate, and play key roles in the phenomena known as El Niño and La Niña. They have also been crucial in navigation and human history; currents helped carry Polynesians to Easter Island, Darwin to the Galápagos, and Europeans to the New World. Currents transport heat, nutrients, pollution, and the larvae of many marine species from place to place.

Some horizontal surface currents are very slow. Others, like the Gulf Stream, are rapid and powerful. From the

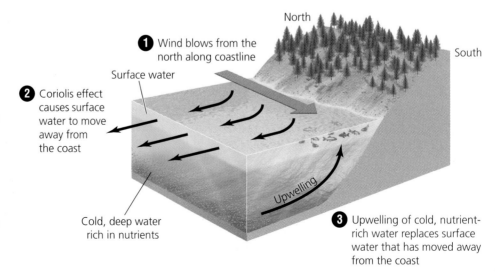

FIGURE 12.5
Upwelling is the movement of bottom waters upward. This type of vertical current often brings nutrients up to the surface, creating rich areas for marine life. For example, north winds blow along the western coast of Vancouver Island (1), while the Coriolis effect draws wind and water away from the coast (2). Water is then pulled up from the bottom (3) to replace the water that moves away from shore.

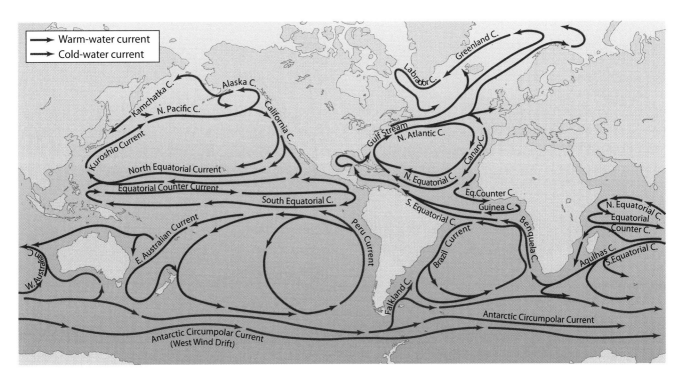

FIGURE 12.6
The upper waters of the oceans flow in currents, which are long-lasting and predictable global patterns of water movement. Warm- and cold-water currents interact with the planet's climate system, and people have used them for centuries to navigate the oceans.
Source: Garrison, T. 2005. Oceanography, 5th ed. Belmont, CA: Brooks/Cole.

Gulf of Mexico, the Gulf Stream moves up the U.S. Atlantic coast and flows past the eastern edges of Georges Bank and the Grand Banks at a rate of 160 km per day (nearly 2 m/s). Averaging 70 km across, the Gulf Stream continues across the North Atlantic, bringing warm water to Europe and moderating that continent's climate, which otherwise would be much colder.

La Niña and El Niño demonstrate the atmosphere–ocean connection

Like the thermohaline circulation, horizontal ocean currents can have far-reaching impacts on climate. An example is the La Niña–El Niño cycle, which demonstrates the linkages between the oceanic and atmospheric systems. Under normal conditions, prevailing winds blow from east to west along the equator, from a region of high pressure in the eastern Pacific to one of low pressure in the western Pacific, forming a large-scale convective loop, or atmospheric circulation pattern (**FIGURE 12.7A**). The winds push surface waters westward, causing water to "pile up" in the western Pacific. As a result, water near Indonesia can be 50 cm higher and 8°C warmer than water near South America. The westward-moving surface waters allow cold water to rise up from the deep in a nutrient-rich upwelling along the coast of Peru and Ecuador.

El Niño conditions are triggered when air pressure increases in the western Pacific and decreases in the eastern Pacific, causing the equatorial winds to weaken. Without these winds, the warm water that collects in the western Pacific flows eastward (**FIGURE 12.7B**), suppressing upwellings along the Pacific coast of South, Central, and North America, and shutting down the delivery of nutrients that support marine life and fisheries. Coastal industries, such as Peru's anchovy fisheries, are devastated by each El Niño event; the 1982–1983 event caused more than $8 billion in economic losses worldwide. El Niños alter weather patterns around the world, creating rainstorms and floods in areas that are generally dry (such as southern California), and causing drought and fire in regions that are typically moist (such as Indonesia). In Canada, El Niño tends to produce weather that is drier and warmer than normal, with more frequent droughts.

La Niña events are the opposite of El Niño; under these conditions, cold surface waters extend far westward in the equatorial Pacific, and weather patterns are affected in opposite ways. La Niña–influenced weather tends to be abnormally cool and wet all the way from British Columbia to southern Quebec. These cycles, called the **El Niño–Southern Oscillation (ENSO)**, are periodic but irregular, occurring every two to eight years. Scientists are getting better at deciphering the triggers for these events, and predicting their impacts on weather. They are also investigating whether globally warming air and sea

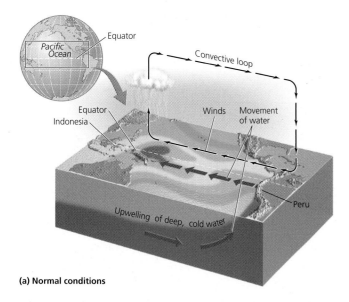

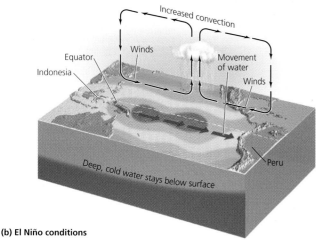

FIGURE 12.7
In these diagrams, red and orange colours denote warmer water, and blue and green colours denote colder water. Under normal conditions **(a)**, prevailing winds push warm surface waters toward the western Pacific. Under El Niño conditions **(b)**, the winds weaken and the warm water flows back across the Pacific toward South America, like water sloshing in a bathtub. This shuts down upwelling along the American coast and alters precipitation patterns regionally and globally. Adapted from National Oceanic and Atmospheric Administration, Tropical Atmospheric Ocean Project.

temperatures may be increasing the frequency and strength of these cycles.

Seafloor topography can be rugged and complex

Although oceans are depicted on most maps and globes as smooth, blue swaths, portions of the ocean floor are complex. Underwater volcanoes shoot forth enough magma to build islands above sea level, such as the Hawaiian Islands. Steep canyons similar in scale to the Grand Canyon lie just offshore of some continents. The lowest spot in the oceans—the Mariana Trench in the South Pacific—is deeper than Mount Everest is high, by more than 2.1 km. Our planet's longest mountain range is under water—the Mid-Atlantic Ridge runs the length of the Atlantic Ocean (**FIGURE 12.8**).

FIGURE 12.8
The seafloor can be rugged. The spreading margin between tectonic plates at the Mid-Atlantic Ridge gives rise to a vast underwater volcanic mountain chain.

Georges Bank and the Grand Banks are essentially huge underwater mounds formed from the debris dumped by glaciers at their southernmost extent. As climate warmed and the glaciers retreated, sea level rose and this hilly terrain became submerged. The 200-nautical-mile fishing limit established by the Convention on the Law of the Sea is particularly problematic here because of the underwater topography—about 10% of the Grand Banks (an area known as the Nose and Tail) lie outside of Canada's Exclusive Economic Zone, which makes it very difficult to regulate the entire fishery as a system.

We can gain an understanding of underwater geographic features by examining a stylized map (**FIGURE 12.9**) that reflects **bathymetry** (the measurement of ocean depths) and **topography** (physical geography, or the shape and arrangement of landforms). In bathymetric profile, gently sloping **continental shelves** underlie the shallow waters bordering the continents. Continental shelves vary in width from 100 m to 1300 km, averaging 70 km wide, with an average slope of 1.9 m/km. These shelves drop off with relative suddenness at the *shelf–slope break*. From there, the **continental slope** angles more steeply downward to the *continental rise*, the gentling slope that connects the continental shelf to the **abyssal plain**, the flat bottom of the deep ocean.

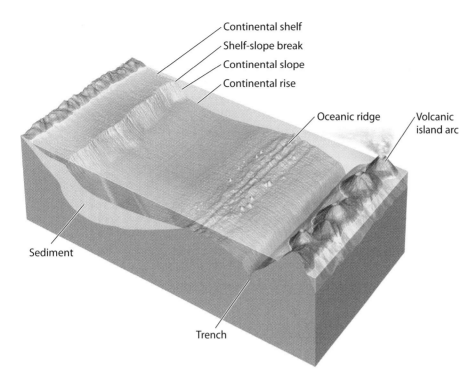

FIGURE 12.9
A stylized bathymetric profile shows key geologic features of the submarine environment. Shallow regions of water exist around the edges of continents over the continental shelf, which drops off at the shelf–slope break. The relatively steep dropoff called the continental slope gives way to the more gradual continental rise, all of which are underlain by sediments from the continents. Vast areas of seafloor are flat abyssal plain. Seafloor spreading occurs at oceanic ridges, and oceanic crust is subducted in trenches. Volcanic activity along trenches often gives rise to island chains, such as the Aleutian Islands. Features on the left side of this diagram are more characteristic of the Atlantic Ocean, and features on the right side of the diagram are more characteristic of the Pacific Ocean. Adapted from Thurman, H. V. 1990. *Essentials of Oceanography*, 4th ed. New York: Macmillan.

Most of the abyssal plain is flat, but volcanic peaks that rise above the ocean floor provide physical structure for marine animals and are often the site of productive fishing grounds. Some island chains, such as the Florida Keys, are formed by the development of reefs and lie atop the continental shelf. Others, such as the Aleutian Islands, which curve across the North Pacific from Alaska toward Russia, are volcanic in origin. The Aleutians are also the site of a deep trench that, like the Mariana Trench, formed at a convergent tectonic plate boundary, where one slab of crust dives beneath another in the process of subduction. These trenches are the deepest places on Earth.

Marine and Coastal Ecosystems

With their variation in topography, temperature, salinity, nutrients, and sunlight, marine and coastal environments feature a variety of ecosystems. Most marine and coastal ecosystems are powered by solar energy, with sunlight driving photosynthesis by phytoplankton in the photic zone. Yet even the darkest ocean depths host life.

Regions of ocean water differ greatly, and some zones support more life than others. The uppermost 10 m of water absorbs 80% of the solar energy that reaches its surface. For this reason, nearly all of the oceans' primary productivity occurs in the well-lit top layer, or **photic zone**. Generally, the warm, shallow waters of continental shelves are most biologically productive and support the greatest species diversity. Habitats and ecosystems occurring between the ocean's surface and floor are termed **pelagic**, whereas those that occur on the ocean floor are called **benthic**.

Open-ocean ecosystems vary in their biological diversity

Biological diversity in pelagic regions of the open ocean is highly variable in its distribution. Primary productivity and animal life near the surface are concentrated in regions of nutrient-rich upwelling. Marine animals that actively swim are referred to as **nekton**. They are contrasted with **plankton**, microscopic organisms that float rather than swim, including *phytoplankton* (plants) and *zooplankton* (animals).

Phytoplankton are particularly important, because they constitute the base of the marine food chain in the pelagic zone (hence concerns about the impacts on marine phytoplankton if the flux of UV radiation to the ocean's surface layer is increased as a result of stratospheric ozone depletion). These photosynthetic algae, protists, and cyanobacteria feed zooplankton, which in turn become food for nektonic fish, jellyfish, whales, and other free-swimming animals (**FIGURE 12.10**). Predators at higher trophic levels include larger fish, sea turtles, and sharks. Many fish-eating birds, such as puffins, petrels, and shearwaters, feed at the surface of the open ocean, returning periodically to nesting sites on islands and coastlines.

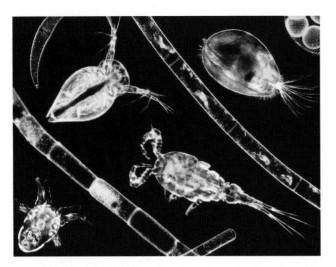

FIGURE 12.10
The uppermost reaches of ocean water contain billions upon billions of phytoplankton—tiny photosynthetic algae, protists, and bacteria that form the base of the marine food chain—as well as zooplankton, small animals and protists that dine on phytoplankton and comprise the next trophic level.

In recent years biologists have been learning more about animals of the very deep ocean, although tantalizing questions remain. In deep-water ecosystems, animals have adapted to deal with extreme water pressures and to live in the dark without food from plants. Some of these often bizarre-looking creatures scavenge carcasses or organic detritus that falls from above. Others are predators, and still others attain food from symbiotic mutualistic bacteria. Some species carry bacteria that produce light chemically by bioluminescence (**FIGURE 12.11**).

As you learned in Chapter 2, some ecosystems form around hydrothermal vents, where heated water spurts from the seafloor, often carrying minerals that precipitate to form large rocky structures. Tubeworms, shrimp, and other creatures in these recently discovered systems use symbiotic bacteria to derive their energy from chemicals in the heated water rather than from sunlight. They manage to thrive within amazingly narrow zones between scalding-hot and icy cold water.

Kelp forests harbour many organisms

Large brown algae, or **kelp**, grow from the floor of continental shelves, reaching upward toward the sunlit surface. Some kelp reaches 60 m in height and can grow 45 cm in a single day. Dense stands of kelp form underwater "forests" along many temperate coasts (**FIGURE 12.12**). Kelp forests supply shelter and food for invertebrates and fish, which in turn provide food for higher trophic level predators, such as seals and sharks. Kelp forests were the setting for our discussion of keystone species in Chapter 4. Recall that sea otters control sea urchin populations, and when otters disappear, urchins overgraze the kelp, destroying the forests and creating "urchin barrens." Kelp forests also absorb wave energy and protect shorelines from erosion. People eat some types of kelp, and kelp provides compounds known as alginates, which serve as thickeners in a wide range of consumer products, including cosmetics, paints, paper, and soaps.

Coral reefs are treasure troves of biodiversity

In shallow subtropical and tropical waters, coral reefs occur. A reef is an underwater outcrop of rock, sand,

FIGURE 12.11
Life is scarce in the dark depths of the deep ocean, but the creatures that do live there can appear bizarre. The anglerfish lures prey toward its mouth with a bioluminescent (glowing) organ that protrudes from the front of its head.

FIGURE 12.12
"Forests" of tall brown algae known as *kelp* grow from the floor of the continental shelf. Numerous fish and other creatures eat kelp or find refuge among its fronds.

or other material. A **coral reef** is a mass of calcium carbonate composed of the skeletons of tiny colonial marine organisms. A coral reef may occur as an extension of a shoreline, along a **barrier island** paralleling a shoreline, or as an **atoll**, a ring around a submerged island.

Corals themselves are tiny invertebrate animals related to sea anemones and jellyfish. They remain attached to rock or existing reef and capture passing food with stinging tentacles. Corals also derive nourishment from microscopic symbiotic algae, known as *zooxanthellae*, which inhabit their bodies and produce food through photosynthesis. Most corals are colonial, and the colourful surface of a coral reef consists of millions of densely packed individuals. As the corals die, their skeletons remain part of the reef while new corals grow atop them, increasing the reef's size.

Like kelp forests, coral reefs protect shorelines by absorbing wave energy. They also host tremendous biodiversity (**FIGURE 12.13A**). The likely reason is that coral reefs provide complex physical structure (and thus many varied habitats) in shallow nearshore waters, which are regions of high primary productivity. Besides the staggering diversity of anemones, sponges, hydroids, tubeworms, and other sessile (stationary) invertebrates, innumerable molluscs, flatworms, sea stars, and urchins patrol the reefs, while thousands of fish species find food and shelter in reef nooks and crannies.

Coral reefs are experiencing worldwide declines, however. Many have undergone **coral bleaching**, a process that occurs when the coloured symbiotic zooxanthellae leave the coral, depriving it of nutrition. Corals lacking zooxanthellae lose colour and frequently die, leaving behind ghostly white patches in the reef (**FIGURE 12.13B**). Coral bleaching is not entirely understood. For example, it is not known exactly why the zooxanthellae leave during coral bleaching—or even *if* they leave; it has been hypothesized, alternatively, that they may lose their pigmentation (hence the "bleaching"), possibly as a result of *photodamage* (*photo* = "light") caused by exposure to ultraviolet radiation.[1]

Coral bleaching may result from stress caused by increased sea surface temperatures associated with global climate change, from changes in light levels in some shallow-water areas, from an influx of pollutants, or from some combination of these and other unknown factors, both natural and anthropogenic. In addition, coral reefs sustain significant damage when divers stun fish with cyanide or by throwing explosives over the side of the boat, a common fishing and fish collection practice (for the pet trade) in the waters of Indonesia and the Philippines. Tourism also has been a significant burden on coral reefs globally; each scuba diver who breaks off a small piece of a reef contributes to its demise. Even with the advent of ecotourism, which of course advocates taking only photographs, the traffic of divers' feet and hands

(a) Coral reef community

(b) Bleached coral

FIGURE 12.13
Corals reefs provide food and shelter for a tremendous diversity **(a)** of fish and other creatures. However, these reefs face multiple environmental stresses from human impacts, and many corals have died as a result of coral bleaching **(b)**, in which corals lose their zooxanthellae. Such bleaching is evident in the whitened portion of this coral.

on the reefs continues to take a toll. Finally, as global climate change proceeds, the oceans are becoming more acidic, as excess carbon dioxide from the atmosphere reacts with seawater to form carbonic acid. Acidification threatens to deprive corals of the carbonate ions they need to produce their structural parts.

A few coral species thrive in waters outside the tropics and build reefs on the ocean floor at depths of 200–500 m. These little-known reefs, which occur in cold-water areas off the coasts of Norway, Spain, the British Isles, and elsewhere, are only now beginning to be studied by scientists. Already, however, many have been badly damaged by trawling—the same practice that has so degraded the benthic habitats of groundfish, such as the Atlantic cod. Norway and other countries are now beginning to protect some of these deep-water reefs.

Intertidal zones undergo constant change

Where the ocean meets the land, **intertidal** or **littoral** ecosystems (**FIGURE 12.14**) spread between the uppermost reach of the high tide and the lowest limit of the low tide. **Tides** are the periodic rising and falling of the ocean's height at a given location, caused by the gravitational pull of the Moon and Sun. High and low tides occur roughly six hours apart, so intertidal organisms spend part of each day submerged in water, part of the day exposed to the air and sunlight, and part of the day being lashed by waves. Subject to tremendous extremes in temperature, moisture, light exposure, and salinity, these creatures must also protect themselves from marine predators at high tide and terrestrial predators—even opportunistic predators like crows, seagulls, and raccoons—at low tide.

The intertidal environment is a tough place to make a living, but it is home to a remarkable diversity of organisms. Rocky shorelines can be full of life among the crevices, which provide shelter and pools of water (*tidepools*) during low tides. Sessile animals, such as anemones, mussels, and barnacles, live attached to rocks, filter-feeding on plankton in the water that washes over them. Urchins, sea slugs, chitons, and limpets eat intertidal algae or scrape food from the rocks. Sea stars (starfish) creep slowly along, preying on the filter-feeders and herbivores at high tide. Crabs clamber around the rocks, scavenging detritus.

The rocky intertidal zone is so diverse because environmental conditions, such as temperature, salinity, and moisture, change dramatically from the high to the low reaches. This environmental variation gives rise to horizontal bands formed by dominant organisms as they array themselves according to their habitat needs. Sandy intertidal areas, such as those of Cape Cod, host less biodiversity, yet plenty of organisms burrow into the sand at low tide to await the return of high tide, when they emerge to feed.

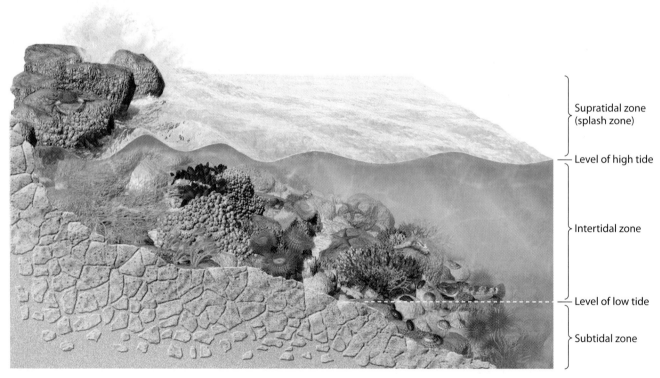

Tidal zones

FIGURE 12.14
The rocky intertidal zone is the swath of a rocky shoreline between the lowest and highest reaches of the tides. This is an ecosystem rich in biodiversity, typically containing large invertebrates, such as sea stars (starfish), barnacles, crabs, sea anemones, corals, bryozoans, snails, limpets, chitons, mussels, nudibranchs (sea slugs), and sea urchins. Fish swim in tidal pools, and many types of algae cover the rocks. Areas higher on the shoreline are exposed to the air more frequently and for longer periods, so organisms that can best tolerate exposure specialize in the upper intertidal zone. The lower intertidal zone is exposed less frequently and for shorter periods, so organisms less able to tolerate exposure thrive in this zone.

FIGURE 12.15
Salt marshes, like the Malbay salt marsh in the Gaspé Peninsula, shown here, occur in temperate intertidal zones where the substrate is muddy, allowing salt-adapted grasses to grow. Tidal waters generally flow through marshes in channels called *tidal creeks*, amid flat areas called *benches*, sometimes partially submerging the grasses.

Salt marshes occur widely along temperate shorelines

Along many of the world's coasts at temperate latitudes, **salt marshes** occur where the tides wash over gently sloping sandy or silty substrates. Rising and falling tides flow into and out of channels called *tidal creeks* and at highest tide spill over onto elevated marsh flats (**FIGURE 12.15**). Marsh flats grow thick with grasses, rushes, shrubs, and other herbaceous plants. Grasses, such as those in the genera *Spartina* and *Distichlis*, are the dominant vegetation in most salt marshes.

Salt marshes boast very high primary productivity and provide critical habitat for shorebirds, waterfowl, and the adults and young of many commercially important fish and shellfish species. Salt marshes also filter out pollution (hence the use of constructed or artificial wetlands for wastewater management, as discussed in Chapter 11). Coastal marshes stabilize shorelines against **storm surges**, the very large waves that crash onto the shore during a major storm, which are caused by a combination of low atmospheric pressure and high winds. However, people like to live along coasts, and coastal sites are desirable for commerce. As a result, people have altered or destroyed vast expanses of salt marshes worldwide to make way for coastal development. When salt marshes are destroyed, we lose the ecosystem services they provide. When Hurricane Katrina struck the Gulf Coast of Louisiana, for instance, the flooding was made worse because vast areas of salt marshes had vanished because of development, subsidence from oil and gas drilling, and dams that had held back marsh-building sediment.

Mangrove forests line coasts in the tropics and subtropics

In tropical and subtropical latitudes, mangrove forests replace salt marshes along gently sloping sandy and silty coasts. The **mangrove** is a tree with a unique type of root system that curves upward (like a snorkel) to obtain oxygen lacking in the mud in which the tree grows, and that curves downward like stilts to support the tree in changing water levels (**FIGURE 12.16**). Fish, shellfish, crabs, snakes, and other organisms thrive among the root networks, and

FIGURE 12.16
Mangrove forests are important ecosystems along tropical and subtropical coastlines throughout the world. Mangrove trees, such as these at Lizard Island, Australia, show specialized adaptations for growing in saltwater and provide habitat for many types of fish, birds, crabs, and other animals.

birds feed and nest in the dense foliage of these coastal forests. Besides serving as nurseries for fish and shellfish that people harvest, mangroves also provide materials that people use for food, medicine, tools, and construction.

From Florida to Mexico to the Philippines, mangrove forests in tropical areas have been destroyed as people have developed coastal areas for residential, commercial, and recreational uses. Shrimp farming in particular has driven the conversion of large areas of mangroves. We have eliminated half the world's mangrove forests, and their area continues to decline by 2%–8% per year. When mangroves are removed, coastal areas lose the ability to slow runoff, filter pollutants, and retain soil. As a result, offshore systems, such as eelgrass beds and coral reefs, are more readily degraded.

Moreover, mangrove forests protect coastal communities against storm surges and tsunamis (tidal waves), as was shown when the 2004 Indian Ocean tsunami devastated areas where mangroves had been removed but caused less damage where mangroves were intact. The loss of coastal mangroves may also have played a role in the scale of devastation from Hurricane Nargus in Burma (or Myanmar), a disaster that was greatly exacerbated by political stubbornness after the fact. Despite these important ecosystem services provided by mangroves, we have granted only about 1% of the world's remaining mangroves protection against development.

Freshwater meets saltwater in estuaries

Many salt marshes and mangrove forests occur in or near **estuaries**, water bodies where rivers flow into the ocean, mixing freshwater with saltwater. Estuaries are biologically productive ecosystems that experience fluctuations in salinity as tides and freshwater runoff vary daily and seasonally. For shorebirds and for many commercially important shellfish species, estuaries provide critical habitat. For anadromous fishes (fishes, such as salmon, that spawn in freshwater and mature in saltwater), estuaries provide a transitional zone where young fish make the passage from freshwater to saltwater.

Estuaries around the world have been affected by urban and coastal development, water pollution, habitat alteration, and overfishing. The Gaspé Peninsula of Quebec—where the St. Lawrence River flows into the Gulf of St. Lawrence and from there into the Atlantic Ocean—is one area in Canada where estuaries and salt marshes are important ecosystems. Florida Bay, where freshwater from the Everglades system mixes with saltwater, provides another example. This estuary has suffered pollution and a reduction in freshwater flow caused by irrigation and fertilizer use by sugarcane farmers, housing development, septic tank leakage, and other human impacts. Coastal ecosystems have borne the brunt of human impact because two-thirds of Earth's people choose to live within 160 km of the ocean.

---

weighing the issues

COASTAL DEVELOPMENT 12–1

A developer wants to build a large marina on an estuary in your coastal town. The marina would boost the town's economy but eliminate its salt marshes. As a homeowner living adjacent to the marshes, how would you respond? Do you think that developers or town officials should offer homeowners insurance against damage from storm surges in such a situation?

Human Use and Impact

Our species has a long history of interacting with the oceans. We have long travelled across their waters, clustered our settlements along coastlines, and been fascinated by the beauty, power, and vastness of the seas. We have also left our mark upon them by exploiting oceans for their resources and polluting them with our waste.

Oceans provide transportation routes

We have used the oceans for transportation for thousands of years, and the oceans continue to provide affordable means of moving people and products over vast distances. Ocean shipping has accelerated the global reach of some cultures and has promoted interaction among long-isolated peoples. It has had substantial impacts on the environment as well. The thousands of ships plying the world's oceans today carry everything from cod to cargo containers to crude oil. Ships transport ballast water, which, when discharged at ports of destination, may transplant aquatic organisms picked up at ports of departure. Some of these species—such as the zebra mussel (Chapter 4)—establish themselves and become invasive.

We extract energy and minerals

We mine the oceans for sources of commercially valuable energy. Worldwide, about 30% of our crude oil and nearly half of our natural gas comes from seafloor deposits. Most offshore oil and gas is concentrated in petroleum-rich regions, such as the North Sea and the Gulf of Mexico, but energy companies extract smaller amounts of oil and gas from diverse locations, among them the Grand Banks and adjacent Canadian waters. Proposals to drill for oil and gas in Georges Bank, however, have been stalled until recently by the Canadian government, in large part because of fears that spilled oil could damage the region's valuable fisheries.

Ocean sediments also contain a novel potential source of fossil fuel energy. **Methane hydrate** is an icelike solid consisting of molecules of methane (CH_4, the main component of natural gas) embedded in a crystal lattice of water molecules. Methane hydrates are stable at temperature and pressure conditions found in many sediments on the Arctic seafloor and the continental shelves. It is estimated that the world's deposits of methane hydrates may hold twice as much carbon as all known deposits of oil, coal, and natural gas combined.

Could methane hydrates be developed as an energy source to power our civilization through the twenty-first century and beyond? Perhaps, but a great deal of research remains before scientists and engineers can be sure how to extract these energy sources safely. Destabilizing a methane hydrate deposit could lead to a catastrophic release of gas. This could cause a massive landslide and tsunami and would also release huge amounts of methane, a potent greenhouse gas, into the atmosphere, exacerbating global climate change.

Fortunately, the oceans also hold potential for providing renewable energy sources that do not emit greenhouse gases. Engineers have developed ways of harnessing energy from waves, tides, and the heat of ocean water. These promising energy sources await further research, development, and investment.

We can extract minerals from the ocean floor, as well. By using large vacuum cleaner–like hydraulic dredges, miners collect sand and gravel from beneath the sea. Also extracted are sulphur from salt deposits in the Gulf of Mexico and phosphorite from offshore areas near the California coast and elsewhere. Other valuable minerals found on or beneath the seafloor include calcium carbonate (used in making cement) and silica (used as fire-resistant insulation and in manufacturing glass), as well as rich deposits of copper, zinc, silver, and gold ore. Many minerals are concentrated in manganese nodules, small ball-shaped accretions that are scattered across parts of the ocean floor. More than 1.5 trillion tons of manganese nodules may exist in the Pacific Ocean alone, and their reserves of metal may exceed all terrestrial reserves.

The logistical difficulty of mining the ocean floor has kept it only marginally economical or uneconomical so far, but that is changing very rapidly. An Australian company called Neptune Mining Company, in conjunction with a Canadian company called Nautilus Minerals, has developed technologies and approaches that make use of existing offshore oil exploration technologies that will make it much cheaper and more efficient to mine on the seafloor. As of 2008 they are set to open the world's first underwater gold, silver, copper, and zinc mine, offshore from Papua New Guinea.

Seafloor mineral development has been one of the most controversial factors in the very long history of attempts to convene a comprehensive, binding Law of the Sea, an international legal discussion that actually began as early as the 1600s. To this day, the United States abstains from participating in some aspects of the Law of the Sea, primarily because of proposed restrictions on seafloor mining. The existing Law of the Sea, or **United Nations Convention on the Law of the Sea (UNCLOS)**, is based on a series of international conferences that took place between 1973 and 1982. This version of UNCLOS established the 200-nautical-mile Exclusive Economic Zones of nations (replacing the previous 12-mile zone), which greatly enhanced the ability of nations to control and manage their own coastal zones.

Marine pollution threatens resources

People have long used oceans as a sink for waste and pollution. Even into the mid-twentieth century, it was common for coastal cities to dump trash and untreated sewage along their shores. Halifax, Nova Scotia, only began to treat municipal sewage outflow into Halifax Harbour— approximately 181 million litres per day[2]—in the early 2000s. A surprising number of Canadian towns continue to discharge raw sewage into coastal and inland waters, even today. Fort Bragg, a bustling town on the northern Californian coast, boasts of its Glass Beach, an area where beachcombers collect sea glass, the colourful surf-polished glass sometimes found on beaches after storms. But Glass Beach is in fact the site of the former town dump, and besides well-polished glass, the perceptive visitor may also spot old batteries, rusting car frames, and other trash protruding from the bluffs above the beach.

Oil, plastic, industrial chemicals, and excess nutrients all eventually make their way from land into the oceans. Raw sewage and trash from cruise ships and abandoned fishing gear from fishing boats add to the input. The scope of trash in the sea can be gauged by the amount picked up each September by volunteers who trek beaches in the Ocean Conservancy's annual International Coastal Cleanup. In this nonprofit organization's 2006 cleanup, 359 000 people from 66 nations picked up 3.2 million kilograms of trash from 55 600 km of shoreline.

Nets and plastic debris endanger marine life

Plastic bags and bottles, discarded fishing nets, gloves, fishing line, buckets, floats, abandoned cargo, and much else that people transport on the sea or deposit into it can harm marine organisms. Because most plastic is not biodegradable, it can drift for decades before washing up on beaches. Marine mammals, seabirds, fish, and sea turtles may mistake floating plastic debris for food and can die as a result of ingesting material they cannot digest or expel. Fishing nets that are lost or intentionally discarded can continue snaring animals for decades (**FIGURE 12.17**).

Of 115 marine mammal species, 49 are known to have eaten or become entangled in marine debris, and 111 of 312 species of seabirds are known to ingest plastic. All five species of sea turtle in the Gulf of Mexico have died from consuming or contacting marine debris. Marine debris affects people, as well. Surveys of fishers have shown that more than half have encountered equipment damage and other problems from plastic debris.

Oil pollution comes from spills of all sizes

Major oil spills, such as the *Exxon Valdez* spill in Prince William Sound, Alaska (Chapter 2), make headlines and cause serious environmental problems. Yet it is important to put such accidents into perspective. The majority of oil pollution in the oceans comes not from large spills in a few particular locations but from the accumulation of innumerable, widely spread small sources (non-point sources), including leakage from small boats and runoff from human activities on land. Moreover, the amount of petroleum spilled into the oceans in recent years is equalled by the amount that seeps into the water from naturally occurring seafloor deposits (**FIGURE 12.18A**).

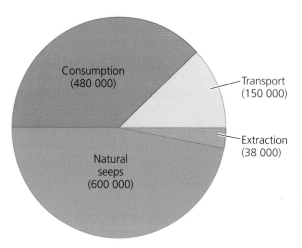

(a) Sources of petroleum input into oceans (metric tons)

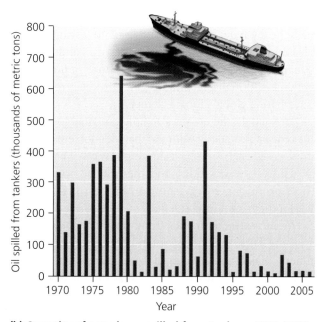

(b) Quantity of petroleum spilled from tankers, 1970–2006

FIGURE 12.18
Of the 1.3 million metric tons of petroleum entering the world's oceans each year, nearly half is from natural seeps. Petroleum consumption by people accounts for 38% of total input, and this includes numerous diffuse non-point sources, especially runoff from rivers and coastal communities and leakage from two-stroke engines. Spills during petroleum transport account for 12%, and leakage during petroleum extraction accounts for 3% **(a)**. Less oil is being spilled into ocean waters today in large tanker spills, thanks in part to regulations on the oil shipping industry and improved spill response techniques **(b)**. The figure shows cumulative quantities of oil spilled worldwide from nonmilitary spills of over seven metric tons. Data in (a) from National Research Council. 2003. *Oil in the Sea III. Inputs, Fates, and Effects.* Washington, DC: National Academies Press. Data in (b) from International Tanker Owners Pollution Federation Ltd. (ITOPF). 2006.

FIGURE 12.17
This northern fur seal became entangled in a discarded fishing net. Each year, many thousands of marine mammals, birds, and turtles are killed by plastic debris, abandoned nets, and other trash that people have dumped in the ocean.

Nonetheless, minimizing the amount of oil we release into coastal waters is important, because petroleum pollution is detrimental to marine life and to our economies. Petroleum can physically coat and kill marine organisms and can poison them when ingested. In response to headline-grabbing oil spills, governments worldwide have begun to implement more stringent safety standards for tankers, such as requiring industry to pay for tugboat escorts in sensitive and hazardous coastal waters, double hulls to preclude punctures, and the development of prevention and response plans for major spills.

The oil industry has resisted many such safeguards. Today, the ship that oiled Prince William Sound is still plying the world's oceans, renamed the *Sea River Mediterranean* and still featuring only a single hull. However, over the past three decades, the amount of oil spilled in global waters has decreased (**FIGURE 12.18B**), in part because of an increased emphasis on spill prevention and response.

Pollutants can contaminate seafood

Marine pollution can make some fish and shellfish unsafe for people to eat. One prime concern today is mercury contamination. Mercury is a toxic heavy metal that is emitted in coal combustion and from other sources. After settling onto land and water, mercury bioaccumulates in animals' tissues and biomagnifies as it makes its way up the food chain. As a result, fish and shellfish at high trophic levels can contain substantial levels of mercury. Eating seafood high in mercury is particularly dangerous for young children and for pregnant or nursing mothers, because the fetus, baby, or child can suffer neurological damage as a result.

Because seafood is a vital part of a healthy diet, nutritionists do not advocate avoiding seafood entirely. However, people in at-risk groups should avoid fish high in mercury (such as swordfish, shark, and albacore tuna) while continuing to eat seafood low in mercury (such as catfish, salmon, and canned light tuna). We should also be careful not to eat seafood from local areas where health advisories have been issued.

Excess nutrients cause algal blooms

Pollution from fertilizer runoff or other nutrient inputs can have dire effects on marine ecosystems. The release of excess nutrients into surface waters can spur unusually rapid growth of phytoplankton, causing *eutrophication* in freshwater and saltwater systems.

Excessive nutrient concentrations sometimes give rise to population explosions among several species of marine algae that produce powerful toxins that attack the nervous systems of vertebrates. Blooms of these algae occur periodically on both the east and west coasts of Canada. Some algal species produce reddish pigments that discolour surface waters, and blooms of these species are nicknamed **red tides** (**FIGURE 12.19**). Harmful algal blooms can cause illness and death among zooplankton, birds, fish, marine mammals, and people as their toxins

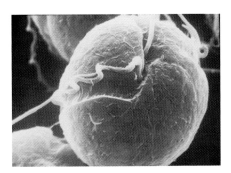

(a) Dinoflagellate (*Gymnodinium*)

FIGURE 12.19
In a harmful algal bloom, certain types of algae multiply to great densities in surface waters, producing toxins that can bioaccumulate and harm organisms. Red tides are a type of harmful algal bloom in which the algae, such as dinoflagellates of the genus *Gymnodinium* **(a)**, produce pigment that turns the water red **(b)**.

(b) Red tide, Gulf of Carpentaria, Australia

are passed up the food chain. They also cause economic loss for communities dependent on fishing or beach tourism. Reducing nutrient runoff into coastal waters can lessen the frequency of these outbreaks. When they occur, we can minimize their health impacts by monitoring to prevent human consumption of affected organisms.

As severe as the impacts of marine pollution can be, however, most marine scientists concur that the more worrisome dilemma is overharvesting. Unfortunately, the old cliché that "there are always more fish in the sea" is not true; the oceans today have been overfished, and like the groundfish of the Northwest Atlantic, many stocks have been largely depleted.

Emptying the Oceans

The oceans and their biological resources have provided for human needs for thousands of years, but today we are placing unprecedented pressure on marine resources. Half the world's marine fish populations are fully exploited, meaning that we cannot harvest them more intensively without depleting them, according to the U.N. Food and Agriculture Organization (FAO). An additional one-quarter of marine fish populations are overexploited and already being driven toward extinction. Thus only one-quarter of the world's marine fish populations can yield more than they are already yielding without being driven into decline.

The total global fisheries catch, after decades of increases, levelled off after about 1988 (**FIGURE 12.20**), despite increased fishing effort. Fishery collapses, such as those off Newfoundland and Labrador and New England, are ecologically devastating and also take a severe economic toll on human communities that depend on fishing. If current trends continue, predicted a comprehensive 2006 study in the journal *Science*, populations of *all* ocean species that we fish for today will collapse by the year 2048.

As our population grows, we will become even more dependent on the oceans' bounty. This makes it vital, say many scientists and fisheries managers, that we turn immediately to more sustainable fishing practices.

We have long overfished

People have always harvested fish, shellfish, turtles, seals, and other animals from the oceans. Although much of this harvesting was sustainable, scientists are learning that people began depleting some marine species centuries or millennia ago. Overfishing then accelerated during the colonial period of European expansion and intensified further in the twentieth century.

A recent synthesis of historical evidence reveals that ancient overharvesting likely affected ecosystems in astounding ways we only partially understand today. Several large animals, including the Caribbean monk seal, Steller's sea cow, and Atlantic grey whale, were hunted to extinction prior to the twentieth century—before scientists were able to study them or the ecological roles they played. Overharvesting of the vast oyster beds of Chesapeake Bay led to the collapse of its oyster fishery in the late nineteenth century. Eutrophication and hypoxia similar to that of the Gulf of Mexico (Chapter 5) resulted, because there are no longer oysters to filter algae and bacteria from the water. In the Caribbean, green sea turtles ate sea grass and likely kept it cropped low, like a lawn. But with today's turtle population a fraction of what it was, sea grass grows thickly, dies, and rots, giving rise to disease, such as sea grass wasting disease, which ravaged sea grass in the 1980s. The best-known case of historical overharvesting is the near-extinction of many species of whales. This resulted from commercial whaling that began centuries ago and was curtailed only in 1986. Since then, some species have been recovering, but others have not.

Groundfish in the Northwest Atlantic historically were so abundant that the people who harvested them

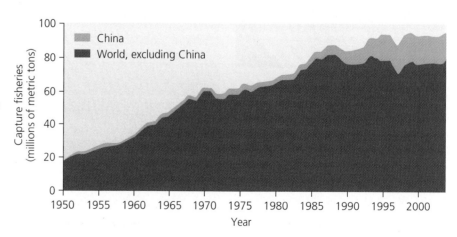

FIGURE 12.20
The total global fisheries catch has increased over the past half-century, but in recent years growth has stalled, and many fear that a global catch decline is imminent if conservation measures are not taken soon. The figure shows trends with and without data on China's substantial fishing industry, which had been withheld from international scrutiny for many years (see "The Science Behind the Story: China's Fisheries Data"). With China's data, global catch has levelled off since the mid-1990s. Without China's data, the catch has decreased slightly since 1988. Data from U.N. Food and Agricultural Organization (FAO).

THE SCIENCE BEHIND THE STORY

China's Fisheries Data

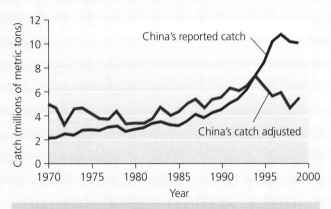

Dr. Daniel Pauly is with the University of British Columbia.

Models estimated China's fish harvests to be much lower than reported. Data from Watson, R., and D. Pauly. 2001. Systematic distortions in world fisheries catch trends. *Nature* 414:536–538.

China is responsible for a larger share of the world's fisheries catch than any other nation, according to the United Nations Food and Agriculture Organization (FAO). Thus, China's catch data has a major impact on the FAO's attempts to assess the health of global fisheries.

In 2001, two fisheries scientists published a paper in the journal *Nature* suggesting that China had exaggerated its catch data by as much as 100% during the 1990s. Because local officials are rewarded for meeting production targets set by the government, they have economic incentive to falsely inflate their catch numbers, these researchers argued, and China's national data ends up inflated as a result. The high catch data had led to complacency among the world's fisheries managers and policy makers, they maintained, because it led the FAO to overestimate the true amount of fish left in the ocean.

The authors of the paper, University of British Columbia researchers Reg Watson and Daniel Pauly, had become suspicious of China's data for several reasons. First, China's coastal fisheries had been overexploited for decades, yet reported catches continued to increase.

Second, China's "catch per unit effort" remained unchanged from 1980 to 1995, even though the abundance of fish had decreased. Finally, China's catch statistics suggested that its waters were far more productive than ecologically similar areas fished by other nations.

By using a number of databases, including FAO catch data collected since the 1950s, Watson and Pauly built a statistical model to predict global fisheries catch. They divided the world's oceans into approximately 180 000 "cells," each spanning half a degree longitudinally and latitudinally. The cells were characterized by factors that affect catch size, such as depth, primary productivity, fishing rights, and species distribution. The researchers used these factors to predict annual catch for each cell.

For most cells, the catch predicted by the model was similar to the catch reported by fleets that fished those waters. In large swaths of China's coastal waters, however, predicted catches were far smaller than reported catches. For instance, the model predicted that in 1999 China's total catch would be 5.5 million metric tons, whereas the figure reported by Chinese fisheries authorities was 10.1 million metric tons.

When Watson and Pauly analyzed global fisheries trends by using the model rather than China's official numbers, they found that the annual global catch had decreased by 360 000 metric tons since 1988, instead of increasing by 330 000 metric tons, as reported by the FAO. Their finding suggested that the global catch, rather than remaining stable through the 1990s, actually had begun to decline in the 1980s.

Fisheries managers and policy makers depend on the FAO for information about the state of the world's fisheries. By using China's apparently inflated numbers, Watson later wrote, the FAO had encouraged a global "mopping-up operation" of the last of the world's fish.

The Chinese government and the FAO both criticized the study. China's reported catch was "basically correct," said Yang Jian, director-general of the Chinese Bureau of Fisheries. Watson and Pauly's model, he suggested, failed to take into account unique aspects of China's fisheries, such as its large catch of crabs and jellyfish.

The FAO noted that it was already treating China's statistics with caution. In its recent publications, global fisheries catches were being reported with and without China's data (see **FIGURE 12.20**). The FAO had also held meetings with Chinese officials to discuss ways of reducing overreporting, which Yang agreed did exist.

---

never imagined they could be depleted. Yet careful historical analysis of fishing records has revealed that even in the nineteenth century, fishers repeatedly experienced locally dwindling catches and each time needed to introduce some new approach or technology to extend their reach and restore their catch rate.

Fishing has become industrialized

Today's commercial fishing fleets are highly industrialized, employing fossil fuels, huge vessels, and powerful new technologies to capture fish in volumes never dreamed of by nineteenth-century mariners. So-called *factory fishing* vessels even process and freeze their catches while at sea. The global reach of today's fleets makes our impacts much more rapid and intensive than in past centuries.

The modern fishing industry uses a number of methods to capture fish at sea. Some vessels set out long **driftnets** that span large expanses of water (**FIGURE 12.21A**). These strings of nets are arrayed strategically to drift with currents so as to capture passing fish, and are held vertical by floats at the top and weights at the bottom. Driftnetting usually targets species that traverse the open water in immense schools (flocks), such as herring, sardines, and mackerel. Specialized forms of driftnetting are used for sharks, shrimp, and other animals.

Longline fishing (**FIGURE 12.21B**) involves setting out extremely long lines with up to several thousand baited hooks spaced along their lengths. Tuna and swordfish are among the species targeted by longline fishing.

Trawling entails dragging immense cone-shaped nets through the water, with weights at the bottom and floats at the top to keep the nets open. Trawling in open water captures pelagic fish, whereas **bottom-trawling** (**FIGURE 12.21C**) involves dragging weighted nets across the floor of the continental shelf to catch groundfish and other benthic organisms, such as scallops.

Fishing practices kill nontarget animals and damage ecosystems

Unfortunately, these fishing practices catch more than just the species they target. **By-catch** refers to the accidental capture of animals, and it accounts for the deaths of many thousands of fish, sharks, marine mammals, and birds each year. Driftnetting captures substantial numbers of dolphins, seals, and sea turtles, as well as countless nontarget fish. Most of these end up drowning (mammals and turtles need to surface to breathe) or dying from air exposure on deck (fish breathe through gills in the water). Many nations have banned or restricted driftnetting because of excessive by-catch. The widespread death of dolphins in driftnets motivated consumer efforts to label tuna as "dolphin-safe" if its capture uses methods designed to avoid dolphin by-catch. Such measures helped reduce dolphin deaths from an estimated 133 000 per year in 1986 to fewer than 2000 per year since 1998.

Similar by-catch problems exist with longline fishing, which kills turtles, sharks, and albatrosses, magnificent seabirds with wingspans up to 3.6 m. Several methods are being developed to limit by-catch from longline fishing, but an estimated 300 000 seabirds of various species die each year when they become caught on hooks while trying to ingest bait.

(a) Driftnetting

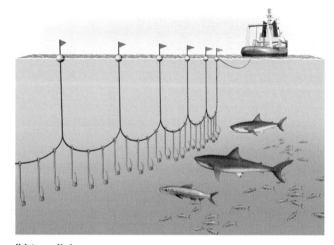

(b) Longlining

(c) Bottom-trawling

FIGURE 12.21
Commercial fishing fleets use several main methods of capture. In driftnetting **(a)**, huge nets are set out to drift through the open water to capture schools of fish. In longlining **(b)**, lines with numerous baited hooks are set out in open water. In bottom-trawling **(c)**, weighted nets are dragged along the floor of the continental shelf. All methods result in the capture of nontarget animals. The illustrations above are schematic for clarity and do not capture the immense scale that these technologies can attain; for instance, industrial trawling nets can be large enough to engulf multiple Boeing 747 jumbo jets.

(a) Before trawling

(b) After trawling

FIGURE 12.22
Bottom-trawling causes severe structural damage to reefs and benthic habitats, and it can decimate underwater communities and ecosystems. A photo of an untrawled location **(a)** on the seafloor of Indonesia shows a vibrant and diverse coral reef community. A photo of a trawled location **(b)** nearby shows a flattened and lifeless expanse of broken coral draped with an abandoned stretch of trawling net.

Bottom-trawling can destroy entire communities and ecosystems. The weighted nets crush organisms in their path and leave long swaths of damaged sea bottom. Trawling is especially destructive to structurally complex areas, such as reefs, that provide shelter and habitat for many animals. In recent years, underwater photography has begun to reveal the extent of structural and ecological disturbance done by trawling (**FIGURE 12.22**). Trawling is often likened to clear-cutting and strip-mining, and in heavily fished areas, the bottom may be damaged more than once. At Georges Bank, it is estimated that the average expanse of bottom has been trawled three times.

Modern fishing fleets deplete marine life rapidly

We can see the effects of large-scale industrialized fishing in the catch records of groundfish from the Northwest Atlantic. Although cod had been harvested since the 1500s on the Grand Banks, catches more than doubled once immense industrial trawlers from Europe, Japan, and the United States appeared in the 1960s (**FIGURE 12.23A**). These record-high catches lasted only a decade; the industrialized approach removed so many fish that the stock has not recovered. Likewise, on Georges Bank, cod catches rose greatly in the 1960s, remained high for 30 years, then collapsed (**FIGURE 12.23B**).

Throughout the world's oceans, today's **industrialized fishing** fleets are depleting marine populations quickly. In a 2003 study, Canadian fisheries biologists Ransom Myers and Boris Worm analyzed fisheries data from FAO archives, looking for changes in the catch rates of fish in various regions of ocean since they were first exploited by industrialized fishing. For one region after another, they found the same pattern: Catch rates dropped precipitously, with 90% of large-bodied fish and sharks eliminated within only a decade (**FIGURE 12.24**). Following that, populations stabilized at 10% of their former levels. This means, Myers and Worm concluded, that the oceans today contain only one-tenth of the large-bodied animals they once did.

As we have seen, when animals at high trophic levels are removed from a food web, the proliferation of their prey can alter the nature of the entire community. Many scientists now conclude that most marine communities may have been very different prior to industrial fishing.

Several factors mask declines

Although industrialized fishing has depleted fish stocks in region after region, the overall global catch has remained roughly stable for two decades (see **FIGURE 12.20**). The seeming stability of the total global catch can be explained by several factors that mask population declines. One is that fishing fleets have been travelling longer distances to reach less-fished portions of the ocean. They also have been fishing in deeper waters; average depth of catches was 150 m in 1970 and 250 m in 2000. Moreover, fishing fleets have been spending more time fishing and have been setting out more nets and lines—expending increasing effort just to catch the same number of fish.

Improved technology also helps explain large catches despite declining stocks. Today's Japanese, European,

FIGURE 12.23
In the North Atlantic off the coast of Newfoundland and Labrador, commercial catches of Atlantic cod **(a)** increased with intensified fishing by industrial trawlers in the 1960s and 1970s. The fishery subsequently crashed, and moratoria imposed in 1992 and 2003 have not brought it back. A similar pattern is seen in the cod catches at Georges Bank **(b)**; industrial fishing produced 30 years of high catches, followed by a collapse and the closure of some areas to fishing. Note also that in each case, there is one peak before 1977 and one after 1977. The first peak and decline resulted from foreign fishing fleets, whereas the second peak and decline resulted from Canadian and U.S. fleets, respectively, after they laid claim to their 200-nautical-mile Exclusive Economic Zones. Data in (a) from Millennium Ecosystem Assessment. 2005. Data in (b) from O'Brien, et al. 2005. Georges Bank Atlantic Cod. In Mayo, R. K. and M. Terceiro, eds. *Assessment of 19 Northeast Groundfish Stocks Through 2004*. Woods Hole, MA: Northeast Fisheries Science Center.

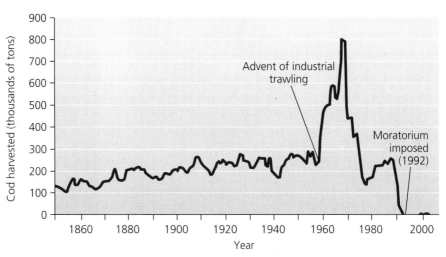

(a) Cod harvested from Grand Banks

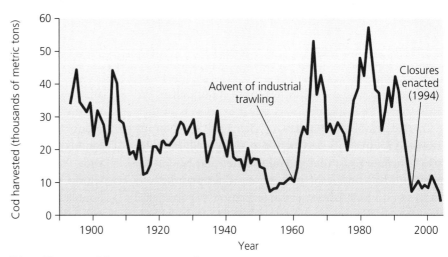

(b) Cod harvested from Georges Bank

Canadian, and U.S. fleets can reach almost any spot on the globe with vessels that attain speeds of 80 km/h. They have access to an array of technologies that militaries have developed for spying and for chasing enemy submarines, including advanced sonar mapping equipment, satellite navigation, and thermal sensing systems. Some fleets rely on aerial spotters to find schools of commercially valuable fish, such as bluefin tuna. One final cause of misleading stability in global catch numbers is that not all data supplied to international monitoring agencies may be accurate, for a variety of reasons.

We are "fishing down the food chain"

Overall figures on total global catch tell only part of the story, because they do not relate the species, age, and size of fish harvested. Careful analyses of fisheries data have revealed in case after case that as fishing increases, the size and age of fish caught decline. Cod caught in the Northwest Atlantic today are on average *much* smaller than they were decades ago, and it is now rare to find a cod more than 10 years of age, although cod of this age formerly were common. The reproductive potential of today's smaller cod is much less—an order of magnitude less—than that of the giant cod of previous decades.

In addition, as particular species become too rare to fish profitably, fleets begin targeting other species that are in greater abundance. Generally this means shifting from large, desirable species to smaller, less desirable ones. Fleets have time and again depleted popular food fish (such as cod) and shifted to species of lower value (such as capelin, smaller fish that cod eat). Because this often entails catching species at lower trophic levels, this phenomenon has been termed "fishing down the food chain," a concept proposed by UBC fisheries scientist Daniel Pauly and colleagues in 1998.

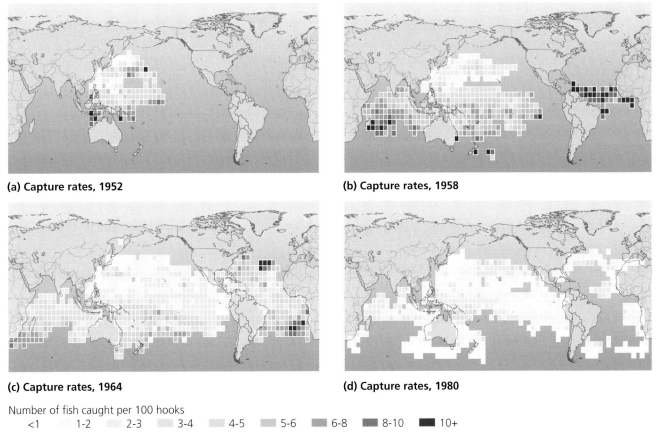

FIGURE 12.24
As industrial fishing fleets reached each new region of the world's oceans, capture rates of large predatory fishes were initially high and then within a decade declined markedly. In the figure, darker colours signify high capture rates, and lighter colours signify low capture rates. High capture rates in the southwestern Pacific in 1952 **(a)** gave way to low ones in later years. Excellent fishing success in the tropical Atlantic and Indian Oceans in 1958 **(b)** had turned mediocre by 1964 **(c)** and poor by 1980 **(d)**. High capture rates in the north and south Atlantic in 1964 **(c)** gave way to low capture rates there in 1980 **(d)**. Adapted from Myers, R. A. and B. Worm. 2003. Rapid worldwide depletion of predatory fish communities. *Nature* 423:280–283.

Aquaculture can partly make up for the loss of capture fisheries

As you learned in Chapter 8, there has been a dramatic increase in farm fisheries, or **aquaculture**, over the past few decades. From a small percentage of the total fish harvest only 20 years ago, aquaculture now accounts for more than 30% of world production. Freshwater fish species—including rainbow trout, brook trout, arctic char, and tilapia, as well as Atlantic, coho, and chinook salmon—are currently farmed in Canada. Canada is the fourth-largest producer of farmed salmon in the world[3] and the economic value of the aquaculture industry is now well over a half-billion dollars annually. Shellfish—including clams, oysters, scallops, and mussels, as well as some types of marine plants, algae, and oysters for pearl production—are also raised in controlled environments.

Aquaculture has both benefits and drawbacks

Aquaculture can help improve a region's or nation's food security by increasing overall amounts of fish available. It also reduces pressure on overharvested and declining wild stocks, as well as reducing by-catch and providing employment for fishers who can no longer fish from depleted natural stocks. Aquaculture also relies far less on fossil fuels than do fishing vessels, and provides a safer work environment. Fish farming can also be remarkably energy-efficient, producing as much as 10 times as much fish per unit area as is harvested from oceanic waters on the continental shelf, and up to 1000 times as much as is harvested from the open ocean.

Along with its benefits, aquaculture has some disadvantages. Dense concentrations of farmed animals can increase the incidence of disease, which necessitates antibiotic treatment and results in additional expense.

If farmed aquatic organisms escape into ecosystems where they are not native, they may spread disease to native stocks or may outcompete native organisms for food or habitat. The opposite has also occurred—recent research suggests that wild Pacific salmon swimming near aquaculture pens may pass parasites, which then spread rapidly as a result of the high population densities in the pens.

The high-density fish populations involved in aquaculture also produce a significant amount of waste, both from the farmed organisms and from the feed that goes uneaten and decomposes in the water column. Farmed fish often are fed grain, and as we have discussed (Chapter 8), growing grain to feed animals that we then eat can reduce the energy efficiency of food production and consumption. In other cases, farmed fish are fed fishmeal made from wild ocean fish, such as herring and anchovies, whose harvest may place additional stress on wild fish populations.

Aquaculture has also led to damaging landscape changes in some coastal ecosystems, including the removal of protective mangrove forests. Coastlines evolve geologically as natural barriers to the battering of storm waves and winds. When these barriers are removed or disrupted, shorelines are left defenceless. The world saw the effects of shoreline modification graphically illustrated during the devastating Sumatra–Andaman tsunami of December 24, 2004, a seismic sea wave generated by a major submarine earthquake off the coast of Indonesia. Many of the coastlines in this part of South Asia and the Pacific had been significantly altered for aquaculture pen construction. The tsunami met with little natural resistance as it rushed onshore.

Consumer choice can influence fishing practices

To most of us, marine fishing practices may seem a distant phenomenon over which we have no control. Yet by exercising careful choice when we buy seafood, consumers can influence the ways in which fisheries function. Purchasing ecolabelled seafood, such as dolphin-safe tuna, is one way to exercise choice, but in most cases consumers have no readily available information about how their seafood was caught.

Several nonprofit organizations have recently devised concise guides to help consumers make informed choices. These guides differentiate fish and shellfish that are overfished or whose capture is ecologically damaging from those that are harvested more sustainably. Table 12.1 has some examples from such a consumer guide, prepared by the Monterey Bay Aquarium, which also provides a wealth of information about sustainable fisheries for the public on its website.

Table 12.1 Seafood Choices for Consumers

Best choices*	Seafood to avoid†
Catfish (farmed)	Atlantic cod
Caviar (farmed)	Caviar (wild-caught)
Sardines	Chilean seabass/toothfish
Snow crab (Canada)	King crab (imported)
Pacific halibut	Atlantic halibut
Striped bass (farmed)	Orange roughy
Oysters (farmed)	Sharks
Shrimp (trap-caught)	Shrimp (imported, farmed, or trawl-caught)
Sturgeon (farmed)	Sturgeon (imported wild-caught)
Pollock (wild-caught from Alaska)	Swordfish (imported)
Tuna: albacore (troll/pole-caught)	Tuna: bluefin

*Fish or shellfish that are abundant, well managed, and fished or farmed in environmentally friendly ways.
†Fish or shellfish from sources that are overfished and/or fished or farmed in ways that harm other marine life or the environment.
Based on Monterey Bay Aquarium. 2005. Seafood Watch national seafood guide.

weighing the issues 12-2
EATING SEAFOOD

After reading this chapter, do you plan to alter your decisions about eating seafood in any way? If so, how? If not, why not? Do you think consumer buying choices can exert an influence on fishing practices? On mercury contamination in seafood?

Marine biodiversity loss erodes ecosystem services

Overfishing, pollution, habitat change, and other factors that deplete biodiversity can threaten the ecosystem services we derive from the oceans. In the 2006 study that predicted global fisheries collapse by 2048, the study's 14 authors analyzed all existing scientific literature to summarize the effects of biodiversity loss on ecosystem function and ecosystem services. They found that across 32 different controlled experiments conducted by various researchers, systems with less species diversity or genetic diversity showed less primary and secondary production and were less able to withstand disturbance.

The team also found that when biodiversity was reduced, so were habitats that serve as nurseries for fish and shellfish. Moreover, biodiversity loss was correlated with reduced filtering and detoxification (as from wetland vegetation and oyster beds), which can lead to harmful algal blooms, dead zones, fish kills, and beach closures.

Marine Conservation

Because we bear responsibility and stand to lose a great deal if valuable ecological systems collapse, marine scientists have been working to develop solutions to the problems that threaten the oceans. Many have begun by taking a hard look at the strategies used traditionally in fisheries management.

Fisheries management has been based on maximum sustainable yield

Fisheries managers conduct surveys, study fish population biology, and monitor catches. They then use that knowledge to regulate the timing of harvests, the techniques used to catch fish, and the scale of harvests. The goal is to allow for maximal harvests of particular populations while keeping fish available for the future—the concept of **maximum sustainable yield**. If data indicate that current yields are unsustainable, managers might limit the number or total biomass of that species that can be harvested, or they might restrict the type of gear fishers can use.

Despite such efforts, several fish and shellfish stocks have plummeted, and many scientists and managers now feel it is time to rethink fisheries management. One key change these reformers suggest is to shift the focus away from individual species and toward viewing marine resources as elements of larger ecological systems. This means considering the impacts of fishing practices on habitat quality, on species interactions, and on other factors that may have indirect or long-term effects on populations. One key aspect of such an *ecosystem-based management* approach is to set aside areas of ocean where systems can function without human interference.

We can protect areas in the ocean

Hundreds of **marine protected areas (MPAs)** have been established, most of them along the coastlines of developed countries[4] (**FIGURE 12.25**). However, despite their name, marine protected areas do not necessarily protect all their natural resources, because nearly all MPAs allow fishing or other extractive activities. As a recent report from an environmental advocacy group put it, even national marine sanctuaries "are dredged, trawled, mowed for kelp, crisscrossed with oil pipelines and fibre-optic cables, and swept through with fishing nets."

Because of the lack of true refuges from fishing pressure, many scientists—and some fishers—want to establish areas where fishing is prohibited. Such "no-take" areas have come to be called **marine reserves**. Designed to preserve entire ecosystems intact without human interference, marine reserves are also intended to improve

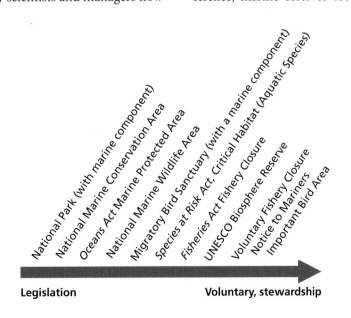

FIGURE 12.25
There are numerous tools that can be used to protect marine areas in Canada, which range from legislated to voluntary measures. Canada's first official marine protected area, designated in 2003, was the Endeavour Hydrothermal Vents area on the Juan da Fuca Ridge off the west coast of Vancouver Island.

THE SCIENCE BEHIND THE STORY

Do Marine Reserves Work?

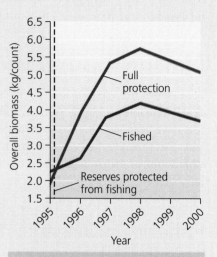

Dr. Callum M. Roberts, University of York, monitors reef fish in Belize.

In November 2001, a team of fisheries scientists published a paper in the journal *Science*, providing some of the first clear evidence that marine reserves can benefit nearby fisheries. The team, led by British researcher Callum Roberts, focused on reserves off the coasts of Florida and the Caribbean island of St. Lucia.

Following the establishment in 1995 of the Soufrière Marine Management Area (SMMA), a network of reserves intended to help restore St. Lucia's severely depleted coral reef fishery, Roberts and his colleague, Julie Hawkins, conducted annual visual surveys of fish abundance in the reserves and nearby areas. Within three years, they found that the biomass of five commercially important families of fish—surgeonfishes, parrot fishes, groupers, grunts, and snappers—had tripled inside the reserves and doubled outside them (see the first figure).

Roberts and Hawkins also interviewed local fishers and found that those with large traps were catching 46% more fish per trip in 2000–2001 than they had in 1995–1996 and that fishers with small traps were catching 90% more (see the second figure). Roberts and his colleagues concluded that *"in five years, reserves have led to improvement in the SMMA fishery, despite the 35% decrease in area of fishing grounds."*

Roberts and his co-workers also studied the oldest fully protected marine reserve in the United States, the Merritt Island National Wildlife Refuge (MINWR), established in 1962 as a buffer around what is today the Kennedy Space Center on Cape Canaveral, Florida. In a previous study, Darlene Johnson and James Bohnsack of the National Oceanic and Atmospheric Administration and Nicholas Funicelli of the United States Geological Service had found that the reserve

Established in 1995, the Soufrière Marine Management Area (SMMA) along the coast of St. Lucia had a rapid impact. By 1998, fish biomass within the five reserves tripled, and in adjacent fished areas it doubled. Data from Roberts, C., et al. 2001. Effects of marine reserves on adjacent fisheries. *Science* 294:1920–1923.

---

fisheries. Scientists argue that marine reserves can act as production factories for fish for surrounding areas, because fish larvae produced inside reserves will disperse outside and stock other parts of the ocean (see "The Science Behind the Story: Do Marine Reserves Work?"). Proponents maintain that by serving both purposes, marine reserves are a win–win proposition for environmentalists and fishers alike.

Many fishers dislike the idea of no-take reserves, however, just as most were opposed to the Canadian groundfish moratoria and the Georges Bank closures. Nearly every marine reserve that has been established or proposed has met with pockets of intense opposition from people and businesses that use the area for fishing or recreation. Opposition comes from commercial fishing fleets as well as from individuals who fish recreationally. Both types of fishers are concerned that marine reserves will simply put more areas off-limits to fishing. In some parts of the world, protests have become violent. For instance, to protest fishing restrictions, fishers in the Galápagos Islands destroyed offices at Galápagos National Park and threatened researchers and park managers with death.

Reserves can work for both fish and fishers

In the past decade, data synthesized from marine reserves around the world have been indicating that reserves *do* work as win–win solutions that benefit ecosystems, fish populations, and fishing economies. In 2001, 161 prominent marine scientists signed a "consensus statement" summarizing the effects of marine reserves. Besides boosting fish biomass, total catch, and record-sized fish, the report stated, marine reserves yield several benefits. Within reserve boundaries, they

- Produce rapid and long-term increases in abundance, diversity, and productivity of marine organisms
- Decrease mortality and habitat destruction
- Lessen the likelihood of extirpation of species

Outside the reserve boundaries, marine reserves

- Can create a "spillover effect" when individuals of protected species spread outside reserves
- Allow larvae of species protected within reserves to "seed the seas" outside reserves

contained more and larger fish than did nearby unprotected areas. This team also found that some of the reserve's fish appeared to be migrating to nearby fishing areas.

Bohnsack, Roberts, and their colleagues corroborated the evidence for migration by analyzing trophy records from the International Game Fish Association. They found that the proportion of Florida's record-sized fish caught near Merritt Island increased significantly after 1962. Nine years after the refuge was established, for instance, the number of spotted sea trout records from the Merritt Island area jumped dramatically. The researchers hypothesized that the reserve was providing a protected zone in which fish could grow to trophy size before migrating to nearby areas, where they were caught by recreational fishers.

Not everyone saw the St. Lucia and Merritt Island cases as proof that marine reserves could rescue depleted fisheries. In February 2002, several alternative interpretations of the evidence were published as letters in *Science*. Mark Tupper, a fisheries scientist at the University of Guam, suggested that the St. Lucia results were relevant only to coral reef fisheries in developing nations, whereas Florida's boost in fish populations was due primarily to limits on recreational fishing. Karl Wickstrom, editor-in-chief of *Florida Sportsman* magazine, suggested that the increase in trophy fish near MINWR was caused by commercial fishing regulations and changes in how trophies were recorded and promoted. And Ray Hilborn, a fisheries scientist at the University of Washington, challenged the study's scientific methods. In the St. Lucia case, he pointed out, there had been no control condition.

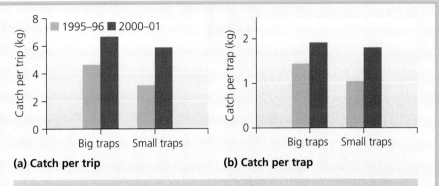

(a) Catch per trip   (b) Catch per trap

Roberts and his colleagues studied biomass of fish caught at the SMMA over two five-month periods in 1995–1996 and 2000–2001. Per fishing trip (a), the catch for fishers with big traps increased by 46%, and the catch for fishers with small traps increased by 90%. Per trap (b), fishers with big traps caught 36% more fish, and fishers with small traps caught 80% more fish. Data from Roberts, C., et al. 2001. Effects of marine reserves on adjacent fisheries. *Science* 294:1920–1923.

In response, Roberts and his colleagues reaffirmed the validity of their results while acknowledging some limitations. They agreed with Tupper that marine reserves are not always effective and often need to be complemented by other management tools, such as size limits. "We agree that inadequately protected reserves are useless," they wrote, "but our study shows that well-enforced reserves can be extremely effective and can play a critical role in achieving sustainable fisheries."

The consensus statement was backed up by research into reserves worldwide. At Apo Island in the Philippines, biomass of large predators increased eightfold inside a marine reserve, and fishing improved outside the reserve. At two coral reef sites in Kenya, commercially fished and keystone species were up to 10 times as abundant in the protected area as in the fished area. At Leigh Marine Reserve in New Zealand, snapper increased fortyfold, and spiny lobsters were increasing by 5%–11% yearly. Spillover from this reserve improved fishing and ecotourism, and local residents who once opposed the reserve now support it.

The review of data from existing marine reserves as of 2001 revealed that just one to two years after their establishment, marine reserves

- Increased densities of organisms on average by 91%
- Increased biomass of organisms on average by 192%
- Increased average size of organisms by 31%
- Increased species diversity by 23%

Since that time, further research has shown that reserves create a fourfold increase in catch per unit effort in fished areas surrounding reserves and that they can greatly increase ecotourism by divers and snorkelers.

On Georges Bank, once commercial trawling was halted in 1994, populations of many organisms began to recover. As benthic invertebrates began to come back, numbers of groundfish, such as haddock and yellowtail flounder, rose inside the closed areas, and scallops increased by 14 times. Moreover, fish from the closure areas appear to be spilling over into adjacent waters, because fishers have been catching more and more groundfish from Georges Bank as a whole since the late 1990s. From these and other data sets, increasing numbers of scientists, fishers, and policy makers are advocating the establishment of fully protected marine reserves as a central management tool.

How should reserves be designed?

If marine reserves work in principle, the question becomes how best to design reserves and arrange them into networks. Scientists today are asking how large reserves

CANADIAN ENVIRONMENTAL PERSPECTIVES

Farley Mowat

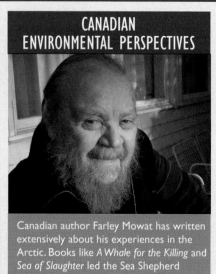

Canadian author Farley Mowat has written extensively about his experiences in the Arctic. Books like *A Whale for the Killing* and *Sea of Slaughter* led the Sea Shepherd Conservation Society to name one of its fleet of "eco-piracy" boats after him.

- **Author**
- **Conservationist** and **naturalist**
- **Animal welfare activist**

He is a "natural storyteller," a gifted writer, and a Canadian icon. His books—more than 18 million of them sold, in many languages—have helped familiarize readers with wildlife and life in wild places. They have raised awareness and changed attitudes, and have even influenced Canadian government policies. Books like *Never Cry Wolf* (1963) are fixtures in Canadian classrooms. Some of his books have brought controversy and criticism, and one, *Sea of Slaughter* (1984), apparently contributed to his being barred from entering the United States.[5]

Farley Mowat was born in Ontario but grew up in Saskatoon during the Great Depression. As a child he was fascinated with nature and kept numerous wild animals and even insects as household pets. Mowat's first formal writing as a young teenager was an article on birds. Several of his later books, such as *Owls in the Family* (1961), were based on reminiscences of animals and nature experiences from his childhood. Mowat served in active combat duty during the Second World War. His intense battlefield experiences eventually provided material and inspiration for some of his later writings, such as *My Father's Son: Memories of War and Peace* (1992).

After returning from the war he studied biology at the University of Toronto, which eventually led him to a two-year sojourn in the Arctic. His first book, *People of the Deer* (1952), was based on his frustration upon becoming aware of the destitute situation of the Ihalmiut, an Inuit band who, he felt, had been misunderstood and exploited by white people.

A number of Mowat's subsequent books also were based on his experiences in the Arctic. He won the Governor General's Award for *Lost in the Barrens* (1956), which tells the story of two young people—one white, the other Cree—who are lost in the Arctic wilderness. They manage to survive for part of the winter, but ultimately are rescued by an Inuit boy whose extensive knowledge of the Arctic environment saves them.

Mowat's relationship with the people of Newfoundland and Labrador, where

---

need to be, how many there need to be, and where they need to be placed. Involving fishers directly in the planning process is crucial for coming up with answers to such questions. In Canada, marine reserves are managed by Parks Canada through the National Marine Conservation Areas Program, and their management plans are designed as partnerships among coastal communities, Aboriginal people, provincial, territorial, and federal government agencies, and other stakeholders.

Of several dozen studies that have estimated how much area of the ocean should be protected in no-take reserves, estimates range from 10% to 65%, with most falling between 20% and 50%. Other studies are modelling how to optimize the size and spacing of individual reserves so that ecosystems are protected, fisheries are sustained, and people are not overly excluded from marine areas (**FIGURE 12.26**). If marine reserves are designed strategically to take advantage of ocean currents,

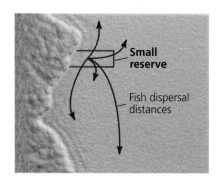 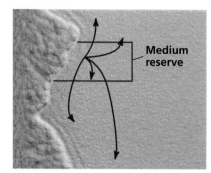 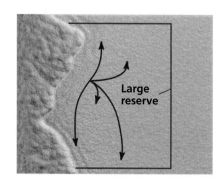

FIGURE 12.26
Marine reserves of different sizes may have varying effects on ecological communities and fisheries. Young and adult fish and shellfish of different species can disperse different distances, as indicated by the red arrows in the figure. A small reserve (left panel) may fail to protect animals because too many disperse out of the reserve. A large reserve (right panel) may protect fish and shellfish very well but will provide relatively less "spillover" into areas where people can legally fish. Thus medium-sized reserves (middle panel) may offer the best hope of preserving species and ecological communities while also providing adequate fish to fishers and human communities.
Source: Halpern, B. S., and R. R. Warner. 2003. Matching marine reserve design to reserve objectives. Proceedings of the Royal Society of London B 270:1871–1878.

he lived for eight years, is complex. His 1968 book *This Rock Within the Sea: A Heritage Lost* tells of a noble people with a calling to the sea. By 1972, however, *A Whale for the Killing* dramatically illustrated his disillusionment after an unfortunate beached whale was inhumanely shot to death. The book is also a plea for action to save whales from extinction.

In 1984, *Sea of Slaughter* was published, providing an overview of animal life in the North Atlantic since the early days of European fishing and whaling. Roger Tory Peterson compared the environmental significance of *Sea of Slaughter* to that of Rachel Carson's *Silent Spring*, saying, "In this masterpiece, Canada's most beloved naturalist-author is as angry about the assault on the living sea as Rachel Carson was about the land in *Silent Spring*."[6] On a promotional tour for *Sea of Slaughter*, Mowat attempted to enter the United States but was barred. He explored the possible reasons for this in his next book, *My Discovery of America* (1985).

Mowat has the distinction—perhaps dubious—of having a ship named after him. The *R/V Farley Mowat* is one of a small fleet of vessels that belong to the Sea Shepherd Conservation Society, of which Mowat is the international chair. Sea Shepherd, founded by Paul Watson (a co-founder of Greenpeace), is an eco-activist organization that has pledged to sink or sabotage any vessels it believes to have violated international whaling laws. Greenpeace, also an activist group, has repeatedly distanced itself from Sea Shepherd because of the extreme nature of its approach.

On April 12, 2008, the Department of Fisheries and Oceans, Transport Canada, Canadian Coast Guard, and RCMP took possession of the *Farley Mowat* for alleged violations of the *Marine Mammal Regulations*.[7] The Sea Shepherd Conservation Society, which claims the charges to be false, is invoicing the Canadian Government $1000 per day for holding the vessel. Sea Shepherd says of the *Farley Mowat*, "She is a protector, and a symbol of hope for a better, more humane, and more ecologically conscious future."[8]

"So ends the story of how the Sea of Whales became a Sea of Slaughter as, one by one, from the greatest to the least, each in turn according to its monetary worth, the several cetacean nations perished in a roaring holocaust fuelled by human avarice." —Farley Mowat, Sea of Slaughter (1984)

Thinking About Environmental Perspectives

The Sea Shepherd Conservation Society, of which Farley Mowat is the international chair, has been criticized for extremism. Its anti-whaling actions have included ramming, disabling, and otherwise confronting whaling vessels from the Faeroe Islands (in the North Atlantic) to Antarctica. Sea Shepherd maintains the position that standing by and observing illegal activities that endanger ocean life is not an acceptable pathway, and it pledges to take whatever actions are necessary to protect sea life and habitats. What do you think? Are extreme and potentially dangerous actions, like ramming a vessel, acceptable, or perhaps even necessary, in some circumstances?

many scientists say, then they may well seed the seas and help lead us toward solutions to one of our most pressing environmental problems.

Conclusion

Oceans cover most of our planet and contain diverse topography and ecosystems, some of which we are only now beginning to explore and understand. We are learning more about the oceans and coastal environments while we are intensifying our use of their resources and causing these areas more severe impacts. In so doing, we are coming to understand better how to use these resources without depleting them or causing undue ecological harm to the marine and coastal systems on which we depend.

Today, scientists are demonstrating that setting aside protected areas of the ocean can serve to maintain natural systems and also to enhance fisheries. This is vital at a time when we are depleting many of the world's marine fish stocks. As historical studies reveal more information on how much biodiversity our oceans formerly contained and have now lost, we may increasingly look beyond simply making fisheries stable and instead consider restoring the ecological systems that once flourished in our waters.

REVIEWING OBJECTIVES

You should now be able to:

Identify physical, geographic, chemical, and biological aspects of the marine environment

- Oceans cover 71% of Earth's surface and contain more than 97% of its surface water.
- Ocean water contains 96.5% H_2O by mass and various dissolved salts.
- Colder, saltier water is denser and sinks. Water temperatures vary with latitude, and temperature variation is greater in surface layers.

- Persistent currents move horizontally through the oceans, driven by density differences, sunlight, and wind.
- Vertical water movement includes upwelling and downwelling, which affect the distribution of nutrients and life.
- Seafloor topography can be complex.

Describe major types of marine ecosystems

- Major types of marine and coastal ecosystems include pelagic and deep-water open ocean systems, kelp forests, coral reefs, intertidal zones, salt marshes, mangrove forests, and estuaries.
- Many of these systems are highly productive and rich in biodiversity. Many also suffer heavy impacts from human influence.

Outline historic and current human uses of marine resources

- For millennia, people have fished the oceans and used ocean waters for transportation.
- Today we extract energy and minerals from the oceans.

Assess human impacts on marine environments

- People pollute ocean waters with trash, including plastic and nets that harm marine life.
- Marine oil pollution results from non-point sources on land, as well as from tanker spills at sea.
- Heavy metal contaminants in seafood affect human health, and nutrient pollution can lead to harmful algal blooms.
- Overharvesting is perhaps the major human impact on marine systems.

Review the current state of ocean fisheries and reasons for their decline

- Half the world's marine fish populations are fully exploited, 25% are already overexploited, and only 25% can yield more without declining.
- Global fish catches have stopped growing since the late 1980s, despite increased fishing effort and improved technologies.
- People began depleting marine resources long ago, but impacts have intensified in recent decades.
- Commercial fishing practices include driftnetting, longline fishing, and trawling, all of which capture nontarget organisms, called by-catch.
- Today's oceans hold only one-tenth the number of large animals that they did before the advent of industrialized commercial fishing.
- As fishing intensity increases, the fish available become smaller.
- Marine biodiversity loss affects ecosystem services.
- Consumers can encourage good fishery practices by shopping for sustainable seafood.
- Traditional fisheries management has not stopped declines, so many scientists feel that ecosystem-based management is needed.

Evaluate marine protected areas and reserves as innovative solutions

- We have established far fewer protected areas in the oceans than we have on land, and most marine protected areas allow many extractive activities.
- Marine reserves can protect ecosystems while also boosting fish populations and making fisheries sustainable.

TESTING YOUR COMPREHENSION

1. What proportion of Earth's surface do oceans cover? What is the average salinity of ocean water? How are density, salinity, and temperature related in each layer of ocean water?
2. What factors drive the system of ocean currents? In what ways do these movements affect conditions for life in the oceans?
3. Where in the oceans are productive areas of biological activity likely to be found?
4. Describe three kinds of ecosystems found near coastal areas and the kinds of life they support.
5. Why are coral reefs biologically valuable? How are they being degraded by human impact?
6. What is causing the disappearance of mangrove forests and salt marshes?
7. Discuss three ways in which people are combating pollution in the oceans and on our coasts.
8. Describe an example of how overfishing can lead to ecological damage and fishery collapse.
9. Name three industrial fishing practices, and explain how they create by-catch and harm marine life.
10. How does a marine reserve differ from a marine protected area? Why do many fishers oppose marine reserves? Explain why many scientists say no-take reserves will be good for fishers.

SEEKING SOLUTIONS

1. What benefits do you derive from the oceans? How does your behaviour affect the oceans? Give specific examples.
2. We have been able to reduce the amount of oil we spill into the oceans, but petroleum-based products, such as plastic, continue to litter our oceans and shorelines. Discuss some ways that we can reduce this threat to the marine environment.
3. Describe the trends in global fish capture over the past 50 years, and explain several factors that account for these trends.
4. Consider what you know about biological productivity in the oceans, about the scientific data on marine reserves, and about the social and political issues surrounding the establishment of marine reserves. What ocean regions do you think it would be particularly appropriate to establish as marine reserves? Why?
5. **THINK IT THROUGH** You make your living fishing on the ocean, just as your father and grandfather did, and as most of your neighbours do in your small coastal village. Your region's fishery has just collapsed, however, and everyone is blaming it on overfishing. The government has closed the fishery for three years, and scientists are pushing for a permanent marine reserve to be established on your former fishing grounds. You have no desire to move away from your village, so what steps will you take now? Will you protest the closure? What compensation will you ask of the government if it prevents you from fishing? Will you work with scientists to establish a reserve that improves fishing in the future, or will you oppose their attempts to create a reserve? What data and what assurances will you ask of them?
6. **THINK IT THROUGH** You are mayor of a coastal town where some residents are employed as commercial fishers and others make a living serving ecotourists who come to snorkel and scuba dive at the nearby coral reef. In recent years, several fish stocks have crashed, and ecotourism is dropping off as fish disappear from the increasingly degraded reef. Scientists are urging you to help establish a marine reserve around portions of the reef, but most commercial and recreational fishers are opposed to this idea. What steps would you take to restore your community's economy and environment?

INTERPRETING GRAPHS AND DATA

The accompanying graph presents trends in the status of North Atlantic swordfish, a highly migratory species managed directly by the National Marine Fishery Service. The solid red line shows the mortality rate from fishing. The solid blue line indicates the biomass of the stock. The dotted lines of corresponding colours indicate the reference levels used to determine whether the stock is overfished or recovered. The graph also indicates the date when an international recovery plan was implemented.

1. Describe the trends in swordfish stocks (1) before the adoption of an international management plan and (2) since the plan was adopted. Describe the interactions between fishing mortality and biomass as illustrated by the graph.
2. Based on the data in the graph, predict the likely trend in swordfish production over the next 10 years, assuming no change to the status quo.
3. This graph illustrates an effort that is succeeding, but not all rebuilding plans lead to stock recovery. Beyond the existence of a plan, what actions might play a role in supporting stock recovery efforts?

Trends in fishing mortality and stock biomass for North Atlantic swordfish, 1978–2000. Adapted from Rosenberg, A., et al. 2006. Rebuilding U.S. fisheries: Progress and problems. *Frontiers in Ecology and the Environment* 4(6).

CALCULATING FOOTPRINTS

The relationship between the ecological goods and services used by individuals and the amount of *land* area needed to provide those goods and services is relatively well developed. People also use goods and services from Earth's oceans, where the concept of *area* is less useful. It is clear, however, that our removal of fish from the oceans has an impact, or an ecological footprint.

The table shows data on the mean annual per capita consumption from ocean fisheries for North America, China, and the world as a whole. By using the data provided, calculate the amount of fish each consumer group would consume per year, given the annual per capita consumption rates for each of these three regions. Record your results in the table.

1. Calculate the ratio of North America's per capita fish consumption rate to that of the world. Compare this ratio to the ratio of the per capita ecological footprints for the United States, Canada, and Mexico (see Chapter 1, Calculating Footprints exercise) versus the world average footprint of 2.23 ha/person/year. Can you account for similarities and differences between these ratios?
2. The population of China has grown at an annual rate of 1.1% since 1987, while over the same period fish consumption in China has grown at an annual rate of 8.9%. Speculate on the reasons behind China's rapidly increasing consumption of fish.
3. What ecological concerns do the combined trends of human population growth and increasing per capita fish consumption raise for you? What role might you play in contributing to these concerns or to their solutions?

Consumer group	Annual Consumption		
	North America (21.6 kg per capita)	China (27.7 kg per capita)	World (16.2 kg per capita)
You			
Your class			
Your province			
Canada	6.91×10^8 kg	8.31×10^9 kg	4.86×10^9 kg
World			

Data from U.N. Food and Agriculture Organization (FAO), Fisheries Department. 2004. The state of world fisheries and aquaculture: 2004. Data are for 2002, the most recent year for which comparative data are available.

TAKE IT FURTHER

Go to www.myenvironmentplace.ca where you will find

- Suggested answers to end-of-chapter questions
- Quizzes, animations, and flashcards to help you study
- *Research Navigator*™ database of credible and reliable sources to assist you with your research projects
- Tutorials to help you master how to interpret graphs
- Current news articles that link the topics that you study to case studies from your region and around the world

- **ECO Occupational Profiles:** If you found this chapter especially interesting, you might want to learn more about the following jobs by visiting the Occupational Profiles website of the Environmental Careers Organization. Go to www.eco.ca and check out the following careers:
  - Aquaculture support worker
  - Aquaculturist
  - Fisheries technician
  - Marine biologist
  - Oceanographer

CHAPTER ENDNOTES

1. NOAA Coral Health and Monitoring Program, *Coral Bleaching*, www.coral.noaa.gov/cleo/coral_bleaching.shtml.
2. Halifax Regional Municipality, Halifax Harbour Project, www.halifax.ca/harboursol/WhatistheHarbourSolutionsProject.html.
3. Fisheries and Oceans Canada, Aquaculture Fact Sheets, *Fin Fish*, www.dfo-mpo.gc.ca/aquaculture/finfish_e.htm.
4. Parks Canada, Fisheries and Oceans Canada, and Environment Canada (2005) *Canada's Marine Protected Areas Strategy*, www.dfo-mpo.gc.ca/oceans-habitat/oceans/mpa-zpm/fedmpa-zpmfed/pdf/mpa_e.pdf.
5. The Canadian Encyclopedia *Historica*, Farley Mowat, www.thecanadianencyclopedia.com/index.cfm?PgNm=TCE&ArticleId=A0005502.
6. Amazon.com, *Sea of Slaughter* by Farley Mowat (Reviews), www.amazon.com/Sea-Slaughter-Farley-Mowat/dp/1576300196.
7. Department of Fisheries and Oceans, News Releases 2008, *Enforcement Actions Taken Against Farley Mowat—Update*, www.dfo-mpo.gc.ca/media/npress-communique/2008/20080414-eng.htm; the vessel remains in custody as of late 2008.
8. Sea Shepherd Conservation Society, *Neptune's Navy*, www.seashepherd.org/fleet/fleet.html.

Atmospheric Science and Air Pollution

13

This is Earth's atmosphere, from space.

Upon successfully completing this chapter, you will be able to

- Describe the composition, structure, and function of Earth's atmosphere
- Outline the scope of outdoor air pollution and assess potential solutions
- Explain stratospheric ozone depletion and identify steps taken to address it
- Define acidic deposition and illustrate its consequences
- Characterize the scope of indoor air pollution and assess potential solutions

The Inco Superstack, at 380 m tall, is the tallest smokestack in the Western Hemisphere.

CENTRAL CASE:
THE RAIN AND THE BIG NICKEL

"Despite Canada's success at reducing acid-causing emissions, acid deposition is still affecting our environment."
—ENVIRONMENT CANADA, *2004 CANADIAN ACID DEPOSITION SCIENCE ASSESSMENT*

"The problem [of acid rain] has not gone away."
—GENE LIKENS, DIRECTOR OF THE INSTITUTE OF ECOSYSTEM STUDIES IN MILLBROOK, NEW YORK

Sudbury, Ontario, is home to some of the world's largest nickel and copper deposits. These are not ordinary mineral deposits—they were formed by rock melting resulting from the impact of a meteorite 15 km in diameter more than 2 billion years ago. The impact left a crater almost 600 km in diameter and 40 km deep, in which the massive ore deposits were localized. A "sister" meteorite impact left a smaller crater, now filled by the deep, clear waters of Lake Wanapitei.

Sudbury also claims to be the "blueberry capital of the world," but here a downside becomes apparent. Blueberries thrive in highly acidic soils, but the soils of the Sudbury area are not naturally so acidic. What happened to create this blueberry-friendly environment?

Nickel mining and refining happened. Mining started in the Sudbury area in the late 1800s and continues today. The mining and, particularly, refining of ores (through *smelting*, in which the ore is crushed and heated to a very high temperature to segregate the metals of interest) generates emissions high in sulphur dioxide (SO_2). These sulphur-rich emissions, if released to the atmosphere, combine with water vapour to form sulphuric acid, which ultimately falls as *acidic deposition*. The rocks and soils of the Sudbury area are naturally susceptible to acidification because of their chemical composition.

As early as the 1920s, the degradation of the natural environment around Sudbury was recognized as a problem. Acid rain devastated the area's forests and water bodies. Pollution stained the treeless soil black. In 1969—quite early in the modern environmental movement (recall that the first Earth Day was in 1970)—the government of Ontario informed Inco, the principal mining company and main polluter in the area at the time, that it would be required to substantially decrease emissions from its facilities. Inco's response, cutting-edge in its day, was to build a "Superstack," 380 m tall, to carry emissions from the smelter far away from the immediate Sudbury area (see photo). The Superstack was completed in 1972, and to this day it remains the tallest smokestack in the Western Hemisphere.

Today we know that the Superstack was only a partial solution. It did disperse the sulphurous emissions, but rather than ending the acidic deposition it just spread the problem farther afield. Environment Canada estimates that 7000 lakes in northern Ontario and Quebec were damaged by acid-causing emissions from smelters in Sudbury.[1] The sensitivity of the underlying granitic rocks contributed to the acidification problem, which affected not only forests and soils, but also fish and the sport fishing industry in an area of approximately 17 000 km^2.[2] Sulphur emissions from the Superstack were dispersed over a much broader area than they would have been, had the stack height been less.

Beginning in the early 1980s, Inco and Falconbridge (the other major producer of smelter emissions in the area) undertook vigorous efforts to clean their emissions prior to releasing them to the atmosphere. The result is that SO_2 emissions today have been reduced by as much as 90%. Lakes, forests, and soils in the region have shown significant biological and chemical improvements in the more than 30 years since the Superstack was constructed. However, many of the damaged lakes are still acidic and contaminated with metals.[3]

The Experimental Lakes Area in northwestern Ontario, set up in 1968 and maintained by Fisheries and Oceans Canada, is one facility where scientific research on lake acidification and recovery takes place. The area consists of 58 small lakes and their watersheds, with a permanent field research station. Here scientists carry out research on the impacts of acidification, investigate the process of recovery, and test a variety of remediation approaches. For example, many of the fish in a sulphur-acidified lake died of starvation as a result of

In the Experimental Lakes Area of northwestern Ontario, many of the trout in an experimentally acidified lake died of starvation as a result of the loss of their food sources (top). When the lake was remediated, the fish population was able to recover (bottom).

the impacts of acidification on their food sources (see photo). When the lake was remediated, the fish population was able to recover. This kind of research gives scientists a better idea of the recovery process and the robustness of aquatic systems undergoing chemical and biological changes.[4]

The most important legacy of the Superstack is that it ushered in an era of ecological awareness, recovery, and restoration in Ontario, and of pride in the natural environment in the Sudbury area. The Regional Land Reclamation Program, launched in 1978, celebrated its 25th anniversary in 2003. The partners continue to carry out environmental research on acidification and the recovery process, which may prove useful in other parts of the world where acidification has caused ecological damage.

The Atmosphere and Weather

Every breath we take reaffirms our connection to the **atmosphere**, the thin layer of gases that surrounds Earth. We live at the bottom of this layer, which provides us with oxygen, absorbs hazardous solar radiation, burns up incoming meteors, transports and recycles water and nutrients, and moderates climate.

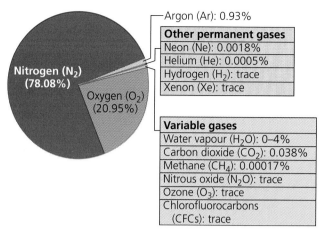

FIGURE 13.1
Earth's atmosphere consists mostly of nitrogen, secondarily of oxygen, and lastly of a mix of gases at dilute concentrations. Permanent gases are fixed in concentration. Variable gases vary in concentration as a result of either natural processes or human activities. Data from Ahrens, C. D. 2007. *Meteorology Today*, 8th ed. Belmont, CA: Brooks/Cole.

The atmosphere consists of roughly 78% nitrogen gas (N_2) and 21% oxygen gas (O_2). The remaining 1% is composed of argon gas (Ar) and minute concentrations of several other gases (**FIGURE 13.1**). These include *permanent gases* that remain at stable concentrations and *variable gases* that vary in concentration from time to time or place to place as a result of natural processes or human activities.

Over Earth's long history, the atmosphere's chemical composition has changed. Oxygen gas began to build up in an atmosphere dominated by carbon dioxide (CO_2), nitrogen, carbon monoxide (CO), and hydrogen (H_2) about 2.7 billion years ago, with the emergence of autotrophic microbes that emitted oxygen as a by-product of photosynthesis. Today, human activity is altering the quantities of some atmospheric gases, such as carbon dioxide, methane (CH_4), and ozone (O_3). In this chapter and in Chapter 14, we will explore the atmospheric changes brought about by artificial pollutants, but we must first begin with an overview of Earth's atmosphere.

The atmosphere is layered

The atmosphere that stretches so high above us and seems so vast is actually just a thin coating about 1/100 of Earth's diameter, like the fuzzy skin of a peach. This coating consists of four layers that atmospheric scientists recognize by measuring differences in temperature, density, and composition (**FIGURE 13.2**).

The bottommost layer, the **troposphere**, blankets Earth's surface and provides us with the air we need to live. The movement of air within the troposphere is also largely responsible for the planet's weather. Although it is

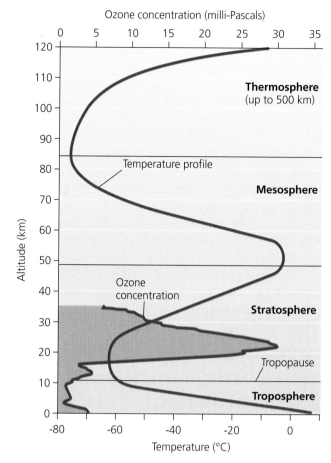

FIGURE 13.2
Temperature drops with altitude in the troposphere, rises with altitude in the stratosphere, drops in the mesosphere, then rises again in the thermosphere. The tropopause separates the troposphere from the stratosphere. Ozone reaches a peak in a portion of the stratosphere, giving rise to the term *ozone layer*. Adapted from Jacobson, M. Z. 2002. *Atmospheric Pollution: History, Science, and Regulation*. Cambridge: Cambridge University Press; and Parson, E. A. 2003. *Protecting the Ozone Layer: Science and Strategy*. Oxford: Oxford University Press.

thin (averaging 11 km high) relative to the atmosphere's other layers, the troposphere contains three-quarters of the atmosphere's mass, because air is denser near Earth's surface. On average, tropospheric air temperature declines by about 6°C for each kilometre in altitude, dropping to roughly –52°C at its highest point. At the top of the troposphere, however, temperatures cease to decline with altitude, marking a boundary called the *tropopause*. The tropopause acts like a cap, limiting mixing between the troposphere and the atmospheric layer above it, the stratosphere.

The **stratosphere** extends from 11 km to 50 km above sea level. Although similar in composition to the troposphere, the stratosphere is 1000 times as dry and less dense. Its gases experience little vertical mixing, so once substances (including pollutants) enter it, they tend to remain for a long time. The stratosphere attains a maximum temperature of –3°C at its highest altitude but is colder in its lower reaches. The reason is that ozone and oxygen absorb and scatter the Sun's ultraviolet (UV) radiation, so that much of the UV radiation penetrating the upper stratosphere fails to reach the lower stratosphere. Most of the atmosphere's minute amount of ozone concentrates in a portion of the stratosphere roughly from 17 km to 30 km above sea level, a region that has come to be called Earth's **ozone layer**. The ozone layer greatly reduces the amount of UV radiation that reaches Earth's surface. Because UV light can damage living tissue and induce mutations in DNA, the ozone layer's protective effects are vital for life on Earth.

Above the stratosphere lies the **mesosphere**, which extends from 50 km to 80 km above sea level. Air pressure is extremely low here, and temperatures decrease with altitude, reaching their lowest point at the top of the mesosphere. From here, the **thermosphere**, our atmosphere's top layer, extends upward to an altitude of 500 km.

Atmospheric properties include temperature, pressure, and humidity

Although the lower atmosphere is stable in its chemical composition, it is dynamic in its movement; air movement within it is due to differences in the physical properties of air masses. Among these properties are pressure and density, relative humidity, and temperature.

Gravity pulls gas molecules toward Earth's surface, causing air to be most dense near the surface and less so as altitude increases. **Atmospheric pressure**, the force per unit area produced by a column of air, also decreases with altitude, because at higher altitudes fewer molecules are pulled down by gravity (**FIGURE 13.3**). At sea level, atmospheric pressure is 1013 millibars (mb). Mountain climbers trekking to Mount Everest, the world's highest mountain, can look up and view their destination from Kala Patthar, a nearby peak, at roughly 5.5 km in altitude. At this altitude, pressure is 500 mb—half the atmosphere's air molecules are above the climber, and half are below. A climber who reaches Everest's peak (8.85 km), where the "thin air" is just more than 300 mb, stands above two-thirds of the molecules in the atmosphere. When we fly on a commercial jet airliner at a cruising altitude of 11 km, we are above roughly 80% of the atmosphere's molecules.

Another property of air is **relative humidity**, the ratio of water vapour a given volume of air contains to the maximum amount it *could* contain at a given temperature. Average daytime relative humidity in June in the desert at Phoenix, Arizona, is only 31% (meaning that the air contains less than a third of the water vapour it possibly can at its temperature), whereas on the tropical island

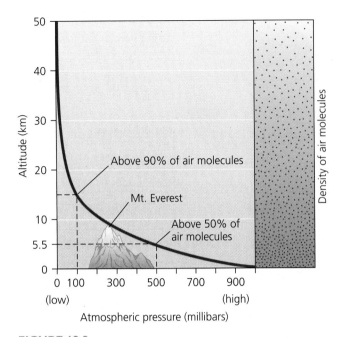

FIGURE 13.3
As one climbs higher through the atmosphere, gas molecules become less densely packed. As density decreases, so does atmospheric pressure. Because most air molecules lie low in the atmosphere, one needs to be only 5.5 km high to be above half the planet's air molecules. Adapted from Ahrens, C. D. 2007. *Meteorology Today*, 8th ed. Belmont, CA: Brooks/Cole.

of Guam, relative humidity rarely drops below 88%. People are sensitive to changes in relative humidity because we perspire to cool our bodies. When humidity is high, the air is already holding nearly as much water vapour as it can, so sweat evaporates slowly and the body cannot cool itself efficiently. This is why high humidity makes it feel hotter than it really is. Low humidity speeds evaporation and makes it feel cooler.

The temperature of air also varies with location and time. At the global scale, temperature varies over Earth's surface because the Sun's rays strike some areas more directly than others. At more local scales, temperature varies because of topography, plant cover, proximity of land to water, and many other factors. Sometimes these local variations are striking—the side of a hill that is sheltered from wind or direct sunlight can have a totally different weather pattern, or **microclimate**, from the side facing into the wind or sunlight.

Solar energy heats the atmosphere, helps create seasons, and causes air to circulate

Energy from the Sun heats air in the atmosphere, drives air movement, helps create seasons, and influences weather and climate. An enormous amount of solar energy continuously bombards the upper atmosphere—more than

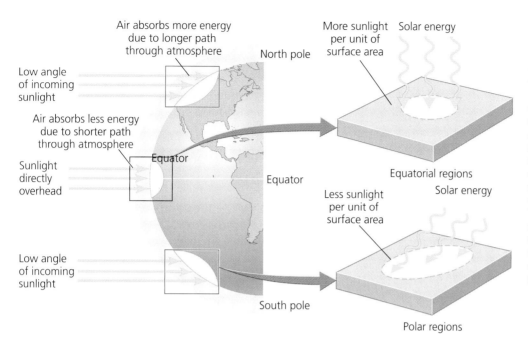

FIGURE 13.4
Because of Earth's curvature, polar regions receive, on average, less solar energy than equatorial regions. One reason is that sunlight gets spread over a larger area when striking the surface at an angle. Another reason is that sunlight approaching at a lower angle near the poles must traverse a longer distance through the atmosphere, during which more energy is absorbed or reflected. These patterns represent year-round averages; the latitude at which radiation approaches the surface perpendicularly varies with the seasons (see **FIGURE 13.5**).

1000 watts/m², many thousands of times greater than the total output of electricity generated by human society. Of that solar energy, about 70% is absorbed by the atmosphere and planetary surface, while the rest is reflected back into space (see **FIGURE 14.1**).

The spatial relationship between Earth and the Sun determines how much solar radiation strikes each point on Earth's surface. Sunlight is most intense when it shines directly overhead and meets the planet's surface at a perpendicular angle. At this angle, sunlight passes through a minimum of energy-absorbing atmosphere, and Earth's surface receives a maximum of solar energy per unit surface area. Conversely, solar energy that approaches Earth's surface at an oblique angle loses intensity as it traverses a longer distance through the atmosphere, and it is less intense when it reaches the surface. This is why, on average, solar radiation intensity is highest near the equator and weakest near the poles (**FIGURE 13.4**).

Because Earth is tilted on its *axis* (an imaginary line connecting the poles) by about 23.5°, the Northern and Southern Hemispheres each tilt toward the Sun for half the year, resulting in the change in seasons (**FIGURE 13.5**).

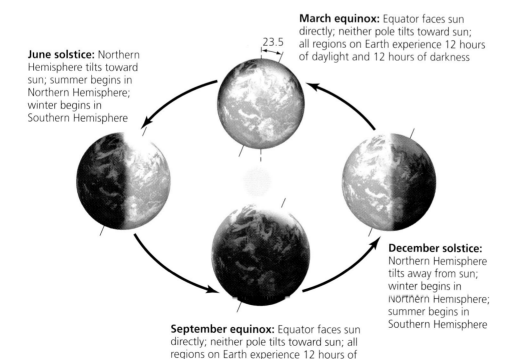

FIGURE 13.5
The seasons occur because Earth is tilted on its axis by 23.5 degrees. As Earth revolves around the Sun, the Northern Hemisphere tilts toward the Sun for one half of the year, and the Southern Hemisphere tilts toward the Sun for the other half of the year. In each hemisphere, summer occurs during the period in which the hemisphere receives the most solar energy because of its tilt toward the Sun.

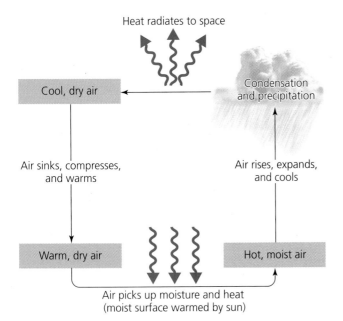

FIGURE 13.6
Weather is driven in part by the convective circulation of air in the atmosphere. Air being heated near Earth's surface picks up moisture and rises. Once aloft, this air cools, and moisture condenses, forming clouds and precipitation. Cool, drying air begins to descend, compressing and warming in the process. Warm, dry air near the surface begins the cycle anew.

Regions near the equator are largely unaffected by this tilt; they experience about 12 hours each of sunlight and darkness every day throughout the year. Near the poles, however, the effect is strong, and seasonality is pronounced.

Land and surface water absorb solar energy, re-radiating heat and causing water to evaporate. Air near Earth's surface therefore tends to be warmer and moister than air at higher altitudes. These differences set into motion a process of **convective circulation** (**FIGURE 13.6**). Warm air, being less dense, rises and creates vertical currents. As air rises into regions of lower atmospheric pressure, it expands and cools. Once the air cools, it descends and becomes denser, replacing warm air that is rising. The air picks up heat and moisture near ground level and prepares to rise again, continuing the process. Similar convective circulation patterns occur in ocean waters, in magma beneath Earth's surface, and even in a simmering pot of soup. Convective circulation influences both weather and climate.

The atmosphere drives weather and climate

Weather and climate involve the physical properties of the troposphere, such as temperature, pressure, humidity, cloudiness, and wind. **Weather** specifies atmospheric conditions over short time periods, typically hours or days, and within relatively small geographic areas. **Climate** describes the pattern of atmospheric conditions found across large geographic regions over long periods—seasons, years, or millennia. Writer Mark Twain once noted the distinction between climate and weather by saying, "Climate is what we expect; weather is what we get."

Air masses interact to produce weather

Weather can change quickly when air masses with different physical properties meet. The boundary between air masses that differ in temperature and moisture (and therefore density) is called a **front**. The boundary along which a mass of warmer, moister air replaces a mass of colder, drier air is termed a **warm front** (**FIGURE 13.7A**). Some of the warm, moist air behind a warm front rises

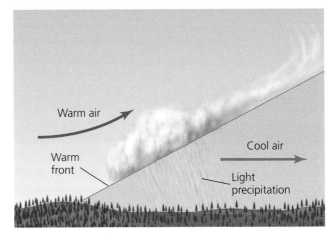

(a) Warm front

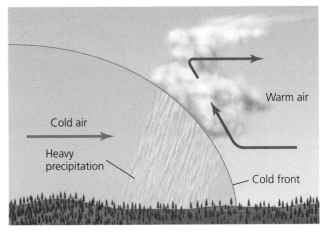

(b) Cold front

FIGURE 13.7
When a warm front approaches **(a)**, warmer air rises over cooler air, causing light or moderate precipitation as moisture in the warmer air condenses. When a cold front approaches **(b)**, colder air pushes beneath warmer air, and the warmer air rises, resulting in condensation and heavy precipitation.

over the cold air mass and then cools and condenses to form clouds that may produce light rain. A **cold front** (**FIGURE 13.7B**) is the boundary along which a colder, drier air mass displaces a warmer, moister air mass. The colder air, being denser, tends to wedge beneath the warmer air. The warmer air rises, expands, then cools to form clouds that can produce thunderstorms. Once a cold front passes through, the sky usually clears, and the temperature and humidity drop.

Adjacent air masses may also differ in atmospheric pressure. A **high-pressure system** contains air that moves outward away from a centre of high pressure as it descends. High-pressure systems typically bring fair weather. In a **low-pressure system**, air moves toward the low atmospheric pressure at the centre of the system and spirals upward. The air expands and cools, and clouds and precipitation often result.

Under most conditions, air in the troposphere decreases in temperature as altitude increases. Because warm air rises, vertical mixing results. Occasionally, however, a layer of cool air occurs beneath a layer of warmer air. This departure from the normal temperature profile is known as a **temperature inversion**, or **thermal inversion** (**FIGURE 13.8**), because the normal direction of temperature change is inverted. The cooler air at the bottom of the inversion layer is denser than the warmer air at the top, so it resists vertical mixing and remains stable. Thermal inversions can occur in different ways, sometimes involving cool air at ground level and sometimes producing an inversion layer higher above the ground (as shown in **FIGURE 13.8B**). One common type of inversion occurs in mountain valleys where slopes block morning sunlight, keeping ground-level air within the valley shaded and cool.

Vertical mixing normally allows air pollution to be diluted upward, but thermal inversions trap pollutants near the ground. A thermal inversion sparked a "killer smog" crisis in London, England, in 1952. A high-pressure system settled over the city, acting like a cap on the pollution; at least 4000 people—possibly as many as 12 000—died as a result of this event. Inversions regularly cause smog buildups in large metropolitan areas in valleys ringed by mountains, such as Los Angeles, Mexico City, Seoul, and Rio de Janeiro.

Large-scale circulation systems produce global climate patterns

At larger geographic scales, convective air currents contribute to broad climatic patterns (**FIGURE 13.9A**). Near the equator, solar radiation sets in motion a pair of convective cells known as **Hadley cells**. Here, where sunlight

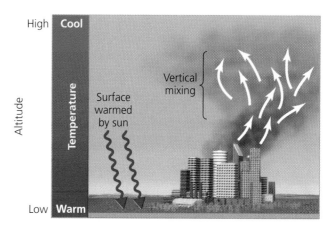

(a) Normal conditions

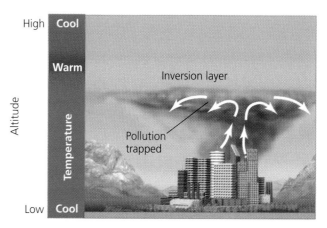

(b) Thermal inversion

FIGURE 13.8
A thermal inversion is a natural atmospheric occurrence that can worsen air pollution locally. Under normal conditions **(a)**, tropospheric temperature decreases with altitude. Air of different altitudes mixes, dispersing pollutants upward and outward from their sources. During a thermal inversion **(b)**, cool air remains near the ground underneath a layer of air that warms with altitude. Little mixing occurs, and pollutants are trapped near the surface.

is most intense, surface air warms, rises, and expands. As it does so, it releases moisture, producing the heavy rainfall that gives rise to tropical rainforests near the equator.

After releasing much of its moisture, this air diverges and moves in currents heading northward and southward. The air in these currents cools and descends back to Earth at about 30° latitude north and south. Because the descending air has low relative humidity, the regions around 30° latitude are quite arid, giving rise to deserts. Two pairs of similar but less intense convective cells, called **Ferrel cells** and **polar cells**, lift air and create precipitation around 60° latitude north and south and cause air to descend at around 30° latitude and in the polar regions.

These three pairs of cells account for the latitudinal distribution of moisture across Earth's surface: warm, wet climates near the equator; arid climates and major deserts near 30° latitude; moist, temperate regions near

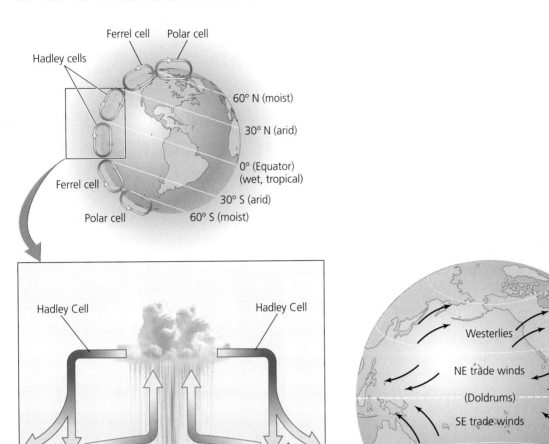

FIGURE 13.9
A series of large-scale convective cells **(a)** helps determine global patterns of humidity and aridity. Warm air near the equator rises, expands, and cools, and moisture condenses, giving rise to a wet climate in tropical regions. Air travels toward the poles and descends around 30° latitude. This air, which loses its moisture in the tropics, causes regions around 30° latitude to be arid. This convective circulation, a Hadley cell, occurs on both sides of the equator. Between roughly 30° and 60° latitude north and south, Ferrel cells occur; and between 60° and 90° latitude, polar cells occur. Air rises around 60° latitude, creating a moist climate, and falls around 90°, creating a dry climate. Global wind currents **(b)** show latitudinal patterns as well. Trade winds between the equator and 30° latitude blow westward, whereas westerlies between 30° and 60° latitude blow eastward.

60° latitude; and dry, cold conditions near the poles. These patterns, combined with temperature variation, help explain why biomes tend to be arrayed in latitudinal bands (**FIGURE 4.16**).

The Hadley, Ferrel, and polar cells interact with Earth's rotation to produce the global wind patterns shown in **FIGURE 13.9B**. As Earth rotates on its axis, locations on the equator spin faster than locations near the poles. As a result, the north–south air currents of the convective cells are deflected from a straight path as some portions of the globe move beneath them more quickly than others. This deflection is called the **Coriolis effect**, and it results in the curving global wind patterns evident in **FIGURE 13.9B**. The Coriolis effect influences the circulation of any freely moving fluid on Earth's surface, including ocean water, but its influence is not noticeable unless the scale of the circulation is quite large.

Between the equator and 30° latitude, the **trade winds** blow from east to west. Where the trade winds meet and are deflected toward the west, just north and south of the equator, lies a region with few winds known as the **doldrums**. Farther from the equator, between 30° and 60° north and south latitude, the **westerlies** originate from the west and blow east. People used these global circulation patterns for centuries to facilitate ocean travel by wind-powered sailing ships.

The atmosphere interacts with the oceans to affect weather, climate, and the distribution of biomes. For instance, winds and convective circulation in ocean water together maintain ocean currents. Trade winds weaken periodically, leading to El Niño conditions. The atmosphere's interactions with other systems of the planet are complex, but even a basic understanding of how the atmosphere functions can help us comprehend how our pollution of the atmosphere can affect ecological systems, economies, and human health.

Outdoor Air Pollution

Throughout human history, we have made the atmosphere a dumping ground for our airborne wastes. Whether from primitive wood fires or modern coal-burning power plants, people have generated **air pollutants**, gases and particulate material added to the atmosphere that can affect climate or harm people or other organisms. **Air pollution** refers to the release of air pollutants. In recent decades, government policy and improved technologies have helped us to substantially diminish **outdoor air pollution**, usually called **ambient air pollution**, in countries of the developed world. However, outdoor air pollution remains a problem, particularly in developing nations and in urban areas.

Natural sources can pollute

When we think of outdoor air pollution, we tend to envision smokestacks belching black smoke from industrial plants. However, natural processes produce a great deal of the world's air pollution. Some of these natural impacts can be exacerbated by human activity and land use policies.

Winds sweeping over arid terrain can send huge amounts of dust aloft. In 2001, strong westerlies lifted soil from deserts in Mongolia and China. The dust blanketed Chinese towns, spread to Japan and Korea, travelled eastward across the Pacific Ocean to the United States, then crossed the Atlantic and left evidence atop the French Alps. Every year, hundreds of millions of tons of dust are blown westward by trade winds across the Atlantic Ocean from northern Africa to the Americas (**FIGURE 13.10A**). These dust storms bring nutrients to the Amazon basin, as well as fungal and bacterial spores that have been linked to die-offs in Caribbean coral reef systems. Although dust storms are natural, the immense scale of these events is exacerbated by unsustainable farming and grazing practices that strip vegetation from the soil, promote wind erosion, and lead to desertification.

Volcanic eruptions release large quantities of particulate matter, as well as sulphur dioxide and other gases, into the troposphere (**FIGURE 13.10B**). Major eruptions may blow matter into the stratosphere, where it can circle the globe and remain aloft for months or years. Sulphur dioxide reacts with water and oxygen, and then condenses into fine droplets called **aerosols**, which reflect sunlight back into space and thereby cool the atmosphere and surface. The 1991 eruption of Mount Pinatubo in the Philippines ejected nearly 20 million tonnes of ash and aerosols and cooled global temperatures by roughly 0.5°C.

Burning vegetation also pollutes the atmosphere with soot and gases. More than 60 million hectares of forest and grassland burn in a typical year (**FIGURE 13.10C**). Fires occur naturally, but many are made more severe by human action. In North America, fuel buildup from decades of fire suppression has caused damaging forest fires in recent years. In the tropics, many fires result from the clearing of forests for farming and grazing by "slash-and-burn". In 1997, a severe drought brought on by the twentieth century's strongest El Niño event caused forest fires in Indonesia to rage out of control. Their smoke sickened 20 million Indonesians and caused a plane to crash and ships to collide. Along with tens of thousands of fires in drought-plagued Mexico, Central America, and Africa, these fires released more carbon monoxide into the atmosphere during 1997–1998 than did our worldwide combustion of fossil fuels.

We create various types of outdoor air pollution

Since the onset of industrialization, human activity has introduced a variety of sources of air pollution. As with water pollution, air pollution can emanate from mobile or stationary sources, and from *point sources* or *non-point sources*. A point source describes a specific spot where large quantities of pollutants are discharged. Non-point sources are more diffuse, often consisting of many small sources. Power plants and factories act as stationary point sources, whereas millions of automobiles on the roadways—each one a tiny point source—together create a massive, mobile non-point source of pollutants.

Once pollutants are in the atmosphere in sufficient concentrations, they may do harm directly, or they may induce chemical reactions that produce harmful compounds. **Primary pollutants**, such as soot and carbon monoxide, are pollutants emitted into the troposphere in a form that can be directly harmful or that can react to form harmful substances. Harmful substances produced

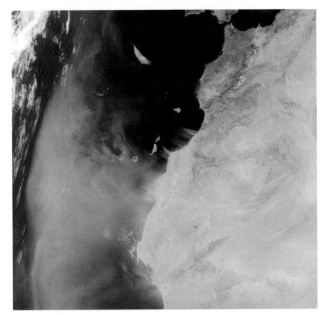
(a) Dust storm off west coast of Africa

(b) Mount St. Helens eruption, 1980

(c) Forest fire in Alberta

FIGURE 13.10
Massive dust storms, such as this one blowing across the Atlantic Ocean from Africa to the Americas **(a)**, are one type of natural air pollution. Volcanic eruptions result in another, as shown by Mount St. Helens **(b)**, which erupted in the state of Washington in 1980. A third natural cause of air pollution is fires in forests and grasslands **(c)**. Often, pollution from natural sources is made worse by human influence, such as when poor agricultural practices lead to soil erosion by wind, or when fire suppression leads to more devastating fires.

when primary pollutants interact or react with constituents of the atmosphere are called **secondary pollutants**. Secondary pollutants include tropospheric ozone, sulphuric acid, and other examples we will explore below.

Arguably the greatest human-induced air pollution problem today is our emission of greenhouse gases that contribute to global climate change. Addressing our release of excess carbon dioxide, methane, and other gases that warm the atmosphere stands as one of our civilization's primary challenges. We will discuss this issue separately and in depth in Chapter 14.

CEPA identifies harmful airborne substances

The *Canadian Environmental Protection Act (CEPA)* (1999) provides a list of air pollutants that are subject to legislative control and management. These pollutants differ widely in their chemical composition, chemical reactivity, emission sources, residence time (how long they remain in various environmental reservoirs, including organisms), persistence (how long they last before breaking down), transportability (their ability to be moved long or short distances), and impacts on the natural and built environments, and on human and ecosystem health. Environment Canada groups the pollutants of greatest concern into four categories:

1. Criteria air contaminants
2. Persistent organic pollutants
3. Heavy metals
4. Toxic air pollutants

We will look at each of the four pollutant categories in turn.[5]

Criteria air contaminants Criteria air contaminants (**CACs**) are produced in varying quantities by a number of processes, including the burning of fossil fuels. In Canada, the list of CACs includes sulphur oxides (SO_x), nitrogen oxides (NO_x), particulate matter (PM), volatile organic compounds (VOCs or VOX), carbon monoxide (CO), ammonia (NH_3), and tropospheric ozone (O_3), as follows.

Sulphur dioxide (SO_2) is a colourless gas with a strong odour. The vast majority of SO_2 and other *sulphur oxide* (SO_x) pollution results from the combustion of coal for electricity generation and industry. During combustion, elemental sulphur (S) in coal reacts with oxygen gas (O_2) to form SO_2. Once in the atmosphere, SO_2 may react to form sulphuric acid (H_2SO_4), which may then fall back to Earth as acid precipitation.

Nitrogen dioxide (NO_2) is a highly reactive, foul-smelling reddish brown gas that contributes to smog and acid precipitation. Along with *nitric oxide* (NO), NO_2 belongs to a family of compounds called *nitrogen oxides* (NO_x). Nitrogen oxides result when atmospheric nitrogen and oxygen react at the high temperatures created by combustion engines. More than half of NO_x emissions result from combustion in motor vehicle engines; electrical utility and industrial combustion account for most of the rest.

Particulate matter (PM) is composed of solid or liquid particles small enough to be suspended in the atmosphere. References to PM often include a number, specifying the size of the particles. PM_{10}, for example, refers to particles less than 10 μm (microns) in diameter; $PM_{2.5}$ refers to extremely fine particles, with diameters less than 2.5 μm; and so on. Particulate matter (which can also be called *suspended particulates, SP*) includes primary pollutants, such as dust and soot, as well as secondary pollutants, such as sulphates and nitrates. Particulates can damage respiratory tissues when inhaled. Most particulate matter (60%) in the atmosphere is wind-blown dust; human activity accounts for much of the rest. Along with SO_2, it was largely the emission of particulate matter from coal combustion that produced London's 1952 killer smog.

A **volatile organic compound (VOC or VOX)** is a carbon-containing chemical used in and emitted by vehicle engines and a wide variety of solvents and industrial processes, as well as by many household chemicals and consumer items. One group of VOCs consists of hydrocarbons, such as methane (CH_4, the primary component of natural gas), propane (C_3H_8, used as a portable fuel), butane (C_4H_{10}, found in cigarette lighters), and octane (C_8H_{18}, a component of gasoline). Human activities account for about half of VOC emissions, and the remainder comes from natural sources. For example, plants produce isoprene (C_5H_8) and terpene ($C_{10}H_{15}$).

Carbon monoxide (CO) is a colourless, odourless gas produced by the incomplete combustion of fuel. Vehicles and engines are the main source, but others include industrial processes, combustion of waste, and residential wood burning. Carbon monoxide poses risk to humans and other animals, even in low concentrations. It can bind to hemoglobin in red blood cells, preventing it from becoming oxygenated.

Ammonia (NH_3) is a colourless gas with a pungent odour—it is the smell associated with urine. Most NH_3 is generated from livestock waste and fertilizer production. Ammonia is poisonous if inhaled in great quantities and is irritating to the eyes, nose, and throat in lesser concentrations. In the atmosphere it combines with sulphates and nitrates to form secondary fine particulate matter ($PM_{2.5}$). NH_3 can also contribute to the nitrification and eutrophication of aquatic systems.

Tropospheric ozone (O_3) is also called ground-level ozone to distinguish it from the ozone in the stratosphere, which shields us from the dangers of UV radiation. In contrast to stratospheric ozone, O_3 from human activity forms and accumulates at ground level as a pollutant. In the troposphere, this colourless gas results from the interaction of sunlight, heat, nitrogen oxides, and carbon-containing chemicals; it is therefore a secondary pollutant. A major component of smog, O_3 can pose health risks because of its

---

weighing the issues 13–1
BAD AIR DAYS

Are you sensitive to smog? Do you suffer from itchy eyes, burning lungs, or other symptoms on "bad air days"? Do you think there is a smog problem in your area?

If you visit the website of the National Air Pollution Surveillance Network (NAPS), maintained by Environment Canada, you can watch animations of severe smog events in which ground-level ozone exceeded the accepted air quality standard levels over very large areas of both Canada and the United States (www.etc-cte.ec.gc.ca/NAPS/naps_smog_e.html).

instability as a molecule; this triplet of oxygen atoms will readily release one, leaving a molecule of oxygen gas (O_2) and a free oxygen atom. The free oxygen atom may then participate in reactions that can injure living tissues and cause respiratory problems. Tropospheric ozone is the pollutant that most frequently exceeds its air quality standard.

Persistent organic pollutants   Persistent organic pollutants (POPs) can last in the environment for long periods of time. They are capable of travelling great distances by air because they are *volatile*, which means that they evaporate readily. The term **persistent** refers to substances that have long residence times, either because they remain in environmental reservoirs for a long time, or because they take a long time to degrade or break down, or both.

POPs are of particular concern because they can enter the food supply, bioaccumulate in body tissues, and have significant impacts on human health and the environment, even in low concentrations. They have few natural sources and come primarily from human activity. Examples include industrial chemicals, such as PCBs (polychlorinated biphenyls); pesticides, such as DDT (dichloro-diphenyl-trichloroethane, FIGURE 13.11), chlordane, and toxaphene; and contaminants and by-products, such as dioxins and furans, which come from incomplete combustion processes.

The indiscriminate spraying and persistent buildup of DDT caused widespread deaths of birds and other fauna in the 1950s and 1960s, leading Rachel Carson to write her famous book *Silent Spring*, one of the pivotal events in the modern environmental movement. DDT was banned for agricultural use in most developed nations in the 1970s and 1980s, but its use continues even today in some parts of the developing world. As recently as 2002, DDT and its by-products were still widely detectable in human blood, tissue, and breast milk samples from subjects in North America.[6]

Heavy metals   **Heavy metals** can be transported by the air, enter our water and food supply, and reside for long periods in sediments. Metals tend to be associated with particulate matter, either occurring in particulate form, or attaching to small particles that can then be transported atmospherically. Heavy metals are poisonous, even in low concentrations, and can bioaccumulate in body tissues. These pollutants occur even in Canada's far north—they are carried from the industrial south by continent-scale atmospheric currents—where they are deposited on land and water surfaces. This is called **long-range transport of atmospheric pollutants (LRTAP)**, and it has been a concern in Canada since at least the early 1970s.

Mercury is a heavy metal of considerable concern in Canada. Mercury is *volatile* (evaporates readily) and occurs in a number of different chemical forms, some more toxic than others. It has natural as well as human sources. It has been used for a variety of industrial purposes, partly because of its unusual property of remaining liquid at surface temperatures and pressures. Like other heavy metals, mercury can enter the food chain, accumulate in body tissues, and cause central nervous system malfunction and other ailments. Mercury is *lipophilic*, which can be translated as "fat-loving"; it is chemically capable of binding to fatty tissues in organisms. Mercury is also of concern as a pollutant in surface water bodies.

Lead enters the atmosphere as a particulate pollutant. The lead-containing compounds tetraethyl lead and tetramethyl lead, when added to gasoline, improve engine performance. However, exhaust from the combustion of leaded gasoline emits lead into the atmosphere, from which it can be inhaled or deposited on land or water. Like mercury, lead is bioaccumulative and can cause damage to the central nervous system. Once people recognized the dangers of lead, leaded gasoline began to be phased out in most industrialized nations in the 1970s. The use of leaded gas ended in Canada in 1993.[7] Today the greatest source of atmospheric lead pollution in developed nations is industrial metal smelting. However, many developing nations still add lead to gasoline and experience significant lead pollution.

Toxic air pollutants are a broad category of "other" pollutants identified by *CEPA* as being harmful or toxic, and therefore subject to regulation, control, and monitoring. They include substances known to cause cancer, reproductive defects, or neurological, developmental,

FIGURE 13.11
Learning about the persistence and negative ecological impacts of some chemical pesticides catalyzed the public and helped launch the modern environmental movement in the late 1960s and early 1970s. Here, a farmer sprays pesticide on his field without adequate personal protection against chemical exposure.

immune system, or respiratory problems in people. Some also negatively affect the health of animals and plants. This category overlaps with the other types of air pollutants (for example, lead, mercury, dioxins, furans, and ozone all appear on the list of toxic pollutants, as well as in their other categories), but it includes additional substances that have been determined to be toxic. One example is asbestos, which we will look at in greater detail in Chapter 19. Chlorofluorocarbons (CFCs) are also on the list. Most toxic air pollutants are produced by human activities, such as metal smelting, sewage treatment, and industrial processes.

---

weighing the issues — INVESTIGATING YOUR REGION'S AIR QUALITY — 13-2

How polluted is the air near where you live? Go to the National Pollutant Release Inventory (NPRI) website at www.ec.gc.ca/pdb/npri/npri_home_e.cfm. Use the *Google Earth* mapping tool to check on the amounts of pollutants released in your own province or local area. Are there any specific facilities in your area that are major emitters of atmospheric pollutants? Were you aware of the existence of these emitters, previously?

Government agencies share in dealing with air pollution

In Canada the management of air-related issues is the responsibility of the federal government—primarily, but not exclusively, Environment Canada—and the provincial and territorial governments, through their environment ministries. For the most part, municipal governments do not have direct regulatory control over activities that affect air quality. However, municipal governments manage so many activities that influence air quality that their role is central to the collaborative effort.

Federal The principal federal legislation under which air quality is regulated is the *Canadian Environmental Protection Act* (1999), described officially as "an Act respecting pollution prevention and the protection of the environment and human health in order to contribute to sustainable development".[8] Although *CEPA* gives the lead responsibility for air quality to Environment Canada, it also defines an important role for Health Canada, which we will investigate in greater detail in Chapter 19. Federal agencies, such as Transport Canada and Natural Resources Canada, also have programs and activities with important linkages to air quality issues.

The federal government is also responsible for entering Canada into international agreements concerning air quality. The *Montreal Protocol* and the *Kyoto Protocol* are examples of multilateral agreements that address air pollution issues of global concern (stratospheric ozone depletion and global warming, respectively).

Canada has a long history of international agreements with the United States to control transboundary pollution. For air quality, these date back to agreements made in the early 1900s, although the modern era began in 1979 with agreements concerning the long-range transport of pollutants related to acid deposition. The present bilateral international agreement on transboundary air pollution, coordinated primarily through the International Joint Commission, is the *Canada–United States Air Quality Agreement*, signed in 1991. This agreement has three annexes, the first dealing with acid rain precursors, the second with coordination of international scientific research, and the third, most recently (in 2000), with ground-level ozone.

Provincial/territorial Each provincial/territorial government approaches air quality issues with its own agenda and set of rules, through its environment ministry. This makes sense—not just politically but scientifically, too—since issues vary dramatically from one region to another. For example, as discussed in "Central Case: The Rain and the Big Nickel," acid deposition caused by both local pollution and pollutants transported from the industrial midwestern states is the major air pollution issue in Ontario. In Saskatchewan, in contrast, air quality issues associated with the handling of grain, feed, and livestock are of broader concern.

Regional differences in the handling of air quality issues can lead to problems, though. For example, in the past the standards for acceptable levels and the protocols of measurement for pollutants have varied significantly from one jurisdiction to another. The government has been trying to bring these standards into conformity across the nation, by working through the Canadian Council of Ministers of the Environment (CCME). In 1998 the federal government, provinces, and territories signed the *Harmonization Accord* and *Canada-Wide Standards Sub-Agreement*, to enhance effectiveness, accountability, and clarity in the management of environmental issues throughout Canada. Roles and responsibilities are assigned, on a case-by-case basis, to the government agency best situated to deal with the matter, but each level of government still retains its legal authority in the matter. The CCME has signed several agreements on harmonized standards for water quality, but those for air quality, the *National Ambient Air Quality Objectives*, have yet to be enacted.[9]

Municipal   Only two municipalities in Canada—Montreal and Greater Vancouver—have been given direct regulatory authority over sources of air pollution by their respective provincial governments. However, all municipalities manage programs and activities that directly influence air quality on a daily basis—public transportation and land use zoning come to mind, for example.

Therefore, most municipalities have programs aimed at improving air quality and raising public awareness of air quality issues. The top concerns differ from one location to another, and air quality may or may not be the top issue in a given municipality. For example, in Mississauga, Ontario—a geographically spread-out, largely suburban city—transportation and related air quality concerns are the central environmental issue, both for government decision makers and for the general public. In Halifax, with a smaller population, shorter travel distances, and fewer cars, wastewater management takes priority over air quality concerns. Pollution types and sources differ, too, from one locality to another. In Sudbury, smelters are a central concern; in Mississauga, the issue is cars and coal-fired power plants. Elsewhere, pulp-and-paper mills are of primary concern, or grain-handling operations, or large feedlots, or dust from cement manufacturers.

Monitoring shows that many forms of air pollution have decreased

CEPA not only lists the pollutants of interest but also requires that any releases of these pollutants be reported to the National Pollutant Release Inventory (NPRI), which is maintained by Environment Canada. The NPRI is thus one important vehicle that can be used to keep track of air quality in Canada. The data on NPRI are submitted, under law, directly by those who emit harmful substances into the atmosphere.

Pollutants also are measured and monitored through a nationwide network of monitoring stations that compose the National Air Pollution Surveillance (NAPS) Network, coordinated by Environment Canada (**FIGURE 13.12**). Many of the stations in the network are NAPS-designated sites; however, stations managed by other federal agencies (such as the Meteorological Service of Canada), as well as provincial, territorial, municipal, or other types of agencies (such as universities) also contribute data to the network.

The main focus of air quality monitoring is on the criteria air contaminants, but other categories of pollutants are monitored at some stations. Specialized and regional networks often contribute to the monitoring of these

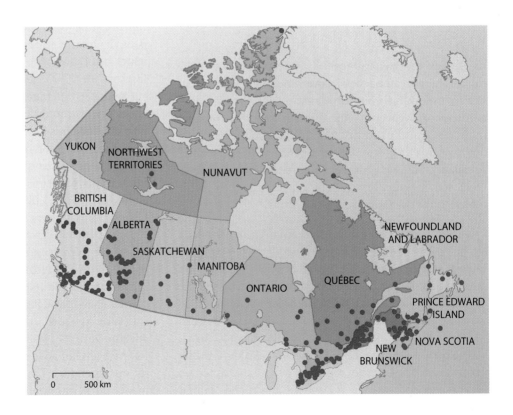

FIGURE 13.12
This map shows the network of real-time air quality monitoring stations across Canada.

other pollutants. Examples include the Canadian Air and Precipitation Monitoring Network (non-urban air quality); Surface Ozone Monitoring Network; Mercury Deposition Network; Air Toxics Monitoring Network; and Particulate Matter Monitoring Network.

In the decades since the first modern anti-pollution actions in North America in the early 1970s, emissions of some criteria air contaminants have decreased substantially. This has resulted in declining levels of these pollutants, as measured at air quality monitoring stations throughout Canada (**FIGURE 13.13**). The most dramatic decrease can be seen in atmospheric lead (**FIGURE 13.13F**). Even though Canada did not phase out leaded gas until 1993, the cessation of leaded gas use in

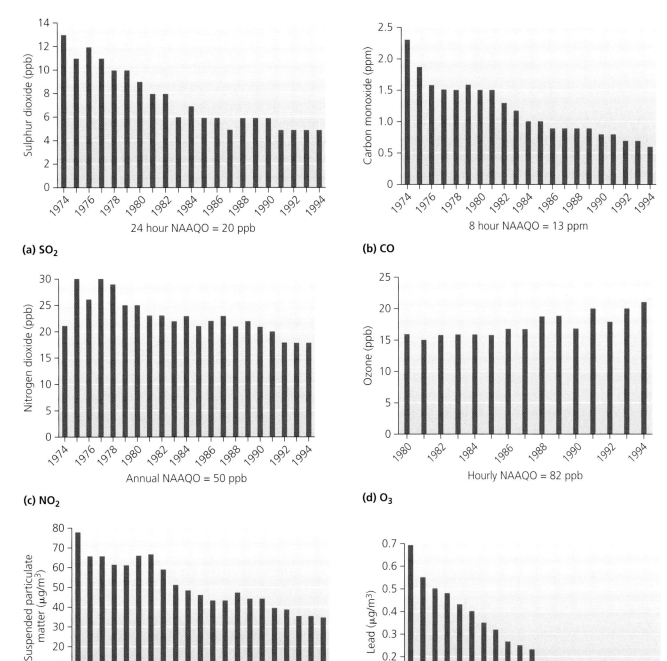

FIGURE 13.13
Mean annual levels of some criteria air contaminants have declined since the early 1970s; others show little to no improvement.

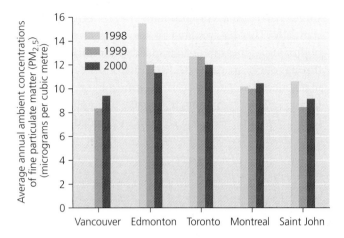

FIGURE 13.14
Levels of fine particulate matter are still a problem in Canadian cities. Based on data from National Air Pollution Surveillance Network, Environment Canada.

the United States in the early 1970s had a clear impact on atmospheric lead in Canada. In contrast, O_3 (**FIGURE 13.13D**) shows an increase over the same period, whereas NO_2 (**FIGURE 13.13C**) and PM (**FIGURE 13.13E**) show only minimal decreases (**FIGURE 13.14**). Not surprisingly, these three pollutants are the core indicators for Canada's Air Quality Health Index.

There are several reasons for the declines in some pollutants, which have occurred in spite of increases in population, energy use, vehicle use, and economic productivity in North America. Cleaner-burning motor vehicle engines and automotive technologies such as catalytic converters have played a large part in decreasing emissions of carbon monoxide and several other pollutants. Sulphur dioxide permit-trading and clean coal technologies have reduced SO_2 emissions. Technologies, such as electrostatic precipitators and **scrubbers** (**FIGURE 13.15**), which chemically convert or physically remove airborne pollutants before they are emitted from smokestacks, have allowed factories, power plants, and refineries to decrease emissions of several pollutants.

Other industrialized nations have also succeeded in reducing emissions and improving air quality, thanks to improved technologies and targeted federal policies. Between 1996 and 2005, London, England, achieved a 56% reduction in CO emissions, a 41% drop in NO_x emissions, a 28% decline in particulate matter release,

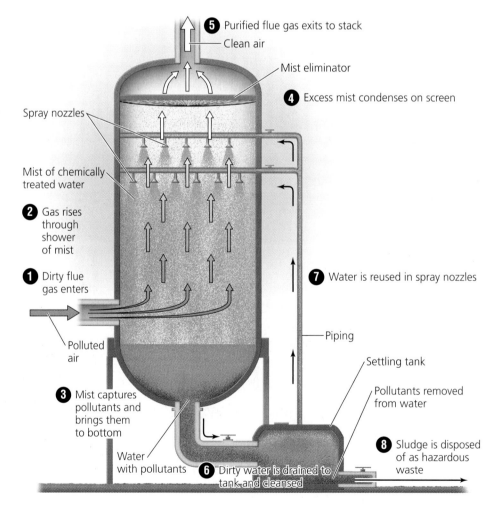

FIGURE 13.15
In this spray-tower wet scrubber, polluted air (1) rises through a chamber while arrays of nozzles spray a mist of water mixed with lime or other active chemicals (2). The falling mist captures pollutants and carries them to the bottom of the chamber (3), essentially washing them out of the air. Excess mist is captured on a screen (4), and air emitted from the scrubber has largely been cleansed (5). Periodically, the dirty water is drained from the chamber (6), cleansed in a settling tank, and recirculated (7) through the spray nozzles. The resulting sludge must be disposed of (8) as hazardous waste. Scrubbers and other pollution control devices come in many designs; the type shown here typically removes at least 90% of particulate matter and gases, such as sulphur dioxide.

and an impressive 73% decrease in SO_2 emissions. Only tropospheric ozone showed an increase, rising 33%.

Canada is attempting to "turn the corner" on air pollution

In 2007, the federal environment minister released a document entitled *Turning the Corner: An Action Plan to Reduce Greenhouse Gases and Air Pollution*,[10] which includes specific targets for the reduction of greenhouse gas (GHG) and criteria air contaminant (CAC) emissions from industrial sources. This represents the first time the federal government will require industry to reduce GHG and CAC emissions. The specific goals are to reduce GHG emissions from industrial sources by 150 megatonnes by 2020, and to halve air pollution emissions from industry by 2015 (**FIGURE 13.16**).

The document offers several alternatives for industries interested in reducing their emissions, including equipment and process upgrades; fuel switching; cap-and-trade, offset, and emissions trading systems; and control technologies. The possibility of federal funding for some of these programs is mentioned, as is increased funding for research and development of new technologies to facilitate emission reductions. The document also offers financial rewards for "early action," for companies that took steps to reduce their emissions prior to the adoption of the new policy. The report includes calculations of the health benefits and other social and economic impacts of the policy.[11]

We will complete our look at outdoor air pollution with an examination of three specific issues: photochemical smog; acidic deposition; and stratospheric ozone depletion, along with a brief consideration of air quality issues in rural areas and in the rapidly industrializing nations of the developing world.

Smog is the most common, widespread air quality problem

In response to the increasing incidence of fogs polluted by the smoke of Britain's Industrial Revolution, an early British scientist coined the term **smog**. Today the term is used worldwide to describe unhealthy mixtures of air pollutants that often form over urban areas.

The deadly smog that enveloped London in 1952 was what we today call **industrial smog**, or grey-air smog. When coal or oil is burned, some portion is completely combusted, forming CO_2; some is partially combusted, producing CO; and some remains unburned and is released as soot, or particles of carbon. Moreover, coal contains varying amounts of contaminants, including

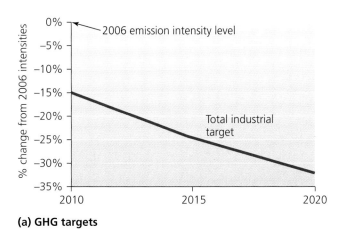

(a) GHG targets

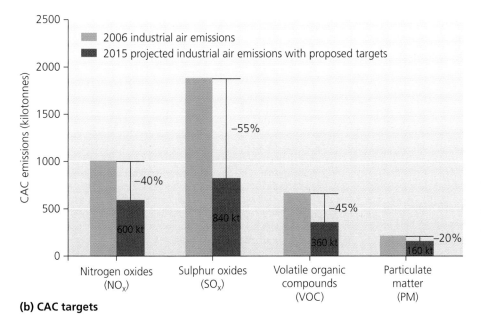

(b) CAC targets

FIGURE 13.16
The federal government's 2007 policy statement *Turning the Corner* includes targets for reduction of emissions of greenhouse gases **(a)** and some criteria air contaminants **(b)** from industrial sources. Source: Environment Canada. 2007. *Turning the Corner: Clean Air Regulatory Agenda*, www.ec.gc.ca/doc/media/m_124/ppt/tech_eng.htm.

mercury and sulphur. Sulphur reacts with oxygen to form sulphur dioxide, which can undergo a series of reactions to form sulphuric acid and ammonium sulphate (**FIGURE 13.17A**). These chemicals and others produced by further reactions, along with soot, are the main components of industrial smog and give the smog its characteristic grey colour.

Industrial smog is far less common today in developed nations than it was 50–100 years ago. In the wake of the 1952 London episode and others, the governments of most developed nations began regulating industrial emissions to minimize the external costs they impose on citizens. However, in regions that are industrializing today, such as China, India, and Eastern Europe, heavy reliance on coal burning (by industry and by citizens heating and cooking in their homes), combined with lax pollution controls, produces industrial smog that poses significant health risks in many areas.

Although coal combustion supplies the chemical constituents for industrial smog, weather also plays a role, as it did in London in 1952. A similar event occurred four years earlier in Donora, Pennsylvania. Here, air near the ground cooled during the night, and because Donora is located in hilly terrain, too little morning sun reached the valley floor to warm and disperse the cold air. The resulting thermal inversion trapped smog containing particulate matter emissions from a steel and wire factory; 21 people were killed, and more than 6000 people—nearly half the town—became ill (**FIGURE 13.17B**).

Hilly topography, such as Donora's, is a factor in the air pollution of many other cities where surrounding mountains trap air and create inversions. This is true for Mexico City, which has long symbolized the smog problems of modern cities. Modern-day Mexico City, however, suffers from a different type of smog, called photochemical smog.

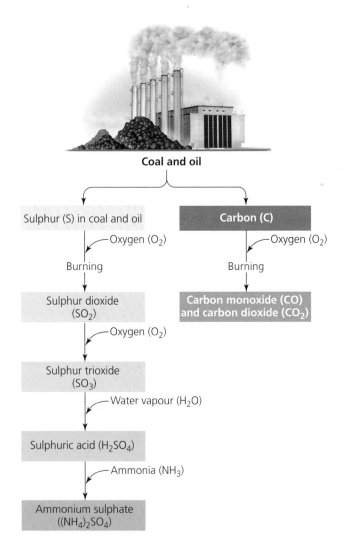

(a) Burning sulphur-rich oil or coal without adequate pollution control technologies

(b) Smog event of 1948, midday in Donora, Pennsylvania

FIGURE 13.17
Emissions from the combustion of coal and oil in manufacturing plants and utilities without pollution-control technologies can create industrial smog. Industrial smog consists primarily of sulphur dioxide and particulate matter, as well as carbon monoxide and carbon dioxide from the carbon component of fossil fuels. When fossil fuels are combusted, sulphur contaminants give rise to sulphur dioxide, which in the presence of other chemicals in the atmosphere can produce several other sulphur compounds **(a)**. Under certain weather conditions, industrial smog can blanket whole towns or regions, as it did in Donora, Pennsylvania, shown here in the daytime during its deadly 1948 smog episode **(b)**.

Photochemical smog is produced by a complex series of reactions

A photochemical process is one whose activation requires light. **Photochemical smog**, or brown-air smog, is formed through light-driven chemical reactions of primary pollutants and normal atmospheric compounds that produce a mix of more than 100 different chemicals, tropospheric ozone often being the most abundant among them (**FIGURE 13.18A**). High levels of NO_2 cause photochemical smog to form a brownish haze over cities (**FIGURE 13.18B**). Hot, sunny, windless days in urban areas provide perfect conditions for the formation of photochemical smog. Exhaust from morning traffic releases large amounts of NO and VOCs into a city's air. Sunlight then promotes the production of ozone and other constituents of photochemical smog. Levels of photochemical pollutants in urban areas typically peak in midafternoon and can irritate people's eyes, noses, and throats.

The cities most afflicted by photochemical smog are those with weather and topography that promote it. The geographic area associated with a particular air mass is called an **airshed**. People who live within the same airshed tend to experience similar weather and "bad air" days. Airsheds that are topographically constrained, such as those that occupy topographical basins or valleys, allow less natural circulation and renewal of the air, and are more prone to inversions and prolonged smog events.

Some provinces have cut emissions leading to photochemical smog through vehicle inspection programs, such as AirCare in British Columbia or Drive Clean in Ontario, where drivers are required to have their vehicle exhaust inspected regularly at check stations to maintain their registrations. Although a failed "smog check" means inconvenience for the car owner, these programs help maintain vehicle condition and make the air measurably cleaner for all of us.

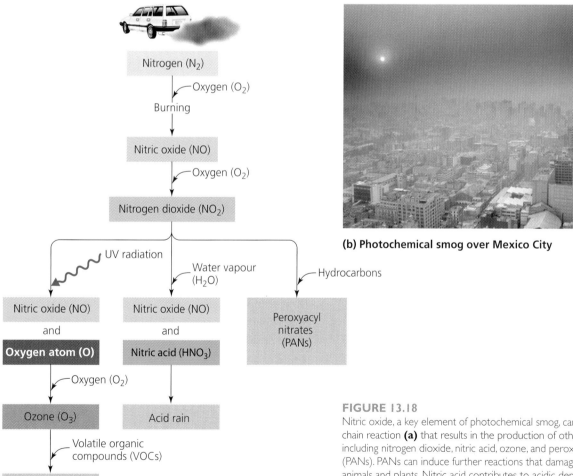

(a) Formation of photochemical smog

(b) Photochemical smog over Mexico City

FIGURE 13.18
Nitric oxide, a key element of photochemical smog, can start a chemical chain reaction **(a)** that results in the production of other compounds, including nitrogen dioxide, nitric acid, ozone, and peroxyacyl nitrates (PANs). PANs can induce further reactions that damage living tissues in animals and plants. Nitric acid contributes to acidic deposition as well as photochemical smog. Photochemical smog is common today over many urban areas, especially those with hilly topography or frequent inversion layers. Mexico City **(b)** frequently experiences photochemical smog.

THE SCIENCE BEHIND THE STORY

Identifying CFCs as the Main Cause of Ozone Depletion

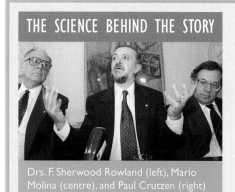

Drs. F. Sherwood Rowland (left), Mario Molina (centre), and Paul Crutzen (right) jointly received the 1995 Nobel Prize in chemistry.

Ozone was discovered in 1839, and its presence in the upper atmosphere was first proposed in the 1880s. In 1924, British scientist G. M. B. Dobson built an instrument that could measure ozone concentrations by sampling incoming sunlight at ground level and comparing the intensities of wavelengths that ozone does and does not absorb. By the 1970s, the Dobson ozone spectrophotometer was being used by a global network of observation stations.

Research on CFCs has also had a long history. First invented in 1928, CFCs were useful as refrigerants, fire extinguishers, and propellants for aerosol spray cans. Starting in the 1960s, CFCs also found wide use as cleaners for electronics and as a part of the process of manufacturing rigid polystyrene foams. Research on the chemical properties of CFCs showed that they were almost completely inert; that is, they rarely reacted with other chemicals. Therefore, scientists surmised that, at trace levels, CFCs would be harmless to both people and the environment.

However, in June 1974, chemists F. Sherwood Rowland and Mario Molina published a paper in the journal *Nature*, arguing that the inertness that made CFCs so ideal for industrial purposes could also have disastrous consequences for the ozone layer.

Whereas reactive chemicals are broken down in the lower atmosphere, CFCs reach the stratosphere unchanged. Once CFCs reach the stratosphere, intense ultraviolet radiation from the Sun breaks them into their constituent chlorine and carbon atoms. In a two-step chemical reaction (see the figure), a chlorine atom can split an ozone molecule, and then ready itself to split another one. Over its lifetime, each free chlorine atom, it was calculated, can catalyze the destruction of as many as 100 000 ozone molecules.

Rowland and Molina were the first to assemble a complete picture of the threat posed by CFCs, but they could not have reached their conclusions without the contributions of other scientists. British researcher James Lovelock had developed an instrument to measure extremely low concentrations of atmospheric gases. American scientists Richard Stolarski and Ralph Cicerone had shown that chlorine atoms can catalyze the destruction of ozone. Dutch meteorologist Paul Crutzen had shown that naturally produced nitrous oxide breaks down ozone. And American researcher James McDonald had predicted that ozone loss, by allowing more UV radiation to reach the surface, would result in thousands more skin cancer cases each year.

Rowland and Molina's analysis earned them the 1995 Nobel Prize in chemistry jointly with Crutzen. It also helped spark discussion among scientists, policy makers, and industry leaders over limits on CFC production. As a result, several nations banned the use of CFCs in aerosol spray cans in 1979. Other uses continued, however, and by

---

In 2003, London, England, instituted a "congestion-charging" program. People driving into central London during weekdays were required to pay £8 (about $15) per day. The money was to be used to enhance bus service and encourage transport by rail, taxi, bicycle, and foot. Many citizens were outraged, arguing that the fees were too high or that the system discriminated against poor people. Others complained that it would not work, or that the promised improvements in public transport were too slow in coming. However, many Londoners supported the program, and business support grew as the benefits became clearer. Traffic congestion in the zone decreased by nearly 30%, and there were 40–70 fewer injuries from traffic accidents per year. The air became cleaner, as well. In the first year, particulate matter in the charging zone declined by 15.5%, nitrogen oxide emissions decreased by 13.4%, and carbon dioxide emissions fell by 16.4%. Similar schemes have been successfully implemented in other cities, such as Singapore. Even so, as of 2008, the new mayor of London has made it a priority to dismantle the policy because of the unpopularity of the fees.

weighing the issues — CONGESTION CHARGING — 13–3

Does your city, or the nearest major city to you, suffer from air pollution? Do you think this city should adopt a congestion-charging program like London's? What benefits would your city enjoy from such a program, and what problems might it bring? What other steps should this city take to tackle pollution?

Air quality is a rural issue, too

Air quality is not only an urban issue. In rural areas, people suffer from drift of airborne pesticides from farms, as well as industrial pollutants transported from cities, factories, and power plants. A great deal of rural air pollution emanates from feedlots, where cattle,

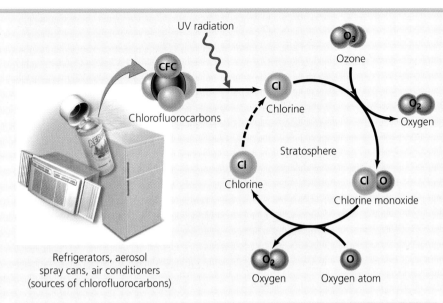

A chlorine atom released from a CFC molecule in the presence of UV radiation reacts with an ozone molecule, forming one molecule of oxygen gas and one chlorine monoxide (ClO) molecule. The oxygen atom in the ClO molecule will then bind with a stray oxygen atom to form oxygen gas, leaving the chlorine atom to begin the destructive cycle anew. In this way, any given chlorine atom may destroy up to 100 000 ozone molecules.

the early 1980s global production of CFCs was increasing.

Then, a new finding shocked scientists and spurred the international community to take further action. Scientists at a British research station in Antarctica had been recording ozone concentrations since the 1950s. In May 1985, Joseph Farman and colleagues reported in *Nature* that Antarctic ozone concentrations had declined dramatically since the 1970s. The decline exceeded even the worst-case predictions.

To determine what was causing the "ozone hole" over Antarctica, expeditions were mounted in 1986 and 1987 to measure trace amounts of atmospheric gases by using ground stations and high-altitude balloons and aircraft. Together with other researchers, Crutzen analyzed data collected on the expeditions and concluded that the ozone hole resulted from a combination of Antarctic weather conditions and human-made chemicals.

In the frigid Antarctic winter, high-altitude—or polar stratospheric—clouds form. In the spring, those clouds provide ideal conditions for CFC-derived chlorine and other chemicals to catalyze the destruction of massive amounts of ozone. The problem is made worse by the fact that prevailing air currents largely isolate Antarctica's atmosphere from the rest of Earth's atmosphere.

In subsequent years, scientists used data from ground stations and satellites to show that ozone levels were declining globally. In 1987, those findings helped convince the world's nations to agree on the Montreal Protocol, which aimed to cut CFC production in half by 1998. Within two years, however, further scientific evidence and computer modelling showed that more drastic measures would be needed if serious damage to the ozone layer was to be avoided. In 1990, the Montreal Protocol was strengthened to include a complete phase-out of CFCs by 2000. By 1998, the amount of chlorine in the atmosphere appeared to be levelling off.

hogs, or chickens are raised in dense concentrations. The huge numbers of animals at feedlots and the voluminous amounts of waste they produce release dust as well as methane, hydrogen sulphide, and ammonia. These gases create objectionable odours, and the ammonia contributes to nitrogen deposition across wide areas. Studies have shown that people working at and living near feedlots have high rates of respiratory problems.

Industrializing nations are suffering increasing air pollution

Although industrialized nations have been improving their air quality, outdoor air pollution is growing worse in many industrializing countries. In these societies, rapidly proliferating factories and power plants are releasing emissions with little effort to control pollution, and citizens continue to burn traditional sources of fuel, such as wood and charcoal, for cooking and home heating. Thus, just as occurred in England during its period of industrialization, new pollution sources are added to traditional sources while populations rise.

China has some of the world's worst air pollution. Four out of five Chinese cities surveyed by the World Bank in 2000 experienced SO_2 or NO_2 emissions above the threshold set by the World Health Organization. Air pollution became a serious concern for both the health and the performance of athletes during the Beijing Olympics of 2008. Together, China and India suffer 58% of the 1.8 million premature deaths that the World Bank estimates occur each year globally as a result of outdoor air pollution.

Southern Asia has a persistent 3 km-thick layer of pollution that hangs over the subcontinent throughout the dry season each December through April. Dubbed the *Asian Brown Cloud*, this massive layer of pollution is thought to reduce the sunlight reaching Earth's surface in that region by 10%–15%, influence climate, decrease rice productivity by 5%–10%, and account for many thousands of deaths each year.

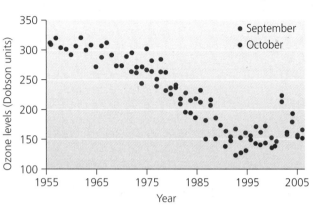
(a) Monthly mean ozone levels at Halley, Antarctica

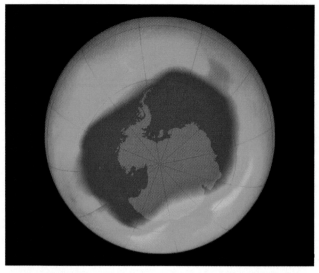
(b) The "ozone hole" over Antarctica, September 24, 2006

FIGURE 13.19
The "ozone hole" consists of a region of thinned ozone density in the stratosphere over Antarctica and the southernmost ocean regions. It has reappeared seasonally each September in recent decades. Data from Halley, Antarctica (a), show a steady decrease in stratospheric ozone concentrations from the 1960s to 1990. Ozone-depleting CFCs began to be regulated under the Montreal Protocol in 1987, and ozone concentrations stopped declining. Satellite imagery from September 24, 2006 (b), shows the "ozone hole" (blue) at its maximal recorded extent to date. Data in (a) from British Antarctic Survey.

Synthetic chemicals deplete stratospheric ozone

A pollutant in the troposphere, ozone is a highly beneficial gas at altitudes of about 25 km in the lower stratosphere, where it is concentrated in the stratospheric *ozone layer* (see **FIGURE 13.2**). Here, concentrations of ozone are only about 12 parts per million. However, ozone molecules are so effective at absorbing incoming ultraviolet radiation from the Sun that this concentration helps to protect life on Earth's surface from the damaging effects of ultraviolet (UV) radiation.

In the 1960s, atmospheric scientists began wondering why their measurements of stratospheric ozone were lower than theoretical models predicted. Researchers hypothesizing that natural or artificial chemicals were depleting ozone finally pinpointed a group of human-made compounds derived from simple hydrocarbons, such as ethane and methane, in which hydrogen atoms are replaced by chlorine, bromine, or fluorine. One class of such compounds, **chlorofluorocarbons (CFCs)**, was being mass-produced by industry at a rate of 1 million metric tons per year in the early 1970s, and this rate was growing by 20% a year.

Soon researchers showed that CFCs could deplete stratospheric ozone by releasing chlorine atoms that split ozone molecules, creating from each of them an O_2 molecule and a ClO molecule (see "The Science Behind the Story: Identifying CFCs as the Main Cause of Ozone Depletion"). Then in 1985, scientists announced that stratospheric ozone levels over Antarctica had declined by 40%–60% in the previous decade, leaving a thinned ozone concentration that was soon dubbed the **ozone hole** (**FIGURE 13.19**).

Research over the next few years confirmed the link between CFCs and ozone loss in the Antarctic and indicated that depletion was also occurring in the Arctic and perhaps globally. The depletion was shown to be growing both in severity and in areal extent, almost without exception, year to year. Already concerned that increased UV radiation would lead to more skin cancer, scientists were becoming anxious over possible ecological effects as well, including harm to crops and to the productivity of ocean phytoplankton, the base of the marine food chain.

There are still many questions to be resolved about ozone depletion

Although significant progress has been made in understanding stratospheric ozone depletion since the initial discovery of the role of CFCs, many questions remain. For example, will ozone depletion spread from the polar regions, where it is most severe, to encompass mid-latitude and even low-latitude regions (and if so, how quickly)? What is the actual relationship between ozone depletion and human health impacts, such as skin cancer? What are the other potential impacts of ozone depletion, for example, on marine and terrestrial ecosystems and on other types of materials? And—importantly, for policy decisions—are the substitute chemicals that are being proposed in international agreements definitely less damaging to the stratospheric ozone layer, or do they raise concerns of their own?

The question of polar ozone depletion is of particular interest, especially to Canadian scientists working in the Arctic. The answer to why ozone depletion is most severe over the poles probably lies in a more refined understanding of the physical processes involved in ozone-depleting reactions. For example, scientists now believe that the ozone-depleting chemical reactions may find ideal launching sites on tiny ice crystals that are found only where the air is extremely cold. These conditions are optimal over Antarctica, where a circular wind pattern called the *polar vortex* traps extremely cold air over the pole.

There are many other interesting questions waiting to be answered. For example, Dr. Ralf Staebler and colleagues from the Air Quality Research Division of Environment Canada are investigating a unique form of ozone depletion which occurs near the ground in the polar regions, as the sun rises in the spring after the long darkness of winter. Here, a sudden loss of ozone is caused by reactions with bromine, another type of halogen, which can deplete ozone at ground level in much the same manner as the chlorine from the CFCs in the upper atmosphere. The scientists measure this ozone loss using a mobile measurement platform—basically, a temperature-controlled box full of instruments and a meteorological mast mounted on a sled—which they call the OOTI ("Out On The Ice") sled (**FIGURE 13.20**). They wanted a platform that would be self-contained, self-sufficient, and mobile, with which to measure ozone, the chemicals implicated in ozone-destroying chemistry, and the associated *micrometeorology*. As part of the Canadian contribution to the International Polar Year, this sled has been deployed near Alert, NU, Barrow, AK, Kuujjuarapik (Hudson Bay), and from the Amundsen Icebreaker on the frozen ocean south of Banks Island.

The researchers are particularly interested in finding out whether the ozone-depleting chemistry is occurring near the ground or directly at the ice-atmosphere interface. The OOTI sleds, combined with instruments operated by other researchers on the Amundsen Icebreaker, have allowed them to measure ozone depletion processes. Ozone-depleting reactions are of further interest to these researchers because they have an impact on the fate of mercury (Hg) in the Arctic. Ozone controls the oxidation

FIGURE 13.20
Researchers from the Air Quality Research Division of Environment Canada employ a mobile lab-on-a-sled called OOTI to measure ozone and ozone-depleting chemicals in the Arctic environment.

state of the atmosphere. Once the ozone is gone, a new chemical regime is entered in which halogens emitted from the ocean take over this role. In an ozone-controlled atmosphere, the residence time of Hg vapour is quite long (on the order of a year). In a halogen-controlled regime, Hg is more efficiently converted to more chemically reactive forms, which can then enter the cryosphere and possibly the biosphere, with ensuing negative consequences.

The Montreal Protocol addressed ozone depletion

Scientists all over the world continue to address problems like these in an attempt to refine our understanding of the complex chemical process of ozone depletion. In the meantime, though, the international community has been unwilling to wait for catastrophe and has moved forward to strike an agreement to reduce and eventually eliminate the production of known ozone-depleting substances, also know as *ODS*.

In response to the scientific concerns, international policy efforts to restrict CFC production finally bore fruit in 1987 with the **Montreal Protocol**. In this treaty, signatory nations (eventually numbering 180) agreed to cut CFC production in half. Five follow-up agreements strengthened the pact by deepening the cuts, advancing timetables for compliance, and addressing related ozone-depleting chemicals. Today the production and use of ozone-depleting compounds has fallen by 95% since the late 1980s, and scientists can discern the beginnings of long-term recovery of the stratospheric ozone layer. Industry has been able to shift to alternative chemicals that have largely turned out to be cheaper and more efficient.

There are still challenges to overcome. Much of the 5 billion kilograms of CFCs emitted into the troposphere has yet to diffuse up into the stratosphere, and CFCs are slow to dissipate or break down. Thus, we can expect a considerable lag time between implementation of policy and the desired environmental effect. (This is one reason scientists often argue for proactive policy guided by the precautionary principle, rather than reactive policy that may respond too late.) Moreover, nations can plead for some ozone-depleting chemicals to be exempted from the ban; for example, the United States recently was allowed to continue using methyl bromide, a fumigant used to control pests on strawberries.

Despite the remaining challenges, the Montreal Protocol and its follow-up amendments are widely considered the biggest success story so far in addressing any global environmental problem. Environmental scientists have attributed this success primarily to two factors:

1. Policy makers engaged industry in helping to solve the problem, and government and industry worked together on developing replacement chemicals. This cooperation reduced the battles that typically erupt between environmentalists and industry.
2. Implementation of the Montreal Protocol after 1987 followed an adaptive management approach, altering strategies midstream in response to new scientific data, technological advances, or economic figures.

Because of its success in addressing ozone depletion, the Montreal Protocol is widely seen as a model for international cooperation in addressing other pressing global problems, such as persistent organic pollutants, climate change, and biodiversity loss.

Acidic deposition is another transboundary pollution problem

As discussed in "Central Case: The Rain and the Big Nickel," **acidic deposition** refers to the settling, or deposition, of acidic or acid-forming pollutants from the atmosphere onto Earth's surface. This can take place either by **acidic precipitation** (commonly referred to as **acid rain**, but also including acid snow, sleet, and hail), by fog, by gases, or by the deposition of dry particles. Acidic deposition is one type of **atmospheric deposition**, which refers more broadly to the wet or dry deposition on land of a wide variety of pollutants, including mercury, lead, nitrates, organochlorides, and others.

Acidic deposition originates primarily with the emission of sulphur dioxide and nitrogen oxides, largely through fossil fuel combustion by automobiles, electric utilities, and industrial facilities like the Inco and Falconbridge smelters in Sudbury. Once airborne, these primary pollutants can react with water, oxygen, and oxidants to produce secondary compounds of low pH, primarily sulphuric acid and nitric acid. Suspended in the troposphere, droplets of these acids may travel for days or weeks, sometimes covering hundreds or thousands of kilometres in a particular airshed before falling in precipitation (**FIGURE 13.21**).

Natural rainwater is not neutral; in fact, it is slightly acidic, with a typical pH of around 5.6. This is mainly because rainwater reacts with naturally occurring carbon dioxide in the air, forming carbonic acid. Rain and other forms of precipitation with pH less than about 5.1 are considered to be acidified. Acidification can occur as a result of natural processes, such as sulphur-rich volcanic eruptions, but the main cause of acid precipitation is human-generated air pollution.

Acidic deposition can have wide-ranging, cumulative detrimental effects on ecosystems and on our built environment (Table 13.1). Acids leach nutrients, such as calcium, magnesium, and potassium, from the topsoil, altering soil chemistry and harming plants and soil organisms. This occurs because hydrogen ions from acidic precipitation take the place of calcium, magnesium, and potassium ions in soil compounds, and these valuable nutrients leach into the subsoil, where they become inaccessible to plant roots.

Acidic precipitation also "mobilizes" toxic metal ions, such as aluminum, zinc, mercury, and copper, by chemically converting them from insoluble forms to soluble forms. Elevated soil concentrations of metals, such as aluminum, hinder water and nutrient uptake by plants. In some regions of Britain and the United States, acid fog with a pH of 2.3 (equivalent to vinegar) has enveloped forests for extended periods, leading to widespread tree mortality.

When acidic water runs off from land, it affects streams, rivers, and lakes. In fact, thousands of lakes in Canada, Scandinavia, the United States, and elsewhere now have lost their fish, because acid precipitation leaches aluminum out of soil and rock and into waterways in a

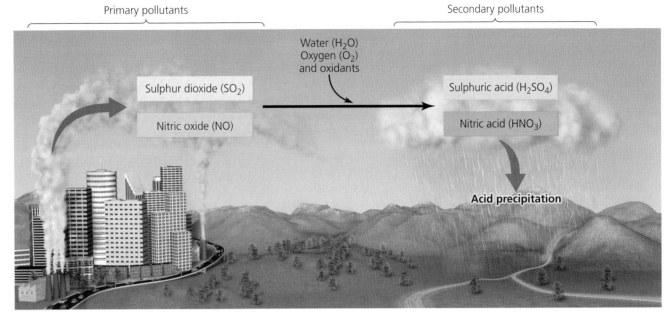

FIGURE 13.21
Acidic deposition can have consequences long distances downwind from its source. Sulphur dioxide and nitric oxide emitted by industries and utilities can be transformed into sulphuric acid and nitric acid through chemical reactions in the atmosphere. These acidic compounds then descend to Earth's surface in rain, snow, fog, and dry deposition.

Table 13.1 Effects of Acidic Deposition on Ecosystems in Northeastern North America
Acidic deposition in northeastern forests has
■ Accelerated leaching of base cations (ions that counteract acidic deposition) from soil
■ Allowed sulphur and nitrogen to accumulate in soil
■ Increased dissolved inorganic aluminum in soil, hindering plant uptake of water and nutrients
■ Caused calcium to leach from needles of red spruce, leading to tree mortality from wintertime freezing
■ Increased mortality of sugar maples because of leaching of base cations from soil and leaves
■ Acidified many lakes, especially those situated on soils and bedrock of granitic composition
■ Lowered lakes' capacity to neutralize further acids
■ Elevated aluminum levels in surface waters
■ Reduced species diversity and abundance of aquatic life, and negatively affected entire food webs

Source: Adapted from Driscoll, C.T., et al. 2001. Acid rain revisited. Hubbard Brook Research Foundation.

form that can be deadly to aquatic life. Aluminum can kill fish by damaging their gills and disrupting their salt balance, water balance, breathing, and circulation.

Besides altering natural ecosystems, acid precipitation also damages agricultural crops. Moreover, it erodes stone buildings, corrodes cars, and erases the writing from tombstones. Ancient cathedrals, monuments, temples, and stone statues in many parts of the world are experiencing irreversible damage as their features gradually wear away.

Because the pollutants leading to acid deposition can travel long distances, their effects may be felt far from their sources—a situation that has led to political bickering among the leaders of states and nations. For instance, much of the pollution from power plants and factories in Pennsylvania, Ohio, and Illinois falls out in southeastern Canada, as well as in states to the east, including New York, Vermont, and New Hampshire. The bedrock geology and soil chemistry of the area that is receiving the acidic deposition also plays a large role in the acid tolerance and the ecological response to acidification. As **FIGURE 13.22** shows, many regions in northeastern United States and southeastern Canada have experienced acid deposition in exceedance of their critical loads.

Acid deposition has not been reduced as much as scientists had hoped

Reducing acid precipitation involves reducing the pollution that contributes to it. New technologies, such as scrubbers, have helped. As a result of declining emissions of SO_2, average sulphate precipitation has decreased in northeastern North America since the early 1980s

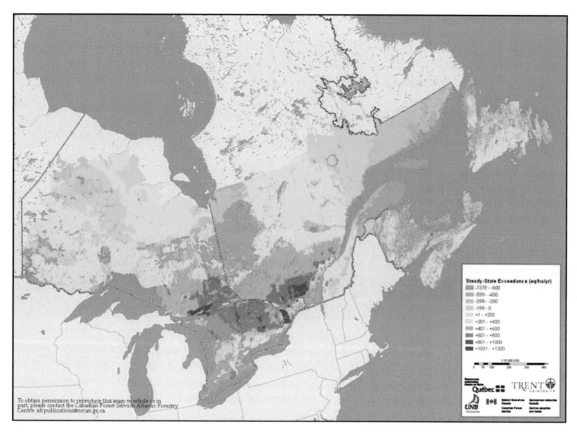

FIGURE 13.22
The maximum amount of acid deposition that a region can receive without damage to its ecosystems—its critical load—depends on the acid rain neutralizing capacity of water, rocks, and soils. This map shows areas of eastern Canada where the levels of acid deposition have exceeded the capacity of the soils to neutralize the acid. From Environment Canada, *Acid Rain and Forests*, www.ec.gc.ca/acidrain/images/Exceedance_E.jpg.

(**FIGURE 13.23A**). However, because of increasing NO_x emissions, average nitrate precipitation has changed little in the same period (**FIGURE 13.23B**). This may partly account for the trends in **FIGURE 13.24**, which suggest that although sulphate levels in many lakes in Ontario, Quebec, and Atlantic Canada have declined, acidification has continued or even deteriorated in some of the lakes.

A recent report by scientists at Hubbard Brook Experimental Forest has disputed the notion that the problem of acid deposition is being solved (see "The Science Behind the Story: Acid Rain at Hubbard Brook Research Forest"). Instead, the report said, the effects are worse than first predicted, and existing clean air legislation and limits on sulphate and nitrate emissions will be insufficient to solve the problem.

Indoor Air Pollution

Indoor air generally contains higher concentrations of pollutants than does outdoor air. As a result, the health effects from **indoor air pollution** in workplaces, schools, and homes outweigh those from outdoor air pollution. One estimate, from the U.N. Development Programme in 1998, attributed 2.2 million deaths worldwide to indoor air pollution and 500 000 deaths to outdoor air pollution. Indoor air pollution alone, then, takes roughly 6000 lives each day.

If the impact of indoor air pollution seems surprising, consider that the average person in North America is indoors at least 90% of the time. Then consider that in the past half-century a dizzying array of consumer products have been manufactured and sold, many of which we keep in our homes and offices and use extensively in our daily lives. Many of these products are made of synthetic materials, and, as we will see in Chapter 19, novel synthetic substances are not comprehensively tested for health effects before being brought to market. Products as diverse as insecticides, cleaning fluids, plastics, and chemically treated wood can all exude volatile chemicals into the air.

In an ironic twist, some attempts to be environmentally prudent during the "energy crisis" of 1973–1974 worsened indoor air pollution in developed countries. To reduce heat loss and improve energy efficiency, building managers sealed off most ventilation in existing buildings, and building designers constructed new buildings

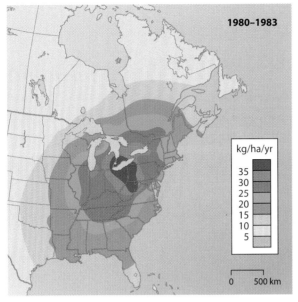
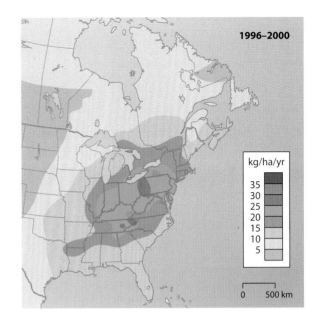

(a) Wet sulphate deposition down since early 1980s

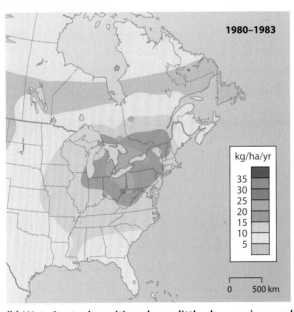
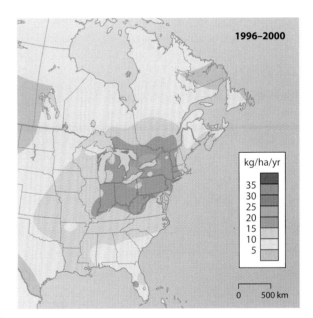

(b) Wet nitrate deposition shows little change since early 1980s

FIGURE 13.23
Acidic deposition of wet sulphates (a) and wet nitrates (b) has declined in recent years in eastern North America. From Canadian National Atmospheric Chemistry Database, Meteorological Service of Canada, Environment Canada.

with limited ventilation and with windows that did not open. These steps may have saved energy, but they also worsened indoor air pollution by trapping stable, unmixed air—and its pollutants—inside.

Indoor air pollution in the developing world arises from fuelwood burning

Indoor air pollution has the greatest impact in the developing world. Millions of people in developing nations burn wood, charcoal, animal dung, or crop waste inside their homes for cooking and heating with little or no ventilation (**FIGURE 13.25**). In the process, they inhale dangerous amounts of soot and carbon monoxide. In the air of such homes, concentrations of particulate matter are commonly 20 times as high as World Health Organization standards. Poverty forces fully half the population and 90% of rural residents of developing countries to heat and cook with indoor fires.

Indoor air pollution from fuelwood burning, the WHO estimates, kills 1.6 million people each year, causing more than 5% of all deaths in some developing nations

THE SCIENCE BEHIND THE STORY

Acid Rain at Hubbard Brook Research Forest

Dr. Gene E. Likens is with the Institute of Ecosystem Studies.

Acidic deposition involves subtle and incremental changes in pH levels that take place over long periods, so no single experiment can give us a complete picture of acidic deposition's effects. Nonetheless, one long-term study conducted in the Hubbard Brook Experimental Forest in New Hampshire's White Mountains has been critically important to our understanding of acidic deposition.

Established by the U.S. Forest Service in 1955, Hubbard Brook was initially devoted to research on hydrology, the study of water flow through forests and streams. In 1963, Hubbard Brook researchers broadened their focus to include a long-term study of nutrient cycling in forest ecosystems. Since then, they have collected and analyzed weekly samples of precipitation. The measurements make up the longest-running North American record of acid precipitation.

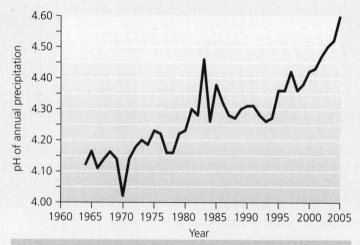

Over the past 40 years, precipitation at the Hubbard Brook Experimental Forest has become slightly less acidic. However, it is still far more acidic than natural precipitation. Data from Likens, G. E. 2004. Ecology 85:2355–2362.

Throughout Hubbard Brook's 3160 ha, small plastic collecting funnels channel precipitation into clean bottles, which researchers retrieve and replace each week. Hubbard Brook's laboratory measures acidity and conductivity, which indicates the amounts of salts and other electrolytic contaminants dissolved in the water. Concentrations of sulphuric acid, nitrates, ammonia, and other compounds are measured elsewhere.

By the late 1960s, ecologists Gene Likens, F. Herbert Bormann, and others had found that precipitation at Hubbard

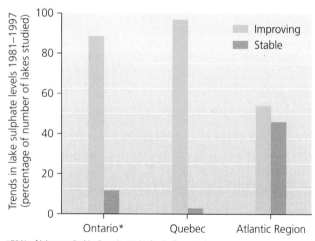

*73% of lakes studied in Ontario are in the Sudbury region.

(a) Trends in lake sulphate levels

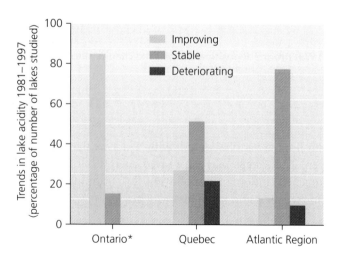

(b) Trends in lake acidity

FIGURE 13.24
Although sulphate levels in many lakes in Ontario, Quebec, and Atlantic Canada have shown improvement (a), acid conditions in many of the lakes have not improved, and some have even continued to deteriorate (b). Source: State of the Environment Infobase Acid Rain, Environment Canada, Ontario, Atlantic, and Quebec Regions, adapted by National Indicators and Reporting Office, www.ec.gc.ca/soer-ree/English/Indicator_series/new_issues.cfm?issue_id=3&tech_id=10#bio_pic.

Brook was several hundred times as acidic as natural rainwater. By the early 1970s, a number of other studies had corroborated their findings. Together, these studies indicated that the precipitation had pH values averaging around 4 and that individual rainstorms showed values as low as 2.1—almost 10 000 times as acidic as ordinary rainwater.

The most severe problems were found to be in the northeastern United States, where prevailing west-to-east winds were blowing emissions from fossil fuel-burning power plants in the Midwest. Scientists hypothesized that when sulphur dioxide, nitrogen oxides, and other pollutants arrived in the Northeast, they were absorbed by water droplets in clouds, converted to acidic compounds, such as sulphuric acid, and deposited on farms, forests, and cities in rain or snow.

Clean air legislation has helped reduce acidic deposition. At an area of Hubbard Brook known as Watershed 6, average pH increased slightly between 1965 and 1995, from about 4.15 to about 4.35 (see the first figure). Nonetheless, acidic deposition continues to be a serious problem in the northeastern United States and southeastern Canada.

Some long-term consequences of acidic deposition are now becoming clear. In 1996, researchers reported that approximately 50% of the calcium and magnesium in Hubbard Brook's soils had leached out. Meanwhile, acidic deposition

Acidic deposition killed these trees on Mount Mitchell in western North Carolina.

had increased the concentration of aluminum in the soil, which can prevent tree roots from absorbing nutrients. The resulting nutrient deficiency slows forest growth and weakens trees, making them more vulnerable to drought and insects (see the second figure). It also reduces the ability of soil and water to neutralize acidity, making the ecosystem increasingly vulnerable to further inputs of acid.

In October 1999, researchers used a helicopter to distribute 50 tons of a calcium-containing mineral called wollastonite over one of Hubbard Brook's watersheds. Their objective was to raise the concentration of base cations to estimated historical levels. Over the next 50 years, scientists plan to evaluate the impact of calcium addition on the watershed's soil, water, and life. By providing a comparison with watersheds in which calcium remains depleted, the results should provide new insights into the consequences of acid rain and the possibilities for reversing its negative effects.

FIGURE 13.25

In the developing world, many people build fires inside their homes for cooking and heating, as seen here in a South African kitchen. Indoor fires expose family members to particulate matter and carbon monoxide. In most regions of the developing world, indoor air pollution is estimated to cause upward of 3% of all health risks.

and 2.7% of the entire global disease burden. Many people who tend indoor fires are not aware of the health risks. They do not have access to the statistics showing that chemicals and soot released by burning coal, plastic, and other materials indoors can increase risks of pneumonia, bronchitis, allergies, sinus infections, cataracts, asthma, emphysema, heart disease, cancer, and premature death. Many who are aware of the health risks are too poor to have viable alternatives.

Even in the developed world, recognizing indoor air pollution as a problem is still quite novel. Fortunately, scientists have identified the most deadly indoor threats. Particulate matter and chemicals from wood and charcoal smoke are the primary health risks in the developing world. In developed nations, the top risks are cigarette smoke and radon, a naturally occurring radioactive gas.

Tobacco smoke and radon are the most dangerous indoor pollutants in the developed world

The health effects of smoking cigarettes are well known, but only recently have scientists quantified the risks of inhaling secondhand smoke. *Secondhand smoke*, or environmental tobacco smoke, is smoke inhaled by a nonsmoker who is nearby or shares an enclosed airspace with a smoker. Secondhand smoke has been found to cause many of the same problems as directly inhaled cigarette smoke, ranging from irritation of the eyes, nose, and throat, to exacerbation of asthma and other respiratory ailments, to lung cancer. This hardly seems surprising when one considers that environmental tobacco smoke consists of a brew of more than 4000 chemical compounds, many of which are known or suspected to be toxic or carcinogenic.

Although smoking remains common in many parts of the developing world, its popularity has declined greatly in developed nations in recent years. Many public and private venues now ban smoking. In Canada, smoking is now prohibited in all indoor workplaces and public places, such as airports. The last of the provinces or territories to implement the ban was Yukon, in 2008.

After cigarette smoke, radon gas is the second-leading cause of lung cancer in the developed world. The WHO estimates that radon may account for 15% of lung cancer cases worldwide. As we will see in Chapter 19, radon is a radioactive gas resulting from the natural decay of uranium in soil, rock, or water, which seeps up from the ground and can infiltrate buildings. Radon is colourless and odourless, and it can be impossible to predict where it will occur without knowing details of an area's underlying geology (**FIGURE 13.26**). As a result, the only way to determine whether radon is entering a building is to measure radon with a test kit.

Indoor radon is taken very seriously in the United States, where it is difficult to purchase a home without tests showing that radon is not a problem. The U.S. Environmental Protection Agency has recommended a maximum acceptable level of radon that is essentially the same as the ambient outdoor level. This has brought thousands of houses into the unacceptable range; the EPA has estimated that 6% of U.S. homes exceed the maximum recommended level.

Historically, Health Canada has taken the approach that some indoor concentration of naturally occurring soil gases, such as radon, is unavoidable. The maximum acceptable, or "actionable," level for indoor radon in homes and other non-occupational settings recommended by Health Canada is 0.02 WL (a Working Level is a unit of measurement that is specific to radon and its daughter by-products), which is roughly equivalent to 200 Bq/m^3 air (a Becquerel is a unit that measures the number of radioactive disintegrations per second), or 5.4 pCi/L air (picoCurie also measures the number of radioactive disintegrations per second). This is the same as the maximum permissible annual average concentration of radon daughters from the operation of a nuclear facility. The actionable level defined by the U.S. EPA is equivalent to 0.016 WL (about 4 pCi/L, or 150 Bq/m^3 of air).

Many VOCs pollute indoor air

In our daily lives at home, we are exposed to many indoor air pollutants (**FIGURE 13.27**). The most diverse indoor pollutants are volatile organic compounds (VOCs). These airborne carbon-containing compounds are released by everything from plastics to oils to perfumes to paints to cleaning fluids to adhesives to pesticides. VOCs evaporate from furnishings, building materials, colour film, carpets, laser printers, fax machines, and sheets of paper. Some products, such as chemically treated furniture, release large amounts of VOCs when new and progressively less as they age. Other items, such as photocopying machines, emit VOCs each time they are used. Seemingly innocent indoor activities, like cooking, can even contribute pollutants; dioxins and furans, the toxic by-products of incomplete combustion that are problematic outdoor pollutants, can come from burning foods during meal preparation.

Although we are surrounded by products that emit VOCs, they are released in very small amounts. Studies have found overall levels of VOCs in buildings nearly

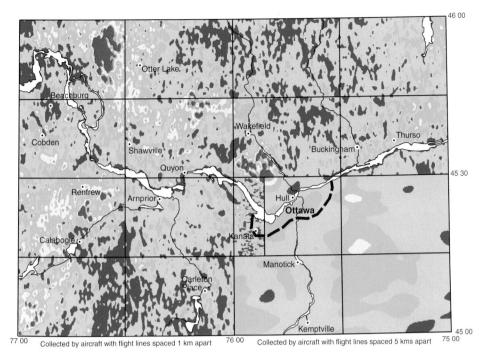

FIGURE 13.26
One's risk from radon depends largely on underground geology. This map shows how the potential for elevated levels of radon can be estimated on the basis of the amount of naturally occurring uranium in rocks and soils. However, there is much fine-scale geographic variation from place to place, and construction materials and techniques also have an impact on the processes that can concentrate radon indoors. Testing is the only sure means of determining whether this colourless, odourless gas could be a problem in your home.
Source: Natural Resources Canada, www.geoscape.nrcan.gc.ca/ottawa/radon_e.php?p=1.

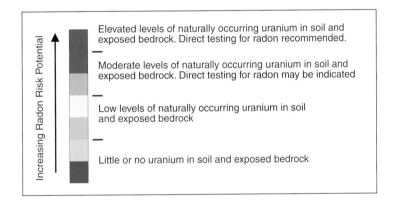

always to be less than 0.1 parts per million. This is, however, a substantially greater concentration than is generally found outdoors, and indoor sources are suspected for most of these compounds.

The implications for human health of chronic exposure to VOCs are far from clear. Because they exist in such low concentrations and because individuals are regularly exposed to mixtures of many different types, it is extremely difficult to study the effects of any one pollutant. An exception is formaldehyde, which does have clear and known health impacts. This VOC, one of the most common synthetically produced chemicals, irritates mucous membranes, induces skin allergies, and causes other ailments. Formaldehyde is used in numerous products, but health complaints have mainly resulted from its leakage from pressed wood and insulation. The use of plywood has decreased in the last decade because of health concerns over formaldehyde.

VOCs also include pesticides, which we examine in Chapters 8 and 19. Three-quarters of U.S. homes use at least one pesticide indoors during an average year, but most are used outdoors. Thus it may seem surprising that the U.S. Environmental Protection Agency found in a 1990 study that 90% of people's pesticide exposure came from indoor sources. Households that the agency tested had multiple pesticide volatiles in their air, at levels 10 times above levels measured outside. Some of the pesticides had apparently been used years earlier against termites and then seeped into the houses through floors and walls. DDT, banned 15 years before the study, was found in five of eight homes, probably having been brought in from outdoors on the soles of occupants' shoes.

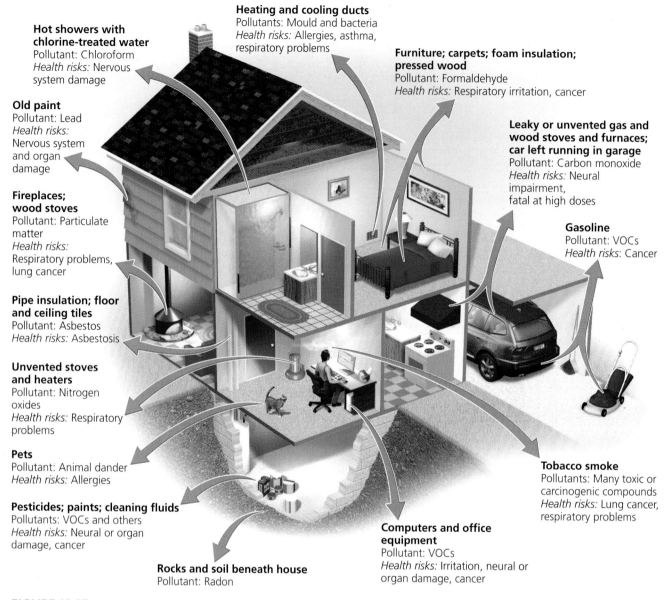

FIGURE 13.27
The typical North American home contains a variety of potential sources of indoor air pollution. Shown are some of the most common sources, the major pollutants they emit, and some of the health risks they pose.

Living organisms can pollute indoor spaces

Tiny living organisms can pollute. In fact, they may be the most widespread source of indoor air pollution in the developed world. Dust mites and animal dander can worsen asthma in children. Some fungi, mould, and mildew (in particular, their airborne spores) can cause severe health problems, including allergies, asthma, and other respiratory ailments. Some airborne bacteria can cause infectious disease. One example is the bacterium that causes Legionnaires' disease. Heating and cooling systems in buildings make ideal breeding grounds for microbes, providing moisture, dust, and foam insulation as substrates, as

HOW SAFE IS YOUR INDOOR ENVIRONMENT?

Think about the amount of time you spend indoors. Name the potential indoor air quality hazards in your home, work, or school environment. Are these spaces well ventilated? What could you do to make the indoor spaces you use safer?

well as air currents to carry the organisms aloft. Microbes that induce allergic responses are thought to be a major

CANADIAN ENVIRONMENTAL PERSPECTIVES

David Phillips

David Phillips has become a spokesperson on climatic change.

- Senior **climatologist** for Environment Canada
- **Meteorologist, weather researcher**, and Canada's "weather ambassador"
- Author and **science writer**

We all know his face and his voice. He is the go-to person for answers to any and all questions about the weather in Canada. He has a fascination with extreme weather trivia and especially for the role that weather has played in shaping the Canadian psyche. David Phillips is the official spokesperson for Environment Canada's Meteorological Service and Canada's unofficial "weather ambassador."

As a senior climatologist, Phillips is responsible for activities related to the study, promotion, and understanding of Canadian weather and climate.[12] Phillips claims that his interest in the weather wasn't sparked by any significant weather event, like a dramatic thunderstorm or lightning display; he says he grew up, like everyone else, as "just someone who hated and loved the weather."[13]

Phillips started as a researcher at Environment Canada, working on a Climatic Severity Index for Canadians. The study assessed 150 locations in Canada to determine where the worst weather occurs. The index ranges from 1 to 100, with 1 being the least severe and 100 the most severe. It combines weather-related stressors, such as extremes of heat and cold, wetness or dryness, windiness, poor air quality, darkness, fog, restricted visibility, lightning, thunderstorms, blowing snow, and freezing precipitation. It also accounts for factors that are responsible for human discomfort caused by the weather, such as mobility limitations, psychological factors, safety and hazardous conditions, and discomfort factors, such as wind chill.

When Phillips was interviewed on national TV about this study, his career as a spokesperson got its start. Phillips has been a voice of reason in the sometimes confusing public discourse about global climatic change. He often reminds people that just because we might be having an unusually hot (or cold) day, month, season, year, or even decade, it doesn't necessarily mean that this is an indicator of long-term climatic change.

"The atmosphere has ... a beauty and fragility that requires us to better understand, enjoy, and protect it." —David Phillips

Thinking About Environmental Perspectives

1. When he first started working at Environment Canada, David Phillips was a researcher working on the Climatic Severity Index. That was back in the 1980s. Today there are numerous versions of such weather-, climate-, and atmosphere-related indexes. Examples include the Air Quality Index, the Palmer Drought Severity Index, and the Winter Severity Index, but there are many others. Choose one example of a weather-related index and find out (1) what are the components of the index (in other words, what weather-related factors are accounted for in the index, and how are they combined and weighted), and (2) how this index has been used in Canada in the past (give some examples of extreme events to which the index has been applied).
2. David Phillips is often seen on television during extreme weather events, calmly reminding people that a single extreme weather event (or season) does not mean that the effects of climate change are necessarily upon us. How *do* scientists tell whether an extreme weather event—take Hurricane Katrina, for example—is actually evidence of longer-term climatic change? (We will discuss this question in greater detail in Chapter 14.)

---

cause of building-related illness, a sickness produced by indoor pollution. When the cause of such an illness is a mystery, and when symptoms are general and nonspecific, the illness is often called **sick-building syndrome**.

We can reduce indoor air pollution

Using low-toxicity material, monitoring air quality, keeping rooms clean, and providing adequate ventilation are the keys to alleviating indoor air pollution in most situations. In the developed world, we can try to limit our use of plastics and treated wood where possible and to limit our exposure to pesticides, cleaning fluids, and other known toxicants by keeping them in a garage or outdoor shed rather than in the house. Health Canada recommends that we test our homes and offices for mould and radon, and monitor continuously for carbon monoxide. Because carbon monoxide is so deadly and so hard to detect, many homes are equipped with detectors that sound an alarm if incomplete combustion produces dangerous levels of CO. In addition, keeping rooms and air ducts clean and free of mildew and other biological pollutants will reduce potential irritants and allergens. Finally, it is important to keep our indoor spaces as well ventilated as possible to minimize concentrations of the pollutants among which we live.

Remedies for fuelwood pollution in the developing world include drying wood before burning (which reduces the amount of smoke produced), cooking outside, shifting to less-polluting fuels (such as natural gas), and replacing inefficient fires with cleaner stoves that burn fuel more efficiently. For example, the Chinese government has invested in a program that has placed more fuel-efficient stoves in millions of homes in China. Installing hoods, chimneys, or cooking windows can increase ventilation for little cost, alleviating the majority of indoor smoke pollution.

Conclusion

Indoor air pollution is a potentially serious health threat. However, by keeping informed of the latest scientific findings and taking appropriate precautions, we as individuals can significantly minimize the risks to our families and ourselves. Outdoor air pollution has been addressed more effectively by government legislation and regulation. In fact, reductions in outdoor air pollution in Canada, the United States, Great Britain, and other developed nations represent some of the greatest strides made in environmental protection to date. Much room for improvement remains, however, particularly in reducing acidic deposition and the photochemical smog that results from urban congestion. Avoiding unhealthy pollutant levels in the developing world will continue to pose a challenge as less-wealthy nations industrialize.

REVIEWING OBJECTIVES

You should now be able to:

Describe the composition, structure, and function of Earth's atmosphere

- The atmosphere consists of 78% nitrogen gas, 21% oxygen gas, and a variety of other gases in minute concentrations.
- The atmosphere includes four principal layers: the troposphere, stratosphere, mesosphere, and thermosphere. Temperature and other characteristics vary across these layers. Ozone is concentrated in the stratosphere.
- The Sun's energy heats the atmosphere, drives air circulation, and helps determine weather, climate, and the seasons.
- Weather is a short-term phenomenon, whereas climate is a long-term phenomenon. Fronts, pressure systems, and the interactions among air masses influence weather.
- Global convective cells called Hadley, Ferrel, and polar cells create latitudinal climate zones.

Outline the scope of outdoor air pollution and assess potential solutions

- Natural sources, such as windblown dust, volcanoes, and fires, account for much atmospheric pollution, but human activity can worsen some of these phenomena.
- Human-emitted pollutants include primary and secondary pollutants from point and non-point sources.
- To safeguard public health, Environment Canada and various other federal, provincial, and territorial agencies monitor a number of air contaminants, including sulphur oxides, nitrogen oxides, particulate matter, volatile organic compounds, carbon monoxide, ammonia, and tropospheric ozone.

- The principal legislation under which air quality and emissions are regulated in Canada is the *Canadian Environmental Protection Act* (1999).
- Industrial smog is produced by fossil fuel combustion and is still a problem in urban and industrial areas of many developing nations.
- Photochemical smog is created by chemical reactions of pollutants in the presence of sunlight. It impairs visibility and human health in urban areas.
- Rural areas suffer air pollution from feedlots and other sources.
- Industrializing nations, such as China and India, are experiencing some of the world's worst air pollution today.

Explain stratospheric ozone depletion and identify steps taken to address it

- CFCs destroy stratospheric ozone, and thinning ozone concentrations pose dangers to life because they allow more ultraviolet radiation to reach Earth's surface.
- The Montreal Protocol and its follow-up agreements have proven remarkably successful in reducing emissions of ozone-depleting compounds.
- The long residence time of CFCs in the atmosphere means a time lag between the protocol and the actual restoration of stratospheric ozone.

Define acidic deposition and illustrate its consequences

- Acidic deposition results when pollutants, such as SO_2 and NO_x react with water in the atmosphere to produce strong acids that are deposited on Earth's surface.
- Acidic deposition may occur a long distance from the source of pollution.

- Water bodies, soils, trees, animals, and ecosystems all experience negative impacts from acidic deposition.

Characterize the scope of indoor air pollution and assess potential solutions

- Indoor air pollution causes far more deaths and health problems worldwide than outdoor air pollution.
- Indoor burning of fuelwood is the developing world's primary indoor air pollution risk.
- Tobacco smoke and radon are the deadliest indoor pollutants in the developed world.
- Volatile organic compounds and living organisms can pollute indoor air.
- Using low-toxicity building materials, keeping spaces clean, monitoring air quality, and maximizing ventilation are some of the steps we can take to reduce indoor air pollution.

TESTING YOUR COMPREHENSION

1. About how thick is Earth's atmosphere? Name one characteristic of each of the four atmospheric layers.
2. Where is the "ozone layer" located? How and why is stratospheric ozone beneficial for people and tropospheric ozone harmful?
3. How does solar energy influence weather and climate? How do Hadley, Ferrel, and polar cells help to determine long-term climatic patterns and the location of biomes?
4. Describe a thermal inversion. How can thermal inversions contribute to the severity of pollution episodes?
5. Name three natural sources of outdoor air pollution and three sources caused by human activity.
6. What is the difference between a primary and a secondary pollutant? Give an example of each.
7. What is smog? How is smog formation influenced by the weather? By topography? How does photochemical smog differ from industrial smog?
8. How do chlorofluorocarbons (CFCs) deplete stratospheric ozone? Why is this depletion considered a long-term international problem? What has been done to address this problem?
9. Why are the effects of acidic deposition often felt in areas far from where the primary pollutants are produced? List three impacts of acidic deposition.
10. Name five common sources of indoor pollution. For each, describe one way to reduce one's exposure to this source.

SEEKING SOLUTIONS

1. Describe several factors that make it difficult to determine the specific causes of air pollution and to develop solutions. How did London try to improve its air quality? Are there any cities near you that have tried different approaches to mitigating air pollution?
2. Name one type of natural air pollution, and discuss how human activity can sometimes worsen it. What potential solutions can you think of to minimize this human impact?
3. Describe how and why emissions of some major pollutants have been reduced in North America since the 1970s, despite increases in population and economic activity.
4. International regulatory action has produced reductions in CFCs, but other transboundary pollution issues, including acidic deposition, have not yet been addressed as effectively. What types of actions do you feel are appropriate for pollutants that cross political boundaries?
5. **THINK IT THROUGH** You have just been elected mayor of the largest city in your province. Your city's residents are complaining about photochemical smog and traffic congestion. Traffic engineers and city planners project that population and traffic will grow by 20% in the next decade. A citizens' group is urging you to implement a congestion-charging program like London's, but businesses are fearful of losing money if shoppers are discouraged from visiting. Consider the particulars of your city, and then decide whether you will pursue a congestion-charging program. If so, how would you do it? If not, why not, and what other steps would you take to address your city's problems?

6. **THINK IT THROUGH** You have just become the head of your region's office of public health and the environment ministry has informed you that your jurisdiction has failed to meet recommended air quality standards for ozone, sulphur dioxide, and nitrogen dioxide. The area is partly rural but is home to a city of 200 000 people, with sprawling suburbs. There are several large and aging coal-fired power plants, a number of factories with advanced pollution control technology, and no public transportation system. What steps would you urge the municipal or provincial government to take to meet the air quality standards? Explain how you would prioritize these steps.

INTERPRETING GRAPHS AND DATA

Visit the NAPS website (www.etc-cte.ec.gc.ca/napsstations/main.aspx). Use the interactive mapping tool to locate and investigate the Air Quality Station closest to your home.

1. Which pollutants are monitored at this station? Is it a NAPS-designated station, or is it a provincial, territorial, or municipal monitoring site?
2. Use the graphing tool to produce a graph showing the trends in several different air pollutants at this station over the past five years. Do you notice any interesting trends?
3. Over the same period, were there any pollutants that exceeded the acceptable levels? Which pollutants? You can determine this by choosing to plot the results from that station against the National Air Quality Objectives, NAQO, and you will obtain a graph that shows the number of *exceedances*, that is, the number of times that pollutant exceeded the desirable or acceptable level at that station. Did exceedances happen on a regular basis? The graph here is an example of such a plot, based on data from the Air Quality Station at Vickers Road and East 18th Street in Hamilton, Ontario, for ground-level ozone in the year 2005. How many times did ozone exceed the maximum desirable level at this station in 2005? How many times did it exceed the maximum acceptable level?
4. Now select an Air Quality Station in a location that has quite different characteristics from where you live—if you live in the far north, choose a station in an urban or industrial or agricultural area; if you live in downtown Toronto or Windsor or Hamilton, choose a station in rural Saskatchewan or Prince Edward Island. Compare your findings for the two stations.

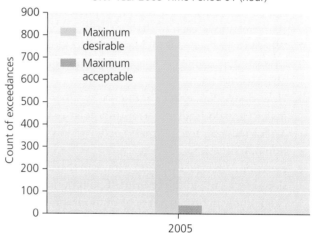

This bar graph shows how many times the maximum desirable and maximum acceptable levels of ozone were exceeded at the air quality station located at Vickers Road and East 18th Street in Hamilton, Ontario. Source: Environment Canada National Air Pollution Surveillance Network, www.etc-cte.ec.gc.ca/napsstations/main.aspx.

CALCULATING FOOTPRINTS

Environment Canada provides online calculators that allow you to determine the quantities of air pollutants that you, personally, produce as a result of various activities, including driving a car, mowing a lawn, or using a snowblower. The pollution calculators are accessible through the Environment Canada website, at www.etc-cte.ec.gc.ca/databases/fuelcalc_e.html.

Use the calculators to investigate lawn mowing, and answer the following questions:

1. What are the main types of emissions associated with mowing your lawn?
2. Are there differences among different models of lawnmowers?
3. If you can mow a lawn of 25 m² in 10 minutes, how big is a lawn that would take 30 minutes to mow? How big is the "60-minute lawn"? How do the quantities of emissions change if you are mowing the 10-minute lawn, as opposed to the 30-minute or 60-minute lawn? Does your house have a lawn? Is it a 10-minute, 30-minute, or 60-minute lawn?
4. How do the quantities of emissions change if the lawnmower is one year old, as opposed to two, five, or 10 years old? Calculate the emissions over a summer of mowing a 30-minute lawn 10 times with a relatively new lawnmower, as opposed to mowing the same lawn 10 times with a 10-year-old mower.

TAKE IT FURTHER

 Go to www.myenvironmentplace.ca where you will find

- Suggested answers to end-of-chapter questions
- Quizzes, animations, and flashcards to help you study
- *Research Navigator*™ database of credible and reliable sources to assist you with your research projects
- Tutorials to help you master how to interpret graphs
- Current news articles that link the topics that you study to case studies from your region and around the world

- **ECO Occupational Profiles:** If you found this chapter especially interesting, you might want to learn more about the following jobs by visiting the Occupational Profiles website of the Environmental Careers Organization. Go to www.eco.ca and check out the following careers:
    - Air quality engineer
    - Air quality specialist
    - Air quality technician
    - Environmental monitoring technician
    - Environmental technician
    - Pollution control technologist

CHAPTER ENDNOTES

1. Environment Canada Green Lane, *Acid Rain and ... Case Studies*, www.ec.gc.ca/acidrain/acidcase.html.
2. Environment Canada Green Lane, *Acid Rain and ... Case Studies*, www.ec.gc.ca/acidrain/acidcase.html.
3. Environment Canada National Water Research Institute, www.nwri.ca/sande/jul_aug_2001-e.html.
4. Department of Fisheries and Oceans, *Experimental Lakes Area*, www.dfo-mpo.gc.ca/regions/central/pub/ela-rle/index_e.htm.
5. This section partially based on information from Environment Canada, *Clean Air Online*, www.ec.gc.ca/cleanair-airpur/Pollutants-WSBCC0B44A-1_En.htm.
6. Centers for Disease Control, Agency for Toxic Substances and Disease Registry, *Division of Toxicology ToxFAQs*, September 2002, www.atsdr.cdc.gov/tfacts35.pdf.
7. UNEP, *Phasing Lead Out of Gasoline: An Examination of Policy Approaches in Different Countries*, 1999.
8. Environment Canada, *CEPA Environmental Registry*, www.ec.gc.ca/CEPARegistry/the_act/.
9. Environment Canada, *Clean Air Online*, www.ec.gc.ca/cleanair-airpur/Taking_Action/Canadian_Governments-WS3067CF5B-1_En.htm.
10. Environment Canada, Clean Air Regulatory Agenda, *Regulatory Framework for Industrial Air Emissions*, www.ec.gc.ca/doc/media/m_124/ppt/tech_eng.htm.
11. Environment Canada, Clean Air Regulatory Agenda, *Regulatory Framework for Industrial Air Emissions*, www.ec.gc.ca/doc/media/m_124/ppt/tech_eng.htm.
12. Meteorological Service of Canada, Environment Canada, David Phillips www.msc-smc.ec.gc.ca/cd/biographies/david_phillips_e.html.
13. Environment Canada, Canada's Weather Guru: A Chat With David Phillips, *Envirozine*, Issue 78, www.ec.gc.ca/EnviroZine/english/issues/78/feature_1_e.cfm.

Global Climate Change

14

This is Muratti Island, one of the low-lying islands of the Maldives.

Upon successfully completing this chapter, you will be able to

- Describe Earth's climate system and explain the many factors influencing global climate
- Characterize human influences on the atmosphere and global climate
- Summarize modern methods of climate research
- Outline current and future trends and impacts of global climate change
- Suggest ways we can respond to climate change

Fishers of the Maldives ply the waters near their island home.

CENTRAL CASE:
RISING SEAS MAY FLOOD THE MALDIVES

"The impact of global warming and climate change can effectively kill us off, make us refugees ..."
—ISMAIL SHAFEEU, MINISTER OF ENVIRONMENT, MALDIVES, 2000

"We're in a giant car heading towards a brick wall and everyone's arguing over where they're going to sit."
—DAVID SUZUKI

A nation of low-lying islands in the Indian Ocean, the Maldives is known for its spectacular tropical setting, colourful coral reefs, and sun-drenched beaches. For visiting tourists it seems to be paradise, and for 370 000 residents of the Maldives it is home. But residents and tourists alike now fear that the Maldives could soon be submerged by the rising seas that are accompanying global climate change.

Nearly 80% of the Maldives' land area lies less than 1 m above sea level. In a nation of 1200 islands whose highest point is just 2.4 m above sea level, rising seas are a matter of life or death. The world's oceans rose 10–20 cm during the twentieth century as warming temperatures expanded ocean water and as melting polar ice discharged water into the ocean. According to current projections, sea level will rise another 18–59 cm by the year 2100.

Higher seas are expected to flood large areas of the Maldives and cause salt water to contaminate drinking water supplies. Storms intensified by warmer water temperatures will erode beaches, cause flooding, and damage the coral reefs that are so vital to the tourism and fishing industries that drive the nation's economy. Because of such concerns, the Maldivian government recently evacuated residents from several of the lowest-lying islands.

On December 26, 2004, the nation got a taste of what could be in store in the future, when a massive *tsunami*, or *seismic sea wave*, devastated coastal areas throughout the Indian Ocean. The tsunami killed 100 Maldivians and left 20 000 homeless. Schools, boats, tourist resorts, hospitals, and transportation and communication infrastructure were damaged or destroyed. The World Bank estimates that direct damage in the Maldives totalled $470 million, an astounding 62% of the nation's gross domestic product (GDP). Soil erosion, saltwater contamination of aquifers, and other environmental damage will result in still greater long-term economic losses.

The 2004 tsunami was caused *not* by climate change, but by an earthquake. Yet as sea levels rise and coastal ecosystems are compromised, the damage that such natural events—or ordinary storm waves—can inflict increases considerably. Maldives islanders are not alone in their predicament. Other island nations, from the Galápagos to Fiji to the Seychelles, also fear a future of constant vigilance against encroaching seawater. Not surprisingly, island nations were the first among the developing nations of the world to organize themselves to make their position on climate change known to the global community, through *AOSIS*, the *Association of Small Island States*. But mainland coastal areas around the world—from the low-lying marshes of the St. Lawrence estuary to the hurricane-battered coast of Louisiana to the cyclone-ravaged shores of Burma—will face similar challenges. One way or another, global climate change will affect each and every one of us.

Our Dynamic Climate

Climate influences virtually everything around us, from the day's weather to major storms, from crop success to human health, from national security to the ecosystems that support our economies. If you are a student in your twenties, climate change may well be *the* major event of your lifetime and the phenomenon that most shapes your future.

The 2007 release of the *Fourth Assessment Report* of the Intergovernmental Panel on Climate Change (IPCC) made clear to the world the scientific consensus that climate is changing, that we are the cause, and that climate change is already exerting impacts that will become increasingly severe if we do not take action. In 2006–2007, millions of people watched former U.S. vice-president Al Gore's Oscar-winning film on climate change, *An Inconvenient Truth*, or read his best-selling book of the same title. In 2007 thousands of students and citizens staged rallies around the world, calling for action against climate change.

Climate change is the fastest-moving area of environmental science today. New scientific papers that refine our understanding of climate are published every week, and policy makers and businesspeople make headlines with decisions and announcements just as quickly. By the time you read this chapter, new issues will have developed, and new problems and solutions may have been found. We urge you to explore further, with your instructor and on your own, the most recent information on climate change and the impacts it will have on your future.

What is climate change?

Climate describes an area's long-term atmospheric conditions, including temperature, moisture content, wind, precipitation, barometric pressure, solar radiation, and other characteristics. As you have learned, *climate* differs from *weather* in that weather specifies conditions at localized sites over hours or days, whereas climate describes conditions across broader regions over seasons, years, or millennia. **Global climate change** describes trends and variations in Earth's climate, involving such aspects as temperature, precipitation, and storm frequency and intensity. People often use the term **global warming** synonymously in casual conversation, but global warming refers specifically to an increase in Earth's average surface temperature; it is thus only one aspect of global climate change.

Our planet's climate varies naturally through time, but the climatic changes taking place today are unfolding at an exceedingly rapid rate. Moreover, scientists agree that human activities, notably fossil fuel combustion and deforestation, are largely responsible. Understanding how and why climate is changing requires understanding how our planet's climate functions. Thus, we first will survey the fundamentals of Earth's climate system—a complex and finely tuned system that has nurtured life for billions of years.

The Sun and atmosphere keep Earth warm

Three factors exert more influence on Earth's climate than all others combined. The first is the Sun; without it, Earth would be dark and frozen. The second is the atmosphere; without it, Earth would be as much as 33°C colder on average, and temperature differences between night and day would be far greater than they are. The third is the oceans, which shape climate by storing and transporting heat and moisture.

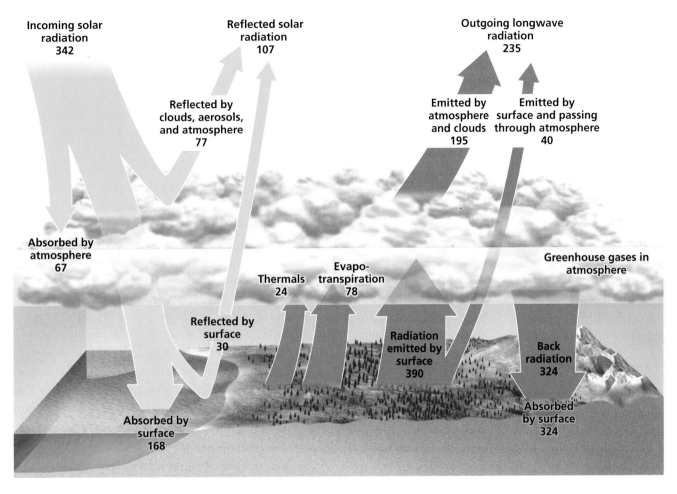

FIGURE 14.1
Our planet absorbs nearly 70% of the solar radiation it receives from the Sun and reflects the rest back into space (yellow arrows). About 30% is reflected off of the top of the atmosphere, clouds, and the surfaces of icecaps, oceans, and land. Absorbed radiation is then reemitted (orange arrows) as heat. Greenhouse gases in the atmosphere absorb some of this radiation and reemit it again, sending some downward to warm the atmosphere and the surface through the greenhouse effect. This illustration, showing major pathways of energy flow in watts per square metre, indicates that our planet naturally emits and reflects 342 watts/m², the same amount it receives from the Sun. Arrow thicknesses in the diagram are proportional to flows of energy in each pathway. Data from Kiehl, J. T., and K. E. Trenberth. 1997. Earth's annual global mean energy budget. *Bulletin of the American Meteorological Society* 78:197–208.

The Sun supplies most of our planet's energy. Earth's atmosphere, clouds, land, ice, and water together absorb about 70% of incoming solar radiation and reflect the remaining 30% back into space (**FIGURE 14.1**); the reflectivity of a surface is called **albedo**. The 70% that is absorbed into the system powers a wide variety of Earth's processes, from photosynthesis to winds, waves, and evaporation.

Greenhouse gases warm the lower atmosphere

As Earth's surface absorbs the incoming short-wavelength solar radiation, surface materials increase in temperature and emit **infrared radiation**, radiation with longer wavelengths than visible light. In this longer-wavelength form, the radiation emitted by Earth's surface begins to make its way back to outer space.

However, some gases that are naturally present in the lower part of the atmosphere (the *troposphere*) absorb this infrared radiation very effectively. These include water vapour, ozone (O_3), carbon dioxide (CO_2), nitrous oxide (N_2O), and methane (CH_4), as well as halocarbons, a diverse group that includes chlorofluorocarbons (CFCs). Such gases are known as **greenhouse gases (GHGs)** or, technically, as **radiatively active gases**. After absorbing radiation emitted from the surface, greenhouse gases subsequently reemit infrared energy of slightly different wavelengths. Some of this reemitted energy is lost to space, but some travels back downward, warming the troposphere and the planet's surface in a phenomenon known as the **greenhouse effect**.

The greenhouse effect is a *natural* phenomenon, and greenhouse gases have been present in our atmosphere for all of Earth's history. It's a good thing, too; without the natural greenhouse effect, our planet would have a

Table 14.1	Global Warming Potentials of Four Greenhouse Gases
Greenhouse gas	**Relative heat-trapping ability (in CO_2 equivalents)**
Carbon dioxide	1
Methane	23
Nitrous oxide	296
HFC-23	12 000

Data from Intergovernmental Panel on Climate Change. 2001. Third assessment report. Climate change 2001: The scientific basis.

much colder surface temperature—probably an average of around −18°C—and life on Earth would be impossible, or very, very different. Thus, it is not the natural greenhouse effect that is the cause of current concerns, but the **anthropogenic**—human-generated—contribution to the greenhouse effect.

There are both natural and anthropogenic sources for almost all greenhouse gases—with the exception of chlorofluorocarbons (CFCs) and other halocarbons, such as HFC-23, which are wholly anthropogenic. However, human activities have increased the concentrations of many greenhouse gases in the past 250–300 years, thereby enhancing the greenhouse effect.

Greenhouse gases differ not only in their concentrations in the atmosphere but also in their ability to warm the troposphere and the surface. **Global warming potential** refers to the relative ability of one molecule of a given greenhouse gas to contribute to warming. Table 14.1 shows the global warming potentials for several greenhouse gases. Values are expressed in relation to carbon dioxide, which is assigned a global warming potential of 1. Thus, a molecule of methane is 23 times as potent as a molecule of carbon dioxide, and a molecule of nitrous oxide is 296 times as potent as a CO_2 molecule.

Carbon dioxide is the anthropogenic greenhouse gas of primary concern

Carbon dioxide (CO_2) is not the most potent greenhouse gas on a per-molecule basis (**TABLE 14.1**), but it is far more abundant in the atmosphere than the other GHGs. Emissions of greenhouse gases from human activity in Canada and other economically developed countries consist mostly of carbon dioxide. Even after accounting for the greater global warming potential of molecules of the other gases, carbon dioxide's abundance in our emissions makes it the major anthropogenic contributor to global warming.

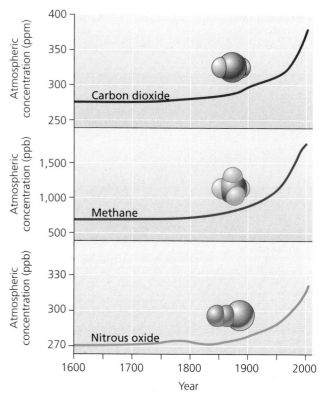

FIGURE 14.2
Since the start of the Industrial Revolution, global atmospheric concentrations of carbon dioxide, methane, and nitrous oxide have increased markedly. Data from Intergovernmental Panel on Climate Change. 2007. *Fourth assessment report.*

The main natural **source** of carbon dioxide moving into the atmosphere is the decay of organic material; volcanoes also emit a significant amount of CO_2. Natural sources greatly outweigh the human contribution; however, human activities have boosted the atmospheric concentration of carbon dioxide from around 280 parts per million (ppm), as recently as the late 1700s, to 387 ppm in 2008[1] (**FIGURE 14.2**). The atmospheric CO_2 concentration is now, by far, at its highest level in more than 650 000 years, and likely the highest in the last 20 million years. More importantly, researchers have established a very strong link between atmospheric carbon dioxide and temperature, as you will see. When atmospheric GHGs are high, so is temperature.

Human activity has released carbon from sequestration in long-term reservoirs

Why has atmospheric CO_2 increased by so much, if the natural sources still outweigh the human sources? The simple answer is that human inputs have shifted the balance of fluxes in the carbon cycle.

Recall our discussion of the carbon cycle and other biogeochemical cycles in Chapter 5, where we examined the idea that human activities often accelerate the movements or **fluxes** of material from one **reservoir** to another in biogeochemical cycles. As we will see in Chapter 15, carbon is stored or **sequestered** for long periods in the lithosphere. The deposition, partial decay, and compression of organic matter (mostly plants) that grew in wetland or marine areas led to the formation of coal, oil, and natural gas in sediments. It takes millions of years for the organic precursors of fossil fuels to be buried, chemically altered, and trapped deep underground in their rock hosts; the main chapter of fossil fuel formation occurred during the Carboniferous Period, 290 million–354 million years ago (please see Appendix D on the Companion Website at www.myenvironmentplace.ca).

In the absence of human activity, these lithospheric carbon reservoirs would be practically permanent. However, over the past two centuries we have extracted fossil fuels and burned them in our homes, factories, and automobiles, transferring large amounts of carbon from one reservoir (the long-term underground deposits) to another (the atmosphere). This human-modified flux of carbon from lithospheric reservoirs into the atmosphere is *much* faster than the natural flux. It is also faster than the combined total of all the fluxes of carbon *out* of the atmosphere and into carbon **sinks** (reservoirs that accept more of the material than they release), such as, notably, the ocean. The release of carbon from long-term reservoirs and the acceleration of the carbon flux from the lithospheric reservoir to the atmospheric reservoir is the main reason atmospheric carbon dioxide concentrations have increased so dramatically since the Industrial Revolution.

At the same time, people have cleared and burned forests to make room for crops, pastures, villages, and cities. Forests, soils, and crops also serve as storage reservoirs and sinks for carbon—in this case, for recently active carbon in the short-term carbon cycle. Their removal reduces the biosphere's ability to absorb carbon dioxide from the atmosphere. In this way, deforestation also has modified the flux of carbon from terrestrial reservoirs to the atmospheric reservoir and has contributed to rising atmospheric CO_2 concentrations.

Other greenhouse gases contribute to warming

Carbon dioxide is not the only greenhouse gas increasing in concentration in the atmosphere as a result of our activities. We release **methane** (CH_4) by tapping into fossil fuel deposits, by raising livestock that emit methane as a metabolic waste product, by disposing of organic matter in landfills, and by growing certain crops, such as rice. Since 1750, atmospheric methane concentrations have risen 250%, and today's concentration is the highest by far in more than 650 000 years. Buried organic matter and natural gas deposits are the main natural sources of methane.

Human activities have also augmented atmospheric concentrations of **nitrous oxides**. These greenhouse gases, by-products of feedlots, chemical manufacturing plants, auto emissions, and synthetic nitrogen fertilizers, have risen by 18% since 1750 (**FIGURE 14.2**). The atmosphere is the largest nitrogen reservoir, but soils and soil-forming processes are the principal drivers of the nitrogen biogeochemical cycle. This is one reason that environmental scientists are taking a great interest in the impacts of climatic warming on soils (Chapter 7).

Ozone—so important for life on Earth because of its function in the stratosphere as a UV filter—is also a radiatively active gas, contributing to warming both near the surface and up in the stratosphere. Concentrations of ozone in the troposphere have risen roughly 36% since 1750, also contributing to photochemical smog. In a confusing twist of chemistry, CFCs and HFCs—the anthropogenic group of chemicals known as **halocarbons**, which have been implicated in stratospheric ozone depletion—are also radiatively active. The contribution of halocarbons to global warming has begun to slow as a result of the Montreal Protocol and subsequent controls on their production and release.

Water vapour is by far the most abundant naturally occurring greenhouse gas in our atmosphere, and contributes most to the natural greenhouse effect. Its concentrations vary locally, but its global concentration has not changed over recent centuries, so it is not viewed as having driven industrial-age climate change. However, as tropospheric temperatures continue to increase, Earth's water bodies should transfer more water vapour into the atmosphere, contributing to higher atmospheric concentrations. This could have a number of effects on climate, weather, and ecosystems, in addition to contributing to further greenhouse warming.

There are many feedback cycles in the climate system

If global warming leads to an increase in the concentration of water vapour in the atmosphere, this could cause further warming because water is a radiatively active gas. This additional warming, in turn, could cause still more evaporation, leading to further increases in water vapour in the atmosphere; and so on. This is called a **positive feedback loop** or *positive feedback cycle*, and the climate system is loaded with them.

The climate system also has lots of **negative feedback loops**. For example, if global warming leads to increased evaporation and more water vapour in the atmosphere, it could give rise to increased cloudiness. Increased cloudiness could slow global warming by reflecting more solar radiation back into space. But the complexity of cloud cover doesn't end there; depending on whether low- or high-elevation clouds resulted, they might shade and cool Earth's surface (a negative feedback), or contribute to warming, thus accelerating evaporation and further cloud formation (a positive feedback).

Water vapour is not the only natural constituent involved in feedbacks in the climate system. For example, warming of soils could cause an accelerated flux of soil gases to the atmosphere; some of these are GHGs, and their release could lead to further warming (a positive feedback). On the other hand, soil formation is accelerated by warmer, wetter weather, and soils function as a major sink for organic matter, removing carbon from the atmospheric reservoir (a negative feedback).

Aerosols—microscopic droplets and particles suspended in the air—also can have either a warming or cooling effect. Soot, or black carbon aerosols, can cause warming by absorbing solar energy, but most tropospheric aerosols are whiter and cool the atmosphere by reflecting the Sun's rays. When sulphur dioxide enters the atmosphere, it undergoes various reactions, some of which lead to acid precipitation. These reactions, along with volcanic eruptions, can contribute to the formation of a sulphur-rich aerosol haze in the upper atmosphere, which reduces the amount of sunlight that reaches Earth's surface. Aerosols released by major volcanic eruptions also can exert short-term cooling effects on Earth's climate over periods of up to several years.

Because of the complexity of these competing effects and feedbacks in the climate system, and the interactions among the various processes, minor modifications of components of the atmosphere can potentially lead to major effects on climate. Sorting out feedbacks and their relative importance and influence on one another is one of the greatest challenges in modern climate science.

Radiative forcing expresses change in energy input over time

Scientists have made quantitative estimates of the degree of influence that aerosols, greenhouse gases, and other factors exert over Earth's energy balance (**FIGURE 14.3**). The amount of change in energy that a given factor causes is called its **radiative forcing**. Positive forcing warms the surface, whereas negative forcing cools it. Scientists' best estimate is that Earth today compared with the pre-industrial Earth of the year 1750 is experiencing

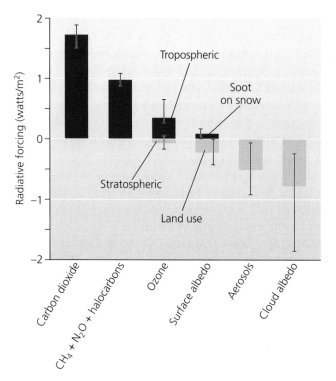

FIGURE 14.3
For each emitted gas or other human impact on the atmosphere since the Industrial Revolution, we can estimate the warming or cooling effect on Earth's climate. We express this as *radiative forcing*, which in this graph is shown as the amount of influence on climate today relative to 1750, in watts per square metre. Red bars indicate positive forcing (warming), and blue bars indicate negative forcing (cooling). A number of more minor influences are not shown. In total, scientists estimate that human impacts on the atmosphere exert a cumulative radiative forcing of 1.6 watts/m^2.

overall radiative forcing of about 1.6 watts/m^2. For context, look back at **FIGURE 14.1** and note that Earth is estimated to receive and give off 342 watts/m^2 of energy. Although 1.6 may seem like a small proportion of 342, it is enough to alter climate significantly.

The atmosphere is not the only factor that influences climate

Our climate is influenced by factors other than atmospheric composition. Among these are cyclic changes in Earth's rotation and orbit, variation in energy released by the Sun, absorption of carbon dioxide by the oceans, and oceanic circulation patterns. We will look at each of these factors in turn.

Milankovitch cycles   During the 1920s, Serbian mathematician Milutin Milankovitch described the influence of periodic changes in Earth's rotation and orbit around the Sun on **insolation**, the amount of solar energy that reaches Earth's surface. These variations, which include wobbling (or *precession*) of Earth's rotational axis, tilt of the axis, and change in the shape of Earth's orbit around

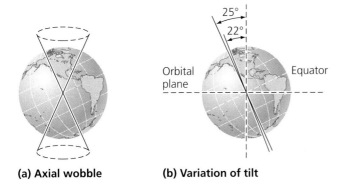

(a) Axial wobble (b) Variation of tilt

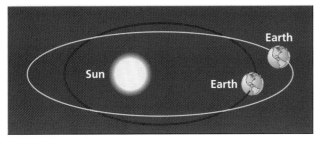

(c) Variation of orbit

FIGURE 14.4
There are three orbital factors whose variations have a significant influence on climate. The first is an axial wobble **(a)** that occurs on a 19 000- to 23 000-year cycle. The second is a 3° shift in the tilt of Earth's axis **(b)** that occurs on a 41 000-year cycle. The third is a variation in Earth's orbit from almost circular to more elliptical **(c)**, which repeats itself every 100 000 years. These variations affect the intensity of solar radiation that reaches portions of Earth at different times, contributing to long-term changes in global climate.

the Sun, alter the way solar radiation is distributed over Earth's surface (**FIGURE 14.4**). The collective impact of the variations causes cyclical changes in insolation and, therefore, in atmospheric heating. These **Milankovitch cycles** (**FIGURE 14.5**) involve variations that are sufficient to trigger long-term climate variations, such as periodic episodes of **glaciation** during which global surface temperatures drop and ice sheets advance from the poles toward the mid-latitudes, and the intervening warm **interglaciations**.

Solar output The Sun varies in the amount of radiation it emits (its *luminosity*), over both short and long timescales. For example, at each peak of its 11-year sunspot cycle, the Sun may emit *solar flares*, bursts of energy strong enough to disrupt satellite communications. However, scientists are concluding that the variation in solar energy reaching our planet in recent centuries has simply not been great enough to drive significant temperature change on Earth's surface. Estimates place the radiative forcing of natural changes in solar output at only about 0.12 watts/m^2—less than any of the anthropogenic causes shown in **FIGURE 14.3**.

Ocean absorption The ocean, acting as a sink, holds 50 times as much carbon as the atmosphere holds and absorbs carbon dioxide from the atmosphere, both through direct solubility of gas in water, and through uptake by marine phytoplankton for photosynthesis. However, the oceans absorb CO_2 more slowly than we are adding CO_2 to the atmosphere (see **FIGURE 5.15**). Thus, carbon absorption by the oceans is slowing global warming but not preventing it. Moreover, recent evidence indicates that this absorption is now decreasing. As ocean water warms, it absorbs less CO_2 because gases are less soluble in warmer water—a positive feedback effect that accelerates warming.

Ocean circulation Ocean water exchanges tremendous amounts of heat with the atmosphere, and ocean currents move energy from place to place. In equatorial regions, such as the area around the Maldives, the oceans receive more heat from the Sun and atmosphere than they emit. Near the poles, the oceans emit more heat than they receive. Because cooler water is denser than warmer water, the cool water at the poles tends to sink, and the warmer surface water from the equator moves poleward to take its place. This is one of the principles underlying global ocean circulation patterns.

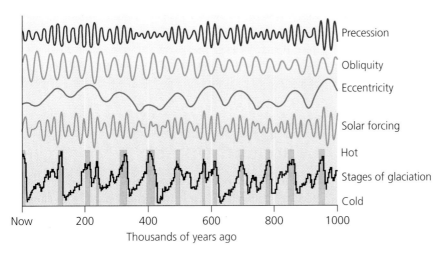

FIGURE 14.5
The collective influence of variations in axial tilt, wobble, and orbital shape cause cyclical variations in solar insolation and heating of Earth's atmosphere and surface, which forces cooling and warming trends in global climate on a variety of timescales and initiates major climatic shifts, such as glaciations.

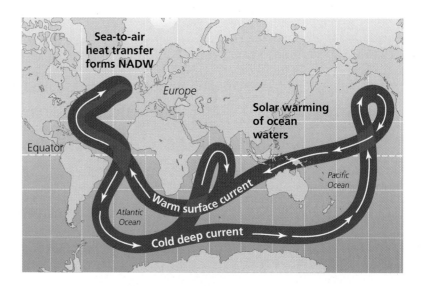

FIGURE 14.6
Warm surface currents carry heat from equatorial waters of the ocean north toward Europe and Greenland, where they release heat into the atmosphere and then cool and sink, forming the North Atlantic Deep Water. Scientists debate whether rapid melting of Greenland's ice sheet could interrupt the flow of heat from equatorial regions and cause Europe to cool dramatically.

One ocean–atmosphere interaction that influences climate is the *El Niño–Southern Oscillation* (*ENSO*), a systematic shift in atmospheric pressure, sea surface temperature, and ocean circulation in the tropical Pacific Ocean. El Niño conditions are triggered when air pressure increases in the western Pacific and decreases in the eastern Pacific, causing the equatorial winds to weaken. La Niña events are the opposite; under these conditions, cold surface waters extend far westward in the equatorial Pacific. Both El Niño and La Niña have dramatic influences on global weather patterns. ENSO cycles are periodic but irregular, occurring every two to eight years. Scientists are getting better at deciphering the triggers for these events and predicting their impacts on weather. They are also investigating whether globally warming air and sea temperatures may be increasing the frequency and strength of these cycles.

Ocean currents and climate also interact through the **thermohaline circulation**, a worldwide current system in which warmer, fresher water moves along the surface and colder, saltier water (which is more dense) moves deep beneath the surface (**FIGURE 14.6**). In the Atlantic Ocean, warm surface water flows northward from the equator in the Gulf Stream, carrying heat to high latitudes and keeping Europe warmer than it would otherwise be. As the surface water of this conveyor belt system releases heat energy and cools, it becomes denser and sinks, creating the *North Atlantic Deep Water* (*NADW*).

Scientists hypothesize that interruptions in the thermohaline circulation could trigger rapid climate change. If global warming causes much of Greenland's ice sheet to melt, freshwater runoff into the North Atlantic would dilute surface waters, making them less dense (because freshwater is less dense than salt water). This could potentially stop NADW formation and shut down the thermohaline circulation and the associated northward flow of warm equatorial water (that is, the Gulf Stream), causing Europe to cool rapidly. Such a change may have had a role in the "Little Ice Age," a period of unusually cold climate that began in the fifteenth century. This scenario also inspired the 2004 blockbuster film *The Day After Tomorrow*, although the filmmakers chose entertainment value over science and grossly exaggerated the potential impacts and the speed with which the effect might occur.

Some data suggest that the thermohaline circulation and formation of NADW in the North Atlantic is already slowing; other researchers argue that Greenland's runoff will not be enough to cause a shutdown this century. However, the possibility is worrisome; there is some evidence that interruptions of the thermohaline circulation have historically been responsible for extremely rapid shifts in climate in the areas bordering the North Atlantic. These climatic "flip-flops" are now thought to have occurred in the geologic past in periods as short as a decade or so.

The Science of Climate Change

To comprehend any phenomenon that is changing, we must study its past, present, and future. Climate scientists monitor present-day climatic conditions, but they also have devised clever means of inferring past change and have developed sophisticated methods to predict future change.

Proxy indicators tell us about the past

To understand how climate is changing today, and to predict future change, scientists must learn what climatic conditions were like thousands or millions of years ago. Evidence about climate in the geologic past—

paleoclimate—is extremely important, because it gives us a baseline against which to measure the changes that we see happening in the climate system on a shorter time scale. The record of actual measurements of temperature, precipitation, and other indicators of climate is not very long—only a few hundred years' worth of data. Environmental scientists have developed a number of methods to decipher clues from the past. **Proxy indicators** are types of indirect evidence that serve as proxies, or substitutes, for direct measurement and that shed light on past climate.

Some human records of historical events can contribute to a longer database of weather information. For example, fishers have recorded the timing of sea ice formation for hundreds of years, and wine makers have kept meticulous records of precipitation and the length of the growing season.

To go back farther in time, we must begin to rely on the record-keeping ability of the natural world. For instance, growth rings in trees can give information about conditions of temperature and precipitation—they act as a proxy, or a stand-in, for actual measurements. The width of each ring of a tree trunk cut in cross-section reveals how much the tree grew in a particular growing season; a wide ring means more growth, generally indicating a wetter year. Long-lived trees, such as bristlecone pines, can provide records of precipitation and drought going back hundreds or thousands of years. Tree rings are also used to study fire history, since a charred ring indicates that a fire took place in that year.

The significance of annual growth rings is simple: Without the rings, scientists would have no way to know *when* a particularly cold or warm or dry or wet growth season occurred. With growth rings, scientists can count back to determine the age of the tree (or any other organism that accumulates material in annual growth cycles). Researchers can gather data on past ocean conditions from corals, for example, which accumulate material in annual rings. As they grow and build their reefs, the living corals take in trace elements from ocean water, incorporating these chemical clues into their rings.

Scientists continue to search for ways to extend the climate record back even farther. In arid regions, packrat middens are a valuable source of climate data. Packrats are rodents that carry seeds and plant parts back to their middens, or dens, in caves and rock crevices sheltered from rain. In an arid location, plant parts may last for centuries, allowing researchers to study the past flora of the region. Researchers also drill cores into sediments that lie beneath bodies of water. Sediments often preserve pollen grains and other remnants from plants that grew in the past, and as we saw with the study of Easter Island, analyzing these materials can illuminate the history of past vegetation. Because climate influences the types of plants that grow in an area, knowing what plants lived in a location at a given time can tell us much about the climate at that place and time.

Earth's icecaps, ice sheets, glaciers, and sediments hold clues to the much longer-term climate history. Over the ages, these huge expanses of snow and ice have accumulated to great depths, preserving within their layers tiny bubbles of the ancient atmosphere (**FIGURE 14.7**). Scientists can examine these trapped air bubbles by drilling into the ice and extracting long columns, or cores. The layered ice, accumulating season after season over thousands of years, provides a timescale, something like the growth rings of the tree. From these ice cores, scientists can determine atmospheric composition, greenhouse gas concentrations, temperature trends, snowfall, solar activity, and even (from trapped soot particles) frequency of forest fires and volcanic eruptions.

By extracting ice cores from Greenland and Antarctica, scientists have now been able to go back in time more than 740 000 years, reading Earth's global climatic history across eight glacial cycles (see "The Science Behind the Story:

(a) Ice core

(b) Micrograph of ice core

FIGURE 14.7
In Greenland and Antarctica, scientists have drilled deep into ancient ice sheets and removed cores of ice like this one **(a)** to extract information about past climates. Bubbles (black shapes) trapped in the ice **(b)** contain small samples of the ancient atmosphere.

Reading History in the World's Longest Ice Core".) During 2007–2009, such research is being funded and promoted as part of the *International Polar Year*, a large international scientific program coordinating research in the Arctic and Antarctic. The main approach to these studies makes use of the fact that the chemistry of the H_2O in the ice itself holds clues about climatic conditions at the time when it first formed as precipitation.

Stable isotope geochemistry is a powerful tool for the study of paleoclimate

Clues about paleoclimates can be partly uncovered through the study of the **stable isotope geochemistry** of the ice. Stable isotopes are naturally occurring variations of elements, which vary just slightly from one another in mass but not in other chemical characteristics. This means, for example, that deuterium—a naturally occurring isotope of hydrogen that is just slightly heavier than simple hydrogen—behaves chemically like hydrogen. However, because of its slight difference in mass, deuterium is separated from hydrogen and becomes differently concentrated in Earth materials (like ice) when acted upon by any process that is influenced by mass.

Consider, for example, the process of precipitation. Water that falls as precipitation (a gravity-influenced process) is naturally enriched in the *heavier* isotopes of its components, hydrogen and oxygen. Water that remains behind in cloud form, on the other hand, will tend to be enriched in the *lighter* isotopes of its chemical constituents. The scientific term for the separation and differential concentration of isotopes of slightly different mass is **fractionation**. Many natural fractionation processes are temperature dependent—that is, they are controlled by variations in temperature. Sampling and analysis of Earth materials—not just ice but any natural material that incorporates isotopes and is affected by temperature-dependent fractionation processes—can reveal the past temperature history of those materials.

Proxy indicators like stable isotopes often give us information about local or regional areas, but to get a global perspective scientists need to combine multiple records from various areas. In fact, one of the most interesting challenges in stable isotope studies of paleoclimate has been to explain the slight mismatch in timing (*asynchronicity*) between major climatic events recorded in ice core records from Antarctic and Greenland ice cores. Because the number of available indicators decreases the farther back in time we go, estimates of global climate conditions for the recent past tend to be more reliable than those for the distant past.

Direct atmospheric sampling tells us about the present

Studying present-day climate is much more straightforward, because scientists can measure atmospheric conditions directly. As mentioned above, our records of direct measurements of temperature and precipitation date back several hundred years, although the more recent records are obviously the most reliable. Atmospheric carbon dioxide concentrations have been measured continuously at the Mauna Loa Observatory in Hawaii, starting in 1958 (**FIGURE 14.8**). These data show that atmospheric CO_2 concentrations have increased from 315 ppm in 1958 to 387 ppm in 2008. Today scientists at Mauna Loa continue to make these measurements, building upon the best long-term data set we have of direct atmospheric sampling of any greenhouse gas.

Models help us understand climate

To understand how climate systems function and to predict future climate change, scientists simulate climate

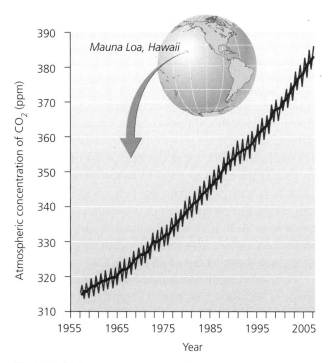

FIGURE 14.8
Atmospheric concentrations of carbon dioxide have risen steeply since these measurements began to be taken at the Mauna Loa Observatory in 1958. The jaggedness of the upward trend reflects seasonal variations in photosynthetic uptake. The Northern Hemisphere has more land area and thus more vegetation than the Southern Hemisphere. Thus, more carbon dioxide is absorbed during the northern summer, when Northern Hemisphere plants are more photosynthetically active. Data from National Oceanic and Atmospheric Administration, Earth System Research Laboratory, Global Monitoring Division, 2007.

processes with sophisticated computer programs. **Coupled general circulation models** (or simply **climate models**) are programs that combine what is known about atmospheric circulation, ocean circulation, atmosphere–ocean interactions, and feedback mechanisms to simulate climate processes (**FIGURE 14.9**). They couple, or combine, the climate influences of the atmosphere and oceans into a single simulation. This requires manipulating vast amounts of data and complex mathematical equations—not possible until the advent of modern computers.

Canadian scientists are internationally recognized as leaders in the development of such models, also called *general circulation models* or *global climate models* (GCMs). Research utilizing climate models developed at the Canadian Centre for Climate Modelling and Analysis (CCCma), part of Environment Canada, has contributed to reports by the Intergovernmental Panel on Climate Change (IPCC). Scenarios based on Canadian models have also been used by the government of the United States in its *National Assessment of the Potential Consequences of Climate Variability and Change*, providing a basis for strategic planning for managing the consequences of climate change.

Researchers develop different climate models for different purposes. You can think of these models as being sort of like very sophisticated versions of computer simulation games, like *SimEarth*, *SimCity*, *SimAnt*, and so on.

If you have played any of these games (especially the earlier versions, which required more hands-on intervention by the gamer, before the advent of high-tech graphics), then you know something about how they work. You provide some starting information to the game (or computer model), setting up the preconditions for the simulation, and then you let it run. You can tweak it as the game goes on—constructing more buildings, taking away food sources, or adding more predators to the ant colony—but eventually the model will run to its natural completion.

That is something like how climate models work. Researchers construct models that are as realistic as possible, building in as much information as they can from what is understood about the functioning of the climate system. They can test the effectiveness of the models by entering *past* climate data and running the model toward the present. If a model produces accurate reconstructions of our *current* climate, based on well-established data from the past, then we have reason to believe that it simulates climate mechanisms realistically and that it may accurately predict *future* climate.

FIGURE 14.10 shows temperature results from three such simulations, as reported by the IPCC. Results in **FIGURE 14.10A** are based on natural climate-changing factors alone (such as volcanic activity and variation in solar energy). Results in **FIGURE 14.10B** are based on

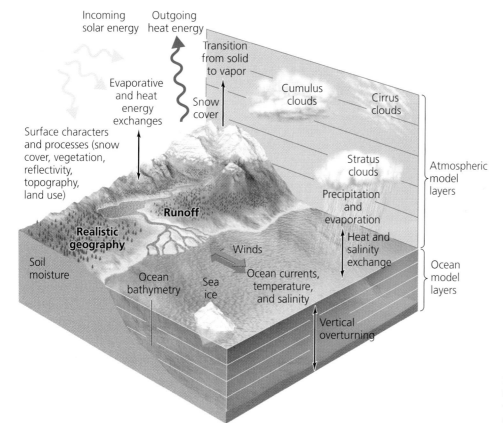

FIGURE 14.9
Modern climate models incorporate many factors, including processes involving the atmosphere, land, oceans, ice, and biosphere. Such factors are shown graphically here, but the actual models deal with them as mathematical equations in computer simulations.

THE SCIENCE BEHIND THE STORY

Reading History in the World's Longest Ice Core

An EPICA researcher prepares a Dome C ice core sample for analysis.

Snow falling year after year compresses into ice and stacks up into immense sheets that scientists can mine for clues to Earth's climate history. The ice sheets of Antarctica and Greenland trap tiny air bubbles, dust particles, and other proxy indicators of past conditions. By drilling boreholes and extracting ice cores, researchers can tap into these valuable archives.

Recently, researchers drilled and analyzed the deepest core ever. At a remote, pristine site in Antarctica named Dome C, they drilled down 3270 m to bedrock and pulled out more than 800 000 years' worth of ice. The longest previous ice core (from Antarctica's Vostok station) had gone back 420 000 years. Ice near the top of these cores was laid down recently; ice at the bottom is oldest. By analyzing ice at intervals along the core's length, researchers can generate a timeline of environmental change.

The Dome C core was drilled by the European Project for Ice Coring in Antarctica (EPICA), a consortium of researchers from 10 European nations. Antarctic operations are expensive and logistically complicated: Ice drilling requires powerful technology, and the analysis requires a diverse assemblage of experts. When the team published its results in the journal *Nature* in 2004, the landmark paper had 56 authors.

The researchers obtained data on surface air temperature going back 740 000 years by measuring the ratio of deuterium isotopes to normal hydrogen in the ice. This ratio is temperature dependent (top panel of figure).

By examining dust particles, they could tell when arid and/or windy climates sent more dust aloft. By analyzing air bubbles trapped in the ice, the researchers were later able to quantify atmospheric concentrations of carbon dioxide, methane, and nitrous oxide, across 650 000 years.

One finding was expected—yet crucial. The researchers documented that temperature swings in the past were tightly correlated with concentrations of carbon dioxide (middle panel of figure), as well as methane and nitrous oxide. Also clear and expected from the data was that temperature varied with swings in solar radiation caused by Milankovitch cycles; the Dome C core spanned eight glacial cycles.

A tight correlation between greenhouse gas concentrations and temperature does not prove *causality*: It doesn't prove that high atmospheric GHG concentrations caused warming, nor does it prove that warming caused GHG concentrations to increase.

What the correlation *does* demonstrate is that there is a close relationship between atmospheric GHGs and climatic warming: when atmospheric GHG concentrations are high, so is temperature, and vice versa. This is crucially important.

The EPICA data also demonstrate that by increasing greenhouse gas concentrations since the Industrial Revolution, we have brought them well above the highest levels they reached naturally in 650 000 years. Today's carbon dioxide concentration (387 ppm in 2008) is far above previous maximum values. Present-day concentrations of methane and nitrous oxide are likewise the highest in 650 000 years. These data show that we as a society have brought ourselves deep into uncharted territory.

Other findings from the ice core are not as easily explained. Intriguingly, earlier glacial cycles differ from recent cycles. For recent cycles, the Dome C core showed that glacial periods were long and interglacial periods were brief, with rapid rise and fall of temperature. Interglacials thus appear on a graph of temperature through time as tall thin spikes. However, in older cycles the glacial and interglacial periods were of more equal duration, and the warm extremes of interglacials were not as great. This change in the nature of glacial cycles through time had been noted by researchers working with oxygen isotope data from the fossils of marine organisms (bottom panel of figure). But why glacial cycles should be different before and after the 450 000-year mark, no one knows.

Today scientists are searching for a site that might provide an ice core stretching back more than 1 million years. Data from marine isotopes tells us that glacial cycles at that time switched from a periodicity of roughly 41 000 years (conforming to the influence of planetary tilt), to about 100 000 years (more similar to orbital changes). An ice core that captures cycles on both sides of the 1 million-year divide might help clarify the influence of Milankovitch cycles.

The Dome C ice core research shows that we still have plenty to learn about our complex climate history. The close correlation between greenhouse gases and temperature evident in the EPICA data provides a strong indication that we would do well to bring greenhouse emissions under control before it is too late.

---

anthropogenic factors only (such as human emissions of greenhouse gases and sulphate aerosols). Results in **FIGURE 14.10C** are based on natural and anthropogenic factors combined, and this produces the closest match between predictions and actual climate.

Such results as those in **FIGURE 14.10C** clearly support the hypothesis that both natural and human factors contribute to climate dynamics, and they also indicate that global climate models can produce reliable predictions. As computing power increases and we glean more and better data from proxies, these models become increasingly reliable. They also are improving in resolution and are beginning to predict climate change region by region for various areas of the world.

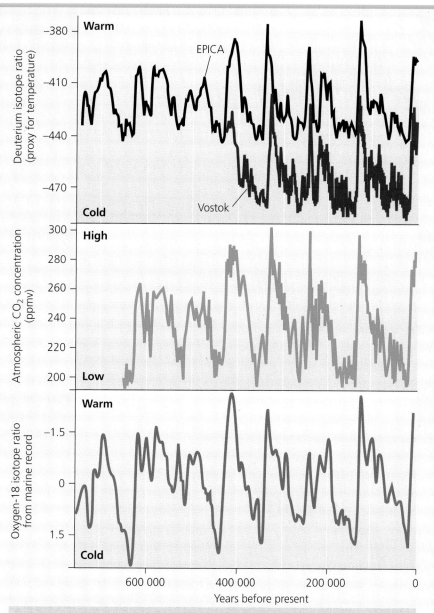

Deuterium isotope data from the EPICA ice core reveals changes in surface temperature across 740 000 years (top panel). High peaks indicate warm interglacial periods, and low troughs indicate cold glacial periods. Data from the Vostok ice core are shown for comparison. Atmospheric carbon dioxide concentrations (middle panel) from the EPICA ice core rise and fall in tight correlation with temperature. These data sets are consistent with oxygen18 isotope data from the marine record (bottom panel), an independent proxy indicator for global temperature. Adapted from EPICA community members. 2004. Eight glacial cycles from an Antarctic ice core. *Nature* 429:623–628; and Siegenthaler, U. 2005. Stable carbon cycle–climate relationship during the late Pleistocene. *Science* 310:1313–1317.

Current and Future Trends and Impacts

Evidence that climate conditions have changed worldwide since industrialization is now overwhelming and indisputable. Climate change in recent years has already had numerous effects on the physical properties of our planet, on organisms and ecosystems, and on human well-being. If we continue to emit greenhouse gases into the atmosphere, the impacts of climate change will only grow more severe.

Future impacts of climate change are subject to regional variation, so the way each of us experiences these impacts over the coming decades will vary tremendously,

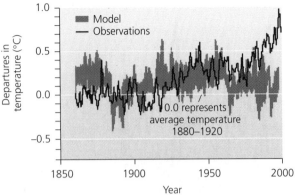

(a) Natural factors only

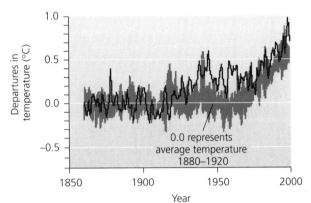

(b) Anthropogenic factors only

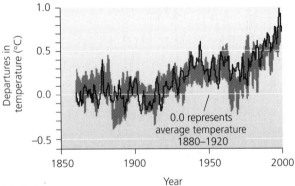

(c) All factors

FIGURE 14.10
Scientists test climate models by entering climate data from past years and comparing model predictions (blue areas) with actual observed data (red lines). Models that incorporate only natural factors **(a)** or only human-induced factors **(b)** do not predict real climate trends as well as models **(c)** that incorporate both natural and anthropogenic factors. Data from the Intergovernmental Panel on Climate Change. 2001. *Fourth assessment report.*

depending on where we live. The impacts on Canada could be particularly severe. According to the CCCma,

> Over the next century, the Canadian climate model indicates that the most northerly regions of the Earth will experience the greatest warming, with potentially serious impacts on Arctic communities and ecosystems. Major changes in climate are also expected in the Prairies, the east and west coasts and the Great Lakes basin. Many communities and climate-sensitive industries will be profoundly affected, including forestry, agriculture, marine transportation, fishing and oil and gas development . . . The magnitude and rate of change projected for the twenty-first century will put enormous pressure on our environment, our infrastructure and our social fabric. Climate modelling can help us avoid aggravating the situation by providing a glimpse of the future.[2]

The IPCC summarizes evidence of climate change and predicts future impacts

In recent years, it seems that virtually everyone is detecting climatic changes around us. A fisher in the Maldives notes the seas encroaching on his home island. A rancher in Alberta suffers a multi-year drought. A homeowner in Florida finds it impossible to obtain insurance against the hurricanes and storm surges that increasingly threaten. Residents of Montreal marvel over freakish weather events.

But can we conclude that all of these impressions are part of a real pattern, or are they just ephemeral fluctuations in the data? A true **trend** is a pattern that persists within a data set, even after short-term fluctuations, background noise, and local anomalies have been removed or accounted for. Is there solid scientific evidence to confirm that we are seeing a trend, and that the global climate is indeed already changing? A wide variety of data sets do appear to show significant trends in climate conditions over the past century and particularly in recent years. The most thoroughly reviewed and widely accepted synthesis of scientific information concerning climate change is a series of reports issued by the **Intergovernmental Panel on Climate Change (IPCC)**. This international panel of scientists and government officials was established in 1988 by the United Nations Environment Programme (UNEP) and the World Meteorological Organization.

In 2007 the IPCC released its *Fourth Assessment Report*, which represents the consensus of scientific climate research from around the world. This report summarizes many thousands of scientific studies, and it documents observed trends in surface temperature, precipitation patterns, snow and ice cover, sea levels, storm intensity, and other factors. It also predicts future changes in these phenomena after considering a range of potential scenarios for future greenhouse gas emissions. The report addresses impacts of current and future climate change on wildlife, ecosystems, and human societies. Finally, it discusses possible strategies we might pursue in response to climate change. **FIGURE 14.11** summarizes a selection

Major Trends and Impacts of Climate Change, from IPCC Fourth Assessment Report, 2007

Global physical indicators

Earth's average surface temperature increased 0.74° C (1.33 °F) in the past 100 years, and will rise 1.8–4.0° C (3.2–7.2° F) in the 21st century.

Eleven of the years from 1995 to 2006 were among the 12 warmest on record.

Atmospheric water vapour increased since at least the 1980s.

Oceans absorbed >80% of heat added to the climate system, and warmed to depths of at least 3000 m (9800 ft).

Glaciers, snow cover, ice caps, ice sheets, and sea ice will continue melting, contributing to sea-level rise.

Sea level rose by an average of 17 cm (7 in.) in the 20th century, and will rise 18–59 cm (7–23 in.) in the 21st century.

Ocean water became more acidic by about 0.1 pH unit, and will decrease in pH by 0.14–0.35 units more by century's end.

Storm surges increased, and will increase further.[1]

Sea level rise will worsen coastal erosion and degrade wetlands.

Carbon uptake by terrestrial ecosystems will likely peak by the mid-21st century and then weaken or reverse, amplifying climate change.[2]

Social indicators

Farmers and foresters have had to adapt to altered growing seasons and disturbance regimes.

Temperate-zone crop yields will rise until temperature warms beyond 3° C (5.4° F), but in the dry tropics and subtropics, crop productivity will fall and lead to hunger.[5]

Timber production may rise slightly in the near-term, but will vary by region.[5]

Impacts on biodiversity will cause losses of food, water, and other ecosystem goods and services.[2]

Sea-level rise will displace people from islands and coastal regions.[3]

Melting of mountain glaciers will reduce water supplies to millions of people.[2]

Economic costs will outweigh benefits as climate change worsens;[2] costs could average 1–5% of GDP globally for 4° C (7.2° F) of warming.

Poorer nations and communities suffer more from climate change, because they rely more on climate-sensitive resources and have less capacity to adapt.[2]

Human health will suffer as increased warm-weather health hazards outweigh decreased cold-weather health hazards.[2]

Regional physical indicators

Arctic areas warmed fastest. Future warming will be greatest in the Arctic and greater over land than over water.

Summer Arctic sea ice thinned by 7.4% per decade since 1978.

Average Northern Hemisphere temperatures of the past 50 years were the highest in at least 1300 years.[1]

Thawing decreased area of Arctic permafrost in spring by 15% since 1900.

Precipitation increased in e. North America, e. South America, n. Europe, and n. and c. Asia since 1900.

Precipitation decreased in the Sahel, the Mediterranean, s. Africa, and parts of s. Asia since 1900.

Precipitation will generally increase at high latitudes and decrease at subtropical latitudes, often making wet areas wetter and dry ones drier.[1]

Heavy precipitation events increased over most land areas.[1]

Droughts became longer, more intense, and more widespread since the 1970s, especially in the tropics and subtropics.[1]

Droughts and flooding will increase, leading to agricultural losses.[2]

Over most land areas, cold and frost days decreased[3] and will continue to decrease[4] while hot days and heat waves increased[3] and will continue to increase.[3]

Hurricanes intensified in the North Atlantic since 1970[1], and will continue to intensify.[1]

The thermohaline circulation will slow, but will not shut down and chill Europe in the 21st century.[3]

Antarctica will continue accumulating snow, but may also continue losing ice around its edges.

Biological indicators

Species ranges are shifting toward the poles and upward in elevation, and will continue to shift.

The timing of seasonal phenomena (such as migration and breeding) is shifting, and will continue to shift.

About 20–30% of species studed so far will face extinction risk if temperature rises more than 1.5–2.5° C (2.7–4.5° F).[5]

Species interactions and ecosystem structure and function could change greatly, resulting in biodiversity loss.

Corals will experience further mortality from bleaching and ocean acidification.[5,4]

FIGURE 14.11

Climate change has had consequences already and is predicted to have many more. Listed here are some of the main observed and predicted trends and impacts described in the Intergovernmental Panel on Climate Change's *Fourth Assessment Report*. For simplicity, this table expresses mean estimates only; the IPCC report provides ranges of estimates as well. [1]Certainty level 5, 66%–90% probability of being correct. [2]Certainty level 5, ~80% probability of being correct. [3]Certainty level 5, 90%–99% probability of being correct. [4]Certainty level 5, > 99% probability of being correct. [5]Certainty level 5, ~50% probability of being correct. Data from the Intergovernmental Panel on Climate Change (IPCC). 2007. *Fourth assessment report*.

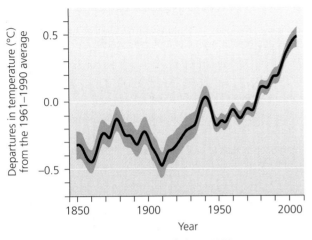

(a) Global temperature measured since 1850

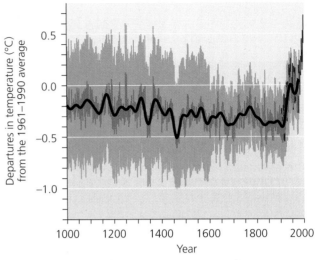

(b) Northern Hemisphere temperature over the past 1000 years

FIGURE 14.12
Data from thermometers **(a)** show changes in Earth's average surface temperature since 1850. Grey shaded area indicates range of uncertainty. In **(b)**, proxy indicators (blue line) and thermometer data (red line) together show average temperature changes in the Northern Hemisphere over the past 1000 years. The grey-shaded zone represents the 95% confidence range. This record shows that twentieth-century warming has eclipsed the magnitude of change during both the "Medieval Warm Period" (tenth to fourteenth centuries) and the "Little Ice Age" (fifteenth to nineteenth centuries). Data in (a) from the Intergovernmental Panel on Climate Change. 2007. *Fourth assessment report.* Data in (b) from and IPCC. 2001. *Third assessment report.*

of the IPCC report's major observed and predicted trends and impacts.

The IPCC report is authoritative, but—like all science—it deals in uncertainties. Its authors have therefore taken great care to assign statistical probabilities to its conclusions and predictions. In addition, its estimates regarding impacts of change on human societies are conservative, because its scientific conclusions had to be approved by representatives of the world's national governments, some of which are reluctant to move away from a fossil-fuel-based economy.

Temperature increases will continue

The IPCC report concludes that average surface temperatures on Earth increased by an estimated 0.74°C in the century from 1906 to 2005 (**FIGURE 14.12**), with most of this increase occurring in the last few decades. Eleven of the years, from 1995 to 2006, were among the 12 warmest on record since global measurements began 150 years earlier. The numbers of extremely hot days and heat waves have increased, whereas the number of cold days has decreased.

Temperature changes are greatest in the Arctic, and climate scientists anticipate that this will likely continue to be the case throughout the rest of this century (**FIGURE 14.13**). Here, ice sheets are melting, sea ice is thinning, storms are increasing, and altered conditions are posing challenges for people and wildlife. As sea ice melts earlier, freezes later, and recedes from shore, it

  CLIMATE CHANGE AND HUMAN RIGHTS 14-1

In December 2005, a group representing North America's Inuit sent a legal petition to the Inter-American Commission on Human Rights, demanding that the United States restrict its greenhouse gas emissions, which the Inuit maintained were destroying their way of life in the Arctic. After a year, the commission dismissed the petition with a terse three-sentence letter.

Do you think Arctic-living people deserve compensation from industrialized nations whose emissions have caused climate change that has disproportionately affected the Arctic? Do you think climate change can be viewed as a human rights issue? What ethical issues, if any, do you think climate change presents? How could these best be resolved?

becomes harder for Inuit and for polar bears alike to hunt the seals they each rely on for food. Thin sea ice is dangerous for people to travel and hunt upon, and in recent years, polar bears have been dying of exhaustion and starvation as they try to swim long distances between ice floes. **Permafrost** (perennially frozen ground) is thawing in the Arctic, destabilizing countless buildings. The strong Arctic warming is contributing to sea-level rise by melting icecaps and ice sheets.

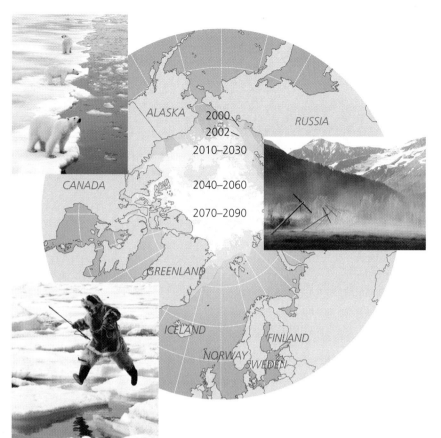

FIGURE 14.
The Arctic has b[...]
change's impacts so [...]
recedes from large are[...]
indicating the mean minim[...]
of sea ice for the recent pas[...]
Inuit find it difficult to hunt and [...]
traditional ways, and polar bears s[...]
they are less able to hunt seals. As pe[...]
thaws beneath them, human-made stru[...]
damaged: phone poles topple and building[...]
lean, buckle, crack, and fall. Map data from [...]
National Center for Atmospheric Research and [...]
National Snow and Ice Data Center.

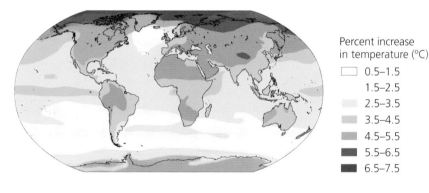

Percent increase in temperature (°C)
- 0.5–1.5
- 1.5–2.5
- 2.5–3.5
- 3.5–4.5
- 4.5–5.5
- 5.5–6.5
- 6.5–7.5

FIGURE 14.14
This map shows projected increases in surface temperature for the decade 2090–2099, relative to temperatures in 1980–1999. Land masses will warm more than oceans, and the Arctic will warm the most, up to 7.5°C. The IPCC uses multiple climate models when predicting regional variation in how temperature will change, and this map was generated by using an emission scenario that is intermediate in its assumptions, involving an average global temperature rise of 2.8°C by 2100. Data from Intergovernmental Panel on Climate Change. 2007. *Fourth assessment report*.

In the future, we can expect average surface temperatures on Earth to rise roughly 0.2°C per decade for the next 20 years, according to IPCC analysis. If we were to cease greenhouse gas emissions today, temperatures would still rise 0.1°C per decade because of the time lag from gases already in the atmosphere that have yet to exert their full influence. At the end of the twenty-first century, the IPCC predicts global temperatures will be 1.8°C–4.0°C higher than today's, depending upon the emission scenario. Unusually hot days and heat waves will become more frequent. Temperature change is predicted to vary from region to region in ways that parallel existing regional differences (**FIGURE 14.14**).

Sea surface temperatures are also increasing as the oceans absorb heat. The record number of hurricanes and tropical storms in 2005—Katrina and 27 others—left many people wondering whether global warming was to blame. Are warmer ocean temperatures spawning more hurricanes, or hurricanes that are more powerful or long lasting? Scientists are not yet sure, but recent analyses of storm data suggest that warmer seas may not be increasing the number of storms but likely are increasing the power of storms and possibly their duration.

Changes in precipitation will vary by region

A warmer atmosphere holds more water vapour, but changes in precipitation patterns have been complex, with some regions of the world receiving more precipitation than usual and others receiving less. In regions from the African Sahel to the Western Prairies, droughts have

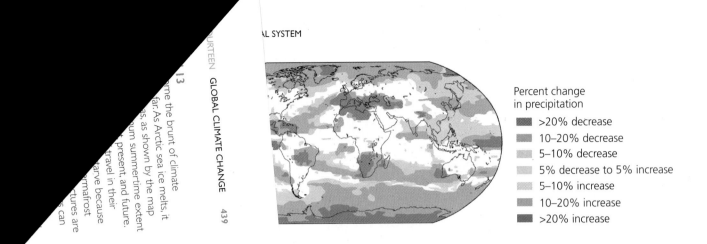

become more frequent and severe, harming agriculture, promoting soil erosion, reducing drinking water supplies, and encouraging forest fires. Meanwhile, in both dry and humid regions, heavy rain events have increased, contributing to damaging floods.

Precipitation changes are predicted to vary among different regions (**FIGURE 14.15**). In general, precipitation will increase at high latitudes and decrease at low and middle latitudes, magnifying differences in rainfall that already exist and worsening water shortages in many developing countries of the arid subtropics. In many areas, heavy precipitation events will become more frequent, increasing the risk of flooding.

Melting ice and snow have far-reaching effects

As the world warms, mountaintop glaciers are disappearing (**FIGURE 14.16**). Since 1980, the World Glacier Monitoring Service estimates, major glaciers have lost an average of 9.6 m in vertical thickness. Many glaciers on tropical mountaintops have disappeared altogether. The Athabasca Glacier, one of the outlet glaciers of the great Columbia Icefield in the Canadian Rockies, has retreated 1.5 km since the late nineteenth century, with the rate of retreat accelerating dramatically after 1980. Peyto Glacier in Banff National Park, Alberta, retreated rapidly during the first half of the twentieth century, stabilized briefly in 1966, and then resumed its retreat in 1979. A comparison of photos taken in 1898 and recently shows that Illecillewaet Glacier in Glacier National Park, British Columbia, has retreated 2 km in the past century (see **FIGURE 14.16**).

Mountains accumulate snow in the winter and release meltwater gradually during the summer. Throughout high-elevation areas of the world, however, warming temperatures will continue to melt mountain glaciers, posing risks of sudden floods as ice dams burst, and reducing summertime water supplies to millions of people. More than one-sixth of the world's people live in regions supplied by mountain meltwater, and some of these people are already beginning to face water shortages. If this water vanishes during drier months, whole communities will be forced to look elsewhere for water, or to move.

Warming of temperatures will have an impact not only on surface snow and ice (**FIGURE 14.17**) but also on subsurface ice and permafrost, as mentioned above. Permafrost is the most characteristic feature of Arctic soils, and its expected behaviour in a changing climate regime is not thoroughly understood from a scientific perspective. The foundations of roads and buildings in the north would be at risk if permafrost undergoes major changes as a result of warming. Permafrost also plays a major role in slope stability, with greatly increased chances

(a) Illecillewaet Glacier in 1898

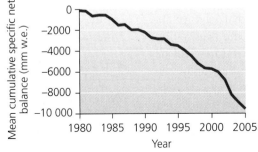

(b) Illecillewaet Glacier in 2008

FIGURE 14.16
Glaciers are melting rapidly around the world as global warming proceeds. The Illecillewaet Glacier in Glacier National Park, British Columbia, has retreated substantially between the photograph taken in 1887 **(a)** and the present **(b)**. The graph shows declines in mass in 30 major glaciers monitored since 1980 by the World Glacier Monitoring Service. Data from World Glacier Monitoring Service.

Decrease in winter snow by 2050s

FIGURE 14.17
The Canadian Regional Climate Model, developed by Environment Canada and the University of Quebec at Montreal, shows the projected change, to the middle of this century, in snow cover. As the climate warms, winter snow could decrease by 50% in much of southern Canada and increase slightly in the high Arctic. *Source: CCCma Environment Canada.*

for landslides as permafrost begins to melt. There are also concerns that the warming of permanently frozen soils would lead to the accelerated release of soil gases, such as methane, which could contribute to a positive feedback cycle in the climate system, leading to further warming.

The Arctic is changing dramatically

Warming temperatures are already reducing snow cover, affecting permafrost stability at high latitudes, and melting the immense ice sheets of the Arctic. Recent research reveals that melting of the Greenland ice sheet is accelerating (see "The Science Behind the Story: Greenland's Glaciers Race to the Sea"). At the other end of the world, in Antarctica, ice shelves almost the size of Prince Edward Island have disintegrated as a result of contact with warmer ocean water, but increased precipitation has so far supplied the continent with enough extra snow to compensate for the loss of ice around its edges.

According to Dr. Luke Copland of the University of Ottawa, the area of Canada's ice shelves has shrunk by approximately 90% over the past 100 years. One major event that contributed to this loss was the collapse of the Ayles Ice Shelf in 2005 (**FIGURE 14.18**). The Ayles Ice Shelf was one of Canada's six major ice shelves, located off the northern coast of Ellesmere Island, Nunavut, about 800 km south of the North Pole. On August 13, 2005, it broke off, creating an ice island 14 km by 5 km, and 37 m in thickness. The oldest ice in the ice island is more than 3000 years old. This ice breakup was so large that it was detected by seismometers. The breakup happened very

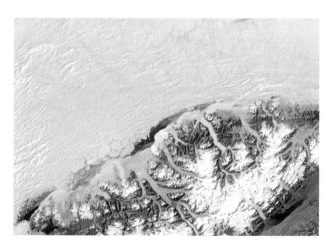

(a) Ayles ice sheet before breakup

(b) Ayles ice island after breakup

FIGURE 14.18
As seen in these satellite images, the Ayles ice sheet **(a)**, one of Canada's major ice sheets north of Ellesmere Island, collapsed in 2005 **(b)**, forming a huge ice island 14 km by 5 km in size.

quickly, lasting less than an hour according to reconstructions based on satellite images. Since the breakup, the ice island has drifted, becoming temporarily locked in place during winter freeze-ups, then breaking free and drifting again, and it has broken into two large parts. The large ice island fragments could pose a risk to oil rigs and drilling in the Beaufort Sea.

One reason warming is accelerating in the Arctic is that as snow and ice melt, darker, less-reflective surfaces are exposed, and Earth's albedo, or capacity to reflect light, decreases. As a result, more of the Sun's rays are absorbed at the surface, fewer reflect back into space, and the surface warms. In a positive feedback, this warming causes more ice and snow to melt, which in turn causes more absorption of radiation and more warming.

Near the poles, snow cover, permafrost, and ice sheets are projected to decrease, and sea ice will continue to shrink in both the Arctic and the Antarctic (**FIGURE 14.19**). Some climate scenarios show Arctic sea ice disappearing completely from the Northwest Passage by the late twenty-first century, creating new shipping lanes for commerce (and likely a rush to exploit underwater oil and mineral reserves that may exist in Arctic waters).

Rising sea levels will affect hundreds of millions of people and coastal zones

As glaciers and ice sheets melt, increased runoff into the oceans causes sea levels to rise. Sea levels also are rising because ocean water is warming, and water expands

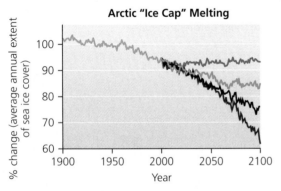

FIGURE 14.19
The various curves on this graph show climate model simulations of several of the future scenarios described by the IPCC. Scenario A2, in which population growth is rapid and technological advancement to lower-emission energy technologies is slow, yields the most dramatic decrease in areal extent of Arctic sea ice.

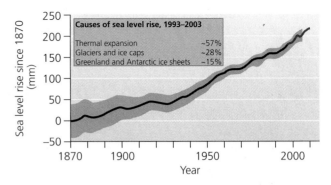

FIGURE 14.20
Data from tide gauges (black line) and satellite observations (red line) show that global average sea level has risen about 200 mm since 1870. Grey shaded area indicates range of uncertainty. Thermal expansion of water accounts for most sea-level rise. Data from Intergovernmental Panel on Climate Change. 2007. *Fourth assessment report.*

in volume as its temperature increases. In fact, recent sea-level rise has resulted primarily from the thermal expansion of seawater. Worldwide, average sea levels rose an estimated 17 cm during the twentieth century (**FIGURE 14.20**). Seas rose by an estimated 1.8 mm/year during 1961–2003 and 3.1 mm/year during 1993–2006. Note that these numbers represent vertical rises in water level, and on most coastlines a vertical rise of centimetres means metres of horizontal incursion inland.

weighing the issues — AN ICE-FREE NORTHWEST PASSAGE AND CANADA'S ARCTIC SOVEREIGNTY 14–2

Now that the Northwest Passage—the long-coveted sea route from the Atlantic to the Pacific via the Arctic Ocean—is almost ice-free, northern nations Canada, Russia, Norway, and Denmark are anxious to confirm their territorial influence in the region. Canada reaffirmed its sovereignty over Arctic waters at an international conference held in Greenland in 2008, and Prime Minister Stephen Harper pledged to spend billions of dollars defending Canada's interests in the Arctic if necessary. In addition to the sea route, the Arctic subsurface may hold as much as 25% of the world's undiscovered oil and gas reserves. The U.S. argues that the High Arctic does not belong to Canada, or to anyone else. What do you think? Who should control Arctic Ocean waters, and for what purposes? How much of Canadian taxpayers' money should be spent to assert and maintain Canada's sovereignty in the Arctic?

Higher sea levels lead to beach erosion, coastal flooding, intrusion of saltwater into aquifers, and other impacts. In 1987, unusually high waves struck the

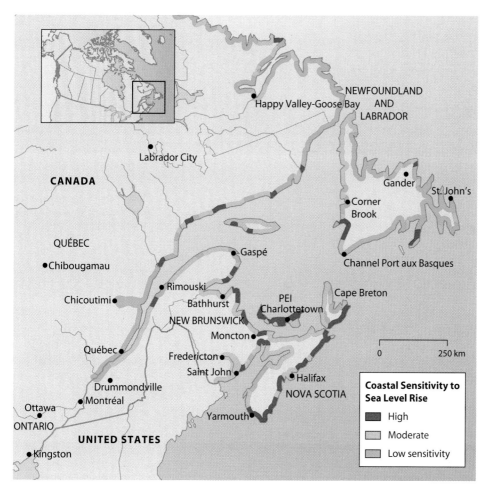

FIGURE 14.21
This map shows the sensitivity of the coastlines of Canada to the expected rise in sea level as a result of climate change. Sensitivity is the degree to which a coastline may experience physical changes, such as flooding, erosion, beach migration, and coastal dune destabilization. Two major regions of high sensitivity are identified: Atlantic Canada and parts of the Beaufort Sea coast.
Source: The Atlas of Canada, NRCAN, Coastal Sensitivity to Sea-Level Rise, http://atlas.nrcan.gc.ca/site/english/index.html.

Maldives and triggered a campaign to build a large seawall around Male, the nation's capital. Known as "The Great Wall of Male," the seawall is intended to protect buildings and roads by dissipating the energy of incoming waves during storm surges. A *storm surge* is a temporary and localized rise in sea level brought on by the high tides and winds associated with storms. The higher that sea level is to begin with, the farther inland a destructive storm surge can reach. "With a mere one-metre rise [in sea level]," Maldivian president Maumoon Abdul Gayoom warned in 2001, "a storm surge would be catastrophic, and possibly fatal to the nation."

The Maldives is not the only nation with such concerns. In fact, among island nations, the Maldives has fared better than many others. It saw a sea-level rise of about 2.5 mm per year throughout the 1990s, but most Pacific islands are experiencing greater changes. Regions experience differing amounts of sea-level change because land elevations may be rising or subsiding naturally, depending on local geologic conditions.

At the end of the twenty-first century, the IPCC predicts mean sea level to be 18–59 cm higher than today's, depending upon the emission scenario. However, these estimates do not take into account findings on accelerated ice melting in Greenland, because that research is so new that it has not yet been incorporated into climate models. If Greenland's melting continues to accelerate, then sea levels will rise more quickly.

Rising sea levels will force hundreds of millions of people to choose between moving upland or investing in costly protections against high tides and storm surges. Canada's coastal regions will be variably affected, as shown in **FIGURE 14.21**. Low-lying coastal locations, such as parts of the St. Lawrence estuary, Nova Scotia, Cape Breton, and Prince Edward Island, are particularly vulnerable to flooding. Some areas of the far north, including the coasts of the Beaufort Sea, and low-lying areas of the B.C. coast are also at risk of both flooding and accelerated coastal erosion.

Worldwide, densely populated regions on low-lying river deltas, such as the Ganges–Brahmaputra Delta in Bangladesh and the Irrawaddy Delta of Burma, will be most affected. So will storm-prone regions, coastal cities, and areas around the world where land is subsiding. Many Pacific islands will need to be evacuated, and some nations, such as Tuvalu and the Maldives, fear for their very existence. In the meantime, the Maldives is likely to suffer from shortages of freshwater because rising seas threaten to bring saltwater into the nation's wells, as the 2004 tsunami did. The contamination of groundwater and soils by seawater threatens not only island nations like the Maldives but all coastal areas that depend on small

THE SCIENCE BEHIND THE STORY

Greenland's Glaciers Race to the Sea

Outlet glaciers melt into Scoresby Sund, Greenland's largest fjord.

Scientists have known for years that the Arctic is bearing the brunt of global warming and that the massive ice sheet covering Greenland is melting around its edges. But data from 1993 to 2003 showed Greenland's ice loss accounting for only 4%–12% of global sea-level rise, about 0.21 mm/yr. And as authors of the IPCC's *Fourth Assessment Report* used results from climate models to predict Greenland's future contributions to sea-level rise, the models told them to expect more of the same.

However, some brand-new research hadn't made it into the models. Scientists studying how ice moves were learning that ice sheets can collapse more quickly than expected and that Greenland's ice loss is accelerating. As a result, they said, the IPCC report underestimates the likely speed and extent of future sea-level rise.

Greenland's ice sheet is massive, nearly 3 km at its thickest point, and covers approximately 1.7 million km²—an area larger than the entire province of Quebec. If the entire ice sheet were to melt, global sea level would rise by a whopping 7 m. The ice sheet, like all perennial ice masses, gains mass by accumulating snow during cold weather, which becomes packed into ice over time. It loses mass as surface ice melts in warm weather, generally at the periphery, where ice is thinnest or contacts seawater. If melting and runoff outpace accumulation, the ice sheet shrinks.

Researchers are now learning that the internal physical dynamics of how ice moves may be even more important. These dynamics can speed the flow of immense amounts of ice in *outlet glaciers* downhill toward the coast, where eroding ice sloughs off and melts into the sea. The first good indication of this process came in 2002 when a team of researchers noted that ice in outlet glaciers flows more quickly during warm months, when pools of meltwater form on the surface and leak down through crevasses and vertical tunnels called *moulins* to the bottom of the glacier. There, the water runs downhill in a layer between bedrock and ice, lubricating the bedrock surface and enabling the ice to slide downhill like a car hydroplaning on a wet road. In addition, the meltwater weakens ice on its way down and warms the base of the glacier, melting some of it to create more water (see the figure).

Other scientists have proposed that warming ocean water melts ice shelves along the coast, depriving outlet glaciers of the buttressing support that holds them in place. Without a floating ice tongue at its terminus, a glacier slides into the ocean more readily. These physical dynamics represent positive feedback, researchers said; once global warming initiates these processes, they encourage further melting. As such, it is likely that Greenland's melting will accelerate.

Researchers have now determined that Greenland lost a grand total of 91 km³ of ice in 1996, 138 km³ in 2000, and 224 km³ in 2005. This last amount exceeds all the water consumed in Canada in almost two decades! Clearly, Greenland's ice losses are accelerating. Scientists will be keeping a close eye on the ice sheet to see whether it continues to discharge water more and more quickly. If it does, climate modellers will have several years in which to incorporate into their models the new and evolving understanding of the physical dynamics of ice—just in time for the next IPCC report, scheduled for 2013.

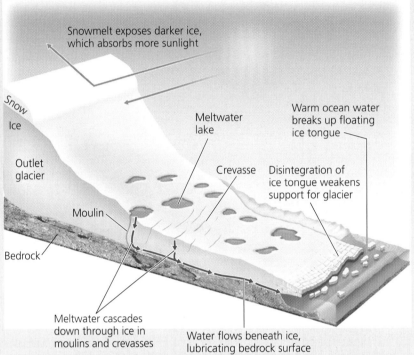

Outlet glaciers from Greenland's ice sheet are accelerating their slide into the ocean because meltwater is descending through moulins and crevasses, lubricating the bedrock surface. Moreover, melting snow exposes darker ice, which absorbs sunlight and speeds melting. Finally, breakup of floating ice tongues weakens support for the glacier.

lenses of freshwater floating atop saline groundwater in coastal aquifers.

Maldives residents also worry about damage to the marine ecosystems that are critical for their economy, including coral reefs. Coral reefs provide habitat for important food fish that are consumed locally and exported; offer snorkelling and scuba diving sites for tourism; and reduce wave intensity, protecting coastlines from erosion. Around the world, rising seas will eat away at the coral reefs, mangrove forests, and salt marshes that serve as barriers protecting our coasts.

Climate change poses two additional threats to coral reefs: warmer waters are causing coral bleaching, and enhanced CO_2 concentrations in the atmosphere are changing ocean chemistry. As ocean water absorbs atmospheric CO_2, it becomes more acidic, which impairs the growth of coral and other organisms whose exoskeletons consist of calcium carbonate. The oceans have already decreased by 0.1 pH unit (a fairly large amount), and are predicted to decline in pH by 0.14–0.35 more units over the next 100 years.

Climate change affects organisms and ecosystems

The many changes in Earth's physical systems have direct consequences for life on our planet. Organisms are adapted to their environments, so they are affected when those environments are altered. As global warming proceeds, it modifies temperature-dependent biological phenomena. For instance, in the spring, birds are migrating earlier, insects are hatching earlier, and animals are breeding earlier. Plants are leafing out earlier, too—an effect confirmed by satellite photography that records whole landscapes "greening up" each year.

---

weighing the issues

ENVIRONMENTAL REFUGEES?

14-3

Citizens of the Maldives see an omen of their future in the Pacific island nation of Tuvalu, which has been losing 9 cm of elevation per decade to rising seas. Appeals from Tuvalu's 11 000 citizens were heard by New Zealand, which began accepting them in small numbers as of 2003, although the government has not officially categorized them as **environmental refugees**—people who have been driven from their homelands as a result of environmental change or natural disaster. Tuvaluans have been particularly vocal about global warming, but several other small Pacific island nations have joined in voicing their concerns.

Will there come a time when neighbouring countries should begin to treat people who leave small island nations as environmental refugees? Should they be doing it now? What will happen to these people after relocation—do you think a national culture can survive if its entire population is relocated?

These changes in seasonal timing are expected to continue, and they are having complex effects. For instance, European birds known as great tits time their breeding cycle so that their young hatch and grow at the time of peak caterpillar abundance. However, as plants leaf out earlier and insects emerge earlier, research shows that great tits are not breeding earlier. As a result of the mismatch in timing, fewer caterpillars are available when young birds need them, and fewer birds survive. Although some organisms will no doubt adapt to such changes in seasonal timing, research so far shows that in most cases mismatches occur.

A similar situation provided an unexpected benefit and is helping to prevent outbreaks of the invasive parasitic spruce budworm on Vancouver Island. Budworm development is triggered by temperature, while the development of the Douglas fir, the preferred food of the budworm larva, is triggered more by light and day length. As a result of a 90-year increase in winter temperatures on southern Vancouver Island, the budworm larvae are emerging from their shelters earlier, at a time when the Douglas fir, their main food source, is not yet available.

Biologists have recorded spatial shifts in the ranges of organisms, with plants and animals moving toward the poles or upward in elevation (toward cooler regions) as temperatures warm. As these trends continue, some organisms will not be able to cope, and as many as 20%–30% of all plant and animal species could be threatened with extinction, the IPCC estimates. Trees may not be able to shift their distributions fast enough. Animals adapted to montane environments may be forced uphill until there is nowhere left to go (as we saw with Monteverde's organisms in Chapter 3). Rare species finding refuge in protected preserves may be forced out of preserves into developed areas, making such refuges far less effective tools for conservation.

The American pika (*Ochotona princeps*), a small mammal that lives on rocky slopes near mountain glaciers in southern Canada, may be the first animal in North America to become extinct as a result of climatic warming. The pika (**FIGURE 14.22**), while clearly not the only animal likely to suffer from warming (polar bears come to mind), is extremely sensitive to environmental conditions, especially temperature. Pikas cannot survive if temperatures are too high. Their favoured habitat, in the vicinity of glaciers, is also threatened as a result of rapidly changing conditions in the glacial environment.

These changes will greatly affect species interactions, and scientists foresee major modifications in the structure and function of communities and ecosystems. In regions where precipitation and stream flow increase, erosion and flooding will pollute and alter aquatic systems.

FIGURE 14.22
Pikas live near glaciers in southern Canada. They are extremely sensitive to temperature, and may become one of the first animals to be driven to extinction by climatic warming.

In regions where precipitation decreases, lakes, ponds, wetlands, and streams will diminish, affecting aquatic organisms, as well as human health and well-being. Acidification of the oceans may pose major threats for corals and other marine animals. Given that corals will likely also suffer increased bleaching from thermal stress, the world's coral reefs are expected to decline substantially. This would reduce marine biodiversity significantly, because so many other organisms depend on living coral reefs for food and shelter.

Effects on plant communities are an important component of climate change (**FIGURE 14.23**). By drawing in CO_2 for photosynthesis, plants act as sinks for carbon. If climate change increases vegetative growth, this could help mitigate carbon emissions, in a process of negative feedback. However, if climate change decreases plant growth (through drought or fire, for instance), then positive feedback could increase carbon flux to the atmosphere. The many impacts on ecological systems will reduce the ecosystem goods and services we receive from nature and that our societies depend on, from food to clean air to drinking water.

Climate change exerts societal impacts—and vice versa

Human society has begun to feel the impacts of climate change. Damage from drought, flooding, hurricanes, storm surges, and sea-level rise, as discussed above, has already taken a toll on the lives and livelihoods of millions of people. However, climate change will have additional consequences for humans, including impacts on agriculture, forestry, economics, and health. **FIGURE 14.24** gives an overview of the complex interactions between human systems and the systems of the natural world in determining and reacting to climatic change.

Let us look in a bit more detail at some of the most important human drivers and impacts of climate change.

Agriculture For farmers, earlier springs require earlier crop planting. For some crops in the temperate zones, production may increase slightly with moderate warming, because growing seasons become longer and because more carbon dioxide is available to plants for photosynthesis. However, rainfall will shift in space and time, and in areas where droughts and floods become more severe, these will cut into agricultural productivity. Overall, global

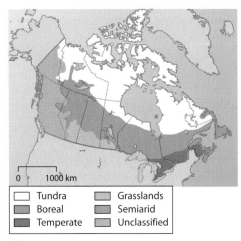

(a) **Present day**

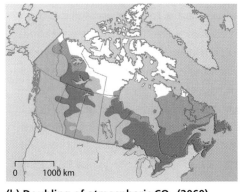

(b) **Doubling of atmospheric CO_2 (2060)**

FIGURE 14.23
These maps, based on the Canadian Climate Model, show how forests and grasslands may be expected to shift in response to a doubling of carbon dioxide in the atmosphere by the year 2060 **(b)**, as compared with today **(a)**.
Source: Government of Canada, National Animal Health Strategy.[3]

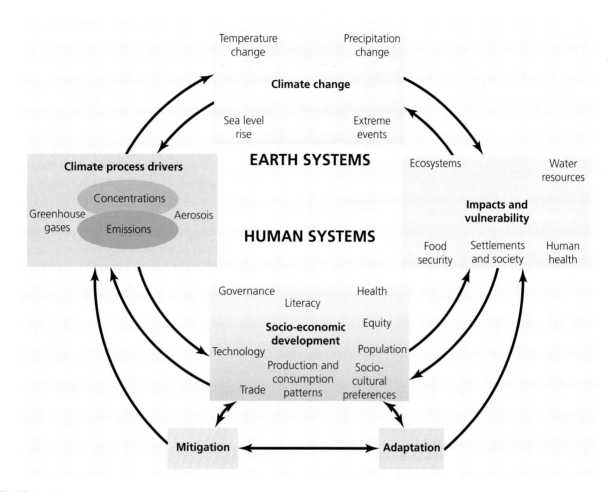

FIGURE 14.24
Human and natural systems interact in complex ways to both influence and react to climatic change, as shown in this diagram from the Fourth Assessment Report of the IPCC.
Source: Intergovernmental Panel on Climate Change. 2007. Fourth assessment report.

crop yields are predicted to increase somewhat, but beyond a rise of 3°C, the IPCC expects crop yields to decline. In seasonally dry tropical and subtropical regions, growing seasons may be shortened, and harvests may be more susceptible to drought and crop failure. Thus, scientists predict that crop production will fall in these regions even with minor warming. This would worsen hunger in many of the world's developing nations.

It is tempting to think that a warmer North would mean that Canada's agricultural region would expand, or, at worst, shift toward the north. This is unlikely to happen. Recall from Chapter 7 that it takes a very long time for weathering and the accumulation of organic material to lead to the formation of soil profiles. Much of Canada's North is bare rock, covered by a thin, rocky, immature soil. The rock of the Canadian Shield was scraped clean by advancing glaciers during the last major glaciation in the Pleistocene Epoch (please see Appendix D on the Companion Website at www.myenvironmentplace.ca). The Pleistocene glaciers retreated from Canada's North more than 10 000 years ago, but in all that time there has been no significant development of a tillable soil horizon over much of the land area.

Forestry Forest managers increasingly find themselves having to battle insect and disease outbreaks, invasive species, and catastrophic fires, which are mostly caused by decades of fire suppression but are also promoted by longer, warmer, drier fire seasons (**FIGURE 14.25**). For timber and forest products, enriched atmospheric CO_2 may spur greater growth in the near term, but this will vary substantially from region to region. Other climatic effects, such as drought, may eliminate these gains. For instance, droughts brought about by a strong El Niño in 1997–1998 allowed immense forest fires to destroy millions of hectares of rainforest in Indonesia, Brazil, Mexico, and elsewhere.

Health As a result of climate change, we will face more heat waves—and heat stress can cause death, especially among older adults. A 2003 heat wave killed at least

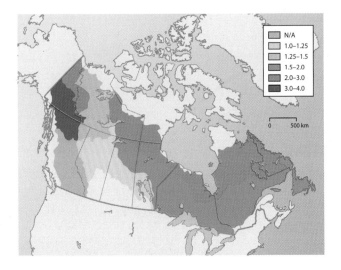

FIGURE 14.25
This map shows the ratio of fire activity at the end of the twenty-first century to current fire activity, derived from the Canadian Climate Model. Values greater than 1.0 mean there will be more fire in the future, and values less than 1.0 mean less fire in the future.
Source: Canadian Forestry Service, NRCAN, http://cfs.nrcan.gc.ca/news/460.

weighing the issues: AGRICULTURE IN A WARMER WORLD 14-4

The IPCC predicts that slight warming could shift agricultural belts toward the poles and marginally increase global crop production. How might this shift affect Canada's North? Locate the nations of Russia and Argentina on a world map, and hypothesize how such a poleward shift of agriculture might affect these nations. Now locate India, Nigeria, and Ethiopia, and hypothesize how the shift might affect them. Given that many developing nations near the equator are already suffering from food shortages and soil degradation, what effects do you think that global warming may have? If climate change magnifies inequities between developed and developing nations, what could we do to alleviate this problem?

---

35 000 people in Europe, for example. In addition, a warmer climate exposes us to other health problems:

- Respiratory ailments that result from air pollution, as hotter temperatures promote formation of photochemical smog
- Expansion of tropical diseases, such as dengue fever, into temperate regions as vectors of infectious disease (such as mosquitoes) move toward the poles; the advent of mosquito-borne West Nile Virus in southern Canada may be a preview of this scenario
- Disease and sanitation problems that occur when floods overcome sewage treatment systems
- Injuries and drownings that will likely increase if storms become more frequent or intense
- Hunger-related ailments that will worsen as the human population grows and climate-related stresses on agricultural systems increase
- Health hazards from cold weather will likely decrease, but researchers feel that the increase from warm-weather hazards will more than offset these gains

Economics People will experience a variety of economic costs—and some benefits, too—from the many impacts of climate change. On the whole, researchers predict that costs will outweigh benefits and that this gap will widen as climate change grows more severe. Climate change will also widen the gap between rich and poor, both within and among nations. Poorer people have less wealth and technology with which to adapt to climate change, and poorer people rely more on resources (such as local food and water) that are particularly sensitive to climatic conditions.

From a wide variety of economic studies, the IPCC estimated that climate change will cost 1%–5% of GDP on average globally, although poor nations would lose proportionally more than rich nations. Economists trying to quantify damages from climate change by measuring the "social cost of carbon" (i.e., external costs) have proposed costs of anywhere from $10 to $350 per ton of carbon. The highest-profile economic study to date has been the *Stern* Review commissioned by the British government. This study maintained that climate change could cost roughly 5%–20% of World GDP by the year 2200 but that investing just 1% of GDP starting now could enable us to avoid these future costs. Regardless of the precise numbers, many economists and policy makers are concluding that spending money now to mitigate climate change will save us a great deal more in the future.

All these physical, biological, and social impacts of climate change are consequences of the warming effect of our greenhouse gas emissions. We are bound to experience further consequences, but by addressing the root causes of anthropogenic climate change we can still prevent the most severe future impacts.

Are we responsible for climate change?

The IPCC's 2007 report concluded conservatively that it is more than 90% likely that most of the global warming recorded over the past half-century is due to the well-documented increase in greenhouse gas concentrations in our atmosphere. Scientists agree that this increase in

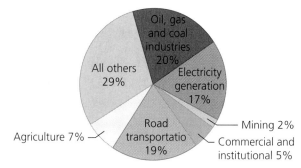

FIGURE 14.26
Greenhouse gas emissions in Canada come primarily from the use of fossil fuels for both energy and transportation.
Source: Environment Canada.[4]

greenhouse gases results primarily from our combustion of fossil fuels for energy and transportation, and secondarily from land use changes, including deforestation and agriculture (**FIGURE 14.26**).

By the time the IPCC's *Fourth Assessment Report* came out, many scientists had already become concerned enough to put themselves on record urging governments to address climate change. In June 2005, as the leaders of the G8 industrialized nations met, the national academies of science from 11 nations (Canada, Brazil, China, France, Germany, India, Italy, Japan, Russia, the United Kingdom, and the United States) issued a joint statement urging these political leaders to take action. Such a broad consensus statement from the world's scientists was virtually unprecedented, on any issue. The statement read, in part,

> *The scientific understanding of climate change is now sufficiently clear to justify nations taking prompt action. It is vital that all nations identify cost-effective steps that they can take now, to contribute to substantial and long-term reduction in net global greenhouse gas emissions . . . A lack of full scientific certainty about some aspects of climate change is not a reason for delaying an immediate response that will, at a reasonable cost, prevent dangerous anthropogenic interference with the climate system.*

Today the debate concerning the human role in climate change is largely over, and most North Americans accept that our fossil fuel consumption is altering the planet that our children will inherit. The tide turned with the 2007 release of the IPCC report, the broad popularity of Al Gore's 2006 movie and book, *An Inconvenient Truth*, grassroots activism, and action by political leaders at all levels.

As a result of this shift in public perception, and in response to demand from their shareholders, many corporations and industries began offering support for reductions in greenhouse gas emissions. These corporate leaders joined ranks with the insurance industry, which many years earlier had grown concerned with climate change as it foresaw increased payouts for damage caused by coastal storms, drought, and floods.

Responding to Climate Change

Today we possess a broadened consensus that climate change is a clear and present challenge to our society. Precisely how we should respond to climate change is a difficult question, however, and one we will likely be wrestling with for decades. The strategy that we choose, as a society, will determine how successful we are at curbing climate change, and how quickly (or *if*) we can make it happen. The IPCC bases its climate predictions on four groups of potential scenarios of human response to climate change (these are called the *SRES* scenarios, which stands for *Special Report on Emissions Scenarios*):

1. Scenario "A1" assumes a world of very rapid economic growth, a global population that peaks in mid-century, and rapid introduction of new and more efficient technologies. A1 is divided into three groups that describe alternative directions of technological change: fossil-intensive energy resources, non-fossil energy resources, and a mix of energy sources.
2. Scenario "B1" describes a world with the same global population as A1 but with more rapid changes in economic structures toward a service- and information-based economy.
3. Scenario "B2" describes a world with intermediate population and economic growth, emphasizing local solutions to economic, social, and environmental sustainability.
4. Scenario "A2" describes a world with high population growth, slow economic development, and slow advancement to lower-emission energy technologies.[5]

These are the same four scenarios that are represented in **FIGURE 14.19**, above, showing the potential for major changes in Arctic ice. As shown in **FIGURE 14.27**, each set of scenarios represents a different set of human responses to climate change, with different implications for the rapidity and vigour—and thus the specific outcomes and success—of the response.

Shall we pursue mitigation, or adaptation?

We can respond to climate change in two fundamental ways: mitigation or adaptation.

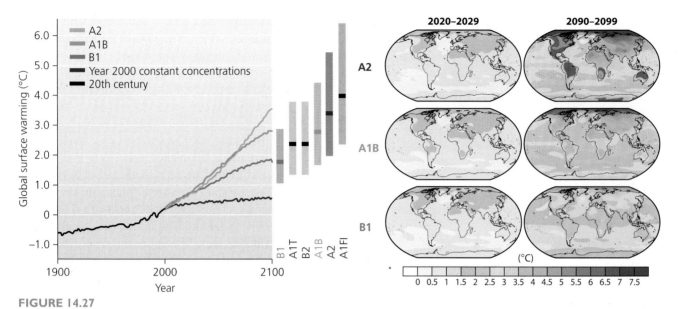

FIGURE 14.27
Here (a) are three scenarios studied by the IPCC for emissions of carbon dioxide and related warming. The pink line shows projected temperature change if carbon dioxide emissions could be held at year 2000 levels, rather than being allowed to increase with population and economic growth. The bars (b) show the range of temperature increases predicted for each of the scenarios. The maps (c) show the timing and distribution of temperature changes resulting from the various scenarios. These predictions are based on a large number of climate models. Data from Intergovernmental Panel on Climate Change. 2007. *Fourth assessment report.*

For **mitigation**, the aim is to mitigate, or alleviate, the problem. In this case we would choose to pursue actions that reduce greenhouse gas emissions, in order to lessen the severity of future climate change. For example, mitigation strategies can focus on reducing greenhouse gas emissions by improving energy efficiency, switching to clean and renewable energy sources, encouraging farm practices that protect soil quality, recovering landfill gas, and preventing deforestation.

The second type of response is to accept that climate change is happening and to pursue strategies to minimize its impacts on us. This strategy is called **adaptation** because the goal is to adapt to change by finding ways to cushion oneself from its blows. Erecting a seawall, like Maldives residents did with the Great Wall of Male, is an example of adaptation using technology and engineering. The people of Tuvalu also adapted, but with a behavioural choice—some chose to leave their island and make a new life in New Zealand. Other examples of adaptation include restricting coastal development; adjusting farming practices to cope with drought; and modifying water management practices to deal with reduced river flows, glacial outburst floods, or salt contamination of groundwater.

James Ford from McGill University, along with other Canadian researchers, has studied Inuit adaptation to climate change in Canada's North. Ford is a geographer, with an interest in global environmental change and natural hazards. He focuses his research specifically on the Arctic, where his doctoral thesis work investigated the vulnerability of a particular community in Nunavut to the effects of climate change. This included, for example, working with hunters to determine how they have been affected by changes in the extent and thickness of sea ice, and in the migratory behaviour of wildlife.[6] This research has contributed to adaptation plans aimed at reducing the vulnerability of communities in Nunavut and elsewhere in the Arctic to the effects of climate change.

Many environmental advocates have criticized adaptation strategies, because they view them as escapist—a way of sidestepping the hard work that must be done to protect future generations from climate change. However, adaptation and mitigation are not mutually exclusive approaches; a person or a nation can pursue both. Indeed, both approaches are necessary.

Adaptation strategies are needed, however. Even if we were to halt all our emissions now, global warming would continue until the planet's systems reached a new equilibrium, with temperature rising an additional 0.6°C, and considerably more in such areas as Canada's North, by the end of this century. Because we will face this change no matter what we do, it is wise to develop ways to minimize its impacts. Canada has been a world leader in assessing and developing adaptation strategies and produced a national assessment of climate change impacts and adaptation, *The Canada Country Study*, in 1998.

Mitigation is also necessary. If we do nothing to slow climate change, it will eventually overwhelm any efforts at adaptation we could make. To leave a sustainable future for our civilization and to safeguard the living planet that we know, we will need to pursue mitigation. Moreover, the faster we begin reducing our emissions, the lower the level at which they will peak, and the less we will alter climate. We will spend the remainder of this chapter examining approaches to the mitigation of climate change.

We can look more closely at our lifestyle

In Canada, major sources of GHG emissions include power generation facilities that produce electricity, heat, or steam by using fossil fuels; oil and gas extraction; mining, smelting, and refining of metals, and steel production; pulp, paper, and saw mills; petroleum refineries; and chemical producers. Other activities that contribute significantly to GHG emissions include transportation, waste disposal, and agriculture-related activities.[7] Some of these are large-scale industrial causes, but many can be influenced by the simple choices we all make in our everyday lives.

The generation of electricity is one significant source of carbon dioxide emissions. From cooking and heating to the clothes we wear, much of what we own and do depends on electricity. Although Canada's electricity comes mainly from hydroelectric sources (58%), 25% comes from fossil fuel–related processes. Therefore, reducing the volume of fossil fuels we burn to generate electricity would lessen greenhouse gas emissions, as would decreasing electricity consumption. There are two ways to reduce the amount of fossil fuels we use: (1) encouraging conservation and efficiency and (2) switching to cleaner and renewable energy sources (Chapters 16 and 17).

Conservation and efficiency Conservation and efficiency in energy use can arise from new technologies, such as high-efficiency light bulbs and appliances, or from individual ethical choices to reduce electricity consumption. For instance, replacing an old washing machine with an energy-efficient washer can cut your CO_2 emissions by 200 kg annually. Replacing standard light bulbs with compact fluorescent lights reduces energy use for lighting by 40%. Such technological solutions are popular, and they can be profitable for manufacturers while also saving consumers money.

Consumers can also opt for lifestyle choices. For nearly all of human history, people managed without the electrical appliances that most of us take for granted today. It is possible for each of us to choose to use fewer greenhouse gas-producing appliances and technologies and to take practical steps to use electricity more efficiently.

Sources of electricity We can reduce greenhouse gas emissions by altering the types of energy we use. Among fossil fuels, natural gas burns more cleanly than oil, and oil is cleaner-burning than coal. Using natural gas instead of coal produces the same amount of energy with roughly one-half the emissions. Moreover, approaches to boosting the efficiency of fossil fuel use, such as cogeneration, produce fewer emissions per unit of energy generated.

Currently, interest in carbon capture and **carbon sequestration** or **carbon storage** is intensifying. **Carbon capture and storage (CCS)** refers to technologies or approaches that remove carbon dioxide from power plant emissions. Successful carbon capture technology would allow power plants to continue using fossil fuels while cutting greenhouse gas pollution. The carbon would then be stored somewhere—perhaps underground, under pressure, in locations where it would not seep out. However, we are still a long way from developing adequate technology and secure storage space to accomplish this, and some experts doubt that we will ever be able to sequester enough carbon to make a dent in our emissions.

---

weighing the issues   **NUCLEAR POWER OR FOSSIL FUELS?**   14-5

Some environmentalists support the use of nuclear power for the generation of electricity, since it is a "clean" energy source that does not add to greenhouse gas emissions or other types of air pollution. Others oppose it fiercely, citing problems with the disposal of highly toxic and radioactive waste, among other reasons. What do you think about nuclear energy? Is it an "environmental" choice? We will return to this question in Chapter 16.

Technologies and energy sources that generate electricity without using fossil fuels represent another means of reducing greenhouse gas emissions. These include nuclear power, hydroelectric power, geothermal energy, photovoltaic cells, wind power, and ocean energy sources. These energy sources give off no emissions during their use (but some in the production of their infrastructure). We will examine these clean and renewable energy sources in detail in Chapters 16 and 17.

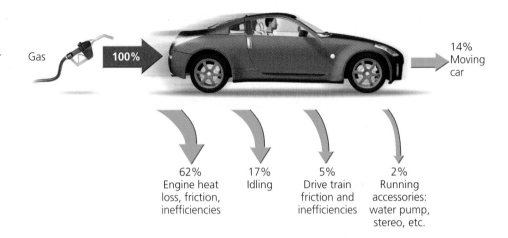

FIGURE 14.28
Conventional automobiles are extremely inefficient. Almost 85% of useful energy is lost, and only 14% actually moves the car down the road.

Transportation is a significant source of greenhouse gases

The typical automobile is highly inefficient. Close to 85% of the fuel you pump into your gas tank does something other than move your car down the road (**FIGURE 14.28**). Although more aerodynamic designs, increased engine efficiency, and improved tire design could help reduce these losses, gasoline-fuelled automobiles may always remain somewhat inefficient.

Automotive technology The technology exists to make our vehicles more fuel-efficient than they currently are. Raising fuel efficiency will require government mandate and/or consumer demand. As gasoline prices rise, demand for more fuel-efficient automobiles will intensify. Advances in technology are also bringing us alternatives to the traditional combustion-engine automobile. These include hybrid vehicles that combine electric motors and gasoline-powered engines for greater efficiency. They also include fully electric vehicles, alternative fuels, such as compressed natural gas and biodiesel, and hydrogen fuel cells that use oxygen and hydrogen and produce only water as a waste product.

Driving less and using public transportation
People can also opt to make lifestyle choices that reduce their reliance on cars. For example, some people are choosing to live nearer to their workplaces. Others use mass transit, such as buses, subway trains, and light rail. Still others bike or walk to work or for their errands (**FIGURE 14.29**). Canadians use public transportation for approximately 7% of daily transportation needs. Increasing this level of use would be an extremely efficient way to reduce energy use and pollution. Unfortunately, reliable and convenient public transit is not available in many communities. Making automobile-based cities and suburbs more friendly to pedestrian and bicycle traffic and improving people's access to public transportation are central challenges for city and regional planners.

FIGURE 14.29
By choosing human-powered transportation methods, such as bicycles, we can greatly reduce our transportation-related greenhouse gas emissions. More people are choosing to live closer to their workplaces and to enjoy the dual benefits of exercise and reduced emissions by walking or cycling to work or school.

We can reduce emissions in other ways as well

Other pathways toward mitigating climate change include advances in agriculture, forestry, and waste management.

In agriculture, sustainable land management that protects the integrity of soil on cropland and rangeland enables soil to store more carbon. Techniques have also been developed to reduce the emission of methane from rice cultivation and from cattle and their manure, and to reduce nitrous oxide emissions from fertilizer. We can also grow renewable biofuels, and this is an active area of current research.

In forest management, the rapid reforestation of cleared areas helps restore forests, which act as reservoirs that pull carbon from the air. Sustainable forestry practices and the preservation of existing forests can help to reverse the carbon dioxide emissions resulting from deforestation.

Waste managers are doing their part to cut greenhouse emissions by recovering methane seeping from landfills, treating wastewater, and generating energy from waste in incinerators. Individuals, communities, and waste haulers also help reduce emissions by encouraging recycling, composting, and the reduction and reuse of materials and products.

We will need to follow multiple strategies to reduce emissions

We should not expect to find a single magic bullet for mitigating climate change. Instead, reducing emissions will require many steps by many people and institutions across many sectors of our economy. The good news is that most reductions can be achieved by using current technology and that we can begin implementing these changes right away.

In the long term, any one approach will not be enough. To stop climate change, we will need to reduce emissions (as opposed to stabilizing them), and this may require us to develop new technology, further change our lifestyles, and/or reverse our population growth. However, there is plenty we can do in the meantime to mitigate climate change simply by scaling up the technologies and approaches for emissions reductions that we already have developed.

How quickly and successfully we translate science and technology into practical solutions for reducing emissions depends largely on the policies we urge our leaders to pursue and on how government and the market economy interact. As we will see in Chapter 22 and in numerous instances throughout this book, governmental command-and-control policy has been vital in safeguarding environmental quality and promoting human well-being. However, government mandates are often resisted by industry, and market incentives can sometimes be more effective in driving change.

With climate change policy, we are in the midst of a dynamic period of debate and experimentation. At all levels—international, national, provincial/territorial, regional, and local—policy makers, industry, commerce, and citizens are searching for ways to employ government and the market to reduce emissions in ways that are fair, economically palatable, effective, and enforceable.

We began tackling climate change by international treaty

In 1992, the United Nations convened the U.N. Conference on Environment and Development Earth Summit in Rio de Janeiro, Brazil. Nations represented at the Earth Summit signed the *U.N. Framework Convention on Climate Change* (*FCCC*), which outlined a plan for reducing greenhouse gas emissions to 1990 levels by the year 2000 through a voluntary, nation-by-nation approach. By the late 1990s, it was already clear that the voluntary approach was not likely to succeed. After watching the seas rise and observing the failure of most industrialized nations to cut their emissions, nations of the developing world—the Maldives among them—helped initiate an effort to create a binding international treaty that would *require* all signatory nations to reduce their emissions. This effort led to the *Kyoto Protocol*.

The Kyoto Protocol seeks to limit emissions

An outgrowth of the FCCC drafted in 1997 in Kyoto, Japan, the **Kyoto Protocol** mandates signatory nations, by the period 2008–2012, to reduce emissions of six greenhouse gases to levels below those of 1990 (Table 14.2). Canada was the 99th country to ratify the agreement, signing in 2002. The treaty took effect in 2005 after Russia became the 127th nation to ratify it.

The United States has continued to refuse to ratify the Kyoto Protocol. U.S. leaders have called the treaty unfair because it requires industrialized nations to reduce emissions but does not require the same of rapidly industrializing nations, such as China and India, whose greenhouse emissions have risen more than 50% in the past 15 years. Proponents of the Kyoto Protocol say the differential requirements are justified because industrialized nations created the current problem and therefore should take the

Table 14.2	Emissions Reductions Required and Achieved	
Nation	Required change 1990–2008/2012*	Observed change 1990–2006†
Russia	0.0%	−32.0%‡
Germany	−21.0%	−17.2%
United Kingdom	−12.5%	−14.3%
France	0.0%	−0.8%
Italy	−6.5%	+12.1%
Japan	−6.0%	+6.5%
United States	−7.0%	+15.8%
Canada	−6.0%	+29.1%

*Percentage decrease in emissions (carbon-equivalents of six greenhouse gases) from 1990 to period 2008–2012, as mandated under Kyoto Protocol.
†Actual percentage change in emissions (carbon-equivalents of six greenhouse gases) from 1990 to 2004. Negative values indicate decreases; positive values indicate increases. Values do not include influences of land use and forest cover.
‡Russia's substantial decrease was due mainly to economic contraction following the breakup of the Soviet Union.
Data from U.N. Framework Convention on Climate Change, National Greenhouse Gas Inventory Reports, 2007.

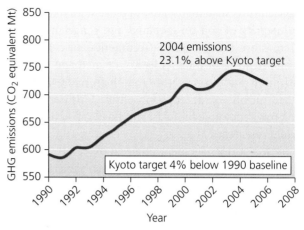

FIGURE 14.30
Total carbon dioxide emissions in Canada show a decrease over the three to four years before 2006; even so, emissions are still considerably above Canada's Kyoto target.
Source: Environment Canada, National GHG Inventory.[8]

lead in resolving it. The refusal of the United States to join international efforts to curb greenhouse emissions has generated resentment among its allies and has undercut the effectiveness of these efforts.

Because resource use and per capita emissions are far greater in the industrialized world, governments and industries there often feel they have more to lose economically from restrictions on emissions. Ironically, industrialized nations are also the ones most likely to gain economically, because they are best positioned to invent, develop, and market new technologies to power the world in a post–fossil fuel era.

As of 2004 (the most recent year with full international data), nations that signed the Kyoto Protocol had decreased their emissions overall by 3.3% from 1990 levels. However, much of this reduction was due to economic contraction in Russia and nations of the former Soviet Bloc following the breakup of the Soviet Union. When these nations are factored out, the remaining signatories showed an 11.0% *increase* in emissions from 1990 to 2004.

As a signatory to the Kyoto Protocol, Canada is obliged to submit an inventory report of its GHG emissions on an annual basis, using an internationally agreed-to format. The National Inventory monitors six gases—carbon dioxide (CO_2), methane (CH_4), nitrous oxide (N_2O), sulphur hexafluoride (SF_6), perfluorocarbons (PFCs), and hydrofluorocarbons (HFCs)—and provides an analysis of the factors underlying the trends in emissions since 1990. Total greenhouse gas emissions in Canada actually decreased in 2006, by 1.9% from 2005 levels, and 2.8% from 2003 levels (**FIGURE 14.30**). The decrease is due primarily to a change in the mix of sources used for electricity production (reduced coal and increased hydro and nuclear generation), lower emissions from fossil fuel production (as a result of fuel switching and a smaller volume of oil refined) and reduced demand for heating fuels because of warmer winters in 2004, 2005, and 2006.

Overall, however, emissions in 2006 were about 22% above the 1990 total, which is still about 29.1% above Canada's Kyoto target.[9] At this point, Canada is not just off-target; it would take radical technological and economic changes to fulfill the commitment we have made, as a nation, in signing the Kyoto Protocol. Kyoto Protocol critics and supporters alike acknowledge that even if every nation complied with the treaty, greenhouse gas emissions would continue to increase—albeit more slowly than they would without the treaty. Nations are now looking ahead and negotiating over what will come next to supersede Kyoto.

Market mechanisms are being used to address climate change

As you will learn in Chapter 22, *permit trading* programs represent a means of harnessing the economic efficiency of the free market to achieve policy goals while allowing business, industry, or utilities flexibility in how they meet those goals. Supporters of permit trading programs argue that they provide the fairest, least expensive, and most effective method of achieving emissions reductions. Polluters get to choose how to reduce their emissions and are given financial incentives for reducing emissions below the legally required amount. We will likely discover how successful these ventures are over the next decade as various carbon trading programs are implemented around the world.

The world's first emissions trading program for greenhouse gas reduction (operating since 2003) is the Chicago Climate Exchange, which now boasts hundreds of member corporations, institutions, and municipalities, mainly in the United States, Canada, and Brazil, but recently beginning to include other countries as well. This legally binding trading system imposes a 6% reduction on overall emissions by 2010.

The world's largest cap-and-trade program is the European Union Emission Trading Scheme, which began on January 1, 2005. All EU member states participate, and each submits for approval a national allocation plan that conforms to the nation's obligations under the Kyoto Protocol. This market got off to a successful start, and carbon prices reached €30 per ton in 2006. However, when investors determined that national governments had allocated too many emissions permits to their industries, the price of carbon fell. The overallocation gave companies little incentive to reduce emissions, so permits lost their value. By early 2007, prices in the market had fallen below €0.30. This drop is roughly equivalent to a stock valued at $40 falling to less than 40 cents.

Proponents of emissions trading chalk up the freefall in the European market as a learning experience. Europeans will have the chance to correct their allocations and revive their market beginning in 2008 as the program enters its next phase, when it expands to include more greenhouse gases, more emissions sources, and additional members.

Carbon offsets are in vogue

Emissions trading programs have allowed participants who cannot or will not adequately reduce their own emissions to use *carbon offsets* instead. A **carbon offset** is a voluntary payment to another entity intended to enable that entity to reduce the greenhouse emissions that one is unable or unwilling to reduce oneself. The payment thus offsets one's own emissions. For instance, a coal-burning power plant could pay a reforestation project to plant trees that will soak up as much carbon as the coal plant emits. Or a university could fund the development of clean and renewable energy projects to make up for fossil fuel energy the university uses.

Carbon offsets are becoming popular among utilities, businesses, universities, governments, and individuals trying to achieve **carbon neutrality**, a state in which no net carbon is emitted. For time-stressed people with enough wealth, offsets represent a simple and convenient way to reduce emissions without investing in efforts to change one's habits. For example, you can go to the website of the company TerraPass, calculate your emissions for travel by car or plane, or for your residence, and purchase offsets for those emissions. Your money funds renewable energy and efficiency projects, and you can advertise your donation with bumper stickers and decals.

In principle, carbon offsets seem a great idea, but in practice they often fall short. Without rigorous oversight to make sure that the offset money actually accomplishes what it is intended for, carbon offsets risk being no more than a way for wealthy consumers to assuage a guilty conscience. Efforts to create a transparent and enforceable offset infrastructure are ongoing. If these efforts succeed, then carbon offsets could become an effective and important key to mitigating climate change.

You can reduce your own carbon footprint

Carbon offsets, emissions trading schemes, national policies, international treaties, and technological innovations will all play roles in mitigating climate change. But in the end the most influential factor may be the collective decisions of millions of regular people. In our everyday lives, each one of us can take steps to approach a carbon neutral lifestyle by reducing greenhouse emissions that result from our decisions and activities. Just as we each have an ecological footprint, we each have a **carbon footprint** that expresses the amount of carbon we are responsible for emitting.

You can apply many of the strategies discussed in this chapter in your everyday life—from deciding where to live and how to get to work to choosing appliances. You will encounter still more solutions in our discussions of energy sources, conservation, and renewable energy in Chapters 15 to 17 and elsewhere throughout this book.

Global climate change may be the biggest challenge facing us and our children. Fortunately, it is still early enough that we may be able to avert the most severe impacts. Taking immediate, resolute action is the most important thing that we, personally and as a society, can do.

CANADIAN ENVIRONMENTAL PERSPECTIVES

Sheila Watt-Cloutier

Sheila Watt-Cloutier, former chair of the Inuit Circumpolar Council, has been widely honoured around the world for her work in bringing attention to climate change and its impacts in the Arctic.

- **Inuit educational and health care advocate**
- **Environmental activist** and **world leader on climate change and human rights**
- Former **chair of the Inuit Circumpolar Council**

Sheila Watt-Cloutier was born in the remote town of Kuujjuaq in northern Quebec (the region known in Inuktitut as *Nunavik*). Even today there are no road connections between Kuujjuaq and the south. She grew up with a traditional Inuit way of life, travelling only by dogsled as a child, eating foods gathered from the wild, and participating in hunts with her family and other community members.

Through McGill University, Watt-Cloutier studied psychology and sociology, and subsequently worked as an educational and health care advocate in Nunavik. She was the Inuk advisor to a widely read report on Nunavik's education system, called *Silatunirmut: The Pathway to Wisdom*, and co-wrote, co-produced, and co-directed a video for Inuit youth, *Capturing Spirit: The Inuit Journey*. On the importance of raising youth awareness of traditional ways, she says that "the Inuit strength is our culture, and our young people need to stay in touch with tradition."[10]

In 1995, Watt-Cloutier helped oversee the administration of Inuit land claims under the James Bay and Northern Quebec Agreement. She was elected president of the Inuit Circumpolar Council Canada in 1995, and through that post she became a spokesperson on the issue of *persistent organic pollutants*—chemicals that remain in the environment for a long time, become widely distributed geographically and accumulate in the fatty tissues of people and wildlife. She describes the contamination of the Arctic food chain by POPs as "a sign of the intrusive effects of globalization."[11] She later helped negotiate the Stockholm Convention on POPs, a global treaty intended to protect human health and the environment from persistent chemical contaminants.[12]

In 2002, Watt-Cloutier became the chair of the International ICC, which represents 150 000 Inuit in Canada, the United States, Russia, and Greenland.[13] In 2005, Watt-Cloutier and 62 other Inuit from Canada and Alaska filed a petition with the Inter-American Commission on Human Rights, seeking changes to policy for the impacts of climate change on the traditional Inuit way of life. Watt-Cloutier is listed on the petition as the principal plaintiff. The petition squarely blames "actions and omissions" of the United States, and provides a detailed scientific analysis of the damaging effects of climate change on temperature, wildlife, human health, and cultural well-being in the Arctic. "For every one degree the temperature increases per year globally, it's more like three to five degrees in the Arctic," she says. "For us, this is a monumental change."[14]

Watt-Cloutier feels that her efforts to make the rest of the world aware of the dramatic changes that northerners are already experiencing have put a human face on the impacts of environmental change and may ultimately help connect people to the environment. Watt-Cloutier, who now lives in Iqaluit, Nunavut, still draws her strength from the northern landscape and her deeply ingrained Inuit culture.

She has been widely recognized for her work. She is an Officer of the Order of Canada and a recipient of a Lifetime Achievement Award for Human Development from the United Nations. She has also received the Rachel Carson Prize, the Global Green Award, and an Aboriginal Achievement Award. She was nominated for the Nobel Peace Prize for her work on climate change, an honour many thought she might share with the eventual winner, former U.S. vice-president Al Gore.

"We are the early warning system for the entire planet." —Sheila Watt-Cloutier

Thinking About Environmental Perspectives

Sheila Watt-Cloutier has said that people in the Arctic are deeply and internally connected to the environment because they depend on it for their survival and that—in contrast—people from the south in Canada tend to be disconnected from nature. What do you think? Is it possible to experience a deep, almost spiritual, connection with the environment if you are not living and surviving in the wilderness?

Conclusion

Many factors influence Earth's climate, and human activities have come to play a major role. Climate change is well underway, and further greenhouse gas emissions will increase global warming and cause increasingly severe and diverse impacts. Sea-level rise and other consequences of global climate change will affect locations worldwide from the Maldives to Bangladesh to Ellesmere Island to Florida. As scientists and policy makers come to better understand anthropogenic climate change and its environmental, economic, and social consequences, more and more of them are urging immediate action. Reducing greenhouse gas emissions and taking other actions to mitigate and adapt to climate change represents the foremost challenge for our society in the coming years.

REVIEWING OBJECTIVES

You should now be able to:

Describe Earth's climate system and explain the many factors influencing global climate

- Earth's climate changes naturally over time, but it is now changing rapidly because of human influence.
- The Sun provides most of Earth's energy and interacts with the atmosphere, land, and oceans to drive climate processes.
- Earth absorbs about 70% of incoming solar radiation and reflects about 30% back into space.
- Greenhouse gases, such as carbon dioxide, methane, water vapour, nitrous oxide, ozone, and halocarbons, warm the atmosphere by absorbing infrared radiation and reemitting infrared radiation of different wavelengths.
- Milankovitch cycles influence climate in the long term.

Characterize human influences on the atmosphere and global climate

- Increased greenhouse gas emissions enhance the greenhouse effect.
- By burning fossil fuels, clearing forests, and manufacturing halocarbons, humans are increasing atmospheric concentrations of many greenhouse gases.
- Human input of aerosols into the atmosphere exerts a variable but slight cooling effect.

Summarize modern methods of climate research

- Geologic records, such as cores through ice or sediments, reveal information about past climatic conditions.
- Direct atmospheric sampling tells us about current composition of the atmosphere.
- Coupled general circulation models serve to predict future changes in climate.

Outline current and future trends and impacts of global climate change

- The IPCC has comprehensively synthesized current climate research, and its periodic reports represent the consensus of the scientific community.
- Temperatures on Earth have warmed by an average of 0.74°C over the past century and are predicted to rise 1.8°C–4.0°C over the next century.
- Changes in precipitation vary by region.
- Sea level has risen an average of 17 cm over the past century.
- Other impacts include melting of glaciers and polar ice, frequency of extreme weather events, impacts on agriculture, forestry, and health, and effects on plants and animals.
- Climate change and its impacts will vary regionally.
- Despite some remaining uncertainties, the scientific community feels that evidence for humans' role in influencing climate is strong enough to justify governments' taking action to reduce greenhouse emissions.

Suggest ways we can respond to climate change

- Both adaptation and mitigation are necessary for responding to climate change.
- Conserving electricity, improving efficiency of energy use, and switching to clean and renewable energy sources will help reduce fossil fuel consumption and greenhouse emissions.
- Encouraging new automotive technologies and investment in public transportation will help reduce greenhouse emissions.
- Solving the climate problem will require the deployment of multiple strategies.
- The Kyoto Protocol has provided a first step for nations to begin addressing climate change.
- Emissions trading programs are providing a way to harness the free market and engage industry in reducing emissions.
- Individuals are increasingly exploring carbon offsets and other means of reducing personal carbon footprints.

TESTING YOUR COMPREHENSION

1. What happens to solar radiation after it reaches Earth? How do greenhouse gases warm the lower atmosphere?

2. Why is carbon dioxide considered the main greenhouse gas? How could an increase in water vapour create either a positive or negative feedback effect?

3. How do scientists study the ancient atmosphere?
4. Has simulating climate change with computer programs been effective in helping us predict climate? How do these programs work?
5. List five major trends in climate that scientists have documented so far. Now list five future trends or impacts that they are predicting.
6. Describe how rising sea levels, caused by global warming, can create problems for people. How may climate change affect marine ecosystems?
7. How might a warmer climate affect agriculture? How is it affecting distributions of plants and animals? How might it affect human health?
8. What are the main sources of greenhouse gas emissions in Canada? In what ways can we reduce these emissions?
9. What roles have international treaties played in addressing climate change? Give two specific examples.
10. Describe one market-based approach for reducing greenhouse emissions. Explain one reason it may work well and one reason it may not work well.

SEEKING SOLUTIONS

1. To determine to what extent current climate change is the result of human activity versus natural processes, which type of scientific research do you think is most helpful? Why?
2. Some people argue that we need "more proof," or "better science" before we commit to substantial changes in our energy economy. How much "science," or certainty, do you think we need before we should take action regarding climate change? How much certainty do you need in your own life before you make a major decision? Should nations and elected officials follow a different standard? Do you believe that the precautionary principle is an appropriate standard in the case of global climate change? Why or why not?
3. Describe several ways in which we can reduce greenhouse gas emissions from transportation. Which approach do you think is most realistic, which approach do you think is least realistic, and why?
4. Imagine that you would like to make your own lifestyle carbon neutral and that you aim to begin by reducing the emissions you are responsible for by 25%. What actions would you take first to achieve this reduction?
5. **THINK IT THROUGH** You have been appointed as the Canadian representative to negotiate the terms of a treaty to take hold after the Kyoto Protocol ends. All nations recognize that the Kyoto Protocol was not fully effective, and most are committed to creating a stronger agreement. The Canadian government has instructed you to take a leading role in designing the new treaty and to engage constructively with other nations' representatives, while protecting your nation's economic and political interests. What type of agreement will you try to shape? Describe at least three components that you would propose or agree to, and at least one that you would oppose.
6. **THINK IT THROUGH** You have just been elected premier of a province. Polls show that the public wants you to take bold action to reduce greenhouse gas emissions. However, polls also show that the public does not want prices of gasoline or electricity to rise more than they already have. Carbon-emitting industries in your province are wary of emissions reductions being required of them but are willing to explore ideas with you. Your parliament will support you in your efforts as long as you remain popular with voters. The province to the west has just passed ambitious legislation mandating steep greenhouse gas emissions reductions. The province to the east has joined a new regional emissions-trading consortium. What actions will you take in your first year as premier?

INTERPRETING GRAPHS AND DATA

The graph illustrates Canada's GHG emissions (CO_2) between 1990 and 2002, with projections to 2010 according to a "business-as-usual" scenario and a "Kyoto-based" scenario. From the information provided on the graph, answer the following questions.

1. Calculate the approximate percentage change in CO_2 emissions between 1990 and 2006.
2. Between 1990 and 2006, the population of Canada increased from 27.7 million to 33.0 million and the inflation-adjusted gross domestic product (GDP)

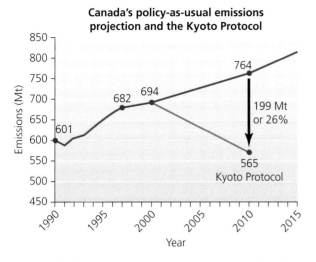

Based on the National GHG Inventory, this graph shows Canada's GHG emissions (CO_2) between 1990 and 2002, with projections to 2010 according to a "business-as-usual" scenario and a "Kyoto-based" scenario.

went from $27 631 to $36 463 per capita (inflation-adjusted, reported in 1997 dollars). Calculate the percentage increases in both population and GDP per capita over this time period. What quantitative conclusions can you draw from these data about CO_2 emissions per capita? About CO_2 emissions per unit of total economic activity? Create a graph and sketch a trend line of CO_2 emissions per capita from 1990 to 2006. Now sketch a trend line of CO_2 emissions per unit of total economic activity from 1990 to 2006.

3. Imagine you are put in charge of designing a strategy to reduce Canada's emissions of CO_2 from fossil fuel combustion. Based on the data presented here, what approaches would you recommend, and how would you prioritize these? Explain your answers.

CALCULATING FOOTPRINTS

Global climate change is something to which we all contribute, because fossil fuel combustion plays such a large role in supporting the lifestyles we lead. As individuals, each one of us can contribute to mitigating global climate change through personal decisions and actions that affect the way we live our lives. Several online calculators enable you to calculate your own personal *carbon footprint*, the amount of carbon emissions for which you are responsible. Go to one of these at www.carbonfootprint.com, take the quiz, and enter the relevant data in the table.

1. How does your personal carbon footprint compare with that of the average Canadian? How does it compare to that of the average person in the world? Why do you think your footprint differs from these in the ways it does?
2. Think of three changes you could make in your lifestyle that would lower your carbon footprint. Now take the footprint quiz again, incorporating these three changes. Enter your resulting footprint in the table. By how much did you reduce your yearly emissions?

	Carbon footprint (kilograms per person per year)
World average	
Average for industrialized nations	
Canadian average	
Your footprint	
Average needed to halt climate change	
Your footprint with three changes	

3. Now take the quiz again, trying to make enough changes to reduce your footprint to the level at which we could halt climate change. Do you think you could achieve such a footprint? What do you think would be an admirable yet realistic goal for you to set as a target value for your own footprint? Would you choose to purchase carbon offsets to help reduce your impact? Why or why not?

TAKE IT FURTHER

 Go to www.myenvironmentplace.ca where you will find

- Suggested answers to end-of-chapter questions
- Quizzes, animations, and flashcards to help you study
- *Research Navigator*™ database of credible and reliable sources to assist you with your research projects
- Tutorials to help you master how to interpret graphs
- Current news articles that link the topics that you study to case studies from your region and around the world

- **ECO Occupational Profiles:** If you found this chapter especially interesting, you might want to learn more about the following jobs by visiting the Occupational Profiles website of the Environmental Careers Organization. Go to www.eco.ca and check out the following careers:
  - Avalanche forecaster
  - Climatologist
  - Glaciologist
  - Meteorologist

CHAPTER ENDNOTES

1. U.S. National Oceanic and Atmospheric Administration (NOAA), *Trends in Atmospheric Carbon Dioxide–Mauna Loa*, www.esrl.noaa.gov/gmd/ccgg/trends/.
2. As Canada's climate changes, and weather patterns shift, Canadian climate models provide guidance in an uncertain future. CCCma Environment Canada, www.cccma.ec.gc.ca/20051116_brochure_e_pgs.pdf.
3. Government of Canada, *National Animal Health Strategy*, www.animauxsains.ca/english/doc/presentation/a_e.shtml.
4. Environment Canada, www.ec.gc.ca/pdb/ghg/newsrelease2006_e.cfm.
5. Intergovernmental Panel on Climate Change, *Fourth Assessment Report*, 2007.
6. Dr. James Ford, *Geography—Global Environmental Change, Area of Specialization*, www.arctic-north.com/JamesPersonalWebsite/specialization.html; and Ford. J. (2008) Emerging trends in climate change policy: The role of adaptation, *International Public Policy Review*, pp. 5–16.
7. Environment Canada, *National GHG Inventory*, www.ec.gc.ca/pdb/ghg/onlinedata/downloadDB_e.cfm#s5.
8. Environment Canada, *National GHG Inventory*, www.ec.gc.ca/pdb/ghg/inventory_report/2006/somsum_eng.cfm.
9. Environment Canada, *National GHG Inventory*, www.ec.gc.ca/pdb/ghg/inventory_report/2006/somsum_eng.cfm.
10. *Canadian Geographic*, Sheila Watt-Cloutier: Citation of Lifetime Achievement, from www.canadiangeographic.ca/cea/archives/archives_lifetime.asp?id=159.
11. *Canadian Geographic*, Sheila Watt-Cloutier: Citation of Lifetime Achievement, from www.canadiangeographic.ca/cea/archives/archives_lifetime.asp?id =159.
12. Stockholm Convention on Persistent Organic Pollutants (POPs), http://chm.pops.int/Default.aspx.
13. Inuit Circumpolar Council, http://inuitcircumpolar.com/index.php?ID=1&Lang=En.
14. *Canadian Geographic*, Sheila Watt-Cloutier: Citation of Lifetime Achievement, from www.canadiangeographic.ca/cea/archives/archives_lifetime.asp?id= 159.

15 Fossil Fuels: Energy and Impacts

This is an aerial view of the Mackenzie River Delta, in the Northwest Territories.

Upon successfully completing this chapter, you will be able to

- Identify the principal energy sources that we use
- Describe the nature and origin of coal and evaluate its extraction and use
- Describe the nature and origin of natural gas and evaluate its extraction and use
- Describe the nature and origin of oil and evaluate its extraction, use, and future availability
- Describe the nature, origin, and potential of alternative fossil fuel types and technologies
- Outline and assess environmental impacts of fossil fuel use
- Evaluate political, social, and economic impacts of fossil fuel use
- Specify strategies for conserving energy and enhancing efficiency

The wildlife of the Mackenzie delta and environs is sure to be affected by the proposed pipeline.

CENTRAL CASE:
ON, OFF, ON AGAIN? THE MACKENZIE VALLEY NATURAL GAS PIPELINE

"We've embarked on the beginning of the last days of the age of oil."
—MIKE BOWLIN, CHAIR, ARCO

"I listened to a brief by northern businessmen in Yellowknife who favour a pipeline through the North. Later, in a native village far away, I heard virtually the whole community express vehement opposition to such a pipeline. Both were talking about the same pipeline; both were talking about the same region—but for one group it is a frontier, for the other a homeland."
—JUSTICE THOMAS BERGER, NORTHERN FRONTIER, NORTHERN HOMELAND, 1977

The Mackenzie Valley Gas Project is a proposal to develop three major natural gas fields located in the Mackenzie Delta, Northwest Territories, and deliver the natural gas to southern markets through a 1220 km pipeline system to be constructed along the Mackenzie Valley (see maps).[1]

The idea of a major pipeline running from the Beaufort Sea down the Mackenzie Valley is not new. A pipeline was proposed and examined seriously in the early 1970s; it was called, at the time, "the biggest project in the history of free enterprise."[2]

In 1974 the federal government appointed Justice Thomas Berger to carry out an inquiry into the potential impacts of the project on the people of Canada's North. Berger travelled throughout the North, meeting with Dene, Inuit, Métis, and white residents and leaders. The report of the inquiry (released in 1977) was called *Northern Frontier, Northern Homeland*. It recommended that pipeline development in the North be delayed by at least 10 years because of deep opposition by native

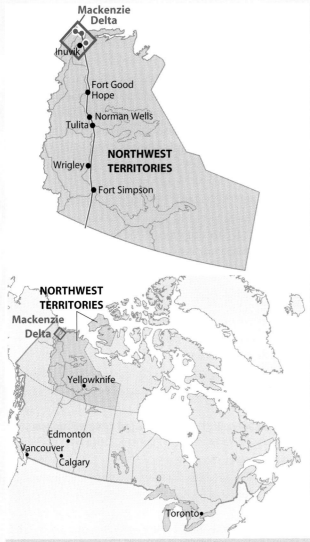

These maps show the location of the proposed Mackenzie River Valley natural gas pipeline.

twenty-first century. Interestingly, many of the native leaders who vehemently opposed the pipeline in the 1970s had by that time become supporters, realizing that it would bring much-needed jobs and revenue to the North. Chief Frank T'Seleie of Fort Good Hope, Northwest Territories, was a vocal opponent of the project as a young man in the 1970s. He now supports the pipeline, since many of the land claims in the area have been settled. He even participated in project negotiations on behalf of native communities.[4]

The partners in the current project are four major Canadian oil and gas companies—Imperial Oil Resources Ventures, ConocoPhillips Canada, Shell Canada, and ExxonMobil Canada, known collectively as the "Producer Group"—and Aboriginal Pipeline Group, representing the interests of Aboriginal people in the project.[5]

The project has been equally controversial on its second go-around. In October 2006, Alternatives North (a social justice group based in Yellowknife) estimated that Imperial Oil and its partners will earn billions of dollars from the project and should not receive federal subsidization. However, principal partner Imperial Oil claims that the project is only marginally economic and will not proceed without assurance of at least $1.2 billion from the federal government. According to Imperial Oil, the cost of the project has increased to $16.2 billion, and the scheduled start date for production (originally 2010) has been delayed until 2014.[6]

According to Nature Canada, the pipeline has the potential to transform the Valley into an industrial landscape; fragment habitat for bears, caribou, and wolves; harm fish and fish habitat by increasing sediment deposition into rivers; permanently damage breeding areas for millions of geese, tundra swans, and other migratory birds; require forests to be cut and heavy machinery deployed for infrastructure; trigger a rush of oil and gas development in the Mackenzie Valley (this was also predicted by the Berger report); and increase greenhouse gas emissions.[7] It remains to be scientifically proven or observed whether these impacts will occur, and to what degree.

An interesting side issue concerns the ultimate fate of the gas itself. Much of the natural gas produced in the Mackenzie Delta gas fields may never make it to the southern Canadian and American consumers for whom it was originally intended. This is because of the

leaders and the potential for negative impacts. The report further recommended that such development not proceed until native land claims in the area had been successfully resolved.

Among the main concerns were the potential impacts on people and animals that would result from the infrastructure (roads, airports, towns) likely to accompany the construction of the pipeline. As reported by CBC News, "Some dismissed the impact of a pipeline, saying it would be like a thread stretched across a football field. Those close to the land said the impact would be more like a razor slash across the Mona Lisa."[3]

In the end, the pipeline project was delayed for much longer than 10 years. The idea was on-again, off-again for many years, but was revitalized at the turn of the enormous acceleration in development of tar sand deposits in Alberta. Tar sand production requires significant inputs of energy, and the complete, successful development of these deposits is dependent on the production and delivery of natural gas from the North. The first 20 years' worth of natural gas from the Mackenzie Delta may go straight into the production of oil from Alberta's tar sands.

Sources of Energy

Humanity has devised many ways to harness the renewable and nonrenewable forms of energy available on our planet (Table 15.1). We use these energy sources to heat and light our homes, power our machinery, fuel our vehicles, and provide the comforts and conveniences to which we've grown accustomed in the modern industrial age.

We use a variety of energy sources

A great deal of energy emanates from Earth's core, making geothermal power available for our use. Energy also results from the gravitational pull of the Moon and Sun, and we are just beginning to harness the power from the ocean tides that these forces generate. An immense amount of energy resides within the bonds among protons and neutrons in atoms, and this energy provides us with nuclear power.

Most of our energy, however, comes ultimately from the Sun. We can harness energy from the Sun's radiation directly in a number of ways. Solar radiation also helps drive wind patterns and the hydrologic cycle, making possible such forms of energy as wind power and hydroelectric power. And of course, sunlight drives photosynthesis and the growth of plants, from which we take wood and other biomass as a fuel source. Finally, when plants die and are preserved in sediments under particular conditions, they may impart their stored chemical energy to **fossil fuels**, highly combustible substances formed from the remains of organisms from past geologic ages. The three fossil fuels we use widely today are oil, coal, and natural gas.

Since the Industrial Revolution, fossil fuels have replaced biomass as our society's dominant source of energy. Global consumption of the three main fossil fuels has risen steadily for years and is now at its highest level ever (**FIGURE 15.1A**). The high energy content of fossil fuels makes them efficient to burn, ship, and store. Besides providing for transportation, heating, and cooking, these fuels are used to generate *electricity*, a secondary form of energy that is easier to transfer over long distances and apply to a variety of uses.

Canada's energy stream is complex; we use energy in a variety of forms for different purposes (**FIGURE 15.1B**). We also both import and export energy—in some cases, different grades of the same form of energy (such as coal, which is both imported and exported). But overall Canada is a net exporter of energy.

As we first noted in Chapter 1, energy such sources as sunlight, geothermal energy, and tidal energy are considered perpetually *renewable* or inexhaustible because their supplies will not be depleted by our use. Other sources, such as timber, are renewable only if we do not harvest them at too great a rate. In contrast, such energy sources as oil, coal, and natural gas are considered *nonrenewable*, because at our current rates of consumption we will use up Earth's accessible store of them in a matter

Table 15.1 Energy Sources We Use Today

Energy source	Description	Type of energy	Chapter
Crude oil	Fossil fuel extracted from ground (liquid)	Nonrenewable	15
Natural gas	Fossil fuel extracted from ground (gas)	Nonrenewable	15
Coal	Fossil fuel extracted from ground (solid)	Nonrenewable	15
Nuclear energy	Energy from atomic nuclei of uranium	Nonrenewable	16
Biomass energy	Energy stored in plant matter from photosynthesis	Renewable	16
Hydropower	Energy from running water	Renewable	16
Solar energy	Energy from sunlight directly	Renewable	17
Wind energy	Energy from wind	Renewable	17
Geothermal energy	Earth's internal heat rising from core	Renewable	17
Tidal and wave energy	Energy from tidal forces and ocean waves	Renewable	17
Chemical fuels	Chemical energy in batteries and fuel cells	Renewable	17

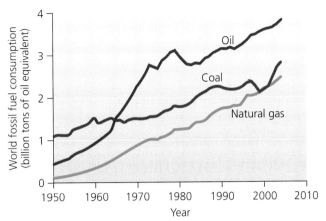

FIGURE 15.1
Global consumption of fossil fuels **(a)** has risen greatly over the past half-century. Oil use rose steeply during the 1960s to overtake coal, and today it remains our leading energy source. Canada's energy mix **(b)** illustrates our fundamental dependence on fossil fuels, as well as the importance of uranium (for nuclear energy) and hydropower in our energy mix. Data for (a) from Worldwatch Institute, 2006. *Vital signs 2006–2007*. Graphic in (b) from Natural Resources Canada. 2006. *Report of the National Advisory Panel on Sustainable Energy Science and Technology*, Office of Energy Research and Development.

(a) Global consumption of fossil fuels

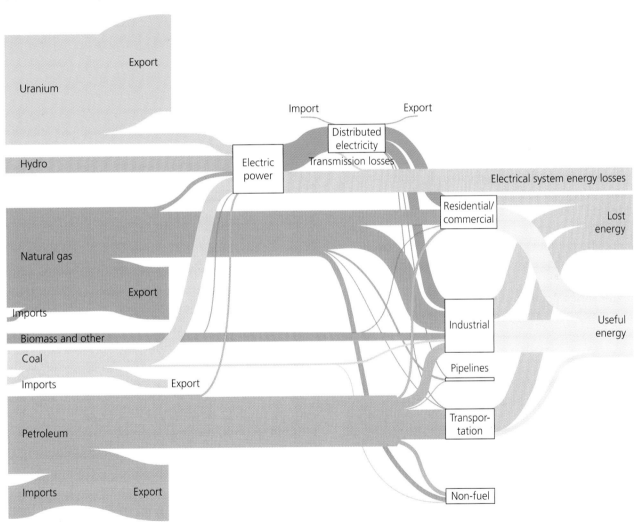

(b) Canada's energy stream

of decades to centuries. Nuclear power as currently harnessed through fission of uranium can be considered nonrenewable to the extent that uranium ore is in limited supply.

Although these nonrenewable fuels result from ongoing natural processes, the timescales on which they are created are so long that once the fuels are depleted, they cannot be replaced within any time span useful to our civilization. It takes a thousand years for the biosphere to generate the amount of organic matter that must be buried to produce a single day's worth of fossil fuels for our society. To replenish the fossil fuels we have depleted so far would take many millions of years. For this reason, and because fossil fuels exert severe environmental impacts, renewable energy sources increasingly are being developed as alternatives to fossil fuels, as we will see in Chapters 16 and 17.

Fossil fuels are indeed fuels created from "fossils"

The fossil fuels we burn today in our vehicles, homes, industries, and power plants were formed from the tissues of organisms that lived 100 million to 500 million years ago. The energy these fuels contain came originally from the Sun and was converted to chemical-bond energy as a result of photosynthesis. The chemical energy in these organisms' tissues then became concentrated as these tissues decomposed and their hydrocarbon compounds were altered and compressed.

Most organisms, after death, do not end up as part of a coal, gas, or oil deposit. A tree that falls and decays as a rotting log undergoes mostly **aerobic** decomposition; in the presence of air, bacteria and other organisms that use oxygen break down plant and animal remains into simpler carbon molecules that are recycled through the ecosystem. Fossil fuels are produced only when organic material is broken down in an **anaerobic** environment, one that has little or no oxygen. Such environments include the bottoms of shallow seas, deep lakes, and swamps (**FIGURE 15.2**).

Over millions of years, organic matter that accumulates at the bottoms of such water bodies undergoes decomposition, forming an oil precursor called **kerogen**. Geothermal heating then acts on the kerogen to create crude oil and natural gas. Natural gas can also be produced nearer the surface by anaerobic bacterial decomposition of organic matter. Oil and gas come to reside in porous rock layers beneath dense, impervious layers. Coal is formed when plant matter is compacted so tightly that there is little decomposition. The fuel that forms in any given place is dependent on the chemical composition of the starting material, the temperatures and pressures to which the material is subjected, the presence or absence of anaerobic decomposers, and the passage of time.

Fossil fuel reserves are unevenly distributed

Fossil fuel deposits are localized and unevenly distributed over Earth's surface, so some regions have substantial reserves of fossil fuels whereas others have very few. How long each nation's fossil fuel reserves will last depends on how much the nation extracts, how much it consumes, and how much it imports from and exports to other nations. Nearly two-thirds of the world's proven reserves of crude oil lie in the Middle East. The Middle East is also rich in natural gas, but Russia contains more than twice as much natural gas as any other country. Russia is also rich in coal, as is China, but the United States possesses more coal than any other nation (Table 15.2).

Developed nations consume more energy than developing nations

Citizens of developed regions generally consume far more energy than do those of developing regions (**FIGURE 15.3**). Per person, the most-industrialized nations use up to 100 times as much energy as do the least-industrialized

FIGURE 15.2
Tropical swamps, like the Okeefenokee Swamp in Florida, shown here, are one type of environment in which the formation of fossil fuels would have begun 100 million to 150 million years ago. Fossil fuels begin to form when organisms die and end up in oxygen-poor conditions, such as when trees fall into bogs and are buried by sediment, or when marine phytoplankton and zooplankton drift to the seafloor and are buried.

Table 15.2 Nations With the Largest Proven Reserves of Fossil Fuels		
Oil (% world reserves)	**Natural gas** (% world reserves)	**Coal** (% world reserves)
Saudi Arabia, 21.9	Russia, 26.3	United States, 27.1
Iran, 11.4	Iran, 15.5	Russia, 17.3
Iraq, 9.5	Qatar, 14.0	China, 12.6
Kuwait, 8.4	Saudi Arabia, 3.9	India, 10.2
United Arab Emirates, 8.1	United Arab Emirates, 3.3	Australia, 8.6
Venezuela, 6.6	United States, 3.3	South Africa, 5.4
Russia, 6.6	Nigeria, 2.9	Ukraine, 3.8
Libya, 3.4	Algeria, 2.5	Kazakhstan, 3.4
Kazakhstan, 3.3	Venezuela, 2.4	Poland, 1.5
Nigeria, 3.0	Iraq, 1.7	Brazil, 1.1

Data from British Petroleum. 2007. Statistical review of world energy 2007.

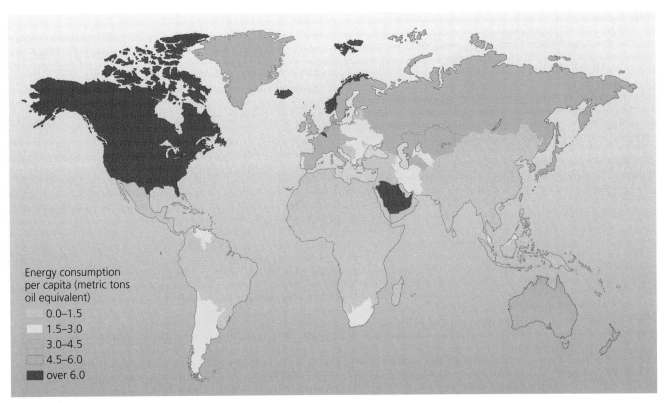

(a) Map of energy consumption per person

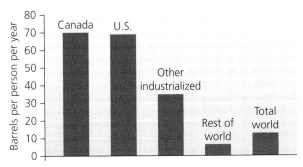

(b) Energy use per capita, in oil equivalents

FIGURE 15.3

Regions vary greatly in their consumption of energy per person **(a)**. People in industrialized nations consume the most. The map combines all types of energy, standardized to metric tons of "oil equivalent," that is, the amount of fuel needed to produce the energy gained from combusting one metric ton of crude oil. The bar graph **(b)** compares energy use in oil equivalents in Canada and the United States—world "leaders" in energy use—to other nations of the world. Data from British Petroleum. 2007. *Statistical review of world energy 2007.*

nations. Although Canada has only about 0.005% of the world's population, it consumes 2.5% of the energy. The United States, with only 4.6% of the world's population, accounts for 22.5% of the world's energy use. Even so, Canada's per capita energy use is higher. This is partly a result of the cold climate and long distances that characterize our nation, but careless use patterns are part of the story, too.

Moreover, developed and developing nations tend to apportion their energy use differently. Industrialized nations use roughly one-third of their energy on transportation, one-third on industry, and one-third on all other uses. Developing nations devote a greater proportion of their energy to subsistence activities, such as agriculture, food preparation, and home heating, and substantially less on transportation. In addition, people in developing countries often rely on manual or animal energy sources instead of automated ones. For instance, rice farmers in Bali plant rice by hand, but industrial rice growers in California use airplanes. Because industrialized nations rely more on equipment and technology, they use more fossil fuels. In Canada, where hydroelectric resources are particularly abundant (Chapter 16), oil, coal, and natural gas still supply 67% of energy needs (**FIGURE 15.4**).

It takes energy to make energy

We don't simply get energy for free. To harness, extract, process, and deliver the energy that we use requires that we invest substantial inputs of energy. Drilling and extracting oil and natural gas requires the construction of an immense infrastructure of roads, wells, vehicles, storage tanks, pipelines, housing for workers, and more—which necessitates the use of energy. Piping and shipping the natural gas out of the Mackenzie River Delta and delivering it to the market for use would require further

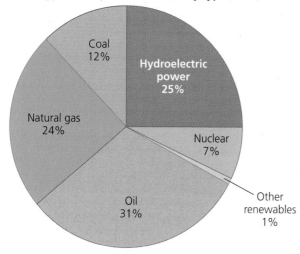

FIGURE 15.4
Fossil fuels dominate energy consumption in Canada, as in other industrialized nations.

means that the amount of energy invested is the same as the amount of energy extracted. Fossil fuels are widely used because their EROI ratios have historically been high. However, EROI ratios can change over time. For instance, those for oil and natural gas declined from more than 100:1 in the 1940s to about 30:1 in the 1970s, and today they are less than 15:1 globally. The EROI ratios have declined because we extracted the easiest deposits first and now must work harder and harder to extract the remaining amounts. Canadian tar sands, for comparison, have EROI ratios of 5:1 to 3:1; ethanol produced from corn has an EROI of less than 2:1; and solar panels have an EROI of approximately 1:1.

Coal, Natural Gas, and Oil

The three principal fossil fuels, on which our modern industrial society has been founded, are coal, natural gas, and oil. Let us consider each of them, in turn.

Coal is the world's most abundant fossil fuel

The proliferation 300 million to 400 million years ago of swampy environments where organic material could be buried has resulted in substantial coal deposits throughout the world. The precursor to coal is *peat*, a moist soil composed of compressed organic matter. **Coal** is organic matter (generally woody plant material) that was compressed under very high pressure to form a dense, carbon-rich solid material (**FIGURE 15.5**). Coal is a rock, whereas peat is a soil. Coal typically results when

energy inputs. Thus, when evaluating the value of an energy source, it is important to subtract costs in energy invested from benefits in energy received. **Net energy** expresses the difference between energy returned and energy invested:

Net Energy = Energy Returned − Energy Invested

When comparing energy sources, it is useful to use a ratio often denoted as **EROI**, or **energy returned on investment**. EROI ratios are calculated as follows:

EROI = Energy Returned/Energy Invested

Higher ratios mean that we receive more energy from each unit of energy that we invest; when EROI = 1, it

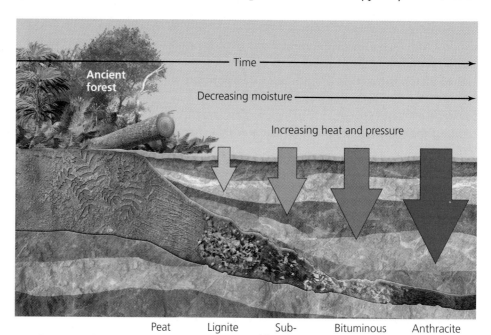

FIGURE 15.5
Coal forms as a result of the compaction of ancient plant matter underground. Scientists categorize coal into several types, depending on the amount of heat, pressure, and moisture involved in its formation. Anthracite coal is formed under greatest pressure, where temperatures are high and moisture content is low. Lignite coal is formed under conditions of much less pressure and heat but more moisture. Peat is also part of this continuum, representing plant matter that is minimally compacted.

little decomposition takes place within the parent organic material because it cannot be digested, or because appropriate decomposers are not present. Coal provides one-quarter of the world's commercial energy consumption.

Coal use has a long history

People have used coal longer than any other fossil fuel. The Romans used coal for heating in the second and third centuries in Britain, as have people in parts of China for 2000–3000 years. Native Americans of the Hopi Nation still follow ancestral traditions by using coal to fire pottery, cook food, and heat their homes. Once commercial mining began in Europe in the 1700s, people began using coal widely as a heating source. Coal found an expanded market after the invention of the steam engine, because it was used to boil water to produce steam. Coal-fired steam engines helped drive the industrial revolution, powering factories, agriculture, trains, and ships. The birth of the steel industry in 1875 increased demand still further because coal fuelled the furnaces used to produce steel.

In the 1880s, people began to use coal to generate electricity. In coal-fired power plants, coal combustion converts water to steam, which turns a turbine to create electricity (see "The Science Behind the Story: Clean Coal for Electricity Generation"). Today coal provides more than 30% of the electrical generating capacity of Canada. Canada hosts approximately 10 billion tonnes of coal reserves—more than our oil and natural gas reserves combined. China and the United States are the primary producers and consumers of coal (Table 15.3).

Table 15.3 Top Producers and Consumers of Coal	
Production (% world production)	**Consumption** (% world consumption)
China, 39.4	China, 38.6
United States, 19.3	United States, 18.4
India, 6.8	India, 7.7
Australia, 6.6	Japan, 3.9
South Africa, 4.7	Russia, 3.6
Russia, 4.7	South Africa, 3.0
Indonesia, 3.9	Germany, 2.7
Poland, 2.2	Poland, 1.9
Germany, 1.6	South Korea, 1.8
Kazakhstan, 1.6	Australia, 1.7

Data from British Petroleum. 2007. Statistical review of world energy 2007.

Coal is mined from the surface and from below ground

We extract coal by using two major methods (**FIGURE 15.6**). We reach underground deposits with **subsurface mining**. Shafts are dug deep into the ground, and networks of tunnels are dug or blasted out to follow coal seams. The coal is removed systematically and

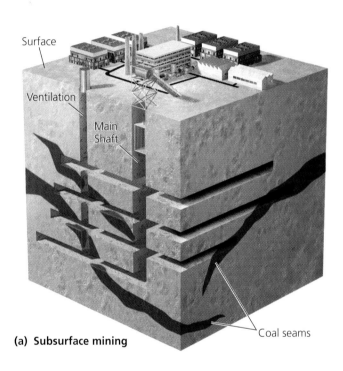

(a) **Subsurface mining**

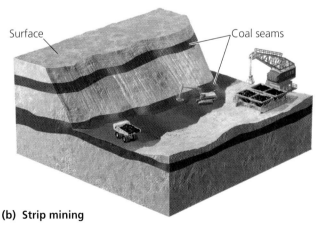

(b) **Strip mining**

FIGURE 15.6
Coal is mined in two major ways. In subsurface mining **(a)**, miners work below ground in shafts and tunnels blasted through the rock; these passageways provide access to underground seams of coal. This type of mining poses dangers and long-term health risks to miners. In open-pit and strip-mining **(b)**, soil is removed from the surface, exposing coal seams from which coal is mined. This type of mining can cause substantial environmental impacts.

THE SCIENCE BEHIND THE STORY

Clean Coal for Electricity Generation

A worker checks a furnace at a coal-fired power plant.

We use coal—lots of it—to generate electricity, in a process that dates back more than a century (see the figure). Once mined, coal is hauled to power plants, where it is pulverized. The crushed coal is blown into a boiler furnace on a superheated stream of air and burned in a blaze of intense heat—typical furnace temperatures often flare at 815°C.

Water circulating around the boiler absorbs the heat and is converted to high-pressure steam. This steam is injected into a **turbine**, a rotary device that converts the kinetic energy of a moving substance, such as steam, into mechanical energy. As steam from the boiler exerts pressure on the blades of the turbine, they spin, turning the turbine's drive shaft.

The drive shaft is connected to a generator, which features a rotor that rotates and a stator that remains stationary. Generators make use of a phenomenon that you may have experimented with yourself in your high school physics class: Moving magnets adjacent to coils of copper wire cause electrons in the copper wires to move, generating alternating electric current.

In some generators, the rotor consists of magnets and spins within a stator of coiled copper wire. As the turbine's drive shaft rotates, it causes the rotor to revolve, creating a magnetic field and causing electrons in the stator's copper wires to move, thereby creating an electrical current. This current flows into transmission lines that travel from the power plant out to the customers who use the plant's electricity.

As we try to balance growing demand for electricity with rising concerns about environmental and health impacts of coal combustion, power plants continue to rely heavily on coal while scientists work to limit the pollution that use of this fuel creates. **Clean coal** technologies largely focus on approaches to rid the generation process of toxic chemicals before or after the coal is burned.

There are two principal clean-coal pathways being investigated by researchers internationally. These involve technologies that are aimed at (1) a cleaner combustion process, and (2) *gasification* and the production of clean synthetic fuels from coal.[8]

Combustion-focused technologies basically start with the present pulverized-coal process and apply improvements aimed at making combustion more efficient and more complete, and thus cleaner. One example is *fluidized bed* technologies, which bathe the finely pulverized coal in jets of air during combustion. This leads to a turbulent, fluid-like environment, which allows the temperature to be increased and raises the efficiency of the chemical reactions that occur during combustion. Gasification technologies are less well developed; they involve creating clean synthetic fuels, including hydrogen, from coal.

Either of these approaches can be combined with corollary technologies aimed at cleaning the coal prior to combustion; cleaning emissions after burning and before they leave the smokestack; co-generation technologies; and carbon capture and storage technologies. For example, some pre-combustion technologies utilize sulphur-metabolizing bacteria to remove sulphur from the coal prior to burning. Technologies that clean emissions before they leave the stack include *scrubbers*, which utilize calcium- or sodium-based materials to absorb and remove sulphur dioxide (SO_2) from the emissions. Other types of scrubbers use chemical reactions to strip away nitrogen oxides (NO_x), breaking them down into elemental nitrogen and water. Multilayered filtering devices can be used to capture tiny ash particles before they leave the stack.

---

shipped to the surface. Today less than 2% of Canada's coal is extracted by underground mining. The history of underground mining in Nova Scotia demonstrates the hazards posed to miners who go deep under the ground to mine coal.

When coal deposits are at or near the surface, open-pit or strip-mining methods are used. Open-pit mining involves large excavations, which are deepened and widened as mining proceeds. This type of coal mining is common in Alberta and British Columbia. In **strip-mining**, heavy machinery removes earth in long, horizontal strips to expose the layers, or *seams*, and extract the coal. The pits are subsequently refilled with the soil that had been removed. Strip-mining of coal is most common in the Prairies and in New Brunswick.[9] Strip-mining operations can occur on immense scales; in some cases entire mountaintops are lopped off. This environmentally destructive process is called *mountaintop removal*. We will explore some of coal's significant environmental impacts later in this chapter.

Coal varies in its qualities

Coal varies from deposit to deposit in many ways, including in its water content and the amount of potential energy it contains. Peat is essentially organic material that is broken down anaerobically but remains wet, near the surface, and not well compressed. Peat has been widely used as a fuel in Britain and other locations. Canada is home to some of the most extensive peat deposits in the world, and 90% of our exported peat goes to the United States. As you learned in Chapter 7, peat is also of interest as a northern soil, and scientists are closely tracking its response to global climate change.

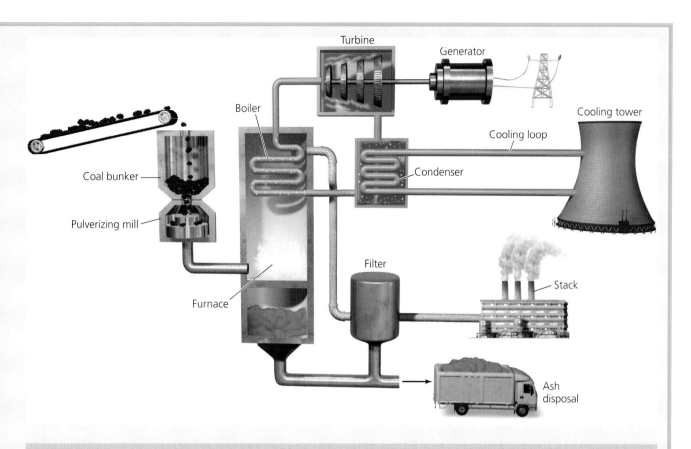

Coal is used as a fuel source for heat, to generate electricity. Pieces of coal are pulverized and blown into a high-temperature furnace. Heat from the combustion boils water, and the resulting steam turns a turbine, generating electricity by passing magnets past copper coils. The steam is then cooled and condensed in a cooling loop and returned to the furnace. "Clean coal" technologies help make the combustion process more efficient, clean the coal prior to combustion, or filter out pollutants after combustion. Toxic ash residue is disposed of at hazardous waste disposal sites.

Some energy analysts and environmental advocates question a policy emphasis on clean coal. Coal, they maintain, is an inherently dirty means of generating power and should be replaced outright with cleaner energy sources. However, with coal-fired power plants still generating a significant proportion of Canada's electricity—as well as Canada's greenhouse gas emissions—the push to clean up coal-based technologies makes sense in today's world.

As peat decomposes further, as it becomes buried more deeply under sediments, as pressure and heat increase, and as time passes, water is squeezed out of the material, and carbon compounds are packed more tightly together, forming coal. Scientists classify coal into four types: lignite, sub-bituminous, bituminous, and anthracite. Lignite is the least-compressed type of coal, and anthracite is the most-compressed type. The greater the compression, the greater is the energy content per unit volume.

Most coal contains impurities, including sulphur, mercury, arsenic, and other trace metals, and coal deposits vary in the amounts of impurities they contain. Sulphur content depends in part on whether the coal was formed in freshwater or saltwater sediments. Coal from eastern provinces of Canada tends to be relatively high in sulphur (e.g., 2%) because it was formed in marine sediments, where sulphur from seawater was present. In comparison, coal from Alberta and British Columbia is typically lower in sulphur content, ranging down to about 0.5%, whereas coal from China can be very sulphur-rich (and thus more polluting), ranging up to about 3%.

When high-sulphur coal is burned, it produces sulphate air pollutants, which contribute to industrial smog and acidic deposition; China's use of high-sulphur coal deposits has contributed to severe air-quality problems in that country. Combustion of coal high in mercury content emits mercury that bioaccumulates in organisms' tissues, poisoning animals as it moves up food chains. Such pollution problems commonly occur downwind of coal-fired power plants. Scientists and engineers are seeking ways to cleanse coal of its impurities so that it can continue to be used as an energy source while minimizing impact on health and the environment. Reducing pollution from coal is important because society's demand for this relatively abundant fossil fuel may soon rise as supplies of oil and natural gas decline.

Natural gas is the fastest-growing fossil fuel in use today

Natural gas consists primarily of methane, CH_4, and typically includes varying amounts of other volatile hydrocarbons. (Natural "gas" is actually something of a misnomer, as the material can be liquid at the ambient pressures and temperatures in subsurface reservoirs.) Natural gas provides one-quarter of global commercial energy consumption. It is a much cleaner-burning fuel than coal or oil, so it produces less pollution. World supplies of natural gas are projected to last perhaps 60 more years.

Natural gas is formed in two ways

Natural gas can arise from either of two processes. *Biogenic* gas is created at shallow depths by the anaerobic decomposition of organic matter by bacteria. An example is the "swamp gas" you can sometimes smell when stepping into the muck of a swamp.

In contrast, *thermogenic* gas results from compression of organic material, accompanied by heating deep underground. The organic precursor materials come most commonly from animal and plant matter, such as zooplankton and phytoplankton in shallow marine waters. As the organic matter is buried more and more deeply under sediments, the pressure exerted by the overlying sediments grows, and temperatures increase; this process is called *maturation*. Carbon bonds in the organic matter begin breaking, and the organic matter turns into kerogen, which acts as a source material for both natural gas and crude oil. Further heat and pressure act on the kerogen to degrade complex organic molecules into simpler hydrocarbon molecules. At very deep levels—below about 3 km—the high temperatures and pressures tend to form natural gas. Whereas biogenic gas is nearly pure methane, thermogenic gas contains small amounts of other gases as well as methane.

Thermogenic gas may be formed directly, along with coal or crude oil, or from coal or oil that is altered by heating. Most gas extracted commercially is thermogenic and is found above deposits of crude oil or seams of coal, so its extraction often accompanies the extraction of those fossil fuels. The natural gas found in the Mackenzie River Delta, discussed in "Central Case: On, Off, On Again? The Mackenzie Valley Natural Gas Pipeline," originated from the thermogenic decomposition of organic matter in shallow marine sediments.

Often, natural gas goes to waste as it escapes from coal mines or oil wells. Methane from coal seams, called *coalbed methane*, commonly leaks to the atmosphere during mining. To avoid this waste, and because methane is a potent greenhouse gas that contributes to climate change, mining engineers are now trying to capture more of this gas for energy. Likewise, in most remote oil-drilling areas, where the transport of natural gas remains prohibitively expensive, natural gas is flared—wasted by simply being burned off. Gas captured during oil drilling is expensive to export, but in some cases it can be reinjected into the ground for potential future extraction.

One source of biogenic natural gas is the decay process in landfills, and many landfill operators are now capturing this gas to sell as fuel. This practice decreases energy waste, can be profitable for the operator, and helps reduce the atmospheric release of methane.

Natural gas has only recently been widely used

Throughout history, naturally occurring seeps of natural gas would occasionally be ignited by lightning and could be seen burning in parts of what is now Iraq, inspiring the Greek essayist Plutarch around 100 C.E. to describe their "eternal fires." The first commercial extraction of natural gas took place in 1821, but until recently its use was localized because technology did not exist to pipe gas safely over long distances. Natural gas was used to fuel streetlamps, but when electric lights replaced most gas lamps in the 1890s, gas companies began marketing gas for heating and cooking. The first major commercial natural gas development in Canada was at Bow Lake, Alberta, southwest of Medicine Hat, in 1908.[10] After the Second World War, wartime improvements in welding and pipe building made gas transport safer and more economical, and during the 1950s and 1960s, thousands of kilometres of underground pipelines were laid throughout North America.

Today natural gas is increasingly favoured because it is versatile and clean-burning, emitting just half as much carbon dioxide per unit of energy produced as coal and two-thirds as much as oil. Converted to a liquid at low temperatures (*liquefied natural gas*, or *LNG*), it can be shipped long distances in refrigerated tankers, although this poses risks of catastrophic explosions. Natural gas deposits are greatest in Russia and the Middle East, and Russia and the United States lead the world in gas production and gas consumption, respectively (Table 15.4); Canada is the world's third-largest producer of natural gas.

Table 15.4 Top Producers and Consumers of Natural Gas

Production (% world production)	Consumption (% world consumption)
Russia, 21.3	United States, 22.0
United States, 18.5	Russia, 15.1
Canada, 6.5	Iran, 3.7
Iran, 3.7	Canada, 3.4
Norway, 3.0	United Kingdom, 3.2
Algeria, 2.9	Germany, 3.0
United Kingdom, 2.8	Japan, 3.0
Indonesia, 2.6	Italy, 2.7
Saudi Arabia, 2.6	Saudi Arabia, 2.6
Netherlands, 2.2	Ukraine, 2.3

Data from British Petroleum. 2007. Statistical review of world energy 2007.

Natural gas extraction becomes more challenging with time

To access some natural gas deposits, prospectors need only drill an opening, because pressure and low molecular weight drive the gas upward naturally. The first gas fields to be tapped were of this type. Most fields remaining today, however, require that gas be pumped to Earth's surface. In Alberta as well as parts of the United States, it is common to see a device called a *horsehead pump* (**FIGURE 15.7**). This pump moves a rod in and out of a shaft, creating pressure to pull both natural gas and crude oil to the surface.

As with oil and coal, many of the most accessible natural gas reserves have already been exhausted, causing their production to decline. Thus, deposits located in more remote areas, such as the Mackenzie River Delta, are becoming more attractive economically. Much extraction today also makes use of sophisticated techniques to break into rock formations and pump gas to the surface. One such "fracturing technique" is to pump salt water under high pressure into the rocks to crack them. Sand or small glass beads are inserted to hold the cracks open once the water is withdrawn. This type of extraction has extensive environmental impacts and it can be very water-intensive.

Offshore drilling produces much of our gas and oil

Drilling for natural gas, as well as for oil, takes place not just on land but also in the seafloor on the continental shelves. Offshore drilling has required developing technology that can withstand the forces of wind, waves, and ocean currents. Some drilling platforms are fixed standing platforms built with unusual strength. Others are resilient floating platforms anchored in place above the drilling site. Most of the offshore gas and oil development in Canada is located in the Beaufort Sea and in the North Atlantic Ocean off the coasts of Newfoundland and Labrador and Nova Scotia (**FIGURE 15.8**).

FIGURE 15.7
Horsehead pumps, like this one in Drayton Valley, Alberta, are used to extract natural gas as well as oil. They are a common feature of the landscape in Alberta and parts of the United States. The pumping motion of the machinery draws gas and oil upward from below ground.

FIGURE 15.8
The Hibernia Offshore Drilling Platform, shown here, is located in the North Atlantic about 300 km off the coast of Newfoundland in the Grand Banks. It is the world's largest offshore platform, and began production in 1997.

Oil is the world's most-used fuel

Oil has dominated world energy use since the 1960s, when it eclipsed coal. It now accounts for 37% of the world's commercial energy consumption. Its use worldwide over the past decade has risen more than 17%.

People have used solid forms of oil (such as tar and asphalt) from deposits easily accessible at Earth's surface for at least 6000 years. The modern extraction and use of petroleum for energy began in the 1850s, when miners drilling for groundwater or salt occasionally encountered oily rocks instead. At first, entrepreneurs bottled the crude oil from these deposits and sold it as a healing aid, unaware that crude oil is carcinogenic when applied to the skin and poisonous when ingested. Soon, however, it was realized that this "rock oil" could be used to light lamps and lubricate machinery. Edwin Drake is generally credited with drilling the world's first oil well, in Titusville, Pennsylvania, in 1859. In fact, however, the first oil well was drilled a full year earlier, in 1858, at Oil Springs, Ontario, by James Miller Williams, who struck free liquid oil only 20 m below the surface while attempting to drill a water well.[11]

Today our global society produces and consumes nearly 750 L of oil each year for every man, woman, and child. The United States consumes nearly one-fourth of the world's oil and shows little sign of abating. For our part, Canadians—less than 0.005% of the world's population—consume 2.5% of the oil. Table 15.5 shows the top oil-producing and oil-consuming nations.

Heat and pressure underground form petroleum

The sludgelike liquid we know as **oil**, **crude oil**, or **petroleum** (a term that includes both oil and natural gas),

Table 15.5 Top Producers and Consumers of Oil

Production (% world production)	Consumption (% world consumption)
Saudi Arabia, 13.1	United States, 24.1
Russia, 12.3	China, 9.0
United States, 8.0	Japan, 6.0
Iran, 5.4	Russia, 3.3
China, 4.7	Germany, 3.2
Mexico, 4.7	India, 3.1
Canada, 3.9	South Korea, 2.7
Venezuela, 3.7	Canada, 2.5*
Kuwait, 3.4	France, 2.4
Norway, 3.3	Saudi Arabia, 2.4

*Canada's population is approximately 0.005% of the global population.
Data from British Petroleum. 2007. Statistical review of world energy 2007.

tends to form within a window of temperature and pressure conditions often found 1.5–3 km below the surface. Like natural gas, most of the crude oil we now extract was formed when dead plant material (and small amounts of animal material) drifted down through shallow coastal marine waters millions of years ago and was buried in sediments on the ocean floor.

Crude oil is a mixture of hundreds of different types of hydrocarbon molecules characterized by carbon chains of different lengths. The specific properties of the oil depend on the chemistry of the organic starting materials, the characteristics of the geologic environment of formation, and the details of the maturation process. A hydrocarbon chain's length affects its chemical properties, which has consequences for human use, such as whether a given fuel burns cleanly in a car engine. Oil refineries sort the various hydrocarbons of crude oil, separating those intended for use in gasoline engines from those, such as tar and asphalt, used for other purposes.

Petroleum geologists infer the location and size of deposits

Because petroleum forms only under certain conditions, it occurs in isolated deposits. Once geothermal heating separates hydrocarbons from their source material and produces crude oil, this liquid migrates upward through rock pores, sometimes assisted by seismic faulting. It tends to collect in porous layers beneath dense, impermeable layers.

Geologists searching for oil (or other fossil fuels) drill rock cores and conduct ground, air, and seismic surveys to map underground rock formations, understand geologic history, and predict where fossil fuel deposits might lie. One method is to create powerful vibrations (by exploding dynamite, thumping the ground with a large weight, or using an electric vibrating machine) at the surface in one location and then measure how long it takes the seismic waves to reach receivers at other surface locations. Density differences in the substrate cause waves to reflect off layers, refract, or bend. Scientists and engineers interpret the patterns of wave reception to infer the densities, thicknesses, and location of underlying geologic layers—which in turn provide clues about the location and size of fuel deposits.

Over the past few decades, geologists have greatly improved their methods for locating new deposits; however, with their scientific understanding of Earth processes, geologists are generally quick to acknowledge that petroleum is ultimately a finite and nonrenewable resource.

Some portion of oil that is located by geologists will be impossible to extract by using current technology and will need to wait for future advances in extraction equipment or methods. Thus, estimates are generally made of *technically recoverable* oil. However, oil companies will not be willing to extract these entire amounts. Some oil would be so difficult to extract that the expense of doing so would exceed the income the company would receive from the oil's sale. Thus, the amount a company chooses to drill for will be determined by the costs of extraction (and transportation), together with the current price of oil on the world market. Because the price of oil fluctuates, the portion of oil from a given deposit that is *economically recoverable* fluctuates as well.

Thus, technology sets a limit on the amount that *can* be extracted, whereas economics determines how much *will* be extracted. The amount of oil, or any other fossil fuel, in a deposit that is technologically and economically feasible to remove under current conditions is the *proven recoverable reserve* of that fuel.

We drill to extract oil

Once geologists have identified an oil deposit, an oil company will typically conduct exploratory drilling. Holes drilled during this phase are usually small in circumference and descend to great depths. If enough oil is encountered, extraction begins. Just as you would squeeze a sponge to remove its liquid, pressure is required to extract oil from porous rock. Oil is typically already under pressure—from above by rock or trapped gas, from below by groundwater, or internally from natural gas dissolved in the oil. All these forces are held in place by surrounding rock until drilling reaches the deposit, whereupon oil will often rise to the surface of its own accord.

Once pressure is relieved, however, both oil and natural gas become more difficult to extract and may need to be pumped out. Even after pumping, a great deal of oil remains stuck to rock surfaces. As much as two-thirds of a deposit may remain in the ground after **primary extraction**, the initial drilling and pumping of available oil (**FIGURE 15.9A**). Companies may then begin **secondary extraction**, in which solvents are used or underground rocks are flushed with water or steam to remove additional oil (**FIGURE 15.9B**).

Even after secondary extraction, quite a bit of oil can remain; we lack the technology to remove every last drop. Secondary extraction is more expensive than primary extraction, so many oil deposits did not undergo secondary extraction when they were first drilled because the price of oil was too low to make the procedure economical. When oil prices rose in the 1970s, many drilling sites were reopened for secondary extraction. Still more are being reopened today, as prices rise again. As mentioned above, secondary extraction also can be harder on the environment than primary extraction.

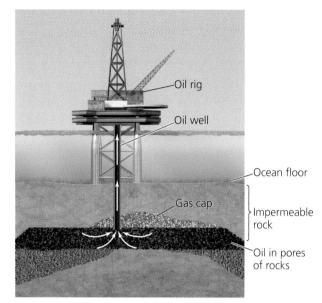

(a) Primary extraction of oil

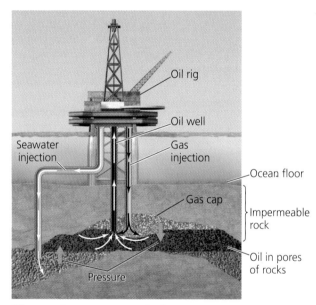

(b) Secondary extraction of oil

FIGURE 15.9
In primary extraction **(a)**, oil is drawn up through the well by keeping pressure at the top lower than pressure at the level of the oil deposit. Once the pressure in the deposit drops, however, material must be injected into the deposit to increase the pressure. Thus, secondary extraction **(b)** involves injecting seawater beneath the oil and/or gases just above the oil to force more oil up and out of the deposit.

Petroleum products have many uses

Once crude oil is extracted, it is put through **refining** processes (see "The Science Behind the Story: How Crude Oil Is Refined"). Because crude oil is a complex mix of hydrocarbons, we can create many types of petroleum products by separating its various components. Since the

> ### THE SCIENCE BEHIND THE STORY
>
> *These workers are at a Citgo oil refinery in Lemont, Illinois.*
>
> ## How Crude Oil Is Refined
>
> Crude oil is a complex mixture of thousands of kinds of hydrocarbon molecules. Through the process of *refining*, these hydrocarbon molecules are separated into classes of different sizes and chemically transformed to create specialized fuels for heating, cooking, and transportation and to create lubricating oils, asphalts, and the precursors of plastics and other petrochemical products. To maximize the production of marketable products while minimizing negative environmental impacts, petroleum engineers have developed a variety of refining techniques.
>
> The first step in processing crude oil is *distillation*, or *fractionation*. This process is based on the fact that different components of crude oil boil at different temperatures. In refineries, the distillation process takes place in tall columns filled with perforated horizontal trays (see figure). The columns are cooler at the top than at the bottom. When heated crude oil is introduced into the column, lighter components rise as vapour to the upper trays, condensing into liquid as they cool, while heavier components sink to the lower trays. Light gases, such as butane, boil at less than 32°C, and heavier oils, such as industrial fuel oil, boil only at temperatures above 343°C.
>
> Since the early twentieth century, light gasoline, used in automobiles, has been in much higher demand than most other derivatives of crude oil. The demand for high-performance, clean-burning gasoline has also risen. To meet these demands, refiners have developed several techniques to convert heavy hydrocarbons into gasoline.
>
> The general name for processes that convert heavy oil into lighter oil is *cracking*. One of the simplest methods is thermal cracking, in which long-chained molecules are broken into smaller chains by heating in the absence of oxygen. (The oil would ignite if oxygen were present.) Catalytic cracking, a related method, uses catalysts—substances that promote chemical reactions without being consumed by them—to control the cracking process. The result is an increase in the amount of a desired lighter product from a given amount of heavy oil. Today, the most widely used form of cracking is fluidized catalytic cracking, in which a finely powdered catalyst that behaves like a fluid is fed continuously into a reaction chamber with heavy oils. The products of cracking are then fed into a distillation column.
>
> Refiners can also change the chemical composition of oil through a process called *catalytic reforming*. Catalytic reforming uses catalysts to promote chemical reactions that transform certain hydrocarbons that are slightly heavier than gasoline so that they can be blended with gasoline to obtain higher octane ratings. The octane rating reflects the amount of compression gasoline can undergo before it spontaneously ignites.
>
> Besides distilling crude oil and altering the chemical structure of some of its components, refineries also remove contaminants. Sulphur and nitrogen compounds, which can be harmful when released into the atmosphere, are the two most common contaminants in crude oil. Government regulations have forced refiners to develop scrubbers and other methods of removing such contaminants, particularly sulphur. Some methods successfully remove up to 98% of sulphur.
>
> As a result of all these approaches, each barrel of crude oil is eventually converted into gasoline and a wide variety of other petroleum products.

1920s, refining techniques and chemical manufacturing have greatly expanded our uses of petroleum to include a wide array of products and applications, from lubricants to plastics to fabrics to pharmaceuticals. Today, petroleum-based products are all around us in our everyday lives (**FIGURE 15.10**).

Because petroleum products have become so central to our lives, many fossil fuel experts today are voicing concern that oil production may soon decline as we continue to deplete the world's recoverable oil reserves.

We may have already depleted half our oil reserves

Some scientists and oil industry analysts calculate that we have already extracted nearly half of the world's oil reserves. So far we have used up about 1.1 trillion barrels of oil, and most estimates hold that somewhat more than 1 trillion barrels remain. (A barrel is not a metric unit of measurement, but it is still commonly used in the oil industry. It is equivalent to 42 U.S. gallons, or 0.158987 cubic metres, or 117 litres.)

To estimate how long this remaining oil will last, analysts calculate the *reserves-to-production ratio*, or *R/P ratio*, by dividing the amount of total remaining reserves by the annual rate of production (i.e., extraction and processing). At current levels of production (30 billion barrels globally per year), 1.2 trillion barrels would last about 40 more years.

Unfortunately, this does not mean that we have a full 40 years in which to figure out what to do once the oil runs out. A growing number of scientists and analysts insist that we will face a crisis not when the last drop of oil is pumped, but when the rate of production first begins to decline. They point out that when production declines as demand continues to increase (because of rising global population and consumption), we will experience an oil shortage immediately. Because production tends to decline once reserves are depleted halfway, most of these experts calculate that this crisis will likely begin within the next several years.

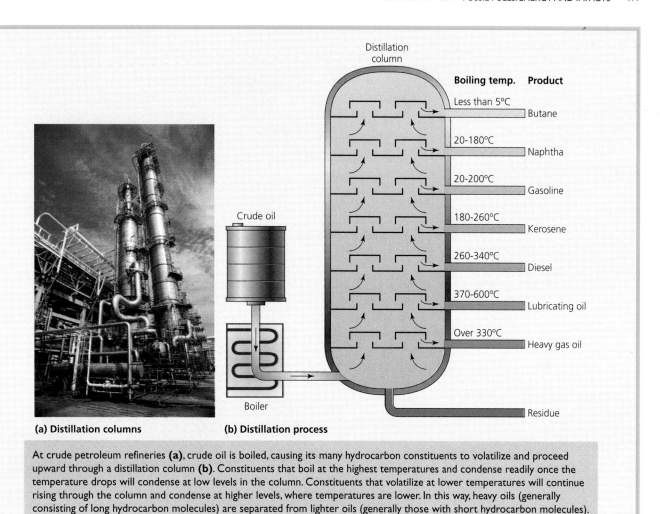

At crude petroleum refineries **(a)**, crude oil is boiled, causing its many hydrocarbon constituents to volatilize and proceed upward through a distillation column **(b)**. Constituents that boil at the highest temperatures and condense readily once the temperature drops will condense at low levels in the column. Constituents that volatilize at lower temperatures will continue rising through the column and condense at higher levels, where temperatures are lower. In this way, heavy oils (generally consisting of long hydrocarbon molecules) are separated from lighter oils (generally those with short hydrocarbon molecules).

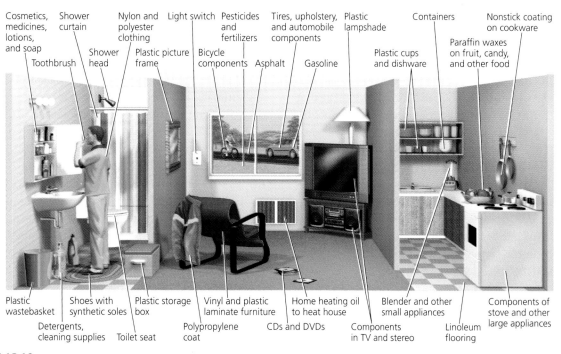

FIGURE 15.10

Petroleum products are everywhere in our daily lives. Besides the gasoline and other fuels we use for transportation and heating, petroleum products include many of the fabrics that we wear and most of the plastics that help make up countless items we use every day.

FIGURE 15.11
Because fossil fuels are nonrenewable resources, supplies at some point pass the midway point of their depletion, and annual production begins to decline. U.S. oil production peaked in 1970, just as geologist M. King Hubbert predicted decades previously; this high point is referred to as Hubbert's peak **(a)**. Today many analysts believe global oil production is about to peak. Shown is the latest projection **(b)**, from a 2007 analysis by scientists at the Association for the Study of Peak Oil. Data in (a) from Deffeyes, K. S. 2001. *Hubbert's Peak: The Impending World Oil Shortage.* Princeton, NJ: Princeton University Press; and U.S. Energy Information Administration. Data in (b) from Campbell, C. J., and the Association for the Study of Peak Oil and Gas. 2007.

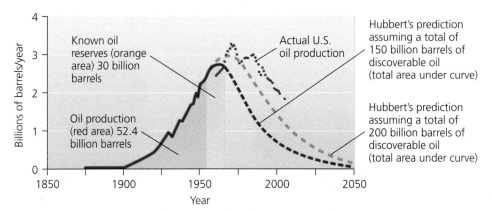

(a) Hubbert's prediction of peak in U.S. oil production, with actual data

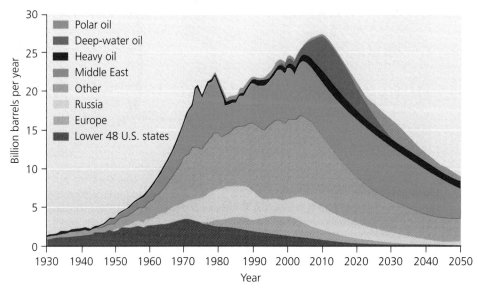

(b) Modern prediction of peak in global oil production

To understand the basis of these concerns, we need to turn back the clock to 1956. In that year, Shell Oil geologist M. King Hubbert calculated that U.S. oil production would peak around 1970. His prediction was ridiculed at the time, but it proved to be accurate; U.S. production peaked in that very year and has continued to fall since then (**FIGURE 15.11A**). The peak in production came to be known as **Hubbert's peak**.

In 1974, Hubbert analyzed data on technology, economics, and geology, predicting that global oil production would peak in 1995. It grew past 1995, but many scientists using newer, better data today predict that at some point in the coming decade, production will begin to decline (**FIGURE 15.11B**). Oil geologist Kenneth Deffeyes even contends that we have already passed the peak—that we did so in December 2005—and he is not alone in this belief. Indeed, because of year-to-year variability in production, we will be able to recognize that we have passed the peak of oil production only several years after it has happened.

Predicting an exact date for **"peak oil"** and the coming decline in production is difficult. Many companies and governments do not reveal their true data on oil reserves, and estimates differ as to how much oil we can extract secondarily from existing deposits. Moreover, a recent U.S. Geological Survey report estimated 2 trillion barrels remaining in the world, rather than 1 trillion, and some estimates predict still greater amounts. A 2007 report by the United States General Accounting Office reviewed 21 studies and found that estimates for the timing of the oil production peak ranged from now through 2040. Regardless of the exact timing, it seems certain that a peak in global oil production will occur. Discoveries of new oil fields peaked 30 years ago, and since then we have been extracting and consuming more oil than we have been discovering. Meanwhile, global demand continues to rise, particularly as China and India industrialize rapidly.

The coming divergence of demand and supply will likely have momentous economic, social, and political consequences that will profoundly affect the lives of each and every one of us. Pessimists predict the collapse of modern industrial society, as fossil fuel supplies become increasingly insufficient. More optimistic observers argue that as oil supplies dwindle, rising prices will create powerful incentives for businesses, governments, and individuals to conserve energy and develop alternative energy sources (Chapters 16 and 17)—and that these developments will save us from major disruptions caused by the coming oil peak.

Indeed, to achieve a sustainable society, we will need to switch to renewable energy sources. Energy conservation can extend the time we have in which to make this transition. However, the research and development needed to construct the infrastructure for a new energy economy depend on having cheap oil, and the time we will have to make this enormous transition will be quite limited.

weighing the issues 15–1
THE END OF OIL

How do you think your life would be affected if our society were to suffer a 50% decrease in oil availability over the next 10 years, as some observers have predicted? What steps would you take to adapt to these changes? What steps should our society take to deal with the coming depletion of oil? Do you think the recent surges in the price of oil and gasoline are an indication that such changes are beginning?

"Unconventional" Fossil Fuels

As oil production declines, we will rely more on natural gas and coal—yet these in turn will also peak and decline in future years. Are there other fossil fuels that can replace them and stave off our day of reckoning? At least three types of alternative fossil fuels exist in large amounts: oil sands, oil shale, and methane hydrates.

Canada owns massive deposits of oil sands

Oil sands (also called **tar sands**) are deposits of moist sand and clay containing 1%–20% **bitumen**, a thick and heavy form of petroleum that is rich in carbon and poor in hydrogen. Oil sands represent crude oil deposits that have been degraded and chemically altered by water erosion and bacterial decomposition.

FIGURE 15.12
In Alberta, companies strip-mine oil sands with the world's largest dump trucks and power shovels. On average, two metric tons of oil sands are required to produce one barrel of synthetic crude oil.

Because bitumen is too thick to extract by conventional oil drilling, oil sands are generally removed by strip-mining (**FIGURE 15.12**), using methods similar to coal strip-mining. For deposits 75 m or more below ground, a variety of *in situ* extraction techniques are being devised. Most of these involve injecting steam or chemical solvents to liquefy the bitumen so it can be extracted through conventional wells. After extraction, bitumen may be sent to specialized refineries, where several types of chemical reactions that add hydrogen or remove carbon can upgrade it into more valuable synthetic crude oil.

Three-quarters of the world's oil sands lie in two areas: eastern Venezuela and northeastern Alberta. Oil sands in each region hold at least 175 billion barrels of oil. In Alberta, strip-mining began in 1967, but as rising crude oil prices make oil sands more profitable, dozens of companies are now angling to begin 100 or more mining projects in the region. In 2005, oil sands produced 966 000 barrels of oil per day, contributing 39% of

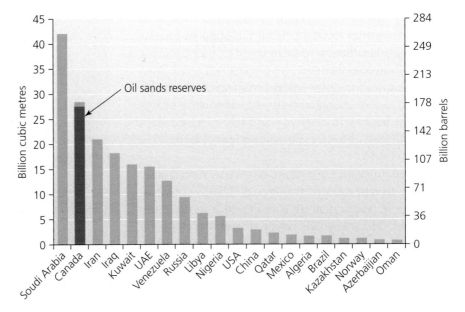

FIGURE 15.13
Canada's tar sands are a significant proven oil reserve, even in an international context. Based on BP Statistical Review of World Energy, 2005.

Canada's petroleum production. If all planned projects go through, production could reach 5 million barrels per day by 2030. The tar sands move Canada into a strong position for proven oil reserves in the international context (**FIGURE 15.13**).

Oil shale is abundant in the American West

Oil shale is sedimentary rock that contains abundant kerogen, which can be processed to produce liquid petroleum. Oil shale is formed by the same processes that form crude oil but occurs when kerogen was not buried deeply enough or subjected to enough heat and pressure to form oil.

We mine oil shale by using strip-mines or subsurface mines. Once mined, oil shale can be burned directly like coal, or it can be baked in the presence of hydrogen and in the absence of air to extract liquid petroleum (a process called *pyrolysis*). Currently, industry is developing *in situ* extraction processes in which rock is heated underground to liquefy and release oil into conventional wells.

The world's known deposits of oil shale may be able to produce more than 600 billion barrels of oil (roughly half as much as the crude oil remaining in the world). About 40% of global oil shale reserves are in the United States. Low prices for crude oil have kept investors away from the more costly oil shale, but as crude prices rise, oil shale is again attracting attention.

Methane hydrate shows potential

In Chapter 12, we discussed **methane hydrate** (also called *methane clathrate* or *methane ice*). This solid consisting of molecules of methane within a crystal lattice of water ice molecules occurs underground in some Arctic locations and more widely under the seafloor on the continental shelves. Most methane in these gas hydrates was formed by bacterial decomposition in anaerobic environments, but some results from thermogenic formation deeper below the surface. Scientists believe there to be immense amounts of methane hydrate on Earth—from 2 to 20 times the amount of natural gas from all other sources.

However, we still need to develop technology to extract methane hydrates safely, avoiding the risk of destabilizing these deposits during extraction. Such destabilization could lead to underwater landslides, tsunamis, and the release of large amounts of methane, a potent greenhouse gas.

These alternative fossil fuels have downsides

These three alternative fossil fuels are abundant, but they are no panacea for our energy challenges. For one thing, their net energy values are low, because they are expensive to extract and process. Thus the energy returned on energy invested (EROI) ratio is low. For instance, at least 40% of the energy content of oil shale is consumed in its production, and oil shale's EROI is only about 2:1 or 3:1, compared with a 5:1 or greater ratio for conventional crude oil. Natural gas extracted from the gas fields of the Mackenzie River Delta might never make it all the way to southern consumers if it must be sidetracked to support the extraction of oil from the Athabasca tar sands.

Second, these fuels exert severe environmental impacts. Oil sands and oil shale require extensive strip-mining, which utterly devastates landscapes over large areas and pollutes waterways that run into other areas. Although most governments require mining companies to restore mined areas to their original condition, regions denuded by the very first oil sand mine in Alberta 30 years ago have still not recovered. *In situ* extraction methods exert less environmental impact, but these techniques are not yet well developed.

Canadian environmentalists are worried about the intensive water use that typically accompanies the extraction of unconventional fossil fuels, as well as the impacts on water quality in surrounding regions. The impacts on wildlife are also of concern; this concern was brought into stark focus in 2008 when hundreds of migratory birds died when their feathers became fouled with oil after landing on Syncrude's massive tailings pond near Fort McMurray, Alberta (FIGURE 15.14). To give you an idea of the magnitude of these ponds, which can hold up to 540000000 m³ of oily sludge, consider that this is the largest dam in Canada, and second in the world only to the Three Gorges hydroelectric dam in China. Canada's largest hydroelectric dam, the Gardiner, has just slightly more than one-tenth the capacity of the Syncrude tailings dam.[12]

Besides impacts from their extraction, our combustion of alternative fossil fuels would emit at least as much carbon dioxide, methane, and other air pollutants as does our use of coal, oil, and gas. Thus, they will worsen the effects that fossil fuels are already causing, including air pollution and global climate change.

Environmental Impacts of Fossil Fuel Use

Our society's love affair with fossil fuels and the many petrochemical products we have developed from them has boosted our material standard of living beyond what our ancestors could have dreamed, has eased constraints on travel, and has helped lengthen our life spans. It has, however, also caused harm to the environment and human health. Concern over these impacts is a prime reason many scientists, environmental advocates, businesspeople, and policy makers are increasingly looking toward renewable sources of energy that exert less impact on natural systems.

Fossil fuel emissions cause pollution and drive climate change

When we burn fossil fuels, we alter flux rates in Earth's carbon cycle. We essentially take carbon that has been retired into a long-term reservoir underground and release it into the air. This occurs as carbon from within the hydrocarbon molecules of fossil fuels unites with oxygen from the atmosphere during combustion, producing carbon dioxide (CO_2). Carbon dioxide is a greenhouse gas, and CO_2 released from fossil fuel combustion warms our planet and drives changes in global climate (Chapter 14). Because global climate change may have diverse, severe, and widespread ecological and socioeconomic impacts, carbon dioxide pollution (FIGURE 15.15) is becoming recognized as the greatest environmental impact of fossil fuel use.

Fossil fuels release more than carbon dioxide when they burn. Methane is a potent greenhouse gas, and other air pollutants resulting from fossil fuel combustion can have serious consequences for human health and the environment. Deposition of mercury and other pollutants from coal-fired power plants is increasingly recognized as a substantial health risk. The burning of fossil

(a) An oil sand tailings pond in Alberta

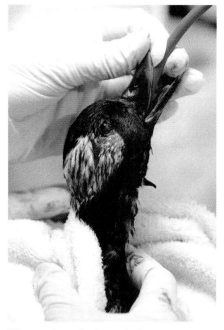

(b) Rescuing oiled sea ducks

FIGURE 15.14
Concerns about the impacts of oil sand exploitation on wildlife came to a head in 2008 when hundreds of migratory birds died after landing in an oily, toxic tailings pond at Fort McMurray, Alberta, similar to the one shown here **(a)**. Only a few of the oiled sea ducks managed to survive **(b)**.

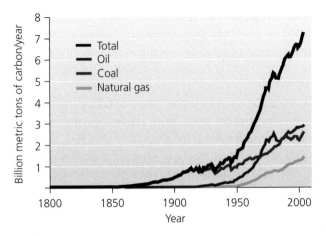

FIGURE 15.15
As industrialization has proceeded, and as population and consumption have grown, emissions from fossil fuel combustion have risen dramatically. Here, worldwide emissions of carbon from carbon dioxide are subdivided by their source (oil, coal, or natural gas). Some other minor sources (such as cement production) are not shown but are included in the graphed total. Data from Marland, G., et al. 2006. Global, regional, and national fossil fuel CO_2 emissions. In *Trends: A Compendium of data on global change*. Carbon Dioxide Information Analysis Center, Oak Ridge National Laboratory, U.S. Department of Energy, Oak Ridge, TN.

fuels in power plants and vehicles releases sulphur dioxide and nitrogen oxides, which contribute to industrial and photochemical smog and to acidic deposition.

We have already employed technologies, such as catalytic converters, to cut down on vehicle exhaust pollution. Gasoline combustion in automobiles releases pollutants that irritate the nose, throat, and lungs. Some hydrocarbons, such as benzene and toluene, are carcinogenic to laboratory animals and likely also to people. In addition, gases, such as hydrogen sulphide, can evaporate from crude oil, irritate the eyes and throat, and cause asphyxiation. Crude oil also often contains trace amounts of known poisons, such as lead and arsenic. As a result, workers at drilling operations, refineries, and other jobs that entail frequent exposure to oil or its products can develop serious health problems, including cancer.

Fossil fuels can pollute water as well as air. Atmospheric deposition of pollutants exerts many impacts on freshwater ecosystems. Moreover, oil from non-point sources—such as industries, homes, automobiles, gas stations, and businesses—runs off roadways and enters rivers and sewage treatment facilities to be discharged eventually into the ocean. Although most spilled oil results from these non-point sources, large catastrophic oil spills can have significant impacts on the marine environment. Crude oil's toxicity to most plants and animals can lead to high mortality. This was the case with the *Exxon Valdez* spill in 1989 (Chapter 2), in which oil from Alaska's North Slope, piped to the port of Valdez through the trans-Alaska pipeline, caused long-term damage to ecosystems and economies in Alaska's Prince William Sound when a ship ran aground.

Oil can also contaminate groundwater supplies, such as when leaks from oil operations penetrate deeply into soil. Of greater concern, as we saw in Chapter 11, are the thousands of underground storage tanks containing petroleum products that have leaked, threatening drinking water supplies.

Some emissions from fossil fuel burning can be "captured"

One relatively new technology for "cleaning up" carbon-based fuel sources is **carbon capture and storage (CCS)** or *carbon capture and sequestration*. Recall from our discussion of biogeochemical cycles that **sequestration** is a term that refers to the storage of materials in geologic reservoirs on a long timescale. In this case, the material of interest is carbon—primarily in the form of CO_2—and the goal is to prevent some of the carbon generated by the burning of fossil fuels from entering the atmosphere and contributing to global warming.

In a nutshell, what would happen is that the CO_2 emitted by, for example, a traditional coal-burning power plant would be captured before it reached the atmosphere and then diverted to a storage reservoir. The most likely reservoirs are the deep ocean, which already acts as a reservoir for atmospheric carbon dioxide, and geologic formations deep underground (**FIGURE 15.16A**). Many of Canada's largest emitters of carbon dioxide are located in reasonable proximity to appropriate underground reservoirs (**FIGURE 15.16B**).

Industry analysts predict that 80%–90% of the carbon dioxide emissions from large emitters like coal-fired generators could be captured and diverted, as compared with a power plant without CCS technology. This would go a long way toward helping Canada meet its commitments in the Kyoto Protocol—at least in the short term, while we attempt to transition to more renewable energy sources.

Many environmentalists are skeptical about CCS, though, arguing that the true environmental impacts of reinjecting carbon dioxide into the ground or into ocean water are not known. Some point, for example, to the possibility of acidification of ocean water (because carbon dioxide mixed with water yields carbonic acid). Others maintain that the technology is still too unproven to be the central focus in Canada's strategy for cutting carbon emissions. Still others argue that the approach is fundamentally flawed, because it takes the burden off large emitters and serves only to prolong our dependence on fossil fuels rather than facilitating a shift to renewables.

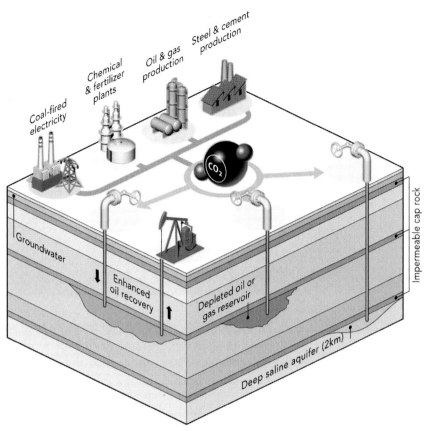

(a) Carbon capture and storage

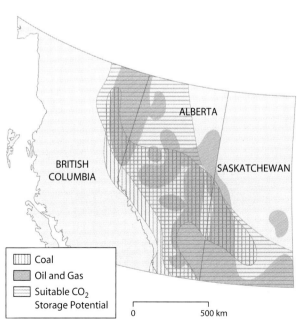

(b) Potential CCS reservoirs

FIGURE 15.16
CCS **(a)** provides a possible mechanism for reducing the harmful environmental impacts of fossil fuel use. The map **(b)** shows the locations of potential reservoirs for CCS in Western Canada, relative to the locations of major fossil fuel deposits.

Coal mining affects the environment

The mining of coal also has substantial impacts on natural systems and human well-being. Surface strip-mining can destroy large swaths of habitat and cause extensive soil erosion. It also can cause chemical runoff into waterways through the process of **acid drainage**. This occurs when sulphide minerals in newly exposed rock surfaces react with oxygen and rainwater to produce sulphuric acid. As the sulphuric acid runs off, it leaches metals from the rocks, many of which are toxic to organisms in high concentrations. Acid drainage is a natural phenomenon, but its rate accelerates greatly when mining exposes many new rock surfaces at once.

Government regulations require mining companies to restore strip-mined land following mining, but impacts can be severe and long lasting just the same. Mountaintop removal (**FIGURE 15.17**) can have even greater impacts than conventional strip-mining. When countless tons of rock and soil are removed from the top of a mountain, it is difficult to keep material from sliding downhill, where immense areas of habitat can be degraded or destroyed and creek beds can be polluted and clogged.

Whereas mountaintop removal threatens the welfare of nearby residents, subsurface mining raises health concerns for miners. Underground coal mining is one of

FIGURE 15.17
Strip-mining in some areas is taking place on massive scales, such that entire mountain peaks are levelled, as at this site in West Virginia. Such "mountaintop removal" can cause enormous amounts of erosion into waterways that flow into surrounding valleys, affecting ecosystems over large areas, as well as the people who live there.

our society's most dangerous occupations. Besides risking injury or death from collapsing shafts and tunnels and from dynamite blasts, miners constantly inhale coal dust in the enclosed spaces of mines, which can lead to respiratory diseases, including fatal black lung disease.

The costs of alleviating all these health and environmental impacts are high, and the public eventually pays them in an inefficient manner. The reason is that the costs are generally not internalized in the market prices of fossil fuels, which are kept inexpensive through government subsidies to extraction companies.

Oil and gas extraction can alter the environment

Much more than drilling is involved in the development of an oil or gas field. Road networks must be constructed, and many sites may be explored in the course of prospecting. The extensive infrastructure needed to support a full-scale drilling operation typically includes housing for workers, access roads, transport pipelines, and waste piles for removed soil. Ponds may be constructed for collecting the toxic sludge that remains after the useful components of oil have been removed. At extraction sites for coalbed methane, groundwater is pumped out to free gas to rise, but salty groundwater dumped on the surface can contaminate soil and kill vegetation over large areas.

Many onshore North American oil reserves are located in Arctic or semi-arid areas. Plants grow slowly in tundra and semi-desert ecosystems, so even minor impacts can have long-lasting repercussions. Tundra vegetation at some northern oil developments, such as Prudhoe Bay in Alaska, still has not fully recovered from temporary roads last used 30 years ago during the exploratory phase of development. Studies at Prudhoe Bay also show that female caribou and their calves avoid all parts of the oil complex, including its roads, sometimes detouring many kilometres to do so. These studies also show that the reproductive rate of female caribou in the Prudhoe Bay region is lower than for those in undeveloped areas in Alaska. As a result, although the herd near Prudhoe Bay has increased over the past 25 years, it has not increased as much as have herds in some other parts of Alaska.

There is no way of knowing how the Prudhoe Bay herd would have performed in the absence of development; that is, there is no control, as there would be for a manipulative experiment. It is difficult, therefore, to draw conclusions about the impacts of oil development on caribou and other wild animals, such as the grizzly bear (**FIGURE 15.18**), at remote oil developments in

FIGURE 15.18
Like the Mackenzie River Valley, Alaska's North Slope is home to a variety of large mammals, including grizzly bears, polar bears, wolves, Arctic foxes, and large herds of caribou. How oil development may affect these animals is a controversial issue, and scientific studies are ongoing. The caribou herd near Prudhoe Bay has increased since oil extraction began there, but less than herds in other parts of Alaska. Grizzly bears, such as the ones shown here, have been found near, or even walking atop, the trans-Alaska pipeline.

places like Prudhoe Bay or the Mackenzie River Valley. It can be anticipated that activities like road building, oil pad construction, worker presence, oil spills, accidental fires, trash buildup, permafrost melting, off-road vehicle trails, and dust from roads would have a significant impact on both vegetation and wildlife.

Political, Social, and Economic Aspects

The political, social, and economic consequences of fossil fuel use are numerous, varied, and far-reaching. Our discussion focuses on several negative consequences of fossil fuel use and dependence, but it is important to bear in mind that their use has enabled much of the world's population to achieve a higher material standard of living than ever before. It is also important to ask in each case whether switching to more renewable sources of energy would solve existing problems.

Nations can become dependent on foreign energy

Virtually all our modern technologies and services depend in some way on fossil fuels. Putting all of one's eggs in one basket is always a risky strategy. The fact that our economies are utterly tied to fossil fuels means that we are vulnerable to supplies' becoming suddenly unavailable or extremely costly. Nations that lack adequate fossil fuel reserves of their own are especially vulnerable. For instance, Germany, France, South Korea, and Japan consume far more energy than they produce and thus rely almost entirely on imports for their continued economic well-being (**FIGURE 15.19**). Canada is both an importer and an exporter of fossil fuels in different forms, but imports are outweighed by exports.

Reliance on foreign oil means that seller nations can control energy prices, forcing buyer nations to pay more and more as supplies dwindle. This became clear in

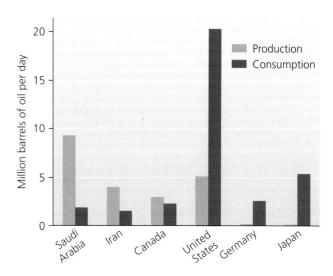

FIGURE 15.19
Japan, Germany, and the United States are among nations that consume far more oil than they produce. Iran, Saudi Arabia, and Canada are among countries that produce more oil than they consume and are able to export oil to high-consumption countries. Data from U.S. Energy Information Administration. 2007. *Annual Energy Review 2006.* Washington, DC.

1973, when the *Organization of the Petroleum Exporting Countries (OPEC)* resolved to stop selling oil to the United States as a consequence of U.S. support of Israel. The embargo created panic in the West and caused oil prices to skyrocket (**FIGURE 15.20**), spurring inflation.

Oil supply and prices affect the economies of nations

More recently, when Hurricanes Katrina and Rita slammed into the Gulf Coast in 2005, they damaged offshore platforms and refineries, causing oil and gas prices to spike significantly. The economic ripple effects served to remind us yet again how much we rely on a steady and ever-increasing supply of petroleum.

With the majority of world oil reserves located in the politically volatile Middle East, crises, such as the 1973–1974 embargo, the Iranian Revolution, the

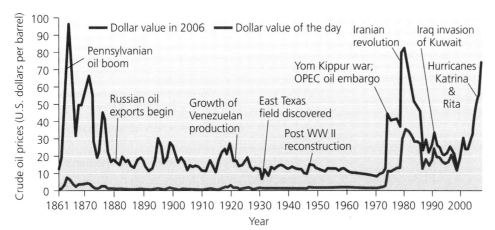

FIGURE 15.20
World oil prices have fluctuated greatly over the decades, often because of political and economic events in oil-producing countries. The greatest price hikes in recent times have resulted from wars and unrest in the oil-rich Middle East. Data from U.S. Energy Information Administration.

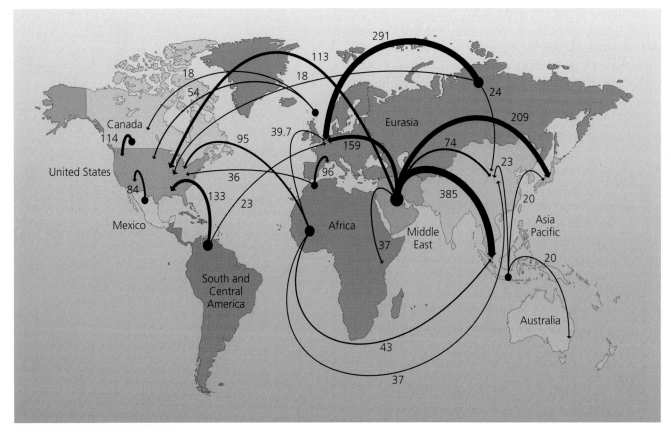

FIGURE 15.21
The global trade in oil is lopsided; relatively few nations account for most exports, and some nations are highly dependent on others for energy. Canada imports some North Sea oil while exporting more to the United States. Numbers in the figure represent millions of metric tons. Data from British Petroleum. 2007. *Statistical review of world energy 2007*.

Iran–Iraq War, and recent events in Iraq, are a constant concern for U.S. policy makers. In response to the 1973 embargo, the U.S. government enacted a series of policies designed to reduce reliance on foreign oil, including the development of domestic sources in remote locations, such as Prudhoe Bay, Alaska. The United States has cultivated a close relationship with Saudi Arabia, the owner of 22% of world oil reserves, despite the fact that Saudi Arabia's political system allows for little of the democracy that U.S. leaders claim to cherish and promote. The world's third-largest holder of oil reserves, at 10%, is Iraq, which is why many people around the world believe that the American-led invasion of that nation in 2003 was motivated primarily to secure access to oil. Major petroleum trade relations among nations and regions of the world are depicted in **FIGURE 15.21**.

Residents may or may not benefit from their fossil fuel reserves

The extraction of fossil fuels can be extremely lucrative; many of the world's wealthiest corporations deal in fossil fuel energy or related industries. These industries provide jobs to millions of employees and supply dividends to millions of investors. Development can potentially yield economic benefits for people who live in petroleum-bearing areas, as well. In addition to the potential for jobs from the Mackenzie River Valley gas pipeline project, for example, the federal government has promised at least $500 million in assistance to residents of the area. Supporters argue that income and federal assistance related to the project could pay for health care, police and fire protection, and other services that are currently scarce in this remote region.

In most parts of the world where fossil fuels have been extracted, local residents have not seen great benefits but instead have frequently suffered. When multinational corporations have extracted oil or gas in developing countries, paying those countries' governments for access, the money often has not trickled down to residents of the regions where the extraction takes place. Moreover, oil-rich developing countries, such as Ecuador, Venezuela, and Nigeria, tend to have few environmental regulations, and existing regulations may go unenforced if a government does not want to risk losing the large sums of money associated with oil development.

In Nigeria, oil was discovered in 1958 in the territory of the Ogoni, one of Nigeria's native peoples, and the Shell Oil Company moved in to develop oil fields. Although Shell extracted $30 billion of oil from Ogoni land over the years, the Ogoni still live in poverty, with no running water or electricity. The profits from oil extraction on Ogoni land went to Shell and to the military dictatorships of Nigeria. The development resulted in oil spills, noise, and constantly burning gas flares, all of which caused illness among people living nearby. From 1962 until his death in 1995, Ogoni activist and leader Ken Saro-Wiwa worked for fair compensation to the Ogoni for oil extraction and environmental degradation on their land. After years of persecution by the Nigerian government, Saro-Wiwa was arrested in 1994, given a trial universally regarded as a sham, and put to death by a military tribunal.

How will we convert to renewable energy?

Fossil fuel supplies are limited, and their use and our continued dependence on them have health, environmental, political, and socioeconomic consequences. Given this, the world's nations have several policy options for guiding future energy use. One option is to continue relying on fossil fuels until they are no longer economically practical and to develop other energy sources only after supplies have dwindled. A second option is to increase funding to develop alternative energy sources dramatically and immediately and to hasten a rapid shift to them. Third, we could steer a middle course and attempt to reduce our reliance on fossil fuels gradually.

Regardless of which course we take, it will benefit us to prolong the availability of fossil fuels as we make the transition to renewable sources. We can prolong our access to fossil fuels by instituting measures to conserve energy, primarily through lifestyle changes that reduce energy use and technological advances that improve efficiency.

Energy conservation has followed economic need

Until our society makes the transition to renewable energy sources, we will need to find ways to minimize the expenditure of energy from our dwindling fossil fuel resources. **Energy conservation** is the practice of reducing energy use to extend the lifetimes of our nonrenewable energy supplies, to be less wasteful, and to reduce our environmental impact.

Many people first saw the value of conserving energy following the OPEC embargo of 1973–1974. Over the past three decades, however, many of the conservation initiatives that followed the oil crisis were abandoned. Without high market prices and an immediate threat of shortages, people lacked economic motivation to conserve. Government funding for research into alternative energy sources decreased, speed limits increased, and countless proposals to raise the mandated average fuel efficiency of vehicles failed. The average fuel efficiency of new vehicles worsened, from 940 km/100 L (22.1 mpg) in 1988 to 892 km/100 L in 2006, primarily as a result of increased sales of light trucks and sport utility vehicles (averaging 782 km/100 L) relative to cars (averaging 1045 km/100 L).

All of this is changing, though, particularly in light of steadily increasing oil prices. As 2008 dawned, oil touched the symbolic $100/barrel landmark price, and increases were translated directly to gas station pumps. A 2008 survey of Canadian spending habits conducted by Investors Group, the largest mutual group company in Canada, revealed that 83% of Canadians planned to buy a more fuel-efficient car next time around, 51% had been cutting down on driving, and 44% planned to change their holiday plans in response to high fuel prices.[13] These types of consumer changes are happening with equal fury in the United States, and as a result the light truck and SUV industry in North America underwent a sudden, severe contraction, with a number of production facilities closing in 2008.

Policy makers have repeatedly failed to raise the *corporate average fuel efficiency (CAFE) standards*, which set benchmarks for auto manufacturers to meet. This is despite a 2001 U.S. National Academy of Sciences report concluding that fuel efficiency could feasibly be raised by 40% without adverse economic effect, and despite the fact that most other developed nations boast autos with considerably better fuel efficiency. Transportation accounts for two-thirds of oil use and more than a quarter of energy use in Canada, and passenger vehicles consume more than half this energy. The vast distances of the Canadian landscape add to the problem. Thus, the failure to improve vehicular fuel economy over the past 20 years, despite the existence of technology to do so, has added greatly to oil consumption. This is unfortunate because the inefficient use of gasoline in auto engines wastes oil that we could put to better use in manufacturing countless products that enhance our lives. Transportation also accounts for about 34% of Canada's greenhouse gas emissions, an increasing concern as we strive to meet our commitments under the Kyoto Protocol.[14]

weighing the issues 15-2

MORE KILOMETRES, LESS GAS

If you drive an automobile, how many kilometres does it travel per 100 L of gasoline? If you drove 2400 km in a car with a fuel efficiency of 1200 km/100 L, instead of making the trip in an SUV with a fuel efficiency of 800 km/100 L, how much less gasoline would you have to buy? How much money would you save on the trip? How much would you save on the amount you typically drive in a year? Do you think that the government should raise taxes on gasoline sales as an incentive to consumers to conserve energy?

Personal choice and increased efficiency are two routes to conservation

Energy conservation can be accomplished in two primary ways. As individuals, we can make conscious choices to reduce our own energy consumption. Examples include driving less, turning off lights when rooms are not being used, turning down thermostats, and investing in more efficient machines and appliances. For any given individual or business, reducing energy consumption can save money while also helping to conserve resources.

As a society, we can conserve energy by making our energy-consuming devices and processes more efficient. Currently, more than two-thirds of the fossil fuel energy we use is simply lost, as waste heat, in automobiles and power plants. In the case of automobiles, we already possess the technology to increase fuel efficiency far above the current North American average of 900 km/100 L. We could accomplish this with more efficient gasoline engines, lightweight materials, continuously variable transmissions, alternative technology vehicles, such as electric/gasoline hybrids or vehicles that use hydrogen fuel cells.

We can also vastly improve the efficiency of our power plants. One way is to use **cogeneration**, in which excess heat produced during the generation of electricity is captured and used to heat workplaces and homes and to produce other kinds of power. Cogeneration can almost double the efficiency of a power plant. The same is true of *coal gasification* and *combined cycle* generation. In this process, coal is treated to create hot gases that turn a gas turbine, while the hot exhaust of this turbine heats water to drive a conventional steam turbine.

weighing the issues 15-3

MACKENZIE VALLEY NATURAL GAS PIPELINE

Do you think Canada should subsidize the development of the Mackenzie Valley Natural Gas Pipeline? Why or why not? What would be gained? What would be lost?

In homes and public buildings, a significant amount of heat is lost in winter and gained in summer because of inadequate insulation (**FIGURE 15.22**). Improvements in the design of homes and offices can reduce the energy required to heat and cool them. Such design changes can involve the building's location, the colour of its roof (light colours keep buildings cooler by reflecting the Sun's rays), and its insulation.

Among consumer products, scores of appliances, from refrigerators to light bulbs, have been reengineered through the years to increase energy efficiency. Energy-efficient lighting, for example, can reduce energy use by 80%, and new energy-efficient appliances have already reduced per-person home electricity use below what it was in the 1970s. Even so, there remains room for further improvement.

FIGURE 15.22
Many of our homes and offices could be made more energy-efficient. One way to determine how much heat a building is losing is to take a photograph that records energy in the infrared portion of the electromagnetic spectrum. In such a photograph, or *thermogram* (shown here), white, yellow, and red signify hot and warm temperatures at the surface of the house, whereas blue and green shades signify cold and cool temperatures. The white, yellow, and red colours indicate areas where heat is escaping.

CANADIAN ENVIRONMENTAL PERSPECTIVES

Mary Griffiths

Mary Griffiths looks for ways to reduce the environmental impacts of fossil fuels and raise awareness about the need for wise management to protect the environment in the long term.

- **Environmental policy analyst**
- **Energy researcher**
- **Author**

Circumstances and opportunities have led Mary Griffiths to "reinvent" herself and her career a number of times in her life. Throughout it all, however, she has maintained her childhood love of nature and her commitment to raising awareness about the interrelationships between people and the natural environment.

In the 1960s—before there was such a thing as a degree in environmental science—Griffiths earned a Ph.D. in geography and subsequently taught the subject for a number of years at the University of Exeter, England. She remembers teaching about the decline of the Ogallala Aquifer beneath the High Plains of the United States and thinking, at the time, how shortsighted it was to be depleting this critical resource. Many years later when she learned about the enormous quantities of freshwater that were being used in Alberta to get oil out of the ground, she felt that it was very important to make sure that the aquifers were not being adversely affected.

Moving with her young family to the Netherlands, Griffiths helped set up a small environmental group to educate individuals and municipalities on the harmful effects of pesticides. "I have always believed in sound research as the basis for action, as well as in the power of leading by example. If one municipal authority could successfully manage its public open spaces without using pesticides, others could do it, too," she says.

The family's next move, to Canada this time, found Griffiths working with the environment and other portfolios for the Liberal Caucus at the Alberta Legislature. Occasionally she sought advice from staff at the Pembina Institute, whose mandate "is to advance sustainable energy solutions through innovative research, education, consulting and advocacy."[15] An opening at Pembina for a new policy analyst gave Griffiths the opportunity to focus entirely on environmental issues once more and to devote her research efforts to finding sustainable energy solutions.

Now a senior policy analyst in the Energy Solutions group at Pembina, Griffiths has become an expert on the environmental impacts of fossil fuel use, especially the impacts on water. She has contributed to evaluations of the environmental impacts of energy projects, including oil sands developments and coal-fired power plants. She has written several books, including *When the Oilpatch Comes to Your Backyard: A Citizens' Guide* (2004), and has co-authored numerous reports on topics ranging from landowner rights to coalbed methane development, the use of water by the oil industry, and carbon capture and storage. She has served on several government committees, including the Alberta Minister of the Environment's Environmental Protection Advisory Committee, and Alberta Environment's Advisory Committee on Water Use Practice and Policy. In 2002, Griffiths received a Canadian Environment Award for her work on clean air issues and was awarded the Alberta Centennial Medal in 2005.[16]

Griffiths' current research focuses on climate change, energy, and water. She is convinced that water resources will come under increasing pressure in the future, and hopes to encourage more and better research, and the formulation of strong policies to ensure that freshwater aquifers are protected and managed in a sustainable manner. Her overall goal is to raise awareness about the need for wise management—in all sectors—to protect the environment. "We must scrutinize all our corporate, government, and individual planning decisions, to see if they meet our long-term goal of sustainability. And we should start planning for 2015 today."[17]

"Knowledge is the foundation for environmental action."[18] —Mary Griffiths

Thinking About Environmental Perspectives

Do you agree with Mary Griffiths that all environmental action should be founded on research and knowledge? Why is research so important in making environmental decisions and formulating environmental policies? How can researchers in science and social science communicate their findings to policy makers in understandable ways?

While manufacturers can improve the energy efficiency of appliances, consumers need to "vote with their wallets" by purchasing these energy-efficient appliances. Decisions by consumers to purchase energy-efficient products are crucial in keeping those products commercially available. For the individual consumer, studies show that the slightly higher cost of buying energy-efficient washing machines is rapidly offset by savings on water and electricity bills. On the national level, France, Great Britain, and many other developed countries have standards of living equal to that of Canada, but they use much less energy per capita. This disparity indicates that Canadian citizens could significantly reduce their energy consumption without decreasing their quality of life.

Both conservation and renewable energy are needed

It is often said that reducing our energy use is equivalent to finding a new oil reserve. Indeed, conserving energy is better than finding a new reserve, because it lessens impacts on the environment while extending our access to fossil fuels.

However, energy conservation does not add to our supply of available fuel. Regardless of how much we conserve, we will still need energy, and it will need to come from somewhere. The only sustainable way of guaranteeing ourselves a reliable long-term supply of energy is to ensure sufficiently rapid development of renewable energy sources, which we will consider in greater detail in Chapters 16 and 17.

Conclusion

Over the past 200 years, fossil fuels have helped us build the complex industrialized societies we enjoy today. However, we are now approaching a turning point in history: Our production of fossil fuels will begin to decline. We can respond to this new challenge in creative ways, encouraging conservation and developing alternative energy sources. Or we can continue our current dependence on fossil fuels and wait until they near depletion before we try to develop new technologies and ways of life. The path we choose will have far-reaching consequences for human health and well-being, for Earth's climate, and for our environment.

The ongoing debates over projects like the Mackenzie Valley Natural Gas Pipeline are microcosms of this debate over our energy future. Fortunately, there is not simply a trade-off between benefits of energy for us and harm to the environment, climate, and health. Instead, as evidence builds that renewable energy sources are becoming increasingly feasible and economical, it becomes easier to envision giving up our reliance on fossil fuels and charting a win–win future for humanity and the environment.

REVIEWING OBJECTIVES

You should now be able to:

Identify the principal energy sources that we use

- A variety of renewable and nonrenewable energy sources are available to us.
- Since the Industrial Revolution, nonrenewable fossil fuels—including oil, natural gas, and coal—have become our primary sources of energy.
- Fossil fuels are formed very slowly as buried organic matter is chemically transformed by heat, pressure, and/or anaerobic decomposition.
- In evaluating energy sources, it is important to compare the amount of energy obtained from them with the amount invested in their extraction and production.

Describe the nature and origin of coal and evaluate its extraction and use

- Coal is our most abundant fossil fuel. It results from organic matter that undergoes compression but little decomposition.
- The first fossil fuel to be widely used for heating homes and powering industry, coal is used today principally to generate electricity.
- Coal is mined underground and strip-mined from the land surface.
- Coal comes in different types and varies in its composition. Combustion of coal that is high in contaminants emits toxic air pollution.

Describe the nature and origin of natural gas and evaluate its extraction and use

- Natural gas consists mostly of methane and can be formed in two ways.
- Use of natural gas is growing rapidly, and it is cleaner burning than coal or oil.
- Natural gas often occurs with oil and coal deposits, is extracted in similar ways, and becomes depleted in similar ways.

Describe the nature and origin of oil and evaluate its extraction, use, and future availability

- Crude oil is a thick, liquid mixture of hydrocarbons that is formed underground under certain temperature and pressure conditions.
- Scientists locate fossil fuel deposits by analyzing subterranean geology. Geologists estimate total reserves, as well as the technically and economically recoverable portions of those reserves.
- Oil drilling often involves primary extraction followed by secondary extraction, in which gas or liquid is injected into the ground to help force up additional oil.
- Petroleum-based products, from gasoline to clothing to plastics, are everywhere in our daily lives.
- Components of crude oil are separated in refineries to produce a wide variety of fuel types.

- We have depleted nearly half the world's oil. Once we pass the peak and production slows, the gap between rising demand and falling supply may pose immense economic and social challenges for our society.

Describe the nature, origin, and potential of alternative fossil fuel types and technologies

- Oils sands, abundant in Canada's West, can be mined and processed into synthetic oil.
- Oil shale is abundant in the western United States.
- Methane hydrate could provide a source of methane gas.

Outline and assess environmental impacts of fossil fuel use

- Emissions from fossil fuel combustion pollute air, pose human health risks, and drive global climate change.
- Oil is a major contributor to water pollution.
- Strip-mining and mountaintop removal can devastate ecosystems locally or regionally, and acid drainage from coalmines pollutes waterways.
- Development for oil and gas extraction exerts various environmental impacts.

Evaluate political, social, and economic impacts of fossil fuel use

- Today's societies are so reliant on fossil fuel energy that sudden restrictions in oil supplies can have major economic consequences.
- Nations that consume more fossil fuels than they produce are especially vulnerable to supply restrictions.
- People living in areas of fossil fuel extraction do not always benefit from their extraction.

Specify strategies for conserving energy and enhancing efficiency

- Energy conservation involves both personal choices and efficient technologies. These two forces interact through the market power of consumer choice.
- Increases in automotive fuel efficiency and efficiency in power plant combustion could help us conserve immense amounts of oil.
- Conservation helps lengthen our access to fossil fuels and reduce environmental impact, but to build a sustainable society we will also need to shift to renewable energy sources.

TESTING YOUR COMPREHENSION

1. Why are fossil fuels our most prevalent source of energy today? Why are they considered nonrenewable sources of energy?
2. How are fossil fuels formed? How do environmental conditions determine what type of fossil fuel is formed in a given location? Why are fossil fuels often concentrated in localized deposits?
3. Describe how net energy differs from energy returned on investment (EROI). Why are these concepts important when evaluating energy sources?
4. Describe how coal is used to generate electricity.
5. Why is natural gas often extracted simultaneously with other fossil fuels? What constraints on its extraction does it share with oil?
6. How do geologists estimate the total amount of oil reserves that remain underground? How is the "technically recoverable" different from the "economically recoverable" oil?
7. How do we create petroleum products? Provide examples of several of these products.
8. What is Hubbert's peak? Why do many experts think we are about to pass the global production peak for oil? What consequences could there be for our society if we do not transition soon to renewable energy sources?
9. List three environmental impacts of fossil fuel production and consumption. Compare some of the contrasting views of scientists regarding the environmental impacts of the Mackenzie Valley Natural Gas Pipeline.
10. Describe two main approaches to energy conservation; give specific examples of each.

SEEKING SOLUTIONS

1. Roughly how much oil is left in the world, and how much longer can we expect to use it? What steps should we take to avoid energy shortages in the future?
2. Compare the effects of coal and oil consumption on the environment. Which process do you think has ultimately been more detrimental to the environment, oil extraction or coal mining, and why? What steps

could governments, industries, and individuals take to reduce environmental impacts?

3. If Canada and other developed countries reduced dependence on foreign oil and on fossil fuels in general, do you think that their economies would benefit or suffer? Might your answer be different for the short term and the long term? What factors come into play in trying to make such a judgment?

4. Contrast the experiences of the Ogoni people of Nigeria with those of the citizens of the Northwest Territories. How have they been similar and different? Do you think businesses or governments should take steps to ensure that local people benefit from oil drilling operations? How could they do so?

5. **THINK IT THROUGH** You have been elected to be a negotiator on behalf of Aboriginal interests in the ongoing discussions about the development of the Mackenzie River Natural Gas Pipeline. What will your position be? Explain it to another negotiator who disagrees with you.

6. **THINK IT THROUGH** Throughout this book in these questions, we have asked you to imagine yourself in various roles. This time we ask you simply to be yourself. Given the information in this chapter on petroleum supplies, consumption, and depletion, what actions, if any, do you plan to take to prepare yourself for changes in our society that may come about as oil production declines? Describe in detail how you think your life may change, and suggest one thing you could do to help reduce negative impacts of oil depletion on our society.

INTERPRETING GRAPHS AND DATA

The fossil fuels that we burn today were formed long ago from buried organic matter. However, only a small fraction of the original organic carbon remains in the coal, oil, or natural gas that is formed. Thus, it requires approximately 90 metric tons of ancient organic matter—so-called paleoproduction—to result in just 3.8 L of gasoline. The graph presents estimates of the amount of paleoproduction required to produce the fossil fuels humans have used each year over the past 250 years.

1. Estimate in what year the annual consumption of paleoproduction, represented by our combustion of fossil fuels, surpassed Earth's current annual net primary production.
2. In 2000, approximately how many times greater than global net primary production was our consumption of paleoproduction?
3. If on average it takes 7000 units of paleoproduction to produce 1 unit of fossil fuel, estimate the total carbon content of the fossil fuel consumed in 2000. How does this amount compare to global NPP?

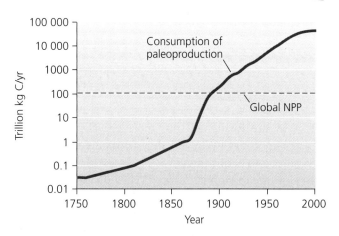

Annual human consumption of paleoproduction by fossil fuel combustion (red line), 1750–2000. The dashed line indicates current annual net primary production (NPP) for the entire planet. Data from Dukes, J. 2003 Burning buried sunshine: Human consumption of ancient solar energy. *Climatic Change* 61:31–44.

CALCULATING FOOTPRINTS

Wackernagel and Rees calculated the energy component of our ecological footprint by estimating the amount of ecologically productive land required to absorb the carbon released from fossil fuel combustion. For the average Canadian, this translates into 3 ha of his or her total ecological footprint. Another way to think about our footprint is to estimate how much land would be needed to grow biomass with an energy content equal to that of the fossil fuel we burn.

Assume that you are an average Canadian who burns about 300 gigajoules of fossil fuels per year and that average terrestrial net primary productivity can be expressed as 160 megajoules/ha/year. Calculate how many hectares of land it would take to supply your fuel use by present-day photosynthetic production. A gigajoule is 10^9 joules; a megajoule is 10^6 joules.

1. Compare the energy component of your ecological footprint calculated in this way with the 3 ha calculated by using the method of Wackernagel and Rees. Explain why results from the two methods may differ.
2. Earth's total land area is approximately 1.5×10^{10} (15 billion) hectares. Compare this with the hectares of land for fuel production from the table.
3. How large a human population could Earth support at the level of consumption of the average North American, if all of Earth's land were devoted to fuel production? Do you consider this realistic? Provide two reasons why or why not.

Hectares of Land for Fuel Production	
You	1794
Your class	
Your province	
Canada	

Data from Wackernagel, M., and W. Rees. 1996. *Our Ecological Footprint: Reducing Human Impact on the Earth.* Gabriola Island, BC: New Society Publishers.

TAKE IT FURTHER

 Go to www.myenvironmentplace.ca where you will find

- Suggested answers to end-of-chapter questions
- Quizzes, animations, and flashcards to help you study
- *Research Navigator*™ database of credible and reliable sources to assist you with your research projects
- Tutorials to help you master how to interpret graphs
- Current news articles that link the topics that you study to case studies from your region and around the world

- **ECO Occupational Profiles**: If you found this chapter especially interesting, you might want to learn more about the following jobs by visiting the Occupational Profiles website of the Environmental Careers Organization. Go to www.eco.ca and check out the following careers:
    - Geologic and geophysical technician
    - Process engineer
    - Reclamation specialist
    - Seismologist

CHAPTER ENDNOTES

1. Mackenzie Gas Project, www.mackenziegasproject.com/theProject/index.html#.
2. CBC News In-depth, *The Mackenzie Valley Pipeline*, www.cbc.ca/news/background/mackenzievalley_pipeline/index.html, updated March 12, 2007.
3. CBC News In-depth, *The Mackenzie Valley Pipeline*, www.cbc.ca/news/background/mackenzievalley_pipeline/index.html, updated March 12, 2007.
4. CBC News In-depth, *The Mackenzie Valley Pipeline*, www.cbc.ca/news/background/mackenzievalley_pipeline/index.html, updated March 12, 2007.
5. Mackenzie Gas Project, www.mackenziegasproject.com/whoWeAre/index.htm.
6. CBC News In-depth, *The Mackenzie Valley Pipeline*, www.cbc.ca/news/background/mackenzievalley_pipeline/index.html, updated March 12, 2007.
7. Nature Canada: Take Action! The Mackenzie River Gas Project, www.naturecanada.ca/take_action_raise_voice_protect.asp.
8. CCTRM Canada's Clean Coal Technology Roadmap, Natural Resources Canada, 2005, www.cleancoaltrm.gc.ca.
9. Coal Association of Canada, www.coal.ca/content/index.php?option=com_content&task=view&id=43&Itemid=40.
10. Petroleum History Society, *Six Historical Events in the First 100 Years of Canada's Petroleum Industry*, www.petroleumhistory.ca/history/wells.html#springs.
11. Petroleum History Society, *Six Historical Events in the First 100 Years of Canada's Petroleum Industry*, www.petroleumhistory.ca/history/wells.html# springs.

12. Tariq Piracha, Natural Resources Canada, *Natural Elements: Squeezing Water From Oil Sands—Resources Management in Petroleum Development*, modified 2008-05-06, www.nrcan-rncan.gc.ca/com/elements/issues/22/wateau-eng.php.
13. Canwest News Service, June 17, 2008, as referenced in the *Montreal Gazette*.
14. The State of Energy Efficiency in Canada, *Office of Energy Efficiency Report 2006*, Natural Resources Canada www.oee.nrcan.gc.ca/publications/statistics/see06/transportation.cfm?attr=0.
15. The Pembina Institute, *About Pembina: Our Mission*, www.pembina.org.
16. Pembina Institute, *Bio: Mary Griffiths*, http://re.pembina.org/author/43.
17. *The Edmonton Journal*, p. 1, 2005, "What Will Edmonton Look Like in 2015?" posted on Pembina Institute Renewable Energy Op-Ed page, http://re.pembina.org/op-ed/1152.
18. *Canadian Geographic*, from the citation for the Canadian Environment Award, Mary Griffiths, http://www.canadiangeographic.ca/cea/archives/archives_individual.asp?id=54.

PART THREE

Canola, seen growing in the foreground, is one of several crops that can be used to produce biofuels. Wind turbines like this one are becoming common, especially in the agricultural regions of Canada.

THE SEARCH FOR SOLUTIONS

Conventional Energy Alternatives

16

Nuclear energy accounts for about one-sixth of the world's electrical power generation. This is a nuclear plant in Pickering, Ontario.

Upon successfully completing this chapter, you will be able to

- Discuss the reasons for seeking alternatives to fossil fuels
- Summarize the contributions to world energy supplies of conventional alternatives to fossil fuels
- Describe the scale, methods, and impacts of hydroelectric power
- Describe nuclear energy and how it is harnessed
- Outline the societal debate over nuclear power
- Describe the major sources, scale, and impacts of biomass energy

This biomass power plant in Skellefteå is part of Sweden's search for sustainable energy alternatives.

CENTRAL CASE:
SWEDEN'S SEARCH FOR ALTERNATIVE ENERGY

"Nowhere has the public debate over nuclear power plants been more severely contested than Sweden."
—WRITER AND ENERGY ANALYST MICHAEL VALENTI

"If [Sweden] phases out nuclear power, then it will be virtually impossible for the country to keep its climate-change commitments."
—YALE UNIVERSITY ECONOMIST WILLIAM NORDHAUS

On the morning of April 28, 1986, workers at a nuclear power plant in Sweden detected suspiciously high radiation levels. Their concern turned to confusion when they determined that the radioactivity was coming not from their own plant, but through the atmosphere from the direction of what was, at the time, the Soviet Union.

They had, in fact, discovered evidence of the disaster at Chernobyl, more than 1200 km away in what is now the nation of Ukraine. Chernobyl's nuclear reactor had exploded two days earlier, but the Soviet government had not yet admitted it to the world.

As low levels of radioactive fallout rained down on the Swedish countryside in the days that followed, contaminating crops and cows' milk, many Swedes felt more certain than ever about the decision they had made collectively six years earlier. In a 1980 referendum, Sweden's electorate had voted to phase out their country's nuclear power program, shutting down all nuclear plants by the year 2010.

But trying to phase out nuclear power has proven difficult. Nuclear power today provides Sweden with one-third of its overall energy supply and nearly half its electricity. If nuclear plants are shut down, something will have to take their place. Aware of the environmental impacts of fossil fuels, Sweden's government and citizens do not favour expanding fossil fuel use. In fact, Sweden is one of the few nations that have managed to decrease use of fossil fuels since the 1970s—and it has done so largely by replacing them with nuclear power.

To fill the gap that would be left by a nuclear phaseout, Sweden's government has promoted research and development of renewable energy sources. Hydroelectric power from running water was already supplying most of the other half of the nation's electricity, but it could not be expanded much more. The government hoped that energy from biomass sources and wind power could fill the gap.

Sweden has made itself an international leader in renewable energy alternatives, but because renewables have taken longer to develop than hoped, the government has repeatedly postponed the nuclear phaseout. Only one of the 12 reactors operating in 1980 has been shut down so far, and efforts to close a second one have generated sustained controversy.

Proponents of nuclear power say it would be fiscally and socially irresponsible to dismantle the nation's nuclear program without a ready replacement. And environmental advocates worry that if nuclear power is simply replaced by fossil fuel combustion, or if converting to biomass energy means cutting down more forests, the nuclear phaseout would be bad news for the environment. Moreover, Sweden has international obligations to hold down its carbon emissions under the Kyoto Protocol, so its incentive to keep nuclear power is strong. Nuclear energy is free of atmospheric pollution and seems, to many, the most effective means of minimizing carbon emissions in the short term.

In 2003, a poll showed 55% of the Swedish public in favour of maintaining or increasing nuclear power, and 41% in favour of abandoning it. But despite the mixed feelings over nuclear power, Swedes have little desire to return to an energy economy dominated by fossil fuels. A concurrent poll showed that 80% of Swedes supported boosting research on renewable energy sources—a higher percentage than in any other European country.

Alternatives to Fossil Fuels

Fossil fuels helped to drive the Industrial Revolution and to create the unprecedented material prosperity we enjoy today. Our global economy is largely powered by fossil fuels; 81% of the world's energy comes from oil, coal, and natural gas (**FIGURE 16.1A**), and these three fuels also power two-thirds of the world's electricity generation (**FIGURE 16.1B**). However, these nonrenewable energy sources will not last forever. As we saw in Chapter 15, oil production is thought to be peaking, and easily extractable supplies of oil and natural gas may not last half a century more. Moreover, the use of coal, oil, and natural gas entails substantial environmental impacts, as described in Chapters 13, 14, and 15.

For these reasons, most scientists and energy experts, as well as many economists and policy makers, accept that we will need to shift from fossil fuels to energy sources that are less easily depleted and gentler on our environment. Developing alternatives to fossil fuels has the added benefit of helping to diversify an economy's mix of energy, thus lessening price volatility and dependence on foreign fuel imports.

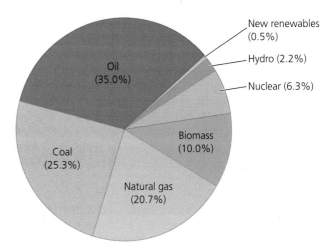

(a) World energy production, by source

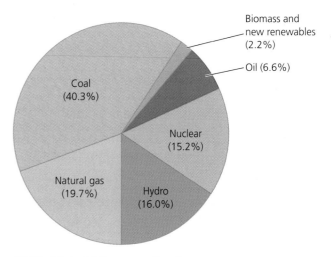

(b) World electricity generation, by source

FIGURE 16.1 Fossil fuels account for 81% of the world's energy production **(a)**. Nuclear and hydroelectric power sources contribute substantially to global electricity generation, but fossil fuels still power two-thirds of our electricity **(b)**. Data are for 2005, from the International Energy Agency (IEA). 2007. *Key World Energy Statistics 2007*. Paris: IEA.

In this chapter we will begin our search for solutions to ever-increasing human impacts on the natural system by examining why we need to find effective alternatives to fossil fuels and what some of those alternatives might be. There is a wide range of alternatives to fossil fuels. Most of these energy sources are renewable, and most have less impact on the environment than oil, coal, or natural gas. However, currently, most remain more costly than fossil fuels, at least in the short term, and external costs are not included in market prices. Moreover, many depend on technologies that are not yet fully developed.

Hydropower, nuclear power, and biomass energy are conventional alternatives

Three alternative energy sources are currently the most developed and most widely used: hydroelectric power, nuclear energy, and energy from biomass. Each of these well-established energy sources plays substantial roles in the energy and electricity budgets of nations today. We can therefore call hydropower, nuclear energy, and biomass energy "conventional alternatives" to fossil fuels.

In some respects, this trio of conventional energy alternatives makes for an odd collection. They are generally considered to exert less environmental impact than fossil fuels but more impact than the "new renewable" alternatives we will discuss in Chapter 17. Yet, as we will see, each involves a unique and complex mix of benefits and drawbacks for human well-being and the environment. Nuclear energy is commonly considered a non-renewable energy source, and hydropower and biomass are generally described as renewable; the reality, however, is more complicated. They are perhaps best viewed as intermediates along a continuum of renewability.

Conventional alternatives provide some of our energy and much of our electricity

Fuelwood and other biomass sources provide 10.0% of the world's energy, nuclear power provides 6.3%, and hydropower provides 2.2%. The less established renewable energy sources account for only 0.5% (see **FIGURE 16.1A**). Although their global contributions to overall energy supply are still minor, alternatives to fossil fuels do contribute greatly to our generation of electricity. Nuclear energy and hydropower each account for nearly one-sixth of the world's electricity generation (see **FIGURE 16.1B**).

Energy consumption patterns in Canada differ from global patterns, primarily because of our heavy reliance on hydropower, especially for electricity generation

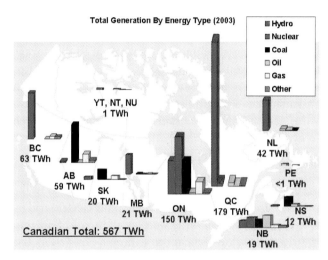

FIGURE 16.2 Energy use patterns in Canada differ from those in the rest of the world, primarily because of our abundant hydropower sources. This map shows regional variations in the type of fuel used to generate electricity in Canada.
Source: Energy Information Administration, U.S. Department of Energy, Country Analysis Briefs: Canada, www.eia.doe.gov/emeu/cabs/Canada/Full.html.

(**FIGURE 16.2**). Canada, like most other industrialized nations, also relies less on fuelwood (biomass) and more on fossil fuels than other countries do. Sweden and some other nations have shown that it is possible to replace fossil fuels gradually with alternative sources. Since 1970, Sweden has decreased its fossil fuel use by 36%, and today nuclear power, biomass, and hydropower together provide Sweden with 63% of its total energy and virtually all of its electricity.

Hydroelectric Power

Next to biomass, we draw more renewable energy from the motion of water than from any other resource. In **hydroelectric power**, or **hydropower**, we use the kinetic energy of moving water to turn turbines and generate electricity. We examined hydropower and its environmental impacts in our discussion of freshwater resources in Chapter 11. Here we will take a closer look at hydropower as an important energy source, particularly in Canada.

Modern hydropower uses two approaches

Just as people have long burned fuelwood, we have long harnessed the power of moving water. Waterwheels spun by river water powered mills in past centuries. Today we harness water's kinetic energy in two major ways: with storage impoundments and by using "run-of-river" approaches.

Most of our hydroelectric power today comes from impounding water in reservoirs behind concrete dams that block the flow of river water, and then letting that water pass through the dam. Because immense amounts of water are stored behind dams, this is called the **storage** technique. If you have ever seen Canada's largest hydroelectric dam, the Gardiner Dam on the South Saskatchewan River, or any other large dam, you have witnessed an impressive example of the storage approach to hydroelectric power (**FIGURE 16.3**).

As reservoir water passes through a dam, it turns the blades of turbines, which cause a generator to generate electricity (**FIGURE 16.4**). Electricity generated in the powerhouse of a dam is transmitted to the electric grid by transmission lines, and the water is allowed to flow into the riverbed below the dam to continue downriver. The amount of power generated depends on the distance the water falls and the volume of water released. By storing water in reservoirs, dam operators can ensure a steady and predictable supply of electricity at all times, even during seasons of naturally low river flow.

An alternative to large dams is the **run-of-river** approach, which generates electricity without greatly disrupting the flow of river water. This approach sacrifices the reliability of water flow across seasons that the storage approach guarantees, but it minimizes many of the impacts of large dams. The run-of-river approach can use several methods, one of which is to divert a portion of a river's flow through a pipe or channel, passing it through a powerhouse and then returning it to the river (**FIGURE 16.5**). This can be done with or without a small reservoir that pools water temporarily, and the pipe or channel can be run along the surface or underground. Another method is to flow river water over a dam small enough not to impede fish passage, siphoning off water to turn turbines, and then returning the water to the river.

Run-of-river systems are particularly useful in areas remote from established electrical grids and in regions without the economic resources to build and maintain large dams. In Chapter 17, where we will look at some of the "new renewable" sources of energy, we will refer to run-of-river power as *small hydro*, in contrast to the *large hydro* of traditional hydroelectric dams. Some environmentalists worry that the impacts of run-of-river systems on water flow and other aspects of the aquatic system are not sufficiently understood as yet.

Hydroelectric power is widely used

Hydropower accounts for 2.2% of the world's energy supply and 16.0% of the world's electricity production. For nations with large amounts of river water and the economic resources to build dams, hydroelectric power has been a keystone of their development and wealth. In addition to Canada, many other nations today obtain large amounts of their energy from hydropower (Table 16.1), including Brazil, Norway, Sweden, and Venezuela.

Sweden receives 11.5% of its total energy and nearly half its electricity from hydropower. In the wake of the nation's decision to phase out nuclear power, many people had hoped hydropower could play a still larger role and compensate for the electrical capacity that would be lost. However, Sweden has already dammed so many of its rivers that it cannot gain much additional hydropower by erecting more dams. Moreover, Swedish citizens have made it clear that they want some rivers to remain undammed, preserved in their natural state.

The great age of dam building for hydroelectric power (as well as for flood control and irrigation) began in the 1930s, when the federal government in the United States

(a) Gardiner Dam, South Saskatchewan River

(b) Satellite image of Gardiner Dam and Lake Diefenbaker

FIGURE 16.3
The Gardiner Dam **(a)** is the largest hydroelectric dam in Canada, with a reservoir capacity of about 65 400 m³. With its partner dam, the smaller Qu'Appelle River Dam, the Gardiner has created Lake Diefenbaker, as seen in this satellite image **(b)**.

(a) Ottawa-Holden Dam

(b) Turbine generator

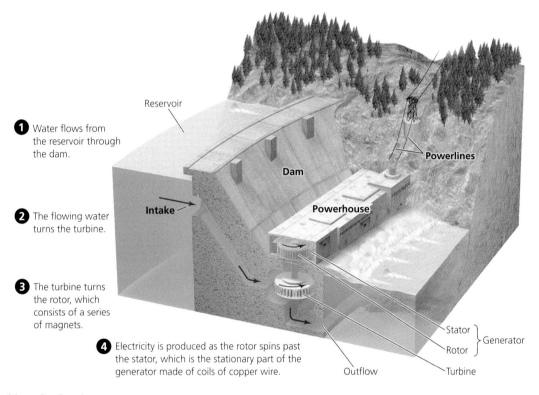

1. Water flows from the reservoir through the dam.
2. The flowing water turns the turbine.
3. The turbine turns the rotor, which consists of a series of magnets.
4. Electricity is produced as the rotor spins past the stator, which is the stationary part of the generator made of coils of copper wire.

(c) Hydroelectric power

FIGURE 16.4
Large dams, such as the Ottawa-Holden Dam on the Ottawa River between Ontario and Quebec **(a)**, generate substantial amounts of hydroelectric power. Inside these dams, flowing water is used to turn turbines similar to the one shown here **(b)**, and generate electricity. Water is funnelled from the reservoir through a portion of the dam **(c)** to rotate turbines, which turn rotors containing magnets. The spinning rotors generate electricity as their magnets pass coils of copper wire. Electrical current is transmitted away through power lines, and the river's water flows out through the base of the dam.

constructed dams as public projects, partly to employ people and help end the economic depression of the time. U.S. dam construction peaked in 1960, when 3123 dams were completed in a single year. American engineers subsequently exported the large dam building technologies elsewhere, notably to the countries of the developing world. In India, Prime Minister Pandit Nehru commented in 1963 that "dams are the temples of modern India," referring to their central importance in the development of energy capacity in that nation.

Hydropower is clean and renewable

For producing electricity, hydropower has two clear advantages over fossil fuels. First, it is renewable: as long as precipitation falls from the sky and fills rivers and reservoirs, we can use water to turn turbines.

The second advantage of hydropower over fossil fuels is its cleanliness. No carbon compounds are burned in the

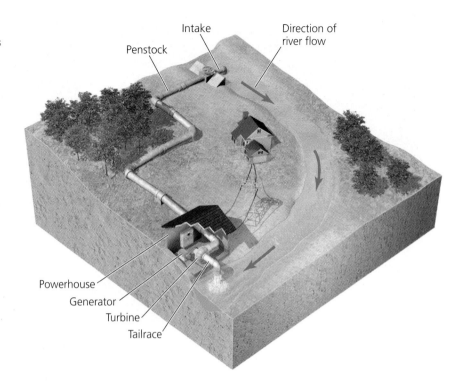

FIGURE 16.5
Run-of-river systems divert a portion of a river's water and can be designed in various ways. Some designs involve piping water downhill through a powerhouse and releasing it downriver, and some involve using water as it flows over shallow dams.

Table 16.1 Top Producers of Hydropower		
Nation	Hydropower produced (terawatt-hours)	Percentage of electricity generation from hydropower
China	397	15.9
Canada	364	59.1
Brazil	337	83.7
United States	290	6.8
Russia	175	18.3
Norway	137	98.9
India	100	14.3
Japan	86	7.8
Venezuela	75	73.9
Sweden	73	46.0
Rest of world	960	13.9

Data are for 2005, from the International Energy Agency.

production of hydropower, so no carbon dioxide or other pollutants are emitted into the atmosphere, which helps safeguard air quality, climate, and human health. Of course, fossil fuels *are* used in constructing and maintaining dams—and recent evidence indicates that large reservoirs release the greenhouse gas methane as a result of anaerobic decay in deep water. But overall, hydropower accounts for only a small fraction of the greenhouse gas emissions typical of fossil fuel combustion.

In addition, hydropower is efficient, for it is thought to have an EROI (*energy returned on investment*) value of 10:1 or more, at least as high as any other modern-day energy source. Fossil fuels had higher EROI values in the past, but as it has become more expensive to reach remaining deposits, their EROI values have dipped below that of hydropower.

Hydropower has negative impacts, too

Although it is renewable, efficient, and produces little air pollution, hydropower does create other impacts. Damming rivers destroys habitat for wildlife as riparian areas above dam sites are submerged and those below dam sites often are starved of water. Because water discharge is regulated to optimize electricity generation, the natural flooding cycles of rivers are disrupted. Suppressing flooding prevents river floodplains from receiving fresh nutrient-laden sediments. Instead, sediments become trapped behind dams, where they begin filling the reservoir.

Dams also cause **thermal pollution**, because water downstream may become unusually warm if water levels are kept unnaturally shallow. Moreover, periodic flushes of cold water occur from the release of reservoir water. Such thermal shocks, together with habitat alteration, have diminished or eliminated many native fish populations in dammed waterways throughout the world. In addition, dams generally block the passage of fish and other aquatic creatures, effectively fragmenting the river and reducing biodiversity in each stretch.

The weight of water in a large reservoir also has been known to cause geologic impacts, such as earthquakes, particularly where water has seeped into fractures in the bedrock underlying the reservoir. Dam collapses—whether as a result of earthquakes or landslides, or from degradation of the construction materials—are regrettably not uncommon and have resulted in many deaths over the past few decades since the great surge in large dam construction.

All these ecological impacts generally translate into negative social and economic impacts on local communities. We discussed the environmental, economic, and social impacts of dams, and their advantages and disadvantages, more fully in Chapter 11.

Hydropower may not expand much more

Today the world is witnessing some gargantuan hydroelectric projects. China's recently completed Three Gorges Dam is the world's largest. Its reservoir displaced more than 1 million people, and the dam should soon be generating as much electricity as dozens of coal-fired or nuclear plants.

However, unlike other renewable energy sources, hydropower is not likely to expand much more. One reason is that, as in Sweden, most of the world's large rivers that offer excellent opportunities for hydropower are already dammed. Another reason is that people have grown more aware of the ecological impacts of dams, and in some regions residents are resisting dam construction. Indeed, in Sweden, hydropower's contribution to the national energy budget has remained virtually unchanged for more than 35 years. In some parts of the world, notably in the United States, many people would like to dismantle some dams and restore river habitats.

Overall, hydropower will likely continue to increase in developing nations that have yet to dam their rivers, but in developed nations hydropower growth will likely slow or stop. The International Energy Agency (IEA) forecasts that hydropower's share of electricity generation will decline between now and 2030, whereas the share of other renewable energy sources will triple, from 2% to 6%.

DAM THE FORESTS 16-1

One of the most significant causes of deforestation in the Amazon rainforest is the construction of large hydroelectric dam projects. Does this serious environmental impact influence your reaction to hydroelectric power as a clean, emission-free source of energy?

Nuclear Power

Nuclear power—usable energy extracted from the force that binds atomic nuclei together—occupies an odd and conflicted position in our modern debate over energy. It is free of the air pollution produced by fossil fuel combustion, so it has long been put forth as an environmentally friendly alternative to fossil fuels. Yet nuclear power's great promise has been clouded by nuclear weaponry, the dilemma of radioactive waste disposal, and the long shadow of Chernobyl and other power plant accidents. As such, public safety concerns and the costs of addressing them have constrained the development and spread of nuclear power in Canada, Sweden, and many other nations.

First developed commercially in the 1950s, nuclear power has expanded fifteenfold worldwide since 1970, experiencing most of its growth during the 1970s and 1980s. Of all nations, the United States generates the most electricity from nuclear power—nearly a third of the world's production—and is followed by France and Japan. Although it is the leader in the quantity of electricity generated, only about 20% of U.S. electricity comes from nuclear sources. Several other nations rely more heavily on nuclear power (Table 16.2). France leads the list, receiving 78% of its electricity from nuclear power. Canada generates approximately 16% of its electricity with nuclear power. The province of Ontario leads the nuclear industry in Canada, generating approximately half of the electricity used in the province from its 16 nuclear generators; Quebec and New Brunswick follow, with one operating reactor each.

Table 16.2 Top Producers of Nuclear Power

Nation	Nuclear power produced*	Number of plants†	Percentage of electricity frrom nuclear power‡
United States	98.4	103	19.4
France	63.4	59	78.1
Japan	47.6	55	30.0
Russia	21.7	31	15.9
Germany	20.4	17	31.8
South Korea	17.5	20	38.6
Ukraine	13.1	15	47.5
Canada	12.6	18	15.8
United Kingdom	11.0	19	18.4
Sweden	9.1	10	48.0
Rest of world	54.2	89	8.0

*In gigawatts, 2007 data, from the European Nuclear Society.
†2007 data, from the International Atomic Energy Agency.
‡2006 data, from the International Atomic Energy Agency.

Fission releases nuclear energy

Strictly defined, **nuclear energy** is the energy that holds together protons and neutrons within the nucleus of an atom. We harness this energy by converting it to thermal energy, which can then be used to generate electricity. Several processes can convert the energy within an atom's nucleus into thermal energy, releasing it and making it available for use. Each process involves transforming isotopes of one element into isotopes of other elements by the addition or loss of neutrons.

The reaction that drives the release of nuclear energy in power plants is **nuclear fission**, the splitting apart of atomic nuclei (**FIGURE 16.6**). In fission, the nuclei of large, heavy atoms, such as uranium or plutonium, are bombarded with neutrons. Ordinarily neutrons move too quickly to split nuclei when they collide with them, but if neutrons are slowed down they can break nuclei apart. Each split nucleus emits heat, radiation, and multiple neutrons. These neutrons (two to three in the case of fissile isotopes of uranium-235) can in turn bombard other nearby uranium-235 (^{235}U) atoms, resulting in a self-sustaining chain reaction.

If not controlled, this chain reaction becomes a runaway process of positive feedback that releases enormous amounts of energy. It is this process that creates the explosive power of a nuclear bomb. Inside a nuclear power plant, however, fission is controlled so that on average only one of the two or three neutrons emitted with each fission event goes on to induce another fission event. In this way, the chain reaction maintains a constant output of energy at a controlled rate.

Nuclear energy comes from processed and enriched uranium

We generate electricity from nuclear power by controlling fission in **nuclear reactors**, facilities contained within nuclear power plants. But this is just one step in a longer process sometimes called the *nuclear fuel cycle*. This process begins when the naturally occurring element uranium is mined from underground deposits, as we will see with the mines on Canadian and Australian Aboriginal lands in Chapter 21.

Uranium minerals are uncommon, and uranium ore is in finite supply, which is why nuclear power is generally considered a nonrenewable energy source. However, Canada is particularly rich in uranium resources, and currently produces about one-third of the world's uranium (**FIGURE 16.7**). Uranium exploration and mining in Canada dates back to the early 1940s. As well, Canada has developed its own indigenous reactor technologies, which are in use in many countries of the world.

Uranium is used for nuclear power because it is radioactive. Radioactive isotopes, or *radioisotopes*, emit subatomic particles and high-energy radiation as they decay into lighter radioisotopes until they ultimately become stable isotopes. The isotope uranium-235 decays into a series of daughter isotopes, eventually forming lead-207. Each radioisotope decays at a rate determined by that isotope's *half-life*, the time it takes for half of the atoms to give off radiation and decay. The half-life of ^{235}U is about 700 million years.

More than 99% of the uranium in nature occurs as the isotope uranium-238. Uranium-235 (with three fewer neutrons) makes up less than 1% of the total. Because ^{238}U does not emit enough neutrons to maintain a chain reaction when fissioned, we use ^{235}U for commercial nuclear power. Therefore, mined uranium ore must be processed to enrich the concentration of ^{235}U to at least 3%. The enriched uranium is formed into pellets of uranium dioxide (UO_2), which are incorporated into metallic tubes called *fuel rods* (**FIGURE 16.8**) that are used in nuclear reactors. After several years in a reactor, enough uranium has decayed that the fuel cannot generate adequate energy, and it must be replaced with new fuel. In some countries, the spent fuel is reprocessed to recover what usable energy may be left. Most spent fuel, however, is disposed of as radioactive waste.

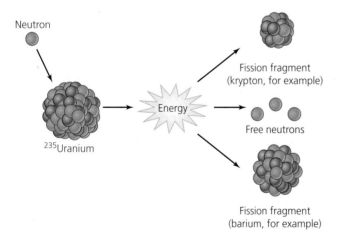

FIGURE 16.6

In nuclear fission, an atom of uranium-235 is bombarded with a neutron. The collision splits the uranium atom into smaller atoms and releases two or three neutrons, along with energy and radiation. The neutrons can continue to split other uranium atoms and set in motion a runaway chain reaction, so engineers at nuclear plants must use control rods to absorb excess neutrons and thereby regulate the rate of the reaction.

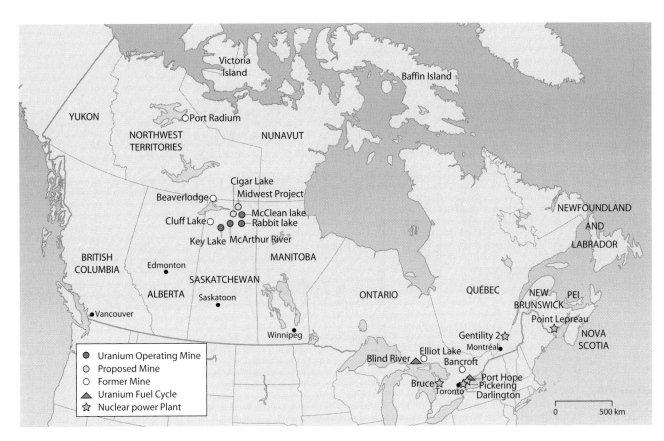

FIGURE 16.7
Canada produces about one-third of the world's uranium; Canada and Australia together produce more than half. This map shows the locations of old and currently operating uranium mines in Canada.
Source: World Nuclear Association, Information Papers, "Canada's Uranium Production & Nuclear Power" (November 2008). Map of Canada accompanying section entitled "Uranium Production." © World Nuclear Association. All rights reserved. Reprinted with permission. www.world-nuclear.org/info/inf49.html/

Fission in reactors generates electricity in nuclear power plants

For fission to begin in a nuclear reactor, the neutrons bombarding uranium are slowed down with a substance called a *moderator*, most often water or graphite. As fission proceeds, it becomes necessary to soak up the excess neutrons produced when uranium nuclei divide, so that on average only a single neutron from each nucleus goes on to split another nucleus. For this purpose, *control rods*, made of a metallic alloy that absorbs neutrons, are placed into the reactor among the water-bathed fuel rods. Engineers move these control rods into and out of the water to maintain the fission reaction at the desired rate.

All this takes place within the reactor core and is the first step in the electricity-generating process of a nuclear power plant (**FIGURE 16.9**). The reactor core is housed within a reactor vessel, and the vessel, steam generator, and associated plumbing are protected within a *containment building*. Containment buildings, with their metre-thick concrete and steel walls, are constructed to prevent leaks of radioactivity caused by accidents or natural catastrophes, such as earthquakes.

FIGURE 16.8
Enriched uranium fuel is packaged into fuel rods, which are encased in metal and used to power fission inside the cores of nuclear reactors. In this photo, the fuel rods are visible, arrayed in a circle within the blue-glowing water.

Breeder reactors make better use of fuel but have raised safety concerns

Using ^{235}U as fuel for fission is only one potential way to harness nuclear energy. **Breeder reactors** like the Canadian Candu reactor make use of ^{238}U, which in

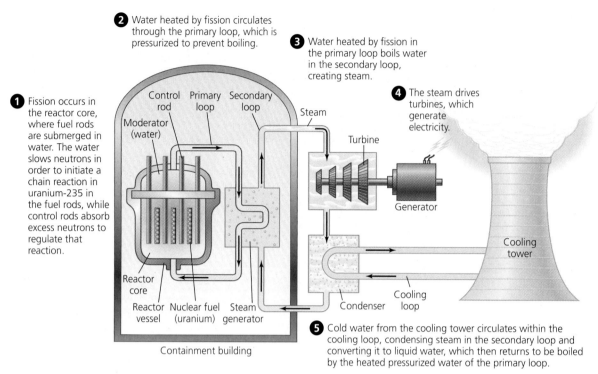

FIGURE 16.9
In a pressurized light water reactor, the most common type of nuclear reactor, uranium fuel rods are placed in water, which slows neutrons so that fission can occur (1). Control rods that can be moved into and out of the reactor core absorb excess neutrons to regulate the chain reaction. Water heated by fission circulates through the primary loop (2) and warms water in the secondary loop, which turns to steam (3). Steam drives turbines, which generate electricity (4). The steam is then cooled in the cooling tower by water from an adjacent river or lake and returns to the containment building (5), to be heated again by heat from the primary loop.

conventional fission goes unused as a waste product. In breeder reactors, the addition of a neutron to ^{238}U and its subsequent electron loss forms plutonium (^{239}Pu). When plutonium is bombarded by a neutron, it splits into fission products and releases more neutrons, which convert more of the remaining ^{238}U fuel into ^{239}Pu, continuing the process. Because 99% of all uranium is ^{238}U, fission in breeder reactors makes much better use of fuel, generates far more power, and produces far less waste. However, breeder reactors also can be dangerous because highly reactive liquid sodium is used as a coolant, raising the risk of explosive accidents, and they are more expensive than conventional reactors.

Fusion remains a dream

For as long as scientists and engineers have generated power from nuclear fission, they have tried to figure out how they might use nuclear fusion instead. **Nuclear fusion**—the process that drives our Sun's vast output of energy and the force behind hydrogen or thermonuclear bombs—involves forcing together the small nuclei of lightweight elements under extremely high temperature and pressure. The hydrogen isotopes deuterium and tritium can be fused together to create helium, releasing a neutron and a tremendous amount of energy (**FIGURE 16.10**).

Overcoming the mutually repulsive forces of protons in a controlled manner is difficult, and fusion requires temperatures of many millions of degrees Celsius. Researchers have not yet developed the process of "cold" fusion for commercial power generation. Despite billions of dollars of funding and decades of research, fusion experiments in the lab still require scientists to put more energy in than they produce from the process. Fusion's potentially huge payoffs, though, make many scientists eager to keep trying.

If one day we were to find a way to control fusion in a reactor, we could produce vast amounts of energy using water as a fuel. The process would create only low-level radioactive wastes, without pollutant emissions or the risk of dangerous accidents, sabotage, or weapons proliferation. A consortium of industrialized nations, including Canada,

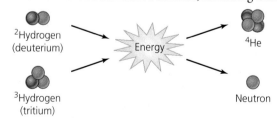

FIGURE 16.10
In nuclear fusion, two small atoms, such as the hydrogen isotopes deuterium and tritium, are fused together, releasing energy along with a helium nucleus and a free neutron. So far, however, scientists have not been able to fuse atoms without supplying far more energy than the reaction produces, so this process is not used commercially.

is collaborating to build a prototype fusion reactor called the International Thermonuclear Experimental Reactor (ITER) in southern France. Even if this multibillion-dollar effort succeeds, however, power from fusion seems likely to remain many years in the future.

> **weighing the issues**
>
> **CHOOSE YOUR RISK** 16–2
>
> Given the choice of living next to a nuclear power plant or a coal-fired power plant, which would you choose? What would concern you most about each option?

Nuclear power delivers energy more cleanly than fossil fuels

Using fission, nuclear power plants generate electricity without creating air pollution from stack emissions. In contrast, combusting coal, oil, or natural gas emits sulphur dioxide that contributes to acidic deposition, particulate matter that threatens human health, and carbon dioxide and other greenhouse gases that drive global climate change. Even considering all the steps involved in building plants and generating power, researchers from the International Atomic Energy Agency (IAEA) have calculated that nuclear power reduces emissions 4–150 times below fossil fuel combustion (see "The Science Behind the Story: Assessing Emissions From Power Sources"). IAEA scientists estimate that at current global levels of use, nuclear power helps us avoid emitting 600 million metric tons of carbon each year, equivalent to 8% of global greenhouse gas emissions.

Nuclear power has additional environmental advantages over fossil fuels—coal in particular. Because uranium generates far more power than coal by weight or volume, less of it needs to be mined, so uranium mining causes less damage to landscapes and generates less solid waste than coal mining. Moreover, in the course of normal operation, nuclear power plants are safer for workers than coal-fired plants.

Nuclear power also has serious drawbacks. One is that the waste it produces is radioactive. Radioactive waste must be handled with great care and must be disposed of in a way that minimizes danger to present and future generations. The second main drawback is that if an accident occurs at a power plant, or if a plant is sabotaged, the consequences could potentially be catastrophic. Given this mix of advantages and disadvantages (**FIGURE 16.11**),

Environmental Impacts of Coal-Fired and Nuclear Power		
Type of Impact	**Coal**	**Nuclear**
Land and ecosystem disturbance from mining	Extensive, on surface or underground	Less extensive
Greenhouse gas emissions	Considerable emissions	None from plant operation; much less than coal over the entire life cycle
Other air pollutants	Sulphur dioxide, nitrogen oxides, particulate matter, and other pollutants	No pollutant emissions
Radioactive emissions	No appreciable emissions	No appreciable emissions during normal operation; possibility of emissions during severe accident
Occupational health among workers	More known health problems and fatalities	Fewer known health problems and fatalities
Health impacts on nearby residents	Air pollution impairs health	No appreciable known health impacts under normal operation
Effects of accident or sabotage	No widespread effects	Potentially catastrophic widespread effects
Solid waste	More generated	Less generated
Radioactive waste	None	Radioactive waste generated
Fuel supplies remaining	Should last several hundred more years	Uncertain; supplies could last for a longer or shorter time than coal supplies

FIGURE 16.11
Coal-fired power plants and nuclear power plants pose very different risks and impacts to human health and the environment. This chart compares the major impacts of each mode of electricity generation. The more severe impacts are indicated by red boxes.

THE SCIENCE BEHIND THE STORY

Assessing Emissions from Power Sources

This power plant in Hebei Province is one of many coal-burning plants in China.

Combusting coal, oil, or natural gas emits carbon dioxide and other greenhouse gases into the atmosphere, where they contribute to global climate change (Chapter 14). Reducing greenhouse emissions is one of the main reasons so many people want to replace fossil fuels with alternative energy sources. But determining how different energy alternatives compare in emissions is a complex process. A number of studies have tried to quantify and compare emission rates of different energy types, but the varied methods used have made it hard to synthesize this information into a coherent picture.

Researchers from the International Atomic Energy Agency (IAEA) attempted such a synthesis for the generation of electricity. Experts met at six meetings between 1994 and 1998 and reviewed the scientific literature, together with data from industry and government. Their goal was to come up with a range of estimates of greenhouse gas emissions for nuclear energy, each major fossil fuel type, and each major renewable energy source. IAEA scientists Joseph Spadaro, Lucille Langlois, and Bruce Hamilton then published the results in the *IAEA Bulletin* in 2000.

The researchers had to decide how much of the total life cycle of electric power production to include in their estimates. Simply comparing the rotation of turbines at a wind farm to the operation of a coal-fired power plant might not be fair, because it would not reveal that greenhouse gases were emitted as a result of manufacturing the turbines, transporting them to the site, and erecting them there. Similarly, because uranium mining is part of the nuclear fuel cycle, and because we use oil-fuelled machinery to mine uranium, perhaps emissions from this process should be included in the estimate for nuclear power.

The researchers decided to conduct a "cradle-to-grave" analysis and include all sources of emissions throughout the entire life cycle of each energy source. This included not just power generation but also the mining of fuel, preparation and transport of fuel, manufacturing of equipment, construction of power plants, disposal of wastes, and decommissioning of plants. They did, however, separate stack emissions from all other sources of emissions in the chain of steps so that these data could be analyzed independently.

Different greenhouse gases were then standardized to a unit of "carbon equivalence" according to their global warming potential. For instance, because methane is 21 times as powerful a greenhouse gas as carbon, each unit of methane emitted was counted as 21 units of carbon-equivalence. The researchers then calculated rates of emission per unit of power produced. They presented figures in grams of carbon-equivalent emitted per kilowatt-hour (gC_{eq}/kWh) of electric power produced.

The overall pattern they found was clear:

- Fossil fuels produce much higher emission rates than renewable energy sources and nuclear energy (see the figure).
- The maximum emission rate for fossil fuels (357 gC_{eq}/kWh for coal) was 4.7 times as high as the maximum emission rate for any renewable energy source (76.4 gC_{eq}/kWh for solar power).
- The minimum emission rate for any fossil fuel (120 gC_{eq}/kWh for natural gas) was nearly 100 times as great as the minimum rate for renewables (1.1 gC_{eq}/kWh for one form of hydropower).

---

many governments (although not necessarily most citizens) have judged the good to outweigh the bad, and today the world has 436 operating nuclear plants in 30 nations.

Nuclear power poses small risks of large accidents

Although scientists calculate that nuclear power poses fewer chronic health risks than does fossil fuel combustion, the possibility of catastrophic accidents has spawned a great deal of public anxiety over nuclear power. Two events were influential in shaping public opinion about nuclear energy.

The first—and closest to home, for North Americans—took place at the Three Mile Island plant in Pennsylvania (**FIGURE 16.12**), where in 1979 the United

FIGURE 16.12
The Three Mile Island nuclear power plant near Harrisburg, Pennsylvania, was the site of a partial meltdown in 1979. This emergency was a "near-miss"—radiation was released but was mostly contained, and no health impacts were confirmed. The incident put the world on notice, however, that a major accident could potentially occur.

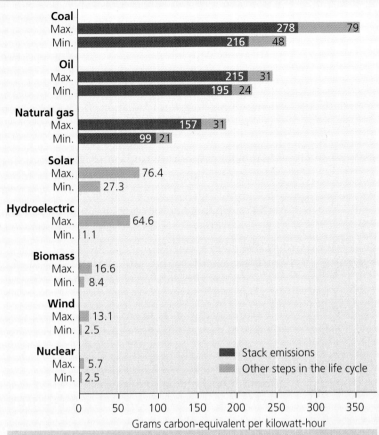

Coal, oil, and natural gas emit far more greenhouse gases than do renewable energy sources and nuclear energy. Red portions of bars represent stack emissions, and orange portions show emissions from other steps in the life cycle. Maximum and minimum values are given for each energy source. Data from Spadaro, J.V., et al. 2000. Greenhouse gas emissions of electricity generation chains: Assessing the difference. IAEA Bulletin 42(2).

Overall, emissions decreased in the following order: coal, oil, natural gas, photovoltaic solar, hydroelectric, biomass, wind, and nuclear. Most fossil fuel emissions were stack emissions directly from power generation, and the amounts in the other steps in the life cycle were roughly comparable to those from renewable sources.

Within each category, emissions values varied considerably. This variation was due to many factors, including the type of technology used, the geographic location and transport costs, the carbon content of the fuel, and the efficiency with which fuel was converted to electricity.

However, technology was expected to improve in the future, creating greater fuel-to-electricity conversion efficiency and lowering emissions rates. Thus, the researchers devised separate emissions estimates for newer technologies expected between 2005 and 2020. These estimates suggested that fossil fuels will improve but still will not approach the cleanliness of nuclear energy and most renewable sources.

Because the IAEA is charged with promoting nuclear energy, critics point out that the agency has clear motivation for conducting a study that shows nuclear power in a favourable light. However, few experts would quibble with the overall trend in the study's data: Nuclear and renewable energy sources are demonstrably cleaner than fossil fuels.

States experienced its most serious nuclear power plant accident. Through a combination of mechanical failure and human error, coolant water drained from the reactor vessel, temperatures rose inside the reactor core, and metal surrounding the uranium fuel rods began to melt, releasing radiation. This process is termed a **meltdown**, and it proceeded through half of one reactor core at Three Mile Island. Area residents stood ready to be evacuated as the nation held its breath, but fortunately most radiation remained trapped inside the containment building.

The accident was brought under control within days, the damaged reactor was shut down, and multibillion-dollar cleanup efforts stretched on for years. Three Mile Island is best regarded as a near-miss; the emergency could have been far worse had the meltdown proceeded through the entire stock of uranium fuel or had the containment building not contained the radiation.

Although residents have shown no significant health impacts in the years since, the event put safety concerns squarely on the map for both citizens and policy makers.

Chernobyl saw the worst nuclear accident yet

In 1986 an explosion at the Chernobyl plant in Ukraine (part of the Soviet Union at the time) caused the most severe nuclear power plant accident the world has yet seen. Engineers had turned off safety systems to conduct tests, and human error, combined with unsafe reactor design, led to explosions that destroyed the reactor and sent clouds of radioactive debris billowing into the atmosphere. For 10 days radiation escaped from the plant while emergency crews risked their lives (some later died from radiation exposure) putting out fires. Most residents of the surrounding countryside remained at home for these 10 days, exposed to radiation, before the Soviet government belatedly began evacuating more than 100 000 people.

In the months and years afterward, workers erected a gigantic concrete sarcophagus around the demolished reactor, scrubbed buildings and roads, and removed irradiated materials (FIGURE 16.13). However, the landscape for at least 30 km around the plant remains contaminated today, and an international team plans to build a larger sarcophagus around the original one, which is deteriorating.

The accident killed 31 people directly and sickened or caused cancer in thousands more. Exact numbers are uncertain because of inadequate data and the difficulty of determining long-term radiation effects (see "The Science Behind the Story: Health Impacts of Chernobyl").

Health authorities estimate that most of the more than 4000 cases of thyroid cancer diagnosed in people who were children at the time resulted from radioactive iodine spread by the accident. Estimates for the total number of cancer cases attributable to Chernobyl, past and future, vary widely, but an international consensus effort 20 years after the event estimated an increase in the cancer rate among exposed people of "up to a few percent," resulting in "up to several thousand fatal cancers."

Atmospheric currents carried radioactive fallout from Chernobyl across much of the Northern Hemisphere, particularly Ukraine, Belarus, and parts of Russia and Europe (FIGURE 16.14). Fallout was greatest where rainstorms brought radioisotopes down from the radioactive cloud. Parts of Sweden received high amounts of fallout. The accident reinforced the Swedish public's fears about nuclear power. A survey taken after the event asked, "Do you think it was good or bad for the country to invest in nuclear energy?" The proportion of respondents answering "bad" jumped from 25% before Chernobyl to 47% afterward.

Fortunately, the world has not experienced another accident on the scale of Chernobyl in the more than two decades since. Moreover, the design of most reactors in Canada and other Western nations is far safer than that of Chernobyl's reactor. Yet, smaller-scale incidents have occurred; for instance, a 1999 accident at a plant in Tokaimura, Japan, killed two workers and exposed more than 400 others to leaked radiation. And Sweden experienced a near-miss in 2006, when the Forsmark plant north of Stockholm narrowly avoided a meltdown after only two of four generators started up following a power outage.

As plants around the world age, they require more maintenance and are therefore less safe. New concerns have also surfaced. The September 11, 2001, terrorist attacks on the World Trade Center and the Pentagon raised fears that similar airplane attacks could be carried out against nuclear plants. Moreover, radioactive material could be stolen from plants and used in terrorist attacks. This possibility is especially worrisome in the cash-strapped nations of the former Soviet Union, where hundreds of former nuclear sites have gone without adequate security for years.

(a) The Chernobyl sarcophagus

(b) Technicians measuring radiation

FIGURE 16.13
The world's worst nuclear power plant accident unfolded in 1986 at Chernobyl, in present-day Ukraine (then part of the Soviet Union). As part of the extensive cleanup operation, the destroyed reactor was encased in a massive concrete sarcophagus (a) to contain further radiation leakage. Technicians scoured the landscape surrounding the plant (b), measuring radiation levels, removing soil, and scrubbing roads and buildings.

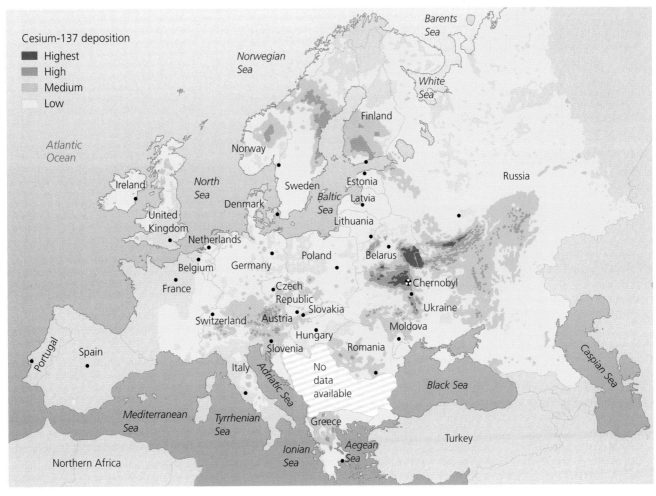

FIGURE 16.14
Radioactive fallout from the Chernobyl disaster was deposited across Europe in complex patterns resulting from atmospheric currents and rainstorms in the days following the accident. Darker colours in this map of cesium-137 deposition indicate higher levels of radioactivity. Although Chernobyl produced 100 times as much fallout as the U.S. bombs dropped on Hiroshima and Nagasaki in the Second World War, it was distributed over a much wider area. Thus, levels of contamination in any given place outside of Ukraine, Belarus, and western Russia were relatively low; the average European received less than the amount of radiation a person receives naturally in a year. Data from Chernobyl information, Swiss Agency for Development and Cooperation, Bern, 2005.

Radioactive waste disposal remains a problem

Even if nuclear power generation could be made completely safe, we would still be left with the conundrum of what to do with spent fuel rods and other radioactive waste. Recall that fission utilizes ^{235}U as fuel, leaving as waste the 97% of uranium that is ^{238}U. This ^{238}U, as well as all irradiated material and equipment that is no longer being used, must be disposed of in a location where radiation will not escape. Because the half-lives of uranium, plutonium, and many other radioisotopes are far longer than human lifetimes, this *high-level waste* will continue emitting radiation for thousands of years. Thus, radioactive waste must be placed in unusually stable and secure locations where radioactivity will not harm future generations.

Currently, nuclear waste from power generation is being held in temporary storage at nuclear power plants in Canada and other places around the world. Spent fuel rods are sunk in pools of cooling water to minimize radiation leakage (**FIGURE 16.15A**). However, plants have only a limited capacity for this type of on-site storage. Many plants are now expanding their storage capacity by storing waste in thick casks of steel, lead, and concrete (**FIGURE 16.15B**).

Because storing waste at many dispersed sites creates a large number of potential hazards, nuclear waste managers want to send all waste to a central repository that can be heavily guarded. In Sweden, that nation's nuclear industry has established a single repository for low-level waste near one power plant and is searching for a single disposal site deep within bedrock for spent fuel rods and

THE SCIENCE BEHIND THE STORY

Health Impacts of Chernobyl

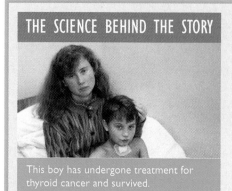
This boy has undergone treatment for thyroid cancer and survived.

In the wake of the nuclear power plant accident at Chernobyl in 1986, medical scientists from around the world rushed to study how the release of radiation would affect human health. Yet determining long-term health impacts of an event is difficult, so it is not surprising that the hundreds of researchers trying to pin down Chernobyl's impacts sometimes came up with very different conclusions.

In an effort to reach consensus, the World Health Organization (WHO) engaged 100 experts from various nations to review all studies through 2006 and issue a report summarizing what scientists had learned in the 20 years since the accident. The WHO report was part of a collaborative 20-year review by an array of international agencies and the governments of Ukraine, Russia, and Belarus.

Doctors documented the most severe effects among emergency workers who battled to contain the incident in its initial hours and days. Medical staff treated and recorded the progress of 134 workers hospitalized with acute radiation sickness (ARS). Radiation destroys cells in the body, and if the destruction outpaces the body's abilities to repair the damage, the person will soon die. Symptoms of ARS include vomiting, fever, diarrhea, thermal burns, mucous membrane damage, and weakening of the immune system by the depletion of white blood cells. In total, 28 people died from acute effects soon after the accident, and those dying had the greatest estimated exposure to radiation.

The major health impact of Chernobyl's radiation, however, has been thyroid cancer. Studies have documented a clear excess of cases among Chernobyl-area residents, particularly children. The thyroid gland is where the human body concentrates iodine, and one of the most common radioactive isotopes released early in the disaster was iodine-131 (^{131}I). Children have large and active thyroid glands, so they are especially vulnerable to thyroid cancer induced by radioisotopes of iodine.

Realizing that thyroid cancer might be a problem, medical workers took measurements of iodine activity from the thyroid glands of several hundred thousand people in Russia, Ukraine, and Belarus in the months following the accident. They also measured food contamination and had people fill out questionnaires on their food consumption. These data showed that drinking milk from cows that had grazed on contaminated grass was the main route of exposure to ^{131}I for most people, although fresh vegetables also contributed.

As doctors had feared, rates of thyroid cancer began rising among children in the regions of highest exposure (see the graph). The yearly number of thyroid cancer cases in the 1990s, particularly in Belarus, far exceeded numbers from years before Chernobyl. Multiple studies found linear dose–response relationships in data from Ukraine and Belarus.

Fortunately, treatment of thyroid cancer has a high success rate, and as of 2002, only 15 of the 4000 children diagnosed with thyroid cancer had died from it. By 2006, medical professionals were estimating that the number of cases had increased to 5000 and was still rising.

Critics pointed out that any targeted search tends to turn up more of whatever medical problem is being looked for. But experts now agree that the increase in childhood thyroid cancer was undeniably attributable to Chernobyl. Furthermore, thyroid cancer also

---

other high-level waste. In the United States, the multiyear search homed in on Yucca Mountain, a remote site in the desert of southern Nevada, 160 km from Las Vegas. No site has yet been chosen in Canada, but at a research facility in Lac du Bonnet, Manitoba (northeast of Winnipeg) scientists are testing proposals for long-term storage deep underground in the stable, ancient crystalline rocks of the Canadian Shield. This approach is called **geologic isolation** (**FIGURE 16.16**), and it is the disposal method of choice among nations that are currently seeking permanent repositories for high-level waste.

In addition to keeping the waste dry, isolated from groundwater and the biosphere, and safe from geologic disruptions, such as earthquakes, another concern is that nuclear waste will need to be transported to its permanent repository. Because this could involve many thousands of shipments by rail and truck across hundreds of public highways through almost every part of the country, many people worry that the risk of an accident or of sabotage is unacceptably high.

weighing the issues
HOW TO STORE WASTE?
16–3

Which do you think is a better option: to transport nuclear waste cross-country to a single repository or to store it permanently at numerous power plants scattered across the nation? Would your opinion be affected if you lived near the repository site? Near a power plant? On a highway route along which waste is transported?

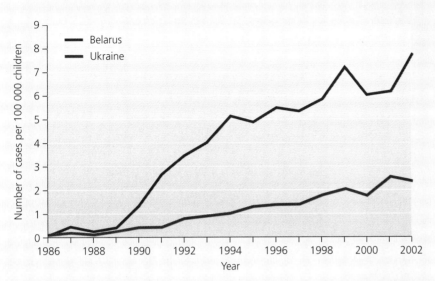

Rates of child thyroid cancer

The incidence of thyroid cancer jumped in Belarus and Ukraine starting four years after the Chernobyl accident released radioactive iodine isotopes. Many babies and young children at the time of the accident developed thyroid cancer in later years. Most have undergone treatment and survived. Data from The Chernobyl Forum. 2006. *Chernobyl's legacy: Health, environmental, and socio-economic impacts*. The Chernobyl Forum: 2003–2005, 2nd rev. version. Vienna, Austria: IAEA.

appears to have risen markedly in adults in Belarus and the most contaminated regions of Russia.

Studies addressing other health impacts have turned up varying results. Some research has shown an increase in cataracts caused by radiation. Other data have revealed that there was apparently no rise in reproductive problems. Studies have also shown psychological effects of the accident among exposed people, including heightened anxiety, particularly about health. However, the WHO report noted that much of this anxiety may result from constant portrayals of exposed people as victims and sufferers rather than as survivors—causing those people to think of themselves in that way.

Researchers have conducted many studies on leukemia and other cancers, but the WHO report concluded that there is "no convincing evidence" so far that rates of any cancer (aside from thyroid cancer) have increased among people exposed to Chernobyl's radiation.

Based on previous studies on survivors of the atomic bombs dropped on Hiroshima and Nagasaki, researchers had expected increases in cancer mortality among the 600 000 people with significant exposure from Chernobyl to be "up to a few percent." This would translate into adding perhaps 4000 cancer fatalities to a normally expected 100 000 cancer deaths in the population from all other causes. However, conducting epidemiological studies with enough statistical power to detect such minor increases in already-rare events is difficult, requiring that huge numbers of people be observed over many years.

Thus, the jury is still out on leukemia and other cancers. Moreover, many cancers do not generally appear for at least 10–15 years after exposure, so it is possible that many illnesses have yet to be diagnosed. For all these reasons, the WHO recommends continued monitoring. As future cancer cases accumulate, continued research will be needed to measure the full scope of health effects from Chernobyl.

Multiple dilemmas have slowed nuclear power's growth

Dogged by concerns over waste disposal, safety, and expensive cost overruns, nuclear power's growth has slowed. Public anxiety in the wake of Chernobyl made utilities less willing to invest in new plants. So did the enormous expense of building, maintaining, operating, and ensuring the safety of nuclear facilities. Almost every nuclear plant has turned out to be more expensive than expected. In addition, plants have aged more quickly than expected because of problems that were underestimated, such as corrosion in coolant pipes. The plants that have been shut down—well over 100 around the world to date—have served on average less than half their expected lifetimes. Moreover, shutting down, or decommissioning, a plant can sometimes be more expensive than the original construction.

As a result of these economic issues, electricity from nuclear power today remains more expensive than electricity from coal and other sources. Governments are still subsidizing nuclear power to keep consumer costs down, but many private investors lost interest long ago. Nonetheless, nuclear power remains one of the few currently viable alternatives to fossil fuels with which we can generate large amounts of electricity in short order.

Many experts predict nuclear power will decrease because three-quarters of Western Europe's capacity is scheduled to be retired by 2030. In Western Europe, not a

(a) Wet storage

(b) Dry storage

FIGURE 16.15
Spent uranium fuel rods are currently stored on-site at nuclear power plants and will likely remain at these scattered sites until a central repository for commercial radioactive waste is fully developed. Spent fuel rods are most often kept in "wet storage" in pools of water **(a)**, which keep them cool and reduce radiation release. Alternatively, the rods may be kept in "dry storage" in thick-walled casks layered with lead, concrete, and steel **(b)**.

single reactor is under construction today, and Germany and Belgium, like Sweden, have declared an intention to phase out nuclear power altogether. Asian nations, in contrast, are adding nuclear capacity. China, India, and South Korea are expanding their nuclear programs to help power their rapidly growing economies. Japan is so reliant on imported oil that it is eager to diversify its energy options. Altogether, Asia hosts two-thirds of the most recent nuclear plants to go into operation and 15 of the 26 plants now under construction.

> **weighing the issues**
>
> **MORE NUCLEAR POWER?** 16–4
>
> Do you think Canada as a whole, or your province or territory in particular, should expand its nuclear power program? Why or why not?

With little or no growth predicted for nuclear power, the era of the large dam drawing to a close, and fossil fuels in limited supply, where will our growing human population turn for additional energy? Increasingly, people are turning to renewable sources of energy: energy sources that cannot be depleted by our use. Although many renewable sources are still early in their stages of development, biomass energy is already well developed and widely used.

Biomass Energy

When people use the term *biomass energy* these days, they can mean very different things. To a poor farmer in Africa, biomass energy means cutting wood from trees or collecting livestock manure by hand and burning it to heat and cook for her family. To an industrialized farmer in Saskatchewan, biomass energy means shipping his grain to a high-tech refinery that converts it to liquid fuel to run automobiles. The diversity of sources and approaches involved in biomass energy (Table 16.3) provides great potential for addressing our energy challenges, but it also means we must be careful in selecting the paths we follow.

(a) Scientific testing

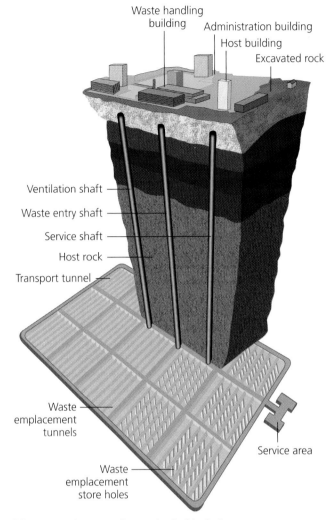

(b) Proposed storage by geological isolation

FIGURE 16.16
There is, as yet, no identified site for the development of a permanent repository for high-level radioactive waste in Canada. Scientific testing of various proposals is ongoing. Here **(a)**, technicians are testing the effects of extreme heat from radioactive decay on the stability of rock. At a site similar to the Lac du Bonnet research facility in Manitoba, waste could be buried in a network of tunnels deep underground in the stable, crystalline rocks of the Canadian Shield **(b)**.

Biomass consists of the organic material that makes up living (and recently deceased) organisms. People harness **biomass energy** from many types of plant matter, including wood from trees, charcoal from burned wood, and matter from agricultural crops, as well as from combustible animal waste products, such as cattle manure. Fossil fuels are not considered biomass energy sources because their organic matter has not been part of living organisms for millions of years and has undergone considerable chemical alteration.

Traditional biomass sources are widely used in the developing world

More than 1 billion people still use wood from trees as their principal energy source. In developing nations, especially in rural areas, families gather fuelwood to burn in their homes for heating, cooking, and lighting (**FIGURE 16.17**; also see **FIGURE 13.25**). In these nations, fuelwood, charcoal, and manure account for fully 35% of energy use—in the poorest nations, up to 90%.

Fuelwood and other traditional biomass sources constitute nearly 80% of all renewable energy used worldwide. However, considering what we have learned about the loss of forests, it is fair to ask whether biomass should truly be considered a renewable resource. In reality, biomass is renewable only if it is not overharvested. At moderate rates of use, trees and other plants can replenish themselves over months to decades. However, when forests are cut too quickly, or when overharvesting leads to soil erosion and forests fail to grow back, then biomass is not effectively replenished. The potential for deforestation makes biomass energy

Table 16.3 Major Sources of Biomass Energy
Direct combustion for heating
■ Wood cut from trees (fuelwood)
■ Charcoal*
■ Manure from farm animals
Biofuels for powering vehicles
■ Corn grown for ethanol
■ Bagasse (sugarcane residue) grown for ethanol
■ Soybeans, rapeseed (or canola), and other crops grown for biodiesel
■ Used cooking oil for biodiesel
■ Plant matter treated with enzymes to produce cellulosic ethanol
Biopower for generating electricity
■ Crop residues (such as cornstalks) burned at power plants
■ Forestry residues (such as wood waste from logging) burned at power plants
■ Processing wastes (such as solid or liquid waste from sawmills, pulp mills, and paper mills) burned at power plants
■ Landfill gas burned at power plants
■ Livestock waste from feedlots for gas from anaerobic digesters
■ Organic components of municipal solid waste from landfills

Charcoal is not "coal" in any sense; it is actually made from charred wood.

less sustainable than other renewable sources, particularly as human population continues to increase.

As developing nations industrialize, fossil fuels are replacing traditional energy sources (**FIGURE 16.18**). As a result, biomass use is growing more slowly worldwide than overall energy use.

New biomass strategies are being developed in industrialized countries

Besides the fuelwood, charcoal, and manure traditionally used in direct combustion for heating, biomass energy sources in today's world include a variety of materials that can be made into several innovative types of energy (see Table 16.3). Some of these sources can be burned in power plants to produce **biopower**, generating heat and electricity in the same way that coal is burned for power. Other new biomass sources can be converted into fuels used primarily to power automobiles; these are termed **biofuels**. Because many of these novel biofuels and biopower strategies depend on technologies resulting from extensive research and development, they are being developed primarily in wealthier industrialized nations.

Biofuels can power automobiles

Liquid fuels from biomass sources are helping to power millions of vehicles on today's roads. The two primary biofuels developed so far are ethanol (for gasoline engines) and biodiesel (for diesel engines).

Ethanol is the alcohol that is in beer, wine, and liquor. It is produced as a biofuel by fermenting biomass, generally from carbohydrate-rich crops, in a process similar to brewing beer. In fermentation, the carbohydrates contained in plants are converted to sugars and then to ethanol. Ethanol is now widely added to gasoline in the United States to reduce automotive emissions, spurred by the 1990 *U.S.*

FIGURE 16.17
More than 1 billion people in developing countries rely on fuelwood for heating and cooking. Wood cut from trees remains the major source of biomass energy used in the world today. In theory, biomass is renewable, but in practice it may not be if forests are overharvested.

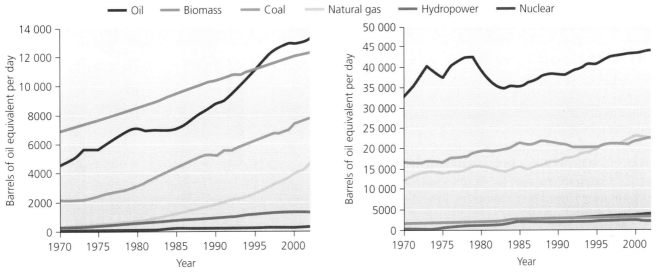

(a) Energy consumption in developing nations

(b) Energy consumption in industrialized nations

FIGURE 16.18
Energy consumption patterns vary greatly between developing nations (a) and wealthier industrialized nations, here represented by nations of the Organisation for Economic Co-operation and Development (OECD) (b). Note the large role that biomass (primarily fuelwood) plays in supplying energy to developing countries. Note also that the y-axes differ; people in developing nations consume far less energy than those in industrialized nations. Data from Energy Information Administration, U.S. Department of Energy.

Clean Air Act. In 2006 in the United States, 18.4 billion litres of ethanol were produced, mostly from corn (**FIGURE 16.19**). This amount is growing rapidly, and the number of U.S. ethanol plants has soared past 100, with 70–80 more now under construction.

The total production capacity for ethanol in Canada, as of 2008, is approaching 1 billion litres per year, up dramatically from just a few years ago, and production could increase further to 1.4 billion litres by 2009. At that point there will be 11 operating ethanol and biodiesel facilities in Canada (in Alberta, Saskatchewan, Manitoba, and Ontario), with an additional five facilities under construction.[1]

Any vehicle with a gasoline engine runs well on gasoline blended with up to 10% ethanol, but more and more vehicles are being produced that can run primarily on

(a) Corn grown for ethanol

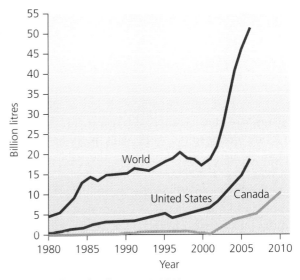

(b) Ethanol production, 1980–2010

FIGURE 16.19
An increasing proportion of the corn crop in Canada, as in the United States, is used to produce ethanol. Brazil produces most of the rest of the world's ethanol, from *bagasse* (sugarcane residue). Ethanol production in the United States, Canada, and elsewhere in the world has grown rapidly in the last several years.

ethanol. Sweden has many such public buses, and the "big three" North American auto makers are now producing *flexible fuel vehicles* that run on E-85, a mix of 85% ethanol and 15% gasoline. More than 5 million such cars are on the road today, but so few gas stations offer E-85 that drivers generally are forced to fill these cars with conventional gasoline. However, increasing infrastructure for ethanol will change this.

In Brazil, sugarcane residue is crushed to make *bagasse*, a material that is then used to make ethanol. Half of all new Brazilian cars are flexible-fuel vehicles, and ethanol from sugarcane accounts for 40% of all automotive fuel that Brazil's drivers use.

Because growing crops to produce ethanol may not be sustainable, as we shall see shortly, researchers are refining techniques to use enzymes to produce ethanol from the cellulose that gives structure to all plant material. If we can produce *cellulosic ethanol* in commercially feasible ways, then ethanol could be made from low-value crop waste (residues, such as corn stalks and husks), rather than from high-value crops. There are other research projects underway in Canada and elsewhere that aim to produce biofuels from algae, rather than utilizing agricultural land that could be producing food crops.

Vehicles with diesel engines have their own biofuel that is growing even faster than ethanol. **Biodiesel** is produced from vegetable oils. Oil is mixed with small amounts of ethanol or methanol (wood alcohol) in the presence of a chemical catalyst. In Europe, where most biodiesel is used, rapeseed, or canola, oil is the oil of choice, whereas U.S. biodiesel producers use mostly soybean oil. Canadian producers also typically rely on canola. In addition, biodiesel producers can utilize animal fats and used grease and cooking oil from restaurants.

Vehicles with diesel engines can run on 100% biodiesel. In fact, when Rudolf Diesel invented the diesel engine in 1895, he designed it to run on a variety of fuels, and he showcased his invention at the 1900 World's Fair by using peanut oil. Since that time, we have mainly used petroleum-based fuel because it is cheaper. Today's diesel engines are designed to work with petrodiesel, and although biodiesel will also power the vehicle, some engine parts wear out more quickly with its use. Most frequently, biodiesel is mixed with conventional petrodiesel; a 20% biodiesel mix (called B20) is common today.

To run on straight vegetable oil, a diesel engine needs to be modified. Extra parts need to be added, so that there are tanks for both the oil and for petrodiesel, which is often needed to start the engine in cooler weather. Although these parts can be bought for as little as $800, it remains to be seen whether using straight vegetable oil might entail further costs, such as reduced longevity or greater engine maintenance.

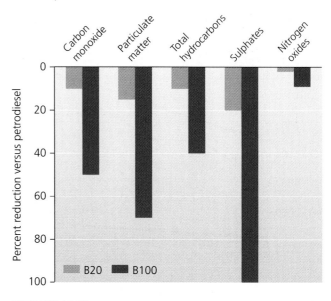

FIGURE 16.20
Burning biodiesel in a diesel engine emits less pollution than burning conventional petroleum-based diesel. Shown are the percentage reductions in several major automotive pollutants that one can attain by using B20 (a mix of 20% biodiesel and 80% petrodiesel) and B100 (pure biodiesel). Data from U.S. Environmental Protection Agency.

Biodiesel cuts down on emissions compared with petrodiesel (**FIGURE 16.20**). Its fuel economy is almost as good, and it costs just slightly more at today's oil prices. It is also nontoxic and biodegradable. Increasing numbers of environmentally conscious individuals in North America and Europe are fuelling their cars with biodiesel from waste oils, and some buses and recycling trucks are now running on biodiesel.

Some enthusiasts have taken biofuel use further. Eliminating the processing step that biodiesel requires, they use straight vegetable oil in their diesel engines. One notable effort is the BIO Tour, in which a group of students, environmentalists, and artists drives across North America in a bus fuelled entirely by waste oil from restaurants. Each summer since 2003, a group has gone on tour with the bus, hosting festive events that combine music and dancing with seminars on environmental sustainability and spreading the word about nonpetroleum fuels.

With biopower we generate electricity from biomass

We harness biopower by combusting biomass to generate electricity. This can be done by using a variety of sources and techniques. Many of the sources used for biopower are the waste products of existing industries or processes. For instance, the forest products industry generates large

FIGURE 16.21
Forestry residues (here from a Swedish logging operation) are a major source material for biopower in some regions.

amounts of woody debris in logging operations and at sawmills, pulp mills, and paper mills (**FIGURE 16.21**). Sweden's efforts to promote biomass energy have focused largely on using forestry residues. Because so much of the nation is forested and the timber industry is a major part of the national economy, plenty of forestry waste is available. Other waste sources include organic waste from municipal landfills, animal waste from agricultural feedlots, and residue from agricultural crops (such as cornstalks and corn husks).

Besides using waste, we also grow certain crops specifically to produce biopower. These include fast-growing trees, such as specially bred willows and poplars, and various fast-growing grasses, such as bamboo, fescue, and switchgrass (**FIGURE 16.22**).

FIGURE 16.22
Switchgrass is one of several fast-growing plants that are being grown as fuel for biopower.

Power plants built to combust biomass operate similarly to those fired by fossil fuels; the combustion heats water, creating steam to turn turbines and generators, thereby generating electricity. Much of the biopower produced so far comes from power plants that generate both electricity and heating through cogeneration. These plants are often located where they can take advantage of forestry waste.

Biomass is also increasingly being combined with coal in coal-fired power plants in a process called *co-firing*. Wood chips, wood pellets, or other biomass is introduced with coal into a high-efficiency boiler that uses one of several technologies. We can substitute biomass for up to 15% of the coal with only minor equipment modification and no appreciable loss of efficiency. Cofiring can be a relatively easy and inexpensive way for fossil-fuel-based utilities to expand their use of renewable energy.

The decomposition of biomass by microbes produces gas that can be used to generate electricity. The anaerobic bacterial breakdown of waste in landfills produces methane and other components, and this **landfill gas** is now being captured at many solid waste landfills and sold as fuel see also Chapter 17, "The Science Behind the Story: Energy From Landfill Gas at Beare Road"). Methane and other gases can also be produced in a more controlled way in anaerobic digestion facilities. The resulting *biogas* can then be burned in a power plant's boiler to generate electricity.

We also harness biopower through **gasification**, a process in which biomass is vaporized at extremely high temperatures in the absence of oxygen, creating a gaseous mixture including hydrogen, carbon monoxide, carbon dioxide, and methane. This mixture can generate electricity when used in power plants to turn a gas turbine to propel a generator. Gas from gasification can also be treated in various ways to produce methanol, synthesize a type of diesel fuel, or isolate hydrogen for use in hydrogen fuel cells. An alternative method of heating biomass in the absence of oxygen results in **pyrolysis**, which produces a mix of solids, gases, and liquids. This includes a liquid fuel called pyrolysis oil, which can be burned to generate electricity.

At small scales, farmers, ranchers, or villages can operate modular biopower systems that use livestock manure to generate electricity. Small household biodigesters now provide portable and decentralized energy production for remote rural areas.

At large scales, industries, such as the forest products industry, are using their waste to generate power, and industrialized farmers are growing crops for biopower. In Sweden, one-sixth of the nation's energy supply now comes from biomass, and biomass provides more fuel for electricity generation than coal, oil, or natural gas. Pulp mill liquors are the main source, but solid wood waste, municipal solid waste, and biogas from digestion are all increasingly used. In the United States, several dozen biomass-fuelled power plants are now operating, and several dozen coal-fired plants are experimenting with co-firing.

Biomass energy brings environmental and economic benefits

Biomass energy has one overarching environmental benefit: It is essentially carbon neutral, releasing no net carbon into the atmosphere. Although burning biomass emits plenty of carbon, the carbon released is simply the carbon that photosynthesis had pulled from the atmosphere to create the biomass in the first place. That is, the carbon that biomass combustion emits is balanced by the carbon that photosynthesis had sequestered within the biomass just years, months, or weeks before. Therefore, when we replace fossil fuels with bioenergy, we reduce net carbon flux to the atmosphere, helping to mitigate global climate change.

However, this holds only if biomass sources are not overharvested. Deforestation will increase carbon flux to the atmosphere because less vegetation means less carbon uptake by plants for photosynthesis. In addition, biomass energy use is not carbon neutral if we need to input fossil fuel energy to produce the biomass (for instance, by driving tractors, using fertilizers, and applying pesticides to produce crops used for biofuel).

Biofuels can reduce greenhouse gas emissions that contribute to climate change in additional ways. Capturing landfill gas reduces emissions of methane, a potent greenhouse gas. And biofuels, such as ethanol and biodiesel, contain oxygen, which when added to gasoline or diesel helps those fuels to combust more completely, reducing pollution.

Shifting from fossil fuels to biomass energy also can have economic benefits. As a resource, biomass tends to be well spread geographically, so using it should help support rural economies and reduce many nations' dependence on imported fuels. Biomass also tends to be the least expensive type of fuel for burning in power plants, and improved energy efficiency brings lower prices for consumers. By enhancing energy efficiency and recycling waste products, the use of biomass energy helps move our industrial systems toward greater sustainability. The forest products industry in North America now obtains much of its energy by combusting the waste it recycles, including woody waste and liquor from pulp mill processing.

Relative to fossil fuels, biomass also benefits human health. By replacing coal in cofiring and direct combustion, biopower reduces emissions of sulphur dioxide because plant matter, unlike coal, contains no appreciable sulphur content. By replacing gasoline and petrodiesel and by burning more cleanly in vehicles, biofuels reduce emissions of nitrogen oxides and other air pollutants.

Biomass energy also brings drawbacks

Biomass energy has negative environmental impacts, too. Burning fuelwood and other biomass in traditional ways for cooking and heating leads to health hazards from indoor air pollution. Harvesting fuelwood at an unsustainably rapid rate leads to deforestation, soil erosion, and desertification, damaging landscapes, diminishing biodiversity, and impoverishing human societies. In arid regions that are heavily populated and that support meagre woodlands, fuelwood harvesting can have enormous impacts. Such is the case with many regions of Africa and Asia. In contrast, moister, well-forested areas with lower population densities, such as Sweden, stand less chance of deforestation.

Growing crops to produce biofuels exerts tremendous impacts on ecosystems. Although crops grown for energy typically receive lower inputs of pesticides and fertilizers than those grown for food, cultivating biofuel crops is land-intensive and brings with it all the impacts of monocultural agriculture. Biofuel crops take up precious land that might otherwise be left in its natural condition or developed for other purposes (**FIGURE 16.23**).

FIGURE 16.23
These hybrid poplars are specially bred for fast growth in dense plantations, and they are harvested for use in biopower. This is renewable energy, but it also has environmental impacts: The vast monocultural plantations, which may replace natural systems over large areas, do not function ecologically as forests.

If we were to try to produce all the automotive fuel currently used in North America with ethanol from corn, we would need to expand the already immense corn acreage by more than 60%, with no loss of productivity and without producing any corn for food. Even at current levels of production, biofuel is competing with food production. As farmers shifted more corn crops to ethanol in 2007 and 2008, corn supplies for food dropped, and international corn prices skyrocketed. In Mexico, where corn tortillas are emblematic of the national cuisine, average citizens found themselves struggling to buy their staple food, and protests erupted across the country over the inflated prices.

Growing bioenergy crops also requires substantial inputs of energy. We currently operate farm equipment by using fossil fuels, and farmers apply petroleum-based pesticides and fertilizers to increase yields. Moreover, fossil fuels are used in refineries to heat water so that we can distill pure ethanol. Thus, shifting from gasoline to ethanol for our transportation needs would not eliminate our reliance on fossil fuels.

Furthermore, growing corn for ethanol yields only a modest amount of energy relative to the energy that needs to be input. Recall our discussion of *energy return on investment* (EROI). The EROI, or ratio of energy

CANADIAN ENVIRONMENTAL PERSPECTIVES

Gráinne Ryder

Gráinne Ryder, a water and energy policy analyst, has spent much of her career working to raise awareness of the huge environmental and social costs of large hydroelectric dams.

- **Water resources engineer**
- **International environmental policy advisor**
- **River protector**

Gráinne Ryder trained as a water resources engineer. She graduated from the University of Guelph and then went to Southeast Asia as a CUSO volunteer—a posting that has stayed with her to this day, both geographically and philosophically. She says of her early work in Thailand, "I got my education from villagers who taught me that water resources problems are 90% social–political and 10% technological. More than this, they taught me that the best water experts are the people who use and depend upon water resources for their livelihoods. Unfortunately, they are usually the last people to be consulted and involved in decision making."

These realizations have followed Ryder throughout her career. Today she is policy director at Probe International, an independent research group (part of Energy Probe Research Foundation) that investigates the environmental and economic consequences of Canadian government and corporate development activities around the world.

Ryder is still intensively involved in Southeast Asia. Among other things, she serves as an advisor to the 3S Rivers Protection Network, a community-run environmental organization in Cambodia, and is co-founder of the Bangkok-based group Towards Ecological Recovery and Regional Alliance.

A typical day for Ryder might involve corresponding with groups in Asia; responding to requests for information from journalists and filmmakers; providing clarification about river science issues, like algal blooms or sedimentation; or pointing out the flaws in developers' environmental assessments that citizens can use in advocacy campaigns. She also works on letters, articles, and in-depth reports for Probe International. Her research interests focus on negotiated river flow management, power sector reform, and decentralized generating technologies, particularly as they relate to internationally financed projects in Asia. She also spends time keeping up with the latest energy trends, such as biofuels and carbon credits, always with the goal of analyzing the potential impacts of new energy approaches on citizens and ratepayers in developing countries.

One of Ryder's best-known publications is *Damming the Three Gorges: What Dam Builders Don't Want You to Know* (1993 and 1990),[2] which she edited. Since the book's first edition, she has become widely known as an expert on hydro development issues in Asia. She has written extensively on the impacts of energy developments in the six-country Greater Mekong Subregion of Southeast Asia, where she works with citizen groups to promote the rights of the public to participate in decisions about the power sector.

Ryder's goal is to stop the use of international aid funds for environmentally destructive and uneconomic large-scale hydro development. Mainly, though, she works to ensure that local stakeholders are empowered to participate in the energy decision-making process.

"Sustainable energy policy has to be based on respect for the rights of local resource users." In 1999, Ryder was identified by *TIME* magazine (Canada) as one of 25 young Canadians likely to make a difference in the next century.[3]

"Dam builders steadfastly refuse to grasp that human suffering and environmental destruction is inevitable with large dams."
—Gráinne Ryder

Thinking About Environmental Perspectives

What do you think about the use of Canadian foreign aid funds to sponsor large energy developments, such as hydroelectric dams, in developing countries? Is this a good way to support economic development in poorer countries, or is it a bad investment because of the potential for negative environmental and social impacts? To what extent should Canadian taxpayers be made aware of the impacts of such projects? (Note that Probe International had to rely on right-to-information legislation to obtain the details of a Canadian environmental impact assessment that supported the Three Gorges Dam project in China—a project that necessitated the relocation of some 1.2 million people.)

returned to energy invested, for corn-based ethanol is controversial, but the best recent estimates place it around 1.5:1. This means that to gain 1.5 units of energy from ethanol, we need to expend 1 unit of energy. The EROI of Brazilian *bagasse* ethanol is much higher, but the low ratio for corn-based ethanol makes this fuel quite inefficient. For this reason many critics do not view ethanol as an effective path to sustainable energy use.

Future advances in cellulosic ethanol may ease the environmental impacts of biofuel crops considerably, and researchers are studying how to obtain biofuel from algae, which could take up far less space than growing corn. Although biomass energy in industrialized nations currently revolves around a few easily grown crops and the efficient use of waste products, the U.S. government envisions a future of specialized crops serving as the basis for a wide variety of fuels and products.

Meanwhile, however, use of fuelwood in the developing world is expected to increase, and the International Energy Agency estimates that in the year 2030, 2.6 billion people will be using traditional fuels for heating and cooking in unsustainable ways. Like nuclear power and hydropower, biomass energy use involves a complex mix of advantages and disadvantages for the environment and human society.

Conclusion

Given limited supplies of fossil fuels and their considerable environmental impacts, many nations have sought to diversify their energy portfolios with alternative energy sources. The three most developed and widely used alternatives so far are hydropower, nuclear power, and biomass energy.

Hydropower is a renewable, pollution-free alternative, but it is not without its own negative ecological impacts. Nuclear power showed promise at the outset to be a pollution-free and highly efficient form of energy. But high costs and public fears over safety in the wake of accidents at Chernobyl and Three Mile Island stalled its growth, and some nations are attempting to phase it out completely. Biomass energy sources include traditional fuelwood, as well as newer biofuels and various means of generating biopower. These sources can be carbon neutral but are not all strictly renewable. Although some nations, such as Sweden, already rely heavily on these three conventional alternatives, it appears that we will need further renewable sources of energy. We will examine some possibilities in Chapter 17.

---

weighing the issues — CORN ETHANOL: FOOD OR FUEL? (16-5)

Do you think producing and using ethanol from corn or other crops is a good idea? Do the benefits outweigh the drawbacks? Can you suggest ways of using biofuels that would minimize environmental impacts? Do you think it is ethical to put a material that people might have eaten as food into the gas tank of an automobile?

REVIEWING OBJECTIVES

You should now be able to:

Discuss the reasons for seeking alternatives to fossil fuels

- Fossil fuels are nonrenewable resources, and we are gradually depleting them.
- Fossil fuel combustion causes air pollution that results in many environmental and health impacts and contributes to global climate change.

Summarize the contributions to world energy supplies of conventional alternatives to fossil fuels

- Biomass provides 10.0% of global primary energy use, nuclear power provides 6.3%, and hydropower provides 2.2%.
- Nuclear power generates 15.2% of the world's electricity, and hydropower generates 16.0%. In Canada, however, hydro is responsible for almost 60% of electricity generation.
- "Conventional energy alternatives" are the alternatives to fossil fuels that are most widely used. They include hydroelectric power, nuclear energy, and biomass energy. In their renewability and environmental impact, these sources fall between fossil fuels and the less widely used "new renewable" sources.

Describe the scale, methods, and impacts of hydroelectric power

- Hydroelectric power is generated when water from a river runs through a powerhouse and turns turbines.
- Dams and reservoirs and run-of-river systems are alternative approaches.

- Hydropower produces little air pollution, but dams and reservoirs can greatly alter riverine ecology and local economies.

Describe nuclear energy and how it is harnessed

- Nuclear power comes from converting the energy of subatomic bonds into thermal energy by using uranium isotopes.
- Uranium is mined, enriched, processed into pellets and fuel rods, and used in nuclear reactors.
- By controlling the reaction rate of nuclear fission, nuclear power plant engineers produce heat that powers electricity generation.

Outline the societal debate over nuclear power

- Many advocates of "clean" energy support nuclear power because it lacks the pollutant emissions of fossil fuels.
- For many people, the risk of a major power plant accident, such as the one at Chernobyl, outweighs the benefits of clean energy.
- The disposal of nuclear waste remains a major dilemma. Temporary storage and single-repository plans each involve health, security, and environmental risks.
- Economic factors and cost overruns have slowed the nuclear industry's growth.

Describe the major sources, scale, and impacts of biomass energy

- Fuelwood remains the major source of biomass energy today, especially in developing nations.
- Biofuels, including ethanol and biodiesel, are used to power automobiles. Some crops are grown specifically for this purpose, and waste oils are also used.
- There are concerns about the impacts on the world's food supply, food prices, and the status of hunger of using food crops, such as corn, to produce ethanol.
- We use biomass to generate electrical power (biopower) in several ways. Sources include special crops as well as waste products from agriculture and forestry.
- Biomass energy theoretically adds no net carbon to the atmosphere and can use waste efficiently. But overharvesting of wood can lead to deforestation, and growing crops solely for fuel production is inefficient and ecologically damaging.

TESTING YOUR COMPREHENSION

1. How much of our global energy supply do nuclear power, biomass energy, and hydroelectric power contribute? How much of our global electricity do these three conventional energy alternatives generate?
2. Describe how nuclear fission works. How do nuclear plant engineers control fission and prevent a runaway chain reaction?
3. In terms of greenhouse gas emissions, how does nuclear power compare with coal, oil, and natural gas? How do hydropower and biomass energy compare?
4. In what ways did the incident at Three Mile Island differ from that at Chernobyl? What consequences resulted from each of these incidents?
5. What has been done so far about disposing of radioactive waste?
6. List five sources of biomass energy. What is the world's most-used source of biomass energy? How does biomass energy use differ between developed and developing nations?
7. Describe two biofuels, where each comes from, and how each is used.
8. Evaluate two potential benefits and two potential drawbacks of biomass energy.
9. Contrast two major approaches to generating hydroelectric power.
10. Assess two benefits and two negative environmental impacts of hydroelectric power.

SEEKING SOLUTIONS

1. Given what you learned about fossil fuels in Chapter 15 and about some of the conventional alternatives discussed in this chapter, do you think it is important for us to minimize our use of fossil fuels and maximize our use of alternatives? If such a shift were to require massive public subsidies, how much should

we subsidize this? What challenges or obstacles would we need to overcome to shift to such alternatives?

2. Nuclear power has by now been widely used for more than three decades, and the world has experienced only one major accident (Chernobyl) responsible for a significant number of deaths. Would you call this a good safety record? Should we maintain, decrease, or increase our reliance on nuclear power? Why might safety at nuclear power plants be better in the future? Why might it be worse?

3. How serious a problem do you think the disposal of radioactive waste represents? How would you like to see this issue addressed?

4. There are many different sources of biomass and many ways of harnessing energy from biomass. Discuss one that seems particularly beneficial to you, and one with which you see problems. In what biomass energy sources and strategies do you think our society should focus on investing?

5. **THINK IT THROUGH** Imagine that you are the head of the national department of energy in a country that has just experienced a minor accident at one of its nuclear plants. A partial meltdown released radiation, but the radiation was fully contained inside the containment building, and there were no health impacts on area residents. However, citizens are terrified, and the media are playing up the dangers of nuclear power. Your country relies on its 10 nuclear plants for 25% of its energy and 50% of its electricity needs. It has no fossil fuel deposits and recently began a promising but still-young program to develop renewable energy options. What will you tell the public at your next press conference, and what policy steps will you recommend taking to ensure a safe and reliable national energy supply?

6. **THINK IT THROUGH** You are an investor and would like to invest in alternative energy. You are considering buying stock in companies that (1) construct nuclear reactors, (2) construct turbines for hydroelectric dams, (3) operate pulp mills, (4) build ethanol refineries, and (5) supply farm waste to co-fired power plants. What questions would you research about each of these companies before deciding how to invest your money? How do you expect you might apportion your investments, and why? Do you expect there may be ways in which you will feel torn between doing what seems financially wise and what seems best for the environment or for energy security? Explain.

INTERPRETING GRAPHS AND DATA

It is not clear that growing crops, such as corn, for the purpose of producing ethanol is the most efficient use of crop lands. Indeed, David Pimentel and Tad Patzek have estimated that replacing just one-third of the gasoline used in North America with ethanol would require more cropland than is needed to feed the population![4] They calculate that 0.6 ha of corn will yield enough ethanol to displace one-third of the gasoline needed to run one average North American car for one year; by comparison, it would require 0.5 ha of corn to feed one person for one year.

As shown on the graph, the area of corn planted in Canada peaked at about 1 315 000 ha in 2001 (because of unusually high demand from french fry manufacturers, in fact). Given that Pimentel and Patzek estimate that it would take 1.8 ha of corn ethanol to displace the gasoline needed to run one car for a year, and given that there are approximately 12 650 000 cars in Canada, how much would the hectares planted in

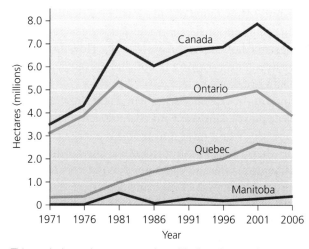

This graph shows the corn area planted in Canada, over time.

corn need to be increased in order to run all of the cars in Canada on ethanol? One-third of the cars? One-fourth of the cars?

CALCULATING FOOTPRINTS

Each of the conventional energy alternatives releases considerably less net carbon dioxide (CO_2) to the atmosphere than do any of the fossil fuels.

For each fuel source in the table, calculate the net greenhouse gas emissions it produces while providing electricity to a typical household that uses 30 kilowatt-hours per day. Use data on emissions rates from each of the energy sources as provided in the figure from "The Science Behind the Story: Assessing Emissions From Power Sources". For each energy source, use the average of the maximum and minimum values given.

1. What is the ratio of greenhouse emissions from fossil fuels, on average, to greenhouse emissions from alternative energy sources, on average?
2. Why are there significant emissions from hydroelectric power, which is commonly touted as being a non-polluting energy source?
3. Which energy source has the lowest emission rate? Would you advocate that your nation further develop that source? Why or why not?

Energy source	Greenhouse gas emission rate (gC_{eq}/kWh)	Period		
		1 Day	1 Year	30 Years
Coal	311	9 330	3 405 450	102 163 500
Oil				
Natural gas				
Photovoltaic solar				
Hydroelectric				
Biomass				
Wind				
Nuclear				

TAKE IT FURTHER

 Go to www.myenvironmentplace.ca where you will find

- Suggested answers to end-of-chapter questions
- Quizzes, animations, and flashcards to help you study
- Research Navigator™ database of credible and reliable sources to assist you with your research projects
- Tutorials to help you master how to interpret graphs
- Current news articles that link the topics that you study to case studies from your region and around the world

- **ECO Occupational Profiles:** If you found this chapter especially interesting, you might want to learn more about the following jobs by visiting the Occupational Profiles website of the Environmental Careers Organization. Go to www.eco.ca and check out the following careers:
  - Energy auditor
  - Environmental assessment analyst
  - Process engineer

CHAPTER ENDNOTES

1. Dow Jones Newswires; resnews@shawbiz.ca.
2. Barber, Margaret, and Gráinne Ryder (eds.) (1993) *Damming The Three Gorges: What Dam Builders Don't Want You To Know, A Critique of the Three Gorges Water Control Project Feasibility Study*, 2nd ed., with introduction by Gráinne Ryder, www.threegorgesprobe.org/pi/documents/three_gorges/damming3g/intro.html.
3. Probe International, *Our Staff*, www.probeinternational.org/our-staff.
4. Pimentel, David, and Tad W. Patzek (2005) Ethanol production using corn, switchgrass, and wood; biodiesel production using soybean and sunflower. *Natural Resources Research*, Vol. 14, No. 1, pp. 65–76.

New Renewable Energy Alternatives

17

(a) Low tide   (b) High tide

The Bay of Fundy is shown here at low tide **(a)** and high tide **(b)**.

Upon successfully completing this chapter, you will be able to

- Outline the major sources of renewable energy and assess their potential for growth
- Describe solar energy and the ways it is harnessed, and evaluate its advantages and disadvantages
- Describe wind energy and the ways it is harnessed, and evaluate its advantages and disadvantages
- Describe geothermal energy and the ways it is harnessed, and evaluate its advantages and disadvantages
- Describe ocean energy sources and the ways they could be harnessed
- Explain hydrogen fuel cells and assess future options for energy storage and transportation

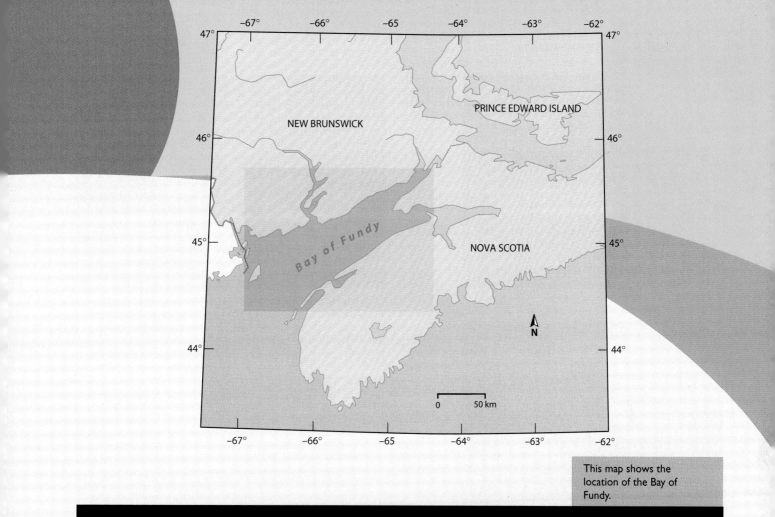

This map shows the location of the Bay of Fundy.

CENTRAL CASE:
HARNESSING TIDAL ENERGY AT THE BAY OF FUNDY

"I believe that water will one day be employed as fuel, that hydrogen and oxygen which constitute it, used singly or together, will furnish an inexhaustible source of heat and light ... Water will be the coal of the future."
—JULES VERNE, IN *THE MYSTERIOUS ISLAND*, 1874

"Not only will atomic power be released, but someday we will harness the rise and fall of the tides and imprison the rays of the sun."
—THOMAS A. EDISON, 1921

"EXTREME HAZARD," reads the posted sign. "Incoming tide rises 5 feet [1.5 m] per hour and may leave you stranded for 8 hours." Not only do ocean tidewaters rise quickly, but nowhere else in the world is the difference between low tide and high tide as great as here, in the Bay of Fundy on Canada's Atlantic coast (see chapter opening photo).

The Bay of Fundy is a long, narrow bay that separates the provinces of Nova Scotia and New Brunswick, and just touches the U.S. state of Maine (see map). The name "fundy" is thought to have come from early Portuguese explorers, who called it the "deep river" or *rio fundo*.

The long, narrow, deep configuration of the bay is responsible for its extreme tidal variation. When the tide comes in, a large volume of ocean water (about 100 billion tonnes) rushes up the length of the narrowly constricted bay in a short period of time (see satellite photos). This phenomenon is called a *tidal bore*. The result of the tidal bore is a very large vertical difference between high and low tide (measured by the Canadian Hydrographic Service at 17 m near the extreme head of the bay, but more typically up to 12 m in the main part of the bay), which is particularly important in making this one of the most suitable locations in the world for the generation of power from ocean tides.

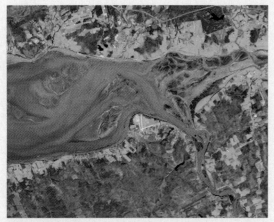
(a) Low tide

(b) High tide
Satellite images show low tide (a) and high tide (b) in the Bay of Fundy.

Water rushes from the sluice gates at the Annapolis Tidal Generating Station in Nova Scotia.

The Bay of Fundy is the site of one of only three operating tidal power plants in the world and the only one in the Western Hemisphere. The Annapolis Tidal Generating Station (see photo), owned by Nova Scotia Power, is located on the Annapolis Basin, a small side-basin on the Nova Scotia side of the main bay, near the town of Annapolis Royal.[1]

The Annapolis power station, which opened in 1984, relies on a relatively early version of tidal power technologies. It works on the same principle as normal hydroelectric power generation—the movement of rushing water turns the blades of a turbine, which, in turn, runs a generator that produces electricity.

Most "normal" hydroelectric power generation extracts energy from rushing river water to turn the turbine. In the case of tidal power, a *sluice*—like a dam with lots of movable openings, or *gates*—is constructed across a narrow part of the water body. When the tide comes in, the sluice gates are opened and the water flows through. At high tide, the gates are closed, trapping the water. When the tide starts to go out, the water above the sluice is held behind the closed gates. After there is a sufficient difference in water level (1 m, in the case of the Annapolis power station), the gates are opened. The water rushes out from behind the sluice to join the rest of the outgoing tide (see photo), turning the turbine blades in the process. At this station, from tidal energy, Nova Scotia Power currently generates as much power as is needed to run about 4000 homes.[2]

Nova Scotia Power and its partners are among those who are working on a new generation of so-called *in-stream* tidal power technologies, some of which they hope to implement in this particularly well suited location by 2010 or so. These technologies involve underwater turbines—like underwater windmills—that would eliminate the need to construct a visible dam or sluice across the waterway.

As you learned in Chapter 16, hydroelectric power is relatively clean and emission free but not entirely without environmental impacts. Many of the negative impacts of traditional hydroelectric power generation are associated with the creation of large reservoirs of standing water behind dams; these impacts are avoided in the case of tidal power, as the tidal water has to be retained behind the sluice gates only for brief periods. The few negative environmental impacts of tidal power generation are mainly associated with interference in the normal currents of the water body. In one case, a whale is thought to have died after following some fish through the sluice gates and becoming trapped. As the new in-stream tidal power technologies become less and less intrusive, these remaining environmental impacts will become less significant.

"New" Renewable Energy Sources

The economic costs, security risks, and environmental impacts of fossil fuel dependence are all growing, and nations across the world are searching for ways to move away from fossil fuels while ensuring a continued supply of energy for their economies.

In Chapter 16 we explored the two renewable energy sources that are most developed and widely used: biomass energy, the energy from combustion of wood and other plant matter, and hydropower, the energy from running water, in addition to our examination of nuclear power. These "conventional" energy sources are renewable or nearly inexhaustible alternatives to fossil fuels, but they still entail a number of undesirable environmental impacts.

In this chapter we explore a group of alternative energy sources that are often called "new renewables." These include energy from the Sun, from wind, from Earth's geothermal heat, and from the movement of ocean water. These energy sources are not truly new; they are as old as our planet, and people have used them for millennia. They are commonly referred to as "new" because (1) they are not yet used on a wide scale in our modern industrial society; (2) they are harnessed by using technologies that are still in a rapid phase of development; and (3) they will likely come to play a much larger role in our energy use in the future.

There are three major categories of applications for renewables in the world's energy market today:

1. Power generation (using wind, solar, tidal, and other energy sources to generate electricity)
2. Space heating (using solar and terrestrial energy sources to heat buildings, factories, and communities)
3. Fuel (using alternatives, such as hydrogen, or using crops, crop residues, or waste materials to manufacture ethanol and biodiesel for use in transportation)[3]

All these categories are very important in Canada, and promising alternative energy resources are available, though they differ from one region of the country to another. Electricity generation powers our modern lifestyle and industry; space heating is important in our northern climate; and fuel for transportation is vital, given the size of the country and the need for mobility. In this chapter we will look at the fundamentals of the new renewable energy sources, and how they can be adapted to applications within these three categories in the Canadian context.

The "new" renewables currently provide little of our power

As a global community, we obtain only half of 1 percent of our energy from the new renewable energy sources. Fossil fuels provide 81% of the world's energy, nuclear power provides 6.3%, and renewable energy sources account for 12.7%, nearly all of which is provided by biomass and hydropower (see **FIGURE 16.1A**). The new renewables make a similarly small contribution to our global generation of electricity (see **FIGURE 16.1B**). Only 18% of our electricity worldwide comes from renewable energy, and of this amount, hydropower accounts for nearly 90%.

Canada's energy mix is somewhat different from that of the rest of the world, primarily because almost 60% of our electricity generation comes from traditional hydroelectric power (which we will refer to as "large hydro" in this chapter). In the United States, by comparison, it is much more common for electricity to be generated by coal-fired power plants or nuclear energy than by running water. The result of the predominance of large hydro in our electricity generation is that approximately 25% of Canada's total primary energy use comes from this source (**FIGURE 17.1**).

This is good news for air quality, at least, since hydroelectric power is largely emission-free, in addition to being renewable. Still, in a world that needs desperately to develop new, environmentally benign energy sources, only about 6% of electricity generation in Canada comes from renewable sources *other* than large hydro (see Table 17.1). The availability of large rivers for traditional hydroelectric installations is limited, even in Canada, and the cost for

Electricity generation in Canada, by fuel type

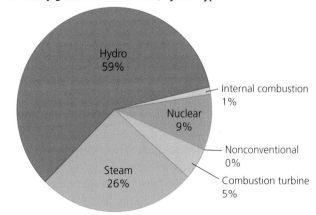

FIGURE 17.1
Electricity generation in Canada is dominated by large hydro, with only about 6% of electricity generation coming from nonhydro "new" renewables. The categories labelled "steam" and "combustion" both refer primarily to coal-fired electricity generation.

Table 17.1	Nonlarge Hydro Renewable Generation Capacity in Canada
Technology	Installed capacity (MW)
Onshore wind	340
Offshore wind	0
Small hydro (run-of-river)	1 800
Solar PV	10
Biomass	1 628
Landfill gas	5
Geothermal and Earth energy	0
Wave energy	0
Tidal energy	20

Source: Pollution Probe. 2004. Pollution Probe Report of the Green Power Workshop Series, www.pollutionprobe.org/Reports/gpworkshopfinalreport2.pdf

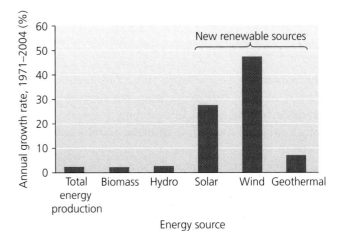

FIGURE 17.2
Globally, the "new renewable" energy sources are growing substantially more quickly than the total primary energy supply. Solar power has grown by 28% each year since 1971, and wind power has grown by 48% each year. Because these sources had such low starting levels, however, their overall contribution to our energy supply is still small. Data from International Energy Agency Statistics, 2007.

new nuclear power facilities is prohibitive. As a result, electricity generation relies more and more on fossil fuels. According to the Pembina Institute in Alberta, "Canada is lagging behind the rest of the world on the use of renewable energy, yet it has more renewable energy resources than most other countries."[4] Even in the United States, according to the Energy Information Administration, the overall consumption of nonhydro renewables was 4.3% of total energy use in 2007, compared with only 1% in Canada for the same period.[5]

Nations and regions vary in the renewable sources they use. In Canada, most nonlarge hydro renewable electricity generation comes from small hydro (run-of-river) installations and biomass (see Table 17.1), followed distantly by wind. However, as you will learn in this chapter, new applications for renewable energy are increasing rapidly, and the contribution of these emerging technologies to Canada's overall energy mix is growing dramatically every year.

The new renewables are growing quickly

Although they compose only a minuscule proportion of our energy budget, the new renewable energy sources are growing at much faster rates than are conventional energy sources. Over the past three decades, solar, wind, and geothermal energy sources have grown far more quickly than has the overall energy supply (**FIGURE 17.2**). The leader in growth is wind power, which has expanded by nearly 50% *each year* since the 1970s. Because these sources started from such low levels of use, however, it will take them some time to catch up to conventional sources. For instance, the absolute amount of energy added by a 50% increase in wind power is still far less than the amount added by just a 1% increase in oil, coal, or natural gas.

Use of new renewables has been expanding quickly because of growing concerns over diminishing fossil fuel supplies (which cause both high prices and national security concerns) and the environmental impacts of fossil fuel combustion (Chapter 15). Advances in technology are also making it easier and less expensive to harness renewable energy sources. The new renewables promise several benefits over fossil fuels. For one, they help alleviate air pollution and the greenhouse gas emissions that drive global climate change (Chapter 14). Unlike fossil fuels, renewable sources are inexhaustible on timescales relevant to human societies. Developing renewables can also diversify an economy's mix of energy, lowering price volatility and protecting against supply restrictions.

In some cases there are additional environmental benefits. For example, using the gases generated by the decomposition of organic wastes in old landfills can be an efficient way to generate both electricity and profits while diverting harmful greenhouse gas emissions from entering the atmosphere. (Methane, the principal component of landfill gases, is a highly effective greenhouse gas.) Landfill gas utilization also solves local environmental problems, cutting down on odours and limiting the possibility of fires and explosions caused by methane gas. Although this continues to be only a tiny source of electricity generation in Canada, it will no doubt grow quickly as Canada strives to meet its objectives for the control of

THE SCIENCE BEHIND THE STORY

Energy from Landfill Gas at Beare Road

Many municipalities utilize passive flaring systems to dispose of landfill gases, preventing the gas from collecting and exploding catastrophically.

Beare Road Landfill (see Chapter 18, "Central Case: The Beare Road Landfill: Making Good Use of Old Garbage"), as at many other sites, the gas was allowed to leak from the landfill via pipes called "candlesticks." These are basically just vertical pipes that allow the gas to flow passively out of the ground, with flames occasionally flaring from the tops of the pipes (see photo).

The passive flaring system at Beare Road has now been replaced by more than 80 vertical wells designed to extract approximately 40 m^3 of landfill gas per minute from the ground.[7] The gas—almost 50% methane—is then brought to the Beare Landfill Power Plant, where particulate matter and moisture are removed (see second photo). The gas then goes to a series of reciprocating gas furnaces, where it is burned to generate electricity. Electricity generation began at this site in 1996.[8]

The Beare Landfill Power Plant currently produces enough electricity to service 4000 homes. The total cost of this LFGTE (landfill gas-to-electricity) facility was about $8.5 million. It has a 10-year anticipated life span, based on the amount of garbage in the landfill, and generates approximately $2 million in revenues each year for its owner, E.S. Fox, which has an agreement with Ontario Power Generation to purchase the electricity. Other LFGTE projects in Canada—including the Keele Valley Landfill in Toronto, Saint Michel Landfill near Montreal, and Clover Bar Landfill near Edmonton—have significantly greater gas and electricity outputs, longer projected life spans, and higher potential profits for the owners of the projects, as well as greater GHG emission savings for Canada.

This is the landfill gas cleanup room at the Beare Landfill Power Plant, where particulates and moisture are removed from the methane gas prior to bringing it to the furnaces.

Methane (CH_4) is the main component of **landfill gas**, gas generated by the decomposition of waste in landfills. (Carbon dioxide, CO_2, is typically a close second.) The capture, recovery, and use of landfill gas for electricity generation can yield 50%–70% of the energy of natural gas, in addition to cutting down significantly on greenhouse gas emissions that would otherwise result from its release into the atmosphere.[6] Capture is particularly important in the case of methane, which is approximately 25 times as effective as carbon dioxide as a radiatively active gas.

Municipalities around the world have long captured and burned off the gases that are naturally generated by garbage accumulating in landfills. In the past, this has been undertaken mainly to avoid odours, fires, and the potential for devastating explosions. For years at

Although LFGTE sounds like a win–win idea (reusing waste materials, cutting back on greenhouse gas emissions and air pollution), some environmentalists are fundamentally opposed to its development. They fear that generating something positive from a pile of garbage will derail attempts to get people to cut down on the amount of waste that they generate. This is a reasonable concern. However, in the case of old sites like the Beare Road Landfill, which hasn't received any new garbage since 1983, the pollution-reducing benefits seem to make it a clear winner.

As of 2003 there were 44 LFGTE systems operating in Canada,[9] with the majority of them in Ontario, British Columbia, and Quebec. These facilities will play an increasingly important role as Canada struggles to meet Kyoto Protocol targets and to keep methane, a highly effective greenhouse gas, from adding to the annual GHG emissions inventory.

---

greenhouse gas emissions. (See "The Science Behind the Story: Energy From Landfill Gas at Beare Road").

New energy sources also can create new employment opportunities and sources of income and property tax for communities, especially in rural areas passed over by other economic development. New and developing technologies generally require more labour per unit of energy output than do established technologies, so shifting to renewable energy should generate more employment than remaining with a fossil fuel economy (**FIGURE 17.3**).

Rapid growth in renewable energy sectors seems likely to continue as population and consumption grow, global energy demand expands, fossil fuel supplies decline, and people demand cleaner environments. More governments, utilities, corporations, and consumers are now promoting and using renewable energy, and, as a result, the prices of renewables are falling.

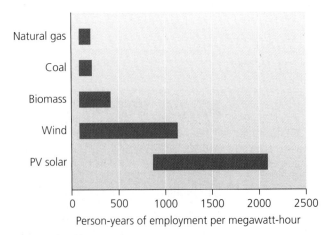

FIGURE 17.3
Manufacturing, installing, and servicing young renewable energy technologies is more labour-intensive than for established fossil fuel technologies, so a shift to renewable energy should increase employment. Renewable energy sources, such as photovoltaic (PV) solar, wind, and biomass tend to support more jobs than natural gas and coal. Data from Worldwatch Institute and Center for American Progress. 2006. *American Energy: The Renewable Path to Energy Security.* Washington, DC.

Our transition must begin soon

If our civilization is to persist in the long term, we will need to shift to renewable energy sources. A key question is whether we will be able to shift soon enough and smoothly enough to avoid widespread war, economic depression, social unrest, and further damage to the environment. The answer to this question will largely determine the quality of life for all of us in the coming decades.

We cannot switch completely to renewable energy sources overnight, because there are technological and economic barriers. Currently, most renewables lack adequate technological development and lack the infrastructure required to transfer power on the required scale. However, rapid advances in science, technology, and infrastructure in recent years suggest that most remaining barriers are political. Renewable energy sources have received far less in subsidies, tax breaks, and other incentives from governments than have conventional sources. For example, in the United States, by one estimate, of the $150 billion in federal government subsidies provided to nuclear, solar, and wind power in the past half-century, the nuclear industry received 96%, solar received 3%, and wind less than 1%. For decades, research and development of renewable sources have gone underfunded because of the continuing availability of fossil fuels made inexpensive in part by government policy.

The funding situation has not been very much different in Canada (**FIGURE 17.4**). In 2006, the *Report of the National Advisory Panel on Sustainable Energy Science and Technology*[10] included the somewhat surprising statement that, ". . . renewable and nuclear technologies were not considered by the Panel to be key priorities for a national

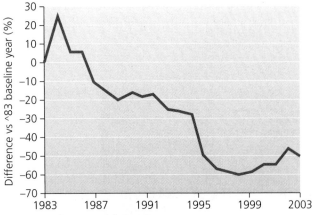

(a) **Federal investment in energy research and development**

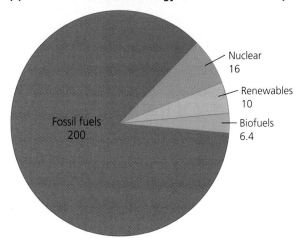

(b) **Annual subsidies worldwide (in U.S. $ billions)**

FIGURE 17.4
Federal funding for energy research and development has plummeted in Canada over the past 20 years **(a)**. Some of the gap has been filled by increases in funding from provinces and the private sector. Of the federal energy R&D funds, a very small proportion has gone directly to the development of alternative energy technologies, primarily nuclear. Internationally, governments are subsidizing the development of fossil fuel resources and technologies much more heavily than for alternative energy sources **(b)**.

energy S&T effort." The eight identified priority areas for federal funding as of 2008, in the context of research and development of new technologies, are[11]

1. Cleaner fossil fuels focusing on the environmental aspects of oil sands development
2. Clean coal and carbon capture and storage
3. Distributed electricity generation from renewable energy and other clean energy sources, including grid management
4. Next-generation nuclear energy technologies
5. Bio-based energy systems
6. Low-emission industrial systems
7. Clean transportation systems; and
8. Built environment, focusing on the integration of renewable energy technologies into buildings and community systems

Of these identified funding priorities, only two (#3, distributed energy generation from renewable energy, and #8, renewable energy technologies in the built environment) are focused more or less directly on new renewable energy technologies.

Failing to make the transition to renewable energy sources would have severe societal repercussions. Most corporations in the fossil fuel and automobile industries understand that they will eventually need to switch to renewable sources. They also know that when the time comes, they will need to act quickly to stay ahead of their competitors. However, in light of continuing short-term profits and unclear policy signals, companies have not been eager to invest in the transition from fossil fuels to renewable energy. Under these circumstances, our best hope may be for a gradual shift driven largely by economic supply and demand. However, if the transition proceeds too slowly—if we wait solely for the market to do its work, without government encouragement—then fossil fuel supplies could dwindle faster than we are able to develop new sources, and we could find our economies disrupted and our environment highly degraded. Thus, encouraging the speedy development of renewable energy alternatives holds promise for bringing us a vigorous and sustainable energy economy without the environmental impacts of fossil fuels.

Solar Energy

The Sun provides energy for almost all biological activity on Earth by converting hydrogen to helium in nuclear fusion. On average, each square metre of Earth's surface receives about one kilowatt of solar energy—17 times the energy of a light bulb. As a result, a typical house has enough roof area to generate all its power needs with rooftop panels that harness **solar energy**. The amount of energy Earth receives from the Sun each day, if it could be collected in full for our use, would be enough to power human consumption for a quarter-century. Clearly, the potential for using sunlight to meet our energy needs is tremendous. However, we are still in the process of developing solar technologies and learning the most effective and cost-efficient ways to put the Sun's energy to use.

The principal uses for solar energy today are as follows:

- Heating and cooling air
- Heating water for a variety of uses, including aquaculture, swimming pools, and home and industrial uses
- Drying crops, such as tea, coffee, timber, fruit, and others
- Generating electricity for off-grid and distributed energy applications
- Detoxifying water and air
- Cooking food
- Daylighting (of course)

Most solar technologies rely on the collection and, in some cases, concentration, of the Sun's rays. The most commonly used way to harness solar energy is through **passive solar energy collection**. In this approach, buildings are designed and building materials are chosen to maximize direct absorption of sunlight in winter, even as they keep the interior cool in the heat of summer. This approach contrasts with **active solar energy collection**, which makes use of technological devices to focus, move, or store solar energy.

We tend to think of using solar energy as a novel phenomenon, but people have chosen and designed their living sites with passive solar collection in mind for millennia. Moreover, solar energy was first harnessed with active solar technology in 1767, when Swiss scientist Horace de Saussure built a thermal solar collector to heat water and cook food. In 1891, U.S. inventor Clarence Kemp claimed the first commercial patent for a solar-powered water heater. Two California entrepreneurs bought the patent rights and outfitted one-third of the homes in Pasadena, California, with solar water heaters by 1897.

Passive solar heating is simple and effective

One passive solar design technique involves installing low, south-facing windows to maximize sunlight capture in the winter (in the Northern Hemisphere; north-facing windows are used in the Southern Hemisphere). Overhangs block light from above, shading these windows in the summer, when the Sun is high in the sky and when cooling, not heating, is desired. Passive solar techniques also include the use of heat-absorbing construction materials (often called **thermal mass**) that absorb heat, store it, and release it later. Thermal mass (of straw, brick, concrete, or other materials) most often makes up floors, roofs, and walls, but it also can be portable blocks.

Thermal mass may be strategically located to capture sunlight in cold weather and radiate heat in the interior of the building. In warm weather, the mass should be located away from sunlight so that it absorbs warmed air in the interior to cool the building. Passive solar design can also involve planting vegetation in particular locations around a building. By heating buildings in cold weather and cooling them in warm weather, passive solar methods conserve energy and reduce energy costs.

Active solar energy collection can heat air and water in buildings

One active method for harnessing solar energy involves the use of **solar panels** or **flat-plate solar collectors**, most often installed on rooftops. These panels generally consist of dark-coloured heat-absorbing metal plates mounted in flat boxes covered with glass panes. Water, air, or antifreeze solutions are run through tubes that pass through the collectors, transferring heat throughout a building. Heated water can be pumped to tanks to store the heat for later use and through pipes designed to release the heat into the building. Such systems have proven especially effective for heating water for residences.

Today only a small fraction of Canada's energy use comes from active solar power, but some estimates suggest that it could account for as much as 5% by 2025.[12] Active solar heating is used more commonly in China and Europe, even in isolated locations. In Gaviotas, a remote town in the high plains of Colombia far from any electrical grid, active solar technology is used for heating, cooling, and water purification (**FIGURE 17.5**). Engineers developed inexpensive solar panels that harvest enough solar energy, even under cloudy skies, to boil drinking water for a family of four. They also designed, built, and installed a unique solar refrigerator in a rural hospital, along with a large-scale solar collector to boil and sterilize water. Their innovations show that solar power does not need to be expensive or confined to regions that are always sunny.

Concentrating solar rays magnifies energy

We can magnify the strength of solar energy by gathering sunlight from a wide area and focusing it on a single point. This is the principle behind *solar cookers*, simple portable ovens that use reflectors to focus sunlight onto food and cook it. Such cookers are proving extremely useful in parts of the developing world.

Utilities have put the solar cooker principle to work in large-scale, high-tech approaches to generating electricity. In one approach, mirrors concentrate sunlight onto a receiver atop a tall "power-tower" (**FIGURE 17.6**). From the receiver, heat is transported by fluids (often molten salts), which are piped to a steam-driven generator to create electricity. These solar power plants can harness light from large mirrors spread across many hectares of land. The world's largest such plant so far—a collaboration among government, industry, and utility companies in the California desert—produces power for 10 000 households. Another approach is the use of solar-trough collection systems, which consist of mirrors that gather sunlight and focus it on oil in troughs. The superheated oil creates steam that drives turbines, as in conventional power plants.

Photovoltaic cells generate electricity directly

A more direct approach to producing electricity from sunlight involves photovoltaic (PV) systems. **Photovoltaic (PV) cells** collect sunlight and convert it to electrical energy by

FIGURE 17.5
Engineers in Gaviotas, a remote highland town in Colombia, developed inexpensive solar panels that provide residences and businesses with active solar power for heating, cooling, and water purification.

FIGURE 17.6
At the Solar Two facility in the desert of southern California, the largest such facility in the world, mirrors are spread across wide expanses of land to concentrate sunlight onto a receiver atop a "power-tower." Heat is then transported through fluid-filled pipes to a steam-driven generator that produces electricity.

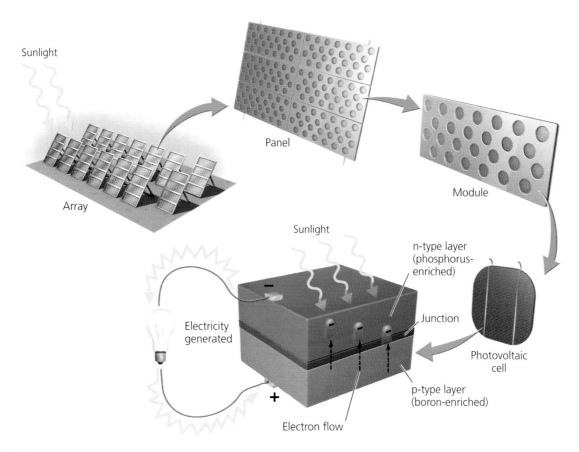

FIGURE 17.7
A photovoltaic (PV) cell converts sunlight to electrical energy. When sunlight hits the silicon layers of the cell, electrons are knocked loose and tend to move from the boron-enriched p-type layer toward the phosphorus enriched n-type layer. Connecting the two layers with wiring remedies this imbalance as electrical current flows from the n-type layer back to the p-type layer. This direct current (DC) is converted to alternating current (AC) to produce usable electricity. PV cells are grouped in modules, which compose panels, which can be erected in arrays.

making use of the *photovoltaic effect*, or *photoelectric effect*. This effect occurs when light strikes one of a pair of metal plates in a PV cell, causing the release of electrons, which are attracted by electrostatic forces to the opposing plate. The flow of electrons from one plate to the other creates an electrical current (direct current, DC), which can be converted into alternating current (AC) and used for residential and commercial electrical power (**FIGURE 17.7**).

The plates of a typical PV cell are made primarily of silicon, a semiconductor that conducts electricity. One silicon plate (the *n-type layer*) is enriched with phosphorus and is rich in electrons. The other plate (the *p-type layer*) is enriched with boron and is electron-poor. When sunlight strikes the PV cell, it knocks electrons loose from some of the silicon atoms, and they tend to flow toward the n-type layer. Connecting the two plates with wires generates electricity as electrons flow from the n-type layer back to the p-type layer. Photovoltaic cells can be connected to batteries that store the accumulated charge until it is needed.

You may be familiar with small PV cells that power your watch or your calculator. Atop the roofs of homes and other buildings, PV cells are arranged in modules, which compose panels, which can be gathered together in arrays. Increasingly, PV roofing tiles are being used instead of these arrays. In some remote areas—such as Xcalak, Mexico—PV systems are used in combination with wind turbines and a diesel generator to power entire villages.

Solar power is little used but fast growing

Although active solar technology dates from the eighteenth century, it was pushed to the sidelines as fossil fuels came to dominate our energy economy. Funding for research and development of solar technology has been erratic. Largely because of the lack of investment, solar energy contributes only a minuscule portion of today's energy production, in Canada and worldwide. However, solar energy use has grown by 28% annually worldwide since 1971, a growth rate second only to that of wind power. Since 1995, solar energy use in Canada has grown by an estimated 25% per year, but we still lag behind solar energy use in Germany, Japan, and elsewhere.[13] The federal government and some provincial and territorial governments provide financial incentives for homeowners and business owners to switch to solar technologies.

Solar power is proving especially attractive in developing countries, many of which are rich in sun but poor in power infrastructure, and where hundreds of millions of people are still without electricity. Some multinational corporations that built themselves on fossil fuels are now investing in alternative energy as well. BP Solar, British Petroleum's solar energy wing, recently completed $30 million projects in the Philippines and Indonesia and is working on a $48 million project to supply electricity to 400 000 people in 150 villages.

Use of PV cells is growing quickly (**FIGURE 17.8**), with Japan leading the world in production and Germany leading the world in installation. Use of solar technology should continue to increase as prices fall, technologies improve, and governments enact economic incentives to spur investment.

Solar power offers many benefits

The fact that the Sun will continue burning for another 4 billion to 5 billion years makes it practically inexhaustible as an energy source for human civilization.

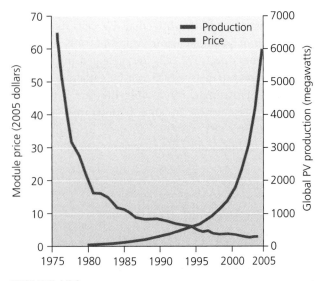

FIGURE 17.8
Global production of PV cells grew sixfold just between 2000 and 2005, and prices have fallen rapidly. Data from Worldwatch Institute and Center for American Progress. 2006. *American Energy: The Renewable Path to Energy Security.* Washington, DC.

Moreover, the amount of solar energy reaching Earth's surface should be enough to power our civilization once solar technology is adequately developed. These primary benefits of solar energy are clear, but the technologies themselves also provide benefits. PV cells and other solar technologies use no fuel, are quiet and safe, contain no moving parts, require little maintenance, and do not even require a turbine or generator to create electricity. An average unit can produce energy for 20–30 years.

Another advantage of solar systems is that they allow for local, decentralized control over power. Homes, businesses, and isolated communities can use solar power to produce their own electricity without being near a power plant or connected to the grid of a city. The challenges of developing and delivering power through a distributed energy system are addressed by priority #3 in the National Advisory Panel's list of energy science and technology funding priorities.

In developing nations, solar cookers enable families to cook food without gathering fuelwood; as a result, they lessen people's daily workload and help reduce deforestation. In such locations as refugee camps, solar cookers are helping relieve social and environmental stress. The low cost of solar cookers—many can be built locally for $2 to $10 each—has made them available for purchase or donation in many impoverished areas.

In the developed world, most PV systems are connected to the regional electric grid. This may enable owners of houses with PV systems to sell their excess solar energy to their local power utility. In this process, called **net metering**, the value of the power the consumer sells to the utility is subtracted from the consumer's monthly utility bill. Net metering is still not common, but as of this writing it is available to consumers in British Columbia, Manitoba, New Brunswick, Ontario, Nova Scotia, Prince Edward Island, Quebec, and Saskatchewan.

Another advantage of solar power is that its development is producing many new jobs. Currently, among major energy sources, PV technology employs the most people per unit energy output (see **FIGURE 17.3**).

Finally, a major advantage of solar power over fossil fuels is that it does not pollute the air with greenhouse gas emissions and other air pollutants. The manufacture of photovoltaic cells *does* currently require fossil fuel use, but once a PV system is up and running, it produces no emissions.

Location and cost can be drawbacks

Solar energy currently has two major disadvantages. One is that not all regions are sunny enough to provide adequate power, given current technology. Although Earth

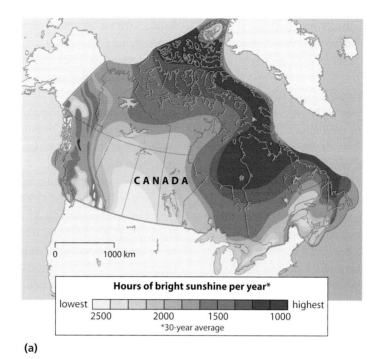

FIGURE 17.9
Because some locations receive more sunlight than others, harnessing solar energy is more profitable in some areas than in others. In spite of the northern location (a), the yearly solar average of Canada's populated areas exceeds that in both Germany and Japan, the world's solar leaders. Solar energy can be used to power remote applications, as in this photograph of a solar installation in Nunavut (b). Data from Canadian Geographic, *The Canadian Atlas Online*, www.canadiangeographic.ca/atlas/themes.aspx?id=weather&sub=weather_power_solarpower&lang=En.

as a whole receives vast amounts of sunlight, not every location on Earth does (**FIGURE 17.9**). People in such cities as Vancouver might find it difficult to harness enough sunlight most of the year to rely on solar power. Daily or seasonal variation in sunlight can also pose problems for stand-alone solar systems if storage capacity in batteries or fuel cells is not adequate or if backup power is not available from a municipal electricity grid. In far northern locations, for example, the seasonal variation in number of hours of daylight makes solar power difficult to utilize.

The primary disadvantage of current solar technology is the upfront cost of investing in the equipment. The investment cost for solar is higher than that for fossil fuels, and indeed, PV solar power remains the most expensive means of producing electricity. Proponents of solar power argue that decades of government promotion of fossil fuels and nuclear power—which have received many financial breaks that solar power has not—have made solar power unable to compete. Because the external costs of nonrenewable energy are not included in market prices, these energy sources have remained relatively cheap. Governments, businesses, and consumers thus have had little economic incentive to switch to solar and other renewables.

However, decreases in price and improvements in energy efficiency of solar technologies so far are encouraging, even in the absence of significant financial commitment from government and industry. At their advent in the 1950s, solar technologies had efficiencies of around 6% while costing $600 per watt. Recent single-crystal silicon PV cells are showing 15% efficiency commercially and 24% efficiency in lab research, suggesting that future solar technologies may be more efficient than any energy technologies we have today. Solar systems have become much less expensive over the years and now can often pay for themselves in 10–20 years. After that time, they provide energy virtually for free as long as the equipment lasts. With future technological advances, some experts believe that the time to recoup investment could fall to one to three years.

Wind Energy

Wind energy—energy derived from moving air masses—is really an indirect form of solar energy, because it is the Sun's differential heating of air masses on Earth that causes wind to blow. We can harness power from wind by using devices called **wind turbines**, mechanical assemblies that convert wind's kinetic energy, or energy of motion, into electrical energy.

Wind has long been used for energy

Today's wind turbines have their historical roots in Europe, where wooden windmills have been used for 800 years to pump water to drain wetlands and irrigate crops and to grind grain into flour. In each application, wind causes a windmill's blades to turn, driving a shaft connected to several cogs that turn wheels, which either grind grain or pull buckets from a well. In North America, countless ranches in the Prairies and the Great Plains use windmills to draw groundwater up for thirsty cattle. The largest wind power producer in Canada is the Le Nordais project in the Gaspé Peninsula, but the first was Cowley Ridge Wind Plant in Alberta, the first phase of which opened in 1993.

The first wind turbine built to generate electricity was constructed in the late 1800s by inventor Charles Brush, who designed a turbine 17 m tall with 144 rotor blades made of cedar wood. But it was not until after the 1973 oil embargo that governments in North America and Europe began funding research and development for wind power. This moderate infusion of funding boosted technological progress, and the cost of wind power was cut in half in less than 10 years. Today wind power at favourable locations generates electricity for nearly as little cost per kilowatt-hour as do conventional sources, and modern wind turbines look more like airplane propellers or sleek new helicopters than romantic old Dutch paintings.

Modern wind turbines convert kinetic energy to electrical energy

Wind blowing into a turbine turns the blades of the rotor, which rotate machinery inside a compartment called a *nacelle*, which sits atop a tall tower (**FIGURE 17.10**). Inside the nacelle are a gearbox and a generator, as well as equipment to monitor and control the turbine's activity. Most of today's towers are 40–100 m tall, so the largest are taller than a football field is long. Higher is generally better, to minimize turbulence (and potential damage) and to maximize wind speed. Most rotors consist of three blades and measure 42–80 m across. Turbines are designed to yaw, or rotate back and forth in response to changes in

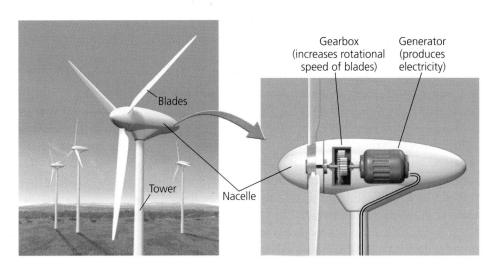

FIGURE 17.10
A wind turbine converts wind's energy of motion into electrical energy. Wind causes the blades of a wind turbine to spin, turning a shaft that extends into the nacelle that is perched atop the tower. Inside the nacelle, a gearbox converts the rotational speed of the blades, which can be up to 20 revolutions per minute (rpm) or more, into much higher rotational speeds (more than 1500 rpm). These high speeds provide adequate motion for a generator inside the nacelle to produce electricity.

wind direction, ensuring that the motor faces into the wind at all times. Turbines can be erected singly, but they are most often erected in groups called wind parks or **wind farms**. The world's largest wind farms contain hundreds of turbines spread across the landscape.

Engineers have designed turbines to begin turning at specific wind speeds to harness wind energy as efficiently as possible. Some turbines create low levels of electricity by turning in light breezes. Others are programmed to rotate only in strong winds, operating less frequently but generating large amounts of electricity in short periods. Slight differences in wind speed yield substantial differences in power output, for two reasons. First, the energy content of a given amount of wind increases as the square of its velocity; thus if wind velocity doubles, energy quadruples. Second, an increase in wind speed causes more air molecules to pass through the wind turbine per unit time, making power output equal to wind velocity cubed. Thus a doubled wind velocity actually results in an eightfold increase in power output.

Wind power is the fastest-growing energy sector

Like solar energy, wind provides only a minuscule proportion of the world's power needs, but wind power is growing quickly—26% per year globally between 2000 and 2005. Germany was the world leader in installed wind capacity by the end of 2006, by quite a large margin (Table 17.2). In Denmark, a series of wind farms supplies more than 20% of the nation's electricity needs (**FIGURE 17.11**). California and Texas account for close to half the wind power generated in the United States. Experts agree that wind power's rapid growth will continue, because only a very small portion of this resource is currently being tapped. Meteorological evidence suggests that wind power could be expanded in Canada to about 30 000 MW, sufficient to meet 15% of the nation's electrical needs. To

Table 17.2 International Rankings of Wind Power Capacity, 2006

Cumulative capacity (end of 2006, MW)		Incremental capacity (2006, MW)	
Germany	20 652	United States	2 454
Spain	11 614	Germany	2 233
United States	11 575	India	1 840
India	6 228	Spain	1 587
Denmark	3 101	China	1 334
China	2 588	France	810
Italy	2 118	Canada	776
United Kingdom	1 967	United Kingdom	631
Portugal	1 716	Portugal	629
France	1 585	Italy	417
Rest of world	11 102	Rest of world	2 305
Total	74 246	Total	15 016

date, Canada's wind power leaders are Quebec, Alberta, and Saskatchewan.[14]

Offshore and high-elevation sites can be promising

Wind speeds on average are roughly 20% greater over water than over land. There is also less air turbulence over water than over land. You can see this clearly in **FIGURE 17.12A**, where the wind conditions over Hudson Bay are noticeably stronger than anywhere over land in Canada. For these reasons, offshore wind turbines are becoming popular. Costs to erect and maintain turbines in water are higher, but the stronger, less turbulent winds produce more power and make offshore wind potentially more profitable. Currently, offshore wind farms are limited to shallow water, where towers are sunk into sediments singly or by using a tripod configuration to stabilize them. However, in the future, towers

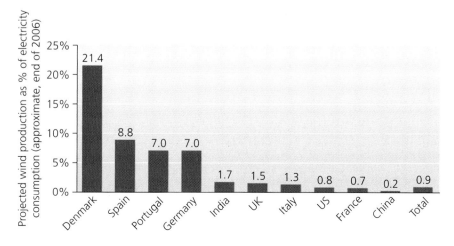

FIGURE 17.11
Tiny Denmark obtains the highest percentage of its energy from wind, but Germany, Spain, the United States, and India have so far developed more total wind capacity. Data from The U.S. Department of Energy's Energy Efficiency and Renewable Energy (EERE) centre, *Annual Report on U.S. Wind Power Installation, Cost, and Performance Trends: 2006*.

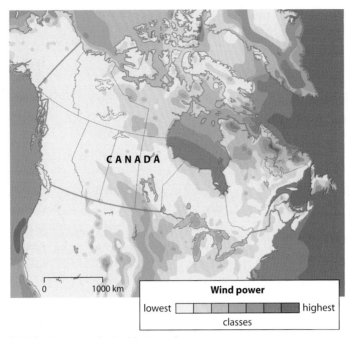

(a) Distribution of wind in Canada

(b) Cowley Ridge Wind Plant, Alberta

FIGURE 17.12
Southern Alberta and Saskatchewan enjoy some of Canada's best conditions for wind power (a). Cowley Ridge Wind Plant in southwestern Alberta (b), Canada's first commercial wind power plant, opened in 1993. Data from Canadian Geographic, *The Canadian Atlas Online*, Wind Power, www.canadiangeographic.ca/atlas/themes.aspx?id=WEATHER&sub=WEATHER_POWER_WINDPOWER &lang=En.

FIGURE 17.13
Winter winds and high elevations combine to make conditions that are appropriate for wind power generation in Yukon. Wind power can be an appropriate energy solution for remote locations, as shown here, but icing of the turbine blades can be a challenge.

may be placed on floating pads anchored to the seafloor in deep water. At great distances from land, it may be best to store the generated electricity in hydrogen fuel and then ship or pipe this to land (instead of building submarine cables to carry electricity to shore), but further research is needed.

Denmark erected the first offshore wind farm in 1991. Over the next decade, nine more came into operation across northern Europe, where the North and Baltic Seas offer strong winds. The power output of these farms increased by 43% annually as larger turbines were erected. In Iceland, wind advocates are considering developing 240 offshore turbines in the nation's waters to meet future electricity demand for its hydrogen economy. There are as yet no operational offshore wind installations in Canada.

Winter weather also promotes stronger wind conditions, as do high elevations. Studies in Yukon, for example, have found many elevated sites suitable for wind power development, and the territory has recently invested in new installations (**FIGURE 17.13**). Wind turbines in cold climates face specific challenges, however, such as icing of the turbine blades.

Wind power has many benefits

Like solar power, wind produces no emissions once the necessary equipment is manufactured and installed. The graph in **FIGURE 17.14** shows emissions per kilowatt-hour of electricity produced over the entire installed lifetime of various technologies. Emissions of CO_2 (the main greenhouse gas associated with global warming), SO_2 (the main precursor of acid rain), and NO_x (the main precursor of photochemical smog) are significantly lower for wind and other renewables, as well as nuclear and hydro, compared with fossil fuel–based electricity generation. Other types of harmful emissions also can be avoided through the use of wind power; for example, the U.S. Environmental Protection Agency (EPA) calculates that running a one-megawatt wind turbine for one year prevents the release of approximately 30 kg of mercury, in comparison with the generation of electricity by a typical coal-fired power plant.

Wind power, under optimal conditions, appears considerably more efficient than conventional power sources in its energy returned on investment. One study found that wind turbines produce 23 times as much energy as they consume. For nuclear energy, the ratio was 16:1; for coal it was 11:1; and for natural gas it was 5:1. Wind farms also use less water

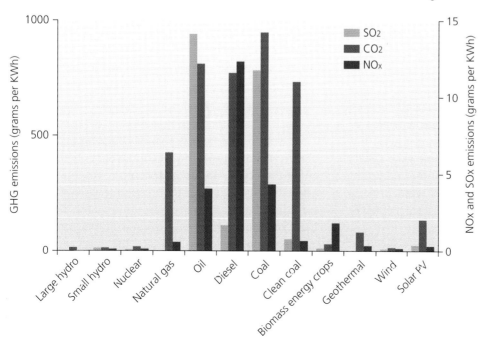

FIGURE 17.14
The lifetime emissions of CO_2, SO_2, and NO_x from wind, solar PV, hydro, nuclear, and even biomass-sourced electricity generation are dramatically lower per kilowatt-hour than for all of the fossil fuel-based sources. Data from Canadian Electricity Association. 2006. *Power Generation in Canada: A Guide.*

than do conventional power plants. Wind turbine technology can be used on many scales, from a single tower for local use to fields of hundreds that supply large regions. Small-scale turbine development can help make local areas more self-sufficient, just as solar energy can.

Another societal benefit of wind power is that farmers and ranchers can lease their land for wind development, which provides them with extra revenue while also increasing property tax income for rural communities. A single large turbine can bring in $2000 to $4500 in annual royalties while occupying just one-tenth of a hectare of land. Because each turbine takes up only a small area, most of the land can still be used for farming, ranching, or other uses.

Wind power has some downsides—but not many

Wind is an intermittent resource; we have no control over when wind will occur, and this unpredictability is a major limitation in relying on it as an electricity source. However, this poses little problem if wind is only one of several sources contributing to a utility's power generation. Moreover, several technologies—for example, batteries or hydrogen fuel—can store energy generated by wind and release it later when needed.

Just as wind varies from time to time, it varies from place to place. Some areas are simply windier than others. Global wind patterns combine with local topography—mountains, hills, water bodies, forests, cities—to create local wind patterns, and companies study these patterns closely before investing in a wind farm. Meteorological research has given us information with which to judge prime areas for locating wind farms (see **FIGURE 17.12A**).

Good wind resources, however, are not always near population centres that need the energy. Transmission networks would need to be greatly expanded to get wind power to where people live. Furthermore, when wind farms *are* proposed near population centres, local residents often oppose them. Turbines are generally located in exposed, conspicuous sites, and many people object to wind farms for esthetic reasons, feeling that the structures clutter the landscape. Although polls show wide public approval of existing wind projects and of the concept of wind power in general, newly proposed wind projects often elicit the so-called **not-in-my-backyard (NIMBY) syndrome** among people living nearby.

Wind turbines also pose a threat to birds and bats, which can be killed when they fly into the rotating blades. Studies at several sites have suggested that bird deaths may be a less severe problem than was initially feared. For instance, one European study indicated that migrating difficulty seabirds fly past offshore turbines without difficulty.

However, other data show that resident seabird densities decline near turbines. On land, an estimated one to two birds are killed per turbine per year—far fewer than the millions already being killed annually by television, radio, and cell phone towers; tall buildings with lighted windows; automobiles; pesticides; and domestic cats. Bat mortality appears to be a more severe problem, but more research is needed. The key for protecting birds and bats may be selecting sites that are not on flyways or in the midst of prime habitat for species that are likely to fly into the blades.

Geothermal Energy

Geothermal energy is one form of renewable energy that does not originate from the Sun; it is generated deep within Earth. The radioactive decay of elements amid the extremely high pressures deep in the interior of our planet generates heat that rises to the surface through magma (molten rock) and through fissures and cracks. Where this energy heats groundwater, spurts of heated water and steam are sent up from below. Terrestrial geysers and submarine hydrothermal vents are the surface manifestations of these processes.

---

weighing the issues   **WIND AND NIMBY**   17–1

If you could choose to get your electricity from either a wind farm or a coal-fired power plant, which would you choose? How would you react if the electric utility proposed to build the wind farm that would generate your electricity atop a ridge running behind your neighbourhood, such that the turbines would be clearly visible from your living room window? Would you support or oppose the development? Why? If you would oppose it, where would you suggest the farm be located? Do you think anyone might oppose it in that location?

Iceland is one country that has abundant geothermal resources. It is an island built from magma that extruded above the ocean's surface and cooled—magma from the Mid-Atlantic Ridge, the area of volcanic activity along the spreading boundary of two tectonic plates. Because of the geothermal heat in this region, volcanoes and geysers are numerous in Iceland. In fact, the word *geyser* originated from the Icelandic *Geysir*, the name for the island's largest geyser, which recently resumed periodic eruptions after many years in dormancy.

Geothermal power plants use the energy of naturally heated underground water and steam for direct heating

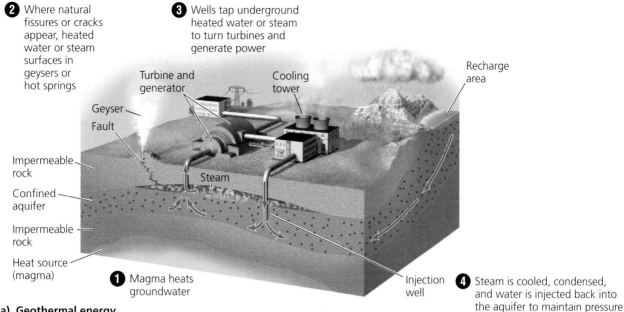

FIGURE 17.15
With geothermal energy (a), magma heats groundwater deep in Earth (1), some of which is let off naturally through surface vents, such as geysers (2). Geothermal facilities tap into heated water below ground and channel steam through turbines in buildings to generate electricity (3). After being used, the steam is often condensed, and the water is pumped back into the aquifer to maintain pressure (4). At the Nesjavellir geothermal power station in Iceland (b), steam is piped from four wells to a condenser at the plant, where cold water pumped from lakeshore wells 6 km away is heated. The water, heated to 83°C, is sent through an insulated 270 km pipeline to Reykjavik and environs, where residents use it for washing and space heating.

and to turn turbines and generate electricity (**FIGURE 17.15**). Geothermal energy is renewable in principle (its use does not affect the amount of heat produced in Earth's interior), but the power plants we build to use this energy may not all be capable of operating indefinitely. If a geothermal plant uses heated water more quickly than groundwater is recharged, the plant will eventually run out of water. This is occurring at The Geysers, in Napa Valley, California, a very early geothermal application where the first generator was built in 1960. In response, operators have begun injecting municipal wastewater into the ground to replenish the supply. More and more geothermal power plants throughout the world are now injecting water, after it is used, back into aquifers to help maintain pressure and thereby sustain the resource.

A second reason geothermal energy may not always be renewable is that patterns of geothermal activity in Earth's crust shift naturally over time. This means that an area that produces hot groundwater now may not always do so.

We can harness geothermal energy for heating and electricity

Geothermal energy can be harnessed directly from geysers at the surface, but most often wells must be drilled down hundreds or thousands of metres toward heated groundwater. Generally, water at temperatures of 150°C–370°C or more is brought to the surface and converted to steam by

THE SCIENCE BEHIND THE STORY

Water and Earth Energy for Heating and Cooling in Toronto and Ottawa

Enwave Energy Corporation's Deep Lake Water Cooling Plant in Toronto is part of a district cooling technology that takes advantage of the cold, deep waters of Lake Ontario.

One application that has been particularly amenable to emerging energy technologies, depending on the specific details of the geographic location, is *district heating* and *district cooling*—that is, distributing centrally generated heating and cooling to many buildings throughout an area.

In Toronto, a forward-looking plan makes use of the cool, deep waters of Lake Ontario to provide air conditioning to more than 2.5 million square metres of downtown office and living spaces. The Deep Lake Water Cooling project is a district-cooling technology that takes advantage of low-temperature lake water lying immediately adjacent to the downtown area. The project is coordinated by EnWave Energy, which was set up (originally as a non-profit corporation) to develop the application more than 20 years ago.

The system works by taking cold water in through three large intake pipes that lie 83 m below the surface of the lake and extend 5.1 km from the shore. Each day, millions of litres of water at 4°C are pumped to the Toronto Island Filtration Plant, where it is treated and circulated for normal distribution into the city's drinking water supply, via a pumping station. Before leaving the pumping station the cold water is diverted through a series of heat exchangers. The coldness of the water is used to remove the thermal energy from warm water passing through pipes on the other side of the heat exchangers. The newly cooled water moves through a system of many kilometres of underground pipes, providing cooling to customers in more than 140 buildings. The water is then returned to the pumping station to be recooled, and the circuit is repeated.

The chilled water and the city's drinking water always remain on opposite sides of the heat exchanger and thus never come into contact with each other. The chilled water is recirculated and recooled through a closed loop system. By using the deep lake water to provide district cooling, the project saves approximately 128 million kilowatt-hours annually in electricity, reducing CO_2 emissions by 79 000 tons and reducing electricity consumption by 90% compared with conventional air conditioning technologies.[15]

Geothermal or terrestrial (*Earth energy*) heat pumps also utilize fluids to store, transport, and supply thermal energy. During the winter months, when the temperature just a few metres under the surface is considerably warmer than the air temperature, heat pumps are used to extract this near-surface thermal energy from groundwater travelling through underground pipes. Note that this heat is primarily solar energy that was absorbed and stored in subsurface soil layers, rather than "true" geothermal energy originating from the hot inner layers of Earth.

Once extracted, the heat is distributed. The cooled fluid is then returned to its underground reservoir to be reheated. Heat pumps are reversible, so the same process can provide cooling. During the summer months, groundwater is used to cool buildings (in a process similar to the Deep Lake Water Cooling process) and is then returned to its underground reservoir to be recooled. Both heating and cooling by this method are cyclic processes that can be repeated indefinitely.

Closed loop systems continuously recirculate the fluid through a pipeline circuit (see diagram) without discharging it back into the aquifer or water body from which it was obtained. In district heating systems in Canada, this fluid is typically a mixture of antifreeze and water, but plain water or air can work in some circumstances. In an **open loop** system, in contrast, heat energy is acquired from the water via heat exchangers, and the water is then discharged back into the aquifer or water body. Open loop systems can be less expensive, but they require a suitable water supply and must be designed to mitigate environmental impacts.[16]

Geothermal heat pump systems

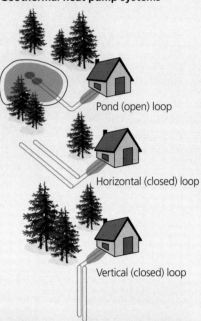

Geothermal heat pump systems operate on either closed loop or open loop systems, in which fluids are used to carry, supply, and store thermal energy. Heat pumps extract energy from the fluids, which are then returned to the underground reservoir.

Injecting water into an aquatic system at a temperature that differs from its original temperature may damage organisms in the ecosystem.

There are tens of thousands of heat pump installations throughout Canada, some for individual buildings and others used in district heating or cooling. One example of a geothermal heat pump system in use in Canada is the Underground Thermal Energy Storage (UTES), which relies on underground reservoirs to store heat during the summer and provide heat during the winter. The UTES technology has been in use at Carleton University in Ottawa since 1990. Environment Canada estimates that the UTES system can reduce cooling costs by 80% and heating costs by 40% or more, in addition to significant reductions in the polluting emissions that would otherwise be associated with space heating and cooling.[17]

lowering the pressure in specialized compartments. The steam is then employed in turning turbines to generate electricity.

Hot groundwater can also be used directly for heating homes, offices, and greenhouses; for driving industrial processes; and for drying crops. Iceland heats most of its homes through direct heating with piped hot water. Iceland began putting geothermal energy to use in the 1940s, and today 30 municipal district heating systems and 200 small private rural networks supply heat to 86% of the nation's residences.

Geothermal **ground source heat pumps** (GSHPs) use thermal energy from near-surface sources of earth and water (see "The Science Behind the Story: Water and Earth Energy for Heating and Cooling in Toronto and Ottawa").

Soil varies in temperature from season to season less than air does, so the pumps heat buildings in the winter by transferring heat from the ground into buildings, and they cool buildings in the summer by transferring heat from buildings into the ground. Both types of heat transfer are accomplished by a network of underground plastic pipes that circulate water. Because heat is simply moved from place to place rather than being produced using outside energy inputs, heat pumps can be highly energy efficient.

Natural Resources Canada estimates that at least 30 000 GSHPs make use of Earth's energy to heat offices, institutions (such as hospitals and universities), factories, and residences in Canada. This is particularly important in Canada, where more than half of the energy demand in institutional and commercial settings is for space heating. Compared with conventional electric heating and cooling systems, GSHPs heat spaces 50%–70% more efficiently, cool them 20%–40% more efficiently, can reduce electricity use by 25%–60%, and can reduce emissions by up to 70%.

Use of geothermal power is growing

Geothermal energy provides less than 0.5% of the total energy used worldwide and remains largely unexploited in Canada. Worldwide it provides more power than solar and wind combined, but only a small fraction of the power derived from hydropower and biomass. In Canada the geologic settings are such that true geothermal energy is commercially viable only in British Columbia because of its proximity to the boundary of the Juan da Fuca tectonic plate.

In the right setting, geothermal power can be among the cheapest electricity to generate. However, at the world's largest geothermal power plants, The Geysers in northern California, generating capacity has declined by more than 50% since 1989 as steam pressure has declined, but The Geysers still provide enough electricity to supply 750 000 homes. Currently Japan, China, and the United States lead the world in use of geothermal power.

Geothermal power has benefits and limitations

Like other renewable sources, geothermal power greatly reduces emissions relative to fossil fuel combustion. Geothermal sources can release variable amounts of gases dissolved in their water, including carbon dioxide, methane, ammonia, and hydrogen sulphide. However, these gases are generally in very small quantities, and geothermal facilities that use the latest filtering technologies produce even fewer emissions. By one estimate, each megawatt of geothermal power prevents the emission of 7.0 million kilograms of carbon dioxide each year.

On the negative side, geothermal sources, as we have seen, may not always be truly sustainable. In addition, the water of many hot springs is laced with salts and minerals that corrode equipment and pollute the air. These factors may shorten the lifetime of plants, increase maintenance costs, and add to pollution.

Moreover, use of geothermal energy is limited to areas where the energy can be tapped. Unless technology is developed to penetrate far more deeply into the ground, geothermal energy use will remain more localized than solar, wind, biomass, or hydropower. Such places as Iceland are rich in geothermal sources, but most of the world is not. Nonetheless, many hydrothermal resources remain unexploited, awaiting improved technology and governmental support for their development.

Ocean Energy

The oceans are home to several underexploited energy sources. Each involves continuous natural processes that could potentially provide us with sustainable energy predictably through time and in substantial amounts. Of the four approaches being developed, three involve motion, and one involves temperature.

We can harness energy from tides, waves, and currents

Just as dams on rivers use flowing freshwater to generate hydroelectric power, some scientists, engineers, businesses, and governments are developing ways to use the kinetic energy from the natural motion of ocean water to generate electrical power.

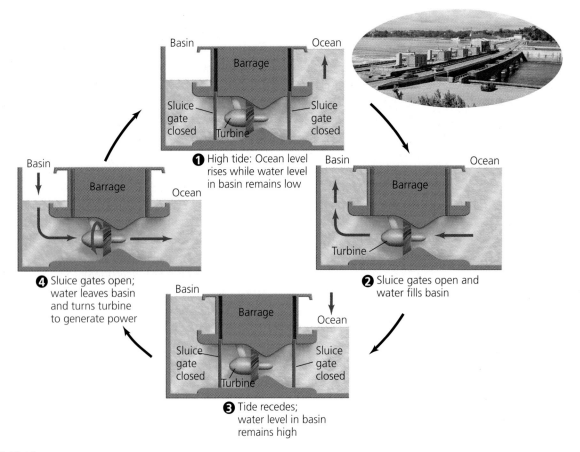

FIGURE 17.16
Energy can be extracted from the movement of the tides at coastal sites where tidal flux is great enough. One way of doing so involves the use of bulb turbines in concert with the outgoing tide. At high tide (1), ocean water is let through the sluice gates, filling an interior basin (2). At low tide (3), the basin water is let out into the ocean, spinning turbines to generate electricity (4). This technology is similar to what has been used at the Annapolis Tidal Generating Station since 1984, and at the La Rance facility in France (photo) for more than four decades.

The rising and falling of ocean tides twice each day at coastal sites throughout the world moves large amounts of water past any given point on the world's coastlines (see "Central Case: Harnessing Tidal Energy at the Bay of Fundy"). Differences in height between low and high tides are especially great in long, narrow bays, such as Alaska's Cook Inlet or the Bay of Fundy. Such locations are best for harnessing **tidal energy**, which is accomplished by erecting sluices across the outlets of tidal basins. The incoming tide flows through the sluice gates and is trapped behind them. Then, as the outgoing tide passes through the gates, it turns turbines to generate electricity (**FIGURE 17.16**). Some designs allow for generating electricity from water moving in each direction.

The world's largest tidal generating facility is the La Rance facility in France (see **FIGURE 17.16**, inset), which has operated for more than 40 years. Smaller facilities operate in China, in Russia, and in Canada at the Bay of Fundy. Tidal stations release few or no pollutant emissions, but they can have impacts on the ecology of estuaries and tidal basins.

Wave energy—energy harnessed from wind-driven waves at the ocean's surface—could be developed at a greater variety of sites than could tidal energy. The principle is to harness the motion of the waves and convert this mechanical energy into electricity. Many designs for machinery to harness wave energy exist, but few have been adequately tested. Some designs are for offshore facilities and involve floating devices that move up and down with the waves. Wave energy is greater at deep-ocean sites, but transmitting the electricity produced to shore would be expensive.

Other designs are for coastal onshore facilities. Some of these designs funnel waves from large areas into narrow channels and elevated reservoirs, from which water is then allowed to flow out, generating electricity as hydroelectric dams do. Other coastal designs use rising and falling waves to push air into and out of chambers, turning turbines to generate electricity (**FIGURE 17.17**). No commercial wave energy facilities are operating yet, but demonstration projects exist in Europe and Japan.

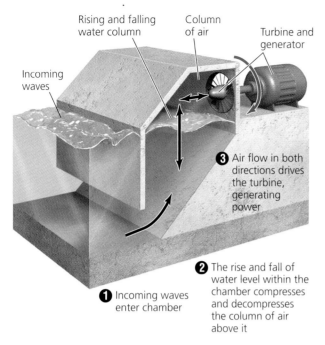

FIGURE 17.17
Coastal facilities can make use of energy from the motion of ocean waves. As waves are let into and out of a tightly sealed chamber, the air inside is alternately compressed and decompressed, creating airflow that rotates turbines to generate electricity.

A third way of harnessing marine kinetic energy is to use the motion of ocean currents, such as the Gulf Stream. Devices that look essentially like underwater wind turbines have been erected in European waters to test this idea.

The ocean stores thermal energy

Each day the tropical oceans absorb an amount of solar radiation equivalent to the heat content of 250 billion barrels of oil—enough to provide about 100 000 times the electricity used daily in Canada. The ocean's sun-warmed surface is higher in temperature than its deep water, and **ocean thermal energy conversion (OTEC)** is based on this gradient in temperature.

In the *closed cycle* approach (similar to the systems discussed in "The Science Behind the Story: Water and Earth Energy for Heating and Cooling in Toronto and Ottawa"), warm surface water is piped into a facility to evaporate chemicals, such as ammonia, that boil at low temperatures. These evaporated gases spin turbines to generate electricity. Cold water piped in from ocean depths then condenses the gases so they can be reused. In the *open cycle* approach, warm surface water is evaporated in a vacuum, and its steam turns turbines and then is condensed by cold water. Because ocean water loses its salts as it evaporates, the water can be recovered, condensed, and sold as desalinized freshwater for drinking or agriculture.

OTEC systems require not only a large temperature difference between the surface and deeper waters but also a rapid drop-off of underwater topography near the coast, so that sufficiently cold temperatures can be accessed within a reasonable distance of the shore. Research on OTEC systems has been conducted in Hawaii and Japan, where conditions are optimal, but costs remain high, and as of yet no facility is commercially operational.

> **weighing the issues**
>
> **YOUR ISLAND'S ENERGY?**
>
> Imagine that you have been elected president of an island nation and that your nation's parliament is calling on you to propose a national energy policy. Your geologists do not yet know whether there are fossil fuel deposits or geothermal resources under your land, but your country gets a lot of sunlight and a fair amount of wind, and broad, shallow shelf regions surround its coasts. Your island's population is moderately wealthy but is growing fast, and importing fossil fuels from mainland nations is becoming increasingly expensive.
>
> What approaches would you propose in your energy policy? What specific steps would you urge your congress to fund immediately? What trade relationships would you seek to establish with other countries? What questions would you ask of your economic advisors? What questions would you fund your country's scientists to research?

Hydrogen

At the beginning of the chapter we mentioned that there are three main categories of applications for renewables: (1) electricity generation, (2) space heating (and cooling), and (3) fuels. All the renewable energy sources we have discussed so far can be used to generate electricity more cleanly than can fossil fuels. Many of them can be applied locally, to provide space heating or cooling for buildings and districts. As useful as these applications are to us, however, the energy and the electricity generated by them cannot be stored and transported easily in large quantities for use when and where they are needed. This is why vehicles still rely on fossil fuels for power. The development of

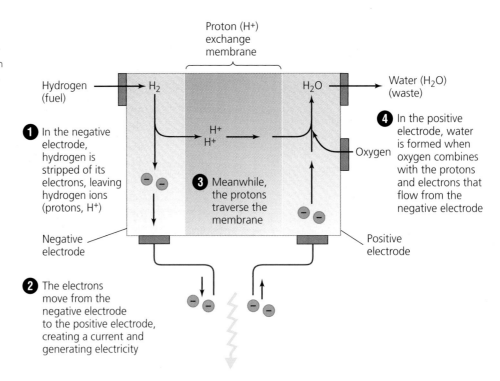

FIGURE 17.18
Hydrogen fuel drives electricity generation in a fuel cell, creating water as a waste product. Atoms of hydrogen are first stripped of their electrons (1). The electrons move from a negative electrode to a positive one, creating current and generating electricity (2). Meanwhile, the hydrogen ions pass through a proton exchange membrane (3) and combine with oxygen to form water molecules (4).

hydrogen fuel shows promise for storing energy conveniently and in considerable quantities and to produce electricity at least as cleanly and efficiently as renewable energy sources.

In the "hydrogen economy" that many energy experts worldwide envision, hydrogen fuel, together with electricity, would serve as the basis for a clean, safe, and efficient energy system. This system would use as a fuel the universe's simplest and most abundant element. In this system, electricity generated from renewable sources that are intermittent, such as wind or solar energy, could be used to produce hydrogen. **Fuel cells**—essentially, hydrogen batteries—could then employ hydrogen to produce electrical energy as needed to power vehicles, computers, cell phones, home heating, and countless other applications (**FIGURE 17.18**).

Basing an energy system on hydrogen could alleviate dependence on foreign fuels and help fight climate change. For these reasons, many governments, including the federal and provincial governments in Canada, are funding research into hydrogen and fuel cell technology. Automobile companies also are investing significant amounts in research and development to produce vehicles that run on hydrogen (**FIGURE 17.19**).

Bringing a hydrogen economy to reality would require collaboration and partnerships. In 2004 the Hydrogen Village (H2V) was launched in the Greater Toronto Area by founding partners Stuart Energy, the University of Toronto Mississauga, Hydrogenics, and the City of Toronto. H2V is thus a public–private–university partnership, now with more than 40 member organizations. H2V is not a "bricks and mortar" village; it is a "virtual" village, whose mandate is to promote the development of community-based applications of hydrogen and fuel cell technologies. The goal is to accelerate the sustainable commercialization of hydrogen and fuel cell technologies—in other words, to develop the marketplace to the point of readiness for hydrogen technologies, through awareness raising, education, and pilot projects for hydrogen technologies.

Hydrogen fuel may be produced from water or from other matter

Hydrogen gas (H_2) does not tend to exist freely on Earth; rather, hydrogen atoms bind to other molecules, becoming incorporated in everything from water to organic molecules. To obtain hydrogen gas for fuel, we must force these substances to release their hydrogen atoms, and this requires an input of energy. Several potential ways of producing hydrogen are being studied. In **electrolysis**, electricity is input to split hydrogen atoms from the oxygen atoms of water molecules:

$$2H_2O \rightarrow 2H_2 + O_2$$

Electrolysis produces pure hydrogen, and it does so without emitting the carbon- or nitrogen-based pollutants of fossil fuel combustion. However, whether this strategy for producing hydrogen will cause pollution over

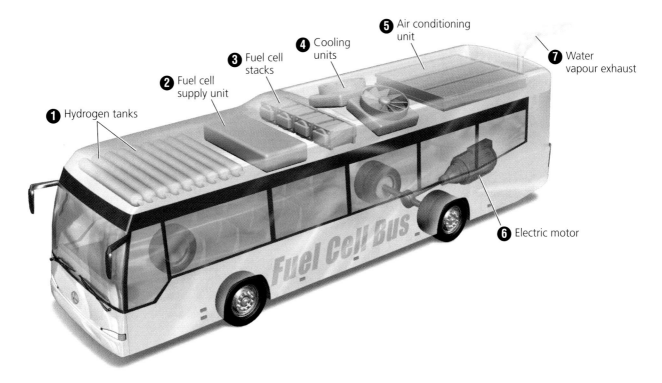

FIGURE 17.19
In the hydrogen-fuelled Citaro buses operating in Reykjavik and other European capitals, hydrogen is stored in nine fuel tanks (1). The fuel cell supply unit (2) controls the flow of hydrogen, air, and cooling water into the fuel cell stacks (3). Cooling units (4) and the air conditioning unit (5) dissipate waste heat produced by the fuel cells. Electricity generated by the fuel cells is changed from direct current (DC) to alternating current (AC) by an inverter, and it is transmitted to the electric motor (6), which powers the operation of the bus. The vehicle's exhaust (7) consists simply of water vapour.

its entire life cycle depends on the source of the electricity used for the electrolysis. If coal is burned to create the electricity, then the entire process will not reduce emissions compared with reliance on fossil fuels. If, however, the electricity is produced by some less-polluting renewable source, then hydrogen production by electrolysis would create much less pollution and greenhouse warming than reliance on fossil fuels. The "cleanliness" of a future hydrogen economy would, therefore, depend largely on the source of electricity used in electrolysis.

The environmental impact of hydrogen production will also depend on the source material for the hydrogen. Besides water, hydrogen can be obtained from biomass and fossil fuels. Obtaining hydrogen from these sources generally requires less energy input but results in emissions of carbon-based pollutants. For instance, extracting hydrogen from the methane (CH_4) in natural gas entails producing one molecule of the greenhouse gas carbon dioxide for every four molecules of hydrogen gas:

$$CH_4 + 2H_2O \rightarrow 4H_2 + CO_2$$

Thus, whether a hydrogen-based energy system is environmentally cleaner than a fossil fuel system depends on how the hydrogen is extracted.

In addition, some new research suggests that leakage of hydrogen from its production, transport, and use could potentially deplete stratospheric ozone and lengthen the atmospheric lifetime of the greenhouse gas methane.

weighing the issues: PRECAUTION OVER HYDROGEN?

Some environmental scientists have recently warned that we do not yet know enough about the environmental consequences of replacing fossil fuels with hydrogen fuel. An increase in tropospheric hydrogen gas would deplete hydroxyl (OH) radicals, they hypothesize, possibly leading to stratospheric ozone depletion and global warming from increased concentrations of methane. Some scientists say such effects will be small; others say there could be further effects that are currently unknown. Do you think we should apply the precautionary principle to the development of hydrogen fuel and fuel cells? Or should we embark on pursuing a hydrogen economy before knowing all the scientific answers? What factors inform your view?

CANADIAN ENVIRONMENTAL PERSPECTIVES

David Keith

David Keith has made a lifetime commitment to solving problems related to climate change.

- **Sustainable energy advisor** and **energy systems expert**
- **Physicist** and **Canada Research Chair in Energy and Environment** at the University of Calgary
- **Canadian Geographic Environmental Scientist of the Year** for 2006

In 2001, *Canadian Geographic* created the Canadian Environment Awards, recognizing a variety of contributions to better understanding and management of the environment, from a wide spectrum of people. In 2006 the magazine also began to recognize the contributions of Canadian scientists, and the very first Canadian Environmental Scientist of the Year was David Keith.

As an undergraduate, Keith won first prize in Canada's National Physics Prize exam. As a graduate student, he won a departmental prize for excellence in experimental physics at the Massachusetts Institute of Technology (MIT). He is now a professor appointed to both the Department of Chemical and Petroleum Engineering and the Department of Economics at the University of Calgary. These seemingly incongruent postings reflect the interdisciplinary nature of Keith's research interests, which lie at the intersections between climate science, energy technology, and public policy. His technical and policy work focuses on technologies for the capture and storage of CO_2, the economics and climatic impacts of large-scale wind power, the use of hydrogen as a transportation fuel, and the technology and implications of *geoengineering*[18]—that is, large-scale interventions in or modifications of Earth's climate system.

Keith, who chairs the Energy and Environmental Systems Group at the University of Calgary, has a reputation for being somewhat blunt, and for challenging entrenched beliefs. He has been known to reject some widely accepted ideas about energy and the environment—both those of the traditional energy sectors and those of the environmental movement. This prompted *Canadian Geographic* to call him a "contrarian" in his citation for Environmental Scientist of the Year.[19] For example, Keith spends a lot of time working out the economics of alternative, non–carbon-based energy technologies, such as wind power. "When it comes to climate change, I'm interested in technologies and social solutions that could take a big bite out of the problem," he says; but he also stresses that, "Economically, it's not even clear that global warming is all negative. It's just not true that the climate we have is automatically the best one—and you have to be straight about this."[20]

Much of Keith's commitment to solving problems related to climate change arises from his deep love of the Arctic. "One of the most fundamental reasons for working on the climate problem is to protect some remaining natural wilderness in the world . . . There are places, like the Arctic and Central Australia, we haven't yet messed with much. These are important, not just biologically but spiritually."[21] Keith comes by this commitment to environmental science naturally—his father, Anthony Keith, began his career as a Canadian Wildlife Service scientist in the 1960s, and was instrumental in convincing then-prime minister Pierre Trudeau to support a widespread ban on DDT use in Canada.

"We don't yet know how to manage the planet, and we don't have the institutions and governing structures for doing so." —David Keith[22]

Thinking About Environmental Perspectives

David Keith's career has combined a number of seemingly unrelated disciplines (physics, economics, and engineering) into a research focus that makes a lot of sense in today's complicated environmental landscape. Try to identify some of your own cross-disciplinary interests. What about psychology and energy conservation (How do you get people to change their energy-using behaviour?)? Or ethics and waste disposal (Is it right to transport someone's garbage a long distance, to dispose of it in someone else's backyard?)? Or economics, political science, and water resources (Can you use laws or pricing structures to effect a change in the way we use and manage our water resources?)?

Research into these questions is ongoing, because scientists do not want society to switch from fossil fuels to hydrogen without first knowing the possible risks from hydrogen.

Fuel cells produce electricity by joining hydrogen and oxygen

Once isolated, hydrogen gas can be used as a fuel to produce electricity within fuel cells. The chemical reaction involved in a fuel cell is simply the reverse of that shown for electrolysis; an oxygen molecule and two hydrogen molecules each split so that their atoms can bind and form two water molecules:

$$2H_2 + O_2 \rightarrow 2H_2O$$

The way this occurs within one common type of fuel cell is shown in **FIGURE 17.19**. Hydrogen gas (usually compressed and stored in an attached fuel tank) is allowed into one side of the cell, whose middle consists of two electrodes that sandwich a membrane that only protons (hydrogen ions) can move across. One electrode, helped by a chemical catalyst, strips the hydrogen gas of its electrons, creating two hydrogen ions that begin moving across the membrane. Meanwhile, on the other side of the cell, oxygen molecules from the open air are split into their component atoms along the other electrode. These oxygen ions soon bind to pairs of hydrogen ions travelling across the membrane, forming molecules of water that are expelled as waste, along with heat. While this is occurring, the electrons from the hydrogen atoms have travelled to a device that completes an electric current between the two electrodes. The movement of the hydrogen's electrons from one electrode to the other creates the output of electricity.

Hydrogen and fuel cells have many benefits

As a fuel, hydrogen offers a number of benefits. We will never run out of hydrogen; it is the most abundant element in the universe. It can be clean and nontoxic to use, and—depending on the source of the hydrogen and the source of electricity for its extraction—it may produce few greenhouse gases and other pollutants. Pure water and heat may be the only waste products from a hydrogen fuel cell, along with negligible traces of other compounds. In terms of safety for transport and storage, hydrogen can catch fire, but if it is kept under pressure, it is probably no more dangerous than gasoline in tanks.

Hydrogen fuel cells are energy efficient. Depending on the type of fuel cell, 35%–70% of the energy released in the reaction can be used. If the system is designed to capture heat as well as electricity, then the energy efficiency of fuel cells can rise to 90%. These rates are comparable or superior to most nonrenewable alternatives.

Fuel cells are also silent and nonpolluting. Unlike batteries (which also produce electricity through chemical reactions), fuel cells will generate electricity whenever hydrogen fuel is supplied, without ever needing recharging. For all these reasons, hydrogen fuel cells are being used to power vehicles, including the buses now operating on the streets of many European, North American, and Asian cities.

Conclusion

The coming decline of fossil fuel supplies and the increasing concern over air pollution and global climate change have convinced many people that we will need to shift to renewable energy sources that will not run out and will pollute far less. Renewable sources with promise for sustaining our civilization far into the future without greatly degrading our environment include solar energy, wind energy, geothermal energy, and ocean energy sources. Moreover, by using electricity from renewable sources to produce hydrogen fuel, we may be able to use fuel cells to produce electricity when and where it is needed, helping to convert our transportation sector to a nonpolluting, renewable basis.

Most renewable energy sources have been held back by inadequate funding for research and development and by artificially cheap market prices for nonrenewable resources that do not include external costs. Despite these obstacles, renewable technologies have progressed far enough to offer hope that we can shift from fossil fuels to renewable energy with a minimum of economic and social disruption. Whether we can also limit environmental impact will depend on how soon and how quickly we make the transition and to what extent we put efficiency and conservation measures into place.

REVIEWING OBJECTIVES

You should now be able to:

Outline the major sources of renewable energy and assess their potential for growth

- The "new renewable" energy sources include solar, wind, geothermal, and ocean energy sources. They are not truly "new"—in fact, many of them are ancient—but they are currently in a stage of rapid development of modern technologies.
- The new renewables currently provide far less energy and electricity than we obtain from fossil fuels or other conventional energy sources.
- Use of new renewables is growing quickly, and this growth is expected to continue as people seek to move away from fossil fuels.

Describe solar energy and the ways it is harnessed, and evaluate its advantages and disadvantages

- Energy from the Sun's radiation can be harnessed by using passive methods or by active methods involving powered technology.
- Solar technologies include solar panels for heating, mirrors to concentrate solar rays, and photovoltaic cells to generate electricity.
- Solar energy is perpetually renewable, creates no emissions, and enables decentralized power.
- Solar radiation varies in intensity from place to place and time to time, and harnessing solar energy remains expensive.

Describe wind energy and the ways it is harnessed, and evaluate its advantages and disadvantages

- Energy from wind is harnessed by using wind turbines mounted on towers.
- Turbines are often erected in arrays at wind farms located on land or offshore, in locations with optimal wind conditions.
- Wind energy is renewable, turbine operation creates no emissions, wind farms can generate economic benefits, and the cost of wind power is competitive with that of electricity from fossil fuels.
- Wind is an intermittent resource and is adequate only in some locations. Turbines kill some birds and bats, and wind farms can face opposition from local residents.

Describe geothermal energy and the ways it is harnessed, and evaluate its advantages and disadvantages

- Energy from radioactive decay in Earth's core rises toward the surface and heats groundwater. This energy is harnessed at the surface or by drilling at geothermal power plants.
- The use of geothermal energy and Earth energy for direct heating of water, electricity generation, and in heat pumps for space heating and cooling can be efficient, clean, and renewable.
- Geothermal sources occur only in certain areas and may be exhausted if water is overpumped.

Describe ocean energy sources and the ways they could be harnessed

- Major ocean energy sources include the motion of tides, waves, and currents, and the thermal heat of ocean water.
- Ocean energy is perpetually renewable and holds much promise, but so far technologies have seen only limited development.

Explain hydrogen fuel cells and assess future options for energy storage and transportation

- Hydrogen can serve as a fuel to store and transport energy, so that electricity generated by renewable sources can be made portable and used to power vehicles.
- Hydrogen can be produced through electrolysis but also by using fossil fuels—in which case its environmental benefits are reduced.
- There is concern that releasing excess hydrogen could have negative impacts on the atmosphere.
- Fuel cells create electricity by controlling an interaction between hydrogen and oxygen, and they produce only water as a waste product.
- Hydrogen can be clean, safe, and efficient. Fuel cells are silent, are nonpolluting, and do not need recharging.

TESTING YOUR COMPREHENSION

1. About how much of our energy now comes from renewable sources? What is the most prevalent form of renewable energy we use? What form of renewable energy is most used to generate electricity?
2. What factors and concerns are causing renewable energy sectors to expand? Which renewable source is experiencing the most rapid growth?
3. Contrast passive and active solar heating. Describe how each works, and give examples of each.
4. Define the photoelectric effect. Explain how photovoltaic (PV) cells function and are used.
5. What are the environmental and economic advantages of solar power? What are its disadvantages?
6. How do modern wind turbines generate electricity? How does wind speed affect the process? What factors affect where wind turbines are placed?
7. What are the environmental and economic benefits of wind power? What are its drawbacks?
8. Define *geothermal energy*, and explain how it is obtained and used. In what ways is it renewable, and in what way is it not renewable? How does it differ from what Natural Resources Canada (NRCan) calls *Earth energy*?
9. List and describe four approaches to obtaining energy from ocean water.
10. How is hydrogen fuel produced? Is this a clean process? What factors determine the amount of pollutants hydrogen production will emit?

SEEKING SOLUTIONS

1. Why might a hydrogen economy be closer than we think? Why might it instead not come to pass? Do you think water could be "the coal of the future"? Why or why not?
2. For each source of renewable energy discussed in this chapter, what factors are standing in the way of an expedient transition from fossil fuel use? What could be done in each case to ease a shift to these renewable sources?
3. Do you think we can develop and implement renewable energy resources to replace fossil fuels without great social, economic, and environmental disruption? What steps would we need to take? Will market forces alone suffice to bring about this transition? Do you think such a shift will be good for our economy?
4. Do you think Canada could make the transition to a hydrogen economy? What steps could or should Canada take to accelerate such a change?
5. **THINK IT THROUGH** You have just graduated from college or university, gotten married, landed a good job, and purchased your first home. You and your spouse plan to stay in this home for the foreseeable future and are considering installing solar panels and/or PV tiles on your roof. What factors will you consider, and what questions will you ask before deciding whether to make the investment in solar energy?
6. **THINK IT THROUGH** You are the CEO of a company that develops wind farms. Your staff is presenting you with three options, listed below, for sites for your next development. Describe at least one likely advantage and at least one likely disadvantage you would expect to encounter with each option. What further information would you like to know before deciding which to pursue?
   - Option A: A remote rural site in Yukon
   - Option B: A ridge-top site among the suburbs of Saskatoon
   - Option C: An offshore site off the Nova Scotia coast

INTERPRETING GRAPHS AND DATA

Of the new renewable energy alternatives discussed in this chapter, photovoltaic conversion of solar energy is one that many areas of Canada could easily adopt. However, the influx of solar radiation varies with time of day, time of year, and location, so all areas are not equally well suited. Today's photovoltaic technology is approximately 10% efficient at converting the energy of sunlight into electricity, but new technologies under development may increase that efficiency to as much as 40%.

1. Given a 10% efficiency for photovoltaic conversion of solar energy, approximately how many square metres of photovoltaic cells would be needed to supply one person's residential electrical needs for a year, based on the yearly average values? How many square metres would be needed if efficiency were improved to 40%?
2. Given the same 10% conversion efficiency, approximately how many square metres of photovoltaic cells would be required to supply one person's residential electrical needs during the month of April? During July? How many square metres would be required to supply the average Canadian household of four people for each of those months?

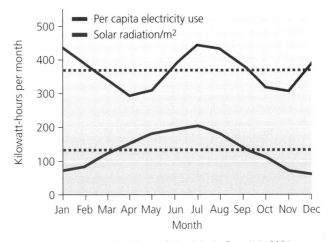

Average per capita residential use of electricity in Canada in 2004 (red line) and average influx of solar radiation per square metre for Winnipeg, Manitoba (blue line). The dashed lines represent hypothetical yearly average values for each.

3. Commercially available photovoltaic systems of this capacity cost approximately $20 000. Assume that the average cost of electricity in Canada is approximately nine cents per kilowatt-hour (this price varies dramatically with location and time). At these prices, how long would it take for the PV system to generate $20 000 worth of electricity? Calculate a combination of PV system cost and electricity cost at which the system would pay for itself in 10 years.

CALCULATING FOOTPRINTS

Assume that average per capita residential consumption of electricity is 12 kWh per day, that photovoltaic cells have an electrical output of 10% incident solar radiation, and that PV cells cost $800 per square metre. Estimate the area and cost of the PV cells needed to provide all the residential electricity used by each group in the table.

1. What additional information do you need to increase the accuracy of your estimates for the specific areas in the table?
2. Considering the distribution of solar radiation in Canada, as shown in **FIGURE 17.9**, where do you think it will be most feasible to greatly increase the percentage of electricity generated from photovoltaic solar cells?
3. The purchase price of a photovoltaic system is considerable. What other costs and benefits should you consider, in addition to the purchase price, when contemplating "going solar"?

	Area of photovoltaic cells(m²)	Cost of photovoltaic cells
You	25	$20 000
Your class		
Your province		
Canada		

TAKE IT FURTHER

Go to www.myenvironmentplace.ca where you will find

- Suggested answers to end-of-chapter questions
- Quizzes, animations, and flashcards to help you study
- *Research Navigator*™ database of credible and reliable sources to assist you with your research projects
- Tutorials to help you master how to interpret graphs
- Current news articles that link the topics that you study to case studies from your region and around the world

- **ECO Occupational Profiles:** If you found this chapter especially interesting, you might want to learn more about the following jobs by visiting the Occupational Profiles website of the Environmental Careers Organization. Go to www.eco.ca and check out the following careers:
  - Emerging energy researcher
  - Environmental technical salesperson
  - Wind energy developer

CHAPTER ENDNOTES

1. Nova Scotia Power Environment, Renewable Energy, *Tidal*, www.nspower.ca/environment/green_power/tidal/technology.shtml.
2. Nova Scotia Power Environment, Renewable Energy, *Tidal*, www.nspower.ca/environment/green_power/tidal/technology.shtml.
3. Pembina Institute, *Renewable Energy*, http://re.pembina.org/global/support.
4. Washington International Renewable Energy Conference (WIREC), March 4–6, 2008, Washington D.C., Draft Report by Roger Peters, the Pembina Institute http://pubs.pembina.org/reports/WIREC-conference-summary.pdf.
5. U.S. Energy Information Administration, *Annual Energy Outlook 2007*, www.eia.doe.gov/cneaf/solar.renewables/page/prelim_trends/rea_prereport.html;

and Canadian Association for Renewable Energies, www.renewables.ca/main/main.php.

6. *Canadian Biogas Industry Overview*, Ken Hogg, October 27, 2005, New Era Renewable Energy Solutions, www.ptac.org/eea/dl/eeaf0501p04.pdf.

7. *Canadian Biogas Industry Overview*, Ken Hogg, October 27, 2005, New Era Renewable Energy Solutions, www.ptac.org/eea/dl/eeaf0501p04.pdf.

8. Environment Canada, Municipal Solid Waste, Landfill Gas, *Beare Road Project*, www.ec.gc.ca/wmd-dgd/default.asp?lang=En&n=3438B2E7-1.

9. Environment Canada, Methane to Markets Partnership Landfill Subcommittee, *Landfill Gas Management in Canada*, October 2005, www.methanetomarkets.org/resources/landfills/docs/canada_lf_profile.pdf.

10. *Powerful Connections: Priorities and Directions in Energy Science and Technology in Canada* (2006) The Report of the National Advisory Committee on Sustainable Energy Science and Technology, Office of Energy Research and Development, Natural Resources Canada, www.nrcan.gc.ca/eps/oerd-brde/report-rapport/toc_e.htm.

11. NRCan Office of Energy Research and Development, *eco-ENERGY Technology Initiative*, www2.nrcan.gc.ca/ES/OERD/english/View.asp?x=1640.

12. Solar Energy, *The Canadian Encyclopedia*, Historica, www.thecanadianencyclopedia.com/index.cfm?PgNm=TCE&Params=A1ARTA0007549.

13. Canadian Geographic, *The Canadian Atlas Online*, Solar Power, www.canadiangeographic.ca/atlas/themes.aspx?id=weather&sub=weather_power_solarpower&lang=En.

14. Canadian Geographic, *The Canadian Atlas Online*, Wind Power, www.canadiangeographic.ca/atlas/themes.aspx?id=WEATHER&sub=WEATHER_POWER_WINDPOWER&lang=En.

15. C40 Large Cities Climate Summit 2007 Case Studies: *Energy*, www.nycclimatesummit.com/casestudies/energy/energy_CANADA.html.

16. Canadian Centre for Energy, www.centreforenergy.com/silos/geothermal/geothermalOverview03.asp.

17. Environment Canada, *Science and the Environment: Earth for Storing Energy*, www.ec.gc.ca/science/sandesept99/article5_e.html.

18. Keith, David, *Homepage*, University of Calgary, www.ucalgary.ca/~keith/.

19. Bergman, Brian (2006) Environmental scientist of the year: Climate contrarian, *Canadian Geographic* May/June, pp. 74–82.

20. Bergman, Brian (2006) Environmental scientist of the year: Climate contrarian, *Canadian Geographic* May/June, pp. 74–82.

21. Bergman, Brian (2006) Environmental scientist of the year: Climate contrarian, *Canadian Geographic* May/June, pp. 74–82.

22. Bergman, Brian (2006) Environmental scientist of the year: Climate contrarian, *Canadian Geographic* May/June, pp. 74–82.

Managing Our Waste

18

These containers are en route to a recycling facility

Upon successfully completing this chapter, you will be able to

- Summarize and compare the types of waste we generate
- List the major approaches to managing waste
- Delineate the scale of the waste dilemma
- Describe conventional waste disposal methods: landfills and incineration
- Evaluate approaches for reducing waste: source reduction, reuse, composting, and recycling
- Discuss industrial solid waste management and principles of industrial ecology
- Assess issues in managing hazardous waste

After only a couple of years the newly established North Garden at Beare Road, planted by community volunteers, was flourishing. Source: Friends of the Rouge, *About the Rouge Watershed: Geology,* www.frw.ca/rouge.php?ID=105.

CENTRAL CASE:
THE BEARE ROAD LANDFILL: MAKING GOOD USE OF OLD GARBAGE

"An extraterrestrial observer might conclude that conversion of raw materials to wastes is the real purpose of human economic activity."
—GARY GARDNER AND PAYAL SAMPAT, WORLDWATCH INSTITUTE

"We can't have an economy that uses our air, water, and soil as a garbage can."
—DAVID SUZUKI

"The issue will never go away. It's going to be a crisis if we can't find another place for our trash."
—JANE PITFIELD, TORONTO CITY COUNCILLOR

In the eastern part of Toronto, not far from the University of Toronto Scarborough campus, there is a park with a grassy hill. On the hill there are trees, bike trails, and fields of wildflowers. This is the highest spot in the neighbourhood, overlooking the Rouge River. Visitors stroll, chat, admire the view, walk their dogs. As you hike toward the top, there are few clues that beneath this grassy hill lie 9.6 million tonnes of garbage.[1] This is the old Beare Road Landfill.

The Beare Road site is located on gravel that was deposited some 12 000 years ago. At that time, when the last glaciation was drawing to a close, meltwater from the Laurentide Ice Sheet flowed into the Lake Ontario Basin, creating a glacial lake much larger than the present-day Lake Ontario. Geologists refer to this ancient lake as Lake Iroquois (see map).

The coarse sandy and gravelly deposits of the former shoreline of Lake Iroquois are now bluffs, stranded high above the current lake level. These porous and permeable units host significant aquifers, including the aquifer of the Oak Ridges Moraine, as well as the headwaters for hundreds of streams and rivers. The deposits have also been profitable for producers of *aggregates*—gravel,

The location of the ancient shoreline of glacial Lake Iroquois is now marked by a series of gravel deposits and bluffs. Old gravel pits show the extent of aggregate extraction from these deposits over the past 100+ years. Many of the worked-out gravel pits were later utilized as dumpsites for municipal solid waste.

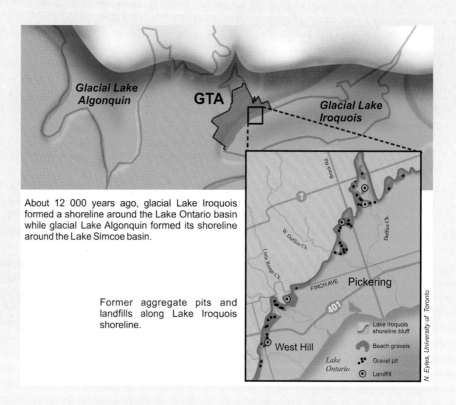

About 12 000 years ago, glacial Lake Iroquois formed a shoreline around the Lake Ontario basin while glacial Lake Algonquin formed its shoreline around the Lake Simcoe basin.

Former aggregate pits and landfills along Lake Iroquois shoreline.

sand, and crushed stone used for various construction purposes, including the building of roads and production of concrete. Many of the small towns scattered along the ancient shoreline in Ontario owe their economic beginnings to the exploitation of gravel and sand from these deposits.

Fast-forward to the latter half of the 1900s, when worked-out gravel pits scattered along the ancient shoreline sat empty. Some of the pits filled with water, serving as recreational lakes for fishing and swimming. In other cases, they beckoned to local residents—what better place to dispose of municipal solid waste? In that era, the negative impacts of tossing waste into such a porous and permeable medium were little known, and a number of the pits were used for this purpose (see maps), including the Beare Road gravel pit.

The Beare Road pit officially began receiving municipal garbage in 1968. It began with a capacity of 3 million tonnes, but this was increased several times over the years as the urgency grew for places to put Toronto's ever-increasing garbage. The final increase in capacity was accompanied by a promise of funding from the government, to be used toward the rehabilitation of the landscape, ultimately for recreational use by the community.[2]

The landfill was eventually closed down in 1983.[3] At that time a system for passive flaring of landfill gases was installed (see "The Science Behind the Story: Energy From Landfill Gas at Beare Road," Chapter 17). Some landscape restoration was undertaken, by the government and by local residents (see photo), and the site was opened as a park.

In 1996, E. S. Fox, in agreement with the City of Toronto (owner of the site) and Ontario Power Generation, began to collect the methane-rich gas being generated by the decomposing garbage at Beare Road. This type of operation is called LFGTE (landfill gas-to-electricity), and it makes use of what would otherwise be a harmful by-product of the garbage. Methane gas smells bad, corrodes vegetation, and is explosive and flammable. It is also a highly effective greenhouse gas, which must be actively managed if Canada hopes to minimize GHG emissions to fulfill its commitment to control global warming in the future.

Some environmental problems persist at the site; they are typical of old dumpsites and will require active management for years. For example, early engineering installations designed to control the collection and movement of leachate failed years ago. The impermeable liner that had been installed to prevent leakage filled with leachate, which then began to seep from the side of the hill at the level where the liner topped out—a classic demonstration of the "bathtub effect." The possibility persists that leachate may one day threaten community developments immediately downstream from the site. The exact content and composition of the waste also are unknown.

In spite of these problems, the Beare Road project provides a hope-inspiring model for the management of old landfill sites. Gas collection and utilization has helped to resolve a number of local environmental problems (such as odour and corrosion of vegetation) and is contributing to the reduction of GHG emissions for Canada.

Approaches to Waste Management

As the world's human population rises, and as we produce and consume more material goods, we generate more waste. **Waste** refers to any unwanted material or substance that results from a human activity or process. The federal government has adopted a definition that states, in part, that waste is "any substance for which the owner/generator has no further use."[4] Another popular definition suggests that waste is "resources out of place," emphasizing the fact that most waste still contains a significant proportion of useful materials. These definitions represent a changing perception of waste—that there is much of value that can be recovered from our waste stream.

For management purposes, waste is divided into several main categories. *Municipal solid waste* is nonliquid waste that comes from homes, institutions, and small businesses. *Industrial solid waste* includes waste from production of consumer goods, mining, agriculture, and petroleum extraction and refining. *Hazardous waste* refers to solid or liquid waste that is toxic, chemically reactive, flammable, corrosive, or radioactive. It can include everything from paint and household cleaners to medical waste to industrial solvents. Another type of waste is **wastewater**, water we use in our households, businesses, industries, or public facilities and drain or flush down our pipes, as well as the polluted runoff from our streets and storm drains. We discussed wastewater in Chapter 11.

We have several aims in managing waste

Waste can degrade water quality, soil quality, and air quality, thereby degrading human health and the environment. Waste is also a measure of inefficiency, so reducing waste can potentially save industry, municipalities, and consumers both money and resources. In addition, waste is unpleasant esthetically. For these and other reasons, waste management has become a vital pursuit.

There are three main components of **waste management**:

1. Minimizing the amount of waste we generate
2. Recovering waste materials and finding ways to recycle them
3. Disposing of waste safely and effectively

Minimizing waste at its source—called *source reduction*—is the preferred approach. There are several ways to reduce the amount of waste that enters the **waste stream**, the flow of waste as it moves from its sources toward disposal destinations (**FIGURE 18.1**). Manufacturers can use materials more efficiently. Consumers can buy fewer goods, buy goods with less packaging, and use those goods longer. Reusing goods you already own, purchasing used items, and donating your used items for others also help reduce the amount of material entering the waste stream.

Recovery (recycling and composting) is widely viewed as the next best strategy in waste management. *Recycling* involves sending used goods to facilities that extract and reprocess raw materials to manufacture new goods. Newspapers, white paper, cardboard, glass, metal cans, appliances, and some plastic containers have all become increasingly recyclable as new technologies have been developed and as markets for recycled materials have grown. Organic waste can be recovered through *composting*, or biological decomposition. Recycling is not a concept that humans invented; recall that all materials are recycled in ecosystems. Recycling is a fundamental feature of the way natural systems function.

Regardless of how effectively we reduce our waste stream, there will always be some waste left to dispose of. Disposal methods include burying waste in landfills and burning waste in incinerators. In this chapter we first examine how these approaches are used to manage municipal solid waste, and then we address industrial solid waste and hazardous waste.

Municipal Solid Waste

Municipal solid waste is waste produced by consumers, public facilities, and small businesses. It is what we commonly refer to as "trash" or "garbage." Everything from paper to food scraps to roadside litter to old appliances and furniture is considered municipal solid waste.

FIGURE 18.1
The most effective way to manage waste is to minimize the amount of material that enters the waste stream. To do this, manufacturers can increase efficiency, and consumers can buy "green" products that have minimal packaging or are produced in ways that minimize waste. Individuals can compost food scraps and yard waste at home and can reuse items rather than buying new ones. Many of us can recycle materials and compost yard waste through municipal recycling and composting programs. For all remaining waste, waste managers attempt to find disposal methods that minimize impact to human health and environmental quality.

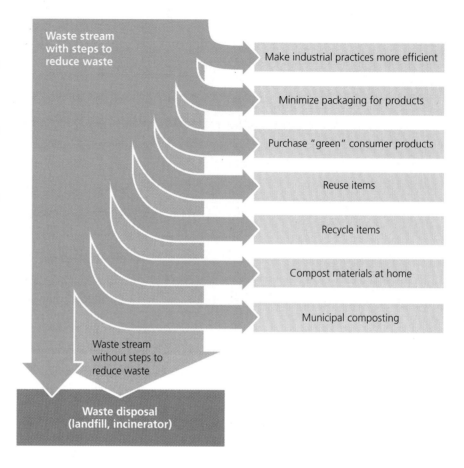

Patterns in the municipal solid waste stream vary from place to place

In Canada, paper, organics (mainly yard debris and food scraps), and plastics are the principal components of municipal solid waste, together accounting for more than 66% of the waste stream (**FIGURE 18.2**). Even after recycling, paper is the largest component of municipal solid waste. Patterns differ in developing countries; there, food scraps are often the primary contributor to solid waste, and paper makes up a smaller proportion.

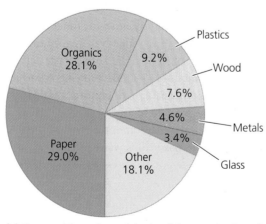

(a) Composition of municipal solid waste in Canada

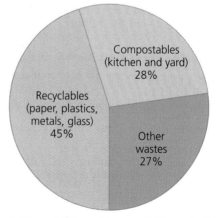

(b) Recyclable and compostable materials in municipal solid waste

FIGURE 18.2
Paper products are the largest component of the municipal solid waste stream in Canada (a), followed closely by organic wastes (yard trimmings and food scraps), and plastics. In total, each Canadian citizen generates more than one tonne (actually, 1037 kg in 2004, including both residential and industrial components)[5] of solid waste each year. Much of the waste generated by Canadians is, in principle, recyclable or compostable (b). Data are for 2004, from Environment Canada, Waste Management, *Municipal Solid Waste*, www.ec.gc.ca/wmd-dgd/Default.asp?lang=En&n=7623F633-1.

Most municipal solid waste comes from packaging and nondurable goods (products meant to be discarded after a short period of use). In addition, consumers throw away old durable goods and outdated equipment as they purchase new products. As we acquire more goods, we generate more waste. According to Statistics Canada, which tracks a variety of social, economic, and environmental indicators, Canadian citizens produced more than 33 million tonnes of municipal solid waste in 2004, for a population of 32 million—more than one tonne per person for that year. This means that Canadians generated about 1000 kg of trash per person that year, or 2.8 kg per person per day.[6]

Surpassing Canada in per capita solid waste generation is the United States, with about 3.3 kg per person per day.[7] Trailing behind is the Netherlands, with 1.4 kg per person per day. Among developed nations, Germany and Sweden produce the least waste per capita, generating just under 0.9 kg per person per day. Differences among nations result in part from differences in the cost of waste disposal; where disposal is expensive, people have incentive to waste less. The wastefulness of the North American lifestyle, with its excess packaging and reliance on nondurable goods, has caused critics to label this as "the throwaway society."

In developing nations, people consume less and generate considerably less waste. One study (see "Interpreting Graphs and Data") found that people of high-income nations waste more than twice as much as people of low-income nations. However, wealthier nations also invest more in waste collection and disposal, so they are often better able to manage their waste proliferation and minimize impacts on human health and the environment.

Waste generation is rising in all nations

In North America since 1960, waste generation has increased by almost 300%, and per capita waste generation has risen by about 70%. Plastics, which came into wide consumer use only after 1970, have accounted for the greatest relative increase in the waste stream during the last several decades. In the past decade or so, waste generation in Canada has kept pace with the growth rate of the population but has lagged slightly behind the growth in real gross domestic product (GDP) (**FIGURE 18.3**). This suggests a promising trend of waste diversion and producing more for less, perhaps because of recycling and a shift to more efficient waste management processes.

The intensive consumption that has long characterized wealthy nations is now increasing rapidly in developing

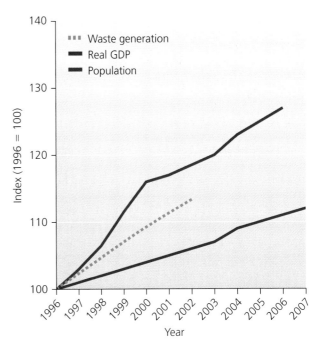

FIGURE 18.3
Waste generation in Canada has increased since 1996, keeping pace with population growth. The increase in waste generation has lagged slightly behind the growth in real GDP, probably mainly because of higher waste diversion rates. Source: Statistics Canada, Human Impacts on the Environment, Feature Article: *Solid Waste in Canada,* 2005.

nations. To some extent, this trend reflects rising material standards of living, but an increase in packaging is also to blame. Items made for temporary use and poor-quality goods designed to be inexpensive wear out and pile up quickly as trash, littering the landscapes of countries from Mexico to Kenya to Indonesia. Over the past three decades, per capita waste generation rates have more than doubled in Latin American nations and have increased more than fivefold in the Middle East. Like consumers in the "throwaway society," wealthy consumers in developing nations often discard items that can still be used. At many dumps and **landfills** throughout the developing world, in fact, poor people still support themselves by selling items they scavenge (**FIGURE 18.4**).

In many industrialized nations, per capita waste generation rates have levelled off or decreased in recent years. Note in **FIGURE 18.3** that waste generation has essentially kept pace with population in Canada since 1996, which means that per capita waste production has been essentially stable over this period (see "Calculating Footprints"). This is due largely to the increased popularity of recycling, composting, reduction, and reuse. We will examine these nondisposal approaches to waste management shortly, but let us first assess how we dispose of waste.

FIGURE 18.4
Tens of thousands of people used to scavenge each day from the dump at Payatas, outside Manila in the Philippines, finding items for themselves and selling material to junk dealers for 100–200 pesos ($2–$4) per day. That so many people could support themselves this way testifies to the immense amount of usable material needlessly discarded by wealthier portions of the population. The dump was closed in 2000 after an avalanche of trash killed hundreds of people.

Open dumping of the past has given way to improved disposal methods

Historically, people dumped their garbage wherever it suited them. Until the mid-nineteenth century, New York City's official method of garbage disposal was to dump it off piers into the East River. As population densities increased, municipalities took on the task of consolidating trash into open dumps at specified locations to keep other areas clean. This is how the worked-out gravel pits of the ancient Lake Iroquois shoreline in southern Ontario came to be used as dumpsites. To decrease the volume of trash, the dumps would be burned from time to time. Open dumping and burning still occur throughout much of the world.

As population and consumption rose in developed nations, more packaging and the use of nondegradable materials increased, waste production increased, and dumps accordingly grew larger. At the same time, expanding cities and suburbs forced more people into the vicinity of dumps and exposed them to the noxious smoke of dump burning. Reacting to opposition from residents living near dumps, and to a rising awareness of health and environmental threats posed by unregulated dumping and burning, many nations improved their methods of waste disposal. Most industrialized nations now bury waste in lined and covered landfills and burn waste in incineration facilities.

Since the late 1980s, the recovery of materials for recycling has expanded, slightly decreasing the pressure on landfills (**FIGURE 18.5**). The total rate of diversion of municipal solid waste in Canada (that is, diversion away from disposal or incineration, to recycling or composting) increased from 21% in 2000 to 24% in 2004. Over the

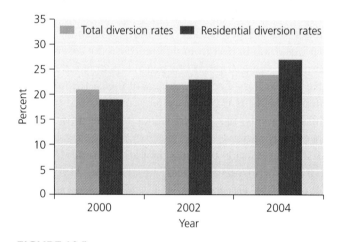

FIGURE 18.5
Waste generation grew by about 19% between 2000 and 2004 in Canada. However, over the same period the diversion rate (to recycling and composting) increased overall from 21% to 24%, and diversion in the residential context increased even more, from 19% to 27%.
Source: Environment Canada, Waste Management, *Municipal Solid Waste*, www.ec.gc.ca/wmd-dgd/default.asp?lang=En&n=7623F633-1.

same period, the rate of diversion of solid waste from residential sources increased more—from 19% to 27%. Change has been slow, but the rate of diversion is now beginning to overcome the increase in rate of waste generation. Even so, an estimated $1.5 billion worth of recyclable materials were disposed of in 2004.[8]

Waste disposal is regulated by three levels of government

In Canada, municipal governments are responsible for the collection, diversion, and disposal of solid waste from residential and small commercial and industrial sources. If you put waste by the side of the road on garbage collection day and it is picked up by city garbage trucks, then it is municipal solid waste. Many municipalities now also provide drop-off facilities for special categories of waste, including household hazardous wastes, such as leftover paint.

Provincial and territorial governments have control over the movement of waste materials within the jurisdiction and over the licensing of treatment facilities and waste generators. Thus, each province or territory has its own legislation and guidelines regulating the design, siting, licensing, operations, and expansion of landfill sites. The federal government, meanwhile, is responsible for looking after international agreements about waste, and regulating transboundary movements of waste materials. Federal involvement in waste management occurs mainly through the *Canadian Environmental Protection Act* and the *Fisheries Act*.

Sanitary landfills are engineered to minimize leakage of contaminants

In a modern **sanitary landfill**, waste is buried in the ground or piled up in large, carefully engineered mounds. In contrast to open dumps, sanitary landfills are designed to prevent waste from contaminating the environment and threatening public health (**FIGURE 18.6**). In a sanitary landfill, waste is partially decomposed by bacteria and compresses under its own weight to take up less space. Waste is typically interlayered with soil, a method that speeds decomposition, reduces odours, and reduces infestation by pests. Limited infiltration of rainwater allows for biodegradation by aerobic and anaerobic bacteria.

To protect against environmental contamination, landfills must be located away from wetlands and situated well above the local water table. The bottoms and sides of modern sanitary landfills are lined with heavy-duty plastic or other high-tech geo-engineered fabrics. They are typically underlain by a thick layer (a metre or more) of impermeable clay to help prevent contaminants from seeping into aquifers. Sanitary landfills also have systems of pipes, collection ponds, and treatment facilities to collect and treat **leachate**, the liquid that results when substances from the trash dissolve in water as rainwater percolates downward. Leachate collection systems must be maintained throughout the lifetime of the landfill, and for many years after closure. Regulations also require that area groundwater and soils be monitored regularly for contamination.

After a landfill is closed, it is capped with an engineered cover that must be maintained. The cap usually consists of a hydraulic barrier that prevents water from seeping down and gas from seeping up; a gravel layer above the hydraulic barrier that drains water, lessening pressure on the hydraulic barrier; a soil barrier that stores water and protects the hydraulic layer from weather extremes; and a topsoil layer that encourages plant growth, helping to prevent erosion.

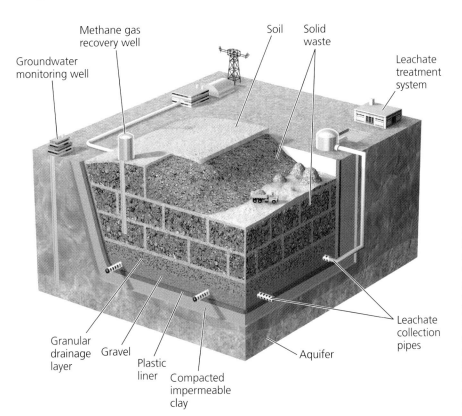

FIGURE 18.6
Sanitary landfills are engineered to prevent waste from contaminating soil and groundwater. Waste is laid in a large, lined depression, underlain by an impervious clay layer designed to prevent liquids from leaching out. Pipes of a leachate collection system draw out these liquids from the bottom of the landfill. Waste is layered along with soil until the depression is filled, and it continues to be built up until the landfill is capped. Landfill gas produced by anaerobic bacteria may be recovered, and waste managers monitor groundwater for contamination.

FIGURE 18.7
Residents and members of Friends of the Rouge plant wildflowers at the Beare Road North Garden. Source: Friends of the Rouge, *About the Rouge Watershed: Geology*, www.frw.ca/rouge.php?ID=105.

At the Beare Road Landfill, some underlying rock and sediment units are naturally clay-rich shales, creating a barrier that appears to slow the subsurface movement of leachate from the site. However, the Iroquois sand and gravel deposits—removed at an earlier stage for aggregate production, forming the pit itself—are much more permeable, providing a pathway for leachate migration.[9] Although it met provincial standards for landfill technology at the time of its construction, the Beare Road Landfill predated most current regulations and guidelines. As a result, it has caused some environmental contamination, and residents of adjacent areas continue to express some concerns about the possibility of gas and leachate migration. There is no functional engineering to control leachate migration at the site, aside from a collection ditch around the perimeter and the gas collection technologies. The refuse mound is surrounded by a number of wells installed for the purpose of monitoring groundwater for any signs of leachate migration into the surrounding areas.[10] The mound was covered after the closure of the site, to inhibit the infiltration of rainwater and the production and migration of leachate.

Landfills can be transformed after closure

Today many landfills lie abandoned. One reason is that waste managers have closed many smaller landfills and consolidated the trash stream into fewer, much larger, landfills. Meanwhile, growing numbers of cities have converted closed landfills into public parks (**FIGURE 18.7**), like the Rouge Park in Toronto, which includes the Beare Road Landfill site. Such efforts date back at least to 1938, when an ash landfill at Flushing Meadows, in Queens, New York, was redeveloped for the 1939 World's Fair, and subsequently the 1964–1965 World's Fair. Designated a park in 1967, today the site hosts Shea Stadium, the Queens Museum of Art, the New York Hall of Science, and the Queens Botanical Garden. Shutting down an industrial site and getting it ready for cleanup and repurposing is called **decommissioning**.

The Fresh Kills redevelopment endeavour in New York will be the world's largest landfill conversion project. The largest landfill in the world, Fresh Kills was the primary repository of New York City's garbage for half a century. On March 22, 2001, New York City Mayor Rudolph Giuliani and New York Governor George Pataki were on hand to celebrate as a barge arrived on the western shore of New York City's Staten Island and dumped the final load of trash at Fresh Kills. The landfill's closure was a welcome event for Staten Island's 450 000 residents, who had long viewed the landfill as a bad-smelling eyesore, health threat, and civic blemish. The 890 ha landfill featured six gigantic mounds of trash and soil. The highest, at 69 m, was higher than the nearby Statue of Liberty.

New York City planned to transform the old landfill into a world-class public park—a verdant landscape of rolling hills and wetlands teeming with wildlife, a mecca for recreation for New York's residents. Later in 2001, however, the landfill had to be reopened temporarily. After the September 11, 2001, terrorist attacks, the 1.8 million tons of rubble from the collapsed World Trade Center towers, including unrecoverable human remains, was taken by barge to Fresh Kills, where it was sorted and buried. A monument will be erected at the site as part of the new park. Today, plans for the park are forging ahead. A draft master plan that incorporated suggestions from the public was released in 2006, and an environmental impact statement was released in the spring of 2008. The master plan involves everything from ecological restoration of the wetlands to construction of roads, ball fields, sculptures, and roller-blading rinks. People will be able to bicycle on trails paralleling tidal creeks of the region's largest estuary and reach stunning vistas atop the hills.

Landfills have drawbacks

Despite improvements in liner and cover technology and landfill siting, many experts believe that leachate will eventually escape even from well-lined landfills. Liners can be punctured, and leachate collection systems eventually cease to be maintained. Moreover, landfills are kept dry to reduce leachate, but the bacteria that break down material thrive in wet conditions. Dryness, therefore, slows waste decomposition. In fact, it is surprising how slowly some materials biodegrade when they are tightly compressed in a landfill. Innovative archeological research has revealed that landfills often contain food that has not decomposed and 40-year-old newspapers that are still legible (see "The Science Behind the Story: Digging Garbage: The Archeology of Solid Waste").

Another problem is finding suitable areas to locate landfills, because most communities do not want them nearby. This *not-in-my-backyard (NIMBY) syndrome* is one reason why Toronto, New York, and many other cities export their waste, and why residents of areas that are receiving the waste are increasingly protesting. The quote from a Toronto city councillor at the beginning of this chapter refers to the ongoing struggle to find somewhere to dispose of Toronto's waste. The past practice—to export Toronto's trash to Michigan—is becoming increasingly unpopular among residents there. In 2007 an average of 74 truckloads per day of solid waste (approximately 441 363 tonnes) went to a Michigan landfill from Toronto (down from 142 daily truckloads in 2003). This amount continued to decline in 2008, as the City surpassed its waste export reduction targets. Toronto's agreement with the receiving landfill in Michigan (by which the City is contractually obligated to continue to deliver garbage to the privately-maintained landfill) expires at the end of 2010. In 2007, the City of Toronto acquired the Green Lane Landfill Site located southwest of London. This landfill is still about 200 km from downtown Toronto, but it provides an alternative to the Michigan exports. The site features the latest landfill engineering technology, including onsite leachate treatment and methane gas collection and flaring systems.[11]

As a result of the NIMBY syndrome, landfills are rarely sited in neighbourhoods that are home to wealthy and educated people with the political clout to keep them out. Instead, they are disproportionately sited in poor and minority communities, as environmental justice advocates have frequently made clear.

The unwillingness of most communities to accept waste became apparent with the famed case of the "garbage barge," the *Mobro 4000*. In Islip, New York, in 1987, the town's landfills were full, prompting town administrators to ship waste by barge to a methane production plant in North Carolina. Prior to the barge's arrival, it became known that the shipment was contaminated with medical waste, including syringes, hospital gowns, and diapers. Because of the medical waste, the methane plant rejected the entire load. The barge sat in a North Carolina harbour for 11 days before heading for Louisiana. However, Louisiana would not permit the barge to dock. The barge travelled toward Mexico, but the Mexican navy prevented it from entering that nation's waters. In the end, the barge travelled 9700 km before eventually returning to New York, where, after several court battles, the waste was finally incinerated at a facility in Queens.

Incinerating trash reduces pressure on landfills

Incineration, or combustion, is a controlled process in which mixed garbage is burned at very high temperatures (**FIGURE 18.8**). At incineration facilities, waste is generally sorted and metals removed. Metal-free waste is chopped into small pieces to aid combustion and then is burned in a furnace. Incinerating waste reduces its weight by up to 75% and its volume by up to 90%.

However, simply reducing the volume and weight does not rid trash of components that are toxic. The ash remaining after trash is incinerated therefore must be disposed of in special landfills for hazardous waste. Moreover, when trash is burned, hazardous chemicals—including dioxins, heavy metals, and PCBs (Chapters 13 and 19)—can be created and released into the atmosphere. Such releases caused a backlash against incineration from citizens concerned about health hazards. Opponents also feel that incineration is incompatible with the more sustainable path of reducing consumption, producing less waste, and diverting more of the waste we produce into recycling and composting.

Most developed nations now regulate incinerator emissions, and some have banned incineration outright. In Canada the ability to ban incineration rests with the provinces and territories. In some provinces where incineration is allowed, such as Ontario (where a previous ban was lifted was lifted in 1995), some municipalities continue to ban incineration in their own jurisdiction, or not include incineration as part of their waste reduction and diversion plan. For example, the City of Toronto's aggressive plan for reducing the amount of waste produced and shipped to Michigan does not include incineration.

THE SCIENCE BEHIND THE STORY

Digging Garbage: The Archeology of Solid Waste

This is Fresh Kills Landfill, before its closure.

"Garbologist" William Rathje has pioneered the study of our culture through the waste we generate.

Garbage and *knowledge* are two words rarely put together. But when scientist William Rathje dons trash-flecked clothes and burrows into a city dump, he gleans valuable information about how we live. By pulling tons of trash out of disposal sites over the course of decades, Rathje has turned dumpster diving into a noteworthy field of scientific inquiry that he calls *garbology*. An archeologist by training, Rathje has brought exacting archeological techniques to the contents of trashcans.

As a professor at the University of Arizona in the early 1970s, Rathje wanted his students to learn a technique common among archeologists—sorting through ancient trash mounds to understand past cultures. With few ancient civilizations or their trash close at hand, he arranged for his students to dig through their neighbours' garbage.

In 1973, he gave that effort a name, "The Garbage Project," and began a methodical study of the contents of modern trash. With rakes and notebooks, the researchers sorted, weighed, itemized, and analyzed the refuse. They then visited the homes of the people who had generated the trash and asked residents about their shopping and consumption habits.

Then in 1987, amid growing debates about how quickly landfills were filling up, Rathje decided to see what was taking up space in them. The Garbage Project headed to landfills with a truck-mounted bucket auger—a large drill commonly employed by geologists and construction crews to handle everything from excavating soil samples to creating new water wells. Rathje and his researchers dug into landfills around North America, boring as far as 30 m down in 15 to 20 garbage "wells" at each site, with each well yielding up to 25 tons of trash.

Once excavated, landfill contents were sorted, weighed, and identified. Rathje's teams sometimes froze the trash before they worked with it to make the garbage easier to separate and to limit odour and flies. Smaller bits of trash were put through sieves and sometimes washed with water to make them easier to label. Rathje has excavated at least 21 dumps, uncovering a host of interesting facts in the process:

- *Not much rot.* Trash doesn't decay much in closed landfills, Rathje has found. In the low-oxygen conditions inside most closed dumps, trash turns into a sort of time capsule. Rathje's teams have found whole hot dogs in most digs, intact pastries that are decades old, and grass clippings that are still green. Decades-old newspapers are legible and can be used to date layers of trash.
- *Paper rules.* Paper-based products make up more than 40% of most landfill content, and construction debris makes up about 20%. Newspapers are often a high-volume item, averaging about 14% of landfill space.
- *Plastic packaging no problem.* Rathje says plastic packaging is not the landfill problem many believe it to be. Plastic packaging makes up only about 4.5% of landfill content, and that figure has not increased substantially since the 1970s, Rathje reported in 1997. Fast-food packaging, polystyrene foam, and disposable diapers also aren't a major problem, making up only about 3% of landfill content. If all plastic packaging were to be replaced by containers made of glass, paper, steel, or similar materials, Rathje maintains, the packaging load to landfills would more than double.
- *Poison in small bottles.* Toxic waste comes in all sizes. If nail polish were sold in large drums, its chemical composition would make it illegal to throw out in a regular dump. Nail polish, however, is discarded in small bottles—hundreds of thousands of them per year. Luckily, however, the potentially toxic ingredients in nail polish don't always spread far, he found. Paper, diapers, and other nontoxic garbage often absorb toxic materials in landfills and keep the poisons from leaching out.

Through garbology, Rathje has gleaned unique insights into how we can change our often-wasteful habits. Now a consulting professor at Stanford University, Rathje has emerged as a leading expert on how to reduce waste.

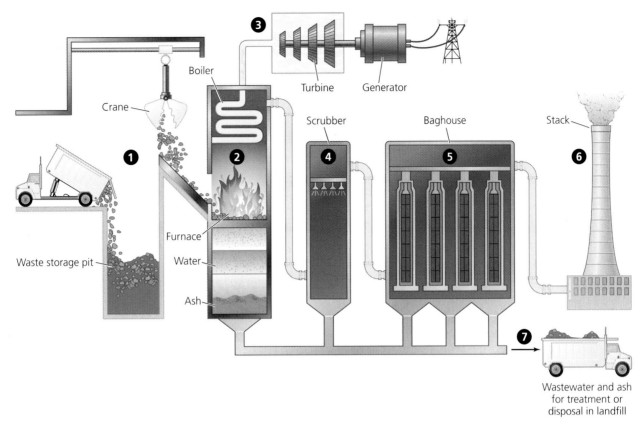

FIGURE 18.8
Incinerators reduce the volume of solid waste by burning it but may emit toxic compounds into the air. Many incinerators are waste-to-energy (WTE) facilities that use the heat of combustion to generate electricity. In a WTE facility, solid waste (1) is burned at extremely high temperatures (2), heating water, which turns to steam. The steam turns a turbine (3), which powers a generator to create electricity. In an incinerator outfitted with pollution-control technology, toxic gases produced by combustion are mitigated chemically by a scrubber (4), and airborne particulate matter is filtered physically in a baghouse (5) before air is emitted from the stack (6). Ash remaining from the combustion process is disposed of (7) in a landfill.

As a result of real and perceived health threats from incinerator emissions—and of community opposition to these plants—engineers have developed several technologies to mitigate emissions. *Scrubbers* chemically treat the gases produced in combustion to remove hazardous components and neutralize acidic gases, such as sulphur dioxide and hydrochloric acid, turning them into water and salt. Scrubbers generally do this either by spraying liquids formulated to neutralize the gases or by passing the gases through dry lime.

Particulate matter is physically removed from incinerator emissions in a system of huge filters known as a *baghouse*. These tiny particles, called fly ash, often contain some of the worst dioxin and heavy metal pollutants. In addition, burning garbage at especially high temperatures can destroy certain pollutants, such as PCBs. Even all these measures, however, do not fully eliminate toxic emissions.

Many incinerators burn waste to create energy

Incineration was initially practised simply to reduce the volume of waste, but today it often serves to generate electricity as well. Most North American incinerators today are **waste-to-energy (WTE) facilities** that use the heat produced by waste combustion to boil water, creating steam that drives electricity generation or fuels heating systems. When burned, waste generates about 35% of the energy generated by burning coal.

Revenues from power generation, however, are usually not enough to offset the considerable financial cost of building and running incinerators. Because it can take many years for a WTE facility to become profitable, many companies that build and operate these facilities require communities contracting with them to guarantee the facility a minimum amount of garbage. In a number of cases, such long-term commitments have interfered with communities' later efforts to reduce their waste through recycling and other waste-reduction strategies.

Landfills can produce gas for energy

Combustion in WTE plants is not the only means of gaining energy from waste. Deep inside landfills, bacteria decompose waste in an oxygen-deficient environment.

> **weighing the issues**
>
> **GARBAGE JUSTICE?** 18–1
>
> Do you know where your trash goes? Where is your landfill or incinerator located? Are the people who live closest to the facility wealthy, poor, or middle class? What race or ethnicity are they? Do you know whether the people of this neighbourhood protested against the introduction of the landfill or incinerator?

This anaerobic decomposition produces *landfill gas*, a mix of gases that consists of roughly half methane. Landfill gas can be collected, processed, and used in the same way as natural gas.

Today more than 40 operational projects in Canada, like the one at Beare Road, collect landfill gas and convert it into energy. Other countries take advantage of this resource as well. In Chile, four facilities in Valparaiso and Santiago supply 40% of the region's demand for natural gas. In the United States, more than 300 facilities convert landfill gas to energy. At landfill sites where gas is not collected for commercial use, it is typically allowed to flow out passively through candlestick pipes, where it is burned off in flares to reduce odours.

Reducing waste is a better option

Reducing the amount of material entering the waste stream avoids costs of disposal and recycling, helps conserve resources, minimizes pollution, and can often save consumers and businesses money. Preventing waste generation in this way is known as **source reduction**.

Much of our waste stream consists of materials used to package goods. Packaging serves worthwhile purposes—preserving freshness, preventing breakage, protecting against tampering, and providing information—but much packaging is extraneous. Consumers can give manufacturers incentive to reduce packaging by choosing minimally packaged goods, buying unwrapped fruit and vegetables, and buying food in bulk. In addition, manufacturers can use packaging that is more recyclable. They can also reduce the size or weight of goods and materials, as they already have with many items, such as aluminum cans, plastic soft drink bottles, and personal computers.

Some governments have recently taken aim at a major source of waste and litter—plastic grocery bags. These lightweight polyethylene bags can persist for centuries in the environment, choking and entangling wildlife and littering the landscape. Several nations have now banned their use. When Ireland began taxing these bags, their use dropped 90%. The IKEA Company began charging for them and saw similar drops in usage. In 2007 the small Manitoba town of Leaf Rapids became the first municipality in Canada to ban plastic bags, and the City of Toronto has approved a per-bag charge of five cents for new plastic bags, scheduled to take effect in 2009.

Increasing the longevity of goods also helps reduce waste. Consumers generally choose goods that last longer, all else being equal. To maximize sales, however, companies often produce short-lived goods that need to be replaced frequently. Thus, increasing the longevity of goods is largely up to the consumer. If demand is great enough, manufacturers will respond.

Reuse is one main strategy for waste reduction

To reduce waste, you can save items to use again or substitute disposable goods with durable ones. Habits as simple as bringing your own coffee cup to coffee shops or bringing sturdy reusable cloth bags to the grocery store can, over time, have substantial impact. You can also donate unwanted items and shop for used items yourself at yard sales and resale centres. Besides doing good for the environment, reusing items is often economically advantageous. Used items are quite often every bit as functional as new ones, and much cheaper. Table 18.1 presents a sampling of actions that we all can take to reduce the waste we generate.

> **weighing the issues**
>
> **REDUCING PACKAGING: IS IT A WRAP?** 18–2
>
> Reducing packaging cuts down on the waste stream, but how, when, and how much should we reduce? Packaging can serve very worthwhile purposes, such as safeguarding consumer health and safety. Can you think of three products for which you would *not* want to see less packaging? Can you name three products for which packaging could easily be reduced without ill effect to the consumer? Would you be any more or less likely to buy these products if they had less packaging?

Table 18.1	Some Everyday Things You Can Do to Reduce and Reuse

Donate used items to charity.

Reuse boxes, paper, plastic wrap, plastic containers, aluminum foil, bags, wrapping paper, fabric, packing material, and so on.

Rent or borrow items instead of buying them, when possible … and lend your items to friends.

Buy groceries in bulk.

Decline bags at stores when you don't need them.

Bring reusable cloth bags shopping.

Make double-sided photocopies.

Bring your own coffee cup to coffee shops.

Pay a bit extra for durable, long-lasting, reusable goods rather than disposable ones.

Buy rechargeable batteries.

Select goods with less packaging.

Compost kitchen and yard wastes in a compost bin or worm bin (often available from your community or waste hauler).

Buy clothing and other items at resale stores and garage sales.

Use cloth napkins and rags rather than paper napkins and towels.

Write to companies to tell them what you think about their packaging and products.

When solid waste policy is being debated, let your government representatives know your thoughts.

Support organizations that promote waste reduction.

Composting recovers organic waste

Composting is the conversion of organic waste into mulch or humus through natural biological processes of decomposition. The resulting compost can then be used to enrich soil. Householders can place waste in compost piles, underground pits, or specially constructed containers. As wastes are added, heat from microbial action builds in the interior, and decomposition proceeds. Banana peels, coffee grounds, grass clippings, autumn leaves, and countless other organic items can be converted into rich, high-quality compost through the actions of earthworms, bacteria, soil mites, sow bugs, and other detritivores and decomposers. Home composting is a prime example of how we can live more sustainably by mimicking natural cycles and incorporating them into our daily lives.

Centralized composting programs—there are now more than 350 of them in Canada—divert food and yard waste from the waste stream to composting facilities, where they decompose into mulch that community residents can use for gardens and landscaping. Some municipalities now ban yard waste from the municipal waste stream, helping accelerate the drive toward composting. Approximately 28% of the Canadian solid waste stream is made up of materials that can easily be composted (see **FIGURE 18.2**). Composting reduces landfill waste, enriches soil and helps it resist erosion, encourages soil biodiversity, makes for healthier plants and more pleasing gardens, and reduces the need for chemical fertilizers.

Recycling consists of three steps

Recycling, too, offers many benefits. **Recycling** consists of collecting materials that can be broken down and reprocessed to manufacture new items. In 2002, 6.6 million tonnes of materials were prepared for recycling by waste management organizations and companies. Of this, paper was the main component (23% of the total).[12]

The recycling loop contains three basic steps (**FIGURE 18.9**). The first step is collecting and processing used recyclable goods and materials. Communities may designate locations where residents can drop off recyclables or receive money for them. Many of these have now been replaced by the more convenient option of curbside recycling, in which trucks pick up recyclable items in front of houses, usually in conjunction with municipal trash pickup. Curbside recycling has grown rapidly, and its convenience has helped boost household recycling rates across Canada.

FIGURE 18.9
The familiar recycling symbol consists of three arrows to represent the three components of a sustainable recycling strategy: collection and processing of recyclable materials, use of the materials in making new products, and consumer purchase of these products.

Items collected are taken to **materials recovery facilities (MRFs)**, where workers and machines sort items, using automated processes including magnetic pulleys, optical sensors, water currents, and air classifiers that separate items by weight and size. The facilities clean the materials, shred them, and prepare them for reprocessing.

Once readied, these materials are used in manufacturing new goods. Newspapers and many other paper products use recycled paper, many glass and metal containers are now made from recycled materials, and some plastic containers are of recycled origin. Some large objects, such as benches and bridges in city parks, are now made from recycled plastics, and glass is sometimes mixed with asphalt (creating "glassphalt") for paving roads and paths. The pages in this textbook are made from recycled paper that is up to 20% post-consumer waste.

If the recycling loop is to function, consumers and businesses must complete the third step in the cycle by purchasing products made from recycled materials. Buying recycled goods provides economic incentive for industries to recycle materials and for new recycling facilities to open or existing ones to expand. In this arena, individual consumers have the power to encourage environmentally friendly options through the free market. Many businesses now advertise their use of recycled materials, a widespread instance of *ecolabelling*. As markets for products made with recycled materials expand, prices continue to fall.

Recycling has grown rapidly and can expand further

The thousands of curbside recycling programs and MRFs in operation today have sprung up only in the last 20 years. For example, paper recycling in Canada rose from 26% in 1990 to almost 45% in 2002, according to Statistics Canada data (**FIGURE 18.10**).[13]

Recycling rates vary greatly from one product or material type to another and from one location to another. Rates for different types of materials and products range from nearly zero to almost 100%. The increase in recycling has been propelled in part by economic forces as established businesses see opportunities to save money and as entrepreneurs see opportunities to start new businesses. It has also been driven by the desire of municipalities to reduce waste and by the satisfaction people take in recycling. These two forces have driven recycling's rise even though it has often not been financially profitable. In fact, many of the increasingly popular municipal recycling programs are run at an economic loss. The expense required to collect, sort, and process recycled goods is often more than recyclables are worth in the market. Furthermore, the more people recycle, the more glass, paper, and plastic is available to manufacturers for purchase, driving down prices.

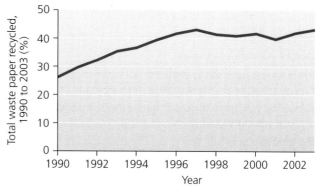

Source: Statistics Canada, Environment Accounts and Statistics Division

FIGURE 18.10
Recycling of paper has risen steadily in Canada over the past 20 years. As of 2002 almost 45% of mixed paper waste was being recovered through recycling.

Recycling advocates, however, point out that market prices do not take into account external costs—in particular, the environmental and health impacts of *not* recycling. For instance, it has been estimated that globally, recycling saves enough energy to power 6 million households per year. And recycling aluminum cans saves 95% of the energy required to make the same amount of aluminum from mined virgin bauxite, its source material.

As more manufacturers use recycled products and as more technologies and methods are developed to use recycled materials in new ways, markets should continue to expand, and new business opportunities may arise. We are still at an early stage in the shift from an economy that moves linearly from raw materials to products to waste, to an economy that moves circularly, using waste products

weighing the issues
COSTS OF RECYCLING AND NOT RECYCLING
18–3

Should recycling programs be subsidized by governments even if they are run at an economic loss? What types of external costs—costs not reflected in market prices—do you think would be involved in not recycling, say, aluminum cans? Do you feel these costs justify sponsoring recycling programs even when they are not financially self-supporting? Why or why not?

as raw materials for new manufacturing processes. The steps we have taken in recycling so far are central to this transition, which many analysts view as key to building a sustainable economy.

Financial incentives can help address waste

Waste managers have employed economic incentives to reduce the waste stream. The "pay-as-you-throw" approach to garbage collection uses a financial incentive to influence consumer behaviour. In these programs, municipalities charge residents for home trash pickup according to the amount of trash they put out. The less waste the household generates, the less the resident has to pay.

Return-for-refund schemes ("bottle bills" in the United States) represent another approach that hinges on financial incentive. To date, all provinces and territories except Nunavut have such programs. Consumers pay a deposit, return bottles and cans to stores after use, and receive a refund—generally $0.05 to $0.20 per bottle or can. The first bottle bills were passed in the 1970s to cut down on litter, but they have also served to decrease the waste stream. Research by Melissa Felder, Clarissa Morawski, and others has shown that where they have been enacted, these laws have proven profoundly effective and resoundingly popular; they are recognized as among the most successful recycling programs of recent decades (**FIGURE 18.11**). Jurisdictions with bottle and can refund programs have reported that their beverage container litter has decreased by 69%–84%, their total litter has decreased by 30%–64%, and their per capita container recycling rate has risen 260%.

Edmonton showcases reduction and recycling

Edmonton, Alberta, has created one of the world's most advanced waste management programs. As recently as 1998, fully 85% of the city's waste was being landfilled, and space was running out. Today, just 35% goes to the new sanitary landfill, whereas 15% is recycled, and an impressive 50% is composted. Edmonton's citizens are proud of the program, and 88% of them participate in its curbside recycling program. Where blue recycling bins are available at apartments and condominiums, the participation rate of residents is 91%. The goal is to divert 90% of the city's waste from landfill by 2012.[14]

When Edmonton's residents put out their trash, city trucks take it to their new *co-composting* plant, the largest in North America (**FIGURE 18.12A**). The waste is

(a) Composting facility, Edmonton, Alberta

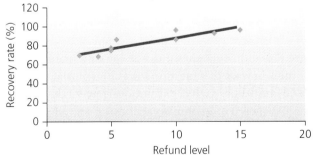

(b) Aeration building, Edmonton composting facility

FIGURE 18.11
Data suggest that bottle and can return-for-refund programs increase recycling rates and that higher redemption amounts boost recycling rates further. The strong correlation between recovery rates and refund levels shown on this diagram indicates that the economic incentive to return containers is a key driver. Data from Melissa Felder and Clarissa Morawski, CM Consulting. 2003. *Evaluating the Relationship Between Refund Values and Beverage Container Recovery.* For the Beverage Container Management Board, BC.

FIGURE 18.12
Edmonton boasts one of North America's most successful waste management programs. Edmonton's gigantic composting facility (a) is the size of eight football fields. Inside the aeration building (b), which is the size of 14 professional hockey rinks, mixtures of solid waste and sewage sludge are exposed to oxygen and composted for 14–21 days.

dumped on the floor of the facility, and large items, such as furniture, are removed and landfilled. The bulk of the waste is mixed with dried sewage sludge for one to two days in five large rotating drums, each the length of six buses. The resulting mix travels on a conveyor to a screen that removes nonbiodegradable items. It is aerated for several weeks in the largest stainless steel building in North America (**FIGURE 18.12B**). The mix is then passed through a finer screen and finally is left outside for four to six months. The resulting compost—80 000 tons annually—is made available to area farmers and residents. The facility even filters the air it emits with a 1 m layer of compost, bark, and wood chips, which eliminates the release of unpleasant odours into the community. Christmas tree composting and "grasscycling" programs are now included, as well.

Edmonton's program also includes a state-of-the-art MRF that handles 40 000 tonnes of waste annually, a leachate treatment plant, a research centre, public education programs, and a wetland and landfill revegetation program. In addition, 100 pipes collect enough landfill gas to power 4000 homes, bringing thousands of dollars to the city and helping power the new waste management centre. Five area businesses reprocess the city's recycled items, including ewastes. Newsprint and magazines are turned into new newsprint and cellulose insulation, and cardboard and paper are converted into building paper and shingles. Household metal is made into rebar and blades for tractors and graders, and recycled glass is used for reflective paint and signs.

Industrial Solid Waste

In Canada, disposal of wastes from nonresidential sources (industrial, commercial, and institutional) increased from 14.6 million to 15.5 million tonnes between 2002 and 2004.[15] (Compare this with the more than 30 million tonnes of solid waste generated by Canadian households each year, and the magnitude of the residential waste problem becomes clear.)

Industrial solid waste includes waste from factories, mining activities, agriculture, petroleum extraction, and more. Waste is generated at various points along the process from raw materials extraction to manufacturing to sale and distribution (**FIGURE 18.13**).

Regulation and economics each influence industrial waste generation

Most methods and strategies of waste disposal, reduction, and recycling by industry are similar to those for

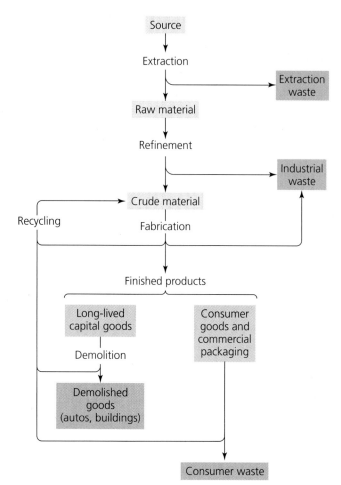

FIGURE 18.13
Industrial and municipal waste is generated at a number of stages throughout the life cycles of products. Waste is first generated when raw materials needed for production are extracted. Further industrial waste is produced as raw materials are processed and as products are manufactured. Waste results from the demolition or disposal of products by businesses and individuals. At each stage, there are opportunities for efficiency improvements, waste reduction, or recycling.

municipal solid waste. For instance, businesses that manage their own waste onsite most often dispose of it in landfills, and companies must design and manage their landfills in ways that meet provincial/territorial, local, or tribal guidelines. Other businesses pay to have their waste disposed of at municipal disposal sites. Regulation and enforcement vary across provinces and territories and from municipality to municipality.

The amount of waste generated by a manufacturing process is one measure of its efficiency; the less waste produced per unit or volume of product, the more efficient that process is, from a physical standpoint. However, physical efficiency is not always equivalent to economic efficiency. Often it is cheaper for industry to manufacture its products or perform its services quickly but messily. That is, it can be cheaper to generate waste than to avoid generating waste. In such cases, economic efficiency is maximized, but physical efficiency is not. The frequent mismatch between these two types of efficiency is a major reason that the output of industrial waste is so great.

Rising costs of waste disposal, however, enhance the financial incentive to decrease waste and increase physical efficiency. Once either government or the market makes the physically efficient use of raw materials also economically efficient, businesses have financial incentives to reduce their own waste.

Industrial ecology seeks to make industry more sustainable

To reduce waste, growing numbers of industries today are experimenting with industrial ecology. A holistic approach that integrates principles from engineering, chemistry, ecology, and economics, **industrial ecology** seeks to redesign industrial systems to reduce resource inputs and to minimize physical inefficiency while maximizing economic efficiency. Industrial ecologists would reshape industry so that nearly everything produced in a manufacturing process is used, either within that process or in a different one.

The larger idea behind industrial ecology is that industrial systems should function more like ecological systems, in which almost everything produced is used by some organism, with very little being wasted. This principle brings industry closer to the ideal of ecological economists, in which human economies attain sustainability by functioning in a circular fashion rather than a linear one.

Industrial ecologists pursue their goals in several ways. For one, they examine the entire life cycle of a given product—from its origins in raw materials, through its manufacturing, to its use, and finally its disposal—and look for ways to make the process more ecologically efficient. This strategy is called **life-cycle analysis**. In addition, industrial ecologists examine industrial processes with an eye toward eliminating environmentally harmful products and materials. Finally, they study the flow of materials through industrial systems to look for ways to create products that are more durable, recyclable, or reusable. Goods that are currently thrown away when they become obsolete, such as computers, automobiles, and some appliances, could be designed to be more easily disassembled and their component parts reused or recycled. In this way, industrial ecology also helps to close the loop by minimizing wastes at both the industry end and the consumer end of the process.

By applying strategies aimed at reducing waste and preventing pollution at its source—commonly referred to as **pollution prevention (P2) strategies**—companies can significantly reduce their waste output, as shown in **FIGURE 18.14**.

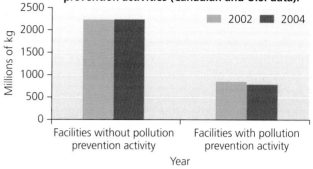

FIGURE 18.14
Pollution prevention or "P2" activities can greatly reduce industrial waste releases. This bar graph shows comparative data for 2000 and 2002, based on Canadian and U.S. data. Source: Commission for Environmental Cooperation, 2004

Businesses are adopting industrial ecology

Businesses are taking advantage of the insights of industrial ecology to reduce waste and lessen their impact on health and the environment while saving money. A good example is the carpet tile company Interface, which asks customers to return used tiles for recycling and reuse as backing for new carpet. Interface also modified its tile design and its production methods to reduce waste. It adapted its boilers to use landfill gas for its energy needs. Through such steps, the company has cut its waste generation by 80%, its fossil fuel use by 45%, and its water use by 70%—all while saving $30 million per year, holding prices steady for its customers, and raising profits by 49%.

Among many other initiatives are such programs as Canadian Tire's auto parts return initiatives, and Xerox's take-back/lease programs for its used photocopiers. Programs like these are founded in good business principles, and they have the added benefit of building customer loyalty. An interesting variation of the take-back concept is the ENVIRx program run by the Pharmacists Association of Alberta, in which consumers are able to return unused medications to participating pharmacies for proper disposal. This helps prevent pharmaceuticals from entering the municipal waste stream.

The Swiss Zero Emissions Research and Initiatives (ZERI) Foundation sponsors dozens of innovative projects worldwide that attempt to create goods and services without generating waste. One example involving breweries is currently being pursued in Canada, Sweden, Japan, and Namibia. Brewers in these projects take waste from the beer-brewing process and use it to fuel other processes. Traditional breweries produce only beer while generating much waste, some of which goes toward animal feed. ZERI-sponsored breweries use their waste grain to make bread and to farm mushrooms. Waste from the mushroom farming, along with brewery wastewater, goes to feed pigs. The pigs' waste is digested in containers that capture natural gas and collect nutrients used to nourish algae for growing fish in fish farms. The brewer derives income from bread, mushrooms, pigs, gas, and fish, as well as beer, while producing little waste. Although most ZERI projects are not fully closed-loop systems, they attempt to approach this ideal.

An international initiative that has been adopted in Canada as an overriding strategy for industrial waste minimization is the Extended Producer Responsibility and Stewardship (or EPR) program. The central goal of the program is to transfer a large part of the responsibility for waste minimization, both physical and financial, to producers.[16] This gives producers the incentive to design more environmentally efficient products and processes, and to take greater responsibility for the product at the end of its life cycle. It also encourages producers to build the environmental costs of a product into its market price—one of the basic tenets of ecological economics. As of 2005 there were more than 50 active EPR initiatives in Canada.

Waste exchanges are an offshoot of industrial ecology

The concept of industrial ecology is based on a "closed loop" in which wastes are recycled back through the system. Following the definition of wastes as "resources out of place," industrial ecologists strive to find practical, economical uses for waste materials. To achieve this goal, they try to identify how waste products from one manufacturing process can be used as raw materials for a different process. For instance, used plastic beverage containers cannot be refilled because of the potential for contamination, but they can be shredded and reprocessed to make other plastic items, such as benches, tables, and decks.

Many network services have emerged with the goal of linking producers of waste with industries or individuals that can make use of the waste as raw materials. Such a network is called a **waste exchange**. You can check out an example of a nationwide waste exchange by visiting the website of The Waste Exchange of Canada at www.recyclexchange.net. Other waste exchanges operate locally or internationally.

For businesses, governments, and individuals alike, there are plenty of ways to reduce waste and mitigate the impacts of our waste generation—and quite often, doing so brings economic benefits. This is true both for solid waste and for hazardous waste.

Hazardous Waste

Hazardous wastes are diverse in their chemical composition and may be liquid, solid, or gaseous. In Canada, according to the *Canadian Environmental Protection Act (CEPA)* (1999), **hazardous waste** is waste that has one or more of the following properties:

- **Flammable**. Substances that easily catch fire (for example, natural gas or alcohol)
- **Corrosive**. Substances that corrode metals in storage tanks or equipment
- **Reactive**. Substances that are chemically unstable and readily react with other compounds, often explosively or by producing noxious fumes
- **Toxic**. Substances that harm human health when they are inhaled, are ingested, or contact human skin (see Chapter 19)

Materials with these characteristics can harm human health and degrade environmental quality. Flammable and explosive materials can cause ecological damage and atmospheric pollution. For instance, fires at tire dumps, such as the one in Hagersville, Ontario, in 1990 (**FIGURE 18.15**), have caused air pollution and highway closures. Toxic wastes in lakes and rivers have caused fish die-offs, endangered aquatic mammals (see Chapter 5,

FIGURE 18.15
In 1990 a pile of 15 million discarded tires in Hagersville, Ontario, caught fire and burned for 17 days, releasing thousands of litres of toxic chemicals, such as benzene and toluene, into the air and water.

"Central Case: The Plight of the St. Lawrence Belugas"), and closed important domestic fisheries.

Certain categories of materials that are clearly "dangerous" are nevertheless not included in the official definition of hazardous waste. An example is *biomedical waste*, which includes things like human tissues and fluids, and discarded medical sharps. These materials are excluded from the definition of hazardous waste not because they are without risk but because they require specialized handling, treatment, and disposal methods and are therefore controlled under different legislation. Similarly, *radioactive waste* requires a special set of management approaches.

Hazardous wastes have diverse sources

Industry, mining, households, small businesses, agriculture, utilities, and building demolition all create hazardous waste. Industry produces the largest amounts of hazardous waste, but in most developed nations industrial waste generation and disposal is highly regulated. This regulation has reduced the amount of hazardous waste entering the environment from industrial activities. As a result, households currently are the largest source of unregulated hazardous waste.

Household hazardous waste (HHW) includes a wide range of items, such as paints, batteries, oils, solvents, cleaning agents, lubricants, and pesticides. There are 13 categories of hazardous materials that are commonly used by municipalities for the purpose of sorting and disposing of HHW (Table 18.2).[17]

Canadians improperly dispose of approximately 27 000 tonnes of household hazardous waste each year,[18] and the average home contains close to 45 kg of it in sheds, basements, closets, and garages. Although many hazardous substances become less hazardous over time as they degrade chemically, two classes of chemicals are particularly hazardous because their toxicity persists over time: organic compounds and heavy metals.

Organic compounds and heavy metals can be hazardous

In our day-to-day lives, we rely on the capacity of synthetic organic compounds and petroleum-derived compounds to resist bacterial, fungal, and insect activity. Such items as plastic containers, rubber tires, pesticides, solvents, and wood preservatives are useful to us precisely because they resist decomposition. We use these substances to protect our buildings from decay, kill pests that attack crops, and

Table 18.2 Categories of Household Hazardous Waste

- *Antifreeze* (ethylene or propylene glycol used or intended for use as a vehicle engine coolant; materials and their containers)
- *Fertilizers* (materials registered under the *Fertilizers Act*, packaged in 30-kg quantities or less, including their containers)
- *Lubricating oils* (petroleum-derived or synthetic oils, Crankcase, engine and gear oils, and hydraulic, transmission and heat transfer fluids, and lubricating fluids used in machinery; containers of 30 L and less)
- *Paints and Coatings* (household and industrial use, including their containers)
- *Pesticides, fungicides, herbicides, insecticides* (and their containers, including domestic, commercial, agricultural, and restricted pesticides)
- *Pressurized containers* (such as propane tanks and cylinders, and oxygen tanks)
- *Single-use dry cell batteries* (alkaline and carbon zinc, mercuric-oxide, silver-oxide and zinc-air, and lithium, including cylindrical, regular and button batteries)
- *Solvents* (and their containers, including turpentine, isopropanol, ethanol, ketones, xylene, toluene, mineral spirits, linseed oils, naptha, and methylene chloride; these are better known as paint thinners, lacquer thinners, automotive body resin solvents, contact cement thinners, paint strippers, and degreasers)
- *Used oil filters* (from hydraulic, transmission or internal combustion engine applications, including diesel fuel filters, household furnace fuel filter, coolant filter, storage tank diesel fuel filter, sump-type automatic transmission filter, and others)

Source: Stewardship Ontario, Municipal Hazardous or Special Waste, Do What You Can Program; for more information see "What's Included,".

keep stored goods intact. However, the resistance of these compounds to decay is a double-edged sword, for it also makes them persistent pollutants. Many synthetic organic compounds are toxic because they can be absorbed readily through the skin of humans and other animals and can act as mutagens, carcinogens, teratogens, and endocrine disruptors.

Heavy metals, such as lead, chromium, mercury, arsenic, cadmium, tin, and copper, are used widely in industry for wiring, electronics, metal plating, metal fabrication, pigments, and dyes. Heavy metals enter the environment when paints, electronic devices, batteries, and other materials are disposed of improperly. Lead from fishing weights and from hunters' lead shot has accumulated in many rivers, lakes, and forests. In older homes, lead from pipes contaminates drinking water, and lead paint remains a problem, especially for infants. Heavy metals that are fat-soluble and break down slowly are prone to bioaccumulating. In California's Coast Range, for instance, mercury washed downstream from abandoned mercury mines enters low-elevation lakes and rivers, is consumed by bacteria and invertebrates, and

THE SCIENCE BEHIND THE STORY

Testing the Toxicity of "e-Waste"

Discarded electronic waste can leach heavy metals and should be considered hazardous waste, researchers say.

Most electronic waste, or "e-waste," is disposed of in conventional sanitary landfills. However, most electronic appliances contain heavy metals that can cause environmental contamination and public health risks. For instance, more than 6% of a typical computer is composed of lead.

Timothy Townsend and colleagues from the University of Florida determined that cathode ray tubes (CRTs) from computer monitors and colour televisions leach an average of 18.5 mg/L of lead, far above the regulatory threshold of 5 mg/L for lead in leachate. (This threshold is different from the allowable concentration of lead in drinking water in Canada, which is 0.010 mg/L.) Following this research, the U.S. Environmental Protection Agency (EPA) proposed classifying CRTs as hazardous waste, and several U.S. states banned them from conventional landfills.

Then in 2004, Townsend's lab group completed experiments on 12 other types of electronic devices. To measure their toxicity, the group used a standard test, the Toxicity Characteristic Leaching Procedure (TCLP), designed to mimic the process by which chemicals leach out of solid waste in landfills. In the TCLP, waste is ground up into fine pieces, and 100 g of it is put in a container with 2 L of an acidic leaching fluid. The container is rotated for 18 hours, after which the leachate is analyzed for its chemical content. Researchers look for eight heavy metals—arsenic, barium, cadmium, chromium, lead, mercury, selenium, and silver—and determine their concentrations in the leachate.

To conduct the standard TCLP, Townsend's team ground up the central processing units (CPUs) of personal computers, creating a mix made up by weight of 15.8% circuit board, 7.5% plastic, 68.2% ferrous metal, 5.4% nonferrous metal, and 3.1% wire and cable. However, grinding up a computer into small bits is no easy task, and it is hard to obtain a sample that accurately represents all components and materials. So the researchers also designed a modified TCLP test in which they placed whole CPUs—with the parts disassembled but not ground up—in a rotating 208 L drum full of leaching liquid. Then they tested their 12 types of devices by using a combination of the standard and modified TCLP methods.

The team's results are summarized in the accompanying bar chart. Lead was the only heavy metal found to exceed the U.S. EPA's regulatory threshold, but this threshold (5 mg/L) was exceeded in the majority of trials. Computer monitors

---

accumulates in increasingly larger quantities up the food chain, poisoning organisms at higher trophic levels and making fish unsafe to eat.

"E-waste" is a new and growing problem

When we first began to conduct much of our business, learning, and communication with computers and other electronic devices, many people predicted that our paper waste would decrease. Instead, the proliferation of computers, printers, VCRs, fax machines, cell phones, GPS devices, MP3 players, and other gadgets has created a substantial new source of waste. These products have short lifetimes before people judge them obsolete, and most are discarded after only a few years.

The amount of **electronic waste**—often called **e-waste**—is growing rapidly. Canadians discarded 74 000 tonnes of computer waste in 2002, including 1.7 million desktop computers, 1.9 million cell phones, 2 million television sets and 1.1 million VCRs. The U.S. Environmental Protection Agency reports that 70% of the heavy metals found in U.S. landfills came from discarded electronic products.[19]

Most e-waste is disposed of in landfills as conventional solid waste. However, most electronic products contain heavy metals and toxic flame retardants, and recent research suggests that ewaste should instead be treated as hazardous waste (see "The Science Behind the Story: Testing the Toxicity of 'e-Waste'". In Canada there are no federal programs or legislation aimed specifically at dealing with e-waste, although initiatives have been started in most provinces.

More and more electronics are now being recycled. The devices are taken apart, and parts are either reused or disposed of more safely. Roughly one-fifth of the nearly 2 million tons of electronics discarded in 2005 in the United States was recycled, the EPA estimates.

There are serious concerns about the health risks that recycling may pose to workers doing the disassembly, and wealthy nations ship much of their e-waste to developing

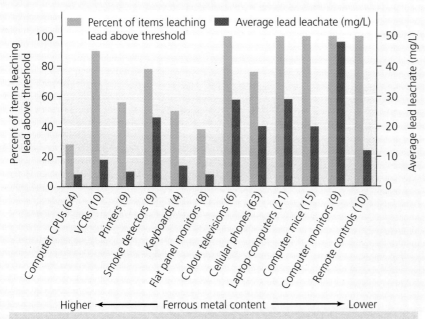

Some proportion of all 12 devices tested exceeded 5 mg/L, the EPA's regulatory threshold for lead leachate. Devices with higher ferrous metal content tended to leach less lead. Where both standard and modified TCLPs were used, results are averaged. Data from Townsend, T. G., et al. 2004. *RCRA toxicity characterization of computer CPUs and other discarded electronic devices.* July 15, 2004, report to the U.S. EPA.

leached the most lead (47.7 mg/L on average), as expected, because monitors include the cathode ray tubes were already known to be a problem. However, laptops, colour TVs, smoke detectors, cell phones, and computer mice also leached high levels of lead. Next came remote controls, VCRs, keyboards, and printers, all of which leached more lead on average than the threshold and did so in 50% or more of the trials. Whole CPUs and flat panel monitors were the only devices to leach less than 5 mg/L of lead on average, but even these exceeded the threshold more than one-quarter of the time.

The researchers found that items containing more ferrous metals (such as iron) tended to leach less lead. For instance, CPUs contain 68% ferrous metals (compared with only 7% in laptops), and laptops leached seven times as much lead as CPUs. Further experiments confirmed that ferrous metals were chemically reacting with lead and stopping it from leaching.

The work suggests that many electronic devices have the potential to be classified as hazardous waste because they frequently surpass the toxicity criterion for lead. However, scientists must decide how to judge results from the modified TCLP methods and must evaluate other research, before determining whether to alter regulatory standards.

Lab tests may or may not accurately reflect what actually happens in landfills. So Townsend's team is filling columns measuring 24 cm wide by 4.9 m long with e-waste and municipal solid waste, burying them in a landfill, and then testing the leachate that results. The results from such research should help regulators decide how best to dispose of e-waste that is not reused or recycled.

countries, where the disassembly is done by poor workers with minimal safety regulations. These environmental justice concerns need to be resolved, but if electronics recycling can be done responsibly, it seems likely to be the way of the future.

TOXIC COMPUTERS? 18–4

The cathode ray tubes in televisions and computer screens can hold up to 5 kg of heavy metals, such as lead and cadmium. These represent the second-largest source of lead in landfills today, behind auto batteries. With more computer screens being purchased, the transition to high-definition television, and the rapid turnover of computers, what future waste problems and environmental health issues might you expect? How do you think we should handle the reuse, recycling, and disposal of these products?

In many North American cities, businesses, nonprofit organizations, or municipal services now collect used electronics for reuse or recycling. The next time you upgrade to a new computer, TV, DVD player, VCR, or cell phone, find out about opportunities to recycle your old ones.

Several steps precede the disposal of hazardous waste

For many years we discarded hazardous waste without special treatment. In many cases, people did not know that certain substances were harmful to human health. In other cases, the danger posed by these substances was known or suspected, but it was assumed that the substances would disappear or be sufficiently diluted in the environment. The resurfacing of toxic chemicals in a residential area years after their burial at Love Canal in upstate New York provided a dramatic demonstration to the North American public that hazardous waste deserves special attention and treatment.

FIGURE 18.16
Many communities designate collection sites or collection days for household hazardous waste. Here, workers handle waste from an Earth Day collection event.

FIGURE 18.17
Unscrupulous individuals or businesses sometimes dump hazardous waste illegally to avoid disposal costs.

Since the 1980s, many communities have designated sites or special collection days to gather household hazardous waste, or have designated facilities for the exchange and reuse of substances (**FIGURE 18.16**). Once consolidated in such sites, the waste is transported for treatment and ultimate disposal.

As for municipal solid waste, the management and control of hazardous waste and hazardous recyclable materials is a shared responsibility in Canada. The federal government regulates international agreements and transport. Provincial and territorial governments regulate intraprovincial transport and are responsible for licensing hazardous waste generators, carriers, and treatment facilities.[20] As hazardous waste is generated, transported, and disposed of, the producer, carrier, and disposal facility must each report the type and amount of material generated, its location, origin, and destination, and its handling. This is intended to prevent illegal dumping and encourage the use of reputable waste carriers and disposal facilities. Because it can be quite costly to dispose of hazardous waste, irresponsible companies sometimes illegally and anonymously dump waste, creating health risks for residents and financial headaches for local governments forced to deal with the mess (**FIGURE 18.17**).

Hazardous waste from industrialized nations is also sometimes dumped illegally in developing nations—a major environmental justice issue. This practice occurs despite the *Basel Convention*, an international treaty to prevent such acts. In 2006, a ship secretly dumped toxic wastes in Abidjan, the capital of Ivory Coast, after being told by Dutch authorities that the Netherlands would charge money to dispose of the waste in Amsterdam. The waste caused several deaths and thousands of illnesses in Abidjan, and street protests forced the government to resign over the scandal. Some jail sentences and fines were eventually handed down in this case, although thousands of victims still have received no compensation for their suffering.

Fortunately, high costs of disposal have also encouraged conscientious businesses to invest in reducing their hazardous waste. Many biologically hazardous materials can be broken down by incineration at high temperatures in cement kilns. Some hazardous materials can be treated by exposure to bacteria that break down harmful components and synthesize them into new compounds. Besides bacterial bioremediation, phytoremediation is also used. Various plants have now been bred or engineered to take up specific contaminants from soil and then break down organic contaminants into safer compounds or concentrate heavy metals in their tissues. The plants are eventually harvested and disposed of.

There are three disposal methods for hazardous waste

There are three primary means of hazardous waste disposal: secure landfills, surface impoundments, and injection wells. These do nothing to lessen the hazards of the substances, but they do help keep the waste isolated from people, wildlife, and ecosystems.

Secure landfills Design and construction standards for landfills that receive hazardous waste are much stricter than those for ordinary sanitary landfills. Hazardous waste landfills, also called **secure landfills**, must have several impervious liners, leachate removal systems, and extensive monitoring wells, and they must be located far from aquifers. Dumping of hazardous waste in ordinary landfills is particularly problematic in closed-down landfills that received wastes prior to the advent of more secure disposal options for hazardous materials.

Surface impoundments Liquid hazardous waste, or waste in dissolved form, may be stored in ponds or **surface impoundments**, shallow depressions lined with plastic and an impervious material, such as clay. Water containing dilute hazardous waste is placed in the pond and allowed to evaporate, leaving a residue of solid hazardous waste on the bottom (**FIGURE 18.18**). This process is repeated until the dry material is removed and transported elsewhere for permanent disposal. Impoundments are not ideal. The underlying layer can crack and leak waste. Some material may evaporate or blow into surrounding areas. Rainstorms may cause waste to overflow and contaminate nearby areas. For these reasons, surface impoundments are used only for temporary storage.

The potential for problems with surface impoundment of hazardous wastes became abundantly clear in the small rural town of Elmira, Ontario, in the late 1980s. UniRoyal (now the Crompton Company) had operated a chemical production facility at Elmira since 1942. One of the substances produced at the plant was the so-called Agent Orange, an extremely powerful herbicide used by the United States during the Vietnam War to defoliate large areas of forest. (U.S. and Vietnamese soldiers and civilians later suffered serious health impacts as a result of exposure to this chemical.) Rubber and agrichemicals were also produced at the plant. The waste by-products of chemical production were disposed of—legally, and within Ontario Ministry of Environment regulations for the day—in a clay-lined surface impoundment pit onsite, starting in the 1960s.

Then, in 1989, traces of chemical markers, notably the carcinogen N-nitrosodimethylamine (NDMA), began to appear in municipal drinking water wells, downslope of the impoundment site. Apparently the clay liner of the impoundment pit had failed, perhaps because it had become saturated, and contaminants were leaking out of the pit and joining the groundwater. The company undertook remediation of both the impoundment pit and the surrounding aquifers in the early 1990s. However, the town of Elmira no longer withdraws its drinking water from these aquifers; instead, it pipes water in from another municipality.

Deep-well injection The third method is intended for long-term disposal. In **deep-well injection**, a well is drilled deep beneath the water table into porous rock, and wastes are injected into it (**FIGURE 18.19**). The waste is meant to remain deep underground, isolated from groundwater and human contact. This idea seems attractive in principle, but in practice wells become corroded and can leak wastes into soil, allowing them to enter aquifers. Alberta accounts for approximately 90% of all deep-well injection of hazardous wastes in Canada, including 40 000 m³ of oilfield waste and 14 000 tonnes of industrial chemical waste.[21]

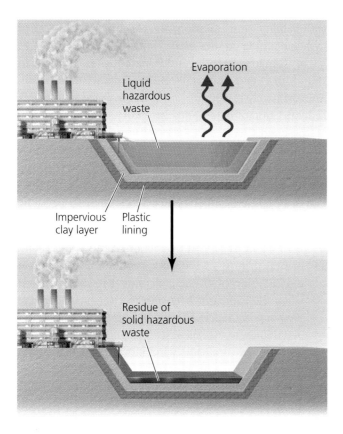

FIGURE 18.18
Surface impoundments are a strategy for temporarily disposing of liquid hazardous waste. The waste, mixed with water, is poured into a shallow depression lined with plastic and clay to prevent leakage. When the water evaporates, leaving a crust of the hazardous substance, new liquid is poured in and the process repeated. This method alone is not satisfactory, because waste can potentially leak, overflow, evaporate, or blow away.

Radioactive waste is especially hazardous

Radioactive waste is particularly dangerous to human health. It is, by definition, **radioactive** (that is, giving off energetic particles and radiation in the process of spontaneous radioactive decay) and thus potentially hazardous

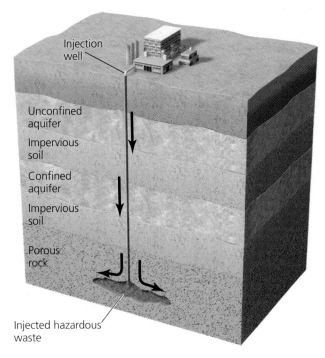

FIGURE 18.19
Liquid hazardous waste may be pumped deep underground, by deep-well injection. The well must be drilled below any aquifers, into porous rock separated by impervious clay. The technique is expensive, and waste may leak from the well shaft into groundwater.

to human and animal health. In addition, some of it is highly toxic, and some of it is extremely persistent. The dilemma of disposal has dogged the nuclear energy industry for decades. As we saw in our discussion of radioactive waste disposal in Chapter 16, there is as yet no identified permanent repository for radioactive waste in Canada. Yucca Mountain in Nevada is now designated as the single-site repository for all nuclear waste in the United States; other countries, including Germany also have designated permanent disposal sites for radioactive waste.

Most proposals for permanent disposal of radioactive waste involve some form of **geologic isolation**, that is, using the absorptive capacity and impermeability of some naturally occurring rock units to block the movement of contaminants away from the disposal site. Canadian proposals focus on disposal in automated facilities located deep underground in the stable, ancient, plutonic igneous rocks of the Canadian Shield. Other geologic settings that are amenable to geologic isolation include salt domes and thick shale units.

Geologic isolation would be combined with the chemical immobilization of the waste, sophisticated engineering design of the facility itself, and multiple-layered impervious containers for storage of the waste. In Canada, this is referred to as a **multiple-barrier approach**—that is, engineering the entire facility to place as many barriers as possible, both physical and chemical, in the pathway of any escaping contaminants.

Contaminated sites are being cleaned up, slowly

Many thousands of former military and industrial sites remain contaminated with hazardous waste in Canada and virtually every other nation on Earth. For most nations, dealing with these messes is simply too difficult, time-consuming, and expensive. Some contaminated sites, especially in the United States, have reached iconic status for the roles they played in raising public awareness, spurring local residents to action, and kick-starting the modern environmental movement. In Love Canal, a residential neighbourhood in Niagara Falls, New York, families were evacuated after toxic chemicals buried by a company and the city in past decades rose to the surface, contaminating homes and an elementary school. In Missouri, the entire town of Times Beach—another name that is practically synonymous with poorly managed toxic waste—was evacuated and its buildings demolished after being contaminated by dioxin from waste oil sprayed on its roads. The beneficial outcome is that these horrific cases led to the establishment of the first legislation to deal with liability, compensation, and cleanup costs associated with contaminated sites in the United States.

In Canada, many contaminated sites have been abandoned and thus fall under federal jurisdiction. Approximately 11 000 sites are currently listed in Canada's Federal Contaminated Sites Inventory. More than 6000 of these sites have been assessed or classified under a National Classification System for contaminated sites, developed by the Canadian Council of Ministers of the Environment.[22] Examples of the nearly 300 priority sites that received funds for cleanup activities in 2007 from the federal government include the following:[23]

- *Faro Mine, Yukon:* $14.6 million for the largest and highest priority of the federal contaminated sites, an old lead, zinc, silver, and gold mine that was shut down when the company went into receivership. There are now approximately 70 million tonnes of tailings and 320 million tonnes of waste rock at the site, with consequent acid drainage, wind-blown particulates, and other environmental hazards.[24]
- *Canadian Forces Base Esquimalt, British Columbia:* $4.56 million for remediation of hydrocarbon- and heavy metal-contaminated soil, assessment, and risk management.
- *Port Radium Mine, Northwest Territories:* $7.1 million funding for sealing of mine openings, covering of areas of elevated radiation levels, stabilization of tailings areas, demolition, and hazardous waste disposal (see Chapter 21, "Central Case: Mining Denendeh").

CANADIAN ENVIRONMENTAL PERSPECTIVES

Brennain Lloyd

Brennain Lloyd defends the North against the irresponsible dumping of garbage and other harmful waste materials.

- **Coordinator** of Northwatch
- **Peace and environmental activist**
- **Advisor** on natural resource management

Brennain Lloyd came to environmental activism from the peace movement, where she confronted issues such as poverty, injustice, and violence. As she explains it, "peace work in northeastern Ontario inevitably led to working with environmental groups" on such issues as the impacts of uranium mining and refining, the threat of a nuclear waste dumpsite in the area, and the establishment of a military testing and training corridor for the delivery of air-launched cruise missiles. "These issues had huge peace implications, but also impacts on forest health, wildlife, and human communities."

Lloyd—who is from North Bay, Ontario—is the coordinator of Northwatch, a coalition of citizen organizations and individual members. Founded in 1988, Northwatch brings a northern perspective to regional environmental issues, including waste dumping; natural resource management, especially mining and logging; and water and air quality concerns. Northwatch also provides support for local citizen groups working to address environmental issues in their communities. Three current focus areas are better forest management, greater community involvement in mine monitoring and management, and preventing northeastern Ontario from becoming a dumping ground.[25]

Today, Lloyd is ubiquitous—speaking about nuclear waste disposal in the Canadian Shield to the Canadian Environmental Assessment Agency; presenting a talk at the University of Waterloo on garbage dumping in the North, as part of the university's Environmentalist-in-Residence series; participating in a task force on mining sector sustainability or running a workshop on forest management for a local citizens' group or advisory committee to the Ministry of Natural Resources; reviewing and analyzing a proposal for a new biodiesel plant.

A few years ago, she was central to a successful intervention in the City of Toronto's plan to ship municipal garbage hundreds of kilometres for disposal in the old Adams Mine site in Kirkland Lake, Ontario. Northwatch coordinated community opposition to this plan, through the Adams Mine Intervention Coalition. The plan, which would have endangered regional water bodies and groundwater, was eventually overturned by Toronto City Council in 2000. In 2004, the Province of Ontario passed the *Adams Mine Lake Act*, which revoked all existing approvals for the use of the old mine site as a landfill, prevented the site from ever being used for that purpose, and extended the same protection to all large water bodies in the province.

In addition to the Adams Mine struggle, Lloyd and Northwatch have intervened in a number of other plans to import waste into Northern Ontario, including Ontario's biomedical waste, and PCBs from around the world. Radioactive waste has been a long-term struggle; the coalition opposes any plans to develop a deep geologic isolation disposal site for nuclear waste in the Canadian Shield. It also works to ensure appropriate decommissioning of old mine sites in the North, such as the uranium mines of Elliot Lake (see Chapter 21, "Central Case: Mining Denendeh").

Northwatch often finds itself in a reactive position, such as opposing plans to use the North as a dumping ground or withdraw resources in a harmful or negligent manner. But Lloyd and her coalition also take a proactive stance on northern development, stating that "the North must realize a long-term objective of diversifying the economy while maintaining the natural resource base and making best use of those resources which are extracted. To this end, economic and social decisions must be made with the priority of creating and contributing to a 'sustainable' North."

"My work on waste management has been of necessity, in response to repeated efforts to use northeastern Ontario as a dumping ground for 'foreign' wastes."[26] —Brennain Lloyd

Thinking About Environmental Perspectives

Toronto reversed its original plan to ship garbage to a dumpsite located in the old Adams Mine near Kirkland Lake, Ontario, partly in response to the enormous opposition from organizations like the Adams Mine Intervention Coalition and Northwatch. What do you think about a city like Toronto shipping its garbage far away? If Toronto or other large cities simply run out of room to dispose of their garbage within their own borders, what realistic alternatives do they have?

- *Belleville Small Craft Harbour, Ontario:* $6.8 million to treat contaminated soil and prevent contaminants in groundwater from discharging into the adjacent Bay of Quinte. Belleville Harbour was used for more than 50 years for the storage of coal and fuel products, which led to petroleum hydrocarbon and heavy metal contamination.

You can examine the complete Federal Contaminated Sites Inventory for yourself. It is maintained (interestingly, perhaps because of the enormous costs associated with remediation of these sites) by the Treasury Board of Canada Secretariat, and the website is www.tbs-sct.gc.ca/fcsi-rscf.

Sites that have been contaminated but have the potential to be cleaned up and remediated for other purposes are called **brownfields**. However, many sites are contaminated with hazardous chemicals we have no effective way to deal with. In such cases, cleanups simply involve trying to isolate waste from human contact, either by building trenches and clay or concrete barriers around a site or by excavating contaminated material, placing it in industrial-strength containers, and shipping it to a hazardous waste disposal facility. For all these reasons, the current emphasis is on preventing hazardous waste contamination in the first place.

All three North American countries monitor industrial pollutants by using Pollutant Release and Transfer Registers (PRTRs), which combine reports from industrial facilities with information about transfers, off-site treatments, and disposal or recycling of pollutants. In Canada the PRTR is the National Pollutant Release Inventory (NPRI), established in 1992, which covers more than 300 chemicals plus the criteria air contaminants. In Mexico, it is the *Registro de Emisiones y Transferencia de Contaminantes (RETC)*, which covers 100 chemicals, and in the United States it is the Toxics Release Inventory (TRI), started in 1987, which tracks data for more than 600 chemicals.[27] By learning where these pollutants come from, where they end up, and how they are transferred, we may ultimately be in a better position to control them.

Conclusion

Our societies have made great strides in addressing our waste problems. Modern methods of waste management are far safer for people and gentler on the environment than past practices of open dumping and open burning. In many countries, recycling and composting efforts are making rapid progress. Canada has changed in a few decades from a country that did virtually no recycling to a nation in which nearly one-quarter of all solid waste is diverted from disposal. The continuing growth of recycling, composting, and pollution-prevention initiatives, driven by market forces, government policy, and consumer behaviour, shows potential to further alleviate our waste problems.

Despite these advances, our prodigious consumption habits have created more waste than ever before. Our waste management efforts are marked by a number of difficult dilemmas, including the cleanup of highly contaminated sites, safe disposal of hazardous and radioactive waste, and frequent local opposition to disposal sites. These dilemmas make clear that the best solution to our waste problem is to reduce our generation of waste. Finding ways to reduce, reuse, and efficiently recycle the materials and goods that we use stands as a key challenge for this century.

REVIEWING OBJECTIVES

You should now be able to:

Summarize and compare the types of waste we generate

- Municipal and industrial solid waste, hazardous waste, and wastewater are major types of waste.

List the major approaches to managing waste

- Source reduction, recovery, and disposal are the three main components of waste management.

Delineate the scale of the waste dilemma

- Developed nations generate far more waste than developing nations do.
- Waste everywhere is increasing as a result of growth in population and consumption.

Describe conventional waste disposal methods: landfills and incineration

- Sanitary landfills guard against contamination of groundwater, air, and soil. Nonetheless, such contamination can occur.
- Incinerators reduce waste volume by burning it. Pollution control technology removes most pollutants from emissions, but some escape, and highly toxic ash needs to be disposed of in landfills.
- We are harnessing energy from landfill gas and generating electricity from incineration.

Evaluate approaches for reducing waste: source reduction, reuse, composting, and recycling

- Reducing waste before it is generated is the best waste management approach. Recovery is the next-best option.
- Consumers can take simple steps to reduce their waste output.
- Composting reduces waste while creating organic matter for gardening and agriculture.
- Recycling has grown in recent years and now removes nearly 24% of the Canadian waste stream.

Discuss industrial solid waste management and principles of industrial ecology

- Regulations differ, but industrial waste management is similar to that for municipal solid waste.
- Industrial ecology urges industrial systems to mimic ecological systems and provides ways for industry to increase its efficiency.

Assess issues in managing hazardous waste

- Hazardous waste is flammable, corrosive, reactive, or toxic.
- Electronic waste may be considered hazardous.
- Hazardous waste is strictly regulated, yet illegal dumping remains a problem.
- No fully satisfactory method of disposing of hazardous waste has yet been devised.
- Cleanup of hazardous waste sites is a long and expensive process.

TESTING YOUR COMPREHENSION

1. Describe five major methods of managing waste. Why do we practise waste management?
2. Why have some people labelled modern North America as "the throwaway society"? How much solid waste do Canadians generate, and how does this amount compare with that of people from other countries?
3. Name several technologies designed to make sanitary landfills safe places for the disposal of waste. Describe three problems with landfills.
4. Describe the process of incineration or combustion. What happens to the resulting ash? What is one drawback of incineration?
5. What is composting, and how does it help reduce input to the waste stream?
6. What are the three elements of a sustainable process of recycling?
7. What are the goals of industrial ecology?
8. What four criteria are used to define hazardous waste? Why are heavy metals and synthetic organic compounds particularly hazardous?
9. What are the largest sources of hazardous waste? Describe three ways to dispose of hazardous waste.
10. How is waste regulated in Canada? What are some of the similarities and differences between the regulation of nonhazardous wastes and that of hazardous wastes?

SEEKING SOLUTIONS

1. How much waste do you generate? Look into your waste bin at the end of the day, and categorize and measure the waste there. List all other waste you may have generated in other places throughout the day. How much of this waste could you have avoided generating? How much could have been reused or recycled?
2. Some people have criticized current waste management practices as merely moving waste from one medium to another. How might this criticism apply to the methods now in practice? What are some potential solutions?
3. Of the various waste management approaches covered in this chapter, which ones are your community and campus pursuing, and which are they not pursuing? Would you suggest that your community or campus start pursuing any new approaches? If so, which ones, and why?
4. Could manufacturers and businesses benefit from source reduction if consumers were to buy fewer products as a result? How? Given what you know about industrial ecology, what do you think the future of sustainable manufacturing may look like?
5. **THINK IT THROUGH** You are the CEO of a major corporation that produces containers for soft drinks and a wide variety of other consumer products. Your company's shareholders are asking that you improve the company's image—while not cutting into profits—by taking steps to reduce waste. What steps would you consider taking?
6. **THINK IT THROUGH** You are the president of your college or university. Your trustees want you to engage with local businesses and industries in ways that benefit both the school and the community. Your faculty and students want you to make the school a

leader in waste reduction and industrial ecology. Consider the industries and businesses in your community and the ways they interact with facilities on your campus. Bearing in mind the principles of industrial ecology, can you think of any novel ways in which your school and local businesses might mutually benefit from one another's services, products, or waste materials? Are there waste products from one business, industry, or campus facility that another might put to good use? Can you design an eco-industrial park that might work on your campus? What steps would you propose to take as president?

INTERPRETING GRAPHS AND DATA

By using 1990 data from 149 countries, researchers for the World Bank examined global patterns in the generation and management of municipal solid waste (MSW). The researchers were particularly interested in the relationships among wealth, population size (in 1990), and per capita generation of MSW. Their results are presented in the accompanying table.

1. Create a bar chart with the four income categories of nations as entries on the *x*-axis, and with percentages from 0 to 60 on the *y*-axis. For each category of nation, plot as paired bars the values of MSW generation and population size as percentages of the world total.
2. By using the data for total MSW generation and for population size (remembering that there are 1000 kg in a tonne), calculate the number of kilograms of MSW per capita per day for each category of nation, and enter these values in the table.
3. Now add a second *y*-axis to your graph, on the right side, ranging from 0 to 1.5, divided into increments of 0.10. Plot the values you calculated for per capita waste generation for each of the four income categories of nations, placing them as data points connected by a line. Describe in general terms the relationship between wealth and per capita generation of MSW. Can you offer at least one possible reason for the trend that you see?
4. Do you think it is possible for wealthy nations to reduce their per capita MSW generation to the rates of poorer nations? Why or why not?

Income category of nations	Total MSW generation		Population size		Kilograms MSW per capita per day
	Millions of tonnes/year	Percentage of world total	Millions of people	Percentage of world total	
Low	596	46.3	3 091	58.5	0.53
Lower-middle	145	11.2	629	11.9	
Upper-middle	192	14.9	748	14.2	
High	357	27.6	816	15.4	
All countries	**1 290**	**100.0**	**5 284**	**100.0**	

Data from Beede, D. N., and D. E. Bloom. 1995. The economics of municipal solid waste. World Bank Research Observer 10:113–150.

CALCULATING FOOTPRINTS

According to Statistics Canada, on a per capita basis, Nova Scotians generate the least solid waste (1.6 kg/day), and Alberta residents generate the most (3.1 kg/day). The average for the entire country is 2.8 kg of solid waste per person per day. Compare this number to the data in the "Interpreting Graphs and Data" table. Now calculate the amount of solid waste generated in one day and in one year by each of the groups of different population sizes indicated below, at each of the rates shown in the accompanying table.

1. Suppose your town of 50 000 people has just approved construction of a landfill nearby. Estimates are that it will accommodate 1 million tonnes of MSW. Assuming the landfill is serving only your

town, for how many years will it accept waste before filling up? How much longer would a landfill of the same capacity serve a town of the same size in another industrialized ("high-income") country?

2. Why do you think Canadians and Americans generate so much more MSW than people in other "high-income" countries, when standards of living in those countries are comparable?

Groups generating MSW	Per capita MSW generation rates				
	Canadian avg. rate (2.8 kg/day)		Nova Scotian avg. rate (1.6 kg/day)	Albertan avg. rate (3.1 kg/day)	World avg. rate (0.668 kg/day)
You	2.8 kg	1 037 kg			
Your class					
Your town					
Your province					
Canada					
World					

Source: Statistics Canada. 2005 (2004 data; some 2002 data projected to 2004 levels with a 2% increase, matching the observed increase from 2000 to 2002).

TAKE IT FURTHER

Go to www.myenvironmentplace.ca where you will find

- Suggested answers to end-of-chapter questions
- Quizzes, animations, and flashcards to help you study
- *Research Navigator*™ database of credible and reliable sources to assist you with your research projects
- Tutorials to help you master how to interpret graphs
- Current news articles that link the topics that you study to case studies from your region and around the world

- **ECO Occupational Profiles:** If you found this chapter especially interesting, you might want to learn more about the following jobs by visiting the Occupational Profiles website of the Environmental Careers Organization. Go to www.eco.ca and check out the following careers:
  - Hazardous waste technician
  - Industrial waste inspector
  - Landfill engineer
  - Recycling coordinator
  - Waste management specialist

CHAPTER ENDNOTES

1. Canadian Biogas Industry Overview, Ken Hogg, October 27, 2005, *New Era Renewable Energy Solutions,* www.ptac.org/eea/dl/eeaf0501p04.pdf.
2. City of Toronto Council and Committees, June 1, 1999, *Ontario Hydro Corridor Lands and Beare Road Ski Facility Trust Fund,* www.toronto.ca/legdocs/1999/agendas/committees/sc/sc990622/it035b.htm.
3. Environment Canada, Waste Management, *Beare Road Project,* www.ec.gc.ca/wmd-dgd/default.asp?lang=En&n=3438B2E7-1.
4. Government of Canada Depository Services Program, *Hazardous Waste Management: Canadian Directions, 1992,* http://dsp-psd.tpsgc.gc.ca/Collection-R/LoPBdP/BP/bp323-e.htm#A.%20Definitions%20and%20Classification(txt).
5. Environment Canada, Waste Management, *Municipal Solid Waste,* www.ec.gc.ca/wmd-dgd/Default.asp?lang=En&n=7623F633-1.
6. Statistics Canada, Human Activity and the Environment 2005, *Feature Article: Solid Waste in Canada,* www.statcan.ca/english/freepub/16-201-XIE/0000516-201-XIE.pdf.
7. Simmons, P., et al. (2006) The state of garbage in America. *BioCycle,* Vol. 47, p. 26.

8. Environment Canada, Waste Management, *Municipal Solid Waste,* www.ec.gc.ca/wmd-dgd/default.asp?lang=En&n=7623F633-1.
9. Desrocher, S., and B. Sherwood-Lollar (1998) Isotopic constraints on off-site migration of landfill CH4, *Ground Water,* Vol. 36, No. 5, pp. 801–809. Research Library Core.
10. Desrocher, S., and B. Sherwood-Lollar (1998) Isotopic constraints on off-site migration of landfill CH4, *Ground Water,* Vol. 36, No. 5, pp. 801–809. Research Library Core.
11. City of Toronto Solid Waste Management, *Facts About Toronto's Trash,* updated November 1, 2007, www.toronto.ca/garbage/facts.htm.
12. Statistics Canada, Human Activity and the Environment 2005, *Feature Article: Solid Waste in Canada,* www.statcan.ca/english/freepub/16-201-XIE/0000516-201-XIE.pdf.
13. Statistics Canada, Human Activity and the Environment 2005, *Feature Article: Solid Waste in Canada,* www.statcan.ca/english/freepub/16-201-XIE/0000516-201-XIE.pdf.
14. City of Edmonton, 2007, *Recycling Fact Sheet,* www.edmonton.ca/Environment/WasteManagement/PDF/2007%20Recycling%20Fact%20Sheet/City_recycle_factsheet(web).pdf.
15. Commission for Environmental Cooperation (2008) *The North American Mosaic: An Overview of Key Environmental Issues: Industrial Pollution and Waste, Commission for Environmental Cooperation,* 2008, www.cec.org/soe/files/en/SOE_IndustrialPollution_en.pdf.
16. Environment Canada, *EPR,* www.ec.gc.ca/epr/default.asp?lang=En&n=EEBCC813-1.
17. Recycling Council of Ontario, *Household Hazardous Waste Fact Sheet, 1996,* www.rco.on.ca/RCO_files/HHW.pdf.
18. Canadian Institute for Environmental Law and Policy, *Understanding Hazardous Waste in Ontario, 2006,* www.cielap.org/pdf/HazWaste2007.pdf.
19. Canadian Institute for Environmental Law and Policy, *Understanding Hazardous Waste in Ontario, 2006,* www.cielap.org/pdf/HazWaste2007.pdf.
20. Environment Canada, Waste Management, *Hazardous Waste,* www.ec.gc.ca/wmd-dgd/default.asp?lang=En&n=FDC36D83-1.
21. Pembina Institute for Appropriate Development, *Alberta GPI: Hazardous Waste, 2005,* http://pubs.pembina.org/reports/49.Hazardous%20Waste.pdf.
22. Environment Canada, News Release, July 26, 2007, and Federal Contaminated Sites Inventory, www.tbs-sct.gc.ca/fcsi-rscf/home-accueil.aspx?Language=EN&sid=wu91133413870.
23. Environment Canada, *Backgrounder: Federal Contaminated Sites Receiving Funds, 2007,* www.ec.gc.ca/default.asp?lang=En&n=714D9AAE-1&news=81941DCD-F8FA-4012-9266-CEC4F186B0F7.
24. Faro Mine Closure, *Challenges,* www.faromineclosure.yk.ca/project/challenges.html.
25. Northwatch, www.web.net/~nwatch/.
26. Brennain Lloyd, from a delegation to the Canadian Environmental Assessment Agency, 2003, www.ceaa.gc.ca/010/0001/0001/0012/0002/0012/s6_e.htm.
27. Commission for Environmental Cooperation (2008) *The North American Mosaic: An Overview of Key Environmental Issues: Industrial Pollution and Waste,* www.cec.org/soe/files/en/SOE_IndustrialPollution_en.pdf.

19 Environmental Health and Risk Management

This is an open-pit asbestos mine in Thetford Mines, Quebec.

Upon successfully completing this chapter, you will be able to

- Identify the major types of environmental health hazards and explain the goals of environmental health
- Describe the types, abundance, distribution, and movement of toxicants in the environment
- Discuss the study of hazards and their effects, including case histories, epidemiology, animal testing, and dose–response analysis
- Assess risk assessment and risk management
- Compare philosophical approaches to risk
- Describe policy and regulation in Canada and internationally

For many years, asbestos was valued and used widely in the built environment for its insulation, fire-resistance, and sound-muffling qualities. Here, asbestos was used as insulation around hot-water pipes.

CENTRAL CASE:
THE ASBESTOS DILEMMA

"Over the past 50 years we have all been unwitting participants in a vast, uncontrolled, worldwide chemistry experiment involving the oceans, air, soils, plants, animals, and human beings."
—UNITED NATIONS ENVIRONMENT PROGRAMME, IN *A GUIDE TO THE STOCKHOLM CONVENTION ON PERSISTENT ORGANIC POLLUTANTS*

"Canada should hang its head in shame. We use tax dollars and foreign missions to host pro-asbestos conferences. We undermine the efforts of other countries to ban asbestos."
—PAT MARTIN, CANADIAN MEMBER OF PARLIAMENT

Some hazardous materials occur naturally but cause problems when they become concentrated in buildings where people are exposed to them on a daily basis. Asbestos is one of these. Therein lies a dilemma for Canada, which hosts some of the world's largest asbestos deposits. **Asbestos** is an industrial and commercial term encompassing several groups of minerals that occur in long, hairlike microscopic fibres (see photo). These fibres give asbestos some very useful characteristics, including thermal insulation, sound muffling, tensile strength, and fire resistance. Because of these qualities, asbestos has been used widely in buildings and in many products, beginning as early as the 1870s, when Canada was the world's largest producer and exporter.[1]

Two groups of minerals produce asbestos fibres—the amphibole group and the serpentine group. Canada produces asbestos only from the serpentine group, which includes only one asbestos-forming mineral, called *chrysotile*. Chrysotile accounts for about 99% of the world's asbestos trade today.[2]

Canada's asbestos deposits are clustered in the Eastern Townships of southern Quebec, where the material is extracted from open-pit mines (see opening

CHAPTER NINETEEN **ENVIRONMENTAL HEALTH AND RISK MANAGEMENT** 589

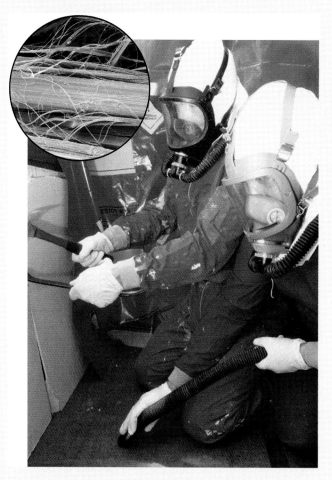

Asbestos has been widely used in insulation and other products. The substance has now been removed from many buildings in which it was used because of human health risks related to exposure. Its removal poses risks as well, however, and removal workers must wear protective clothing and respirators.

photo). People in this area spend their lives living and working on top of chrysotile-bearing rocks. The town of Asbestos is the location of the world's largest asbestos deposit. In Asbestos, Thetford Mines, and other small towns of southeastern Quebec, the asbestos mining and processing industry has been the principal employer.

Unfortunately, the fibrous structure of asbestos minerals that gives it its tensile strength and insulating properties also makes it dangerous when inhaled. When lodged in lung tissue, asbestos induces the body to produce acid to combat it. The acid scars the lung tissue but doesn't dislodge or dissolve the asbestos. Within a few decades the scarred lungs may cease to function, a disorder called *asbestosis*.

As early as the late 1800s, reports of illnesses and fatalities associated with asbestos exposure were reported among British and European miners. In 1912, the link between asbestos and disease in miners was officially denied by the Canadian Department of Labour.[3] Then, in 1955, a medical officer from Thetford Mines reported that 128 of 4000 miners were confirmed to have asbestosis.[4]

In subsequent decades, many scientific studies confirmed the links between asbestos and asbestosis, as well as establishing connections with lung cancer and mesothelioma, a rare cancer of the lining of the chest cavity. One by one, international agencies and governments began to limit and regulate the use of asbestos-containing products.

However, controversy rages about the role of serpentine asbestos (i.e., chrysotile) in causing negative health impacts. There is widespread agreement that exposure to amphibole asbestos is extremely hazardous, but the government of Canada, supported by the asbestos industry, maintains that chrysotile asbestos is significantly less hazardous to human health. To some, this is purely a scientific discussion, requiring further studies to sort out the relative risks. To others, it is a clear case of the politicization of science, with federal and provincial government agencies ignoring and effectively overruling scientific consensus on the matter.

Risks from asbestos exposure are greatest for miners and those involved in the production of asbestos products. In the asbestos processing industry in India and other developing countries, exposure in the workplace with inadequate protection may worsen the problem. There are also concerns for residents of towns like Thetford Mines and Asbestos, where everyday, lifelong exposures have occurred. Asbestos-related health risks arise from inhalation of the fibres, so the risk depends on

- The concentration of fibres in the air
- How long your exposure lasts
- How often you are exposed
- The size of fibres inhaled
- The amount of time since the initial exposure (because some asbestos-related diseases are dormant for many years after the initial exposure)[5]

Because of these risks, asbestos has been removed from many schools and offices (see photo) throughout North America. In some cases, the dangers of exposure from asbestos removal may exceed the dangers of leaving it in place. This can occur, for example, if the asbestos would be exposed, disturbed, or displaced if removal were attempted. Once asbestos enters the air circulation system of a building it can be almost impossible to eliminate. To prevent this, the decision is sometimes taken to leave the material in place.

Today, Canada exports chrysotile asbestos to 60 countries, mainly developing nations in Asia, Africa, and Latin America. The federal and Quebec provincial governments strongly support the asbestos industry, and have spent more than $40 million in the last few years to protect the asbestos market.[6] The value of Canadian asbestos exports in 2006–2007 was more than $1 billion.[7]

Health Canada still does not support the conclusion that exposure to chrysotile asbestos is hazardous to human health or causes cancer, although the World Health Organization and even the Canadian Cancer Society do. Health Canada recommends wearing a respirator when dealing with asbestos-containing products and refers consumers to the *Hazardous Products Act* and the *Canadian Environmental Protection Act*.[8] In 2008, Health Canada began a study of the health risks of exposure to chrysotile asbestos, which presumably will contribute to policy development, once completed.

Canada is a party to the United Nations' Rotterdam Convention, which is intended to enforce the prior informed consent procedure in the context of international trade. *Prior informed consent* is based on the ethical principle that consent to an activity is not legally valid unless the consenting group has been adequately informed and can be shown to have a reasonable understanding of all potential impacts before giving consent. The application of prior informed consent to international trade involves the listing, by consensus of the parties, of certain hazardous substances, with the result that producers would have to obtain prior informed consent from receiving nations before exporting these substances.

Five types of asbestos are listed under the Rotterdam Convention. Chrysotile is not on the list; its inclusion has been blocked numerous times by a small group of nations, led by Canada. In October 2008, at the fourth meeting of the parties to the convention, Canada again blocked the inclusion of chrysotile on the prior informed consent list, in spite of widespread support for its inclusion from health and environmental organizations and experts around the world. According to an editorial in the *Canadian Medical Association Journal* in October 2008, "Canada is the only Western democracy to have consistently opposed international efforts to regulate the global trade in asbestos. And the government of Canada has done so with shameful political manipulation of science."[9]

Environmental Health

Examining the impacts of human-made chemicals on wildlife and people is one aspect of the broad field of environmental health. The study and practice of **environmental health** assesses environmental factors that influence human health and quality of life. Those environmental factors include wholly natural aspects of the environment over which we have little or no control, as well as *anthropogenic* (human-caused) factors for which we ourselves are responsible. Practitioners of environmental health seek to prevent adverse effects on human health and on the ecological systems that are essential to environmental quality and human well-being.

Environmental hazards can be physical, chemical, biological, or cultural

There are innumerable environmental health threats, or hazards, in the world around us. People are exposed to environmental hazards every day. Some—like earthquakes, heat waves, poison ivy, and venomous snakes—occur naturally. Others arise from our exposure to activities and materials that are wholly anthropogenic, such as smoking or airplane flight, or the synthetic chemicals DDT, BPA, and PDBE.

For ease of understanding, we can categorize hazards into four main types: physical, chemical, biological, and cultural. For each of these four types of hazards, there is some amount of risk that we cannot avoid—but there is also some amount of risk that we *can* avoid by taking precautions. Much of environmental health consists of taking steps to minimize the impacts of hazards and the risks of encountering them.

Physical hazards **Physical hazards** can arise from processes that occur naturally and pose risks to human life, health, and property (and to the natural environment, as well). These include earthquakes, volcanic eruptions, fires, floods, blizzards, landslides, hurricanes, and droughts (**FIGURE 19.1**). We can do little to predict the

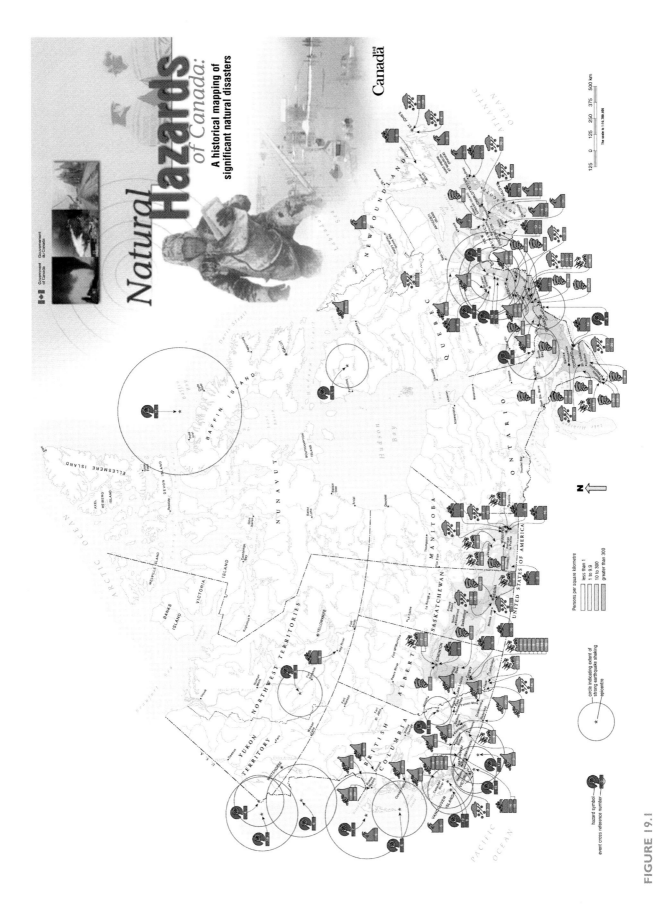

FIGURE 19.1

This map shows the geographic distribution of significant natural disasters in Canada. The map, from Public Safety Canada, was prepared collaboratively with Natural Resources Canada, the Geological Survey of Canada, *The Atlas of Canada*, Environment Canada, and Statistics Canada.

FIGURE 19.2
Environmental health hazards can be divided into four types. The sun's ultraviolet radiation is an example of a physical hazard (a). Excessive exposure increases the risk of skin cancer. Chemical hazards (b) include both artificial and natural chemicals. Much of our exposure comes from household chemical products, such as pesticides. Biological hazards (c) include exposure to organisms that transmit disease. Some mosquitoes are vectors for certain pathogenic microbes, including those that cause malaria. Cultural or lifestyle hazards (d) include decisions we make about how to behave, as well as constraints forced on us by socioeconomic factors. Smoking is a lifestyle choice that raises one's risk of lung cancer and other diseases.

(a) Physical hazard

(b) Chemical hazard

(c) Biological hazard

(d) Cultural hazard

timing of a natural disaster, such as an earthquake, and nothing to prevent one. However, scientists can map geologic faults to determine areas at risk of earthquakes, engineers can design buildings in ways that help them resist damage, and citizens and governments can take steps to prepare for the aftermath of a severe quake. Other types of natural disasters are more amenable to prediction, forecasting, and early warning. Meteorologists have combined their skills and knowledge with sophisticated GIS-based terrain models to improve the forecasting of major storms and floods, for example.

Some regions are naturally prone to certain types of hazards, by virtue of their geology, topography, climate, and other physical characteristics. For example, you can see from **FIGURE 19.1** that the West Coast, southern Ontario, and south-central Quebec are the most vulnerable to earthquakes, whereas the Rockies and West Coast are susceptible to landslides, and the Prairies and southern Ontario to tornadoes. However, some common practices increase our vulnerability to physical hazards. Deforesting slopes makes landslides more likely, for instance, and channelizing rivers makes flooding more likely in some areas while preventing flooding in others.

We can reduce risk from such hazards by improving forestry and flood control practices and by choosing not to build in areas prone to floods, landslides, fires, and coastal waves.

Other physical hazards arise from our exposure to ongoing natural processes, such as ultraviolet (UV) radiation from sunlight (**FIGURE 19.2A**). Excessive exposure to UV radiation damages DNA and has been tied to skin cancer, cataracts, and immune suppression in humans. We can reduce UV exposure and risk by using clothing and sunscreen to shield our skin from intense sunlight.

Chemical hazards **Chemical hazards** include spills and other exposures to the many synthetic chemicals that our society produces, such as disinfectants, pesticides (**FIGURE 19.2B**), and pharmaceuticals. Chemicals produced naturally by organisms also can be hazardous. Following our overview of environmental health, much of this chapter will focus on chemical health hazards and the ways people study and regulate them.

Biological hazards **Biological hazards** result from ecological interactions among organisms (**FIGURE 19.2C**).

weighing the issues 19-1
RISKY BUSINESS

Do you live in a part of Canada that is "naturally" hazardous? Check your region on **FIGURE 19.1** to see if the natural disasters shown on the map match your experience of life in that part of the country. Can you determine which aspects of the natural environment make your region vulnerable to that particular type of hazard? Do you know what kinds of preparations have been undertaken by governments in your province or territory to minimize risk to the public?

When we become sick from a virus, bacterial infection, or other **pathogen**, we are suffering parasitism by other species that are simply fulfilling their ecological roles. This is **infectious disease**, also called *communicable* or *transmissible disease*. Infectious diseases, such as malaria, cholera, tuberculosis, and influenza (flu), all are considered environmental health hazards. As with physical and chemical hazards, it is impossible for us to avoid risk from biological agents completely, but we can take steps to reduce the likelihood of infection.

Cultural hazards Hazards that result from the place we live, our socioeconomic status, occupation, or behavioural choices can be thought of as **cultural hazards** or **lifestyle hazards**. For instance, choosing to smoke cigarettes, or living or working with people who smoke, greatly increases our risk of lung cancer (**FIGURE 19.2D**). Choosing to smoke is a personal behavioural decision, but exposure to secondhand smoke in the home or workplace may not be under one's control. Much the same might be said for drug use, diet and nutrition, crime, and mode of transportation. As advocates of environmental justice argue, such health factors as living in proximity to toxic waste sites or working unprotected with pesticides are often correlated with lack of education and awareness, and with socioeconomic deprivation.

Disease is a major focus of environmental health

Among the hazards people face, disease stands preeminent. Despite all our technological advances, we still find ourselves battling disease, which causes the vast majority of human deaths worldwide (**FIGURE 19.3A**). Many major killers, such as cancer, heart disease, and respiratory disorders, have genetic bases but are also influenced by environmental factors. For instance, whether a person develops asthma is influenced not only by genes but also by environmental conditions. Pollutants from fossil

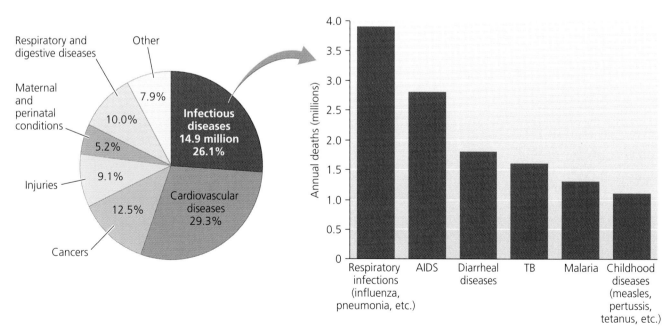

(a) Leading causes of death across the world   (b) Leading causes of death by infectious disease

FIGURE 19.3
Infectious diseases are the second-leading cause of death worldwide, accounting for more than one-quarter of all deaths per year (a). Six types of diseases—respiratory infections, AIDS, diarrhea, tuberculosis (TB), malaria, and childhood diseases, such as measles—account for 80% of all deaths from infectious disease (b). Data are for 2004, from World Health Organization.

fuel combustion worsen asthma, and fewer children raised on farms suffer asthma than children raised in cities, studies have shown. Malnutrition can foster a wide variety of illnesses, as can poverty and poor hygiene. Moreover, lifestyle choices can affect risks of acquiring some noninfectious diseases: Smoking can lead to lung cancer and lack of exercise to heart disease, for example.

Infectious diseases account for 26% of deaths that occur worldwide each year—nearly 15 million people (**FIGURE 19.3B**). Some pathogenic microbes attack us directly, and sometimes infection occurs through a **vector**, an organism that transfers the pathogen to the host. Mosquitoes are common disease vectors; so are rats. Many infectious agents are transported by water, or spend part or all of their life cycle in or near the water. Infectious diseases account for close to half of all deaths in developing countries but for very few deaths in developed nations. This discrepancy is due to differences in hygiene conditions and access to nutrition, health care, and medicine, which are tightly correlated with wealth.

Public health efforts have lessened the impact of infectious disease in developed nations and even have eradicated some diseases. Nevertheless, other diseases—among them tuberculosis, acquired immunodeficiency syndrome (AIDS), and West Nile virus—are increasing (**FIGURE 19.4**). Still others, such as the avian flu that began spreading worldwide in 2005–2006, remain as threats for a possible global epidemic.

Many diseases are spreading because of the mobility we have achieved in our era of globalization; a virus for influenza or SARS (Severe Acute Respiratory Syndrome), which paralyzed Toronto in 2003 after being imported from Hong Kong, can now hop continents in a matter of hours by airplane in its human host. Other diseases, such as tuberculosis and strains of malaria, are evolving resistance to antibiotics, in the same way as pests evolve resistance to pesticides. Tropical diseases, such as malaria, West Nile virus, dengue, cholera, and yellow fever, threaten to expand into the temperate zones with climatic warming. And habitat alteration can affect the abundance, distribution, and movement of certain disease vectors.

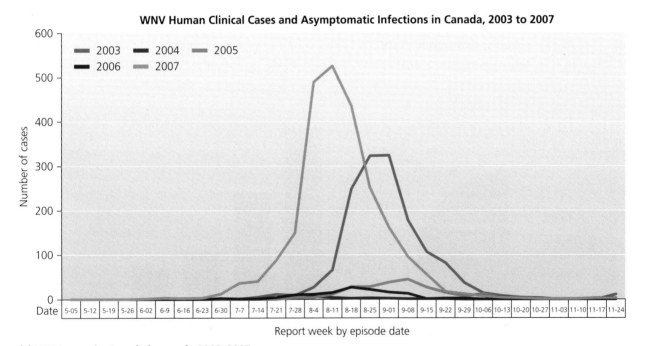

(a) WNV cases in Canada by week, 2002–2007

FIGURE 19.4
Some infectious diseases are on the rise. One example is West Nile virus, which spread rapidly through the Western Hemisphere after first being detected in the New York City area in 1999. Birds are most affected by this mosquito-transmitted virus, but more than 5000 human cases were recorded in Canada by the end of 2007, since 2002 when the first cases were confirmed. The WNV season corresponds to mosquito season, roughly May to October, but the peak of reported cases **(a)** varies according to geographic location and the weather for that year. Manitoba, Alberta, and Ontario reported the highest number of WNV cases in 2006 **(b)**; however, Saskatchewan surpassed all of them in WNV cases in 2007 **(c)**. Only a small proportion of people exposed to WNV (perhaps as low as 1%) contract a serious clinical case. The fatality rate among those who do contract the disease is between 3% and 15%, influenced by age and overall health.[10] Source: Public Health Agency of Canada, *West Nile Virus MONITOR*, www.phac-aspc.gc.ca/wnv-vwn/index-eng.php.

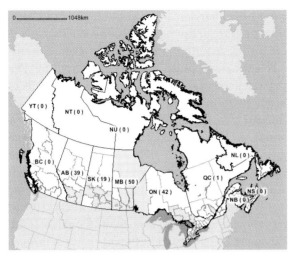

(b) Geographic distribution of WNV cases in Canada, 2006

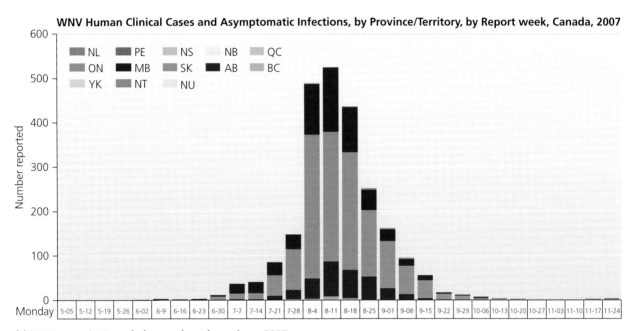

(c) WNV cases in Canada by week and province, 2007

FIGURE 19.4 (CONTINUED)

To predict and prevent infectious disease, environmental health experts deal with the often complicated interrelationships among technology, land use, and ecology. One of the world's leading infectious diseases, malaria (which takes nearly 1.3 million lives each year), provides an example. The microscopic organisms (four species of *Plasmodium*) that cause malaria depend on mosquitoes as a vector. These microorganisms can sexually reproduce only within a mosquito, and it is the mosquito that injects them into a human or other host. Thus the primary mode of malaria control has been to use insecticides, such as DDT, to kill mosquitoes. Large-scale eradication projects involving insecticide use and draining of wetlands have removed malaria from large areas of the temperate world where it used to occur, such as throughout the southern United States. However, types throughout the southern United States. However, types of land use that create pools of standing water in formerly well drained areas can boost mosquito populations, potentially allowing malaria to reinvade.

Environmental health hazards exist indoors as well as outdoors

Outdoor hazards are generally more familiar to us, but we spend most of our lives indoors. Therefore, we must consider the spaces inside our homes and workplaces to be part of our environment and, as such, the sources of potential environmental hazards (Table 19.1). Asbestos, discussed in the Central Case, is one example of a naturally occurring material that is particularly hazardous when concentrated in products in the built environment. Radon is another indoor hazard. **Radon** is a highly toxic radioactive gas that is colourless and undetectable without specialized kits. Radon is naturally occurring; it seeps up from the ground in areas with certain types of bedrock and can build up inside basements and homes with poor air circulation. Exposure to radon over long periods is thought to be the second most important cause of lung cancer, after smoking.

Table 19.1 Selected Environmental Hazards

Air

Smoking and secondhand smoke

Chemicals from automotive exhaust

Chemicals from industrial pollution

Tropospheric ozone

Pesticide drift

Dust and particulate matter

Water

Pesticide and herbicide runoff

Nitrates and fertilizer runoff

Mercury, arsenic, and other heavy metals in groundwater and surface water

Food

Natural toxins

Pesticide and herbicide residues

Indoors

Asbestos

Radon

Lead in paint and pipes

Toxicants in paints, plastics, and consumer products

Smoking—a lifestyle choice—exacerbates the effects of radon exposure. This type of combined factor also makes it complicated to estimate the number of deaths attributable to a single cause, such as radon exposure. The World Health Organization estimates that for a lifetime of exposure at 20 Bq/m^3 of air,[11] the risk of developing cancer is about 3 people per 1000.[12] According to U.S. EPA calculations, if 1000 nonsmokers were exposed in the home over their lifetime to radon at the concentration of the EPA's "actionable" limit (the limit above which you should undertake remedial action, which is 4 pCi/L, or 150 Bq/m^3),[13] about seven of them would be expected to get lung cancer as a result; this is about the same as the risk of dying in a car crash. On the other hand, if those 1000 people were smokers, 62 of them would be expected to get lung cancer as a result; this is about seven times the risk of dying in a car crash.[14] This translates, according to the EPA and the Centers for Disease Control, into about 21 000 deaths per year in the United States from lung cancer caused by radon exposure.

Lead poisoning represents another indoor health hazard. When ingested, **lead**, a heavy metal, can cause damage to the brain, liver, kidney, and stomach; learning problems and behavioural abnormalities; anemia; hearing loss; and even death. It has been suggested that the downfall of ancient Rome was caused in part by chronic lead poisoning.

Today, lead poisoning can result from drinking water that has passed through the lead pipes common in older houses, especially those built before about 1950. Even in newer pipes, lead solder was widely used into the 1980s and is still sold in stores. Lead is perhaps most dangerous, however, to children through its presence in paint. Until 1976, most paints contained lead, and interiors in most houses were painted with lead-based paint. If you strip layers of paint from woodwork in an older home, you are likely to be exposed to airborne lead and should take appropriate precautions. Babies and young children often take an interest in peeling paint from walls and may ingest or inhale some of it. About 5% of children in North America are thought to have been affected by lead poisoning; the number exposed may be much higher, as the health impacts from casual exposure may not be immediately apparent or easily diagnosed.

Lead in paint has been strictly limited in Canada since 1976 (and since 1972 in the United States). The government of Canada has responded to lead paint in imported toys and other items intended for children through legislation, such as the *Children's Jewellery Regulations*, which came into force in 2005. This law made it illegal to import, advertise, or sell children's jewellery or accessories that contain more than 600 mg/kg of total lead, or more than 90 mg/kg of migratable lead.[15] Lead that is *migratable* will leach from the item under certain circumstances; for children's toys, jewellery, and other accessories, the circumstance of greatest concern would occur when the child puts the item into his or her mouth.

In 2007 concerns were heightened when it was discovered that millions of children's toys, pieces of jewellery, baby bibs, and other items manufactured in China still contained lead-based paint. This included many toys sold under the popular Fisher-Price and Mattel brand names. Near the end of 2007, China agreed to limit the lead content of paint and to closely monitor the use of lead-based paints in manufacturing. Health Canada is working on further reducing the risk of children's exposure by limiting the lead contents of consumer products through the *Hazardous Products Act*. These limits will focus mainly on categories of products with which children are most likely to come into close contact, such as cribs and other children's furniture; toys; cooking utensils; candles; and other items.

One recently recognized hazard is a group of chemicals known as *polybrominated diphenyl ethers* (*PBDEs*). These compounds provide fire-retardant properties and are used in a diverse array of consumer products, including computers, televisions, plastics, and furniture. They appear to be released during production and disposal of products and also to evaporate at very slow rates throughout the lifetime of products. These chemicals persist and accumulate in living tissue, and their abundance in the environment and in people is doubling every few years. PBDEs appear to be *endocrine disruptors*, which means that they interfere with hormone levels; lab testing with animals shows them to affect thyroid hormones.

Animal testing also shows limited evidence that PBDEs may affect brain and nervous system development and might possibly cause cancer. Concern about PBDEs rose after a study showed that concentrations in the breast milk of Swedish mothers had increased exponentially from 1972 to 1997. The European Union decided in 2003 to ban PBDEs, and industries in Europe had already begun to phase them out. As a result, concentrations in breast milk of European mothers have fallen substantially. So far, Health Canada does not consider that the amounts of PDBEs consumed by Canadians pose a health risk.

Another common indoor air pollutant is formaldehyde (CH_2O), a colourless gas with a sharp, irritating odour. Health Canada states that some formaldehyde is present in the air in all Canadian homes, in varying concentrations.[16] It is widely used as a disinfectant and preservative. In the home, it can come from a wide variety of plastics and adhesives, textiles (such as permanent press clothing), and wood products—that "new furniture" or "new carpet" smell is often related to the presence of formaldehyde. Cigarette smoke also contributes formaldehyde to the indoor environment.

Toxicology is the study of poisonous substances

Studying the health effects of chemical agents suspected to be harmful, such as PBDEs, is the focus of the field of **toxicology**, the science that examines the effects of poisonous substances on humans and other organisms. Toxicologists assess and compare substances to determine their **toxicity**, the degree of harm a chemical substance can inflict. The concept of toxicity among chemical hazards is analogous to that of *pathogenicity* or *virulence* of the biological hazards that spread infectious disease. Just as types of microbes differ in their ability to cause disease, chemical hazards differ in their capacity to endanger us.

However, any chemical substance may exert negative effects if it is ingested in great enough quantities or if exposure is extensive enough. Conversely, a **toxic** agent (also known as a **toxin** or **toxicant**) in a minute enough quantity may pose no health risk at all. These facts are often summarized in the catchphrase, "The dose makes the poison." In other words, a substance's toxicity depends not only on its chemical identity but also on its quantity.

During the past century, our ability to produce new chemicals has expanded, concentrations of chemical contaminants in the environment have increased, and public concern for health and the environment have grown. These trends have driven the rise of **environmental toxicology**, which deals specifically with toxic substances that come from or are discharged into the environment. Environmental toxicology includes the study of health effects on humans, other animals, and ecosystems, and it represents one approach within the broader scope of environmental health.

Toxicologists generally focus on human health, using other organisms as models and test subjects. In environmental toxicology, animals are also studied out of concern for their welfare and because—like canaries in a coal mine—animals can serve as indicators of health threats that could soon affect humans.

As we review the effects of human-made chemicals throughout this chapter, it is important to keep in mind that artificially produced chemicals have played a crucial role in giving us the standard of living we enjoy today. These chemicals have helped create the industrial agriculture that produces our food, the medical advances that protect our health and prolong our lives, and many of the modern materials and conveniences we use every day. It is important to remember these benefits as we examine some of the unfortunate side effects of these advances, and look for better alternatives.

Toxic Agents in the Environment

The environment contains countless natural chemical substances that may pose health risks. These substances include oil oozing naturally from the ground; radon gas seeping up from bedrock; and toxic chemicals stored or manufactured in the tissues of living organisms—for example, toxins that plants use to ward off herbivores and toxins that insects use to defend themselves from predators. In addition, we are exposed to many synthetic (artificial, or human-made) chemicals.

Synthetic chemicals are ubiquitous in our environment

Synthetic chemicals are all around us in our daily lives (**FIGURE 19.5**). Hundreds of thousands of synthetic chemicals have been manufactured (Table 19.2), about 100 000 are in common use, and many have found their way into soil, air, and water. Synthetic chemicals that have been identified in Canadian lakes and streams include

Table 19.2 Estimated Numbers of Chemicals in Commercial Substances during the 1990s	
Type of chemical	**Estimated number**
Chemicals in commerce	100 000
Industrial chemicals	72 000
New chemicals introduced per year	2 000
Pesticides (21 000 products)	600
Food additives	8 700
Cosmetic ingredients (40 000 products)	7 500
Human pharmaceuticals	3 300

Data from Harrison, P., and F. Pearce. 2000. AAAS Atlas of Population and Environment. Berkeley, CA: University of California Press.

antibiotics, detergents, drugs, steroids, plasticizers, disinfectants, solvents, perfumes, and many other substances. The pesticides we use to kill insects and weeds on farms, lawns, and golf courses are some of the most widespread synthetic chemicals. As a result of all this exposure, every one of us carries traces of numerous industrial chemicals in our body.

This should not *necessarily* be cause for alarm. Not all synthetic chemicals pose health risks, and relatively few are known with certainty to be toxicants. However, of the roughly 100 000 synthetic chemicals on the market today, very few (perhaps as low as 10%) have been thoroughly tested for harmful effects. For the vast majority, we simply do not know what effects, if any, they may have.

Why are there so many synthetic chemicals around us? Let us consider pesticides and herbicides, made widespread by advances in chemistry and production capacity during and following the Second World War. As material prosperity grew in Westernized nations in the decades following the war, people began using pesticides not only for agriculture but also to improve the look of their lawns and golf courses and to fight termites, ants, and other insects inside their homes and offices. Pesticides were viewed as means toward a better quality of life.

It was not until the 1960s that people began to learn about the risks of exposure to pesticides. A key event was the publication of Rachel Carson's 1962 book, *Silent Spring*, which brought the pesticide dichloro-diphenyl-trichloroethane (DDT) to the public's attention.

FIGURE 19.5
Synthetic chemicals, such as those in household products, are everywhere around us in our everyday lives. Some of these compounds may potentially pose environmental or human health risks.

Silent Spring began the public debate over synthetic chemicals

Rachel Carson was a naturalist, author, and government scientist. In *Silent Spring*, she brought together a diverse

FIGURE 19.6
Before the 1960s, the environmental and health effects of potent pesticides, such as DDT, were not widely studied or publicly known. Public areas, such as parks, neighbourhoods, and beaches, were regularly sprayed for insect control without safeguards against excessive human exposure. Here children on a beach are fogged with DDT from a pesticide spray machine in 1945.

collection of scientific studies, medical case histories, and other data that no one had previously synthesized and presented to the general public. Her message was that DDT in particular and artificial pesticides in general were hazardous to people's health, the health of wildlife, and the well-being of ecosystems. Carson wrote at a time when large amounts of pesticides virtually untested for health effects were indiscriminately sprayed over residential neighbourhoods and public areas, on the assumption that the chemicals would do no harm to people (FIGURE 19.6). Most consumers had no idea that the store-bought chemicals they used in their houses, gardens, and fields might be toxic.

Although challenged vigorously by spokespeople for the chemical industry, who attempted to discredit both the author's science and her personal reputation, Carson's book was a bestseller. Carson suffered from cancer as she finished *Silent Spring*, and she lived only briefly after its publication. However, the book helped generate significant social change in views and actions toward the environment. The use of DDT is now illegal in a number of nations and was banned in Canada in 1985. The United States banned DDT for almost all uses (except public emergencies) in 1972 but still manufactures and exports DDT to countries that do use it. Many developing countries with tropical climates use DDT to control human disease vectors, such as mosquitoes that transmit malaria. In these countries, malaria represents a far greater health threat than do the toxic effects of the pesticide.

Toxicants come in several different types

Toxicants, whether they are natural or synthetic, can be classified into different types based on their particular effects on health. The best known are the **carcinogens**, which include chemicals and types of radiation that cause cancer. In cancer, malignant cells grow uncontrollably, creating tumours, damaging the body's functioning, and often leading to death. In our society today, the greatest number of cancer cases is thought to result from carcinogens contained in cigarette smoke. Carcinogens can be difficult to identify because there may be a long lag time between exposure to the agent and the detectable onset of cancer. Historically, much toxicological work focused on carcinogens. Now, however, we know that toxicants can produce many different types of effects, so scientists have many more endpoints, or health impacts, to look for.

Mutagens are chemicals that cause mutations in the DNA of organisms. Although most mutations have little

---

weighing the issues
THE CIRCLE OF POISON
19–2

It has been called the "circle of poison." Although the United States banned the use of DDT in 1972, U.S. companies still manufacture and export the compound to many developing nations. Thus, pesticide-laden food can be imported back into the United States and to other countries, including Canada. How do you feel about this? Is it unethical for one country to sell to others a substance that it has deemed toxic? Are there factors or circumstances that might change the view you take? Compare this with Canada's exportation of asbestos, discussed in the Central Case. What are the similarities and differences?

or no effect, some can lead to severe problems, including cancer and other disorders. If mutations occur in an individual's sperm or egg cells, then the individual's offspring suffer the effects.

Chemicals that cause harm to the unborn are called **teratogens**. Teratogens that affect the development of human embryos in the womb can cause birth defects. One example involves the drug thalidomide, developed in the 1950s as a sleeping pill and to prevent nausea during pregnancy. Tragically, the drug turned out to be a powerful teratogen, and its use caused birth defects in thousands of babies (**FIGURE 19.7A**). Even a single exposure during pregnancy could result in limb deformities and organ defects. Thalidomide was banned in the 1960s once scientists recognized its connection with birth defects. Ironically, today the drug shows promise in treating a wide range of diseases, including Alzheimer's disease, AIDS, and various types of cancer.

The human immune system protects our bodies from disease. Some toxicants weaken the immune system, reducing the body's ability to defend itself against bacteria, viruses, allergy-causing agents, and other attackers. Others, called **allergens**, overactivate the immune system, causing an immune response when one is not necessary. One hypothesis for the increase in asthma in recent years is that allergenic synthetic chemicals are more prevalent in our environment.

Still other chemical toxicants, **neurotoxins**, assault the nervous system. Neurotoxins include various heavy metals, such as lead, mercury, and cadmium, as well as pesticides and some chemical weapons developed for use in war. A famous case of neurotoxin poisoning occurred in Japan, where a chemical factory dumped waste laden with **mercury**, a heavy metal, into Minamata Bay between the 1930s and 1960s. Thousands of people in and around the town on the bay were poisoned by eating fish contaminated with mercury (**FIGURE 19.7B**). People began to show symptoms that included slurred speech, loss of muscle control, disfiguring birth defects, and even death. The company and the government eventually paid about $5000 in compensation to each affected resident.

Endocrine disruption may be widespread

Scientists have recently begun to recognize the importance of **endocrine disruptors**, toxicants that interfere with the endocrine system. The **endocrine system** consists of chemical messengers (**hormones**) that travel through the body. Sent through the bloodstream at extremely low concentrations, these messenger molecules have many vital functions. They stimulate growth, development, and

(a) Thalamide poisioning

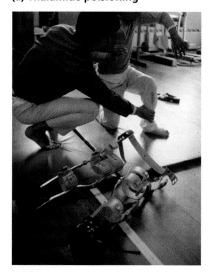

(b) Mercury poisioning

FIGURE 19.7
Two episodes of almost iconic status in the history of synthetic toxicants are the use of thalidomide by pregnant women in the late 1950s to early 1960s, and the mercury poisoning of the residents of Minamata, Japan. The drug thalidomide turned out to be a potent teratogen **(a)**. It was banned in the 1960s but not before causing thousands of birth defects. Butch Lumpkin was an exceptional "thalidomide baby" who learned to overcome his short arms and deformed fingers, becoming a professional tennis instructor. To this day, the name of Minamata, Japan, is synonymous with the mercury poisoning that occurred there between the 1930s and 1960s **(b)**. The horrific impacts of mercury exposure are illustrated in this classic photograph of a victim, taken in 1972 by Eugene Smith.

sexual maturity, and they regulate brain function, appetite, sex drive, and many other aspects of our physiology and behaviour.

Many endocrine disruptors are so similar to hormones in their molecular structure and chemistry that they

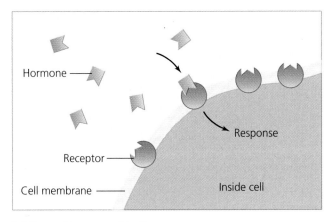

(a) Normal hormone binding

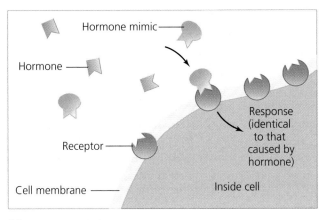

(b) Hormone mimicry

FIGURE 19.8
Many endocrine-disrupting substances mimic the structure of hormone molecules. Like a key similar enough to fit into another key's lock, the hormone mimic binds to a cellular receptor for the hormone, causing the cell to react as though it had encountered the hormone.

"mimic" the hormone by interacting with receptor molecules just as the actual hormone would (**FIGURE 19.8**). Such effects were first noted as far back as the 1960s, with the pesticide DDT. Hormone-disrupting toxicants can affect an animal's endocrine system by blocking the action of hormones or accelerating their breakdown.

One common type of endocrine disruption involves the feminization of male animals, as shown by studies on alligators, fish, frogs, and other organisms. Feminization may be widespread because a number of chemicals appear to mimic the female sex hormone estrogen and bind to estrogen receptors. For example, when biologist Louis Guillette began studying the reproductive biology of alligators in Florida lakes in 1985, he discovered that alligators from one lake in particular, Lake Apopka northwest of Orlando, showed a number of bizarre reproductive problems. Females were having trouble producing viable eggs. Male hatchling alligators had severely depressed levels of the male sex hormone testosterone, and female hatchlings showed greatly elevated levels of the female sex hormone estrogen. The young animals had abnormal gonads, too, and smaller penises. Testes of the males produced sperm at a premature age, and ovaries of the females contained multiple eggs per follicle instead of the expected one egg.

Guillette and his co-workers grew to suspect that environmental contaminants might be somehow responsible for the reproductive abnormalities. Lake Apopka had suffered a major spill of the pesticides dicofol and DDT in 1980, and yearly surveys thereafter showed a precipitous decline in the number of juvenile alligators in the lake. In addition, the lake received high levels of chemical runoff from agriculture and was experiencing eutrophication from nutrient input from fertilizers. Comparing alligators from heavily polluted Lake Apopka with those from cleaner lakes nearby, Guillette's team found that Lake Apopka alligators had abnormally low hatching rates in the years after the pesticide spill. Even as hatching rates recovered in the 1990s, the alligators continued to show aberrant hormone levels and bizarre gonad abnormalities.

Similar problems began cropping up in other lakes that experienced runoff of chemical pesticides. In the lab, researchers found that several contaminants detected in alligator eggs and young could bind to receptors for estrogen and exist in concentrations great enough to cause sex reversal of male embryos. One chemical in particular, atrazine—a widely used herbicide—appeared to disrupt hormones by inducing production of aromatase, an enzyme that converts testosterone to estrogen. In 2003, Guillette reported preliminary findings that nitrate from fertilizer runoff may also act as an endocrine disruptor; when nitrate concentrations in lakes are above the standard for drinking water, juvenile male alligators have smaller penises and 50% lower testosterone levels.

To date, endocrine effects have been found most widely in nonhuman animals, but the results of Guillette and others have raised concern not only for alligator health but also for human health. Endocrine-disrupting chemical contaminants could be affecting people, just as they have affected alligators. For example, some scientists attribute a striking drop in sperm counts among men worldwide to endocrine disruptors. Danish researchers reported in 1992 that the number and motility of sperm in men's semen had declined by 50% since 1938 (**FIGURE 19.9A**). The research involved a review of 61 studies that included 15 000 men from 20 nations on six continents. Subsequent studies by other researchers—including some who set out to disprove the findings—have largely confirmed the results by using other methods and other populations (although there is tremendous geographic variation that remains unexplained).

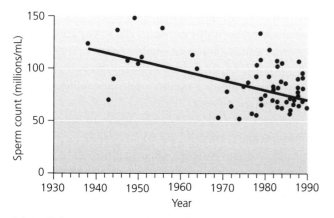

(a) Declining sperm count in men, based on 61 studies

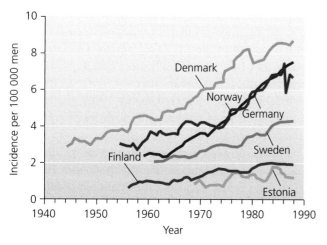

(b) Increasing incidence of testicular cancer

FIGURE 19.9
Research in 1992 synthesized the results of 61 studies of sperm counts in men from various localities since 1938. The data were highly variable but showed a significant decrease in human sperm counts over time (a). Many scientists have hypothesized that this decrease may result from exposure to endocrine-disrupting chemicals in the environment. Some have also hypothesized that endocrine disruptors could be behind an increased incidence of testicular cancer (b). (a) Data from Carlsen, E., et al. 1992. Evidence for decreasing quality of semen during the last 50 years. British Medical Journal 305:609–613, as adapted by Toppari, J., et al. 1996. Male reproductive health and environmental xenoestrogens. Environmental Health Perspectives 104 (Suppl. 4): 741–803. (b) Adami, H. O., et al. 1994. Testicular cancer in nine northern European countries. International Journal of Cancer 50:33–38

Some researchers have also voiced concerns about rising rates of testicular cancer (FIGURE 19.9B), undescended testicles, and genital birth defects in men. Other scientists have proposed that the rise in breast cancer rates (one in nine Canadian women alive today will develop breast cancer) may also be due to hormone disruption, because an excess of estrogen appears to feed tumour development in older women.

Endocrine disruptors can affect more than just the reproductive system. Some impair the brain and nervous system. North American studies have shown neurological problems associated with PCB contamination. In one study, mothers who ate Great Lakes fish contaminated with PCBs had babies with lower birth weights and smaller heads, compared with mothers who did not eat fish. These babies grew into children who showed weak and jerky reflexes and tested poorly in intelligence tests.

Endocrine disruption research has generated debate

Much of the research into hormone disruption has brought about strident debate. This is partly because a great deal of scientific uncertainty is inherent in any young and developing field. Another reason is that negative findings about chemicals pose an economic threat. Our society has invested heavily in some chemicals that now are suspect. One example is asbestos, the chrysotile form of which is still exported by Canada to the developing world, even as its health impacts are under increasing scrutiny, as discussed in the Central Case. Another example is bisphenol-A (see "The Science Behind the Story: Is Bisphenol-A Safe?"). A building block of polycarbonate plastic, bisphenol-A (BPA) occurs in a wide variety of plastic products we use daily, from drink containers to eating utensils to auto parts to CDs and DVDs. It is used in epoxy resins, including those that coat metal food and drink cans. It is even used in dental sealants.

BPA leaches out of many products into water and food, and recent experimental evidence ties it to birth defects and other abnormalities in lab animals. The plastics industry vehemently protests that the chemical is safe, pointing to other research backing its contention. However, by one count through the end of 2006, 151 of the 178 published studies with lab animals report harm from even small amounts of bisphenol-A. Almost without exception, the studies reporting harm received public government funding, whereas those reporting no harm were industry funded. In 2008, Canada became the first country to declare bisphenol-A a dangerous substance. Many retailers responded by pulling BPA-containing products off their store shelves.[17]

Research results with bisphenol-A and mice, and those with atrazine and frogs, have both shown effects at extremely low levels of the chemical. This is also the case with research on other known or purported endocrine disruptors. The apparent reason is that the endocrine system is geared to respond to minute concentrations of substances (normally, hormones in the bloodstream). Because the endocrine system responds to minuscule amounts of chemicals, it is especially vulnerable to effects from environmental contaminants that are dispersed and diluted through the environment and that reach our bodies in very low concentrations.

THE SCIENCE BEHIND THE STORY

Is Bisphenol-A Safe?

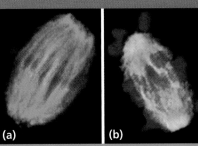

During normal cell division (a), chromosomes align properly. Exposure to bisphenol-A causes abnormal cell division (b), whereby chromosomes scatter and are distributed improperly and unevenly between daughter cells.

Can a compound in everyday products damage the most basic processes necessary for healthy pregnancies and births? Canada has been a leader in the international effort to determine whether bisphenol-A—found in plastic baby bottles, reusable water bottles, pitchers, tableware, storage containers, and other products—might be damaging the health of Canadians, especially children and infants.

Canada's *Chemicals Management Plan* was introduced in 2006 to review the safety of approximately 200 chemicals that have been on the market for many years without adequate scientific knowledge of their impacts. Bisphenol-A, or BPA, is one of these chemicals, and its potential for reproductive impacts made it a high priority for investigation.

The plastics industry produces more than 2.5 billion kilograms of BPA every year—more than a third of a kilogram for every person on the planet. BPA has been known to be an estrogen mimic since the 1930s and more recently has been linked to reproductive abnormalities in mice. Research has shown that BPA can leach out of plastic into water and food when the plastic is treated with extreme heat, acidity, or harsh soap. One study conducted by researchers for the magazine *Consumer Reports* found that BPA seeped out of the plastic walls of heated baby bottles into infant formula.

One early investigation took place almost by accident, at a laboratory at Case Western Reserve University in 1998. The study was initiated when geneticist Patricia Hunt was making a routine check of female lab mice, which included extracting and examining developing eggs from the ovaries. The results on this occasion showed chromosome problems in about 40% of the eggs.

A bit of sleuthing revealed that a lab assistant had mistakenly washed the plastic mouse cages and water bottles with an especially harsh soap. The soap damaged the cages so badly that parts of them melted. The cages were made from plastic containing BPA. To recreate the accidental BPA exposure from the cage-washing incident, Hunt and other researchers washed the polycarbonate cages and water bottles by using varying levels of the harsh soap. They then compared mice kept in damaged cages with plastic water bottles with mice kept in undamaged cages with glass water bottles. The developing eggs of mice exposed to BPA through the damaged plastic showed significant problems during meiosis, the division of chromosomes during egg formation—just as they had in the original incident (see the photo). In contrast, the eggs of mice in the control cages were normal.

In additional tests, three sets of female mice were given oral doses of BPA over three, five, and seven days, and the same abnormalities were observed, although at lower levels (see the graph). The mice given BPA for seven days were most severely affected.

Published in 2003 in the journal *Current Biology*, Hunt's findings set off a wave of concern over the safety of bisphenol-A. Dozens of other studies have since come out; most have shown harmful effects in lab animals. A diversity of reproductive and other effects has been shown, and findings with obese mice have led some to suggest that bisphenol-A may be contributing to the current obesity epidemic. Health Canada's assessment of BPA did not support a link to obesity but did conclude that even low levels of exposure can affect neural development and behaviour if exposure occurs early in life.[18]

The result was a recommendation by Health Canada that BPA be declared a "toxic" substance as defined by the *Canadian Environmental Protection Act* (1999). Health Canada maintains that typical BPA exposures for Canadians are far below harmful levels but urges caution for children and infants under 18 months. The assessment also concluded that BPA can harm fish and other organisms, and confirmed that the chemical has been identified in municipal wastewater.

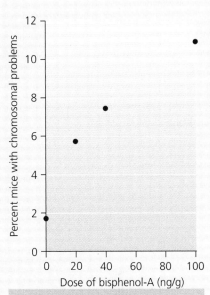

In this dose–response experiment, the percentage of mice showing chromosomal problems during cell division rose with increasing dose of bisphenol-A. Data from Hunt, P.A., et al. 2003. Bisphenol-A exposure causes meiotic aneuploidy in the female mouse. *Current Biology* 13:546–553.

In the United States and Europe, regulators have set safe intake levels for people at doses of 50 ng/g of body weight per day. The acceptable level in Canada has been set at half that amount, 25 ng/g of body weight per day. For comparison, lab tests show concentrations of 5–8 ng/mL of liquid in heated plastic baby bottles, 95% of which still are manufactured with BPA (as of 2007), suggesting significant potential for exposure at harmful levels.[19] Studies in the United States have found BPA in the urine of 95% of people tested.[20] With a chemical that is so widely present in our lives, the scientific and social debate over bisphenol-A seems set to continue building.

Toxicants may concentrate in water

Toxicants are not evenly distributed in the environment, and they move about in specific ways (**FIGURE 19.10**). For instance, water, in the form of runoff, often carries toxicants from large areas of land and concentrates them in small volumes of surface water. If chemicals persist in soil, they can leach into groundwater and contaminate drinking water supplies.

Many chemicals are soluble in water and enter organisms' tissues through drinking or absorption. For this reason, aquatic animals, such as fish, frogs, and stream invertebrates, are effective indicators of pollution. When aquatic organisms become sick, we can take it as an early warning that something is amiss. This is why many scientists see findings that show impacts of low concentrations of pesticides on frogs, fish, and invertebrates as a warning that humans could be next. The contaminants that wash into streams and rivers also

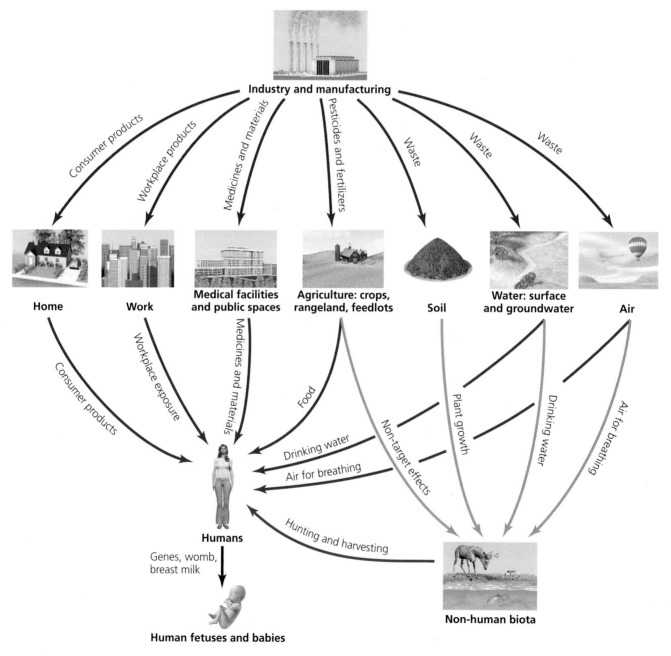

FIGURE 19.10
Synthetic chemicals take many routes in travelling through the environment. Although humans take in only a tiny proportion of these compounds, and although many compounds are harmless, humans—particularly babies—receive small amounts of toxicants from many sources.

flow and seep into the water we drink and drift through the air we breathe.

Airborne toxicants can travel widely

Because many chemical substances can be transported by air (Chapter 13), the toxicological effects of chemical use can occur far from the site of direct chemical use. For instance, airborne transport of pesticides is sometimes termed *pesticide drift*. The Central Valley of California is widely considered the most productive agricultural region in the world. Because it is a naturally arid area, food production depends on intensive use of irrigation, fertilizers, and pesticides. Roughly 143 million kilograms of pesticide active ingredients are used in California each year, mostly in the Central Valley. The region's frequent winds often blow the airborne spray and dust particles containing pesticide residue for long distances. In the Sierra Nevada Mountains, research has associated pesticide drift from the Central Valley with population declines in four species of frogs. Families living in towns in the Central Valley have suffered health impacts, and activists for farm workers maintain that hundreds of thousands of residents are at risk.

Synthetic chemical contaminants are ubiquitous worldwide. Despite being manufactured and applied mainly in the temperate and tropical zones, contaminants appear in substantial quantities in the tissues of Arctic polar bears, Antarctic penguins, and people living in Greenland. Scientists can travel to the most remote and seemingly pristine alpine lakes in British Columbia and find them contaminated with foreign toxicants, such as PCBs. The surprisingly high concentrations in polar regions result from patterns of global atmospheric circulation that move airborne chemicals systematically toward the poles (**FIGURE 19.11**).

Some toxicants persist for a long time

Once toxic agents arrive somewhere, they may degrade quickly and become harmless, or they may remain unaltered and persist for many months, years, or decades. Large numbers of such **persistent** synthetic chemicals exist in our environment today because we design them to persist. The synthetic chemicals used in plastics, for instance, are used precisely because they resist breakdown.

The rate at which chemicals degrade depends on such factors as temperature, moisture, and sun exposure and on how these factors interact with the chemistry of the

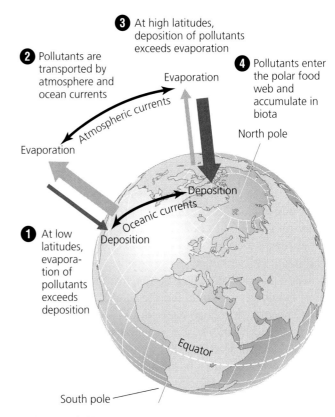

FIGURE 19.11

In a process called *global distillation*, pollutants that evaporate and rise high into the atmosphere at lower latitudes, or are deposited in the ocean, are carried toward the poles by atmospheric currents of air and oceanic currents of water. For this reason, polar organisms take in more than their share of toxicants, despite the fact that relatively few synthetic chemicals are manufactured or used near the poles.

toxicant. Toxicants that persist in the environment have the greatest potential to harm many organisms over long periods of time. A major reason people have been so concerned about toxic chemicals, such as DDT and PCBs, is that they have long persistence times. In contrast, the Bt toxin used in biocontrol and in genetically modified (GM) crops (**FIGURE 8.12**) has a very short persistence time.

Most toxicants eventually degrade into simpler compounds called **breakdown products**. Often these are less harmful than the original substance, but sometimes they are just as toxic as the original chemical, or more so. For instance, DDT breaks down into DDE, a highly persistent and toxic compound in its own right. A large number of chemical breakdown products have not been fully studied.

Toxicants may accumulate over time and up the food chain

Of the toxicants that organisms absorb, breathe, or consume, some are quickly excreted, and some are degraded into harmless breakdown products. However, others remain intact in the body. Toxicants that are fat-soluble or oil-soluble (often organic compounds, such as DDT and DDE) are absorbed and stored in fatty tissues. Others, such as methylmercury, may be stored in muscle tissue. If the rate of ingestion of such toxicants is greater than the rate of excretion, they may build up in an animal over time, in a process termed **bioaccumulation**.

Toxicants that bioaccumulate in the tissues of one organism may be transferred to other organisms when predators consume prey. When one organism consumes another, it takes in any stored toxicants and stores them itself, along with the toxicants it has received from eating other prey. Thus with each step up the food chain, from producer to primary consumer to secondary consumer and so on, concentrations of toxicants can be greatly magnified. This process, called **biomagnification**, occurred most famously with DDT. Top predators, such as birds of prey, ended up with high concentrations of the pesticide because concentrations were magnified as DDT moved from water to algae to plankton to small fish to bigger fish and finally to fish-eating birds (**FIGURE 19.12**).

Biomagnification caused populations of many North American birds of prey, such as the peregrine falcon, the bald eagle, the osprey, and the brown pelican, to decline precipitously from the 1950s to the 1970s. Eventually scientists determined that DDT was causing the birds' eggshells to grow thinner, so that eggs were breaking while in the nest.

All these birds' populations have rebounded since DDT was banned in North America, but such scenarios are by no means a thing of the past. The polar bears of Svalbard Island in Arctic Norway are at the top of the food chain and feed on seals that have biomagnified toxicants. Despite their remote Arctic location, Svalbard Island's polar bears show some of the highest levels of PCB contamination of any wild animals tested, as a result of biomagnification and the process of global distillation shown in **FIGURE 19.11**. The contaminants are likely responsible for the immune suppression, hormone disruption, and high cub mortality that the bears seem to be suffering. Cubs that survive receive PCBs in their mothers' milk, so the contamination persists and accumulates over generations.

Not all toxicants are synthetic

Although we have focused mainly on synthetic chemicals, chemical toxicants also exist naturally in the environment around us and in the foods we eat. We have good reason as citizens and consumers to insist on being informed about risks synthetic chemicals may pose, but it is a mistake to assume that all artificial chemicals are unhealthful and that all natural chemicals are healthful. In fact, the

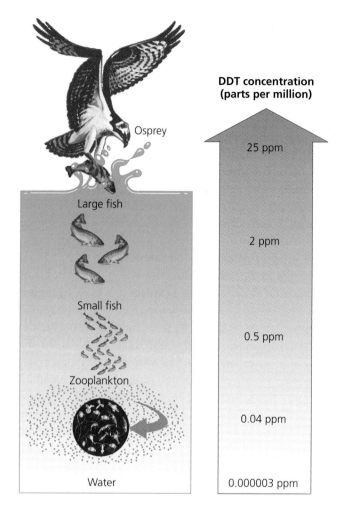

FIGURE 19.12
Fat-soluble compounds, such as DDT, bioaccumulate in the tissues of organisms. As animals at higher trophic levels eat organisms lower on the food chain, their load of toxicants passes up to each consumer. DDT moves from zooplankton through various types of fish, finally becoming highly concentrated in fish-eating birds, such as ospreys.

plants and animals we eat contain many chemicals that can cause us harm. Recall that plants produce toxins to ward off animals that eat them. In domesticating crop plants, we have selected for strains with reduced toxin content, but we have not eliminated these dangers. Furthermore, when we consume animal meat, we take in toxins the animals have ingested from plants or animals they have eaten.

Scientists are actively debating just how much risk natural toxicants pose. Some even maintain that the amounts of synthetic chemicals in our food from pesticide residues are dwarfed by the quantities of natural toxicants. However, others argue that natural toxicants are more readily metabolized and excreted by the body than synthetic ones, that synthetic toxicants persist and accumulate in the environment, and that synthetic chemicals expose people (such as farm workers and factory workers) to risks in ways other than the ingestion of food. What is clear is that more research is required in this area.

Studying the Effects of Hazards

Determining health effects of particular environmental hazards is a challenging job, especially because any given person or organism has a complex history of exposure to many hazards throughout life. Scientists rely on several different methods, including correlative surveys and manipulative experiments.

Wildlife studies use careful observations in the field and lab

When scientists were zeroing in on the impacts of DDT, one key piece of evidence came from museum collections of wild birds' eggs from the decades before synthetic pesticides were manufactured. Eggs from museum collections had measurably thicker shells than the eggs scientists were studying in the field from present-day birds. Scientists have pieced together the puzzle of toxicant effects on alligators by taking measurements from animals in the wild, then doing controlled experiments in the lab to test hypotheses. With frogs and atrazine, scientists first measured toxicological effects in lab experiments, then sought to demonstrate correlations with herbicide use in the wild.

Often the study of wildlife advances in the wake of some conspicuous mortality event. Off the California coast in 1998–2001, populations of sea otters fell noticeably, and many dead otters washed ashore. Field biologists documented the population decline, and specialists went to work in the lab performing autopsies to determine causes of death. The most common cause of death was found to be infection with the parasite *Toxoplasma*, which killed otters directly and also made them vulnerable to shark attack. *Toxoplasma* occurs in the feces of cats, so scientists hypothesized that sewage runoff containing waste from litter boxes was entering the ocean from urban areas and infecting the otters.

Human studies rely on case histories, epidemiology, and animal testing

In studies of human health, we gain much knowledge by studying sickened individuals directly. Medical professionals have long treated victims of poisonings, so the effects of common poisons are well known. Autopsies help us understand what constitutes a lethal dose. This process of observation and analysis of individual patients is known as a *case history* approach. Case histories have advanced our understanding of human illness, but they do not always help us infer the effects of rare hazards, newly manufactured compounds, or chemicals that exist at low environmental concentrations and exert minor long-term effects. Case histories also tell us little about probability and risk, such as how many extra deaths we might expect in a population because of a particular cause.

For such situations, which are common in environmental toxicology, epidemiological studies are necessary. Studies in the field of **epidemiology** involve large-scale comparisons among groups of people, usually contrasting a group known to have been exposed to some hazard and a group that has not. Epidemiologists track the fate of all people in the study, generally for a long period of time (often years or decades), and measure the rate at which deaths, cancers, or other health problems occur in each group. The epidemiologist then analyzes the data, looking for observable differences between the groups, and statistically tests hypotheses accounting for differences. When a group exposed to a hazard shows a significantly greater degree of harm, it suggests that the hazard may be responsible. For example, asbestos miners have been tracked for asbestosis, lung cancer, and mesothelioma rates. Survivors of the Chernobyl disaster in Ukraine have been monitored for thyroid and other cancers. Currently, levels of BPA in humans are being tracked as part of the Canadian Health Measures Survey, with expected completion of the BPA study in late 2009.

This type of human tracking is part of an ongoing approach taken by Health Canada to the management and tracking of toxic chemicals, called *Human Biomonitoring of Environmental Chemical Substances*. The process is akin to a natural experiment, in which the experimenter takes advantage of the presence of groups of subjects made possible by some event or long-term exposure that has already occurred. A slightly different type of natural experiment was conducted by anthropologist Elizabeth Guillette (see "The Science Behind the Story: Pesticides and Child Development in Mexico's Yaqui Valley").

The advantages of epidemiological studies are their realism and their ability to yield relatively accurate predictions about risk. The drawbacks include the need to wait a long time for results and the inability to address future effects of new products just coming to market. In addition, participants in epidemiological studies encounter many factors that affect their health besides the one under study. Epidemiological studies measure a statistical association between a health hazard and an effect, but they do not confirm that the hazard *causes* the effect. It can also be difficult to disentangle the contributions of various factors to any observed negative health impacts. In cases where a number of factors are present, they may interact, affecting results.

THE SCIENCE BEHIND THE STORY

Pesticides and Child Development in Mexico's Yaqui Valley

These Yaqui people live in a farming area in Mexico, where agricultural chemicals are used.

With spindly arms and big, round eyes, one set of pictures resembles the stick figures drawn by young children everywhere. Next to them is another group of drawings, mostly disconnected squiggles and lines. Both sets of pictures are intended to depict people. The main difference between the two groups of young artists: long-term pesticide exposure.

Children's drawings are not a typical tool of toxicology, but anthropologist Elizabeth Guillette wanted to try new methods. Guillette was interested in the effects of pesticides on children. She devised tests to measure childhood development based on techniques from anthropology and medicine. Guillette used the Yaqui Valley of northwestern Mexico as her study site.

The Yaqui Valley is farming country, worked for generations by the indigenous group that gives the region its name. Synthetic pesticides arrived in the area in the 1940s. Some Yaqui embraced the agricultural innovations, spraying their farms in the valley to increase their yields. Yaqui farmers in the surrounding foothills, however, generally chose to bypass the chemicals and to continue following more traditional farming practices. Although differing in farming techniques, Yaqui in the valley and foothills continued to share the same culture, diet, education system, income levels, and family structure.

At the time of the study, in 1994, valley farmers planted crops twice a year, applying pesticides up to 45 times from planting to harvest. A previous study conducted in the valley in 1990 had indicated high levels of multiple pesticides in the breast milk of mothers and in the umbilical cord blood of newborn babies. In contrast, foothill families avoided chemical pesticides in their gardens and homes.

To understand how pesticide exposure affects childhood development, Guillette and fellow researchers studied 50 preschoolers aged four to five—33 from the valley and 17 from the foothills. Each child underwent a half-hour exam, during which researchers showed a red balloon, promising to give the balloon later as a gift, and using the promise to evaluate long-term memory. Each child was put through a series of physical and mental tests:

- Catching a ball from up to 3 m away, to test overall coordination
- Jumping in place for as long as possible, to assess endurance
- Drawing a picture of a person, as a measure of perception
- Repeating a short string of numbers, to test short-term memory
- Dropping raisins into a bottle cap from a height of about 13 cm, to gauge fine-motor skills

The researchers also measured each child's height and weight. When all tests were completed, each child was asked what he or she had been promised and then received the red balloon.

Although the two groups of children were not significantly different in height and weight, they differed markedly in other areas of development. Valley children were far behind the foothill children developmentally in coordination, physical endurance, long-term memory, and fine motor skills:

- From a distance of 3 m, valley children had great difficulty catching the ball.
- Valley children could jump for an average of 52 seconds, compared with 88 seconds for foothill children.
- Most valley children missed the bottle cap when dropping raisins, whereas foothill children dropped them into the caps far more often.

An example is smoking, which complicates the interpretation of epidemiological studies on the linkages between cancer and radon exposure. Smoking acts not only as an additional factor that may cause cancer in subjects but also as a reinforcing or *synergistic* factor.

Manipulative experiments are needed to establish causation. However, subjecting people to massive amounts of toxicants in a lab experiment would clearly be unethical. So researchers have traditionally used other animals as subjects to test toxicity. Foremost among these animal models have been laboratory strains of rats, mice, and other mammals. Because of our shared evolutionary history, the bodies of other mammals function similarly to ours. The extent to which results from animal lab tests apply to humans varies from one study to the next.

Some people feel the use of rats and mice for testing is unethical, but animal testing enables scientific and medical advances that would be impossible or far more difficult otherwise. However, new techniques (with human cell cultures, bacteria, or tissue from chicken eggs) are being devised that may one day replace some live-animal testing.

Dose–response analysis is a mainstay of toxicology

The standard method of testing with lab animals in toxicology is dose–response analysis. Scientists quantify the toxicity of a given substance by measuring how much effect a toxicant produces at different doses or how many animals are affected by different doses of the toxic agent. The **dose** is the amount of toxicant the test animal receives or is exposed to and absorbs; the **response** is the

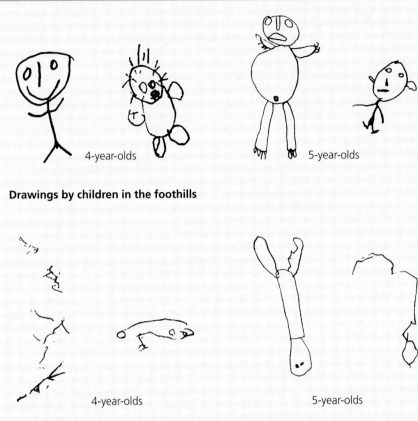

Drawings by children in the foothills

Drawings by children in the valley

Elizabeth Guillette's study in Mexico's Yaqui Valley offers a startling example of apparent neurological effects of pesticide poisoning. Young children from foothills areas where pesticides were not commonly used drew recognizable figures of people. Children the same age from valley areas where pesticides were used heavily in industrialized agriculture could draw only scribbles. Adapted from Guillette, E.A., et al. 1998. *Environmental Health Perspectives* 106:347–353

■ Each group did fairly well repeating numbers, but valley children showed poor long-term memory. At the end of the test, all but one of the foothill children remembered that they had been promised a balloon, and 59% remembered it was red. Of the valley children only 55% remembered they'd be getting a balloon, only 27% remembered the colour of the balloon, and 18% were unable to remember anything about the balloon.

The children's drawings exhibited the most dramatic differences (see the figure). The foothill children drew pictures that looked like people; in contrast, the scribbles of the valley children resembled little that looked like a person. By the standards of developmental medicine, the four- and five-year-old valley children drew at the level of a two-year-old.

Some scientists greeted Guillette's study with skepticism, pointing out that its sample size was too small to be meaningful. Others said that such factors as different parenting styles or unknown health problems could be to blame. Toxicologists argued that without blood or tissue tests on the children, the study results couldn't be tied to agrichemicals. Regardless of these criticisms, Guillette maintains that her findings show that nontraditional study methods are a valid way to track the effects of environmental toxins and that pesticides present a complex long-term risk to human growth and health.

type or magnitude of negative effects the animal exhibits as a result. The response is generally quantified by measuring the proportion of animals exhibiting negative effects. The data are plotted on a graph, with dose on the *x*-axis and response on the *y*-axis (**FIGURE 19.13A**). The resulting curve is called a **dose–response curve**.

Once they have plotted a dose–response curve, toxicologists can calculate a convenient shorthand gauge of a substance's toxicity: the amount of toxicant it takes to kill half the population of study animals used. This lethal dose for 50% of individuals is termed the **lethal-dose-50%** or **LD$_{50}$**. A high LD$_{50}$ indicates low toxicity, and a low LD$_{50}$ indicates high toxicity.

If the experimenter is instead interested in nonlethal health effects, he or she may want to document the level of toxicant at which 50% of a population of test animals is affected in some other way (for instance, what level of toxicant causes 50% of lab mice to lose their hair?). Such a level is called the **effective-dose-50%**, or **ED50**.

Sometimes responses occur only above a certain dose. Such a **threshold dose** (**FIGURE 19.13B**) might be expected if the body's organs can fully metabolize or excrete a toxicant at low doses but become overwhelmed at higher concentrations. It might also occur if cells can repair damage to their DNA only up to a certain point.

Sometimes responses *decrease* with increased dose. For example, the effectiveness of some vitamins, such as vitamin A, increases with increasing dose until a threshold is reached, beyond which the benefits cease to accrue. With still higher doses, however, a turning point is eventually reached at which negative health effects begin to occur as dose is increased (**FIGURE 19.13C**). This type of dose–response curve is typical of materials that are

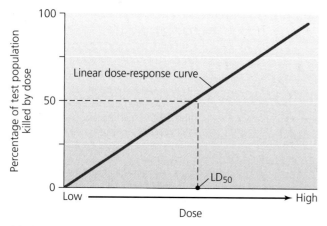

(a) Linear dose-response curve

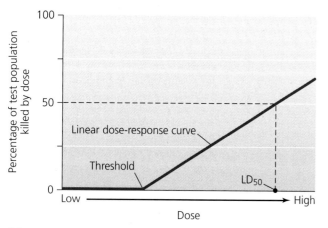

(b) Dose-response curve with threshold

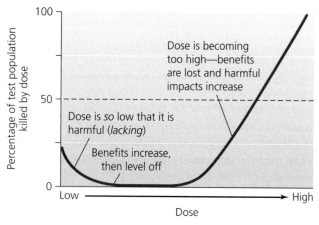

(c) U-shaped dose-response curve

FIGURE 19.13
In a classic linear dose–response curve (a), the percentage of animals killed or otherwise affected by a substance rises with the dose. The point at which 50% of the animals are killed is labelled the lethal-dose-50, or LD_{50}. For some toxic agents, a threshold dose (b) exists, below which doses have no measurable effect. Some substances have dose–response curves shaped like a U. For example (c), at low doses the benefits of vitamin A increase with dose, until a threshold is reached beyond which no further benefits accrue. With further increases, a dose is eventually reached beyond which negative impacts will begin to occur.

harmful to health if they are missing altogether, but toxic if the dose is too high. Toxicologists are finding that some dose–response curves are U-shaped, J-shaped, or shaped like an inverted U. Such counterintuitive curves often occur with endocrine disruptors, likely because the hormone system is geared to function with extremely low concentrations of hormones and so is vulnerable to disruption by toxicants at extremely low concentrations. Inverted dose–response curves present a challenge for policy makers attempting to set safe environmental levels for toxicants. We may have underestimated the dangers of compounds that behave in these ways, because many such chemicals exist in very low concentrations over wide areas.

Knowing the shape of dose–response curves is crucial if one is planning to extrapolate from them to predict responses at doses below those that have been tested. Scientists generally give lab animals much higher doses relative to body mass than humans would receive in the environment. This is so that the response is great enough to be measured and so that differences between the effects of small and large doses are evident. Data from a range of doses help give shape to the dose–response curve. Once the data from animal tests are plotted, scientists extrapolate to estimate the effect of still-lower doses on a hypothetically large population of animals. This way, they can come up with an estimate of, say, what dose causes cancer in 1 mouse in 1 million. A second extrapolation is then required to estimate the effect on humans, with our greater body mass. Because these two extrapolations go beyond the actual data obtained, they introduce uncertainty into the interpretation of what doses are acceptable for humans. As a result, to be on the safe side, regulatory agencies set standards for maximum allowable levels of toxicants that are well below the minimum toxicity levels estimated from lab studies.

Individuals vary in their responses to hazards

Different individuals may respond quite differently to identical exposures to hazards. These differences can be genetically based or can be due to a person's current condition. People in poorer health are often more sensitive to biological and chemical hazards. Sensitivity also can vary with sex, age, and weight. Because of their smaller size and rapidly developing organ systems, fetuses, infants, and young children tend to be much more sensitive to toxicants than are adults. Regulatory agencies, such as Health Canada, set standards for adults and then extrapolate downward for infants and children. However, in many cases the linear extrapolations still do not protect babies adequately. Many critics today contend that despite improvements, regulatory agencies still do not account explicitly enough for risks to fetuses, infants, children, older adults, and the immunocompromised.

Health Canada and Environment Canada collaborate on the **Air Quality Health Index (AQHI)**, a newly standardized national-level index that provides a measure of the level of air pollution at a given time and location, and translates this into the expected health impacts of the pollution. Air pollution is reported on a scale of 1 (very low) to 10+ (very high), based on measured concentrations of ozone (O_3), particulate matter ($PM_{2.5}$, PM_{10}), and nitrogen dioxide (NO_2).[21] All these common air pollutants contribute to the occurrence of respiratory problems, especially in children, older adults, and those with allergies or other kinds of health impairment. AQHI readings of 1 to 3 correspond to a low health risk, 4 to 6 a moderate health risk, 7 to 10 a high health risk, and 10+ a very high health risk.[22] High AQHI readings can trigger policy responses, such as warnings for those who are most susceptible to remain indoors.

The type of exposure can affect the response

The risk posed by a hazard often varies according to whether a person experiences high exposure for short periods of time, known as **acute exposure**, or lower exposure over long periods of time, known as **chronic exposure**. Incidences of acute exposure are easier to recognize, because they often stem from discrete events, such as accidental ingestion, an oil spill, a chemical spill, or a nuclear accident. Lab tests and LD_{50} values generally reflect acute toxicity effects. However, chronic exposure is more common—and more difficult to detect and diagnose. Chronic exposure often affects organs gradually, as when smoking causes lung cancer, or when alcohol abuse induces liver or kidney damage. Pesticide residues on food or low levels of arsenic in drinking water also pose chronic risk. Because of the long time periods involved, relationships between cause and effect may not be readily apparent.

Mixes may be more than the sum of their parts

It is difficult enough to determine the impact of a single hazard on an organism, but the task becomes astronomically more difficult when multiple hazards interact. For instance, chemical substances, when mixed, may act in concert in ways that cannot be predicted from the effects of each in isolation. Mixed toxicants may sum each other's effects, cancel out each other's effects, or multiply each other's effects. Whole new types of impacts may arise when toxicants are mixed together.

Such interactive impacts—those that are more than or different from the simple sum of their constituent effects—are called **synergistic effects**. This creates one of the main problems associated with epidemiological studies that follow groups of people with potential exposure to hazards over a period of time. Most people are routinely exposed to a complex mixture of hazards from the home, the workplace, the environment, and the lifestyle choices they make, including such activities as smoking. In addition to its own negative health impacts, for example, smoking can synergistically reinforce the impacts of other factors. Disentangling the health effects of these various hazards is challenging and always brings a certain degree of uncertainty to the conclusions of epidemiological studies.

Lab experiments with alligators have indicated that DDE can either help cause or inhibit sex reversal, depending on the presence of other chemicals. Mice exposed to a mixture of nitrate, atrazine, and aldicarb have been found to show immune, hormone, and nervous system effects that were not evident from exposure to each of these chemicals alone. Wood frogs in the wild are increasingly suffering limb deformities, apparently the result of being parasitized by trematode flatworms. A frog's being near an agricultural field with pesticide runoff increases the rate of parasitic infection, because, as lab studies have shown, pesticides suppress the frog's immune response, making it more vulnerable to parasites. We saw examples of experiments designed to investigate the synergistic impacts of pesticides combined with other environmental stresses on frogs, in Chapter 9 "The Science Behind the Story: Amphibian Diversity and Decline".

Traditionally, environmental health has tackled effects of single hazards one at a time. In toxicology, the complex experimental designs required to test interactions, and the sheer number of chemical combinations, have meant that single-substance tests have received priority. This approach is changing, but scientists in environmental health and toxicology will never be able to test all possible combinations. There are simply too many hazards in the environment.

Risk Assessment and Risk Management

Policy decisions on whether to ban chemicals or restrict their use generally follow years of rigorous testing for toxicity. Likewise, strategies for combating disease and other health threats are often based on extensive research.

Policy and management decisions reach beyond the scientific results on health to incorporate considerations about economics and ethics. And all too often, they are influenced by political pressure from powerful interests. The steps between the collection and interpretation of scientific data and the formulation of policy involve assessing and managing risk.

Risk is expressed in terms of probability

Exposure to an environmental health threat does not invariably produce some given effect. Rather, it causes some probability of harm, some statistical chance that damage will result. To understand the impact of a health threat, a scientist must know more than just its identity and strength. He or she must also know the chance that an organism will encounter it, the frequency at which the organism may encounter it, the amount of substance or degree of threat to which the organism is exposed, and the organism's sensitivity to the threat. Such factors help determine the overall risk posed.

Risk can be measured in terms of *probability*, a quantitative description of the likelihood of a certain outcome. The probability that some harmful outcome (for instance, injury, death, environmental damage, or economic loss) will result from a given action, event, or substance expresses the **risk** posed by that phenomenon.

Our perception of risk may not match reality

Every action we take and every decision we make involves some element of risk, some (generally small) probability that things will go wrong. We try in everyday life to behave in ways that minimize risk, but our perceptions of risk do not always match statistical reality (**FIGURE 19.14**). People often worry unduly about negligibly small risks but happily engage in other activities that pose high risks. For instance, most people perceive flying in an airplane as a riskier activity than driving a car, but driving a car is statistically far more dangerous.

Psychologists agree that this difference between risk perception and reality stems from the fact that we feel more at risk when we are not controlling a situation and more safe when we are "at the wheel"—regardless of the actual risk involved. When we drive a car, we feel we are in control, even though statistics show we are at greater risk than as a passenger in an airplane. This psychology can account for people's great fear of nuclear power, toxic waste, and pesticide residues on foods—environmental hazards that are invisible or little understood and whose presence in their lives is largely outside their personal control. In contrast, people are more ready to accept and ignore the risks of smoking cigarettes, overeating, and not exercising, all voluntary activities statistically shown to pose far greater risks to health.

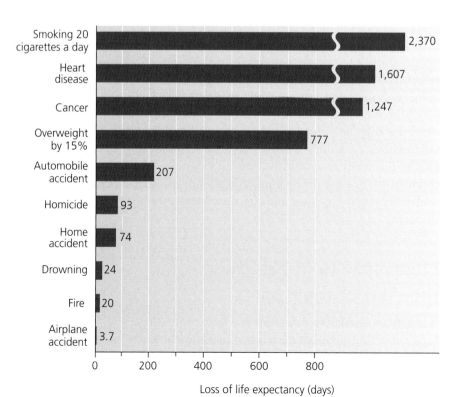

FIGURE 19.14
Our perceptions of risk do not always match the reality of risk. Listed here are several leading causes of death, along with a measure of the risk each poses. Risk is measured in days of lost life expectancy, that is, the number of days of life lost by people suffering the hazard, spread across the entire population—a measure commonly used by insurance companies. By this measure, one common source of anxiety, airplane accidents, poses 20 times less risk than home accidents, more than 50 times less risk than auto accidents, and more than 200 times less risk than being overweight by 15%. Data from Cohen, B. 1991. Catalogue of risks extended and updated. *Health Physics* 61:317–335.

Risk assessment analyzes risk quantitatively

The quantitative measurement of risk and the comparison of risks involved in different activities or substances together are termed **risk assessment**. Risk assessment is a way of identifying and outlining problems. In environmental health, it helps ascertain which substances and activities pose health threats to people or wildlife and which are largely safe.

Assessing risk for a chemical substance involves several steps. The first steps involve the scientific study of toxicity outlined above—determining whether a given substance has toxic effects and, through dose–response analysis, measuring how effects on an organism vary with the degree of toxicant exposure. Subsequent steps involve assessing the individual's or population's likely extent of exposure to the substance, including the frequency of contact, the concentrations likely encountered, and the length of time the substance is expected to be encountered. Risk assessment studies may be performed by scientists associated with the same industries that manufacture toxicants. This, in many people's minds, undermines the objectivity of the process.

Risk management combines science and other social factors

Accurate risk assessment is a vital step toward effective **risk management**, which consists of decisions and strategies to minimize risk (**FIGURE 19.15**). In most developed nations,

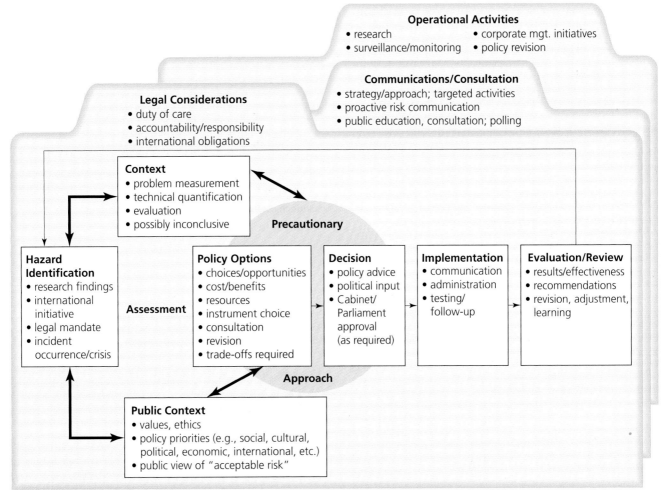

Source: *Risk Management for Canada and Canadians: Report of the ADM Working Group on Risk Management* (PCO), Annex A.

FIGURE 19.15
The first step in addressing the risk of an environmental hazard is quantifying the hazard and its potential impacts. The precautionary approach is at the core of this integrated risk management strategy, from the report of the government of Canada's Working Group on Risk Management (2000). Once science identifies and measures the hazard, then risk management can proceed. In this process, economic accountability, political effectiveness, social responsiveness, and ethical issues and values are considered in light of the scientific data. The consideration of all these types of information is designed to result in policy decisions that minimize the risk of the environmental hazard. Citizen involvement is intended to be a central focus of this framework.
Source: Treasury Board of Canada Secretariat. 2000. *Results for Canadians—A Management Framework for the Government of Canada*. www.tbs-sct.gc.ca/report/res_can/rc-eng.asp.

risk management is handled largely by federal agencies, such as Environment Canada, Health Canada, Public Safety Canada, and Public Health Canada, as well as the Canadian Council of Ministers of the Environment (CCME), which comprises the federal, territorial, and provincial environment ministers. In risk management, scientific assessments of risk are considered in light of economic, social, and political needs and values. The costs and benefits of addressing risk in various ways are assessed with regard to both scientific and nonscientific concerns. Decisions whether to reduce or eliminate risk are then made.

In environmental health and toxicology, comparing costs and benefits can be difficult because the benefits are often economic, whereas the costs often pertain to health. Moreover, economic benefits are generally known, easily quantified, and of a discrete and stable amount, whereas health risks are hard-to-measure probabilities, often involving a very small percentage of people likely to suffer greatly and a large majority likely to experience little effect. When a government agency bans a pesticide, it may mean measurable economic loss for the manufacturer and the farmer, whereas the benefits accrue less predictably over the long term to some percentage of factory workers, farmers, and the public. Because of the lack of equivalence in the way costs and benefits are measured, risk management frequently stirs up debate.

Philosophical and Policy Approaches

Because we cannot know a substance's toxicity until we measure and test it, and because there are so many untested chemicals and combinations, science will never eliminate the many uncertainties that accompany risk assessment. In such a world of uncertainty, there are two basic philosophical approaches to categorizing substances as safe or dangerous.

Two approaches exist for determining safety

One approach is to assume that substances are harmless until shown to be harmful. We might nickname this the *innocent-until-proven-guilty approach*. Because thoroughly testing every existing substance (and combination of substances) for its effects is a hopelessly long, complicated, and expensive pursuit, the innocent-until-proven-guilty approach has the benefit of not slowing down technological innovation and economic advancement. However, it has the disadvantage of putting into wide use some substances that may later turn out to be dangerous.

The other approach is to assume that substances are harmful until they are shown to be harmless. This approach follows the **precautionary principle**. This more cautious approach should enable us to identify troublesome toxicants before they are released into the environment, but it may also significantly impede the pace of technological and economic advance.

These two approaches are actually two ends of a continuum of possible approaches. The two endpoints differ mainly in where they lay the burden of proof—specifically, whether product manufacturers are required to prove safety or whether government, scientists, or citizens are required to prove danger.

---

weighing the issues | THE PRECAUTIONARY PRINCIPLE | 19–3

Industry's critics say chemical manufacturers should bear the burden of proof for the safety of their products before they hit the market. Industry's supporters say that mandating more safety research will hamper the introduction of products that consumers want, increase the price of products as research costs are passed on to consumers, and cause companies to move to nations where standards are more lax. What do you think? Should government follow the precautionary principle and require proof of safety prior to a chemical's introduction to the market?

Philosophical approaches are reflected in policy

Because of the health and environmental consequences of hazardous substances, governments of developed nations have viewed regulatory oversight over manufactured substances as one solution to environmental health threats. One's philosophical approach has implications for policy, affecting what materials are allowed into our environment. Most nations follow a blend of the two approaches, but there is marked variation among countries. At the present time, European nations are embarking on a new policy course that largely incorporates the precautionary principle regarding the regulation of synthetic chemicals.

The precautionary principle has long been a foundation of Canada's approach to environmental management, perhaps more so in Environment Canada than in Health Canada, although Health Canada's move to declare BPA a toxic substance is a clear example of the use of the precautionary principle. The same is not true of the United States, though, where environmental and consumer advocates criticize policy makers and regulatory agencies for largely following the innocent-until-proven-guilty approach. For instance, compounds involved in cosmetics require no FDA review or approval before being sold to the public.

In Canada, several federal agencies are jointly responsible for tracking and regulating synthetic chemicals, emissions, and effluents. Table 19.3 shows the structure of laws that govern the environment and environmental health issues in Canada; the centrepiece (in bold) is the *Canadian Environmental Protection Act* (1999). All the acts ultimately have an impact on the health and welfare of the environment in general and humans specifically, but those in the shaded boxes are the most directly related to human health concerns. The *Fisheries Act* is not

Table 19.3 Some Major Federal Laws Covering Environmental Health Issues

Regulation of products	Regulation of emissions and effluents	Habitat protection, land use, and natural resource management
Canadian Environmental Protection Act, 1999		
		Species at Risk Act
Transportation of Dangerous Goods Act, 1992	Canada Shipping Act, 2001	Wild Animal and Plant Protection and Regulation of International and Interprovincial Trade Act
Food and Drugs Act	Fisheries Act	Migratory Birds Convention Act, 1994
Pest Control Products Act	Canadian Environmental Assessment Act	Canada Wildlife Act
Feeds Act	Canada Water Act	International River Improvements Act
Seeds Act	Indian Act	International Boundary Waters Treaty Act
Fertilizer Act	Arctic Waters Pollution Prevention Act	Oceans Act
Health of Animals Act	Territorial Lands Act	
Hazardous Products Act		
Canada Agricultural Products Act, including regulations such as *Dairy Products Regulations*		

shaded in Table 19.3 because the protection of human health is not its central goal or purpose. Interestingly, however, the *Fisheries Act*, which prohibits the introduction of deleterious substances into any water body that provides habitat for fish, has proven to be one of the most powerful tools available for the protection of water quality in Canada.

Toxicants are also regulated internationally

Nations have sought to address chemical pollution with international treaties. One example is the *Basel Convention [of the United Nations] on the Control of Transboundary Movements of Hazardous Wastes and Their Disposal*. The Basel Convention (mentioned in the context of waste management) is probably the most comprehensive international agreement on hazardous materials. The convention came into effect in 1992 and has 170 parties (i.e., signatory nations that have ratified the agreement). Its central goal is to protect human health and the environment against the adverse effects of hazardous and other wastes.[23]

Another example of a U.N.-moderated vehicle is the *International Code of Conduct on the Distribution and Use of Pesticides*. This is not an international law but a set of voluntary guidelines and best practices for both private and public entities. The World Health Organization also plays a role in establishing these guidelines and monitoring adherence to them. The *Rotterdam Convention on the Prior Informed Consent Procedure for Certain Hazardous Chemicals and Pesticides in International Trade*, discussed in the Central Case, is an example of an international legal instrument that operates according to rules established by consensus of the parties.

The *Stockholm Convention on Persistent Organic Pollutants* (POPs) came into force in 2004 and has been ratified by roughly 140 nations. POPs are toxic chemicals that persist in the environment, bioaccumulate in the food chain, and often can travel long distances. The PCBs and other contaminants found in polar bears are a prime example. Because contaminants often cross international boundaries, an international treaty seemed the best way of dealing fairly with such transboundary pollution. The Stockholm Convention aims first to end the use and release of 12 of the POPs shown to be most dangerous, a group nicknamed the "dirty dozen" (Table 19.4).

It sets guidelines for phasing out these chemicals and encourages transition to safer alternatives. In September 2006, Canada completed the initial assessments of the

Table 19.4 The "Dirty Dozen" Persistent Organic Pollutants (POPs) Targeted by the Stockholm Convention

Toxicant	Type	Description
Aldrin	Pesticide	Kills termites, grasshoppers, corn rootworm, and other soil insects
Chlordane	Pesticide	Kills termites and is a broad-spectrum insecticide on various crops
DDT	Pesticide	Widely used in the past to protect against insect-spread diseases; continues to be applied in several countries to control malaria
Dieldrin	Pesticide	Controls termites and textile pests; also used against insect-borne diseases and insects in agricultural soil
Dioxins	Unintentional by-product	Produced by incomplete combustion and in chemical manufacturing; released in some kinds of metal recycling, pulp and paper bleaching, automobile exhaust, tobacco smoke, and wood and coal smoke
Endrin	Pesticide	Kills insects on cotton and grains; also used against rodents
Furans	Unintentional by-product	Result from the same processes that release dioxins; also are found in commercial mixtures of PCBs
Heptachlor	Pesticide	Kills soil insects, termites, cotton insects, grasshoppers, and mosquitoes
Hexachlorobenzene	Fungicide; unintentional by-product	Kills fungi that affect crops; released during chemical manufacture and from processes that give rise to dioxins and furans
Mirex	Pesticide	Combats ants and termites; also is a fire retardant in plastics, rubber, and electronics
PCBs	Industrial chemical	Used in industry as heat-exchange fluids, in electrical transformers and capacitors, and as additives in paint, sealants, and plastics
Toxaphene	Pesticide	Kills insects on crops; kills ticks and mites on livestock

Data from United Nations Environment Programme (UNEP), 2001.

23 000 chemicals on its Domestic Substances List and derived a list of 2600 medium-priority substances, thereby becoming the first country to prioritize domestic chemicals as per the Stockholm agreement.[24]

The European Union is taking the world's boldest step toward testing and regulating manufactured chemicals. In 2007, the EU's REACH program went into effect (REACH stands for Registration, Evaluation, Authorization, and Restriction of Chemicals). REACH shifts the burden of proof for testing chemical safety from national governments to industry and requires that chemical substances produced or imported in amounts of over one metric ton per year be registered with a new European Chemicals Agency. This agency will evaluate

CANADIAN ENVIRONMENTAL PERSPECTIVES

Wendy Mesley

Journalist Wendy Mesley became an environmental health crusader when she investigated the environmental causes of cancer for CBC's Marketplace.

- **Journalist** with strong interest in environmental issues
- **Cancer survivor**
- **Co-host** of CBC's consumer program, *Marketplace*

Wendy Mesley isn't an obvious choice of a person to profile in a book about environmental science. She isn't a scientist at all, in fact—she is a journalist. She started out in high school answering the phone at CHUM Radio in Toronto and eventually moved on to earn a journalism degree at Ryerson. In recent years she has been a regular correspondent for CBC's evening news centrepiece, *The National*, as well as the anchor for *Sunday Report*. In 2002 she became the co-host of *Marketplace*, CBC's consumer watchdog show, and she has received several Gemini Awards for her work over the years.

Mesley has always taken an intelligent, investigative approach to her reporting. That is probably why, when she was diagnosed with breast cancer in 2005, she began to ask questions—lots of questions, some of them very difficult ones.

Of course she asked, "Why me?" Says Mesley, "I have been a health nut all my adult life, watching what I eat and exercising ... and there is no history of it in my family."[25]

But more than that, she was determined to make some kind of sense about all of the *other* people who are ending up with cancer. Mesley was stunned to discover that the Canadian Cancer Society now predicts that almost one in every two people in Canada will get some form of cancer in their lifetimes. "The biggest shock was ... to find out how common a club it really was. That was the big tragedy. Mine and everyone else's."[26]

Mesley eventually translated her investigative curiosity into a *Marketplace* series called *Chasing the Cancer Answer*. In the series, Mesley addressed what she calls "the elephant in the room"[27] that no one is talking about—the role of the myriad cancer-causing contaminants in the environment.

In the series, Mesley directed some pointed questions at the Canadian Cancer Society. She wondered, for example, why the society places such a huge financial emphasis on behavioural causes, like smoking and diet, while devoting virtually no attention, investigation, or funding to the environmental causes of cancer. She blasted the society for focusing so much attention on finding a cure and so little on prevention.[28] She questioned why there is so little awareness of common cancer-causing agents. For example, after taking birth control pills for most of her adult life, thinking she would only be at risk if she were also a smoker, Mesley was surprised to find out that in 2005 the World Health Organization had reclassified the synthetic hormones used in birth control pills as powerful (Class I) carcinogens.

The response to the series from some groups was immediate and vehement. The Physicians for a Smoke-Free Environment published an open letter to her, suggesting that highlighting environmental contaminants as a cause of cancer may have diverted valuable attention away from the cancer-causing role of human behaviour, especially smoking and secondhand smoke.

Today, more than four years after being diagnosed with malignant tumours in her breast, Wendy Mesley is doing well. She is back at work, cancer free—at least for now—and maintaining the positive attitude that she has had from the very beginning of her cancer journey. She now does her best to avoid exposure to toxic chemicals, such as those found in cleaning products, for example, and is looking forward to spending many more years with her husband and young daughter.

"Talking about things in our environment that are making us sick means you have to do something about them. But it's easier to get out a message that will sell products than to take products off the shelf. There's no money in it."[29] —Wendy Mesley

Thinking About Environmental Perspectives

Wendy Mesley was criticized after her documentary on cancer aired on CBC television, for drawing attention away from the role of smoking and diet in causing cancer. What do you think? Has the Canadian Cancer Society focused too much of its efforts on quitting smoking, eating veggies, wearing sunblock, and living a healthy lifestyle, while ignoring the role of environmental exposure to carcinogenic chemicals? Do we need to consider both the behavioural and the environmental factors in the fight against cancer?

industry research and decide whether the chemical seems safe and should be approved, whether it is unsafe and should be restricted, or whether more testing is needed.

Previous policy had required industry to test chemicals brought to market after 1981 (there are 4300 such chemicals) but required no such testing for chemicals already on the market in 1981 (which number more than 100 000). It is expected that REACH will require 30 000 substances to be registered. The REACH policy also aims to help industry by giving it a single streamlined regulatory system and by exempting it from having to file paperwork on substances under one metric ton. By requiring stricter review of major chemicals already in use, exempting chemicals made only in small amounts, and providing financial incentives for innovating new chemicals, the EU hopes to help European industries research and develop safer new chemicals and products while safeguarding human health and the environment.

The EU expects that roughly 1500 substances may be judged harmful enough to be replaced with safer substances. These include carcinogens, mutagens, teratogens, persistent substances, and chemicals that bioaccumulate. (EU commissioners held off on including endocrine disruptors.) The EU also expects that 1%–2% of substances may cease to be manufactured because their production will no longer be profitable.

In an impacts assessment in 2003, EU commissioners estimated that REACH will cost the chemical industry and chemical users 2.8 billion to 5.2 billion euros over 11 years but that the health benefits to the public would be roughly 50 billion euros over 30 years. Changes in the program since then have made the predicted cost–benefit ratio even better.

Conclusion

International agreements, such as the Rotterdam and Stockholm Conventions, inspire hope that governments will act to protect the world's people, wildlife, and ecosystems from toxic chemicals and other environmental hazards. At the same time, solutions often come more easily when they do not arise from government regulation alone. To many minds, consumer choice, exercised through the market, may be the best way to influence industry's decision making. Consumers of products can make decisions that influence industry when they have full information from scientific research regarding the risks involved. Once scientific results are in, a society's philosophical approach to risk management will determine policy decisions.

Whether the burden of proof is laid at the door of industry or of government, it is important to realize that we will never attain complete scientific knowledge of any risk. Rather, we must make choices based on the information available. Synthetic chemicals have brought us innumerable modern conveniences, a larger food supply, and medical advances that save and extend human lives. Human society would be very different without these chemicals. Yet a safer and happier future, one that safeguards the well-being of both humans and the environment, depends on knowing the risks that some hazards pose and on having means in place to phase out harmful substances and replace them with safer ones.

REVIEWING OBJECTIVES

You should now be able to:

Identify the major types of environmental health hazards and explain the goals of environmental health

- Environmental health seeks to assess and mitigate environmental factors that adversely affect human health and ecological systems.
- Environmental health threats include physical, chemical, biological, and cultural hazards.
- Disease is a major focus of environmental health.
- Environmental hazards exist indoors as well as outdoors.
- Toxicology is the study of poisonous substances.

Describe the types, abundance, distribution, and movement of toxicants in the environment

- Thousands of potentially toxic substances exist around us.
- Toxicants may be of human or natural origin. They include carcinogens, mutagens, teratogens, allergens, neurotoxins, and endocrine disruptors.
- Toxicants may enter and move through surface and groundwater reservoirs, or they may travel long distances through the atmosphere.
- Some chemicals break down very slowly and thus persist in the environment.

- Some organic poisons bioaccumulate and move up the food chain, poisoning consumers at high trophic levels through the process of biomagnification.

Discuss the study of hazards and their effects, including case histories, epidemiology, animal testing, and dose–response analysis

- In case histories, researchers study health problems in individual people.
- Epidemiology involves gathering data from large groups of people over long periods of time and comparing groups with and without exposure to the environmental health threat being assessed.
- In dose–response analysis, scientists measure the response of test animals to various doses of the suspected toxicant.
- Toxicity or strength of response may be influenced by the dose or amount of exposure, the nature of exposure (acute or chronic), individual variation, and synergistic interactions with other hazards.

Assess risk assessment and risk management

- Risk assessment involves quantifying and comparing risks involved in different activities or substances.
- Risk management integrates science with political, social, and economic concerns, in order to design strategies to minimize risk.

Compare philosophical approaches to risk

- An innocent-until-proven-guilty approach assumes that a substance is not harmful unless it is shown to be so.
- A precautionary approach entails assuming that a substance may be harmful unless proven otherwise.

Describe policy and regulation in Canada and internationally

- Health Canada, Environment Canada, and other federal and provincial/territorial agencies are jointly responsible for regulating environmental health threats under Canadian policy.
- Canada and European nations tend to take a more precautionary approach to environmental hazards and the testing of chemical products, as compared with the United States.

TESTING YOUR COMPREHENSION

1. What four major types of health hazards does research in the field of environmental health encompass?
2. In what way is disease the greatest hazard that people face? What kinds of interrelationships must environmental health experts study to learn about how diseases affect human health?
3. Where does most exposure to lead, asbestos, radon, and PBDEs occur? How has each been addressed?
4. When did concern over the effects of pesticides start to emerge? Describe the argument presented by Rachel Carson in *Silent Spring*. What impact did it have on public perception of risk from synthetic chemicals? Is DDT still used?
5. List and describe the six types or general categories of toxicants described in this chapter.
6. How do toxicants travel through the environment, and where are they most likely to be found? What are the life spans of toxic agents? Describe the processes of bioaccumulation and biomagnification.
7. What are epidemiological studies, and how are they most often conducted?
8. Why are animals used in laboratory experiments in toxicology? Explain the dose–response curve. Why is a substance with a high LD_{50} considered safer than one with a low LD_{50}?
9. What factors may affect an individual's response to a toxic substance? Why is chronic exposure to toxic agents often more difficult to measure and diagnose than acute exposure? What are synergistic effects, and why are they difficult to measure and diagnose?
10. How do scientists identify and assess risks from substances or activities that may pose health threats?

SEEKING SOLUTIONS

1. Describe some environmental hazards that you think you may be living with indoors. How do you think you may have been affected by indoor or outdoor environmental hazards in the past? What philosophical

approach do you plan to take in dealing with these toxicants in your own life?

2. Why is it that research on endocrine disruption has spurred so much debate? What steps do you think could be taken to help establish more consensus among scientists, industry, regulators, policy makers, and the public?

3. Do you feel that laboratory-bred animals should be used in experiments in toxicology? Why or why not?

4. Describe differences in the policies of Canada, the United States, and the European Union toward the study and management of the risks of synthetic chemicals. Which approach do you believe is better, and why?

5. **THINK IT THROUGH** You are the parent of two young children, and you want to minimize the environmental health risks to your family as your kids grow up. Name five steps that you could take in your household and in your daily life that would reduce your children's exposure to environmental health hazards.

6. **THINK IT THROUGH** You have just been hired as the office manager for a high-tech startup company that employs bright and motivated young people but is located in an old, dilapidated building. Despite their youth and vigour, the company's employees seem perpetually sick with colds, headaches, respiratory ailments, and other unexplained illnesses. Looking into the building's history, you discover that the water pipes and ventilation system are many decades old, that there have been repeated termite infestations, and that part of the building was remodelled just before your company moved in but there are no records of what was done in the remodelling. Your company has all the latest furniture, computers, and other electronics. Most windows are sealed shut.

You want to figure out what is making the employees sick, and you want to convince your boss to give you a budget to hire professionals to examine the building for hazards. What hazards might you expect? What arguments will you use to convince your employer to fund tests and inspections? What questions will you ask employees to help focus and prioritize any funds you are granted for testing and inspections?

INTERPRETING GRAPHS AND DATA

To minimize exposure to ultraviolet (UV) radiation and thus the risk of skin cancer, people have increased their use of sunscreen lotions in recent decades. Recently, however, some research has shown that chemicals in sunscreens may themselves pose some risk to human health. The compounds commonly used as UV protectants are fat soluble, environmentally persistent, and prone to bioaccumulation. Moreover, they exhibit estrogenic effects in laboratory rats (see Schlumpf et al., 2001, cited in the source note to the graph). Although the benefits of sunscreen use are substantial, the risks are not yet well understood. A hypothetical trade-off between the risk factors of UV exposure and those of sunscreen use illustrates the balancing act of risk management.

1. What dosage of applied sunscreen on the graph corresponds to the greatest risk caused by UV exposure? What dosage corresponds to the greatest risk caused by chemicals in the sunscreen? Which of these two points on the graph is associated with the greater risk?

2. What dosage of applied sunscreen on the graph corresponds to the least risk caused by UV exposure? What dosage corresponds to the least risk caused by

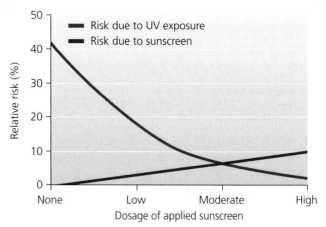

Hypothetical risk distributions for individuals using an estrogenic sunscreen to prevent skin cancer. Schlumpf, M., et al. 2001. *In vitro* and *in vivo* estrogenicity of UV screens. *Environmental Health Perspectives* 109:239–244.

chemicals in the sunscreen? Which of these two points is associated with the greater risk?

3. The total risk to the individual is the sum of the two individual risks. What point on the graph corresponds to the greatest total risk? What sunscreen

dosage corresponds to the least total risk? Based on the data shown here, how much sunscreen would you choose to apply the next time you go to the beach? Is there any other information you would like to know before you change the way you use sunscreen? Can you think of any other cases that illustrate this sort of trade-off between dose-dependent risk factors?

CALCULATING FOOTPRINTS

In 2001, the population of Canada was 30 million, the population of the United States was approximately 285 million, and the world's population totalled 6.16 billion. In that same year, pesticide use in the United States was approximately 545 million kilograms of active ingredient, and world pesticide use totalled 2295 million kilograms of active ingredient. Statistics on total agricultural pesticide use in Canada are very difficult to obtain; the government of Canada has admitted this to the Food and Agricultural Organization of the United Nations and has pledged to improve on the monitoring and data collection of pesticide use. Pesticides include hundreds of chemicals used as insecticides, fungicides, herbicides, rodenticides, repellents, and disinfectants. They are used by farmers, governments, industries, and individuals. Pesticide use for agricultural purposes in Canada tends to be heavily weighted toward the use of insecticides, rather than the other categories of pesticides. Agricultural pesticide use in Canada tends to be lower, per hectare, than in much of the rest of the world, possibly even half as much as in the United States (roughly 0.9 kg/ha, compared with 1.6 kg/ha in the United States as of 2001), probably mainly because of shorter growing seasons. On this basis, we can calculate that the total pesticide use in Canada in 2001 was probably on the order of 29 million kilograms.

In the table, calculate your share of pesticide use as a Canadian citizen in 2001 and the amount used by (or on behalf of) the average citizen of the world and the average U.S. citizen.

	Annual pesticide use (kilograms of active ingredient)
Canada (total)	29 million
Canada (per capita)	
United States (total)	545 million
United States (per capita)	1.9
World (total)	2295 million
World (per capita)	

1. What is the ratio of your annual pesticide use to the world's per capita average?
2. Refer to the Calculating Footprints section in Chapter 1, and find the ecological footprints of the average Canadian citizen, average U.S. citizen, and average world citizen. Compare the ratio of pesticide usage with the ratio of the overall ecological footprints. What are the differences, and how would you account for them?
3. Does the figure for per capita pesticide use for you as a Canadian citizen seem reasonable for you personally? Why or why not? Do you find this figure alarming or of little concern? What else would you like to know to assess the risk associated with this level of pesticide use?

TAKE IT FURTHER

Go to www.myenvironmentplace.ca where you will find

- Suggested answers to end-of-chapter questions
- Quizzes, animations, and flashcards to help you study
- *Research Navigator*™ database of credible and reliable sources to assist you with your research projects
- Tutorials to help you master how to interpret graphs
- Current news articles that link the topics that you study to case studies from your region and around the world
- **ECO Occupational Profiles:** If you found this chapter especially interesting, you might want to learn more about the following jobs by visiting the Occupational Profiles website of the Environmental

Careers Organization. Go to www.eco.ca and check out the following careers:
- Ecotoxicologist
- Emergency manager
- Environmental epidemiologist
- Environmental health officer
- Environmental monitoring technician
- Hazardous materials specialist
- Laboratory assessor
- Microbiologist
- Occupational hygienist

CHAPTER ENDNOTES

1. Greenberg, M. (2008) The defence of chrysotile, 1912–2007. *International Journal of Occupational Environmental Health*, Vol. 14, No. 1, pp. 57–66.
2. NRCAN, *Main Minerals and Metals Produced in Canada*, www.nrcan.gc.ca/mms/scho-ecol/main_e.htm#asbestos.
3. Greenberg, M. (2008) The defence of chrysotile, 1912–2007. *International Journal of Occupational Environmental Health*, Vol. 14, No. 1, pp. 57–66.
4. Greenberg, M. (2008) The defence of chrysotile, 1912–2007. *International Journal of Occupational Environmental Health*, Vol. 14, No. 1, pp. 57–66.
5. Health Canada, *It's Your Health: Health Risks of Asbestos*, www.hc-sc.gc.ca/hl-vs/iyh-vsv/environ/asbestos-amiante-eng.php.
6. Greenberg, M. (2008) The defence of chrysotile, 1912–2007. *International Journal of Occupational Environmental Health*, Vol. 14, No. 1, pp. 57–66.
7. Canada Asbestos Update, released Nov. 15, 2007, in response to a question posed in Parliament by MP Pat Martin, www.bwint.org/default.asp?index=1265&Language=EN.
8. Health Canada, *It's Your Health: Health Risks of Asbestos*, www.hc-sc.gc.ca/hl-vs/iyh-vsv/environ/asbestos-amiante-eng.php.
9. Attaran, Amir, David R. Boyd, and Matthew B. Stanbrook (2008) Asbestos mortality: a Canadian export, *CMAJ*, Vol. 179, No. 9, pp. 871–872.
10. West Nile Virus, *Canadian Family Physician*, www.cfpc.ca/cfp/2005/Jun/vol51-jun-cme-1.asp.
11. There are several units of measurement in common use for radon, and they are not directly convertible. WL (Working Level) is a measure of the concentration of the daughter products of radon; pCi (picoCurie, usually measured per litre of air) and Bq (Becquerel, usually measured per cubic metre of air) are measures of the number of radioactive disintegrations or transformations per second. 1 Bq is equivalent to 27 pCi. 1 WL is approximately equivalent to 200 pCi/L.
12. World Health Organization, 2004, *Radon and Health Information Sheet*, www.who.int/phe/radiation/en/2004Radon.pdf.
13. Health Canada's actionable limit is currently 200 Bq/m^3, or 5.4 pCi/L.
14. U.S. Environmental Protection Agency (2007) *A Citizen's Guide to Radon: The Guide to Protecting Yourself and Your Family from Radon*, www.epa.gov/radon/pubs/citguide.html.
15. Health Canada Advises Canadians About Children's Potential Exposure to Lead, September 6, 2007, www.hc-sc.gc.ca/ahc-asc/media/advisories-avis/_2007/2007_114-eng.php.
16. Health Canada, *It's Your Health: Formaldehyde and Indoor Air*, www.hc-sc.gc.ca/hl-vs/iyh-vsv/environ/formaldehyde-eng.php.
17. CBC News, *Health Canada, Bisphenol A: Announcement Imminent*, April 2008, www.cbc.ca/consumer/story/2008/04/15/bisphenol.html.
18. Government of Canada, *Chemical Substances: An Eco-Action Initiative, Questions and Answers for Action on Bisphenol A Under the Chemicals Management Plan*, updated June 2008, www.chemicalsubstanceschimiques.gc.ca/faq/bisphenol_a_qa-qr_e.html#14.
19. Environmental Defence, *Toxic Nation: Toxic Baby Bottles in Canada*, 2008, www.toxicnation.ca/files/toxicnation/report/ToxicBabyBottleReport.pdf.
20. Environmental Defence, *Toxic Nation: Toxic Baby Bottles in Canada*, 2008, www.toxicnation.ca/files/toxicnation/report/ToxicBabyBottleReport.pdf
21. Environment Canada, *About the Air Quality Health Index*, www.ec.gc.ca/cas-aqhi/default.asp?Lang=En&n=065BE995-1.
22. Environment Canada, *About the Air Quality Health Index*, www.ec.gc.ca/cas-aqhi/default.asp?Lang=En&n=065BE995-1.
23. Basel Convention on the Transboundary Movement of Hazardous Wastes and Their Disposal, www.basel.int.

24. Chatterjee, Rhitu (2008) Hunting for persistent chemicals that might pollute the Arctic, *Environmental Science and Technology,* Vol. 42, No. 14, p. 5034, http://pubs.acs.org/subscribe/journals/esthag-w/2008/jun/science/rc_pops.html.
25. Harvey, Robin (2008) Wendy Mesley: Cancer diagnosis compelled journalist to search for answers, *Being Well Magazine,* Winter, www.yorkregion.com/Health/beingwell%20magazine/article/67934.
26. Harvey, Robin (2008) Wendy Mesley: Cancer diagnosis compelled journalist to search for answers, *Being Well Magazine,* Winter, www.yorkregion.com/Health/beingwell%20magazine/article/67934.
27. Yates, Dana (2007) Moving On, *Ryerson University Alumni Magazine,* January, pp. 28–29, http://danayates.ca/samples/Wendy%20Mesley%20-%20winter%202007.pdf.
28. Harvey, Robin (2008) Wendy Mesley: Cancer diagnosis compelled journalist to search for answers, *Being Well Magazine,* Winter, www.yorkregion.com/Health/beingwell%20magazine/article/67934.
29. Yates, Dana (2007) Moving On, *Ryerson University Alumni Magazine,* January, pp. 28–29, http://danayates.ca/samples/Wendy%20Mesley%20-%20winter%202007.pdf.

The Urban Environment: Creating Liveable Cities

20

Vancouver is said (by some) to be the world's most liveable city.

Upon successfully completing this chapter, you will be able to

- Describe the scale of urbanization
- Assess urban and suburban sprawl
- Outline city and regional planning and land use strategies
- Evaluate transportation options
- Describe the roles of urban parks
- Analyze environmental impacts and advantages of urban centres
- Assess the pursuit of sustainable cities

The roof of the new addition to the Vancouver Convention and Exhibition Centre will be the largest green roof in Canada.

CENTRAL CASE:
PLANNING FOR LONG-TERM URBAN SUSTAINABILITY IN VANCOUVER

"We have planning boards. We have zoning regulations. We have urban growth boundaries and 'smart growth' and sprawl conferences. And we still have sprawl."
—ENVIRONMENTAL SCIENTIST DONELLA MEADOWS, 1999

"No one will find out what works for our cities by looking at [them] ... you've got to get out and walk."
—URBAN ACTIVIST JANE JACOBS

According to The Economist, Vancouver is "the world's most liveable city."[1] As it turns out, Vancouver has some credentials with which to back this up.

In 2002–2003, nine cities from Canada, Japan, Russia, Germany, India, Argentina, China, the United States, and Mexico participated in the International Sustainable Urban Systems Design competition. The competition was designed to encourage cities to plan for the transition from a fossil fuel–based economy to one based on alternative energy sources. The assigned task was to develop 100-year plans for urban sustainability.

The project that formed the basis for the Canadian entry was called citiesPLUS (Planning for Long-term Urban Sustainability). Focusing on the Greater Vancouver area, the project involved sustainable transportation and urban greening, among other things. The Canadian team consisted of the Sheltair Group, a Vancouver-based urban planning and resource management company, the International Centre for Sustainable Cities (ICSC), Greater Vancouver Regional District (GVRD), and the Liu Institute for Global Issues at UBC.

Some of the design teams came from industrialized nations; others came from the developing world. The projects were all different in scope, focus, and approach, but they all came to similar conclusions: Within 30 to 50 years, without major changes, the energy, food, water, transportation, and other systems that support urban centres will begin to break down.[2]

The citiesPLUS team produced Canada's first 100-year sustainability plan for a metropolitan area.[3] At the competition's final event in Tokyo in 2003, the team was awarded the Grand Prize. According to ICSC, one of the project partners, part of the reason the Canadian entry won is because it was participatory and collaborative, involving community residents, elected officials, academics, and professionals.[4]

Western Economic Diversification Canada (WEDC), a federal ministry with the mandate to promote the development and diversification of Western Canada, agreed with the consensus of teams at the International Sustainable Urban Systems Design competition. In a discussion paper, the Vancouver Working Group of WEDC concluded that "business as usual" is not an acceptable scenario and that changes need to be put in place now if we hope to keep the urban environment liveable. The Working Group pointed to a projection of population distribution in the Vancouver area to the year 2040 (see maps), concluding that "unless urban systems are transformed, expected population growth in the region will lead to massive loss in natural amenities, and will undermine the region's prosperity and quality of life, long before 100 years from now."[5]

One outgrowth of the citiesPLUS project was the Sustainable Cities PLUS Network, a resource-sharing network whose members are cities and regions engaged in long-term planning for urban sustainability. The purpose of the network is to transform the way cities approach planning decisions by placing them in a much longer timeframe—three generations.[6] Members share experiences, expertise, tools, and learning about long-term planning with other member cities. The network gives member communities an opportunity to learn from other cities' successes and setbacks.[7] There are now 35 member cities from all over the world, including Calgary, Canmore, Edmonton, Halifax Regional Municipality, Iqaluit, Metro Vancouver, Niagara Region, Ottawa, Regina, Saint John, Vernon/Okanagan Indian Band, and Whistler.[8]

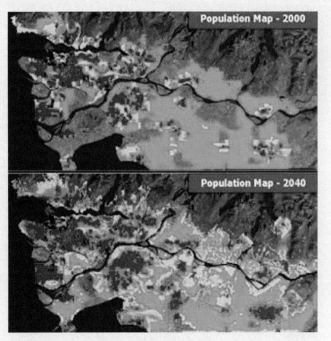

These maps, from the Vancouver Working Group Discussion Paper, compare the distribution of population in the Greater Vancouver area in 2000 with a "business-as-usual" projection to 2040. The paper concluded that unless the urban system is transformed, population spread will continue to encroach on green space, ultimately degrading both prosperity and quality of life. Source: Vancouver Working Group of Western Economic Diversification Canada Discussion Paper, *The Liveable City*, www.wd.gc.ca/rpts/research/livable/2a_e.asp.

Meanwhile, Vancouver's city planners continue to try to incorporate the principles of sustainability into new urban developments. An interesting example (see photo) is the new addition to the Vancouver Convention and Exhibition Centre, which will undertake its first duties during the 2010 Olympics. The roof of the building will be the largest green roof in Canada. The rest of the addition is being built to meet high environmental standards, with efficient lighting, heating and cooling by thermal exchange with seawater, and an onsite desalinization plant.[9]

Will citiesPLUS, the PLUS Network, and other similar urban sustainability initiatives change the world dramatically? Probably not; they likely will not even lead to any of the member cities' laying claim to the title of "Sustainable City." But taking steps to move the world's cities toward sustainability is more urgent than ever before, as the world's population becomes increasingly urbanized. If we want the cities of the future to provide a safe, healthy, liveable environment, then initiatives like those of citiesPLUS will need to become more common, more effective, and more widely adopted.

Our Urbanizing World

In 2008, a milestone was reached when 3.3 billion people—half the world's population—were city dwellers. This number will likely increase to 5 billion or more within the next two decades.[10] This shift from the countryside into towns and cities, or **urbanization**, is arguably the single greatest change our society has undergone since its transition from a nomadic hunter–gatherer lifestyle to a sedentary agricultural one. Even though we commonly associate cities with noise, smog, garbage, and other negative environmental impacts, the truth is that cities offer many features that are beneficial—perhaps even crucial—to the environment. We will explore those features in this chapter.

Industrialization has driven the move to urban centres

Since 1950, the world's urban population has more than quadrupled. Urban populations are growing for two reasons: (1) the human population overall is growing (Chapter 6), and (2) more people are moving from farms to cities than are moving from cities to farms.

The shift of population from country to city began long ago. Agricultural harvests that produced surplus food freed a proportion of citizens from farm life and allowed the rise of specialized manufacturing professions, class structure, political hierarchies, and urban centres. The earliest Canadian towns were administrative, military, and/or trading centres, and were located for strategic purposes and for ease of access to waterways (**FIGURE 20.1A**). The process of urbanization in Canada thus began in 1608, with the founding of Quebec City.

In Canada, as elsewhere, the establishment of transportation routes for the shipment of raw materials and products was a central characteristic of the next stage of urban development (**FIGURE 20.1B**). Starting in the early 1800s, the Industrial Revolution spawned technological innovations that created jobs and opportunities in urban centres for people who were no longer needed on farms. Industrialization and urbanization bred further technological advances that increased production efficiencies, both on the farm and in the city.

Worldwide, the proportion of the population that is urban rose from 30% half a century ago to 49% in 2005. Between 1950 and 2005, the global urban population increased by 2.65% each year, whereas the rural population rose only by 1.12% annually. From 2005 to 2030, the United Nations projects that the urban population will grow by 1.78% annually, whereas the rural population will decline by 0.03% each year.

(a)

(b)

FIGURE 20.1
The city of Montreal **(a)**, originally an Iroquoian settlement, flourished as an early trade route because of its strategic port location at the confluence of the St. Lawrence and Ottawa rivers. Although there was a settlement at Fort Calgary prior to 1883 **(b)**, the actual town site was laid out by the Canadian Pacific Railway Company and was incorporated in 1884 as the first town in Alberta.

Trends differ between developed and developing nations, however (**FIGURE 20.2**). In 1851, only 13% of Canadian citizens were urban dwellers; the percentage passed 50% shortly after 1920.[11] Following the industrialization of the 1800s and early 1900s, urban centres in both Canada and the United States continued to grow very rapidly through the early twentieth century and into the 1960s. Urban population growth then slowed dramatically—even reversing, in some cases—starting in the 1970s. In the United States, in particular, this led to stagnancy and deterioration in the urban cores of many larger cities.

More recently there has been some recovery in city centres, but urban population growth remains slow throughout North America. This is partly because four of every five people already live in cities, towns, and suburbs. The proportion of Canada's population that is classified as "urban" now stands at more than 80% (**FIGURE 20.3**).

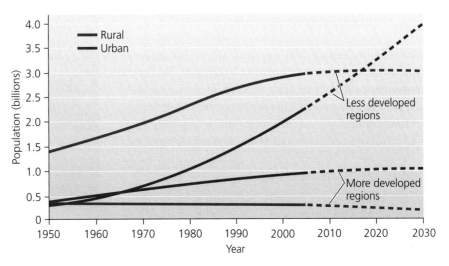

FIGURE 20.2
In developing countries today, urban populations are growing quickly, whereas rural populations are levelling off and may soon begin to decline. Developed countries are already largely urbanized, so in these countries urban populations are growing more slowly, whereas rural populations are falling. Solid lines in the graph indicate past data, and dashed lines indicate projections of future trends. Data from United Nations Population Division (UNPD). 2007. *World Urbanization Prospects: The 2005 Revision.* New York: UNPD.

Although the population of Canada is as urbanized as that of the United States (also 80% urban), Canada has never been as heavily *sub*urban as the United States. Most U.S. urban dwellers reside in suburbs; fully 50% of the American population today is suburban. In Canada, 72% of the urban population lives in cities, with the remaining 28% in suburban areas.[12] In other words, only 22% of the total Canadian population is suburban. Statistics Canada defines **suburbs** as areas that are peripheral to—and strongly "influenced by"—cities or Census Metropolitan Areas. (For more complete and precise definitions of this and other terms related to population and urbanization, you can consult the *2006 Census Dictionary*, which is made available online by Statistics Canada.[13])

Today Ontario, Quebec, and British Columbia are the most urbanized provinces in Canada. About 45% of Canada's population is housed in just six large urban centres, each with a population of more than 1 million: Toronto, Calgary, Edmonton, Montreal, Vancouver, and Ottawa-Gatineau. Canada's population is also highly concentrated in the south; according to the 2006 census, two out of three Canadians live within 100 km of the southern border, occupying only about 4% of the land area of the country.

Today's developing nations, where many people still reside on farms, are now urbanizing rapidly. In China, India, Nigeria, and other nations, rural people are streaming to cities in search of jobs and urban lifestyles. U.N. demographers estimate that virtually all the world's population growth over the next 25 years will be absorbed by urban areas of developing nations.

Today's urban centres are unprecedented in scale and rate of growth

Cities in themselves are nothing novel. Urban centres where population, cultural activities, and political power are concentrated have been part of human culture for several thousand years. Ancient Mediterranean civilizations, the great Chinese dynasties, and the Mayan and Incan empires all featured sophisticated and powerful urban centres.

What is new in the urban setting of today is the sheer scale of today's metropolitan areas. Human population growth (Chapter 6) has placed greater numbers of people in towns and cities than ever before. Today, 25 cities are home to more than 10 million residents (Table 20.1). The greater metropolitan area of the world's most populous city, Tokyo, Japan (defined as the Tokyo-Osaka corridor, an almost continuous urban belt), is home to 35 million people. North America's largest metropolises, Mexico City and New York City, each hold about 19 million. However, the majority of urban dwellers in North America live in much smaller cities.

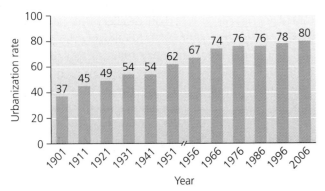

Sources: Statistics Canada, censuses of population, 1901 to 2006

FIGURE 20.3
Canada's population has become increasingly urbanized, as shown in this graph based on census data from 1921 through 2006. Today, more than 80% of Canada's population lives in cities, towns, and suburbs. Statistics Canada. 2006. Portrait of the Canadian Population in 2006: Subprovincial population dynamics, *Census Analytical Series*, www12.statcan.ca/english/census06/analysis/popdwell/Subprov1.cfm.

Table 20.1 Metropolitan "Mega-Cities" With 10 Million Inhabitants or More, as of 2008*	
City, country	Millions of people
Tokyo, Japan	35.2
New York, United States	19.7
Seoul, South Korea	19.5
São Paulo, Brazil	18.7
Jakarta, Indonesia	18.2
Mexico City, Mexico	18.1
Osaka-Kobe-Kyoto, Japan	17.3
Mumbai, India	17.0
Manila, Philippines	16.3
Cairo, Egypt	15.8
Delhi, India	15.3
Moscow, Russian Federation	14.0
Los Angeles Area, United States	13.8
Shanghai, China	13.6
Kolkata, India	13.2
Dhaka, Bangladesh	12.4
Buenos Aires, Argentina	12.0
Karachi, Pakistan	11.6
Beijing, China	11.3
Tianjin, China	11.0
Rio de Janeiro, Brazil	10.9
Lagos, Nigeria	10.9
Istanbul, Turkey	10.5
Guangzhou, China	10.5
Paris, France	10.4

* Note that the definition of "city" can vary dramatically from one analysis to another, depending on where and how one draws the boundaries of the city. Furthermore, reliable, recent census data is not available for some cities in the developing world.
Source: United Nations Population Division. 2007. World Urbanization Prospects: The 2005 Revision. New York: UNPD; and Demographia: World Urban Areas (World Agglomerations) and Population Projections, 4th Comprehensive Edition, June 2008.

The **"mega-cities"** of 10 million people or more that now characterize much of the developing world are of a scale that is unfamiliar to most North Americans. For example, by using the broadest possible definition of "metropolitan area," which bundles everything from Hamilton to Oshawa into one large Toronto-centred economic area called the Golden Horseshoe, the population is still only just over 5.6 million. (This is based on a definition that considers the continuity of urbanized or built-up areas, economic integration, and the location of the workforce in a region.) Some of the large urban centres of the developing world now host homeless and transient populations that are larger than the entire population of the Golden Horseshoe area.

Most fast-growing cities today are in the developing world. Some of this growth and rural–urban migration is occurring because industrialization is decreasing the need for farm labour and promoting commerce and jobs in cities. Sadly, another reason is that wars, conflict, and ecological degradation are driving millions of people out of the countryside and into cities. Cities like Mumbai (India), Lagos (Nigeria), and Cairo (Egypt) are growing in population even more quickly than North American cities did prior to the 1970s.

Many cities in the developing world are growing at rates of 3%–5% per year and even higher. These rates have been matched by some North American cities during their fastest growth, but only for short periods. (For example, Calgary—currently Canada's fastest-growing city—grew by 3.6% between July 2005 and June 2006.) Recall how exponential growth works; on a sustained basis, even these seemingly small growth rates imply doubling times of just 14 to 23 years for many cities in the developing world.

All too often, this is happening without the economic growth and infrastructure development needed to match the population growth. As a result, many of these cities are facing overcrowding, pollution, and poverty. Nearly three of every four governments of developing nations have enacted policies to discourage the movement of people from the countryside into cities.

weighing the issues

DEFINING HOMELESSNESS   20–1

Definitions of "homelessness" vary dramatically. Are you homeless if your lack of shelter is "voluntary"? What if you are unable to maintain a job and home because of a mental illness or an addiction? Are you homeless if you live in wholly inadequate housing? (What if the "housing" is a cardboard box or a shipping crate?) Are you homeless if you are a transient worker, sleeping each night, for example, at the construction site where you will work the next day?

Are there homeless people in the town or city where you are living? If you think not, look again. Many homeless people are "invisible," either because they prefer to keep a low profile, or because we choose not to see them.

Can you think of some ways in which environmental change could cause homelessness, perhaps on a temporary basis? How do environmental change and environmental hazards make life difficult for homeless people? Try to think beyond the borders of the familiar, and consider what life is like for homeless people in the developing world.

Various factors influence the geography of urban areas

Location is vitally important for urban centres. Environmental variables, such as climate, topography, and the configuration of waterways, go a long way toward determining whether a small settlement will become a large city—and successful cities historically have been located in places that give them economic advantages. Think of any major city, and chances are that it is situated along a major river, seacoast, railroad, or highway—some corridor for trade that has driven economic growth (see **FIGURE 20.1**).

Many well-located cities have acted as linchpins in trading networks, funnelling resources from agricultural regions, processing them, manufacturing products, and shipping those products to other markets. Montreal's location at the confluence of two rivers cemented its strategic location for trade and commerce as early as the 1700s. For Calgary and other Western towns, it took the arrival of the main line of the Canadian Pacific Railway in the late 1800s to initiate major growth.

All cities, from ancient times to the present day, have supported themselves by drawing in resources from outlying rural areas through trade, persuasion, or conquest. In turn, cities have historically influenced how people use land in surrounding areas. Although city life and country life may seem very different, cities and the rural regions surrounding them have always been linked by tight economic relationships.

Spatial patterns of urbanization can change with changing times. Today, several factors are causing population centres to decentralize in developed nations. For one thing, people now are globally interconnected to an unprecedented degree. Being located on a river or seacoast is no longer as vital to a city's success in our age of global commerce, jet travel, diplomacy, television, cell phones, and the internet. Globalization has connected distant societies, and businesses and individuals can more easily communicate from locations away from major city centres. Moreover, fossil fuels have enabled the outward spread of cities. By easing long-distance transport, fossil fuels and the proliferation of highway networks have made it easier to commute into and out of cities and to import and export resources, goods, and waste. Such factors have enabled a shift of population from cities to suburbs, particularly in North America.

People have moved to suburbs

By the mid-twentieth century, many cities in North America had accumulated more people than these cities had jobs to offer. Unemployment rose, and crowded inner-city areas began to suffer increasing poverty and crime. As inner cities declined economically from the 1960s onward, many affluent city dwellers chose to move outward to the cleaner, less crowded, and more parklike suburban communities beginning to surround the cities (**FIGURE 20.4**). These people were pursuing more space, better economic opportunities, cheaper real estate, less crime, and better schools for their children.

The development of highway systems and reliable transportation options allowed millions of people to commute by car to their downtown workplaces from new homes in suburban "bedroom communities." The exodus to the suburbs, in turn, hastened the economic decline of central cities, especially in the United States; for example, in the 1960s and 1970s, Chicago's population declined to 80% of its peak because so many residents moved to the suburbs.

In most ways, suburbs have delivered the qualities people sought in them. The wide spacing of houses, with each house on its own plot of land, gives families room and privacy. However, by allotting more space to each person, suburban growth has spread human impact across the landscape. Natural areas have disappeared as housing developments are constructed. We have built extensive

weighing the issues

WHAT MADE YOUR CITY? 20–2

Consider the town or city in which you live, or the major urban centre located nearest you. Why do you think it developed into an urban area? What physical, social, or environmental factors may have initiated settlement of the site and aided the growth of the town during different periods?

FIGURE 20.4
Suburbs, as illustrated in this aerial photograph of Markham, Ontario, have grown because they allow each person to have a bit more individual space and privacy, and a cleaner environment than in the inner city. In recent years, suburbs may have become the victim of their own success, as crowding and sprawl have made the suburbs more and more like the cities their inhabitants sought to avoid in the first place.

road networks to ease travel, but suburbanites now find themselves needing to drive everywhere. They commute longer distances to work and spend more time in congested traffic. The expanding rings of suburbs surrounding cities have grown larger than the cities themselves, and towns are running into one another. These aspects of suburban growth have inspired a new term: *sprawl*.

Sprawl

The term *sprawl* has become laden with meanings and connotes different things to different people. To some, sprawl is esthetically ugly, environmentally harmful, and economically inefficient. To others, it is the collective outgrowth of reasonable individual desires and decisions in a world of growing human population. We can begin our discussion by giving **sprawl** a simple, nonjudgmental definition: the spread of low-density urban or suburban development outward from an urban centre.

Today's urban areas spread outward

As urban and suburban areas have grown in population, they have also grown spatially. This growth is obvious from maps and satellite images of rapidly spreading cities, such as Las Vegas (**FIGURE 20.5**).

Because suburban growth entails allotting more space per person than does city dwelling, in most cases this outward spatial growth across the landscape has outpaced the growth in numbers of people. According to the 2006 census, the urbanized land area in Canada increased by 2% between 2001 and 2006, whereas the urban population only increased by 0.3% over the same time period. Between 1950 and 1990, the population of 58 major U.S. metropolitan areas rose by 80%, but the land area they covered rose by 305%.

Many researchers thus define *sprawl* simply as the physical spread of development at a rate greater than the rate of population growth. This phenomenon of the geographic spread of population can be seen in maps of Vancouver and Calgary (**FIGURE 20.6**). These maps, based on 2006 census data, show the highest population growth rates between 2001 and 2006 in areas peripheral to the urban cores.

Several types of standard development approaches can result in sprawl (**FIGURE 20.7**). However, to determine whether the growth patterns shown in **FIGURE 20.6** actually represent "sprawl" would require examining the nature and density of development within the high-growth areas. The high rate of growth might be indicative of the phenomenon of "in-fill," in which suburban areas fill in and become more densely populated as people move out of areas that have already reached their optimal densities. Some of the questions a researcher might ask

(a) Las Vegas, Nevada, 1972

(b) Las Vegas, Nevada, 2002

FIGURE 20.5
Satellite images show the type of rapid urban and suburban expansion that many people have dubbed *sprawl*. Las Vegas, Nevada, is currently one of the fastest-growing cities in North America. Between 1972 **(a)** and 2002 **(b)**, the population increased more than fivefold, and the quantity of developed area rose more than threefold.

FIGURE 20.6
This map shows areas of growth in the Vancouver area **(a)** and the Calgary area **(b)** between 2001 and 2006. In general, in both cases, areas peripheral to the urban "core" have experienced the fastest growth rates, shown in dark purple.

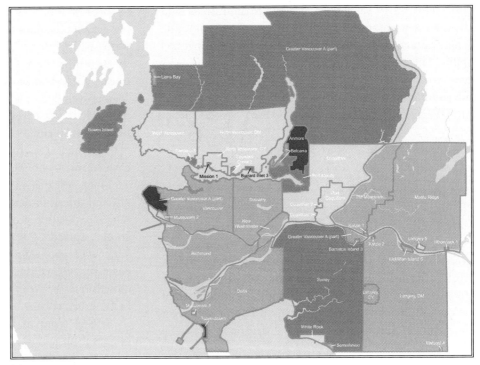

(a) Vancouver

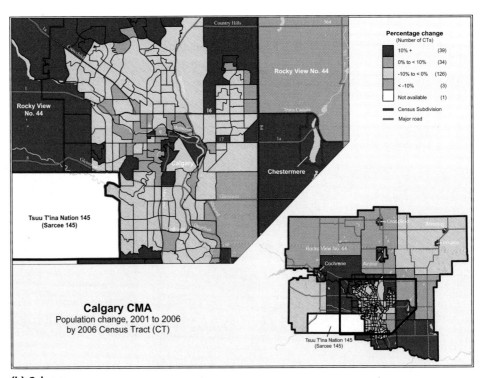

(b) Calgary

about the maps in **FIGURE 20.6** and about sprawl in general include the following:

- Are the peripheral growth areas characterized by low-density development throughout, or are there pockets of higher-density development?
- Does the spread of urbanized land area relative to the growth in urban population necessarily imply that each resident is occupying more land?
- If low-density development is occurring and spreading, will this lead necessarily to a qualitative experience of "sprawl" by residents of urban areas?
- What characterizes the daily experience of sprawl—longer commuting times, more air pollution, less parking?
- What, if any, will be the environmental impacts?
- What are the causes of sprawl?

(a) Uncentred commercial strip development

(b) Low-density single-use development

(c) Scattered, or leap-frog, development

(d) Sparse street network

FIGURE 20.7
Several conventional approaches to suburban development can result in sprawl. In uncentred commercial strip development **(a)**, businesses are arrayed in a long strip along a roadway, and no attempt is made to create a centralized community with easy access for consumers. In low-density, single-use residential development **(b)**, homes are located on large lots in residential tracts far away from commercial amenities. In scattered or leapfrog development **(c)**, developments are created at great distances from a city centre and are not integrated. In developments with a sparse street network **(d)**, roads are far enough apart that moderate-sized areas go undeveloped, but not far enough apart for these areas to function as natural areas or sites for recreation. All these development approaches necessitate frequent automobile use.

- What are the factors that control the rate and distribution of sprawl in different cities?
- And finally, can—or should—anything be done to curb sprawl?

Sprawl has several causes

From the preceding discussion it should be clear that there are two main components of sprawl. One is human population growth—there are simply more people alive each year. The other is per capita land consumption—each person takes up more land. In a basic sense, the amount of sprawl is a function of the number of people added to an area times the amount of land the average person occupies.

In Chapter 6, we discussed reasons for human population growth. As for the increase in per capita land consumption, there are numerous reasons. Highways, widely available automobiles, and technologies, such as telecommunications and the internet, have fostered movement away from city centres because they free businesses from dependence on the centralized infrastructure a major city provides, and they give workers greater flexibility to live wherever they desire.

The primary reasons for greater per capita land consumption, however, are that most people simply like having some space and privacy, and dislike congestion. Furthermore, in the consumption-oriented North American lifestyle that promotes bigger houses, bigger cars, and bigger TVs, having more space to house one's possessions becomes important. Unless there are overriding economic or social disadvantages, most people prefer living in a less congested, more spacious, more affluent community.

Economists, politicians, and city boosters have almost universally encouraged the unbridled spatial expansion of cities and suburbs. The conventional assumption has been that growth is good and that attracting business, industry,

THE SCIENCE BEHIND THE STORY

Measuring the Causes and Impacts of Sprawl

Dr. Reid Ewing, Rutgers University.

Critics of sprawl have blamed it for so many societal ills that a person can be made to feel guilty just for having been born in the suburbs. But what does scientific research tell us are the actual causes and consequences of sprawl?

University of Toronto researcher Matthew Turner and colleagues used satellite imagery to investigate the factors that drive and control sprawl. They found that sprawl varies greatly from one metropolitan area to another, even during periods when the overall extent of sprawl does not change. They also determined that a variety of factors are statistically correlated with sprawl, including temperate climate, decentralized employment, early public transport infrastructure, and the availability of unincorporated land in the urban fringe.[14]

When Reid Ewing of Rutgers University and his team set out to measure the impacts of sprawl, they discovered that it was difficult even to agree on a definition of the term. Surveying the literature, Ewing's team found that researchers using different criteria ranked cities in very different ways. Turner and colleagues also concluded that many U.S. municipalities widely thought to suffer from sprawl, in fact, did not.[15] For example, they found that Miami—thought by many people to be a classic example of urban sprawl—is about a third more compact than New York, which is conventionally regarded as compact and dense as a result of historical growth patterns.

So Ewing's team tried to define *sprawl* as simply as possible, without mixing the consequences of sprawl into the definition. They decided that sprawl occurs when the spread of development across the landscape far outpaces population growth. They devised four criteria by which to rank metropolitan areas. Sprawling cities would show the following:

1. Low residential density
2. Distant separation of homes, employment, shopping, and schools
3. Lack of "centredness," that is, lack of activity in community centres and downtown areas
4. Street networks that make many streets hard to access

For each criterion, the researchers measured multiple factors (22 variables in all), analyzed them, and arrived at a cumulative index of sprawl. They correlated their sprawl scores with a number of transportation-related variables. They

---

and residents will unfailingly increase a community's economic well-being, political power, and cultural influence. Today, however, this assumption is increasingly being challenged. As the negative effects of sprawl on citizens' lifestyles accumulate, growing numbers of people have begun to question the mantra that all growth is good.

What is wrong with sprawl?

Sprawl means different things to different people. To some, the word evokes strip malls, homogenous commercial development, and tracts of cookie-cutter houses encroaching on farmland and ranchland. It may suggest traffic jams, destruction of wildlife habitat, and loss of natural land around cities.

However, for other people, sprawl represents the collective result of choices made by millions of well-meaning individuals trying to make a better life for themselves and their families. In this view, those who decry sprawl are being elitist and fail to appreciate the good things about suburban life. Let us try, then, to leave the emotional debate aside and assess the impacts of sprawl (see "The Science Behind the Story: Measuring the Causes and Impacts of Sprawl").

Transportation Most studies show that sprawl constrains transportation options, essentially forcing people to drive cars. These constraints include the need to own a vehicle and to drive it most places, the need to drive greater distances or to spend more time in vehicles, a lack of mass transit options, and more traffic accidents. An automobile-oriented culture also increases dependence on nonrenewable petroleum, with its attendant economic and environmental consequences.

Pollution Sprawl's effects on transportation give rise, in turn, to increased pollution. Carbon dioxide emissions from vehicles cause global climate change (Chapter 14), while nitrogen- and sulphur-containing air pollutants contribute to tropospheric ozone, urban smog, and acid precipitation. Motor oil and road salt from roads and parking lots pollute waterways, posing risks to ecosystems and human health. Runoff of polluted water from paved areas is estimated to be about 16 times as great as from naturally vegetated areas.

found that people in the most sprawling metros owned more cars and drove an average of 9 km farther per day than people in the least sprawling metros. They also determined that people in the most sprawling metros used public transit far less and suffered 67% more traffic fatalities than those in the least sprawling metros.

Strikingly, the study found no significant difference in commute time from home to work for people of sprawling versus less sprawling metro areas. Critics of sprawl have long blamed it for traffic congestion and commute delays (see photo). Advocates of suburban spread argue that more streets ease commutes and that regions can sprawl their way out of congestion. The Ewing team's results seem to suggest that each side may have a point.

None of these results prove that sprawl *causes* these impacts, because statistical correlation alone does not imply causation. However, taken together, the results suggest that spatial patterns of development may influence people's transportation options, impacts, and behaviour. The Ewing team's results were published in 2003 in the *Transportation Research Record* and in a 2002 report published by Smart Growth America.[16]

Ewing and his colleagues are examining other aspects of sprawl, including health impacts. Because studies have found that residents of sprawling areas depend more on cars and walk less, the researchers hypothesized that they would find more obesity and poorer health in people living in sprawling areas. Indeed, they found that people from sprawling metros are on average heavier for their height and show increased instances of high blood pressure.

However, Matthew Turner and UBC colleague Lawrence Frank have challenged the conclusion that sprawl causes people to gain weight, although they do agree on the basic fact that people who live in sprawling neighbourhoods are heavier overall. The researchers undertook a study in which they followed nearly 6000 individuals over a period of six years, during which time about 80% of the study participants changed residences. The researchers found that even though people moved to different types of neighbourhoods, the change in environment didn't necessarily affect their weight.[17] These findings appear to contradict the now

Traffic congestion is one of the most widely recognized impacts of sprawl, but the traffic implications of sprawl require further investigation.

"conventional" idea that sprawl causes people to become lazier and therefore fatter.

As more studies on the impacts of sprawl accumulate, we will have a better idea of the causes and consequences of urban and suburban development patterns, and the benefits and costs of our choices in urban design.

Health Aside from the health impacts of pollution, some research suggests that sprawl promotes physical inactivity because driving cars largely takes the place of walking during daily errands. Physical inactivity increases obesity and high blood pressure, which can in turn lead to other ailments. A 2003 study found that people from the most sprawling areas weigh 2.7 kg more for their height than people from the least-sprawling metropolitan areas, although a cause-and-effect relationship has yet to be established.

Land use The spread of low-density development means that more land is developed while less is left as forests, fields, farmland, or ranchland. At the current rate of development around the city of Toronto, for example, 1070 km² of rural land will be newly urbanized by 2021. Most of this land—about 92% of it—is prime agricultural land. When farmland, forests, and grasslands are converted to suburban development, their ecosystem services are diminished or lost. These services include resource production, esthetic beauty, habitat for wildlife, cleansing of water, places for recreation, and many others.

Economics Sprawl drains tax dollars from existing communities and funnels them into infrastructure for new development on the fringes of those communities. Money that could be spent maintaining and improving downtown centres is instead spent on extending the road system, water and sewer system, electricity grid, telephone lines, police and fire service, schools, and libraries. Advocates for sprawling development argue that taxes on new development eventually pay back the investment made in infrastructure, but studies have found that in most cases taxpayers continue to subsidize new development if municipalities do not pass infrastructure costs along to developers.

SPRAWL NEAR YOU — 20–3

Is there sprawl in the area where you live? Are you bothered by it, or not? Has development in your area had any of the impacts described above? Do you think your city or town should use its resources to encourage outward growth, or should densification be encouraged?

Creating Liveable Cities

To respond to the challenges that urban and suburban sprawl present, architects, planners, developers, and policy makers across North America today are trying to restore the vitality of city centres and to plan and manage how urbanizing areas develop.

City and regional planning are means for creating liveable urban areas

Planning is the professional pursuit that attempts to design cities and other human settlements so as to maximize their efficiency, functionality, and beauty. City planners advise policy makers on development options, transportation needs, public parks, and other matters.

City planning in North America came into its own in the early twentieth century (**FIGURE 20.8**) and grew in importance as urban populations expanded and wealthier residents fled to the suburbs. In today's world of sprawling metropolitan areas, regional planning has become just as important. Regional planners deal with the same issues as city planners, but they work on broader geographic scales and must coordinate their work with multiple municipal governments. In some places, regional planning has been institutionalized in formal governmental bodies; the Greater Vancouver Regional District is an example of regional entity.

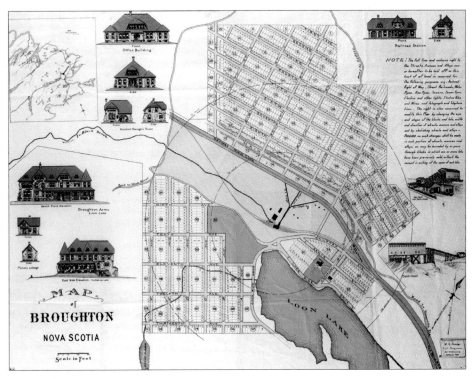

(a) Broughton, Nova Scotia (1905)

FIGURE 20.8
Most cities in Canada were "planned," especially the newer towns of the West, although many originated on the sites of preexisting settlements. The town of Broughton, Nova Scotia **(a)**, was a planned community that failed when the nearby mining operation was abandoned. Mount Royal, Quebec **(b)**, was planned by the Canadian Northern Railway Company, which purchased land for the settlement and constructed an underground rail tunnel to the site from Montreal in 1912.

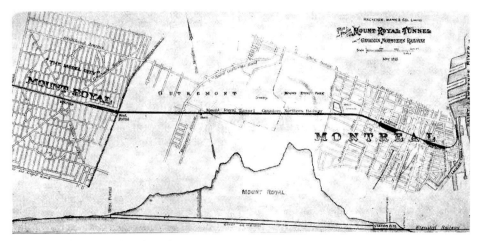

(b) Mount Royal, Québec (1913)

weighing the issues: YOUR URBAN AREA | 20–4

Think of your favourite parts of the city you know best. What aspects do you like about them? What do you dislike about some of your least favourite parts of the city? What could this city do to improve the quality of life for its inhabitants?

Zoning is a key tool for planning

One tool that planners use is **zoning**, the practice of classifying areas for different types of development and land use (**FIGURE 20.9**). For instance, to preserve the cleanliness and tranquility of residential neighbourhoods, industrial plants may be kept out of districts zoned for residential use. The specification of zones for different types of development gives planners a powerful means of guiding what gets built where. Zoning can restrict areas to a single use, as is often done with suburban residential tracts in so-called bedroom communities. Or zoning can allow the type of mixed use—residential and commercial, for instance—that some planners say can reinvigorate urban neighbourhoods. Zoning also gives homebuyers and business owners security; they know in advance what types of development can and cannot be located nearby.

Zoning involves government restriction on the use of private land and represents a top-down constraint on personal property rights. However, many people feel that government has a proper role in setting certain limitations on property rights for the good of the community. Similar debates arise with endangered species management and

weighing the issues: ZONING AND DEVELOPMENT | 20–5

Imagine you own a 5 ha parcel of land that you want to sell for housing development—but the local zoning board rezones the land so as to prohibit the development. How would you respond?

Now imagine that you live next to someone else's undeveloped 5 ha parcel. You enjoy the privacy it provides—but the local zoning board rezones the land so that it can be developed into a dense housing subdivision. How would you respond?

What factors do you think members of a zoning board should take into consideration when deciding how to zone or rezone land in a community?

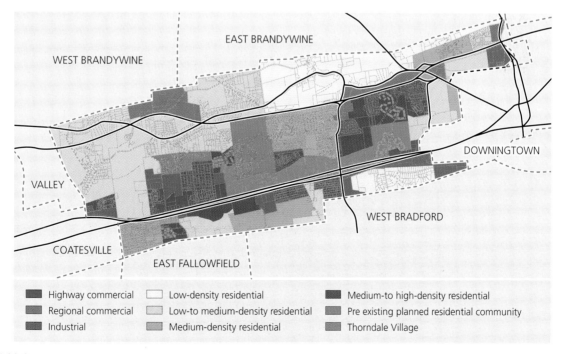

FIGURE 20.9
By zoning areas for different uses, planners guide how a community develops. Zoning restricts what landowners can do with their land, but it is intended to maximize prosperity, efficiency, and quality of life for the community. This sample zoning map shows several patterns common to modern zoning practice. Public and institutional uses are clustered together in a downtown area. Industrial uses are clustered together, away from most residential areas. Commercial uses are clustered along major roadways, and residential zones generally are higher in density toward the centre of town.

other environmental issues. For the most part, people have supported zoning over the years because the common good it produces for communities is widely felt to outweigh the restrictions on private use.

Urban growth boundaries and greenbelts are now widely used

The state of Oregon in the northwestern United States took a forward-thinking step in the early 1970s, requiring every city and county to draw up a comprehensive land use plan in line with statewide guidelines. As part of the land use plan, each metropolitan area had to establish an **urban growth boundary** (**UGB**), a line intended to separate areas designated to be urban from areas desired to remain rural. Development for housing, commerce, and industry would be encouraged within these urban growth boundaries but severely restricted beyond them. The intent was to revitalize city centres, prevent suburban sprawl, and protect farmland, forests, and open landscapes around the edges of urbanized areas.

A number of other cities in the United States and Canada have adopted some form of UGB. In their own ways, all UGBs aim to concentrate development, prevent sprawl, and preserve working farms, orchards, ranches, and forests. UGBs also appear to reduce the amounts municipalities have to pay for infrastructure, compared with sprawl. However, UGBs also seem to increase housing prices within their boundaries.

In the Portland, Oregon, area, the UGB has restricted development outside the UGB. It has increased the density of new housing inside the UGB by more than 50% as homes are built on smaller lots and as multi-storey apartments fulfill a vision of "building up, not out." Downtown employment has risen, and Portland has been able to absorb considerable immigration while avoiding rampant sprawl. However, urbanized area still increased by 101 km^2 in the decade after the UGB was established, because 146 000 people were added to the population. This suggests that relentless population growth may thwart even the best antisprawl efforts. Indeed, the Portland-area UGB has been enlarged three dozen times, and population projections for the region suggest there will be pressure for still more expansion. Many other locations that have instituted UGBs have chosen to expand them later. Housing and land prices also have risen dramatically within the UGB while declining in areas outside of the boundary. Given the reality of UGB expansions, it must be asked whether UGBs will truly limit sprawl in the long run.

In Canada, UGBs often take the form of greenbelts, rather than a "line on a map." A **greenbelt**, like an urban growth boundary, is a land use or zoning designation that is intended to contain urban development while protecting natural or agricultural lands in surrounding areas. Greenbelts provide additional benefits, including access to natural areas for city dwellers, better air and water quality, and protection for plant and animal habitats. Cities in Canada that have instituted some form of greenbelt policy include Ottawa, Toronto, and Vancouver.

Toronto's greenbelt (**FIGURE 20.10**), established by the provincial government of Ontario, is particularly

FIGURE 20.10
Toronto's greenbelt is thought to be one of the "most successful and most useful" greenbelts in the world. It stretches around the eastern half of Lake Ontario, encompassing the entire urban–commercial Golden Horseshoe district.

extensive. It includes much of the land area that overlies the Oak Ridges Moraine, an environmentally important area and a major aquifer that serves as the source region for many area streams. It also includes the Niagara Escarpment, designated as a UNESCO Biosphere Reserve. According to the Canadian Institute for Environmental Law and Policy, Ontario's greenbelt is "positioned to be the most useful and successful greenbelt in the world," in a report comparing international examples of urban development containment boundaries.[18] As with UGBs, however, greenbelts can be controversial. Opponents argue, for example, that homeowners in the "green" area are acting solely on their own behalf, hoping to increase housing prices and maintain their rural lifestyle close to the city.

"Smart growth" aims to counter sprawl

As more people have begun to feel negative effects of sprawl on their everyday lives, efforts to control growth have sprung up throughout North America. Urban growth boundaries and many other ideas from these policies have coalesced under the concept of **smart growth**. Smart growth principles (Table 20.2) vary in their details from place to place, but they always include a combination of environmental, economic, and social development goals.

Proponents of smart growth want municipalities to manage the rate, placement, and style of development so as to promote healthy neighbourhoods and communities, jobs and economic development, transportation options, and environmental quality. They aim to rejuvenate the older existing communities that so often are drained and impoverished by sprawl. Smart growth means "building up, not out"—focusing development and economic investment in existing urban centres and favouring multi-storey live/work spaces and high-rises. Many initiatives and experiments in smart growth are going on now in municipalities across Canada.

The "new urbanism" and "liveable cities" are now in vogue

Greenbelts and smart growth are efforts to envision the city of the future—a more sustainable, more environmentally friendly urban environment. A related movement among many architects, planners, and developers is labelled the **new urbanism**. This approach seeks to design neighbourhoods on a walkable scale, with homes, businesses, schools, and other amenities all close together for convenience. The aim is to create functional neighbourhoods in which most of a family's needs can be met close to home without the use of a car. Greenspaces, trees, a mix of architectural styles, and creative street layouts add to the visual interest and pleasantness of new urbanist developments. By aiming to accommodate diversity in age, ethnicity, and socioeconomic status, these developments mimic the traditional urban neighbourhoods that existed until the advent of suburbs.

One of the greatest advocates for liveable communities and walkable, people-friendly cities was Toronto resident Jane Jacobs (**FIGURE 20.11**), who died in April 2006. An activist and co-founder of the Energy Probe Research Foundation, Jacobs was widely known as an "urbanist."

Table 20.2 Ten Principles of "Smart Growth"

1	Mix land uses
2	Take advantage of compact building design and green buildings
3	Create a range of housing opportunities and choices
4	Create walkable neighbourhoods
5	Foster distinctive, attractive communities with a strong sense of place
6	Preserve open space, farmland, natural beauty, and critical environmental areas
7	Strengthen and direct development toward existing communities
8	Provide a variety of transportation choices
9	Make development decisions predictable, fair, and cost effective
10	Encourage community and stakeholder collaboration in development decisions

Source: Sierra Club, Ontario Chapter (based on U.S. Environmental Protection Agency, 2005).

FIGURE 20.11
One of the great advocates of the "liveable city" was Jane Jacobs, one of the most important figures in the history of urban planning.

She was a grassroots activist and an outspoken critic of urban renewal policies that ignored the needs of neighbourhood residents. Although she had no formal training as a planner, Jacobs' various books on cities and urban renewal are highly regarded for their insightful observations on what works and what doesn't work in North American cities. Jacobs' writings and her work in Toronto and New York City (where she spent a significant part of her career) inspired the founders of the Project for Public Spaces, a New York–based nonprofit organization that assists city planners in creating vibrant, people-friendly gathering places in downtown areas.

In 2007, Torontonians and New Yorkers honoured Jacobs with "Jane's Walk," a series of coordinated walks through urban neighbourhoods. By 2008, Jane's Walk had already grown to include walks through 152 different neighbourhoods across Canada. As Jacobs herself said about how to make our cities more liveable and people-friendly, "you've got to get out and walk."

Transportation options are vital to liveable cities

A key ingredient in any planner's recipe for improving the quality of urban life is making multiple transportation options available to citizens. These options include public buses, trains and subways, and light rail (smaller rail systems powered by electricity). As long as an urban centre is large enough to support the infrastructure necessary, these mass transit options are cheaper, more energy-efficient, and cleaner than roadways choked with cars (**FIGURE 20.12**). They also ease traffic congestion by carrying passengers who would otherwise be driving cars. Mass transit rail systems take up less space than road networks and emit less pollution than cars. The fuel and productivity lost on roadways to traffic jams have been estimated to cost billions of dollars each year.

In Canada, subway train systems in Montreal and Toronto enjoy heavy use on a daily basis. The most-used train systems in the United States are the extensive heavy rail subway systems in New York, Washington, D.C., Boston, and San Francisco. Major cities internationally from Moscow to Paris to Tokyo have large and heavily utilized subway systems. Some cities with severe traffic problems—such as Bangkok, Shanghai, and Athens—have recently opened new rail systems that carry hundreds of thousands of commuters a day. Light rail use is increasing in Europe, and ridership is now rising faster than is the rate of new car drivers. Most countries have bus systems that are far more accessible to citizens than are those of Canada or the United States.

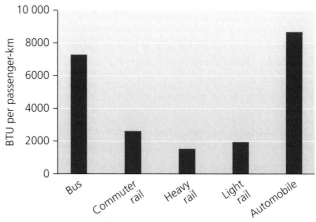

(a) Energy consumption for different modes of transit

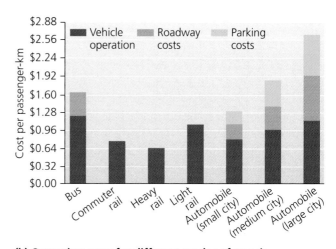

(b) Operating costs for different modes of transit

FIGURE 20.12
Rail transit consumes far less energy per passenger kilometre **(a)** than bus or automobile transit. Rail transit involves fewer costs per passenger kilometre **(b)** than bus or automobile transit. Based on data from Litman, T. 2005. *Rail Transit in America: A Comprehensive Evaluation of Benefits.* Victoria, BC: Victoria Transport Policy Institute.

One city famous for its efficient transportation system is Curitiba, Brazil (**FIGURE 20.13**). Faced with a heavy influx of immigrants from outlying farms in the 1970s, visionary city leaders led by Mayor Jaime Lerner decided to pursue an aggressive planning process so that they could direct growth rather than being overwhelmed by it. They reconfigured Curitiba's road system to maximize the efficiency of a large fleet of public buses. Today, this metropolis of 2.5 million people has an outstanding bus system that is used each day by three-quarters of the population. The 340 bus routes, 250 terminals, and 1900 buses accompany measures to encourage bicycles and pedestrians. All of this has resulted in a steep drop in car use, despite the city's rapidly growing population.

Establishing mass transit is not always easy, however. Once a road system has been developed and businesses

FIGURE 20.13
The rail and bus transit system in Curitiba, a planned community in Brazil, is one of the most efficient urban transportation systems in the world.

and homes are built alongside roads, it can be difficult and expensive to replace or complement the road system with a mass transit system. In addition, modes of mass transit differ in their effectiveness (see "The Science Behind the Story: Assessing the Benefits of Rail Travel"), depending on city size, size of the transit system, and other factors. To make urban transportation more efficient, governments can raise fuel taxes, tax inefficient modes of transport, reward carpoolers with carpool lanes, encourage bicycle use and bus ridership, and charge trucks for road damage. They can choose to minimize investment in infrastructure that encourages sprawl and to stimulate investment in renewed urban centres.

Parks and open space are key elements of liveable cities

City dwellers often desire some sense of escape from the noise, commotion, and stress of urban life. Natural lands, public parks, and open space provide greenery, scenic beauty, freedom of movement, and places for recreation. These lands also keep ecological processes functioning by regulating climate, producing oxygen, filtering air and water pollutants, and providing habitat for wildlife. The animals and plants of urban parks and natural lands also serve to satisfy *biophilia*, our natural affinity for contact with other organisms.

Protecting natural lands and establishing public parks become more important as our societies become more urbanized, because many urban dwellers come to feel increasingly isolated and disconnected from nature. In the wake of urbanization and sprawl, people of every industrialized society in the world today have, to some degree, chosen to conserve land in public parks.

City parks were widely established at the turn of the last century

At the turn of the twentieth century, civic improvement was garnering interest and support as politicians and citizens alike yearned for ways to make their crowded and dirty cities more liveable. In the late 1800s, urban public parks began to be established, using esthetic ideals borrowed from European parks, gardens, and royal hunting grounds. The lawns, shaded groves, curved pathways, and pastoral vistas we see today in many city parks and cemeteries originated with these European ideals. One of the preeminent park designers of the era was the American landscape architect Frederick Law Olmsted, who designed New York's Central Park in 1853 and a host of urban park systems afterward (**FIGURE 20.14**). Olmsted also designed Mount Royal Park in Montreal,

FIGURE 20.14
City parks were developed in many urban areas in the late nineteenth century to provide citizens esthetic pleasure, recreation, and relief from the stresses of the city. Central Park in New York City, shown here, was one of the first, designed by the famous landscape architect Frederick Law Olmsted.

THE SCIENCE BEHIND THE STORY

Assessing the Benefits of Rail Transit

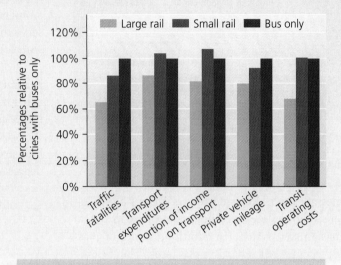

These subway riders are on a busy commute.

Large-rail cities outperform bus-only cities in transport-oriented variables, but the benefits of rail are not so clear-cut for small-rail cities. Adapted from Litman, T. 2005. *Rail Transit in America: A Comprehensive Evaluation of Benefits*. Victoria, BC: Victoria Transport Policy Institute.

Most urban planning experts see benefits in public mass transit and in making transit systems readily accessible to citizens. But building a mass transit system—and then maintaining and expanding it—can be an expensive undertaking. Thus, planners and policy makers value quantitative information on the costs and benefits of different types of transit systems and on how extensive a given transit system should be to produce the benefits they desire.

For this reason, researcher Todd Litman of the nonprofit Victoria Transport Policy Institute in British Columbia conducted a comprehensive evaluation of the benefits of rail transit. Focusing on cities in the United States, Litman first divided the cities into three categories. "Bus-only" cities have bus service but no rail service; "small-rail" cities have rail systems serving fewer than 12% of daily commuters; "large-rail" cities feature rail service as a major component of the transportation system, serving up to 48% of daily commuters.

Litman compared these three groups of cities for a number of variables. Several key results are summarized in the accompanying figure. Compared with bus-only cities, large-rail cities had 36% fewer per capita traffic deaths each year. Residents of large-rail cities drove 21% fewer kilometres yearly than those of bus-only cities. Large-rail city residents also saved money on transportation, spending 14% less than bus-only city residents on transportation—equivalent to annual savings of $448. Transportation costs took up only 12% of their household budgets, compared with 14.9% of those of bus-only city residents.

The analysis also revealed benefits of large-rail systems for municipalities running the systems. For example, large-rail cities paid less in operating costs per kilometre travelled and recovered far more of their costs than bus-only cities (38% cost recovery versus 24%).

Small-rail cities were intermediate in traffic deaths and vehicle mileage, as might be expected. However, they performed no better than bus-only cities in terms of operating costs per passenger kilometre or transit cost recovery. And residents of small-rail cities fared slightly worse than residents of bus-only cities in annual transport expenditures and proportion of income spent on transportation. These results suggest that not all benefits of rail transit begin to accrue with small systems. Rather, for many factors, a rail system has to be large enough or accommodate enough riders to gain the economy of scale needed to provide benefits.

A similar pattern was found for annual per capita costs caused by traffic congestion. For bus-only and small-rail cities, congestion costs increased with city size. But for large-rail cities, congestion costs did not vary with size and were lower than in comparably sized cities.

Litman also found that each year governments spend billions of dollars on rail transit systems that they do not get back in revenues from fares, but this is outweighed by the monetary benefits of rail systems. Each year, rail systems save billions of dollars in congestion costs, consumer transportation costs, roadway maintenance costs, parking costs, and accident costs, not to mention indirect savings because of enhanced environmental quality.

A number of studies have critiqued rail transit and portrayed it as ineffective. The Litman study indicates that although not all rail systems are cost-effective, rail systems become more beneficial as they become larger and carry a greater proportion of a city's commuters.

FIGURE 20.15
Riverwood Park is a 60 ha, multiuse urban park in Mississauga, Ontario. Two-thirds of the property is designated to remain in its natural state (or be restored as such); the remainder is designated for a variety of uses by artists, hikers, fishers, bicyclists, school groups, and others.

although the park that was eventually built followed few of his suggestions.

Two sometimes conflicting goals motivated the establishment and design of early city parks in North America. On the one hand, the parks were meant to be "pleasure grounds" for the wealthy, who helped support their establishment financially and who would ride the parks' winding roadways in carriages. On the other hand, parks were meant to alleviate congestion for poverty-stricken immigrants, and these park users were more interested in active recreation, such as ballgames, than in carriage rides.

At times during the historical development of urban parks, the esthetic interests of the educated elite, the recreational interests of the broader citizenry, and the ecological interests of urban wildlands came into conflict—a friction that survives today in debates over recreation in city parks. For example, Riverwood Park is a 60 ha urban park located in the middle of Mississauga, Ontario (**FIGURE 20.15**). Park planners had to balance many competing interests in designing the park, including the interests of naturalists, artists, mountain bikers, fishers, historians (interested in the heritage buildings on site), schoolchildren, seniors, residents, students and teachers, and the park's wildlife and natural vegetation.

Smaller public spaces are also important

Large city parks are a key component of a healthy urban environment, but even small spaces can make a big difference. Playgrounds provide places where children can be active outdoors and interact with their peers. Community gardens allow people to grow their own vegetables and flowers in a neighbourhood setting (**FIGURE 20.16**). Vancouver offers gardeners access to 10 community gardens scattered throughout the city, and many other cities across North America now feature thriving community gardens.

Greenways or *corridors*, strips of land that connect parks or neighbourhoods, are often located along rivers, streams, or canals, and they may provide access to networks of walking trails. (Compare this use of "corridor" with the concept of habitat corridors, discussed in Chapter 10.) They can protect water quality, boost property values, and serve as corridors for the movement of birds and wildlife. The Rails-to-Trails Conservancy has spearheaded the conversion of abandoned railroad rights-of-way into trails for walking, jogging, and biking. To date, nearly 24 000 km of 1200 rail lines have been converted across North America.

Besides creating new types of urban spaces, many cities are working to enhance the "naturalness" of their parks through ecological restoration, the practice of restoring native communities. In Mississauga's Riverwood Park, for example, volunteer teams gather periodically to remove garlic mustard, an invasive plant that smothers native plants on the forest floor. In Vancouver's Hastings Park,

FIGURE 20.16
Urban community gardens like this one provide city residents with a place to grow vegetables. They also serve as greenspaces that beautify cities.

restoration activities have included tearing down a number of old park buildings and establishing a 4 ha natural area with a pond. On Vancouver Island, extensive restoration activities have been undertaken to reestablish degraded Garry oak ecosystems. At some Chicago-area forest preserves, scientists and volunteers have used prescribed burns to restore prairie grasses native to the region.

Urban Sustainability

Urbanization and urban centres exert both positive and negative environmental impacts. These impacts depend strongly on how we utilize resources, produce goods, transport materials, and deal with waste.

Urban resource consumption brings a mix of environmental impacts

Most of us might guess that urban living has a greater environmental impact than rural living. However, the picture is not so simple; instead, urbanization brings a complex mix of consequences.

Resource sinks Cities and towns are sinks for resources, having to import from beyond their borders nearly everything they need to feed, clothe, and house their inhabitants. In other words, the ecological footprints of cities tend to be very large. Urban and suburban areas rely on large expanses of land elsewhere to supply food and other crops, as well as natural resources, such as water, timber, metal ores, and mined fuels. Urban centres also need areas of natural land to provide ecosystem services, including purification of water and air, nutrient cycling, and waste treatment. Major cities, such as New York, depend for their day-to-day survival on water they pump in from faraway watersheds. The city has gone to lengths to acquire, protect, and manage watershed land to minimize pollution of these water sources. When confronted in 1989 with an order by the United States Environmental Protection Agency to build a $6 billion filtration plant to protect its citizens against waterborne disease, the city opted instead to purchase and better protect watershed land, for a fraction of the cost.

As cities have grown, and as the material wealth of most societies has risen, the inexorable pull of resources from the countryside to the cities has become stronger. And as urban areas extend their reach, it becomes increasingly hard for urban residents isolated from natural lands to have a tangible demonstration of the environmental impacts of their choices.

The long-distance transportation of resources and goods requires a great deal of fossil fuel use, which has significant environmental impacts (Chapter 15). This is of particular concern in Canada because of the great distances from shore to shore to shore. Because of pollution, waste, and other environmental impacts, the centralization of resource use that urbanization entails may seem a bad thing for the environment. However, imagine that all the world's 3.3 billion urban residents were instead spread evenly across the landscape. What would the transportation requirements be, then, to move all those resources and goods around to all those people? A world without cities would likely require *more* transportation to provide people with the same level of access to resources and goods.

Efficiency Once resources have arrived at an urban centre where people are densely concentrated, however, cities can help minimize per capita consumption by maximizing the efficiency of resource use and delivery of goods and services. For instance, providing electricity from a power plant for urban houses close together is more efficient than providing electricity to far-flung homes in the countryside. The density of cities facilitates the provision of many social services that improve quality of life, including medical services, education, water and sewer systems, waste disposal, and public transportation. This is called *economy of proximity*—it is simpler and cheaper to deliver goods and services to people who are clustered together, and it generates fewer environmental impacts, as well.

More consumption Because cities draw resources from afar, as mentioned above, their ecological footprints are much greater than their actual land areas. For instance, urban scholar Herbert Girardet calculated that the ecological footprint of London, England, extends 125 times as large as the city's actual area. By another estimate, cities take up only 2% of the world's land surface but consume more than 75% of its resources.

In 2005, the City of Calgary participated in an urban ecological footprint study by the Federation of Canadian Municipalities. The city's footprint team identified energy as the main component of Calgary's ecological footprint, at 62% (**FIGURE 20.17A**). They calculated an ecological footprint of 9.86 *global hectares* (*gha*) per capita, the highest of any city in Canada; this compares with a Canadian urban average of 7.25 gha, and 1.9 gha available globally (**FIGURE 20.17B**). The global hectare (gha) is a unit that can be used in ecological footprint calculations to describe both the area of biologically productive land

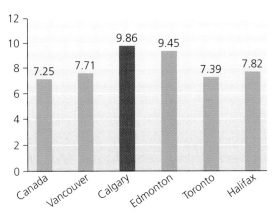

(a) Ecological footprints of some Canadian cities

FIGURE 20.17
A 2005 study found that Calgary had the largest ecological footprint of any Canadian city **(a)**, 9.86 gha, compared with the Canadian average of 7.25 gha and global availability of 1.9 gha. Calgary's ecological footprint team found that energy use accounted for 62% of the city's ecological footprint **(b)**.

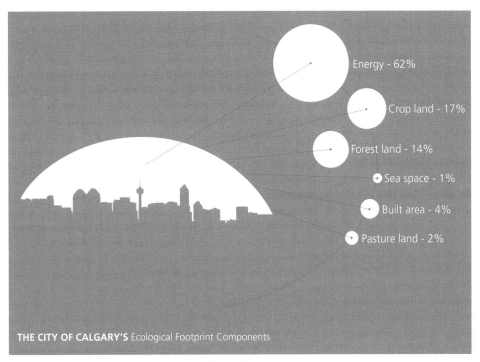

(b) Components of Calgary's ecological footprint

or water and the demand on it. Different land types have different productivities. This means that a global hectare of, for example, agricultural land, would be physically smaller than a global hectare of pasture land, which is much less biologically productive. In other words, more pasture would be needed to provide the same productivity as one hectare of agricultural land.

The City of Calgary is using the ecological footprint exercise as a guide for both citizens and city planners in making decisions that will affect the city's future. They have made reducing the city's ecological footprint a priority through initiatives, such as ImagineCALGARY, a 100-year vision for the city, which includes 30-year targets and strategies for urban sustainability.[19]

However, the ecological footprint concept is most meaningful when used on a per capita basis. So, in asking whether urbanization causes increased resource consumption, we must ask whether the average urban dweller has a larger footprint than the average rural dweller. The answer is yes, but urban and suburban residents also tend to be wealthier than rural residents, and wealth correlates with resource consumption. Thus, although urban citizens tend to consume more than rural ones, the reason could be simply that they are wealthier.

Urban intensification preserves land

The ecological footprints of urban areas are large, but because people are packed densely together in cities, more land outside cities is left undeveloped. Indeed, this is the idea behind urban growth boundaries. If cities did not exist, and if instead all 6.7 billion of us were evenly spread across the planet's land area, we would have much less room for agriculture, wilderness, biodiversity, or privacy. There would be no large blocks of land left uninhabited by people or of unfragmented habitat for wildlife. The fact that half the human population is concentrated in discrete locations helps allow room for natural ecosystems to maintain themselves, continue functioning, and provide the ecosystem services on which all of us, urban and rural, depend.

Urban centres suffer and export pollution

Just as cities import resources, they export wastes, either passively through pollution or actively through trade. In so doing, urban centres transfer the costs of their activities to other regions—and mask the costs from their own residents. Citizens of Toronto may not recognize that pollution from coal-fired power plants in their region worsens acid precipitation hundreds of kilometres to the east. Citizens of New York City may not realize how much garbage their city produces if it is shipped to other states or nations for disposal.

However, not all waste and pollution leaves the city. Urban residents are exposed to heavy metals, industrial compounds, and chemicals from manufactured products that accumulate in soil and water. Airborne pollutants cause photochemical smog, industrial smog, and acid precipitation. Fossil fuel combustion releases carbon dioxide and other pollutants, leading to climate change.

Cities often have ambient temperatures that are several degrees higher than the surrounding suburbs and rural areas. This is called the **urban heat island effect**, and it results from the concentration of heat-generating buildings, cars, factories, and people in the city centre. Tall buildings and paved surfaces are important contributors to the urban heat island effect, by absorbing heat and releasing it slowly, and by interfering with the convective circulation of air that would otherwise cool the city. When heated air becomes trapped over the city, the smog and particulate air pollution it carries become trapped as well; this can lead to a phenomenon called a **dust dome**. If you drive toward a large city on a hot, hazy summer day—especially if the city is located in a basin or a valley—you will likely see the trapped smog hovering like a brown or blue-grey cloud over the city skyline.

Urban residents also suffer noise pollution and light pollution. **Noise pollution** consists of undesired ambient sound. Excess noise degrades one's surroundings esthetically, can induce stress, and at intense levels (such as with prolonged exposure to the sounds of leaf blowers, lawnmowers, and jackhammers) can harm hearing. The glow of **light pollution** from the city lights obscures the night sky, impeding the visibility of stars.

These various forms of pollution and the health threats they pose are not evenly shared among urban residents. Those who bear the brunt of the pollution are often those who are too poor to live in cleaner areas. Environmental justice concerns centre on the fact that a disproportionate number of people living near, downstream from, or downwind from factories, power plants, and other polluting facilities are people who are poor and, often, people of racial minorities.

Urban centres foster innovation and offer cultural resources

One of the greatest impacts of urbanization on environmental quality is also one of the most indirect and intangible. Cities promote a flourishing cultural life and, by mixing together diverse people and influences, spark innovation and creativity. The urban environment can promote education and scientific research, and cities have long been viewed as engines of technological and artistic inventiveness. This inventiveness can lead to solutions to societal problems, including ways to reduce environmental impacts.

For instance, research into renewable energy sources is helping us develop ways to replace fossil fuels. Technological advances have helped us reduce pollution. Wealthy and educated urban populations provide markets for low-impact goods, such as organic produce. Recycling programs help reduce the solid waste stream. Environmental education is helping people choose their own ways to live cleaner, healthier, lower-impact lives.

All these phenomena grow from the education, innovation, science, and technology that are part of urban culture. Citizens' Environment Watch (CEW) is an example of an urban-based program in Toronto that offers activities for those who are interested in learning about and protecting the natural environment in the urban setting (**FIGURE 20.18**). For example, the Toronto Lichen Count (TLC) is a CEW program in which participants are trained to count lichen colonies, which can be indicators of changes in air quality. The TLC monitoring sites are all located next to subway stations in the downtown Toronto area, for easy access by public transit.

Some seek sustainability for cities

Modern cities that import all their resources and export all their wastes have a linear, one-way metabolism.

FIGURE 20.18
Urban and suburban children take part in environmental education programs that foster awareness and practical training for children and adults interested in learning about and protecting nature in the urban environment. Although natural land is rare in urbanized areas, the education and innovation that urbanization often promotes can lead to solutions that reduce environmental impact.

Such linear models of production and consumption tend to destabilize environmental systems and are not sustainable. Proponents of sustainability for cities stress the need to develop circular systems, akin to systems found in nature, which recycle materials and use renewable sources of energy.

Researchers in the field of **urban ecology** hold that cities can be viewed explicitly as ecosystems and that the fundamentals of ecosystem ecology and systems science (Chapter 5) apply to urban areas. To help cities improve their standards of living while reducing their environmental impacts, urban sustainability advocates suggest that cities follow an ecosystem-centred model by striving to

- Maximize efficient use of resources
- Recycle as much as possible
- Develop environmentally friendly technologies
- Account fully for external costs
- Offer tax incentives to encourage sustainable practices
- Use locally produced resources
- Use organic waste and wastewater to restore soil fertility
- Encourage urban agriculture

More and more cities are adopting these strategies. For instance, urban agriculture is a growing pursuit in many urban areas, from Cuba to Japan. Singapore produces all of its meat and 25% of its vegetable needs within its city limits. In Berlin, Germany, 80 000 people grow food in community gardens, and 16 000 more are on waiting lists.

Curitiba, Brazil, shows the kind of success that can result when a city invests in well planned infrastructure. Besides the highly effective bus transportation network described earlier, the city provides recycling, environmental education, job training for the poor, and free health care. Surveys show that its citizens are unusually happy and better off economically than people living in other Brazilian cities.

In general, experts advise that developed countries should invest in resource-efficient technologies to reduce their impacts and enhance their economies, whereas developing countries should invest in basic infrastructure to improve health and living conditions. Successes in places from Vancouver to Curitiba suggest that cities need not be unsustainable. Indeed, because they affect the environment in some positive ways and have the potential for efficient resource use, cities can and should be a key element in achieving progress toward global sustainability.

Conclusion

As half the human population has shifted from rural to urban lifestyles in the relatively recent past, the nature of our impact on the environment has changed. As urban and suburban dwellers, our impacts are less direct but often more far-reaching. Resources must be delivered to us over long distances, requiring the use of still more resources. Limiting the waste of those resources by making our urban and suburban areas more sustainable will be vital for the future. Fortunately, the innovative cultural environment that cities foster has helped us develop solutions to alleviate impact and promote sustainability.

Part of seeking urban sustainability lies in making urban areas better places to live. One key component of these efforts involves expanding transportation options to relieve congestion and make cities run more efficiently. Another lies in ensuring access to adequate parklands and greenspaces near and within our urban centres, to keep us from becoming wholly isolated from nature. Accomplishments in city and regional planning have made many cities more liveable than they once were, and we should be encouraged by such progress. Proponents of smart growth and the new urbanism believe they have solutions to the challenges posed by urban and suburban sprawl.

Continuing experimentation in cities from British Columbia to Brazil will help us determine how best to ensure that urban growth improves our quality of life and does not degrade the quality of our environment. With the continued urbanization of the world's population, it is important that we learn how to make cities work for both people and the environment.

CANADIAN ENVIRONMENTAL PERSPECTIVES

Nola-Kate Seymoar

Nola-Kate Seymoar has devoted her life and career to supporting sustainable communities.

- **President** and **CEO**, International Centre for Sustainable Cities
- **Advisor** on community economic development, empowerment, and capacity-building
- **Lead developer**, Sustainable Cities PLUS Network

"We are a *do-tank*, not a *think tank*,"[20] says Nola-Kate Seymoar, CEO of the International Centre for Sustainable Cities. The ICSC helps cities resolve problems related to urban growth. The participatory, consultative approach that characterizes the organization may be a holdover from Seymoar's very first job as a recreation director for underprivileged children in Edmonton, a job that she credits with teaching her the value of working cooperatively to achieve shared goals.[21]

Seymoar's academic background is as varied as her career has been since those early days on the playground. She has earned three interdisciplinary degrees: a B.A. in recreation administration, an M.A. in community development, and a Ph.D. in social psychology, with research interests focusing on communities and empowerment. She has since applied these credentials across a spectrum of private, public, and nonprofit organizations, and in the academic sector. She was even an entrepreneur early in her career, starting three successful businesses in film, video, and interior design.

Throughout it all, Seymoar's main focus has been on communities, sustainability, and advocacy for youth and local-level empowerment. This is reflected, for example, in her work for the Alberta Department of Youth, the Arctic Children and Youth Foundation, and ICSC's new Youth Led Development Program. She has served on advisory committees and boards for many organizations and events, including the advisory committee for the World Urban Forum and the World Peace Forum, both held in Vancouver in 2006. She received a Global Citizen Award in 1995 and the Queen's Golden Jubilee Medal in 2002.

Seymoar was part of the multidisciplinary Vancouver team that won the Grand Prize in the 2003 International Gas Union's Cities PLUS competition, profiled in the Central Case. After the competition, determined to maintain the momentum, Seymoar conceived and launched the Sustainable Cities PLUS Network, which connects cities around the world that are engaged in long-term planning for sustainability. The network promotes city-to-city peer dialogue, enabling member communities to learn from one another's successes and setbacks. Seymoar continues to lead the PLUS Network.

Meanwhile, her ICSC and PLUS Network connections with cities around the world continue to lead Seymoar into new project areas. A current example is her role as senior adviser to a two-year capacity-building project in Sri Lanka, called *Centering Women in Reconstruction and Governance* working with a local organization, Sevanatha. The project is funded by the Canadian International Development Agency, GROOTS (Grass Roots Organizations Operating Together in Sisterhood), Beedie Construction, and local municipalities. It involves the participatory design and construction of women-led community resource centres in Matara and Moratuwa, Sri Lanka.

She has worked on urban greening, urban waste management, community resiliency, and neighbourhood rehabilitation projects from Thailand to Turkey, China to South Africa, Poland to the Philippines. Nola-Kate Seymoar seems tireless in her commitment to community-level sustainability in the urban centres of the world. Our chances for urban sustainability and quality of life in cities are no doubt greater because of her efforts.

"Sustainability is about the future—the quality of life that will be possible for our grandchildren." —Nola-Kate Seymoar

Thinking About Environmental Perspectives

Thinking about the city nearest to where you live, what advice do you think this city could offer to, say, Beijing or Bangkok or Dar es Salaam with regard to sustainability, quality of life, and urban liveability? What kinds of advice from cities elsewhere in the world might be of benefit to your home city?

REVIEWING OBJECTIVES

You should now be able to:

Describe the scale of urbanization

- The world's population is becoming predominantly urban.
- The shift from rural to urban living is driven largely by industrialization and is proceeding fastest now in the developing world.
- Nearly all future population growth will be in cities of the developing world.
- The geography of urban areas is changing as cities decentralize and suburbs grow and expand.

Assess urban and suburban sprawl

- Sprawl covers large areas of land with low-density development. Both population growth and increased per capita land use contribute to sprawl.
- Sprawl has resulted from the home-buying choices of individuals who prefer suburbs to cities, and it has

been facilitated by government policy and technological developments.
- Sprawl may lead to negative impacts involving transportation, pollution, health, land use, natural habitat, and economics.

Outline city and regional planning and land use strategies

- City and regional planning and zoning are key tools for improving the quality of urban life.
- "Smart growth," urban growth boundaries, greenbelts, and the "new urbanism" attempt to recreate compact and vibrant urban spaces.

Evaluate transportation options

- Mass transit systems can enhance the efficiency of urban areas, but bus and train systems of different sizes bring different benefits.

Describe the roles of urban parks

- Urban parklands are vital for active recreation, soothing the stress of urban life, and keeping people in touch with natural areas.

Analyze environmental impacts and advantages of urban centres

- Cities are resource sinks with high per capita resource consumption. However, cities also allow natural lands to be preserved.
- Urban centres can maximize efficiency and help foster innovation that can lead to solutions for environmental problems.

Assess the pursuit of sustainable cities

- The linear mode of consumption and production is unsustainable, and more circular modes will be needed to create sustainable cities.

TESTING YOUR COMPREHENSION

1. What factors lie behind the shift of population from rural areas to urban areas? What types of cities and countries are experiencing the fastest urban growth today, and why?
2. Why have so many city dwellers in Canada, the United States, and other developed nations moved into suburbs?
3. Give two definitions of *sprawl*. Describe five negative impacts that have been suggested to result from sprawl.
4. What are city planning and regional planning? Contrast planning with zoning.
5. How are some people trying to prevent or slow sprawl? Describe some key elements of "smart growth." What effects, positive and negative, do urban growth boundaries tend to have?
6. Describe several apparent benefits of rail transit systems. What is a potential drawback?
7. How are city parks thought to make urban areas more liveable? What types of smaller spaces in cities can serve some of the functions of parks?
8. Why do urban dwellers tend to consume more resources per capita than rural dwellers?
9. Describe the connection between urban ecology and sustainable cities. List three actions a city can take to enhance its sustainability.
10. Name two positive effects of urban centres and intensification on the natural environment.

SEEKING SOLUTIONS

1. Assess the reasons that urban populations are increasing and why rural populations are stable or declining. Do you think these trends will continue in the future, or might they change for some reason?
2. Evaluate the causes of the spread of suburbs and of the environmental, social, and economic impacts of sprawl. Overall, do you think the spread of urban and suburban development that many people label *sprawl* is predominantly a good thing or a bad thing? Do you think it is inevitable? Give reasons for your answers.
3. Would you personally want to live in a neighbourhood developed in the style of the new urbanism? Would you like to live in a city or region with an urban growth boundary? Why or why not?
4. All things considered, do you feel that cities have a positive or a negative effect on environmental quality? How much do you feel we may be able to improve the sustainability of our urban areas?
5. **THINK IT THROUGH** You are a person who aims to live in the most ecologically sustainable way you can.

Which of the following places would you choose to live: in a high-rise apartment in a big city, or on a 30 ha ranch abutting a national forest? Why? What considerations will you factor into your decision?

6. **THINK IT THROUGH** Let us say that after you graduate you are offered three equally desirable jobs, in three very different locations. If you take the first, you will live in the midst of a highly diverse but densely populated city. If you accept the second, you will live in a suburb where you have more space but where development and sprawl may soon surround you. If you select the third, you will live in a rural area with plenty of space and a beautiful natural environment, but a long commute and few cultural amenities. Where would you choose to live? Why?

INTERPRETING GRAPHS AND DATA

In the accompanying graph, urban population density is used as an indicator of sprawl (lower density = more sprawl), and carbon emissions per capita provide some measure of the environmental impact of the transportation system or preferences for each of the cities represented.

1. Describe the relationship between urban density and carbon emissions, as shown in the graph.
2. Assuming that the standard of living is similar in these cities, to what might you attribute the relationship described in your answer to Question 1?
3. If zoning ordinances slowed urban sprawl and resulted in a doubling of urban population density in a city like Houston, Texas, how would you predict that carbon emissions per capita in that city might change? What about in Vancouver?

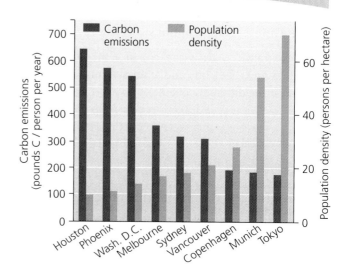

Population density versus carbon emissions from transportation in 1990. Data from Kenworthy, J., et al. 1999. *An International Sourcebook of Automobile Dependence in Cities.* Boulder, CO: University Press of Colorado, as cited by Sheehan, M. O. 2002. *What Will It Take to Halt Sprawl?* Washington DC: Worldwatch Institute.

CALCULATING FOOTPRINTS

One way of altering your ecological footprint is to consider transportation alternatives. Each litre of gasoline is converted to approximately 34 kg of carbon dioxide (CO_2) during combustion, and this CO_2 is then released into the atmosphere. The table lists typical amounts of CO_2 released for each person per kilometre, through various forms of transportation, assuming typical fuel efficiencies.

For an average North American person who travels about 19 000 kilometres per year, calculate and record in the table the CO_2 emitted yearly for each transportation option, and the reduction in CO_2 emission that one could achieve by relying solely on each option.

1. What transportation option will give you the most kilometres travelled per unit of carbon dioxide emitted?

2. Clearly, it is unlikely that any of us will walk or bicycle 19 000 km per year or travel only in vanpools of eight people. In the last two columns, estimate what proportion of the 19 000 annual kilometres you think that you actually travel by each method, and then calculate the CO_2 emissions that you are responsible for generating over the course of a year. Which transportation option accounts for the most emissions for you?

3. How could you reduce your CO_2 emissions? How many kilograms of emissions do you think you could realistically eliminate over the course of the next year?

4. As discussed in the chapter, energy use and CO_2 emissions constitute the main portion of Calgary's

ecological footprint (see Figure 20.17). Given that the city is interested in reducing its ecological footprint, what transportation strategies would you recommend to council to help them achieve this goal? What do you think are the other contributors to the energy component of Calgary's ecological footprint, besides transportation?

	CO_2 per person per kilometre	CO_2 per person per year (19 000 km avg. travel)	CO_2 emission reduction	Your estimated kilometres per year	Your CO_2 emissions per year
Automobile (driver only)	0.60 kg	11 400 kg	0		
Automobile (2 persons)	0.23 kg				
Automobile (4 persons)	0.15 kg				
Vanpool (8 persons)	0.08 kg				
Bus	0.19 kg				
Walking	0.06 kg				
Bicycle	0.04 kg				
				Total = 19 000	

TAKE IT FURTHER

 Go to www.myenvironmentplace.ca where you will find

- Suggested answers to end-of-chapter questions
- Quizzes, animations, and flashcards to help you study
- *Research Navigator*™ database of credible and reliable sources to assist you with your research projects
- Tutorials to help you master how to interpret graphs
- Current news articles that link the topics that you study to case studies from your region and around the world

- **ECO Occupational Profiles:** If you found this chapter especially interesting, you might want to learn more about the following jobs by visiting the Occupational Profiles website of the Environmental Careers Organization. Go to www.eco.ca and check out the following careers:
    - Environmental engineer
    - Environmental planner
    - GIS analyst
    - Landscape architect

CHAPTER ENDNOTES

1. *The Economist*, Where the grass is greener: Cities in Australia and Canada are rated the most liveable in the world, August 22, 2007, www.economist.com/markets/rankings/displaystory.cfm?story_id=8908454&CFID=16415879&CFTOKEN=94552766.
2. cities*PLUS*, www.citiesplus.ca
3. cities*PLUS*, www.citiesplus.ca
4. International Centre for Sustainable Cities, Projects, Cities PLUS Network, http://sustainablecities.net/plusnetwork.
5. Vancouver Working Group of Western Economic Diversification Canada Discussion Paper, www.wdgc.ca/rpts/research/livable/2a_e.asp.

6. Sustainable Cities: PLUS Network, http://sustainablecities.net/plusnetwork.
7. International Centre for Sustainable Cities, Projects, Cities PLUS Network, http://sustainablecities.net/plusnetwork.
8. Sustainable Cities: PLUS Network, http://sustainablecities.net/plusnetwork.
9. Vancouver Convention and Exhibition Centre, www.vcec.ca.
10. UNFPA, *State of world population: Unleashing the potential of urban growth,* 2007, www.unfpa.org/swp/2007/english/introduction.html.
11. Statistics Canada (2006) Census, *Summary Tables,* www40.statcan.ca/l01/cst01/demo62a.htm.
12. Statistics Canada (2006) Census, *Summary Tables,* www40.statcan.ca/l01/cst01/demo05a.htm.
13. Statistics Canada (2006) *Census Dictionary,* www12.statcan.ca/english/census06/reference/dictionary/index.cfm.
14. Burchfield, Marcy, Henry G. Overman, Diego Puga, and Matthew A. Turner (2005) *Causes of sprawl: A portrait from space,* Working Paper, http://diegopuga.org/papers/sprawl.pdf.
15. Casselman, Anne (2006) Is urban sprawl an urban myth? *Discover Magazine,* September 1, http://discovermagazine.com/2006/sep/urbanmyth.
16. Ewing, R., et al. (2002) *Measuring Sprawl and Its Impact.* Washington, DC: Smart Growth America.
17. Eid, Jean, Henry G. Overman, Diego Puga, and Matthew Turner (2007) Fat City: The relationship between urban sprawl and obesity. Centre for Economic Performance, LSE Discussion Papers, no. dp0758, http://cep.lse.ac.uk/pubs/download/dp0758.pdf.
18. Carter-Whitney, Maureen (2008) *Ontario's Greenbelt in an International Context: Comparing Ontario's Greenbelt to its Counterparts in Europe and North America.* Friends of the Greenbelt Foundation Occasional Paper Series. Toronto: Friends of the Greenbelt Foundation, p. 1, www.ourgreenbelt.ca/sites/ourgreenbelt.ca/files/CIELAPsmallerfile.pdf.
19. City of Calgary (2008) *Toward a Preferred Future: Understanding Calgary's Ecological Footprint,* www.calgary.ca/docgallery/bu/environmental_management/ecological_footprint/towards_preferred_future.pdf.
20. From Kowal, Jaime (2006) *Waking Up the West: Healers and Visionaries.* Vancouver: Catalyst Publications.
21. Tse, Catherine (2006) Swing Shift: Nola-Kate Seymoar, International Centre for Sustainable Cities, *Shared Vision: Dialogue for Change,* August, www.shared-vision.com/20060817/swing_shift.

21 Environmental Ethics and Economics: Values and Choices

This is an aerial view of Diavik Diamond Mine, Northwest Territories.

Upon successfully completing this chapter, you will be able to

- Characterize the influences of culture and world view on the choices people make
- Outline the nature, evolution, and expansion of environmental ethics in Western cultures
- Describe precepts of classical and neoclassical economic theory, and summarize their implications for the environment
- Compare the concepts of economic growth, economic health, and sustainability
- Explain the fundamentals of environmental economics and ecological economics

This is the mining settlement at Port Radium, Northwest Territories, in 1930.

CENTRAL CASE:
MINING DENENDEH[1]

"This forest is where our people have fished and trapped for generations—so we have this responsibility to take care of it, and that responsibility came from the elders, our ancestors, who told us to do whatever we have to do to protect the land."
—SOPHIA RABLIAUSKAS, GOLDMAN ENVIRONMENT PRIZE WINNER; POPLAR RIVER FIRST NATION COMMUNITY

"Let us not forget that, in the end, all economies, whether new or old, are built on foundations of access to land, natural wealth and resources."
—MATTHEW COON COME, FORMER GRAND CHIEF OF THE CREE NATION AND NATIONAL CHIEF OF THE ASSEMBLY OF FIRST NATIONS

The traditional lands of the Dene and Métis form the Northwest Territories of Canada, known as *Denendeh*. The people of Denendeh have mined, manufactured, and traded metals—especially copper—since long before the arrival of Europeans.

Industrial-scale mining has played a significant role in the history and development of the Northwest Territories since 1930, when the first modern mining operation, Eldorado uranium mine, was established at Port Radium on Great Bear Lake (*Sahtu*, in the Dene language; see map and photo). The Con and Giant gold mines, which began production at Yellowknife in

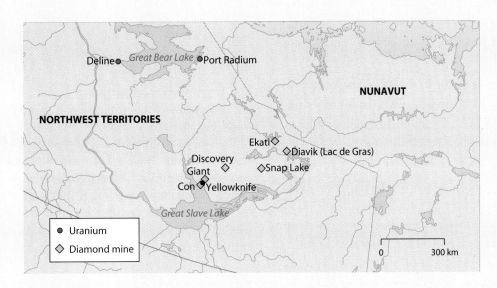

Great Bear Lake in the Northwest Territories was the site of uranium mining at the Eldorado Mine in Port Radium from 1930 until the 1960s. More recent diamond mining centres on an area to the northeast of Yellowknife.

1938 and 1948, respectively, were important to the economy of the area for more than 50 years. Mining continues today in the North but the emphasis has shifted to diamonds, with the development of new mines like Ekati and Diavik, the first diamond mines in Canada.

Recent staking rushes in the Northwest Territories, particularly for diamonds, have generated concerns over unsettled land claims. For example, the public hearings of the Mackenzie Valley Environmental Impact Review Board for Wool Bay and Drybones Bay in 2003 centred on conflicts between unresolved Aboriginal land claims and mineral claims made by exploration companies. The environmental impact assessments halted exploration. The Dene claimed that their spiritual and cultural use of the land was in jeopardy. This included many practical issues, such as the locations of gravesites, as well as improperly managed waste materials from exploration. In this case, the Aboriginal rights and title to the land were given precedence over mining exploration.

The environmental, social, and economic impacts of past mining development in the North have been considerable, including contamination of land and water; devastating economic boom-and-bust cycles; influxes of large numbers of non-native temporary workers; and persistent health issues for workers and residents. An early case in the Aboriginal rights movement involved Eldorado's alleged discrimination against Dene workers, discussed later in this chapter, which came to light when former mine workers developed health problems that may have resulted from inadequate protection against radiation.

Mines can leave a legacy of contamination, and Canada's North is strewn with closed and abandoned mines. Discovery gold mine, 80 km northeast of Yellowknife (see map), began production in 1944 and closed in 1969. Left behind were uncontrolled tailings and waste rock piles, mercury and cyanide contamination of the nearby lake, and an abandoned and crumbling town site that was finally demolished in 2005.

Today, greater care is legally required when new mines are developed, to ensure that they will have minimal environmental, social, and health impacts. Mitigation, closure, and restoration plans, and financing for those plans, must be in place before a mine even begins production. However, the legacy of closures, bankruptcies, and abandonment of mine sites in the North persists.

Mining also can cause profound social and economic changes. The inability to maintain traditional food systems and cultural practices can undermine community health and wellness; impacts can include substance abuse and family breakdown. When Giant closed, the economic impacts on Yellowknife were devastating, causing housing prices to plummet. The recent diamond rush has created a new economic boom in Yellowknife and a construction push to house incoming workers, with the result that rents and housing prices are now some of the highest in Canada.

Today the life cycle of a mine, and the involvement of a mining company with a site and its people, no longer ends when the supply/demand for metal and minerals is depleted. The decision to transform an area into a mine now comes with the responsibility of post-operational management and remediation, and a commitment to ongoing quality of life for workers and the community. The new diamond mines at Ekati, Diavik, and Snap Lake have been designed so that, once mining is finished, the land will be reclaimed and restored. There will be no town, no permanent roads, no waste rock or tailings left behind. The new mines also distinguish themselves in the consideration they give to traditional knowledge, including learning from elders about environmental conditions, and applying this knowledge to construction, aquatic monitoring, and other aspects of mine management.

There is no single perspective or consensus on mining among Dene people today. Some call for greater involvement of Dene leaders in mine development. Some wish for a greater share of economic benefits and reduced risks. Others are concerned about the preservation of their land and culture. Decisions about mining in Canada's North must be based not only on scientific assessments but on social and cultural concerns as well. Mining in Denendeh provides an example of how values, beliefs, lifestyles, and traditions interact with economic interests to influence decisions about how to live within our environment.

To read the full version of the article on which this case is based, see www.pimatisiwin.com/index.php?search_val=Denendeh&act=search.

Culture, World View, and the Environment

The Dene have an opportunity to gain substantial economic benefits by allowing mining to proceed on their lands, but they are cognizant of the long history of environmental and cultural damage associated with mining in Canada's North. Trade-offs in which economic benefits and social or ethical concerns appear to be in conflict crop up frequently in environmental issues. In this chapter we will examine some of the underlying causes of such conflicts, how they may be resolved, and how they influence the environmental choices we all make every day.

Ethics and economics involve values

As discussed in Chapter 1 and revisited throughout this book, environmental science examines Earth's natural systems, how they affect humans, and how humans affect them. To address environmental problems, however, requires input from disciplines beyond the natural sciences. It is necessary to understand how people perceive their environment, how they relate to it philosophically and pragmatically, and how they value its elements. Ethics and economics are quite different disciplines, but each deals with questions of what we value and how those values influence our decisions and actions. Anyone trying to address an environmental problem must try to understand not only how natural systems work but also how values shape human behaviour.

Culture and world view influence our perception of the environment

Every action we take affects our environment. Growing food requires soil, cultivation, and irrigation. Building homes requires land, lumber, and metal. Manufacturing and fuelling vehicles require metal, plastic, glass, and petroleum. From nutrition to housing to transportation, we meet our needs by withdrawing resources and altering our surroundings. Decisions about how we manipulate and exploit our environment to meet our needs depend in part on rational assessments of costs and benefits.

Our decisions are also heavily influenced by the culture of which we are a part and by our particular world view. **Culture** can be defined as the ensemble of knowledge, beliefs, values, and learned ways of life shared by a group of people. Culture, together with personal experience, influences each person's perception of the

THE SCIENCE BEHIND THE STORY

The Mirrar Clan Confronts the Jabiluka Uranium Mine

These protestors rallied against the proposed Jabiluka uranium mine.

"The Jabiluka uranium mine will improve the quality of the environment. The uranium resource there has been polluting the river system naturally, probably for thousands of years... With the uranium resource removed and put to good use, the level of radioactivity will fall." —Michael Darby, Australian political commentator

"My country is in danger."—Yvonne Margarula, senior traditional land owner, Mirrar Clan and Goldman Environment Prize winner; to the U.N. World Heritage Committee, 1998

The remote Kakadu region of Australia's Northern Territory is home to several groups of **Aborigines**, the native or **indigenous** people who lived there before British colonization. The region features Kakadu National Park (see photo), a World Heritage Site recognized by the United Nations for its irreplaceable natural and cultural resources. The land also holds uranium, a naturally occurring radioactive metal valued for its use in nuclear power plants, nuclear weapons, and medical and industrial tools. Uranium mining is a key contributor to the Australian national economy, accounting for 7% of Australia's economic output.

The occurrence of uranium deposits on Aboriginal land has led to conflicts between corporations seeking to develop mining operations and Aboriginal people trying to maintain their traditional culture. One such group is the Mirrar Clan, an extended family of Kakadu-area Aborigines. The Mirrar have been living with the region's first uranium mine, the Ranger mine, since the Australian government approved its development on their land in 1978.

When the corporate owners of Ranger proposed to open another uranium mine, Jabiluka, on their land, the Mirrar launched into an environmental battle that would rage over a number of years and several continents. The Mirrar viewed Jabiluka as a threat to their health and to the integrity of their environment, particularly given repeated radioactive spills at the Ranger mine. Many feared that contaminated water would be released into area creeks and that radioactive radon gas would emanate from stored waste materials. Moreover, mindful of geologic faults that exist in the area, the Mirrar worried that dams holding mine waste could fail catastrophically in an earthquake. "We are talking about a uranium mine inside our largest national park," said Peter Robertson, coordinator of the Environment Centre of the Northern Territory at the time. "This is not a place to cut corners."[2]

Environmental activists worldwide joined the Mirrar's struggle. In 1998 nearly 3000 people travelled to the Kakadu region to protest Jabiluka. In late 2002 their efforts finally succeeded. Sir Robert Wilson, chief executive officer of Rio Tinto, the corporation holding rights to the ore body, announced the cancellation of mining plans at Jabiluka, citing economic factors (declining world uranium prices) and ethical factors (concerns about developing the mine without Mirrar consent). Wilson added that the company planned to rehabilitate the site and restore damage done there during exploration and assessment.

Since that time, the price of uranium has risen on the world market. The corporation's plans are now in a holding pattern as it waits and hopes that the Mirrar will one day give their consent. In 2007 the new CEO of Rio Tinto, Tom Albanese, reiterated the company's intention not to pressure the government for approval to develop Jabiluka without the prior informed consent of the traditional landowners.[3] **Prior informed consent**, one of the hallmark principles of modern environmental ethics, means that consent or acceptance of an activity (such as land development or waste disposal) is not legally valid unless the consenting person or group has been properly and adequately informed *and* can be shown to have a reasonable understanding of all potential impacts before giving consent.

The Mirrar opposed the mine development despite the economic benefits promised to them in the form of jobs, income, development, and a higher material standard of living. The decision to bypass these economic incentives was not easy; indeed, other Aboriginal groups in the Kakadu region supported the mine development. In formulating their approaches to the mining proposal, the Mirrar and others had to weigh economic, social, cultural, and philosophical questions as well as scientific ones.

This is Kakadu National Park, Australia, where Aboriginal residents, environmentalists, and mining companies have battled over mining rights.

world and his or her place within it, something described as the person's **world view.** A world view reflects a person's (or group's) beliefs about the meaning, operation, and essence of the world.

People with different world views can study the same situation and review identical data yet draw dramatically different conclusions. For example, many well-meaning people support mining in Canada's North, while many other well-meaning people oppose it. The officers, employees, and shareholders of the mining companies, and the government officials who support mining, view it as a source of jobs, income, energy, and economic growth. They believe mining will benefit the North in general and the Dene in particular. Opponents, in contrast, foresee environmental problems and negative social consequences. They recognize that mining disturbs the landscape and can pollute air and water, while community disruption, substance abuse, and crime can accompany mining booms.

In Australia, too, there have been conflicts between corporations seeking to develop mining operations and Aboriginal people trying to maintain their traditional culture. One such group is the Mirrar Clan, who have been living with the region's first uranium mine, the Ranger mine, on their land since 1978 (see "The Science Behind the Story: The Mirrar Clan Confronts the Jabiluka Uranium Mine").

Uranium mining is a key contributor to Australia's economy; however, many of Australia's uranium deposits occur on Aboriginal lands in the remote region of Kakadu. When the corporate owners of the Ranger mine proposed to open another uranium mine on their land, the Mirrar fought back. They saw the mine as a threat not only to their health and the integrity of the environment but also to their culture and religion, which are deeply tied to the landscape. The proposed mine site is near traditional hunting and gathering sites, in the floodplain of a river that provides the clan with food and water. Like many other Aborigines, the Mirrar hold the landscape to be sacred, and they depend on its resources for their daily needs.

Many factors shape our world views and perception of the environment

The traditional culture and world view of the Mirrar Clan helped shape its response to the proposed Jabiluka mine. Australian Aborigines view the landscape around them as the physical embodiment of stories that express the beliefs and values central to their culture. The landscape to them is a sacred text, analogous to the Bible in Christianity, the Koran in Islam, or the Torah in Judaism. Australian Aborigines believe that spirit ancestors possessing human and animal features travelled routes called "dreaming tracks," leaving signs and lessons in the landscape. By explaining the origins of specific landscape features, dreaming-track stories assign meaning to notable landmarks and help Aborigines construct detailed mental maps of their surroundings. The stories also teach lessons concerning family relations, hunting, food gathering, and conflict resolution. The Mirrar who opposed the Jabiluka uranium mine believed that it would desecrate sacred sites and compromise their culture (**FIGURE 21.1**).

Similarly, many Aboriginal people of Canada's North, including the Dene and Innu, believe that the landscape is inhabited by spirits, both benevolent and malevolent, and thus must be honoured and protected.[4] They worry that mining may have negative impacts on their lands, traditional hunting routes, water sources, or resource-gathering sites. There is also concern that noise, disruption, and emissions from such operations might cause harm to sacred animals or to sites that have spiritual or cultural significance.

Religion and spiritual beliefs are among many factors that can shape people's world views and perception of the environment. A community may also share a particular view of the environment if its members have lived through similar experiences. For example, early European settlers in both Australia and North America viewed their environment as a hostile force because inclement weather,

---

weighing the issues

ECOIMPERIALISM? 21-1

The Mirrar Clan opposed the development of the Jabiluka uranium mine on their land, in spite of potential economic benefits. They were supported in this effort by an extensive international network of environmentalists. How do you think those environmentalists would have reacted if the Mirrar, having gained self-determination over their land, had changed their position and opted to approve the mine? The term **ecoimperialism** has been used to describe the imposition of Western environmental priorities and values on Aborigines and people in developing nations. Do you believe that the right to self-determination *obliges* Aboriginal people to retain their traditional way of life and preserve the integrity of their ancestral lands? Does your view extend to Aborigines in the tropical rainforests of the Brazilian Amazon? What about Aboriginal people in Canada's North?

(a) Ranger Uranium Mine, Kakadu region, Australia

(b) Voisey's Bay Nickel Mine, northern Labrador, Canada

FIGURE 21.1
The Ranger mine (a), located on Australian Aboriginal lands amid sacred sites, caused enough environmental impacts to spark fierce opposition to the proposed Jabiluka mine nearby. (b). The Voisey's Bay mine and concentrator site in northern Labrador faced similar challenges from both Innu Nation and Nunatsiavut Government The mine's owner, Vale Inco negotiated special agreements with the two Aboriginal groups prior to the opening of the mine, dealing with land rights, social and economic considerations, and numerous environmental concerns.

weighing the issues

MINING IN MECCA . . . ?

21–2

Suppose a mining company discovered uranium near the Sacred Mosque at Mecca—or the site in Bethlehem believed to be the birthplace of Jesus or the Wailing Wall in Jerusalem. What do you think would happen if the company announced plans to develop a mine close to one of these sacred locations, assuring the public that environmental impacts would be minimal and that the mine would create jobs and stimulate economic growth? Which aspects of these unlikely situations resemble that of the Jabiluka case, and which are different?

wild animals, and other natural forces frequently destroyed crops, killed livestock, and took settlers' lives. Such experiences were shared in stories and in songs and helped shape prevailing social attitudes in many frontier communities. The view of nature as a hostile force and an adversary to be overcome has passed from one generation to the next and still influences the way many North Americans and Australians view their surroundings.

Political ideology also shapes a person's attitude toward the environment. For instance, one's views on the role of government will influence whether or not one wants government to intervene in a market economy to protect environmental quality. Economic factors also sway how people perceive their environment and make decisions. An individual with a strong interest in the outcome of a decision that may result in his or her private gain or loss is said to have a *vested interest*. Mining company executives and shareholders have a vested interest in a decision to open an area to mining because a new mine can increase profits. Vested interests may lead people to view a proposed mine as a source of economic gain, while minimizing (whether consciously or subconsciously) the potential for negative environmental impacts.

There are many ways to understand the environment

An interesting aspect of the relationship of indigenous peoples with their local environment is **traditional ecological knowledge (TEK)** (or *indigenous ecological knowledge*), the intimate knowledge of a particular environment possessed and passed along by those who have inhabited an area for many generations.[6] Examples include knowledge of the medicinal properties of local plants; wintering-over or migration habits of local animals; local geographic and microclimatic variations; or the sequence of tasks required to carry out a traditional task, such as trapping and butchering a large animal. Such a deep understanding is gained through generations of hunters, fishers, gatherers, and harvesters passing along their knowledge of the natural world, usually by way of oral teachings, songs, and storytelling.

In some circumstances, TEK can be assigned a market value. For example, indigenous knowledge of local plants might be extremely valuable to a pharmaceutical company searching for plants with modern medicinal applications. In recent years the value of TEK has become more widely recognized, acknowledged, and remunerated by governments and industry. For example, the Nunavut Wildlife Management Board, which meets annually to set limits on the annual polar bear hunt for Inuit traditional hunters, receives and weighs information from government scientists and from the hunters themselves as part of the decision-making process.

Throughout this book you have encountered scientific data regarding the environmental impacts of our choices (where to make our homes, how to make a living, what to wear, what to eat, how to travel, how to spend our leisure time, and so on). Culture, world views, and values play critical roles in such choices and even can influence the interpretation of scientific data. Thus, acquiring a foundation of scientific understanding is only one part of the search for solutions to environmental problems. Attention to ethics and economics helps us understand why and how we value those things we value.

Environmental Ethics

The field of **ethics** is a branch of philosophy that involves the study of good and bad, right and wrong. The term *ethics* can also refer to the set of moral principles or values held by a person or a society. Ethicists help clarify how people judge right from wrong by elucidating the criteria, standards, or rules that people use in making these judgments. Such criteria are grounded in values—for instance, promoting human welfare, maximizing individual freedom, or minimizing pain and suffering.

weighing the issues 21-3

THE ATLANTIC SEAL HUNT

No environmental issue identified with Canada is more emotionally charged than the Atlantic seal hunt. Each year environmentalists and animal activists around the world mobilize to try to stop the hunt, arguing that too many seals are killed, that the practice of killing seals is barbaric, and that the methods used are inhumane. The hunters and their supporters counter that they are continuing a way of life that has been practised by Aboriginal people in Canada for at least 4000 years, that it is their right to practise their traditional ways, and that the hunt is vital for the economic well-being and survival of their communities.

What do you think? Who should decide which of these sets of values—animal rights or Aboriginal self-determination—should take precedence in this case?

People of different cultures or with different world views may differ in their fundamental values and thus may differ in the specific actions they consider to be right or wrong. This is why some ethicists are **relativists**; that

weighing the issues 21-4

VALUING TEK

Let us say that you are a researcher working for a large pharmaceutical company. You are doing botanical fieldwork, searching for a plant that may offer a new cure for cancer. Some of the indigenous people in the area have a deep understanding of the medicinal properties of local plants, and you would like to ask them some questions. Under what circumstances should you do this? What if you were to make a major discovery on the basis of something you learned from them, with potential earnings of billions of dollars for your company—would the people who passed along the crucial information have a legitimate claim to part of those earnings?

---

is, they believe that ethics do and should vary with social context. However, many ethicists are **universalists**; that is, they maintain that there exist some fundamental, objective notions of right and wrong, good and bad, that hold across cultures and situations. For both relativists and universalists, ethics is a *normative* or *prescriptive* pursuit; it tells us how we *ought to* behave.

Ethical standards are the criteria that help differentiate right from wrong. One classic ethical standard is the *categorical imperative* proposed by philosopher Immanuel Kant, which roughly approximates the "golden rule" common to many of the world's great religions. For example, Hindus learn that they should "not do to others what would cause pain if done to you"; a central tenet of Buddhism is to "hurt not others in ways that you yourself would find hurtful"; and Christians are encouraged to "do unto others as you would have others do unto you." The universality of the golden rule or categorical imperative makes it a fundamental ethical standard.

Another ethical standard is the principle of *utility*, elaborated by British philosopher John Stuart Mill, among others. The **utilitarian principle** holds that something is right when it produces the greatest practical benefits for the most people. For example, a utilitarian might argue that forest biodiversity should be conserved because the possibility exists that a cure for cancer might be found there among the naturally occurring biological compounds. The argument that forest species should be preserved because they have an *intrinsic value*, or an inherent right to exist, would be much less convincing to a utilitarian. We all employ such ethical standards as tools for making decisions, consciously or unconsciously, in our everyday lives.

Environmental ethics pertains to humans and the environment

The application of ethical standards to relationships between humans and nonhuman entities is known as **environmental ethics.** This relatively new branch of ethics arose once people began to perceive environmental changes brought about by industrialization. Human interactions with the environment frequently give rise to ethical questions that can be difficult to resolve. Consider some examples:

1. Does the present generation have an obligation to conserve resources for future generations? If so, how should this influence our decision making, and how much are we obligated to sacrifice?
2. Are there situations that justify exposing some communities to a disproportionate share of pollution? If not, what actions are warranted in preventing this problem? By extension, if a certain community stands to gain the most from a particular activity, should that community be expected to take on most of the risk associated with the activity?
3. Are humans justified in driving species to extinction? Are we justified in causing other permanent changes in ecological systems? If destroying a forest would drive extinct an insect species few people have heard of, but would create jobs for 10 000 people, would that action be ethically admissible? What if it were an owl species, or an ape, or a whale? What if only 100 jobs would be created? What if it were a species that is harmful to humans, such as mosquitoes? What about a bacterium or a virus?

The intergenerational question—whether we owe consideration to those who will live on this planet and make use of its resources years from now—is of particular interest. The most common definition of *sustainable development* (Chapter 1) says that we must meet our current needs without compromising the availability of natural resources or the quality of life for future generations. But how can we tell what future generations may need or want, or what they will value or hold sacred? In 2007 and 2008 construction crews began digging trenches for the laying of an oil pipeline extension through pristine wilderness areas of Jasper National Park and Mount Robson Provincial Park in the Canadian Rockies. The extension, which will allow for the movement of an extra 40 000 barrels of oil each day from Alberta to markets in the United States, was approved in 1952. Although the construction work is being carefully monitored, there is necessarily some environmental disruption associated with this activity. If the pipeline extension had been requested today, it is highly unlikely that it would have been approved. Would it have been possible to know, in 1952, how this decision would be viewed more than 50 years later? And how can we determine how the environmental decisions that we make today will be viewed 50 years into the future?

We have extended ethical consideration to more entities through time

Answers to questions like those above depend partly on what ethical standard(s) a person chooses to use. They also depend on the breadth and inclusiveness of the person's domain of ethical concern. A person who feels responsibility for the welfare of insects would answer the third question very differently from a person whose domain of ethical concern ends with humans. Most of us feel moral obligations to some entities in the world but by no means to all.

Throughout the history of Western cultures (i.e., European and European-derived societies), people have gradually enlarged the array of entities they feel deserve ethical consideration. The enslavement of human beings by other human beings was common in many societies until recently, for instance. Women were not allowed to vote in Canada until 1916 (even then, only in Manitoba), and many still receive lower pay for equal work. Consider, too, how little ethical consideration citizens of one nation generally extend to those of another on which their government has declared war. Human societies are only now beginning to embrace the principle that all people should be granted equal ethical consideration.

Our expanding domain of ethical concern has begun to include nonhuman entities as well. Mahatma Gandhi reportedly said, "The greatness of a nation and its moral progress can be judged by the way its animals are treated." Concern for the welfare of domesticated animals is evident today in humane societies and in the way many people provide for their pets. Animal-rights activists voice concern for animals that are hunted or used in laboratory testing. Most people now accept that wild animals (at least obviously sentient animals, such as primates and other large vertebrates, with which we share similarities) merit ethical consideration.

Today many environmentalists are concerned not only with certain animals but also with the well-being of whole natural communities. Some have gone still further, suggesting that all of nature—living and nonliving things, even rocks—should be ethically represented (**FIGURE 21.2**). If you think this is a silly idea, consider how you might react if someone put a fast-food restaurant on the top of Mt. Everest, or if a multinational corporation

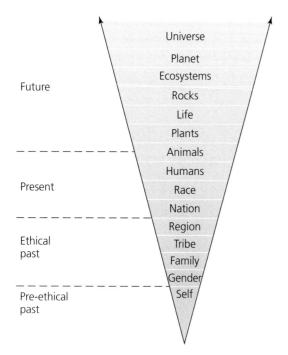

FIGURE 21.2
Through time, people in Western cultures have broadened the scope of their ethical consideration for others. We can view ethics progressing through time in a generalized way outward from the self.

decided to paint a gigantic corporate logo on the Moon. Do the unique landmarks of the natural environment deserve ethical consideration? Do they have any inherent value, or the "right" to exist unaltered? Is their value distinct from the services they may render to humans, or inseparable from human interests? These questions are all worth considering.

What is behind this ongoing expansion of ethical consideration? Rising economic prosperity in Western cultures, as people became less anxious about their day-to-day survival, has helped enlarge our ethical domain. Science has also played a role in demonstrating that humans do not stand apart from nature but rather are part of it. Ecology, as it has developed over the past 75 years, has made clear that all organisms are interconnected, and what affects plants, animals, and ecosystems can in turn affect humans. Evolutionary biology has shown that humans are merely one species out of millions and have evolved subject to the same pressures as other organisms.

For many non-Western cultures, expansive ethical domains are nothing new. Many traditional cultures have long granted ethical standing to nonhuman entities. Aborigines, like the Mirrar, Innu, and Dene, who view their landscape as sacred and alive, are a case in point. However, it is worthwhile to examine Western ethical expansion because it underlies so many of modern secular society's beliefs and actions regarding the environment.

We can simplify the continuum of attitudes toward the natural world by dividing it into three ethical perspectives: anthropocentrism, biocentrism, and ecocentrism.

Anthropocentrism Anthropocentrism takes a human-centred view of our relationship with the environment. An anthropocentrist denies or ignores the notion that nonhuman entities can have rights and measures the costs and benefits of actions solely according to their impact on people (**FIGURE 21.3**). To evaluate an action that affects the environment, an anthropocentrist might use such criteria as impacts on human health, economic costs and benefits, and esthetic concerns.

For example, if a mine provides a net economic benefit while doing no harm to human health and having little esthetic impact, the anthropocentrist would conclude it was a worthwhile venture, even if it might drive some native species extinct. If protecting the area would provide spiritual, economic, or other benefits to humans now or in the future, an anthropocentrist might favour its protection. In the anthropocentric perspective, anything not providing benefit to people is considered to be of negligible value.

Biocentrism In contrast to anthropocentrism, **biocentrism** ascribes values to actions, entities, or properties on the basis of their effects on all living things or on the integrity of the biotic realm in general (see **FIGURE 21.3**). In this perspective, all life has ethical standing. A biocentrist evaluates actions in terms of their overall impact on living things, including—but not exclusively focusing on—human beings. In the case of a mine proposal, a biocentrist might oppose the mine if it posed a serious threat to the abundance and variety of living things in the area, even if it would create jobs, generate economic growth, and pose no threat to human health. Some biocentrists advocate equal consideration of all living things, whereas others advocate that some types of organisms should receive more than others.

Ecocentrism Ecocentrism judges actions in terms of their benefit or harm to the integrity of whole ecological systems, which consist of biotic and abiotic elements and the relationships among them (see **FIGURE 21.3**). An ecocentrist would value the well-being of entire species, communities, or ecosystems over the welfare of a given individual. Implicit in this view is that the preservation of larger systems generally protects their components, whereas selective protection of the components may not always safeguard the entire system. Ecocentrism is a *holistic* perspective. Not only does it encompass a wide variety of entities, but it also stresses preserving the connections that tie the entities together into functional systems.

FIGURE 21.3
An anthropocentrist extends ethical standing only to humans and judges actions in terms of their effects on humans. A biocentrist values and considers all living things, human and otherwise. An ecocentrist extends ethical consideration to living and nonliving components of the environment. The ecocentrist also takes a holistic view of the connections among these components, valuing the larger functional systems of which they are a part.

Environmental ethics has ancient roots

Environmental ethics arose as an academic discipline in the early 1970s, but people have contemplated our relationship with nature for thousands of years. Ancient Aboriginal oral traditions and dreaming-track stories treat the environment as a source of sacred teachings, so boulders, caves, patches of lichen, and other entities are felt to have moral significance worthy of contemplation and protection. In the Jain Dharma, one of the oldest religious traditions in the world, compassion for all life—human and nonhuman—is a core belief. Jains are typically vegetarian or vegan and try to avoid foods obtained with unnecessary cruelty; for some, this even involves refusing to eat root vegetables (such as potatoes and onions) to avoid killing the plant from which they were obtained. In the Western tradition, the ancient Greek philosopher Plato expressed what he considered humans' moral obligation to the environment, writing, "The land is our ancestral home and we must cherish it even more than children cherish their mother."

Some ethicists and theologians have pointed to the religious traditions of Christianity, Judaism, and Islam as sources of anthropocentric hostility toward the environment. They point out biblical passages such as, "Be fruitful and multiply, and fill the earth and subdue it; and have dominion over the fish of the sea and over the birds of the air and over every living thing that moves upon the earth." Such wording has justified and encouraged separation from and animosity toward nature over the centuries, some scholars say. Others interpret sacred texts of these religions to encourage benevolent human stewardship over nature. Consider the directive, "You shall not defile the land in which you live. . . ." Although people have held differing views of their ethical relationship with the environment for millennia, environmental impacts that became apparent during the Industrial Revolution intensified debate about our relationship with the environment.

The Industrial Revolution inspired environmental philosophers

As the Industrial Revolution (Chapter 6) spread from Great Britain throughout Europe and elsewhere, its technological advances and resultant population growth amplified human impacts on the environment. In this period of social and economic transformation, agricultural economies became industrialized, machines enhanced or replaced human and animal labour, and much of the rural population moved into cities. Consumption of natural resources accelerated, and pollution increased dramatically as coal combustion fuelled railroads, steamships, ironworks, and factories.

Many writers and philosophers of the time criticized the drawbacks of industrialization. British critic John Ruskin (1819–1900) called cities "little more than laboratories for the distillation into heaven of venomous smokes and smells." Ruskin also complained that people prized the material benefits that nature could provide, but no longer appreciated its spiritual and esthetic benefits. Motivated by similar concerns, a number of citizens' groups sprang up in nineteenth-century England that could be considered some of the first environmental organizations. These included the Commons Preservation Society, the Coal Smoke Abatement Society, and the Selborne League, dedicated to the protection of rare plants, birds, and landscapes.

During the 1840s, a philosophical movement called *transcendentalism* flourished, espoused primarily by American philosophers Ralph Waldo Emerson and Henry David Thoreau and poet Walt Whitman. The transcendentalists viewed nature as a direct manifestation of the divine, emphasizing the soul's oneness with nature and God. They objected to what they saw as their fellow citizens' obsession with material things. Through their writings the transcendentalists promoted a holistic view of nature. They identified a need to experience wild nature and portrayed natural entities as symbols or messengers of a deeper truth. Thoreau viewed nature as divine, but he also observed the natural world closely, in the manner of a scientist; he was in many ways one of the first ecologists. His book *Walden*, in which he recorded his observations and thoughts while he lived at Walden Pond away from the bustle of urban Massachusetts, is a classic of philosophical and environmental literature.

Conservation and preservation arose at the start of the twentieth century

One admirer of Emerson and Thoreau was John Muir (1838–1914), a Scottish immigrant who eventually settled in California and made the Yosemite Valley his wilderness home. Although Muir chose to live in isolation in his beloved Sierra Nevada for long stretches of time, he nonetheless became politically active and won fame as a tireless advocate for the preservation of wilderness (**FIGURE 21.4**).

Muir was motivated by the rapid deforestation and environmental degradation he witnessed throughout North America and by his belief that the natural world should be treated with the same respect that cathedrals receive. Today he is associated with the **preservation ethic**, which holds that we should protect the natural environment in a pristine, unaltered state. Muir argued

FIGURE 21.4
A pioneering advocate of the preservation ethic, John Muir is remembered for his efforts to protect the Sierra Nevada from development and for his role in founding the Sierra Club. Muir (right) is shown with President Theodore Roosevelt in Yosemite National Park. After his 1903 wilderness camping trip with Muir, the president instructed his interior secretary to increase protected areas in the Sierra Nevada.

that nature deserved protection for its own inherent value (an ecocentrist argument), but he also maintained that nature played a large role in human happiness and fulfillment (an anthropocentrist argument). "Everybody needs beauty as well as bread," he wrote in 1912, "places to play in and pray in, where nature may heal and give strength to body and soul alike."

Canadian James Bernard Harkin (1875–1955) also believed in preserving the beauty of nature. He was strongly influenced by the writings of Muir, his contemporary. Harkin was the first commissioner of Dominion Parks (which eventually became Parks Canada) and is credited with saving vast areas of Canadian wilderness from development (**FIGURE 21.5**). He believed in the spiritual, healing, and restorative power of nature. He was drawn to mountains, which, he said, "elevate the mind and purify the spirit." Harkin believed that setting aside land as national parks was only the beginning of preservation; the real challenge would be to maintain them as wilderness. Resource extraction and even some vehicle use were limited in national parks during his service, and he came to be known as the "Father of National Parks" of Canada.[7]

Some of the factors that motivated Muir and Harkin also inspired American forester Gifford Pinchot (1865–1946), who opposed rapid deforestation and unregulated economic development of land. However, Pinchot took a more anthropocentric view of how and why nature should be valued. He is today the person most closely associated with the **conservation ethic**, which

FIGURE 21.5
Kootenay National Park **(a)** (shown here, Dog Lake and the Mitchell Mountain Range) was designated as a national park in 1920. It was one of 11 national parks designated during the tenure of James Harkin **(b)** as Canada's commissioner of Dominion Parks. Mount Harkin in the Mitchell Range is named in his honour.

(b) James Harkin

(a) Kootenay National Park

holds that humans should put natural resources to use but also that we have a responsibility to manage them wisely. Whereas preservation aims to preserve nature for its own sake and for the esthetic, spiritual, symbolic, and recreational benefit of people, conservation promotes the prudent, efficient, and sustainable extraction and use of natural resources for the benefit of present and future generations. The conservation ethic uses a utilitarian standard, stating that in using resources, humans should attempt to provide the greatest good to the greatest number of people for the longest time.

Pinchot's counterpart in Canada was Clifford Sifton (1861–1929), a controversial politician and conservationist. As minister of the interior, Sifton aggressively lured immigrants to settle and farm in the West; he was deeply committed to the agricultural development of land as Canada's principal natural resource—not an obvious conservationist position. However, Sifton was also a "champion" of Canada's natural resources, and was particularly devoted to forest conservation and reforestation. He was the first chairman of the Commission for the Conservation of Natural Resources, which undertook detailed inventories of Canadian natural resources.[8]

Conservation and preservation are rooted in fundamentally different ethical approaches, which often meant that advocates were pitted against one another on policy issues of the day. Nonetheless, both branches represented reactions against the prevailing "frontier development ethic," which held that humans should be masters of nature, and which promoted economic development without regard to its negative consequences. Those who led the conservation and preservation movements in the nineteenth and early twentieth centuries left legacies that reverberate today in our ethical approaches to the environment.

> **weighing the issues** 21-5
>
> **PRESERVATION AND CONSERVATION**
>
> With which ethic do you most identify—preservation or conservation? Think of a forest or other important natural resource in your region. Give an example of a situation in which you might adopt a preservation ethic and an example of one in which you might adopt a conservation ethic. Are there conditions under which you would follow neither, but instead adopt a "development ethic"? What was your reaction when you learned about the extension of the oil pipeline through wilderness areas in Jasper National Park? Would you classify your reaction as preservationist, conservationist, or utilitarian?

The land ethic and deep ecology enlarged the boundaries of the ethical community

As a young forester and wildlife manager, Aldo Leopold (1887–1949) (**FIGURE 21.6**) began his career as a conservationist. At first, he embraced the government policy

FIGURE 21.6
Aldo Leopold, wildlife manager and pioneering environmental philosopher, articulated a new relationship between people and the environment. In his essay "The Land Ethic," he called on people to include the environment in their ethical framework.

of shooting predators, such as wolves, to increase populations of deer and other game animals. At the same time, however, Leopold was following the development of ecological science. He eventually ceased to view certain species as "good" or "bad" and instead came to see that healthy ecological systems depend on the protection of all their interacting parts, including predators as well as prey. Drawing an analogy to mechanical maintenance, he wrote, "to keep every cog and wheel is the first precaution of intelligent tinkering," the quotation with which we opened Chapter 3.

Leopold argued that humans should view themselves and "the land" as members of the same community, and that people are obliged to treat the land in an ethical manner based on mutual respect. In his 1949 essay "The Land Ethic," he wrote:

> *All ethics so far evolved rest upon a single premise: that the individual is a member of a community of interdependent parts . . . The land ethic simply enlarges the boundaries of the community to include soils, waters, plants, and animals, or collectively: the land . . . A land ethic changes the role of Homo sapiens from conqueror of the land-community to plain member and citizen of it . . . It implies respect for his fellow-members, and also respect for the community as such.*

Leopold intended that the land ethic would help guide decision making. "A thing is right," he wrote, "when it tends to preserve the integrity, stability, and beauty of the biotic community. It is wrong when it tends otherwise." Many today view Aldo Leopold as the most eloquent and important philosopher of environmental ethics.

One philosophical perspective that goes beyond even ecocentrism is **deep ecology**, which emerged in the 1970s. Proponents describe the movement as resting on principles of "self-realization" and biocentric equality. They define self-realization as the awareness that humans are inseparable from nature and that the air we breathe, the water we drink, and the foods we consume are both products of the environment and integral parts of us. Biocentric equality is the concept that all living beings have equal value and that because we are truly inseparable from our environment, we should protect all living things as we would protect ourselves.

Ecofeminism recognizes connections between the oppression of nature and of women

As deep ecology and mainstream environmentalism were extending people's ethical domains during the 1960s and 1970s, major social movements, such as the civil rights movement and the feminist movement, were gaining prominence. A number of feminist scholars saw parallels in human behaviour toward nature and men's behaviour toward women. The degradation of nature and the social oppression of women shared common roots, these scholars asserted.

Ecofeminism argues that the patriarchal (male-dominated) structure of society—which traditionally grants more power and prestige to men than to women—is a root cause of both social and environmental problems. Ecofeminists hold that a world view traditionally associated with women, which interprets the world in terms of interrelationships and cooperation, is more compatible with nature than a world view traditionally associated with men, which interprets the world in terms of hierarchies and competition. Ecofeminists maintain that a male tendency to try to dominate and conquer what men hate, fear, or do not understand has historically been exercised against both women and the natural environment.

One of the most interesting environmental movements of our time is Chipko Andolan, which had its philosophical grounding in the principles of Gandhian nonviolent resistance, grassroots social activism, and ecofeminism.[9] Chipko (which literally means "hug" or "stick to" in Hindi, and probably led to the use of the term "tree hugger" in reference to environmentalists) emerged in the early 1970s in the northern Uttarakhand region

(a) Villagers placed their bodies between the trees and the contractor's axes.

(b) Wangari Maathai

FIGURE 21.7
In the early 1970s, grassroots resistance efforts by village women led to the establishment of the Chipko Movement, dedicated to preventing deforestation in northern India's Himalayan foothills **(a)**. The founder of Kenya's Green Belt Movement, Professor Wangari Maathai **(b)**, was awarded the Nobel Peace Prize in 2004 for her work to empower women and fight deforestation.

of India, as an effort to stop clear-cutting from decimating the vast forests of northern India. The movement came to a climax in 1973, when government workers turned up unannounced to cut trees but were met by a group of village women who refused to allow the work to proceed (**FIGURE 21.7A**). Leader Gaura Devi reportedly stated, "The forest is like our mother. You will have to shoot us before you can cut it down." The women stood watch over the forest, wrapping their arms around the trees. Eventually, after considerable negotiation, the government declared the region to be an environmentally "sensitive" area. Today, Chipko is an international icon of grassroots environmentalism.

Another movement rooted in ecofeminism and the empowerment of the poor is the Green Belt Movement of Kenya. This organization, which began by paying impoverished village women to plant tree seedlings, was founded in 1977. The Nobel Peace Prize was awarded to Green Belt's founder, Wangari Maathai, in 2004 (**FIGURE 21.7B**).[10]

Environmental justice seeks equitable access to resources and protection from environmental degradation

Our society's domain of ethical concern has been expanding from rich to poor and from majority races and ethnic groups to minority ones. This ethical expansion involves applying a standard of fairness and equality and has given rise to the environmental justice movement. **Environmental justice** is based on the principle that all people—regardless of race, colour, national origin, or income—have the right to live and work in a clean, healthy environment; to receive protection from the risks and impacts of environmental degradation, and to be compensated for having suffered such impacts; and to have equitable access to environmental resources of high quality.

The environmental justice movement is fuelled by the fact that the poor and minorities tend to be exposed to a greater share of pollution, hazards, and environmental degradation than are richer people and whites. This has been supported by scientific research (**FIGURE 21.8**). For example, studies have found the percentage of minorities in areas with toxic waste sites to be twice that in areas without toxic waste sites. Researchers in many parts of the world who study air pollution, lead poisoning, pesticide exposure, and workplace hazards have found similar patterns.

A protest in the early 1980s by black people in Warren County, North Carolina, against a toxic waste dump in their community is widely seen as the beginning of the movement in North America. The state had chosen to site the dump in the county with the highest percentage of blacks, prompting residents to suspect "environmental racism." Environmental justice grew in prominence as more people began fighting environmental hazards in their communities.

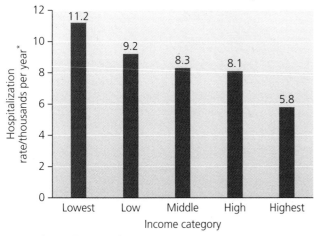

FIGURE 21.8
Many studies have demonstrated links between exposure to pollution and socioeconomic status. This is demonstrated by a graph comparing incidences of hospitalization for respiratory illnesses—a common health impact of exposure to environmental pollutants—among children of different income levels in Toronto.

The early environmental justice movement—like the Chipko and Green Belt movements internationally but in contrast to early environmental movements in North America—was made up largely of low-income people and minorities. Today the movement has broadened to encompass worker health and safety; education and financial services for preparation for, response to, and recovery from natural disasters; and access to land, water, and other environmental resources worldwide.

There are two basic ways that environmental injustice is manifested. First, a community or group can be denied equitable access to environmental resources. This often occurs where poverty is common, or where land of high quality is scarce. For example, in many developing countries rich landowners control the best agricultural land, while poor subsistence farmers are left to scratch out a living on small patches of unsuitable land. In North America we may see this inequity expressed as expensive beachfront or mountain-view properties that are available only to the wealthiest few, or as limited access to an expensive diet rich in vitamins and nutrients.

Second, a community or group can be subjected to environmental injustice by having disproportionate risks or costs of pollution or degradation transferred to them, as in the example (above) of toxic waste dumps sited in communities of racial or ethnic minorities. In less economically developed parts of the world, the poor and other marginalized groups are directly dependent on the environment for survival and therefore suffer the harshest and most immediate impacts of environmental degradation. Cultural differences also can lead to disproportionate exposure of specific groups to the effects of environmental degradation. For example, the traditional diet of many of Canada's Aboriginal people is heavily reliant on fish and marine mammals, which tend to have high concentrations of harmful pollutants, such as mercury.

Aboriginal groups, like the Mirrar and others, struggling to maintain a traditional lifestyle have been linked with the environmental justice movement. Many such struggles originate not as environmental causes but as *social justice* causes. The rubbertappers or *seringueiros* of the Brazilian Amazon are representative of causes that originate at this intersection of environmental justice and social justice. The *seringueiros* and their leader, Chico Mendes (**FIGURE 21.9**), caught the world's attention when it became apparent that the battle to save their traditional economy—based on the sustainable extraction of latex and other forest resources—was intimately connected with efforts to stop deforestation of the Amazonian rainforest for ranching.

The *seringueiros* forged a partnership with indigenous Amazonians, who face many of the same challenges to their traditional lands, resources, and lifestyle. Chico Mendes thought of himself first as a social activist, fighting on behalf of his people and their way of life. Would the international environmental movement have embraced his cause if it had *not* been intimately connected with the rainforest? Sadly, Chico Mendes' iconic status and effectiveness as an activist on behalf of his people and their beloved forests and traditions were

FIGURE 21.9
Chico Mendes, seen here shortly before his assassination in 1988, was a social activist on behalf of the traditional lifestyle of his people, but he caught the eye of the world as an environmental activist struggling to preserve the Amazon rainforest.

> ### weighing the issues
>
> **ENVIRONMENTAL JUSTICE AT HOME** 21-6
>
> Consider the place where you live. Where are the factories, waste dumps, and polluting facilities located? Who lives closest to them? Are there certain groups that live in substandard housing, where they might be exposed to environmental hazards? What could be done to ensure that poor communities are no more polluted than wealthy ones? Do all groups have access to the same level of nutrition and quality of food? If your community is prone to environmental hazards, such as earthquakes, floods, or tornadoes, do people from all walks of life have equal access to education, preparation, insurance, and recovery services to deal with these hazards?

dramatically heightened by his assassination, in 1988, by Brazilian ranchers.

Critics have characterized the attempts of the predominantly white Australian government and uranium mining companies to open mines on traditional lands of the Mirrar as environmental injustice. In North America, as well, uranium mining has been a focus of environmental justice concerns. As discussed in the Central Case, Dene mineworkers from the Northwest Territories and Saskatchewan, and Navajo mineworkers from the United States, suffered delayed health effects that may have been caused by working in uranium mines with minimal safeguards.

Dene workers were hired as early as 1930 to haul radioactive uranium ore wrapped in cloth bags at Eldorado. The governments of Canada (then the world's largest supplier of uranium) and the United States (main purchaser of the ore) were aware of the health hazards of dealing with radioactive ores, as documented in government publications from the time.[11] However, these concerns were not communicated to Dene workers, many of whom did not speak English, and protective equipment was not provided. For decades, until the mine was shut down in the 1960s, Dene miners "slept on the ore, ate fish from water contaminated by radioactive tailings and breathed radioactive dust while on the barges, docks and portages," according to the *Calgary Herald*.[12]

The later deaths, from lung cancer, of many Port Radium miners led to the nickname "Village of Widows" for the settlement of Deline, and the Dene Nation called for an official response from the federal government. In 2005 the government of Canada issued the results of its study, which concluded that there was insufficient evidence to link the deaths of the miners with exposure to radiation during mining work at Port Radium. The report acknowledged that the mine had had an impact on soil and water quality in the immediate vicinity of Great Bear Lake and called for the proper decommissioning of the site, with provisions for continued environmental monitoring. Compensation for residents and the remaining miners was not addressed in the report.[13]

Although cases of lung cancer began to appear among uranium mineworkers in the early 1960s, scientific studies of radiation's effects on miners at the time specifically excluded Aboriginal mineworkers. The decision to include only white miners in those studies was attributed to the researchers' desire to study a "homogeneous population." A later generation would perceive this as negligence and discrimination.

In the United States, the *Radiation Exposure Compensation Act* of 1990 compensated Navajo miners who suffered health effects from unprotected work in the mines. Even the compensation process, though, has been controversial; as of 2006, approximately 80% of the $300 million allocated to compensate these miners and their families had been allocated to non-native miners. Many of the Navajo miners lacked the extensive documentation required to apply for compensation, including medical documentation and records of the exact dates and conditions in which they worked. Some Navajo miners also have been denied compensation because they fail to qualify as "nonsmokers" as a result of having participated in traditional native ceremonies involving smoke.[14]

Cases like these illustrate the interplay between changing ethical values and resultant policy making, which we will examine further in Chapter 22. First, we will explore economics, which, like ethics, addresses people's values, influences behaviour, and widely informs policy.

Economics: Approaches and Environmental Implications

People who oppose mining and other developments in wilderness areas or on traditional lands typically do so on the basis of ethical concerns and worries over environmental impacts. Few challenge such activities on economic grounds; even opponents generally recognize them as lucrative activities that generate jobs and income. On the other hand, support for mining and other development and resource extraction activities is primarily founded in economic reasoning. Conflict between ethical and economic motivations is a recurrent theme in environmental issues.

Is there a trade-off between economics and the environment?

Although measures to safeguard the environment may frequently mesh well with ethical considerations, we often hear it said that environmental protection works in opposition to economic progress. Arguments are made that environmental protection costs too much money, interferes with progress, and leads to job loss. But is this necessarily the case? Growing numbers of economists assert that there need be no such trade-off—that, in fact, environmental protection can be *good* for the economy. The position one takes often depends on whether one thinks in the short term or the long term, and whether one holds to traditional economic schools of thought or to newer ones that view human economies as coupled to the natural environment.

Economics studies the allocation of scarce resources

Like ethics, economics examines factors that guide human behaviour. **Economics** is the study of how people decide to use scarce resources to provide goods and services in the face of demand for them. By this definition, environmental problems are economic problems that can intensify as population and resource consumption increase. For example, pollution may be viewed as depleting the scarce resources of clean air, water, or soil. Indeed, the words *economics* and *ecology* come from the same Greek root, *oikos*, meaning "household." In its broadest context, the human "household" is Earth itself. Economists traditionally have studied the household of human society, and ecologists the broader household of all life.

Several types of economies exist today

An **economy** is a social system that converts resources into **goods**, material commodities manufactured for and bought by individuals and businesses; and **services**, work done for others as a form of business. The oldest type of economy is the *subsistence economy*. People in subsistence economies—who still compose much of the human population—meet most or all of their daily needs directly from nature and do not purchase or trade for most of life's necessities.

A second type of economy is the *capitalist market economy*. In this system, buyers and sellers interact to determine which goods and services to produce, how much to produce, and how these should be produced and distributed. Capitalist economies are often contrasted with state socialist economies, or *centrally planned economies*, in which government determines in a top-down manner how to allocate resources. In today's world, capitalism predominates over socialism.

A pure market economy would operate without any government intervention. In reality, all capitalist market economies today, including that of Canada, are hybrid systems (often called *mixed economies*). In modern market economies, governments intervene for several reasons: (1) to eliminate unfair advantages held by single buyers or sellers; (2) to provide social services, such as national defence, medical care, and education; (3) to provide "safety nets" (for older adults, the unemployed, those with chronic illnesses or disabilities, victims of natural disasters, and so on); (4) to manage commonly owned resources (the "commons"); and (5) to mitigate pollution and other types of environmental damage.

Environment and economy are intricately linked

All human economies exist within the larger environment and depend on it in important ways. Economies receive inputs from the environment, process them in complex ways that enable human society to function, then discharge outputs of waste from this process into the environment. Economies are thus *open systems* integrated with the larger environmental system of which they are a part. Earth, in turn, is a *closed system*. This means that the material inputs Earth can provide to economies are ultimately finite and so is the waste-absorbing capacity of the planet.

Although the interactions between human economies and the nonhuman environment are readily apparent, traditional economic schools of thought have long overlooked the importance of these connections. Indeed, most conventional economists today still adhere to a world view that largely ignores the environment (**FIGURE 21.10A**), and this world view continues to drive most policy decisions. A conventional economic world view essentially holds that environmental resources (the inputs into the economy) are limitless and free and that wastes (outputs) can be endlessly exported and absorbed by the environment, at no cost. However, modern economists belonging to the fast-growing fields of environmental economics, ecological economics, and natural resource economics explicitly accept that human economies are subsets of the environment and depend crucially on the environment (**FIGURE 21.10B**).

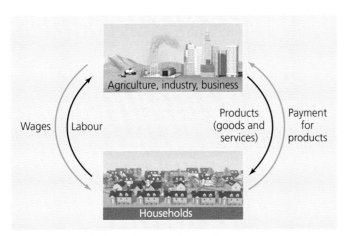

(a) Conventional view of economic activity

FIGURE 21.10
Standard neoclassical economics focuses on processes of production and consumption between households and businesses **(a)**, viewing the environment only as a "factor of production" that helps enable the production of goods. Environmental and ecological economists view the human economy as existing within the natural environment **(b)**, receiving resources from it, discharging waste into it, and interacting with it through ecosystem services.

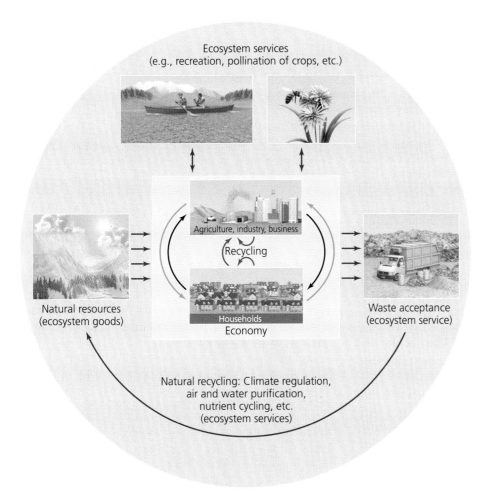

(b) Economic activity as viewed by environmental and ecological economists

Economic activity uses resources from the environment. Natural resources are the substances and forces we need to survive: the Sun's energy, freshwater, trees that provide lumber, rocks that provide metals, and fossil fuels and other energy sources that power our machines. We can think of natural resources as "goods" produced by nature. Without Earth's natural resources, there would be no human economies and no human beings.

Environmental systems also function in a manner that naturally supports economies. Earth's ecological systems purify air and water, cycle nutrients, provide for the pollination of plants by animals, and serve as receptacles and recycling systems for the wastes generated by our

Table 21.1 Ecosystem Services

Type of ecosystem service*	Example(s)
Regulation of atmospheric gases	Maintaining the ozone layer; balancing oxygen, carbon dioxide, and other gases
Regulation of climate	Controlling global temperature and precipitation through oceanic and atmospheric currents, greenhouse gases, cloud formation, and so on
Protection and buffering	Providing storm protection, flood control, and drought recovery, mainly through vegetation and shoreline structure
Regulation of water flow	Providing water for agriculture, industry, transportation
Storage of water	Providing water through watersheds, reservoirs, aquifers
Control of erosion	Preventing soil loss from wind or runoff; storing silt in lakes and wetlands
Formation of soil	Weathering rock; accumulating organic material
Cycling of nutrients	Cycling carbon, nitrogen, phosphorus, sulphur, and other nutrients through ecosystems
Waste treatment	Removing toxins, recovering nutrients, controlling pollution
Pollination of plants	Transporting floral gametes by wind or pollinating animals, enabling crops and wild plants to reproduce
Population control	Controlling prey with predators; controlling hosts with parasites; controlling herbivory on crops with predators and parasites
Provision of habitat	Providing ecological settings in which creatures can breed, feed, rest, migrate, winter
Provision of food	Producing fish, game, crops, nuts, and fruits that humans obtain by hunting, gathering, fishing, subsistence farming
Supply of raw materials	Producing lumber, fuel, metals, fodder
Genetic resources	Providing unique biological sources for medicine, materials science, genes for resistance to plant pathogens and crop pests, ornamental species (pets and horticultural plant varieties)
Recreational opportunities	Ecotourism, sport fishing, hiking, birding, kayaking, other outdoor recreation
Noncommercial services	Esthetic, artistic, educational, spiritual, and/or scientific values of ecosystems

*Ecosystem "goods" are here included in ecosystem services.
Source: Adapted with permission from Costanza, R., et al. 1997. The value of the world's ecosystem services and natural capital. Nature 387:253–260.

economic activity. Such essential services, often called **ecosystem services** (Table 21.1), sustain the life that makes our economic activity possible. Some ecosystem services represent the very nuts-and-bolts of our survival; others enhance our quality of life.

Although the environment allows economic activity to occur by providing ecosystem goods and services, that economic activity can affect the environment in return. When we deplete natural resources or produce pollution, we degrade the ability of ecological systems to function. The *Millennium Ecosystem Assessment* concluded in 2005 that 15 of 24 ecosystem services surveyed globally were being degraded or used unsustainably. The degradation of ecosystem services can in turn negatively affect economies. Ecological degradation harms poor people more than wealthy people, the Millennium Ecosystem Assessment found. As a result, restoring ecosystem services is a prime objective for alleviating poverty in much of the world.

These interrelationships have only recently become widely recognized. Let us briefly examine how economic thought has changed over the years, tracing the path that is now beginning to lead economies to become more compatible with natural systems.

Classical economics promoted the free market

Economics shares a common intellectual heritage with ethics, and practitioners of both have long been interested in the relationship between individual action and societal well-being. Some philosophers argued that individuals acting in their own self-interest would harm society (as in the tragedy of the commons). Others believed that such behaviour could benefit society, as long as the behaviour was constrained by the rule of law and private property rights and operated within fairly competitive markets.

The latter view was articulated by Scottish philosopher Adam Smith (1723–1790). Known today as the father of **classical economics**, Smith believed that when people are free to pursue their own economic self-interest in a competitive marketplace, the marketplace will behave as if guided by "an invisible hand" that ensures their actions will benefit society as a whole. In his 1776 book *Inquiry into the Nature and Causes of the Wealth of Nations*, Smith wrote:

> [Each individual] intends only his own security, only his own gain. And he is led in this by an invisible hand to promote an end which was no part of intention. By pursuing his own interests he frequently promotes that of society more effectually than when he really intends to.

Smith's philosophy remains a pillar of free market thought today, and many credit it for the tremendous gains in material prosperity that industrialized nations have experienced in the past few centuries. Others argue that the policies spawned by free-market thought worsen inequalities between rich and poor and contribute to environmental degradation. Market capitalism, these critics assert, should be constrained and regulated by democratic government.

Neoclassical economics incorporates human psychology

Economists subsequently took more quantitative approaches and incorporated human psychology into their work. Modern **neoclassical economics** examines the psychological factors underlying consumer choices, explaining market prices in terms of consumer preferences for units of particular commodities. In neoclassical economic theory, buyers desire the lowest possible price, whereas sellers desire the highest possible price. This conflict between buyers and sellers results in a compromise price being reached and the "right" quantity of commodities being bought and sold. This is often phrased in terms of *supply*, the amount of a product offered for sale at a given price, and *demand*, the amount of a product people will buy at a given price if free to do so. Theoretically, when prices go up, demand drops and supply increases; when prices fall, demand rises and supply decreases. In theory, the market automatically moves toward an equilibrium point, a price at which supply equals demand (**FIGURE 21.11A**). Similar reasoning can be applied to environmental issues, such that economists can determine "optimal" levels of resource use or pollution control (**FIGURE 21.11B**).

Cost–benefit analysis is a useful tool

Neoclassical economists commonly use a method referred to as **cost–benefit analysis**. In this approach, estimated costs for a proposed action are totalled and compared with

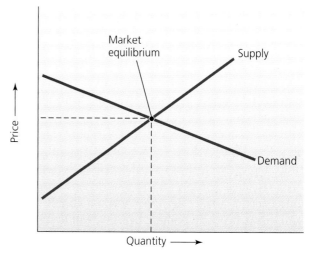

(a) Classic supply-demand curve

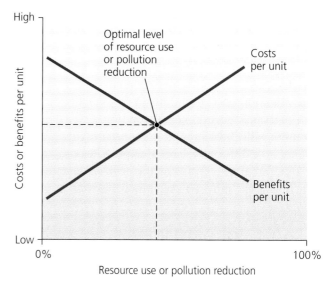

(b) Marginal benefit and cost curves

FIGURE 21.11
This basic supply-and-demand curve **(a)** illustrates the relationship between supply, demand, and market equilibrium, the "balance point" at which demand is equal to supply. We can use a similar graph **(b)** to determine an "optimal" level of resource use or pollution mitigation. In this graph, the cost per unit of resource use or pollution cleanup (blue line) rises as the resource use or pollution cleanup proceeds and it becomes expensive to extract or clean up the remaining amounts. Meanwhile, the benefits per unit of resource use or pollution cleanup (red line) decrease. The point where the lines intersect gives the optimal level.

the sum of benefits anticipated to result from the action. If total benefits exceed costs, the action should be pursued; if costs exceed benefits, it should not. When choosing among multiple alternative actions, the one with the greatest excess of benefits over costs should be chosen.

This reasoning seems eminently logical; however, problems often arise because not all costs and benefits are easily quantified, or even easily identified. It may be simple to quantify wages paid to uranium miners or the market value of uranium extracted from a mine or even the cost of measures to minimize health risks for miners. But it is much more difficult to assess the cost of a landscape's being scarred by mine development or the cost of radioactive contamination of a stream or the costs (not just in terms of money, but in health and emotional well-being) to the families of miners who die from mining-related cancers.

Because some costs and benefits cannot easily be assigned monetary values—and because it is difficult to identify and agree on all costs and benefits—cost–benefit analysis is often controversial. Moreover, because economic benefits usually are more easily quantified than environmental costs, economic benefits tend to be over-represented in traditional cost–benefit analyses. As a result, environmental advocates often feel these analyses are biased in favour of economic development and against environmental protection.

A corollary to cost–benefit analysis has been the development of a variety of mechanisms for understanding, calculating, and defining the value, to society and to individuals, of environmental costs and benefits. In general, **valuation** refers to the attempt to quantify the value of a particular environmental good or service—even if it cannot easily be expressed in monetary terms. We will discuss how different valuation approaches can be used to more accurately represent the environment in cost–benefit analysis and asset accounting, below.

Aspects of neoclassical economics have profound implications for the environment

Today's capitalist market systems operate largely in accord with the precepts of neoclassical economics. These systems have generated unprecedented material wealth, employment, and other desirable outcomes, but they have also contributed to environmental problems. Four of the fundamental assumptions of neoclassical economics have implications for the environment:

1. Resources are infinite or substitutable.
2. Long-term effects should be discounted.
3. Costs and benefits are internal.
4. Growth is good.

Let us critically examine each of these assumptions in turn.

1. Are resources infinite or substitutable?
Neoclassical economic models generally treat the supply of workers and other resources as being either infinite or largely "substitutable and interchangeable." This implies that once we have depleted a resource—natural, human, or otherwise—we should be able to find a replacement for it. Human resources can substitute for financial resources, for instance, or manufactured resources can substitute for natural resources. Theory allows that the substituted resource may be less efficient or more costly, but some degree of substitutability is generally assumed. In other words, traditional economists have considered environmental goods and services to be so-called "free gifts of nature"—infinitely abundant and resilient, and ultimately substitutable by human technological ingenuity.

It is true that many resources can be replaced; societies have transitioned from manual labour to animal labour to steam-driven power to fossil fuel power and may yet transition to renewable power sources, like solar energy. However, Earth's material resources are ultimately limited. Nonrenewable resources, such as fossil fuels, can be depleted, and even renewable resources can be used up if we exploit them more quickly than they can be replenished. This was experienced by the inhabitants of Rapa Nui, who harvested wood faster than their forests could regrow (see "The Science Behind the Story: The Lesson of Rapa Nui").

2. Should long-term effects be discounted?
Although few people would dispute that resources are *ultimately* limited, many assume that their depletion will take place so far in the future that there is no need for current generations to worry. For economists in the neoclassical tradition, an event far in the future counts much less than one in the present; in economic terminology, we say that these future effects are "discounted." In discounting, short-term costs and benefits are granted more importance than long-term costs and benefits. This encourages policy makers to play down long-term consequences of decisions we make today. (See "The Science Behind the Story: Ethics in Economics: Discounting and Global Climate Change".)

For example, a stock of uranium ore in the ground or living trees standing in the forest or fish in a stream

SUBSTITUTABILITY AND THE ENVIRONMENT
21–7

Can you think of a natural resource that would be difficult to replace with a substitute? What problems might arise from the assumption that *all* resources, including clean air and water, are substitutable and interchangeable? Can you think of examples that contradict any of the other three assumptions of neoclassical economics?

represent potential commodities that would have monetary or **market values** if they were extracted or harvested and sold. In cost–benefit analysis, however, such commodities are typically discounted for future use to the extent that they appear to be of economic value only if they can be used up as quickly as possible. Some governments and businesses use a 10% annual discount rate for decisions on resource use. This means that the long-term value of a stand of ancient trees worth $500 000 for the timber it contains would drop by 10% each year; after 10 years of discounting, it would be worth only $174 339.22. By this logic, the more quickly the trees are cut down, the more they are worth.

A related problem is that accounting procedures at the national level typically do not assign an *asset value* to intact natural resources as they would, for example, for a physical asset, such as a factory or a system of roads. This means that an unexploited natural resource has *no discernible value* to the nation that possesses it, providing a powerful incentive for nations to exploit their natural resources to the greatest extent and as quickly as possible.

This was stated eloquently by Robert Repetto, one of the pioneers of **natural resource accounting**, which seeks mechanisms by which to incorporate the economic asset values of natural resources into national accounting systems. According to Repetto:

> *A country can cut down its forests, erode its soils, pollute its aquifers and hunt its wildlife and fisheries to extinction, but its measured income is not affected as its assets disappear . . . By failing to recognize the asset value of natural resources, the accounting framework that underlies the principal tools of economic analysis misrepresents the policy choices nations face.*[15]

To further complicate matters, national accounting systems also fail to adequately incorporate certain categories of activities and expenses into the calculations. For example, an individual or a nation might spend money to prevent future environmental degradation, or to restore a degraded natural environment. This could include, for example, launching a program of pollution abatement, constructing windbreaks to prevent soil erosion, or installing heavy-walled containers in oil tankers to prevent spills. Such costs, which economists call *defensive expenditures*, are calculated only as current expenses; the future value of the resource these activities may be helping to preserve or restore is of no economic value to the system.

3. Are costs and benefits internal? A third assumption of neoclassical economics is that all costs and benefits associated with a particular exchange of goods or services are borne by individuals engaging directly in the transaction. In other words, it is assumed that the costs

FIGURE 21.12
An Indonesian boy wading in a polluted river suffers external costs, costs that are not borne by the buyer or seller. External costs may include water pollution, esthetic harm, human health problems, property damage, harm to aquatic life, esthetic degradation, declining real estate values, and other impacts.

and benefits of a transaction are "internal" to the transaction, experienced by the buyer and seller alone, and do not affect other members of society.

However, in many situations this is simply not the case. Pollution from a factory can harm people living nearby. In such cases, someone—often taxpayers not involved in producing the pollution—ends up paying the costs of alleviating it. Market prices do not take the social, environmental, or economic costs of this pollution into account. Costs or benefits of a transaction that involve people other than the buyer or seller are known as **externalities**. A positive externality is a benefit enjoyed by someone not involved in a transaction, and a negative externality, or *external cost*, is a cost borne by someone not involved in a transaction (**FIGURE 21.12**). Negative externalities often harm groups of people or society as a whole, while allowing certain individuals private gain.

External costs commonly include the following:

- Property damage
- Declines in desirable elements of the environment, such as poorer air quality, or fewer fish in a stream
- Esthetic damage, such as that resulting from air pollution or clear-cutting
- Stress and anxiety experienced by people downstream or downwind from a pollution source
- Declining real estate values resulting from these problems

THE SCIENCE BEHIND THE STORY

Ethics in Economics: Discounting and Global Climate Change

Sir Nicholas Stern was head of the Government Economic Service, United Kingdom.

To help decide how to respond to global climate change, the British government commissioned esteemed economist Sir Nicholas Stern to assess the economic costs that a changing climate may impose on society. When Stern's report was published in November 2006, a debate ensued among economists that had more to do with ethics than with economics. The dispute centred on discounting. Some economists argued that Stern had used an unusually low discount rate, leading to unrealistically high cost estimates.

To produce the *Stern Review on the Economics of Climate Change*, Stern and his research team surveyed the burgeoning literature on rising global temperatures, changing regional rainfall patterns, and increasing storminess (see Chapter 14). They then estimated the economic consequences of these climatic changes and tried to put a price tag on the global cost. The report concluded that without action to forestall it, climate change would cause losses in annual global gross domestic product (GDP) of 5%–20% by the year 2200 (see figure).

Stern's team calculated that by paying just 1% of GDP annually starting now, society could stabilize atmospheric greenhouse gas concentrations and prevent most of these monetary losses. The bottom line: *Spending a relatively small amount of money now will save us much larger expenses in the future.* This conclusion caught the attention of governments worldwide; for the first time, economists were advancing a strong economic argument for tackling climate change immediately.

However, the *Stern Review*'s numbers depend significantly on how one chooses to weigh impacts in the future versus impacts happening now. Stern used two discount factors. One accounted for the likelihood that people in the future will be richer than we are today and thus better able to handle economic costs. The other considered whether the future should be discounted simply because it is the future. This latter discount factor (called a "pure time discount") is an ethical issue because it places explicit values on the welfare of future generations.

The *Stern Review* used a pure time discount rate of 0.1%. This means that an impact occurring next year is judged to be 9% as important as one occurring this year, and the welfare of a person born 10 years from now is 99.0% as valuable as the welfare of one born today. This discount rate treats current and future generations *nearly* equally. Future generations are slightly down-weighted only because of the (very small) possibility that our species could go extinct (in which case, there would be no future generations to be concerned about).

The Mirrar experienced external costs in the form of pollution from the Ranger mine. In March 2002, radioactive material from the Ranger mine contaminated a stream on Mirrar land with uranium concentrations 4000 times as high as allowed by law. According to an Aboriginal representative, this was the fourth such violation in less than three months. In 2004 the government temporarily shut down the mine and brought the owners to court for these violations.[16] For the Dene of Great Bear Lake the external costs of uranium mining have come in the form of lingering contamination of soil, water, and fish, as well as health impacts on workers and, by extension, economic and psychological impacts on their families.

By ignoring external costs, economies create a false idea of the true and complete costs of particular choices and unjustly subject people to the consequences of transactions in which they did not participate. External costs are one reason governments develop environmental legislation and regulations. Unfortunately, external costs are difficult to account for and eliminate. It is tough to assign a monetary value to illness, premature death, or degradation of an esthetically or spiritually significant site.

weighing the issues

PRIVATE VERSUS SOCIAL COSTS AND BENEFITS

21–8

A central concern of modern environmental economics is to identify mechanisms through which governments and markets can balance the optimization of benefits from a *private* (i.e., corporate) perspective versus a *social* perspective. This often requires convincing (or coercing) private corporations to internalize some production externalities, such as the cost to society of pollution, environmental degradation, or resource depletion resulting from production. Can you think of mechanisms that governments and markets might use to accomplish this? These could be legal mechanisms, or financial regulatory, or ethical measures. Some are "carrots" (intended to encourage and reward socially responsible behaviour), and some are "sticks" (intended to punish behaviour that transfers too much of the environmental cost of production to society). We will consider the use of such mechanisms in greater detail in Chapter 22.

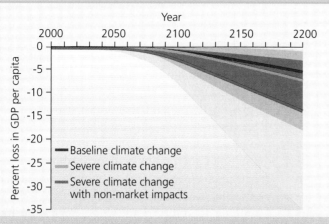

The *Stern Review* estimated that global climate change could cause substantial losses in future wealth. Baseline climate change as predicted by the IPCC *Third Assessment Report* could bring losses of 5.3% of per capita GDP annually by the year 2200. Severe climate change, as predicted by research since that report was issued, could bring annual losses of 7.3%, and adding nonmarket values raises this figure to 13.8%. The grey shaded areas around the three data lines represent ranges of values judged statistically to be 95% likely. Data from HM Treasury, 2007. *Stern Review on the Economics of Climate Change*. London, U.K.

Some economists viewed this discount rate as too low. Economist William Nordhaus critiqued the *Stern Review* in scientific journals and proposed a discount rate starting at 3% and falling to 1% in 300 years. Nordhaus maintained that such numbers were more objective because they are more in line with how people's values are revealed by the prices they pay for things in the free market. He argued that Stern's near-zero discount rate overweights the future, forcing people today to pay too much to address hypothetical future impacts. Indeed, when economists use discounting in standard ways to assess capital investments or construction projects (development of a railroad, dam, or highway) they typically choose discount rates closer to those Nordhaus suggested.

In their response, Stern and his team argued that a discount rate of 3% or even 1% is too high. Such rates may be useful for assessing development projects, they maintained, but they are not appropriate for long-term environmental problems that directly affect human well-being. A 3% rate means that a person born in 1985 is valued only half as much as a person born in 1960. It means a grandchild is worth far less than a grandparent simply because of when they were born. Stern and his supporters argued that the market should not be used to guide ethical decisions.

Stern's group also published sensitivity analyses that analyzed how their conclusions would change if different discount rates were used. These showed that the report's main message—that it's cheaper to prevent climate change now than to pay to deal with it later—was robust to a fair amount of change in the discount rate, up to at least a rate of 1.5%.

The choice of a discount rate is an ethical decision on which well-intentioned people may differ. As governments, businesses, and individuals begin to invest in addressing climate change, the debate over the *Stern Review* reveals how ethics and economics remain intertwined.

4. Is growth good? A fourth assumption of the neoclassical economic approach is that economic growth is required to keep employment high and maintain social order. The argument goes something like this: If the poor view the wealthy as the source of their suffering, they may revolt. Promoting economic growth can defuse this situation by creating opportunities for the poor to become wealthier themselves. By making the overall economic pie larger, everyone's slice becomes larger, even if some people still have much smaller slices than others.

The idea that economic growth is good has been encouraged over the centuries by the concept of material progress, espoused by Western cultures since the Enlightenment. Everywhere, every day in the modern industrialized world we see expressions of the view that "more and bigger" is always better. We hear constantly in business news of increases in an industry's output or percentage growth in a country's economy, with increases touted as good news and decreases, stability, or even a minor drop in the *rate* of growth presented as bad news. Economic growth has become the quantitative ruler by which progress is measured.

Is the growth paradigm good for us?

The rate of economic growth in recent decades is unprecedented in human history. As a result, the world economy is seven times the size it was half a century ago. All measures of economic activity—trade, rates of production, amount and value of goods manufactured—are higher than they have ever been and are still increasing. This growth has brought many people much greater material wealth (although gaps between rich and poor are immense, and growing).

To the extent that economic growth is a means to an end—a tool with which we can achieve greater human happiness—it can be a good thing. However, many observers today worry that growth has become an end in itself and is no longer necessarily the best tool with which to pursue happiness. Critics of the growth paradigm often note that runaway growth resembles the multiplication of cancer cells, which eventually overwhelm and destroy the organism in which they grow. These critics fear that runaway economic growth will likewise destroy the economic system on which we all depend. Resources for growth are ultimately limited, they argue, so nonstop growth is not sustainable and will fail as a long-term strategy.

Defenders of traditional economic approaches reply that critics have been saying for decades that limited resources would doom growth-oriented economies, yet most of these economies are still expanding dramatically. If resources are dwindling, why are we witnessing the most rapid growth of material wealth in human history?

One prime reason is technological innovation. In case after case, improved technology has enabled us to push back the limits on growth, effectively expanding the carrying capacity of the environment. More powerful technology for extracting minerals, fossil fuels, and groundwater has expanded the amounts of these natural resources available to us. Technological developments, such as automated farm machinery, fertilizers, and chemical pesticides, have allowed us to grow more food per unit area of land, boosting agricultural output. Faster, more powerful machines in our factories have enabled us to translate our enhanced resource extraction and agricultural production into faster rates of manufacturing.

Economists disagree on whether economic growth is sustainable

Can we conclude, then, that endless improvements in technology are possible and that we will never run into shortages of resources? At one end of the spectrum are those who believe that technology can solve everything—a philosophy that has greatly influenced economic policy in market economies over the past century.

At the other end of the spectrum, **ecological economists** argue that a couple of centuries is not a very long period of time and that history suggests that civilizations do not, in the long run, overcome their environmental limitations. Ecological economics, which has emerged as a discipline only in the past decade or two, applies the principles of ecology and systems science (Chapters 3 through 5) to the analysis of economic systems. Earth's natural systems generally operate in self-renewing cycles, not in a linear or progressive manner. Ecological economists advocate sustainability in economies and see natural systems as good models.

To evaluate an economy's sustainability, ecological economists take a long-term perspective and ask, "Could we continue this activity forever and be happy with the outcome?" Most ecological economists argue that the growth paradigm will eventually fail and that if nothing is done to rein in population growth and resource consumption, depleted natural systems could plunge our economies into ruin. Many advocate economies that do not grow and do not shrink but rather are stable. Such **steady-state economies** are intended to mirror natural ecological systems.

Environmental economists tend to agree that economies are unsustainable if population growth is not reduced and resource use is not made more efficient. However, they maintain that we can accomplish these changes and attain sustainability within our current economic systems. By retaining the principles of neoclassical economics but modifying them to address environmental challenges, environmental economists argue that we can keep our economies growing and that technology can continue to improve efficiency. Environmental economists were the first to develop ways to tackle the problems of external costs and discounting, and to weigh the true costs and benefits associated with resource use. They blazed a trail, and ecological economists then went farther, proposing that sustainability requires far-reaching changes leading ultimately to a steady-state economy.

A steady-state economy is a revolutionary alternative to growth

The idea of a steady-state economy did not originate with the rise of ecological economics. Back in the nineteenth century, British economist John Stuart Mill (1806–1873) hypothesized that as resources became harder to find and extract, economic growth would slow and eventually stabilize. Economies would carry on in a state in which individuals and society subsist on steady flows of natural resources and on savings accrued during occasional productive but finite periods of growth. Such a model appears to match the economies of many traditional societies throughout the world before the global expansion of colonialism and European culture.

Modern proponents of a steady-state global economy, such as the pioneering economist Herman Daly, are not so sanguine as to expect that a steady state will evolve on its own from a capitalist market system. Instead, most believe we will need to rethink our assumptions and fundamentally change the way we conduct economic transactions. Critics often assume that an end to growth will mean an end to a rising quality of life. Ecological economists, however, argue that quality of life can continue to rise under a steady-state economy and, in fact, may be more likely to do so. Technological advances will not cease just because growth stabilizes, they argue, and neither will behavioural changes (such as greater use of recycling) that enhance sustainability. Instead, wealth and human happiness can continue to rise after economic growth has levelled off.

Attaining sustainability will certainly require reforms and may well require fundamental shifts in thinking, values, and behaviour. How can these goals be attained in a world whose economic policies are still largely swayed by a cornucopian world view that barely takes the environment into account? While keeping in mind that ecological and environmental economic approaches are still actively being developed, we will now survey a few strategies for sustainability that have been offered so far.

We can measure economic progress differently

For decades, economists have assessed the economic robustness of a nation by calculating its **gross domestic product (GDP)**, the total monetary value of final goods and services produced in a country each year (**FIGURE 21.13**). GDP is an extremely powerful indicator, used to make financial policy decisions by federal governments worldwide, with fundamental impacts on quality of life and well-being for billions of people.

However, there are problems with using this measure of economic activity to represent a nation's economic well-being. For one, GDP does not account for the non-market values of ecosystem goods and services, nor is GDP necessarily an expression of *desirable* economic activity. In fact, GDP can increase, even if the economic activities driving it hurt the environment or society.

For example, a large oil spill would increase GDP because oil spills require cleanups, which cost money and, as a result, increase the production of goods and services. Such activities generate income, and are therefore reflected by a positive change in GDP, but the negative health and ecological costs of disasters like these are typically *not* reflected in the GDP—unless they generate jobs or cash transactions, in which case they may show up as positive changes. A radiation leak at a uranium mine on Mirrar homelands would likely add to the Australian GDP because of the many monetary transactions required for cleanup and medical care. Similarly the $6.7 million in contracts offered by the government of Canada for the cleanup of abandoned uranium mine sites in the Northwest Territories will add to the GDP of Canada. Even Hurricane Katrina—which caused the deaths of almost 2000 people, the loss of tens of thousands of homes and jobs, and the contamination or permanent flooding of millions of square kilometres of coastal marshlands and forests in the United States—generated positive growth in GDP in the third quarter of 2005.

Some economists have attempted to develop economic indicators that differentiate between desirable and undesirable economic activity. Such indicators can function as more accurate guides to nations' welfare. One alternative to the GDP is the **genuine progress indicator (GPI)**, introduced in 1995 by Redefining Progress, a nonprofit organization that develops economic and policy tools to promote

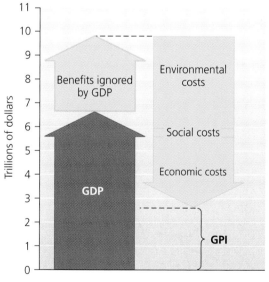

(a) Components of GDP vs. GPI

(b) Change in GDP vs. GPI

FIGURE 21.13
Gross domestic product (GDP; red arrow) sums together all economic activity, whether good or bad. As a result, many researchers today believe that GDP is lacking as an indicator of overall well-being. The genuine progress indicator (GPI) is one example of an alternative indicator of progress and well-being. GPI adds to GDP benefits by accounting for the value of activities such as volunteering and parenting (upward-pointing gold arrow). The GPI also subtracts external environmental costs, such as pollution; social costs, such as divorce and crime; and economic costs, such as borrowing and the gap between rich and poor (downward-pointing gold arrow). Data from Venetoulis, J., and C. Cobb. 2004. *The Genuine Progress Indicator, 1950–2002 (2004 Update)*. Redefining Progress.

accurate market prices and sustainability (see **FIGURE 21.13**). The GPI has not yet gained widespread acceptance, partly because it is tremendously data-intensive, but it has generated a great deal of discussion that has drawn attention to the weaknesses of the GDP.

To calculate GPI, economists begin with conventional economic activity and then add to it all *positive* contributions to the economy that do not have to be paid for with money, such as volunteer work and parenting. They then subtract the *negative* impacts, such as crime, pollution, gaps between rich and poor, and other detrimental social, environmental, and economic factors. The GPI thereby summarizes many more forms of economic activity than does GDP and differentiates between economic activity that increases societal well-being and economic activity that decreases it.

Thus, whereas GDP increases when fossil fuel use increases, GPI declines because of the adverse environmental and social impacts of such consumption, including air and water pollution, increased road congestion and traffic accidents, and global climate change. **FIGURE 21.14** compares changes in per capita GPI and GDP in Alberta from 1961 to 2003. The Pembina Institute in Alberta is a leader in the calculation and application of alternative indicators of economic and overall well-being. The province's GDP has increased greatly as a result of increased economic activity; however, GPI has declined over the same period.

GPI is not the only alternative to GDP. The Index of Sustainable Economic Welfare (ISEW) is based on income, wealth distribution, the value of volunteerism, and natural resource depletion. The Net Economic Welfare (NEW) index adjusts GDP by deducting the costs of environmental degradation. The United Nations uses a tool called the Human Development Index, calculated on the basis of a nation's standard of living, life expectancy, and education. Any of these indices, ecological economists maintain, should give a more accurate portrait of a nation's welfare than GDP, which policy makers currently use so widely.

We can give ecosystem goods and services monetary values

Economies receive from the environment vital resources and ecosystem services. However, any survey of environmental problems today—deforestation, biodiversity loss, pollution, collapsed fisheries, climate change, and so on—makes it immediately apparent that our society often mistreats the very systems that keep it alive and healthy. Furthermore, the values of environmental goods and services are routinely underrepresented in cost–benefit analysis, one of the most powerful tools of economic decision making. Why is this? From the economist's perspective, humans overexploit natural resources and systems because the market assigns them no quantitative monetary value or, at best, assigns values that underestimate their true worth.

Think for a minute about the nature of some of these services. The esthetic and recreational pleasure we obtain from natural landscapes, whether wildernesses or city parks, is something of real value. Yet this value is hard to quantify and appears in no traditional measures of economic worth. Or consider Earth's water cycle, by which rain fills our reservoirs with drinking water, rivers give us hydropower and flush away our waste, and water evaporates, purifying itself of contaminants and readying itself to fall again as rain. This natural cycle is absolutely vital to our existence, yet because its value is not quantified, markets impose no financial penalties when we interfere with it.

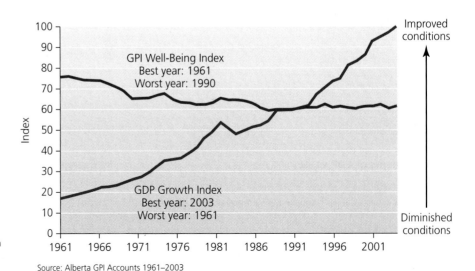

FIGURE 21.14
Although the GDP of Alberta has increased dramatically since 1961, the GPI actually shows a decline over the same period. GPI advocates suggest that this discrepancy means that we are spending more money than ever but that our lives are not that much better. Data from Pembina Institute, *Alberta GPI Accounts, 1961–2003*.

Source: Alberta GPI Accounts 1961–2003

Ecosystem services are said to have **nonmarket values**, values not usually included in the price of a good or service (Table 21.2 and **FIGURE 21.15**). Because the market does not assign value to ecosystem services, debates, such as that over the Jabiluka mine, often involve comparing apples and oranges—in this case, the intangible cultural, ecological, and spiritual arguments of the Mirrar versus the hard numbers of mine proponents.

To partially resolve this dilemma, environmental and ecological economists have sought ways to assign values to ecosystem goods and services. One technique, **contingent valuation**, uses surveys to determine how much people are willing to pay to protect a resource or to restore it after damage has been done.

Such an exercise was conducted with a mining proposal in the Kakadu region in the early 1990s that preceded the Jabiluka proposal. The Kakadu Conservation Zone, a government-owned 50 km² plot of land surrounded by Kakadu National Park, was either to be developed for mining, or to be preserved and added to the park. To determine the degree of public support for environmental protection versus mining, a government

Table 21.2	Values That Modern Market Economies Generally Do Not Address
Nonmarket value	**Is the worth we ascribe to things that . . .**
Use value	We use directly
Option value	We do not use now but might use or find a use for at a later time
Esthetic value	We appreciate for their beauty or emotional appeal
Cultural value	Sustain or help define our culture
Spiritual value	Are sacred to certain groups, or evoke spiritual, religious, or philosophical responses in us
Scientific value	May be significant as subjects of scientific research
Educational value	May teach us about ourselves and the world
Existence value	Are important or have value simply because they exist, even though we may never experience them directly (e.g., remote wilderness, or endangered species in a far-off place)

(a) Existence values

(b) Use values

(c) Option values

(d) Esthetic values

(e) Scientific values

(f) Educational values

(g) Cultural values

FIGURE 21.15
Accounting for nonmarket values, such as those shown here, may help us to make better environmental and economic decisions.

commission sponsored a contingent valuation study to determine how much Australian citizens valued keeping the Kakadu Conservation Zone preserved and undeveloped. Researchers interviewed 2034 citizens, asking them how much money they would be willing to pay to stop mine development.

The interviewers presented two scenarios: (1) a "major-impact" scenario based on predictions of environmentalists who held that mining would cause great harm, and (2) a "minor-impact" scenario based on predictions of mining executives who held that development would have few downsides. After presenting both scenarios in detail, complete with photographs, the interviewers asked the respondents how much their households would pay if each scenario, in turn, were to occur. Respondents on average said their households would pay $80 per year to prevent the minor-impact scenario and $143 per year to prevent the major-impact scenario. Multiplying these figures by the number of households in Australia (5.4 million at the time), the researchers found that preservation was "worth" $435 million annually to the Australian population under the minor-impact scenario, and $777 million under the major-impact scenario. Because both of these numbers significantly exceeded the $102 million in annual economic benefits expected from mine development, the researchers concluded that preserving the land in its undeveloped state was worth more than mining it.

Because contingent valuation relies on survey questions, critics complain that in such cases people will volunteer idealistic (inflated) values rather than realistic ones, knowing that they will not actually have to pay the price they name. In part because of such concerns, the Australian government commission decided not to use the Kakadu contingent valuation study's results. (The mine was stopped, ultimately, but mainly as a result of Aboriginal opposition.)

Whereas contingent valuation measures people's *expressed* preferences, other methods aim to measure people's *revealed* preferences—preferences as revealed by data on actual behaviour. For example, the amount of money, time, or effort people expend to travel to parks for recreation has been used to measure the value people place on parks. Economists have also analyzed housing prices, comparing homes with similar characteristics but different environmental settings to infer the dollar value of landscapes, views, greenspace, and peace and quiet. Another approach assigns environmental amenities value by measuring the cost required to restore natural systems that have been damaged or to mitigate harm from pollution.

In 1997 a research team led by Robert Costanza reviewed ecosystem valuation studies, with the goal of calculating the global economic value of all the services that ecosystems provide (FIGURE 21.16). The team identified more than 100 studies that estimated the worth of such ecosystem services as water purification, greenhouse gas regulation, plant pollination, and pollution cleanup. The studies used such methods as contingent valuation to estimate the values of such aspects of natural systems as biodiversity and esthetics.

To estimate the worth of ecosystem services more accurately, the team reevaluated the data by using alternative

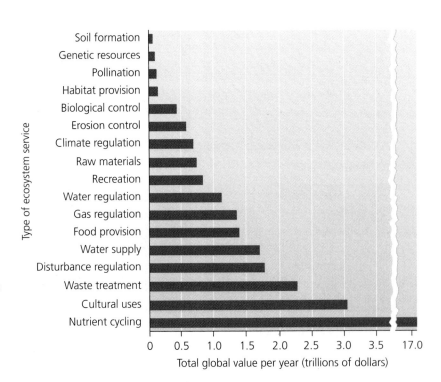

FIGURE 21.16
Costanza and colleagues estimated the total value of the world's ecosystem services at approximately $33 trillion. Shown are subtotals for each major class of ecosystem service. The $33 trillion figure does not include values from some ecosystems, such as deserts and tundra, for which adequate data were unavailable. Data from Costanza, R., et al. 1997. The value of the world's ecosystem services and natural capital. *Nature* 387:253–260.

valuation techniques. One method was to calculate the cost of replacing ecosystem services with technology. For example, marshes protect people from floods and filter out water pollutants. If a marsh were destroyed, the researchers would calculate the value of the services it had provided by measuring the cost of the levees and water-purification technology that would be needed to assume those tasks. The researchers then calculated the global monetary value of such wetlands by multiplying those totals by the global area occupied by the ecosystem. By calculating similar totals from coral reefs, deserts, tundra, and other ecosystems, they arrived at a global value for ecosystem services.[17] The total figure was $33 trillion per year (1997 dollars)—greater than the combined gross domestic products of all nations in the world. A follow-up study in 2002 concluded that the economic benefits of preserving the world's remaining natural areas outweighed the benefits of exploiting them by a factor of 100 to 1.

Markets can fail

When they do not reflect the full costs and benefits of actions, markets are said to fail. **Market failure** occurs when markets do not take into account the environment's positive effects on economies (such as ecosystem services) or when they do not reflect the negative effects of economic activity on the environment or on people (external costs).

Traditionally, market failure has been countered by government intervention. Governments can dictate limits on corporate behaviour through laws and regulations. They can institute *green taxes,* which penalize environmentally harmful activities. Or they can design economic incentives that put market mechanisms to work to promote fairness, resource conservation, and economic sustainability. We will examine legislation, regulation, green taxation, and market incentives in Chapter 22 as part of our policy discussion.

Corporations are responding to sustainability concerns

As more consumers and investors express preferences for sustainable products and services, more industries, businesses and corporations are finding that they can make money by "greening" their operations.

Some companies, such as the Body Shop, cultivate eco-conscious images; others donate a portion of their proceeds to environmental and other progressive non-profit groups. Today, some newer businesses are trying to go even further than these pioneers. The outdoor apparel company Nau manufactures items from materials made of corn biomass and recycled bottles, and helps fund environmental nonprofits. Entrepreneurs are starting thousands of local sustainability-oriented businesses across the world.

In the past few years, corporate sustainability has gone mainstream, and some of the world's largest corporations have joined in, including McDonald's, Starbucks, IKEA, Dow, and British Petroleum. Nike collects millions of used sneakers each year and recycles the materials to create synthetic surfaces for basketball courts, tennis courts, and running tracks. Nike also uses more organic cotton and has developed less-toxic rubber and adhesives. In response to media attention and consumer concern, such corporations as Nike and the Gap are also working to improve labour conditions in their factories overseas.

Of course, corporations exist to make money for their shareholders, so they cannot be expected to pursue goals that are not profitable. Moreover, some corporate greening efforts are more rhetoric than reality, and corporate *greenwashing* may mislead some consumers into thinking that companies are acting more sustainably than they are. However, as consumer preferences turn increasingly to sustainable products and practices, many corporations are seeing the economic wisdom of moving toward a more sustainable model of operation.

Perhaps the most celebrated recent corporate greening is that of Wal-Mart. Environmentalists have long criticized the world's largest retailer for its environmental and social impacts. In 2006 the company began a quest to sell organic and sustainable products, reduce packaging and use recycled materials, enhance fuel efficiency in its truck fleets (**FIGURE 21.17**), reduce energy use in its stores, cut carbon dioxide emissions, and preserve an equivalent area of natural land for every parcel of land developed. Many observers remain skeptical of Wal-Mart's commitment,

FIGURE 21.17
Giant retail corporation Wal-Mart has recently launched an environmental campaign that includes a fleet of "green" fuel-efficient hybrid trucks.

CANADIAN ENVIRONMENTAL PERSPECTIVES

Matthew Coon Come

Matthew Coon Come is an advocate for the environmental, social, and political rights of Aboriginal people in Canada.

- **Environmental** and **social activist**
- Former **grand chief** of the Grand Council of the Cree
- Former **national chief** of the Assembly of First Nations

Matthew Coon Come is a politician and activist of Cree descent. He was born in a bush tent on his father's trapline and was taken away at the age of six to attend La Tuque Indian Residential School almost 500 km from his home. He studied law at McGill University and as a young man was elected grand chief of the Grand Council of the Cree. He has since served in many leadership positions on behalf of Aboriginal communities in Canada and around the world. He is a past winner of the Equinox Environmental Award and the Condé Nast Environment Award, among others.[18]

Coon Come is an advocate for his people in a world where environmental rights and social justice collide with economic development. He led the Cree in the late 1980s and early 1990s in opposition to a massive hydroelectric development project, the Great Whale River Phase of Hydro Quebec's James Bay project, which began in 1971. This phase of the project would have comprised more than 30 dams and 600 dikes, blocking nine major rivers,[19] affecting watersheds with an area the size of France, and creating reservoirs the size of Lake Erie.[20]

Coon Come gathered support from local, national, and international environmental, human rights, and tribal communities, creating a coalition that vocally and visibly opposed the project. As a result of the coalition's efforts, the state of New York cancelled major contracts to purchase electricity from Hydro Quebec,[21] putting the viability of the project in jeopardy. In 1994, the Supreme Court of Canada ruled that federal environmental assessments were required for all Hydro Quebec electricity exports. Shortly after that decision, the premier of Quebec announced the indefinite suspension of work on the Great Whale Project.[22]

For his role in this struggle, Matthew Coon Come was awarded the Goldman Environmental Prize, a prestigious international prize that honours grassroots environmental leaders around the world.

By 2002, though, many Cree had changed their position on the James Bay Project, citing the desperate need for economic development in the North. Cree leaders signed an agreement with the government of Quebec that cleared the way for the project to proceed. The next phase, which passed its environmental assessment in 2006, will involve the diversion of about 50% of the water flow of the Rupert River, as well as the construction of two new power stations and four large dams, flooding for a new reservoir, new roads, and worker settlements.

After all these years, the project continues to polarize the Cree Nation. Its supporters argue that the participation of the Cree in this project is a crucial step toward modernization and economic and social stability. Opponents—including residents from the three communities nearest to the project, as well as current grand chief Matthew Mukash, who is pressing instead for the development of wind energy—are concerned that the negative environmental impacts will outweigh any economic benefits.

Meanwhile, Matthew Coon Come became the national chief of the Assembly of First Nations, a position he held until 2003. In 2008 he issued a statement in response to the apology by the government of Canada for the years of abuse suffered by Native children, including Coon Come, in church-run residential schools. The statement, while accepting the government's apology, detailed the damage done to him by the residential school. It said in part, "They told me that my culture, and my people's ways of life, would never sustain me. They lied. I am a son of a hunter, fisherman, and a trapper. My father taught me how to walk the land, and to love and respect the animals and all of creation. I have not lost my culture. Our way of life is thriving."[23]

"Business as usual will not ensure the protection of the land of our children, yet unborn." —Matthew Coon Come

Thinking About Environmental Perspectives

Put yourself in the place of a Cree community leader like Matthew Coon Come. How would you handle a controversy like the Hydro Quebec James Bay project? How would you weigh the importance of much-needed jobs and other economic development against the potential for negative environmental impacts like mercury contamination, loss of habitat, or variations in water flow? What do you think of Cree leaders who have changed their positions on this and other projects, first opposing and then supporting them—do you think they are indecisive, or fickle, or are they courageous and forward-thinking?

---

calling it superficial "greenwashing"; yet, if the company achieves only a fraction of its stated goals, the environmental benefits could be substantial because of the corporation's vast global reach.

The bottom line is that corporate actions hinge on consumer behaviour. It is up to all of us as consumers to encourage trends in sustainability by rewarding corporations that truly promote sustainable solutions.

Conclusion

Corporate sustainability, alternative ways of measuring growth, and the valuation of ecosystem goods and services are a few of the recent developments that have brought economic approaches to bear on environmental protection and resource conservation. As economics becomes more environmentally friendly, it renews some of its historic ties to ethics.

Environmental ethics has expanded people's sphere of ethical consideration outward to encompass other societies and cultures, other creatures, and even nonliving entities that were formerly outside the realm of ethical concern. This ethical expansion involves the concept of *distributional equity*, or equal treatment for all, which is the aim of environmental justice. One type of distributional equity is equity among generations. Such concern by current generations for the welfare of future generations is the basis for the notion of sustainability.

Although we tend to think of sustainability as a "modern" idea, it is actually inherent to some of the most basic concepts of neoclassical economics. Sir John Hicks (1904–1989), one of the most influential economists of the twentieth century, defined income as "the maximum value [a person] can consume during a week, and still expect to be as well off at the end of the week as he was at the beginning." The implication is that *true* income is *sustainable* income—if your spending compromises your resource base and reduces your future ability to produce, then you are depleting your capital. As former World Bank economist Herman E. Daly explains,

> *Why all the fuss about sustainability? Because, contrary to the theoretical definition of income, we are in fact consuming productive capacity and counting it as income in our national accounts... Depletion of natural capital and consequent reduction of its life-sustaining services is the meaning of unsustainability.*[24]

Is sustainability a pragmatic pursuit for us? The answer largely depends on whether we believe that economic well-being and environmental well-being are opposed to each other, or whether we accept that they can work in tandem. Equating economic well-being with economic growth, as most economists traditionally have, suggests that economic welfare entails a trade-off with environmental quality. However, if economic welfare can be enhanced in the absence of growth, we can envision economies and environmental quality benefiting from one another.

REVIEWING OBJECTIVES

You should now be able to:

Characterize the influences of culture and world view on the choices people make

- A person's culture strongly influences his or her world view. Such factors as religion and political ideology are especially influential.

Outline the nature, evolution, and expansion of environmental ethics in Western cultures

- Our society's domain of ethical concern has been expanding, such that we have granted ethical consideration to more and more entities.
- Anthropocentrism values humans above all else, whereas biocentrism values all life and ecocentrism values ecological systems.
- The preservation ethic (preserving natural systems intact) and the conservation ethic (promoting responsible long-term use of resources) have guided branches of the environmental movement during the past century.
- The environmental justice movement, seeking equal treatment for people of all races and income levels, is a recent outgrowth of environmental ethics.

Describe precepts of classical and neoclassical economic theory, and summarize their implications for the environment

- Classical economic theory proposes that individuals acting for their own economic good can benefit society as a whole. This view has provided a philosophical basis for free-market capitalism.
- Neoclassical economics focuses on consumer behaviour and supply and demand as forces that drive economic activity.
- Several assumptions of neoclassical economic theory contribute to environmental impact.

Compare the concepts of economic growth, economic health, and sustainability

- Conventional economic theory has promoted never-ending economic growth, with little regard to possible environmental impact.

- Economic growth is not necessarily required for overall economic well-being.
- In the long run, some economists believe that a steady-state economy will be necessary to achieve sustainability.

Explain the fundamentals of environmental economics and ecological economics

- Environmental economists advocate reforming economic practices to promote sustainability. Key approaches are to identify external costs, assign value to nonmonetary items, find new approaches to measuring growth, and attempt to make market prices reflect real costs and benefits.
- Ecological economists support these efforts and others. Many support developing a steady-state economy.
- Consumer choice in the marketplace can help drive businesses and corporations to pursue sustainability goals.

TESTING YOUR COMPREHENSION

1. What does the study of ethics encompass? Describe the three classic ethical standards. What is environmental ethics?
2. Why in Western cultures have ethical considerations expanded to include nonhuman entities?
3. Describe the philosophical perspectives of anthropocentrism, biocentrism, and ecocentrism. How would you characterize the perspective of the Mirrar Clan?
4. Differentiate between the preservation ethic and the conservation ethic. Explain the contributions of John Muir, James Harkin, Gifford Pinchot, and Clifford Sifton in the history of environmental ethics.
5. Describe Aldo Leopold's "land ethic." How did Leopold define the "community" to which ethical standards should be applied?
6. Name four key contributions the environment makes to the economy.
7. For each of these basic tenets of neoclassical economics, explain the potential impacts on the environment and provide a hypothetical example:
   - Resources are infinite or substitutable.
   - Long-term effects should be discounted.
   - Costs and benefits are internal.
   - Growth is good.
8. Describe Adam Smith's metaphor of the "invisible hand." How did neoclassical economists refine classical economics? Neoclassical economists have moved away from Smith's original definition of income as "economic gains made with no negative impacts on the resource base." What are the environmental implications of straying from this definition of income?
9. Compare and contrast the views of neoclassical economists, environmental economists, and ecological economists.
10. What is contingent valuation, and what is one of its weaknesses? Describe an alternative method that addresses this weakness.

SEEKING SOLUTIONS

1. Do you feel that an introduction to environmental ethics and world views is an important part of a course in environmental science? Should ethics and world views be a component of other science courses? Explain your answers.
2. Describe your world view as it pertains to your relationship with the environment. How do you think your culture has influenced your world view? How do you think your personal experience has influenced it, including your gender and race? Do you feel that you fit into any particular category discussed in this chapter? Why or why not?
3. How would you analyze the case of the Mirrar Clan and the proposed Jabiluka uranium mine from each of the following perspectives? In your description, list two questions that a person of each perspective would likely ask when attempting to decide whether the mine should be developed. Be as specific as possible,

and be sure to identify similarities and differences in approaches:
- Preservationist
- Conservationist
- Deep ecologist
- Environmental justice advocate
- Indigenous land rights activist
- Ecofeminist
- Neoclassical economist
- Ecological economist

4. What is a steady-state economy? Do you think this model is a practical alternative to the growth paradigm? Why or why not?
5. Do you think we should attempt to quantify and assign market values to ecosystem services and other entities that have only nonmarket values? Why or why not?
6. **THINK IT THROUGH** A manufacturing facility on a river near your home provides jobs for 200 people in your community and pays $2 million in taxes to the local government each year. Sales taxes from purchases made by plant employees and their families contribute an additional $1 million to local government coffers. However, a recent peer-reviewed study in a well-respected scientific journal revealed that the plant has been discharging large amounts of waste into the river, causing a 25% increase in cancer rates, a 30% reduction in riverfront property values, and a 75% decrease in native fish populations.

The plant owner says the facility can stay in business only because there are no regulations mandating expensive treatment of waste from the plant. If any such regulations were imposed, he says he would close the plant, lay off its employees, and relocate to a more business-friendly community. How would you recommend resolving this situation? What further information would you want to know before making a recommendation? In arriving at your recommendation, how did you weigh the costs and benefits associated with each of the plant's impacts?

INTERPRETING GRAPHS AND DATA

Economists use various indicators of economic well-being. One that has been used for decades is the gross domestic product (GDP), the total monetary value of final goods and services produced each year. An alternative measure called the genuine progress indicator (GPI) is calculated as follows:

GPI = GDP + (Benefits Ignored by GDP) − (Environmental Costs) − (Social and Economic Costs)

Benefits include such things as the value of parenting and volunteer work. Environmental costs include the costs of water, air, and noise pollution; loss of wetlands; depletion of nonrenewable resources; and other environmental damage. Social and economic costs include investment, lending, and borrowing costs; costs of crime, family breakdown, underemployment, commuting, pollution abatement, and automobile accidents; and loss of leisure time.

1. Describe economic growth as measured by GDP for the province of Alberta from 1961 to 2006. Now describe economic growth as measured by GPI over the same time period. To what factors would you attribute the growing difference between these measures?

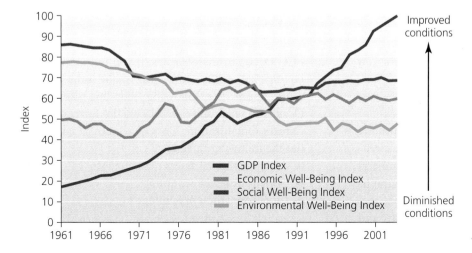

Components of GPI and GDP in Alberta, 1961–2003. Data from *Pembina Institute, Alberta GPI Accounts, 1961–2003.*

2. For GPI to grow, one or more things must happen: Either GDP must grow faster, benefits must grow faster, or social, economic, and environmental costs must shrink relative to the other terms. How would you explain the changes in GDP and GPI over the past few decades for the province of Alberta? How do the data in the graph support your answer?
3. Even with regulations for air and water pollution control, hazardous waste disposal, solid waste management, forestry practices, and species protection, environmental costs continue to increase. Why do you suppose the trend is still in that direction?
4. Alberta in the 2000s is in the midst of a huge economic boom, related largely to the development of the Athabasca and other fossil fuel deposits. How do you think this ongoing resource development will affect Alberta's GDP and GPI for the province over the next few decades? Think about all of the factors—economic, environmental, and social—that will play a role, and whether the impacts will likely be positive or negative.

CALCULATING FOOTPRINTS

Although the gross domestic product (GDP) of Alberta grew impressively between 1961 and 2003 (see the graph in "Interpreting Graphs and Data"), so did the province's population. According to Statistics Canada, the population of the province was 1 332 000 in 1961, and 3 375 000 in 2006—an increase of more than 150%.

Estimate the values of the components of the genuine progress indicator (GPI) for Alberta in 1961 and 2003 from the graph (above), and enter your estimates into the table below. Then, by using the population figures, calculate per capita values for each component in 1961 and 2003.

1. Consider your own life. What would you estimate is the value of the benefits in which you participate? Compare your personal estimate with the average values for the province of Alberta in 2003.
2. What would you estimate are the values of the environmental, social, and economic costs for which you are personally responsible? How do they compare with the averages for Alberta?

	Alberta Total 1961	Per Capita 1961	Alberta Total 2003	Per Capita 2003
GDP				
Environmental				
Social				
Economic				
GPI				

3. In 2003, social and economic costs were proportionally larger relative to GDP than they were in 1961, and environmental costs were roughly the same in proportion to GDP. How would you account for these trends? What could you do to help improve these trends in your own personal accounting?

TAKE IT FURTHER

Go to www.myenvironmentplace.ca where you will find

- Suggested answers to end-of-chapter questions
- Quizzes, animations, and flashcards to help you study
- Research Navigator™ database of credible and reliable sources to assist you with your research projects
- Tutorials to help you master how to interpret graphs
- Current news articles that link the topics that you study to case studies from your region and around the world

- **ECO Occupational Profiles:** If you found this chapter especially interesting, you might want to learn more about the following jobs by visiting the Occupational Profiles website of the Environmental Careers Organization. Go to www.eco.ca and check out the following careers:
  - Environmental assessment analyst
  - Environmental economist
  - Environmental psychologist

CHAPTER ENDNOTES

1. This piece is based on a summary of *Mining Denendeh: A Dene Nation Perspective on Community Health Impacts of Mining,* by Chris Paci (Lands and Environment Department, Dene Nation) and Noeline Villebrun, Dene national chief, prepared for the 2004 Mining Ministers Conference, Coppermine, Nunavut. Dr. Chris Paci, Lands and Environment Department, Dene Nation, Box 2338, Yellowknife, NT, Canada, X1A 2P7, cpaci@denenation.com Prepared for the 2004 Mining Ministers Conference, Coppermine, Nunavut.
2. Regarding the incident of contamination that led to the mine's closure in 2004, as reported by Friends of the Environment, Australia, www.foe.org.au/media-releases/2004-media-releases/mr_24_3_04.htm.
3. *The Age*, July 27, 2007, as cited by World Information Service on Energy, www.wise-uranium.org/upjab.html.
4. Based partly on information from *Living with the Land: A Manual for Documenting Cultural Landscapes*, NWT Cultural Places Program, Government of NWT, 2007.
5. Voisey's Bay Nickel Mine, Ltd., www.vbnc.com/iba.asp.
6. Parts of this paragraph are based on information from *Mining Denendeh: A Dene Nation Perspective on Community Health Impacts of Mining*, by Chris Paci (Lands and Environment Department, Dene Nation) and Noeline Villebrun, Dene national chief, prepared for the 2004 Mining Ministers Conference, Coppermine, Nunavut.
7. Based on information from the Canadian Museum of Civilization, James Bernard Harkin, www.civilization.ca/hist/biography/biographi204e.html.
8. Based on information from Canadian Museum of Civilization, Clifford Sifton, www.civilization.ca/hist/advertis/ads2-06e.html.
9. Rajiv Rawat (1996, May) *Women of Uttarakhand: On the Frontiers of the Environmental Struggle*, http://uttarakhand.prayaga.org/chipko.html.
10. Green Belt Movement, http://eratos.utm.utoronto.ca/research.htm; and other sources.
11. Recall that radioactivity had only been discovered some 30 years earlier, in the late 1890s.
12. Andrew Nikiforuk (1998) "Echoes of the Atomic Age: Cancer kills fourteen aboriginal uranium workers," *Calgary Herald*, Saturday, March 14.
13. World Information Service on Energy, *Decommissioning of Port Radium*, www.wise-uranium.org/uippra.html#MORE.
14. Radiation Exposure Compensation Program www.usdoj.gov/civil/torts/const/reca/; and "Navajo President Joe Shirley, Jr., updates Navajo miners on progress toward getting fair RECA compensation," September 2006.
15. Repetto, Robert (1992) Accounting for environment assets, *Scientific American*, June.
16. Environment Centre Northern Territory (Australia), *Mining Archives*, www.ecnt.org/index.html.
17. Costanza, R., et al. (1997) The value of the world's ecosystem services and natural capital. *Nature* 387: 253–260.
18. Matthew Coon Come, *Goldman Prize*, www.goldmanprize.org/node/93 and other sources.
19. Matthew Coon Come, *Goldman Prize* www.goldmanprize.org/node/93 and other sources.
20. Government of Canada (2007) *Key Economic Events: 1972—The James Bay Project*, www.canadianeconomy.gc.ca/English/economy/1972James_Bay_Project.html.
21. Government of Canada (2007) *Key Economic Events: 1972—The James Bay Project*, www.canadianeconomy.gc.ca/English/economy/1972James_Bay_Project.html.
22. Government of Canada (2007) *Key Economic Events: 1972—The James Bay Project*, www.canadianeconomy.gc.ca/English/economy/1972James_Bay_Project.html.
23. Nation Talk (2008) *General Statement by Matthew Coon Come*, June 13, www.nationtalk.ca/modules/news/article.php?storyid=10522.
24. Daly, Herman E. (2001) Sustainable development and OPEC. Paper invited for the conference OPEC and the Global Energy Balance: Towards a Sustainable Energy Future. Vienna, Austria.

Environmental Policy: Decision Making and Problem Solving

22

Lake Erie has recovered from the worst of its pollution. However, the environmental woes of the lake are far from over.

Upon successfully completing this chapter, you will be able to

- Describe environmental policy and assess its societal context
- Identify the institutions important to Canadian environmental policy and recognize major Canadian environmental laws
- List the institutions involved with international environmental policy and describe how nations handle transboundary issues
- Categorize the different approaches to environmental policy

Lake Erie is the shallowest of the Great Lakes, and the smallest by volume. This increases its susceptibility to eutrophication and algal blooms. In this satellite image, sediment plumes can be seen swirling through the water. Rows of crops can be seen on the land around the lake. The Canada–U.S. international border runs lengthwise down the middle of Lake Erie.

CENTRAL CASE:
THE DEATH AND REBIRTH OF LAKE ERIE

We must ask ourselves seriously whether we really wish some future universal historian on another planet to say about us: "With all their genius and with all their skill, they ran out of foresight and air and food and water and ideas," or, "They went on playing politics until their world collapsed around them."
—U THANT, FORMER SECRETARY-GENERAL OF THE UNITED NATIONS, 1970

"When you get ready to vote, make sure you know what you're doing."
—BOB HUNTER, JOURNALIST, ACTIVIST, AND CO-FOUNDER OF GREENPEACE

During the 1970s, Lake Erie became infamous when it effectively "died" as a result of pollution. Sewage, fertilizers, phosphate-containing detergents, and other chemicals in runoff combined to result in over-fertilization, algal blooms, eutrophication, and oxygen depletion. The water turned a sickly green colour, and the lake was widely declared to be dead. The water subsequently became so depleted in oxygen that portions of it became crystal clear—not because it was "clean," but because it was incapable of supporting aquatic life.

Lake Erie is the smallest of the Great Lakes in volume and depth, though not in surface area. Erie therefore is lacking the very cold, deep waters that characterize the larger lakes (notably Superior). It is slightly warmer, circulates nutrients more effectively, and is the most biologically productive of the Great Lakes. Erie is surrounded by agricultural land, with significant industrial development and a population that surged in the late nineteenth century and into the twentieth century. These factors combined to expose this most sensitive of the Great Lakes to an onslaught of chemicals that overloaded its capacity.

A concentrated international effort brought Lake Erie back from the brink of disaster. After tough legal

Through the *Canada–U.S. Great Lakes Water Quality Agreement (GLWQA)*, Canada and the United States have identified 42 Areas of Concern (AOCs) around the Great Lakes and connecting channels. Each of these has associated with it a Remedial Action Plan, or RAP. There are currently seven AOCs associated with Lake Erie.

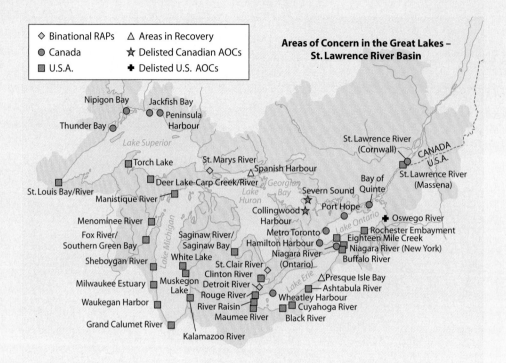

restrictions on nutrient runoff were implemented on both sides of the border, improvements began to be seen. By the 1980s, nutrient levels in the lake had decreased sufficiently that aquatic populations began to rebound. Achieving this required significant cross-border communication and collaboration on policy and decision making.

The principal binational agreement governing the Great Lakes is the *Canada–U.S. Great Lakes Water Quality Agreement (GLWQA)*. Its precursor was the *Boundary Waters Treaty* of 1909, which set up a mechanism for resolving international disputes over transboundary water resources and established the *International Joint Commission (IJC)* to ensure the respective and common interests of Canada and the United States. Given the significance of the Great Lakes, which contain about 22% of the world's freshwater, with a total surface area of 244 100 km^2, it is not surprising that this is where much of the work of the IJC has been focused.

In 1972, in response to the "death" of Lake Erie, other pollution crises of the early 1970s, and the increasing environmental awareness of citizens, Canada and the United States signed the *GLWQA*. Its main objective is to restore and maintain the chemical, physical, and biological integrity of the waters of the Great Lakes Basin ecosystem, including the St. Lawrence River system.

Under the *GLWQA*, Canada and the United States have designated 42 Areas of Concern (AOCs) in the lakes and connecting channels. Sixteen are in Canada; seven are associated with Lake Erie (see figure). By definition, AOCs are areas in which environmental degradation is pronounced, such that restrictions must be placed on swimming, fishing, and drinking water consumption. Some AOCs, especially those associated with connecting channels, are of particular concern because they contribute to the overall degradation of the Great Lakes. For each AOC, an individually tailored *Remedial Action Plan*, or RAP, has been developed.

The RAPs focus on specific locations, but more holistic *Lakewide Management Plans (LaMPs)* have been created for Erie, Ontario, and Superior. The Lake Erie LaMP involves the province of Ontario, Great Lakes states, and a network of other stakeholders. The goal is to describe the lake's problems, identify the source of the problems, and envision the preferred future state of the resource.[1]

Efforts to cooperate on behalf of the Great Lakes have not been restricted to the two federal governments. The government of Canada has reached an agreement with the province of Ontario, called the *Canada–Ontario Agreement Respecting the Great Lakes Basin Ecosystem (COA)*. Its intention is to promote cooperative action between Canada and Ontario to restore and sustain the environmental quality of the Great Lakes. The Great Lakes states and a number of Tribal Nations in the United States participate in similar agreements.

Lake Erie recovered from the worst of its pollution after the *GLWQA* came into effect. However, the environmental woes of the lake are far from over. For example, phosphate levels in the lake—largely under control since the 1980s—began to increase in the mid-1990s. Invasive species, such as the zebra mussel (Chapter 4), are having complex impacts on native populations. Bioaccumulative toxins in the lake are of concern, as well. The Lake Erie LaMP will attempt to delineate these problems and propose solutions for a cleaner, healthier lake.

Who knew that it would take so many decades, so many organizations (and acronyms), and so much effort to get two nations to work together for the good of this crucial resource? The rescue of Lake Erie is one example of how people and organizations work together to achieve environmental goals, and how laws and policies directly influence the environment.

Environmental Policy

When a society reaches broad agreement that a problem exists, it may persuade its leaders to try to resolve the problem through the making of policy. **Policy** is a formal set of general plans and principles intended to address problems and guide decision making in specific instances. **Public policy** is policy made by governments, including those at the local, provincial or territorial, federal, and international levels. Public policy consists of laws, regulations, orders, incentives, and practices intended to advance societal welfare. **Environmental policy** is policy that pertains to human interactions with the environment. It generally aims to regulate resource use or reduce pollution to promote human welfare and/or protect natural systems.

Forging effective policy requires input from science, ethics, and economics. As discussed in Chapter 1, science provides the information and analysis needed to identify and understand environmental problems and devise potential solutions to them. Ethics and economics (Chapter 21) offer criteria to assess the extent and nature of problems and to help clarify how society might like to address them. Government interacts with individual citizens, organizations, and the private sector in a variety of ways to formulate policy (**FIGURE 22.1**).

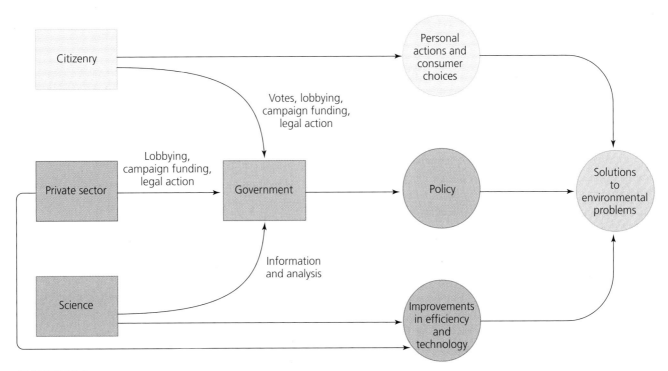

FIGURE 22.1
Policy plays a central role in how we as a society address environmental problems. Voters, the private sector, and groups representing various interests influence government representatives. Scientific research also informs government decisions. Governmental representatives and agencies formulate policy that aims to address social problems, including environmental problems. Public policy—along with improvements in technology and efficiency from the private sector and personal actions and consumer purchasing choices exercised by citizens—can produce lasting solutions to environmental problems.

The pollution problems of Lake Erie show how science, economics, and ethics can inform and motivate the making of policy, both domestically and internationally. Scientific research tells us that agricultural and urban runoff carries potentially harmful pathogens and other contaminants (Chapters 11 and 19). Science also reveals how excess organic material can radically alter conditions for aquatic and marine life, lowering concentrations of dissolved oxygen and increasing mortality for many species (Chapters 5 and 12). Pollution and beach closures and impacts on fisheries inflict economic loss by reducing recreation, tourism, and other economic activity associated with clean water and coastal areas. Ethically, water pollution poses problems because pollution from upstream users degrades water quality for downstream users not responsible for the pollution.

Many environmental problems share this combination of impacts—harming human health, altering ecological systems, inflicting economic damage, and creating inequities among people. In the Great Lakes, as in many other environmental systems, all of this takes place in a transboundary, multijurisdictional context, with political pressures that come from both domestic and international sources.

Environmental policy addresses issues of equity and resource use

The capitalist market economic systems of modern democracies are largely driven by incentives for short-term economic gain rather than long-term social and environmental stability. Market capitalism provides little incentive for businesses or individuals to behave in ways that minimize environmental impact or equalize costs and benefits among parties. Such *market failure* has traditionally been viewed as justification for government intervention. Environmental policy aims to protect environmental quality and the natural resources people use and to promote equity in people's use of resources.

The tragedy of the commons Policy to protect resources held in common by the public is intended to safeguard these resources from depletion or degradation. As Garrett Hardin explained in his essay "The Tragedy of the Commons", a resource held in common that is accessible to all and is unregulated will eventually become overused and degraded. Therefore, he argued, it is in our best interest to develop guidelines for the use of such resources. In Hardin's illustrative example of a commonly owned pasture, guidelines might limit the number of animals each individual can graze or might require pasture users to pay to restore and manage the shared resource. These two concepts—regulation of use and active management—are central to environmental policy today.

The tragedy of the commons does not always play itself out as Hardin predicted. Some traditional societies have devised safeguards against exploitation, and in modern Western societies resource users have occasionally cooperated to prevent overexploitation. Moreover, in many cases Hardin's starting assumptions are not met; resources on public lands may not be equally accessible to everyone but may instead be more accessible to wealthier or more established resource extraction industries. Nonetheless, the threat of overexploiting public resources is real and has been a driving force behind much environmental policy.

Free riders Another reason to develop policy for publicly held resources is the **free rider** problem. Let us say a community on a river suffers from water pollution that emanates from 10 different factories. The problem could in theory be solved if every factory voluntarily agreed to reduce its own pollution. However, once they all begin reducing their pollution, it becomes tempting for any one of them to stop doing so. Such a factory, by avoiding the sacrifices others are making, would in essence get a "free ride" on the efforts of others. If enough factories take a free ride, the whole effort will collapse.

---

weighing the issues

PRIVATE VS. PUBLIC GOOD

22-1

Imagine you have purchased land and plan to clear its forest and build condominiums on it. A local environmental group finds an endangered plant species on the property and petitions the government to prevent development of the land. Should you be allowed to build? If not, should you be financially compensated?

Now imagine that you are a member of the environmental group and also a neighbour of the landowner in question. The land holds the last stand of forest in this region, which has experienced extensive forest loss over the past decade. You feel your quality of life and that of your neighbours will be compromised if the development goes ahead. Should you be allowed to claim damages in civil court? Make a case why the landowner should not be allowed to build.

Which argument do you feel has greater merit? Would it make a difference to you if the land had been zoned as developable (or not) when the purchase was made? What do you think is the best way to balance private property rights with protection of the public good in cases like this?

Because of the free rider problem, private voluntary efforts are often less effective than efforts mandated by public policy. Public policy can prevent free riders and ensure that all parties sacrifice equitably, by enforcing compliance with laws and regulations or by taxing parties to attain funds with which to pursue societal goals.

External costs Environmental policy is also developed to ensure that some parties do not use resources in ways that harm others. One way to promote fairness is by dealing openly with *externalities* and *external costs*, harmful impacts that result from market transactions but are borne by people not involved in the transactions. For example, a factory may reap greater profits by discharging waste freely into a river and avoiding paying for proper waste disposal. Its actions, however, impose external costs (water pollution, decreased fish populations, esthetic degradation, or other problems) on downstream users of the river (FIGURE 22.2).

In Chapter 5, for example, we discussed the problem of downstream pollution in the estuary of the St. Lawrence River. The pollution that ends up in the estuary originates far upstream. It comes from the industries, towns, and farms that line the river, all the way upstream to the shores of the Great Lakes. The costs—both economic and ecological—of downstream pollution are passed along to someone else or to society as a whole and are not directly internalized to the activities that generated the pollution in the first place.

These, then, are the fundamental goals of environmental policy: to protect resources against the tragedy of the commons, and to promote equity by eliminating free riders and addressing external costs.

Many factors hinder implementation of environmental policy

If the goals of environmental policy are seemingly so noble, why is it that environmental laws are so often challenged, environmental regulations frequently derided, and the ideas of environmental activists repeatedly ignored or rejected by citizens and policy makers?

In North America, most environmental policy has come in the form of laws instituted by government regulators. Some businesses and individuals view these regulations as overly restrictive, bureaucratic, or unresponsive to human needs. For instance, many landowners fear that zoning regulations or protections for endangered species will impose restrictions on their activities or on the use of their land. Developers complain of time and money lost to bureaucracy in obtaining permits; reviews by government agencies; surveys for endangered species; and required environmental controls, monitoring, and mitigation. In the eyes of such property owners and businesspeople, environmental regulation all too often means inconvenience and economic loss.

Another reason people sometimes do not see a need for environmental policy stems from the nature of most environmental problems, which often develop gradually. The degradation of ecosystems and public health caused by human impact on the environment is a long-term process. Human behaviour is geared toward addressing short-term needs, and this tendency is reflected in our social institutions. Businesses usually opt for short-term economic gain over long-term considerations. The news

FIGURE 22.2
River pollution raises many issues that have been viewed as justification for environmental policy. This woman washing clothes in the river may suffer upstream pollution from factories, and her use of detergents may cause pollution for people living downstream.

media have a short attention span based on the daily news cycle, whereby new events are given more coverage than slowly developing long-term trends. Politicians often act out of short-term interest because they depend on re-election every few years. For all these reasons, many environmental policy goals that seem admirable and that attract wide public support in theory may be obstructed in their practical implementation.

Canadian Environmental Law and Policy

Canadians have been leaders in the development of environmental management approaches and policies. However, our approach to environmental management, law, and policy has been inevitably influenced by the United States. This is partly because we are next-door neighbours; it is also because we have economies that are closely linked, as well as a shared history of life on this continent.

Canada's environmental policies are influenced by our neighbour

The United States was a leader in the early development of national-level environmental laws, passing the first comprehensive environmental protection law (*National Environmental Protection Act, NEPA*) in 1970. This led, in turn, to the creation of the Environmental Protection Agency (EPA). Canada followed shortly thereafter, with the establishment of Environment Canada, mandated by the *Department of the Environment Act* in 1971. Most countries in the world have now followed the lead of Canada and the United States in this regard.

Canada is also heavily influenced by the United States in its environmental management approach because of their trading relationship, now governed largely by the North American Free Trade Agreement (NAFTA), which regulates how marketable commodities are handled. This includes things like wood, agricultural crops, animal products, and even water. In Chapter 11, for example, we discussed the federal government's approach to maintaining control over Canada's water resources, even though NAFTA identifies water as a marketable commodity that should not be restricted from export (Chapter 11).

Canadian environmental management also is strongly influenced by the sheer extent of the environmental resources we share with our southern neighbour—across the 8891 km of the world's longest undefended border we share countless rivers and watersheds, rock and soil masses, airsheds and weather systems, animal migration paths, and ecosystems. In the past, and still today, human activities on both sides of the border have negatively affected people and ecosystems on the other side. For example, in Chapter 7 we discussed how agriculture in the Fraser River Valley may be causing nitrate contamination of an aquifer that is shared with the residents of upstate Washington. In Chapter 13 we examined how industrial emissions from the central and northeastern states are transported into southern Canada, leading to smog and acid precipitation.

Sometimes Americans and Canadians don't agree on how our shared natural systems and cross-border impacts should be managed. For the most part, however, Canada–U.S. binational management of transboundary pollution and shared environmental resources has been characterized by cooperation and dialogue and serves as a successful model for international environmental management. We have mentioned a number of these collaborations throughout this book; an example is the IJC, which was instrumental in the recovery of Lake Erie.

Several legal instruments are used to ensure that environmental goals are achieved

Government agencies rely on several types of instruments to achieve environmental goals. **Acts** are laws, or statutes, proposed and voted upon by the Parliament of Canada or the provincial and territorial parliaments. **Regulations** are legal instruments too, but they are more specific; they are detailed sets of requirements (such as numerical limits, licensing requirements, performance specifications, and exemption criteria) established by governments to allow them to implement, enforce, and achieve the objectives of environmental acts.

Agreements can be either enforceable or voluntary; they are entered into by agencies of government, often with the goal of streamlining, clarifying, or harmonizing the administration of environmental legislation. **Permits** are documents that grant a group or an individual legal permission to carry out an activity that will have environmental impacts, usually within certain limitations and for a specified period of time. Examples at the federal level include permits governing the disposal of substances at sea, import and export of hazardous wastes, and hunting of migratory birds.[2] We will examine agreements, permits, and other types of nonbinding instruments in greater detail, later in the chapter.

There are four basic ways in which individuals and organizations can be required or compelled to obey environmental laws in Canada, or penalized if they do not obey the law. They include criminal enforcement; penalties or fees; administrative orders to investigate, clean up, or otherwise address an environmental situation of concern; and finally, civil actions, through which an individual or corporation that causes environmental damage to another's property or person may be held responsible for the damage via a lawsuit.[3]

Environmental goals and best practices can be promoted by voluntary initiatives

One alternative to environmental legislation that appeals to many in the private sector is the adoption, by consensus, of **voluntary guidelines** that are sector-based and self-enforced. A familiar example is the self-policing by television and music producers, for content deemed too violent for children. Some sectors have had success with this approach; others have not. The Canadian mining industry, for example, which is heavily regulated by both federal and provincial environmental legislation, has undertaken some voluntary initiatives and self-imposed guidelines on behalf of the environment. Examples include guidelines on the use of traditional ecological knowledge in mining, and best practices for the management of acid drainage at mine sites. The Accelerated Reduction and Elimination of Toxics (ARET) Program was a voluntary initiative that involved industries from a number of sectors, including mining, as well as government agencies and nongovernmental organizations. The program issued a proactive challenge to industry to reduce emissions of toxic substances in advance of legislation and was successful in reaching many of its targets.

An example of voluntary guidelines for environmental practice that have been widely adopted on an international scale are the ISO 14001 standards for environmental management. *ISO is the International Organization for Standardization*, a nongovernmental organization with 157 member nations that is headquartered in Geneva, Switzerland. Probably the best-known set of international standards is ISO 9000, which is aimed at ensuring the consistent use of best practices for quality assurance in industrial processes. The ISO 14000 series was similarly designed to promote consistency and best practices, but in the specific context of environmental management. If a factory, office, government, or other type of organization chooses to become certified under ISO 14001, it voluntarily agrees to meet the standards and follow accepted procedures for environmental management. In turn, it becomes eligible to promote itself as an ISO 14001-certified organization.

It is open to debate whether voluntary initiatives and self-policing are as effective as legislation in achieving environmental goals. Some critics argue that industry is simply seeking to avoid legislation or—more cynically—trying to divert attention from its environmental shortcomings. Even those in industry who favour voluntary initiatives acknowledge that these generally represent the "best practices" of the most proactive, environmentally responsible companies and that legislation is needed in order to avoid free riders that do not follow the voluntary guidelines.

Canadian environmental policy arises from all three levels of government

Even with the influence of such a populous nation and heavyweight economy right next door, Canadians have developed our own approach to environmental management and regulation. This approach has tended to make environmental management collaborative, cooperative, and consultative (**FIGURE 22.3**). It is codified in the *Canadian Environmental Protection Act* (1999) that the administration of the act falls to Environment Canada, but all other aspects of its implementation are collaborative. Public consultation also is part of the wording of the act.

In Canada, the federal government shares responsibility for environmental protection with provincial/territorial, Aboriginal, and municipal/local governments. The principal responsibility for the environment falls to the provincial/territorial governments in most situations. This

---

weighing the issues

GOVERNMENT REGULATION VERSUS VOLUNTARY GUIDELINES 22-2

Should industrial sectors that have a significant impact on the environment, such as mining and forestry, be allowed to police their own activities and impacts via voluntary guidelines? Has this approach worked satisfactorily in the television and music industries? Or would the interests of profit interfere with the successful implementation of such guidelines, necessitating more rigid and comprehensive government regulations? What about voluntary guidelines for the bottled water industry, which has been somewhat less regulated than mining or forestry up to now?

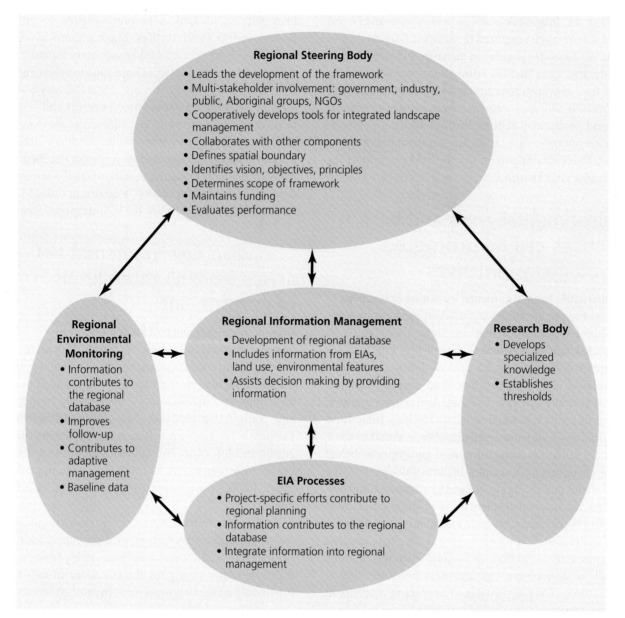

FIGURE 22.3
The Canadian approach to environmental management has always been consultative. The management process shown here, from the Canadian Environmental Assessment Agency, is based on a model for the management of cumulative effects in the region of Cold Lake, Alberta. In this model, stakeholders bring information from a wide variety of sources, to be applied collaboratively to determine a solution to an environmental problem.

multilevel approach necessitates close collaboration among the levels and agencies of government; sometimes it results in overlapping and redundancies, confusion of jurisdictions, and even contradictions between provincial and federal law.

One way the government attempts to overcome the overlapping of environmental jurisdictions is through the work of the Canadian Council of Ministers of the Environment (CCME), which comprises the federal, provincial, and territorial environment ministers. They meet regularly to work on projects of mutual and overlapping interest. For example, the CCME is developing a set of nationwide standards, the Canadian Environmental Quality Guidelines. The first harmonized standards, for water quality, were signed into accord by the ministers in 1987. Since then, several sets of standards and refinements for water quality (for various uses) and for sediment and soil quality have been added. The standards for air quality (National Ambient Air Quality Objectives) are still being developed, as of 2008.

Federal government There are many federal-level laws that affect the environment, some directly and others indirectly. Some of the most important are the following:

- ***Canadian Environmental Protection Act (CEPA) (1999),*** the centrepiece of Canada's federal-level environment legislation. *CEPA* focuses on preventing pollution, and protecting the environment and human health. The act gives the government broad enforcement powers, and mandates public participation and consultation, giving citizens input into environmental policy decisions.[4]

- *Canadian Environmental Assessment Act* (1992), which requires any project involving federal funds and lands, Aboriginal lands, or international effects to undertake an **environmental impact assessment (EIA)**, a study of the potential impacts of the project on the environment. The depth or magnitude of the assessment, the specific requirements, and the mitigation efforts that may be required depend on the extent and type of project.
- *Fisheries Act* (1985), which prohibits the release of "deleterious substances" into any body of water where fish may be present at any time. The *Fisheries Act* has proven to be one of the most powerful laws that can be brought to bear on the protection of water resources.
- *Transportation of Dangerous Goods Act* (1992), which regulates the transport of hazardous materials in Canada by air, road, rail, or ship, no matter what the purpose or point of origin of the transfer. Hazardous waste materials are also covered under this act.
- *Canadian Wildlife Act* (1985), *Migratory Birds Act* (1994), and *Species at Risk Act* (2002), all of which aim, in various ways, to respect the needs, identify risk factors, and protect organisms in Canada from harm or extinction.

Environment Canada is charged with the administration, implementation, and enforcement of *CEPA* and several other acts, collaborating in this effort with a wide variety of other federal-level departments. For example, Environment Canada and Health Canada together decide on the allowable levels for various contaminants in drinking water; the regulations are then published by Health Canada. Some other federal-level agencies that commonly work with Environment Canada to set environmental policy and implement environmental legislation are listed in Table 22.1.

Provincial/territorial governments

The provincial and territorial governments are very active in environmental protection and regulation. The specific details of content and procedure vary from one provincial jurisdiction to the next, but there is some consistency. For example, all of the provincial and territorial governments have some form of legislation that sets limits on the amounts and concentrations of specific harmful or potentially harmful substances that can be released (or *discharged*) into the environment.

Table 22.1 Federal Agencies that Influence and Implement Environmental Policy and Natural Resource Management

- **Environment Canada**, which encompasses a number of agencies, including the Canadian Environmental Assessment Agency, Parks Canada, Canadian Wildlife Service, Meteorological Service of Canada, and a number of other agencies and regional offices
- Health Canada, which participates in scientific research concerning disease and the health risks of exposure to substances, and the setting of regulatory guidelines for potentially harmful substances in water, air, soil, and food
- Fisheries and Oceans Canada, which has the central responsibility for the implementation and enforcement of the *Fisheries Act*.
- Natural Resources Canada (including the Office of Energy Efficiency), which promotes the responsible use of Canada's natural resources
- Agriculture and Agri-Foods Canada, which provides information, research, technology, policies, and programs for security of the food system, soils, and agricultural lands
- Natural Sciences and Engineering Research Council (NSERC) of Canada, which funds scientific research on the environment and related technologies
- Indian and Northern Affairs, which (among other roles) promotes the sustainable development of Aboriginal communities and their natural resources
- Transport Canada, which regulates the transport of hazardous materials by air, rail, road, and sea in Canada, as well as playing a major role in transportation system design, and in Canada's efforts to reduce greenhouse gas emissions
- Statistics Canada, which maintains extensive databases on population, patterns of travel, consumption, trade, and lifestyle, and other human activities that affect the environment
- Many other agencies, some of which come into play primarily or exclusively when it becomes necessary to respond to an emergency, such as a major flood

(The exact limits and the specific lists of controlled substances vary, but the approaches are similar.)

All of the provinces and territories also require organizations and individuals to obtain approvals or permits before undertaking activities that might prove damaging to the environment. For example, the taking of water from aquifers or surface water bodies in amounts over a certain threshold requires a special permit. Construction, dredging, mining, and logging are other examples of activities that are regulated provincially and territorially, and require approval before being undertaken.

In Ontario, citizens are protected by an Environmental Bill of Rights (EBR, 1994), one of the only such laws in the world. The EBR, administered by the appointed Environmental Commissioner of Ontario, establishes that the people of Ontario have as a common goal the protection of the natural environment. It guarantees to all Ontarians the right to a clean, healthy environment, and the right to participate in and be made aware of government decisions regarding the environment. The EBR mandates the establishment and maintenance of an Environmental Registry whereby any citizen may enter a query against an activity that may be damaging the environment, to which the government and/or industry agents responsible for that activity are required to respond.[5]

Aboriginal governments Aboriginal governments in Canada participate in environmental governance through a wide variety of mechanisms. In some cases, this involvement stems directly from land claims. For example, Aboriginal governments are extensively involved in decision making about where, when, and under what conditions to allow resource extraction activities, such as mining, forestry, or oil drilling, to proceed on Indian lands. The settlement of land claims and legal establishment of resource rights are central to such activities.

In other cases, involvement stems from the desire to ensure adequate Aboriginal representation in environmental decision making. For example, Aboriginal communities have been represented in all decision making regarding the Mackenzie Valley Natural Gas Pipeline discussed in Chapter 15 ("Central Case: On, Off, On Again? The Mackenzie Valley Natural Gas Pipeline"). Local Aboriginal groups also participate extensively in stakeholder consultations and scientific efforts to document traditional resource use patterns and traditional ecological knowledge of the environment. As discussed in Chapter 21 ("Central Case: Mining Denedeh"), modern diamond mining companies in the Northwest Territories depend heavily on collaborations with Aboriginal stakeholders.

Much of Aboriginal involvement in environmental management in Canada is based on ensuring equitable access to natural resources and preventing excessive exposure of Aboriginal communities to environmental degradation. These are the two central concerns of the environmental justice movement, as discussed in Chapter 21. For example, there is an ongoing struggle in Canada to ensure adequate drinking-water quality in Aboriginal communities—a resource that most of the rest of Canada takes for granted.

Municipal/local governments The legal landscape is further complicated by the role taken by municipal governments in environmental regulation and enforcement, a role that is growing in importance. Municipalities in Canada have traditionally taken responsibility for the

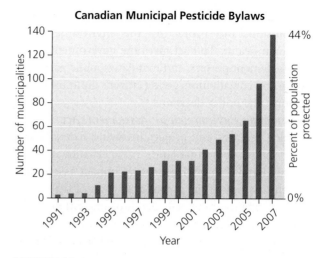

FIGURE 22.4
Many municipalities are expanding their traditional environmental roles in water and waste management to encompass other environmental initiatives, policy, and legislation. This graph shows the increasing number of municipalities in Canada that have enacted by-laws banning the local use of pesticides, effectively bypassing provincial environmental legislation.

management of water and sewage systems, noise issues, waste disposal, land-use zoning to regulate development, and local air quality concerns, such as requiring permits for backyard burning of yard wastes. All of these traditional roles have some environmental influence.

Expanding their environmental role, many municipalities have passed by-laws that restrict or prohibit the use of pesticides and herbicides on lawns in urban and suburban areas, even if their use is still approved by the provincial government. Other municipalities have adopted initiatives to deal with sprawl and transportation issues across municipal boundaries (**FIGURE 22.4**). Many urban municipalities are becoming more involved in the documentation, monitoring, and even regulation of brownfields, contaminated sites that may become available for redevelopment once they have been rehabilitated to provincial standards. Some municipalities have even entered into agreements to limit CO_2 emissions—the type of action that traditionally has fallen within the jurisdiction of the federal government.

International agreements In addition to federal, provincial/territorial, and municipal legislation, Canadians (and anyone operating a business within Canada) are also governed by the various international agreements Canada has entered into. For example, it is illegal to import elephant ivory into Canada because Canada is a party to the United Nations Convention on International Trade in Endangered Species. We will discuss the various approaches to international environmental law in greater detail, below.

Government and nongovernmental agencies work together to resolve environmental issues

The Canadian approach to environmental management is consultative; in fact, as mentioned above, public participation and consultation are mandated by *CEPA*. This means that government typically does not undertake any major revisions to policy without extensive consultation with stakeholders. A **stakeholder** is any person or group that has an interest in, or might be affected by, the outcome of a particular undertaking. This includes both environmental projects and policies (see **FIGURE 22.3**, for example).

Individual citizen stakeholders in environmental policy development are often represented by **environmental nongovernmental organizations (ENGOs)**. These can be activist groups, such as Greenpeace; or political advocacy groups, like the Sierra Club, Riverkeepers, or the Environmental Defence Fund; or groups with specific mandates, such as the Nature Conservancy, which aims to preserve land, habitat, and ecosystems by acquiring and setting aside land in trust. Stakeholders in the industrial sector may be represented by sector-specific nongovernmental agencies. For example, the Protected Areas Initiative in Manitoba employs an extensive process of public consultation on the environmental impacts of mining, in which the mining industry is represented by the Mining Association of Manitoba and the Mineral Exploration Liaison Committee; the government of Manitoba is represented by staff from the Department of Conservation and the Manitoba Geological Survey and Mines Branch; and other stakeholders are represented by World Wildlife Fund Canada.

Other types of agencies are also involved in the environmental management process in some circumstances. For example, as discussed in Chapter 11 ("The Science Behind the Story: When Water Turns Deadly: The Walkerton Tragedy"), a Crown corporation called the Ontario Clean Water Agency has been in charge of cleaning up the municipal water supply in the town of Walkerton, Ontario, since the deadly episode of contamination there in 2000.

Another collaborative and consultative vehicle that is commonly used in Canada is the **round table**, a multi-stakeholder working group established to consult on a particular issue, generally within a particular sector or area of concern. Round tables involve representatives from a variety of governmental, nongovernmental, and private-sector organizations. The National Round Table on the Environment and the Economy (NRTEE) was established in 1994 as an advisory council to the prime minister. The NRTEE has 25 appointed members, representing a variety of regions, sectors, and disciplines, including business, labour, environmental organizations, academic institutions, and First Nations. They develop recommendations on how best to integrate environmental, economic, and social considerations into decision making.[6]

Different environmental media require different regulatory approaches

Environmental law in Canada today is also influenced by history. The laws governing different environmental media—water, land, forests, mineral resources, even plants, animals, and air—have built upon different legal precedents. Let us briefly consider how these historical precedents influence the legal landscape today.

For example, water law in Canada has developed from two different historical/legal concepts. The first is called **riparian** law, in reference to the *riparian zone*, or water's edge. In riparian law, anyone who has legal access to the water's edge (such as by owning property on a river bank) has the legal right to withdraw water from the resource. The second legal concept in water law is **prior appropriation**. This refers to the "first come, first right" principle, by which one's right to withdraw water is established by historical precedent—if you have always withdrawn water from this river, then your right to do so has been established historically.

There are three basic models for water management, depending on whether the resource is public, private, or common property. For public ownership, management of the resource is typically by a government agency. In private ownership models, resource allocations are primarily controlled and reallocated by market transactions. For common property resources, the management model usually requires users to work cooperatively to establish the rules granting access to the resource. In Canada, the Crown owns all water, and water rights are granted by licence; therefore these distinctions of ownership technically do not apply. However, all three management models are used in water management to greater or lesser degrees, varying somewhat in emphasis and the details of application from province to province.[7]

Many of the environmental problems in natural resource management today were not evident when these historical precedents were set. For example, groundwater depletion wasn't a problem a hundred years ago and neither was pollution. These have necessitated some interesting reconsiderations of water law. For example, if you spill a toxic chemical on land that you own, you may not be held liable for any damages; however, if the toxic chemical

FIGURE 22.5
Indians travel to Nelson House, Manitoba, in 1910 for a treaty signing. There is record of the Aboriginal representatives asking many questions at the time of signing, about the continuation of their resource rights. The Treaty Commissioner is reported to have assured them that "not for many years to come, probably not in the lifetime of any of them, would their hunting rights be interfered with." Source (photo): Archives of Manitoba; Source (quote): Frank Tough, *Economic Aspects of Aboriginal Title in Northern Manitoba: Treaty 5 Adhesions and Métis Scrip*, Manitoba History, 15, Spring 1988.

infiltrates, joins the groundwater, and then migrates in the subsurface into the groundwater in your neighbour's property, contaminating the aquifer and wells there, you would be liable for damages.

The right to govern and allocate water (and other natural resources) was granted to the provinces by the federal government, which continues to play a role (in the management of fisheries, for example). For the eastern provinces, the transfer of rights to govern and allocate access to natural resources happened at the time of Confederation; for the western provinces, it didn't happen until the *Natural Resources Transfer Act* of 1930. These transfers of jurisdiction over land and natural resources from the federal to the provincial governments happened with little or no regard for the rights of prior Aboriginal inhabitants of the land.

The history of land law and especially mining law in Canada is intricately connected to Aboriginal land claims. There are two categories of Aboriginal land claims in Canada—those for which treaties exist, and those for which no treaty exists. Resolving these claims often rests on establishing traditional occupancy and continuous use of lands and resources (**FIGURE 22.5**).

Environmental policy has changed with the society and the economy

Environmental policy in North America has evolved, along with social and economic conditions, in tandem with and influenced by the evolution of environmental ethics (Chapter 21). From the 1780s to the late 1800s, environmental law dealt primarily with the management of public lands. It grew from early explorations and accompanied the westward expansion of settlers across the continent. This period is associated with a "frontier ethic," characterized by efforts to tame and conquer the wilderness. Environmental laws of this period were intended to promote settlement and the extraction and use of the continent's abundant natural resources.

In the late 1800s, as the continent became more populated and its resources were increasingly exploited, public perception and government policy toward natural resources began to shift. Laws of this period aimed to regulate resource use and mitigate some of the environmental problems associated with westward expansion. Policies were influenced by the emergence of the conservation and preservation ethics (Chapter 21). This period saw the opening of the first national parks, including Banff National Park in 1885.

Probably no person is more emblematic of this period in Canada than Clifford Sifton, minister of the interior in the late 1800s and early 1900s. Under Sifton's policies, wave upon wave of immigrants were encouraged to move into Canada's West to settle the Prairie grasslands and convert them into farms (**FIGURE 22.6**). Sifton

FIGURE 22.6
Early natural resource laws in North America were intended to encourage westward expansion, settlement, and resource development. Here, loggers in the late 1800s fell large cedars in British Columbia. Source: Library and Archives Canada.

FIGURE 22.7
Scientist, writer, and citizen activist Rachel Carson illuminated the problem of pollution from DDT and other pesticides in her 1962 book, *Silent Spring*.

recognized the value of Canada's natural resources, but unlike some others he also realized that those resources were not unlimited. He saw conservation and reforestation as an economic necessity and created a forestry branch of the department of the interior with the goal of regulating logging and conserving federal forests. Later he commissioned detailed studies of all Canadian natural resources, and came to be known as the "father of conservation" in Canada.

Land management policies continued to develop through the twentieth century, targeting soil conservation in the Dust Bowl years, with initiatives like the Prairie Farm Rehabilitation Administration (1935), and extending through the establishment of the National Soil Conservation Program in 1986.

The next wave of environmental policy responded mainly to pollution and environmental crises, and built upon public awareness of the impacts of environmental degradation. During the 1960s and 1970s, several events triggered increased awareness of environmental problems and brought about a shift in public priorities and important changes in public policy. One landmark event was the 1962 publication of *Silent Spring*, by American scientist and writer Rachel Carson (**FIGURE 22.7**). *Silent Spring* awakened the public to the negative ecological and health effects of pesticides and industrial chemicals. (The book's title refers to Carson's warning that pesticides might kill so many birds that few would be left to sing in springtime.)

Several other books brought to the public consciousness environmental issues, such as the limitations of resources, the impacts of human activities, and the health implications of environmental degradation. Among these were *The Limits to Growth* (1972),[8] one of the very first efforts to use computers to develop quantitative models to explore the interplay between population growth, resource use, and resource depletion; *Small is Beautiful: A Study of Economics as if People Mattered* (1973),[9] a collection of essays that critiqued the "growth is good" mantra of neoclassical economics; *The Population Bomb* (1968),[10] an extreme neo-Malthusian call for action against uncontrolled human population growth; and *Diet For a Small Planet* (1971),[11] probably the first widely read book linking food and vegetarianism with the responsible use of the planet's resources.

The impacts of pollution on surface water bodies also became starkly evident to the average citizen. The "death" of Lake Erie in the early 1970s was a highly publicized environmental disaster. The Cuyahoga River (**FIGURE 22.8**), which flows into Lake Erie, also did its part to bring attention to the hazards of pollution. The Cuyahoga was so polluted with oil and industrial waste that the river actually caught fire near Cleveland, Ohio, more than half a dozen times during the 1950s and 1960s.

FIGURE 22.8
Ohio's Cuyahoga River caught fire several times in the 1950s and 1960s. The Cuyahoga, which flows into Lake Erie, was so polluted with oil and industrial waste that the river would burn for days at a time.

This spectacle moved the public throughout North America to do more to protect the environment.

Other iconic eco-disasters of the 1960s and 1970s included the leakage of hazardous wastes from the old Hooker Chemical waste dump at Love Canal, New York, which led to the first-ever declaration of a federal state of emergency in the United States from an environmental cause; and the complete evacuation and abandonment of the town of Times Beach, Missouri, as a result of dioxin contamination. The Amoco *Cadiz* oil spill off the coast of France in 1978—still one of the largest oil spills ever—also served to raise public awareness.

In response to these events, and armed with unprecedented access to knowledge and information about the environmental and health impacts of pollution, the North American public was moved to action. Several young Canadian activists, including Bob Hunter (quoted at the beginning of this chapter), founded a tiny environmental organization called Greenpeace (**FIGURE 22.9**). Their original intent was to stage daring, high-profile protests against submarine testing of nuclear devices by the United States. They soon branched out into protests against whaling, bottom trawling, and other industrial activities they saw as exploitive and unsustainable. Greenpeace has since grown into one of the most powerful ENGOs in the world, still known for the daring nature of its protests.

Today, largely because of grassroots activism and environmental policies enacted since the 1960s, pesticides are more strictly regulated, and the air and water are considerably cleaner. The public enthusiasm for environmental protection that spurred such advances remains strong today. Polls repeatedly show that an overwhelming majority of Canadians favour environmental protection—even if it means paying more. For example, a 2008 poll showed that the majority of Canadians want our government to take decisive action against climate change, even in the face of rising oil prices reflected in the soaring price of gasoline at the pump.

Such support is evident each year in April, when millions of people worldwide celebrate Earth Day in thousands of locally based events featuring speeches, demonstrations, hikes, bird walks, cleanup parties, and more. Since the first Earth Day, on April 22, 1970, participation in this event has grown and spread to nearly every country in the world (**FIGURE 22.10**).

The social context for environmental policy changes over time

Historians have suggested that major advances in environmental policy occurred in the 1960s and 1970s because three factors converged. First, evidence of environmental problems became widely apparent. Second, people could visualize policies to deal with the problems. Third, the political climate was ripe, with a supportive public and leaders who were willing to act.

There was a fourth reason for the advancement in environmental policy: economic confidence. By the 1960s and 1970s, people in North America had reached a point in their economic development where life was reasonably comfortable, for more people than ever before. The basic necessities for survival were ensured, and people found themselves willing to make sacrifices—notably financial sacrifices but also behavioural changes—to obtain a cleaner, healthier environment for themselves and their children.

Economists have a name for this change in attitude; it is called a *willingness-to-pay transition* (**FIGURE 22.11**). Most environmental problems (such as lack of sanitation or clean water) improve with increasing economic status of societies, sometimes after an initial stage of worsening while industrial development accelerates (as is the case with air pollution). Until recently, waste generation and carbon emissions were two environmental problems for which there seemed to be no willingness-to-pay threshold—they just continued to increase in severity as incomes rose. In recent years, people in Western nations appear to be more willing to make financial sacrifices and modify their behaviour to address these problems.

FIGURE 22.9
Jim Bohlen, co-coordinator for Don't Make a Wave Committee (l), with Greenpeace skipper John Cormack, and Erving Stowe and Paul Cote of the British Columbia branch of the Sierra Club (l-r) at a Vancouver harbour in 1971, prior to setting sail for Amchitka Island where the United States was scheduled to set off an underground nuclear blast.

(a) The first Earth Day, Washington, D.C., 1970

(b) Schoolchildren celebrating Earth Day, Katmandu, Nepal, 2002

(c) Earth Day Canada's familiar logo

FIGURE 22.10
April 22, 1970, saw the first Earth Day celebration **(a)**. This public outpouring of support for environmental protection sparked a wave of environmental policy to address pollution. Decades later, Earth Day is celebrated by millions of people across the globe **(b)**, as shown here in Nepal. The familiar Earth Day Canada© logo **(c)** symbolizes leadership in environmental education and action in addition to traditional awareness-raising Earth Day activities. www.earthday.ca

During the Industrial Revolution in England, and during later industrial development in North America, air pollution from factories and refineries was a huge problem. Financial progress led the older developed nations to invest in technologies to limit air pollution, with the result that air quality has improved dramatically. Developing and rapidly industrializing nations today are facing similar problems, and some of them are reaching their own willingness-to-pay transition.

The concept of sustainable development now guides environmental policy

We may now be embarking on a new wave of environmental policy, focused on sustainable development. The concept gained popularity as a result of the 1987 report of the United Nations Commission on Environment and Development, led by the (then) prime minister of Norway, Gro Harlem Brundtland. As discussed in Chapter 1, the commission's report, entitled *Our Common Future*, defined *sustainable development* as "development that meets the needs of the present without compromising the ability of future generations to meet their own needs."[12] The concept of sustainable development got a further boost at the 1992 Earth Summit at Río de Janeiro, Brazil. This was the largest international diplomatic conference ever held, drawing representatives from 179 nations and unifying these leaders around the idea of sustainable development.

The idea of sustainable development has not been without controversy. Some people find it too vague. Others find it prone to misuse and misinterpretation. (Is it "sustainable" development, or "sustained" development?) But if nothing else, the concept of sustainable

FIGURE 22.11
This figure shows the theoretical relationship between income and various environmental problems. Access to sanitation facilities and clean water generally improves with increasing income in a society undergoing industrial development **(a)**. Air pollution tends to worsen with increasing income early in industrial development **(b)** because early industrial development typically involves increased use of fossil fuels. As incomes increase with development, a willingness-to-pay threshold is eventually reached; pollution control technologies are employed, and air quality improves. Until recently, a willingness-to-pay transition for waste generation and carbon emissions had not been evident **(c)**. These two problems traditionally worsen with increasing income. We now seem to be reaching the point where we are willing to alter our behaviour to see some improvements in these two areas.

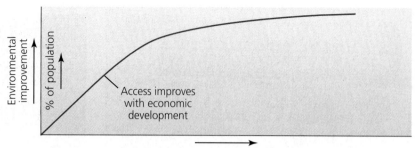
(a) Access to sanitation and clean water

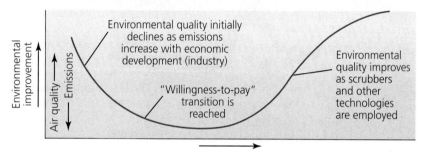
(b) Air quality

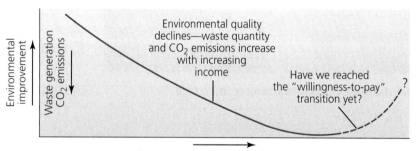
(c) Waste generation and carbon emissions

development has led to enthusiastic discussion and debate among the people of the world.

An alternative definition that has emerged more recently defines sustainable development as maximizing the co-achievement of economic, environmental, and social goals. In Chapter 23 we will explore some of the ways that people in Canada and around the world are working toward this objective.

On an international level, sustainable development as a new policy approach tries to find ways to safeguard the functionality of natural systems while raising living standards for the world's poorer people (**FIGURE 22.12**). As the world's nations continue to feel the social, economic, and ecological effects of environmental degradation, environmental policy will without doubt become a more central part of governance and everyday life in all nations in the years ahead.

Scientific monitoring and reporting helps with environmental policy decisions

All levels of government report to the Canadian public about their activities and any changes, positive or negative, in the condition of the environment within their jurisdiction. **State-of-the-environment reporting (SOER),** refers to the collection, organization, and reporting of information that can be used to measure and monitor changes in the environment and in processes or factors that have impacts on the environment over time. The information is reported by using **indicators,** values that can be measured and in comparison to which changes can be assessed. Some indicators are simple numbers (e.g., number of species on the endangered species list, or

FIGURE 22.12
Many nations are shifting their policies to support sustainable development efforts, trying to increase standards of living while safeguarding the environment. Here, a woman stirs rice on a solar-powered oven at a restaurant made of recycled drink cans, showcased at the U.N. World Summit on Sustainable Development in Johannesburg, South Africa, in 2002.

hectares of forested area). Others are complex, composite, or derived numbers (e.g., the Air Quality Health Index, which combines several indicators into one number).

According to Environment Canada, the purpose of state-of-the-environment reporting is to answer five key questions:[13]

1. What is happening in the environment (i.e., how are environmental conditions and trends changing)?
2. Why is it happening (i.e., how are human activities and other stresses linked to the issue)?
3. Why is it significant (i.e., what are its ecological and socioeconomic effects)?
4. What is being done about it (i.e., how is society responding to the issues)?
5. Is this sustainable (i.e., are human actions depleting environmental capital and causing deterioration of ecosystem health)?

The beginning of formal SOER in Canada dates back to *Our Common Future*, the 1987 Report of the United Nations Commission on Environment and Development. A response was assembled by the Canadian Council of Ministers of the Environment (CCME) in 1987. Around the same time, several provinces and territories were beginning to develop their own sets of indicators and plans for sustainability (Yukon, Alberta, British Columbia, Nova Scotia, and Saskatchewan were the earliest). Not all of the provinces and territories have regular programs for comprehensive state-of-the-environment or sustainability reporting, though all of them collect and report environmental information in one way or another. It is time consuming and expensive to collect information on a comprehensive set of indicators, and keep the information up-to-date.

Environment Canada takes the lead role in a number of SOER activities. Many other federal-level departments also deliver regular state-of-the-environment reports. These include Agriculture and Agri-Food Canada; Fisheries and Oceans Canada; Health Canada; Natural Resources Canada; Parks Canada; Indian and Northern Affairs Canada; Statistics Canada; Senate of Canada; Transport Canada; and the Treasury Board of Canada Secretariat. Some of these efforts are coordinated. In 1996 the so-called "5NR" departments (the five natural resource departments—the first four listed above, plus Environment Canada) produced a co-ordinated plan and signed a Memorandum of Understanding on SOER.

Regions in Canada that have common environmental concerns sometimes collaborate on the production of SOE reports. Some of these represent Canadian participation in international efforts; an example is the Arctic State of the Environment Report, sponsored by the Circum-Polar Council. Others are collaborations among different stakeholders and levels of government; an example is Criteria and Indicators of Sustainable Forest Management in Canada, sponsored by the multistake-holder CCME.

Many municipalities in Canada also produce SOE reports. Municipal-level reports tend to focus more on community-level indicators of well-being, of both ecosystems and people. Many corporations, too, have adopted environmental or sustainability reporting as sections or addenda to their annual business reporting framework.

SOER presents organizational challenges

There is so much environmental information that can be measured and reported; what is the most effective way to organize and analyze all of this information? Environment Canada has been a leader in the development of approaches to SOER and was instrumental in the development and early adoption of the **pressure–state–response (PSR) model** for SOER. This organizational framework is based on establishing linkages and causalities (**FIGURE 22.13**). Human activities, such as mining, water extraction, or logging, exert stresses on the

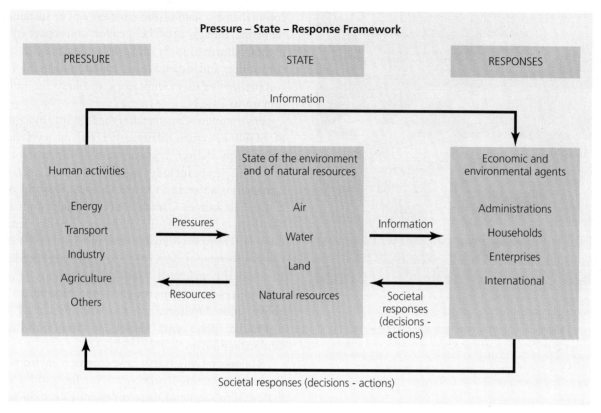

FIGURE 22.13
The pressure–state–response model allows researchers and decision makers to assess linkages and causalities, and to assess the effectiveness of responses to environmental change. Activities that cause environmental stress are shown in the "pressure" box; indicators of environmental condition in the "state" box; and human policies in the "response" box. Source: OECD Environment Monographs, No. 83, *OECD Core Set of Indicators for Environmental Performance Reviews: A Synthesis Report by the Group on the State of the Environment*, 1993.

environment (shown by indicators in the "pressure" box). This affects both the quality of the environment and the quality and quantity of natural resources (shown by indicators in the "state" box). Society responds to the changes by adopting environmental and economic policies (shown by indicators in the "response" box). The whole system is characterized by cause-and-effect feedback loops.[14]

Another organizational issue in SOER is how to subdivide the information. Most reports are based on geographic or geopolitical boundaries, but these often don't correspond to natural ecosystem boundaries. For example, it is more meaningful to produce a report for the St. Lawrence system as a whole than for any of the provinces or states that border it to produce a report based only on its own jurisdictional boundary. Therefore, some SOE reports focus on a biogeographic feature, such as a specific basin or a location with unique characteristics, such as the Arctic. Others focus on a particular sector, such as the forestry sector, a specific current issue, such as climate change, or a particular environmental medium, such as air, water, or soils.

International Environmental Law and Policy

Environmental systems pay no heed to political boundaries, so environmental problems often are not restricted to the confines of particular countries. For instance, most of the world's major rivers straddle or cross international borders, including the Great Lakes and the St. Lawrence, the Red River, and many others. Because Canada's laws have no authority in the United States or any other nation outside Canada, international law is vital to solving transboundary problems.

Often nations make progress on international issues not through legislation but through creative bilateral or multilateral agreements hammered out after a lot of hard work and diplomacy. Such was the case with the effort to develop a long-term plan to manage water quality in the Great Lakes through the International Joint Commission. Other examples of important North

American environmental agreements, some of which we have mentioned elsewhere in this book, include the Agreement on Air Quality; Migratory Birds Convention; North American Agreement on Environmental Cooperation; Canada–U.S. Agreement on the Transboundary Movement of Hazardous Waste; and Canada–U.S. Joint Marine Pollution Contingency Plan.

International law includes conventional and customary law

Because solving transboundary dilemmas requires international cooperation, several principles of international law and a number of international organizations have arisen. International law known as **customary law** arises from long-standing practices, or customs, held in common by most cultures. International law known as **conventional law** arises from **conventions** or **treaties** into which nations enter.

The Office of the Auditor General estimates that Canada has entered into more than 100 international environmental agreements (there are some 500 such agreements worldwide).[15] One example is the Montreal Protocol, a 1987 accord among more than 160 nations to reduce the emission of airborne chemicals that deplete the ozone layer. Another example is the Kyoto Protocol, aimed at reducing fossil fuel emissions that contribute to global climate change. Some examples of important international environmental laws and agreements are listed in Table 22.2.

Several organizations shape international environmental policy

Although there is no real mechanism for enforcing international environmental law, a number of international organizations regularly act to influence the behaviour of nations by providing funding, applying peer pressure, and/or directing media attention.

The United Nations In 1945, representatives of 50 countries founded the United Nations. Headquartered in New York City, this organization's purpose is "to maintain international peace and security; to develop friendly relations among nations; to cooperate in solving international economic, social, cultural and humanitarian problems and in promoting respect for human rights and fundamental freedoms; and to be a centre for harmonizing the actions of nations in attaining these ends."

Table 22.2 Some Important International Environmental Laws and Agreements in Which Canada is a Party

Air
- Stockholm Convention on Persistent Organic Pollutants (POPs)
- United Nations Framework Convention on Climate Change (UNFCCC)–Kyoto Protocol
- Vienna Convention for the Protection of the Ozone Layer–Protocol on Substances that Deplete the Ozone Layer (Montreal Protocol)
- Convention of the World Meteorological Organization (WMO)

Biodiversity
- Agreement on the Conservation of Polar Bears
- Convention on Biological Diversity
- Convention on International Trade in Endangered Species of Wild Fauna and Flora (CITES)
- Convention on Wetlands of International Importance (Ramsar 1971)

Ecosystems
- Antarctic Treaty
- Arctic Council

Environmental Cooperation
- UNECE Convention on Environmental Impact Assessment in a Transboundary Context (Espoo Convention)

Hazardous Materials and Wastes
- Basel Convention on the Control of Transboundary Movements of Hazardous Wastes and Their Disposal
- Rotterdam Convention on the Prior Informed Consent (PIC) Procedure for Certain Hazardous Chemicals and Pesticides in International Trade

Lakes and Rivers
- Canada–U.S. Agreement on Great Lakes Water Quality
- Treaty Relating to the Boundary Waters and Questions Arising Along the Border Between the United States and Canada

Oceans
- Convention on the Prevention of Marine Pollution by Dumping of Waste and Other Matter (LC72)
- International Convention for the Prevention of Pollution From Ships (MARPOL 73/78)
- International Convention on Oil Pollution Preparedness, Response, and Cooperation
- United Nations Fish Stocks Agreement

OECD Environment Monographs, No. 83, OECD Core Set of Indicators for Environmental Performance Reviews: A synthesis report by the Group on the State of the Environment, 1993, www.ens.gu.edu.au/AES I 161/Topic I/Images/gd93179.pdf.

FIGURE 22.14
The United Nations is active in international environmental policy making. For instance, it sponsored the 2002 Earth Summit in Johannesburg, South Africa.

The United Nations has taken an active role in shaping international environmental policy (**FIGURE 22.14**). Of several agencies within it that influence environmental policy, most notable is the **United Nations Environment Programme (UNEP)**, created in 1972, which helps nations understand and solve environmental problems. Based in Nairobi, Kenya, its mission is sustainability, enabling countries and their citizens "to improve their quality of life without compromising that of future generations." UNEP's extensive research and outreach activities provide a wealth of information useful to policy makers and scientists throughout the world.

The European Union The European Union (EU) seeks to promote Europe's unity and its economic and social progress (including environmental protection) and to "assert Europe's role in the world." The EU can sign binding treaties on behalf of its 27 member nations and can enact regulations that have the same authority as national laws in each member nation. It can also issue *directives*, which are more advisory in nature. The EU's European Environment Agency works to address waste management, noise pollution, water pollution, air pollution, habitat degradation, and natural hazards. The EU also seeks to remove trade barriers among member nations. It has classified some nations' environmental regulations as barriers to trade because some northern European nations have traditionally had more stringent environmental laws that prevent the import and sale of environmentally harmful products from other member nations.

The World Trade Organization Based in Geneva, Switzerland, the World Trade Organization (WTO) was established in 1995, having grown from a 50-year-old international trade agreement. The WTO represents multinational corporations and promotes free trade by reducing obstacles to international commerce and enforcing fairness among nations in trading practices. Whereas the United Nations and the European Union have limited influence over nations' internal affairs, the WTO has real authority to impose financial penalties on nations that do not comply with its directives. These penalties can on occasion play major roles in shaping environmental policy.

Like the EU, the WTO has interpreted some national environmental laws as unfair barriers to trade. For instance, in a well-known example, in 1995 the U.S. EPA issued regulations requiring cleaner-burning gasoline in U.S. cities. Brazil and Venezuela filed a complaint with the WTO, saying the new rules unfairly discriminated against the petroleum they exported to the United States, which did not burn as cleanly. The WTO agreed, ruling that even though the South American gasoline posed a threat to human health in the United States, the EPA rules represented an illegal trade barrier. Not surprisingly, critics have frequently charged that the WTO and trade agreements like NAFTA aggravate environmental problems (see "The Science Behind the Story: The Environment in NAFTA and NAAEC").

The World Bank Established in 1944 and based in Washington, D.C., the World Bank is one of the globe's

---

weighing the issues

TRADE BARRIERS AND ENVIRONMENTAL PROTECTION 22–3

In 1999 the state of California voted to ban a gasoline additive called methyl tertiary butyl ether (MTBE), which is made from methanol and boosts gasoline octane, after traces of the chemical were found in the state's drinking water. Vancouver-based Methanex Corporation, the world's largest manufacturer of methanol, concluded that this was trade protectionism. The company contested the ban and claimed $970 million from the United States Treasury in compensation for its losses; it would have been the largest compensation ever awarded under NAFTA. In 2005, NAFTA's arbitration tribunal dismissed the claim.

Some people thought this case would open the floodgates for trade protectionism arguments against any jurisdiction that bans products that may be harmful to health or the environment. Others say that California was using the environment unfairly to protect its domestic methanol producers against foreign competition. What do you think?

THE SCIENCE BEHIND THE STORY

The Environment in NAFTA and NAAEC

The brown cloudy water in this photo is a plume of wastewater being released into the Pacific Ocean, near Tijuana.

In 1992, the United States, Mexico, and Canada signed the North American Free Trade Agreement, better known as NAFTA. NAFTA is a comprehensive trade and investment agreement with aggressive measures to eliminate tariffs and reduce other kinds of barriers to trade. Since January 1, 2003, virtually all trade among the three countries has been tariff free as a result of NAFTA. Since the signing of NAFTA, however, it has been a major concern of ENGOs and other organizations to determine the impacts—positive or negative—of trade in general, and NAFTA in particular, on the environment.

The critical section of the agreement is the infamous "Chapter 11," which aims to prevent the mistreatment of investors by foreign governments. This means, for example, that investors should not have their assets seized (expropriated) by a foreign government and should not be subjected to regulations that give unfair advantages to domestic investors. Chapter 11 provides a dispute settlement mechanism whereby investors may challenge a government's handling of a particular industry or foreign investment.

Canada has participated in a number of disputes; probably the most famous was the softwood lumber dispute, in which the United States claimed that Canada gives unfair advantage to domestic logging companies by subsidizing the forestry industry. The softwood lumber dispute dates back many decades and has been brought to NAFTA and WTO tribunals on numerous occasions. The dispute centred on *stumpage fees* charged for logging on public lands. The U.S. maintained that stumpage fees were set too low by the Canadian government, giving unfair advantage to domestic loggers. The dispute was settled in 2006 to the satisfaction of the two federal and three provincial governments involved (British Columbia, Ontario, and Quebec).

The greatest environmental concern arising from Chapter 11 has been the possibility that domestic environmental legislation could be interpreted as a barrier to international free trade, and thus be overruled by NAFTA. An example was the lawsuit brought by Vancouver-based Methanex, the manufacturers of the gasoline additive MTBE, against the state of California. Methanex sought almost $1 billion in damages, claiming that California's ban on the additive amounted to trade protectionism. The Methanex claim was dismissed by a NAFTA tribunal in 2005, and this seems to have calmed the fears of some environmentalists. Other similar cases remain to be settled, however.

One major environmental concern has arisen from the establishment of factories in Mexico by Canadian and U.S. companies, as a result of free trade. Mexico's Federal *Law on Ecological Equilibrium and Environmental Contamination* (1988) provides strict rules for environmental protection, but the government lacks the financial resources to fully implement it. There has been considerable concern among critics of NAFTA that industries and jobs would migrate from the United States and Canada, where environmental and labour standards are high, to so-called "pollution havens" in Mexico, where labour is less expensive and environmental standards are less rigorously enforced. This has been labelled the "race to the bottom," referring to the potential for international investors to maximize profits by finding the "lowest common denominator" of environmental protection.

A case in point is the Tijuana River, which winds northwestward through the arid landscape of northern Baja California, Mexico, crossing the U.S. border south of San Diego. The river's watershed covers 4500 km^2 and is home to 2 million people. It is a transboundary watershed, with approximately 70% of its area in Mexico (see map and photo). On the Mexican side of the border, the river and the *arroyos*, or creeks, that flow into it are lined with farms, apartments, shanties, and factories, as well as leaky sewage treatment plants and toxic dump sites. Rains wash pollutants from these sources through the arroyos into the Tijuana River and eventually onto U.S. and Mexican beaches. The problem has grown worse in recent years as the region's population has boomed, outstripping the capacity of sewage treatment facilities. Beach closures and pollution advisories have become commonplace, and garbage litters the beaches.

The proliferation of U.S.- and Canadian-owned factories, or *maquiladoras,* on the Mexican side of the border has contributed to the river's pollution, both through direct disposal of industrial waste and by attracting thousands of new workers to the already crowded region. Things are worse on the Mexican side because most Mexican residents of the Tijuana River watershed live in poverty relative to their U.S. neighbours. Close to one-third of Tijuana's homes are not connected to a sewer system, and river pollution directly affects people's day-to-day lives.

To address some of the environmental concerns raised by NAFTA, the three nations reached a corollary agreement called the North American Agreement on Environmental Cooperation (NAAEC, 1993). (It is one of two side-agreements to NAFTA; the other is an agreement about labour standards.) The objectives are to promote sustainable development, encourage pollution prevention policies and practices, and enhance compliance with environmental laws and regulations. The agreement also promotes transparency and public participation in the development of environmental policies.[16] The agreement mandated the international Commission for Environmental Cooperation, provided a quasi-judicial mechanism for the resolution of disputes, and established a cooperative work plan for the environment among the three nations.[17] The ultimate goal is to ensure that the parties do not lower their environmental standards to attract investment.

As we approach the twentieth anniversary of NAFTA, it seems that the

(Continued)

The Tijuana River winds northwestward from Mexico into California just south of San Diego, draining 4500 km² of land in its watershed (coloured green in map). Pollution entering the river affects Mexican residents of the watershed and U.S. citizens on San Diego County beaches, but the situation is much worse on the Mexican side of the border (on the left in the photo), compared with the California side (on the right). During high-water episodes that overwhelm treatment facilities, millions of gallons of raw sewage have flushed into the river and the Pacific Ocean.

agreement has left the North American environment both better and worse off. According to research reported at a symposium on The Environmental Impacts of Free Trade held by the CEC, there has not been one overall or generalized environmental effect of NAFTA; the impacts have been mixed.[18] In his introduction, symposium chair Pierre Marc Johnson commented that:

In the fisheries sector, evidence does not suggest that NAFTA, per se, has had much of an effect, either positive or negative, on sustainable fisheries management. For forest products, on the other hand, the restructuring of the industry has been accompanied by significant changes, including its exposure to contestation from

largest sources of funding for economic development. This institution has shaped environmental policy through its funding of dams, irrigation infrastructure, and other major development projects. In fiscal year 2005, the World Bank provided more than $22.3 billion in loans for projects in the poorest countries around the world.

The World Bank has frequently been criticized for funding unsustainable projects that cause more environmental problems than they solve. Providing for the needs of growing human populations in poor nations while minimizing damage to the environmental systems on which people depend can be a tough balancing act. Environmental scientists today agree that the concept of sustainable development must be the guiding principle for such efforts.

Organization of the Petroleum Exporting Countries (OPEC)

OPEC is an intergovernmental organization of oil-producing and exporting nations, with 13 members.[19] It was originally founded to protect the interests of its members in the international marketplace, which it has accomplished by controlling the world price of oil. The position of OPEC with regard to recent developments in our understanding of global climate change and its relationship to fossil fuel use are understandably wary. According to the OPEC website, OPEC "is concerned that some countries may impose environmental and taxation policies that are harmful to those who rely on fossil fuels for a substantial part of their income." It maintains that OPEC functions to supply a stable petroleum

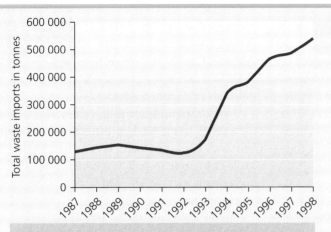

Hazardous waste imports into Canada have increased dramatically since NAFTA was signed. Other areas have seen environmental improvements. Source: Marisa Jacott, Cyrus Reed, and Mark Winfield, The Generation and Management of Hazardous Wastes and Transboundary Hazardous Waste Shipments Between Mexico, Canada, and the United States, 1990–2000, in *North American Symposium on Understanding the Linkages Between Trade and the Environment,* Commission on Environmental Cooperation of North America, 2002.

international competition. For freshwater, concerns persist about the possibility of bulk water exports in the face of dwindling water resources and the effects of investor-state challenges under NAFTA's Chapter 11 ... Some air pollution indicators show increases of carbon monoxide and sulfur dioxide (SO_2) levels in the U.S., and of SO_2 in Mexico, and significant reductions of air pollution in the Canadian and Mexican paper sectors. Data on hazardous waste show a significant increase in hazardous waste generation in some Canadian provinces, along with a decrease in some northern U.S. states. A key finding ... is that the total amount of trade in hazardous wastes—in particular, waste imported into Canada from the United States—has increased dramatically since NAFTA came into effect.[20]

Johnson continued, addressing the question of the relocation of industries to Mexico:

Evidence suggests that ... production relocation can be linked to free trade, and that declines in production bring both environmental "benefits"—from lessened industrial activity—and greater social dislocation linked to unemployment. At the same time, environmental problems understandably increase where production expands ... Is there any evidence of a "pollution haven"? In fact, the evidence suggests that Mexican export specialization has resulted in less and less pollution. In contrast, Canadian export specialization is now much more pollution-prone than Mexico's.[21]

The implications of NAFTA for trade in water are of fundamental importance to Canadians (see Chapter 11, "Central Case: Turning the Tap: The Prospect of Canadian Bulk Water Exports"). In a CEC symposium paper, *NAFTA Effects on Water: Testing for NAFTA Effects in the Great Lakes Basin*, Christine Elwell, senior policy analyst for the Sierra Club of Canada, concluded that NAFTA has had clearly demonstrable negative impacts on both water quality and quantity in the Great Lakes, and that without greater attention, future concerns and environmental stresses would arise from the privatization and commodification of water.[22]

With regard to the *maquiladoras* and the impacts of free trade on the environment in northern Mexico, the jury is still out. NAFTA opponents still maintain that Mexican workers are paid far less, their civil and labour rights are suppressed, and environmental regulations are ignored, to the profit of the companies that do business there. NAFTA supporters, on the other hand, argue that the savings would be insufficient for a company to consider relocating to Mexico simply to avoid the cost of compliance with environmental regulations.

In his introduction to the CEC symposium, Johnson came to a conclusion that many environmentalists had been waiting to hear, when he stated that "environmental concerns need to be addressed during trade negotiations and treaty implementation."[23] Both supporters and opponents of NAFTA should have no trouble agreeing on this point.

market, which is essential for the world economy, and opposes "discriminatory" taxes on fossil fuel use.[24]

International ENGOs A number of ENGOs have become international in scope and exert influence over international environmental policy. The nature of these advocacy groups is diverse. Some, such as the Nature Conservancy, focus on accomplishing conservation objectives without becoming politically involved. Other groups, including Conservation International, the World Wide Fund for Nature, Greenpeace, Population Connection, and many others, attempt to shape policy directly or indirectly through research, education, lobbying, or protest. NGOs apply more funding and expertise to environmental problems—and conduct more research intended to solve them—than do many national governments. In some Communist and post-Communist countries, nongovernmental organizations that have been illegal or severely restricted are now becoming much more visible and are influencing both awareness and public policy.

This is just a small sampling of the very many organizations that play central roles in determining international environmental policy that affects us all. As globalization proceeds, our world is becoming ever more interconnected. As a result, human societies and Earth's ecological systems are being altered at unprecedented rates. Trade and technology have expanded the global reach of all societies, especially those that, like Canada and the United States, consume resources from across the world. Highly consumptive nations that import goods and resources from far and wide exert extensive impacts on the planet's environmental systems. Multinational corporations operate outside the reach of national laws and rarely have incentive to conserve resources or conduct their business sustainably in the nations where they operate. For all these reasons, in today's globalizing world the organizations and institutions that influence international policy are becoming increasingly vital.

Exploring Approaches to Environmental Policy

We have discussed environmental policy so far in this chapter mainly from the regulatory or legislative perspective ("top-down"), and from the perspective of public awareness and grassroots environmental activism ("bottom-up"). There are other important approaches that inform environmental policy development and serve to modify both corporate and consumer behaviour. Let us explore some of these.

Science plays a role in policy, but it can be politicized

Ethical values, economic interests, and political ideology influence most policy decisions. However, environmental policy decisions that are effective are generally those informed by scientific research. For instance, when deciding whether and how to regulate a substance that may pose a public health risk, regulatory agencies, such as Health Canada, comb the scientific literature and may commission new studies or have their own scientists carry out studies, seeking to gain as full an understanding of the health risks as science can reasonably provide. When trying to win support for a bill to reduce pollution, a representative may use data from scientific studies to quantify the cost of the pollution or the predicted benefits of its reduction. The more information a policy maker can glean from science, the better the policy he or she will be able to create. In today's world, a nation's strength depends on its commitment to science, and this is the reason that governments devote a portion of our taxes to fund scientific research.

Unfortunately, sometimes policy makers choose to ignore science and instead allow political ideology to determine policy. In past years, some scientists—particularly those working on politically sensitive issues such as climate change or endangered species protection—have occasionally found their work suppressed or discredited, or their jobs threatened. Examples of cases in which the work or opinions of Canadian scientists have been ignored or overlooked by some politicians, who may have been under social and economic pressure, include the collapse of the cod fishery in Atlantic Canada, the connections between asbestos and disease, and recent work on global warming and the potential impacts of climate change.

When scientists are gagged, and when taxpayer-funded science is suppressed or distorted for political ends, we all lose. We cannot simply take for granted that science will play a role in policy. As scientifically literate citizens of a democracy, we all need to make sure our government representatives are making proper use of the tremendous scientific assets we have at our disposal.

Command-and-control policy has improved our lives, but it is not perfect

A great deal of environmental policy has functioned by setting rules or limits and threatening punishment for violating these rules or limits, in what is often called a **command-and-control** approach. The command-and-control approach has resulted in some major successes. Without doubt, our environment would be in far worse shape today were it not for this type of government regulatory intervention.

Most of the major environmental laws of recent decades, and most regulations enforced by agencies today, use the command-and-control approach. This simple and direct approach to policy making has brought citizens of Canada and many other nations cleaner air, cleaner water, safer workplaces, healthier neighbourhoods, and many other improvements in quality of life. The relatively safe, healthy, comfortable lives most of us enjoy today owe much to the environmental policy of the past few decades.

Despite the successes of command-and-control policy, it is not without its drawbacks. Although policy steps in to respond to market failure, it is clear that government intervention sometimes fails every bit as badly as markets can fail. Sometimes government actions are well intentioned but not well-enough informed, so they can lead to unforeseen consequences. Policy can also fail if a government does not live up to its responsibilities to protect its citizens or treat them equitably.

Economic tools also can be used to achieve environmental goals

The most common critique of command-and-control policy is that it achieves its goals in a more costly and less efficient manner than the free market can. By mandating particular solutions to problems, command-and-control policy fails to take advantage of the fact that private entities competing in the free market can often produce better solutions at lower cost. Many minds that are economically motivated to compete in the market are more likely to innovate and find optimal solutions than a smaller number of policy makers with no such economic incentive.

The most widely developed alternatives to command-and-control policies therefore involve the creative use of economic incentives to encourage desired outcomes, discourage undesired outcomes, and set market dynamics in motion to achieve goals in an economically efficient manner. Policy makers now often try to combine the advantages of government and the private sector. The challenge of crafting economic policy tools is to channel the innovation and economic efficiency of the free market in directions that benefit the public.

Subsidies One set of economic policy tools aims to encourage industries or activities that are deemed desirable. Governments may give *tax breaks* to certain types of businesses or individuals, for instance. Relieving the tax burden lowers costs for the business or individual, thus assisting the desirable industry or activity. A similar economic policy tool is the **subsidy**, a government giveaway of cash or publicly owned resources that is intended to encourage a particular activity.

National governments commonly provide subsidies to industries they judge to benefit the nation in some way. Subsidies can be used to promote environmentally sustainable activities, but all too often they have been used to prop up unsustainable ones. Some studies suggest that subsidies that are harmful to the environment total roughly $1.45 *trillion* yearly across the globe—an amount larger than the economies of all but five nations.

Although there are many examples of government subsidies in Canada that result—either directly or indirectly—in harm to the environment (subsidies that promote logging in old-growth forests come to mind), perhaps the most controversial are those that support the oil and gas industry (**FIGURE 22.15**). The Sierra Club of Canada estimates that since signing the Kyoto Protocol in 1997, Canada has spent $2 in subsidies to support oil and gas industry development for every $1 spent finding ways to meet Kyoto targets.[25]

Advocates of sustainable resource use have long urged governments to subsidize environmentally sustainable activities instead. So far, Canada has lagged behind most other developed nations in subsidizing research into sustainable solutions. For instance, in 2000–2001 less than 5% of federal energy-related investment was devoted to renewable energy research—$12.9 million of a total

FIGURE 22.15
Subsidies designed to support industry can have a negative impact on the environment, either directly or indirectly. The billions of dollars of subsidies to the oil and gas industry in Canada have so far greatly outweighed investment and subsidies for alternative energy sources.

expenditure of $230.2 million.[26] In the same period, billions of dollars in tax deductions were passed along to the oil, gas, and nuclear industries. We have a long way to go in getting subsidies to work for *both* the economy *and* the environment.

Green taxes and "polluter pays" Another economic policy tool—taxation—can be used to discourage undesirable activities. Taxing undesirable activities helps to "internalize" external costs by making them part of the overall cost of doing business. Taxes on environmentally harmful activities and products are called **green taxes**. By taxing activities and products that cause undesirable environmental impacts, a tax becomes a tool for policy as well as simply a way to fund government.

Green taxes have yet to gain widespread support, although similar "sin taxes" on cigarettes and alcohol are common tools of Canadian social policy. The Liberal Party's "green shift" campaign involved a carbon tax that caused confusion and consternation among voters in the 2008 federal election. Taxes on pollution have been widely instituted in Europe, though, where many nations have adopted the **polluter pays principle**. This principle specifies that the price of a good or service should include all its costs, including costs of environmental degradation that would otherwise be passed on as external costs.

Under green taxation, a factory that pollutes a waterway would pay taxes based on the amount of pollution it discharges. The idea is to give companies a financial incentive to reduce pollution, while allowing the polluter the freedom to decide how best to minimize its expenses. One polluter might choose to invest in technologies to reduce its pollution if doing so is less costly than paying the taxes. Another polluter might find abating its pollution more costly and could choose to pay the taxes instead—funds the government might then apply toward mitigating pollution in some other way. Bottle-return programs are an example of the polluter pays principle applied at the consumer end.

Green taxation provides incentive for industry to lower emissions not merely to a level specified in a regulation but to still-lower levels. However, green taxes do have disadvantages. One is that businesses will most likely pass on their tax expenses to consumers, and these increased costs may affect low-income consumers disproportionately more than high-income ones.

Permit trading A different, market-based approach to the management of environmental impacts is permit trading. In a **permit trading** system, the government creates a market in permits for an environmentally harmful activity, and companies, utilities, or industries are allowed to buy, sell, or trade rights to conduct the activity. For instance, to decrease emissions of air pollutants, a government might grant emissions permits and set up an **emissions trading system**. The government first determines the overall amount of pollution it will accept and then issues marketable permits to polluters that allow them each to emit a certain fraction of that amount. Polluters may buy, sell, and trade these permits with other polluters. Each year, the government may reduce the amount of overall emissions allowed.

In such a **cap-and-trade** system, a polluting party that is able to reduce its pollution receives credit for the amount it did not emit and can sell this credit to other parties. Suppose, for example, you are a plant owner with permits to release 10 units of pollution, but you find that you can become more efficient and release only 5 units of pollution instead. You then have a surplus of permits, which might be very valuable to some other plant owner who is having trouble reducing pollution or who wants to expand production. In such a case, you can sell your extra permits. Doing so generates income for you and meets the needs of the other plant, while preventing any increase in the total amount of pollution. Moreover, environmental organizations can buy up surplus permits and "retire" them, thus reducing the overall amount of pollution.

Marketable permits provide companies with an economic incentive to find ways to reduce emissions. If successful, permit trading can end up costing both industry and government much less than a conventional regulatory system. In 2007 the National Round Table on Environment and Economy undertook a major study of emissions reduction goals and market instruments in the Canadian context, in response to a request for advice from the federal government on reducing carbon emissions. They concluded that reaching the government's stated goal of a 65% reduction from 2005 levels by 2050 would require meeting an interim goal of a 20% reduction by 2020, and that this would necessitate the use of an emissions tax, a cap-and-trade system, or a combination of the two (see "Interpreting Graphs and Data").[27]

Ontario has had a capped emissions trading system in place since 2001, covering NO_x and SO_2 emissions. The Ontario Emissions Trading Registry serves as a mechanism for tracking emissions, as well as the transfer and use of allowances and emission-reduction credits, which are issued under the Emissions Trading Regulation. It also provides a forum for public notification and commentary on emissions trading transactions.[28]

Cap-and-trade programs are no panacea. They can reduce pollution overall, but they do allow hotspots of pollution to occur around plants that buy permits to pollute more. Moreover, large firms can hoard permits, deterring smaller new firms from entering the market and thereby suppressing competition. Nevertheless, permit trading has shown promise for safeguarding environmental quality while granting industries the flexibility to lessen their impacts in ways that are economically palatable.

Presently, a market in carbon emissions is operating among European nations as a result of the Kyoto Protocol to address climate change. Under the Kyoto Protocol, nations have targets for reducing their carbon emissions from power plants, automobiles, and other sources that are driving climate change. Each nation participating in the European Union Emission Trading Scheme takes the emissions permits it is allowed and allocates them to its industries according to their emissions at the start of the program. The industries then can trade permits freely, establishing a market whereby the price of a carbon emissions permit fluctuates according to supply and demand.

Unfortunately, European nations allocated too many permits in the program's first phase, destroying industries' financial incentive to cut emissions and causing the permits to become nearly worthless. These nations plan to correct the overallocation when the program enters its next phase.

weighing the issues

22-4 EMISSIONS TRADING

Some environmental activists oppose emissions trading because they view it as giving polluters "a licence to pollute." How do you feel about emissions trading as a means of reducing air pollution? Would you favour command-and-control regulation or market-based permit trading? What advantages and disadvantages do you see in each?

Market incentives are being tried widely on the local level

You may well have already taken part in transactions involving financial incentives as policy tools. For example, many municipalities charge residents for waste disposal according to the amount of waste they generate. Other cities place taxes or disposal fees on items that require costly safe disposal, such as tires and motor oil. Still others give rebates to residents who buy water-efficient toilets and appliances, because the rebates can cost the city less than upgrading its sewage treatment system. Likewise, power companies sometimes offer discounts to customers who buy high-efficiency light bulbs and appliances, because doing so is cheaper for the utilities than expanding the generating capacity of their plants.

ENGOs and even private companies are involved in this process at the local level, too. Many new programs have emerged in the past five years or so, aimed at providing rewards for behavioural changes that will benefit the environment. These include rebate programs for purchasing energy-efficient appliances or light bulbs, vouchers for turning in old, gasoline-guzzling lawnmowers, and financial incentives for businesses that turn down their lighting or air conditioning, among countless other examples.

"Car Heaven" is one example of a program that helps individual Canadians reduce their impact on the environment by giving them incentives and options for disposing of old, energy-inefficient vehicles. The program was started in Toronto in 2000 under the Clean Air Foundation; it then spread to Edmonton, Calgary, and Vancouver, and is now almost nationwide. In addition to retiring high-polluting vehicles from the road, the program ensures that the cars (including all parts and materials,

FIGURE 22.16
Car Heaven provides incentives for people to get their old, polluting clunkers off the road. The program ensures that the vehicles (unlike these cars, stacked in a wrecking yard) are recycled and disposed of in an environmentally safe manner, using a three-step program of pretreatment, reuse of parts, and recycling of remaining materials.

such as mercury switches, tires, residual oil, batteries, and refrigerants) are treated, reused, or recycled, and then disposed of in an environmentally responsible manner (**FIGURE 22.16**).

At all levels, from the local to the international, market-based incentives can reduce environmental impact while minimizing overall costs to industry and easing concerns about the intrusiveness of government regulation. Command-and-control policy is straightforward to implement, easy to monitor, and frequently works. Market-based approaches can be more complicated, but if they work, they can lessen environmental impact at a lower overall cost.

Ecolabelling gives some choice back to the consumer

More and more, not just in response to legislation but voluntarily, manufacturers are designating on their labels how their products were grown, harvested, or manufactured. This method, called **ecolabelling**, serves to tell consumers which brands use environmentally benign processes. It can also raise consumer awareness of environmental issues, as well as encouraging companies

CANADIAN ENVIRONMENTAL PERSPECTIVES

Maude Barlow

Through her Blue Planet Project, Maude Barlow campaigns for global water justice.

- **Water rights activist** and **author**
- **Co-founder** of The Blue Planet Project
- National chairperson **of the Council of Canadians**

Maude Barlow does not shy away from controversy. When she sees an injustice, she says, "I have to tell people . . . I have to do something so that other people will also take action."[29] Barlow comes from a family tradition of concern for social justice issues, and over the years she has taken action in some challenging arenas. She became involved in politics early in her career, serving as Pierre Trudeau's adviser on women's issues in the 1980s. Previously she had worked as the director of equal opportunity for the City of Ottawa and led a campaign to stop violence against women.

In 1985 she resigned from political life to start the Council of Canadians, with a small group of activists that included Farley Mowat (profiled in Chapter 12). The principal mission of the CoC is to protect Canadian independence.[30] This includes, in part, a goal to "work with Canadians and people around the world to reclaim the global and local commons which are the shared heritage of humanity and of the earth."[31] The CoC is now the largest citizens' organization in Canada; to maintain its ideological independence, it accepts no funding from governments or the private sector.[32]

The first significant undertaking of the organization was to assemble a coalition of labour and social justice organizations opposed to the North American Free Trade Agreement (see Chapter 22, "The Science Behind the Story: The Environment in NAFTA and NAAEC"). The organization saw success in its fight against the Multilateral Agreement on Investment (MAI) in the late 1990s. The MAI, an agreement among OECD member countries to standardize investment regulations across international boundaries, would have restricted the ability of governments to limit foreign multinational investments, even if they were seen as potentially harmful. Critics of the agreement felt that it would greatly weaken national sovereignty in the areas of human rights, labour rights, and environmental standards.

More recently, the efforts of both Barlow and the CoC have been focused on fighting the bulk export of Canadian water (Chapter 11, "Central Case: Turning the Tap: The Prospect of Canadian Bulk Water Exports"). Not surprisingly, for Barlow this is more than an environmental issue; it is a social justice issue. In her most recent book, *Blue Covenant*, she pushes for a full U.N.-based international treaty recognizing the right to water. "Finally, the global water justice movement is demanding a change in international law to settle once and for all the question of who controls water. It must be commonly understood that water is not a commercial good, although of course it has an economic dimension, but rather a human right and a common trust."[33]

For all the controversy that surrounds her, Barlow has been widely recognized for her work. She has received honorary doctorates from six Canadian universities, and the Canadian Environment Award Citation of Lifetime Achievement. In her acceptance speech for the latter, Barlow highlighted the impacts of the petrochemical industry on the environment and health in Canada.[34] In 2005 she received the Lannan Foundation Cultural Freedom Fellowship, and the Right Livelihood Award, which she shared with fellow water rights activist Tony Clarke. Barlow is now the national chairperson of the Council of Canadians, and co-founder of the Blue Planet Project, through which she continues to work for international water rights. In October 2008, Barlow accepted an appointment as the first United Nations senior adviser on water issues.

"I am part of a family of activists and environmentalists in Canada and around the world."[35] —Maude Barlow

Thinking About Environmental Perspectives

According to the global water justice movement and Maude Barlow, "What is needed now is binding law to codify that states have the obligation to deliver sufficient, safe, accessible, and affordable water to their citizens as a public service." Do you think access to drinking water is a basic human right, like access to air? Or should water be a marketable commodity to be controlled and protected, at least in part, by the private sector? If it is a basic human right, should it be protected by international treaty?

---

to choose more environmentally friendly production methods. By preferentially buying ecolabelled products, consumers can provide businesses with a powerful incentive to switch to more sustainable processes. One early example was labelling cans of tuna as "dolphin-safe," indicating that the methods used to catch the tuna avoid the accidental capture of dolphins. Other examples include labelling recycled paper, organically grown foods, genetically modified foods (widely done in Europe but not in North America), fair-trade and shade-grown coffee, lumber harvested through sustainable forestry, and clothing made from organically grown cotton (**FIGURE 22.17**).

In a similar vein, individuals who invest their money in the stock market can choose to pursue *socially responsible investing*, which entails investing only in companies that have met certain criteria for environmental or social sustainability.

FIGURE 22.17
Ecolabelling allows businesses to promote products that minimize environmental impacts and gives consumers the opportunity to choose healthier, low-impact products. Roots, which was first started in Canada as a natural footwear company in 1973, was an early advocate of ecolabelling and the use of organically produced cotton in clothing.

Conclusion

Environmental policy is a problem-solving tool that makes use of science, ethics, and economics and that requires an astute understanding of the political process. Conventional command-and-control approaches of legislation and regulation are the most common approaches to policy making, but various innovative economic policy tools are also being developed. As we have seen in the case of Lake Erie, the Great Lakes, and the St. Lawrence system, environmental issues often overlap political boundaries and require international cooperation. Through the hard work of concerned citizens interacting with their government representatives, the political process eventually produced promising solutions through binational agreements and management plans.

The central focus of this book has been the science behind the pressing environmental issues of our day. In this chapter, we have departed somewhat from this central focus to consider the fundamentals of environmental law and policy, and some of the approaches to environmental management that have been used in the past and are currently emerging in Canada. By understanding these fundamentals and combining them with your understanding of environmental science, you will be well equipped to develop your own creative solutions to many of the challenging problems we will encounter.

REVIEWING OBJECTIVES

You should now be able to:

Describe environmental policy and assess its societal context

- Policy is a tool for decision making and problem solving that makes use of information from science and values from ethics and economics.
- Environmental policy is designed to protect natural resources and environmental amenities from degradation or depletion and to promote equitable treatment of people.

Identify the institutions important to Canadian environmental policy and recognize major Canadian environmental laws

- Federal, provincial/territorial, Aboriginal, and municipal/local governments, together with administrative agencies, all play roles in Canadian environmental policy.
- Early environmental policy encouraged frontier expansion and resource extraction. Policies of the next period, in the late 1800s and early 1900s, aimed to regulate resource use and mitigate impacts of the first. More recently, in the 1960s and subsequent years, several high-profile pollution events raised public awareness and gave us many of today's major environmental laws.
- Some major Canadian environmental laws include the *Canadian Environmental Protection Act*, the *Fisheries Act*, and the *Canadian Environmental Assessment Act*.

List the institutions involved with international environmental policy and describe how nations handle transboundary issues

- Many environmental problems cross political boundaries and thus must be addressed internationally.
- Currently, international environmental policy centres on the concept of sustainable development.
- International policy includes customary law (law by shared traditional custom) and conventional law (law by treaty).

- Institutions, such as the United Nations, European Union, World Bank, World Trade Organization, and nongovernmental organizations all play roles in international policy.

Categorize the different approaches to environmental policy

- Science plays a role in policy making, although some policy makers may ignore or distort it for political ends.
- Legislation that comes from a central government agency is referred to as a top-down or command-and-control approach.
- Shortcomings of the command-and-control approach have led many economists to advocate economic policy tools.
- Market-based approaches include subsidies, green taxation, and permit trading.

TESTING YOUR COMPREHENSION

1. Describe two common justifications for environmental policy, and discuss three problems that environmental policy commonly seeks to address.
2. What are some factors that may hinder the implementation or enforcement of environmental policy?
3. Environmental policy and management in Canada is shared among all levels of government. What does this mean, in practice, and what are some of the most important agencies that work on behalf of the environment at all levels of government?
4. Summarize the evolution of environmental law from the early settling of the West to the present day. What is the approach that appears to characterize the present wave of environmental policy?
5. What are the central goals of the *Canadian Environmental Protection Act*?
6. What is the difference between customary law and conventional law? What special difficulties do transboundary environmental problems present?
7. What are some of the important international environmental agreements to which Canada is a party?
8. Why are environmental regulations sometimes considered to be unfair barriers to trade?
9. Differentiate among a green tax, a subsidy, a tax break, and a marketable emissions permit.
10. Many recent environmental initiatives have focused on giving incentives to individuals and corporations to change their environmental behaviour. Explain what this means, and give some examples of programs that do this.

SEEKING SOLUTIONS

1. Many free-market advocates maintain that environmental laws and regulations are an unnecessary government intrusion into private affairs. As you may recall from Chapter 21, Adam Smith argued that individuals can benefit society by pursuing their own self-interest. Do you agree? Can you describe a situation in which an individual acting in his or her self-interest could harm society by causing an environmental problem? Can you describe how environmental policy might rectify the situation? What are some advantages and disadvantages of instituting environmental laws and regulations, versus allowing unfettered exchange of materials and services?
2. Reflect on the causes for the historical transitions from one type of environmental policy to another. Now peer into the future, and think about how life might be different in 25, 50, or 100 years. What would you speculate about the environmental policy of the future? What issues might it address?
3. Compare the roles of the United Nations, the European Union, the World Bank, the World Trade Organization, and nongovernmental organizations. If you could gain the support of just one of these institutions for a policy you favoured, which would you choose? Why?
4. Think of one environmental problem that you would like to see solved. From what you have learned about the policy-making process, describe how you think you could best shepherd your ideas through the process to address this problem.
5. Compare the main approaches to environmental policy—command-and-control and economic or market-based approaches. Can you describe an advantage and a disadvantage of each? Do you think any one approach is most effective? Could we do with just one approach, or does it help to have more than one?
6. **THINK IT THROUGH** You have just been named minister of the environment. New legislation mandates

reductions in water pollution from untreated municipal wastewater, chemical discharges from factories, and oil spillage from commercial and recreational boats. The new law mandates a 25% reduction in these pollution sources over 10 years, but it does not specify how these reductions are to be accomplished. What policy approaches would you choose to pursue to carry out the mandates of the legislation? Give reasons for your choices.

7. **THINK IT THROUGH** You are now the prime minister of Canada. You must represent Canada at a meeting of G8 leaders, and they want to know what Canada's position will be on the future of international agreements to limit climate change. What will you say?

INTERPRETING GRAPHS AND DATA

In 2007 the National Round Table on Environment and Economy (NRTEE) undertook a major study of emissions trading in the Canadian context, in response to a request for advice from the federal government on reducing carbon emissions.[36] The federal government's current stated goal is to reduce greenhouse gas emissions to 20% below 2005 levels by 2020, and 65% below 2005 levels by 2050. The NRTEE produced the following graph in its report, showing a business-as-usual scenario (BAU); a downstream cap-and-trade system scenario (DCT), which would mainly target large industrial emitters; and an upstream cap-and-trade system (UCT), which would limit the carbon content of fuels before they even reach the industrial users.

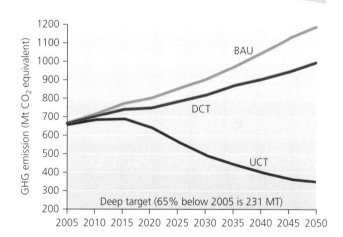

This graph compares emissions reductions over time, based on a business-as-usual scenario (BAU); a downstream cap-and-trade system scenario (DCT), which would mainly target large industrial emitters; and an upstream cap-and-trade (UCT) and/or carbon emissions tax system, which would limit the carbon content of fuels before they even reach the industrial users.

1. The NRTEE concluded that achieving the government's 2050 emissions targets will require reaching an interim goal of 20% reduction below 2005 levels by 2020. Which of the three scenarios presented in this graph achieves this interim goal?
2. Both the DCT and BAU scenarios end up above the target in 2050. By what percentage do they exceed the target? By what percentage do they exceed 2005 levels?
3. Even the UCT scenario doesn't quite achieve the 2050 target. By how much would the UCT scenario reduce emissions by 2050? By what percentage would emissions still exceed the established target?

CALCULATING FOOTPRINTS

Referring to the graph in Interpreting Graphs and Data, the baseline level of greenhouse gas emissions used by the NRTEE for its calculations was 660 Mt CO_2 equivalent in 2005. The federal government's stated goal is to reduce this by 65% by 2050 (i.e., to 231 Mt CO_2 equivalent). Given that there are approximately 32 million people living in Canada at this time (and let us assume, for the moment, that the population will remain stable between now and 2050), how much of this reduction will have to come from you, personally? From your university and your province? What about the interim goal of reducing emissions by 20% by the year 2020?

	Current	20% Reduction	65% Reduction
You			
Your university			
Your province			
Canada	660 Mt	528 Mt	231 Mt

Mt = million tonnes

TAKE IT FURTHER

 Go to www.myenvironmentplace.ca where you will find

- Suggested answers to end-of-chapter questions
- Quizzes, animations, and flashcards to help you study
- *Research Navigator*™ database of credible and reliable sources to assist you with your research projects
- Tutorials to help you master how to interpret graphs
- Current news articles that link the topics that you study to case studies from your region and around the world

- **ECO Occupational Profiles:** If you found this chapter especially interesting, you might want to learn more about the following jobs by visiting the Occupational Profiles website of the Environmental Careers Organization. Go to www.eco.ca and check out the following careers:
    - Compliance promotion specialist
    - Environmental enforcement officer
    - Environmental assessment analyst
    - Environmental lawyer
    - Environmental policy analyst

CHAPTER ENDNOTES

1. Environment Canada, *Great Lakes Portraits: Lake Erie: A Lake in Flux; Canada-Ontario Agreement Respecting the Great Lakes Basin Ecosystem,* www.on.ec.gc.ca/laws/coa/2001/lake-erie-e.html.
2. Environment Canada, *Acts, Regulations, and Agreements,* www.ec.gc.ca/default.asp?lang=En&n=48D356C1-1.
3. Based on information from C. W. Daniel Kirby, Radha Curpen, Shawn Denstedt (2006) *Doing Business in Canada: Environmental Law in Canada,* June, www.osler.com/resources.aspx?id=8745.
4. CEPA Fact Sheet No. 2: *CEPA 1999 At A Glance: What It Is, What It Does, How It Works,* Government of Canada, October 2005.
5. Based on information from Environmental Commissioner of Ontario, *Environmental Bill of Rights, FAQs,* www.eco.on.ca/eng/index.php/environmental-bill-of-rights/about-the-ebr.php.
6. National Round Table on the Environment and the Economy, www.nrtee-trnee.ca.
7. Hurlburt, Margot (2007) *Canada's Water Law,* prepared for the National Council of Women of Canada, www.ncwc.ca/pdf/waterlaw.pdf.
8. Meadows, Donella H., and Dennis L. Meadows, Jørgen Randers, and William W. Behrens III (1972) *Limits to Growth.* New York: Universe Books, ISBN 0-87663-165-0.
9. Schumacher, Edmund (1973) *Small is Beautiful: A Study of Economics as if People Mattered.* Vancouver: Hartley and Marks, ISBN 0-88179-169-5.
10. Ehrlich, Paul R. (1968) *The Population Bomb.* New York: Ballantine Books.
11. Lappé, Frances Moore (1971) *Diet for a Small Planet.* New York: Ballantine Books, ISBN 0345023781.
12. *Our Common Future: The Report of the United Nations Commission on Environment and Development* (1987). Oxford, UK: Oxford University Press.
13. Bond, Wayne, Dennis O'Farrell, Gary Ironside, Barb Buckland, and Risa Smith, Knowledge Integration Strategies Division, Environmental Reporting Branch, Strategic Information Integration Directorate (2005) *Environmental Indicators and State of the Environment Reporting: An Overview for Canada—Background paper to an Environmental Indicators and State of the Environment Reporting Strategy, 2004–2009, Environment Canada.* Ottawa: Environment Canada, www.ec.gc.ca/soer-ree/English/resource_network/bg_paper2_e.cfm.
14. Environment Canada, International Policy and Cooperation Branch, International Relations Directorate (2002) *Compendium of International Environmental Agreements,* 3rd ed. Ottawa: Environment Canada, www.ec.gc.ca/international/multilat/compendium_e.htm.
15. Office of the Auditor General of Canada (2004) Chapter 1: International Environmental Agreements, *Report of the Commissioner of Environment and Sustainable Development,* Ottawa: Office of the Auditor General, www.oag-bvg.gc.ca/internet/English/aud_ch_cesd_2004_1_e_14914.html.
16. North American Agreement on Environmental Cooperation–Canadian Office, *NAAEC Overview,* www.naaec.gc.ca/eng/agreement/agreement_e.htm.

17. Murray, William (1993) *NAFTA and the Environment*. Ottawa: Environment Canada Science and Technology Division, http://dsp-psd.tpsgc.gc.ca/Collection-R/LoPBdP/MR/mr116-e.htm.
18. *The Environmental Effects of Free Trade: Papers Presented at the North American Symposium on Assessing the Linkages Between Trade and the Environment, 2000* (2002). Ottawa: Commission for Environmental Cooperation of North America, www.cec.org/files/pdf/ECONOMY/symposium-e.pdf.
19. OPEC's members are Algeria, Angola, Ecuador, Indonesia, the Islamic Republic of Iran, Iraq, Kuwait, the Socialist People's Libyan Arab Jamahiriya, Nigeria, Qatar, Saudi Arabia, United Arab Emirates, and Venezuela.
20. Johnson, Dr. Pierre Marc (symposium chair) (2002) Introduction. *The Environmental Effects of Free Trade: Papers Presented at the North American Symposium on Assessing the Linkages Between Trade and the Environment*. Ottawa: Commission for Environmental Cooperation of North America, www.cec.org/files/pdf/ECONOMY/symposium-e.pdf.
21. Johnson, Dr. Pierre Marc (symposium chair) (2002) Introduction. *The Environmental Effects of Free Trade: Papers Presented at the North American Symposium on Assessing the Linkages Between Trade and the Environment*. Ottawa: Commission for Environmental Cooperation of North America, www.cec.org/files/pdf/ECONOMY/symposium-e.pdf.
22. Elwell, Christine (2002) NAFTA Effects on Water: Testing for NAFTA Effects in the Great Lakes Basin. *The Environmental Effects of Free Trade: Papers Presented at the North American Symposium on Assessing the Linkages Between Trade and the Environment*. Ottawa: Commission for Environmental Cooperation of North America, www.cec.org/files/pdf/ECONOMY/symposium-e.pdf.
23. Johnson, Dr. Pierre Marc (symposium chair) (2002) Introduction. *The Environmental Effects of Free Trade: Papers Presented at the North American Symposium on Assessing the Linkages Between Trade and the Environment*. Ottawa: Commission for Environmental Cooperation of North America, www.cec.org/files/pdf/ECONOMY/symposium-e.pdf.
24. OPEC, *Frequently Asked Questions: Does OPEC Support Environmental Policies?*, www.opec.org/library/FAQs/aboutOPEC/q17.htm.
25. Ecojustice (2005) *Media Release: Misdirected Spending: Groups Demand Investigation into billions in Federal Subsidies to Canada's Oil and Gas Industry*, www.ecojustice.ca/media-centre/press-releases/pressrelease.2007-12-03.6697873163.
26. Myers, Lynne. *Financial Incentives and Subsidies for Renewable Energy in Canada and the United States*, March 8, 2002.
27. *National Round Table on the Environment and the Economy, Getting to 2050: Canada's Transition to a Low-Emission Future*, www.nrtee-trnee.ca/eng/publications/getting-to-2050/Getting-to-2050-low-res-eng.pdf.
28. Ontario Ministry of the Environment (2005) *Emissions Trading: Fact Sheet*, www.ene.gov.on.ca/programs/4346e02.pdf.
29. CBC (2001) *Life and Times: Maude Barlow*, www.cbc.ca/lifeandtimes/barlow.html.
30. Council of Canadians, *About Us*, www.canadians.org/about/index.html.
31. Council of Canadians, *Vision Statement*, www.canadians.org/about/BOD/vision.html.
32. Council of Canadians, *About Us*, www.canadians.org/about/index.html.
33. Barlow, Maude (2007) *Blue Covenant: The Global Water Crisis and the Coming Battle for the Right to Water*, p. 164, http://canadians.org/about/documents/Blue_Covenant_Excerpt_07.pdf.
34. Council of Canadians, *About Us: Maude Barlow*, www.canadians.org/about/Maude_Barlow/.
35. On the Road with Maude Barlow (2008) *Canadian Perspectives*, Summer, www.canadians.org/publications/CP/2008/summer/CP_summer_08_Maude.html.
36. National Round Table on the Environment and the Economy, *Getting to 2050: Canada's Transition to a Low-Emission Future*, www.nrtee-trnee.ca/eng/publications/getting-to-2050/Getting-to-2050-low-res-eng.pdf.

Sustainable Solutions

23

Sri Lankan children plant tree seedlings on deforested hillsides around their village.

Upon successfully completing this chapter, you will be able to

- List and describe approaches being taken on college and university campuses to promote sustainability
- Explain the concept of sustainable development
- Discuss how protecting the environment can be compatible with promoting economic welfare
- Describe and assess key approaches to designing sustainable solutions
- Discuss the need for action on behalf of the environment and the tremendous human potential to solve problems

Transportation-related initiatives are among the many projects and programs aimed at promoting sustainability on UBC's campuses.

CENTRAL CASE:
THE CAMPUS SUSTAINABILITY MOVEMENT

"We solemnly pledge to the peoples of the world and the generations that will surely inherit this Earth that we are determined to ensure that our collective hope for sustainable development is realized."
—DECLARATION SIGNED BY 193 NATIONS AT THE WORLD SUMMIT, JOHANNESBURG, SOUTH AFRICA, 2002

"We have to move decisively to protect our children's future."
—ELIZABETH MAY, LEADER OF THE GREEN PARTY OF CANADA

The University of British Columbia bills itself as "Canada's Leader in Campus Sustainability." Risking the ire of the many other campuses across Canada that have extremely active environmental and sustainability programs—the University of Waterloo and Vancouver Island University, both early adopters of campus greening and sustainability plans, come to mind—it is nevertheless clear that UBC has been a trailblazer and remains a leader in the international movement for campus sustainability.

UBC may have been Canada's first university to open a sustainability office and adopt a comprehensive sustainable development plan for the campus. In 2003, UBC became Canada's first university to receive Green Campus Recognition from the U.S.-based National Wildlife Federation.[1] To build student engagement and buy-in, UBC's sustainability office created the Sustainability Pledge in 2002. The pledge is an official commitment by students to use the knowledge they gain at UBC to improve the sustainability of the world. The goal is to inspire at least 20% of students to sign by 2010.[2]

The university's president also was one of the very first signers (in 1990) of the **Talloires Declaration**, a commitment on the part of college and university presidents around the world to pursue and foster sustainability. The declaration provides a 10-point action plan for incorporating sustainability and environmental literacy in teaching, research, operations, and outreach at colleges and universities (Table 23.1A). As of mid-2008, the Talloires Declaration had been signed by 375 university presidents and chancellors from 50 nations, including 29 Canadian universities and colleges (Table 23.1B).[3]

Proponents of campus sustainability view colleges and universities as microcosms of society, noting that institutions of learning consume resources, emit pollution, and exert other environmental impacts. These proponents seek to change the ways these institutions operate and to reduce their ecological footprints.

Table 23.1A The 10-Step Plan of the Talloires Declaration

1. Raise public, government, industry, foundation, and university awareness by publicly addressing the urgent need to move toward an environmentally sustainable future.
2. Engage in education, research, policy formation, and information exchange on population, environment, and development to move toward a sustainable future.
3. Establish programs to produce expertise in environmental management, sustainable economic development, population, and related fields to ensure that all university graduates are environmentally literate and responsible citizens.
4. Develop the capability of university faculty to teach environmental literacy.
5. Set an example of environmental responsibility by establishing programs of resource conservation, recycling, and waste reduction at the universities.
6. Encourage the involvement of government (at all levels), foundations, and industry in supporting university research, education, policy formation, and information exchange in environmentally sustainable development. Expand work with nongovernmental organizations to assist in finding solutions to environmental problems.
7. Convene school deans and environmental practitioners to develop research, policy, information exchange programs, and curricula for an environmentally sustainable future.
8. Establish partnerships with primary and secondary schools to help develop the capability of their faculty to teach about population, environment, and sustainable development issues.
9. Work with the U.N. Conference on Environment and Development, the U.N. Environment Programme, and other national and international organizations.
10. Establish a steering committee and a secretariat to continue this momentum and inform and support each other's efforts in carrying out this declaration.

University Leaders for a Sustainable Future, Programs: Talloires Declaration, www.ulsf.org/programs_talloires.html.

Table 23.1B Canadian Signatories to the Talloires Declaration (as of June 2008)

Acadia University, Wolfville, Nova Scotia
Algonquin College, Ottawa, Ontario
Atlantic School of Theology, Halifax, Nova Scotia
Carleton University, Ottawa, Ontario
Concordia University, Montreal, Quebec
Dalhousie University, Halifax, Nova Scotia
Emily Carr Institute of Art and Design, British Columbia
Lakehead University, Thunder Bay, Ontario
McGill University, Montreal, Quebec
Mount Saint Vincent University, Halifax, Nova Scotia
Royal Roads University, Victoria, British Columbia
Ryerson University, Toronto, Ontario
Saint Francis Xavier University, Antigonish, Nova Scotia
Saint Mary's University, Halifax, Nova Scotia
Saint Thomas University, Fredericton, New Brunswick
Simon Fraser University, Burnaby, British Columbia
University College of Cape Breton, Sydney, Nova Scotia
University of British Columbia, Vancouver, British Columbia
University of Guelph, Guelph, Ontario
University of Lethbridge, Lethbridge, Alberta
University of Manitoba, Winnipeg, Manitoba
University of Northern British Columbia, Prince George, British Columbia
University of Ottawa, Ottawa, Ontario
University of Saskatchewan, Saskatoon, Saskatchewan
University of Victoria, Victoria, British Columbia
University of Western Ontario, London, Ontario
University of Windsor, Windsor, Ontario
University of Winnipeg, Winnipeg, Manitoba
York University, Toronto, Ontario

Most colleges and universities have student-run environmental organizations, recycling programs, and courses or majors in environmental science or environmental studies. It is perhaps not surprising that UBC has been one of the leaders in the development of programs and courses focusing on sustainability; after all, the inventors of the ecological footprint concept—Mathis Wackernagel and Bill Rees—were at UBC when they collaborated on the concept. Today the SEEDS Program (Social, Ecological, Economic Development Studies) connects UBC students with staff and faculty advisors to promote collaborative research that contributes to campus sustainability projects.

UBC is in the midst of implementing a campus-wide vision statement, *Inspirations and Aspirations: The Vancouver Campus Sustainability Strategy, 2006–2010*, which outlines nine sets of goals for sustainability:

- Social goals:
  - Improve human health and safety
  - Make UBC a model sustainable community
  - Increase the understanding of sustainability inside and outside the university
- Economic goals:
  - Ensure ongoing economic viability
  - Maintain and enhance the asset base
  - Maintain and maximize the utilization of the physical infrastructure
- Environmental goals:
  - Reduce pollution
  - Conserve resources
  - Protect biodiversity

The strategy identifies specific targets and timelines within each of these categories. A recent campus initiative that is contributing to the achievement of environmental goals, for example, is the mandatory U-Pass (universal transportation pass) program. As of 2006, the program had reduced overall automobile use by 22% below 1997 levels despite a 28% growth in population, representing an estimated total savings of 12 000 tonnes of CO_2 per year.

Many of these sustainability initiatives have cost money, and their proponents have had to argue their case to penny-conscious administrators. But the university's administration has recognized that many short-term costs are actually investments that will save money in the long term. On the UBC sustainability office website (www.sustain.ubc.ca), a calculator clicks off the use of paper, electricity, and water on campus, comparing it in real time with the virgin paper, electricity, water, greenhouse gas emissions, and dollars saved by campus environmental initiatives.

The university's position on sustainability is perhaps best summed up by its mission statement, which articulates the goal that "graduates of UBC will value diversity, work with and for their communities, and be agents for positive change. They will acknowledge their obligations as global citizens and strive to secure a sustainable and equitable future for all."[4]

Sustainability on Campus

If we are to attain a sustainable civilization, we will need to make efforts at every level, from the individual to the household to the community to the nation to the world. Governments, corporations, and organizations must all encourage and pursue sustainable practices. Among the institutions that can contribute to sustainability efforts are colleges and universities.

We tend to think of colleges and universities as enlightened and progressive institutions that generate benefits for society. However, colleges and universities are also centres of lavish resource consumption. Institutions of higher education feature extensive infrastructure, including classrooms, offices, research labs, and residential housing. Most also have dining establishments, sports arenas, vehicle fleets, and road networks. The ecological footprint of a typical college or university is substantial (see "The Science Behind the Story: A Campus Ecological Footprint Calculator").

Reducing the size of this footprint is challenging. Colleges and universities tend to be bastions of tradition, where institutional habits are deeply ingrained and where bureaucratic inertia and financial constraints can often block the best intentions for positive change. Nonetheless, faculty, staff, administrators, and especially students are progressing on a variety of fronts to make the operations of educational institutions more sustainable.

In 2008 the Alma Mater Society (AMS), which represents 44 000 students at UBC Vancouver, created the AMS Lighter Footprint Strategy, based on the ecological footprint concept of Rees and Wackernagel.[5] The intent of the strategy was to coordinate and build upon the successes of student-run sustainability initiatives. The strategy defines internal goals that can be achieved by a student organization working on its own and interactive goals that require communication or collaboration with other agencies, both within the university and outside of it. The goals are divided into such categories as food and beverage, communications, transportation, and campus policies. Each category has suggested actions and timeframes, as well as indicators with which progress can be measured. Student-run organizations at other campuses are beginning to adopt some of the actions and principles of the AMS Lighter Footprint Strategy.

Why strive for campus sustainability?

You enrolled at your college or university to gain an education, not to transform the institution. Why, then, are

THE SCIENCE BEHIND THE STORY

A Campus Ecological Footprint Calculator[6]

UTM's ecological footprint project kicked off under the campus's "Grow Smart, Grow Green" plan.

In the summer of 2004, University of Toronto Mississauga (UTM) undergraduate Greg Bunker worked with professor Tenley Conway to complete an Ecological Footprint Assessment Analysis (EFA) for the UTM campus. The results of the work revealed that the available online calculators, designed for individuals and schools, were ill suited for university campuses. The calculators yielded widely varying results. More importantly, it was discovered that the existing calculators were organized in a manner that made it difficult to collect the required information, and some of the indicators did not accurately reflect the university setting.

For example, when Bunker defined the intercampus shuttle bus as an intercity bus, using the online calculators that were available at the time, fossil fuel consumption was defined as the number of passengers multiplied by the number of kilometres travelled per month. Increasing the number of students using this mode of mass transportation would actually *increase* UTM's footprint, and any improvement in the fuel efficiency of the service would not be considered. This result is counterintuitive, and suggested that the indicator—if calculated in this manner—was not an effective diagnostic tool for resource use at an institution. It became clear that the existing calculators did not accurately reflect the situation of the university campus.

Following up on the recommendations that came out of this early work, undergraduate students Chelsea Dalton and Jennifer Loo worked with Professor Conway in the summer of 2005 to develop the UTM Ecological Footprint Calculator, which is uniquely tailored to the campus setting. The most useful aspect of the calculator is its organizational structure. Rather than being organized by impact categories, as most calculators are, it follows the university's departmental structure. This streamlines the collection of data and allows the user to easily see which departments on campus should be targeted for footprint reduction strategies.

The calculator has a total of six impact categories: materials and waste, built-up land, water, energy, food, and transportation. The calculator breaks down the footprint by impact category, with the colour-coded results displayed at the bottom of the spreadsheet. As a result of the design of the calculator, the impact categories and departments that have the largest contributions to the campus footprint are clearly indicated. The calculator includes on-campus consumption and commuting only; activities that occur off-campus, such as field trips and at-home consumption by commuter students, are not included. However, these have been the subjects of subsequent student research projects at UTM. In addition, materials used to construct new buildings are not included, because of the difficulty of tracking the weights and exact amounts consumed.

When computed by using the new calculator, UTM's total footprint came out to 7827 ha, or 1.04 ha per full-time equivalent campus community member. Of all

---

increasing numbers of students promoting sustainable practices on their campuses? First, reducing the ecological footprint of a campus really can make a difference. The consumptive impact of educating, feeding, and housing hundreds or thousands of students is immense. Second, campus sustainability efforts make students aware of the need to address environmental problems, and students who act to promote campus sustainability serve as models for their peers. Finally, students who engage in sustainability efforts learn and grow as a result. The challenges, successes, and failures that they encounter can serve as valuable preparation for similar efforts in transforming inertia-bound institutions in the broader society.

Support from faculty, staff, and administrators is crucial for success, but students are often the ones who initiate change. Students often feel freer than faculty or staff to express themselves. Students also arrive on campus with new ideas and perspectives, and they generally are less attached to traditional ways of doing things.

Campus efforts may begin with an audit

Campus sustainability efforts often begin with a quantitative assessment of the institution's operations. Such audits provide baseline information on what an institution is doing or how much it is consuming and help set priorities and goals. It is useful in such an **environmental audit** to target items that can lead directly to specific recommendations to reduce impacts and enhance sustainability. Once changes are implemented, the institution can monitor progress by comparing future measurements to the audit's baseline data.

Student auditors at Mount Allison University undertake a comprehensive environmental audit every two years.

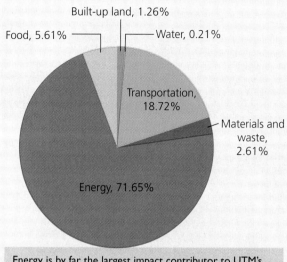

Energy is by far the largest impact contributor to UTM's campus ecological footprint, at almost 72% of the total.

the impact categories, energy contributes the most to the footprint, accounting for about 72%, followed by transportation at 19%, then food, materials and waste, built-up land, and lastly, water (see figure).

UTM was not the first university to undertake an ecological footprint assessment (EFA). Several other universities and colleges have conducted EFAs of varying scope, with widely differing results. For example, the 2004 EFA for UTM, using two online calculators, determined UTM's footprint to be in excess of 40 000 ha—more than five times the current result!

The completeness of the data affected the calculated footprint size, and differences in methodologies were crucial.

Part of the goal of the UTM Ecological Footprint Project has been to compare how various campus EFA studies have been done and to investigate how the methodologies could be improved and customized for the university setting. Additional goals were to consider the implications of EFAs for sustainability at post-secondary institutions and to propose some ways in which campus ecological footprints might be reduced.

The Ecological Footprint Project was partly sponsored by the chief executive officer of UTM, who subsequently used some of the results and recommendations to derive strategies for UTM's *Grow Smart, Grow Green* development plan. One aspect of this plan was the establishment of UTM's Centre for Emerging Energy Technologies, whose director is Chris Cheh. The centre has been responsible for a number of initiatives and partnerships on campus, including the acquisition of solar golf carts for grounds maintenance crews and campus tours; a photovoltaic system for supplementary power supply; a solid oxide fuel cell combined heat and power installation for student residences; and a test venue for hydrogen buses, in partnership with the Hydrogen Village.

Goals for the future of the Ecological Footprint Project include expanding the calculator to make it more widely applicable; developing a calculator that allows the user to see the footprint reductions that would result from making certain improvements on campus; and calculating the footprint impacts of various emerging energy technologies that are being used on the UTM campus. Incorporating the project into related course work is another avenue being pursued. This would provide undergraduates with an opportunity to conduct research while contributing in a meaningful way to an ongoing project.

The results of the 2005 audit—the university's fourth—contributed to the development of a green action plan, which provides ideas for campus groups looking for possible activities and serves as a "tool kit" for the environmental issues committee.

Recycling and waste reduction are common campus efforts

Campus sustainability efforts frequently involve waste reduction, recycling, or composting. Waste management initiatives are relatively easy to start and maintain because they offer many opportunities for small-scale improvements and because people generally enjoy recycling and reducing waste. Waste and recycling audits, battery drop-offs, and residence composting programs are popular on Canadian campuses.

Students, faculty, and staff at Concordia University are diverting 68% of their waste through recycling and composting initiatives. This includes drop-offs for old computers and other electronic wastes, which are becoming an increasingly problematic component of municipal waste. In June 2007, Concordia became the first university certified under the *Ici on recycle* program, sponsored by Recyc-Québec and the government of Quebec, which recognizes institutions that divert at least 65% of their waste from landfills. McGill University students—inspired by what they saw happening at Concordia and other universities—started Gorilla Composting, a student-run initiative to promote composting on campus (**FIGURE 23.1**).

Students at some campuses run events that promote the reuse of items. At the University of Calgary, the U-Bike program was started when 10 discarded bicycles were salvaged, repaired, and pressed into service. The bikes,

FIGURE 23.1
The mascot of McGill University's student-run Gorilla Composting Team (inset shows the team's logo) introduces new students to on-campus composting.

painted with bright red and yellow stripes, are available for use free of charge for any university student. Programs similar to this, which apparently originated in the Netherlands in 1998, are springing up at campuses across North America. The U-Bike program reuses discarded bikes, promotes a healthier lifestyle, and limits the use of cars around the campus.

Administrators are more easily convinced to enact institutional changes if they save money. The EcoTrek Project, Canada's largest university energy and water retrofit project, accomplished this at UBC. The project began in the fall of 2002 and was completed in 2004. The objectives were to generate savings, reduce energy use in university buildings by 20%, and reduce water use by 30%; these goals have been met and exceeded. The project is now saving UBC an estimated $2.6 million annually, making the payback period approximately 17 years for the $35 million project. The work involved retrofits to nearly 300 campus buildings, including installing more efficient lighting; weather-stripping nearly 2000 doors; and replacing old steam boilers with energy-efficient burners.[7]

The Renew Project at UBC provided similar savings. This project was aimed at incorporating the full long-term costs—environmental and social, as well as economic—into renovate-versus-replace decisions on campus. Suzanne Poohkay, director of UBC facilities and capital planning, commenting on the results of one Renew project, said that "the refurbishment cost $7.2 million as opposed to $17.6 million had we built it from scratch. We diverted 220 000 kg of waste from going to landfill, 775 000 kg of CO_2 equivalent emissions were not released, and 3.5 million litres of water were saved."[8]

Green building design is a key to sustainable campuses

Dozens of campuses now boast "green" buildings that are constructed from sustainable building materials and whose design and technologies encourage energy efficiency, water efficiency, renewable energy, and the reduction of pollution. As with any type of ecolabelling, agreed-upon standards are needed, and for sustainable buildings these are the **Leadership in Energy and Environmental Design (LEED)** standards. Developed and maintained by the nonprofit Green Building Council, LEED standards guide the design and certification of new construction and the renovation of existing structures on campuses and elsewhere.

One of the most striking green buildings on a college campus is the Adam Joseph Lewis Center for Environmental Studies at Oberlin College in Ohio (**FIGURE 23.2**). This building was constructed by using materials that were recycled or reused, took little energy to produce, or were locally harvested, produced, or distributed. Some materials, such as carpeting, are leased and then returned to the company for recycling when they wear out. The building contains energy-efficient lighting, heating, and appliances, and it maximizes indoor air quality with a state-of-the-art ventilation system as well as paints, adhesives, and carpeting that emit few volatile organic compounds. The structure is powered largely by

Lewis Center at Oberlin College

FIGURE 23.2
Oberlin College's Adam Joseph Lewis Center for Environmental Studies is one of the best-known green campus buildings.

ate liveable spaces that promote social interaction, where plantings supply shade, prevent soil erosion, create attractive settings, and provide wildlife habitat. It has been said, in fact, that groundskeepers are more vital to schools' recruiting efforts than are vice-presidents.

The University of Victoria, known for the beauty of its 160 ha campus grounds, protects certain areas from development and has made a commitment to restoration and monitoring of naturalized areas. The university has adopted the concept of "naturescaping," a landscaping approach that emphasizes restoring, preserving, and enhancing wildlife habitat in urban and rural areas. The intent is to create new habitats, utilize native plant species, reduce the need for watering, and eliminate chemical pesticides and herbicides. There are several natural landscaping projects on campus, including the Native Plant Study Garden and the Lorene Kennedy Memorial Native Plants Garden (**FIGURE 23.3**).[9]

solar energy from photovoltaic panels on the roof, active solar heating, and passive solar heating from south-facing walls of glass and a tiled slate floor that acts as a thermal mass. More than 150 sensors in the building monitor conditions, such as temperature and air quality; these are reported online in real-time, queryable data formats. A unique component of the building is the "living machine," a solar-powered treatment system that recycles wastewater for nonpotable uses.

Vancouver Island University (formerly Malaspina College–University) was an early leader in green buildings and environmental retrofits in Canada. The university adopted a holistic, environmentally sustainable approach to building design in 1990, when it began planning for the Nanaimo campus. Since then, the university has realized millions of dollars in energy savings and now offers an undergraduate certificate program in green building and renewable energy.

The movement for "green buildings" continues to grow. Pavilions Lassonde at the École Polytechnique de Montréal was the first LEED-certified building in Canada, achieving its gold status in October 2005, followed closely by the Life Sciences Centre at UBC in December 2005 and the University of Victoria Medical Sciences Centre in August 2006. Only a handful of buildings in Canada have earned platinum LEED certification, which requires that the building be retrofitted rather than new; these include the Centre for Child Development at the University of Calgary, certified in 2007.

Sustainable architecture doesn't stop at the walls of a building. Careful design of campus landscaping can cre-

FIGURE 23.3
The Lorene Kennedy Memorial Native Plant Garden is one of a number of "naturescaping" projects at the University of Victoria.

Efficient water use is important

Managing water efficiently is a key element of sustainable campuses. Simon Fraser University was recently recognized with an Earth Award from the Building Owners and Managers Association (BOMA) of British Columbia, for the new Arts and Social Sciences Complex (ASSC1). The building has many energy-saving features, as well as a cistern under the inner courtyard that can store more than 227 000 L of rainwater, to be used for landscape irrigation purposes.[10] ASSC1 has also won engineering and architectural design awards; it is certified under the BOMA-sponsored Go Green program, a green building certification program similar to LEED. Simon Fraser University was the first post-secondary institution in North America to have 26 core campus buildings certified under Go Green.

The Living with the Lakes Centre at Laurentian University demonstrates another approach to water management integrated with green building design. This project, which will house the university's aquatic ecology programs, will be completely integrated into its lakeside setting, featuring a flow-through aquatic laboratory with natural lake water. A green roof will store and filter rainwater, and two constructed wetlands, one outdoors and one indoors, will treat and purify stormwater. The building and site are designed to improve the quality of water entering Ramsay Lake, a drinking water reservoir.[11]

Water conservation is just as important indoors. Water-saving technologies, such as waterless urinals and "living machines" to treat wastewater, are being installed at a number of campuses. UBC's EcoTrek upgrades, mentioned above, reduced water use on campus by 30% each year—saving enough water for 11 785 homes. A study by the sustainability office of the University of Toronto found that urinals in some buildings on the downtown campus were set to flush automatically at timed intervals, 24 hours a day, seven days a week, resulting in what the study called an "astronomical" waste of water. The problem was resolved by installing motion sensor flush systems in 15 washrooms, resulting in estimated savings of 7 million litres of water and $10 000 per year.[12]

Energy conservation is achievable

Students are finding many ways to conserve energy on campus. Campuses can harness large energy savings simply by not powering unused buildings. How many times have you walked through a classroom building at night or in the summer when it was totally empty, yet found that all the lights were on and the heating or air conditioning was running full blast? Large buildings are expensive to heat, cool, and light, so powering them down when not in use saves a great deal of energy, money, and greenhouse gas emissions.

One successful on-campus energy conservation program is the University of Toronto's ReWire Project, which aims to help students, staff, and faculty reduce their energy consumption through behavioural changes. Components of the project include sustainability pledges, environmental education initiatives, and "toolkits" that provide information, resources, and strategies to users. The toolkits contain an assortment of "action tools" that allow users to pass along information to promote sustainable behaviour. The toolkits are designed for use in specific campus settings, such as residence rooms, common areas, and offices. The project has received financial support from a number of sources outside the university, including the Toronto Atmospheric Fund, the Better Buildings Partnership, and the Ontario Power Authority.[13]

Students can promote renewable energy

Campuses can reduce energy consumption and greenhouse emissions by altering the types of energy they use. Student initiatives on campus can also influence the types of energy we use in our society. In October 2009, teams of students from 20 universities competed in the sixth annual Solar Decathlon, which included teams from Germany, Puerto Rico, and Spain, as well as Canada and the United States. Two collaborative teams represented Canada: the University of Calgary, SAIT Polytechnic, and Mount Royal College (all from Calgary, Alberta); and the University of Waterloo (Waterloo, Ontario), Ryerson University (Toronto, Ontario), and Simon Fraser University (Burnaby, British Columbia). In this remarkable event, teams of students travel to Washington, D.C., bringing material for the solar-powered homes they have spent months designing. They erect their homes on the National Mall in Washington, where the buildings stand for three weeks (**FIGURE 23.4**). The homes are judged on 10 criteria, and prizes are awarded to winners in each category.

Solar and wind power and other alternative energy sources play roles on many campuses. Since the 1980s, Carleton University has made use of geothermal energy for space heating, through underground thermal energy storage (UTES) technologies. UTES systems use underground reservoirs to store heat during the summer and provide heat during the winter. Heat is withdrawn from and

FIGURE 23.4
Teams from colleges and universities around the world compete in the biennial Solar Decathlon in Washington, D.C. Each team erects an entire house, of the students' own design, fully powered by solar energy. The 2009 event included two collaborative teams representing six Canadian campuses. This is a rendering of the Alberta Solar Decathlon Team's entry in the 2009 competition.

returned to the reservoir, as needed, by pumping fluids through a system of underground pipes.

Trent University, located on the fast-flowing Otanabe River, has its own hydroelectric power plant, the Stan Adamson Powerhouse, which houses three turbine generators. The plant currently supplies about 75%–80% of the total electricity used on campus in a given year, and excess power from low-use times is exported to the power grid. The power plant technology dates from the original construction in the late 1800s, but Trent plans to start a major upgrade of the facility very soon, which should allow the campus to be 100% self-sufficient on a source of energy that is associated with virtually no pollutants or emissions.[14]

Carbon neutrality is a new goal

Now that global climate change has vaulted to the forefront of society's concerns, reducing greenhouse gas emissions from fossil fuel combustion has become a top priority for campus sustainability proponents. Today many campuses are aiming to become carbon neutral.

By the autumn of 2007, more than 350 university presidents in the United States had signed on to the American College and University Presidents Climate Commitment (Confederation College in Thunder Bay, Ontario, was the only Canadian campus to have signed this agreement, as of 2008). This pledge to reduce campus greenhouse emissions commits presidents to undertake inventory of emissions, set target dates for becoming carbon neutral, and take immediate steps to lower emissions with short-term actions, while also integrating sustainability into the curriculum. In March 2008, a similar Climate Change Statement of Action was signed by university and college presidents across British Columbia. The statement commits each institution to initiate a comprehensive plan to reduce greenhouse gases.

Student pressure and petitions at many campuses have nudged administrators and trustees to set targets for reducing greenhouse emissions, and to strive to do more to counteract climate change than they do to contribute to it. Common Energy is a network of students, faculty, staff, and community stakeholders working to encourage their universities to move beyond climate neutrality, by providing resources and undertaking actions to promote community involvement in climate-related initiatives.[15] Go Beyond is a project of the B.C. Campus Climate Network, a partnership among high school, college, and university sustainability organizations. Go Beyond, which supports students in implementing climate action on their campuses, states as its objective to "move beyond climate-neutral step by step together."[16]

Dining services and campus gardens let students eat sustainably

Campus food service operations can promote sustainable practices by buying organic produce, composting food scraps, and purchasing food in bulk or with less packaging. Buying locally grown or produced food supports local economies and cuts down on fuel use from long-distance transportation. At the University of Waterloo, students and the University Food Services teamed up to start an on-campus Farm Market to promote local foods. The market features 100% local produce, preserves and honeys made in Waterloo County, and fresh baked goods made right on campus. McGill University's Organic Campus is a student-run nonprofit organization that delivers local organic fruits and vegetables on campus.

Some college campuses even have community gardens where students can grow food. McGill's Campus Crops is a student-run urban gardening cooperative

FIGURE 23.5
A number of campuses now include gardens where students can grow organic vegetables that are used for meals in dining halls. Here, a student with McGill University's Campus Crops helps set up a plot for urban agriculture.

(**FIGURE 23.5**). Faced with a shortage of community garden plots, the students are aiming to identify unused spaces on the McGill campus that could support urban agriculture. McGill also has a vegan food collective (The Midnight Kitchen).

Institutional purchasing matters

The kinds of purchasing decisions made in dining halls favouring local food, organic food, and biodegradable products can be applied across the entire spectrum of a campus's needs. When campus purchasing departments buy recycled paper, certified sustainable wood, energy-efficient appliances, goods with less packaging, and other ecolabelled products, they send signals to manufacturers and increase the demand for such items. Students are working with campus bookstores to carry, promote, and sell more environmentally and socially sustainable books, paper products, and school supplies.

The University of Guelph is one of a number of universities that have undertaken green purchasing policies for their computing services departments, with a pledge to take into consideration the entire life cycle of electronic products. "Green computing" considers everything from "cradle" (the materials and processes used in the manufacturing and shipping of the product), through operational use (the energy efficiency of the product), to "grave" (the effective recycling and disposal of the product). The Green Energy Research Institute at the University of Waterloo has contributed much of the background science and engineering research that has provided the foundation for green computing and electronics policies.

At Chatham College in Pennsylvania, students chose to honour their school's best-known alumnus, Rachel Carson, by seeking to eliminate toxic chemicals on campus. Administrators agreed to this effort, provided that alternative products to replace the toxic ones worked just as well and were not more expensive. Students brought in the CEO of a company that produces nontoxic cleaning products, who demonstrated to the janitorial staff that his company's products were superior. The university switched to the nontoxic products, which were also cheaper, and proceeded to save $10 000 per year. Chatham students then found a company offering paint without volatile organic compounds and negotiated with it for a free paint job and discounted prices on later purchases. Students also worked with grounds staff to eliminate herbicides and fertilizers used on campus lawns and to find alternative treatments.

Transportation alternatives are many

Many campuses struggle with traffic congestion, parking shortages, commuting delays, and pollution from vehicle exhaust. Some are addressing these issues by establishing

FIGURE 23.6
The sustainable Concordia *allégo* program holds free bike tune-up events for students, staff, and faculty.

or expanding bus and shuttle systems; encouraging bicycling, walking, and carpooling; and introducing alternative vehicles to university fleets (**FIGURE 23.6**). The *allégo* program at Concordia University makes sustainable transportation choices available to those who are commuting to and from the campus, offering a free rideshare board, bicycle safety and maintenance workshops, and a Guaranteed Ride Home program for carpoolers.[17]

The University of British Columbia is also a leader in transportation efforts. For $20 a month, UBC's U-Pass program provides students with biking programs and facilities, expanded campus bus and shuttle services, unlimited use of some city transit systems, rides home in emergencies, merchant discounts, and priority parking spaces and ride-matching services for carpoolers. The program has boosted transit ridership to campus by 53% and has decreased single-person car use by 20%.

Campuses are restoring native plants, habitats, and landscapes

No campus sustainability program would be complete without some attempt to enhance the campus's natural environment, such as on the University of Victoria campus, described above. Such efforts remove invasive species, restore native plants and communities, improve habitat for wildlife, enhance soil and water quality, reduce pesticide use, and create healthier, more attractive surroundings.

The University of Saskatchewan College of Education has constructed a Prairie Habitat Garden with native grasses, shrubs, and wildflowers. The garden helps preserve the natural heritage and habitat of prairie grasslands while acting as an educational resource (**FIGURE 23.7**).

In a unique partnership, the University of Toronto Mississauga joined with Evergreen Foundation's

FIGURE 23.7
Many schools have embarked on habitat restoration projects to beautify their campuses, provide wildlife habitat, restore native plants, and filter water runoff. Here, students learn about the local environment at the University of Saskatchewan's Prairie Habitat Garden.

Learning Grounds program for a major campus greening initiative. The partnership has won local environmental awards. Regular volunteer plantings of native shrubs and flowers have contributed to the naturalization of many areas of the campus. Students, staff, faculty, and neighbours routinely turn out for the events. Besides providing wildlife habitat, the restored areas reduce runoff, erosion, and maintenance costs, and provide opportunities for research and education. For example, student researchers have assessed Jefferson salamander populations and habitat conditions at the campus. These studies will provide information on the threatened salamander's breeding ponds and terrestrial habitat, educate the university community about the significance of the species, and determine whether habitat restoration actions are required. The university also undertook a controlled burn to restore native vegetation on a small area of the campus, planting experimental plots of native plants to study ecological patterns in the wake of the restoration.

Sustainability efforts include curricular changes

Campus sustainability activities provide students with active, hands-on ways to influence how their campuses function. Sustainability concerns are also transforming academic curricula and course offerings. The course for which you are using this book right now likely did not exist a generation ago. As our society comes to appreciate the looming challenge of sustainability, colleges and universities are attempting to train students to confront this challenge more effectively.

Almost all universities and colleges in Canada now offer courses and programs on environment and sustainability issues. Trent University in Peterborough, Ontario, has offered courses in environmental and resource management since 1975. The university now partners with Fleming College to offer a practical, field-based program in ecological restoration. Faculty at Queen's University have received awards for excellence in education for the promotion of sustainable practices, in recognition of their contributions to sustainability education.[18]

Some universities specialize within the general fields of environment and sustainability. For example, McMaster University is widely recognized for excellence in the field of environmental health. The Institute of Island Studies at Prince Edward Island University emphasizes the study of the culture, environment, and economy of small islands. The institute's associates undertake studies on environmental policy, water quality, land use, and other topics from the particular perspective of island economies, and partner with other island researchers from around the world.

Besides offering new courses and programs, schools are incorporating sustainability issues into established courses across many disciplines. UBC, for example, now offers more than 300 courses with sustainability-related themes. Many of these are also *interdisciplinary* courses and programs, meaning that they integrate knowledge from a variety of different disciplines. An example of an interdisciplinary research program is the National First Nations Environmental Contaminants Program (NFN–ECP), a collaborative program between the First Nations University of Canada, the Assembly of First Nations, and Health Canada. The goal of NFN–ECP is to support research on environmental contaminant exposure and the potential for risks to the health and well-being of First Nations people in Canada.[19]

Sustainability courses and programs often have a practical slant. For example, the University of Waterloo's ERS250 "Greening the Campus" course and the University of Toronto's ENV421 "Environmental Research" course both offer students the opportunity to carry out practical research on campus for course credits. Often the results of student research are used in developing campus policies; examples of past ENV421 projects include "Recommendations for Reducing Energy Consumption," and "The Natural Step and the University of Toronto: A Value Added Approach to Energy Efficiency Planning."[20]

As mentioned in the Central Case, the UBC SEEDS program links students, staff, and faculty to research projects that enhance sustainability on campus, and allows students to get course credits for their projects. SEEDS projects have benefited the campus through a

---

weighing the issues

SUSTAINABILITY ON YOUR CAMPUS 23–1

Find out what sustainability efforts are being made on your campus. What results have these achieved so far? What further efforts would you like to see pursued on your campus? Do you foresee any obstacles to these efforts? How could these obstacles be overcome? How could you become involved? Are there opportunities for you to undertake research or project work for course credits?

number of research projects that have led to policy changes and/or initiatives, including these:

- A biodiesel project that led UBC's Vancouver Plant Operations to use 20% biodiesel fuel in its diesel fleet vehicles
- Research that led to UBC's becoming a pesticide-free campus
- A reassessment of landscape techniques to reduce heavy metal contaminants in stormwater
- Research that led to new seafood purchasing policies[21]

In the best tradition of colleges and universities, classroom learning and real-world learning go hand in hand in SEEDS and similar programs on campuses across Canada.

Organizations are available to assist campus efforts

Many campus sustainability initiatives are supported by organizations, such as Sierra Youth Coalition Campus Sustainability Project; University Leaders for a Sustainable Future; the Association for the Advancement of Sustainability in Higher Education; and the National Wildlife Federation's Campus Ecology program. These organizations act as information clearinghouses for campus sustainability efforts. With the assistance of these organizations, it is easier than ever to start sustainability efforts on your own campus and obtain the support to carry them through to completion.

The Sierra Youth Coalition (www.syc-cjs.org) offers a framework for sustainability assessments, which has provided a starting point for many initiatives on campuses across Canada (FIGURE 23.8). The framework incorporates indicators of both human and ecosystem well-being.

Sustainability and Sustainable Development

Efforts toward sustainability on college and university campuses parallel efforts in the world at large. As more people come to appreciate Earth's limited capacity to accommodate our rising population and consumption, they are voicing concern that we will need to modify our behaviours, institutions, and technologies if we wish to sustain our civilization and the natural environment on which it depends. In the quest for sustainability, the strategies pursued on campuses reflect those pursued in the wider society, and they also can serve as models.

When people speak of **sustainability**, what precisely do they mean to sustain? Generally they mean to sustain human institutions in a healthy and functional state—and also to sustain ecological systems in a healthy and functional state. The contributions of biodiversity and ecosystem goods and services to human welfare are tremendous. Indeed, they are so fundamental (some would say infinitely valuable, thus literally priceless) that we have long taken them for granted.

Sustainable development aims to achieve a triple bottom line

We first explored **sustainable development** in Chapter 1 and offered the United Nations' definition: "Development that meets the needs of the present without compromising the ability of future generations to meet their own needs." Today, it is widely recognized that sustainability does not mean simply protecting the environment against the ravages of human development. Instead, it means finding ways to promote social justice, economic well-being, and environmental quality at the same time. Meeting this **triple bottom line** is the goal of modern

FIGURE 23.8
The Sierra Youth Coalition's Sustainability Assessment Framework has served as a starting point for many campus initiatives across Canada.

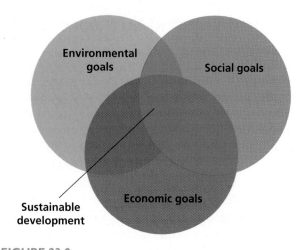

FIGURE 23.9
Modern conceptions of sustainable development hold that sustainability occurs at the intersection between three sets of goals: social, economic, and environmental.

sustainable development (**FIGURE 23.9**). Achieving this goal is most pressing in nations of the developing world, but it is a vital need everywhere. It is our primary challenge for this century and likely for the rest of our species' time on Earth.

"Environmental sustainability" is one of the United Nations' Millennium Development Goals set by the international community at the turn of this century, and the drive for sustainability overall meshes with all eight goals. In the past few years, the Millennium Project and the Millennium Ecosystem Assessment have determined the following:

- Environmental degradation is a major barrier to achieving the Millennium Development Goals.
- Investing in environmental assets and management is vital to relieving poverty, hunger, and disease.
- Reaching environmental goals requires progress in eradicating poverty.

The many actions being taken today on campuses and by governments, businesses, industries, organizations, and individuals across the globe are giving people optimism that achieving these goals and developing in sustainable ways is within reach.

Environmental protection can enhance economic opportunity

Reducing resource consumption and waste often saves money, as many colleges and universities discover when they embark on sustainability initiatives. Sometimes savings accrue immediately, and other times an up-front investment brings long-term savings.

For society as a whole, attention to environmental quality can enhance economic opportunity by providing new types of employment. This reality contrasts with the common perception that environmental protection hurts the economy by costing people jobs. In the controversy over logging of old-growth forest in the Pacific Northwest (Chapter 10), protection for the endangered northern spotted owl (**FIGURE 23.10**), whose natural habitat ranges from Northern California to British Columbia, set limits on timber extraction. Proponents of logging claim that such restrictions cost local loggers their jobs. However, loggers' jobs are far more at risk when timber companies cut trees at unsustainable rates and then leave a region, seeking mature forests elsewhere, as has happened in region after region throughout history.

The jobs-versus-environment debate frequently overlooks the fact that as some industries decline, others spring up to take their place. As jobs in logging, mining, and manufacturing have disappeared in developed nations over the past few decades, jobs have proliferated in service occupations and high-technology sectors. As we decrease our dependence on fossil fuels, jobs and investment opportunities are opening up in renewable energy sectors, such as wind power and fuel cell technology (Chapter 17). Many jobs were lost in the Ontario automobile

FIGURE 23.10
The northern spotted owl (*Strix occidentalis occidentalis*) has become a symbol of the "jobs-versus-environment" debate. This bird of the Pacific rainforest is considered endangered because of the logging of mature forests. Proponents of logging argue that laws protecting endangered species cause economic harm and job loss. Advocates of endangered species protection argue that unsustainable logging practices pose a larger risk of job loss.

industry in 2008, when General Motors downsized in response to a major consumer move away from gas-guzzling trucks. Almost immediately, talks began between the Canadian Auto Workers Union (CAW) and a foreign manufacturer of windmill parts who expressed an interest in the kinds of skills the autoworkers possess.

Will the alternative energy technology industry provide jobs for all of the displaced Ontario autoworkers? Not singlehandedly, and certainly not right away. However, the talks opened up the possibility that long-term sustainable alternatives would emerge. Environmental protection need not lead to economic stagnation, but instead can enhance economic opportunity. This connection is also suggested by the fact that global economies have expanded rapidly in the past 30 years, the very period during which environmental protection measures have proliferated.

Moreover, if we look beyond conventional economic accounting (which measures only private economic gain and loss) and instead include external costs and benefits that affect people at large, then environmental protection becomes still more valuable. Take several studies reviewed by the Millennium Ecosystem Assessment: They each show how overall economic value is maximized by conserving natural resources rather than exploiting them for short-term private gain (**FIGURE 23.11**).

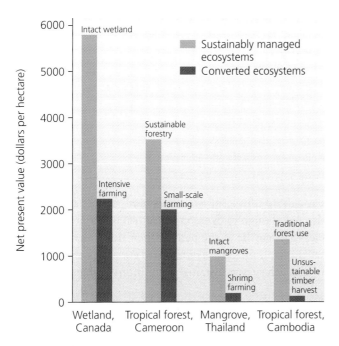

FIGURE 23.11
Once external costs and benefits are factored in, the economic value of sustainably managed ecosystems generally exceeds the economic value of ecosystems that have been converted for intensive private resource harvesting. Shown are land values calculated by researchers in four such comparisons from sites around the world. Data from Millennium Ecosystem Assessment, 2005.

Although people desire private monetary gain, they also desire to live in areas that have clean air, clean water, intact forests, and parks and open space. Environmental protection increases a region's attractiveness, drawing more residents and increasing property values and the tax revenues that help fund social services. As a result, those regions that act to protect their environments are generally the ones that retain and increase their wealth and quality of life.

What accounts for the perceived economy-versus-environment divide?

If environmental protection and economic development are mutually reinforcing, what, then, accounts for the common view that we cannot simultaneously protect the environment and provide for people's needs? One proximate explanation lies in the fact that economic development since the Industrial Revolution has so clearly diminished biodiversity, decreased habitat, and degraded ecological systems. A second proximate explanation lies in the fact that many people believe command-and-control environmental policy poses excessive costs for industry and restricts rights of private citizens.

An ultimate explanation may lie in our species' long history. For the thousands of years that we lived as nomadic hunter–gatherers, population densities were low, and consumption and environmental impact were limited. With natural resources in little danger of running out, people were free to exploit them limitlessly and had little reason to adopt a conservation ethic. Our establishment of sedentary agricultural societies, followed by urbanization, has increased our impact while also causing us to overlook the connections between our economies and our environments. It is common to hear "humans and the environment" or "people and nature" being set in contrast, as though they were separate. Some philosophers venture to say that the perceived dichotomy between humans and nature is the root of all our environmental problems.

Humans are not separate from the environment

On a day-to-day basis, it is easy to feel disconnected from the natural environment, particularly in industrialized nations and large cities. We live inside houses, work in shuttered buildings, travel in enclosed vehicles, and generally know little about the plants and animals around us. Millions of urban citizens have never set foot in an undeveloped area. Just a few centuries or even

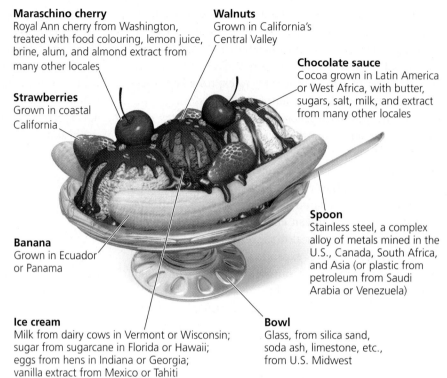

FIGURE 23.12
A banana split eaten at an ice cream shop in a North American town consists of ingredients from around the world, whose production has impacts on the environments of many distant locations. Ice cream requires milk from dairy cows that graze pastures or are raised in feedlots on grain grown in industrial monocultures. Ice cream is sweetened with sugar from sugar beet farms or sugarcane plantations. The banana was shipped thousands of kilometres by oil-fuelled transport from a tropical country, where it grew on a plantation that displaced rainforest and where it was liberally treated with fertilizers and fungicides. Fruits and nuts grown in California's Central Valley were irrigated generously with water piped in from elsewhere. The spoon originated with metal ores mined along with thousands of tons of soil and processed into stainless steel by using energy from fossil fuels.

Maraschino cherry
Royal Ann cherry from Washington, treated with food colouring, lemon juice, brine, alum, and almond extract from many other locales

Walnuts
Grown in California's Central Valley

Chocolate sauce
Cocoa grown in Latin America or West Africa, with butter, sugars, salt, milk, and extract from many other locales

Strawberries
Grown in coastal California

Spoon
Stainless steel, a complex alloy of metals mined in the U.S., Canada, South Africa, and Asia (or plastic from petroleum from Saudi Arabia or Venezuela)

Banana
Grown in Ecuador or Panama

Ice cream
Milk from dairy cows in Vermont or Wisconsin; sugar from sugarcane in Florida or Hawaii; eggs from hens in Indiana or Georgia; vanilla extract from Mexico or Tahiti

Bowl
Glass, from silica sand, soda ash, limestone, etc., from U.S. Midwest

decades ago, most of the world's people were able to name and describe the habits of the plants and animals that lived nearby. They knew exactly where their food, water, and clothing came from. Today it seems that water comes from the faucet, clothing from the mall, and food from the grocery store. It is not surprising that we have lost track of the connections that tie us to our natural environment.

However, this doesn't make our connections to the environment any less real. Consider a thoroughly un-"natural" invention of the human species: the banana split (**FIGURE 23.12**), in which each and every element has ties to the resources of the natural environment, and each exerts environmental impacts.

Once we learn to consider where the things we use and value each day actually come from, it becomes easier to see how people are part of the environment. And once we reestablish this connection, it becomes readily apparent that our own interests are best served by preservation or responsible stewardship of the natural systems around us. Because what is good for the environment can also be good for people, win–win solutions are very much within reach, if we learn from what science can teach us, think creatively, and act on our ideas.

Strategies for Sustainability

Sustainable solutions to environmental problems are numerous, and we have seen specific examples throughout this book. The challenges lie in being imaginative enough to think of solutions and being shrewd and dogged enough to overcome political or economic obstacles that may lie in the path of their implementation. We will now summarize several broad strategies or approaches that can spawn sustainable solutions (Table 23.2).

Table 23.2 Some Major Approaches to Sustainability

- Refine our ideas about economic growth and quality of life.
- Reduce unnecessary consumption.
- Limit population growth.
- Encourage green technologies.
- Mimic natural systems by promoting closed loop industrial processes.
- Think in the long term.
- Enhance local self-sufficiency, and embrace some aspects of globalization.
- Be politically active.
- Vote with our wallets.
- Promote research and education.

We can refine our ideas about economic growth and quality of life

It is conventional among economists and the policy makers who heed their advice to speak of economic growth as an ultimate goal. Many politicians view nurturing an expanding economy as their prime responsibility while in office. Yet economic growth is merely a tool with which we try to attain the real goal of maximizing human happiness. Thus, if economic growth depends on an ever-increasing consumption and depletion of nonrenewable resources, then we will not be able to attain long-term happiness by endlessly expanding the size of our economy.

We may also want to incorporate external costs into the market prices of goods and services. Currently, goods and services are priced as though pollution and resource extraction involved no costs to society. If we can make our accounting practices reflect indirect consequences to the public, then we can provide a clearer view of the full costs and benefits of any given action or product. In that case, the free market could become the optimal tool for improving environmental quality, our economy, and our quality of life. Moreover, implementing green taxes and phasing out harmful subsidies could hasten our attainment of prosperous and sustainable economies. The political obstacles to this are considerable, and such changes will require educated citizens to push for them and courageous policy makers to implement them.

We can consume less

Economic growth is driven by consumption: the purchase of material goods and services (and thus the use of resources involved in their manufacture) by consumers (**FIGURE 23.13**). Our tendency to believe that more, bigger, and faster are always better is reinforced by advertisers seeking to sell more goods more quickly. Consumption has grown tremendously, with the wealthiest nations leading the way. Our houses are larger than ever, sport–utility vehicles are among the most popular automobiles, and many citizens have more material belongings than they know what to do with. We think nothing today of having home computers with high-speed internet access, let alone the televisions, telephones, refrigerators, and dishwashers that were marvels just decades ago.

Because many of Earth's natural resources are limited and nonrenewable, consumption cannot continue growing forever. Eventually, if we do not shift to sustainable resource use, per capita consumption will drop for rich and poor alike as resources dwindle. Cornucopian critics often scoff at the notion that resources are limited, but we must remember that our perspective in time is limited and that our consumption is taking place within an extraordinarily brief slice of time in the long course of history (**FIGURE 23.14**). Our lavishly consumptive lifestyles are a brand-new phenomenon on Earth. We are enjoying the greatest material prosperity in all of history, but if we do not find ways to make our wealth sustainable, the party may not last much longer.

Fortunately, material consumption alone does not reflect a person's quality of life. For many people in industrialized nations, the accumulation of possessions has not brought contentment. Observing how affluent people often fail to find happiness in their material wealth, social critics have given this phenomenon a name like a dread disease: **"affluenza."** Whether or not it is a disease, scientific research does back up the contention that money cannot buy nearly as much happiness as people typically believe (**FIGURE 23.15**). Although economic growth is generally equated with "progress," true progress consists of an increase in human happiness. In the end we are, one would hope, more than just the sum of what we buy.

FIGURE 23.13
Citizens of Canada consume more than the people of any other nation, with the exception (but only marginally) of the United States. Unless we find ways to increase the sustainability of our manufacturing processes, this rate of consumption cannot be maintained in the long run.

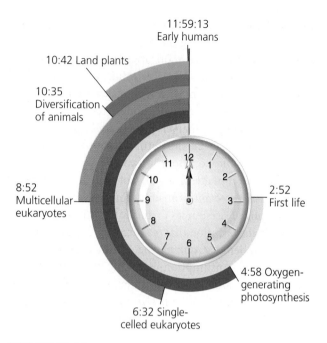

FIGURE 23.14
By viewing Earth's 4.6 billion year history as a 12-hour clock, we can gain a better understanding of relative time scales across the immense span of geologic time. *Homo sapiens*, as a species, has come into existence only during the final one or two seconds, around 11:59:59. The Agricultural and Industrial Revolutions that have so greatly increased our environmental impacts have taken up only a minuscule fraction of a second.

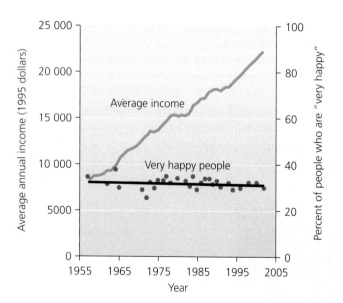

FIGURE 23.15
Although average income of North Americans has risen steadily in the past half-century, the percentage of people reporting themselves as being "very happy" has remained stable or declined slightly. When asked how much a fivefold increase in income improves a person's mood day to day, respondents guessed on average that such an increase would improve mood by 32%, but the actual improvement, from respondents' self-reporting, is only 12%. Data from Myers, D. G. 2000. *The American Paradox: Spiritual Hunger in an Age of Plenty*. New Haven, CT: Yale University Press; Gardner, G., and E. Assadourian, 2004. Rethinking the good life, pp. 164–179 in *State of the World 2004*, Worldwatch Institute; and Kahneman, D., et al. 2006, Would you be happier if you were richer? A focusing illusion. *Science* 312:1908–1910.

FIGURE 23.16
The homegrown, local food, and slow-food movements are growing, all over North America. "Kitchen gardens," like the one shown here, are becoming more and more popular, even in urban areas.

We can reduce our consumption while enhancing our quality of life—squeezing more from less—in at least three ways. One way is to improve the technology of materials and the efficiency of manufacturing processes, so that industry produces goods using fewer natural resources. Another way is to develop a sustainable manufacturing system—one that is circular and based on recycling, in which the waste from a process becomes raw material for input into that process or others. A third way is to modify our behaviour, attitudes, and lifestyles to minimize consumption.

At the outset, such choices may seem like sacrifices, but people who have slowed down the pace of their busy lives and freed themselves of an attachment to material possessions say it can feel tremendously liberating. Fans of local, homegrown foods and the "slow-food" movement (**FIGURE 23.16**) feel that they are helping themselves as much as they are helping the environment.

Population growth must eventually cease

Just as continued growth in consumption is not sustainable, neither is growth in the human population. We have seen that populations may grow exponentially for a time but eventually encounter limiting factors and decline or level off. We have used technology to increase Earth's carrying capacity for our species, but our population cannot continue growing forever; sooner or later, human population growth will end. The question is how: through war, plagues, and famine, or through voluntary means as a result of wealth and education?

The demographic transition is already far along in many industrialized nations thanks to urbanization, wealth, education, and the empowerment of women. If today's developing nations also pass through a demographic transition, then there is hope that humanity may halt its population growth while creating a more prosperous and equitable society.

Technology can help us toward sustainability

It is largely technology—developed with the Agricultural Revolution, the Industrial Revolution, and advances in medicine and health—that has spurred our population increase. Technology has magnified our impact on Earth's environmental systems, yet it can also give us ways to reduce our impact. Recall the IPAT equation, which summarizes human environmental impact (I) as the interaction of population (P), consumption or affluence (A), and technology (T). Technology can exert either a positive or negative value in this equation. The short-sighted use of technology may have gotten us into this mess, but wiser use of environmentally friendly, or "green," technology can help get us out.

In recent years, technology has intensified environmental impact in developing countries as industrial technologies from the developed world have been exported to poorer nations eager to industrialize. In developed nations, meanwhile, green technologies have begun mitigating our environmental impact. Catalytic converters on cars have reduced emissions (**FIGURE 23.17**), as have scrubbers on industrial smokestacks (**FIGURE 13.15**). Recycling technology and advances in wastewater treatment are helping reduce our waste output. Solar, wind, and geothermal energy technologies are producing cleaner renewable energy (Chapter 17). Countless technological advances such as these are one reason that people of North America and Western Europe today enjoy cleaner environments—although they consume far

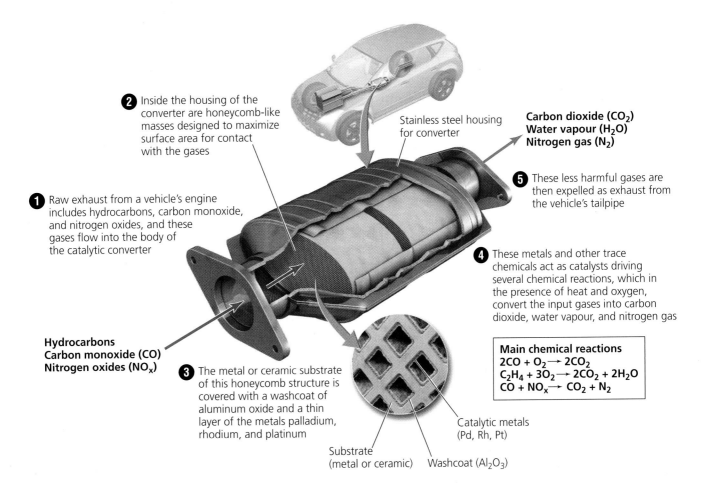

FIGURE 23.17
The catalytic converter is a classic example of green technology. This device filters air pollutants from vehicle exhaust and has helped to improve air quality in Canada and many other nations.

Industry can mimic natural systems

As industries seek to develop green technologies and sustainable practices, they have an excellent model: nature itself. As we saw in Chapter 5 and throughout this book, environmental systems tend to operate in cycles consisting of feedback loops and the circular flow of materials. In natural systems, output is recycled into input.

In contrast, human manufacturing processes traditionally run on a linear model in which raw materials are input and processed to create a product, while by-products and waste are generated and discarded. Some forward-thinking industrialists are making their processes more sustainable by transforming linear pathways into closed cycles in which waste is recovered, recycled, and reused. For instance, several companies now produce carpets that can be retrieved from the consumer when they wear out, and these materials are then recycled to create new carpeting. Some automobile manufacturers are planning cars that can be disassembled and recycled into new cars. Proponents of this industrial model see little reason why virtually all appliances and other products cannot be recycled, given the right technology. Their ultimate vision is to create truly closed loop industrial processes, generating no waste.

We can think in the long term

To be sustainable, a solution must work in the long term. Often the best long-term solution is not the best short-term solution, which explains why much of what we currently do is not sustainable. Policy makers in democracies often act for short-term good because they aim to produce immediate, positive results so that they will be reelected. This poses a major hurdle for addressing environmental dilemmas, many of which are cumulative, worsen gradually, and can be resolved only over long periods. Often the costs of addressing an environmental problem are short-term, whereas the benefits are long-term, giving politicians little incentive to tackle the problem. In such a situation, citizen pressure on policy makers is especially vital, because policy is an essential tool for pursuing sustainability (see "The Science Behind the Story: Rating the Environmental Performance of Nations").

Businesses may act according to either long-term or short-term interests. A business committed to operating in a particular community for a long time has incentive to sustain environmental quality. However, a business merely attempting to make a profit and move on has little incentive to invest in environmental protection measures that involve short-term costs.

We can promote local self-sufficiency and embrace some aspects of globalization

As our societies become more globally interconnected, we experience a diversity of impacts, positive and negative. To many people, encouraging local self-sufficiency is an important element of building sustainable societies. When people feel closely tied to the area in which they live, they tend to value the area and seek to sustain its environment and human communities. This line of reasoning is frequently made in relation to locally based organic or sustainable agriculture.

Many advocates of local self-sufficiency criticize globalization. However, as ecological economist Herman Daly has explained, globalization means different things to different people. Those who view it as a positive phenomenon generally focus on how people of the world's diverse cultures are increasingly communicating and learning about one another. Books, airplanes, television, and the internet have made us more aware of one another's cultures and more likely to respect and celebrate—rather than fear—differences among cultures.

Those who view globalization in a negative light generally cite the homogenization of the world's cultures, by which a few cultures and world views displace many others. For instance, the world's many languages are going extinct with astonishing speed. Traditional ways of life in many areas are being abandoned as more people take up the material and cultural trappings of a few dominant Western cultures.

In recent years, many people have reacted against this homogenization and the growing power of large multinational corporations. In France, farm activist Jose Bove became a popular hero when he wrecked a McDonald's restaurant with his tractor to protest what many French farmers view as a threat to local French cuisine. In Seattle in 1999, thousands of protesters picketed a meeting of the World Trade Organization, which they viewed as a symbol of Western market capitalism. Since then, protesters have picketed every WTO meeting (**FIGURE 23.18**). We explored the complex relationship between free trade and the environment in Chapter 22 ("The Science Behind the Story: The Environment in NAFTA and NAAEC").

Daly and others argue that globalization entails a process in which multinational corporations attain

FIGURE 23.18
Protesters picketed the World Trade Organization's meeting in Quebec in 2003, criticizing the homogenizing effects of globalization, as well as relaxations in labour and environmental protections brought about by free trade.

greater and greater power over global trade while governments retain less and less. Most critics of globalization consider corporations less likely than governments to support environmental protection, so they feel that globalization will hinder progress toward sustainability. Moreover, market capitalism, almost by definition, promotes a high-consumption lifestyle, which does indeed threaten efforts to attain sustainable solutions.

On the positive side, globalization may foster sustainability because Western democracy, as imperfect as it is, serves as a model for people living under repressive governments. Open societies allow for entrepreneurship and the flowering of creativity in business, research, and academia. Millions of free minds thinking about issues are more likely to come up with sustainable solutions than the minds of a few holding authoritarian power.

Citizens exert political influence

Politically open democracies offer a compelling route for pursuing sustainability: the power of the vote. Many of the changes needed to attain sustainable solutions require policy making, and individually and collectively we can guide our political leaders to enact policies for sustainability. Policy makers respond to whoever exerts influence. Corporations and interest groups employ lobbyists to influence politicians all the time. Citizens in a democratic republic have the same power, *if* they choose to exercise it. You can exercise your power at the polling station, by attending public hearings, by donating to advocacy groups that promote ideas you favour, and by writing letters and making phone calls to office-holders. You might be surprised how little input policy makers receive from the public; sometimes a single letter or phone call can make a big difference.

Today's environmental laws came about because citizens pressured their governmental representatives to tackle environmental problems. The raft of environmental legislation enacted in the 1970s in North America (Chapter 22) might never have come about had ordinary citizens not stepped up and demanded action. We owe it to our children to be engaged and to act responsibly now so that they have a better world in which to live.

Consumers vote with their wallets

Expressing one's preferences through the political system is important, but we also wield influence through the choices we make as consumers. When products produced sustainably are ecolabelled, consumers can "vote with their wallets" by purchasing these products. Consumer choice has helped drive sales of everything from recycled paper to organic produce to "dolphin-safe" tuna.

Individuals can multiply their own influence by promoting "green" purchasing habits at their school or workplace. We saw how purchasing power at colleges and universities has spurred sales of certified sustainable wood, organic food, energy-efficient appliances, and more. Employees in businesses and government agencies can promote change within those institutions by voicing their preferences in purchasing decisions. As Dr. Seuss wrote in the children's classic *The Lorax*, "Unless someone like you cares a whole awful lot, nothing is going to get better. It's not."

---

weighing the issues — GLOBALIZATION (23-2)

What advantages and disadvantages do you see in globalization? Have you personally benefited or been hurt by it in any way? In what ways might promoting local self-sufficiency be helpful for the pursuit of global sustainability? In what ways might it not?

Now consider this: Many people enjoy eating at Vietnamese restaurants in Canada, yet many who criticize globalization would frown on the presence of McDonald's restaurants in Vietnam. Do you think this represents a double standard, or are there reasons the two are not comparable?

THE SCIENCE BEHIND THE STORY

Rating the Environmental Performance of Nations

Top-Ranked Nations, 2008 EPI[22]

Nation	EPI Score
1. Switzerland	95.5
2. Norway	93.1
3. Sweden	93.1
4. Finland	91.4
5. Costa Rica	90.5
6. Austria	89.4
7. New Zealand	88.9
8. Latvia	88.8
9. Colombia	88.3
10. France	87.8

Bottom-Ranked Nations, 2008 EPI[23]

Nation	EPI Score
140. Guinea-Bissau	49.7
141. Yemen	49.7
142. Democratic Republic of Congo	47.3
143. Chad	45.9
144. Burkina Faso	44.3
145. Mali	44.3
146. Mauritania	44.2
147. Sierra Leone	40.0
148. Angola	39.5
149. Niger	39.1

The Environmental Performance Index is a weighted composite indicator that can be used to measure the progress of nations and their relative standings in environmental sustainability.

To measure the progress of nations toward environmental sustainability, researchers devised the Environmental Performance Index (EPI), which rates countries using data from 25 indicators of environmental conditions for which governments can be held accountable. A report detailing the results was produced and published online in 2006 and updated in 2008, by Daniel Esty and five colleagues at the Yale Center for Environmental Law and Policy and Columbia University's Center for International Earth Science Information Network, in collaboration with the World Economic Forum and the Joint Research Centre of the European Commission. At http://epi.yale.edu/Home you will find an interactive map showing the details of the EPI for all of the 149 nations included in the 2008 study.

The U.N.'s Millennium Development Goals did not define how to quantify progress on environmental measures. The lack of quantitative measures, these researchers said, had stymied progress on how to implement effective environmental policy. To address this problem, the researchers aimed to track the performance of environmental policy with the same quantitative rigour as statistics are tracked for health, poverty reduction, and other development goals. Giving nations scores and ranking them would reveal "leaders and laggards," showing which nations are on the right track and which are not.

The researchers gathered internationally available data from U.N. and other sources on 16 indicators, which funnelled into six categories: environmental health, air quality, water resources, productive natural resources, sustainable energy, and biodiversity and habitat. The last five categories were considered components of "ecosystem vitality." Nations were scored on their performance in each category, with overall scores based 50% on their environmental health score and 50% on the five ecosystem vitality scores. Scores for each category ranged from zero to 100, with 100 representing a target value established by international consensus.

The researchers included 149 nations in the 2008 report; some nations could not be included because of lack of data.

The main pattern in the results was conspicuous: Nations with the highest scores (see the first table) are wealthy, industrialized nations with the capacity to commit substantial resources toward environmental protection. Nations with the lowest scores (see the second table) are largely developing nations with few resources to invest. They include countries with dense populations and stressed ecosystems that are

Promoting research and education is vital

None of these approaches will succeed fully if the public is not aware of its importance. An individual's decisions to reduce consumption, purchase ecolabelled products, or vote for candidates who support sustainable approaches will have limited impact unless many others do the same. Individuals can influence large numbers of people by educating others and by serving as role models through their actions. The campus sustainability efforts at many colleges and universities accomplish both approaches. Moreover, the discipline of environmental science plays a key role in providing information that people can use to make wise decisions about environmental issues. By promoting scientific research and by educating the public about environmental science, we can all assist in the pursuit of sustainable solutions.

Precious Time

The pace of our lives is getting faster, and life's commotion can make it hard to give attention to problems we don't need to deal with on a daily basis. The world's sheer load of environmental dilemmas can feel overwhelming, and even the best intentioned among us may feel we have little time to devote to saving the planet.

However, the natural systems we depend on are changing quickly. Many human impacts continue to intensify, including deforestation, overfishing, land clearing, wetlands draining, and resource extraction. Our window of opportunity for turning some of these trends around is shrinking. Even if we can visualize sustainable solutions to our many problems, how can we possibly find the time to implement them before we do irreparable damage to our environment and our own future?

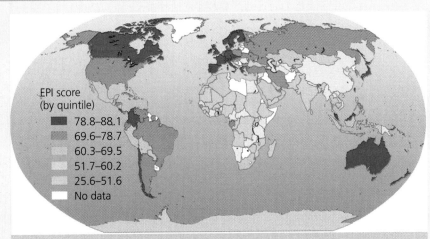

For the Environmental Performance Index, researchers scored nations for their performance in approaching environmental sustainability goals. Higher EPI values (red is highest in this map) indicate better performance. Data from *2008 Environmental Performance Index*, http://epi.yale.edu/CountryScores.

trying to industrialize, those with arid environments and limited natural resources, and those facing extreme poverty.

The correlation between economic vitality and environmental protection was expected, but the researchers were more interested in the variation around this trend. For any given level of income or development, some nations vastly outperformed others. The researchers say this demonstrates that other factors are at work—namely, the political choices national leaders make. Indeed, their data show a strong correlation between EPI scores and "good governance," which involves aspects such as rule of law, open political debate, and lack of corruption.

These interpretations touched sensitive nerves in nations that did not rank as highly as they would have liked. For example, the United States placed 39th—far below most of its economic peers, including Canada, which ranked 12th, with an EPI score of 86.6. Critics say the rankings have little meaning because only certain data sets were measured, and these were then weighted subjectively. Supporters counter that although no one will ever agree on exact numbers, the EPI provides a useful tool for evaluating actions and investments, highlighting policies that work, and identifying future priorities.

Conversely, poorer nations tended to score low on environmental health and higher on indicators that reflected the availability of remaining unexploited natural resources. Overall across all nations, performance was worst for the indicators of renewable energy and wilderness protection.

The EPI scores are only as solid as the data that go into them, and not every nation provides reliably accurate data. Moreover, Esty's group acknowledges that there are ways in which the formulas might be improved in the future. Nonetheless, the Environmental Performance Index provides a unique means of assessing what is working in environmental policy. As the methodology is refined year by year, the EPI promises to become a more accurate performance indicator and a more helpful policy tool.

Whereas controversy generated by the overall rankings drew media attention, other trends were apparent for those who examined the data more deeply. For instance, wealthy industrialized nations tended to score highest in environmental health and lowest in ecosystem vitality indicators, especially biodiversity and habitat. This is presumably because they possess money to invest in protecting the health of their citizens but achieved economic development by exploiting their natural resources and degrading their natural environment.

Today, humanity faces a challenge more important than any previous one—the challenge to achieve sustainability. Attaining sustainability will be a larger and complex process, but it is one to which every single person on Earth can contribute; in which government, industry, and citizens can all cooperate; and toward which all nations can work together. Human ingenuity is capable of it; we merely need to rally public resolve and engage our governments, institutions, and entrepreneurs in the race.

We must pass through the environmental bottleneck

Human ingenuity and compassion give us reason to hope that we may achieve sustainability before doing too much damage to our planet and our prospects, but we must be realistic about the challenges that lie ahead. As we deplete the natural capital we can draw on, we give ourselves, and the rest of the world's creatures, less room to manoeuvre. Until we implement sustainable solutions, we will be squeezing ourselves through a progressively tighter space, like being forced through the neck of a bottle. The key question for the future of our species and our planet is whether we can make it safely through this bottleneck. Biologist Edward O. Wilson has written eloquently of this view:

> At best, an environmental bottleneck is coming in the twenty-first century. It will cause the unfolding of a new kind of history driven by environmental change. Or perhaps an unfolding on a global scale of more of the old kind of history, which saw the collapse of regional civilizations [in] Mesopotamia, and subsequently Egypt, then the Mayan and many others ...

CANADIAN ENVIRONMENTAL PERSPECTIVES

Keleigh Annau

Keleigh Annau, now a student at Mount Allison University, had the idea for Lights Out Canada when she was a high school student in Parksville, B.C.

- **Youth environmental activist**
- **Toyota Earth Day Scholarship** and **Bell Scholarship winner**
- **Founder** of Lights Out Canada
- **Musician** and **dancer**

When you read about Keleigh Annau, three words that keep popping up are "inspiration," "passion," and "tenacity." Annau—now an inspiration to others—was first inspired to act by a conference on climate change that she attended at the age of 16. The conference opened her eyes to the potential impacts of climate change on her generation. She returned home to Parksville, British Columbia, with the idea for Lights Out Canada—a simple way to "turn off the lights and switch on education about global warming."[24]

Instead of thinking about all of the obstacles in her way, Annau shared her idea with several friends and with mentor Dev Aujla from the organization DreamNow. Together they launched a successful pilot event in December 2005. The first full-fledged Lights Out event took place in May 2006, with about 50 000 students participating in schools across Canada. By the third annual event in April 2008, more than 100 000 students in 10 countries participated in what had now become Lights Out World.

Annau and her international network of young environmental activists—many of whom she connected with as a student at United World College of the Adriatic, in Italy—are working to expand Lights Out Canada & Lights Out World[25] in preparation for the 2009 event. Now a student in International Relations at Mount Allison University in New Brunswick, Annau is focusing her attention on the Atlantic Provinces, and creating a Lights Out team at Mount Allison. She is applying for grants, and hopes to expand Lights Out from an annual event into a full-time program. Her plan involves training student representatives to give Lights Out presentations in their schools, acquiring specific commitments from participants, and working with university campuses.

Annau is keenly aware that turning out the lights once a year won't have a significant impact, in itself. However, she recognizes that the impacts on the personal awareness, decisions, and behaviours among youth participants can be much more far-reaching. "The youth of today will be designating policy and managing the industrial, business, and agricultural sectors in the future. It is my hope that by instilling awareness about the importance of living sustainably in the youth that participate in Lights Out Canada, this will plant a seed and encourage them to make green practices a priority when they make life and career choices in the future."

This is where the "passion" and "tenacity" come in. Annau still points to Lights Out as her central passion, and the process of building it as a transformative experience in her life. "My experience with Lights Out has changed my life, not only in how I feel about environmental issues or what my career goals are, but also because I feel empowered." In the meantime, her tenacity has allowed Annau to keep taking the small steps required to move forward, and to build a little idea into a big force for change.

"Small changes add up to make a big difference." —Keleigh Annau

Thinking About Environmental Perspectives

What do *you* think is the most pressing environmental issue facing your generation? Can you think of a way that you might be able to raise the awareness of other students concerning this issue?

Somehow humanity must find a way to squeeze through the bottleneck without destroying the environments on which the rest of life depends.

We must think of Earth as an island

We began this book with the vision of Earth as an island, and indeed that is what it is (**FIGURE 23.19**). Islands can be paradise, as Rapa Nui (Easter Island) likely was when the Polynesians first reached it. But when Europeans arrived at Easter Island, they witnessed the aftermath of a civilization that had depleted its island's resources, degraded its environment, and collapsed as a result. For the few people who remained of the once-mighty culture, life was difficult and unrewarding. They had lost even the knowledge of the history of their ancestors, who had cut trees unsustainably, kicking the base out from beneath their prosperous civilization.

As Easter Island's trees disappeared, some individuals must have spoken out for conservation and for finding ways to live sustainably amid dwindling resources. Likely others ignored those calls and went on extracting more than the land could bear, assuming that somehow things would turn out all right. Indeed, whoever cut the last tree atop the most remote mountaintop could have looked out across the island and seen that it was the last tree. And yet that person cut it down.

in ways that encourage sustainable practices, and by employing science to help us achieve these ends, we may yet be able to live happily and sustainably on our wondrous island, Earth.

Conclusion

In any society facing dwindling resources and environmental degradation, there will be those who raise alarms and those who ignore them. Fortunately, in our global society today we have many thousands of scientists who study Earth's processes and resources. For this reason, we are amassing a detailed knowledge and an ever-developing understanding of our dynamic planet, what it offers us, and what impacts it can bear. The challenge for our global society today, our one-world island of humanity, is to support that science so that we may judge false alarms from real problems and distinguish legitimate concerns from thoughtless denial. This science, this study of Earth and of ourselves, offers us hope for our future.

FIGURE 23.19
We end as we began the book, with a photo of Earth from space. This photograph of Earth and Earth's Moon, taken by the spacecraft *Voyager* as it sped away from home, shows our planet as it truly is—an island in space. Everything we know, need, love, and value comes from and resides on this small sphere, so we had best treat it well.

It would be tragic folly to let such a fate befall our planet as a whole. By recognizing this, by deciding to shift our individual behaviour and our cultural institutions

Never doubt that a small group of thoughtful, committed people can change the world. Indeed, it's the only thing that ever has.—**Margaret Mead**

REVIEWING OBJECTIVES

You should now be able to:

List and describe approaches being taken on college and university campuses to promote sustainability

- Audits produce baseline data on how much a campus consumes and pollutes.
- The most common campus sustainability efforts involve recycling and waste reduction.
- Green buildings are being constructed on a growing number of campuses.
- There are many ways to reduce water use, and these efforts often save money.
- Students have many feasible ways to conserve energy and promote renewable energy sources.
- A current drive is to make campuses carbon neutral and responsive to concerns about climate change.
- Dining services can help sustainability efforts by providing local food and reducing waste.
- Colleges and universities can favour sustainable products in institutional purchasing.
- Campuses can use alternative fuels and vehicles and encourage bicycling, walking, and public transportation.
- Habitat restoration is one of the most popular campus sustainability activities.
- Curricula are adding issues of sustainability and enabling students to earn course credits while working on campus and community sustainability projects.

Explain the concept of sustainable development

- Sustainable development entails environmental protection, economic development, and social justice.
- Proponents of sustainable development feel that economic development and environmental quality can enhance one another.

Discuss how protecting the environment can be compatible with promoting economic welfare

- Environmental protection and green technologies and industries can create rich sources of new jobs.
- Protecting environmental quality enhances a community's desirability and economy.

Describe and assess key approaches to designing sustainable solutions

- Approaches that can inspire sustainable solutions include refining our ideas about quality of life; reducing unnecessary consumption; limiting population growth; encouraging green technologies; mimicking natural systems; thinking long-term; enhancing local self-sufficiency; being politically active; voting with our wallets; and promoting research and education.
- Growth in population and per capita consumption will likely need to be halted if we are to create a sustainable society.
- Technology has traditionally increased environmental impact, but new "green" technologies can help reduce impact.

Discuss the need for action on behalf of the environment and the tremendous human potential to solve problems

- Time for turning around our increasing environmental impacts is running short.
- Canada and other nations have met tremendous challenges before, so we have reason to hope that we will be able to attain a sustainable society.

TESTING YOUR COMPREHENSION

1. In what ways are campus sustainability efforts relevant to sustainability efforts in the broader society?
2. Name one way in which campus sustainability proponents have addressed each of the following areas:
   a. Recycling and waste reduction
   b. "Green" building
   c. Water conservation
   d. Energy efficiency
   e. Renewable energy
   f. Global climate change
3. Name one way in which campus sustainability proponents have addressed each of the following areas:
   a. Dining services
   b. Institutional purchasing
   c. Transportation
   d. Habitat restoration
   e. Curricula
4. What do environmental scientists mean by *sustainable development*?
5. Describe three ways in which environmental protection can enhance economic well-being.
6. Why are many people now living at the highest level of material prosperity in history? Is this level of consumption sustainable? How can it feel good to consume less?
7. In what ways can technology help us achieve sustainability? How do natural processes provide good models of sustainability for manufacturing? Provide examples.
8. Why do many people feel that local self-sufficiency is important? What consequences of globalization may threaten sustainability? How can open democratic societies help to promote sustainability?
9. Explain Edward O. Wilson's metaphor of the "environmental bottleneck."
10. How can thinking of Earth as an island help prevent us from repeating the mistakes of previous civilizations?

SEEKING SOLUTIONS

1. What sustainability initiatives would you like to see attempted on your campus? If you were to take the lead in promoting such initiatives, how would you go about it? What obstacles would you expect to face, and how would you deal with them?
2. Choose one item or product that you enjoy, and consider how it came to be. Think of as many components of the item or product as you can, and determine how each of them was obtained or created. Now refer to **FIGURE 23.12**. What steps were involved in creating your item's components, and where did the raw materials come from? How was your item manufactured? How was it delivered to you?
3. Do you think that we can increase our quality of life through development while also protecting the

integrity of the environment? Discuss examples from your course or from other chapters of this book that illustrate possible win–win solutions. Are you familiar with any cases in your community or at your college that bear on this issue? Describe such a case, and state what lessons you would draw from it.

4. Reflect on the experiences of prior human civilizations and how they came to an end. What is your prognosis for our current human civilization? Do you see a vast world of independent cultures all individually responsible for themselves, or do you consider human civilization to be one great entity? If we accept that all people depend on the same environmental systems for sustenance, what resources and strategies do we have to ensure that the actions of a few do not determine the outcome for all and that sustainable solutions are a common global goal?

5. **THINK IT THROUGH** You have been elected president of your student government, and your school's administrators promise to be responsive to student concerns. Many of your fellow students are asking you to promote sustainability initiatives on your campus. Consider the many approaches and activities pursued by the colleges and universities mentioned in this chapter, and now think about your own school. Which of these approaches and activities are most needed at your school? Which might be most effective? What ideas would you prioritize and promote during your term as president?

6. **THINK IT THROUGH** In our final "Think It Through" question, you are . . . *you!* In this chapter and throughout this book, you have encountered a diversity of ideas for sustainable solutions to environmental problems. Many of these are approaches you can pursue in your own life. Name at least five ways in which you think you can make a difference—and would most like to make a difference—in helping to attain a more sustainable society. For each approach, describe one specific thing you could do today or tomorrow or next week to begin.

INTERPRETING GRAPHS AND DATA

An undergraduate class at Pennsylvania State University conducted an ecological assessment of one of the biology laboratory buildings in response to the question: "How is this building like an ecosystem?" The result of the assessment was a 52-page report outlining ways to reduce the ecological footprint of the Mueller Laboratory Building in the areas of energy use, water use, communications/computing, furnishings/renovation, maintenance, and food. The students found ways to save an estimated $45 500 per year in the cost of energy alone for a building occupied by 123 scientists and support staff. Their data on the current use and potential savings in the energy component of the ecological footprint are shown in the graph.

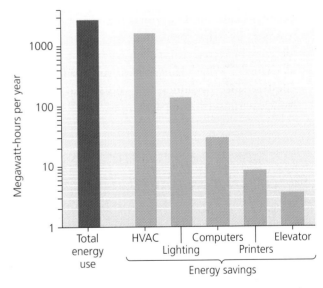

Total annual electrical energy use (red bar) and potential energy savings in five areas (orange bars) at Penn State University's Mueller Laboratory Building. The y-axis is logarithmic, with each tick mark representing an increase equal to the value indicated below it on the axis. Data from Penn State Green Destiny Council. 2001. *The Mueller Report: Moving Beyond Sustainability Indicators to Sustainability Action at Penn State.*

1. From the graph, estimate the amount of potential savings for each of the five areas identified in the Mueller Report. Approximately what percentage of the total electrical energy use of the building do these savings together represent?

2. What was the approximate cost of electricity in cents per kilowatt-hour used to calculate the savings of $45 500? Do you think this cost may have changed since the report was issued in 2001? How and why?

3. How is the building where you take your course like an ecosystem? How is it not like an ecosystem? Is your building operating sustainably? What improvements do you think could be made to it?

CALCULATING FOOTPRINTS

As we have seen throughout this book, individuals can contribute to sustainable solutions for our society and our planet in many ways. Some of these involve advocating for change at high levels of government or business or academia. But plenty of others involve the countless small choices we make in how we live our lives day to day. Where we live, what we buy, how we travel—these types of choices we each make as citizens, consumers, and human beings determine how we affect the environment and the people around us. As you know, such personal choices are summarized (crudely, but usefully) in an ecological footprint.

Turn back to the "Calculating Footprints" exercise in Chapter 1, and recover the numerical value of your own personal ecological footprint that you calculated at the beginning of your course. Enter it in the table to the right. Now return to the same online ecological footprint calculator that you used for Chapter 1's exercise. For many of you, this will have been www.myfootprint.org or http://ecofoot.org. Take the footprint quiz again, and calculate your current footprint.

	Footprint value (hectares per person)
World average	2.23
Canadian average	9.6
Your footprint from Chapter 1	
Your footprint now	
Your footprint with three more changes	

1. Enter your current footprint, as determined by the online calculator, in the table. How does this value compare with your footprint at the beginning of your course? By what percentage did your footprint decrease or increase? If it changed, why do you think it changed? What changes have you made in your lifestyle since beginning this course that influence your environmental impact?
2. How does your personal footprint compare with the average footprint of a Canadian resident? How does it compare with that of the average person in the world? What do you think would be an admirable yet realistic goal for you to set as a target value for your own footprint?
3. Now think of three changes in your lifestyle that would decrease your footprint. These should be changes that you would like to make and that you believe you could reasonably make. Take the footprint quiz again, incorporating these three changes. Enter the resulting footprint in the table.
4. Now set a goal of reducing your footprint by 25%, and experiment by changing various answers in your footprint quiz. What changes would allow you to attain a 25% reduction in your footprint? What changes would be needed to reduce your footprint to the hypothetical target value you set in Question 2?

TAKE IT FURTHER

Go to www.myenvironmentplace.ca where you will find

- Suggested answers to end-of-chapter questions
- Quizzes, animations, and flashcards to help you study
- *Research Navigator*™ database of credible and reliable sources to assist you with your research projects
- Tutorials to help you master how to interpret graphs
- Current news articles that link the topics that you study to case studies from your region and around the world
- **ECO Occupational Profiles:** If you found this chapter especially interesting, you might want to learn more about the following jobs by visiting the Occupational Profiles website of the Environmental Careers Organization. Go to www.eco.ca and check out the following careers:
    - Environmental auditor
    - Ecotourism operator
    - Environmental training specialist
    - Industrial designer
    - Sustainable architect
    - Sustainable interior designer

CHAPTER ENDNOTES

1. University of British Columbia Sustainability Office, www.sustain.ubc.ca.
2. *UBC Sustainability Report 2006–2007,* www.sustain.ubc.ca/pdfs/ar/UBC-Sustainability_Report_2006-2007-final.pdf
3. University Leaders for a Sustainable Future, *Programs: Talloires Declaration,* www.ulsf.org/programs_talloires.html.
4. University of British Columbia Mission Statement, 2005, *Trek 2010.*
5. AMS Lighter Footprint Strategy, 2008, www.amsubc.ca/uploads/government/AMS_Lighter_Footprint_Strategy.pdf.
6. UTM Ecological Footprint and Campus Sustainability Assessment Project, http://eratos.erin.utoronto.ca/conway/ecofootprint/.
7. UBC Campus Sustainability Office, ECOTREK Project Complete, *Sustainable Energy Management Program,* www.ecotrek.ubc.ca.
8. *UBC Sustainability Report 2006–2007,* www.sustain.ubc.ca/pdfs/ar/UBC-Sustainability_Report_2006-2007-final.pdf.
9. University of Victoria, *Sustainability in Planning, Buildings, and Land Use,* http://web.uvic.ca/sustainability/buildings.php.
10. Simon Fraser University, www.sfu.ca/sfunews/Stories/sfunews071008019.shtml.
11. Laurentian University, Cooperative Freshwater Ecology Unit, *Living with the Lakes Centre: The Facility,* www.laurentian.ca/Laurentian/Home/Departments/Cooperative+Freshwater+Ecology+Unit/Living+with+Lakes+Centre/The+Facility.htm.
12. University of Toronto Sustainability Office, *Projects: Buildings and Infrastructure,* www.sustainability.utoronto.ca:81/projects/projects/buildings-infrastructure.
13. University of Toronto Sustainability Office, *ReWire Project,* http://rewire.utoronto.ca.
14. Trent University, *Physical Resources, Powerhouse,* www.trentu.ca/physicalresources/powerhouse.php.
15. Common Energy, www.commonenergy.org.
16. B.C. Campus Climate Network, *Go Beyond Project,* www.campusclimatenetwork.org/wiki/Go_Beyond_Project.
17. Sustainable Concordia, *allégo: Smarter transport for a more sustainable future,* http://sustainable.concordia.ca/ourinitiatives/allego/.
18. Queen's University Faculty of Applied Science, http://appsci.queensu.ca/news/2005-2006/index.php?article=31&view=print.
19. First Nations University of Canada, *Academic Programs, Department of Science,* www.firstnationsuniversity.ca/default.aspx?page=30
20. University of Toronto Sustainability Office, *Education and Awareness,* www.sustainability.utoronto.ca:81/course-research/education-awareness.
21. *UBC Sustainability Report 2006–2007,* www.sustain.ubc.ca/pdfs/ar/UBC-Sustainability_Report_2006-2007-final.pdf.
22. Data from (principal authors) Esty, Daniel C., Christine Kim, and Tanja Srebotnjak, Marc A. Levy, Alex de Sherginin, and Valentina Mara (2008) *2008 Environmental Performance Index,* Yale Center for Environmental Law and Policy and the Center for International Earth Science Information Network of Columbia University in collaboration with the World Economic Forum and the Joint Research Centre of the European Commission, http://sedac.ciesin.columbia.edu/es/epi/papers/2008EPI_rankingsandscores_23Jan08.pdf.
23. Data from (principal authors) Esty, Daniel C., Christine Kim, and Tanja Srebotnjak, Marc A. Levy, Alex de Sherginin, and Valentina Mara (2008) *2008 Environmental Performance Index,* Yale Center for Environmental Law and Policy and the Center for International Earth Science Information Network of Columbia University, in collaboration with the World Economic Forum and the Joint Research Centre of the European Commission, http://sedac.ciesin.columbia.edu/es/epi/papers/2008EPI_rankingsandscores_23Jan08.pdf.
24. Lights Out Canada, *Stuff for You,* www.lightsoutcanada.tpweb.ca/LightsOutCanada-ConceptPage-NEW.pdf.
25. Lights Out Canada & Lights Out World, www.lightsoutcanada.org.

Appendices

Appendix A Some Basics on Graphs

Presenting data in ways that help make trends and patterns visually apparent is a vital part of the scientific endeavour. For scientists, businesspeople, and others, the primary tool for expressing patterns in data is the graph. Thus, the ability to interpret graphs is a skill that you will want to cultivate. This appendix guides you in how to read graphs, introduces a few vital conceptual points, and surveys the most common types of graphs, giving rationales for their use.

Navigating a Graph

A graph is a diagram that shows relationships among *variables,* which are factors that can change in value. The most common types of graphs relate values of a *dependent variable* to those of an *independent variable.* As explained in Chapter 1, a dependent variable is so named because its values "depend on" the values of an independent variable. In other words, as the values of an independent variable change, the values of the dependent variable change in response. In a manipulative experiment, changes that a researcher specifies in the value of the independent variable *cause* changes in the value of the dependent variable. In observational studies, there may be no causal relationship, and scientists may plot a correlation. In either case, the values of the independent variable are known or specified, and the values of the dependent variable are unknown and are what we are interested in observing or measuring.

By convention, independent variables are generally represented on the horizontal axis, or *x-axis,* of a graph, while dependent variables are represented on the vertical axis, or *y-axis.* Numerical values of variables generally become larger as one proceeds rightward on the *x*-axis or upward on the *y*-axis. In many cases, independent variables are not numerical at all, but categorical. For example, in a graph presenting population sizes of several nations, the nations comprise a categorical independent variable, whereas population size is a numerical dependent variable.

As a simple example, **FIGURE A.1** shows data from a scientist who ran an experiment to test the effects on laboratory mice of the chemical bisphenol-A. The *x*-axis shows values of her independent variable, the dose of bisphenol-A given to the mice. The values are expressed in units of nanograms per gram of water (ng/g). The researcher was interested in what proportion of mice developed chromosomal problems as a result.

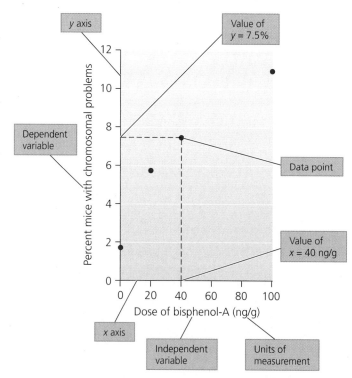

FIGURE A.1 Frequency of chromosomal problems relative to dose of bisphenol-A.

Thus, her dependent variable, presented on the *y*-axis, is the percentage of mice found to develop chromosomal problems. For each of four doses of bisphenol-A she supplied in her experiment, a data point on the graph is plotted to show the corresponding percentage of mice with chromosomal problems.

Now that you're familiar with the basic building blocks of a graph, let's survey the most common types of graphs you'll see, and examine a few vital concepts in graphing.

GRAPH TYPE: Line Graph

A line graph is drawn when a data set involves a sequence of some kind, such as a series of values that occur one by one and change through time or across distance (**FIGURE A.2**). Line graphs are most appropriate when the *y*-axis expresses a continuous numerical variable, and the *x*-axis expresses either continuous numerical data or discrete sequential categories (such as years).

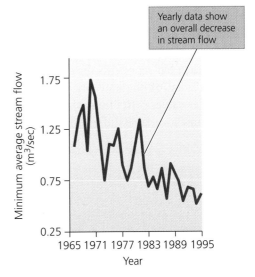

FIGURE A.2 Minimum stream flow.

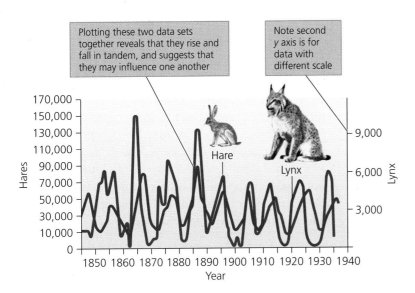

FIGURE A.3 Population fluctuations in hare and lynx.

One useful technique is to plot two or more data sets together on the same graph (**FIGURE A.3**). This allows us to compare trends in the data sets to see whether and how they may be related.

KEY CONCEPT: Projections

Besides showing observed data, we can use graphs to show data that is predicted for the future, based on models, simulations, or extrapolations from past data. Often, projected future data on a line graph are shown with dashed lines, as in **FIGURE A.4**, to indicate that they are less certain than data that has already been observed.

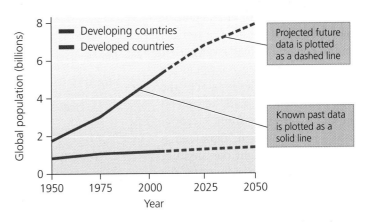

FIGURE A.4 Past and projected population growth for developing and developed countries.

GRAPH TYPE: Bar Chart

A bar chart is most often used when one variable is categorical and the other is numerical. In such a chart, the height (or length) of each bar represents the quantitative value of a given category of the categorical variable; longer bars mean larger values (**FIGURE A.5**). Bar charts allow us to visualize how a variable differs quantitatively among categories.

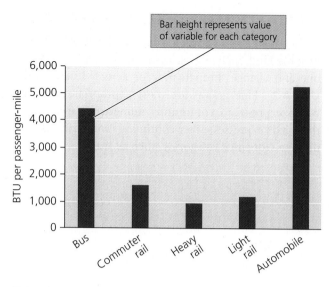

FIGURE A.5 Energy consumption for different modes of transit.

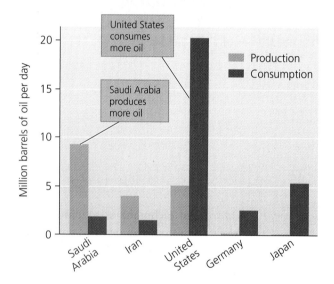

FIGURE A.6 Oil production and consumption by selected nations.

It is often instructive to graph two or more data sets together to reveal patterns and relationships. A bar chart, such as **FIGURE A.6**, allows us to compare two data sets (oil production and oil consumption) both within and among nations. A graph that does double duty in this way allows for higher-level analysis (in this case, suggesting which nations depend on others for petroleum imports). Most bar charts in this book illustrate multiple types of information at once in this manner.

FIGURE A.7 illustrates two ways in which a bar chart can be modified. First, note that the orientation of bars is horizontal instead of vertical. In this configuration, the categorical variable is along the *y*-axis and the numerical variable is along the *x*-axis. Second, the bars can extend in either direction from a central *x*-axis value of zero, representing either positive (right) or negative (left) values. Depending on the nature of one's data and the points one wants to make, sometimes such arrangements can make for a clearer presentation.

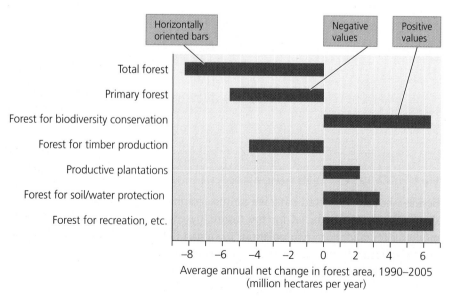

FIGURE A.7 Loss or gain of forests, by type.

One special type of horizontally oriented bar chart is the age pyramid used by demographers (**FIGURE A.8**). Age categories are displayed on the *y*-axis, with bars representing the population size of each age group varying in their horizontal length.

KEY CONCEPT: Statistical Uncertainty

Most data sets involve some degree of uncertainty. Sometimes exact measurements are impossible, so the researcher estimates the likely range of measurement error around the data point. Other times, data points represent the *mean* (average) of many measurements, and the researcher may want to show the degree to which the data vary around this mean. Mathematical techniques are used to obtain precise statistical probabilities for degrees of variation. Results from such analyses are represented in

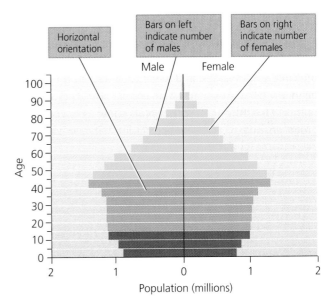

FIGURE A.8 Age structure of Canada, 2005.

bar charts, such as **FIGURE A.9**, as thin black lines called *error bars*. These bars extend past and within the end of each bar, representing the degree of variation around the bar's value. Longer error bars indicate more uncertainty or variation, whereas short error bars mean we can have high confidence in the value.

The statistical analysis of data is critically important in science. In this book we provide a broad and streamlined introduction to many topics, so we often omit error bars from our graphs and details of statistical significance from our discussions. This is for clarity of presentation only; the research we discuss analyzes its data in far more depth than any textbook could possibly cover.

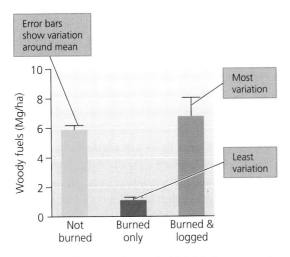

FIGURE A.9 Fine-scale woody debris left after treatments in salvage logging study.

KEY CONCEPT: Logarithmic Scales

When data span a large range of values, it can help to change the scale on a graph from the standard linear scale to a logarithmic scale. In a logarithmic scale, each equal unit of distance on an axis corresponds to a ratio of values rather than an additive increase in values. Most often, logarithmic scales advance by factors of 10. **FIGURE A.10** uses a logarithmic scale on its *y*-axis for a graph in which we also show a linear scale using white horizontal lines crossing the graph space. The choice of scale does not affect data values, but does drastically affect the appearance of lines or bars on a graph. For further discussion, see the "Interpreting Graphs and Data" feature for Chapter 1.

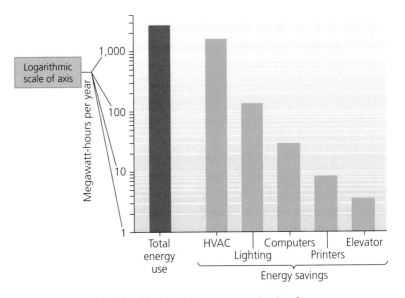

FIGURE A.10 Annual energy use and savings from a campus sustainability study.

GRAPH TYPE: Scatter Plot

A scatter plot is often used when there is no sequential aspect to the data, when a given *x*-axis value could have multiple *y*-axis values, and when each data point is independent, having no particular connection to other data points (**FIGURE A.11**). Scatter plots allow us to visualize a broad positive or negative correlation between variables.

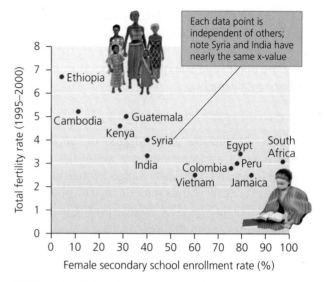

FIGURE A.11 Fertility rate and female education.

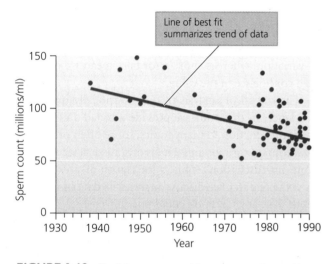

FIGURE A.12 Declining sperm count in men across the world.

A "line of best fit" may be drawn through the data of a scatter plot in order to make a trend in the data clearer to the eye (**FIGURE A.12**). These lines are not drawn casually, however; their placement and slope are determined by precise mathematical analysis of the data through a statistical technique called linear regression.

GRAPH TYPE: Pie Chart

A pie chart is used when we wish to compare the proportions of some whole that are taken up by each of several categories (**FIGURE A.13**). A pie chart is appropriate when one variable is categorical and one is numerical. Each category is represented visually like a slice from a pie, with the size of the slice reflecting the percentage of the whole that is taken up by that category.

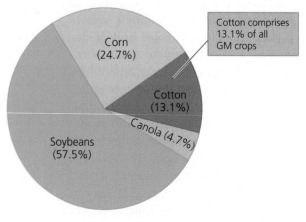

FIGURE A.13 Genetically modified crops grown worldwide, by type.

Appendix B Units and Conversions

Measurement	Unit and abbreviation	Metric equivalent	Metric to english conversion factor	English to metric conversion factor
Length	1 kilometre (km)	= 1000 (10^3) metres	1 km = 0.62 miles	1 mile = 1.61 km
	1 metre (m)	= 100 (10^2) centimetres	1 m = 1.09 yards	1 yard = 0.914 m
		= 1000 millimetres	1 m = 3.28 feet	1 foot = 0.305 m
			1 m = 39.37 inches	
	1 centimetre (cm)	= 0.01 (10^{-2}) metres	1 cm = 0.394 inches	1 foot = 30.5 cm
				1 inch = 2.54 cm
	1 millimetre (mm)	= 0.001 (10^{-3}) metres	1 mm = 0.039 inches	
Area	1 square metre (m^2)	= 10 000 square centimetres	1 m^2 = 1.1960 square yards	1 square yard = 0.8361 m^2
			1 m^2 = 10.764 square feet	1 square foot = 0.0929 m^2
	1 square centimetre (cm^2)	= 100 square millimetres	1 cm^2 = 0.155 square inches	1 square inch = 6.4516 cm^2
	1 hectare (ha)	= 0.01 square kilometres	1 ha = 2.471 acres	1 acre = 0.40469 hectares
Mass	1 metric ton (t)	= 1000 kilograms	1 t = 1.103 ton	1 ton = 0.907 t
	1 kilogram (kg)	= 1000 grams	1 kg = 2.205 pounds	1 pound = 0.4536 kg
	1 gram (g)	= 1000 milligrams	1 g = 0.0353 ounces	1 ounce = 28.35 g
	1 milligram (mg)	= 0.001 gram		
Volume (solids)	1 cubic metre (m^3)	= 1 000 000 cubic centimetres	1 m^3 = 1.3080 cubic yards	1 cubic yard = 0.7646 m^3
			1 m^3 = 35.315 cubic feet	1 cubic foot = 0.0283 m^3
	1 cubic centimetre (cm^3 or cc)	= 0.000 001 cubic metres = 1 millilitre	1 cm^3 = 0.0610 cubic inches	1 cubic inch = 16.387 cm^3
	1 cubic millimetre (mm^3)	= 0.000 000 001 cubic metres		
Volume (liquids and gases)	1 kilolitre (kL)	= 1000 litres	1 kL = 264.17 gallons	1 gallon = 3.785 L
	1 litre (L)	= 1000 millilitres	1 L = 0.264 gallons	1 quart = 0.946 L
			1 L = 1.057 quarts	
	1 millilitre (mL)	= 0.001 litre	1 mL = 0.034 fluid ounce	1 quart = 946 mL
		= 1 cubic centimetre	1 mL = approximately 0.25 teaspoons	1 pint = 473 mL
				1 fluid ounce = 29.57 mL
				1 teaspoon = approx. 5 mL
Time	1 millisecond (ms)	= 0.001 seconds		
Temperature	Degrees Celsius (°C)		°C = $\frac{5}{9}$(°F − 32)	°F = $\frac{9}{5}$°C + 32
Energy and power	1 kilowatt-hour	= 34 113 BTU = 860 421 calories		
	1 watt	= 3.413 BTU/h		
		= 14.34 calories/min		
	1 calorie	= the amount of heat necessary to raise the temperature of 1 g (1 cm^3) of water 1°C		
	1 horsepower	= 7.457 × 10^2 watts		
	1 joule	= 9.481 × 10^{-4} BTU		
		= 0.239 cal		
		= 2.778 × 10^{-7} kilowatt-hours		
Pressure	1 pound per square inch (psi)	= 6894.757 pascals (Pa)		
		= 0.068045961 atmospheres (atm)		
		= 51.71493 millimetres of mercury (mm hg = Torr)		
		= 68.94757 millibars (mbar)		
		= 6.894757 kilopascals (kPa)		
	1 atmosphere (atm)	= 101.325 kilopascals (kPa)		

Appendix C Periodic Table of the Elements

	Representative (main group) elements																	
	IA	IIA											IIIA	IVA	VA	VIA	VIIA	VIIIA
1	1 H 1.0079 Hydrogen																	2 He 4.003 Helium
2	3 Li 6.941 Lithium	4 Be 9.012 Beryllium											5 B 10.811 Boron	6 C 12.011 Carbon	7 N 14.007 Nitrogen	8 O 15.999 Oxygen	9 F 18.998 Fluorine	10 Ne 20.180 Neon
3	11 Na 22.990 Sodium	12 Mg 24.305 Magnesium	IIIB	IVB	VB	VIB	VIIB		VIIIB		IB	IIB	13 Al 26.982 Aluminum	14 Si 28.086 Silicon	15 P 30.974 Phosphorus	16 S 32.066 Sulfur	17 Cl 35.453 Chlorine	18 Ar 39.948 Argon
4	19 K 39.098 Potassium	20 Ca 40.078 Calcium	21 Sc 44.956 Scandium	22 Ti 47.88 Titanium	23 V 50.942 Vanadium	24 Cr 51.996 Chromium	25 Mn 54.938 Manganese	26 Fe 55.845 Iron	27 Co 58.933 Cobalt	28 Ni 58.69 Nickel	29 Cu 63.546 Copper	30 Zn 65.39 Zinc	31 Ga 69.723 Gallium	32 Ge 72.61 Germanium	33 As 74.922 Arsenic	34 Se 78.96 Selenium	35 Br 79.904 Bromine	36 Kr 83.8 Krypton
5	37 Rb 85.468 Rubidium	38 Sr 87.62 Strontium	39 Y 88.906 Yttrium	40 Zr 91.224 Zirconium	41 Nb 92.906 Niobium	42 Mo 95.94 Molybdenum	43 Tc 98 Technetium	44 Ru 101.07 Ruthenium	45 Rh 102.906 Rhodium	46 Pd 106.42 Palladium	47 Ag 107.868 Silver	48 Cd 112.411 Cadmium	49 In 114.82 Indium	50 Sn 118.71 Tin	51 Sb 121.76 Antimony	52 Te 127.60 Tellurium	53 I 126.905 Iodine	54 Xe 131.29 Xenon
6	55 Cs 132.905 Cesium	56 Ba 137.327 Barium	57 La 138.906 Lanthanum	72 Hf 178.49 Hafnium	73 Ta 180.948 Tantalum	74 W 183.84 Tungsten	75 Re 186.207 Rhenium	76 Os 190.23 Osmium	77 Ir 192.22 Iridium	78 Pt 195.08 Platinum	79 Au 196.967 Gold	80 Hg 200.59 Mercury	81 Tl 204.383 Thallium	82 Pb 207.2 Lead	83 Bi 208.980 Bismuth	84 Po 209 Polonium	85 At 210 Astatine	86 Rn 222 Radon
7	87 Fr 223 Francium	88 Ra 226.025 Radium	89 Ac 227.028 Actinium	104 Rf 261 Unnilquadium	105 Db 262 Unnilpentium	106 Sg 263 Unnilhexium	107 Bh 262 Unnilseptium	108 Hs 265 Unniloctium	109 Mt 266 Unnilmennium	110 Uun 269 Ununnilium	111 Uuu 272 Unununium	112 Uub 277 Ununbium		114		116		

— Transition metals —

Rare earth elements

Lanthanides

58 Ce 140.115 Cerium	59 Pr 140.908 Praseodymium	60 Nd 144.24 Neodymium	61 Pm 145 Promethium	62 Sm 150.36 Samarium	63 Eu 151.964 Europium	64 Gd 157.25 Gadolinium	65 Tb 158.925 Terbium	66 Dy 162.5 Dysprosium	67 Ho 164.93 Holmium	68 Er 167.26 Erbium	69 Tm 168.934 Thulium	70 Yb 173.04 Ytterbium	71 Lu 174.967 Lutetium

Actinides

| 90
Th
232.038
Thorium | 91
Pa
231.036
Protactinium | 92
U
238.029
Uranium | 93
Np
237.048
Neptunium | 94
Pu
244
Plutonium | 95
Am
243
Americium | 96
Cm
247
Curium | 97
Bk
247
Berkelium | 98
Cf
251
Californium | 99
Es
252
Einsteinium | 100
Fm
257
Fermium | 101
Md
258
Mendelevium | 102
No
259
Nobelium | 103
Lr
262
Lawrencium |
|---|---|---|---|---|---|---|---|---|---|---|---|---|---|

The periodic table arranges elements according to atomic number and atomic weight into horizontal rows called periods and vertical columns called groups.

Elements of each group in Class A have similar chemical and physical properties. This reflects the fact that members of a particular group have the same number of valence shell electrons, which is indicated by the group's number. For example, group IA elements have one valence shell electron, group IIA elements have two, and group VA elements have five. In contrast, as you progress across a period from left to right, properties of the elements change, varying from the very metallic properties of groups IA and IIA to the nonmetallic properties of group VIIA to the inert elements (noble gases) in group VIIIA. This reflects changes in the number of valence shell electrons.

Class B elements, or transition elements, are metals, and generally have one or two valence shell electrons. In these elements, some electrons occupy more distant electron shells before the deeper shells are filled.

In this periodic table, elements with symbols printed in black exist as solids under standard conditions (25°C and 1 atm of pressure), while elements in red exist as gases, and those in dark blue as liquids. Elements with symbols in green do not exist in nature and must be created by some type of nuclear reaction.

Glossary

abiotic Refers to any nonliving component of the *environment*. Compare *biotic*.

Aborigine (Aboriginal) Used in reference to native people who have a long history of cultural connection with and occupation of a particular geographic area. The First Nations, Inuit, and Métis are the Aboriginal groups (sometimes called *founding* or *native peoples*) of the land that now comprises Canada. Compare *indigenous*.

abyssal plain The flat, mostly topographically featureless ocean floor.

acid drainage A process in which sulphide *minerals* in newly exposed *rock* surfaces react with oxygen and rainwater to produce sulphuric acid, which causes chemical *runoff* as it *leaches* metals from the rocks. Although acid drainage can be a natural phenomenon, mining can greatly accelerate its rate by exposing many new rock surfaces at once.

acidic The property of a solution in which the concentration of hydrogen (H^1) *ions* is greater than the concentration of hydroxide (OH^2) ions. Compare *basic*.

acidic deposition The deposition of *acidic* or acid-forming pollutants from the *atmosphere* onto Earth's surface by *precipitation*, by fog, by gases, or by the deposition of dry particles.

acidic precipitation See *acidic deposition*.

acid rain *Acidic precipitation* deposited in the form of rain.

act A law or statute, proposed and voted upon by the Parliament of Canada or the Provincial and Territorial Parliaments. Compare *regulation, agreement*.

active solar (energy collection) An approach in which technological devices are used to focus, move, or store solar energy. Compare *passive solar energy collection*.

acute exposure Exposure to a *toxicant* occurring in high amounts for short periods of time. Compare *chronic exposure*.

adaptation A responsive strategy in which it is accepted that an event with potential impacts (such as climatic change) is going to occur, and plans are made to adjust to the impacts. Compare *mitigation*.

adaptive management The systematic, scientific testing of different management approaches to improve methods over time.

adaptive radiation A burst of *species* diversification that occurs in response to *natural selection* induced by environmental change.

adaptive trait (adaptation) A trait that confers greater likelihood that an individual will reproduce.

aerobic Occurring in an *environment* where oxygen is present. For example, the decay of a rotting log proceeds by aerobic decomposition. Compare *anaerobic*.

aerosols Very fine liquid droplets or solid particles aloft in the atmosphere.

"affluenza" Term coined by social critics to describe the failure of material goods to bring happiness to people who have the financial means to afford them.

afforestation Planting of trees where the land has not been forested for a long time.

age distribution The relative numbers of organisms of each age within a *population*. Age distributions can have a strong effect on rates of population growth or decline and are often expressed as a ratio of age classes, consisting of organisms (1) not yet mature enough to reproduce, (2) capable of reproduction, and (3) beyond their reproductive years. The age distribution of a population can be displayed on a type of diagram called an *age structure diagram* or *age pyramid*, which shows the breakdown of a population in terms of both gender and age cohorts.

age structure See *age distribution*.

age structure diagram (age pyramid) See *age distribution*.

agreement An instrument for environmental regulatory purposes, that can be either enforceable or voluntary; they are entered into by agencies of government, often with the goal of streamlining, clarifying, or harmonizing the administration of environmental legislation. Compare *act, regulation*.

Agricultural Revolution The shift around 10,000 years ago from a hunter-gatherer lifestyle to an agricultural way of life in which people began to grow their own crops and raise *domesticated* animals; this much more intensive, manipulative way of producing and extracting resources marked a permanent change in the relationship of people to the natural *environment*. Also called *Neolithic Revolution*. Compare *Industrial Revolution; Medical-Technological Revolution*.

agriculture The practice of cultivating *soil*, producing crops, and raising livestock for human use and consumption.

agroforestry Planting of trees in conjunction with crops. The trees benefit the crops by aiding in the biogeochemical cycling of nutrients and water, and by contributing organic material to the soil, and by providing shade. The trees can also provide harvestable products, such as fruits and nuts.

A horizon A layer of *soil* found in a typical *soil profile*. It forms the top layer or lies below the *O horizon* (if one exists). It consists of mostly inorganic mineral components such as *weathered* substrate, with some organic matter and humus from above mixed in. The A horizon is often referred to as *topsoil*. Compare *B horizon; C horizon; R horizon*.

air pollutants Gases and particulate material added to the atmosphere that can affect *climate* or harm people or other organisms.

air pollution The act of polluting the air, or the condition of being polluted by *air pollutants*.

Air Quality Health Index (AQHI) A standardized, national-level index that quantifies the level of *air pollution* (from $1 =$ very low to $10+ =$ very high) and associated health risks.

airshed The geographical area associated with a particular air mass.

albedo The reflectivity of a surface.

allergen A *toxicant* that overactivates the immune system, causing an immune response when one is not necessary.

alley cropping A type of *agroforestry* in which fields planted in rows of mixed crops are surrounded by or interspersed with rows of trees that provide fruit, wood, or protection from wind. Compare *intercropping; shelterbelt*.

allopatric speciation *Species* formation due to the physical separation of *populations* over some geographic distance.

alpine tundra *Tundra* that occurs at high altitudes.

ambient air pollution Outdoor air pollution.

amensalism A relationship between members of different *species* in which one organism is harmed and the other is unaffected. Compare *commensalism*.

amino acids Organic *molecules* that join in long chains to form *proteins*.

anaerobic Occurring in an *environment* that has little or no oxygen. The conversion of *organic matter* to *fossil fuels* (crude oil, coal, natural gas) at the bottom of a deep lake, swamp, or shallow sea is an example of anaerobic decomposition. Compare *aerobic*.

anthropocentrism A human-centred view of our relationship with the *environment*. Compare *biocentrism; ecocentrism*.

anthropogenic Human-generated, manufactured.

aquaculture The raising of aquatic organisms for food in controlled *environments*.

aquifer An underground water reservoir.

area effect In *island biogeography theory*, the pattern that large islands host more *species* than smaller islands, because larger islands provide larger targets for *immigration* and because *extinction* rates are reduced.

artesian aquifer See *confined aquifer*.

artificial selection *Natural selection* conducted under human direction. Examples include the *selective breeding* of crop plants, pets, and livestock.

artifical wetland See *constructed wetland*.

asbestos Any of several types of *mineral* that form long, thin microscopic fibres—a structure that gives asbestos the capacity to insulate for heat, muffle sound, and resist fire. When inhaled and lodged in lung tissue, asbestos scars the tissue and may eventually lead to lung cancer or *asbestosis*.

atmosphere The thin layer of gases surrounding planet Earth. Compare *biosphere; hydrosphere; lithosphere*.

atmospheric deposition The wet or dry deposition on land of a wide variety of pollutants, including mercury, nitrates, organochlorides, and others. *Acidic deposition* is one type of atmospheric deposition.

atmospheric pressure The weight per unit area produced by a column of air.

atoll A ring-shaped island (generally of *coral reef*) surrounding an older submerged island area.

atom The smallest component of an *element* that maintains the chemical properties of that element.

atomic number The number of *protons* in a given *atom*.

autotroph (primary producer) An organism that can use the energy from sunlight to produce its own food. Includes green plants, algae, and cyanobacteria.

***Bacillus thuringiensis* (Bt)** A naturally occurring *soil* bacterium that produces a protein that kills many *pests*, including caterpillars and the larvae of some flies and beetles.

background rate (of extinction) The average rate of *extinction* that occurred before the appearance of humans. For example, the *fossil record* indicates that for both birds and mammals, one *species* in the world typically became extinct every 500–1,000 years. Compare *mass extinction event*.

barrier island A long thin island that parallels a shoreline. Generally of *sand* or *coral reef*, barrier islands protect coasts from storms.

basic The property of a *solution* in which the concentration of hydroxide (OH^2) *ions* is greater than the concentration of hydrogen (H^1) ions. Compare *acidic*.

bathymetry The measurement of ocean depths and the *topography* of the ocean floor.

bedrock The continuous mass of solid *rock* that makes up Earth's crust.

benthic Of, relating to, or living on the bottom of a water body. Compare *pelagic*.

benthic zone The bottom layer of water body. Compare *littoral zone; limnetic zone; profundal zone*.

B horizon The layer of *soil* that lies below the *A horizon* and above the *C horizon*. Minerals that leach out of the A horizon are carried down into the B horizon (or subsoil) and accumulate there. Sometimes called the *subsoil*, or "zone of accumulation" or "zone of deposition." Compare *O horizon; R horizon*.

bioaccumulation The buildup of *toxicants* in the tissues of an animal.

biocentrism A philosophy that ascribes relative values to actions, entities, or properties on the basis of their effects on all living things or on the integrity of the *biotic* realm in general. The biocentrist evaluates an action in terms of its overall impact on living things, including—but not exclusively focusing on—human beings. Compare *anthropocentrism; ecocentrism*.

biodiesel Diesel fuel produced by mixing vegetable oil, used cooking grease, or animal fat with small amounts of *ethanol* or methanol (wood alcohol) in the presence of a chemical catalyst.

biodiversity (biological diversity) The sum total of all organisms in an area, taking into account the diversity of *species,* their *genes,* their *populations,* and their *communities*.

biodiversity hotspot An area that supports an especially great diversity of *species,* particularly species that are *endemic* to the area.

biofuel Fuel produced from *biomass energy* sources and used primarily to power automobiles.

biogeochemical cycle See *nutrient cycle*.

biological control (biocontrol) The attempt to battle *pests* and weeds with organisms that prey on or parasitize them, rather than by using *pesticides*.

biological diversity See *biodiversity*.

biological hazard Human health hazards that result from ecological interactions among organisms. These include *parasitism* by viruses, bacteria, or other *pathogens*. Compare *infectious disease; chemical hazard; cultural hazard; physical hazard*.

biological weathering *Weathering* that occurs when living things break down *parent material* by physical or chemical means. Compare *chemical weathering; physical weathering*.

biomagnification The magnification of the concentration of *toxicants* in an organism caused by its consumption of other organisms in which toxicants have *bioaccumulated*.

biomass Biological material; comprises living and recently deceased organic matter.

biomass energy *Energy* harnessed from plant and animal matter, including wood from trees, charcoal from burned wood, and combustible animal waste products, such as cattle manure. *Fossil fuels* are not considered biomass energy sources because their organic matter has not been part of living organisms for millions of years and has undergone considerable chemical alteration since that time.

biome A major regional complex of similar plant *communities;* a large ecological unit defined by its dominant plant type and vegetation structure. Compare *ecoregion*.

biophilia A hypothetical phenomenon defined as "the connections that human beings subconsciously seek with the rest of life."

biopower The burning of *biomass energy* sources to generate electricity.

bioremediation The attempt to clean up *pollution* by enhancing natural processes of biodegradation by living organisms. Compare *phytoremediation*.

biosphere The sum total of all the planet's living organisms and the *abiotic* portions of the *environment* with which they interact. Compare *atmosphere; hydrosphere; lithosphere*.

biosphere reserve A tract of land with exceptional *biodiversity* that couples preservation with *sustainable development* to benefit local people. Biosphere reserves are designated by UNESCO (the *United Nations* Educational, Scientific, and Cultural Organization), following application by local stakeholders.

biotechnology The material application of biological *science* to create products derived from organisms. The creation of *transgenic* organisms is one type of biotechnology.

biotic Refers to any living component of the *environment*. Compare *abiotic*.

biotic potential An organisms' capacity to produce offspring.

bitumen A thick and heavy form of *petroleum* rich in *carbon* and poor in *hydrogen*.

bog A type of *wetland* in which a pond is thoroughly covered with a thick floating mat of vegetation. Compare *freshwater marsh; swamp*.

bonding The joining together of two or more *atoms*, usually by sharing or exchange or *electrons*.

boreal forest A *biome* of northern coniferous forest that stretches in a broad band across much of Canada, Alaska, Russia, and Scandinavia. Also known as *taiga*, boreal forest consists of a limited number of *species* of evergreen trees, such as black spruce, that dominate large regions of forests interspersed with occasional bogs and lakes.

bottom-trawling Fishing practice that involves dragging weighted nets across the seafloor to catch *benthic* organisms. Bottom-trawling crushes many organisms in its path and leaves long swaths of damaged sea bottom.

breakdown product A *compound* that results from the degradation of a *toxicant*.

breeder reactor A *nuclear reactor* that creates more fissile material than it consumes, and uses primarily uranium-238 and plutonium-239. Although breeder reactors make better use of fuel, generate more power, and produce less waste than conventional reactors, most have been closed because of concerns over nuclear weapons proliferation.

brownfield Sites that have been contaminated by hazardous materials but that have the potential to be cleaned up and remediated for other purposes.

bulk water exports Large-scale diversions of fresh water, with export as a marketable commodity.

by-catch That portion of a commercial fishing catch consisting of animals caught unintentionally. By-catch kills many thousands of fish, sharks, marine mammals, and birds each year.

Calvin cycle In *photosynthesis*, a series of chemical reactions in which *carbon atoms* from *carbon dioxide* are linked together to manufacture sugars.

Canadian Environmental Protection Act (1999) The centrepiece of Canadian environmental legislation.

canopy (of a forest) The more or less continuous upper level of leaves and branches defined by the tree tops. Forests can have *closed* or *open canopies*.

cap-and-trade A *permit-trading* system in which government determines an acceptable level of *pollution* and then issues polluting parties *permits* to pollute. A company receives credit for amounts it does not emit and can then sell this credit to other companies. Compare *emissions trading system*.

captive breeding The practice of capturing members of threatened and endangered *species* so that their young can be bred and raised in controlled *environments* and subsequently reintroduced into the wild.

carbohydrate An *organic compound* consisting of *atoms* of carbon, hydrogen, and oxygen.

carbon The chemical *element* with six protons and six neutrons. A key element in *organic compounds*.

carbon capture and storage (CCS) Technologies or approaches that remove *carbon dioxide* from power plant or other emissions and sequester it, in an effort to mitigate *global climate change*.

carbon cycle A major *nutrient cycle* consisting of the routes that carbon *atoms* take through the nested networks of environmental *systems*.

carbon dioxide (CO_2) A colorless gas used by plants for *photosynthesis*, given off by *respiration*, and released by burning *fossil fuels*. A primary *greenhouse gas* whose buildup contributes to *global climate change*.

carbon footprint The cumulative amount of *carbon*, or *carbon dioxide* equivalent, that a person or institution emits, and is indirectly responsible for emitting, into the *atmosphere*, contributing to *global climate change*. Compare *ecological footprint*.

carbon monoxide (CO) A colorless, odorless gas produced primarily by the incomplete combustion of fuel. A *criteria pollutant*.

carbon neutrality The state in which an individual, business, or institution emits no net *carbon* to the atmosphere. This may be achieved by reducing carbon emissions and/or employing *carbon offsets* to offset emissions.

carbon offset A voluntary payment to another entity intended to enable that entity to reduce the *greenhouse gas* emissions that one is unable or unwilling to reduce oneself. The payment thus offsets one's own emissions.

carbon sequestration Technologies or approaches to sequester, or store, *carbon dioxide* from industrial emissions, e.g., underground under pressure in locations where it will not seep out, in an effort to mitigate *global climate change*.

carbon storage See *carbon sequestration* or *carbon capture and storage (CSS)*.

carcinogen A chemical or type of radiation that causes cancer.

carnivore An organism that consumes animals. Compare *herbivore*; *omnivore*.

carrying capacity The maximum *population size* that a given *environment* can sustain.

cation exchange Process by which plants' roots donate *hydrogen ions* to the *soil* in exchange for cations (positively charged *ions*) such as those of calcium, magnesium, and potassium, which plants use as *nutrients*. The soil particles then replenish these cations by exchange with soil water.

cell The most basic organizational unit of organisms.

cellular respiration The process by which a *cell* uses the chemical reactivity of *oxygen* to split glucose into its constituent parts, water and *carbon dioxide*, and thereby release *chemical energy* that can be used to form chemical bonds or to perform other tasks within the cell. Compare *photosynthesis*.

channelization Modification of a river's channel or banks by straightening, widening, or concrete-lining, usually for the purposes of navigation, flood control, or diversion for irrigation or water supply.

chaparral A *biome* consisting mostly of densely thicketed evergreen shrubs occurring in limited small patches. Its "Mediterranean" *climate* of mild, wet winters and warm, dry summers is induced by oceanic influences. In addition to ringing the Mediterranean Sea, chaparral occurs along the coasts of California, Chile, and southern Australia.

chemical energy *Potential energy* held in the force of *bonding* between *atoms*.

chemical hazard Chemicals that pose human health hazards. These include *toxins* produced naturally, as well as many anthropogenic chemicals, including some disinfectants, *pesticides*, and other synthetic chemicals. Compare *biological hazard; cultural hazard; physical hazard*.

chemical weathering *Weathering* that results when water or other substances chemically interact with *parent material*. Compare *biological weathering; physical weathering*.

chemosynthesis The process by which bacteria in hydrothermal vents use the *chemical energy* of hydrogen sulphide (H_2S) to transform inorganic carbon into *organic compounds*. Compare *photosynthesis*.

chlorofluorocarbon (CFC) One of a group of human-made *organic compounds* derived from simple *hydrocarbons,* such as ethane and methane, in which hydrogen *atoms* are replaced by chlorine, bromine, or fluorine. CFCs deplete the protective *ozone layer* in the *stratosphere*.

chlorophyll The light-absorbing pigment that enables *photosynthesis* and makes plants green.

chloroplast A *cell organelle* containing *chlorophyll*, in which *photosynthesis* occurs.

C horizon The layer of *soil* that lies below the *B horizon* and above the *R horizon*. It contains *rock* particles that are larger and less *weathered* than the layers above. It consists of *parent material* that has been altered only slightly or not at all by the process of *soil* formation. Compare *A horizon; O horizon*.

chronic exposure Exposure for long periods of time to a *toxicant* occurring in low amounts. Compare *acute exposure*.

classical economics The study of the behaviour of buyers and sellers in a free-market *economy*; holds that individuals acting in their own self-interest may benefit society, provided that their behaviour is constrained by the rule of law and by private property rights and operates within competitive markets. See also *neoclassical economics*.

clay *Sediment* consisting of particles less than 0.002 mm in diameter. Compare *sand; silt*.

clean coal Term used to describe technologies and approaches that seek to reduce the generation and release of sulphur and other pollutants before, during, or after coal is burned for power.

clear-cutting The harvesting of timber by cutting all the trees in an area, leaving only stumps. Although it is the most cost-efficient method, clear-cutting is also the most damaging to the *environment*.

climate The pattern of atmospheric conditions found across large geographic regions over long periods of time. Compare *weather*.

climate model See *coupled general circulation model*.

climatograph (climate diagram) A visual representation of a region's average monthly temperature and *precipitation*.

climax community In the traditional view of ecological *succession*, a *community* that remains in place with little modification until disturbance restarts the successional process. Today, ecologists recognize that community change is more variable and less predictable than originally thought, and that assemblages of species may instead form complex mosaics in space and time.

closed canopy (of a forest) A forest canopy that has very few openings where light can penetrate to the forest floor.

closed cycle (or loop) A *cycle* that operates as a *closed system*.

closed system A *system* that is self-contained with regard to exchanges of matter (but not energy) with its surroundings. Scientists may treat a system as closed to simplify some question they are investigating, but no natural system is truly closed. Compare *open system*.

coal A *fossil fuel* composed of *organic matter* that was compressed under high pressure to form a dense, carbon-rich solid material.

cogeneration A practice in which the extra heat generated in the production of electricity is captured and put to use heating workplaces and homes, as well as producing other kinds of power.

cold front The boundary where a mass of cold air displaces a mass of warmer air. Compare *warm front*.

command-and-control An approach to protecting the *environment* that sets strict legal limits and threatens punishment for violations of those limits.

commensalism A relationship between members of different *species* in which one organism benefits and the other is unaffected. Compare *amensalism*.

community A group of *populations* of organisms that live in the same place at the same time.

community ecology The study of the interactions among *species*, from one-to-one interactions to complex interrelationships involving entire *communities*.

community-supported agriculture (CSA) A practice in which consumers pay farmers in advance for a share of their yield, usually in the form of weekly deliveries of produce.

compaction (of soils) A decrease in the volume of *soil* and collapse of the *soil structure*, which can occur when fluids are withdrawn or when the soil is too heavily tilled, dries out as a result of the removal of vegetation, or bears too much weight.

competition A relationship in which multiple organisms seek the same limited resource.

competitive exclusion An outcome of *interspecific competition* in which one *species* excludes another species from resource use entirely.

compost A mixture produced when *decomposers* break down *organic matter*, including food and crop waste, in a controlled environment.

composting The conversion of organic *waste* into mulch or *humus* by encouraging, in a controlled manner, the natural biological processes of decomposition.

compound A *molecule* whose *atoms* are composed of two or more *elements*.

concentrated animal feeding operation (CAFO) A *feedlot* or factory farm.

confined (artesian) aquifer A water-bearing, porous layer of *rock, sand,* or gravel that is trapped between an upper and lower layer of less permeable substrate, such as *clay*. The water in a confined aquifer is under pressure because it is trapped between two impermeable layers. Compare *unconfined aquifer*.

coniferous Refers to trees that are "evergreen," that is, they do not lose their leaves in the fall. Coniferous trees produce cones to host their seeds, and typically have needles rather than broad, flat leaves. Compare *deciduous*.

conservation biology A scientific discipline devoted to understanding the factors, forces, and processes that influence the loss, protection, and restoration of *biological diversity* within and among *ecosystems*.

conservation ethic An *ethic* holding that humans should put *natural resources* to use but also have a responsibility to manage them wisely. Compare *preservation ethic*.

constructed wetland A *wetland* that is built, usually for the purpose of stormwater runoff management or *wastewater management*. Also called *artificial wetland*.

consumer See *heterotroph*.

consumptive use Freshwater use in which water is removed from a particular *aquifer* or surface water body and is not returned to it. *Irrigation* for *agriculture* is an example of consumptive use. Compare *nonconsumptive use*.

continental shelf The very gently sloping underwater edge of a continent, varying in width from 100 m to 1,300 km. At the shelf-slope break, the continental shelf gives way to the steeper *continental slope*, which leads down to the ocean floor or *abyssal plain*.

continental slope The relatively steep offshore slope that extends downward from the *continental shelf* to the continental rise, which connects it to the *abyssal plain* of the ocean.

contingent valuation A technique that uses surveys to determine how much people would be willing to pay to protect a resource or to restore it after damage has been done.

contour farming The practice of ploughing furrows sideways across a hillside, perpendicular to its slope, to help prevent the formation of rills and gullies. The technique is so named because the furrows follow the natural contours of the land.

control The portion of an *experiment* in which a *variable* has been left unmanipulated or untreated.

controlled burn See *prescribed burn*.

controlled experiment An *experiment* in which the effects of all *variables* are held constant, except the one whose effect is being tested by comparison of *treatment* and *control* conditions.

convective circulation A circular *current* (of air, water, magma, etc.) driven by temperature differences. In the atmosphere, warm air rises into regions of lower *atmospheric pressure*, where it expands and cools and then descends and becomes denser, replacing warm air that is rising. The air picks up heat and moisture near ground level and prepares to rise again, continuing the process.

convention A *treaty* or binding agreement among national governments.

conventional law International law that arises from conventions, or treaties, that nations agree to enter into. Compare *customary law*.

(United Nations) Convention on Biological Diversity An international treaty that aims to conserve *biodiversity*, use biodiversity in a *sustainable* manner, and ensure the fair distribution of biodiversity's benefits. Many nations have agreed to the treaty, including Canada (as of 2007, 189 nations and the European Union had become parties to the agreement); several others, including the United States, have not.

(United Nations) Convention on International Trade in Endangered Species of Wild Fauna and Flora (CITES) A 1973 international treaty, facilitated by the *United Nations*, that protects *endangered species* by banning the international transport of their body parts.

convergent plate boundary Area where tectonic plates collide. Can result in *subduction* or mountain range formation.

convergent evolution The process whereby two completely separate and distinct *species* evolve similar traits, generally as a result of *adaptation* to selective pressures from similar *environments* or *habitats*.

coral bleaching The loss of the coloured *symbiotic algae* that normally inhabit corals; possibly caused by stress due to increased water temperature or *turbidity*.

coral reef A mass of calcium carbonate composed of the skeletons of tiny colonial marine organisms.

core The innermost part of the Earth, made up mostly of iron, that lies beneath the *crust* and *mantle*.

core The part of a forest or reserve that is isolated from the surrounding area by a transitional or buffer zone. Compare *edge*.

Coriolis effect The apparent deflection of north-south air *currents* to a partly east-west direction, caused by the faster spin of regions near the equator than of regions near the poles as a result of Earth's rotation.

correlation A relationship among *variables*.

corridor A passageway of protected land established to allow animals to travel between islands of protected *habitat*.

corrosive Substances that corrode (gradually erode, or eat away) metals. One criterion for defining *hazardous waste*.

cost-benefit analysis A method commonly used by *neoclassical economists*, in which estimated costs for a proposed action are totalled and then compared to the sum of benefits estimated to result from the action.

coupled general circulation model A computer program that combines what is known about *weather* patterns, atmospheric circulation, atmosphere-ocean interactions, and feedback mechanisms to simulate *climate* processes.

criteria (air) pollutants Six *air pollutants*—carbon monoxide, sulphur dioxide, nitrogen dioxide, tropospheric ozone, particulate matter, and lead—for which maximum allowable concentrations have been established for ambient outdoor air because of the threats they pose to human health.

cropland Land that humans use to raise plants for food and fibre.

crop rotation The practice of alternating the kind of crop grown in a particular field from one season or year to the next.

crude birth rate The number of births per 1,000 individuals for a given time period. Compare *crude death rate*.

crude death rate The number of deaths per 1,000 individuals for a given time period. Compare *crude birth rate*.

crude oil (petroleum) A *fossil fuel* produced by the conversion of *organic compounds* by heat and pressure. Crude oil is a mixture of hundreds of different types of *hydrocarbon* molecules characterized by carbon chains of different length.

crust The lightweight outer layer of Earth, consisting of *rock* that floats atop the malleable *mantle*, which in turn surrounds a mostly iron *core*.

cultivar Domesticated varieties of crop plants.

cultural hazard Human health hazards that result from the place we live, our socioeconomic status, our occupation, or our behavioral choices. These include choosing to smoke cigarettes, or living or working with people who do. Also called *lifestyle hazard*. Compare *biological hazard; chemical hazard; physical hazard*.

culture The overall ensemble of knowledge, beliefs, values, and learned ways of life shared by a group of people.

current The flow of a liquid (or gas, such as air) in a certain direction.

customary law International law that arises from long-standing practices, or customs, held in common by most *cultures*. Compare *conventional law*.

cycle Flows of elements, compounds, and energy from reservoir to reservoir through the Earth system.

dam Any obstruction placed in a river or stream to block the flow of water so that water can be stored in a reservoir. Dams are built to prevent floods, provide drinking water, facilitate *irrigation*, and generate electricity.

data Information, generally quantitative information.

debt-for-nature swap An approach to conservation, in which a private or nongovernmental organization assumes a portion of the debt of a developing country, in exchange for some form of conservation or land protection action or policy.

deciduous Refers to trees, usually broad-leafed, that lose their leaves each fall and remain dormant during the winter. Compare *coniferous*.

decommissioning The process by which an industrial facility is permanently removed from service and the land on which it resides is reclaimed, remediated, or stabilized.

decomposer Organisms, mainly fungi and bacteria, that break down leaf litter and other nonliving matter into simpler constituents that can then be taken up and used as *nutrients* by plants.

deep ecology A philosophy established in the 1970s based on principles of self-realization (the awareness that humans are inseparable from nature) and *biocentric* equality (the precept that all living beings have equal value). Holds that because we are truly inseparable from our *environment*, we must protect all other living things as we would protect ourselves.

deep-well injection A *hazardous waste* disposal method in which a well is drilled deep beneath an area's *water table* into porous *rock* below an impervious *soil* layer. Wastes are then injected into the well, so that they will be absorbed into the porous rock and remain deep underground, isolated from *groundwater* and human contact. Compare *surface impoundment*.

deforestation The loss of forested land.

demographer A scientist who studies human populations.

demographic transition A theoretical model of economic and cultural change that explains the declining death rates and birth rates that occurred in Western nations as they became industrialized. The model holds that industrialization caused these rates to fall naturally by decreasing mortality and by lessening the need for large families. Parents would thereafter choose to invest in quality of life rather than quantity of children.

demography A *social science* that applies the principles of *population ecology* to the study of statistical change in human *populations*.

denitrifying bacteria Bacteria that convert the nitrates in *soil* or water to gaseous nitrogen and release it back into the *atmosphere*.

deoxyribonucleic acid See *DNA*.

dependent variable The *variable* that is affected by manipulation of the *independent variable*.

deposition The arrival of transported material at a new location. For example, eroded *sediment* is deposited in streams, and wind-borne *particulates* are deposited on the surfaces of land and water bodies.

desalination (desalinization) The removal of salt from seawater.

desert The driest *biome* on Earth, with annual *precipitation* of less than 25 cm. Because deserts have relatively little vegetation to insulate them from temperature extremes, sunlight readily heats them in the daytime, but daytime heat is quickly lost at night, so temperatures vary widely from day to night and in different seasons.

desertification A loss of more than 10% of a land's productivity due to *erosion*, *soil* compaction, forest removal, *overgrazing*, drought, *salination*, *climate change*, depletion of water sources, or other factors. Severe desertification can result in the actual expansion of desert areas or creation of new ones in areas that once supported fertile land.

detritivore Animals such as millipedes and soil insects, which scavenge the waste products or the dead bodies of other *community* members.

development The use of *natural resources* for economic advancement (as opposed to simple subsistence, or survival).

dike A long raised mound of earth erected along a river bank to protect against floods by holding rising water in the main channel.

directional selection Mode of *natural selection* in which selection drives a feature in one direction rather than another—for example, toward larger or smaller, or faster or slower. Compare *disruptive selection*; *stabilizing selection*.

discharge zone An area where *groundwater* emerges from the subsurface and flows out on the surface to become or join a surface water body.

disruptive selection Mode of *natural selection* in which a trait diverges from its starting condition in two or more directions. Compare *directional selection*; *stabilizing selection*.

dissolved oxygen The amount of *oxygen* dissolved in water, and freely available to fish and other aquatic life; an *indicator* of aquatic *ecosystem* health.

distance effect In *island biogeography theory*, the pattern that islands far from a mainland host fewer *species* because fewer species tend to find and colonize it.

divergent evolution The process whereby two *species* with a common genetic ancestor (or two *populations* of the same species, which have become geographically separated or reproductively isolated) evolve different traits over time, generally as a result of *adaptation* to selective pressures from different *environments* or *habitats*. Divergent evolution of two populations may be expected eventually to result in *speciation*—that is, the characteristics of the populations eventually become so different that they are no longer of the same species.

divergent plate boundary Area where *magma* surging upward to the surface divides tectonic plates and pushes them apart, creating new *crust* as it cools and spreads. A prime example is the Mid-Atlantic ridge. Compare *transform plate boundary* and *convergent plate boundary*.

diversion (of water) Removing water from a river system or changing its flow for use in another location.

DNA (deoxyribonucleic acid) A double-stranded *nucleic acid* composed of four nucleotides, each of which contains a sugar (deoxyribose), a phosphate group, and a nitrogenous base. DNA carries the hereditary information for living organisms and is responsible for passing traits from parents to offspring. Compare *RNA*.

doldrums A region with little wind activity, located near the equator.

domesticated (animals) Animals bred and raised in captivity, for human use.

dose The amount of *toxicant* a test animal receives in a dose-response test. Also, the amount of toxicant ingested, absorbed, or otherwise taken in by an organism. Compare *response*.

dose-response curve A curve that plots the response of test animals to different doses of a *toxicant*. The response is generally quantified by measuring the proportion of animals exhibiting negative effects.

downwelling In the ocean, the flow of warm surface water toward the ocean floor. Downwelling occurs where surface *currents* converge. Compare *upwelling*.

drainage basin Land area where all precipitation that falls drains off through a particular river channel. Compare *watershed*.

driftnet An enormous net that stretches across many kilometres of water, used for commercial *industrialized fishing*.

dryland An area with generally low *precipitation* but not so arid as to be classified as a *desert*; grasslands, savannahs, and shrublands are typical dryland *biomes*.

dust dome A phenomenon in which *smog* and *particulate air pollution* become trapped in a layer overlying an urban centre. Can be exacerbated by the *urban heat island effect*.

dynamic equilibrium The state reached when processes within a *system* are moving in opposing directions at equivalent rates so that their effects balance out.

ecocentrism A philosophy that considers actions in terms of their damage or benefit to the integrity of whole ecological *systems*, including both *biotic* and *abiotic* elements. For an ecocentrist, the well-being of an individual organism—human or otherwise—is less important than the long-term well-being of a larger integrated ecological system. Compare *biocentrism*; *anthropocentrism*.

ecofeminism A philosophy holding that the patriarchal (male-dominated) structure of society is a root cause of both social and environmental problems. Ecofeminists hold that a *worldview* traditionally associated with women, which interprets the world in terms of interrelationships and cooperation, is more in tune with nature than a worldview traditionally associated with men, which interprets the world in terms of hierarchies and competition.

ecoimperialism The imposition of Western environmental priorities and values on *Aborigines* and people in developing nations.

ecolabelling The practice of designating on a product's label how the product was grown, harvested, or manufactured, so that consumers buying it are aware of the processes involved and can differentiate between brands that use processes believed to be environmentally beneficial (or less harmful than others) and those that do not.

ecological economics A developing school of *economics* that applies the principles of *ecology* and *systems* thinking to the description and analysis of *economies*. Compare *environmental economics*; *neoclassical economics*.

ecological footprint The cumulative amount of land and water required to provide the raw materials a person or *population* consumes and to dispose of or *recycle* the *waste* that is produced.

ecological restoration Efforts to reverse the effects of human disruption of ecological *systems* and to restore *communities* to their "natural" state. Compare *restoration ecology*.

ecology The *science* that deals with the distribution and abundance of organisms, the interactions among them, and the interactions between organisms and their *abiotic environments*.

economics The study of how we decide to use scarce resources to satisfy the demand for *goods* and *services*.

economy A social *system* that converts resources into *goods* and *services*.

ecoregion A large area of land or water that contains a geographically distinct assemblage of natural *communities* that share a large majority of their *species* and ecological dynamics; share similar environmental conditions; and interact ecologically in ways that are critical for their long-term persistence. Compare *biome*.

ecosystem All organisms and nonliving entities that occur and interact in a particular area at the same time.

ecosystem-based management The attempt to manage the harvesting of resources in ways that minimize impact on the *ecosystems* and ecological processes that provide the resources.

ecosystem diversity The number and variety of ecosystems in a particular area. One way to express *biodiversity*. Related concepts consider the geographic arrangement of *habitats*, *communities*, or *ecosystems* at the landscape level, including the sizes, shapes, and interconnectedness of patches of these entities.

ecosystem ecology The study of how the living and nonliving components of *ecosystems* interact and transfer *energy* among themselves.

ecosystem service An essential service an *ecosystem* provides that supports life and makes economic activity possible. For example, ecosystems naturally purify air and water, cycle *nutrients*, provide for plants to be *pollinated* by animals, and serve as receptacles and *recycling* systems for the *waste* generated by our economic activity.

ecotone A transitional zone where *ecosystems* meet.

ecotourism Visitation of natural areas for tourism and recreation. Most often involves tourism by more-affluent people, which may generate economic benefits for less-affluent communities near natural areas and thus provide economic incentives for conservation of natural areas.

ED$_{50}$ (effective dose–50%) The amount of a *toxicant* it takes to affect 50% of a *population* of test animals. Compare *threshold dose; LD$_{50}$*.

edge The part of a forest that is immediately adjacent to the surrounding area, and is separated from the forest *core* by a transitional or buffer zone.

effluent Water that flows out of a facility such as a *wastewater* treatment plant, mine, or power plant.

electrolysis A process in which electrical current is passed through a *compound* to release *ions*. Electrolysis offers one way to produce *hydrogen* for use as fuel: Electrical current is passed through water, splitting the water *molecules* into hydrogen and *oxygen atoms*.

electromagnetic radiation Radiant energy in the form of radio waves, infrared radiation, visible light, ultraviolet light, X-rays, and gamma rays.

electron A negatively charged particle that surrounds the nucleus of an *atom*.

electronic waste (e-waste) Discarded electronic products such as computers, monitors, printers, DVD players, cell phones, and other devices. *Heavy metals* in these products mean that this waste may require treatment as *hazardous waste*.

element A fundamental type of *matter*; a chemical substance with a given set of properties, which cannot be broken down into substances with other properties. Chemists currently recognize 92 elements that occur in nature, as well as more than 20 others that have been artificially created.

El Niño The exceptionally strong warming of the eastern Pacific Ocean that occurs every 2 to 7 years and depresses local fish and bird *populations* by altering the marine *food web* in the area. Originally, the name that Spanish-speaking fishermen gave to an unusually warm surface *current* that sometimes arrived near the Pacific coast of South America around Christmas time. Compare *La Niña*.

El Niño – Southern Oscillation (ENSO) A systematic shift in atmospheric pressure, sea-surface temperature, and ocean circulation in the tropical Pacific Ocean. ENSO cycles give rise to *El Niño* and *La Niña* conditions.

emergent property A characteristic that is not evident in a *system*'s components individually.

emigration The departure of individuals from a *population*.

emissions trading system A *permit-trading* system for emissions in which a government issues marketable emissions permits to conduct environmentally harmful activities. Under a *cap-and-trade* system, the government determines an acceptable level of *pollution* and then issues permits to pollute. A company receives credit for amounts it does not emit and can then sell this credit to other companies. Compare *cap-and-trade*.

endangered Refers to a *species* in danger of *extirpation* or *extinction*.

endemic Native or restricted to a particular geographic region. An endemic species occurs in one area and nowhere else on Earth.

endocrine disruptor A *toxicant* that interferes with the endocrine (hormone) system.

endocrine system *Hormonal* system.

"end-of-pipe" A response to *pollution* that deals only with *effluents* as they emerge from the "end of the pipe," rather than reducing or eliminating the pollution at its source.

energy An intangible phenomenon that can change the position, physical composition, or temperature of matter.

energy conservation The practice of reducing *energy* use as a way of extending the lifetime of our *fossil fuel* supplies, of being less wasteful, and of reducing our impact on the *environment*.

environment The sum total of our surroundings, including all of the living things and nonliving things with which we interact.

environmental audit An assessment undertaken to determine whether a facility's operations are in compliance with environmental regulations, and/or whether there are changes that could be made to reduce the environmental impacts or enhance the sustainability of the operation.

environmental economics A developing school of *economics* that modifies the principles of *neoclassical economics* to address environmental challenges. An environmental economist believes that we can attain *sustainability* within our current economic *systems*. Compare *ecological economics; neoclassical economics*.

environmental ethics The application of *ethical standards* to environmental questions.

environmental health Environmental factors that influence human health and quality of life and the health of ecological *systems* essential to environmental quality and long-term human well-being.

environmental impact statement (EIS) A report of results from detailed studies that assess the potential effects on the *environment* that would likely result from development projects or other actions undertaken by the government.

environmental justice The principle that all people (sometimes extended to all beings) have the right to a clean, healthy *environment*, and should therefore have equal access to environmental *resources* and protection from the impacts of environmental degradation.

environmental nongovernmental organization (ENGO) An environmental organization that is not a government or a private-sector agency. Well-known examples include the Sierra Club, Conservation International, and Greenpeace.

environmental policy *Public policy* that pertains to human interactions with the *environment*. It generally aims to regulate resource use or reduce *pollution* to promote human welfare and/or protect natural systems.

environmental refugee A person who has been driven from his or her homeland by natural disasters, resource shortages, or environmental change.

environmental resistance The collective force of limiting factors, which together stabilize a *population size* at its *carrying capacity*.

environmental science The study of how the natural world works and how humans and the *environment* interact.

environmental studies An academic program of study about the *environment* that heavily incorporates the *social sciences* as well as the *natural sciences*.

environmental toxicology The study of *toxicants* that come from or are discharged into the *environment*, including the study of health effects on humans, other animals, and *ecosystems*.

environmentalism A social movement dedicated to protecting the natural world.

Environment Canada The department of the federal government that is mandated to preserve and enhance the quality of the natural *environment*; conserve Canada's *renewable natural resources*; conserve and protect Canada's water resources; forecast weather and environmental change; enforce rules relating to boundary waters; and coordinate environmental policies and programs; created in 1971 by the *Department of the Environment Act*.

enzyme A *protein* that catalyzes chemical reactions in organisms.

epidemiology A field of study that involves large-scale comparisons among groups of people, usually contrasting a group known to have been exposed to some *toxicant* and a group that has not.

equilibrium theory of island biogeography A *theory* that was initially applied to oceanic islands to explain how *species* come to be distributed among them. Since its development, researchers have increasingly applied the theory to islands of *habitat* (patches of one type of habitat isolated within vast "seas" of others). Aspects of the theory include *immigration* and *extinction* rates, the effect of island size, and the effect of distance from the mainland.

EROI (Energy Returned on Investment) The ratio determined by dividing the quantity of *energy* returned from a process by the quantity of energy invested in the process. Higher EROI ratios mean that more energy is produced from each unit of energy invested.

erosion The removal of material from one place and its transport to another by the action of wind or water.

estuary An area where a river flows into the ocean, mixing *freshwater* with salt water.

ethanol The alcohol in beer, wine, and liquor, produced as a *biofuel* by fermenting biomass, generally from *carbohydrate*-rich crops such as corn.

ethical standards The criteria that help differentiate right from wrong.

ethics The study of good and bad, right and wrong. The term can also refer to the set of moral principles or values of a group or an individual.

eukaryote A multicellular organism. The *cells* of eukaryotic organisms consist of a membrane-enclosed nucleus that houses *DNA*, an outer membrane of lipids, and an inner fluid-filled chamber containing *organelles*. Compare *prokaryote*.

eutrophic Term describing a water body that has high-nutrient and low-oxygen conditions. Compare *oligotrophic*.

eutrophication The process of *nutrient* enrichment, increased production of organic matter, and subsequent *ecosystem* degradation.

evaporation The conversion of a substance from a liquid to a gaseous form.

even-aged Condition of timber plantations—generally *monocultures* of a single *species*—in which all trees are of the same age. Most ecologists view plantations of even-aged stands more as crop *agriculture* than as ecologically functional forests. Compare *uneven-aged*.

evenness See *relative abundance*.

evolution Genetically based change in the appearance, functioning, and/or behaviour of organisms across generations, often by the process of *natural selection*.

e-waste See *electronic waste*.

Exclusive Economic Zone The area of ocean under the jurisdiction of the bordering nation, which streatches out 200 nautical miles from the shore, as per the *United Nations Convention on the Law of the Sea*.

exotic A *species* that is non-native to an area; alien. Compare *invasive species*.

experiment An activity designed to test the validity of a *hypothesis* by manipulating *variables*. See *manipulative experiment*, *controlled experiment*, and *natural experiment*.

exponential growth The increase of a *population* (or of anything) by a fixed percentage each year; geometric growth. Contrast with *linear growth*.

extensification Increasing resource productivity by bringing more land into production. Compare *intensification*.

externality A cost or benefit of a transaction that affects people other than the buyer or seller. Examples of negative externalities or external costs include harm to citizens from water *pollution* or *air pollution* discharged by nearby factories.

extinction The disappearance of an entire *species* from the face of the Earth. Compare *extirpation*.

extirpation The disappearance of a particular *population* from a given area, but not the entire *species* globally. Compare *extinction*.

extrusive Term for *igneous rock* formed when *magma* is ejected from a volcano and cools quickly (e.g., basalt).

feedback loop (or cycle) A circular process in which a *system*'s output serves as input to that same system. See *negative feedback loop*; *positive feedback loop*.

feedlot A huge barn or outdoor pen designed to deliver *energy*-rich food to animals living at extremely high densities. Also called a "factory farm" or *concentrated animal feeding operation* (*CAFO*).

Ferrel cell One of a pair of cells of *convective circulation* between 308 and 608 northand south latitude that influence global *climate* patterns. Compare *Hadley cell*; *polar cell*.

fertilizer A substance that promotes plant growth by supplying essential *nutrients* such as nitrogen or phosphorus.

first law of thermodynamics Physical law stating that *energy* can change from one form to another but cannot be created or destroyed. The total energy in the universe remains constant and is said to be conserved. Compare *law of conservation of matter*.

flammable Substances that easily catch fire. One criterion for defining *hazardous waste*.

flat-plate solar collectors See *solar panels*.

floodplain The region of land over which a river has historically wandered and periodically floods.

floor (of a forest) The lowest level of a *forest*, consisting of the *topsoil*, organic *litter*, and *humus*.

flux The movement of materials (or *energy*) among pools or *reservoirs* in a *cycle*.

food chain A simple relationship in which *primary producers* are eaten by primary *consumers*, who are, in turn, eaten by secondary consumers, and so on. A significant amount of *energy* is typically lost in moving from the bottom of the food chain to higher levels, in turn.

food security An adequate, reliable, and available food supply to all people at all times.

food web A visual representation of feeding interactions within an ecological *community* that shows an array of relationships between organisms at different *trophic levels*.

forest A densely wooded area.

forestry The scientific study and professional management of forests.

fossil The remains, impression, or trace of an animal or plant of past geological ages that has been preserved in *rock* or *sediments*.

fossil fuel A *nonrenewable natural resource,* such as *crude oil, natural gas,* or *coal*, produced by the decomposition and compression of *organic matter* from ancient life.

fossil record The cumulative body of *fossils* worldwide, which palaeontologists study to infer the history of past life on Earth.

fractionation Any process (both natural and artificial) in which materials become separated from one another, usually on the basis of a physical property such as mass. In *oil refining*, fractionation is part of the distillation process.

free rider A party that fails to invest in controlling *pollution* or carrying out other environmentally responsible activities and instead relies on the efforts of other parties to do so. For example, a factory that fails to control its emissions gets a "free ride" on the efforts of other factories that do make the sacrifices necessary to reduce emissions.

fresh water Water that is relatively pure, holding very few dissolved salts.

freshwater marsh A type of *wetland* in which shallow water allows plants such as cattails to grow above the water surface. Compare *swamp*; *bog*.

front The boundary between air masses that differ in temperature and moisture (and therefore density). See *warm front*; *cold front*.

fuel cell A technology that produces energy from a chemical reaction similar to that in a battery, without combustion of the fuel.

fundamental niche The full *niche* of a *species*. Compare *realized niche*.

fungicide A type of chemical *pesticide* that kills fungi.

gasification A process in which *biomass* is vaporized at extremely high temperatures in the absence of *oxygen*, creating a gaseous mixture including *hydrogen, carbon monoxide,* and *methane,* in order to produce *biopower* or *biofuels*.

GDP See *Gross Domestic Product*.

gene A stretch of *DNA* that represents a unit of hereditary information.

gene bank See *seed bank*.

generalist A *species* that can survive in a wide array of *habitats* or use a wide array of resources. Compare *specialist*.

genetically modified (GM) organism An organism that has been *genetically engineered* using a technique called *recombinant DNA* technology.

genetic diversity A measurement of the differences in *DNA* composition among individuals within a given *species*.

genetic engineering Any process scientists use to manipulate an organism's genetic material in the lab by adding, deleting, or changing segments of its *DNA*.

genome The entirety of an organism's *genes*.

Genuine Progress Indicator (GPI) An *economic* indicator introduced in 1995 that attempts to differentiate between desirable and undesirable economic activity. The GPI accounts for benefits such as volunteerism and for costs such as environmental degradation and social upheaval. Compare *Gross Domestic Product (GDP)*.

geographic information system (GIS) Computer software that takes multiple types of *data* (for instance, on geology, hydrology, vegetation, animal *species*, and human development) and overlays them on a common set of geographic coordinates. The idea is to create a complete picture of a landscape and to analyze how elements of the different datasets are arrayed spatially and how they may be correlated. A common tool of geographers, *landscape ecologists*, resource managers, and *conservation biologists*.

geological isolation An approach to *waste disposal* that makes use of the natural buffering and containment properties of *rocks* and *minerals* to hold, absorb, and isolate the waste from contact with the *hydrosphere* and *biosphere*. Mainly used for *hazardous* or *radioactive wastes*.

geothermal energy Renewable *energy* that is generated deep within Earth. The *radioactive* decay of *elements* amid the extremely high pressures and temperatures at depth generate heat that rises to the surface in *magma* and through fissures and cracks. Where this energy heats *groundwater*, natural eruptions of heated water and steam are sent up from below.

glaciation The extension of ice sheets from the polar regions far into Earth's temperate zones during cold periods of Earth's history.

global climate change Any change in aspects of Earth's *climate*, such as temperature, *precipitation*, and storm intensity. Generally refers today to the current warming trend in global temperatures and associated climatic changes, which appears to be at least partly of anthropogenic origin.

global warming An increase in Earth's average surface temperature. The term is most frequently used in reference to the pronounced warming trend of recent years and decades. Global warming is one aspect of *global climate change*, and in turn drives other components of climate change.

global warming potential A quantity that specifies the ability of one *molecule* of a given *greenhouse gas* to contribute to atmospheric warming, relative to *carbon dioxide*.

good A material commodity manufactured for and bought by individuals and businesses.

grassland An area of land in which grasses are the dominant plant *species*.

green belt A protected natural area that is intended to function like an *urban growth boundary*, restricting the spread of *urban development*.

greenhouse effect The warming of Earth's surface and *atmosphere* (especially the *troposphere*) caused by the *energy* emitted by greenhouse gases.

greenhouse gas A gas that absorbs infrared radiation released by Earth's surface and then warms the surface and *troposphere* by emitting *energy*, thus giving rise to the *greenhouse effect*. Greenhouse gases include *carbon dioxide* (CO_2), water vapour, *ozone* (O_3), *nitrous oxide* (N_2O), *halocarbon* gases, and *methane* (CH_4).

Green Revolution An *intensification* of the industrialization of *agriculture* in the latter half of the 20th century, which has led to dramatically increased crop yields per unit area of farmland. Practices include devoting large areas to *monocultures* of crops specially bred for high yields and rapid growth; heavy use of *fertilizers*, *pesticides*, and *irrigation* water; and sowing and harvesting on the same piece of land more than once per year or per season; compare *Medical-Technological Revolution*.

green tax A levy on environmentally harmful activities and products aimed at providing a market-based incentive to correct for *market failure*. Compare *subsidy*.

Gross Domestic Product (GDP) The total monetary value of final *goods* and *services* produced in a country each year. The GDP sums all *economic* activity, whether good or bad, and does not account for benefits such as volunteerism or for *external costs* such as environmental degradation and social upheaval. Compare *Genuine Progress Indicator (GPI)*.

gross primary production The *energy* that results when *autotrophs* convert *solar energy* (sunlight) to energy of chemical *bonds* in sugars through *photosynthesis*. Autotrophs use a portion of this production to power their own metabolism, which entails oxidizing *organic compounds* by *cellular respiration*. Compare *net primary production*; *secondary production*.

groundfish A fish that feeds near the bottom of the water body, in the *benthic zone*.

ground source heat pump A pump that harnesses *geothermal energy* from near-surface sources of earth and water, and that can help heat residences.

groundwater Water held in *aquifers* underground.

growth rate The net change in a *population's* size, per 1,000 individuals. Calculated by adding the *crude birth rate* to the *immigration* rate and then subtracting the *crude death rate* and the *emigration* rate, each expressed as the number per 1,000 individuals per year.

Haber-Bosch process A process to synthesize ammonia on an industrial scale, which has doubled the natural rate of *nitrogen fixation* on Earth, and thereby increasing *agricultural* productivity, but also dramatically altering the *nitrogen cycle*.

habitat The specific *environment* in which an organism lives, including both *biotic* and *abiotic factors*.

habitat fragmentation The process by which large expanses of habitat are broken up into smaller, isolated pieces. The size of contiguous habitat is an issue for some large animals, but the character of the habitat can be an issue even for smaller animals; smaller habitat fragments have a greater proportion of edge, and the characteristics of *core* and *edge* habitat can differ substantially.

habitat selection The process by which organisms select *habitats* from among the range of options they encounter.

habitat use The process by which organisms select and use *habitats* from among the range of options they encounter.

Hadley cell One of a pair of cells of *convective circulation* between the equator and 308 north and south latitude that influence global *climate* patterns. Compare *Ferrel cell*; *polar cell*.

half-life The amount of time it takes for one-half the atoms of a *radioisotope* to emit radiation and decay. Different radioisotopes have different half-lives, ranging from fractions of a second to billions of years.

halocarbon A *compound* in which *carbon* is bonded to one of the halogens, such as fluorine, chlorine, or bromine.

halocline A zone of the ocean beneath the surface in which *salinity* increases rapidly with depth. Compare *pycnocline*, *thermocline*.

hardwood Wood derived from broad-leafed, deciduous trees. Compare *softwood*.

harvesting Gathering, withdrawal, capture, or other method of removal of product from the *stock* of a resource.

hazardous waste *Waste* that is *toxic*, chemically *reactive*, *flammable*, or *corrosive*. Compare *industrial solid waste*; *municipal solid waste*.

herbicide A type of chemical *pesticide* that kills plants.

herbivore An organism that consumes plants. Compare *carnivore*; *omnivore*.

herbivory The consumption of plants by animals.

heterotroph (consumer) An organism that consumes other organisms. Includes most animals, as well as fungi and microbes that decompose organic matter.

high-pressure system An air mass with elevated *atmospheric pressure*, containing air that descends, typically bringing fair *weather*. Compare *low-pressure system*.

homeostasis The tendency of a *system* to maintain constant or stable internal conditions.

horizon A distinct layer of *soil*. See *A horizon*; *B horizon*; *C horizon*; *E horizon*; *O horizon*; *R horizon*.

hormone A chemical messenger that travels though the bloodstream to stimulate growth, development, and sexual maturity; and regulate brain function, appetite, sexual drive, and many other aspects of physiology and behavior.

host The organism in a parasitic relationship that suffers harm while providing the *parasite* nourishment or some other benefit.

Hubbert's peak The peak in production of *crude oil* in the United States, which occurred in 1970 just as Shell Oil geologist M. King Hubbert had predicted in 1956.

humus A dark, spongy, crumbly material in the *soil*, composed of complex *organic compounds*, resulting from the partial *decomposition* of organic matter.

hydrocarbon An *organic compound* consisting solely of hydrogen and carbon *atoms*.

hydroelectric power (hydropower) The generation of electricity using the *kinetic energy* of moving water.

hydrogen The lightest *element*, consisting of one *proton* and one *electron*.

hydrologic cycle The flow of water—in liquid, gaseous, and solid forms—through our *biotic* and *abiotic environment*.

hydropower See *hydroelectric power*.

hydrosphere Earth's water—salt or fresh, liquid, ice, or vapour—that resides in surface bodies, underground, and in the *atmosphere*. Compare *biosphere; lithosphere*.

hypothesis An educated guess that explains a phenomenon or answers a *scientific* question. Compare *theory*.

hypoxia The condition of extremely low *dissolved oxygen* concentrations in a body of water.

igneous rock One of the three main categories of *rock*. Formed from cooling *magma*. Granite and basalt are examples of igneous rock. Compare *metamorphic rock; sedimentary rock*.

immigration The arrival of individuals from outside a *population*.

incineration A controlled process of burning solid waste for disposal in which mixed garbage is combusted at very high temperatures. Compare *sanitary landfill*.

independent variable The *variable* that the scientist manipulates in a *manipulative experiment*.

indicator A piece of data or information that can be used to measure performance or progress.

indigenous Cultural groups who have maintained a historical connection with a particular geographic area or *environment* for many generations, usually dating back to before any modern colonial contact or occupation of the land. Compare *Aboriginal*.

indoor air pollution *Air pollution* that occurs indoors and can be exacerbated by porr ventilation and the materials of which buildings and their furnishings are composed.

industrial ecology A holistic approach to industry that integrates principles from engineering, chemistry, *ecology, economics,* and other disciplines and seeks to redesign industrial *systems* in order to reduce resource inputs and minimize inefficiency.

industrialized agriculture A form of *agriculture* that uses large-scale mechanization and *fossil fuel* combustion, enabling farmers to replace horses and oxen with faster and more powerful means of cultivating, harvesting, transporting, and processing crops. Other aspects include *irrigation* and the use of *inorganic fertilizers*. Use of chemical herbicides and *pesticides* reduces *competition* from weeds and *herbivory* by insects. Compare *traditional agriculture*.

industrialized fishing Modern commercial fishing, which involves large, mechanized factory trawlers, sophisticated fish-finding equipment, and large inputs of fossil fuels for long-distance travel.

Industrial Revolution The dramatic shift in the mid-1700s from rural life, animal-powered agriculture, and manufacturing by craftsmen to an urban society powered by *fossil fuels* such as *coal* and *crude oil*. The Industrial Revolution led to rapid industrialization urbanization, with related economic and social changes in population, health, transportation, agricultural productivity, and environmental quality. Compare *Agricultural Revolution; Medical-Technological Revolution*.

industrial smog Gray-air smog caused by the incomplete combustion of *coal* or oil when burned. Compare *photochemical smog*.

industrial solid waste Nonliquid *waste* that is not especially hazardous and that comes from production of consumer goods, mining, *petroleum* extraction and *refining*, and agriculture. Compare *hazardous waste; municipal solid waste*.

industrial stage The third stage of the *demographic transition* model, characterized by falling birth rates that close the gap with falling death rates and reduce the rate of *population* growth. Compare *pre-industrial stage; post-industrial stage; transitional stage*.

infectious disease A disease in which a *pathogen* attacks a *host*. Also called communicable or transmissable diseases.

infrared radiation Electromagnetic radiation with wavelengths longer than visible red.

inorganic compound Chemical *compounds* that are of *mineral* (rather than biological) origin. Inorganic compounds may contain *carbon*—the *element* that characterizes *organic compounds*—but lack the carbon-carbon bonds that are typical of organic compounds.

inorganic fertilizer A *fertilizer* that consists of mined or synthetically manufactured *mineral* supplements. Inorganic fertilizers are generally more susceptible than *organic fertilizers* to *leaching* and *runoff* and may be more likely to cause unintended off-site impacts.

insecticide A type of chemical *pesticide* that kills insects.

insolation solar radiation that reaches Earth's surface.

integrated pest management (IPM) The use of multiple techniques in combination to achieve long-term suppression of *pests*, including *biocontrol*, use of *pesticides,* close monitoring of *populations*, habitat alteration, *crop rotation, transgenic* crops, alternative tillage methods, and mechanical pest removal.

intensification Increasing the resource productivity of a given unit of land, usually by applying new technologies to enhance productivity. Compare *extensification*.

interbasin transfer (of water) The diversion of water from one drainage basin or watershed to another.

intercropping Planting different types of crops in alternating bands or other spatially mixed arrangements. Compare *alley cropping, agroforestry*.

interdisciplinary field A field that borrows techniques from several more traditional fields of study and brings together research results from these fields into a broad synthesis.

interglacial A period of *global warming*, between *glaciations*.

Intergovernmental Panel on Climate Change (IPCC) An international panel of *atmospheric* scientists, *climate* experts, and government officials established in 1988 by the *United Nations Environment Programme* and the World Meteorological Organization, whose mission is to assess information relevant to questions of human-induced *global climate change*. The IPCC's 2007 *Fourth Assessment Report* summarizes current and probable future global trends and represents the consensus of atmospheric scientists around the world.

interspecific competition *Competition* that takes place among members of two or more different *species*. Compare *intraspecific competition*.

intertidal Of, relating to, or living along shorelines between the highest reach of the highest *tide* and the lowest reach of the lowest tide.

intraspecific competition *Competition* that takes place among members of the same *species*. Compare *interspecific competition*.

intrusive Term for *igneous rock* that forms when *magma* cools slowly while it is well below Earth's surface (e.g., granite).

invasive species A *species* that spreads widely and rapidly becomes dominant in a *community*, interfering with the community's normal functioning. Compare *exotic*.

ion An electrically charged *atom* or combination of atoms.

IPAT model A formula that represents how humans' total impact (I) on the *environment* results from the interaction among three factors: *population* (P), affluence (A), and technology (T).

irrigation The artificial provision of water to support *agriculture*.

island biogeography See *equilibrium theory of island biogeography*.

isotope One of several forms of an *element* having differing numbers of *neutrons* in the nucleus of its *atoms*. Chemically, isotopes of an element behave almost identically, but they have different physical properties because they differ in mass. Compare *stable isotope geochemistry*.

kelp Large brown algae or seaweed that can form underwater "forests," providing *habitat* for marine organisms.

kerogen A substance derived from deeply buried *organic matter* that acts as a precursor or source material for both *natural gas* and *crude oil*.

keystone species A *species* that has an especially far-reaching effect on a *community*.

kinetic energy *Energy* of motion. Compare *potential energy*.

K–selected (K-strategist) Term denoting a *species* with low *biotic potential* whose members produce a small number of offspring and take a long time to gestate and raise each of their young, but invest heavily in promoting the survival and growth of these few offspring. *Populations* of K–selected species are generally regulated by *density-dependent factors*. Compare *r–selected*.

Kyoto Protocol An agreement drafted in 1997 that calls for reducing, by 2012, emissions of six *greenhouse gases* to levels lower than their levels in 1990. Canada was the 99th country to ratify the agreement, which happened in 2002. Although the United States has refused to ratify the protocol, it came into force in 2005 when Russia ratified it, the 127th nation to do so.

La Niña An exceptionally strong cooling of surface water in the equatorial Pacific Ocean that occurs every 2 to 7 years and has widespread climatic consequences. Compare *El Niño*.

landfill See *sanitary landfill, secure landfill*.

landfill gas A mix of gases that consists of roughly half *methane* produced by *anaerobic* decomposition deep inside *landfills*, and which can be captured and used as a source of *energy*.

land trust Local or regional organization that preserves lands valued by its members. In most cases, land trusts purchase land outright with the aim of preserving it in its natural condition. The Nature Conservancy may be considered the world's largest land trust.

landscape ecology An approach to the study of organisms and their *environments* at the landscape scale, focusing on geographical areas that include multiple *ecosystems*.

La Niña An exceptionally strong cooling of surface water in the equatorial Pacific Ocean that occurs every 2 to 7 years and has widespread climatic consequences. Compare *El Niño*.

latitudinal gradient The increase in *species richness* as one approaches the equator. This pattern of variation with latitude has been one of the most obvious patterns in *ecology*, but one of the most difficult ones for scientists to explain.

lava *Magma* that is released from the *lithosphere* and flows or spatters across Earth's surface.

law of conservation of matter *Matter* may be transformed from one form or type of substance into others, but it cannot be created or destroyed. Compare *first law of thermodynamics*.

LD_{50} (lethal dose–50%) The amount of a *toxicant* it takes to kill 50% of a *population* of test animals. Compare ED_{50}; *threshold dose*.

leachate Liquids that seep through liners of a *sanitary landfill* and leach into the *soil* underneath.

leaching The process by which solid materials such as *minerals* are dissolved in a liquid (usually water) and transported to another location.

lead A *heavy metal* that may be ingested through water or paint, or that may enter the *atmosphere* as a *particulate pollutant* through combustion of leaded gasoline or other processes. Atmospheric lead deposited on land and water can enter the *food chain*, accumulate within body tissues, and cause *lead poisoning* in animals and people. A *criteria pollutant*.

Leadership in Energy and Environmental Design (LEED) A widely-used set of standards for sustainable building.

lichen A *mutualistic* aggregate of fungi and algae in which the algal component provides food and *energy* via *photosynthesis* while the fungal component takes a firm hold on *rock* and captures moisture. Often helps kickstart *primary succession*.

life-cycle analysis In *industrial ecology*, the examination of the entire life cycle of a given product—from its origins in raw materials, through its manufacturing, to its use, and finally its disposal—in an attempt to identify ways to make the process more ecologically efficient.

life expectancy The average number of years that individuals in particular age groups are likely to continue to live.

lifestyle hazard See *cultural hazard*.

light pollution Pollution from city lights that obscures the night sky.

limiting factor A physical, chemical, or biological characteristic of the *environment* that restrains *population* growth.

limnetic zone In a water body, the layer of open water through which sunlight penetrates. Compare *littoral zone*; *benthic zone*; *profundal zone*.

linear growth The increase of a *population* (or of anything) by a fixed amount each year; arithmetic growth. Contrast with *exponential growth*.

lipid One of a chemically diverse group of large, biologically important *molecules* that are classified together because they do not dissolve in water. Lipids include fats, phospholipids, waxes, pigments, and steroids.

lithification The formation of *rock* through the processes of compaction, binding, and crystallization.

lithosphere The solid part of the Earth, including the *rocks*, *sediment*, and *soil* at the surface and extending down many kilometres underground. Compare *atmosphere*; *biosphere*; *hydrosphere*.

littoral See *intertidal*.

littoral zone The region ringing the edge of a water body. Compare *benthic zone*; *limnetic zone*; *profundal zone*.

loam *Soil* with a relatively even mixture of *clay-*, *silt-*, and *sand*-sized particles.

logistic growth curve A plot that shows how the initial *exponential growth* of a *population* is slowed and finally brought to a standstill by *limiting factors*, yielding an S-shaped or sigmoidal growth curve.

longline fishing Fishing practice that involves setting out extremely long lines with up to several thousand baited hooks spaced along their lengths. Kills turtles, sharks, and an estimated 300,000 seabirds each year in *by-catch*.

long-range transport of atmospheric pollutants (LRTAP) Tranportation of *pollutants* in fine *particulate* form via winds and wind systems over great distances; many of the pollutants in the Arctic have been carried there from more populated, industrialized areas via LRTAP.

low-pressure system An air mass in which the air moves toward the low *atmospheric pressure* at the centre of the system and spirals upward, typically bringing clouds and *precipitation*. Compare *high-pressure system*.

macronutrient A *nutrient* that organisms require in relatively large amounts. Compare *micronutrient*.

magma Molten, liquid *rock*.

malnutrition The condition of lacking *nutrients* the body needs, including a complete complement of vitamins and *minerals*.

mangrove A tree with a unique type of roots that curve upward to obtain oxygen, which is lacking in the mud in which they grow, and that serve as stilts to support the tree in changing water levels. Mangrove forests grow on the coastlines of the tropics and subtropics.

manipulative experiment An *experiment* in which the researcher actively chooses and manipulates the *independent variable*. Compare *natural experiment*.

mantle The malleable layer of *rock* that lies beneath Earth's *crust* and surrounds a mostly iron *core*.

marine protected area (MPA) An area of the ocean set aside to protect marine life from fishing pressures. An MPA may be protected from some human activities but be open to others. Compare *marine reserve*.

marine reserve An area of the ocean designated as a "no-fishing" zone, allowing no extractive activities. Compare *marine protected area*.

market failure The failure of markets to take into account the *environment*'s positive effects on *economies* (for example, *ecosystem services*) or to reflect the negative effects (*externalities* or external costs) of economic activity on the environment and thereby on people.

market value (of natural resources) The monetary value that would be achievable by placing a *natural resource* for sale on the open market. It is relatively straightforward to quantify the potential market value of a resource (such as a standing forest, or an ore deposit) that is already represented in the market as a traded commodity (timber, or *minerals*); it is much more challenging to quantify the market value of an environmental *good* or *service* with no existing analogy in the marketplace, such as the value of a particular habitat, or a beautiful view, or the natural filtering of **groundwater**. Compare *valuation*.

mass extinction (event) The extinction of a large proportion of the world's *species* in a very short time period due to some extreme and rapid change or catastrophic event. Earth has seen five mass extinction events in the past half-billion years.

mass number The combined number of *protons* and *neutrons* in an *atom*.

materials recovery facility (MRF) A *recycling* facility where items are sorted, cleaned, shredded, and prepared for reprocessing into new items.

matter Any material that has mass and occupies space.

maximum sustainable yield The maximal harvest of a particular *renewable natural resource* that can be accomplished while still keeping the resource available for the future.

mechanical weathering See *physical weathering*.

Medical-Technological Revolution The modern era of medicine and technological advances, including the shift to modern agriculture through what is known as the Green Revolution. Compare *Agricultural Revolution*; *Industrial Revolution*; *Green Revolution*.

"mega-city" A city of enormous size, generally with a population greater than 10 million.

meltdown The accidental melting of the uranium fuel rods inside the core of a *nuclear reactor*, causing the release of radiation.

mercury A volatile *heavy metal* that has both industrial and natural sources, and occurs in a variety of forms, some more *toxic* than others. Atmospheric mercury deposited on land and water can enter the *food chain*, accumulate within body tissues, and cause *mercury poisoning*. A *criteria pollutant*.

mesosphere The atmospheric layer above the *stratosphere*, extending 50–80 km above sea level.

metamorphic rock One of the three main categories of rock. Formed by great heat and/or pressure that reshapes crystals within the rock and changes its appearance and physical properties. Common metamorphic rocks include marble and slate. Compare *igneous rock; sedimentary rock*.

methane The simplest hydrocarbon (CH_4), the key component of natural gas, and a naturally occurring greenhouse gas.

methane hydrate (methane clathrate or methane ice) An ice-like solid consisting of *molecules* of *methane* (CH_4) embedded in a crystal lattice of water molecules. Methane hydrates are being investigated as a potential new source of *energy* from *fossil fuels*.

microclimate Variations in *weather* and *climate* that are occur on an extremely local scale, such as from one side of a hill to the other.

micronutrient A *nutrient* that organisms require in relatively small amounts. Compare *macronutrient*.

Milankovitch cycle One of three types of variations in Earth's rotation and orbit around the sun that result in slight changes in the relative amount of solar radiation reaching Earth's surface at different latitudes. As the cycles proceed, they change the way solar radiation is distributed over Earth's surface and contribute to changes in *atmospheric* heating and circulation that have triggered the ice ages and other *climate* changes.

Millennium Development Goals A program of targets for *sustainable development* set by the international community through the *United Nations* at the turn of this century.

Millennium Ecosystem Assessment A comprehensive scientific assessment, completed in 2005 by over 2,000 of the world's leading environmental scientists, of the present condition of the world's ecological *systems* and their ability to continue supporting our civilization.

mineral A naturally occurring, solid, crystalline, *inorganic compound*; minerals are the building blocks of *rocks*.

mitigation A responsive strategy in which efforts are made to forestall or minimize the anticipated impacts of environmental change. Compare *adaptation*.

molecule A combination of two or more *atoms*.

monoculture The uniform planting of a single crop over a large area. Characterizes *industrialized agriculture*.

Montreal Protocol International treaty ratified in 1987, in which the (now 191) signatory nations agreed to restrict production of *chlorofluorocarbons (CFCs)* in order to forestall stratospheric ozone depletion. Because of its effectiveness in decreasing global CFC emissions, the Montreal Protocol is considered the most successful effort to date in addressing a global environmental problem.

mortality Deaths within a population.

multiple-barrier approach (to waste containment) An approach to *hazardous waste* disposal and management that places as many impediments as possible in the way of any escaping waste, or *leachate* or *effluents* or other harmful emissions from the waste (such as *radioactivity*), including both manufactured and natural barriers.

multiple use A principle that has nominally guided management policy for national forests over the past half century. The multiple use principle specifies that the forests be managed for recreation, wildlife habitat, mineral extraction, and various other uses.

municipal solid waste Nonliquid *waste* that is not especially hazardous and that comes from homes, institutions, and small businesses. Compare *hazardous waste; industrial solid waste*.

mutagen A *toxicant* that causes *mutations* in the *DNA* of organisms.

mutation An accidental change in *DNA* that may range in magnitude from the deletion, substitution, or addition of a single nucleotide to a change affecting entire sets of chromosomes. Mutations provide the raw material for evolutionary change.

mutualism A relationship in which all participating organisms benefit from their interaction. Compare *parasitism*.

natality Births within a population.

national park A scenic area set aside for recreation and enjoyment by the public. Canada's national park system today includes national parks, marine conservation areas, cultural, historic, and natural heritage sites, and other types of protected areas.

natural experiment An *experiment* in which the researcher cannot directly manipulate the *variables* and therefore must observe nature, comparing conditions in which variables differ, and interpret the results. Compare *manipulative experiment*.

natural gas A *fossil fuel* composed primarily of *methane* (CH_4), produced as a by-product when bacteria decompose *organic material* under *anaerobic* conditions.

natural rate of population change The rate of change in a *population's* size resulting from birth and death rates alone, excluding migration.

natural resource Any of the various substances and *energy* sources we need in order to survive.

natural resource accounting A discipline that has the goal of adjusting national accounts and indicators, such as *GDP*, to reflect the true costs of economic production, including the costs of depleting *natural resources* and damaging the *environment*.

natural science An academic discipline that studies the natural world. Compare *social science*.

natural selection The process by which traits that enhance survival and reproduction are passed on more frequently to future generations of organisms than those that do not, thus altering the *genetic* makeup of *populations* through time. Natural selection acts on genetic variation and is a primary driver of *evolution*.

negative feedback loop (or cycle) A *feedback loop* in which output of one type acts as input that moves the *system* in the opposite direction. The input and output essentially neutralize each other's effects, stabilizing the system. Compare *positive feedback loop*.

nekton Marine animals that actively swim. Compare *plankton*.

neoclassical economics A *theory* of *economics* that explains market prices in terms of consumer preferences for units of particular commodities. Buyers desire the lowest possible price, whereas sellers desire the highest possible price. This conflict between buyers and sellers results in a compromise price being reached and the "right" quantity of commodities being bought and sold. Compare *ecological economics; environmental economics*.

net energy The quantitative difference between *energy* returned from a process and energy invested in the process. Positive net energy values mean that a process produces more energy than is invested. See also *energy conversion efficiency; EROI*.

net metering Process by which owners of houses with *photovoltaic* systems can sell their excess *solar energy* to their local power utility.

net primary production The *energy* or biomass that remains in an ecosystem after *autotrophs* have metabolized enough for their own maintenance through *cellular respiration*. Net primary production is the energy or biomass available for consumption by *heterotrophs*. Compare *gross primary production; secondary production*.

net primary productivity The rate at which *net primary production* is produced. See *productivity; gross primary production; net primary production; secondary production*.

neurotoxin A *toxicant* that assaults the nervous system. Neurotoxins include heavy metals, *pesticides*, and some chemical weapons developed for use in war.

neutron An electrically neutral (uncharged) particle in the nucleus of an *atom*.

new forestry A set of *ecosystem-based management* approaches for harvesting timber that explicitly mimic natural disturbances. For instance, "sloppy clear-cuts" that leave a variety of trees standing mimic the changes a forest might experience if hit by a severe windstorm.

new urbanism A school of thought among architects, *planners*, and developers that seeks to design neighbourhoods in which homes, businesses, schools, and other amenities are within walking distance of one another. In a direct rebuttal to *sprawl*, proponents of new urbanism aim to create functional neighbourhoods in which families can meet most of their needs close to home without the use of a car.

niche The functional role of a *species* in a *community*. See *fundamental niche; realized niche*.

NIMBY syndrome "Not In My Back Yard," refers to the tendency for people to favour an installation such as a toxic waste facility or a refinery only if it is constructed somewhere far removed from their own property.

nitrification The conversion by bacteria of ammonium ions (NH_4^1) first into nitrite ions (NO_2^2) and then into nitrate ions (NO_3^2).

nitrogen The chemical *element* with seven *protons* and seven *neutrons*. The most abundant element in the *atmosphere*, a key element in *macromolecules*, and a crucial plant *nutrient*.

nitrogen cycle A major *nutrient cycle* consisting of the routes that nitrogen *atoms* take through the nested networks of environmental *systems*.

nitrogen dioxide (NO_2) A *criteria air contaminant* and a common by-product of internal combustion engines.

nitrogen fixation The process by which inert nitrogen gas combines with hydrogen to form ammonium ions (NH_4^1), which are chemically and biologically active and can be taken up by plants.

nitrogen-fixing Term describing bacteria that live in a *mutualistic* relationship with many types of plants and provide *nutrients* to the plants by converting nitrogen to a usable form.

nitrous oxides Various gaseous compounds, including NO_2 and NO_3, which commonly result from industrial processes involving combustion.

noise pollution Ambient sound that is undesireable or unhealthy.

nonconsumptive use *Freshwater* use in which the water from a particular *aquifer* or surface water body either is not removed or is removed only temporarily and then returned. The use of water to generate electricity in hydroelectric *dams* is an example. Compare *consumptive use*.

nonmarket value A value that is not usually included in the price of a *good* or *service*.

non-point source A diffuse source of *pollutants*, often consisting of many small sources. Compare *point source*.

nonrenewable natural resource A *natural resource* that is in limited supply and is formed much more slowly than we use it. Compare *renewable natural resource*.

North Atlantic Deep Water (NADW) The deep portion of the *thermohaline circulation* in the northern Atlantic Ocean.

no-till *Agriculture* that does not involve tilling (ploughing, disking, harrowing, or chiseling) the *soil*.

not-in-my-backyard See *NIMBY*.

nuclear energy The *energy* that holds together *protons* and *neutrons* within the nucleus of an *atom*. Several processes, each of which involves transforming *isotopes* of one *element* into isotopes of other elements, can convert nuclear energy into thermal energy, which is then used to generate electricity. See also *nuclear fission; nuclear reactor*.

nuclear fission The conversion of the *energy* within an *atom*'s nucleus to usable thermal energy by splitting apart atomic nuclei. Compare *nuclear fusion*.

nuclear fusion The conversion of the *energy* within an *atom*'s nucleus to usable thermal energy by forcing together the small nuclei of lightweight *elements* under extremely high temperature and pressure. Developing a commercially viable method of nuclear fusion remains an elusive goal.

nuclear reactor A facility within a nuclear power plant that initiates and controls the process of *nuclear fission* in order to generate electricity.

nucleic acid A *molecule* that directs the production of *proteins*. Includes *DNA* and *RNA*.

nutrient An *element* or *compound* that organisms consume and require for survival.

nutrient cycle The comprehensive set of cyclical pathways by which a given *nutrient* moves through the *environment*.

oceanography The study of the physics, chemistry, biology, and geology of the oceans.

ocean thermal energy conversion (OTEC) A potential *energy* source that involves harnessing the solar radiation absorbed by tropical oceans in the tropics.

O horizon The top layer of *soil* in some *soil profiles*, made up of organic matter, such as decomposing branches, leaves, crop residue, and animal waste. Compare *A horizon; B horizon; C horizon; R horizon*.

oil See *crude oil*.

oil sands (tar sands) Deposits that can be mined from the ground, consisting of moist *sand* and *clay* containing 1–20% *bitumen*, that some envision as a replacement for *crude oil* as this resource is depleted. Oil sands represent crude oil deposits that have been degraded and chemically altered by water *erosion* and bacterial decomposition.

oil shale *Sedimentary rock* filled with *kerogen* that can be processed to produce liquid *petroleum*. Oil shale is formed by the same processes that form *crude oil*, but occurs when kerogen was not buried deeply enough or subjected to enough heat and pressure to form oil.

old-growth forest A complex, *primary forest* in which the trees are generally at least 200 years old.

oligotrophic Term describing a water body that has low-nutrient and high-oxygen conditions. Compare *eutrophic*.

omnivore An organism that consumes both plants and animals. Compare *carnivore; herbivore*.

open canopy (of a forest) A *forest canopy* with many openings that allow light to pass through to the forest *floor*.

open cycle (or loop) A *cycle* that operates as an *open system*.

open system A *system* that exchanges *energy*, matter, and information with other systems. Compare *closed system*.

organelle A structure, such as a ribosome or mitochondrion, inside the *cell* that performs specific functions.

organic agriculture *Agriculture* that uses no synthetic *fertilizers* or *pesticides* but instead relies on biological approaches such as *composting* and *biocontrol*.

organic compound A *compound* made up of carbon *atoms* (and, generally, hydrogen atoms) joined by covalent *bonding* and sometimes including other *elements*, such as nitrogen, oxygen, sulphur, or phosphorus. The unusual ability of carbon to build elaborate *molecules* has resulted in millions of different organic compounds showing various degrees of complexity. Compare *inorganic compound*.

organic fertilizer A *fertilizer* made up of natural materials (largely the remains or wastes of organisms), including animal manure, crop residues, fresh vegetation, and compost. Compare *inorganic fertilizer*.

outdoor air pollution *Air pollution* that occurs outdoors. Also called *ambient air pollution*.

overgrazing The consumption by too many animals of plant cover, impeding plant regrowth and the replacement of *biomass*. Overgrazing can exacerbate damage to *soils*, natural *communities*, and the land's productivity for further grazing.

overnutrition A condition of excessive food intake in which people eat more than their daily caloric needs.

oxygen The chemical *element* with eight *protons* and eight *neutrons*. A key element in the *atmosphere* that is produced by *photosynthesis*.

ozone A *molecule* consisting of three atoms of *oxygen*. Absorbs ultraviolet radiation in the *stratosphere*. Compare *ozone layer; tropospheric ozone*.

ozone hole Term popularly used to describe the annual depletion of the stratospheric ozone layer as a result of chemical interactions with *anthropogenic pollutants* such as *chlorofluorocarbons*.

ozone layer A portion of the *stratosphere*, roughly 17–30 km above sea level, that contains most of the ozone in the *atmosphere*.

paleoclimate A climate of Earth's ancient past.

paradigm A dominant philosophical and theoretical framework within a scientific discipline.

parasite The organism in a parasitic relationship that extracts nourishment or some other benefit from the *host*.

parasitism A relationship in which one organism, the parasite, depends on another, the host, for nourishment or some other benefit while simultaneously doing the host harm. Compare *mutualism*.

parent material The base geological material in a particular location.

particulate matter Solid and (sometimes) liquid particles small enough to be suspended in the *atmosphere* and able to damage respiratory tissues when inhaled. Includes *primary pollutants* such as dust and soot, as well as *secondary pollutants* such as suphates and nitrates. A *criteria air contaminant*; abbreviated PM.

passive solar (energy collection) An approach in which buildings are designed and building materials are chosen to maximize their direct absorption of sunlight in winter, even as they keep the interior cool in the summer. Compare *active solar energy collection*.

pathogen A microbe that causes disease.

peak oil Term used to describe the point of maximum production of *petroleum* in the world (or for a given nation), after which *oil* production declines. This is also expected to be roughly the midway point of extraction of the world's oil supplies. The term is generally used in contexts suggesting that our society will face tremendous challenges once the peak has been passed. Compare *Hubbert's peak*.

peat A precursor stage to *coal*, produced when organic material that is broken down by *anaerobic* decomposition remains wet, near the surface, and not well compressed.

peer review The process by which a manuscript submitted for publication in an academic journal is examined by other specialists in the field, who provide comments and criticism (generally anonymously), and judge whether the work merits publication in the journal.

pelagic Of, relating to, or living between the surface and floor of the ocean. Compare *benthic*.

permafrost Permanently frozen ground; a layer of perennially frozen water under the surface, essentially the *W horizon* in an arctic soil.

permeability A measure of the interconnectedness of pore spaces in a *rock*, *sediment*, or *soil*. Compare *porosity*.

permit A document that grants a group or individual legal permission to carry out an activity that will have environmental impacts, usually within certain limitations and for a specified period of time. Compare *act*, *regulation*, *agreement*.

permit-trading The practice of buying and selling government-issued *marketable emissions permits* to conduct environmentally harmful activities. Under such a system, the government determines an acceptable level of *pollution* and then issues permits to pollute. A company receives credit for amounts it does not emit and can then sell this credit to other companies.

persistent A chemical that does not break down, degrade, or decompose easily, and that consequently may have a long *residence time* in a given environmental *reservoir*.

pest A *species*—typically, but not always, a non-native or alien *invasive species*—that has more harmful than beneficial impacts on an *ecosystem* or *community* (or on human interests, such as human health or crop health).

pesticide An artificial chemical used to kill insects (*insecticide*), plants (*herbicide*), or fungi (*fungicide*).

petroleum See *crude oil*.

pH A measure of the concentration of *hydrogen ions* in a solution. The pH scale ranges from 0 to 14: A solution with a pH of 7 is neutral; solutions with a pH below 7 are *acidic*, and those with a pH higher than 7 are *basic*. Because the pH scale is logarithmic, each step on the scale represents a tenfold difference in hydrogen ion concentration.

phosphorus The chemical *element* with 15 *protons* and 15 *neutrons*. An abundant element in the *lithosphere*, a key element in *macromolecules*, and a crucial plant *nutrient*.

phosphorus cycle A major *nutrient cycle* consisting of the routes that phosphorus *atoms* take through the nested networks of environmental *systems*.

photic zone In the ocean or a freshwater body, the well-lit top layer of water where *photosynthesis* occurs.

photochemical smog Brown-air smog caused by light-driven reactions of *primary pollutants* with normal atmospheric *compounds* that produce a mix of over 100 different chemicals, ground-level ozone often being the most abundant among them. Compare *industrial smog*.

photosynthesis The process by which *autotrophs* produce their own food. Sunlight powers a series of chemical reactions that convert carbon dioxide and water into sugar (glucose), thus transforming low-quality *energy* from the sun into high-quality energy the organism can use. Compare *cellular respiration*.

photovoltaic (PV) cell A device designed to collect sunlight and convert it to electrical *energy* directly by making use of the photoelectric effect.

phylogenetic tree A treelike diagram that represents the history of divergence of *species* or other taxonomic groups of organisms.

physical hazard Physical processes that occur naturally in our *environment* and pose human health hazards. These include discrete events such as earthquakes, volcanic eruptions, fires, floods, blizzards, landslides, hurricanes, and droughts, as well as ongoing natural phenomena such as ultraviolet radiation from sunlight. Compare *biological hazard*; *chemical hazard*; *cultural hazard*.

physical weathering *Weathering* that breaks *rocks* down without triggering a chemical change in the *parent material*. Wind and rain are two main forces. Compare *biological weathering*; *chemical weathering*.

phytoplankton Microscopic floating plants, mainly algae. Compare *plankton*, *zooplankton*.

phytoremediation A process for cleaning up contaminated *soil* by allowing or promoting the absorption of *toxin* through the roots or stomata of plants. Compare *bioremediation*.

pioneer species A *species* that arrives earliest, beginning the ecological process of *succession* in a terrestrial or aquatic *community*.

plankton Microscopic aquatic organisms that float (rather than swim). Compare phytoplankton, zooplankton, nekton.

planning The professional pursuit that attempts to design cities in such a way as to maximize their efficiency, functionality, and beauty.

plastic A *synthetic polymer* material, mostly derived from the *hydrocarbons* in *petroleum*.

plate tectonics The process by which Earth's surface is shaped by the extremely slow movement of tectonic plates, or sections of *crust*. Earth's surface includes about 15 major tectonic plates. Their interaction gives rise to processes that build mountains, cause earthquakes, and otherwise influence the landscape.

point source A specific spot—such as a factory's smokestacks—where large quantities of *pollutants* are discharged. Compare *non-point source*.

polar cell One of a pair of cells of *convective circulation* between the poles and 60° north and south latitude that influence global *climate* patterns. Compare *Ferrel cell*; *Hadley cell*.

policy A rule or guideline that directs individual, organizational, or societal behaviour.

pollination An interaction in which one organism (for example, bees) transfers pollen (male sex cells) from one flower to the ova (female cells) of another, fertilizing the female flower, which subsequently grows into a fruit.

polluter-pays principle Principle in which the party that produces *pollution* pays the costs of cleaning up or mitigating the pollution.

pollution Any matter or *energy* released into the *environment* that causes undesirable impacts on the health and well-being of humans or other organisms. Pollution can be physical, chemical, or biological, and can affect water, air, or soil.

pollution-prevention (P2)

polymer A chemical *compound* or mixture of compounds consisting of long chains of repeated *molecules*. Some polymers play key roles in the building blocks of life.

pool See *reservoir*.

population A group of organisms of the same *species* that live in the same area. Species are often composed of multiple populations.

population density The number of individuals within a *population* per unit area. Compare *population size*.

population distribution The spatial arrangement of organisms within a particular area. Also called *population disperson*.

population ecology Study of the quantitative dynamics of how individuals within a *species* interact with one another—in particular, why *populations* of some species decline while others increase.

population size The number of individual organisms present at a given time.

porosity The proportion of "empty" space in a *rock*, *sediment*, or *soil*. Compare *permeability*.

positive feedback loop A *feedback loop* in which output of one type acts as input that moves the *system* in the same direction. The input and output drive the system further toward one extreme or another. Compare *negative feedback loop*.

post-industrial stage The fourth and final stage of the *demographic transition* model, in which both birth and death rates have fallen to a low level and remain stable there, and *populations* may even decline slightly. Compare *industrial stage*; *pre-industrial stage*; *transition stage*.

potential energy *Energy* of position. Compare *kinetic energy*.

precautionary principle The idea that one should not undertake a new action until the ramifications of that action are well understood.

precipitation Water that condenses out of the *atmosphere* and falls to Earth in droplets or crystals.

predation The process in which one *species* (the predator) hunts, tracks, captures, and ultimately kills its prey.

predator An organism that hunts, captures, kills, and consumes individuals of another *species*, the *prey*.

prediction A specific statement, generally arising from a *hypothesis*, that can be tested directly and unequivocally.

pre-industrial stage The first stage of the *demographic transition* model, characterized by conditions that defined most of human history. In pre-industrial societies, both death rates and birth rates are high. Compare *industrial stage; post-industrial stage; transitional stage.*

prescribed (controlled) burns The practice of burning areas of forest or grassland under carefully controlled conditions to improve the health of *ecosystems*, return them to a more natural state, and help prevent uncontrolled catastrophic fires.

preservation ethic An ethic holding that we should protect the natural *environment* in a pristine, unaltered state. Compare *conservation ethic.*

pressure-state-response model A framework for environmental monitoring and reporting, which considers activities and processes that cause pressure or stress on the *environment*, the resulting state or condition of the environment, and the effectiveness of human responses to the management of these stresses.

prey An organism that is killed and consumed by a *predator.*

primary consumer An organism that consumes *producers* and feeds at the second *trophic level.*

primary extraction The initial drilling and pumping of available *crude oil.* Compare *secondary extraction.*

primary forest Forest uncut by people. Compare *second-growth.*

primary pollutant A hazardous substance, such as soot or carbon monoxide, that is emitted into the *troposphere* in a form that is directly harmful. Compare *secondary pollutant.*

primary producer See *autotroph.*

primary production The conversion of solar energy to the energy of chemical bonds in sugars during *photosynthesis*, performed by *autotrophs.* Compare *secondary production.*

primary succession A stereotypical series of changes as an ecological *community* develops over time, beginning with a lifeless substrate. In terrestrial *systems*, primary succession begins when a bare expanse of *rock, sand,* or *sediment* becomes newly exposed to the atmosphere and *pioneer species* arrive. Compare *secondary succession.*

primary treatment A stage of *wastewater* treatment in which contaminants are physically removed. Wastewater flows into tanks in which sewage solids, grit, and particulate matter settle to the bottom. Greases and oils float to the surface and can be skimmed off. Compare *secondary treatment.*

prior appropriation A concept of water law in which a user who can demonstrate a history of earlier is entitled to have access to the *resource.*

prior informed consent The ethical and legal principle that consent or acceptance of an activity (such as land development or waste disposal) is not legally valid unless the consenting person or group has been properly and adequately informed and can be shown to have a reasonable understanding of all potential impacts before giving consent.

producer See *autotroph.*

productivity The rate at which plants convert solar *energy* (sunlight) to *biomass.* Ecosystems whose plants convert solar energy to biomass rapidly are said to have high productivity. See *net primary productivity; gross primary production; net primary production.*

profundal zone In a water body, the volume of open water that sunlight does not reach. Compare *littoral zone; benthic zone; limnetic zone.*

prokaryote A typically unicellular organism. The *cells* of prokaryotic organisms lack *organelles* and a nucleus. All bacteria and archaea are prokaryotes. Compare *eukaryote.*

protein A large *molecule* made up of long chains of amino acids.

proton A positively charged particle in the nucleus of an *atom.*

proxy indicator Indirect evidence, such as pollen from *sediment* cores and air bubbles from ice cores, of the *climate* of the past.

public policy Policy that is made by governments, including those at the local, state, federal, and international levels; it consists of *legislation, regulations*, orders, incentives, and practices intended to advance societal welfare. See also *environmental policy.*

pycnocline A zone of the ocean beneath the surface in which density increases rapidly with depth. Compare *thermocline, halocline.*

pyrolysis The chemical breakdown of *organic matter* (such as *biomass*, oil shale, and so on) by heating in the absence of *oxygen*, which often produces materials that can be more easily converted to useable *energy.* A number of variations on this process exist.

radiative forcing The amount of change in *energy* that a given factor (such as *aerosols, albedo,* or *greenhouse gases*) exerts over Earth's energy balance. Positive radiative forcing warms the surface, whereas negative radiative forcing cools it.

radiatively active gas Greenhouse gas. A gas in the atmosphere that is effective at absorbing infrared radiation.

radioactive The quality by which some *isotopes* decay, changing their chemical identity as they spontaneously shed atomic particles and emit high-energy radiation.

radiocarbon dating A technique to establish the age of preserved organic materials by measuring the percentage of *carbon* that occurs as the *isotope* carbon-14, and matching this against carbon-14's clocklike progression of decay.

radioisotopes Radioactive *isotopes* that emit subatomic particles and high-*energy* radiation as they "decay" into progressively lighter isotopes until becoming stable isotopes.

radon A colourless, *radioactive* gas that is naturally occurring in certain *rocks* and *soils*. It becomes problematic for human health if it becomes concentrated inside basements and homes with poor air circulation.

rangeland Land used for grazing livestock.

reactive Materials that are chemically unstable and readily able to react with other compounds, often explosively or by producing noxious fumes. One criterion for defining *hazardous waste.*

realized niche The portion of the *fundamental niche* that is fully realized (used) by a *species.*

recharge zone An area where *precipitation* falls and infiltrates the ground, percolating downward to eventually join and replenish *groundwater* in an *aquifer.*

recombinant DNA DNA that has been patched together from the DNA of multiple organisms in an attempt to produce desirable traits (such as rapid growth, disease and pest resistance, or higher nutritional content) in organisms lacking those traits.

recovery Waste management strategy composed of *recycling* and *composting.*

recycling The collection of materials that can be broken down and reprocessed to manufacture new items.

Red List An updated list of *species* facing unusually high risks of *extinction.* The list is maintained by the World Conservation Union (IUCN).

red tide A harmful algal bloom consisting of algae that produce reddish pigments that discolour surface waters.

redox An abbreviation for reduction-oxidation reaction, an important type of chemical reaction that involves an exchange of *electrons.*

refining Process of separating the *molecules* of the various *hydrocarbons* in crude oil into different-sized classes and transforming them into various fuels and other petrochemical products.

reforestation Planting or replanting of trees in a previously forested area from which the trees had been removed by logging, fire, or some other cause.

regulation A specific rule or set of criteria issued by an administrative agency, based on the more broadly written statutory law or *act.*

relative abundance The extent to which numbers of individuals of different species are equal or skewed. One way to express *species diversity.* See *evenness;* compare *species richness.*

relative humidity The ratio of the water vapour contained in a given volume of air to the maximum amount the air could contain, for a given temperature.

relativist An ethicist who maintains that *ethics* do and should vary with social context. Compare *universalist.*

renewable natural resource A *natural resource* that is virtually unlimited or that is replenished by the *environment* over relatively short periods of hours to weeks to years. Compare *nonrenewable natural resource.*

replacement fertility The *total fertility rate (TFR)* that maintains a stable *population* size.

reservoir A location where materials in a *cycle* remain for a period of time, before moving to another reservoir. Also called a *pool.* Compare *flux.*

residence time The amount of time a material in a *cycle* remains in a given pool or *reservoir* before moving to another reservoir. Compare *flux, turnover time.*

resilience The ability of an ecological *community* to change in response to disturbance but later return to its original state. Compare *resistance.*

resistance The ability of an ecological *community* to remain stable in the presence of a disturbance. Compare *resilience*.

resource management Strategic decision making about who should extract resources and in what ways, so that resources are used wisely and not wasted.

resource partitioning The process by which *species* adapt to *competition* by evolving to use slightly different resources, or to use their shared resources in different ways, thus minimizing interference with one another.

response The type or magnitude of negative effects an animal exhibits in response to a *dose* of *toxicant* in a dose-response test. Compare *dose*.

restoration ecology The study of the historical conditions of ecological communities as they existed before humans altered them. Compare *ecological restoration*.

R horizon The bottommost layer of *soil* in a typical *soil profile*. Also called bedrock. Compare *A horizon; B horizon; C horizon; O horizon*.

ribonucleic acid See *RNA*.

riparian Relating to a river or the area along a river.

risk The mathematical probability that some harmful outcome (for instance, injury, death, environmental damage, or economic loss) will result from a given action, event, or substance.

risk assessment The quantitative measurement of *risk*, together with the comparison of risks involved in different activities or substances.

risk management The process of considering information from scientific *risk assessment* in light of economic, social, and political needs and values, in order to make decisions and design strategies to minimize *risk*.

RNA (ribonucleic acid) A usually single-stranded *nucleic acid* composed of four nucleotides, each of which contains a sugar (ribose), a phosphate group, and a nitrogenous base. RNA carries the hereditary information for living organisms and is responsible for passing traits from parents to offspring. Compare *DNA*.

rock Durable Earth material made principally of *minerals*.

rock cycle The very slow process in which rocks and the *minerals* that make them up are heated, melted, cooled, broken, and reassembled, forming *igneous*, *sedimentary*, and *metamorphic* rocks.

round table An approach to decision-making in which stakeholders with differing views meet, exchange ideas, and seek consensus. An example is the *National Round Table on the Environment and the Economy (NRTEE)*.

r–selected (r-strategist) Term denoting a *species* with high biotic potential whose members produce a large number of offspring in a relatively short time but do not care for their young after birth. *Populations* of r–selected species are generally regulated by *density-independent factors*. Compare *K–selected*.

runoff The water from *precipitation* that flows into streams, rivers, lakes, and ponds, and (in many cases) eventually to the ocean.

run-of-river Any of several methods used to generate *hydroelectric power* without greatly disrupting the flow of river water. Run-of-river approaches eliminate much of the environmental impact of large *dams*. Compare *storage*.

salination (salinisation) The buildup of salts in surface *soil* layers.

salinity Saltiness; the concentration of dissolved salts in water.

salt marsh Flat land that is intermittently flooded by the ocean where the *tide* reaches inland. Salt marshes occur along temperate coastlines and are thickly vegetated with grasses, rushes, shrubs, and other herbaceous plants.

salvage logging The removal of dead trees following a natural disturbance. Although it may be economically beneficial, salvage logging can be ecologically destructive, because the dead trees provide food and shelter for a variety of insects and wildlife and because removing timber from recently burned land can cause severe *erosion* and damage to *soil*.

sand *Sediment* consisting of particles 0.005–2.0 mm in diameter. Compare *clay; silt*.

sanitary landfill A site at which solid waste is buried in the ground or piled up in large mounds for disposal, designed to prevent the waste from contaminating the *environment*. Compare *incineration, secure landfill*.

savannah A *biome* characterized by grassland interspersed with clusters of acacias and other trees. Savannah is found across parts of Africa (where it was the ancestral home of our *species*), South America, Australia, India, and other dry tropical regions.

science A systematic process for learning about the world and testing our understanding of it.

scientific method A formalized method for testing ideas with observations that involves several assumptions and a more or less consistent series of interrelated steps.

scrubber Technology used to chemically treat gases produced in combustion to remove hazardous components and neutralize *acidic* gases, such as *sulphur dioxide* and hydrochloric acid, turning them into water and salt, in order to reduce smokestack emissions.

secondary consumer An organism that consumes *primary consumers* and feeds at the third *trophic level*.

secondary extraction The extraction of *crude oil* remaining after *primary extraction* by using solvents or by flushing underground *rocks* with water or steam. Compare *primary extraction*.

secondary pollutant A hazardous substance produced through the reaction of substances added to the *atmosphere* with chemicals normally found in the atmosphere. Compare *primary pollutant*.

secondary production The total biomass that *heterotrophs* generate by consuming *autotrophs*. Compare *gross primary production* and *net primary production*.

secondary succession A stereotypical series of changes as an ecological *community* develops over time, beginning when some event disrupts or dramatically alters an existing community. Compare *primary succession*.

secondary treatment A stage of *wastewater* treatment in which biological means are used to remove contaminants remaining after *primary treatment*. Wastewater is stirred up in the presence of *aerobic* bacteria, which degrade organic pollutants in the water. The wastewater then passes to another settling tank, where remaining solids drift to the bottom. Compare *primary treatment*.

second-growth Term describing trees that have sprouted and grown to partial maturity after virgin timber has been cut.

second law of thermodynamics Physical law stating that the nature of *energy* tends to change from a more-ordered state to a less-ordered state.

secure landfill A *landfill* that is especially engineered to receive and effectively isolate *hazardous wastes*. Compare *sanitary landfill*.

sediment The eroded remains of *rocks*.

sedimentary rock One of the three main categories of *rock*. Formed when dissolved minerals seep through *sediment* layers and act as a kind of glue, crystallizing and binding sediment particles together. Sandstone and shale are examples of sedimentary rock. Compare *igneous rock; metamorphic rock*.

seed bank A storehouse for samples of the world's crop diversity.

selection systems (for forest harvesting) Timber harvesting approaches in which some trees are cut and others are selectively allowed to remain standing.

selective breeding See *artificial selection*.

septic system A *wastewater* disposal method, common in rural areas, consisting of an underground tank and series of drainpipes. Wastewater runs from the house to the tank, where solids precipitate out. The water proceeds downhill to a drain field of perforated pipes laid horizontally in gravel-filled trenches, where microbes decompose the remaining waste.

sequestration Isolation; very long-term storage in a *reservoir*.

service Work done for others as a form of business.

sex ratio The proportion of males to females in a *population*.

shelterbelt A row of trees or other tall perennial plants that are planted along the edges of farm fields to break the wind and thereby minimize wind *erosion*.

shrubland An area of open *woodland*, characterized by low, bushy vegetation and occasional taller trees.

sick-building syndrome An illness produced by *indoor air pollution* of which the specific cause is not identifiable.

silt *Sediment* consisting of particles 0.002–0.005 mm in diameter. Compare *clay; sand*.

silviculture See *forestry*.

sink In a *cycle*, a *reservoir* that takes in more material that it releases. Compare *flux, source*.

sinkhole An area where the ground has given way with little warning as a result of subsidence caused by depletion of water from an *aquifer*.

SLOSS (Single Large or Several Small) dilemma The debate over whether it is better to make reserves large in size and few in number or many in number but small in size.

sludge Solid material that is removed during the *wastewater* treatment process.

smart growth A *planning* concept in which a community's growth is managed in ways that limit *sprawl* and maintain or improve residents' quality of life. It involves guiding the rate, placement, and style of development such that it serves the *environment,* the *economy,* and the community.

smog See *industrial smog, photochemical smog.*

social science An academic discipline that studies human interactions and institutions. Compare *natural science.*

softwood Wood derived from coniferous or needle-bearing trees. Compare *hardwood.*

soil A complex plant-supporting *system* consisting of disintegrated *rock, organic matter,* air, water, *nutrients,* and *microorganisms.*

soil degradation Damage to soils (typically through loss of *organic matter* or moisture, or through chemical contamination), or loss of soils (typically through *erosion*).

soil profile The cross-section of a *soil* as a whole, from the surface to the *bedrock.*

soil structure A measure of the organization or "clumpiness" of *soil.*

soil texture A characteristic of *soil,* which is determined by the relative proportions of *clay, sand,* and *silt* particles.

solar collector See *solar panel.*

solar energy Energy from the sun. It is perpetually renewable and may be harnessed in several ways.

solar panels Panels generally consisting of dark-colored, heat-absorbing metal plates mounted in flat boxes covered with glass panes, often installed on rooftops to harness *solar energy.*

solution A chemical mixture; most often used in reference to liquids, it can also be applied to solid and gaseous mixtures.

source In a *cycle,* a *reservoir* that releases more material than it takes in. Compare *sink, flux.*

source reduction The reduction of the amount of material that enters the *waste stream* to avoid the costs of disposal and *recycling,* help conserve resources, minimize *pollution,* and save consumers and businesses money.

specialist A *species* that can survive only in a narrow range of *habitats* that contain very specific resources. Compare *generalist.*

speciation The process by which new *species* are generated.

species A *population* or group of populations of a particular type of organism, whose members share certain characteristics and can breed freely with one another and produce fertile offspring. Different biologists may have different approaches to diagnosing species boundaries.

species-area curve A graph showing how number of *species* varies with the geographic area of a landmass or waterbody. *Species richness* commonly doubles as area increases tenfold.

Species at Risk Act (SARA) Canada's *endangered species* protection law, enacted in 2002.

species coexistence An outcome of *interspecific competition* in which no competing *species* fully excludes others, and the species continue to live side by side.

species diversity The number and variety of *species* in the world or in a particular region.

species richness The number of species in a particular region. One way to express *species diversity.* Compare *evenness; relative abundance.*

sprawl The unrestrained spread of urban or *suburban* development outward from a city centre and across the landscape.

stabilizing selection Mode of *natural selection* in which selection produces intermediate traits, in essence preserving the status quo. Compare *directional selection; disruptive selection.*

stable isotope geochemistry The study of the behaviour of isotopes that are not radioactive. Stable isotopes can act as tracers to help scientists unravel the effects of past environmental processes.

stakeholder A party (an individual or organization) that has a specific interest or stake in a proceeding. Stakeholders have specific, legally defined rights and responsibilities in processes like *environmental impact assessments.*

state-of-the-environment reporting (SOER) Monitoring and analysis of the conditions of the *environment* and periodic reporting on the results of the monitoring, usually undertaken by agencies at various levels of government.

steady state A state of *dynamic equilibrium* or balance in which there is no net change in the *system.* Compare *homeostasis.*

steady-state economy An *economy* that does not grow or shrink but remains stable.

stock The harvestable portion of a resource.

stock-and-flow resources *Renewable natural resources,* whose effective management depends upon balancing the rate of withdrawal from the stock with the rate of renewal or replenishment of the stock.

storage Technique used to generate *hydroelectric power,* in which large amounts of water are impounded in a reservoir behind a concrete *dam* and then passed through the dam to turn *turbines* that generate electricity. Compare *run-of-river.*

storm surge A temporary and localized rise in sea level and associated large waves, brought on by the high *tides,* low *atmospheric pressure,* and winds associated with storms.

stratosphere The layer of the *atmosphere* above the *troposphere* and below the mesosphere; it extends from 11 km to 50 km above sea level.

strip-mining The use of heavy machinery to remove huge amounts of earth to expose *coal* or ore *minerals,* which are mined out directly. Compare *subsurface mining.*

subduction The *plate tectonic* process by which denser ocean *crust* slides beneath lighter continental *crust* at a *convergent plate boundary.*

subsidy A government incentive (a giveaway of cash or publicly owned resources, or a tax break) intended to encourage a particular activity. Compare *green tax.*

subsistence agriculture The oldest form of *traditional agriculture,* in which farming families produce only enough food for themselves.

subsoil B horizon in a *soil profile.*

subsurface mining Method of mining underground *coal* deposits, in which shafts are dug deeply into the ground and networks of tunnels are dug or blasted out to follow coal seams. Compare *strip-mining.*

suburb A smaller community that fringes a city.

succession A stereotypical series of changes in the composition and structure of an ecological *community* through time. See *primary succession; secondary succession.*

sulphur dioxide (SO_2) A *criteria air contaminant* and a common by-product of the burning of *fossil fuels.*

surface impoundment A *hazardous waste* disposal method in which a shallow depression is dug and lined with impervious material, such as *clay.* Water containing small amounts of hazardous waste is placed in the pond and allowed to evaporate, leaving a residue of solid hazardous waste on the bottom. Compare *deep-well injection.*

survivorship curve A graph that shows how the likelihood of death for members of a *population* varies with age.

sustainability A guiding principle of *environmental science* that requires us to live in such a way as to maintain Earth's systems and its *natural resources* for the foreseeable future.

sustainable agriculture *Agriculture* that does not deplete *soils* faster than they form.

sustainable development Development that satisfies our current needs without compromising the future availability of *natural resources* or our future quality of life.

sustainable forestry certification A form of *ecolabeling* that identifies timber products that have been produced using *sustainable* methods. Several organizations issue such certification.

swamp A type of *wetland* consisting of shallow water rich with vegetation, occurring in a forested area. Compare *bog; freshwater marsh.*

swidden The traditional form of *agriculture* in tropical forested areas, in which the farmer cultivates a plot from one to a few years and then moves on to clear another plot, leaving the first to grow back to forest. When the forest is burned to clear the land, this may be called "slash-and-burn" agriculture.

symbiosis A *parasitic* or *mutualistic* relationship between different *species* of organisms that live in close physical proximity.

sympatric speciation *Species* formation that occurs when *populations* become reproductively isolated within the same geographic area. Compare *allopatric speciation.*

synergistic effect An interactive effect (as of *toxicants*) that is more than or different from the simple sum of their constituent effects.

synthetic Human-made; not naturally occurring

system A network of relationships among a group of parts, elements, or components that interact with and influence one another through the exchange of *energy*, matter, and/or information.

taiga The northern *boreal forest biome*.

Talloires Declaration A document composed in Talloires, France, in 1990 that commits university leaders to pursue *sustainability* on their campuses. It has been signed by over 300 university presidents and chancellors from more than 40 nations, including 29 Canadian universities as of mid-2008.

tar sands See *oil sands*.

taxonomist A scientist who classifies *species*.

temperate deciduous forest A *biome* consisting of midlatitude forests characterized by broad-leafed trees that lose their leaves each fall and remain dormant during winter. These forests occur in areas where *precipitation* is spread relatively evenly throughout the year: much of Europe, eastern China, and eastern North America.

temperate grassland A *biome* whose vegetation is dominated by grasses and features more extreme temperature differences between winter and summer, and less *precipitation* than *temperate deciduous forests*; also known as steppe or prairie.

temperate rainforest A *biome* consisting of tall coniferous trees, cooler and less species-rich than *tropical rainforest* and milder and wetter than *temperate deciduous forest*.

temperature (thermal) inversion A departure from the normal temperature distribution in the *atmosphere*, in which a pocket of relatively cold air occurs near the ground, with warmer air above it. The cold air, denser than the air above it, traps *pollutants* near the ground and causes a buildup of smog.

teratogen A *toxicant* that causes harm to the unborn, resulting in birth defects.

terracing The cutting of level platforms, sometimes with raised edges, into steep hillsides to contain water from *irrigation* and *precipitation*. Terracing transforms slopes into series of steps like a staircase, enabling farmers to cultivate hilly land while minimizing their loss of *soil* to water *erosion*.

theory A widely accepted, well-tested explanation of one or more cause-and-effect relationships that has been extensively validated by a great amount of research. Compare *hypothesis*.

thermal inversion See *temperature inversion*.

thermal mass Construction materials that absorb heat, store it, and release it later, for use in *passive solar energy* approaches.

thermal pollution Heat or heated water released into the *environment*, which can damage natural *ecosystems*.

thermocline A zone of the ocean beneath the surface in which temperature decreases rapidly with depth. Compare *pycnocline, halocline*.

thermohaline circulation A worldwide system of ocean currents in which warmer, fresher water moves along the surface and colder, saltier water (which is more dense) moves deep beneath the surface.

thermosphere The *atmosphere*'s top layer, extending upward to an altitude of 500 km.

threatened Refers to a *species* that appears to be on the verge of becoming *endangered*.

threshold dose The *dose* at which a *toxicant* begins to affect a *population* of test animals. Compare ED_{50}; LD_{50}.

tidal energy *Energy* harnessed by erecting a *dam* across the outlet of a tidal basin. Water flowing with the incoming or outgoing *tide* through sluices in the dam turns turbines to generate *electricity*.

tide The periodic rise and fall of the ocean's height at a given location, caused by the gravitational pull of the Moon and Sun.

topography Physical geography; the shape and arrangement of landforms.

topsoil That portion of the *soil* that is most nutritive for plants and is thus of the most direct importance to *ecosystems* and to *agriculture*. Also known as the *A horizon*.

total fertility rate (TFR) The average number of children born per female member of a *population* during her lifetime.

toxic Materials that are poisonous; able to harm health of people or other organisms when a substance is inhaled, ingested, or touched. One criterion for defining *hazardous waste*. Compare *toxicant*.

toxic air pollutant *Air pollutant* that is known to cause cancer, reproductive defects, or neurological, developmental, immune system, or respiratory problems in humans, and/or to cause substantial ecological harm by affecting the health of nonhuman animals and plants.

toxicant (toxin or **toxic agent)** A substance that acts as a poison to humans or wildlife.

toxicity The degree of harm a chemical substance can inflict.

toxicology The scientific field that examines the effects of poisonous chemicals and other agents on humans and other organisms.

toxin See *toxicant*.

trade winds Prevailing winds between the equator and 30° latitude that blow from east to west.

traditional agriculture Biologically powered *agriculture*, in which human and animal muscle power, along with hand tools and simple machines, perform the work of cultivating, harvesting, storing, and distributing crops. Compare *industrialized agriculture*.

traditional ecological knowledge (TEK) (or *indigenous ecological knowledge*) The intimate knowledge of a particular *environment* possessed and passed along by those who have inhabited an area for many generations, typically via an oral tradition of teachings, songs, and storytelling.

tragedy of the commons The scenario in which each individual withdraws whatever benefits are available from an unregulated or poorly regulated common property resource, as quickly as possible, until the resource becomes over-used and depleted

transboundary Crossing or straddling a political boundary such as a national border.

transform plate boundary Area where two tectonic plates meet and slip and grind alongside one another. For example, the Pacific Plate and the North American Plate rub against each other along California's San Andreas Fault.

transgene A *gene* that has been extracted from the *DNA* of one organism and transferred into the DNA of an organism of another *species*.

transgenic Term describing an organism that contains *DNA* from another *species*.

transitional stage The second stage of the *demographic transition* model, which occurs during the transition from the *pre-industrial stage* to the *industrial stage*. It is characterized by declining death rates but continued high birth rates. See also *post-industrial stage*. Compare *industrial stage; post-industrial stage; pre-industrial stage*.

transpiration The release of water vapour by plants through their leaves.

trawling Fishing method that entails dragging immense cone-shaped nets through the water, with weights at the bottom and floats at the top to keep the nets open. Compare *bottom-trawling*.

treaty See *convention*.

tributary A smaller river that flows into a larger one.

triple bottom line An approach to *sustainability* that attempts to meet environmental, economic, and social goals simultaneously.

trophic level Rank in the feeding hierarchy of a *food chain*. Organisms at higher trophic levels consume those at lower trophic levels.

trophic pyramid A diagram showing the *trophic levels* in a *food chain* from bottom to top, with the *autotrophs* at the bottom, moving up through the various levels of *consumers*. The diagram is typically pyramid-shaped because *biomass, energy*, and numbers of individuals decrease upward through the trophic levels.

tropical dry forest A *biome* that consists of deciduous trees and occurs at tropical and subtropical latitudes where wet and dry seasons each span about half the year. Widespread in India, Africa, South America, and northern Australia.

tropical rainforest A *biome* characterized by year-round rain and uniformly warm temperatures. Found in Central America, South America, southeast Asia, west Africa, and other tropical regions. Tropical rainforests have dark, damp interiors; lush vegetation; and highly diverse *biotic communities*.

troposphere The bottommost layer of the *atmosphere*; it extends to 11 km above sea level. See also *stratosphere*.

tropospheric ozone Ground-level ozone, an air pollutant and a component of *smog*.

tundra A *biome* that is nearly as dry as *desert* but is located at very high latitudes along the northern edges of Russia, Canada, and Scandinavia. Extremely cold winters with little daylight and moderately cool summers with lengthy days characterize this landscape of lichens and low, scrubby vegetation.

turbidity Water cloudiness caused by suspended solids.
turbine A rotary device that converts the *kinetic energy* of a moving substance, such as steam, into mechanical energy. Used widely in commercial power generation from various types of energy sources.
turnover time The time it would take for a material to work its way through and out of a *reservoir*, if all of the *sources* or *fluxes* of that material into the reservoir were stopped. Compare *residence time*.
unconfined aquifer A water-bearing, porous layer of *rock*, *sand*, or gravel that lies atop a less-permeable substrate. The water in an unconfined aquifer is not under pressure because there is no impermeable upper layer to confine it. Compare *confined aquifer*.
undernourishment A condition of insufficient *nutrition* in which people receive less than 90% of their daily caloric needs.
understory (of a forest) The floor and the lowest levels of growth within a forest.
uneven-aged Term describing stands of trees in timber plantations that are of different ages. Uneven-aged stands more closely approximate a natural forest than do *even-aged* stands.
United Nations Convention on the Law of the Sea (UNCLOS) The international agreement that established, among other things, the 200-nautical-mile *Exclusive Economic Zones* of nations.
United Nations Environment Programme (UNEP) The department of the United Nations, headquartered in Nairobi, Kenya, established after the Conference on the Human Environment in 1972 to deal with international environmental concerns and encourage nations to adopt environmentally sound approaches to development. One of several agencies within the U.N. that influence international environmental policy and practice.
universalist An *ethicist* who maintains that there exist objective notions of right and wrong that hold across various *cultures* and situations. Compare *relativist*.
upwelling In the ocean, the flow of cold, deep water toward the surface. Upwelling occurs in areas where surface *currents* diverge. Compare *downwelling*.
urban ecology A scientific field that views cities explicitly as *ecosystems*. Researchers in this field seek to apply the fundamentals of *ecosystem ecology* and *systems* science to urban areas.
urban growth boundary (UGB) In *planning*, a geographic boundary intended to separate areas desired to be urban from areas desired to remain rural. Development for housing, commerce, and industry are encouraged within urban growth boundaries, but beyond them such development is severely restricted. Compare *green belt*.
urban heat island effect A phenomenon in which cities are generally several degrees warmer than surrounding suburbs and rural areas, due to tall buildings that interfere with convective cooling, paved surfaces that absorb heat, and concentrated activities that generate waste heat. The urban heat island effect can contribute to the development of *dust domes*.
urbanization The shift from rural to city and *suburban* living.
utilitarian principle The principle that something is right when it produces the greatest practical benefits for the most people.
valuation The attempt to quantify the value of a particular environmental *good* or *service*—even if it cannot easily be expressed in monetary terms.
variable In an *experiment*, a condition that can change. See *dependent variable* and *independent variable*.
variable retention harvesting Approaches to logging that involve harvesting only some of the trees from an area. Compare *clear-cutting*.
vector An organism that transfers a *pathogen* to its *host*. An example is a mosquito that transfers the malaria pathogen to humans.
volatile organic compound (VOC) One of a large group of potentially harmful organic chemicals used in industrial processes.
voluntary guidelines Standards that are undertaken willingly by a particular sector, and self-regulated.

warm front The boundary where a mass of warm air displaces a mass of colder air. Compare *cold front*.
waste Any unwanted product that results from a human activity or process.
waste exchange A system whereby producers of waste materials are matched with organizations in need of the same materials as inputs into other industrial processes.
waste management Strategic decision making to minimize the amount of *waste* generated and to dispose of waste safely and effectively.
waste stream The flow of *waste* as it moves from its sources toward disposal destinations.
waste-to-energy (WTE) facility An *incinerator* that uses heat from its furnace to boil water to create steam that drives electricity generation or fuels heating systems.
wastewater Any water that is used in households, businesses, industries, or public facilities and is drained or flushed down pipes, as well as the polluted *runoff* from streets and storm drains.
waterlogging The saturation of *soil* by water, in which the *water table* is raised to the point that water bathes plant roots. Waterlogging deprives roots of access to gases, essentially suffocating them and eventually damaging or killing the plants.
water mining The withdrawal of water at a rate faster than it can be replenished.
water quality The suitability of water for various purposes (such as drinking or swimming), as determined by comparing the water's physical, chemical, and biological characteristics with a set of predetermined standards.
watershed The entire area of land from which water drains into a given river.
water table The upper limit of *groundwater* held in an *aquifer*.
wave energy Energy harnessed from the motion of wind-driven waves at the ocean's surface. Many designs for machinery to harness wave energy have been invented, but few have been adequately tested.
weather The local physical properties of the *troposphere*, such as temperature, pressure, humidity, cloudiness, and wind, over relatively short time periods. Compare *climate*.
weathering The physical, chemical, and biological processes that break down *rocks* and minerals, turning large particles into smaller particles.
weed A plant that competes with crops. The term is defined by economic interest, and is not biologically meaningful. Compare *pest*.
westerlies Prevailing winds from 30° to 60° latitude that blow from west to east.
wetland A system that combines elements of *fresh water* and dry land. These biologically productive systems include *freshwater marshes*, *swamps*, and *bogs*.
wildlife refuge An area set aside to serve as a haven for wildlife and also sometimes to encourage hunting, fishing, wildlife observation, photography, environmental education, and other public uses.
windbreak See *shelterbelt*.
wind farm A development involving a large group of *wind turbines*.
wind turbine A mechanical assembly that converts the wind's *kinetic energy*, or energy of motion, into electrical energy.
woodland Land that is covered with woody vegetation, including trees and shrubs. Compare *forest*.
world heritage site A location designated by the *United Nations* for its cultural or natural value.
worldview A way of looking at the world that reflects a person's (or a group's) beliefs about the meaning, purpose, operation, and essence of the world.
xeriscaping Landscaping using plants adapted to arid conditions.
zoning The practice of classifying areas for different types of development and land use.
zooplankton Very small or microscopic floating animals. Compare *plankton*, *phytoplankton*.

Photo Credits

Part Opener Photos: Part 1 Wayne Lynch **Part 2** Daryl Benson/Stone/Getty Images **Part 3** All Canada Photos/Alamy

Chapter 1 Opening Photo NASA/Visible Earth **Central Case** Image courtesy of the Image Science & Analysis Laboratory, NASA Johnson Space Center **1.2b** Charles O'Rear/CORBIS **1.4** George Konig/Hulton Archive Photos **1.5** CP Photo/Sean Kilpatrick **1.6** © Barrie Harwood/Alamy **1.11** CORBIS **1.12** © Bryan & Cherry Alexander Photography/Alamy **1.14** Reuters **1.15** Paul Gendell/Alamy Images **Science Behind the Stories: Easter Island** Patricia Jones **The Science behind the Story: Mission to Planet Earth miniphoto** Data made available by NASA/GSFC/MITI/ERSDAC/JAROS, and U.S/Japan ASTER Science Team; **main** Jacques Descloitres, MODIS Land Rapid Response Team NASA/GSFC CP Photo/Sean Kilpatrick

Chapter 2 Opening Photo Donnie Reid **Case Study** Donnie Reid **2.1** Exxon Corporation **2.2** AP Photo/Jack Smith **2.5a** Digital Vision/PictureQuest **2.13 left** Daniel Zheng/CORBIS **2.13 right** Anne-Marie Weber/Taxi **2.16** © NeilRabinowitz/CORBIS **2.17** Ken MacDonald/SPL/Photo Researchers, Inc. **2.17b** Woods Hole Oceanographic Institute **2.18** Chip Clark **2.19a** © Roger Garwood & Trish Ainslie/CORBIS **2.19b** Michael Melford/National Geographic/Getty Images **2.20** Dorling Kindersley **The Science behind the Story: Dirty Work** © Steffen Hauser/botanikfoto/Alamy **The Science behind the Story: Isotopes** Keith Hobson

Chapter 3 Opening Photo Michael & Patricia Fogden/CORBIS **Case Study** Michael & Patricia Fogden/CORBIS **3.1a** Nicholas Athanas/Tropical Birding **3.1b** Michael Fogden/DRK Photo **3.1c** Michael & Patricia Fogden/CORBIS **3.1d** Michael Fogden/Photolibrary **3.6** Painting by Marella J. Sibbick/The National History Museum, London **3.7** Chip Clark/National Museum of Natural History, Smithsonian Institution **3.8 (r)** © Fletcher & Baylis **3.11a** Wisconsin Historical Society **3.11b** G. I. Bernard/PhotoResearchers, Inc. **3.12a** Keenan Ward/CORBIS **3.12b** Art Wolfe/The Image Bank **3.12c** PhotoDisc Green **3.18a** Mike Danzenbaker **3.18b** A.Witte/C.Mahaney/Getty Images **3.18c** John Waters/Nature Picture Library **3.19** Matthias Clamer/Stone **The Science behind the Story: K-T Mass Extinction miniphoto** Roger Ressmeyer/CORBIS/Bettmann **main left** Benjamin Cumming **main right** NASA/GSFC/LaRC/JPL, MISR Team **The Science behind the Story: Climate Change and Monteverde** J. Alan Pounds

Chapter 4 Opening Photo Robert Estall/Robert Harding World Imagery **Case Study** Wolfgang Polzer **4.3** Michael & Patricia Fogden/CORBIS **4.5a** Peter Johnson/CORBIS **4.5b** Getty Images **4.5c** Michael and Patricia Fogden/CORBIS **4.6a** Tom Stack and Associates, Inc. **4.6b** Peter Arnold Inc./Alamy **4.7** Joke Stuurman-Huitema/Foto Nat/Minden Pictures **4.8** Michael & Patricia Fogden/CORBIS **4.12** Charles D.Winters/Photo Researchers, Inc. **4.15** Michael Wheatley © AllCanadaPhotos.com **4.18a** DarrellGulin/CORBIS **4.19a** Liz Hymans/CORBIS **4.20a** Pat O'Hara/CORBIS **4.21a** Philip Gould/CORBIS **4.22a** Charles Mauzy/CORBIS **4.23a** David Samuel Robbins/CORBIS **4.24a** Getty Images/Digital Vision **4.25a** Wolfgang Kaehler/CORBIS **4.26a** Sylvester Adams/Getty Images **4.27a** Charles Mauzy/CORBIS **The Science behind the Story: Zebra Mussels** Fisheries and Oceans Canada **The Science behind the Story: Otters, Urchins, Kelp** Jim Bodkin

Chapter 5 Opening Photo Yves Marcoux/Firstlight **Case Study** Peter Arnold Inc./Alamy **5.3 top** left Megapress/Alamy **5.3 bottom** left Gary Blackeley/Shutterstock **5.3 top right** © Joseph Sohm; Visions of America/CORBIS **5.5a** All Canada Photos/Alamy **5.5b** Glow Images/Alamy **5.5c** dbimages/Alamy **5.5d** All Canada Photos/Alamy **5.5e** All Canada Photos/Alamy **5.10** D.W. Schindler **5.11** NASA/Visible Earth **The Science behind the Story: Gulf of Mexico** Yann Arthus-Bertrand/CORBIS

Chapter 6 Opening Photo Redlink/CORBIS **Case Study** Louise Gibb/The Image Works **6.1** CORBIS **6.4a** Art Resource **6.4b** Bettman/CORBIS **6.12c** SETBOUN/CORBIS **6.12** David Turnley/CORBIS **6.18 inset** AFP/Getty Images **6.17** Tiziana and Gianni Baldizzone/CORBIS **6.19a** Peter Menzel/Peter Menzel Photography **6.19b** Peter Ginter **6.24** Getty Images **The Science behind the Story: Fertility Decline** Mark Edwards/Peter Arnold, Inc.

Chapter 7 Opening Photo Robert Williams **Case Study** Landsat, Peter Lafleur Trent University **7.10** Ron Giling/Peter Arnold, Inc. **7.11a** Barry Runk/Stan/Grant Heilman Photography **7.11b** Grant Heilman/Grant Heilman Photography **7.11c, d** U.S. Department of Agriculture **7.12** CP Photo/Library and Archives Canada PA-139645 **7.13** Ted Spiegel/CORBIS **7.14a** Sylvan Wittwer/Visuals Unlimited **7.14b** Kevin Horan/Stone **7.14c** Keren Su/Stone **7.14d** Ron Giling/Peter Arnold, Inc. **7.14e** Yann Arthus-Bertrand/CORBIS **7.14f** U.S. Department of Agriculture **7.15** China Photos/GettyImages **7.16a** Photo Disc **7.16b** Carol Cohen/CORBIS **7.19** W. Perry Conway/CORBIS **The Science behind the Story: No-Till Agriculture** Joanna B. Pinneo/AURORA **The Science behind the Story: Nitrate Contamination** © Her Majesty the Queen in Right of Canada, Environment Canada, 1990. Reproduced with the permission of the Minister of Public Works and Government Services Canada.

Chapter 8 Opening Photo Paul Austring Photography/Firstlight **Case Study** Macduff Everton/CORBIS, John Schmeiser **8.2** Alexandra Avakian/CORBIS **8.3** Art Rickerby/Time Life Pictures/Getty Images **8.4** Jack Dykinga/ImageBank **8.4 inset** Barry Runk/Stan/Grant Heilman Photography, Inc. **8.6 a, b** Department of Natural Resources, Queensland, Australia **8.8** Photo Disc **8.9** Bob Rowan, Progressive Image/CORBIS **8.14a** Native Seeds/SEARCH **8.14b** Hal Fritts, Native Seeds/SEARCH **8.16** Arthur C. Smith III/Grant Heilman Photography, Inc. **8.20** Aqua Bounty Farms/AP **8.22** Chris Cheadle/Alamy **8.23** Martin Bourque **The Science behind the Story: The Alfalfa and the Leafcutter** Wikipedia **The Science behind the Story: Organic Farming** Mark A. Liebig

Chapter 9 Opening Photo Maurice Hornocker **Case Study** Maurice Hornocker **9.12** Kathy Ferguson-Johnson/Photo Edit **9.13** Getty Images **9.16** Konrad Wothe/Photolibrary **9.17** Joel Sartore/Getty/National Geographic Society/CORBIS **9.21** Natalie Fobes/CORBIS **9.22** Tom & Pat Leeson/Photo Researchers, Inc. **The Science behind the Story: Amphibian Diversity miniphoto** Asanga Ratnaweera and Madhava Meegaskumbura **main** Craig K. Lorenz/Photo Researchers, Inc. **The Science behind the Story: Swift Fox** All Canada Photos/Alamy

Chapter 10 Opening Photo © Dewitt Jones/CORBIS **Case Study** Garth Lenz; © Dewitt Jones/CORBIS **10.3** John E. Marriot **10.4a** Ron Niebrugge/Alamy **10.4b** Tim Ennis/Alamy **10.4c** © Perry Mastrovito/Corbis **10.7** Library and Archives Canada, PA-040927 **10.8** Shutterstock **10.9** grazing Alec Pytlowany/Masterfile **10.11a** A.S. Zain Shutterstock **10.11b** Chris Fredriksson/Alamy **10.12** Weyerhaeuser Company **10.13** Rob Badger/Getty Photos **10.15** CP Photo/Brent Braaten **10.16** Reuters/CORBIS **10.17** David Noton Photography/Alamy **10.20** All Canada Photos/Alamy **10.21a** Global Forest Watch Canada, 2006 (Landsat ETM + Path 44/Row 23, 2000) **10.21b** James Zipp/Photo Researchers, Inc. **The Science behind the Story: Spruce Budworm** john t. fowler/Alamy; All Canada Photos/Alamy **The Science behind the Story: Surveying Earth's Forests** Dan Lamont/Corbis

CR-2 PHOTO CREDITS

Chapter 11 Opening Photo Robert McGouey/Alamy **Case Study** NASA **11.2** Wally Bauman/Alamy **11.3** Frans Lanting/Minden Pictures **11.6** © Department of Natural Resources Canada, 2006. All rights reserved. **11.10a** AP Wide World Photos **11.10b** Ian Berry/Magnum Photos **11.10c** © Hulton-Deutsch Collection/CORBIS **11.11** Canadian Space Agency **11.13a** Gilles Saussier/Liaison **11.13b** NASA Visible Earth **11.17** Mike Grandmaison/Firstlight **11.18** Jim Tuten/Black Star **11.19** © Bettmann/CORBIS **11.20a** Robert T. Zappalorti/Photo Researchers, Inc. **11.20b** David Muench/CORBIS **11.22** © Department of Natural Resources Canada, 2006. All rights reserved. **11.23** David J. Cross/Peter Arnold, Inc. **The Science behind the Story: Arsenic** Shezad Nooran/Woodfin Camp & Associates **The Science behind the Story: Walkerton cover** CP Photo/Frank Gunn **main** © Charles O'Rear/CORBIS

Chapter 12 Chapter Opener Jeffrey L. Rotman/CORBIS **Case Study** Library and Archives Canada, PA-110814 **12.8** Alcoa Technical Center **12.10** Laguna Design/Science Photo Library/Photo Researchers, Inc. **12.11** Bruce Robison/Minden Pictures **12.12** Ralph A. Clevenger/CORBIS **12.13a** Harvey Lloyd/Taxi **12.13b** Stephen Frink/CORBIS **12.14** Carol Cohen/CORBIS **12.15** Robert Williams **12.16** Mark A. Johnson/CORBIS **12.17** Momatiuk Eastcott/Animals Animals/Earth Scenes **12.19a** David M. Phillips/Visuals Unlimited **12.19b** Bill Bachman/Photo Researchers, Inc. **12.22a** Norbert Wu/Norbert Wu Productions **12.22b** Norbert Wu Productions **The Science behind the Story: China's Fisheries** Sherman Lai **The Science behind the Story: Marine Reserves** Rachel Graham/Callum Roberts **Canadian Environmental Perspectives** TedAmsdenPhotography

Chapter 13 Opening Photo Bettmann/CORBIS **Case Study** Wikipedia, The Department of Fisheries and Oceans, Canada. photographer: Dr. Ken Mills **13.10a** NASA Earth Observatory **13.10b** Bettmann/CORBIS **13.10c** Noel Hendrickson/Masterfile **13.11** Frederic J. Brown/AFP/Getty Images **13.15b** Pittsburgh Post-Gazette **13.16b** Allen Russell/Index Stock Imagery **13.19b** NASA **13.20** C. Halsall **13.25** David Turnley/CORBIS **The Science behind the Story: Ozone Depletion** AP Wide World Photos **The Science behind the Story: Acid Rain miniphoto** HO Photography **main** Will & Deni McIntyre/CORBIS

Chapter 14 Opening Photo Rossi, Guido Alberto/Image Bank **Case Study** Adam Woolfitt/CORBIS **14.7a** Ted Spiegel/CORBIS **14.14 top** Lars Johansson/Alamy **14.14 middle** Alaska Stock **14.14 bottom** Brian & Cherry Alexander/Photo Researchers, Inc. **14.16 a** Library and Archives Canada, PA-031956 **14.16b** Dan McCarthy, Department of Earth Sciences, Brock University **14.18a, b** NASA **14.22** Adrian Baras/Shutterstock **14.29** CP Photo/Steve White **The Science behind the Story: World's Longest Ice Core** Pasquale Sorrentino/Photo Researchers, Inc. **The Science behind the Story: Greenland's Glaciers** Rex Stucky/National Geographic Image Collection

Chapter 15 Opening Photo Stocktrek Images, Inc. Alamy **Case Study** National Geographic/Getty Images **15.2** UNEP **15.7** Wikipedia **15.8** Watts/Hall Inc/Firstlight **15.12** Dave Olecko/Bloomberg News/Landov LLC **15.14a** CP Photo/Larry MacDougal **15.14b** CP Photo/Jordan Verlage **15.17** Vivian Stockman **15.18** Steven Kazlowski/Alaska Stock **15.22** AGA Infrared/Photo Researchers Inc. **The Science behind the Story: Clean Coal** Sam Ogden/Photo Researchers, Inc. **The Science behind the Story: How Crude Oil Is Refined miniphoto** John Gress/CORBIS/Reuters America LL **main** CORBIS

Chapter 16 Opening Photo CP Photo/Kevin Frayer **Case Study** Skellefteå Kraft AB **16.3a** 22 DigiTal/Alamy **16.3b** NASA **16.4a** Robert McGouey/Alamy **16.8** argus/Peter Arnold, Inc. **16.12** John S. Zeedick/Getty Images **16.13a** Paul Fusco/Magnum Photos **16.13b** Reuters/CORBIS **16.15a** Roger Ressmeyer/CORBIS **16.15b** U.S. Department of Energy/Photo Researchers, Inc. **16.16a** U.S. Department of Energy **16.17** Chris Steele-Perkins/Magnum Photos **16.19a** Larry Lefever/Grant Heilman Photography **16.21** Lars-Erik Larsson/Svebio **16.22** Wolfgang Hoffmann/USDA/ARS/Agricultural Research Service **16.23** GreenWood Resource, Inc/Jake Eaton **The Science behind the Story: Assessing Emissions** Liu Liqun/CORBIS **The Science behind the Story: Chernobyl** Caroline Penn/CORBIS

Chapter 17 Opening Photo Laszio Podor/Alamy **Case Study** NASA, CP Photo/Mike Dembeck **17.5** Alan Weisman **17.6** CORBIS **17.9b** © Her Majesty the Queen in Right of Canada, as represented by the Minister of Natural Resources, 2000 **17.12b** JUPITER IMAGES/Comstock Images/Alamy **17.13** Yukon Energy **17.15b** Pascal Goetgheluck/Science Photo Library **The Science behind the Story: Beare Road** john t. fowler/Alamy, Aldworth Engineering Inc. **The Science behind the Story: Heating and Cooling** EnWave Energy Corporation

Chapter 18 Opening Photo Mark Gibson/Index Stock **Case Study** Jim Robb, Friends of the Rouge Watershed **18.4** Nigel Dickinson/Stone **18.7** Jim Robb, Friends of the Rouge Watershed **18.12a, b** City of Edmonton **18.15** CP Photo/Bill Sanford **18.16** Joe Sohm/Alamy **18.17** Robert Brook/Photo Researchers, Inc. **The Science behind the Story: Digging Garbage miniphoto** Andrew Holbrooke/CORBIS **main** Louie Psihoyos/©psihoyos.com **The Science behind the Story: Toxicity of "E-Waste"** AP Photo/Keystone, Walter Bieri

Chapter 19 Opening Photo Bill Brooks/Masterfile **Case Study** AFP/Getty Images, Photofusion Picture Library/Alamy **inset** SciMAT/Photo Researchers, Inc. **19.2a** Ingram Publishing/Alamy **19.2b** Spencer Grant/Photo Edit **19.2c** Martin Dohrn, Photo Researchers, Inc. **19.2d** SWA-Dann Tardit/CORBIS **19.5** Joel W. Rogers/CORBIS **19.6** Bettmann/CORBIS **19.7a** Bettmann/CORBIS **19.7b** James L. Stanfield/National Geographic/Getty Images **The Science behind the Story: Bisphenol-A** From Hunt, PA, KE Koehler, M Susiarjo, CA Hodges, A Ilagan, RC Voight, S Thomas, BF Thomas and TJ Hassold. 2003. *Bisphenol A exposure causes meitic aneuploidy in the female mouse.* Current Biology 13: 546–553 **The Science behind the Story: Pesticides and Child Development** Jeff Conant

Chapter 20 Opening Photo Rich Iwaski/age footstock **Case Study** Courtesy of VCEC, Courtesy of Sightline **20.1a** Corbis Premium RF/Alamy **20.1b** Radius Images/All Canada Photos **13.3a** Chad Palmer/Shutterstock **20.4** Wikipedia **20.5 a, b** U.S. EPA **20.6a** Statistics Canada, 2006 Census: Portrait of the Canadian Population in 2006: Findings, Catalogue 97-550-XWE2006001, Released March 13, 2007, URL: http://www12.statcan.ca/census-recensement/2006/as-sa/97-550/index-eng.cfm?CFID=740501&CFTOKEN=57947669 <http://www12.statcan.ca/as-sa/97-550/index-eng.cfm?CFID=740501&CFTOKEN=57947669 **20.6b** Map: "2006 Census, Calgary population change". Source: Statistics Canada, 2006 Census: Portrait of the Canadian Population in 2006: Findings, Catalogue 97-550-XWE2006001, Released March 13, 2007, URL: http://www12.statcan.ca/census-recensement/2006/as-sa/97-550/index-eng.cfm?CFID=740501&CFTOKEN=57947669 <http://www12.statcan.ca/census-recensement/2006/as-sa/97-550/index-eng.cfm?CFID=740501& CFTOKEN=57947669 **20.7a** Lester Lefkowitz/CORBIS **20.7b** Bob Krist/CORBIS **20.7c** David R. Frazier Photolibrary, Inc. **20.7d** Aldo Torelli/Getty Images **20.8a, b** Wikipedia **20.11** Time & Life Pictures/Getty Images **20.13** © Carlos Cazalis/Corbis **20.14** Bettmann/CORBIS **20.15** Dick Hemingway **20.16** Bohemian Nomad Picture makers/CORBIS **20.18** Sally and Richard Greenhill/Alamys **The Science behind the Story: Sprawl miniphoto** Reid Ewing **main** Editorial Fotos/Alamy **The Science behind the Story: Rail Transit** David M. Grossman/The Image Works

Chapter 21 Opening Photo Diavik Diamond Mine **Case Study** Library and Archives Canada, PA-061609 **21.2a** The Gundjehmi Aboriginal Corporation **21.2b** Reprinted with permission from Vale Inco. **21.4** Library of Congress **21.5a** NORMA JOSEPH/Alamy **21.5b** Library and Archives Canada, PA-121371 **21.6** CORBIS **21.7a** © Pallava Bagla/Corbis **21.7b** © Patrick Robert/Corbis **21.9** ANTONIO SCORZA/AFP/Getty Images **21.12** AFP/CORBIS **21.15a** Konrad Wothe/Minden Pictures **21.15b** Bill Hatcher/National Geographic Image Collection **21.15c** Bruce Forster/The Image Bank **21.15d** CORBIS **21.15e** Charles O'Rear/CORBIS **The Science behind the Story: Mirrar Clan** The Gundjehmi Aboriginal Corporation, Reuters New Media Inc./CORBIS **The Science behind the Story: Ethics in Economics** AP Wide World Photos

Chapter 22 Opening Photo © Peter Johnson/CORBIS **Case Study** © NASA/Corbis **3.1** Lori Saldaña **22.2** Annie Griffiths Belt/CORBIS **22.5** Library and Archives Canada, PA094969 **22.6** Library and Archives Canada, PA011632 **22.7** Erich Hartmann/Magnum Photos, Inc. **22.8** Bettmann/CORBIS **22.9** © Bettmann/CORBIS **22.10a** CWH, Associated Press **22.10b** Binod Joshi/Associated Press **22.12** Don Boroughs/The Image

Works **22.14** Jon Hrusa/AP Photo **22.15** Peter Bowater/Alamy **22.16** Philip Rostron/Masterfile **22.17** Realimage/Alamy **The Science behind the Story: NAFTA and NAAEC** Jamie Kum/Ocean Imaging, www.oceani.com, Lori Saldaña

Chapter 23 Opening Photo Mark Edwards/Peter Arnold, Inc. **Case Study** UBC Sustainability Report 2006-2007 **23.1** Gorilla Composting **23.2a** Oberlin College Archives **23.3** Don Pierce/UVIC Photo Services **23.4** Alberta Solar Decathlon Project **23.5** Christopher DeWolf **23.6** Cindy Lopez **23.7** Photo by Jay Wilson **23.10** Galen Rowell/CORBIS **23.13** Costco Wholesale **23.16** John Howard/Riser/Getty Images **23.18** NORMAND BLOUIN/AFP/Getty Images **23.20** NASA/Langley Research Center **The Science behind the Story: Footprint Calculator** Barbara Murck **The Science behind the Story: Environmental Performance of Nations** Daniel Esty

Additional Credits

Chapter 1
Profile and Photograph of David Suzuki. Reprinted by permission of David Suzuki; photo Kent Kalberg

Chapter 2
An excerpt from "Central Case: The Unusual Microbialites of Pavilion Lake" by Darlene Lim & Bernard Laval. Pavillion Lake Research Project: Relevance to Astrobiology and Space Exploration. Http://supercritical.civil.ubc.ca/~pavilion/information/relevance.htm © Lim & Laval. Reprinted by permission of the authors.

Profile and Photograph of Praveen Saxena. Reprinted by permission of Praveen Saxena & an extract from article on "phytoremediation (Lemon Scented Geranums) from http://www.carlton.ca/jmc/cnews/30031997/story1.html. Reprinted by permission of Praveen Saxeena.

Chapter 3
Profile and Photogrpah of Maydianne Andrade. Reprinted by permission of Maydianne Andrade.

Chapter 4
Jessica M. Ward & Anthony Ricciardi, Impacts of Dreissena Invasions on Benthic Macroinvertebrate Communities: A Metaanalysis, Diversity and Distribution (2007) 13, pp. 155-165. Reprinted by permission of the authors. M.E. Palmer & Anthony Ricciardi, Community Interactions Affecting the Relative Abundances of Natives and Invasive Amphipods in the St. Lawrence River, Can. J. Fish. Aquat. Sci. 62: pp.1111-1118 (2005). Reprinted by permission of the authors.

Profile and Photograph of Zoe Lucas. Reprinted by permission of Zoe Lucas.

Excerpt from Boreal Leadership Council, www.borealcanada.ca, Canadian Boreal Initiative. © Canadian Boreal Initiative. Reprinted by permission of the Canadian Boreal Initiative.

Chapter 5
Profile and Photograph of Robert Bateman. Reprinted by permission of Robert Bateman. Photo Norm Lightfoot.

Chapter 6
Profile and Photograph of William Rees. Reprinted by permission of William Rees.

Chapter 7
Profile and Photograph of Loretta Ford. Reprinted by permission of Loretta Ford.

Mer Bleue research project: Based on information from Environment Canada, Clean Air Online, http://www.ec.gc.ca/cleanair-airpur/Pollutants-WSBCC0B44A-1_En.htm; Centers for Disease Control, Agency for Toxic Substances and Disease Registry, Division of Toxicology ToxFAQs, September 2002, http://www.atsdr.cdc.gov/tfacts35.pdf; & UNEP, Phasing Lead Out of Gasoline: An Examination of Policy Approaches in Different Countries, 1999.

Nitrate Contamination in the Abbotsford Aquifer: Environment Canada, Nitrate Levels in the Abbotsford Aquifer: An Indicator of Groundwater Contamination in the Lower Fraser Valley from http://www.ecoinfo.org/env_ind/region/nitrate/nitrate_e.cfm. © Her Majesty the Queen in Right of Canada, Environment Canada, (2009). Reproduced with the permission of the Minister of Public Works and Government Services Canada.

Categories of Pesticides Used in Canada: Appendix 3.2 Description of Major Groups of Chemical Pesticides from http://www2.parl.gc.ca/HousePublications/Publication.aspx?DocID=1031697&Language=E&Mode+1&Parl=36&Ses=2&File=508. This table is excerpted from information available in the Government of Canada report Pesticides: Making the Right Choice for Health and the Environment. Report of the Standing Committee on Environment and Sustainable Development, 2000 http://cmte.parl.gc.ca/cmte/CommitteePublication.aspx?COM=173&Lang=1&SourceId=36396. © Government of Canada. Reprinted by the Government of Canada.

Excerpts from The 100 Mile Diet Website & Blog: Alisa Smith & James MacKinnon. © 2006 Alisa Smith & James MacKinnon. Reprinted by permission of the authors. Profile & Photo of Alisa Smith & James MacKinnon reprinted by permission of Paul Joseph.

Chapter 9
Profile and Photograph of Birute Mary Galdikas. Reprinted by permission of Birute Mary Galdikas.

Chapter 10
Changing Climate and the Spruce Budworm: Changing Climate Stops Pest Outbreaks on Vancouver Island, http://cfs.nrcan.gc.ca/news/585, Natural Resources Canada, Canadian Forest Service. Reproduced with the permission of the Minister of Public Works and Government Services, 2009.

Excerpts from National Forest Strategy 2003-2008. A Sustainable Forest: The Canadian Commitment, National Forest Strategy Coalition, http://nfsc.ca/strategies/strategy5.html. © 2008 Canadian Council of Forest Ministers. Reprinted by permission of the Canadian Council of Forest Ministers.

Profile and Photograph of Tzeporah Berman. Reprinted by permission of Tzeporah Berman.

Chapter 11
Profile and Photograph of David Schindler. Reprinted by permission of David Schindler. Photo Creative Services, University of Alberta.

Chapter 12
Profile and Photograph of Farley Mowat. Profile reprinted by permission of Farley Mowat. Photograph reprinted by permission of Ted Amsden.

Chapter 13
Opening quotation: Excerpted from http://www.ec.gc.ca/acidrain/new.html (among other sites).

Criteria Air Pollutants: Adapted from information from Environment Canada, Clean Air Online, http://www.ec.gc.ca/cleanair-airpur/Pollutants-WSBCC0B44A-1_En.htm

Excerpts from Meteorological Service of Canada, Environment Canada, David Phillips www.msc-smc.ec.gc.ca/cd/biographies/david_phillips_e.html. & Environment Canada Envirozone, Issue 78, Canada's Weather Guru: A Chat with David Phillips, www.ec.gc.ca/EnviorZine/english/issues/78/feature_1_e.cfm. © Government of Canada & David Phillips.

Chapter 14
CCMA: Adapted from Environment Canada, Canadian Centre for Climate Modelling and Analysis (CCMA) http://www.cccma.ec.gc.ca/20051116_brochure_e_pgs.pdf

Profile and Photograph of Sheila Watt-Cloutier. Reprinted by permission of Sheila Watt-Cloutier.

Chapter 15
Profile and Photograph of Mary Griffiths. Reprinted by permission of Mary Griffiths.

Chapter 16
Profile and Photograph of Grainne Ryder. Reprinted by permission of Grainne Ryder.

Chapter 17
Electricity Generation in Canada by types of fuel: The Conference Board of Canada

Federal investment in energy R&D: © Department of Natural Resources Canada, 2006. All rights reserved.

Lifetime air emissions from electricity generation by various energy types: Canadian Electricity Association

Profile and photograph of David Keith. Reprinted by permission of David Keith.

Chapter 18
Extracts from Stewardship Ontario, Municipal Hazardous or Special Waste, "Do What You Can Program." For more information, see "What's Included?" at www.dowhatyoucan.ca/WhatsIncluded.aspx. Reprinted with permission of Stewardship Ontario.

Two Maps showing locations of glacial Lake Iroguois shoreline, nd old aggregate pits and landfill sites: Professor Nick Eyles

Total releases and transfers of controlled industrial wastes for facilities with and without pollution prevention activities (Canadian and US data): Commission for Environmental Cooperation. The North American Mosaic: An Overview of Key Environmental Issues: Industrial Pollution and Waste, Commission for Environmental Cooperation, 2008, http://www.cec.org/soe/files/en/SOE_IndustrialPollution_en.pdf

Profile and photograph of Brennain Lloyd. Reprinted by permission of Brennain Lloyd.

Chapter 19
Graph of Human West Nile Virus Cases in Canada 2003-2007: Public Health Agency of Canada, West Nile Virus MONITOR, Human Surveillance Maps and Statistics http://www.phac-aspc.gc.ca/wnvnr_2007final-eng.pdf. Reproduced with the permission of the Minister of Public Works and Government Services Canada, 2009.

Human West Nile Virus Clinical Cases in Canada, 2006: Public Health Agency of Canada, West Nile Virus MONITOR, Human Surveillance Maps Archive http://www.cnphi-wnv.ca/human2006/index.htm. Reproduced with the Minister of Public Works and Government Services Canada, 2009.

West Nile Virus National Surveillance Report, Summary of 2007 WNV Season: Public Health Agency of Canada, West Nile Virus MONITOR, Human Surveillance Maps Archive http://www.cnphi-wnv.ca/human2006/index.htm. Reproduced with the permission of the Minister of Public Works and Government Services Canada, 2009.

Natural Hazards Map of Canada: Natural Hazards of Canada: A Historical Mapping of Significant Natural Disasters, http://www.publicsafety.gc.ca/res/em/nh/hazardsmap.pdf. Reproduced with the permission of the Minister of Public Works and Government Services 2009.

Profile and photograph of Wendy Mesley. Reprinted by permission of Wendy Mesley and CBC Still Photo Collection.

Chapter 20
Profile and photograph of Nola-Kate Seymoar. Reprinted by permission of Nola-Kate Seymoar.

Chapter 21
Extract from Mining Denendeh: A Dene Nation Perspective on Community Health Impacts of Mining by Chris Paci and Noeline Villebrun. Prepared for the 2004 Mining Ministers Conference, Coppermine, Nunavut. Reprinted with the permission of the authors.

Profile and photograph of Matthew Coon Come. Reprinted by permission of Matthew Coon Come.

Chapter 22
Municipalities in Canada that have banned pesticide use: Coalition for a Healthy Ottawa http://www.flora.org/healthyottawa/news-flash.htm

Total hazardous waste imports to Canada: Marisa Jacott, Cyrus Reed and Mark Winfield, The Generation and Management of Hazardous Wastes and Transboundary Hazardous Waste Shipments Between Mexico, Canada and the United States, 1990-2000, in North American Symposium on Understanding the Linkages Between Trade and the Environment, Commission on Environmental Cooperation of North America, 2002, http://www.cec.org/files/pdf/ECONOMY/symposium-e.pdf

Table 22.2: Environment Canada, International Policy and Cooperation Branch, International Relations Director, Compendium of International Agreements. Third edition. www.ec.gc.ca/international/multi/at/compendium_e.htm. Reprinted with permission of Public Works and Government Services Canada.

Profile and photograph of Maude Barlow. Reprinted by permission of Maude Barlow.

Chapter 23
Taillores Declaration 10 Point Action Plan: Excerpted from the Association of University Leaders for a Sustainable Future, Talloires Declaration 10 Point Action Plan. Reprinted by permission of ULSF. Available online at www.ulsf.org/programs_talloires.html.

UTM Ecological Footprint: Reprinted by permission of Tenley Conway

Profile and photograph of Keleigh Annau. Reprinted by permission of Keleigh Annau.

Selected Sources and References for Further Reading

Chapter 1

Bahn, Paul, and John Flenley. 1992. *Easter Island, Earth island.* Thames and Hudson, London.

Bowler, Peter J. 1993. *The Norton history of the environmental sciences.* W. W. Norton, New York.

Diamond, Jared. 2005. *Collapse: How societies choose to fail or succeed.* Viking, New York.

Edwards, Andres R. 2005. *The sustainability revolution: Portrait of a paradigm shift.* New Society Publishers, Gabriola Island, British Columbia, Canada.

Ehrlich, Paul. 1968. *The population bomb.* 1997 reprint, Buccaneer Books, Cutchogue, New York.

Flenley, John, and Paul Bahn. 2003. *The enigmas of Easter Island.* Oxford University Press, New York.

Goudie, Andrew. 2005. *The human impact on the natural environment,* 6th ed. Blackwell Publishing, London.

Hardin, Garrett. 1968. The tragedy of the commons. *Science* 162: 1243–1248.

Hunt, Terry L., and Carl P. Lipo. 2006. Late colonization of Easter Island. *Science* 311: 1603–1606.

Katzner, Donald W. 2001. *Unmeasured information and the methodology of social scientific inquiry.* Kluwer, Boston.

Kuhn, Thomas S. 1962. *The structure of scientific revolutions,* 2nd ed., 1970. University of Chicago Press, Chicago.

Lomborg, Bjorn. 2001. *The skeptical environmentalist: Measuring the real state of the world.* Cambridge University Press, Cambridge.

Malthus, Thomas R. *An essay on the principle of population.* 1983 ed. Penguin USA, New York.

Millennium Ecosystem Assessment. 2005. *Ecosystems and human well-being: General synthesis.* Millennium Ecosystem Assessment and World Resources Institute.

Musser, George. 2005. The climax of humanity. *Scientific American* 293(3): 44–47.

Ponting, Clive. 1991. *A green history of the world: The environment and the collapse of great civilizations.* Penguin Books, New York.

Popper, Karl R. 1959. *The logic of scientific discovery.* Hutchinson, London.

Porteous, Andrew. 2000. *Dictionary of environmental science and technology,* 3rd ed. John Wiley & Sons, Hoboken, New Jersey.

Redman, Charles R. 1999. *Human impact on ancient environments.* University of Arizona Press, Tucson.

Sagan, Carl. 1997. *The demon-haunted world: Science as a candle in the dark.* Ballantine Books, New York.

Schneiderman, Jill S., ed. 2003. *The Earth around us: Maintaining a livable planet.* Perseus Books, New York.

Siever, Raymond. 1968. Science: Observational, experimental, historical. *American Scientist* 56: 70–77.

Valiela, Ivan. 2001. *Doing science: Design, analysis, and communication of scientific research.* Oxford University Press, Oxford.

Van Tilburg, Jo Anne. 1994. *Easter Island: Archaeology, ecology, and culture.* Smithsonian Institution Press, Washington, D.C.

Venetoulis, Jason, and John Talberth. 2005. *Ecological footprint of nations: 2005 update.* Redefining Progress, Oakland, California.

Wackernagel, Mathis, and William Rees. 1996. *Our ecological footprint: Reducing human impact on the earth.* New Society Publishers, Gabriola Island, British Columbia, Canada.

Wackernagel, Mathis, et al. 2002. Tracking the ecological overshoot of the human economy. *Proceedings of the National Academy of Sciences of the USA* 99: 9266–9271.

World Bank. 2007. *World development indicators 2007.* World Bank, Washington, D.C.

Worldwatch Institute. 2006. *Vital signs 2006–2007.* Worldwatch Institute and W. W. Norton, Washington, D.C., and New York.

WWF–World Wide Fund for Nature. 2006. *Living planet report 2006.* WWF, Gland, Switzerland.

Young, Emma. 2006. A monumental collapse? *New Scientist,* 29 July 2006: 30–34.

Chapter 2

Alaska Department of Environmental Conservation. 1993. *The Exxon Valdez oil spill: Final report, State of Alaska response.* June 1993.

Allen, K. C., and D. E. G. Briggs, eds. 1989. *Evolution and the fossil record.* John Wiley & Sons, Hoboken, New Jersey.

Atlas, Ronald M. 1995. Petroleum biodegradation and oil spill bioremediation. *Marine Pollution Bulletin* 31: 178–182.

Atlas, Ronald M., and Carl E. Cerniglia. 1995. Bioremediation of petroleum pollutants. *Bio Science* 45: 332–338.

Azaizeh, Hassan, et al. 2006. Phytoremediation of selenium using subsurface-flow constructed wetland. *International Journal of Phytoremediation* 8: 187–198.

Berry, R. Stephen. 1991. *Understanding energy: Energy, entropy and thermodynamics for every man.* World Scientific Publishing Co.

Bragg, James R., et al. 1994. Effectiveness of bioremediation for the *Exxon Valdez* oil spill. *Nature* 368: 413–418.

Campbell, Neil A., and Jane B. Reece. 2005. *Biology,* 7th ed. Benjamin Cummings, San Francisco.

Fenchel, Tom. 2003. *Origin and early evolution of life.* Oxford University Press, Oxford.

Fortey, Richard. 1998. *Life: A natural history of the first four billion years of life on Earth.* Alfred Knopf, New York.

Gee, Henry. 1999. *In search of deep time: Beyond the fossil record to a new history of life.* Free Press, New York.

Hall, David O., and Krishna Rao. 1999. *Photosynthesis,* 6th ed. Cambridge University Press, Cambridge.

Kertulis-Tartar, Gina M., et al. 2006. Phytoremediation of an arsenic-contaminated site using *Pteris vittata* L.: A two-year study. *International Journal of Phytoremediation* 8: 311–322.

Lancaster, M. 2002. *Green chemistry.* Royal Society of Chemistry, London.

Manahan, Stanley E. 2004. *Environmental chemistry,* 8th ed. Lewis Publishers, CRC Press, Boca Raton, Florida.

Mc Murry, John E. 2007. *Organic chemistry,* 7th ed. Brooks/Cole, San Francisco.

National Response Team. *NRT fact sheet: Bioremediation in oil spill response.* U.S. EPA. www.epa.gov/oilspill/pdfs/biofact.pdf.

Nealson, Kenneth H. 2003. Harnessing microbial appetites for remediation. *Nature Biotechnology* 21: 243–244.

Ridley, Mark. 2003. *Evolution,* 3rd ed. Blackwell Science, Cambridge, Massachusetts.

Timberlake, Karen C. 2007. *General, organic, and biological chemistry,* 2nd ed. Benjamin Cummings, San Francisco.

United States Environmental Protection Agency. Oil program. www.epa.gov/oilspill.

Van Dover, Cindy Lee. 2000. *The ecology of deep-sea hydrothermal vents*. Princeton University Press, Princeton.

Van Ness, H. C. 1983. *Understanding thermodynamics*. Dover Publications, Mineola, New York.

Ward, Peter D., and Donald Brownlee. 2000. *Rare Earth: Why complex life is uncommon in the universe*. Copernicus, New York.

Wassenaar, Leonard I., and Keith A. Hobson. 1998. Natal origins of migratory monarch butterflies at wintering colonies in Mexico: New isotopic evidence. *Proceedings of the National Academy of Sciences of the USA* 95: 15436–15439.

Chapter 3

Alvarez, Luis W., et al. 1980. Extraterrestrial cause for the Cretaceous-Tertiary extinction. *Science* 208: 1095–1108.

Barbour, Michael G., et al. 1998. *Terrestrial plant ecology*, 3rd ed. Benjamin Cummings, Menlo Park, California.

Begon, Michael, Martin Mortimer, and David J. Thompson. 1996. *Population ecology: A unified study of animals and plants*, 3rd ed. Blackwell Scientific, Oxford.

Begon, Michael, Colin R. Townsend, and John L. Harper. 2005. *Ecology: From individuals to ecosystems*, 4th ed. Blackwell Publishing, London.

Campbell, Neil A., and Jane B. Reece. 2005. *Biology*, 7th ed. Benjamin Cummings, San Francisco.

Clark K. L., et al. 1998. Cloud water and precipitation chemistry in a tropical montane forest, Monteverde, Costa Rica. *Atmospheric Environment* 32: 1595–1603.

Crump, L. Martha, et al. 1992. Apparent decline of the golden toad: Underground or extinct? *Copeia* 1992: 413–420.

Darwin, Charles. 1859. *The origin of species by means of natural selection*. John Murray, London.

Endler, John A. 1986. *Natural selection in the wild*. Monographs in Population Biology 21, Princeton University Press, Princeton.

Freeman, Scott, and Jon C. Herron. 2006. *Evolutionary analysis*, 4th ed. Benjamin Cummings, San Francisco.

Futuyma, Douglas J. 2005. *Evolution*. Sinauer Associates, Sunderland, Massachusetts.

Krebs, Charles J. 2001. *Ecology: The experimental analysis of distribution and abundance*, 5th ed. Benjamin Cummings, San Francisco.

Lawton, Robert O., et al. 2001. Climatic impact of tropical lowland deforestation on nearby montane cloud forests. *Science* 294: 584–587.

Molles, Manuel C., Jr. 2005. *Ecology: Concepts and applications*, 3rd ed. McGraw-Hill, Boston.

Nadkarni, Nalini M., and Nathaniel T. Wheelwright, eds. 2000. *Monteverde: Ecology and conservation of a tropical cloud forest*. Oxford University Press, New York.

Pounds, J. Alan. 2001. Climate and amphibian declines. *Nature* 410: 639.

Pounds, J. Alan, et al. 1997. Tests of null models for amphibian declines on a tropical mountain. *Conservation Biology* 11: 1307–1322.

Pounds, J. Alan, et al. 2006. Widespread amphibian extinctions from epidemic disease driven by global warming. *Nature* 439: 161–167.

Pounds, J. Alan, and Martha L. Crump. 1994. Amphibian declines and climate disturbance: The case of the golden toad and the harlequin frog. *Conservation Biology* 8: 72–85.

Pounds, J. Alan, Michael P. L. Fogden, and John H. Campbell. 1999. Biological response to climate change on a tropical mountain. *Nature* 398: 611–615.

Powell, James L. 1998. *Night comes to the Cretaceous: Dinosaur extinction and the transformation of modern geology*. W. H. Freeman, New York.

Raup, David M. 1991. *Extinction: Bad genes or bad luck?* W. W. Norton, New York.

Ricklefs, Robert E., and Gary L. Miller. 2000. *Ecology*, 4th ed. W. H. Freeman, New York.

Ricklefs, Robert E., and Dolph Schluter, eds. 1993. *Species diversity in ecological communities*. University of Chicago Press, Chicago.

Savage, Jay M. 1966. An extraordinary new toad (*Bufo*) from Costa Rica. *Revista de Biologia Tropical* 14: 153–167.

Savage, Jay M. 1998. The "brilliant toad" was telling us something. *Christian Science Monitor*, 14 September 1998: 19.

Smith, Thomas M., and Robert L. Smith. 2006. *Elements of ecology*, 6th ed. Benjamin Cummings, San Francisco.

Ward, Peter. 1994. *The end of evolution*. Bantam Books, New York.

Williams, George C. 1966. *Adaptation and natural selection*. Princeton University Press, Princeton.

Wilson, Edward O. 1992. *The diversity of life*. Harvard University Press, Cambridge, Massachusetts.

Chapter 4

Breckle, Siegmar-Walter. 2002. *Walter's vegetation of the Earth: The ecological systems of the geo-biosphere*, 4th ed. Springer-Verlag, Berlin.

Bright, Chris. 1998. *Life out of bounds: Bioinvasion in a borderless world*. Worldwatch Institute and W. W. Norton, Washington, D.C., and New York.

Bronstein, Judith L. 1994. Our current understanding of mutualism. *Quarterly Journal of Biology* 69: 31–51.

Chase, Jonathan M., et al., 2002. The interaction between predation and competition: A review and synthesis. *Ecology Letters* 5: 302.

Connell, Joseph H., and Ralph O. Slatyer, 1977. Mechanisms of succession in natural communities. *American Naturalist* 111: 1119–1144.

Drake, John M., and Jonathan M. Bossenbroek. 2004. The potential distribution of zebra mussels in the United States. *Bio Science* 54: 931–941.

Estes, J.A., et al. 1998. Killer whale predation on sea otters linking oceanic and nearshore ecosystems. *Science* 282: 473–476.

Ewald, Paul W., 1987. Transmission modes and evolution of the parasitism-mutualism continuum. *Annals of the New York Academy of Sciences* 503: 295–306.

Gurevitch, Jessica, and Dianna K. Padilla. 2004. Are invasive species a major cause of extinctions? *Trends in Ecology and Evolution* 19: 470–474.

Hobbs, Richard J., et al. 2006. *Foundations of restoration ecology: The science and practice of ecological restoration*. Island Press, Washington, D.C.

Menge, Bruce A., et al. 1994. The keystone species concept: Variation in interaction strength in a rocky intertidal habitat. *Ecological Monographs* 64: 249–286.

Molles, Manuel C. Jr. 2005. *Ecology: Concepts and applications*, 3rd ed. McGraw-Hill, Boston.

Morin, Peter J. 1999. *Community ecology*. Blackwell, London.

Power, Mary E., et al. 1996. Challenges in the quest for keystones. *Bio Science* 46: 609–620.

Ricklefs, Robert E., and Gary L. Miller. 2000. *Ecology*, 4th ed. W. H. Freeman and Co., New York.

Schrope, Mark. 2007. Killer in the kelp. *Nature* 445: 703–705.

Shea, Katriona, and Peter Chesson. 2002. Community ecology theory as a framework for biological invasions. *Trends in Ecology and Evolutionary Biology* 17: 170–176.

Sih, Andrew, et al. 1985. Predation, competition, and prey communities: A review of field experiments. *Annual Review of Ecology and Systematics* 16: 269–311.

Smith, Robert L., and Thomas M. Smith. 2001. *Ecology and field biology*, 6th ed. Benjamin Cummings, San Francisco.

Springer, A. M., et al. 2003. Sequential megafaunal collapse in the North Pacific Ocean: An ongoing legacy of industrial whaling? *Proceedings of the National Academy of Sciences of the USA* 100: 12223–12228.

Stokstad, Erik. 2007. Feared quagga mussel turns up in western United States. *Science* 315: 453.

Strayer, David L., et al. 1999. Transformation of freshwater ecosystems by bivalves: A case study of zebra mussels in the Hudson River. *Bio Science* 49: 19–27.

Strayer, David L., et al. 2004. Effects of an invasive bivalve (*Dreissena polymorpha*) on fish in the Hudson River estuary. *Canadian Journal of Fisheries and Aquatic Sciences* 61: 924–941.

Thompson, John N. 1999. The evolution of species interactions. *Science* 284: 2116–2118.

Van Andel, Jelte, and James Aronson. 2005. *Restoration ecology: The new frontier*. Blackwell Publishing, London.

Weigel, Marlene, ed. 1999. *Encyclopedia of biomes*. UXL, Farmington Hills, Michigan.

Whittaker, Robert H., and William A. Niering. 1965. Vegetation of the Santa Catalina Mountains, Arizona: A gradient analysis of the south slope. *Ecology* 46: 429–452.

Woodward, Susan L. 2003. *Biomes of Earth: Terrestrial, aquatic, and human-dominated*. Greenwood Publishing, Westport, Connecticut.

Chapter 5

Appenzeller, Tim. 2004. The case of the missing carbon. *National Geographic*, February 2004: 88–117.

Capra, Fritjof. 1996. *The web of life: A new scientific understanding of living systems*. Anchor Books Doubleday, New York.

Carpenter, Edward J., and Douglas G. Capone, eds. 1983. *Nitrogen in the marine environment*. Academic Press, New York.

Christopherson, Robert W. 2006. *Geosystems*, 6th ed. Prentice Hall, Upper Saddle River, New Jersey.

Committee on Environment and Natural Resources. 2000. *An integrated assessment: Hypoxia in the northern Gulf of Mexico*. CENR, National Science and Technology Council, Washington, D.C.

Ferber, Dan. 2004. Dead zone fix not a dead issue. *Science* 305: 1557.

Field, Christopher B., et al. 1998. Primary production of the bio-sphere: Integrating terrestrial and oceanic components. *Science* 281: 237–240.

Jacobson, Michael, et al. 2000. *Earth system science from biogeochemical cycles to global changes*. Academic Press.

Keller, Edward A. 2004. *Introduction to environmental geology*, 3rd ed. Prentice Hall, Upper Saddle River, New Jersey.

Larsen, Janet. 2004. Dead zones increasing in world's coastal waters. *Eco-economy Update #41*, 16 June 2004. Earth Policy Institute. www.earth-policy.org/Updates/Update41.htm.

Mississippi River/Gulf of Mexico Watershed Nutrient Task Force. 2001. *Action plan for reducing, mitigating, and controlling hypoxia in the northern Gulf of Mexico*. Washington, D.C.

Mitsch, William J., et al. 2001. Reducing nitrogen loading to the Gulf of Mexico from the Mississippi River Basin: Strategies to counter a persistent ecological problem. *Bio Science* 51: 373–388.

Montgomery, Carla. 2005. *Environmental geology*, 7th ed. Mc Graw-Hill, New York.

National Oceanic and Atmospheric Administration: National Ocean Service. 2000. Hypoxia in the Gulf of Mexico: Progress toward the completion of an integrated assessment. www.nos.noaa.gov/products/pubs_hypox.html.

National Science and Technology Council, Committee on Environment and Natural Resources. 2003. *An assessment of coastal hypoxia and eutrophication in U.S. waters*. National Science and Technology Council, Washington, D.C.

Pima County, Arizona. Sonoran Desert Conservation Plan. www.pima.gov/cmo/sdcp.

Rabalais, Nancy N., R. E. Turner, and D. Scavia. 2002. Beyond science into policy: Gulf of Mexico hypoxia and the Mississippi River. *Bio Science* 52: 129–142.

Rabalais, Nancy N., R. E. Turner, and W. J. Wiseman, Jr. 2002. Hypoxia in the Gulf of Mexico, a.k.a. "The dead zone." *Annual Review of Ecology and Systematics* 33: 235–263.

Raloff, Janet. 2004. Dead waters: Massive oxygen-starved zones are developing along the world's coasts. *Science News* 165: 360–362. June 5, 2004.

Raloff, Janet. 2004. Limiting dead zones: How to curb river pollution and save the Gulf of Mexico. *Science News* 165: 378–380. June 12, 2004.

Ricklefs, Robert E., and Gary L. Miller. 2000. *Ecology*, 4th ed. W. H. Freeman and Co., New York.

Schlesinger, William H. 1997. *Biogeochemistry: An analysis of global change*, 2nd ed. Academic Press, London.

Skinner, Brian J., and Stephen C. Porter. 2003. *The dynamic earth: An introduction to physical geology*, 5th ed. John Wiley and Sons, Hoboken, New Jersey.

Smith, Robert L., and Thomas M. Smith. 2001. *Ecology and field biology*, 6th ed. Benjamin Cummings, San Francisco.

Stiling, Peter. 2002. *Ecology: Theories and applications*, 4th ed. Prentice Hall, Upper Saddle River, New Jersey.

Takahashi, Taro. 2004. The fate of industrial carbon dioxide. *Science* 305: 352–353.

Turner, R. Eugene, and Nancy N. Rabalais. 2003. Linking landscape and water quality in the Mississippi River Basin for 200 years. *Bio Science* 53: 563–572.

Vitousek, Peter M., et al. 1997. Human alteration of the global nitrogen cycle: Sources and consequences. *Ecological Applications* 7: 737–750.

Whittaker, Robert H. 1975. *Communities and ecosystems*, 2nd ed. Macmillan, New York.

Wu, Jianguo, and Richard J. Hobbs, eds. 2007. *Key topics in landscape ecology*. Cambridge University Press, Cambridge, U.K.

Chapter 6

Ausubel, Jesse. H. 1996. Can technology spare the earth? *American Scientist* 84: 166–178.

Balter, Michael. 2006. The baby deficit. *Science* 312: 1894–1897.

Cohen, Joel E. 1995. *How many people can the Earth support?* W. W. Norton, New York.

Cohen, Joel E. 2003. Human population: The next half century. *Science* 302: 1172–1175.

Cohen, Joel E. 2005. Human population grows up. *Scientific American* 293(3): 48–55.

De Souza, Roger-Mark, et al. 2003. Critical links: Population, health, and the environment. *Population Bulletin* 58(3). Population Reference Bureau, Washington, D.C.

Eberstadt, Nicholas. 2000. China's population prospects: Problems ahead. *Problems of Post-Communism* 47: 28.

Ehrlich, Paul R., and John P. Holdren. 1971. Impact of population growth: Complacency concerning this component of man's predicament is unjustified and counterproductive. *Science* 171: 1212–1217.

Ehrlich, Paul R., and Anne H. Ehrlich. 1990. *The population explosion*. Touchstone, New York.

Engelman, Robert, Brian Halweil, and Danielle Nierenberg. 2002. Rethinking population, improving lives. Pp. 127–148 in *State of the world 2002*. Worldwatch Institute and W. W. Norton, Washington, D.C., and New York.

Greenhalgh, Susan. 2001. Fresh winds in Beijing: Chinese feminists speak out on the one-child policy and women's lives. *Signs: Journal of Women in Culture & Society* 26: 847–887.

Harrison, Paul, and Fred Pearce, eds. 2000. *AAAS atlas of population & environment*. University of California Press, Berkeley.

Haub, Carl, and O. P. Sharma. 2006. India's population reality: Reconciling change and tradition. *Population Bulletin* 61(3). Population Reference Bureau, Washington, D.C.

Hesketh, Therese, and Wei Xing Zhu. 1997. Health in China: The one child family policy: The good, the bad, and the ugly. *British Medical Journal* 314: 1685.

Holdren, John P., and Paul R. Ehrlich. 1974. Human population and the global environment. *American Scientist* 62: 282–292.

Imhoff, Marc L., et al. 2004. Global patterns in human consumption of net primary production. *Nature* 429: 870–873.

Kane, Penny. 1987. *The second billion: Population and family planning in China*. Penguin Books, Australia, Ringwood, Victoria.

Kane, Penny, and Ching Y. Choi. 1999. China's one child family policy. *British Medical Journal* 319: 992.

Kent, Mary M., and Carl Haub. 2005. Global demographic divide. *Population Bulletin* 60(4). Population Reference Bureau, Washington, D.C.

Lamptey, Peter R., Jami L. Johnson, and Marya Khan. 2006. The global challenge of HIV and AIDS. *Population Bulletin* 61(1). Population Reference Bureau, Washington, D.C.

Mastny, Lisa, and Richard P. Cincotta. 2005. Examining the connections between population and security. Pp. 22–41 in *State of the world 2005*. Worldwatch Institute and W. W. Norton, Washington, D.C., and New York.

Mc Donald, Mia, with Danielle Nierenberg. 2003. Linking population, women, and biodiversity. Pp. 38–61 in *State of the world 2003*. Worldwatch Institute and W. W. Norton, Washington D.C., and New York.

Meadows, Donella, Jørgen Randers, and Dennis Meadows. 2004. *Limits to growth: The 30-year update.* Chelsea Green Publishing Co., White River Junction, Vermont.

Notestein, Frank. 1953. Economic problems of population change. Pp. 13–31 in *Proceedings of the Eighth International Conference of Agricultural Economists.* Oxford University Press, London.

Population Reference Bureau. 2007. *2007 World Population Data Sheet.* Population Reference Bureau, Washington, D.C., and John Wiley & Sons, Hoboken, New Jersey.

Riley, Nancy E. 2004. *China's population: New trends and challenges. Population Bulletin* 59(2). Population Reference Bureau, Washington, D.C.

UNAIDS and World Health Organization. 2006. *AIDS epidemic update: December 2006.* UNAIDS and WHO, New York.

United Nations Population Division. 2007. *World population prospects: The 2006 revision.* UNPD, New York.

United Nations Population Fund. UNFPA, the 2005 World Summit and the millennium development goals. UNFPA. www.unfpa.org/icpd.

United Nations Population Fund. 2007. *State of world population 2007.* UNFPA, New York.

United States Census Bureau. www.census.gov.

Wackernagel, Mathis, and William Rees. 1996. *Our ecological footprint: Reducing human impact on the earth.* New Society Publishers, Gabriola Island, British Columbia, Canada.

Chapter 7

Ashman, Mark R., and Geeta Puri. 2002. *Essential soil science: A clear and concise introduction to soil science.* Blackwell Publishing, Malden, Massachusetts.

Brown, Lester R. 2002. World's rangelands deteriorating under mounting pressure. *Eco-Economy Update #6,* 5 February 2002. Earth Policy Institute. www.earth-policy.org/Updates/Update6.htm.

Brown, Lester R. 2004. *Outgrowing the Earth: The food security challenge in an age of falling water tables and rising temperatures.* Earth Policy Institute, Washington, D.C.

Charman, P. E. V., and Brian W. Murphy. 2000. *Soils: Their properties and management,* 2nd ed. Oxford University Press, South Melbourne, Australia.

Curtin, Charles G. 2002. Integration of science and community-based conservation in the Mexico/U.S. borderlands. *Conservation Biology* 16: 880–886.

Curtin, Charles. 2007. Integrating landscape and ecosystems app-roaches through science-based collaborative conservation. *Conservation Biology,* in press.

Diamond, Jared. 1999. *Guns, germs, and steel: The fates of human societies.* W. W. Norton, New York.

Diamond, Jared, and Peter Bellwood. 2003. Farmers and their languages: The first expansions. *Science* 300: 597–603.

Food and Agriculture Organization of the United Nations. 2001. Conservation agriculture: Case studies in Latin America and Africa. *FAO Soils Bulletin No. 78.* FAO, Rome.

Fox, Stephen. 1985. *The American conservation movement: John Muir and his legacy.* University of Wisconsin Press, Madison.

Glanz, James. 1995. *Saving our soil: Solutions for sustaining Earth's vital resource.* Johnson Books, Boulder, Colorado.

Goudie, Andrew. 2000. *The human impact on the natural environment,* 5th ed. MIT Press, Cambridge, Massachusetts.

Halweil, Brian. 2002. Farmland quality deteriorating. Pp. 102–103 in *Vital signs 2002.* Worldwatch Institute and W. W. Norton, Washington D.C., and New York.

Harrison, Paul, and Fred Pearce, eds. 2000. *AAAS atlas of population & environment.* University of California Press, Berkeley.

Jenny, Hans. 1941. *Factors of soil formation: A system of quantitative pedology.* Mc Graw-Hill, New York.

Kaiser, Jocelyn. 2004. Wounding Earth's fragile skin. *Science* 304: 1616–1618.

Larsen, Janet. 2003. Deserts advancing, civilization retreating.*Eco-Economy Update #23,* 27 March 2003. Earth Policy Institute. www.earth-policy.org/Updates/Update23.htm.

Malpai Borderlands Group. Malpai Borderlands Group. www.malpaiborderlandsgroup.org.

Millennium Ecosystem Assessment. 2005. *Ecosystems and human well-being: Desertification synthesis.* Millennium Ecosystem Assessment and World Resources Institute.

Morgan, R. P. C. 2005. *Soil erosion and conservation,* 3rd ed. Blackwell, London.

Natural Resources Conservation Service. 2001. *National resources inventory 2001: Soil erosion.* NRCS, USDA, Washington, D.C.

Natural Resources Conservation Service. Soils. NRCS, USDA. www.soils.usda.gov.

Pieri, Christian, et al. 2002. *No-till farming for sustainable rural development.* Agriculture & Rural Development Working Paper. International Bank for Reconstruction and Development, Washington, D.C.

Pierzynski, Gary M., et al. 2005. *Soils and environmental quality,* 3rd ed. CRC Press, Boca Raton, Florida.

Pretty, Jules, and Rachel Hine. 2001. *Reducing food poverty with sustainable agriculture: A summary of new evidence.* Occasional Paper 2001–2. Center for Environment and Society, University of Essex.

Richter, Daniel D. Jr., and Daniel Markewitz. 2001. *Understanding soil change: Soil sustainability over millennia, centuries, and decades.* Cambridge University Press, Cambridge.

Ritchie, Jerry C. 2000. Combining cesium-137 and topographic surveys for measuring soil erosion/deposition patterns in a rapidly accreting area. TEKTRAN, USDA Division of Agricultural Research, January 14, 2000.

Shaxson, T. F. 1999. The roots of sustainability, concepts and practice: Zero tillage in Brazil. *ABLH Newsletter ENABLE; World Association for Soil and Water Conservation (WASWC) Newsletter.*

Soil Science Society of America. 2001. Internet glossary of soil science terms. www.soils.org/sssagloss.

Stocking, M. A. 2003. Tropical soils and food security: The next 50 years. *Science* 302: 1356–1359.

Trimble, Stanley W., and Pierre Crosson. 2000. U.S. soil erosion rates—myth and reality. *Science* 289: 248–250.

Troeh, Frederick R., J. Arthur Hobbs, and Roy L. Donahue. 2004. *Soil and water conservation for productivity and environmental protection,* 4th ed. Prentice Hall, Upper Saddle River, New Jersey.

Troeh, Frederick R., and Louis M. Thompson. 2004. *Soil and soil fertility,* 6th ed. Blackwell Publishing, London.

United Nations Convention to Combat Desertification. 2001. *Global alarm: Dust and sandstorms from the world's drylands.* UNCCD and others, Bangkok, Thailand.

United Nations Environment Programme. 2002. Land. Pp. 62–89 in *Global environment outlook 3 (GEO-3).* UNEP and Earthscan Publications, Nairobi and London.

Uri, Noel D. 2001. The environmental implications of soil erosion in the United States. *Environmental Monitoring and Assessment* 66: 293–312.

Wilkinson, Bruce H. 2005. Humans as geologic agents: A deep-time perspective. *Geology* 33: 161–164.

Chapter 8

Bazzaz, Fakhri A. 2001. Plant biology in the future. *Proceedings of the National Academy of Sciences of the United States of America* 98: 5441–5445.

Brown, Lester R. 2004. *Outgrowing the Earth: The food security challenge in an age of falling water tables and rising temperatures.* Earth Policy Institute, Washington, D.C.

Buchmann, Stephen L., and Gary Paul Nabhan. 1996. *The forgotten pollinators.* Island Press/Shearwater Books, Washington, D.C./Covelo, California.

Center for Food Safety. 2005. *Monsanto vs. U.S. farmers.* Center for Food Safety, Washington, D.C.

Commission for Environmental Cooperation. 2004. *Maize and biodiversity: The effects of transgenic maize in Mexico.* CEC Secretariat.

[Correspondence to *Nature*, various authors]. 2002. *Nature* 416: 600–602, and 417: 897–898.

The Farm Scale Evaluations of spring-sown genetically modified crops. 2003. A themed issue from *Philosophical Transactions of the Royal Society of London B: Biological Sciences* 358(1439), 29 November 2003.

Fedoroff, Nina, and Nancy Marie Brown, 2004. *Mendel in the kitchen: A scientist's view of genetically modified foods*. National Academies Press, Washington, D.C.

Food and Agriculture Organization of the United Nations. 2007. *The state of world fisheries and aquaculture 2006*. FAO Fisheries and Aquaculture Department, Rome.

Gardner, Gary, and Brian Halweil. 2000. *Underfed and overfed: The global epidemic of malnutrition*. Worldwatch Paper #150. Worldwatch Institute, Washington, D.C.

Halweil, Brian. 2004. *Eat here: Reclaiming homegrown pleasures in a global supermarket*. Worldwatch Institute, Washington, D.C.

Halweil, Brian, and Danielle Niereberg. 2004. Watching what we eat. Pp. 68–95 in *State of the world 2004*. Worldwatch Institute and W. W. Norton, Washington, D.C., and New York.

Harrison, Paul, and Fred Pearce, eds. 2000. *AAAS atlas of population & environment*. University of California Press, Berkeley.

International Food Information Council. 2004. Food biotechnology. IFIC, Washington, D.C. www.ific.org/food/biotechnology/index.cfm.

James, Clive. 2006. *Global status of commercialized biotech/GM crops*. International Service for the Acquisition of Agri-biotech Applications.

Kristiansen, P., A. Taji, and J. Reganold, eds. 2006. *Organic agriculture: A global perspective*. CABI Publishing, Oxfordshire, U.K.

Kuiper, Harry A. 2000. Risks of the release of transgenic herbicide-resistant plants with respect to humans, animals, and the environment. *Crop Protection* 19: 773.

Liebig, Mark A., and John W. Doran. 1999. Impact of organic production practices on soil quality indicators. *Journal of Environmental Quality* 28: 1601–1609.

Losey, John E., Linda S. Raynor, and Maureen E. Carter. 1999. Transgenic pollen harms monarch larvae. *Nature* 399: 214.

Maeder, Paul, et al. 2002. Soil fertility and biodiversity in organic farming. *Science* 296: 1694–1697.

Mann, Charles C. 2002. Transgene data deemed unconvincing. *Science* 296: 236–237.

Manning, Richard. 2000. *Food's frontier: The next green revolution*. North Point Press, New York.

Miller, Henry I., and Gregory Conko. 2004. *The frankenfood myth: How protest and politics threaten the biotech revolution*. Praeger Publishers, Westport, Connecticut.

Nestle, Marion. 2002. *Food politics: How the food industry influences nutrition and health*. University of California–Berkeley Press, Berkeley.

Nierenberg, Danielle. 2005. *Happier meals: Rethinking the global meat industry*. Worldwatch Paper #171. Worldwatch Institute, Washington, D.C.

Nierenberg, Danielle. 2006. Meat consumption and output up. Pp. 24–25 in *Vital signs 2006–2007*. Worldwatch Institute and W. W. Norton, Washington, D.C., and New York.

Nierenberg, Danielle, and Brian Halweil. 2005. Cultivating food security. Pp. 62–79 in *State of the world 2005*. Worldwatch Institute and W. W. Norton, Washington, D.C., and New York.

Norris, Robert F., Edward P. Caswell-Chen, and Marcos Kogan. 2002. *Concepts in integrated pest management*. Prentice Hall, Upper Saddle River, New Jersey.

Ortiz-García, Sol, et al. 2005. Absence of detectable transgenes in local landraces of maize in Oaxaca, Mexico (2003–2004). *Proceedings of the National Academy of the United States of America* 102: 12338–12343.

Paoletti, Maurizio G., and David Pimentel. 1996. Genetic engineering in agriculture and the environment: Assessing risks and benefits. *Bio Science* 46: 665–673.

Pearce, Fred. 2002. The great Mexican maize scandal. *New Scientist* 174: 14.

Pedigo, Larry P., and Marlin E. Rice. 2006. *Entomology and pest management*, 5th ed. Prentice Hall, Upper Saddle River, New Jersey.

Pimentel, David. 1999. Population growth, environmental resources, and the global availability of food. *Social Research*, Spring 1999.

Pinstrup-Andersen, Per, and Ebbe Schioler. 2001. *Seeds of contention: World hunger and the global controversy over GM (genetically modified) crops*. International Food Policy Research Institute, Washington, D.C.

Polak, Paul. 2005. The big potential of small farms. *Scientific American* 293(3): 84–91.

Pringle, Peter. 2003. *Food, Inc.: Mendel to Monsanto—The promises and perils of the biotech harvest*. Simon and Schuster, New York.

Quist, David, and Ignacio H. Chapela. 2001. Transgenic DNA introgressed into traditional maize landraces in Oaxaca, Mexico. *Nature* 414: 541–543.

Ruse, Michael, and David Castle, eds. 2002. *Genetically modified foods: Debating technology*. Prometheus Books, Amherst, New York.

Schmeiser, Percy. Monsanto vs. Schmeiser. www. percyschmeiser.com.

Shiva, Vandana. 2000. *Stolen harvest: The hijacking of the global food supply*. South End Press, Cambridge, Massachusetts.

Smil, Vaclav. 2001. *Feeding the world: A challenge for the twenty-first century*. MIT Press, Cambridge, Massachusetts.

Stewart, C. Neal. 2004. *Genetically modified planet: Environmental impacts of genetically engineered plants*. Oxford University Press, Oxford.

Teitel, Martin, and Kimberly Wilson. 2001. *Genetically engineered food: Changing the nature of nature*. Park Street Press.

Tuxill, John. 1999. Appreciating the benefits of plant biodiversity. Pp. 96–114 in *State of the world 1999*. Worldwatch Institute and W. W. Norton, Washington D.C., and New York.

Westra, Lauren. 1998. Biotechnology and transgenics in agriculture and aquaculture: The perspective from ecosystem integrity. *Environmental Values* 7: 79.

Wolfenbarger, L. La Reesa. 2000. The ecological risks and benefits of genetically engineered plants. *Science* 290: 2088.

Chapter 9

Balmford, Andrew, et al. 2002. Economic reasons for conserving wild nature. *Science* 297: 950–953.

Barnosky, Anthony D., et al. 2004. Assessing the causes of late Pleistocene extinctions on the continents. *Science* 306: 70–75.

Baskin, Yvonne. 1997. *The work of nature: How the diversity of life sustains us*. Island Press, Washington, D.C.

Bright, Chris. 1998. *Life out of bounds: Bioinvasion in a borderless world*. Worldwatch Institute and W. W. Norton, Washington D.C., and New York.

CITES Secretariat. Convention on International Trade in Endangered Species of Wild Fauna and Flora. www.cites.org.

Convention on Biological Diversity. www.biodiv.org.

Daily, Gretchen C., ed. 1997. *Nature's services: Societal dependence on natural ecosystems*. Island Press, Washington, D.C.

Ehrenfeld, David W. 1970. *Biological conservation*. International Thomson Publishing, London.

Gaston, Kevin J., and John I. Spicer. 2004. *Biodiversity: An introduction*, 2nd ed. Blackwell, London.

Groom, Martha J., et al. 2005. *Principles of conservation biology*, 3rd ed. Sinauer Associates, Sunderland, Massachusetts.

Groombridge, Brian, and Martin D. Jenkins. 2002. *Global biodiversity: Earth's living resources in the 21st century*. UNEP, World Conservation Monitoring Centre, and Aventis Foundation; World Conservation Press, Cambridge, U.K.

Groombridge, Brian, and Martin D. Jenkins. 2002. *World atlas of biodiversity: Earth's living resources in the 21st century*. University of California Press, Berkeley.

Hanken, James. 1999. Why are there so many new amphibian species when amphibians are declining? *Trends in Ecology and Evolution* 14: 7–8.

Harris, Larry D. 1984. *The fragmented forest: Island biogeography theory and the preservation of biotic diversity*. University of Chicago Press, Chicago.

Harrison, Paul, and Fred Pearce, eds. 2000. *AAAS atlas of population & environment*. University of California Press, Berkeley.

Jenkins, Martin. 2003. Prospects for biodiversity. *Science* 302: 1175–1177.

Louv, Richard. 2005. *Last child in the woods: Saving our children from nature-deficit disorder*. Algonquin Books, Chapel Hill, North Carolina.

Lovejoy, Thomas E., and Lee Hannah, eds. 2006. *Climate change and biodiversity*. Yale University Press, New Haven, Connecticut.

Mac Arthur, Robert H., and Edward O. Wilson. 1967. *The theory of island biogeography*. Princeton University Press, Princeton.

Maehr, David S., Reed F. Noss, and Jeffrey Larkin, eds. 2001. *Large mammal restoration: Ecological and sociological challenges in the 21st century*. Island Press, Washington, D.C.

Matthiessen, Peter. 2000. *Tigers in the snow.* North Point Press, New York.
Meegaskumbura, Madhava, et al. 2002. Sri Lanka: An amphibian hot spot. *Science* 298: 379.
Millennium Ecosystem Assessment. 2005. *Ecosystems and human well-being: Biodiversity synthesis.* Millennium Ecosystem Assessment and World Resources Institute.
Miquelle, Dale, Howard Quigley, and Maurice Hornocker. 1999. *A habitat protection plan for Amur Tiger conservation: A proposal outlining habitat protection measures for the Amur Tiger.* Hornocker Wildlife Institute.
Mooney, Harold A., and Richard J. Hobbs, eds. 2000. *Invasive species in a changing world.* Island Press, Washington, D.C.
Newmark, William D. 1987. A land-bridge perspective on mammal extinctions in western North American parks. *Nature* 325: 430.
Pimm, Stuart L., and Clinton Jenkins. 2005. Sustaining the variety of life. *Scientific American* 293(3): 66–73.
Primack, Richard B. 2006. *Essentials of conservation biology*, 4th ed. Sinauer Associates, Sunderland, Massachusetts.
Quammen, David. 1996. *The song of the dodo: Island biogeography in an age of extinction.* Touchstone, New York.
Relyea, Rick, and Nathan Mills. 2001. Predator-induced stress makes the pesticide carbaryl more deadly to gray treefrog tadpoles. *Proceedings of the National Academy of Sciences, USA* 98: 2491–2496.
Rosenzweig, Michael L. 1995. *Species diversity in space and time.* Cambridge University Press, Cambridge.
Sepkoski, John J. 1984. A kinetic model of Phanerozoic taxonomic diversity. *Paleobiology* 10: 246–267.
Simberloff, Daniel S. 1969. Experimental zoogeography of islands: A model for insular colonization. *Ecology* 50: 296–314.
Simberloff, Daniel. 1998. Flagships, umbrellas, and keystones: Is single-species management passé in the landscape era? *Biological Conservation* 83: 247–257.
Simberloff, Daniel S., and Edward O. Wilson. 1969. Experimental zoogeography of islands: The colonization of empty islands. *Ecology* 50: 278–296.
Simberloff, Daniel S., and Edward O. Wilson. 1970. Experimental zoogeography of islands: A two-year record of colonization. *Ecology* 51: 934–937.
Soulé, Michael E. 1986. *Conservation biology: The science of scarcity and diversity.* Sinauer Associates, Sunderland, Massachusetts.
Takacs, David. 1996. *The idea of biodiversity: Philosophies of paradise.* Johns Hopkins University Press, Baltimore.
United Nations Environment Programme. 2002. Biodiversity. Pp. 120–149 in *Global environment outlook 3 (GEO-3).* UNEP and Earthscan Publications, Nairobi and London.
United Nations Environment Programme. 2003. Sustaining life on Earth: How the Convention on Biological Diversity promotes nature and human well-being. www.biodiv.org/doc/publications/guide.asp.
United States Fish and Wildlife Service. The endangered species act of 1973. Accessible online at www.fws.gov/endangered/esa.html.
Wilson, Edward O. 1984. *Biophilia.* Harvard University Press, Cambridge, Massachusetts.
Wilson, Edward O. 1992. *The diversity of life.* Harvard University Press, Cambridge, Massachusetts.
Wilson, Edward O. 1994. *Naturalist.* Island Press, Shearwater Books, Washington, D.C.
Wilson, Edward O. 2002. *The future of life.* Alfred A. Knopf, New York.
Wilson, Edward O., and Daniel S. Simberloff. 1969. Experimental zoogeography of islands: Defaunation and monitoring techniques. *Ecology* 50: 267–278.
World Conservation Union. IUCN Red List. www.iucnredlist.org.
WWF–World Wide Fund for Nature. 2006. *Living planet report 2006.* WWF, Gland, Switzerland.

Chapter 10

British Columbia Ministry of Forests. Introduction to Silvicultural Systems. www.for.gov.bc.ca/hfd/pubs/SSIntroworkbook/index.htm. British Columbia Ministry of Forests, Victoria, B.C.
Canadian Broadcasting Corporation. 1993. A little place called Clayoquot Sound. CBC broadcast, 13 April 1993. http:// archives.cbc.ca/IDC-1-75-679-3918/Science_technology/clearcutting/clip6.
Cascade Resources Advocacy Group Law Center. Scientific integrity and academic freedom. www.crag.org/fire/academicfreedom.php.
Clary, David. 1986. *Timber and the Forest Service.* University Press of Kansas, Lawrence.
Donato, Daniel C., et al. 2006. Post-wildfire logging hinders regeneration and increases fire risk.*Science* 311: 352.
Food and Agriculture Organization of the United Nations. 2005. *Global forest resources assessment.* FAO Forestry Department, Rome.
Foster, Bryan C., and Peggy Foster. 2002. *Wild logging: A guide to environmentally and economically sustainable forestry.* Mountain Press, Missoula, Montana.
Gardner, Gary. 2006. Deforestation continues. Pp. 102–103 in *Vital signs 2006–2007.* Worldwatch Institute and W. W. Norton, Washington, D.C., and New York.
Harrison, Paul, and Fred Pearce, eds. 2000. *AAAS atlas of population & environment.* University of California Press, Berkeley.
Haynes, Richard W., and Gloria E. Perez, tech. eds. 2001. Northwest Forest Plan research synthesis. *Gen. Tech. Rep. PNW-GTR-498.* USDA Forest Service, Pacific Northwest Research Station, Portland, Oregon.
Myers, Norman, and Jennifer Kent. 2001. *Perverse subsidies: How misused tax dollars harm the environment and the economy.* Island Press, Washington, D.C.
National Forest Management Act of 1976. October 22, 1976 (P.O. 94–588, 90 Stat. 2949, as amended; 16 U.S.C.).
Natural Resources Canada. 2006. *The state of Canada's forests, 2005–2006.* Natural Resources Canada, Ottawa.
Runte, Alfred. 1979. *National parks and the American experience.* University of Nebraska Press, Lincoln.
Sedjo, Robert A. 2000. *A vision for the U.S. Forest Service.* Resources for the Future, Washington, D.C.
Shatford, Jeffrey P. A., David E. Hibbs, and Klaus J. Puettman. 2007. Conifer regeneration after forest fire in the Klamath-Siskiyous: How much, how soon? *Journal of Forestry* 105: 139–146.
Singh, Ashbindu, et al. 2001. An assessment of the status of the world's remaining closed forests. United Nations Environmental Program, UNEP/DEWA/TR 01–2l, August 2001.
Smith, David M., et al. 1996. *The practice of silviculture: Applied forest ecology*, 9th ed. Wiley, New York.
Smith, W. Brad, et al. 2004. Forest resources of the United States, 2002. *Gen. Tech. Rep.* NC-241, North Central Research Station, USDA Forest Service, St. Paul, Minnesota.
Soulé, Michael E., and John Terborgh, eds. 1999. *Continental conservation.* Island Press, Washington, D.C.
Stegner, Wallace. 1954. *Beyond the hundredth meridian: John Wesley Powell and the second opening of the West.* Houghton Mifflin, Boston.
Thompson, Jonathan R., Thomas A. Spies, and Lisa M. Ganio. 2007. Re-burn severity in managed and unmanaged vegetation in a large wildfire. *Proceedings of the National Academy of Sciences of the USA* 104: 10743–10748.
United Nations Environment Programme. 2002. Forests. Pp. 90–119 in *Global environment outlook 3 (GEO-3).* UNEP and Earthscan Publications, Nairobi and London.
USDA Forest Service. 2001. *U.S. forest facts and historical trends.* FS-696, March 2001.
U.S. National Park Service. 2005. *National Park Service statistical abstract 2005.* NPS Public Use Statistics Office, U.S. Department of the Interior, Denver, Colorado.
Woodard, Colin. 2006. The sale of the century. *Nature Conservancy*, Autumn 2006: 20–25.

Chapter 11

American Rivers. 2002. *The ecology of dam removal: A summary of benefits and impacts.* American Rivers, Washington, D.C., February 2002.
British Geographical Society and Bangladesh Department of Public Health Engineering. 2001. *Arsenic contamination of groundwater in Bangladesh.* Technical report WC/00/19, volume 1: Summary.

Dunn, Seth, and Christopher Flavin. 2002. Moving the climate change agenda forward. Pp. 24–50 in *State of the world 2002*. Worldwatch Institute and W. W. Norton, Washington, D.C., and New York.
EPICA community members. 2004. Eight glacial cycles from an Antarctic ice core. *Nature* 429: 623–628.
Flannery, Tim. 2005. *The weather makers: The history and future impact of climate change*. Text Publishing, Melbourne, Australia.
Gelbspan, Ross. 1997. *The heat is on: The climate crisis, the cover-up, the prescription*. Perseus Books, New York.
Gelbspan, Ross. 2004. *Boiling point: How politicians, big oil and coal, journalists, and activists are fueling the climate crisis—and what we can do to avert disaster*. Basic Books, New York.
Gettleman, Jeffrey. 2006. Annan faults "frightening lack of leadership" for global warming. *New York Times*, 16 November 2006.
Gore, Al. 2006. *An inconvenient truth: The planetary emergency of global warming and what we can do about it*. Rodale Press and Melcher Media, New York.
Intergovernmental Panel on Climate Change. 2001. *IPCC third assessment report—Climate change 2001: Synthesis report*. World Meteorological Organization and United Nations Environment Programme.
Intergovernmental Panel on Climate Change. 2007. *Climate change 2007: The physical science basis*. Contribution of Working Group I to the fourth assessment report of the Intergovernmental Panel on Climate Change. World Meteorological Organization and United Nations Environment Programme, Geneva, Switzerland.
Intergovernmental Panel on Climate Change. 2007. *Climate change 2007: Impacts, adaptation, and vulnerability*. Contribution of Working Group II to the fourth assessment report of the Intergovernmental Panel on Climate Change. World Meteorological Organization and United Nations Environment Programme, Geneva, Switzerland.
Intergovernmental Panel on Climate Change. 2007. *Climate change 2007: Mitigation of climate change*. Contribution of Working Group III to the fourth assessment report of the Intergovernmental Panel on Climate Change. World Meteorological Organization and United Nations Environment Programme, Geneva, Switzerland.
Intergovernmental Panel on Climate Change. www.ipcc.ch.
Jonzén, Niclas, et al. 2006. Rapid advance of spring arrival dates in long-distance migratory birds. *Science* 312: 1959–1961.
Karl, Thomas R., and Kevin E. Trenberth. 2003. Modern global climate change. *Science* 302: 1719–1723.
Kerr, Richard A. 2006. A worrying trend of less ice, higher seas. *Science* 311: 1698–1701.
Kerr, Richard A. 2006. A tempestuous birth for hurricane climatology. *Science* 312: 676–678.
Mastny, Lisa. 2005. Global ice melting accelerating. Pp. 88–89 in *Vital signs 2005*. Worldwatch Institute and W. W. Norton, Washington, D.C., and New York.
Mayewski, Paul A., and Frank White. 2002. *The ice chronicles: The quest to understand global climate change*. University Press of New England, Hanover, New Hampshire.
National Assessment Synthesis Team. 2000. *Climate change impacts on the United States: The potential consequences of climate variability and change*. U.S. Global Change Research Program. Cambridge University Press, Cambridge.
National Research Council, Committee on the Science of Climate Change, Division of Earth and Life Studies. 2001. *Climate change science: An analysis of some key questions*. National Academies Press, Washington, D.C.
Pacala, Stephen, and Robert Socolow. 2004. Stabilization wedges: Solving the climate problem for the next 50 years with current technologies. *Science* 305: 968–972.
Parmesan, Camille, and Gary Yohe. 2003. A globally coherent fingerprint of climate change impacts across natural systems. *Nature* 421: 37–42.
Pew Center on Global Climate Change. www.pewclimate.org.
Real Climate. www.realclimate.org.
Rignot, Eric, and Pannir Kanagaratnam. 2006. Changes in the velocity structure of the Greenland Ice Sheet. *Science* 311: 986–990.
Root, Terry L., et al. 2003. Fingerprints of global warming on wild animals and plants. *Nature* 421: 57–60.
The Royal Society. 2005. *Ocean acidification due to increasing atmospheric carbon dioxide*. The Royal Society, London, U.K., June 2005.
Sawin, Janet L. 2005. Climate change indicators on the rise. Pp. 40–41 in *Vital signs 2005*. Worldwatch Institute and W. W. Norton, Washington, D.C., and New York.
Schiermeier, Quirin. 2006. Climate credits. *Nature* 444: 976–977.
Schneider, Stephen H., and Terry L. Root, eds. 2002. *Wildlife responses to climate change: North American case studies*. Island Press, Washington, D.C.
Shapiro, Robert J., Kevin A. Hassett, and Frank S. Arnold. 2002. *Conserving energy and preserving the environment: The role of public transportation*. American Public Transportation Association, July 2002.
Siegenthaler, Urs, et al. 2005. Stable carbon cycle–climate relationship during the Late Pleistocene. *Science* 310: 1313–1317.
Spahni, Renato, et al. 2005. Atmospheric methane and nitrous oxide of the Late Pleistocene from Antarctic ice cores. *Science* 310: 1317–1321.
Speth, James Gustave. 2004. *Red sky at morning: America and the crisis of the global environment*. Yale University Press, New Haven, Connecticut.
Stevens, William K. 1999. *The change in the weather: People, weather and the science of climate*. Delta Trade Paperbacks, New York.
Taylor, David. 2003. Small islands threatened by sea level rise. Pp. 84–85 in *Vital signs 2003*. Worldwatch Institute and W. W. Norton, Washington D.C., and New York.
Time. 2006. Special report: Global warming. *Time*, 3 April 2006: 28–62.
Victor, David G. 2004. *Climate change: Debating America's policy options*. U.S. Council on Foreign Relations Press, Washington, D.C.
Victor, David G., Joshua C. House, and Sarah Joy. 2005. A Madisonian approach to climate policy. *Science* 309: 1820–1821.
United Nations. United Nations Framework Convention on Climate Change. http://unfccc.int/2860.php.
United Nations. Kyoto Protocol. http://unfccc.int/kyoto_protocol/items/2830.php.
Zwally, H. Jay, et al. 2002. Surface melt–induced acceleration of Greenland Ice-Sheet flow. *Science* 297: 218–222.

Chapter 15

Appenzeller, Tim. The end of cheap oil. *National Geographic*, June 2004: 80–109.
Association for the Study of Peak Oil and Gas. www.peakoil.net.
British Petroleum. 2007. *BP statistical review of world energy 2007*. BP, London.
Campbell, Colin J. 1997. *The coming oil crisis*. Multi-Science Publishing Co., Essex, U.K.
Deffeyes, Kenneth S. 2001. *Hubbert's peak: The impending world oil shortage*. Princeton University Press, Princeton, New Jersey.
Deffeyes, Kenneth S. 2005. *Beyond oil: The view from Hubbert's peak*. Farrar, Straus, and Giroux, New York.
Douglas, D. C., P. E. Reynolds, and E. B. Rhode, eds. 2002. *Arctic Refuge coastal plain terrestrial wildlife research summaries. Biological science report*. USGS/BRD/BSR-2002-0001. United States Geological Survey, Washington, D.C.
Dunn, Seth. 2001. Decarbonizing the energy economy. Pp. 83–102 in *State of the world 2001*. Worldwatch Institute and W. W. Norton, Washington, D.C., and New York.
Energy Information Administration, U.S. Department of Energy. www.eia.doe.gov.
Energy Information Administration, U.S. Department of Energy. 1999. *Petroleum: An energy profile, 1999*. DOE/EIA-0545(99).
Energy Information Administration, U.S. Department of Energy. 2006. *International energy annual 2004*. Washington, D.C.
Energy Information Administration, U.S. Department of Energy. 2007. *Annual energy review 2006*. DOE/EIA, Washington, D.C.
Freese, Barbara. 2003. *Coal: A human history*. Perseus Books, New York.
Goodstein, David. 2004. *Out of gas*. W. W. Norton, New York.
Holmes, Bob, and Nicola Jones. 2003. Brace yourself for the end of cheap oil. *New Scientist*, 2 August 2003: 9–11.
International Energy Agency. 2005. *Key world energy statistics 2005*. IEA Publications, Paris.
International Energy Agency. 2005. *World energy outlook 2005*. IEA Publications, Paris.

International Energy Agency. 2005. *Resources to reserves: Oil and gas technologies for the energy markets of the future*. IEA Publications, Paris.

Kunstler, James H. 2005. *The long emergency*. Atlantic Monthly Press, New York.

Lovins, Amory B. 2005. More profit with less carbon. *Scientific American* 293(3): 74–83.

Lovins, Amory B., et al. 2004. *Winning the oil endgame: Innovation for profits, jobs, and security*. Rocky Mountain Institute, Snowmass, Colorado.

Nellemann, Christian, and Raymond D. Cameron. 1998. Cumulative impacts of an evolving oil-field complex on the distribution of calving caribou. *Canadian Journal of Zoology* 76: 1425–1430.

Pelley, Janet. 2001. Will drilling for oil disrupt the Arctic National Wildlife Refuge? *Environmental Science and Technology* 35: 240–247.

Powell, Stephen G. 1990. Arctic National Wildlife Refuge: How much oil can we expect? *Resources Policy*, Sept. 1990: 225–240.

Prugh, Tom, et al. 2005. Changing the oil economy. Pp. 100–121 in *State of the world 2005*. Worldwatch Institute and W. W. Norton, Washington, D.C., and New York.

Ristinen, Robert A., and Jack J. Kraushaar. 2006. *Energy and the environment*, 2nd ed. John Wiley and Sons, New York.

Roberts, Paul. 2004. *The end of oil: On the edge of a perilous new world*. Houghton Mifflin, Boston.

Rottmann, Katja. 2006. Fossil fuel use continues to grow. Pp. 32–33 in *Vital signs 2006–2007*. Worldwatch Institute and W. W. Norton, Washington, D.C., and New York.

Russell, D. E., and P. Mc Neil. 2005. *Summer ecology of the Porcupine caribou herd*. Porcupine Caribou Management Board, Whitehorse, Yukon.

Sawin, Janet L. 2004. Making better energy choices. Pp. 24–45 in *State of the world 2004*. Worldwatch Institute and W. W. Norton, Washington, D.C., and New York.

Skinner, Brian J., and Stephen C. Porter. 2003. *The dynamic earth: An introduction to physical geology*, 5th ed. John Wiley and Sons, Hoboken, New Jersey.

United States Environmental Protection Agency. 2006. *Light-duty automotive technology and fuel economy trends: 1975 through 2006*. EPA Office of Transportation and Air Quality, Washington, D.C.

United States Fish and Wildlife Service. 2001. Potential impacts of proposed oil and gas development on the Arctic Refuge's coastal plain: Historical overview and issues of concern. Web page of the Arctic National Wildlife Refuge, Fairbanks, Alaska. http://arctic.fws.gov/issues1.htm.

United States Geological Survey. 2001. *The National Petroleum Reserve–Alaska (NPRA) data archive*. USGS Fact Sheet FS-024-01, March 2001.

United States Geological Survey. 2001. *Arctic National Wildlife Refuge, 1002 Area, petroleum assessment, 1998, including economic analysis*. USGS Fact Sheet FS-028-01, April 2001.

United States Geological Survey. 2002. *Petroleum resource assessment of the National Petroleum Reserve Alaska (NPRA)*. USGS, Washington, D.C.

United States Government Accountability Office (GAO). 2007. *Crude oil: Uncertainty about future oil supply makes it important to develop a strategy for addressing a peak and decline in oil production*. Report to Congressional Requesters, February 2007.

Walker, Donald A. 1997. Arctic Alaskan vegetation disturbance and recovery. Pp. 457–479 in R. M. M. Crawford, ed., *Disturbance and recovery in Arctic lands*. Kluwer Academic Publishers, Dordrecht, Netherlands.

Chapter 16

British Petroleum. 2007. *BP statistical review of world energy 2007*. BP, London.

Chandler, David. 2003. America steels itself to take the nuclear plunge. *New Scientist*, 9 August 2003: 10–13.

The Chernobyl Forum. 2006. *Chernobyl's legacy: Health, environmental and socio-economic impacts* and *recommendations to the governments of Belarus, the Russian Federation and Ukraine. The Chernobyl Forum: 2003–2005*. Second revised version. World Health Organization and International Atomic Energy Agency, Vienna.

Dunn, Seth. 2001. Decarbonizing the energy economy. Pp. 83–102 in *State of the world 2001*. Worldwatch Institute and W. W. Norton, Washington, D.C., and New York.

Energy Information Administration, U.S. Department of Energy. www.eia.doe.gov.

Energy Information Administration, U.S. Department of Energy. 2006. *International energy annual 2004*. Washington, D.C.

Energy Information Administration, U.S. Department of Energy. 2007. *Annual energy review 2006*. DOE/EIA, Washington, D.C.

Energy Information Administration. 2007. *Annual energy outlook 2007*. Washington, D.C.

European Commission/International Atomic Energy Agency/World Health Organization. 1996. One decade after Chernobyl: Summing up the consequences of the accident. Summary of the conference results. Vienna, Austria, 8–12 April 1996. EC/IAEA/WHO.

Hunt, Suzanne, and Janet L. Sawin, with Peter Stair. 2006. Cultivating renewable alternatives to oil. Pp. 61–77 in *State of the world 2006*. Worldwatch Institute and W. W. Norton, Washington, D.C., and New York.

Hunt, Suzanne, and Peter Stair. 2006. Biofuels hit a gusher. Pp. 40–41 in *Vital signs 2006–2007*. Worldwatch Institute and W. W. Norton, Washington, D.C., and New York.

International Atomic Energy Agency. 2006. *Annual report 2005*. IAEA, Vienna, Austria.

International Atomic Energy Agency. *Nuclear power and sustainable development*. IAEA Information Series 02-01574/FS Series 3/01/E/Rev.1. Vienna, Austria.

International Atomic Energy Agency. 2006. *Environmental consequences of the Chernobyl accident and their remediation: Twenty years of experience*. Report of the U.N. Chernobyl Forum Expert Group "Environment." IAEA, Vienna.

International Energy Agency. 2006. *Key world energy statistics 2006*. IEA Publications, Paris.

International Energy Agency. 2006. *World energy outlook 2006*. IEA Publications, Paris.

International Energy Agency. 2007. *Biomass for power generation and CHP*. IEA Publications, Paris.

Klass, Donald L. 2004. Biomass for renewable energy and fuels. In *The Encyclopedia of Energy*. Elsevier.

Lenssen, Nicholas. 2006. Nuclear power inches up. Pp. 34–35 in *Vital signs 2006–2007*. Worldwatch Institute and W. W. Norton, Washington, D.C., and New York.

Li, Zijun. 2006. Hydropower rebounds slightly. Pp. 46–47 in *Vital signs 2006–2007*. Worldwatch Institute and W. W. Norton, Washington, D.C., and New York.

Lovins, Amory B., et al. 2004. *Winning the oil endgame: Innovation for profits, jobs, and security*. Rocky Mountain Institute, Snowmass, Colorado.

Murray, Danielle. 2005. Ethanol's potential: Looking beyond corn. *Eco-economy Update* #49, 5 June 2005. Earth Policy Institute. www.earth-policy.org/Updates/2005/Update49.htm.

National Renewable Energy Lab, U.S. Department of Energy. www.nrel.gov.

Nature. 2006. Special report: Chernobyl and the future. *Nature* 440: 982–989.

Nuclear Energy Agency. 2002. *Chernobyl: Assessment of radiological and health impacts*. 2002 update of *Chernobyl: Ten years on*. OECD, Paris.

Nuclear Energy Agency. 2005. *NEA annual report 2004*. NEA, Organisation for Economic Co-operation and Development. OECD, Paris.

Office of Energy Efficiency and Renewable Energy, U.S. Department of Energy. www.eere.energy.gov.

Organisation for Economic Co-operation and Development. 2000. *Business as usual and nuclear power*. OECD Publications, Paris.

Pearce, Fred. 2006. Fuels gold: Are biofuels really the greenhouse-busting answer to our energy woes? *New Scientist*, 23 Sept. 2006: 36–41.

REN21 Renewable Energy Policy Network. 2005. *Renewables 2005 global status report*. Worldwatch Institute, Washington, D.C.

Science. 2005. News Focus: Rethinking nuclear power. *Science* 309: 1168–1179.

Spadaro, Joseph V., Lucille Langlois, and Bruce Hamilton. 2000. Greenhouse gas emissions of electricity generation chains: Assessing the difference. *IAEA Bulletin* 42(2).

Swedish Bioenergy Association (SVEBIO). 2003. *Focus: Bioenergy*. Nos. 1–10. SVEBIO, Stockholm.

Swedish Energy Agency. 2004. *Renewable electricity is the future's electricity*. Swedish Energy Agency, Eskilstuna, Sweden.

Swedish Energy Agency. 2006. *Energy in Sweden: Facts and figures 2005*. Swedish Energy Agency, Eskilstuna, Sweden.

Swedish Energy Agency. 2006. *The Swedish Energy Agency 2005*. Swedish Energy Agency, Eskilstuna, Sweden.

Swedish Energy Agency. 2006. *Energy in Sweden 2006*. Swedish Energy Agency, Eskilstuna, Sweden.

U.N. Food and Agriculture Organization. *Biomass energy in ASEAN member countries*. FAO/ASEAN/EC. FAO Regional Wood Energy Development Programme in Asia, Bangkok, Thailand.

U.S. Environmental Protection Agency. Alternative fuels website. www.epa.gov/otaq/consumer/fuels/altfuels/altfuels.htm.

World Health Organization. 2006. *Health effects of the Chernobyl accident and special health care programmes*. Report of the U.N. Chernobyl Forum Expert Group "Health." WHO, Geneva.

Worldwatch Institute and Center for American Progress. 2006. *American energy: The renewable path to energy security*. Worldwatch Institute and Center for American Progress, Washington, D.C.

Chapter 17

American Wind Energy Association. 2005. *Global wind energy market report*. AWEA, Washington, D.C.

Ananthaswamy, Anil. 2003. Reality bites for the dream of a hydrogen economy. *New Scientist*, 15 November 15 2003: 6–7.

Arnason, Bragi, and Thorsteinn I. Sigfusson. 2000. Iceland—a future hydrogen economy. *International Journal of Hydrogen Energy* 25: 389–394.

Ásmundsson, Jón Knútur. 2002. Will fuel cells make Iceland the "Kuwait of the North"? *World Press Review*, 15 February 2002.

Burkett, Elinor. 2003. A mighty wind. *New York Times magazine*. June 15, 2003.

Chow, Jeffrey, et al. 2003. Energy resources and global development. *Science* 302: 1528–1531.

Daimler Chrysler. 2003. *360 DEGREES/Daimler Chrysler Environmental Report 2003*. Daimler Chrysler AG, Stuttgart, Germany.

Dunn, Seth. 2000. The hydrogen experiment. *World Watch* 13: 14–25.

Energy Information Administration, U.S. Department of Energy. www.eia.doe.gov.

Energy Information Administration, U.S. Department of Energy. 2006. *International energy annual 2004*. Washington, D.C.

Energy Information Administration, U.S. Department of Energy. 2007. *Annual energy review 2006*. DOE/EIA, Washington, D.C.

Energy Information Administration, U.S. Department of Energy. 2007. *Annual energy outlook 2007*. Washington, D.C.

Flavin, Christopher, and Seth Dunn. 1999. A new energy paradigm for the 21st century. *Journal of International Affairs* 53: 167–190.

Grant, Paul M., Chauncey Starr, and Thomas J. Overbye. 2006. A power grid for the hydrogen economy. *Scientific American*, July 2006: 77–83.

Hirsch, Tim. 2001. Iceland launches energy revolution. *British Broadcasting Corporation News*, 24 December 2001.

Hogan, Jenny, and Philip Cohen. 2004. Is the green dream doomed to fail? *New Scientist*, 17 July 2004: 6–7.

Hydrogen & Fuel Cell Letter. 2003. World's first commercial hydrogen station opens in Iceland. *Hydrogen & Fuel Cell Letter*, May 2003.

Idaho Wind Power Working Group for the Idaho Department of Water Resources Energy Division. 2002. *Idaho wind power development strategic plan*. Boise, Idaho.

Idaho Wind Power Working Group for the Idaho Department of Water Resources Energy Division. 2002. *Wind power potential in Idaho by county*. Boise, Idaho.

International Energy Agency. 2005. *Renewables information 2005*. IEA Publications, Paris.

International Energy Agency. 2006. *Key world energy statistics 2006*. IEA Publications, Paris.

International Energy Agency. 2006. *World energy outlook 2006*. IEA Publications, Paris.

International Energy Agency. 2007. *Renewables in global energy supply: An IEA fact sheet*. IEA Publications, Paris.

International Energy Agency Renewable Energy Working Party. 2002. *Renewable energy . . . into the mainstream*. SITTARD, The Netherlands.

Jacobson, Mark Z., W. G. Colella, and D. N. Golden. 2005. Cleaning the air and improving health with hydrogen fuel-cell vehicles. *Science* 308: 1901–1905.

Knott, Michelle. 2003. Power from the waves. *New Scientist*, 20 Sept. 2003: 33–35.

Lovins, Amory B., et al. 2004. *Winning the oil endgame: Innovation for profits, jobs, and security*. Rocky Mountain Institute, Snowmass, Colorado.

Martinot, Eric, et al. 2002. Renewable energy markets in developing countries. *Annual Review of Energy and the Environment* 27: 309–48.

Martinot, Eric, Ryan Wiser, and Jan Hamrin. 2005. *Renewable energy markets and policies in the United States*. Center for Resource Solutions, San Francisco. www.martinot.info/Martinot_et_al_CRS.pdf.

Melis, Anastasios, et al. 2000. Sustained photobiological hydrogen gas production upon reversible inactivation of oxygen evolution in the green alga *Chlamydomonas reinhardtii*. *Plant Physiology* 122: 127–135.

National Renewable Energy Lab, U.S. Department of Energy. www.nrel.gov.

Office of Energy Efficiency and Renewable Energy, U.S. Department of Energy. www.eere.energy.gov.

Reeves, Ari, with Fredric Beck. 2003. *Wind energy for electric power: A REPP issue brief*. Renewable Energy Policy Project, Washington, D.C.

REN21 Renewable Energy Policy Network. 2005. *Renewables 2005 global status report*. Worldwatch Institute, Washington, D.C.

Ristinen, Robert A., and Jack J. Kraushaar, 1998. *Energy and the environment*. John Wiley and Sons, New York.

Sawin, Janet. 2004. *Mainstreaming renewable energy in the 21st century*. Worldwatch Paper 169. Worldwatch Institute, Washington, D.C.

Sawin, Janet L. 2006. Wind power blowing strong. Pp. 36–37 in *Vital signs 2006–2007*. Worldwatch Institute and W. W. Norton, Washington, D.C., and New York.

Sawin, Janet L. 2006. Solar industry stays hot. Pp. 38–39 in *Vital signs 2006–2007*. Worldwatch Institute and W. W. Norton, Washington, D.C., and New York.

Wald, Matthew L. 2006. The energy challenge: It's free, plentiful, and fickle. *New York Times*, 28 December 2006.

Weisman, Alan. 1998. *Gaviotas: A village to reinvent the world*. Chelsea Green Publishing Co., White River Junction, Vermont.

World Alliance for Decentralized Energy. 2005. *World survey of decentralized energy 2005*. WADE, Edinburgh, Scotland.

Worldwatch Institute and Center for American Progress. 2006. *American energy: The renewable path to energy security*. Worldwatch Institute and Center for American Progress, Washington, D.C.

Chapter 18

Allen, G. H., and R. A. Gearheart, eds. 1988. *Proceedings of a conference on wetlands for wastewater treatment and resource enhancement*. Humboldt State University, Arcata, California.

Ayres, Robert U., and Leslie W. Ayres. 1996. *Industrial ecology: Towards closing the materials cycle*. Edward Elgar Press, Cheltenham, U.K.

Beede, David N., and David E. Bloom. 1995. The economics of municipal solid waste. *World Bank Research Observer* 10: 113–150.

Diesendorf, Mark, and Clive Hamilton. 1997. *Human ecology, human economy*. Allen and Unwin, St. Leonards.

Douglas, Ed. 2007. Better by design. *New Scientist*, 6 Jan. 2007: 31–35.

Edmonton, Alberta, City of. 2003. Waste management. www.edmonton.ca/portal/server.pt/gateway/PTARGS_0_2_104_0_0_35/http%3B/cmsserver/COEWeb/environment+waste+and+recycling/waste.

Gitlitz, Jenny, and Pat Franklin. 2004. *The 10-cent incentive to recycle*, 3rd ed. Container Recycling Institute, Arlington, Virginia.

Graedel, Thomas E., and Braden R. Allenby. 2002. *Industrial ecology*, 2nd ed. Prentice Hall, Upper Saddle River, New Jersey.

Lilienfeld, Robert, and William Rathje. 1998. *Use less stuff: Environmental solutions for who we really are*. Ballantine, New York.

Manahan, Stanley E. 1999. *Industrial ecology: Environmental chemistry and hazardous waste*. Lewis Publishers, CRC Press, Boca Raton, Florida.

Mc Donough, William, and Michael Braungart. 2002. *Cradle to cradle: Remaking the way we make things*. North Point Press, New York.

Mc Ginn, Anne Platt. 2002. Toxic waste largely unseen. Pp. 112–113 in *Vital signs 2002*. Worldwatch Institute and W. W. Norton, Washington, D.C., and New York.

New York City Department of Parks and Recreation. Fresh Kills Park. www.nycgovparks.org/sub_your_park/fresh_kills_park/html/fresh_kills_park.html.

New York City Department of Planning. Fresh Kills Park Project. www.nyc.gov/html/dcp/html/fkl/fkl3.shtml.

New York City Department of Sanitation. 2000. Closing the Fresh Kills landfill. *The DOS Report*, Feb. 2000.

Rathje, William, and Colleen Murphy. 2001. *Rubbish! The archeology of garbage.* University of Arizona Press.

Simmons, Phil, et al. 2006. The state of garbage in America. *Biocycle* 47(4): 26.

Smith, Ronald S. 1998. *Profit centers in industrial ecology.* Quorum Books, Westport.

Socolow, Robert H., et al., eds. 1994. *Industrial ecology and global change.* Cambridge University Press, Cambridge.

United Nations Environment Programme. 2000. *International source book on environmentally sound technologies (ESTs) for municipal solid waste management (MSWM).* UNEP IETC, Osaka, Japan.

United States Environmental Protection Agency. 2006. *Municipal solid waste generation, recycling, and disposal in the United States: Facts and figures for 2006.* EPA Office of Solid Waste and Emergency Response.

United States Environmental Protection Agency. Municipal solid waste. www.epa.gov/epaoswer/non-hw/muncpl.

Chapter 19

Ames, Bruce N., Margie Profet, and Lois Swirsky Gold. 1990. Nature's chemicals and synthetic chemicals: Comparative toxicology. *Proceedings of the National Academy of Sciences of the USA* 87: 7782–7786.

Bloom, Barry. 2005. Public health in transition. *Scientific American* 293(3): 92–99.

Cagen, S. Z., et al. 1999. Normal reproductive organ development in wistar rats exposed to bisphenol A in the drinking water. *Regulatory Toxicology and Pharmacology* 30: 130–139.

Carlsen, Elisabeth, et al. 1992. Evidence for decreasing quality of semen during past 50 years. *British Medical Journal* 305: 609–613.

Carson, Rachel. 1962. *Silent spring.* Houghton Mifflin, Boston.

Colburn, Theo, Dianne Dumanoski, and John P. Myers. 1996. *Our stolen future.* Penguin USA, New York.

Crain, D. Andrew, and Louis J. Guillette Jr. 1998. Reptiles as models of contaminant-induced endocrine disruption. *Animal Reproduction Science* 53: 77–86.

Gross, Liza. 2007. The toxic origins of disease. *Plo S Biology* 5(7): e193. doi:10.1371/journal.pbio.0050193.

Guillette, Elizabeth A., et al. 1998. An anthropological approach to the evaluation of preschool children exposed to pesticides in Mexico. *Environmental Health Perspectives* 106: 347–353.

Guillette, Louis J. Jr., et al. 1999. Plasma steroid concentrations and male phallus size in juvenile alligators from seven Florida lakes. *General and Comparative Endocrinology* 116: 356–372.

Guillette, Louis J. Jr., et al. 2000. Alligators and endocrine disrupting contaminants: A current perspective. *American Zoologist* 40: 438–452.

Halweil, Brian. 1999. Sperm counts dropping. Pp. 148–149 in *Vital signs 1999.* Worldwatch Institute and W. W. Norton, Washington, D.C., and New York.

Hayes, Tyrone, et al. 2003. Atrazine-induced hermaphroditism at 0.1 PPB in American leopard frogs (*Rana pipiens*): Laboratory and field evidence. *Environmental Health Perspectives* 111: 568–575.

Hunt, Patricia A., et al. 2003. Bisphenol A exposure causes meiotic aneuploidy in the female mouse. *Current Biology* 13: 546–553.

Kolpin, Dana W., et al. 2002. Pharmaceuticals, hormones, and other organic wastewater contaminants in U.S. streams, 1999–2000: A national reconnaissance. *Environmental Science and Technology* 36: 1202–1211.

Landis, Wayne G., and Ming-Ho Yu. 2004. *Introduction to environmental toxicology,* 3rd ed. Lewis Press, Boca Raton, Florida.

Loewenberg, Samuel. 2003. E.U. starts a chemical reaction. *Science* 300: 405.

Manahan, Stanley E. 2004. *Environmental chemistry,* 8th ed. Lewis Publishers, CRC Press, Boca Raton, Florida.

Mc Ginn, Anne Platt. 2000. *Why poison ourselves? A precautionary approach to synthetic chemicals.* Worldwatch Paper #153. Worldwatch Institute, Washington, D.C.

Mc Ginn, Anne Platt. 2002. Reducing our toxic burden. Pp. 75–100 in *State of the world 2002.* Worldwatch Institute and W. W. Norton, Washington, D.C., and New York.

Mc Ginn, Anne Platt. 2003. Combating malaria. Pp. 62–84 in *State of the world 2003.* Worldwatch Institute and W. W. Norton, Washington, D.C., and New York.

Millennium Ecosystem Assessment. 2005. *Ecosystems and human well-being: Health synthesis.* World Health Organization.

Moeller, Dade. 2004. *Environmental health,* 3rd ed. Harvard University Press, 2004.

National Center for Environmental Health; U.S. Centers for Disease Control and Prevention. 2005. *Third national report on human exposure to environmental chemicals.* NCEH Pub. No. 05-0570, Atlanta.

National Center for Health Statistics. 2004. *Health, United States, 2004, with chartbook on trends in the health of Americans.* Hyattsville, Maryland.

Pirages, Dennis. 2005. Containing infectious disease. Pp. 42–61 in *State of the world 2005.* Worldwatch Institute and W. W. Norton, Washington, D.C., and New York.

Renner, Rebecca. 2002. Conflict brewing over herbicide's link to frog deformities. *Science* 298: 938–939.

Rodricks, Joseph V. 1994. *Calculated risks: Understanding the toxicity of chemicals in our environment.* Cambridge University Press, Cambridge.

Salem, Harry, and Eugene Olajos. 1999. *Toxicology in risk assessment.* CRC Press, Boca Raton, Florida.

Spiteri, I. Daniel, Louis J. Guillette Jr., and D. Andrew Crain. 1999. The functional and structural observations of the neonatal reproductive system of alligators exposed *in ozo* to atrazine, 2,4-D, or estradiol. *Toxicology and Industrial Health* 15: 181–186.

Stancel, George, et al. 2001. Report of the bisphenol A sub-panel. Chapter 1 in *National Toxicology Program's report of the endocrine disruptors low-dose peer review.* U.S. EPA and NIEHS, NIH.

Stephens, Carolyn, and Peter Stair. 2007. Charting a new course for urban public health. Pp. 134–151 in *State of the world 2007: Our urban future.* Worldwatch Institute and W. W. Norton, Washington, D.C., and New York.

Stockholm Convention on Persistent Organic Pollutants. www.pops.int.

United States Environmental Protection Agency. 2003. Pesticide registration program. www.epa.gov/pesticides/factsheets/registration.htm.

United States Environmental Protection Agency. 2003. Toxic Substances Control Act. www.epa.gov/region5/defs/html/ tsca.htm.

United States Environmental Protection Agency. 2003. *EPA's draft report on the environment.* EPA 600-R-03-050. EPA, Washington, D.C.

Williams, Phillip L., Robert C. James, and Stephen M. Roberts, eds. 2000. *The principles of toxicology: Environmental and industrial applications,* 2nd ed. Wiley-Interscience, New York.

World Health Organization. 2004. *World health report 2004: Changing history.* WHO, Geneva, Switzerland.

Yu, Ming-Ho. 2004. *Environmental toxicology:Biological and health effects of pollutants,* 2nd ed. CRC Press, Boca Raton, Florida.

Chapter 20

Abbott, Carl. 2001. *Greater Portland: Urban life and landscape in the Pacific Northwest.* University of Pennsylvania Press.

Abbott, Carl. 2002. Planning a sustainable city. Pp. 207–235 in Squires, Gregory D., ed. *Urban sprawl: Causes, consequences, and policy responses.* Urban Institute Press, Washington, D.C.

Beck, Roy, et al. 2003. *Outsmarting smart growth: Population growth, immigration, and the problem of sprawl.* Center for Immigration Studies, Washington, D.C.

Breuste, Jurgen, et al. 1998. *Urban ecology.* Springer-Verlag, Berlin.

Brockerhoff, Martin P. 2000. An urbanizing world. *Population Bulletin* 55(3). Population Reference Bureau, Washington, D.C.

Cronon, William. 1991. *Nature's metropolis: Chicago and the great West.* W. W. Norton, New York.

Duany, Andres, et al. 2001. *Suburban nation: The rise of sprawl and the decline of the American dream*. North Point Press, New York.
Ewing, Reid, et al. 2002. *Measuring sprawl and its impact*. Smart Growth America.
Ewing, Reid, et al. 2003. Measuring sprawl and its transportation impacts. *Transportation Research Record* 1831: 175–183.
Girardet, Herbert. 2004. *Cities people planet: Livable cities for a sustainable world*. Academy Press.
Hall, Kenneth B., and Gerald A. Porterfield. 2001. *Community by design: New urbanism for suburbs and small communities*. Mc Graw-Hill, New York.
Jacobs, Jane. 1992. *The death and life of great American cities*. Vintage.
Kalnay, Eugenia, and Ming Cai. 2003. Impact of urbanization and land-use change on climate. *Nature* 423: 528–531.
Kirdar, Uner, ed. 1997. *Cities fit for people*. United Nations, New York.
Litman, Todd. 2004. *Rail transit in America: A comprehensive evaluation of benefits*. Victoria Transport Policy Institute and American Public Transportation Association.
Logan, Michael F. 1995. *Fighting sprawl and city hall*. University of Arizona Press, Tucson.
Metro. www.metro-region.org.
New Urbanism. www.newurbanism.org.
Northwest Environment Watch. 2004. *The Portland exception: A comparison of sprawl, smart growth, and rural land loss in 15 U.S. cities*. Northwest Environment Watch, Seattle.
Oppenheimer, Laura. 2006. Measure 37 changed state, but how much? *Oregonian*, 3 Dec. 2006: 1 and A15.
Pearce, Fred. 2005. Cities lead the way to a greener world. *New Scientist*, 4 June 2005: 8–9.
Portney, Kent. E. 2003. *Taking sustainable cities seriously: Economic development, the environment, and quality of life in American cities (American and comparative environmental policy)*. MIT Press, Cambridge, Massachusetts.
Pugh, Cedric, ed. 1996. *Sustainability, the environment, and urbanization*. Earthscan Publications, London.
Sheehan, Molly O'Meara. 2001. *City limits: Putting the brakes on sprawl*. Worldwatch Paper #156. Worldwatch Institute, Washington, D.C.
Sheehan, Molly O'Meara. 2002. What will it take to halt sprawl? *World Watch*, Jan/Feb 2002: 12–23.
Sprawl City. www.sprawlcity.org.
Stren, R., et al. 1992. *Sustainable cities: Urbanization and the environment in international perspective*. Westview Press, Boulder, Colorado, and San Francisco.
United Nations Environment Programme. 2002. Urban areas. Pp. 240–269 in *Global environment outlook 3 (GEO-3)*. UNEP and Earthscan Publications, Nairobi and London.
United Nations Population Division. 2006. *World urbanization prospects: The 2005 revision*. UNPD, New York.
United States Environmental Protection Agency. Smart growth. www.epa.gov/smartgrowth.
Wiewel, Wim, and Jospeh J. Persky, eds. 2002. *Suburban sprawl: Private decisions and public policy*. M. E. Sharpe, Armond, New York.
Worldwatch Institute. 2007. *State of the world 2007: Our urban future*. Worldwatch Institute and W. W. Norton, Washington, D.C., and New York.

Chapter 21

Balmford, Andrew, et al. 2002. Economic reasons for conserving wild nature. *Science* 297: 950–953.
Barbour, Ian G. 1992. *Ethics in an age of technology*. Harper Collins, San Francisco.
Brown, Lester. 2001. *Eco-economy: Building an economy for the Earth*. Earth Policy Institute and W. W. Norton, New York.
Carson, Richard T., Leanne Wilks, and David Imber. 1994. Valuing the preservation of Australia's Kakadu Conservation Zone. *Oxford Economic Papers* 46: 727–749.
Cole, Luke W., and Sheila R. Foster. 2001. *From the ground up: Environmental racism and the rise of the environmental justice movement*. New York University Press, New York.
Costanza, Robert, et al. 1997. The value of the world's ecosystem services and natural capital. *Nature* 387: 253–260.
Costanza, Robert, et al. 1997. *An introduction to ecological economics*. St. Lucie Press, Boca Raton, Florida.
Daily, Gretchen. 1997. *Nature's services: Societal dependence on natural ecosystems*. Island Press, Washington, D.C.
Daly, Herman E. 1996. *Beyond growth*. Beacon Press, Boston.
Daly, Herman E. 2005. Economics in a full world. *Scientific American* 293(3): 100–107.
De Graaf, John, David Wann, and Thomas Naylor. 2002. *Affluenza: The all-consuming epidemic*. Berrett-Koehler Publishers, San Francisco.
Elliot, Robert, and Arran Gare, eds. 1983. *Environmental philosophy: A collection of readings*. Pennsylvania State University Press, University Park.
Esty, Daniel C., and Andrew S. Winston. 2006. *Green to gold: How smart companies use environmental strategy to innovate, create value, and build competitive advantage*. Yale University Press, New Haven, Connecticut.
Field, Barry C., and Martha K. Field. 2005. *Environmental economics*, 4th ed. Mc Graw-Hill, New York.
Fox, Stephen. 1985. *The American conservation movement: John Muir and his legacy*. University of Wisconsin Press, Madison.
Gardner, Gary, et al. 2004. The state of consumption today. Pp. 3–23 in *State of the world 2004*. Worldwatch Institute and W. W. Norton, Washington, D.C., and New York.
Gardner, Gary, and Erik Assadourian. 2004. Rethinking the good life. Pp. 164–179 in *State of the world 2004*. Worldwatch Institute and W. W. Norton, Washington, D.C., and New York.
Goodstein, Eban. 1999. *The tradeoff myth: Fact and fiction about jobs and the environment*. Island Press, Washington, D.C.
Goodstein, Eban. 2007. *Economics and the environment*, 5th ed. John Wiley & Sons, Hoboken, New Jersey.
Gundjeihmi Aboriginal Corporation. Welcome to the Mirarr site. www.mirarr.net.
Hawken, Paul, Amory Lovins, and L. Hunter Lovins. 1999. *Natural capitalism*. Little, Brown, and Co., Boston.
HM Treasury. 2006. *Stern review on the economics of climate change*. HM Treasury and Cambridge University Press, U.K.
Kolstad, Charles D. 2000. *Environmental economics*. Oxford University Press, Oxford.
Leopold, Aldo. 1949. *A Sand County almanac, and sketches here and there*. Oxford University Press, New York.
Millennium Ecosystem Assessment. 2005. *Ecosystems and human well-being: Opportunities and challenges for business and industry*. Millennium Ecosystem Assessment and World Resources Institute.
Nash, Roderick F. 1989. *The rights of nature*. University of Wisconsin Press, Madison.
Nash, Roderick F. 1990. *American environmentalism: Readings in conservation history*, 3rd ed. Mc Graw-Hill, New York.
Nordhaus, William. 2007. Critical assumptions in the Stern Review on Climate Change. *Science* 317: 201–202.
O'Neill, John O., R. Kerry Turner, and Ian J. Bateman, eds. 2002. *Environmental ethics and philosophy*. Edward Elgar, Cheltenham, U.K.
Pearson, Charles S. 2000. *Economics and the global environment*. Cambridge University Press, Cambridge.
Ricketts, Taylor, et al. 2004. Economic value of tropical forest to coffee production. *Proceedings of the National Academy of Sciences of the USA* 101: 12579–12582.
Sachs, Jeffrey. 2005. Can extreme poverty be eliminated? *Scientific American* 293(3): 56–65.
Singer, Peter, ed. 1993. *A companion to ethics*. Blackwell Publishers, Oxford.
Smith, Adam. 1776. *An inquiry into the nature and causes of the wealth of nations*. 1993 ed., Oxford University Press, Oxford.
Stone, Christopher D. 1972. Should trees have standing? Towards legal rights for natural objects. *Southern California Law Review* 1972: 450–501.
Tietenberg, Tom. 2006. *Environmental economics and policy*, 5th ed. Addison Wesley, Boston.

Turner, R. Kerry, David Pearce, and Ian Bateman. 1993. *Environmental economics: An elementary introduction*. Johns Hopkins University Press, Baltimore.

Venetoulis, Jason, and Cliff Cobb. 2004. *The genuine progress indicator 1950–2002 (2004 update)*. Redefining Progress, Oakland, California.

Wenz, Peter S. 2001. *Environmental ethics today*. Oxford University Press, Oxford.

White, Lynn. 1967. The historic roots of our ecologic crisis. *Science* 155: 1203–1207.

Chapter 22

Clark, Ray, and Larry Canter. 1997. *Environmental policy and NEPA: Past, present, and future*. St. Lucie Press, Boca Raton, Florida.

Dietz, Thomas, et al. 2003. The struggle to govern the global commons. *Science* 302: 1907–1912.

Fogleman, Valerie M. 1990. *Guide to the National Environmental Policy Act*. Quorum Books, New York.

Fox, Stephen. 1985. *The American conservation movement: John Muir and his legacy*. University of Wisconsin Press, Madison.

French, Hilary. 2000. Environmental treaties gain ground. Pp. 134–135 in *Vital Signs 2000*. Worldwatch Institute and W. W. Norton, Washington D.C., and New York.

Green Scissors. 2004. *Green Scissors 2004: Cutting wasteful and environmentally harmful spending*. Friends of the Earth, Taxpayers for Common Sense, and U.S. Public Interest Research Group.

Herzog, Lawrence A. 1990. *Where north meets south: Cities, space, and politics on the U.S.–Mexico border*. Center for Mexican-American Studies, University of Texas at Austin.

Houck, Oliver. 2003. Tales from a troubled marriage: Science and law in environmental policy. *Science* 302: 1926–1928.

Kraft, Michael E. 2003. *Environmental policy and politics*, 3rd ed. Longman, New York.

Kubasek, Nancy K., and Gary S. Silverman. 2004. *Environmental law*, 5th ed. Prentice Hall, Upper Saddle River, New Jersey.

Myers, Norman, and Jennifer Kent. 2001. *Perverse subsidies: How misused tax dollars harm the environment and the economy*. Island Press, Washington, D.C.

The National Environmental Policy Act of 1969, as amended (Pub. L. 91–190, 42 U.S.C. 4321–4347, January 1, 1970, as amended by Pub. L. 94–52, July 3, 1975, Pub. L. 94–83, August 9, 1975, and Pub. L. 97–258, § 4(b), Sept. 13, 1982). www.nepa.gov/nepa/regs/nepa/nepaeqia.htm.

Office of Management and Budget, Executive Office of the President of the United States, Washington, D.C. 2003. *Informing regulatory decisions: 2003 report to Congress on the costs and benefits of federal and unfunded mandates on state, local, and tribal entities*. Washington, D.C., September 2003.

Percival, Robert V., et al. 2003. *Environmental regulation: Law, science, and policy*, 4th ed. Aspen Publishers, New York.

Shellenberger, Michael, and Ted Nordhaus. 2004. *The death of environmentalism: Global warming politics in a post-environmental world*. Presented at the Environmental Grantmakers Association meeting, October 2004.

Steel, Brent S., Richard L. Clinton, and Nicholas P. Lovrich. 2002. *Environmental politics and policy*. Mc Graw-Hill, New York.

Tietenberg, Tom. 2006. *Environmental economics and policy*, 5th ed. Addison Wesley, Boston.

Union of Concerned Scientists. 2004. Restoring scientific integrity in policymaking. www.ucsusa.org/scientific_integrity/interference/scientists-signon-statement.html.

United States Congress. House. H.R. 3378. 2000. The Tijuana River Valley Estuary and Beach Sewage Cleanup Act of 2000.

United States Environmental Protection Agency. 2000. *Regulatory impact analysis: Heavy-duty engine and vehicle standards and highway diesel fuel sulfur control requirements*. EPA420-R-00-026. EPA, Washington, D.C.

United States Environmental Protection Agency. 2006. *Acid rain program 2005 progress report*. U.S. EPA, Washington, D.C.

Vig, Norman J., and Michael E. Kraft, eds. 2005. *Environmental policy: New directions for the twenty-first century*, 6th ed. CQ Press, Congressional Quarterly, Inc., Washington, D.C.

Wilkinson, Charles F. 1992. *Crossing the next meridian: Land, water, and the future of the West*. Island Press, Washington, D.C.

Chapter 23

Bartlett, Peggy, and Geoffrey W. Chase, eds. 2004. *Sustainability on campus: Stories and strategies for change*. MIT Press, Cambridge, Massachusetts.

Brower, Michael, and Warren Leon. 1999. *The consumer's guide to effective environmental choices: Practical advice from the Union of Concerned Scientists*. Three Rivers Press, New York.

Brown, Lester. 2001. *Eco-economy: Building an economy for the Earth*. Earth Policy Institute and W. W. Norton, New York.

Brown, Lester. 2006. *Plan B 2.0: Rescuing a planet under stress and a civilization in trouble*. Earth Policy Institute and W. W. Norton, New York.

Carlson, Scott. 2006. In search of the sustainable campus: With eyes on the future, universities try to clean up their acts. *Chronicle of Higher Education*, 20 Oct. 2006: vol 53: A10.

Creighton, Sarah Hammond. 1998. *Greening the ivory tower: Improving the environmental track record of universities, colleges, and other institutions*. MIT Press, Cambridge, Massachusetts.

Daly, Herman E. 1996. *Beyond growth*. Beacon Press, Boston.

Dasgupta, Partha, Simon Levin, and Jane Lubchenco. 2000. Economic pathways to ecological sustainability. *Bio Science* 50: 339–345.

Durning, Alan. 1992. *How much is enough? The consumer society and the future of the Earth*. Worldwatch Institute, Washington, D.C.

Erickson, Jon D., and John M. Gowdy. 2002. The strange economics of sustainability. *Bio Science* 52: 212.

Esty, Daniel C., et al. 2006. *Pilot 2006 Environmental Performance Index*. Yale Center for Environmental Law & Policy, New Haven, Connecticut.

French, Hilary. 2004. Linking globalization, consumption, and governance. Pp. 144–163 in *State of the world 2004*. Worldwatch Institute and W. W. Norton, Washington, D.C., and New York.

Gardner, Gary. 2001. Accelerating the shift to sustainability. Pp. 189–206 in *State of the world 2001*. Worldwatch Institute and W. W. Norton, Washington, D.C., and New York.

Gardner, Gary, and Erik Assadourian. 2004. Rethinking the good life. Pp. 164–179 in *State of the world 2004*. Worldwatch Institute and W. W. Norton, Washington, D.C., and New York.

Gibbs, W. Wayt. 2005. How should we set priorities? *Scientific American* 293(3): 108–115.

Hawken, Paul. 1994. *The ecology of commerce: A declaration of sustainability*. Harper Business, New York.

Kahneman, Daniel, et al. 2006. Would you be happier if you were richer? A focusing illusion. *Science* 312: 1908–1910.

Keniry, Julian. 1995. *Ecodemia: Campus environmental stewardship at the turn of the 21st century*. National Wildlife Federation, Washington, D.C.

Mc Michael, A. J., et al. 2003. New visions for addressing sustainability. *Science* 302: 1919–1921.

Mc Intosh, Mary, et al. 2001. *State of the campus environment: A national report card on environmental performance and sustainability in higher education*. National Wildlife Federation Campus Ecology.

Meadows, Donella, Jørgen Randers, and Dennis Meadows. 2004. *Limits to growth: The 30-year update*. Chelsea Green Publishing. Co., White River Junction, Vermont.

Millennium Ecosystem Assessment. 2005. *Ecosystems and human well-being: General synthesis*. Millennium Ecosystem Assessment and World Resources Institute.

National Research Council, Board on Sustainable Development. 1999. *Our common journey: A transition toward sustainability*. National Academies Press, Washington, D.C.

National Wildlife Federation. Campus ecology. www.nwf.org/campusecology.

Sanderson, Eric W., et al. 2002. The human footprint and the last of the wild. *Bio Science* 52: 891–904.

Schor, Juliet B., and Betsy Taylor, eds. 2002. *Sustainable planet: Solutions for the twenty-first century*. The Center for a New American Dream. Beacon Press, Boston.

Toor, Will, and Spenser W. Havlick. 2004. *Transportation and sustainable campus communities: Issues, examples, solutions*. Island Press, Washington, D.C.

United Nations. 2002. *Report of the World Summit on Sustainable Development, Johannesburg, South Africa, 26 August–4 September 2002.* United Nations, New York.

United Nations. 2002. *The road from Johannesburg: What was achieved and the way forward.* United Nations, New York.

United Nations Department of Economic and Social Affairs. 2006. *The Millennium Development Goals Report 2006.* U.N. DESA, New York.

United Nations Development Programme. 2002. *Human development report 2002.* Oxford University Press, Oxford.

United Nations Environment Programme. 2002. Outlook: 2002–2032. Pp. 319–400 in *Global environment outlook 3 (GEO-3).* UNEP and Earthscan Publications, Nairobi and London.

University Leaders for a Sustainable Future. www.ulsf.org.

U.S. Green Building Council. www.usgbc.org.

Wackernagel, Mathis, Lillemor Lewan, and Carina Borgström-Hansson. 1999. Evaluating the use of natural capital with the ecological footprint. *Ambio* 28: 604.

Wilson, Edward O. 1998. *Consilience: The unity of knowledge.* Alfred A. Knopf, New York.

World Commission on Environment and Development. 1987. *Our common future.* Oxford University Press, Oxford.

Index

Note: Page numbers in **bold** indicate definitions; pages numbers followed by *f* indicate figures; page numbers follow by *t* refer to tables.

A

Abbotsford 212
Abidjan 578
abiotic components **4**
Aboriginal peoples
  and campus sustainability 736
  and environmental justice 668
  and environmental law and policy 700
  and land claims 300, 702, 702*f*
  and the Mackenzie Valley Pipeline 462–464
  and mining 654–656
  and National Forest Strategy 302
  perception of environment 658, 663
  and radiation studies 669
Aboriginal Pipeline Group 463
Aborigines **657**–658
abrasion 202
abyssal plain **358**
acacias 118
Acadian forest region 293
Accelerated Reduction and Elimination of Toxics Program 697
Acherontia atropos 98*f*
acid drainage 12, **483**
acidic **40**
acidic deposition
  consequences of 409*f*
  effects on northeastern forests 409*t*
  explained 408–409
  Hubbard Brook Experimental Forest 412–413
  lake acidity 411*f*
  and neutralizing capacity 410*f*
  reduction of 409–410
  sulphate levels 411*f*
  wet nitrate 412*f*
  wet sulphate 411*f*
acidic precipitation **408**
acidity, of soil 195
acid rain 385, **408**
Acinonyx jubatus 257
active solar energy collection **533**, 534
acts **696**

acute exposure **611**
acute radiation sickness 512
Adam Joseph Lewis Center for Environmental Studies 730, 731*f*
Adams Mine Lake Act 581
adaptation 63, 96*f*, 443–**450**
adaptive management **302**–303
adaptive radiation **64**, 65*f*
adaptive trait **63**
adenine 43*f*
aeolian erosion 202
aerobic **466**
aerosols **393**, 428
affluence 179–180
affluenza **741**
afforestation **303**
Afghanistan 203
Africa. *See also* specific countries
  agriculture 198
  contraception use 175
  GM crops 237
  HIV/AIDS 180–181
  parks and reserves 309
  soil erosion 202
African lions 255
age distribution **77**
Agent Orange 579
age pyramids **77**, 170*f*
age structure **77**, 77*f*, 170–171
agreements **696**
Agricultural Revolution 7, **167**
agriculture
  and climate change 446–448, 452–453
  community-supported **245**
  defined **196**
  as environmental challenge 20–22
  feedlot operations 239–240
  history of 196–199, 198*f*, 199*f*
  industrialized **199**
  and land conversion 295, 297, 298
  livestock grazing 297–298
  locally supported 244–245
  low-input 243
  organic 243–244, 243*f*, 245–247, 247*f*
  and population growth 198
  subsistence **198**
  sustainable. *See* sustainable agriculture
  traditional **198**
  urban 245–247, 247*f*
  and water consumption 335

agroforestry **206**
AirCare 368
air pollutants **393**
air pollution. *See also* indoor air pollution; outdoor air pollution
  Air Quality Health Index (AQHI) 611
  defined **393**
  international agreements 709*t*
  willingness-to-pay transition 706*f*
Air Quality Health Index (AQHI) **611**
airshed **403**
Alaska 482, 484*f*
albedo **425**
Alces alces 114
Aleutians 135
alfalfa 231
algal blooms 367–368, 367*f*
Algeria 177
alien **108**
alkalinity, of soil 195
allelopathy 99
allergens **600**
alley cropping **207**
alligators 601
allopatric speciation **66**, 66*f*, 67*t*
alpine biomes 120–121
alpine fir 292
alpine tundra **114**
altitudinal zonation 119*f*, 120
Alvarez, Dr. Luis 72–73
Alvarez, Dr. Walter 72–73
Amazon rainforest 200
ambient air pollution **393**
amensalism **99**
American chestnut 108–109
American pika 445, 446*f*
Ames Research Center 32
amino acids **42**, 42*f*
ammonia **395**
Amoco Cadiz oil spill 704
amphibian diversity and decline 265
amphipod 103
AMS Lighter Footprint Strategy 727
Amur tigers 254–255
anaerobic **466**
Anasazi 10
Anderson, David 278, 352
Andes Mountains 135
Andrade, Maydianne 69–86
anglerfish 360*f*

animal dander 416
Animal Farm (Orwell) 102
animal food production 239–240, 239*f*, 241*f*, 297–298
animal rights 240, 661
animals, domesticated **239**
Annapolis-Cornwallis Valley Aquifers 322
Annapolis Tidal Generating Station 528
Annau, Keleigh 748
Anoplophora glabripennis 109
Antarctica 405, 406*f*, 431*f*, 434, 441
anthracite coal 468*f*, 471
anthropocentrism **662**, 663*f*
anthropogenic **426**
anthropogenic factors 590
Antilocapra americana 116
ants 97*f*
Apis apis 230
Apo Island 377
Appalachian Mountains 135
aquaculture
    benefits of 242, 373
    carp 243
    defined **242**, **373**
    drawbacks of 373–374
    explained 241–242
    growth of 373
    negative impacts 242–243
    world production 242*f*
aquatic systems 121
aqueous solution 40
aquifers **143**, 212, **321–322**, 323*f*
Aral Sea 317*f*, 329
Aransas National Wildlife Refuge 277
arbutus 293
archaea 44
archeology 566
the Arctic
    climate change 439*f*, 441–442
    global warming 269, 444
    human impact 169–170
    polar ozone depletion 406–407
Arctic foxes 484*f*
area effect **276**
Arends-Kuenning, Mary 177
Argentina 207
Ariolilmax columbianus 116
Arizona 238*f*
armyworms 225*f*, 226
arsenic 340
artesian aquifer **322**
artificial selection 64, 197
artificial wetlands **346**
asbestos **588**–590, 602
asbestosis 589
Asian Brown Cloud 405

Asian long-horned beetle 109
aspen 292
Assembly of First Nations 736
Association of Small Island States 424
asynchronicity 432
Atlantic cod 352–353
Atlantic grey whale 267, 368
Atlantic Ocean 129–130
atmosphere
    and climate 391–393
    composition of 387, 387*f*
    defined **130**, **386**
    and Earth temperature 424–425, 426
    layers of 387–388, 387*f*
    properties of 388
    and solar energy 388–390
    and weather 390–391, 390*f*
atmospheric deposition **408**
atmospheric pressure **388**, 388*f*
atmospheric sampling 432
atoll **361**
atomic number **36**
atoms
    bonding 37
    defined **36**
    mixtures and solutions 37
    redox reactions 37
Aujla, Dev 748
Australia
    cactus moths 228, 228*f*
    Kakadu National Park 309, 657, 681–682
    medicinal plants 272
    red tides 367*f*
    uranium mining 658, 659*f*, 676
Australian Plate 134–135
automotive technology 452, 452*f*, 488
autotrophs **48**, 48*f*, 99
avian influenza 240
axial wobble 429, 429*f*
axis 389
Ayles Ice Shelf 441, 441*f*
Azraq Oasis 333

B

babassu palm 271
Bachman's warbler 262
Bacillus thuringiensis **229**
background extinction rate 70
background rate of extinction **262**
bacteria
    and bioremediation 42
    on early Earth 53
    functions of 44
    and oil spills 33–34
    structure of 43
bad air days 403

bagasse 517*f*, 518, 522
baghouse 567
Bahrain 172
Balil tiger 263
balsam fir 293
balsam poplar 292
Baltic Sea 138
banana slug 116
Banff National Park 440
Bangladesh
    family planning 177
    fertility decline 176–177
    water pollution 340
Bardwaj, Michael 92
Barlow, Maude 718
barrier island **361**
basalt 133*f*
Basel Convention on the Control of Transboundary Movements of Hazardous Wastes and Their Disposal 578, 616
BASF 236
basic **40**
Bateman, Robert 152
bathymetry **358**, 359*f*
Bayer CropScience 236
Bay of Fundy 527–528
Beare Road Landfill 531, 557–559, 564, 564*f*, 596
bears 114
Becquerel 414
bedrock **191**
beech trees 115, 293
beer brewing 574
bees 230
beetles 230, 258
Belize 281
Belleville Small Craft Harbour 581
beluga whales 126–127
Bengal tigers 257
benthic habitats **359**
benthic organisms 101, 103
benthic zone **320**, 321*f*
Benton, Ross 297
Berger, Thomas 462
Berman, Tzeporah 311
B horizon **193**
big-leaf maple 293
bioaccumulation 127, **606**
biocentrism **662**, 663*f*
biocontrol **228–229**
biodegrade 34
biodiesel **518**, 518*f*
biodiversity
    amphibian 265
    benefits of 269–274
    biophilia 272–273

and conservation biology. *See* conservation biology
conservation of 82–83, 85
in coral reefs 361
defined **22, 64, 255**
distributional patterns 259–261
ecosystem diversity 258
and ecosystem function 270–271
and ecosystem services 270
and ethics 273–274
explained 64–65
and extinction. *See* extinction
and food security 271, 271*f*
and fragmentation 276, 276*f*
genetic diversity 257
and hierarchy of life 256*f*
hotspots 280, 281*f*
international agreements 709*t*
loss of 261–262, 263–286
marine 374–375
measuring 258–259
and medicine 271–272
in ocean ecosystems 359–360
and parks and reserves 308
and speciation 67–69
species diversity 255–257, 258, 258*f*, 259*f*, 260*f*
and tourism 272
biodiversity hotspots **280**, 281*f*
Biodiversity Portrait of the St. Lawrence 140–141
biofuels
defined **516**
and food availability 225–226, 520–521
to power automobiles 516–518, 518*f*
biogas 519
biogenic gas 472
biogeochemical cycles
carbon cycle 145–147
defined **141**
hydrologic cycle 143–144
nitrogen cycle 146–148, 149
nutrient circulation 141–143
phosphorus cycle 150–153
biological control **228**–229
biological diversity. *See* biodiversity
biological hazards **592–593**, 592*f*
biological weathering **191**
biomagnification 127, **606**
biomass **135**
biomass energy
benefits of 520
biofuels 516–518
defined **515**
in the developing world 515–516, 516*f*

drawbacks of 520–522
electricity generation 518–519, 519*f*
and forestry residues 519*f*
sources of 516*t*
strategies, new 516
traditional 515–516
biomedical waste 575
biomes
alpine 120–121
altitude 119*f*, 120–121
aquatic 121
boreal forest 114–115
chaparral 119
defined **111**
desert 118–119
distribution map 111*f*
explained 111–113
forest 290–291
savannah 118
and species diversity 260
temperate deciduous forest 115
temperate grassland 115–116
temperate rainforest 116
temperature and precipitation 112–113, 112*f*
tropical dry forest 117
tropical rainforest 116–117
tundra 113–114
types of 113–121
biophilia **273**, 641
biopower **516**, 518–519, 519*f*, 520*f*
bioremediation **33**, 34
biosensors 343
biosphere **130**, 309*f*
biosphere reserves **309**
biotechnology **232–233**
biotic components **4**
biotic potential 79, **81–82**
bird flu 240
birth control pills 617
birth rates 164, 172, 174, 175*f*
bison 116
Bison bison 116
bisphenol-A 602, 603
bitumen **479**
bituminous coal 471
black blizzards 204
Black Death 181
black-footed ferret 116
black gum 293
Black Sea 138
black spruce 114
blooms 127
blueberries 189, 385
Blue Covenant (Barlow) 718
Blue Marble 4
bobcats 257

Body Shop 683
bog rosemary 189
bogs **320**
Bohnsack, James 376
bonds **37**, 45
boreal forest
defined **114**
explained 113*f*, 114–115
and latitudinal gradient 260
location 290, 292
trees common to 260, 292
Boreal Forest Conservation Framework 114
Boreal Leadership Council 115
Borlaug, Norman 224, 252*f*
Bormann, F. Herbert 412
Borneo 299, 302*f*
Bosch, Carl 148
Botswana 320*f*
Bottaccione Gorge 72–73
bottled water 333
bottleneck 148
bottom-trawling **370**, 370*f*, 371, 371*f*
Bove, Jose 738
bovine spongiform encephalitis 240
Bowlin, Mike 462
Boza, Mario 83
BPA 602, 603
BP Solar 536
Brander, James 11
Brazil
Earth Summit 705
ethanol 518
GM crops 237
jabiru stork nest 274*f*
land conservation 297
no-till farming 208
rubbertappers 668
sustainable forestry 307*f*
urban sustainability 647
urban transportation 640, 641*f*
breakdown products **605**
breeder reactors **505–506**
breweries 573
Britain 235
British Columbia 289
Broughton, Nova Scotia 636*f*
brownfields **582**
Brundtland, Gro Harlem 25, 705
Brundtland Commission 25
brunisolic soil 194*t*
Brush, Charles 538
Bt crops 234*f*
buffer zone 309
Bufo marinus 109
Bufo periglenes 61
bulk water exports **317**

Bunker, Greg 728
Burgess Shale 68f
Burma 364
burning vegetation 393
butterflies 39, 230
by-catch 242, **370**

C

Cactoblastis cactorum 228, 228f
cactus moths 228, 228f
Cadiz oil spill 704
Cahuita National Park 83
Cairo conference 177–178
calcium carbonate 365
Calgary 627f, 632f, 644, 645f
California 271, 605
California condor 262, 277, 277f
Callosobruchus maculatus 80f
Calvin cycle **48**
Cambrian period 262
campus sustainability. *See also* sustainability
   audits 728
   carbon neutrality 733
   and climate change 733
   curricular changes 736
   energy conservation 732
   food services 734
   footprint calculator 728–729
   green building design 730–731
   institutional purchasing 734
   landscapes 735
   organizations for support 736–737
   reasons for 727–728
   recycling 729–730
   renewable energy 732–733
   Talloires Declaration **725–726**, 726t
   transportation 734–735
   University of British Columbia 725–727
   waste reduction 729–730
   water use 732
Campylobacter jejuni 341
Canada
   carbon dioxide emissions 454f
   climate change 324
   consumption in 741f
   electricity generation 529f
   energy conservation 487
   energy consumption 467, 468f
   energy funding 532–533, 532f
   energy mix 465f, 529
   energy use patterns 499f
   environmental health management 614–615t
   environmental policy. *See* Canadian environmental law and policy
   forestry management 300–302
   forests 292–293, 295, 296f
   genetically modified organisms 221, 237
   hydroelectric power 530t
   natural disasters 591f
   northern mining 654–656
   no-till farming 207
   outdoor air pollution 602–603
   radioactive waste 580–581
   renewable energy 530, 532–533, 532f
   soil conservation 203, 204–205
   solar energy use 537f
   subsidies 715
   sunshine distribution 537f
   urban growth 628
   waste 561f, 562–563, 562f
   water exports 317–318
   wind distribution 540f
Canada-Ontario Agreement Respecting the Great Lakes Basin Ecosystem 692
Canada-U.S. Great Lakes Water Quality Agreement 692
Canada-United States Air Quality Agreement 397
Canada-Wide Standards Sub-Agreement 397
Canadian Agricultural Products Act 243
Canadian Association of Food Banks 223
Canadian Auto Workers Union 731
Canadian Cancer Society 617
Canadian Council of Ministers of the Environment 698
Canadian Environmental Assessment Act 699
Canadian environmental law and policy
   aboriginal governments 700
   Canadian Environmental Assessment Act 699
   Canadian Environmental Protection Act 341, 394–397, 563, 590, 615, 697, 699
   Canadian Environmental Quality Guidelines 698
   Canadian Wildlife Act 699
   Department of the Environment Act 696
   and economics 702–704, 706f
   and ENGOs 701
   federal level 698–699, 699t
   Fisheries Act 341, 563, 615, 699
   historical precedents 701–702
   history of 702–704
   instruments 696–697
   international agreements 700
   management process 698f
   Migratory Birds Act 699
   monitoring 706–708
   municipal/local level 700, 700f
   penalties 697
   provincial/territorial level 699–700
   social context 702–704, 706
   Species at Risk Act 699
   state-of-the-environment reporting 706–708
   Transportation of Dangerous Goods Act 699
   U.S. influence 696
   voluntary guidelines 697
Canadian Environmental Protection Act 341, 394–397, 563, 590, 615, 697, 699
Canadian Environmental Quality Guidelines 698
Canadian Environment Awards 550
Canadian Forces Bases Equimalt 580
Canadian Forest Service 300–301
Canadian Geographic 92
Canadian Regional Climate Model 441f
Canadian Shield 133, 133f
Canadian Tire 573
Canadian Wildlife Act 699
Canadian Wildlife Service 38, 140, 279
cancer
   and asbestos 589
   and carcinogens 599
   and Chernobyl 510, 512–513
   environmental factors 617
   and hormone disruption 602
   and radon 596
   testicular 602f
   and uranium 669
candlesticks 531
cane toad 109
Canis lupus 114
canola 221, 233
canopy **291**
cap-and-trade programs 455–**716**
Cape Breton Highlands National Park 104
Cape Canaveral 376
capitalist market economy 670, 694
Capra, Fritjof 126
captive breeding **276**
carbamates 227t
carbohydrates **43–44**, 43f
carbon. *See also* coal; fossil fuels
   defined **35**, **145**
   emissions 482f
   in lithosphere 427
   in peat 189
   sediment storage 146
carbonate rocks 33

carbon capture and storage **451**, **482**, 483f
carbon cycle **145**–147, 145f
carbon dioxide 401, **426**, 454f, 482f
carbon footprint **455**
carbon monoxide **395**, 417
carbon neutrality **455**, 520, 733
carbon offsets **455**
carbon sequestration 427, **451**, **482**
carbon sinks **427**
carbon storage **451**
carcinogens **599**
Car Heaven 717, 717f
Caribbean monk seal 368
caribou 114, 484, 484f
Carleton University 544, 733
carnivores **99**
Carolina parakeet 262
Carolinian species 293
carrying capacity
   defined **8**
   and human population growth 81, 165, 167–168, 167f
   and population growth 79f, 80
Carson, Rachel 379, 396, 703f, 734
Cartagena Protocol on Biosafety 237
Cascades 135
case history approach 607
Case Western Reserve University 603
Caspian tiger 263
Cassandras 24
Castanea dentata 108–109
catalytic converters 736, 736f
catalytic cracking 476
catalytic reforming 476
categorical imperative 660
Catharanthus roseus 271
Catholic Church 177
cation exchange **196**
cattails 189
cedar trees 116
cells **44**, 45f
cellular respiration **49**
cellulose 43, 43f
cellulosic ethanol 518, 522
cement 132
centrally planned economy 670
Central Park 641f
Central Valley 605
Centre for Biological Diversity 278
Cernan, Eugene 3
channelization 324
chaparral 118f, **119**
character displacement 95
Chatham College 734
Chatterton Hill Park 110
cheetahs 257

Chelonia mydas 83, 83f
chemical energy **45**
chemical hazards **592**, 592f
Chemicals Management Plan 603
chemical weathering **191**
chemistry
   building blocks 34–37
   importance of 33
   and solutions to environmental problems 33–34
chemoautotrophic hypothesis 51–52
chemosynthesis **50**
Chernobyl 497, 509–510, 510f, 511f
chernozemic soil 194t
China
   age pyramid 170f
   age structure 170–171
   agriculture 198
   air pollution 405
   carp aquaculture 243
   desertification 203
   fisheries 369
   and indoor pollution 417–418
   and IPAT model 166–167
   land ownership 9
   lead 597
   overgrazing 213
   population 161–162, 163
   reforestation 209, 209f
   reproductive policy 162, 163, 170
   sensitivity to human disturbance 166
   sex ratios 171–172
   soil erosion 202
   Three Gorges Dam 325–328, 327f, 481, 503
   transgenic crops 237
   wheat varieties 238
Chipko Andolan movement 666–667, 667f
chitin 44
chlordane 396
chlorofluorocarbons **406**
chlorophyll **48**
chloroplasts **48**
C horizon **193**
chromosomes **43**
chronic exposure **611**
chrysotile 588–590, 602
Cicerone, Ralph 404
circle of poison 599
Citaro busses 549f
cities. *See also* urban growth
   cultural resources 646
   efficiency 644
   environmental impacts 644–645
   factors influencing 630
   growth of 628–629

   innovation 646
   liveable 639–640
   mega-cities 629t
   parks and open space 641–644, 641f, 643f
   planning 636, 636f
   pollution 753
   resource consumption 644–645
   sustainability 646–647
   transportation 640–641, 640f, 642
   zoning 637, 637f
Citizens' Environment Watch 646, 647f
civilizations, mistakes of past 10
cladograms 67
classical economics **672**–673
clay **194**
Clayoquot Sound 272, 288–289, 293, 308, 309, 310f, 311
clean coal **470**
clear-cutting 201, 288–289, **304**, 304f, 305f
climate
   and atmosphere 391–393
   defined **390**, **424**
   dynamic nature of 424
   feedback loops 427–428
   and soil formation 192t
climate change
   and agriculture 446–448
   and the Arctic 439f
   the Arctic 441–442
   and automotive technology 452, 452f
   and biodiversity loss 268–269
   in Canada 324
   carbon offsets 455
   and conservation 451
   defined **424**
   direct atmospheric sampling 432
   and economics 448
   and ecosystems 446
   and electricity sources 451
   emission reduction 452–453, 454t
   and energy efficiency 451
   explained 424
   and forest fires 448f
   and forestry 447
   and fossil fuels 481
   future impacts 435–449, 437f
   and health 448
   human responsibility for 448–449
   Intergovernmental Panel on Climate Change 436–438
   international treaties 453
   and lifestyle 451
   market mechanisms 454–455
   melting ice and snow 440–441, 444
   Milankovitch cycles 428–429

climate change (*continued*)
  mitigation vs. adaptation 449–450
  models 432–434, 433*f*, 436*f*, 441*f*
  ocean absorption 429
  ocean circulation 429–430
  organisms, effect on 445
  and pollution 22
  precipitation 439–440, 440*f*
  proxy indicators 430–432
  radiative forcing 428, 428*f*
  response to 449–455
  scenarios 449, 450*f*
  science of 430–434
  sea levels 423–424, 442–445, 442*f*, 443*f*
  societal impacts 446–448, 447*f*
  solar output 429
  and the spruce budworm 297
  stable isotope geochemistry 432
  temperature increases 438–439
  and transportation 452
  trends 436, 437*f*
  and water 323–324
climate models
  Canadian 441*f*, 446*f*
  defined **432**
  examples 436*f*, 442*f*
  explained 432–434
  factors involved 433*f*
climatographs **112**
climax community **107**
cloning 277–279
closed canopy **291**
closed cycle 547
closed loop systems **544**
closed systems **128**, 670
cloud forests 60, 64, 69, 71, 73
clumped distribution 76, 76*f*
clustering 225*f*
coal
  abundance of 468–469
  and air pollution 402
  clean 470–471
  consumers 469*t*
  defined **468**
  for electricity 470–471
  environmental impact 483–484
  environmental impacts 507*f*
  formation of 468*f*
  history of usage 469
  mining 469–470, 469*f*, 483–484
  producers 469*t*
  qualities 470–471
  types of 468*f*, 471
coalbed methane 472
coal gasification 488
Coal Smoke Abatement Society 664

coastal ecosystems
  estuaries 364
  mangrove forests 363–364, 363*f*
  salt marshes 363
coast forest region 293
Coast Range Mountains 133, 133*f*
Cochrane Ecological Institute 279
cocomposting 571
cod fisheries 352–353, 371, 372*f*
coffee 280
cofiring 519
cogeneration 488
cold fronts 390*f*, **391**
Collapse (Diamond) 10
Coloma, Dr. Luis 84
Colombia 205*f*, 534, 534*f*
Colorado River **329**
coloration 96*f*
colour, of soil 193–194, 195*f*
Columbia forest region 293
Columbia Icefield 440
Columbia Wetlands 332*f*
combine cycle generation 488
command-and-control approach **714**
commensalism **99**
Committee on the Status of Endangered Wildlife in Canada 278
Common Energy 733
commons, tragedy of the 8–9, 694
Commons Preservation Society 664
communicable disease 593
communities
  biomass 100–101
  changes in composition of 82
  defined **71**
  energy use 100–101
  feeding relationship 101
  interactions in 99–111
  invasive species 107–110
  keystone species 102–104
  numbers 100–101
  restoration of 110–111
  roles of organisms 101–105
  succession 105, 107
  trophic levels 99–101
Community Baboon Sanctuary 281
community-based conservation 280–281
Community Conservation, Inc. 281
community ecologists 99
community ecology 72
community gardens 643, 643*f*, 734
community-supported agriculture **245**
compaction 213, 333
competition **94**–95, 95*f*
competitive exclusion **94**
compost **211**

composting 559, 569, 571*f*
compounds
  defined **37**
  inorganic 41
  macromolecules 42–44
  organic 41
ConAgra 240
concentrated animal feeding operations **239**, 239*f*
concession 299
Concordia University 729, 734, 735*f*
Confederation College 733
conference presentations 18
confined aquifer **322**
congestion-charging program 404
Congo, Democratic Republic of 172, 172*f*, 309
coniferous **114**
ConocoPhillips Canada 463
conservation. *See also* campus sustainability; conservation biology
  of biodiversity 82–83, 85
  and climate change 451
  community-based 280–281
  conservation concession 281
  Costa Rica 82–83, 85
  debt-for-nature swap 281
  and efficiency 488–489, 488*f*
  of energy 487–490
  ethics of 664–665
  of freshwater 335–336
  innovative strategies 281
  marine 375–379
  national and international efforts 278–280
  personal choice 488–489
  of pollinators 230–231
conservation biology
  advances in 23
  and biodiversity loss 274
  captive breeding 276
  cloning 277–279
  defined **141**, **274**
  island biogeography theory 275–276, 275*f*, 276*f*
  multiple levels 274–275
  reintroduction 277
  umbrella species 278
conservation concession 281
conservation ethic **665**
conservation geneticists 274–275
Conservation International 280, 281
conserved 46
Constitution Act of 1982 300
constructed wetlands **346**
consumer influence 745
consumers **49**, 99

consumption
 environmental impacts 179–180
 and sustainability 20, 741–742
consumptive use of water 325
containment building 505
continental shelves 358
continental slope 358
contingent valuation 681–682
contour farming 205, 206f
contraception 175–187
controlled burns 306
controlled experiments 16
control rods 505, 506f
convection currents 392f
convective circulation 390
conventional law 709
Convention on Biodiversity 237
Convention on Biological
 Diversity 280
Convention on International Trade in
 Endangered Species of Wild Fauna
 and Flora 279
*Convention on Wetlands of International
 Importance, Especially as Waterfowl
 Habitat* 331–332
conventions 709
convergent evolution 64
convergent plate boundaries
 134f, 135
Conway, Tenley 728
Coon Come, Matthew 654, 684
Copernicus, Nicolaus, 19
Copland, Dr. Luke 441
copper 365, 385
coral 98
coral bleaching 361
coral reefs 360–361, 361f, 446
Cordyceps 97f
core 133
core area 309
Coriolis effect 392
cork 272
corn
 conflicting interests 226
 for ethanol 517f
 genetically modified 233
 history of 220
corn ethanol 226
cornucopians 24
corporate average fuel efficiency
 standards 487
corporations 683–684
correlation 16
corridors 310, 643
corrosive 574
cosmology 17
Costanza, Robert 682, 682f

Costa Rica
 biodiversity 64–65, 85f
 bioprospecting 273
 bird study 74
 conservation 82–83, 85
 golden toads 59–61, 62f, 64–65, 69, 73
 harlequin frog 75–76
 immigrant species 82
cost-benefit analysis 673–675
cotton 233, 244
cottongrass 189
cottonwood 293
Council of Canadians 718
coupled general circulation models 433
covalent bond 37
Cowley Ridge Wind Plant 538
cracking 476
Cree Nation 684
Cretaceous period 262
criteria air contaminants 395
critical thinking 14
Crompton Company 579
crop diversity 237–238
cropland 10–11, 196
crop rotation 205, 206f
Crosbie, John 353
crude birth rate 77, 172
crude death rate 77, 172
crude oil 474, 482
Crump, Martha 67
Crutzen, Dr. Paul 404
cryosolic soil 194t
crypsis 96f
cryptic coloration 96f
Cuba 245–247, 247f
cultivars 55, 238
cultural hazards 592f, 593
culture 656–659, 658
Curitiba, Brazil 640, 641f, 647
currents 356
customary law 709
Cuyahoga River 703, 703f
cycles 127
Cytisus scoparius 109
cytosine 43f

D

Daley, Herman 678, 685, 744
Dalton, Chelsea 728
*Damming the Three Gorges: What Dam
 Builders Don't Want You to Know*
 (Ryder) 521
dams
 benefits and costs 315t
 defined 325
 Gardiner Dam 500f
 hydroelectric 326f, 499–500

 number and size of 325
 Ottawa-Holden Dam 501f
 removal of 328
 Three Gorges Dam 325–328, 327f,
  481, 503
Darwin 64
Darwin, Charles 19, 61
data 16
The Day After Tomorrow 430
DDT
 banned 396, 599
 and biomagnification 606
 and endocrine disruptors 601
 exporting 599
 and head lice 13f
 and malaria 12, 13f, 596
 public spraying 599f
dead zone 147–150, 212
death, causes of 593f, 612f
death rates 164, 172, 174
debt-for-nature swap 281
deciduous 115, 117–118
deciduous forest
 food web from 102f
 region 293
 temperate 112–113, 114f, 115
 tropical 117–118, 117f
decommissioning 564
decomposers 99
Decree 795 279
deep ecology 666
Deep Lake Water Cooling project 544
deep-well injection 572, 579f
defensive expenditures 675
Deffeyes, Kenneth 478
deflation 202
deforestation
 causes of 296f
 defined 295
 negative impacts 298
 pace of 298–299
 by region 299f
Delphinapterus leucas 126–127
Democratic Republic of Congo 172,
 172f, 309
democratization of science 14
demographers 77, 168
demographic fatigue 181
demographic transition 173–175,
 174, 174f
demography
 age pyramids 170f
 age structure 170–171
 and carrying capacity 167–168, 167f
 defined 167
 explained 168
 population density 168–169, 169f

demography (*continued*)
   population distribution 168–170
   population size 168
   repercussions 181
   sex ratios 170–172
   total fertility rate 172–173, 173*f*
   transitions in 173–175, 174*f*
The Demon Haunted World (Sagan) 14
Dendroica chrysoparia 83*f*
Dene 462, 654, 658, 669, 676
Denendeh 654–656
denitrifying bacteria **148**
Denmark 541
density-dependent rfactors 81
density-independent factors 81
deoxyribonucleic acid (DNA)
   defined **42**
   and genes 43, 63
   and mutations 63
   structure of 43*f*
Department of the Environment Act 696
dependent variables **16**
deposition **200**
desalination **334–335**, 335*f*
desertification **203–204**, 292
deserts 112, **118–119**, 118*f*
DesGranges, Jean-Luc 140
detritivores **99**
deuterium 432
developed nations 466–467, 517*f*
developing world. *See also* industrializing nations
   biomass energy 515, 516*f*
   energy consumption 466–467, 517*f*
   indoor air pollution 411–414, 413*f*
   urban growth 628*f*
development **24–25**
Devi, Gaura 667
Diamond, Jared 10
diamond mines 654
dicofol 601
Diesel, Rudolph 518
differential extinction 275*f*
dikes **328**
Dinosaur Provincial Park 133*f*
Dioscoreophyllum cumminsii 271
dioxins 396
direct atmospheric sampling 432
directional selection **63**, 63*f*
Dirty Dozen 616*t*
discharge zones **322**
Discover gold mine 655
disease 593–596
disruptive selection **63**, 63*f*
dissociate 40
dissolved oxygen **339**

distance effect **275**, 275*f*
distilling 335, 476
distributional equity 685
district cooling 544
district heating 544
divergent evolution **66**, 67–69
divergent plate boundaries **134–135**, 134*f*
diversions **324**, 329
diversity, and natural selection 61–62
DNA
   defined **42**
   and *E. coli* detection 343
   and genes 43, 63
   and mutations 63
   and nitrogen 146–147
   recombinant **232**, 232*f*
   structure of 43*f*
Dobson, G.M.B. 404
dodo 262
doldrums **392**
Dome C 434
domesticated animals **239**
Donna Creek Biodiversity Project 303
Donora Pennsylvania 402, 402*f*
dose **608**
dose-response analysis 608–610
dose-response curve **609**, 610*f*
Douglas Channel 152
Douglas fir trees 116, 297
Dow 236
downwelling **356**
drainage basin **318**
Drake, Edwin 474
Drayton Valley, Alberta 473*f*
DreamNow 748
Dreissena buensis 93
Dreissena polymorpha 92–93, 103
driftnets **370**, 370*f*
drinking water 333, 341–342
drip irrigation **210**, 210*f*
Drive Clean 368
drugs 271–272
Drybones Bay 655
drylands **291**
dry storage 514*f*
Duboisia leichhardtii 272
Ducks Unlimited Canada 308
Ducruc, Jean-Pierre 140
DuPont 236
Dust Bowl 204
dust dome **646**
dust mites 416
dust pneumonia 204
dust storms 203, 204, 393, 394*f*
dynamic equilibrium **129**

E

E. coli 232, 240, 341
Earth
   curvature of 389*f*
   early Earth 50–51
   history as 12.hour clock 742*f*
   as island 4–10, 748–749
   photography of 3–4, 21, 749
   relationship to Sun 388–390, 389*f*
   rotation of 392
Earth Day 704, 705*f*
Earth energy 544
Earth Summit 453, 705
Earth system science 21
Easter Island 10, 11, 748
eastern hemlock 293
eastern white pine 293
echinacea 294
Echingammarus ischnus 103
ecocentrism **662**, 663*f*
ecofeminism **666**
ecoimperialism **658**
ecolabelling 570, **717–718**, 719*f*
École Polytechnique de Montréal 731
ecological communities. *See* communities
ecological economists **678**
ecological estoration **110–111**
ecological footprint
   affluence and 179
   calculating 9
   on campus 727, 728–729
   of cities 645*f*
   comparison 20*f*
   defined **9**
   global 180*f*
ecology
   defined 17
   levels of organization 71–72, 71*f*
   nutrients 136–138
   organismal 73–74
economically recoverable oil 475
economics
   classical 672–673
   and climate change 448
   corporations 683–684
   cost-benefit analysis **673**
   defined **670**
   economies, types of 670
   and environmental policy 714–717
   and environmental protection 739
   environmental trade-off 670
   ethics in 676–677
   greening business 683
   growth, sustainability of 672–678
   growth paradigm 677–678
   market failure 683

measurement of progress 679–680
monetary values 680–683
neoclassical. *See* neoclassical economics
private vs. social costs and benefits 676
and quality of life 740–741
and sprawl 635
steady state economies **678**–679
and sustainability 670–672, 738–739, 738f
views of 671f
economy **670**
economy of proximity 644
ecoregion **111**–112, **280**
ecosystem-based management **302**, 375
ecosystem diversity **258**
ecosystem ecology **72**
ecosystem engineers 104, 271
ecosystem function 270–271
ecosystems
    defined **71**, **135**
    energy flow 135–136, 136f
    explained 135
    international agreements 709t
    and landscape ecology 139
    marine and coastal 359–364
    matter cycle 136f
    spatial integration 138–139
ecosystem services 270, **672**, 672t, 682–683, 682f
ecotones **139**
ecotourism **84**, 272
EcoTrek Project 729
Ecotrust Canada 289
Ectopistes migratorius 75
ED50 **609**
edge habitats **309**
Edison, Thomas A. 527
Edmonton 571–572, 571f
education 746
effective-dose-50% **609**
effluent **343**
Ehrlich, Paul 164, 165
Eldorado 654, 669
Eldridge, Niles 62
electricity
    from biomass 518–520
    Canadian production 529f
    and climate change 451
    from coal 470–471
    world production 498f
electrolysis 548–549
electromagnetic energy **46**
electronic waste 576–577
electrons **36**
elements **34**–35, 36–36t

Elk Island National Park 14f, 104
Ellesmere Island 441, 441f
Ellesmere Island national Park Reserve 21
Elmira, Ontario 579
elm trees 293
El Niño **357**–358, 358f, 393
El Niño-Southern Oscillation **357**, 430
embryogenesis 35
emergent properties **129**
Emerson, Ralph Waldo 664
emigration **78**
emission reduction 452–453, 454t
emissions 508–509, 541f, 706f
emissions trading programs 455, **716**, 717
employment opportunities 531, 532f, 738–739, 738f
endangered **261**
Endangered Species Act 278
Endeavour Hydrothermal Vents 375f
endemic **69**, **280**
endocrine disruptors 597, **600**–602, 601f
endocrine system **600**
end-of-pipe **342**
energy
    biomass. *See* biomass energy
    Canada 465f, 499f
    changes in 46–47
    coal. *See* coal
    conservation of 46, 487–490
    consumption of 466–467, 467f, 468f, 517f
    defined **45**
    emissions comparison 508–509
    fundamentals of 45–50
    geothermal 49–50
    harnessing of 47
    hydroelectric power. *See* hydroelectric power
    inputs 467–468
    natural gas. *See* natural gas
    nonrenewable 465
    nuclear. *See* nuclear energy
    oil. *See* oil
    renewable 464, 489–490
    solar 47
    sources of 427–429, 464t
    tidal power 527–528
    types of 45–46
    world production 498f
energy conservation **487**–490, 732
energy efficiency 47, 451, 488–489, 488f
energy returned on investment **468**, 502, 521
Energy Solutions Group 489

Engelmann spruce 293
Enhydra lutris 106–107
entropy 46
environment
    culture and 656–658
    defined **4–5**
    and economics 739
    and human connections 739–740, 740f
    ways to understand 659–660
    worldview and 656–658, 659
environmental activisim 13–14, 13f
environmental audits **361**
Environmental Bill of Rights 699
environmental bottleneck 747
environmental challenges 20–22
environmental cooperation 709t
Environmental Defence 152
environmental degradation 179f
environmental economists **678**
environmental ethics
    Atlantic seal hunt 660
    conservation and preservation 664–665
    deep ecology 666
    defined **661**
    ecofeminism 666–667
    environmental justice 667–669
    explained 660
    human-nonhuman relationships 661–662
    and the Industrial Revolution 663–664
    land ethic 666
    progression of 662f
    roots of 663
environmental health
    biological hazards 592–593, 592f
    Canada 614–615, 615t
    chemical hazards 592, 592f
    cultural hazards 592f, 593
    death, causes of 593f
    defined **590**
    disease 593–596
    endocrine disruption 600–602, 601f
    hazards, types of 590–593, 592f
    human studies 588–608
    indoor 596–597, 596t
    international 615–618
    lifestyle hazards 592f, 593
    mixed toxicants 611
    philosophical approaches to 614
    physical hazards 590–592f
    policy 614–615
    research into effects 607–611
    risk management 612–614, 613f
    synthetic chemicals 598–599, 598f, 598t, 604f
    toxicants 599–600

environmental health (*continued*)
  toxic agents 598–606
  toxicants 604–606, 604f
  toxicology 597
  types of exposure 611
  variations in response 610–611
  wildlife studies 607
environmental impact assessment 699
environmental impact statement (EIS) 104
environmentalism 13–14
environmental justice 667–669, 668f
environmental nongovernmental organizations 701, 713
Environmental Performance Index 746–747
environmental policy
  approaches to 714–718
  Canadian. *See* Canadian environmental law and policy
  command-and-control approach 714
  defined 693
  ecolabelling 717–718, 719f
  and economic tools 714–717
  green taxes 715–716
  health 614–615
  hindrances to 695–696
  international. *See* international environmental law and policy
  issues 694–695
  market incentives 717
  participants in 693f
  permit trading 716
  and science 714
  subsidies 715, 715f
  sustainability development 705–706
environmental problems 12–13
Environmental Protection Agency 150, 696
environmental refugees 445
environmental resistance 79f
environmental science
  avoidance of mistakes 10
  defined 5
  vs. environmentalism 13–14
  and human population 7–8, 7f
  interactions of 5–6
  as interdisciplinary pursuit 10–12, 12f
  and natural resources 6–7
  nature of 10–14
  and resource consumption 8–9
environmental studies 10
environmental systems
  categorization of 130
  component parts 130f
  defined 128
  explained 127

  properties of 128–129
  subsystems 129–130
Environment Canada 5, 417
ENVIRx 573
EnWave Energy 544
enzymes 42
Eotetranychus sexmaculatus 80f
epidemiology 607
epiphytes 73
epiphytes 117
equilibrium theory of island biogeography 275
EROI 468, 502, 521–522
erosion
  in Canada 203, 203f
  causes of 200–201
  control practices 209
  defined 192
  explained 200
  global problem 202–203
  types of 201–202, 201f
erosive crops 244
Erwin, Terry 258
Eskimo curlew 262
Esquimalt 580
An Essay on the Principle of Population (Malthus) 163
Estes, James 106
estuaries 364
Esty, Daniel 746
ethane 41–42, 41f
ethanol 226, 516–518, 517f, 521–522
ethical standards 660
ethics
  defined 660
  and economics 656, 676–677
  environmental. *See* environmental ethics
  and other species 273–274
eukaryotes 44, 45f
Eurasian Plate 135
Europe, population 173
European honeybee 230, 231f
European Union
  emissions trading programs 455
  GM foods 237
  and international environmental law 710
  organic farming 244
  regulation of chemicals 618
eutrophic 321
eutrophication
  in coastal ecosystems 367
  defined 127
  and fertilizer 211
  and nutrient pollution 336, 337f
  St. Lawrence estuary 149f

evaporation 143
even-aged 303
evenness 255
Everglades 110, 331, 364
Evergreen Foundation 735
evolution
  and biodiversity 61, 64–65, 67–69
  diagram 60
  divergent 66, 67–69
  and extinction. *See* extinction
  natural selection. *See* natural selection
  and speciation 66–67, 69
ewaste 576–577
Ewing, Reid 634
Exclusive Economic Zone 353
Exclusive Economic Zones 365
exotic 108
Experimental Lakes Area 386
experiments
  defined 16
  manipulative 17–18, 17f
  natural 17
  and testing predictions 16
exploitative interactions 95–98
exponential growth 78–79, 79f
Extended Producer Responsibility and Stewardship program 574
extensification 196, 224
external costs 675, 675f, 695
externalities 675
extinction
  background rate of 262
  causes of 69
  current rates 263, 264f
  defined 69, 261
  and ecosystem function 270–271
  KT boundary 72–73
  mass extinction 53, 70, 72–73, 262, 262f, 263, 263t
  as natural process 262
  North American megafauna 69f
  and overfishing 368
  reasons for concern 71
  Red List 263
  and speciation 69
  vulnerability to 69–70
extirpation 261
extremophiles 52
extrusive rock 131
ExxonMobil Canada 463
Exxon Valdez 33–36, 366–367, 482

F

facilitation 99
factory farms 239, 239f
factory fishing 370

Falconbridge 386
fallow state 297
family planning 176–177, 178, 178f
Farman, Joseph 405
farmers' markets 246f
far Mine 580
fast 44
Federal Contaminated Sites Inventory 581
feedback loops **128**, 128f, 427–428
feedlots **239**–240, 239f
Felder, Melissa 571
Felidae 257
Felis rufus 257
Felus Lynx 114
female literacy 175f
feminization of male animals 601
Ferrel cells **391**, 392f
Fertile Crescent 198, 203
fertility rates
  Bangladesh 176–177
  total fertility rate **172–173**, 173f
fertilizers
  defined **210**
  and Green Revolution **224**
  and marine ecosystems 367
  and soil degradation 210–211, 211f, 212
fire
  and climate change 448f
  and forests 305–306, 306f, 394f, 447
  and indoor pollution 411–414
  and outdoor pollution 393
fire-bellied snake 95f
fire policy 305–306, 306f
first law of thermodynamics **46**
First Nations peoples 289
First Nations University of Canada 736
Fisheries Act 341, 563, 615, 699
fish farms 241–243
fishing
  capture rates 364f
  China 369
  cod fisheries 352–353, 371, 372f
  consumer choice 366f, 374
  "down the food chain" 372
  ecosystem damage 370–371
  global 368, 368f, 371
  industrialized 369–370, 371
  management of 368
  methods of 369–370, 370f
  and nontarget animals 370–371
  North Atlantic 352–353
  overfishing 368–370, 371
fission **504**, 504f, 505
5NR departments 707
flagship species 278

flammable **574**
flat-plate solar collectors **534**
Flavr Savr tomato 234f
Fleming College 736
Flenley, John 11
flexible fuel vehicles 518
flies 230
floodplain **320**
floor, of the forest **291**
Florida Bay 364
Florida Everglades 110
flows 300
fluidized bed technologies 470
Flushing Meadows 564
fluxes **142**, 143f, **427**
Fluxnet Canada-Canada Carbon Project 190
Food and Agriculture Organization 204, 300–301
food chain **101**, 372
food insecurity 223
food labels 244
food production
  animal products 239–241, 239f, 241f
  aquaculture 241–243, 242f
  biofuels 225–226
  crop diversity 237–238
  decline in 222
  and energy use 240–241
  farmers' markets 246f
  feedlot agriculture 239–240
  genetically modified crops. *See* genetically modified food
  global 222f
  Green Revolution 224–225
  increase in 222
  input of animal feed 241, 241f
  labelling 244
  seafood 239f
  seed banks 238–239
food security **222**, 223, 271, 271f
food webs
  defined **101**
  explained 101
  and phosphorus 151–152
  and photosynthesis 145–146
  and respiration 145–146
  in temperate deciduous forest 102f
Ford, James 450
Ford, Loretta 214
ForestEthics 311
forest fires 305–306, 306f, 394f
forestry
  adaptive management 302–303
  and climate change 453
  defined **299**
  ecosystem-based management 302

fire policy 305–306, 306f
forest management 300–302
new forestry **304**
sustainable 301, 306–307f
forests
  biomes, types of 290–291, 290f
  and biopower 519f
  boreal. *See* boreal forest
  in Canada 292–293, 295, 296f
  and climate change 447
  defined **291**
  eastern Canada 293
  ecological value 293
  economic value 293–294
  as ecosystem 291f, 293
  and fire 447
  global inventory 300–301
  land conversion. *See* land conversion
  northern Canada 292
  and nutrient cycle 293, 294f
  parks and reserves 307–310
  plantation forestry 303–304
  products 293–294, 294f
  tropical. *See* tropical dry forest; tropical rainforest
  uses of 294f
  and water cycle 293, 294f
  western Canada 293
Forest Stewardship Council 307
formaldehyde 597
Fort Bragg 365
Fossey, Dian 282
fossil fuels
  alternatives to 498–499
  capture of emissions 482, 483f
  and carbon in atmosphere 146
  and climate change 481
  coal. *See* coal
  defined **464**
  distribution of 466, 466t
  economic impact 485–486
  emissions 508
  environmental impacts of 481–484
  formation of 146, **466**
  global consumption 465f
  impact of 22–23
  methane 480
  natural gas. *See* natural gas
  vs. nuclear energy 451
  in oceans 364–365
  oil. *See* oil
  oil sands 479
  oil shale 480
  petroleum spills 366–367, 366f
  and pollution 481–482
  reliance on foreign energy 485, 485f
  and renewable energy 487

fossil fuels (continued)
  social impact 486–487
  unconventional, disadvantages of 480–481
  unconventional, types of 479–480
  usage amounts 529
fossil record 52–**53**, 53f, 54f
fossils **53**
fractionation **432**, 476–477
fracturing technique 473
fragmentation of habitat **276**, 276f, 310, 310f
France 546
Frank, Lawrence 635
free riders **694**
Fresh Kills redevelopment 564
freshwater. *See also* water
  agricultural demand 335
  bottled water 333
  and climate change 323–324
  conservation 335–336
  depletion, solutions to 334–336
  distribution of 323, 324f
  drinking water treatment 341–342
  exports 317–318
  industrial use 335
  irrigation 329–330
  proportions of 319f
  residential use 335
  supply vs. demand 334
  systems 318–324
  use of 324–334, 325f
  water wars 334
freshwater marshes **320**
frogs 76–265
fronts **390–391**
fuel cells 541f, **548**, 549f, 550–551
fuel rods 504, 505f
fundamental niche **94**, 94f
fungi 416
fungicides **226**
fungus 97f, 98
Funicelli, Nicholas 376
furans 396
fusion 506–506f
Futurama 691

G

Gadus morhua 352–353
Galdikas, Biruté 282
Gammarus fasciatus 103
Gandhi, Mahatma 661
garbage barge 565
garbage justice 568
The Garbage Project 566
garbology 566
gardens 734, 742f

Gardiner Dam 325, 500, 500f
Gardner, Gary 557
Garry oak 109, 110, 110f, 293
gasification 470, **519**
Gaspé Peninsula 364
gates 528
Gates, Bill 161
Gaviotas 534, 534f
Gayoom, Maumoon Abdul 443
gazelles 118
gene banks **238**
generalists **74**
gene revolution 236
genes **43**, 53
genetically modified crops
  explained 232–233
  extent of usage 233
  impact of 234–235
  Mexican maize 221
  by nation 236f
  and selective breeding 233
genetically modified foods
  debate over 235–237
  examples 234f
  labelling 244
  opposition to 234f
  safety 233, 234–235
  salmon 242–243
  Zambia 237, 237f
genetically modified organisms **232**
genetic diversity **257**
genetic engineering **232**
genetic variation 62–64
genital defects 602
genomes **43**, **232**
genuine progress indicator **380**, 680f
geoengineering 550
geographic information system (GIS) 139, **141**, 142f
geologic isolation **512**, 515f, 580
geologic systems
  plate tectonics 133–135
  rock cycle 132
geometric growth 78
Georges Bank 371, 372f, 377
geothermal energy
  benefits of 545
  defined **49**
  explained 49–50, 542–543
  formation of 543f
  harnessing of 543–545
  limitations of 527–528
  thermal pools 50f
  usage, increase of 545
geraniums 35
geyser 542
The Geysers 545

Giant gold mine 654
giant tubeworms 50f
ginseng 294
giraffes 118
Girardet, Herbert 644
Giuliani, Rudolph 564
glaciation **429**
Glacier National Park 440, 440f
glaciers 440–441, 444
Glass Beach 365
Glen Canyon Dam 325
gleysolic soil 194t
Global Amphibian Assessment 70, 265
global climate change. *See* climate change
Global Databank for Animal Genetic Resources for Food and Agriculture 239
global distillation 605f
Global Footprint Network 9
Global Forest Resources Assessment 300–301
global hectares 644
globalization 22, 744–745
global warming **424**
global warming potential **426**, 426t
global wind patterns 392f
Globoruncana 72
glucose 43, 43f
gneiss 133, 133f
Go Beyond 733
gold 365
Golden-cheeked warbler 83f
golden lion tamarin 280
Golden rice 234f
golden rule 660
golden toads 60–61, 64, 75, 84
gold mines 654
Goodall, Jane 282
goods **670**
Gore, Al 424
gorillas 268
Gould, Stephen Jay 62
Grand Banks 352–353, 371, 372f
grand fir 293
Granéli, Edna 138
granite 131–133, 133f
grants and funding 19
graphs, instructions on 652
grasshoppers 204
grasslands **291**, 291f
grazing 212–213
great auk 262
Great Barrier Reef 272
Great Bear Lake 654
Great Britain 235
Great Depression 204

Great Lakes
  Erie, rebirth of 691–693
  invading organisms 92–93, 95–97, 103, 108f
  pollution 152, 341
  as a system 129–130, 131f
Great Lakes-St. Lawrence forest region 293
Great Lakes Water Quality Agreement 152, 341
great tits 445
the Great Wall of Male 443
Greek civilization 10
Green Belt Movement 667
greenbelts **638**–639, 638f
green building design 730–731
green computing 734
Green Energy Research Institute 734
greenhouse effect **425**
greenhouse gases
  Canadian emission targets 401f
  carbon dioxide 426
  defined **425**
  and Earth temperature 425–426, 425f
  and fossil fuels 449f
  global warming potential 426t
  and human activity 426–427
  methane 427
  natural sources 426
  nitrous oxides 427
  ozone 427
  water vapour 427
Greenland 430, 431f, 434, 444
Green Lane Landfill Site 565
green manures 208
Greenpeace 13f, 289, 311, 704, 704f
Green Revolution
  defined **224**
  environmental effects 224–225
  explained 224
  fertilizers 224
  and irrigation 329
  and population 225
green sea turtle 83, 83f
green shift 715
green taxes 683, **715**–716
greenwashing 683
greenways 643
Griffiths, Mary 489
Grizzly Bear Habitat Project 303
grizzly bears 484, 484f
grocery bags 568
gross domestic product **679**, 679f, 680f
gross primary production **136**
groundfish **352**, 370
ground-level ozone 395
ground source heat pumps **545**

groundwater
  aquifer layers 322f
  defined **143**, **321**
  depletion of 332–333
  pollution 339–341, 339f, 341f
  role in hydrologic cycle 321–322
Grow Smart, Grow Green 729
growth rate **78**
Guam 388
Guillette, Elizabeth 607, 608–609
Guillette, Louis 601
Gulf of Carpentaria 367f
Gulf of Mexico 147–150
Gulf of St. Lawrence 128, 135, 138f
gully erosion 201f
Gymnodinium 367f
Gymnogyps califorianus 277f

H

Haber, Fritz 148
Haber-Bosch process **148**
habitat alteration 266
habitat fragmentation **276**, 276f, 310, 310f
habitat heterogeneity 261
habitat island 275
habitats 73–74
habitat selection **74**
habitat use **74**
Hadley cells **391**, 392f
Hagersville, Ontario 574, 574f
Haldane, J. B.S. 258
half-life **36**, 504
Halifax 398
halocarbons 427
halocline **355**
Ha Long Bay, Vietnam 133f
the Hamburger Connection 241
Hamilton, Bruce 508
Hamilton Naturalists Club 308
happiness 742f
Hardin, Garrett 8–9, 694
hardpan 195
hard water 339
hardwood 294
Harkin, James Bernard 664, 665f
harlequin frog 75–76
Harmonization Accord 397
Harper, Stephen 442
harvesting, of forests **299**–300
Harvey, Charles 340
Hawaiian honeycreepers 65f, 133f
Hawkins, Julie 376
Hazardous Products Act 590
hazardous waste **559**
  biomedical 575
  defined **574**

  disposal, methods of 578–579, 579f
  disposal, steps preceding 577–578
  ewaste 576–577
  heavy metals 575–576
  household 575, 575t
  international agreements 709t
  and NAFTA 713
  organic compounds 575
  radioactive 575, 579–582
  regulation of 578
  sources of 575
  types of 574
head hawk moth 98f
head lice 13f
health. *See also* environmental health
  and climate change 448
  and nuclear energy 509–510
  and sprawl 635
heat capacity 355
heavy metals **396**, 575–576
hemlock trees 116
herbal remedies 55
herbicides **226**
herbivores 97–98, **99**
herbivory **97**
heritable 63
heterotrophs **49**
hetertrophic hypothesis 51
Hibernia Offshore Drilling Platform 473f
Hicks, Sir John 685
high-level waste 511
high-pressure system **391**
Himalayas 135
historical sciences 17
HIV/AIDS 180–181, 181f, 594
Hobson, Dr. Keith 38, 39
Hodgkin's disease 271
Holdren, John 165
holistic perspective 662
Home Depot 307
homelessness 629
homeostatis **129**
Homer-Dixon, Thomas 165, 334
honeybees 230, 231f
Hong Kong 170f
Hoover Dam 325
Hopi Nation 469
A horizon **193**
horizons **192**–193, 193f
horizontal currents 356
hormones **600**
Hornocker, Maurice 273
Hornocker Wildlife Institute 255
horsehead pumps 473, 473f
Horwich, Robert 281
host **97**, 97f

household hazardous waste 575, 575t
Hoyle, Sir Frederick 4
Huang He 167
Hubbard Brook Experimental Forest 412–413
Hubbert's peak **478**, 478f
Hudson Bay 539
Hudson River 95–96
Human Biomonitoring of Environmental Chemical Substances 607
Human Development Index 680
human-environment connections 739–740, 740f
human impact
   affluence and 179–180
   on carbon cycle 146
   on hydrologic cycle 144
   on nitrogen cycle 148–149
   on phosphorus cycle 152–153
human population
   in 2006 168f
   and carrying capacity 81
   China 161–162
   growth of. *See* human population growth
   impact of 20–22
   and society 175–182
   and sustainability 20
human population growth
   and agriculture 198
   birth rates 164, 172
   and carrying capacity 167–168, 167f
   China 161–162
   death rates 164, 172
   and environmental scarcity 163–164
   family planning 176–177
   graph 163f
   and HIV/AIDS 180–181, 181f
   immigration 172
   impact on environment 165–167
   Industrial Revolution 163, 164, 164f
   IPAT model 165–167
   perspectives on 163–165
   population control policies 162, 163, 170, 175, 176–177
   and poverty 178, 179t
   as a problem 164–165
   projections 166f, 169f
   rate of 162–163, 164f, 173f
   and resource use 7–8
   and soil 196–199
   and status of women 175–176
   and sustainability 736–743
   transitions in 167–168, 173–175, 174f
   and urban society 7f
   and wealth gap 180, 180f
human studies 588–608

humidity 388
humus **192**
The 100-Mile Diet 246
The 100-Mile Diet (Smith and MacKinnon) 246
hunger 223
Hunt, Patricia 603
Hunter, Bob 691, 704
Hunter Dickinson Services 214
hunting and gathering 196, 199f
Hurricane Katrina 150, 328, 363
Hurricane Nargus 364
hybridize 343
hydrocarbons **41**–42, 41f
hydroelectric dams 326f
hydroelectric power
   advantages 501–502
   approaches 499–500
   Canada 530t
   cleanliness 501–502
   defined **499**
   EROI 502
   limiting factors 503
   negative impacts 502–503
   producers 502t
   renewable 501
   usage amounts 529
   use of 500–501
hydrogen
   and acidity 40–41
   bonding 37
   defined **35**
   fuel. *See* hydrogen fuel
   isotope 37f
hydrogen bond 38, 39f
hydrogen fuel
   benefits of 555
   electricity generation 550–551
   environmental impact 549–550
   explained 547–548
   fuel cells 541f, **548**, 549f, 550–551
   production of 548–549
Hydrogenics 548
Hydrogen Village 548
hydrologic cycle **143**–144, 144f, 319
hydrophilic 44
hydrophobic 44
hydropower **499**
hydrosphere **130**
hydrothermal vents 49, 50f, 52, 135, 360, 375f
hyenas 118
hyoscine 272
hypothesis
   defined **15**
   developing 15
   testing 17–18

hypoxia **127**, 150
hypoxic zones 138

I

ice, melting 440–441, 444
ice cores 431–432, 431f, 434
Ice-minus strawberries 234f
igneous rock **131**, 132f, 133f
Iisaak Forest Resources 289, 307
Iisaak Forest Services 304
Illecillewaet Glacier 440, 440f
immigration **77**, 172
Imperial Oil Resources Ventures 463
in vitro 55
inbreeding depression 257
incineration **565**, 567f
Inco 385, 659f
An Inconvenient Truth (Gore) 424, 449
independent variables **16**
Index of Sustainable Economic Welfare 680
India
   Chipko Andolan movement 666–667
   GM crops 237
   overgrazing 213
   population control policies 176
   population growth 162
indicators **706–707**
indigenous ecological knowledge 659
indigenous peoples **657**
Indonesia 229, 229f, 393
indoor air pollution
   defined 410
   in developing world 411, 413f
   explained 410–411
   living organisms 416
   radon gas 414, 415f
   reduction of 417–418
   sources of 416f
   tobacco smoke 414
   volatile organic compounds 414–415
industrial ecology **573–574**
industrialized agriculture 199, **224**
industrialized nations. *See* developed nations
industrializing nations 405
Industrial Revolution
   and agriculture 199
   as environmental challenge 20
   and environmental philosophers 663–664
   and fossil fuels 22–23
   and greenhouse gases 426f
   and human population growth 163, 164, 164f, 167
   and pollution 8–22

and population growth 8
and urban growth 627–628
industrial smog **401**–402
industrial solid waste **559**
  defined **572**
  exchanges 574
  regulation of 572–573
  sources of 572f
  and sustainability 573–574
industrial stage **174**
industry, and sustainability 744
infant mortality 164, 173
infectious disease **593**, 593f, 594–596
infrared radiation **425**
innocent-until-proven-guilty
    approach 614
Innu Nation 658, 659f
inorganic 211
inorganic compounds **41**
inorganic fertilizers **210**–211
*Inquiry into the Nature and Causes of
    the Wealth of Nations* (Smith) 673
insecticides **226**
insects, number of species of 258
insolation **428**
in-stream tidal power 528
integrated management **229**, 229f
intensification **198**, 224
interactions
  competition 95f
  exploitative 95–98
  herbivores 97–98
  mutualists 98–99
  with no effect 99
  parasites 96–97
  pollination 99
  predation 95–96, 95f, 96f
Inter-American Commission on
    Human Rights 438
interbasin transfers **318**
intercropping **205**–207, 206f
interdisciplinary field **10**, 12f
Interface 573–586
Intergovernmental Panel on Climate
    Change 424, 433, **436**–438
international agreements 700
International Atomic Energy Agency
    507, 508
*International Boundary Waters Treaty
    Act* 152, 318, 341
*International Code of Conduct on
    the Distribution and Use of
    Pesticides* 616
International Conference on
    Population and Development 177
International Crane Recovery
    Team 277

international environmental law
    and policy
  agreements 709t
  conventional law 709
  customary law 709
  NAAEC 711
  NAFTA 711–713
  North American agreements
      708–709
  participants in 709–714
International Joint Commission 92,
    152, 693
International Organization for
    Standardization 307, 697
International Polar Year 432
interspecific competition **94**
intertidal zones **362**, 362f
intraspecific competition **94**
intrinsic value 660
intrusive rock **131**
Inuit 450, 456, 462
Inuit Circumpolar Council 456
Inuvialuit 654
invasive species 107–110, **108**, 266, 267f
ionic bonds 37
ionic compounds 37
ions **36**–37
IPAT model 8–15, **165**–167, 743
Iran 203, 331
irrigation
  conventional irrigation 210f
  defined **209**
  drip irrigation 210f
  and soil degradation 209, 210
  and waste 329–331
  and water consumption 331f, 335
island biogeography theory 275–276,
    275f, 276f
isotopes **36**, 37f, 38–39
isotopic signatures 38–39
ivory-billed woodpecker 262
Ivory Coast 578

J

Jabiluka uranium mine 657–658
jack pine 292
Jacobs, Jane 625, 639, 639f
jaguars 255
Jain Dharma 663
James Bay Project 684
Japan 135
Javan tiger 263
Jeans, Sir James 32
Jian, Yang 369
Johannesburg 725
Johnson, Darlene 376
Johnson, Pierre Marc 712

Jordan 333
Juan da Fuca hydrothermal vents 135
Juan da Fuca Ridge 375f
Juan da Fuca tectonic plate 545
Jubail Desalinization Plant 335f

K

Kakadu National Park 309, 657,
    681–682
Kant, Immanuel 660
Kazakhstan 202
Keith, David 550
kelp 106–107, **360**
kelp forests 360, 360f
Kemp, Clarence 533
Kennedy Space Center 376
Kenya 203, 377, 667
kerogen **466**, 472, 480
keystone species **102**–104, 270
killer smog 391
killer whales 106
Kimmins, Dr. Hamish 288
kinetic energy **45**, 46, 47f
Kirtland's warbler 262
kitchen gardens 742f
Konza Prairie Reserve 213f
Kootenay National Park 665f
Krakatau 72
Krull, Ulrich 343
K-selected **81**–82, 82t
K-strategists 81
K-T mass extinction 72–73
Kuhn, Thomas 19
Kuwait 172
kwashiorkor 223f, 224
Kyoto Protocol 397, **453**–454, 709, 716

L

Labrador duck 262
Labrador Inuit Association 659f
Lac du Bonnet 512, 515f
Lake Apopka 601
Lake Athabasca 21
Lake Baikal 321
Lake Erie 152, 691–693
Lake Iroquois 557
lakes 320–321, 321f, 709t
land conversion. *See also* deforestation
  for agriculture 295, 297
  for livestock 297–298
  for settlement of North America 295
  wetlands 331–332, 332f
land degradation 292
"The Land Ethic" 666–666f
landfill gas **519**, **531**, 567–568
landfill gas-to-electricity 531, 558

landfills
    and archeology 566
    Beare Road 531, 557–559, 596
    closure 596
    drawbacks of 565
    gas capture 472, 531, 567–568
    Green Lane Landfill Site 565
    sanitary 563–564, 563f
    scavenging 561, 562f
    secure 579
landscape 139
landscape ecology 139, 140f, 141
land trusts 308, 308f
Langlois, Lucille 508
La Niña 357, 430
La Rance facility 546
large hydro 500
Las Vegas 631f
Laszlo, Ervin 126
latitudinal gradient 260, 261f
Latrodectus hasselti 86
Laurentian University 732, 732f
law of conservation of matter 34
Law of the Sea 365
LD50 609
leachate 563
leaching 193
lead 396, 596
leaded gas 396, 400
Leadership in Energy and
    Environmental Design 730
lead poisoning 596–597
leafcutter bees 230, 231
Leakey, Louis 282
Legionnaire's disease 416
legumes 205
Leigh Marine Reserve 377
lemon-scented geranium 35
Le Nordais Project 538
Leontopithecus rosalia 280
leopards 255
Leopold, Aldo 60, 271, 665–666, 666f
lethal-dose-50% 609
leukemia 272, 509–510
levees 328
lice 13f
lichens 105
life, origin of
    early Earth 50–51
    fossil record 52–53
    genes 53
    hypotheses 51–52
    life from the depths 51–52
    and Mars 50
    primordial soup 51
    seeds from space 44–46
life-cycle analysis 573

life expectancy 173
life from the depths 51–52
lifestyle hazards 592f, 593
Lighter Footprint Strategy 727
light pollution 646
Lights Out Canada 748
lignite coal 468f, 471
Likens, Dr. Gene E. 412
Likens, Gene 385
lilophilic 396
Lim, Darlene 32, 33, 52
limestone 133f
limiting factors 79, 79f, 137
The Limits to Growth 703
limnetic zone 320, 321f
limnologist 345
linear growth 78
lions 118, 255, 280
Liophis epinephalus 95f
lipids 42, 44
lipophilic 127
liquefied natural gas 472
literacy 175f
lithification 132
lithosols 118
lithosphere 130
Litman, Todd 642
litter 206, 293
Little Ice Age 430
littoral 101, 362
littoral zone 320, 321f
livestock grazing 297–298
Living Planet Index 264, 264f
Lizard Island 363f
Lloyd, Brennain 581
locally supported agriculture
    244–245, 246
locavores 246
lodgepole pine 292–293, 297
logging
    in Borneo 302f
    clear-cutting 201, 288–289
    locations 294
    rate of 300
    and reforestation 303–304
    salvage logging 306
Lomborg, Bjorn 24
London, England 164f, 404, 644
Long Island Sound 138
longline fishing 370, 370f
long-range transport of atmospheric
    pollutants 396
Loo, Jennifer 728
The Lorax 745
Lorene Kennedy Memorial Native
    Plants Garden 731f
Los Angeles 334

Lost in the Barrens (Mowat) 378
Lost Villages Historical Society 328
Lougheed, Peter 317
Louisiana Universities Marine
    Consortium 150
Louv, Richard 273
Love Canal 580, 704
Lovelock, James 404
lower montane rainforests 60
low-input agriculture 243
low-pressure system 391
Lucas, Zoe 120
luminosity 429
Lumpkin, Butch 600f
luvisolic soil 194t
lynx 114

M

Maathai, Wangari 667, 667f
MacArthur, Robert 275
Mackenzie Valley Environmental
    Impact Review Board 655
Mackenzie Valley Gas Project 462–464,
    486, 488
MacKinnon, James 246
MacMillan Bloedel 289, 311
macromolecules
    carbohydrates 43–44
    compartmentalization 44
    defined 42
    lipids 44
    nucleic acids 42–43
    proteins 42, 42f
macronutrients 137
Madagascar 272
"mad cow" disease 240
magma 131, 134f, 135
maize 220–233, 271
maladaptive 63
malaria 12, 13f, 594
Malaysia 299, 302f
Maldives 423–424, 443, 445, 450
malee 119
malnutrition 223
Malthus, Thomas 61, 78, 163, 164f
manganese nodules 365
mangrove forests 363–364, 363f
Manicouagan Crater 73
Manila 562f
manipulative experiments 17–18, 17f
Manitoba 701
mantle 133
Mao Zedong 161
maple trees 115
maquis 119
marasmus 223f, 224
marble 132

Marble Canyon Provincial Park 32
marginal land 203
Mariana Trench 135
marine biodiversity 374–375
marine protected areas **375**–376, 375*f*
marine reserves
  defined **376**
  design of 377–378
  effectiveness of 376–377
  effects of 378*f*
  study of 376–377
market failure **683**, 694
market incentives 717
market values **674–675**
Markham, Ontario 630*f*
Marmota vancouverensis 70, 70*f*
Mars 33
Marshall, Howard 150
Martin, Pat 588
Martineau, Daniel 126–127
mass extinction
  defined **262**
  K-T mass extinction 72–73
  previous episodes 53, 70, 262, 262*f*, 263*t*
  sixth 262–263
mass number **36**
materials recovery facilities 570
Matlab Family Planning and Health Services Project 176
matorral 119
matter **34**, 41–42, 46*f*
maturation 472
Mauna Loa Observatory 432, 432*f*
maximum sustainable yield **303**, **375**
May, Elizabeth 725
Maya 10
McDonald, James 404
McDonald's restaurant 738
McGill University 103, 729, 730*f*, 734, 734*f*
McKay, Chris 32
McMaster University 736
McQuigge, Dr. Murray 342
Mead, Margaret 749
meadowlarks 116
Meadows, Donella 166*f*, 625
meat consumption 239, 239*f*, 241, 335
Mecca 659
mechanical energy 46
mechanical weathering **191**
Medical-Technological Revolution 8, **168**
medicinal plants 271–272, 294
Mediterranean "scrub" woodland 119
Meegaskumbura, Dr. Madhava 265
Megachile rotundata 231
mega-cities **629**, 629*t*

meltdown **509**
Mendes, Chico 668–668*f*
Mer Bleue Conservation Area 189, 190
Merck 273
mercury **396**, 600, 600*f*
mercury contamination 367
Merritt Island National Wildlife Refuge 376
Mesley, Wendy 617
mesosphere **388**
meta-analysis 103
metamorphic rock **132**, 132*f*, 133*f*
metapopulation 141, 275
methane
  Beare Landfill Power Plant 531, 558
  as biogenic gas 472
  chemical structure of 41*f*
  defined **41**
  from landfills 519, 531, 567–568
  sources of **427**
methane clathrate 480
methane hydrate **365**, **480**
methane ice 480
Methanex 710, 711
Métis 462, 654
Mexico
  environmental policy 711
  maize 220–221, 238, 271
  pesticides and child development 608–609
  wheat varieties 224
Mica Dam 325
microbes 33–34, 53, 98
microbialites 32–33
microclimate **388**
micrometeorology 406
micronutrients **137**
Mid-Atlantic Ridge 134, 358*f*
migration 39
Migratory Birds Act 699
Milankovitch cycles **429**
mildew 416
Mill, John Stuart 660, 678
Millennium Declaration 182
Millennium Development Goals **182**, 182*t*, 746
Millennium Ecosystem Assessment **22**, 23–23*t*, 672, 738
Miller, Stanley 51
mimicry 96*f*
Minamata Bay 600, 600*f*
minerals
  defined **131**
  in oceans 365
mining
  Australia 657–658, 659*f*, 676
  contamination 580–582

  diamond 654
  gold 654, 655
  Northwest Territories 654–656
  uranium 658, 659*f*, 669
Mirrar Clan 658, 676
missing carbon sink 146
Mission to Planet earth 21
Mississauga, Ontario 398, 643, 643*f*, 644
mites 80*f*
mitigation **450**–451
mitochondria 44
mixed economy 670
mixtures and solutions 37
moderators 505
molecules **37**
Molina, Dr. Mario 404
monarch butterflies 39
monoculture **199**
monocultures **225**, 225*f*, 226
monosaccharides 43
Monsanto Company 220, 221, 236–237
montaine forest region 292–293
Monteverde
  biodiversity 62*f*
  and climate change 84
  dry days 85
  golden toads 60–61, 69, 73
  harlequin frog 75–76
  immigrant species 82
  ocean conditions 85
Monteverde Cloud Forest Biological Reserve 83
Montreal 627*f*
Montreal Protocol 143, 397, 405, **407**–408, 709
monumentalism 307
Mooers, Arne 278
Moon, as energy source 49
moose 104, 114
Morawski, Clarissa 571
morphology 32
mortality **77**
mosaic 139
moths 228, 228*f*, 230
motile organisms 74
mould 416
moulins 444
mountain pine beetle 297
mountaintop removal 470, 483, 484*f*
Mount Allison University 728
Mount Everest 388
Mount Pinatubo 393
Mount Royal, Quebec 636*f*
Mount Royal College 733
Mount St. Helens 135, 394*f*
Mowat, Farley 378–379

Muir, John 664, 664f
Mukash, Matthew 684
multiple-barrier approach 580
multiple use **301**
municipal regulation 700, 700f
municipal solid waste
   in Canada 561f
   composition of 560f
   composting 569
   by country 561
   defined **559**
   disposal methods 562–567
   financial incentives 571, 571f
   incineration 565–567, 567f
   increase in 561, 561f, 562f
   landfills. *See* landfills
   patterns in 560–561
   reduction of 568, 569t, 571–572, 571f
   regulation of 562–563
Murchison meteorite 51
musk oxen 114
Mustela nigripes 116
mutagens **599**
mutations **63**
mutualism **98**–**99**, 98f
Myanmar 364
mycorrhizae 98
My Discovery of America (Mowat) 379
Myers, Norman 280
Myers, Ransom 371
My Father's Son: Memories of War and Peace (Mowat) 378
The Mysterious Island (Verne) 527

N

nacelle 538
NAFTA 711–713
Nahanni National Park Reserve 50f
Napa Valley 536
napthalene 41f
Nasikabatrachus sahyadrensis 265
natality **77**
National Air Pollution Surveillance Network 395, 398
National Ambient Air Quality Objectives 398
National Biodiversity Institute 273
National Environmental Protection Act 696
National Farmers Union of Canada 221
National Forest Strategy 301–302
National Marine Conservation Areas Program 378
national parks, **308**. *See also* parks and reserves
National Pollutant Release Inventory 397, 398

National Population Health Survey 223
National Round Table on Environment and Economy 716
National Soil Conservation Program 204, 703
Native Americans 238f
Native Seeds/SEARCH 238f
natural disasters 591f
natural experiments **17**
natural gas
   consumers 473t
   defined **472**
   economic impact 485–486
   environmental impact 484
   extraction of 473, 484
   formation of 472
   history of use 472
   offshore drilling 473, 473f
   producers 472t–473t
Naturalist (Wilson) 274
natural rate of population change **173**
natural resource accounting **670**
natural resources
   consumption of 8–10
   continuum 6f
   defined **6**
   management of 6–7
   nonrenewable 6–7
   renewable 6
Natural Resources Canada 300
natural sciences **10**
Natural Sciences and Engineering Research Council 19
natural selection
   defined **61**
   and diversity 61–62
   evidence of 64
   explained 61–62
   and genetic variation 62–64
   logic of 62t
   types of 63, 63f
Nature Conservancy land trust 139, 308
Nautilus Minerals 365
Navajo miners 669
NAWAPA 318
negative feedback loops **128**–**129**, 128f, 428
Nehru, Pandit 501
nekton **359**
neoclassical economics
   cost-benefit analysis 673–675, **674**
   defined **673**
   explained 673
   implications for environment 674–677
neolithic period 7

Neptune Mining Company 365
Nesjavellir 543f
Net Economic Welfare index 680
net energy **468**
Netherlands 578
net metering **536**
net primary production **136**, 137f
net primary productivity **136**
neurotoxins **600**
neutral 40
neutrons **36**
Never Cry Wolf (Mowat) 378
new forestry **304**
Newfoundland 352–353
New Stone Age 167
new urbanism **639**–**640**
New York 564
New York City 641f, 644
New Zealand 377
Niagara Escarpment 639
niches **74**, 94, 94f
Nicholson, Joel 279
nickel mining 385–386
Nickson, Ross 340
Nigeria 487
Nike 683
nitrates 211f, 212
nitric oxide 395
nitrification **147**
nitrogen **35**, 146–147
nitrogen cycle 146–148, **147**, 147f, 149f
nitrogen dioxide 395
nitrogen fixation 148f
nitrogen-fixing bacteria **147**
nitrogen oxides 395
nitrous oxides **427**
noise pollution **646**
nonconsumptive use of water **325**
nondurable goods 561
nonmarket values **681**, 681f
non-point-source pollution **338**, 338f, 393
nonrenewable natural resources **6**
non-timber forest products 294
Nordhaus, William 497, 670
normative **660**
North American Agreement on Environmental Cooperation 711
North American Free Trade Agreement 711–713
North American megafauna 69f
North American Plate 135
North American Water and Power Alliance 318
North Atlantic Deep Water (NADW) **356**, 430, 430f
northern fur seals 366f

northern spotted owl 738f
North Slope 482, 484f
Northwatch 581
Northwest Passage 442
Northwest Territories 462–464, 654
Norway 239, 606
Notestein, Frank 174
no-till farming 206f, **207**–208, 207t
not-in-my-backyard (NIMBY) syndrome **542**, 565
ntype layer 535
nuclear energy
    accident risk 508–509
    advantages of 507
    breeder reactors 505–506
    Chernobyl 497, 509–510, 510f, 511f
    cleanliness 507–508
    defined 46, 504
    drawbacks of 507
    electricity generation 505, 506f
    environmental impacts 507f
    explained 503
    fission 504, 504f, 505
    fusion 506–506f
    health risks 509–510
    producers 503, 503t
    reactors 505
    slow growth of 753
    Sweden 497
    and uranium 504, 505f, 507
    usage amounts 529
    waste 507–508, 511–512, 514f, 515f
nuclear energy vs. fossil fuels 451
nuclear fission **504**, 504f, 505
nuclear fuel cycle 504
nuclear fusion **506**, 506f
nuclear reactors **504**
nucleic acids **42**–43, 43f
null hypothesis 15–16
Nunavik 456
Nunavut 279, 441, 450, 654
nutrient cycles **141**, 293, 294f
nutrient overenrichment 127
nutrient pollution 336, 367
nutrients
    circulation of 141–143
    defined **136**
    and productivity 136–138

O

Oak Ridges Moraine 322, 557
oak trees 115, 293
Oaxaca 220–221, 225
Oberlin College 730, 731f
observations, and scientific method 15
ocean absorption 429
ocean circulation 429–430

ocean energy
    Bay of Fundy 527–528
    currents 547
    harnessing of 545–547, 555f
    thermal energy 753
    tidal energy 545–546
    wave energy 546, 547f
oceanography **354**
oceans
    biodiversity 359–360, 374–375
    and carbon cycle 146
    and climate change 429–430
    climatic influence 356–357f
    conservation 375–379
    content of 354–355, 355f
    coral reefs 360–361, 361f
    coverage of Earth 354, 354f
    currents 356–357f
    ecosystems 359–364
    excess nutrients 367–368
    fishing. *See* fishing
    human use and impact 364–368
    international agreements 709t
    intertidal zones 362, 362f
    kelp forests 360, 360f
    marine protected areas 375–376
    marine reserves. *See* marine reserves
    petroleum spills 366–367, 366f
    pollution 365–368
    resource extraction 364–365
    seafloor topography 358–359, 358f
    temperature 355–356, 356f, 439
    transportation routes 364
    vertical structure 355–356, 356f
ocean thermal energy conversion **547**
Ochotona princeps 445, 446f
Odum, Eugene 74
Ogallala Aquifer 322
O horizon **193**
oikos 670
oil
    consumers 474f
    defined **474**
    depletion of 476–479
    economic impact 485–486
    environmental impact 484
    extraction of 475, 475f, 484
    formation of 474
    global trade 486f
    history of consumption 474
    location of deposits 474–475
    offshore drilling 473f
    producers 474f
    reliance on foreign supplies 485, 485f
    uses of 475–476, 477f
    world prices 485f
oil palm agriculture 299

oils 44
oil sands **479**, 480f, 481f
oil shale **480**, 480f
oil spills
    Amoco *Cadiz* 704
    Exxon Valdez 33–36, 366–367, 482
    sizes of 366–367, 366f
    Torrey Canyon 4
Oil Springs, Ontario 474
Okeefenokee Swamp **466**
old-growth forests **289**
Old Stone Age 7, 167
oligotrophic **321**, 337f
Olmsted, Frederick Law 641
Oman 172
omnivores **99**
one-child policy 170
Ontario 699–700
Ontario Emissions Trading Registry 716
OOTI sled 407, 407f
OPEC 485, 712
open canopy forests **291**
open cycle 547
open loop systems **544**
open systems **128**, 670
orangutans 282, 299
Orbignya phalerata 271
orbit 428–429, 429f
orcas 106
orchids 117
Orcinus orca 106
organelles **44**
organic 41
organic agriculture **243**–244, 243f, 245–247, 247f
organic compounds **41**, 575
organic fertilizers **211**
Organic Products Regulations 243
organic soil 194t
Organisation for Economic Co-operation and Development. *See* developed nations
organismal ecology 73–74
organisms
    and climate change 445
    and soil formation 192t
Organization of the Petroleum Exporting Countries 485, 712
organochloride compounds 127
organochlorines 227t
organophosphates 227t
organs 44
organ systems 44
origin of life. *See* life, origin of
Orwell, George 101–102
Ottawa-Holden Dam 501f, 544

Our Common Future 25, 705, 707
Our Ecological Footprint: Reducing Human Impact on the Earth (Rees) 183
outdoor air pollution
  acidic deposition 408–413
  chlorofluorocarbons 406
  consequences of 22f
  criteria air contaminants 395–396
  decrease in 398–401, 399f
  defined **393**
  federal regulation 397
  heavy metals 396
  in industrializing nations 405
  monitoring stations 398, 398f
  persistent organic pollutants 396
  photochemical smog **403**
  pollutant categories 394–397
  provincial regulation 397–398
  rural issue 404–405
  scrubbers 400f
  smog 391, **401**–402, 403f
  sources, natural 393
  synthetic chemicals 406
  toxic air pollutants 397
  transboundary 408–413
  "turning the corner" 602–603
  types of 393–394
outlet glaciers 444
overcultivating 201
overfishing 352–353, 353f, 368–370
overgrazing 201, **212**–213, 213f
overharvesting 266–268
overnutrition **222**
Ovibos moschatus 114
Owls in the Family (Mowat) 378
oxbow 319, 319f
oxbow lake 319
oxidation 37
oxidizing agents 37
oxygen **35**
oysters 368
ozone
  depletion of 406, 407
  and greenhouse gases 427
  hole 405, 406, 406f
  tropospheric ozone 401
  types of 395
ozone layer **388**

P

Pachycondlyla 97f
Pacific Cordillera 131
Pacific Forestry Centre 297
Pacific Plate 135
Pacific Rim National Park Reserve 308
Pacific yew 272

packaging 560, 566, 568
packrat middens 431
PAHs 42
Paine, Robert 103
Pakistan 177
paleoclimate **431**
paleolithic period 7, 167
paleontology 17
Palmer, Michelle 103
pandas 278
panspermia hypothesis 44–46
Pantera pardus 255
Pantera tigris balica 263
Pantera tigris sondaica 263
Panthera leo 255
Panthera onca 255
Panthera tigris 255, 257f
Panthera tigris altaica 254–255, 257, 266–267, 277, 279
Panthera tigris tigris 257
Panthera tigris virgata 263
paper 566, 570f
paper parks 309
paradigm **19**
paradigm shifts 19–20
parasites 96–**97**, 97f
parasitism **97**
parasitoid wasps 228
parent material **191**, 192t
parks and reserves
  in cities 641–644, 641f
  design of 310
  international 309–310
  land trusts 308, 308f
  non-federal entities 308
  opposition to 308
  reasons for 307–308
particulate matter **395**, 400f
Paskapoo Formation 322
passenger pigeon 75, 75f, 262
passive solar energy collection **533**
Pataki, George 564
patches 139
pathogenicity 597
pathogens 336–337, **593**
Paul, Donald 352
Paul reindeer 80f
Pauly, Dr. Daniel 369, 372
Pavilion First Nations Indian Band 32
Pavilion Lake 32–33
pay-as-you-throw 571
Payto Glacier 440
PCBs 396, 596
peace parks 309
peak oil **478**
peat 189–190, **192**, 468, 468f, 470–471
Peatland Carbon Study 190

peer review **18**
pelagic **359**
People of the Deer (Mowat) 378
periodic table of elements 35
permafrost 113, **193**, **438**, 440–441
permanent gases 387
permeability **194**
Permian period 73, 262
permits **696**
permit trading programs 454–**716**
persistent **396**
persistent organic pollutants 396
persistent synthetic chemicals **605**
Pest Control Products Act 226
pesticide drift 605
pesticides
  categories of 227t
  and child development in Mexico 608–609
  defined **226**
  integrated management 229
  overview 226–227
  and pollution 127, 396f
  resistance to 227
pests
  and biocontrol 228–229
  defined **108**, **226**
  and GM crops 234
  vs. pollinators 230
  resistance to pesticides 227
  Peterson, Roger Tory 379
petroleum, **474**. *See* natural gas; oil
petroleum geology 474–475
petroleum spills. *See* oil spills
Petromyzon marinus 97, 109
Pfiesteria 240
pH
  scale **41**, 41f
  of soil 195–196
pharmaceutical products 271–272
Pharmacists Association of Alberta 573
Pharomachrus mocinno 74
phenoxy 227t
Philippines 177, 377, 562f
Phillips, David 417
Phillips, James 176
philosophy
  and environmental health 614
  and Industrial Revolution 663–664
Phoenix, Arizona 388
phosphates 211f
phospholipids 44
phosphorus 137, 138f, **150**
phosphorus cycle **150**–153, 153f
photic zone **359**
photochemical smog 403–404, 403f
photodamage 361

photoelectric effect 535
photosynthesis
  and cellular respiration 49
  defined **48**
  equation for 49
  explained 48–49, 48*f*
  and food webs 145–146
photovoltaic cells **534–535**, 535*f*, 536*f*
photovoltaic effect 535
phylogenetic trees **67**, 68*f*
physical hazards **590**, 590–592, 591*f*, 592*f*
physical weathering **191**
phytoplankton **93**, 95, 138, 138*f*, 359, 360*f*
phytoremediation **34**, 35
Picea mariana 114
pigeons 75, 75*f*, 262
pikas 445, 446*f*
Pinchot, Gifford 664
Pinus sylvestris 79*f*
pioneer species **105**, 105*f*
Pisaster ochraceus 103
Pitfield, Jane 557
plankton **359**
planning **636**
plantation forestry 303–304, 303*f*
plantio directo 208
plants, and phytoremediation 35
plasmids 232, 232*f*
plastics **44**, 366, 561, 566, 568, 603
plates 134
plate tectonics **133–135**, 134*f*
Plato 10, 663
ploughpan 195
Plutarch 472
plutonic rock **131**
Poás Volcano National Park 83
podzolic soil 194*t*
point-source pollution **338**, 338*f*, 393
Polanyi, John Charles 32
polar bears 23*f*, 114, 269, 269*f*, 484*f*, 606
polar cells **391**
polar covalent bonds 37
polar vortex 407
policy, **693**. *See also* environmental policy
political influence 745
pollination **230**, 230*f*, 238*f*
pollinators
  conservation of 230–231
  importance of 104, 229
Pollutant Release and Transfer Registers 582
polluter pays principle **715**
pollution
  acid rain 41, 385
  air. *See* air pollution
  and biodiversity loss 266
  and climate change 22
  defined **336**
  and fossil fuels 481–482
  and human population growth 166
  light 646
  marine 365–368
  mercury contamination 367
  nitrogen 149
  noise 646
  and plastic 44
  and seafood 367
  and sprawl 634
  St. Lawrence River 127, 131*f*
  thermal 502
  types of 596*t*
  and urban centres 753
  water. *See* water pollution
pollution prevention **573**, 573*f*
polybrominated diphenyl ethers 597
polyculture 199
polycyclic aromatic hydrocarbons (PAHs) 42, 127
polymers
  defined **42**
  synthetic 44
  types of 42–43
polysaccharides 43
ponds 320–321, 321*f*
Poohkay, Suzanne 730
pool **142**
poplar trees 520*f*
The Population Bomb (Ehrlich) 164, 703
population control policies 162, 163, 170, 175, 176–177
population density 75–76, 81, **169**, 169*f*
population dispersion 76
population distribution **76**, 76*f*, 168–170
population ecology **71**, 74–82
population growth
  carrying capacity 80
  differences in 80*f*
  exponential growth 78–79, 79*f*
  growth rate 78
  human. *See* human population growth
  limiting factors 79–80
Population Monster 225
Population Reference Bureau 165
populations. *See also* human population
  age structure 77, 77*f*
  biotic potential 81–82
  birth rates 77
  changes in 77–78
  composition of communities 82
  death rates 77
  defined **64**
  density 75–76, 81
  distribution 76, 76*f*
  dynamics 74–77
  ecology of 74–82
  growth. *See* population growth
  isolation of 67, 67*t*
  limiting factors 79–80
  sex ratios 76
  size 75
  unregulated 78–79
population size **75**, **168**
porosity **194**
Portland, Oregon 638
Port Radium Mine 580, 669
positive feedback loops 128*f*, **129**, 197, **427**
post-industrial stage **175**
potential energy **45**, 46, 47*f*
Potholes region 320
Pounds, J. Alan 67, 84
poverty
  and environmental justice 667–669
  and food insecurity 223
  and human population growth 178, 179*t*
Prairie and Northern Wildlife Research Center 38
prairie dogs 116
Prairie Farm Rehabilitation Administration 204, 703
Prairie Habitat Garden 735, 735*f*
prairies 115
precautionary principle **235**, **614**
precession 428
precipitation **143**, 439–440, 440*f*
predation **95**–96, 95*f*, 96*f*
predator **95**
predator and prey models 11, 95–96, 96*f*
predictions 15–16
pre-industrial stage **174**
prescribed burns 14*f*, **306**
prescriptive 660
preservation 664–665
preservation ethic **664**
pressure-state-response model **707**, 708*f*
prey **95**
prickly pear 228, 228*f*
primary consumers **49**, 99
primary extraction **475**, 475*f*
primary forest **295**
primary pollutants **394**
primary producers **48**
primary production **48**, 136

primary succession **105**, 105*f*
primary treatment **343**
primordial soup 51
Prince Edward Island 193
Prince Edward Island University 736
Prince William Sound 33, 366–367, 482
prior appropriation **701**
prior informed consent 590, **657**
private vs. public good 694
probability 612
Producer Group 463
producers 99
productivity **136**
profundal zone **320**, 321*f*
prokaryotes **44**, 45*f*
pronghorn antelope 116
proteins **42**, 42*f*
proven recoverable reserve 475
provincial regulation 397–398, 699–700
proxy indicators 430–432, **431**
Prudhoe Bay 484, 484*f*
Pseudotsuga menziesii 116
ptype layer 535
public policy, **693**. *See also* environmental policy
punctuated equilibrium 62
pycnocline **355**
pyrethroids 227*t*
pyrolysis 480, **519**

Q

Qatar 172
quagga mussels 93, 109
qualitative data 16
quality of life 740–741, 742*f*
quantitative data 16
Quebec 589
Queen's University 736
Quercus garryana 109
questions, and scientific method 15
quetzel 74
Quttinirpaaq National Park 23*f*

R

Rabalais, Dr. Nancy 150
Rabbitkettle Hot Springs 50*f*
Rabliauskas, Sophia 654
radiation 512–513
radiative forcing **428**, 428*f*
radiatively active gases **425**
radioactive **36**
radioactive waste
   cleanup of 580–582
   defined **579**
   disposal of 507–508, 511–512, 514*f*, 515*f*, 579–580

radioactivity. *See also* radioisotopes
   defined **49**
   and geothermal energy 49
Radioan Exposure Compensation Act 669
radiocarbon dating **38**
radioisotopes **36**, 504
radon gas 414, 415*f*, **596**
Rafflesia arnoldii 299, 302*f*
Rahman, Ziaur 176
Rails-to-Trails Conservancy 643
rail transit 642
rainshadow desert 121
rainwater 408
Ramsar Convention 331
random distribution 76, 76*f*
rangelands **196**, 297
Ranger mine 658–659*f*
Rangifer tarandus 80*f*
Rapa Nui 11, 166, 748
Rathje, William 566
REACH 617–618
reactive **574**
realized niche **94**, 94*f*
recharge zone **322**
recombinant DNA **232**, 232*f*
recovery **559**
recycling 559, 569–570, 569*f*, 571–572
Recyc-Québec 729
red alder 293
red-backed squirrel monkey 83*f*
redback spider 86
Redefining Progress 183
Red List of Threatened Species (Savage) 61, **263**
redox reactions **37**
Redpath Museum 103
red pine 293
Red River flood 328*f*, 329
red tides **367**, 367*f*
reducing agents 37
reduction 37
redwood trees 81
Rees, William 9, 183
refining **475**, 476–477
reforestation **303**
refugees 172, 445
refugia 103
regolith 191
regosolic soil 194*t*
regulations **696**
reintroduction 277
relative abundance **255**
relative humidity **388**
relativists **660**
religion 177, 658–659, 660, 663
Remedial Action Plan 692

remote sensing 21, 141
renewable **464**
renewable energy
   applications of 529
   emissions 541*f*
   employment opportunities 531, 532*f*
   geothermal. *See* geothermal energy
   growth of 530–531, 530*f*
   hydrogen. *See* hydrogen fuel
   new sources 529–533
   ocean energy. *See* ocean energy
   solar. *See* solar energy
   transition to 487, 532–533
   usage amounts 529–530
   wind. *See* wind energy
renewable natural resources **6**
repeatability 19
Repetto, Robert 670
replacement fertility **172–173**
Report of the National Advisory Panel on Sustainable Energy Science and Technology 532
reproductive strategies 81–82
Research Institute of Organic Agriculture 245
reserves-to-production ratio 476
reservoirs **142**, 143*f*, **427**
resident time 142
resilience **104**, 129
resistance **104**, 129
resource consumption 8–9
resource management **6**
resource partitioning **95**
respiration 145–146
response **608**
restoration ecology **110**
return-for-refund schemes 571, 571*f*
reuse 568, 569*t*
reverse osmosis 335
Reykjavik 549*f*
R horizon **193**
ribonucleic acid (RNA) **42**, 43
ribosomes 44
Ricciardi, Dr. Anthony 92, 103
rice 234*f*
Richman, Sheldon 161, 165
Right Livelihood Award 221
rill erosion 201*f*
riparian 289, **320**, **701**
risk **612**
risk assessment **613**
risk management 612–614, 612*f*, **613**, 613*f*
rivers 319–320, 319*f*, 709*t*
Riverwood Park 643, 643*f*, 644
RNA **42**, 43, 146
Roberts, Dr. Callum M. 376

rock cycle **131**, 132, 132*f*
rocks **131**
Roman Catholic Church 177
Roman civilization 10
Roosevelt, Franklin D. 189
Roosevelt, Theodore 664*f*
Roots 719*f*
rosy periwinkle 271
rotation time 303
Rotterdam Convention on the Prior Informed Consent Procedure for Certain Hazardous Chemicals and Pesticides in International Trade 590, 616
Roubik, Dave 230
Rouge Park 564
round table **701**
Roundup Ready crops 221, 236
Rowland, Dr. F. Sherwood 404
Royal Botanic Gardens' Millennium Seed Bank 238
R/P ratio 476
r-selected **81**, 82*t*
r-strategists 81
rubbertappers 294, 668
runoff **143**
run-of-river approach **500**, 502*f*
Runte, Alfred 307
rural air quality 404–405
Ruskin, John 363
Russia 279
Rwanda 165, 172, 172*f*, 309
Ryder, Gráinne 521
Ryerson University 733

S

Sable Island 120
Saccharomyces cerevisiae 80*f*
Safe Drinking Water Act 343
Sagan, Carl 14
Saha, K. C. 340
Sahara Desert 170
Sahel 170, 179*f*
Saimiri oerstedii 83*f*
SAIT Polytechnic 733
salinity **354**
salinization **209**–210
salmon 242–243
salt marshes **363**
salts 37
salvage logging **306**
Sampat, Payal 557
San Andreas Fault 135
sand **194**
sandstone 133*f*
sanitary landfills **563**, 563*f*
Santa Rosa National Park 83

SARA Public Registry 262
Sarawak 299
Saro-Wiwa, Ken 487
SARS 594
Saskatchewan 200
Saudi Arabia 335, 335*f*
Saussure, Horace de 533
Savage, Jay M. 60–61, 69
savannah 117*f*, **118**, **291**, 291*f*
Saxena, Dr. Praveen 35, 55
scavenging 561, 562*f*
Schaller, George 254
Schindler, David 137, 345
Schmeiser, Percy 220, 221
science
  defined **14**
  democratization of 14
  environmental. *See* environmental science
  and environmental policy 714
  historical 17
  natural 10
  nature of 14–20
  scientific method 15–16
  social 10, 17
scientific inquiry 14
scientific method 18*f*
  assumptions of 3–15
  defined **15**
  diagram 15*f*
  elements of 15–16
Scientific Panel for Sustainable Forest Practices on Clayoquot Sound 302
scientific process
  components of 18*f*
  conference presentations 18
  grants and funding 19
  peer review 18
  repeatability 19
  and scientific method 18
  theories 19
Scotch broom 109
Scots pine 79*f*
scrubbers **400**, 400*f*, 470, 567
scrub woodland 119
seafloor topography 358–359, 358*f*, 359*f*
seafood 366*f*, 367
seafood consumption 239*f*
sea lamprey 97, 97*f*, 109
sea levels 423–424, 442–445, 442*f*, 443*f*
Sea of Slaughter (Mowat) 378
sea otters 106–107, 360
Sea River Mediterranean 366–367
Sea Shepherd Conservation Society 379
seasons 389*f*
Seattle, Washington 738
sea turtles 366

sea urchins 106–107, 360
secondary consumers **49**, 99
secondary extraction **475**, 475*f*
secondary pollutants **394**
secondary production **49**, 136
secondary succession **105**–106, 105*f*
secondary treatment **343**
second-growth forest **295**
secondhand smoke 414
second law of thermodynamics **46**–47
secure landfills **579**
sedentary lifestyle 295
sediment 337
sedimentary rock 131–**132**, 132*f*, 133*f*, 146
sediments **131**
seed banks **238**–239
seeds from space 44–46
SEEDS projects 736
seed-tree approach 304, 305*f*
seismic sea wave 423
Selborne League 664
selection systems 304
selective breeding 64, 233
selective tree harvest 305*f*
septic systems **343**
sequestered **427**
sequestration **482**
Sequoia sempervirens 81
Serageldin, Ismail 317
serendipity berry 271
seringeiro rubbertappers 294
services **670**
sessile animals 74
Seuss, Dr. 745
sex ratios **76**, 170–172
sexual reproduction, and genetic variation 63
Seymoar, Nola-Kate 648
Shafeeu, Ismail 423
sharks 268
sheet erosion 201*f*
shelf-slope break 358
Shell Oil 463, 487
Sheltair Group 625
shelterbelts 206*f*, **207**
shelterwood approach 304, 305*f*
shrublands **291**
Siberian Tiger Project 255, 273
Siberian tigers 254–255, 257, 266–267, 277, 279
sick-building syndrome 417
Sierra Club 664*f*
Sierra de Manantlán Biosphere Reserve 238
Sierra Nevada 605, 664*f*
Sierra Youth Coalition 737, 737*f*

Sifton, Clifford 665, 702
Sikhote-Alin Mountains 254–255, 275
Silent Spring (Carson) 396, 598–599, 703
silica 365
silicosis 204
Sillett, T. Scott 74
silt **194**
silver 365
silviculture **299**
Simon Fraser University 278, 732, 733
sinkholes **333**, 333*f*
sinks **142**, 302, **427**
Sistan Basin 203
Sitka spruce 293
The Skeptical Enviromentalist (Lomborg) 24
slash-and-burn agriculture 200*f*, 297, 393
slate 132
SLOSS dilemma **310**
slow food 246
sludge **345**
sluice 528
small hydro 500
Small is Beautiful: A Study of Economics as if People Mattered 703
smart growth 639, 639*t*
Smeeton, Miles and Beryl 279
smelting 385
Smith, Adam 672–673
Smith, Alisa 246
smog 391, **401**, 402*f*, 403*f*
smog check 368
smoking 596
snags 293
snow 440–441
social and economic factors 82–83
social sciences **10**, 17
Social Sciences and Humanities Research Council 19
societal impacts, on climate change 446–448, 447*f*
SOER **706**–708
softwood **294**
soil
    cation exchange 196
    characteristics of 193–196
    classification system 194*t*
    colour of 193–195*f*
    components of 191*f*
    conservation 204–205, 206*f*
    defined **190**
    degradation. *See* soil degradation
    formation of 191–192, 192*t*
    horizons 192–193, 193*f*
    pH scale 195
    and population growth 196–199
    regional differences 199–200
    structure of **195**
    as a system 190
    texture of **194**–195, 195*f*
Soil at Risk: Canada's Eroding Future 204
Soil Conservation Council of Canada 204, 207
soil degradation
    causes of 197*f*
    comparison map 197*f*
    defined **196**
    desertification 203–204
    Dust Bowl 204
    erosion. *See* erosion
    fertilizers 210–211, 212
    grazing 212–213
    and irrigation 209
    and population and consumption 196
    protection against 205–208
    regional differences 199–200
soil profile **192**
solar cookers 534
Solar Decathlon 733, 733*f*
solar energy
    active 533, 534
    and the atmosphere 388–390
    benefits of 536
    Canada, usage in 537*f*
    defined **48**
    drawbacks of 536–538
    electromagnetic spectrum 48*f*
    explained 47, 533
    magnification 534, 534*f*
    passive 533
    and photovoltaic cells 534–535, 535*f*, 536*f*
    usage, increase of 535–536
    uses of 533
solar flares 429
solar output 429
solar panels 24*f*, **534**
Solar Two 534*f*
solonetzic soil 194*t*
solutions **37**
Somalia 223*f*
Soufrière Marine Management Area 376
source reduction 559, **568**
sources **142**
South Africa 181, 725
soybeans 205, 233
Spadaro, Joseph 508
Spartina 363
specialists **74**
Special Report on Emissions Scenarios 449
speciation
    defined **66**
    and diversification 67–69
    explained 66–67
    and extinction 69
species
    biodiversity 64–65
    competition 94–95
    defined **64**
    ethical obligations toward 273–274
    herbivores 97–98
    hierarchy of 100*f*, 256*f*
    interactions 93*t*, 94–101
    mutualists 98–99
    parasites 96–97
    predation 95–96, 95*f*, 96*f*
    resource partitioning 95*f*
    umbrella species 278
species, competition 95*f*
species-area curves **276**, 276*f*
Species-At-Risk Act **261**, 278–279
Species at Risk Act 699
species coexistence **94**
species diversity **255**–257, 258, 258*f*, 259*f*, 260*f*
species richness **255**
sperm counts 601, 602*f*
Sphagnum moss 189
spillover effect 376, 378*f*
splash erosion 201*f*
splash zones 76
spotted owl 116
sprawl
    approaches to 639*f*
    Calgary 632*f*
    causes of 633–634
    defined **631**
    and economics 635
    explained 631–633
    and health 635
    impacts 634–635
    and land use 635
    Las Vegas 631*f*
    and pollution 634
    problems with 634–635
    and transportation 634, 635
    Vancouver 632*f*
spruce budworm 297, 445
spruce trees 116, 293
Sri Lanka 265
St. John's wort 294
St. Lawrence River
    and belugas 126–127
    biodiversity portrait 140–141
    as a system 128, 129–130, 131*f*
St. Lawrence Seaway 328
St. Lucia 377

stabilizing selection **63**, 63f
stable isotope geochemistry **432**
stable isotopes 36, **38**
Staebler, Dr. Ralf 406
stakeholder **701**
starch 43, 43f
state-of-the-environment reporting 706–708
Steadman, David 11
steady state **129**
steady-state economies **678**–679
Steller's sea cow 368
Stelopus varius 75
steppes 115
Stern, Sir Nicholas 676
Stern Review 448, 676–677
steroids 44
stock **6**
stock-and-flow resources **6**
Stockholm Convention on Persistent Organic Ppollutants 616, 616t
stocks 299
Stolarski, Richard 404
stone soils 118
storage impoundments **499**–500
stored product beetle 80f
storm surges **363**, 443
strangler figs 117
stratification 149f
stratosphere **419**
streams 319–320
strip-mining 469f, **470**, 480, 480f, 483, 484f
Strix occidentalis 116
Strix occidentalis occidentalis 738f
stromatolites 52f
Strongylocentrotus spp. 106–107
The Structure of Scientific Revolutions (Kuhn) 19
Stuart Energy 548
stumpage fees 711
subalpine forest region 292–293
sub-bituminous coal 471
subcanopies 293
subduction **135**
subsidies **715**, 715f
subsistence agriculture **198**
subsistence economy 670
subsoil **193**
subspecies 257
subsurface mining **469**, 469f, 480
subsystems 129–130
suburbs 628, 630–631, 630f
succession **105**–107
Sudbury, Ontario 385–386, 398
sugarcane 517f, 518
sugar maple 293

sugars 43
sulphur, in coal 471
sulphur dioxide 385, 393, **395**, 401
Sun. *See also* solar energy
    and Earth temperature 424–425, 425f
    and ecosystems 135–136
    as energy source. *See* solar energy
    relationship to Earth 389, 389f
sunshine distribution, in Canada 537f
superpests 234
Superstack 386
supply-and-demand curve 673f
surface impoundments **579**, 579f
Surinam 200f, 281
survivorship curves **77**, 78f
suspended particulates 395
sustainability
    on campus. *See* campus sustainability
    consumer influence 745
    and consumption 741–742
    defined **3**, **24**, **737**
    and development 24–26
    and economic opportunity 738–739
    and environmental bottleneck 747
    Environmental Performance Index 746–747
    and globalization 744–745
    as goal 24
    and human population growth 736–743
    and industrial solid waste 573
    and industry 744
    and local self-sufficiency 744
    long-term thinking 744
    Millennium Ecosystem Assessment 22, 23–23t
    by nation 746
    and political influence 745
    and population and consumption 20
    research and education 746
    and solutions 23
    strategies for 740t
    and technology 743
    urban 288–290, 644–647
    window of opportunity 741–742
Sustainability Assessment Framework 737f
sustainable agriculture
    defined **243**
    locally supported agriculture 244–245
    no-till farming 206f, **207**–208, 207t
    organic farming 243–244, 243f, 245–247, 247f
sustainable development
    defined **25**, **705**, **737**
    and ethics 661
    explained 24–25, 26

    and public policy 705–706, 707f
    and triple bottom line 737, 738f
sustainable forestry 301, 306–307f
sustainable forestry certification 307
Suzuki, David 14, 25, 385, 557
Svalbard Island 606
swamp gas 472
swamps 320
Sweden 497–498, 510, 511
swidden agriculture **200**, 200f, **297**
swift fox 279
switchgrass 519f
Switzerland 573
symbiosis **98**
sympatric speciation **67**
Syncrude tar sand tailings dam 325, 481
synergistic effects **611**
synergistic factors 608
Syngenta 236
synthetic **44**
synthetic chemicals 598–599, 598f, 598t, 604f
systems. *See* ecosystems; environmental systems; geologic systems

T

table salt 37
taiga **114**, 292
tailings 325, 481, 481f
Talloires Declaration **725**–726, 726t
tamarack 189, 292
Tansley, Arthur 135
target size 275f
tar sands **479**, 480f
tax breaks 715
Taxol 272
taxonomists **255**
taxus 272
Taylor, Scott 11
technically recoverable oil 475
technology 743
temperate deciduous forest 114f, **115**
temperate forest 290
temperate grassland **115**–116, 115f
temperate rainforest **116**, 116f, 293
temperature
    of air 388
    in the Arctic 439f
    climate change 438–439, 439f
    historic measurements 438f
    of oceans 355–356, 356f, 439
temperature inversion **391**, 391f
teratogens **600**
terracing 206f, **207**
tertiary consumers 99
Thailand 177

thalidomide 600, 600f
theory 19
thermal energy 753
thermal inversion **391**, 391f
thermal mass **533**
thermal pollution 337–338, 502
thermocline **355**
thermodynamics, first law of **46**
thermodynamics, second law of **46**–47
thermogenic gas 472
thermogram 488f
thermohaline circulation **356**, **430**
thermosphere **388**
Thetford Mines 589
This Rock Within the Sea: A Heritage Lost (Mowat) 379
Thomson, Dr. Alan 297
Thoreau, Henry David 664
threatened **261**
Three Gorges Dam 325–328, 327f, 481, 503
Three Mile Island 508f, 509
thymine 43f
thyroid cancer 510, 512
tidal bore 527
tidal creeks 363
tidal energy **546**
tidepools 362
tides 362
tigers 254–255, 257, 257f, 263, 269
Tijuana River 711, 712
tilt 389–390, 428, 429f
timber
  in Canada 300, 301
  harvesting 295f, 304–305, 305f
  types of 294
time, and soil formation 192t
Times Beach 580, 704
tires 574f, 575
tissues 44
Titusville, Pennsylvania 474
toads 60–61, 64, 75, 109
tobacco smoke 414
Toogood, J. A. 189
topographical relief 192t
topography **358**
topsoil **193**
Toronto 544, 548, 557–559, 565
Toronto Lichen Count 646
Tortuguero National Park 83
total fertility rate **172**–**173**, 173f
tourism 84, 272, 361
Townsend, Timothy 576
toxaphene 396
toxic **574**, **597**
toxic air pollutants 397

toxicants
  accumulation of 605–606
  airborne 605, 605f
  defined 597
  international regulation 615–618
  mixed 611
  natural 606
  persistence of 605
  regulation of 614–615, 615t
  risk management 612–614
  types of 599–600
  types of exposure to 611
  variations in response 610–611
  in water 604–604f
toxic chemicals 337
toxicity **597**
Toxicity Characteristic Leaching Procedure 576
Toxic Nation 152
toxicology **597**, 608–610
toxic waste 566
toxin **597**
Toxoplasma 607
trade barriers 710
trade winds 392
traditional agriculture **198**–**199**
traditional ecological knowledge 659–660
tragedy of the commons 8–9, 694
transboundary park 309
transboundary waterways **334**
transcendentalism 664
transform plate boundaries 134f, **135**
transgenes **220**, **232**
transgenic organisms **232**
transitional stage **174**
transitional zone 309
transmissible disease 593
transpiration **143**
transportation
  and climate change 452, 452f
  energy consumption 640f
  ocean 364
  rail transit 642
  and sprawl 634, 635
  and sustainability 734–735
  urban 640–641, 640f, 642
Transportation of Dangerous Goods Act 699
trawling **370**, 371f
treaties **709**
tree huggers 667
tree litter 206, 293
trend **436**
Trent University 733, 736
tributary **318**
triple bottom line **737**, 738f

trophic levels **99**–**101**, 100f, 101f
trophic pyramid **101**, 101f
tropical deciduous forest **117**, 117f
tropical dry forest **117**, 117f
tropical forest 290
tropical rainforest 116–**117**, 116f, 260
tropopause 387
troposphere **387**, 391
tropospheric ozone **395**, 401
T'Seleie, Frank 463
tsunami 423
tuberculosis 594
tubeworms, giant 50f
tulip trees 293
tundra 113–**114**, 113f, 169–170, 290, 291
Tunisia 177
Tupper, Mark 377
turbidity **339**
turbine **470**
Turkey 271
Turner, Matthew 634
Turner, Ted 161
Turning the Corner: An Action Plan to Reduce Greenhouse Gases and Air Pollution 401, 401f
turnover time **142**
Tuvalu 445, 450
Twain, mark 390
Tylophora 272

U

U.N. Framework Convention on Climate Change 453
U.S. Clean Air Act 517
UBike program 729
Uganda 280, 309
ultraviolet radiation 88, 388, 404, 406, 592
umbrella species 278
unconfined aquifer **322**
Underground Thermal Energy Storage 544
undernourishment **222**
understory **293**
uneven-aged **304**
uniform distribution **76**, 76f
UniRoyal 579
United Arab Emirates 172
United Nations
  biodiversity protection 279
  Convention on International Trade in Endangered Species of Wild Fauna and Flora 279
  and desertification 203
  and international environmental law 709–710, 710f

parks and reserves 309
and population growth 177–178
soil conservation 204
sustainable development goals 182, 182*t*
United Nations Commission on Environment and Development 705
United Nations Convention on the Law of the Sea 365
United Nations Environment Programme **710**
United States
and biodiversity conservation 280
energy funding 532
nuclear energy 503
universalists **660**
Universal Soil Loss Equation 202, 202*t*
University of British Columbia 32, 183, 725–727, 729–730, 731, 735, 736
University of Calgary 279, 729, 731, 733
University of California 106
University of Guelph 35, 55, 245, 734
University of Montreal 126
University of Saskatchewan 38, 735, 735*f*
University of Toronto 548, 728–729, 732, 735, 736
University of Victoria 731, 731*f*
University of Waterloo 725, 733, 734, 736
upwelling **356**, 356*f*
uranium
Australia 657–658
isotopes 36
locations in Canada 505*f*
nuclear energy 504, 507
urban agriculture 245–247, 247*f*
urban centres. *See* cities
urban ecology **647**
urban growth. *See also* cities
boundaries 638
factors influencing 630
greenbelts 638–639, 638*f*
impacts of 169
increase in 628–629
and industrialization 627–628
land preservation 645–646
mega-cities 629*t*
new urbanism 639–640
population graph 627*f*
and population growth 7*f*
smart growth 639, 639*t*
sprawl. *See* sprawl
urban growth boundary **638**
urban heat island effect **646**
urchin barrens 360
Urey, Harold 51
Ursus maritimus 114, 269*f*

utilitarian principle **660**
utility 660
UTM Ecological Footprint Project 728–729

V

Valenti, Michael 497
valuation **674**
Vancouver 288–290, 632*f*
Vancouver Island 297
Vancouver Island marmot 70, 70*f*
Vancouver Island University 725, 731
variable retention harvesting 289
variables **16**
vector **594**
Verne, Jules 527
vertical current 356, 356*f*
vertisolic soil 194*t*
vested interest 659
Village of Widows 669
virulence 597
Voisey's Bay Nickel Mine 659*f*
volatile organic compounds **395**, 396, 414–415
volcanic eruptions 393
volcanic rock 131
voluntary guidelines **697**
Voroney, Paul 245
Vulpes velox 279

W

Wackernagel, Mathis 3, 9
Walkerton, Ontario 240, 341, 342–343
Wallace, Alfred Russell 61
WalMart 683, 683*f*
warblers 262
Ward, Jessica 103
Wareing, Mark 288
warm fronts **390**, 390*f*
warning coloration 96*f*
Warren County, North Carolina 667
wasps 230
waste
and biogenic natural gas 472, 531
and climate change 453
defined **559**
hazardous. *See* hazardous waste
industrial solid waste. *See* industrial solid waste
municipal solid waste. *See* municipal solid waste
radioactive 507–512, 514*f*, 515*f*
recycling 569–570, 569*f*
reduction of 559, 560*f*, 568, 569*t*
toxic 566
willingness-to-pay transition 706*f*

The Waste Exchange of Canada 574
waste exchanges **574**
waste management **559**
waste stream **559**
waste-to-energy facilities **567**, 567*f*
wastewater
and artificial wetlands 346
defined **343**, **559**
treatment of 343–345, 344*f*
water. *See also* freshwater; groundwater; hydrologic cycle
bonding 37
chemical structure 37, 38–39, 39*f*
conservation 732
desalination 334–335
as ice 40–40*f*
properties for life 40
as solvent 40*f*
waterborne diseases 336–337
water cycle 143, 293, 294*f*
waterlogging 209
water management 701–702
water mining **331**
water pollution
Abbotsford aquifer 212
arsenic 340
groundwater 339–341, 339*f*
legislation and regulation 341
nitrates 212, 341*f*
nutrient pollution 336
pathogens and waterborne diseases 336–337
point and non-point sources 338, 338*f*
prevention 342
quality indicators 338–339
rivers 695, 695*f*
sediment 337
sources of 339–341, 341*f*
thermal 337–338
toxic chemicals 337
Walkerton, Ontario 240, 341, 342–343
willingness-to-pay transition 706*f*
water quality **338–339**
watershed 130–131, **318**
water table **144**, **322**
Waterton-Glacier National Parks 309
water vapour 427, 428
water wars 334
Watson, Jane 106
Watson, Paul 379
Watson, Reg 369
Watt-Cloutier, Sheila 456
wave energy **546**, 547*f*
waxes 44
wealth gap 180, 180*f*

weather **390**–391, 390f
weathering **191**, 191f
weeds **226**
West Arm Demonstration Forest Experiments 303
westerlies **392**
Western Economic Diversification Canada 626
western hemlock 293
western larch 293
western red cedar 293
western white pine 293
West Nile virus 594, 594f
wetlands
  artificial 346
  Botswana 320f
  Columbia Wetlands 332f
  defined **320**
  drainage of 330–332, 332f
  explained 320
  and wastewater treatment 346
wet storage 514f
A Whale for the Killing (Mowat) 379
whales 126–127, 368
When the Oilpatch Comes to Your Backyard: A Citizens' Guide (Griffiths) 489
white birch 292
white spruce 292
white whales 126–127
Whitman, Walt 664
whooping crane 262
whooping cranes 277
W horizon 193
Wildlife Conservation Society 255
wildlife refuges **308**
wildlife studies 607
Wild Species 1005 265
Wilkinson, Bruce 202
Williams, James Miller 474

willingness-to-pay transition 704, 706f
Wilson, Edward O. 22, 254, 273, 274, 275
wind 392, 392f
windbreaks **207**
wind energy
  benefits of 541–542
  defined **538**
  distribution of, in Canada 540f
  drawbacks of 542
  growth of 539
  history of 538
  international capacity 539f, 539t
  sites 539–541
  turbines 538–539, 538f
wind erosion 202, 203f
Wind Erosion Prediction Equation 202, 202t
wind turbines **538**–539, 538f
Winnipeg 328f, 329
wise-use movement 315
wolves 102, 114, 277, 484f
women, and population growth 175–176
Wood Buffalo Park 277
wooded land 290f, 291
woodlands **291**
woodpeckers 262
Wool Bay 655
Working Forest Policy 289
Working Level 414
World Bank 710–712
World Commission on Water 336
world heritage sites **309**
World Summit, Johannesburg 725
World Trade Organization 710, 738, 745f
worldview
  defined **656**
  factors shaping 658–659

  and perception of environment 656–658
World Wide Fund for Nature 280
Worm, Boris 371
Wyeth, Andrew 152

X

Xerox 573

Y

Yaqui Valley 608–609
Yeager, Justin 76
yeast cells 80f
yellow birch 293
yellow cypress 293
Yellowknife 654
Yellow River 167
Yellowstone National Park 277, 308
Yosemite National Park 308, 664f
Yucca Mountain 512, 580
Yukon 541, 541f

Z

Zambia 177, 181
  GM crops 237, 237f
Zea diploperennis 271
zebra mussels 92–93, 96–103, 108f, 109
zebras 118
Zedong, Mao 161
Zero Emissions Research and Initiatives 573
zero-tillage 208
Zimbabwe 177
zinc 365
zone of aeration 322
zone of saturation 322
zoning **637**, 637f
zooplankton **93**
zooxanthellae 361